Investigative and Predictive DNA Testing

Investigative and Predictive DNA Testing

Edited by

Hirak Ranjan Dash
Department of Forensic Science, National Forensic Sciences University, Delhi Campus, New Delhi, India; School of Forensic Sciences, Centurion University of Technology and Management, Bhubaneswar, Odisha, India

ELSEVIER

ACADEMIC PRESS
An imprint of Elsevier

Academic Press is an imprint of Elsevier
125 London Wall, London EC2Y 5AS, United Kingdom
50 Hampshire Street, 5th Floor, Cambridge, MA 02139, United States

ISBN: 978-0-443-30128-5

For Information on all Academic Press publications
visit our website at https://www.elsevier.com/books-and-journals

Publisher: Stacy Masucci
Acquisitions Editor: Ali Afzal-Khan
Editorial Project Manager: Rahul Basak
Production Project Manager: Kiruthigadevi N
Cover Designer: Mark Rogers

Typeset by MPS Limited, Chennai, India

Contents

Part III
Module III: Predictive forensic DNA analysis

Part IV
Module IV: challenges and concerns in forensic DNA analysis

26. Quality assurance and quality control in forensic science: from evidence collection to laboratory analysis

Abdullah Al-Matrouk, Abdullah Aldasem, Mahdi Haidar and Fatema Mousawi

27. Ethical issues in forensic DNA analysis

Prachilekha Sahoo

List of contributors

Joe Adserias-Garriga Department of Applied Forensic Sciences, Mercyhurst University, Erie, PA, United States

Bhavna Ahlawat Birbal Sahni Institute of Palaeosciences, Lucknow, Uttar Pradesh, India

Abdullah Aldasem University of Strathclyde, Glasgow, United Kingdom

Mohammad H. Al-Hasan Environmental and Life Science Research Centre (ELSRC), Kuwait Institute for Scientific Research (KISR), Kuwait City, Kuwait

Arash Alipour Tabrizi Legal Medicine Research Center, Legal Medicine Organization, Tehran, Iran

Sarah J.M. Alley Chemistry and Forensic Science Department, Towson University, Towson, MD, United States

Abdullah Al-Matrouk General Department of Criminal Evidence, Quality Control Department, Ministry of Interior, Kuwait City, Kuwait

Noora R. Al-Snan Forensic Science Laboratory, Directorate of Forensic Science, General Directorate of Criminal Investigation and Forensic Science, Ministry of Interior, Manama, Bahrain

Juan Carlos Álvarez Faculty of Medicine, Laboratory of Genetic Identification & Human Rights (LABIGEN-UGR), Department of Legal Medicine, University of Granada, Granada, Spain

Noemi Argirò Department of Medical and Surgical Specialities, Radiological Sciences and Public Health, Institute of Legal Medicine, University of Brescia, Brescia, Italy

Bhawna Bangwal Department of Forensic Biology and Biotechnology, Delhi Campus, National Forensic Sciences University, Ministry of Home Affairs, Government of India, Delhi, India

Anna Barbaro President Worldwide Association of Women Forensic Experts (WAWFE), Reggio Calabria, Italy

Dora Becher Laboratory of Biomedical Genomics and Oncogenetics, Institut Pasteur de Tunis, Tunis, Tunisia; Directorate of Technical and Scientific Police, Sub-directorate of Forensic and Scientific Laboratories, Tunis, Tunisia

Cairenn Binder Ramapo College of New Jersey, Investigative Genetic Genealogy Center, Mahwah, NJ, United States

Francis Okechukwu Chibueke Department of Biotechnology, School of Health Science and Technology, UPES, India

Venusia Cortellini Department of Medical and Surgical Specialities, Radiological Sciences and Public Health, Institute of Legal Medicine, University of Brescia, Brescia, Italy

Heather Miller Coyle Forensic Science Department, Henry C. Lee College of Criminal Justice & Forensic Sciences, University of New Haven, West Haven, CT, United States

Salem Terwase Daniel Department of Pharmaceutical and Food Manufacturing, Red River College Polytechnic, Winnipeg, MB, Canada

Rutupurna Das Department of Zoology, School of Applied Sciences, Centurion University of Technology and Management, Bhubaneswar, Odisha, India

Hirak Ranjan Dash Department of Forensic Science, National Forensic Sciences University, Delhi Campus, New Delhi, India; School of Forensic Sciences, Centurion University of Technology and Management, Bhubaneswar, Odisha, India

Antonio De Luca Department of Medical and Surgical Specialities, Radiological Sciences and Public Health, Institute of Legal Medicine, University of Brescia, Brescia, Italy

Helena Correia Dias Research Centre for Anthropology and Health (CIAS), University of Coimbra, Coimbra, Portugal; Instituto Nacional de Medicina Legal e Ciências Forenses (INMLCF), I.P, Coimbra, Portugal; Laboratory of Forensic Anthropology (LFA), Centre for Functional Ecology (CFE), University of Coimbra, Coimbra, Portugal

Daniel Díaz-Ruiz Faculty of Medicine, Laboratory of Genetic Identification & Human Rights (LABIGEN-UGR), Department of Legal Medicine, University of Granada, Granada, Spain

George Duncan Halmos College of Natural Sciences and Oceanography, Nova Southeastern University, Dania Beach, FL, United States

Aparna Dwivedi Birbal Sahni Institute of Palaeosciences, Lucknow, Uttar Pradesh, India; Academy of Scientific and Innovative Research (AcSIR), Ghaziabad, Uttar Pradesh, India

Edeaghe Ehikhamenor Center for Forensic Programmes and DNA Studies, University of Benin, Benin, Nigeria

Kelly M. Elkins Chemistry and Forensic Science Department, Towson University, Towson, MD, United States

Aibuedefe Osagie Eremwanarue Faculty of Medicine, Health and Social Care, Canterbury Christ Church University, Canterbury, United Kingdom

Sylvia Etim Department of Chemistry and Biochemistry, Florida International University, Miami, FL, United States

Xiomara Gálvez Faculty of Medicine, Laboratory of Genetic Identification & Human Rights (LABIGEN-UGR), Department of Legal Medicine, University of Granada, Granada, Spain

Shristi Gupta Birbal Sahni Institute of Palaeosciences, Lucknow, Uttar Pradesh, India; Academy of Scientific and Innovative Research (AcSIR), Ghaziabad, Uttar Pradesh, India

David Gurney Ramapo College of New Jersey, Investigative Genetic Genealogy Center, Mahwah, NJ, United States

Christian Haarkötter Faculty of Medicine, Laboratory of Genetic Identification & Human Rights (LABIGEN-UGR), Department of Legal Medicine, University of Granada, Granada, Spain

Mahdi Haidar Kuwait Identification DNA Laboratory (KIDL), General Department of Criminal Evidence, Ministry of Interior, Kuwait City, Kuwait

Mariem Hechmi Laboratory of Biomedical Genomics and Oncogenetics, Institut Pasteur de Tunis, Tunis, Tunisia; University of Tunis El Manar, Tunis, Tunisia; Genetic Typing Service, Institut Pasteur de Tunis, Tunis, Tunisia

Haifa Jmel Laboratory of Biomedical Genomics and Oncogenetics, Institut Pasteur de Tunis, Tunis, Tunisia; University of Tunis El Manar, Tunis, Tunisia

Rym Kefi Laboratory of Biomedical Genomics and Oncogenetics, Institut Pasteur de Tunis, Tunis, Tunisia; University of Tunis El Manar, Tunis, Tunisia; Genetic Typing Service, Institut Pasteur de Tunis, Tunis, Tunisia

Nadia Kheriji Laboratory of Biomedical Genomics and Oncogenetics, Institut Pasteur de Tunis, Tunis, Tunisia; University of Tunis El Manar, Tunis, Tunisia

Amnuay Kleenayoon Samraong, Cambodia

Snigdha Konar Birbal Sahni Institute of Palaeosciences, Lucknow, Uttar Pradesh, India; Academy of Scientific and Innovative Research (AcSIR), Ghaziabad, Uttar Pradesh, India

Abhilash Kumar Das School of Forensic Sciences, Centurion University of Technology and Management, Bhubaneswar, Odisha, India

José Antonio Lorente Faculty of Medicine, Laboratory of Genetic Identification & Human Rights (LABIGEN-UGR), Department of Legal Medicine, University of Granada, Granada, Spain

Sanjoy Majumder School of Biotechnology, Centurion University of Technology and Management, Bhubaneswar, Odisha, India

Danielle Malheiros Departamento de Genética, Laboratório de Genética Molecular Humana, Universidade Federal do Paraná, Curitiba, PR, Brazil

Licínio Manco Research Centre for Anthropology and Health (CIAS), University of Coimbra, Coimbra, Portugal; Department of Life Sciences, University of Coimbra, Coimbra, Portugal

Bruce McCord Department of Chemistry and Biochemistry, Florida International University, Miami, FL, United States

María Isabel Medina-Lozano Faculty of Medicine, Laboratory of Genetic Identification & Human Rights (LABIGEN-UGR), Department of Legal Medicine, University of Granada, Granada, Spain

Celso Teixeira Mendes-Junior Departamento de Química, Laboratório de Pesquisas Forenses e Genômicas, Faculdade de Filosofia, Ciências e Letras de Ribeirão Preto, Universidade de São Paulo, Ribeirão Preto, SP, Brazil

Khadija Abdullahi Mohammed Department of Biochemistry and Molecular Biology, Federal University of Santa Maria, Santa Maria, RS, Brazil

Vítor Matheus Soares Moraes Departamento de Genética, Faculdade de Medicina de Ribeirão Preto, Universidade de São Paulo, Ribeirão Preto, SP, Brazil

Fatema Mousawi Kuwait Identification DNA Laboratory (KIDL), General Department of Criminal Evidence, Ministry of Interior, Kuwait City, Kuwait

Ngozichukwu Francis Ngozichukwu School of Integrated Biological and Chemical Sciences, University of Texas Rio Grande Valley, Edinburg, TX, United States

Stanley Udogadi Nwawuba Department of Biochemistry, Institute of Basic Health Sciences, Federal University of Rio Grande do Sul, Porto Alegre, RS, Brazil

Vidhi Paida Biological and Life Sciences, School of Arts and Sciences, Ahmedabad University, Central Campus, Navrangpura, Ahmedabad, Gujarat, India

Gagan Kumar Panigrahi Department of Zoology, School of Applied Sciences, Centurion University of Technology and Management, Bhubaneswar, Odisha, India

Dipanwita Raha Department of Chemistry and Biochemistry, Florida International University, Miami, FL, United States

Niraj Rai Birbal Sahni Institute of Palaeosciences, Lucknow, Uttar Pradesh, India; Academy of Scientific and Innovative Research (AcSIR), Ghaziabad, Uttar Pradesh, India

Shreya Rastogi National Law University, Delhi, India; LLM New York University School of Law, New York, NY, United States; The Square Cirlce Clinic, NALSAR University of Law, Hyderabad, Telangana, India

Guilherme Schmitt Rieder Department of Biochemistry, Institute of Basic Health Sciences, Federal University of Rio Grande do Sul, Porto Alegre, RS, Brazil

Maria Divya Sahayaselvan The Square Cirlce Clinic, NALSAR University of Law, Hyderabad, Telangana, India; Biotechnology, St. Joseph's College, Bengaluru, Karnataka, India; Forensic Science, University of Strathclyde, Glasgow, Scotland

Annapurna Sahoo Department of Zoology, Jatni College, Jatni, Odisha, India

Prachilekha Sahoo School of Law, Ramaiah University of Applied Sciences, Bengaluru, Karnataka, India

María Saiz Faculty of Medicine, Laboratory of Genetic Identification & Human Rights (LABIGEN-UGR), Department of Legal Medicine, University of Granada, Granada, Spain

Nurdan Sandikci Department of Forensic Sciences, Faculty of Engineering and Natural Sciences, Üsküdar University, İstanbul, Türkiye

Francisco Santisteban-Álvarez Faculty of Medicine, Laboratory of Genetic Identification & Human Rights (LABIGEN-UGR), Department of Legal Medicine, University of Granada, Granada, Spain

Kunja Bihari Satapathy School of Applied Sciences, Centurion University of Technology and Management, Bhubaneswar, Odisha, India

Claudia Marina Schellin-Becker Programa de Pós-Graduação em Genética, Universidade Federal do Paraná, Curitiba, PR, Brazil; Laboratório de Genética Molecular Forense da Polícia Científica do Paraná, Curitiba, PR, Brazil

Andrei Semikhodskii Genecode Diagnostics LLC, Kyiv, Ukraine

Francesco Sessa Department of Medical, Surgical and Advanced Technologies "G.F. Ingrassia," University of Catania, Catania, Italy

Sweety Sharma Department of Forensic Biology and Biotechnology, Delhi Campus, National Forensic Sciences University, Ministry of Home Affairs, Government of India, Delhi, India

Ritesh K. Shukla Biological and Life Sciences, School of Arts and Sciences, Ahmedabad University, Central Campus, Navrangpura, Ahmedabad, Gujarat, India

Oliver M. Tuazon Center for Law and Digital Technologies (eLaw), Institute for the Interdisciplinary Study of the Law, Leiden Law School, Leiden University, Leiden, The Netherlands

Tugba Unsal Sapan Faculty of Engineering and Natural Science, Department of Forensic Sciences, Institute of Addiction and Forensic Sciences, Üsküdar University, Istanbul, Turkey

Kamayani Vajpayee Biological and Life Sciences, School of Arts and Sciences, Ahmedabad University, Central Campus, Navrangpura, Ahmedabad, Gujarat, India

Andrea Verzeletti Department of Medical and Surgical Specialities, Radiological Sciences and Public Health, Institute of Legal Medicine, University of Brescia, Brescia, Italy

Diana C. Vinueza-Espinosa Faculty of Medicine, Laboratory of Genetic Identification & Human Rights (LABIGEN-UGR), Department of Legal Medicine, University of Granada, Granada, Spain

Ray A. Wickenheiser New York State Police Crime Laboratory System, Forensic Investigation Center, (Retired during the course of writing this chapter; Ray Wickenheiser Forensic Consulting, Youngsville, LA) Albany, NY, United States

Graham A Williams Centre for Biomedicine, Hull York Medical School, University of Hull, Hull, United Kingdom

Viroj Wiwanitkit Research Center, Chandigarh University, Chandigarh, Punjab, India

Preface

Humans, the *Homo sapiens*, evolved rapidly and outclassed the other *Homo* species present in the early time. Scientific evolution played a crucial role in establishing the *Homo sapiens* as the wisest creature on the earth. Compared to other approaches of human evolution, such as cognitive revolution and agricultural revolution, the scientific revolution occurred recently in the last 500 years. However, the scientific discoveries taken place in the last five decades outnumber the number of discoveries happened in the last five centuries. One such evolution in the criminal justice system is the discovery of the forensic DNA profiling technique. After the discovery of forensic DNA profiling technique by Sir Alec Jeffreys in 1985, the technique has shown rapid advancements in terms of its sensitivity, robustness, and acceptability in the court of law.

The book *Investigative and Predictive DNA Testing* highlights the important discoveries and scientific advancements that shaped the DNA profiling technique as one of the most irrefutable pieces of evidence in the criminal justice system. Most commonly, the forensic DNA profiling technique is being used for investigative purposes. However, after the inception of massively parallel sequencing (MPS) technique, forensic DNA profiling is capable of generating many investigative leads. Useful investigative leads, such as prediction of biogeographic ancestry, lineage, and phenotype by forensic DNA profiling, are playing a major role in solving various heinous crimes around the world. The adoption of the genetic genealogy approach in forensic investigation has solved many cold cases. Besides, standardized procedures and interpretation guidelines have also been recommended for the predictive forensic DNA profiling by various organizations.

The book *Investigative and Predictive DNA Testing* is a collective effort of the renowned forensic scientists, academicians, and law practitioners around the world to present the advanced techniques of forensic DNA profiling to the readers. This book has been written in four modules. Module I: *Conventional Forensic DNA Analysis* describes the routinely used techniques currently used in most of the forensic DNA laboratories. Module II: *Technological Advancements in Forensic DNA Analysis* emphasizes the technological advancements in the field of forensic DNA profiling in the last four decades. Module III: *Predictive Forensic DNA Analysis* thoroughly describes the tools and techniques used to yield various investigative leads by analyzing DNA. Finally, Module IV: *Challenges and Concerns in Forensic DNA Analysis* describes a general overview of the technological, legal, and ethical challenges faced by the forensic DNA profiling technique. Hope the hard work and expertise of the contributors and editors will make a huge impact on the readers. Wishing a happy reading to all!!!!

Dr. Hirak Ranjan Dash

Part I

Module I: Conventional forensic DNA analysis

Chapter 1

Fundamentals and principles of forensic DNA analysis

Noora R. Al-Snan

Forensic Science Laboratory, Directorate of Forensic Science, General Directorate of Criminal Investigation and Forensic Science, Ministry of Interior, Manama, Bahrain

1.1 Introduction

The advent of forensic DNA analysis has revolutionized the field of criminal investigation, offering unparalleled precision in identifying individuals based on biological evidence. Since its inception in the mid-1980s, DNA profiling has played a pivotal role in both criminal and civil justice systems, serving as an essential tool for identifying perpetrators, exonerating the innocent, and solving cold cases that have long remained unresolved (Bell & Butler, 2022). DNA, the fundamental building block of life, provides a unique genetic code for every individual, excluding identical twins, which makes it an invaluable resource in forensic investigations.

1.1.1 Historical background of forensic DNA analysis

In 1984, while studying hereditary diseases at the University of Leicester in Great Britain, Dr. Alec Jeffreys—later knighted for his contributions—made a groundbreaking discovery. He worked with DNA samples, cutting them into fragments using restriction enzymes. Most of the DNA was broken into small pieces, but some larger sections, known as restriction fragment length polymorphisms, remained intact. By placing these fragments on a gel, adding a radioactive probe, and allowing an exposure to film for several weeks, Jeffreys was able to produce an image resembling a barcode. These "barcodes" varied significantly between individuals, and Jeffreys recalls the moment he realized the potential for using DNA to identify people: "We suddenly realized that we'd essentially stumbled upon a DNA-based method for biological identification. My life completely changed at that point" (Zagorski, 2006).

This discovery, which became known as "DNA fingerprinting," was first applied in England to help solve two murder cases involving teenage girls (Wambaugh, 2016). After the rape and murder of Lynda Mann, the UK Forensic Science Service, using available forensic techniques, was able to determine that semen found on her body came from a male with type A blood and a matching enzyme profile phosphoglucomutase (PGM), which was present in about 10% of males. However, with no further evidence, the case remained unsolved. Two years later, Dawn Ashworth was also raped and murdered similarly. A 17-year-old suspect named Richard Buckland, who had learning difficulties, confessed to one of the murders. However, Jeffreys' new DNA fingerprinting technique exonerated Buckland by showing that the DNA from the semen samples in both cases matched but did not belong to him. This case highlighted one of the most important roles of forensic science: proving innocence.

Following this, authorities conducted a mass screening of local men, and Colin Pitchfork was eventually identified as the perpetrator. Interestingly, he was caught not through DNA but by his attempt to evade detection by convincing another man to provide a DNA sample in his place. This first-ever use of DNA typing in a criminal investigation showcased its potential in exonerating the innocent, conducting mass screenings, and solving crimes, albeit with some elements of luck involved.

One of the most profound impacts of DNA analysis in forensic science has been its role in exonerating individuals wrongfully convicted of crimes. The case of Richard Buckland, who was exonerated through DNA evidence, was the first of many such cases. Since then, organizations like the Innocence Project have used DNA testing to overturn wrongful

Investigative and Predictive DNA Testing. DOI: https://doi.org/10.1016/B978-0-443-30128-5.00018-7

convictions. In many of these cases, individuals had spent years, sometimes decades, in prison for crimes they did not commit, only to be freed once DNA evidence was properly analyzed (Butler, 2011).

DNA's ability to exonerate the innocent has led to significant changes in how criminal justice systems around the world handle evidence. Many jurisdictions now mandate the preservation of DNA evidence in cold cases, and postconviction DNA testing is available in many countries to prevent or correct miscarriages of justice (Bell & Butler, 2022). Since that time, many crimes have been solved using DNA, and now it is considered a pivotal tool for human identification in forensics. Fig. 1.1 illustrates the key milestones in the history of forensic DNA analysis.

1.1.2 Structure and function of DNA in forensic applications

DNA, or deoxyribonucleic acid, is the molecule that serves as the blueprint for life. Found in almost every cell of an organism, DNA carries the genetic information necessary for growth, development, functioning, and reproduction. Structurally, DNA is composed of four nucleotide bases: adenine (A), thymine (T), cytosine (C), and guanine (G). These bases pair in specific ways—adenine with thymine (A-T) and cytosine with guanine (C-G)—to form the familiar double helix structure, which was first elucidated by James Watson and Francis Crick in 1953. The arrangement of these base pairs is critical, as it encodes the instructions necessary for the functioning of every organism (Butler, 2009).

This double helix structure, while visually striking, plays an essential role in genetic expression and inheritance. Each individual's DNA is almost identical to that of others of the same species, but small variations in the sequence of these nucleotide bases account for differences between individuals. These small variations, or polymorphisms, are particularly useful in the field of forensic science, where DNA analysis has become a powerful tool for identifying individuals based on biological evidence left at crime scenes.

In forensic DNA analysis, specific regions of the DNA molecule are examined to create a genetic profile unique to each individual. One of the most important of these regions is known as **short tandem repeats** (STRs). STRs are short sequences of DNA that repeat a variable number of times in different individuals. For example, one person might have a sequence that repeats five times, while another has the same sequence repeated seven times. These differences make STRs

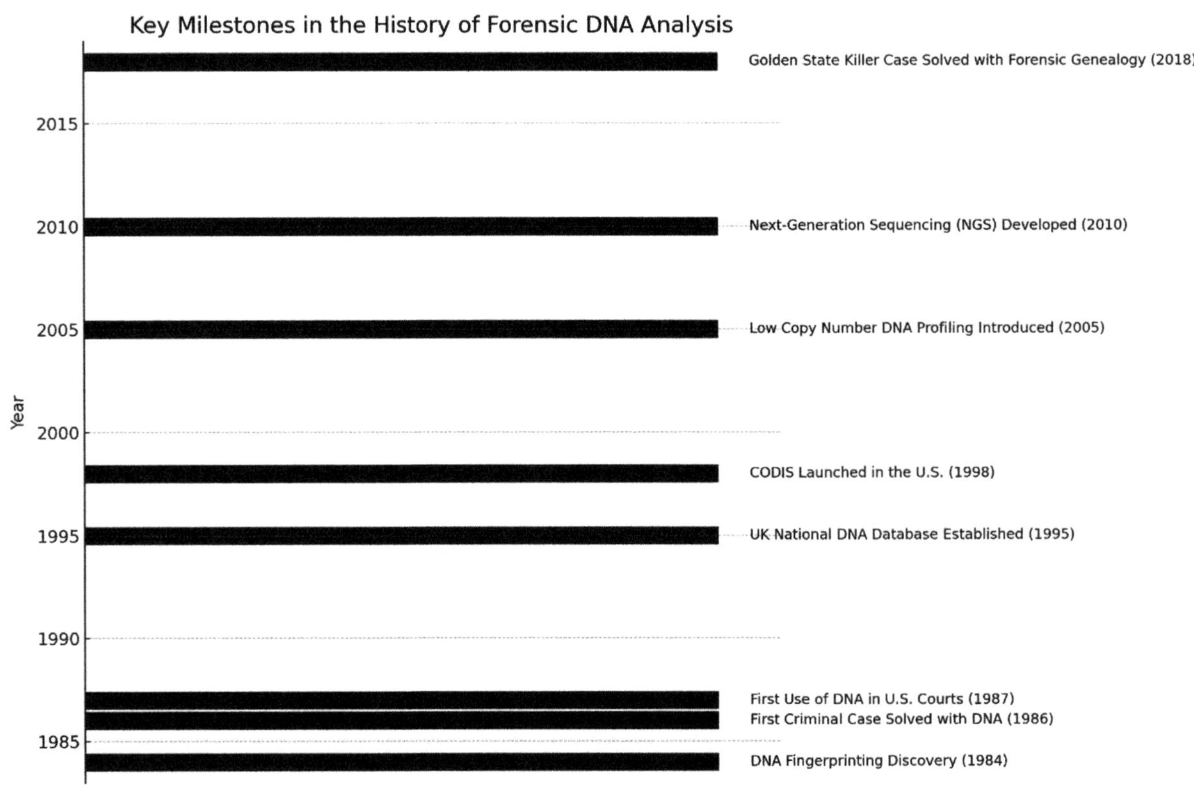

FIGURE 1.1 Key milestones in the history of forensic DNA analysis; *CODIS*: Combined DNA Index System.

highly variable between individuals, even among close relatives, which makes them particularly useful for forensic purposes (Shewale & Liu, 2013).

Forensic scientists analyze these STR regions to generate DNA profiles. Because these regions vary significantly among people, they provide enough information to uniquely identify an individual from a biological sample such as blood, saliva, or hair. Forensic DNA testing typically examines several different STR markers to ensure the accuracy and reliability of a genetic match. This multilocus approach reduces the likelihood of two individuals having the same STR profile to near impossibility, except in the case of identical twins.

As illustrated in Fig. 1.2, the DNA molecule consists of two strands twisted into a double helix, with the nucleotide bases forming rungs on the ladder. These base pairs are held together by hydrogen bonds, which ensure the stability of the DNA structure. In forensic analysis, scientists take advantage of the highly specific nature of these base pairings to isolate and analyze DNA sequences. By focusing on STRs, forensic experts can compare genetic material found at crime scenes with DNA from suspects or DNA databases to make definitive identifications.

The precision of DNA analysis has led to its widespread adoption in forensic investigations, not only for identifying perpetrators but also for exonerating individuals wrongfully accused of crimes. Advances in DNA technology, such as the polymerase chain reaction (PCR), have further enhanced the ability of forensic scientists to work with even tiny or degraded samples, amplifying the amount of DNA available for analysis and ensuring accurate results in a wide variety of cases (Butler, 2011).

1.2 Techniques in forensic DNA analysis

Since its inception in the mid-1980s, forensic DNA analysis has undergone significant advancements, making it one of the most reliable methods in criminal investigations today, as shown in Fig. 1.3. Among the various techniques, **STR profiling** remains the most commonly used and widely accepted method. STR profiling focuses on specific regions of the DNA that exhibit considerable variability between individuals. These short repeating sequences, usually located in noncoding regions of the genome, provide a robust way to differentiate between individuals—even those closely related—due to the unique number of repetitions in each person's DNA. This method is widely used in criminal cases for suspect identification, paternity testing, and even disaster victim identification (Butler, 2011).

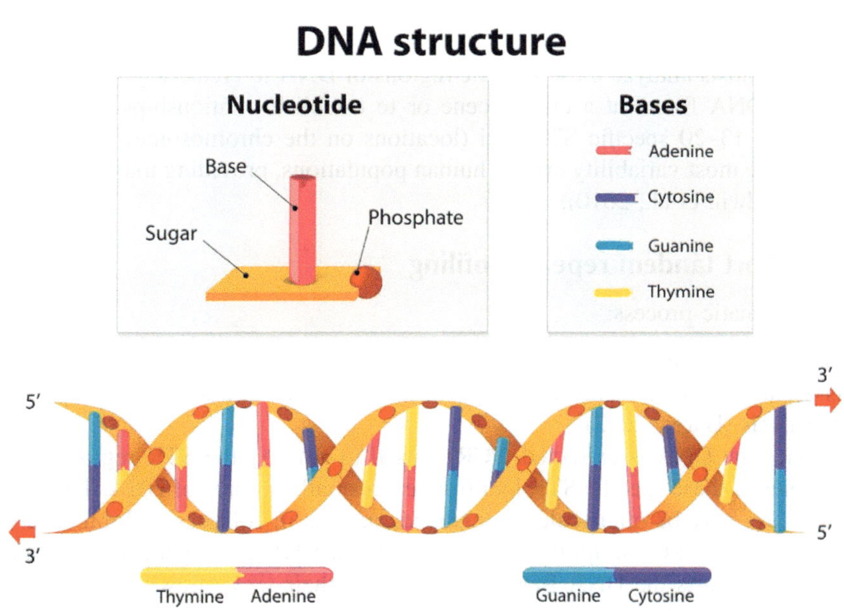

FIGURE 1.2 DNA structure illustrating the bases.
Retrieved with permission from Istock (Credit: ttsz). From DNA structure. (2024). https://www.istockphoto.com/vector/dna-structure-gm480973642-68805397?clarity=false (Accessed 1 October 2024).

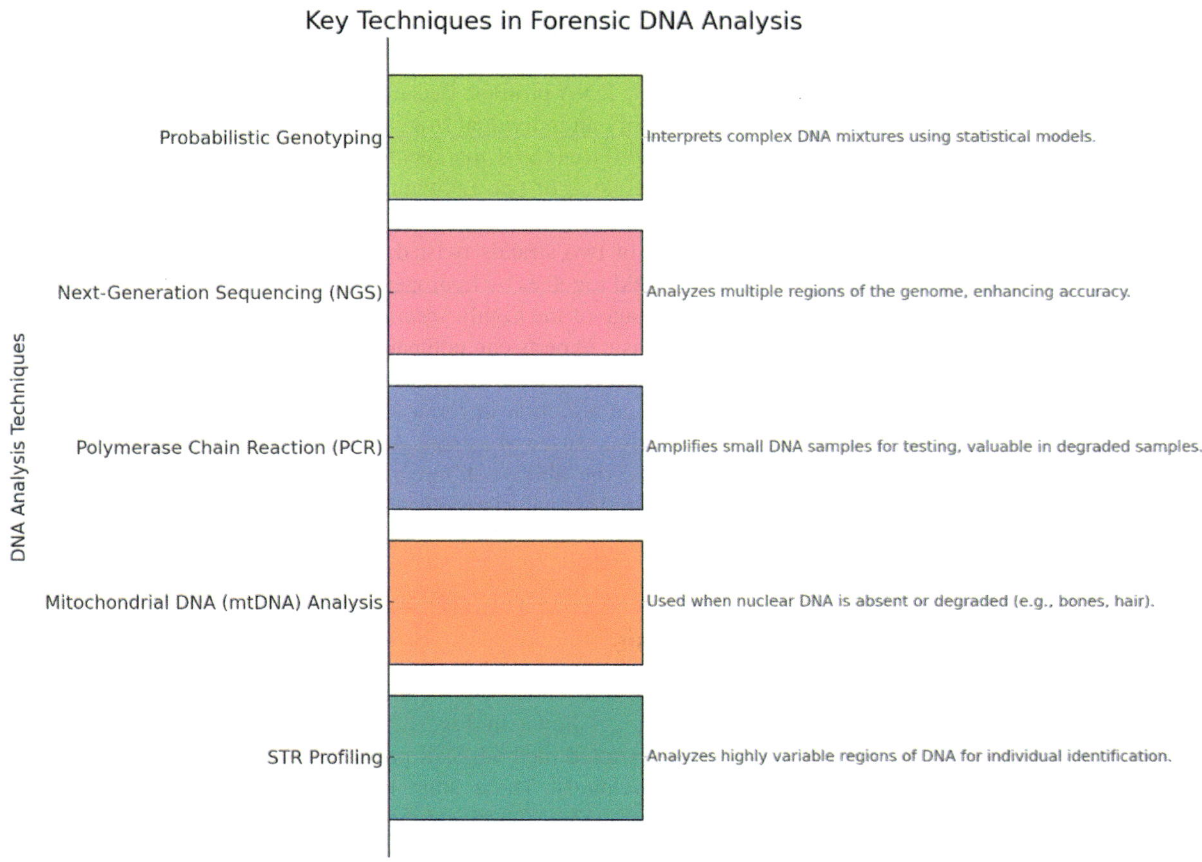

FIGURE 1.3 Key techniques used in forensic DNA analysis, with each technique and its description. It helps to illustrate the essential methods such as short tandem repeat profiling, mitochondrial DNA analysis, polymerase chain reaction, next-generation sequencing, and probabilistic genotyping.

1.2.1 Short tandem repeat profiling

In STR profiling, forensic scientists analyze these specific regions of DNA to create a genetic fingerprint, which can then be used to compare against DNA found at a crime scene or to establish relationships in paternity testing. Typically, forensic laboratories focus on 13–20 specific STR loci (locations on the chromosome) during analysis. These loci are chosen because they exhibit the most variability among human populations, providing maximum discriminatory power for individual identification (Goodwin et al., 2010).

1.2.2 The process of short tandem repeat profiling

STR profiling follows a systematic process:

1. **DNA extraction**: First, DNA is isolated from the biological sample, which could be blood, saliva, semen, or hair. In cases where DNA is highly degraded or in limited quantities, advanced extraction techniques ensure the recovery of enough material for analysis (Hares, 2012).
2. **PCR amplification**: Once the DNA is extracted, PCR is used to amplify the STR regions. PCR allows scientists to replicate millions of copies of these specific STR loci from even small amounts of DNA. This step is crucial in forensic cases where the sample size may be minuscule or degraded, such as in cold cases (Jobling & Gill, 2004).
3. **Separation and detection**: After PCR amplification, the amplified STRs are separated using techniques like capillary electrophoresis. This method sorts the DNA fragments based on their size, allowing forensic scientists to detect the different STR allele patterns. Each STR allele has a different length due to the number of repeats present, which forms the basis for individual identification (Butler, 2011).
4. **Comparison of profiles**: The resulting STR profile from a suspect or biological evidence is compared to other profiles, either from a suspect's reference sample or a database like the Combined DNA Index System (CODIS) in the United

States. If the STR patterns match across multiple loci, it strongly suggests that the two DNA samples originate from the same individual (Goodwin et al., 2010).

The power of STR profiling lies in its high level of discrimination. With the analysis of multiple loci, the chance of two unrelated individuals having the same DNA profile is incredibly small, on the order of one in billions. This precision is why STR profiling is considered the gold standard in forensic DNA analysis for identifying individuals in criminal investigations and legal cases (Hares, 2012). STR profiling is also noninvasive and can be conducted on small, degraded, or mixed samples, making it versatile for a wide range of forensic applications.

Additionally, STR profiles are robust in court, providing concrete genetic evidence that can link a suspect to a crime scene or exonerate an innocent individual. STR analysis is admissible as evidence worldwide, having undergone extensive validation and scrutiny for reliability (Jobling & Gill, 2004).

1.2.3 Limitations and considerations

Despite its reliability, STR profiling is not without its limitations. The method can be less effective in cases where DNA mixtures are present, such as when biological material from multiple individuals is combined. In such scenarios, interpreting the STR profile becomes complex, and additional techniques like probabilistic genotyping may be needed (Goodwin et al., 2010). Furthermore, STR analysis requires a relatively intact DNA sample. While PCR can amplify degraded samples, extremely damaged or compromised DNA may still pose challenges.

1.2.4 Mitochondrial DNA

In cases where nuclear DNA is either absent or too degraded to analyze using traditional methods, forensic scientists turn to **mitochondrial DNA (mtDNA) analysis**. Unlike nuclear DNA, which is inherited from both parents, mtDNA is inherited only from the mother. As a result, mtDNA can remain intact in highly degraded samples, such as hair shafts, teeth, or bones, which might not contain sufficient nuclear DNA for analysis (Butler & Levin, 1998). This technique is particularly valuable in identifying victims of mass disasters or analyzing ancient remains where conventional methods fall short. However, because mtDNA is shared by all maternal relatives, it is less discriminating than STR profiling and cannot provide the level of individual identification offered by nuclear DNA.

mtDNA is found in the mitochondria, the energy-producing structures of cells, and each cell contains hundreds to thousands of mitochondria. This is in contrast to nuclear DNA, which is present in only two copies per cell. Due to the sheer number of mitochondria per cell, mtDNA is much more abundant than nuclear DNA, even in highly degraded biological samples like hair shafts, bones, or teeth, which might lack sufficient nuclear DNA (Amorim et al., 2019). Because mtDNA can often be retrieved from these types of samples, it has become invaluable in situations such as identifying victims of mass disasters, missing persons, or even analyzing ancient remains.

One of the major advantages of mtDNA is that it is more resilient to environmental damage and degradation compared to nuclear DNA. mtDNA can persist in a sample for hundreds or even thousands of years. This makes it particularly valuable in cases involving historical or archaeological investigations. For example, forensic scientists have used mtDNA analysis to examine the remains of ancient human populations or to identify long-lost individuals, such as historical figures (Rudin & Inman, 2001).

1.2.5 The process of mitochondrial DNA analysis

The process of mtDNA analysis involves several key steps that are similar to other forensic DNA analysis techniques but tailored to mtDNA's unique properties:

1. **DNA extraction**: First, DNA is extracted from the biological sample, which can be highly degraded tissue, bones, teeth, or hair. Since mtDNA is present in greater quantities than nuclear DNA, this step is critical for recovering genetic material from these types of samples (Butler & Levin, 1998).
2. **PCR amplification**: Similar to nuclear DNA analysis, **PCR** is used to amplify specific regions of the mtDNA genome. In most forensic analyses, scientists focus on two highly variable regions of mtDNA known as the **hypervariable regions I and II**. These regions exhibit variation between individuals and provide enough genetic information to differentiate between different maternal lineages (Coble et al., 2004).
3. **Sequencing and comparison**: After amplification, the mtDNA is sequenced, and the resulting genetic profile is compared to reference samples, such as samples from living maternal relatives. Because mtDNA is inherited maternally, all individuals in the same maternal line will share the same mtDNA sequence. This means that mtDNA can

be used to establish maternal relationships, but it is less discriminatory than techniques like STR profiling, which provides more individual-specific information (Budowle et al., 2003).

1.2.6 Applications in forensic science

mtDNA analysis has a range of important applications in forensic science, particularly in challenging scenarios where other types of DNA analysis may not be possible. For example, mtDNA is frequently used to identify the remains of individuals in mass disaster situations, such as plane crashes or natural disasters, where the bodies have been exposed to environmental conditions that degrade nuclear DNA. mtDNA analysis was used in the identification of victims from the 9/11 World Trade Center disaster, where severely degraded samples limited the utility of nuclear DNA testing (Butler, 2009).

Additionally, mtDNA analysis is commonly employed in forensic anthropology and archaeology. It allows scientists to study ancient remains and trace the genetic lineage of populations over time. One famous example of mtDNA analysis in archaeology is the identification of the remains of **Nicholas II**, the last Russian Tsar, and his family, whose bodies were identified through mtDNA analysis decades after their deaths (Ivanov et al., 1996).

1.2.7 Limitations of mitochondrial DNA analysis

While mtDNA is highly useful in certain situations, it does come with limitations. The primary disadvantage is that mtDNA cannot distinguish between individuals who share the same maternal lineage. For example, a mother and all her offspring, as well as her siblings and their children, will have identical mtDNA sequences. As a result, mtDNA cannot provide the level of individual specificity that nuclear DNA or STR profiling offers (Budowle et al., 2003).

Another limitation is that mtDNA analysis is more labor-intensive and costly compared to STR profiling. The process of sequencing mtDNA and comparing it to reference samples is time-consuming, which can limit its use in high-throughput forensic laboratories (Butler & Levin, 1998). Despite these limitations, mtDNA remains a critical tool in cases where nuclear DNA analysis is not possible or practical.

1.2.8 Polymerase chain reaction

The **PCR** is one of the most transformative and foundational techniques in modern forensic science. Developed in the 1980s by Dr. Kary Mullis, PCR revolutionized the field by allowing forensic scientists to amplify minute amounts of DNA into quantities that are large enough to be analyzed. This groundbreaking technology has significantly expanded the capacity of forensic laboratories to work with limited, degraded, or old DNA samples—such as those collected from cold crime scenes—thus opening up new possibilities for solving previously unsolvable cases (Saiki et al., 1988).

1.2.9 How polymerase chain reaction works

At its core, PCR is a molecular biology technique that mimics the natural process of DNA replication but in a controlled laboratory environment. PCR consists of a series of repetitive heating and cooling cycles designed to amplify a specific segment of DNA. The process relies on the following key steps:

1. **Denaturation**: The double-stranded DNA is heated to around 94°C–96°C, causing the hydrogen bonds between the base pairs to break, and the DNA strands to separate into single strands. This step is critical in preparing the DNA for replication (Mullis & Faloona, 1987).
2. **Annealing**: The temperature is then lowered to around 50°C–65°C, allowing primers—short sequences of nucleotides designed to bind to the specific DNA sequence of interest—to attach to their complementary sequences on the single-stranded DNA. These primers mark the starting point for DNA replication (Saiki et al., 1988).
3. **Extension**: The reaction mixture is then heated to approximately 72°C, at which point **Taq polymerase**, a heat-resistant enzyme, begins synthesizing new strands of DNA by adding nucleotides to the primer-bound sequences. This replication process results in two new double-stranded DNA molecules, each containing one original strand and one newly synthesized strand (Mullis & Faloona, 1987).

This cycle of denaturation, annealing, and extension is repeated multiple times—typically 30–40 cycles—resulting in exponential amplification of the target DNA region. By the end of the process, the original DNA sample has been replicated billions of times, making it detectable and analyzable, even if the starting sample was only a few molecules of DNA (Butler, 2011).

1.2.10 Applications of polymerase chain reaction in forensic science

PCR has proven invaluable in forensic DNA analysis, particularly in cases where biological evidence is limited, degraded, or compromised. The ability to amplify tiny amounts of DNA has made it possible to analyze samples that would otherwise be too small to work with using traditional methods.

In addition, PCR has enabled forensic scientists to reinvestigate **cold cases**—unsolved cases that have gone dormant due to a lack of evidence. In many cold cases, biological evidence, such as blood, hair, or saliva, has been stored for years, but the samples may have deteriorated over time due to environmental conditions. PCR allows for the amplification of even partially degraded DNA, which can then be used to generate a genetic profile. As a result, PCR has played a crucial role in exonerating wrongfully convicted individuals and identifying perpetrators in cases that were once considered unsolvable (Alketbi, 2023).

1.2.11 Advantages of polymerase chain reaction in forensics

PCR offers numerous advantages that make it an indispensable tool in forensic DNA analysis. One of the most significant advantages is its ability to amplify DNA from a very small starting sample. As long as even a trace amount of DNA is present, PCR can generate enough material for forensic testing. This makes PCR particularly valuable in cases where only a small amount of biological evidence is available, such as a single hair, a drop of blood, or a fragment of skin (Mullis, 1990).

Furthermore, PCR is highly versatile and can be used to analyze different types of biological evidence, including blood, semen, hair, and tissue. It can also be applied to degraded or environmentally damaged samples, making it useful in a wide range of forensic contexts, from recent crime scenes to ancient or historical remains (Jobling & Gill, 2004).

Another key advantage is the speed with which PCR can produce results. Once the DNA sample is extracted, the entire amplification process typically takes only a few hours, allowing forensic scientists to generate DNA profiles quickly and efficiently. This rapid turnaround time is essential in criminal investigations, where timely identification of suspects or exoneration of wrongfully accused individuals can make a significant difference (Butler, 2011).

1.2.12 Limitations of polymerase chain reaction

While PCR is an incredibly powerful technique, it does have limitations that forensic scientists must be mindful of. One of the main challenges with PCR is the potential for **contamination**. Because PCR is so sensitive, even trace amounts of contaminating DNA from the environment, laboratory personnel, or other sources can be amplified along with the target DNA, leading to inaccurate results (Butler, 2011). Strict laboratory protocols are essential to minimize the risk of contamination during PCR analysis.

Another limitation of PCR is that it works best with relatively intact DNA. Although PCR can amplify degraded DNA, there are limits to how much damage the DNA can sustain before amplification becomes unreliable. In cases where the DNA is highly fragmented or chemically altered, additional techniques, such as **next-generation sequencing (NGS)**, may be required to obtain usable genetic information (Butler, 2011).

1.2.13 Next-generation sequencing

In forensic DNA analysis, technological advancements continue to drive improvements in both accuracy and the scope of investigations. **NGS** is one of the most transformative technologies in recent years, allowing scientists to analyze multiple regions of the genome simultaneously. Unlike traditional methods such as STR profiling, which focuses on specific loci within noncoding regions of DNA, NGS provides a broader and deeper examination of both coding and noncoding regions. This gives forensic scientists far more comprehensive genetic data, enhancing the accuracy and resolution of DNA profiles (Bell & Butler, 2022).

1.2.14 How next-generation sequencing works

NGS is a high-throughput technique that enables the sequencing of millions of DNA fragments in parallel. It begins with **DNA extraction** from a biological sample, after which the extracted DNA is broken into smaller fragments. Each fragment is then amplified and sequenced. What sets NGS apart from traditional DNA sequencing techniques is its ability to process multiple DNA regions at once, rather than sequencing one region at a time. This parallel sequencing approach dramatically increases both the speed and volume of data generated (Dijk et al., 2018).

The data obtained from NGS can provide highly detailed insights into an individual's genetic makeup, including single-nucleotide polymorphisms (SNPs) and other types of genetic variations. This can be particularly important in complex forensic cases, where small genetic differences between individuals may hold the key to solving a crime (Schmedes et al., 2019).

1.2.15 Advantages of next-generation sequencing in forensics

NGS has several advantages over traditional methods like STR profiling, particularly in complex forensic investigations:

1. **Simultaneous analysis of multiple regions**: While STR profiling is limited to examining specific loci, NGS allows for the analysis of a much larger portion of the genome, including both coding and noncoding regions. This provides a richer genetic profile, making it possible to identify individuals more precisely (Bell & Butler, 2022).
2. **Detailed genetic variations**: NGS can detect **SNPs**, small genetic variations that occur at a single base pair in the DNA sequence. SNPs are more common in the human genome than STRs, making them an additional valuable tool for distinguishing between individuals, even those with closely related genetic profiles (Schmedes et al., 2019).
3. **Analysis of complex DNA mixtures**: One of the most significant challenges in forensic DNA analysis is the interpretation of **DNA mixtures**—samples that contain genetic material from more than one individual, such as in cases involving multiple perpetrators or victims. Traditional methods like STR profiling often struggle with complex mixtures, but NGS can help forensic scientists deconvolute these samples by providing deeper and more detailed genetic data, making it easier to identify individual contributors (Schmedes et al., 2019).
4. **Degraded and low-quality samples**: NGS is also more robust than traditional techniques when it comes to analyzing **degraded DNA**. Forensic cases often involve samples that have been exposed to environmental damage, resulting in fragmented or chemically modified DNA. NGS is better equipped to handle these types of samples, as it can sequence even small and damaged fragments, extracting more genetic information than traditional STR profiling (Bell & Butler, 2022).

1.2.16 Applications of next-generation sequencing in forensic science

The applications of NGS in forensic science are vast and continue to grow as the technology becomes more accessible. One of the primary uses of NGS in forensics is in **missing persons** investigations. By analyzing both nuclear and mtDNA with NGS, forensic scientists can generate a more complete genetic profile, which can be compared to relatives to confirm identity. This has been especially useful in cold cases or mass disaster scenarios, where DNA samples are often degraded or incomplete (Schmedes et al., 2019).

NGS is also being increasingly used in **ancestry testing** and **phenotypic prediction**. By analyzing a person's genetic markers, forensic scientists can predict traits such as eye color, hair color, and skin tone, which can provide valuable leads in cases where no suspects have been identified. This emerging application, known as **forensic DNA phenotyping**, could greatly enhance investigative efforts by offering physical descriptions of individuals based solely on their genetic data (Dijk et al., 2018).

Furthermore, NGS has opened new avenues in **population genetics**. By analyzing genetic variations across different populations, NGS can help forensic scientists understand the distribution of genetic markers and predict the likely ancestry of an unknown individual. This is particularly useful in cases where traditional methods of identification have failed, and a broader understanding of genetic background is needed.

1.2.17 Challenges and limitations of next-generation sequencing

Despite its many advantages, NGS is not without challenges. One of the primary obstacles is the **cost** associated with implementing and maintaining NGS technologies in forensic laboratories. Although the price of sequencing has decreased dramatically over the past decade, NGS still requires significant financial investment in terms of both equipment and expertise (Bell & Butler, 2022).

Another challenge is the **interpretation of data**. The sheer volume of genetic data generated by NGS can be overwhelming, and forensic scientists must be skilled in bioinformatics to analyze and interpret the results accurately. Unlike traditional methods like STR profiling, which have been thoroughly validated and standardized for use in court, NGS is still relatively new to the forensic field, and its legal admissibility is evolving (Dijk et al., 2018).

Moreover, as NGS enables the sequencing of both coding and noncoding regions of the genome, there is an increased risk of accessing sensitive genetic information unrelated to the case at hand, such as predispositions to certain diseases or other personal traits. This raises **ethical concerns** about privacy and consent, as the use of such information in forensic investigations must be carefully regulated to avoid potential misuse (Schmedes et al., 2019).

1.2.18 Probabilistic genotyping

In the evolving field of forensic DNA analysis, **probabilistic genotyping** has emerged as a powerful tool for interpreting complex DNA evidence. Traditional DNA analysis methods, such as STR profiling, often face challenges when dealing with **DNA mixtures**—samples that contain genetic material from multiple individuals. This is particularly common in cases involving multiple perpetrators or crime scenes where different sources of DNA are mixed. Probabilistic genotyping addresses these challenges by applying advanced statistical models to assess the likelihood that a particular individual contributed to a mixture, even when the DNA evidence is incomplete, partial, or degraded (Perlin et al., 2015).

1.2.19 How probabilistic genotyping works

Probabilistic genotyping moves beyond the binary, deterministic approach of traditional DNA analysis, which often struggles to provide clear conclusions when dealing with complex mixtures. In a traditional approach, forensic scientists rely on visual analysis and straightforward comparisons to determine whether a suspect's DNA matches a sample from a crime scene. However, this method becomes problematic when more than two individuals are involved in a DNA mixture, leading to ambiguous results.

Probabilistic genotyping, in contrast, utilizes **mathematical algorithms** and **likelihood ratios** (LRs) to calculate the probability that a given individual's DNA contributed to the sample in question. By considering various scenarios—such as the number of contributors, potential allelic dropout, and the possibility of DNA degradation—the system generates a range of probabilities rather than a simple yes-or-no answer (Orozco, 2024). This makes it possible to interpret DNA evidence even when only partial or degraded DNA is available, significantly enhancing the scope and reliability of forensic analysis.

Probabilistic genotyping involves calculating LRs to compare two competing hypotheses:

1. **Hypothesis 1 (H1):** The DNA sample contains the suspect's DNA.
2. **Hypothesis 2 (H2):** The DNA sample does not contain the suspect's DNA (i.e., it belongs to someone else).

The LR is calculated as:

$$LR = \frac{P(\text{DNA Evidence}/H1)}{P(\text{DNA Evidence}/H2)}$$

This provides a statistical measure of the strength of the DNA evidence in supporting either hypothesis.

1.2.20 Key advantages of probabilistic genotyping

1. **Handling complex DNA mixtures**: One of the major advantages of probabilistic genotyping is its ability to **deconvolute DNA mixtures**. Traditional methods often reach their limits when more than two contributors are involved, as distinguishing between overlapping DNA profiles becomes difficult. Probabilistic genotyping, however, is designed to work with such mixtures by assigning **probability distributions** to the likelihood that a given DNA profile matches any one of the contributors (Perlin et al., 2015). This capability is particularly useful in cases of gang-related crimes or sexual assaults involving multiple individuals.
2. **Addressing degraded and low-quality samples**: Another challenge in forensic DNA analysis is working with **degraded or low-quality DNA samples**. DNA can degrade over time due to environmental exposure or improper storage, which leads to incomplete or damaged genetic information. In such cases, traditional methods may fail to yield meaningful results. Probabilistic genotyping compensates for this by considering **allelic dropout**—the loss of certain alleles in a sample due to degradation. The system can incorporate this information into its statistical model, producing more reliable results even when the DNA evidence is less than ideal (Buckleton et al., 2018).
3. **Objective interpretation**: Unlike traditional methods that require a certain level of subjective interpretation by forensic analysts, probabilistic genotyping is largely **objective**. The algorithms used in probabilistic genotyping systems, such as **TrueAllele** and **STRmix**, remove much of the human bias that can influence forensic interpretations. This objectivity ensures that the results are consistent and can be validated by other forensic experts, making it more robust for legal proceedings (Perlin et al., 2015).

1.2.21 The role of likelihood ratios in probabilistic genotyping

At the core of probabilistic genotyping is the use of **LRs**, a statistical measure that compares two competing hypotheses: one that assumes the suspect's DNA is part of the mixture and one that assumes it is not. The LR helps forensic scientists determine how strongly the DNA evidence supports the presence of the suspect's DNA in the mixture compared to an alternative hypothesis, such as the DNA belonging to another individual (Taylor et al., 2013).

For example, if a DNA sample from a crime scene contains a mixture of three individuals' DNA, the LR will calculate the probability that the suspect's DNA is included in that mixture versus the probability that it is not. If the LR is high, this indicates that the suspect's DNA is more likely to be part of the mixture. This probabilistic approach provides a **quantitative measure** of the strength of the DNA evidence, making it particularly useful in court, where subjective opinions can be challenged (Orozco, 2024).

1.2.22 Challenges and considerations in probabilistic genotyping

Despite its advantages, probabilistic genotyping also comes with certain challenges. One of the main hurdles is the **computational complexity** of the models used. While these systems are highly accurate, they require extensive computational resources and sophisticated software, which can limit their availability to smaller forensic labs (Perlin et al., 2015).

Another concern is the **admissibility of probabilistic genotyping results in court**. Since the method is relatively new, it is still being evaluated in terms of legal standards and judicial acceptance. Defense attorneys may challenge the use of probabilistic genotyping because jurors may not fully understand the statistical models and could be influenced by the perceived objectivity of the technology. However, as probabilistic genotyping becomes more widespread and its methodologies more thoroughly validated, its use in legal proceedings is expected to grow (Buckleton et al., 2018).

1.3 Case studies and real-world applications

Probabilistic genotyping has already played a pivotal role in numerous high-profile forensic cases. For example, the **TrueAllele** system was used to interpret complex DNA evidence in the **2008 Allegheny County homicide case**, where multiple DNA profiles were found on a firearm. The system helped distinguish the primary suspect's DNA from that of other contributors, providing crucial evidence for the prosecution (Perlin et al., 2015).

Similarly, **STRmix**, another probabilistic genotyping system, was used in several gang-related crime cases, where the DNA mixtures of multiple suspects and victims were analyzed. The ability to deconvolute these mixtures allowed forensic scientists to identify individual contributors with greater certainty, even when traditional methods had failed to provide conclusive results (Taylor et al., 2013).

With these ongoing advancements, forensic DNA analysis continues to become more sophisticated, enabling investigators to solve cases with greater precision. As technologies such as NGS and probabilistic genotyping become more integrated into forensic labs, the scope of DNA evidence will expand, providing law enforcement with more robust tools to solve both current and cold cases.

Table 1.1 shows the different forensic DNA techniques (STR, PCR, mtDNA, and NGS). The table includes the advantages, limitations, and typical applications for each technique.

1.3.1 Application of forensic DNA in criminal justice

Forensic DNA analysis has played a crucial role in the identification of suspects and victims in criminal investigations. In cases of violent crimes such as homicide, sexual assault, and terrorism, DNA profiling serves as a key tool for identifying perpetrators. The utility of DNA analysis is further underscored in its ability to exonerate the wrongfully convicted, contributing to the growing body of work on postconviction DNA testing (Rudin & Inman, 2001).

One of the earliest and most well-known cases where DNA evidence was pivotal in criminal justice was the conviction of Colin Pitchfork in the UK. Pitchfork was the first person convicted of murder based on DNA evidence, setting a precedent for the use of DNA in criminal cases (DNA technology in forensic science, 1992).

1.3.2 The role of DNA databases in modern criminal justice

Since the pioneering days of DNA analysis in the 1980s, technology has evolved, and today DNA databases like CODIS have become invaluable tools in solving both current and cold cases. CODIS was established in the United States in the 1990s to allow law enforcement agencies to store and compare DNA profiles. This system is designed to help investigators

TABLE 1.1 Comparison between different forensic DNA techniques.

Technique	Advantages	Limitations	Applications
Short tandem repeat profiling	High discrimination power; useful in criminal investigations, paternity testing, and disaster victim identification; widely accepted in court	Less effective with DNA mixtures or degraded samples; requires intact DNA for best results	Criminal cases, paternity testing, disaster victim identification
Polymerase chain reaction	Amplifies tiny or degraded DNA samples; fast turnaround time; critical in cold cases with limited evidence	Sensitive to contamination; works best with relatively intact DNA; limited effectiveness with highly degraded samples	Cold cases, degraded samples, rapid DNA profiling
Mitochondrial DNA	Useful when nuclear DNA is degraded or absent (e.g., hair shafts, bones); can be used for ancient or severely degraded samples	Less discriminatory than nuclear DNA; cannot differentiate between individuals sharing the same maternal lineage	Historical remains, mass disaster victim identification, maternal lineage tracing
Next-generation sequencing	Can analyze both coding and noncoding regions simultaneously; provides detailed genetic variations; useful in complex cases and for ancestry testing	Expensive; computationally complex; limited availability in smaller forensic labs; ethical concerns around privacy	Ancestry testing, phenotypic prediction, complex DNA mixture analysis

link DNA evidence from unsolved crimes to known offenders or previously unlinked cases, thereby improving the efficiency and effectiveness of criminal investigations (Bell & Butler, 2022).

CODIS operates at both the state and national levels and contains millions of DNA profiles from individuals convicted of crimes, arrestees, and evidence collected at crime scenes. Through CODIS, investigators can quickly and accurately compare DNA samples collected from a crime scene with profiles stored in the database, which greatly increases the likelihood of solving cases that might otherwise have gone cold (Butler, 2011). This system has been instrumental in closing long-dormant investigations, providing closure for victims' families, and bringing criminals to justice even years after their offenses were committed.

1.3.3 Exoneration and postconviction DNA testing

One of the most significant contributions of forensic DNA analysis in criminal justice is its role in **exonerating individuals who were wrongfully convicted**. DNA evidence has proven to be a powerful tool in reviewing old cases where the technology was not previously available or where other forms of evidence led to wrongful convictions. Since the advent of DNA testing, postconviction reviews have resulted in hundreds of exonerations worldwide, many of them facilitated by organizations like the **Innocence Project** (Garrett, 2011).

One notable case that highlights the impact of DNA in overturning wrongful convictions is the **exoneration of Kirk Bloodsworth** in 1993. Bloodsworth was the first American on death row to be exonerated by DNA evidence. He had been convicted of the rape and murder of a 9-year-old girl based on eyewitness testimony, but later DNA testing of the evidence revealed that he was not the perpetrator. Bloodsworth's case became a landmark example of how forensic DNA analysis can correct miscarriages of justice (Rudin & Inman, 2001).

1.3.4 Cold case investigations

Forensic DNA analysis has also become an essential tool in solving **cold cases**, where traditional investigative methods failed or where there was insufficient evidence to charge a suspect at the time. As DNA technology has advanced, forensic laboratories have been able to reanalyze old evidence, often leading to the resolution of cases that had remained unsolved for decades.

For example, in the case of the **Golden State Killer**, who terrorized California in the 1970s and 1980s, DNA evidence collected from multiple crime scenes sat dormant for years due to the limitations of technology at the time. However, with the advent of **genetic genealogy** and advanced DNA sequencing techniques, investigators were able to link DNA from the crime scenes to the killer's family tree, eventually identifying **Joseph James DeAngelo** as the perpetrator in 2018. This case not only illustrated the power of forensic DNA to solve cold cases but also showcased how new developments in DNA analysis continue to push the boundaries of criminal investigation (Bell & Butler, 2022).

1.4 Ethical and legal considerations in forensic DNA analysis

As forensic DNA analysis has become an indispensable tool in criminal investigations, it has simultaneously raised important ethical and legal concerns. While the technique offers immense potential for solving crimes and exonerating the wrongfully convicted, it also presents challenges regarding privacy, consent, and the possibility of misusing DNA data stored in forensic databases. Striking the right balance between enhancing public safety and protecting individual rights has become an ongoing debate in the field (Oosthuizen & Howes, 2022).

1.4.1 Privacy concerns in DNA databases

One of the foremost ethical concerns associated with forensic DNA analysis is the collection and storage of DNA profiles in national and international databases. Databases such as the CODIS in the United States contain millions of profiles of convicted offenders, arrestees, and sometimes even individuals who have been acquitted. While these databases are instrumental in solving both current and cold cases, they raise important questions about **privacy** and **consent**.

A particularly contentious issue is **familial searching**, where investigators search DNA databases not just for an exact match but also for profiles of family members that may share portions of the suspect's DNA. This method has been highly effective in tracking down perpetrators in cases where the suspect's DNA is not present in the database but the DNA of a close relative is. However, familial searching raises significant privacy concerns for individuals who have not personally consented to having their genetic information used in such investigations. Even though these individuals are not directly involved in the crime, their DNA can indirectly implicate family members, which can lead to invasions of privacy (Goodwin et al., 2010). Ethical questions are raised about whether law enforcement should be allowed to access this type of information without the consent of the individual whose DNA is being scrutinized, particularly in jurisdictions where laws governing DNA use vary.

Furthermore, as DNA databases grow and include more individuals—some of whom may not have been convicted of any crime—the risk of **misuse** of this sensitive genetic information increases. Data breaches, unauthorized access, or the use of DNA information for nonforensic purposes (such as insurance companies or employers) are among the concerns that privacy advocates have raised. The challenge for law enforcement and policymakers is to ensure that the benefits of DNA analysis in solving crimes do not come at the expense of individual privacy rights (Oosthuizen & Howes, 2022).

1.4.2 Familial searching and consent

The use of familial searching in forensic DNA analysis is not without controversy. While it has been instrumental in solving cases like the **Golden State Killer**—where law enforcement tracked down the suspect through a relative's DNA found in a genealogy database—this method raises ethical concerns regarding **informed consent**. In many cases, individuals may have submitted their DNA to genealogy services for personal reasons, such as tracing their ancestry, without fully understanding that their DNA could potentially be accessed by law enforcement for investigative purposes (Greely, 2016). This creates a gray area between voluntary participation in genetic testing and the involuntary use of genetic data in criminal investigations.

Moreover, familial searching has implications for **genetic privacy** within families. Individuals who submit their DNA may unknowingly expose their relatives to police scrutiny, even if those relatives have never consented to the collection or use of their genetic information. This raises questions about the fairness of using familial DNA as a tool for criminal investigations and whether such practices violate individuals' privacy rights (Goodwin et al., 2010). These ethical considerations call for a thoughtful review of policies governing DNA databases and familial searching to ensure that the balance between solving crimes and respecting personal privacy is maintained.

1.4.3 Risk of DNA contamination and misinterpretation

In addition to privacy concerns, forensic DNA analysis carries the risk of **contamination** and **misinterpretation** of DNA evidence, which can have serious consequences for the integrity of investigations. Several high-profile cases have highlighted how the mishandling of DNA evidence can lead to wrongful convictions, underscoring the need for rigorous quality control and oversight in forensic laboratories.

For example, the case of **Adam Scott** in the UK serves as a cautionary tale. In 2011, Scott was wrongfully arrested and charged with a rape that he did not commit after a DNA sample from the crime scene was contaminated with his DNA. Scott had never been to the location where the crime occurred, but due to procedural errors in the forensic lab, his DNA was accidentally transferred onto the evidence. The contamination led to his wrongful arrest before further investigation revealed the mistake. This case illustrates the potential for contamination to undermine the credibility of forensic investigations (Rudin & Inman, 2001).

Misinterpretation of DNA evidence is another significant issue. DNA is often perceived as **infallible** in courtrooms, with juries and judges sometimes viewing DNA matches as definitive proof of guilt. However, DNA evidence can be complex, especially in cases involving **DNA mixtures** or **low-template DNA**, where only small amounts of genetic material are available for analysis. In these cases, the interpretation of the DNA evidence can be subjective, leading to potential errors (Bell & Butler, 2022).

To mitigate these risks, forensic laboratories must adhere to **strict protocols** for handling DNA evidence, and courts must remain vigilant in understanding the limitations of DNA analysis. **Probabilistic genotyping**—an advanced method that uses statistical models to interpret complex DNA evidence—has helped improve the accuracy of DNA analysis in such cases, but it is not immune to challenges (Perlin et al., 2015). Misinterpretation of probabilistic results, combined with overconfidence in DNA evidence, can lead to wrongful convictions if forensic scientists and legal professionals do not approach DNA analysis with the appropriate caution.

1.4.4 Legal and ethical oversight

Given the complexities and challenges associated with forensic DNA analysis, **legal oversight** is essential to ensure that the use of DNA in criminal investigations is both **ethical** and **just**. Laws governing DNA databases, familial searching, and DNA evidence in court vary from country to country, and there is an ongoing need for clear, consistent regulations to protect individuals' rights while allowing law enforcement to utilize DNA as an investigative tool (Greely, 2016).

Many privacy advocates call for stronger **consent requirements** before DNA can be collected or accessed by law enforcement, particularly in cases where familial DNA is being used. Additionally, there is a growing movement to establish **independent oversight bodies** that can audit DNA databases and forensic laboratories to ensure that they are operating within ethical and legal boundaries (Oosthuizen & Howes, 2022).

1.4.5 Recent advancements and future directions

In recent years, advancements in forensic DNA analysis have opened new avenues for research and application. NGS is one such technology that allows for the comprehensive analysis of DNA, providing more detailed information than traditional methods like STR profiling (Butler, 2011). NGS is expected to revolutionize forensic DNA analysis by enabling the examination of a wider array of genetic markers, which could lead to more accurate identifications and stronger links between individuals and crime scenes (Shewale & Liu, 2013). Additionally, probabilistic genotyping is another emerging technology that addresses one of the major challenges in forensic DNA analysis: DNA mixtures. These are samples that contain genetic material from multiple individuals, making it difficult to interpret results (Bright & Coble, 2019).

1.4.6 Automation and increased sensitivity in forensic DNA analysis

Looking ahead, the field of forensic DNA analysis is poised to benefit from increased **automation** and more sensitive **detection methods**, which will streamline (Bell & Butler, 2022) the investigative process and improve the quality of the evidence. Currently, DNA profiling can be a time-consuming process that requires manual handling of samples and interpretation of results. As forensic laboratories adopt more automated systems, the speed at which DNA profiles can be generated is expected to increase, allowing for faster turnaround times in criminal investigations (Bell & Butler, 2022).

Automation will also reduce the risk of human error, which has been a source of concern in some forensic cases. By automating key steps in the DNA profiling process, forensic scientists can ensure that the analysis is more consistent and reliable. Automated systems can handle larger volumes of samples with greater precision, improving the overall efficiency of forensic laboratories.

In addition to automation, the development of **more sensitive detection methods** is expected to enhance the ability to analyze low-template or degraded DNA samples. These advancements will allow forensic scientists to extract usable genetic material from even the smallest traces of evidence, which is particularly important in cold cases or investigations where biological material is limited. This increased sensitivity will broaden the scope of DNA evidence, enabling investigators to draw conclusions from samples that would have been too compromised to analyze using older techniques (Schmedes et al., 2019).

1.4.7 Integration with other forensic disciplines

Another exciting development in the future of forensic DNA analysis is the potential for greater **integration with other forensic disciplines**, such as **forensic anthropology**, **forensic toxicology**, and **digital forensics**. By combining DNA evidence with other types of forensic data, investigators can build a more comprehensive picture of a crime. For example, in cases involving unidentified remains, DNA analysis can be used alongside forensic anthropology to determine not only the identity of the deceased but also their ancestry, physical characteristics, and the likely cause of death (Buckleton et al., 2018).

As technology advances, forensic DNA analysis will become increasingly interconnected with other fields of forensic science, creating a more holistic approach to criminal investigations. This integration will enable forensic scientists to tackle complex cases more effectively, using multiple lines of evidence to arrive at more accurate conclusions.

1.4.8 Future directions in forensic DNA analysis

The future of forensic DNA analysis is full of promise. As technologies like **NGS**, **probabilistic genotyping**, and **automation** become more widespread, they will continue to push the boundaries of what forensic science can achieve. These advancements are expected to improve the **speed, accuracy, and scope** of forensic investigations, helping to solve more cases, both current and cold. Additionally, the integration of **phenotypic prediction** and **ancestry testing** into forensic DNA analysis could provide investigators with new tools for identifying unknown suspects and closing unsolved cases (Perlin et al., 2015).

However, as the field evolves, it will also be important to address the **ethical and legal challenges** that come with these advancements, such as concerns about privacy, consent, and the potential misuse of genetic information. Ensuring that forensic DNA analysis is conducted responsibly and transparently will be key to maintaining public trust in these technologies.

1.5 Conclusion

The principles of forensic DNA analysis form the cornerstone of modern forensic science, enabling law enforcement agencies to solve complex cases and deliver justice. With continued technological advancements and a growing understanding of DNA's role in forensic science, the future of this field looks promising. However, alongside these scientific developments, there must be a careful consideration of the ethical and legal implications to ensure that DNA analysis is used responsibly and justly.

AI disclosure

During the preparation of this work the authors used ScholarGPT in order to assist in language development and editing. After using this tool/service the authors reviewed and edited the content as needed and take full responsibility for the content of the publication.

References

Alketbi, S. K. (2023). The role of DNA in forensic science: A comprehensive review. *SSRN, United Arab Emirates SSRN*. https://doi.org/10.2139/ssrn.4550489, https://www.ssrn.com/index.cfm/en/.

Amorim, A., Fernandes, T., & Taveira, N. (2019). Mitochondrial DNA in human identification: A review. *PeerJ, 7*, 29. https://doi.org/10.7717/peerj.7314.

Bell, & Butler (2022). *Understanding forensic DNA*. Cambridge University Press.

Bright, J. A., & Coble, M. (2019). *Forensic DNA profiling: A practical guide to assigning likelihood ratios*. https://doi.org/10.4324/9780429001017/forensic-dna-profiling-jo-anne-bright-michael-coble.

Buckleton, J., Bright, J. A., & Taylor, D. (2018). *Forensic DNA evidence interpretation: The probabilistic approach*. CRC Press.

Budowle, B., Allard, M. W., Wilson, M. R., & Chakraborty, R. (2003). Forensics and mitochondrial DNA: Applications, debates, and foundations. *Annual Review of Genomics and Human Genetics, 4*, 119–141. https://doi.org/10.1146/annurev.genom.4.070802.110352.

Butler, J. (2009). *Fundamentals of forensic DNA typing*. Academic Press.

Butler, J. (2011). *Advanced topics in forensic DNA typing: Methodology*. Academic Press.

Butler, J. M., & Levin, B. C. (1998). Forensic applications of mitochondrial DNA. *Trends in Biotechnology, 16*(4), 158–162. https://doi.org/10.1016/S0167-7799(98)01173-1.

Coble, M. D., Just, R. S., O'Callaghan, J. E., Letmanyi, I. H., Peterson, C. T., Irwin, J. A., & Parsons, T. J. (2004). Single nucleotide polymorphisms over the entire mtDNA genome that increase the power of forensic testing in Caucasians. *International Journal of Legal Medicine, 118*(3), 137–146. https://doi.org/10.1007/s00414-004-0427-6.

Dijk, Y. J., Naquin, D., & Thermes, C. (2018). The third revolution in sequencing technology. *Trends in Genetics, 30*(9), 418–426.

DNA technology in forensic science. National Academy Press (1992).

Garrett, B. L. (2011). *Convicting the innocent: Where criminal prosecutions go wrong*. Harvard University Press.

Goodwin, W., Linacre, A., & Hadi, S. (2010). *An introduction to forensic genetics*. Wiley.

Greely, H. T. (2016). *The end of sex and the future of human reproduction*. Harvard University Press.

Hares, D. R. (2012). Expanding the CODIS core loci in the United States. *Forensic Science International: Genetics, 6*(1), e52. https://doi.org/10.1016/j.fsigen.2011.04.012.

Ivanov, P. L., Wadhams, M. J., Roby, R. K., Holland, M. M., Weedn, V. W., & Parsons, T. J. (1996). Mitochondrial DNA sequence heteroplasmy in the Grand Duke of Russia Georgij Romanov establishes the authenticity of the remains of Tsar Nicholas II. *Nature Genetics, 12*(4), 417–420. https://doi.org/10.1038/ng0496-417.

Jobling, M. A., & Gill, P. (2004). Encoded evidence: DNA in forensic analysis. *Nature Reviews. Genetics, 5*(10), 739–751. https://doi.org/10.1038/nrg1455, http://www.nature.com/reviews/genetics.

Mullis, K. B. (1990). The unusual origin of the polymerase chain reaction. *Scientific American, 262*(4), 56–65. https://doi.org/10.1038/scientificamerican0490-56.

Mullis, K. B., & Faloona, F. A. (1987). Specific synthesis of DNA in vitro via a polymerase-catalyzed chain reaction. *Methods in Enzymology, 155*(C), 335–350. https://doi.org/10.1016/0076-6879(87)55023-6.

Oosthuizen, T., & Howes, L. M. (2022). The development of forensic DNA analysis: New debates on the issue of fundamental human rights. *Forensic Science International: Genetics, 56*, 18780326. https://doi.org/10.1016/j.fsigen.2021.102606, http://www.elsevier.com.

Orozco, D. (2024). UC Davis TrueAllele® and STRmix: A comparison of two probabilistic genotyping software programs in forensic DNA profile analysis. Doctoral dissertation, UC Davis 121 Doctoral dissertation.

Perlin, M. W., Hornyak, J. M., Sugimoto, G., & Miller, K. W. P. (2015). TrueAllele® genotype identification on DNA mixtures containing up to five unknown contributors. *Journal of Forensic Sciences, 60*(4), 857–868. https://doi.org/10.1111/1556-4029.12788, http://onlinelibrary.wiley.com/journal/10.1111/(ISSN)1556-4029.

Rudin, N., & Inman, K. (2001). *An introduction to forensic DNA analysis*. CRC Press.

Saiki, R. K., Gelfand, D. H., Stoffel, S., Scharf, S. J., Higuchi, R., Horn, G. T., Mullis, K. B., & Erlich, H. A. (1988). Primer-directed enzymatic amplification of DNA with a thermostable DNA polymerase. *Science, 239*(4839), 487–491. https://doi.org/10.1126/science.2448875.

Schmedes, S. E., Sajantila, A., & Budowle, B. (2019). Expansion of forensic DNA typing: Application of next generation sequencing technology. *Forensic Science International: Genetics, 43*, 100–105.

Shewale, J. G., & Liu, R. H. (2013). *Forensic DNA analysis: Current practices and emerging technologies*. United States: Taylor and Francis, 1–414. http://doi.org/10.1201/b15361, https://www.routledge.com/products/9781466571266.

Taylor, D., Bright, J. A., & Buckleton, J. (2013). The interpretation of single source and mixed DNA profiles. *Forensic Science International: Genetics, 7*(5), 516–528. https://doi.org/10.1016/j.fsigen.2013.05.011.

Wambaugh, J. (2016). The blooding: The dramatic true story of the first murder case solved by genetic. *Fingerprinting*.

Zagorski, N. (2006). Profile of Alec J. Jeffreys. *Proceedings of the National Academy of Sciences, 103*(24), 8918–8920. https://doi.org/10.1073/pnas.0603953103.

Chapter 2

Autosomal short tandem repeat analysis and its usefulness in forensic DNA analysis

Stanley Udogadi Nwawuba[1], Guilherme Schmitt Rieder[1], Khadija Abdullahi Mohammed[2], Ngozichukwu Francis Ngozichukwu[3], Salem Terwase Daniel[4], Francis Okechukwu Chibueke[5], Aibuedefe Osagie Eremwanarue[6], and Edeaghe Ehikhamenor[7]

[1]*Department of Biochemistry, Institute of Basic Health Sciences, Federal University of Rio Grande do Sul, Porto Alegre, RS, Brazil,* [2]*Department of Biochemistry and Molecular Biology, Federal University of Santa Maria, Santa Maria, RS, Brazil,* [3]*School of Integrated Biological and Chemical Sciences, University of Texas Rio Grande Valley, Edinburg, TX, United States,* [4]*Department of Pharmaceutical and Food Manufacturing, Red River College Polytechnic, Winnipeg, MB, Canada,* [5]*Department of Biotechnology, School of Health Science and Technology, UPES, India,* [6]*Faculty of Medicine, Health and Social Care, Canterbury Christ Church University, Canterbury, United Kingdom,* [7]*Center for Forensic Programmes and DNA Studies, University of Benin, Benin, Nigeria*

2.1 Introduction

Autosomal short tandem repeat (STR) analysis has been demonstrated to have revolutionized forensic deoxyribonucleic acid (DNA) analysis, and it is presently considered the gold standard within the context of forensic DNA analysis, covering key areas like human identification in criminal investigations, paternity testing, and disaster victim identification (Keerti & Ninave, 2022; Nwawuba et al., 2020). STRs are short repeating motifs of 2–6 base pairs found scattered throughout the human genome, exhibiting a high degree of polymorphism due to variations in the number of repeat units at specific loci (Tanudisastro et al., 2024). These variations provide a powerful means to differentiate individuals, as no two people, except identical twins, will have the same combination of STR alleles at multiple loci (Nwawuba et al., 2020; Tanudisastro et al., 2024). The unique inheritance patterns of STRs, which follow Mendelian genetics, further enhance their utility in resolving familial relationships (Tanudisastro et al., 2024).

The utilization of STRs in forensic DNA analysis has been demonstrated to be fundamental to modern criminal justice systems, aiding authorities to link suspects to crime scenes, exonerate the innocent, and identify missing persons (Budowle & Sajantila, 2024; Nwawuba et al., 2020). A large body of studies has shown that the robustness and reliability of STR analysis anchored on its ability to generate reproducible and statistically significant results, even from degraded and minute samples (Budowle and Sajantila, 2024; Nwawuba et al., 2020). By using capillary electrophoresis and the polymerase chain reaction (PCR) approaches to STR loci amplification, forensic practitioners can generate unique DNA profiles (Budowle & Sajantila, 2024; Tanudisastro et al., 2024). This has enabled these profiles to also be accessible with the Combined DNA Index System (CODIS) databases to promote faster investigations as well as serve as informative evidence during trials (Nwawuba Stanley & Akpata Chinyere, 2020). The scientific basis of the autosomal STR analysis is fundamentally provided by molecular biology and the science of population genetics (Keerti and Ninave, 2022; Nwawuba et al., 2020). STR loci are selected for forensic use on the basis of considerable allelic variation and amenability to multiplex PCR systems, which can analyze several loci at the same time (Hölzl-Müller et al., 2021; Liu et al., 2019).

This chapter explores the principles and applications of autosomal STR analysis in forensic DNA investigations. It includes an overview of DNA, forensic evidence, STR selection criteria, and the relevance of STRs in forensic contexts. Case studies are presented to demonstrate the practical utility of autosomal STRs in solving criminal cases, identifying victims of disasters, and resolving paternity disputes. Beyond criminal investigations, other applications of autosomal STR analysis are also discussed. In cases of mass causality, STR analysis serves the purpose of aiding in the identification of victims in instances where visual identification and other related techniques are not feasible (Marrone et al., 2021; Soniya & Suresh, 2022). Also, the inheritance of autosomal markers is critical in the paternity and kinship disputes resolution,

Investigative and Predictive DNA Testing. DOI: https://doi.org/10.1016/B978-0-443-30128-5.00014-X

utilizing the inheritance of paternally inherited and maternally inherited genetically inherited alleles (Cho et al., 2024). Nevertheless, like all methods, forensic DNA analysis using STR as a marker has its limitations. The same processes are useful for interpreting more complex scenarios, including mixed DNA samples, allelic drop-out, and the presence of stutters require careful interpretation and adherence to stringent guidelines to ensure accuracy and minimize the potential for erroneous conclusions (Haddrill, 2021). Next-generation sequencing (NGS) technology in the past few years has also developed rapidly and has further augmented the effectiveness of STR analysis, which has brought improvement in the scope of the techniques by allowing for high coverage and targeted DNA sequencing (Satam et al., 2023). Integration of new technologies is likely to increase the range of forensic DNA analysis by improving sensitivity, reducing turnaround times, and enhancing high precision in analysis of degraded samples (Abdi et al., 2024; Satam et al., 2023). This chapter hence systematically discusses fundamentals of autosomal STR analysis, including the scientific principles of its application in crime scene investigations.

2.2 Overview of DNA analysis in forensics

The study of DNA has made a great impact on forensic evidence and the art of identifying people, as well as finding the criminals responsible for the crime. Its usage in criminal investigations began in the 1980s; however, its development has advanced so far that it is arguably the most dependable and accurate method of associating biological evidence with specific people (Nwawuba et al., 2020). DNA is defined as the genetic material that is found in almost every living organism. Individual polymorphisms make it possible to identify someone, making it a great help to forensic cases where the aim is to identify a person using biological samples collected from a scene of the crime (Nwawuba et al., 2020).

The first attempts to apply DNA to forensic science started in the middle of the 1980s, and they were initiated by the introduction of the idea of DNA profiling by Sir Alec Jeffreys. His approach, utilizing noncoding DNA sequences known as variable number tandem repeats (VNTR), changed the way criminal and paternity cases were being conducted (Butler, 2015). An important event occurred in the case of Colin Pitchfork in 1987, where DNA was used to successfully link a suspect to a crime, and an innocent person was exonerated; this actively implanted DNA methodology into the forensic practice (Butler, 2015). Since then, there has been a considerable shift in the methods employed to conduct forensic DNA analysis, moving from the VNTR methods to the STR methods to enhance the ability to produce good quality DNA profiles even from limited or degraded samples. PCR technology also greatly transformed DNA testing as it made it possible to make copies of target DNA regions that have enhanced sensitivity that could be used in cases with minimal or trace biological samples. To take it further, forensic analysis has undergone a sea change with the commencement of DNA analysis, as it has dramatically progressed in terms of identification and solving of criminal cases in comparison with the past. There has been a progression in the field starting with DNA fingerprinting and the development of NGS, which increased speed, sensitivity, and scope of DNA analysis. The growth of these technologies further indicates that forensic DNA analysis will be a central science in forensic investigations (Nwawuba et al., 2020; Satam et al., 2023) and will be relevant for contemporary investigative works (Uguen et al., 2024).

2.3 Forensic DNA profiling

DNA profiling, also referred to as DNA fingerprinting, is a sequence of tests and methodology employed to examine and identify the genetic information contained within an individual's cell (Nwawuba et al., 2020; Wet et al., 2011). In the context of Forensics, Forensic DNA, a fortuitous byproduct of some of the most highly regarded, cutting-edge research in the scientific community, is regarded as the comparison of genetic information or profile of a person to a profile of biological evidence found at the scene of crime, or profile stored in the DNA database for the sole purpose of administration of justice (Butler, 2022; Wet et al., 2011). One of the core bases for DNA profiling is the fact that about 99.9% sequence of DNA is found to be common among humans, and the other 0.1% is unique and varies from person to person (Nwawuba et al., 2020). The likelihood of two individuals having exactly the same DNA profile is about 1 in 594.1 trillion individuals (Nizami et al., 2018), and as a result, forensic DNA profiling has outstandingly contributed to exonerating the innocent and convicting the guilty (Pyrek, 2007).

Further, the greater part of the human genome is predominated by the "noncoding region" with well over 90% of which do not carry the genetic information necessary for the gene expression, and as a result, it is found to be very suitable for forensic DNA profiling because they are informative for individual variability and identification (Carracedo et al., 2018; Vitsios et al., 2021). A large body of scientific findings has demonstrated that genetic variation is rather limited in the coding region, with the exception of the human leukocyte antigen region (Carracedo et al., 2018; Nwawuba et al., 2020). This is as a result of the fact that expressed genes are subject to a selection pressure through the course of evolution in order to maintain their specialized function. Whereas, the noncoding region is not mainly influenced by selection pressure, and as such, there is a generational transmission of mutation leading to a great increase in genetic variability (Carracedo et al., 2018; Prahlow et al., 2017). Additionally, a significant part of the noncoding region of about 50% contains repetitive sequences broadly divided into two classes, including; tandem repeat sequences and interspersed elements. In the forensic settings, most of the DNA fingerprinting techniques target the tandem repetitive sequences (Carracedo et al., 2018). Forensic DNA profiling approach has been corroborated to be reliable, reproducible, accurate, and founded on validated methods for equally generating DNA profile and the interpretation of that profile (Nwawuba et al., 2020; Panneerchelvam & Norazmi, 2003), and as such, forensic DNA evidence has been granted a transcendent and epistemic status in many courts around the world (Lynch, 2013; O'Brien et al., 2015).

2.4 Forensic DNA evidence within the criminal justice system

Forensic DNA evidence assumes an important position with respect to abating criminal activities in the criminal justice system (Obafunwa et al., 2018). Evidences have emerged demonstrating that forensic science and the law epitomize interdisciplinary epistemologies, and as such quite several empirical studies, reported high-profile cases (El-Alfy & Abd El-Hafez, 2012; O'Brien et al., 2015; Robertson, 2013), and evidences from around the world have expressly revealed that progress in forensic science expedites the administration of justice (Nwawuba, Momoh, et al., 2020).

The adoption of a forensic science approach toward criminal investigation is routine in developed countries, and the use of forensic evidence to aid a gamut of activities in the criminal justice system has been recognized worldwide (Walsh, 2023). In England and Wales, to a certain degree, a laissez-faire approach to the admissibility of forensic evidence has been demonstrated, and it is reported that forensic evidence is rarely considered inadmissible except for a stark inappropriateness (O'Brien et al., 2015). In the US courts have positioned "DNA" evidence in an exceptional epistemic space (Lynch, 2013), and in Belgium, the only justification for refusal of forensic evidence is on the account that the criminal dossier already holds enough evidence to make an appropriate pronouncement; in consequence, any additional forensic expert evidence is considered superfluous (Vermylen, 2010).

Despite the transcendent, epistemic status granted to DNA evidence, pitfalls and challenges have also been recognized vis-à-vis the introduction of forensic evidence in the courtrooms (Lynch, 2013; O'Brien et al., 2015). Some of the central challenges range from an inappropriate chain of custody, the investigative methodology in the laboratory setting, the manner in which the evidence is presented at trials, its interpretation and evaluation within a case context, and its contribution in convictions or acquittals once admitted (O'Brien et al., 2015; Robertson, 2013).

2.5 DNA profiling markers

Various forensic DNA profiling methods and tools are capable of individual differentiation to varying degrees of effectiveness, speed, and sensitivity. Recently, notable progress has been made in DNA profiling, shortening the time from more than a week to a few hours (Carracedo et al., 2018; Nwawuba et al., 2020). In historical terms, such polymorphic protein biomarkers and enzymes were not only used rarely but also provided very a few identification information (McKiernan & Danielson, 2017). This prompted the shift to more sophisticated ones for DNA profiling. Ever since Sir Alec Jeffreys created an effective method of genetic profiling that provides accurate identification of an individual during criminal investigations, there has been an exponential growth in technologies for DNA profiling. This has reduced the time taken to a greater degree for the production of the specific and discriminating profiles (Børsting & Morling, 2015; Van-Oorschot et al., 2010). DNA profiling past and current histories utilizes various molecular techniques and markers among them: restriction fragment length polymorphisms, VNTR typing, and most recently STR typing, which still remains the gold standard of the profiles in DNA databases.

2.6 Short tandem repeat

In the human genome, various genes contain repeated sequences (Depienne & Mandel, 2021). STRs are present in the genomes of prokaryotes and eukaryotes, including humans (Chistiakov et al., 2006; Ellegren, 2004; Fan & Chu, 2007). They are relatively evenly distributed throughout the human genome, constituting approximately 7% of the total genome (Chintalaphani et al., 2021). Their distribution varies, and they can be found in areas close to telomeres, intergenic regions, introns, regulatory regions, and, to a lesser extent, in exons (Ellegren, 2004; Fan & Chu, 2007; Verbiest et al., 2023, 2024). Most STRs are located in noncoding regions of DNA, particularly within the intron regions (Fan & Chu, 2007).

Largely, genomic sequence variation generally occurs in two forms: mutations and polymorphisms (Depienne & Mandel, 2021; Zhang & Yu, 2007). Among polymorphisms, two major types can be distinguished: single-nucleotide polymorphisms (SNPs) and STRs (Gymrek, 2017). STRs, also referred to as microsatellites, are prone to alterations such as deletions or expansions, which can lead to microsatellite instability (Tanudisastro et al., 2024; Zhang & Yu, 2007). STRs, also known as microsatellites or simple sequence repeats, represent one of the most variable classes of DNA sequences within the genome (Verbiest et al., 2024). STRs consist of short DNA motifs, typically 1–6 base pairs in length, repeated in tandem to form arrays of up to several hundred nucleotides (Chintalaphani et al., 2021; Chistiakov et al., 2006; Fan & Chu, 2007).

STRs can be classified according to the motif (e.g., AC, ATC, ATCG), the number of repeats (e.g., in ATCATCATC, the motif ATC is repeated three times), and length, as the number of repeats can vary between individuals (Ellegren, 2004; Fan & Chu, 2007). Regarding the types of STRs, nucleotide repeats containing two, three, and four nucleotides (di-, tri-, and tetranucleotides, respectively) are the most frequent STRs found in living organisms, but there are also repeats known as penta- and hexanucleotides (Chistiakov et al., 2006; Ellegren, 2004) (Fig. 2.1). Among the main nucleotide repeats mentioned above, dinucleotide repeats are the most frequent, followed by mononucleotide and tetranucleotide repeats (Ellegren, 2004; Verbiest et al., 2023). Trinucleotide repeats are less frequent; however, it is important to note that although they are less common than the others, they are associated with several neurological diseases (Chintalaphani et al., 2021; Fan & Chu, 2007). It is also worth noting that the most common STRs in humans contain a high proportion of the nucleotide adenine (A). Among the types of STRs, dinucleotides are the most frequent, with the dinucleotide repeat (CA)n being the most common, followed by (AT)n, (GA)n, and (GC)n (Ellegren, 2004; Fan & Chu, 2007) (Fig. 2.1).

Due to their repetitive structure, STRs often lead to frequent mutations during cell division (Fan & Chu, 2007; Verbiest et al., 2023). In comparison, the mutation rate associated with STR repeats is an order of magnitude higher than that caused by SNPs (Chintalaphani et al., 2021; Verbiest et al., 2023). Specifically, unique DNA sequences in the genome have a mutation rate considered very low, approximately 10^{-9} nucleotides per generation, while mutations inside the STRs occur at a mutation rate ranging from 10^{-6} to 10^{-2} nucleotides per generation (Fan & Chu, 2007; Verbiest et al., 2023). Consequently, the rate of mutation inside the STRs is about 3–7 orders of magnitude higher than that in other regions of the genome. As a consequence of these mutations, STRs exhibit diverse polymorphisms within members of the same species (Budowle & Sajantila, 2024; Verbiest et al., 2023). Once STRs reach larger sizes because of frequent mutations, they acquire high instability (microsatellite instability), which promotes more mutation processes and STR expansion. This can result in STR repeat expansions that are involved in several neurological diseases (pathogenic repeat expansions) (Fig. 2.2) (Chintalaphani et al., 2021; Fan & Chu, 2007; Verbiest et al., 2023).

The main diseases related to the high rate of mutation and expansion of STRs (pathogenic repeat expansions) are Huntington's disease, Fragile X syndrome in humans, myotonic dystrophies, hereditary cerebellar ataxias, amyotrophic lateral sclerosis, and frontotemporal dementia. Additionally, there are studies that link STR expansions with complex polygenic diseases such as heart disease, bipolar disorder, major depressive disorder, and schizophrenia (Chintalaphani et al., 2021; Ellegren, 2004; Fan & Chu, 2007; Verbiest et al., 2023). Although these mutations are frequently associated

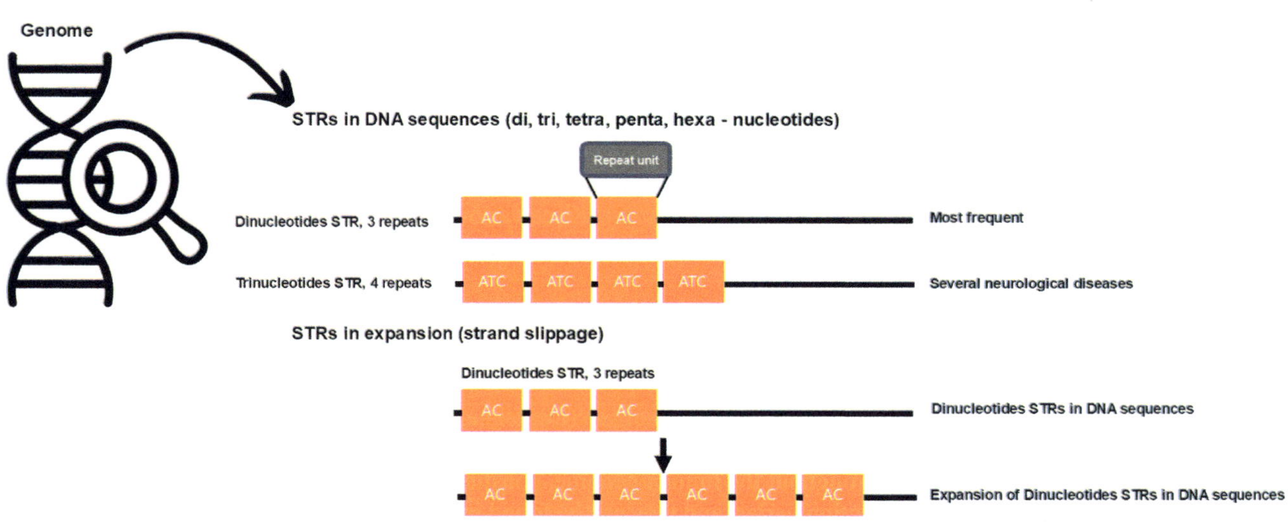

FIGURE 2.1 Types of short tandem repeats (STRs) in DNA sequences, in relation to quantities of nucleotides, repeats, length, and their expansion.

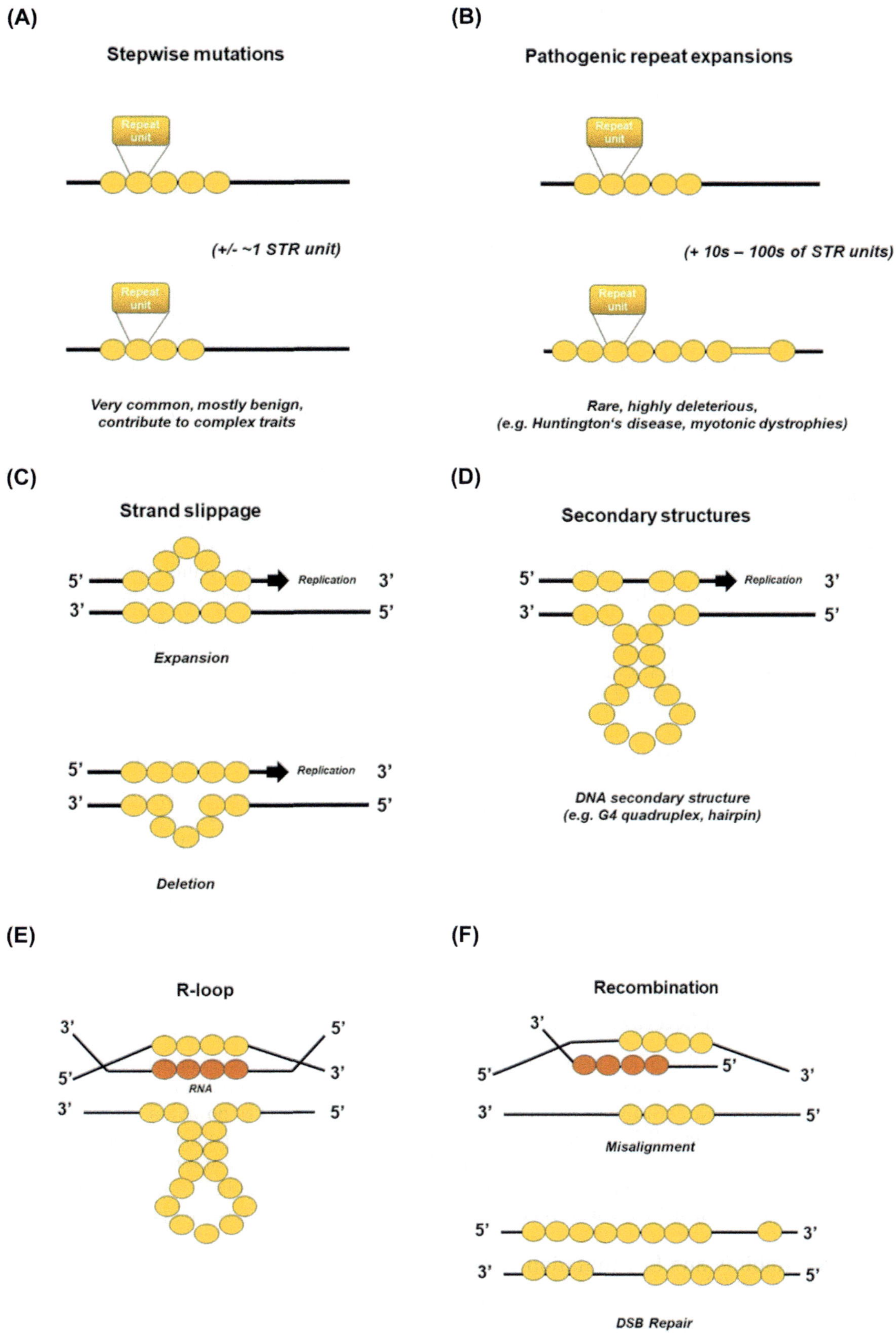

FIGURE 2.2 Types of mutations in repeats. (A) Stepwise mutations. (B) Pathogenic repeat expansions. Multiple processes caused by different short tandem repeat (STR) sequences. (C) Strand slippage. (D) Secondary structures. (E) R-loop. (F) Recombination/double-strand break (DSB)-repair .
Based on Verbiest, M., Maksimov, M., Jin, Y., Anisimova, M., Gymrek, M., & Bilgin Sonay, T. (2023). Mutation and selection processes regulating short tandem repeats give rise to genetic and phenotypic diversity across species. *Journal of Evolutionary Biology, 36*(2), 321–336. https://doi.org/10.1111/jeb.14106.

with rare diseases, the frequent mutations that occur are typically called stepwise changes. These mutations promote the deletion or addition of a repeat unit. These modest mutations practically do not cause any phenotypic effect (Fan & Chu, 2007; Verbiest et al., 2023, 2024) (Fig. 2.2). Although these mutations are frequently associated with rare diseases, the frequent mutations that occur are typically called stepwise changes. These mutations promote the deletion or addition of a repeat unit. These modest mutations practically do not cause any phenotypic effect (Fan & Chu, 2007; Verbiest et al., 2023, 2024) (Fig. 2.2).

Typically, most STR mutations occur through the process of strand slippage, where this slippage results in gradual changes in the repeat bands and consequently in the length of the strand (Ellegren, 2004; Verbiest et al., 2023, 2024). This is the most widely accepted model to explain the STR mutation process and may also be called DNA slippage, polymerase slippage, or slipped-strand mispairing (Ellegren, 2004; Fan & Chu, 2007; Verbiest et al., 2023). As described above, most STRs result in gradual copy number changes (Verbiest et al., 2023).

Strand misalignment can occur through multiple processes during replication or transcription and contribute to STR mutations, which affect biology and disease risk in multiple ways through structural and functional changes in the surrounding DNA (Verbiest et al., 2023, 2024). Among these multiple processes caused by different STR sequences (Figs. 2.2 and 2.3), we can highlight that STR sequences can promote the formation of secondary structures, such as hairpins, which can interrupt DNA synthesis and promote chromosome breakage. Furthermore, GC-rich STR sequences in promoter regions can form stable G4 quadruplexes that regulate transcriptional activity. Furthermore, they can also promote the formation of R-loops during transcription, during homologous recombination, or during double-strand break (DSB) repair. In addition, STRs can also act as cis-regulatory factors, promoting a greater binding affinity of the transcription factor. STRs, due to their variation, also exert important effects on biological processes involving RNA and proteins in the organism, deregulating splicing through different mechanisms and sequestering RNA-binding proteins, as well as preventing the formation of a viable protein, or promoting the translation of toxic proteins (Fig. 2.3) (Ellegren, 2004; Fan & Chu, 2007; Tanudisastro et al., 2024; Verbiest et al., 2023). Therefore, due to this high rate of polymorphism found in microsatellite sequences, STRs have proven to be extremely useful and effective tools in a variety of practical applications, such as the construction of genetic maps, disease diagnosis, genetic localization, paternity testing, genetic linkage analysis, identification of individuals, and currently, one of their main applications is in the area of forensic genetics (Budowle & Sajantila, 2024; Chintalaphani et al., 2021; Chistiakov et al., 2006; Ellegren, 2004; Fan & Chu, 2007; Verbiest et al., 2023; 2024) (Fig. 2.4).

2.7 Criteria for selecting short tandem repeat markers in forensic analysis

The use of STR markers is significant in forensic DNA analysis because it assists in genetic descent and individualization in a forensic context. STRs, which are comprised of a series of nucleotide sequences present in specific regions of the genome, display variation among individuals and are thus appropriate for use in forensics (Fig. 2.5) (Tanudisastro et al., 2024). Nevertheless, STR loci are only useful if an appropriate locus has been chosen with predetermined criteria for optimal discriminatory power, reproducibility, and applicability in forensic science (Nwawuba, Momoh, et al., 2020). Meeting these expectations allows forensic experts to ascertain the suitability of specific STR markers insofar as the results can be used in the resolution of law enforcement issues, the identification of missing individuals, and other forensic purposes.

2.7.1 Polymorphism and heterozygosity

A primary criterion for the selection of any marker, in this case an STR marker, is the level of polymorphism, which refers to the variability in the number of repeats at a given locus across individuals in a population (Uguen et al., 2024). It is always preferred to use markers with high polymorphism since they increase the power of discrimination between individuals. Further, heterozygous STR markers "heterozygous" indicating the proportion of individuals possessing different alleles at a locus, are ideal because they also increase the discriminatory power of forensic examinations (Bukyya et al., 2021). The reason for the very high frequency of heterozygosity of such markers is due to the high number of different alleles in a population, and these markers are therefore quite useful for forensic purposes (Uguen et al., 2024).

2.7.2 Genomic stability and low mutation rates

The forensic uses of STR markers require genomic stability. This is because markers with higher mutation rates can introduce inconsistencies in the genetic profiles of closely related individuals (i.e., parent-child and sibling comparisons), as well as skew long-term population frequency estimations (Omar et al., 2024). Also, one of the most efficacious

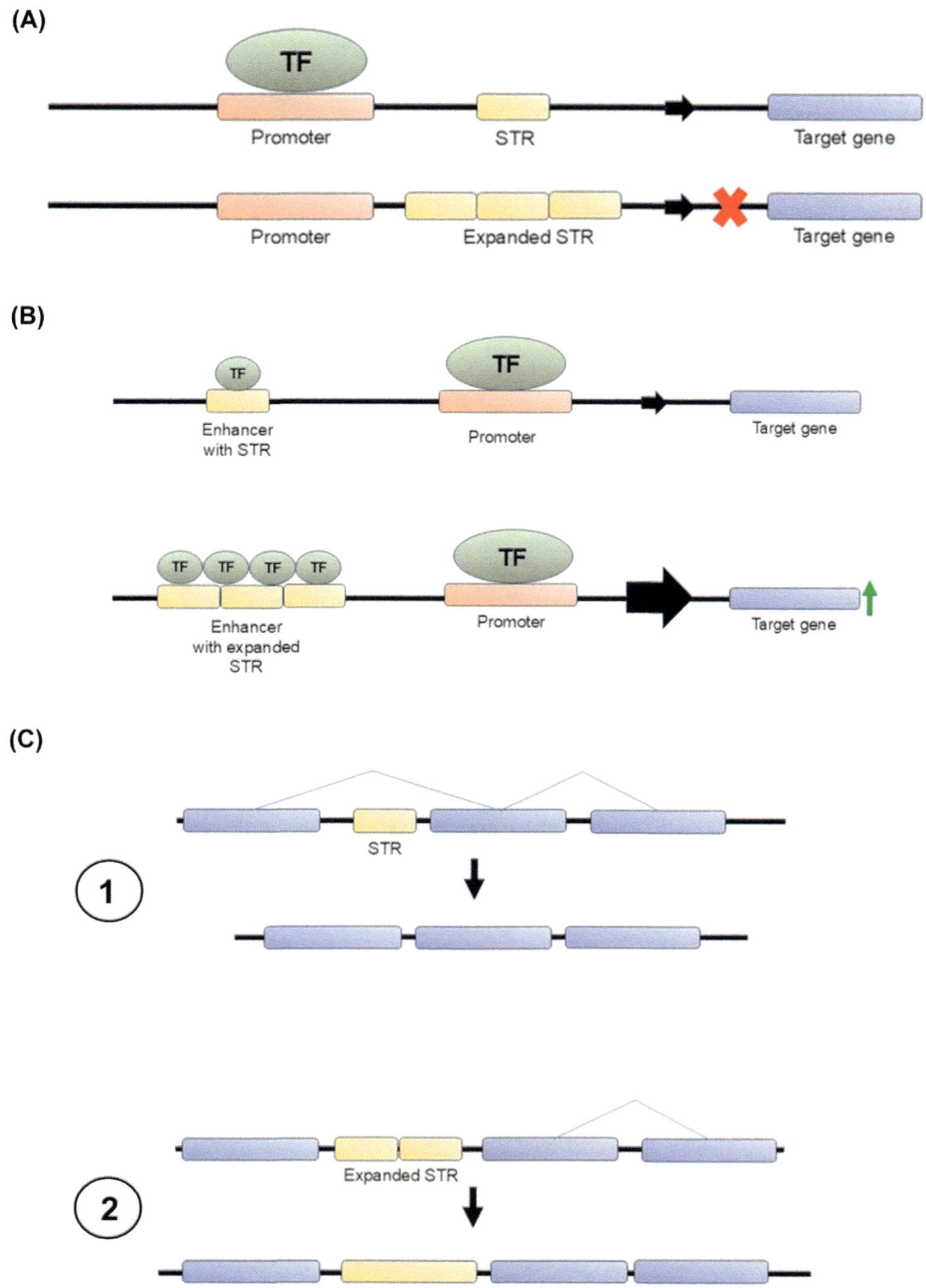

FIGURE 2.3 Multiple processes caused by different short tandem repeat (STR) sequences. (A) Expanded STRs can promote a shift of a promoter region away from its target gene, thereby silencing gene expression. (B) STRs can act as cis-regulatory elements that increase the affinity of enhancer sequences for transcription factors. (C) STRs can dysregulate mRNA processing. Exons are represented in blue. In 1, the STR is not expanded and does not interfere with mRNA processing. In 2, the expanded STR remains in the mRNA after processing.
Based on Tanudisastro, H.A., Deveson, I.W., Dashnow, H., & MacArthur, D.G. (2024). Sequencing and characterizing short tandem repeats in the human genome. *Nature Reviews. Genetics, 25*(7), 460–475. https://doi.org/10.1038/s41576-024-00692-3, https://www.nature.com/nrg/.

approaches for minimizing inconsistency across generations would be avoiding STR markers sensitive to replication slippage and high mutation segment regions. Forensic STR markers are typically selected from genomic regions known to have low mutation rates, ensuring the reliability of DNA comparisons in both criminal and civil cases (Uguen et al., 2024).

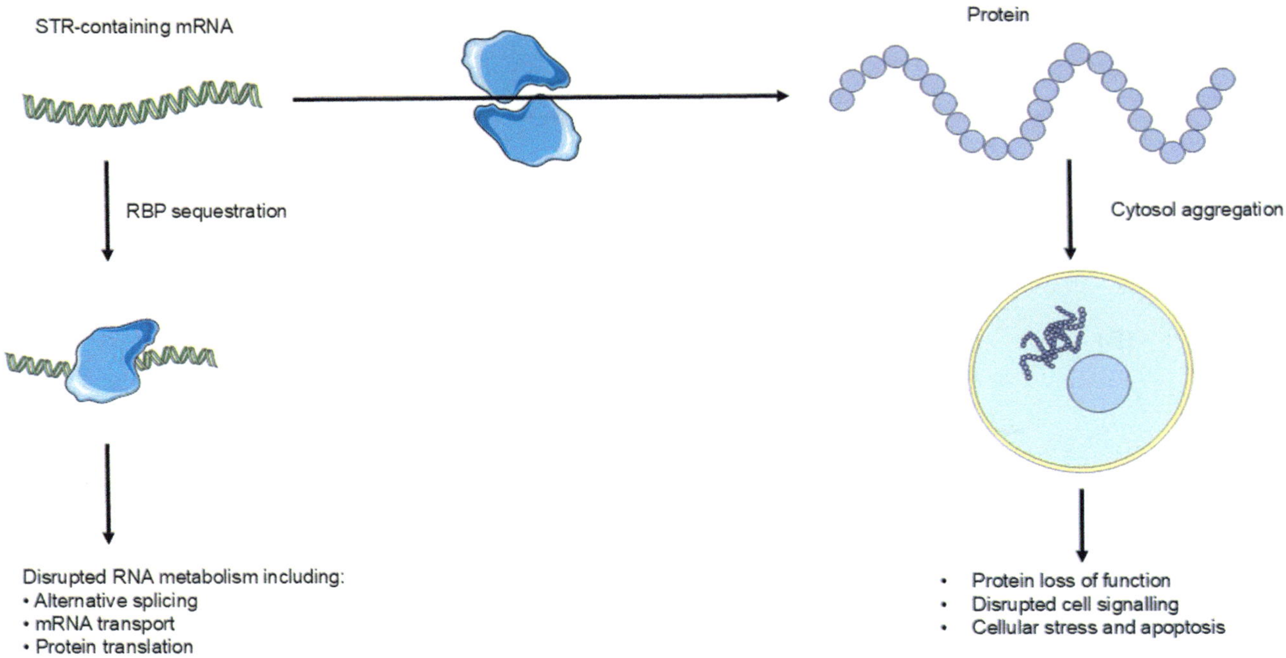

FIGURE 2.4 Short tandem repeat (STR)-containing mRNA can cause dysfunction by binding to RNA-binding proteins (RBPs), while proteins translated from STR-containing transcripts can form toxic aggregates that cluster in the cell cytosol.
Based on Tanudisastro, H.A., Deveson, I.W., Dashnow, H., & MacArthur, D.G. (2024). Sequencing and characterizing short tandem repeats in the human genome. *Nature Reviews. Genetics, 25*(7), 460–475. https://doi.org/10.1038/s41576-024-00692-3, https://www.nature.com/nrg/.

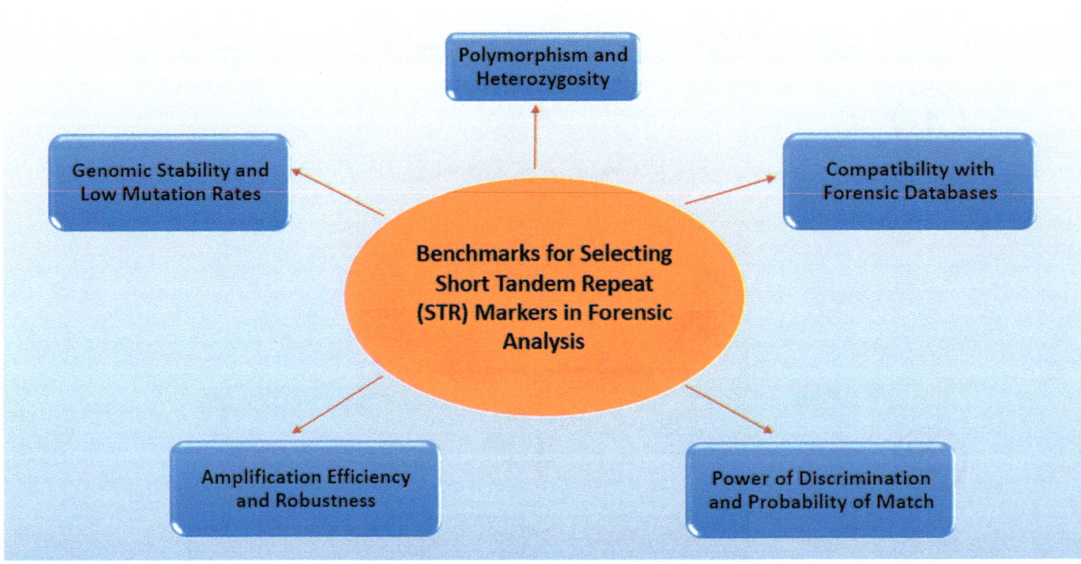

FIGURE 2.5 Benchmarks for selecting short tandem repeat (STR) markers in forensic analysis.

2.7.3 Amplification efficiency and robustness

Amplification efficiency and robustness are some of the criterion features in ascertaining the appropriate STR markers for forensic analysis, due to the fact that there are a number of forensic samples that are quite difficult to work with, as they are degraded, limited, or have the likelihood of contamination. In order to provide accurate and reproducible DNA profiles reliably, it is important to have the ability to optimize STR loci (Nwawuba et al., 2020). For STR amplification in forensic applications, the PCR is going to be the most handy technology, and its application is dependent on the selection of certain markers that can perform under certain conditions (McDonald et al., 2024). As such, it is the duty of forensic scientists to ensure that the STR markers selected mainly for the current analysis can be amplified well, even if they are in low amounts or degraded. Robustness also includes the capacity of the markers to ensure similar outcomes regardless of the conditions employed in terms of quality and quantity of the DNA, as well as the presence of PCR inhibitors. Even when such inhibitors are present, strong STR markers still work so that the overall pool of DNA profiles generated is accurate and reproducible (McDonald et al., 2024; Nwawuba et al., 2020).

2.7.4 Compatibility with forensic databases

STR markers need to be compatible with established forensic DNA databases, such as the CODIS in the US or European Standard Set in Europe. STR markers in these standardized forensic panel profiles make the comparison of genetic profiles across jurisdictions easier and therefore more usable for the solution of cross-border criminal cases (Hölzl-Müller et al., 2021). Their use of established STR markers results in profiles that are easily uploaded and searched against international databases, supporting global forensic cooperation and cold case resolution.

2.7.5 Strength of discrimination and likelihood of match

The overall power of discrimination of a set of STR markers is a key aspect in forensic identification. It signifies the likelihood that two individuals from different backgrounds will differ in their alleles in any particular locus, which goes hand in hand with the extent of diversity and heterozygosity in the markers (Hölzl-Müller et al., 2021). STR markers, which have been reported with high discrimination power, are preferred since they greatly in most cases make a greater sense to the overall population insofar as the population identifying one individual is concerned (Nwawuba et al., 2020). A typical forensic panel would consist of STR markers that give a combined probability of matching equal to less than 1 in one billion. This helps to diminish the odds of accidental matches involving unrelated individuals, which is very important in enhancing the credibility of forensic evidence (Hölzl-Müller et al., 2021).

2.8 The usefulness of autosomal short tandem repeat in forensic DNA analysis

The autosomal STRs have become the cornerstone in recent times in forensic DNA analysis because of their immense variability, robustness, and reliability with regard to individual identification and kinship testing (Nwawuba et al., 2020). Further, the analysis of autosomal STRs represents one of the most powerful and versatile tools in forensic science. In this way, it is clear that DNA profiling with its wide applicability for most types of biological evidence, high discriminatory power, and compatibility with national and international databases make it indispensable in solving crimes, identification of victims, and establishment of biological relationships (Kumar, 2022; Nwawuba et al., 2020; Shrivastava et al., 2020). That this technique can deal with degraded or mixed samples, along with the advances in probabilistic genotyping, means that even the most challenging forensic cases are solvable with a great degree of confidence. With the broad acceptance and legal defensibility of autosomal STR analysis, this further cements autosomal STR profiling as a cornerstone in forensic DNA analysis, hence making it indispensable in modern forensic investigations (Budowle & Sajantila, 2024; Wyner et al., 2020).

2.8.1 Paternity and relationship testing

Autosomal STR analysis is an efficient and one of the most frequently used methods in solving biological relationships through paternity and family testing (Uguen et al., 2024). This technique makes use of the principle of inheritance displayed by autosomal STR markers transmitted from parents to offspring, therefore allowing a very accurate assessment of genetic relatedness (Keerti & Ninave, 2022). Since it can enable the possible detection of the inheritance of STR alleles from both parents, autosomal STR analysis is a powerful tool in determining paternity, but also sibling, grandparent, and more complex forms of kinship testing (El-Alfy & Abd El-Hafez, 2012; Mahalinga et al., 2021).

Empirical studies have substantiated the fact that the paternity test using autosomal STR analysis is based upon the basic characteristic of Mendelian inheritance, where the offspring receives half of the genetic material from both parents (Shrivastava et al., 2021). In this context, at every locus, an individual inherits one STR allele from the mother and one from the father. Based on this principle, in a comparison of the alleles at several loci of STRs in a child and the putative father, forensic analysts can conclude that the latter would have contributed the expected paternal alleles (El-Alfy & Abd El-Hafez, 2012; Mahalinga et al., 2021). These tests commonly analyze 15–24 autosomal STR markers, which turn out to be very good indicators for confirming or excluding the putative father at a high level of certainty. In cases of paternity confirmation, the paternity probability computation will be mostly in excess of 99.99%, which is quite strong evidence of biological paternity. Conversely, whenever the alleged father does not share the expected alleles at more than one locus, a case of paternity exclusion can definitely be established. It is for this reason that this high degree of precision makes autosomal STR testing the gold standard in legal paternity cases, from child support, custody, and down to inheritance disputes (El-Alfy & Abd El-Hafez, 2012; Nadia et al., 2013; Mahalinga et al., 2021).

While paternity testing is the most common application of autosomal STR analysis, testing for many other familial relationships also values the method. The STR markers are capable of providing evidence of genetic relatedness among siblings (full or half), grandparents and grandchildren, aunts and uncles, and even more distant relatives (Keerti & Ninave, 2022; Mahalinga et al., 2021). Such tests are at times necessary when the direct testing between a parent and child is not possible, for example, a deceased, unavailable, or uncooperative parent. For instance, in sibling testing, full siblings would be expected to share approximately 50% of their autosomal alleles, whereas half-siblings share approximately 25% (El-Alfy & Abd El-Hafez, 2012). By studying the STR profiles of individuals in question, a forensic scientist can quantify the amount of allele sharing that provides the likelihood pertaining to whether the individuals are full siblings, half-siblings, or unrelated individuals. Similarly, grand-parentage testing is based upon the same principle—a child receives one-quarter of his or her genetic material from each grandparent, which in turn is detectable by analyzing autosomal STR loci. Such relationship tests are very helpful in immigration cases, where the establishment of a biological relationship is necessary to satisfy legal requirements; they are also helpful in cases involving adoption, donor conception, or missing person investigations (El-Alfy & Abd El-Hafez, 2012; Keerti & Ninave, 2022; Wyner et al., 2020).

Moreover, one of the major strengths of autosomal STR analysis in relationship testing involves the high degree of accuracy and statistical confidence that it provides (Nwawuba, Momoh, et al., 2020). In the case of paternity testing, a sufficient number of loci would be able to guarantee that the probability of a chance match between an unrelated individual and the child is very low (El-Alfy & Abd El-Hafez, 2012; Wyner et al., 2020). This accumulated discriminatory power makes it possible for the forensic scientist to give results with near-certain conclusions. In those cases where other relationships are tested, additional STR loci usually form part of the analysis, and supplementary statistical calculations will be provided, such as the kinship index or likelihood ratio (LR) (Keerti & Ninave, 2022; Mahalinga et al., 2021). These values are the numerical estimate of the likelihood that tested individuals possess a certain relationship, compared with unrelated individuals. According to Ge and Budowle (2020) and Thelingwani et al. (2024), the greater the LR is, the more supportive the genetic evidence is of the hypothesized relationship. When the values are low, relatedness may be indicated. Although the autosomal STR analysis technique applied to the testing of relationships has proved highly sensitive, there are specific challenges and considerations. For instance, there are cases of high relatedness between people-say, incestuous relationships or inbred populations, in which the sharing of alleles becomes high and, thus hard to interpret. In this sense, further genetic markers would include the Y-STRs or mtDNA used complementarily to autosomal STR analysis (Nadia et al., 2013; Wyner et al., 2020; Mahalinga et al., 2021). Nevertheless, the robustness of STR analysis and modern computational methods is such that most of these problems can be managed. Autosomal STRs are used in direct paternity testing, sibling testing, or more complex relative relationships to define critical personal, legal, and social problems concerning family-related relationships (El-Alfy & Abd El-Hafez, 2012; Keerti & Ninave, 2022; Shrivastava et al., 2020; Mahalinga et al., 2021). And as technological improvements continue to be made in the analysis of STRs, their applicability to both established and new uses within forensic genetics will increase, further integrating their worth into modern science and law.

2.8.2 Genealogy

Autosomal STR analysis is one of the most efficient methods used in tracing genealogical history in several generations, both in a closer and distant mode of ancestry (Uguen et al., 2024). In contrast to either Y-chromosome or mitochondrial DNA (mtDNA) analysis that focuses on either the paternal or maternal lineage effectively, analysis of autosomal STRs processes the DNA inherited from both parents for a complete genetic lineage within an individual (Connell et al., 2024). For this reason, the analysis of autosomal STRs becomes an excellent resource for creating family trees as well as for understanding how people genetically relate to each other in different generations and branches of their family tree (Uguen et al., 2024). In genealogical research, the analysis of autosomal STRs, as applied to law, concerns the analysis of shared alleles between two or more people who show evidence of a genetic relationship. The frequency of occurrence of such matching genetic markers corresponds to the closeness of the kinship of the two individuals involved (Kling et al., 2021). Determining the relatedness of individuals is accomplished by analysis of their autosomal DNA profiles: genealogists can tell if the individuals are closely related (e.g., siblings or cousins) or if they share a more remote common ancestor several generations earlier (Kling et al., 2021; Mengge et al., 2024). The fact that it is possible to establish genetic connections over a wide spectrum of genealogical distance makes autosomal STR analysis an excellent method of resolving complicated family trees.

Aside from familiar relationships, autosomal STR analysis can also help in determining the individual's ethnic background and population origins (Uguen et al., 2024). By comparing one's STR profile to worldwide reference populations, it could indicate the geographic and ethnic origins of an individual's ancestors (Mengge et al., 2024). This approach has been important for many in investigating their ancestral genetic background and finding those ancestral regions that contribute to one's total genetic makeup. Genealogical research often comprises the use of autosomal STR data for inferring the proportion of ancestry from different populations, such as African, European, Native American, or Asian ancestry (Kling et al., 2021; Mengge et al., 2024). The broad patterns of ancestry, which this population-based analysis gives, help individuals to make sense of their family lineages in regard to historical migrations and genetic diversity. These findings with regard to population origins are important to genealogists insofar as they work at building a detailed and accurate picture of ancestral background (Kling et al., 2021; Mengge et al., 2024).

These large genealogical databases have greatly increased the utility of autosomal STR analysis in genealogy (Zhang et al., 2023). Many of these databases allow people to upload their genetic profiles in order to compare their profiles with others, finding shared markers indicative of common ancestry. Services such as 23andMe, Ancestry DNA, and GED match allow autosomal DNA testing by analyzing STRs (Kling et al., 2021; Zhang et al., 2023; Mengge et al., 2024). More recently, this collaborative aspect of genetic research, which is made possible by these databases, has brought new dimensions into the process of discovery. Such genetic markers would also provide a means for individuals to come together with distant relatives in filling gaps in family trees, completing the missing pieces of family history, and even finding new branches of the lineage (Zhang et al., 2023; Mengge et al., 2024). This further enhances the prospect of finding previously unknown relatives, extending genealogical research beyond what was conventionally achievable by making use of traditional historical records (Uguen et al., 2024). Using genetic diversity at STR loci, genealogists can map relationships across generations, solve cases of unknown parentage, and collaborate with far-flung relatives in piecing together family trees. Although there are some limiting factors, such as lower resolution when confirming more distant relationships, the now-widespread utilization of autosomal STR analysis in genealogical databases revolutionizes the field and has placed such a valuable resource in the hands of people curious about their deeper genetic heritage (Kling et al., 2021; Zhang et al., 2023; Mengge et al., 2024).

2.8.3 Missing persons and disaster victim identification

The main strength of the autosomal STR marker, regarding missing persons and disaster victim identification, is that it has very high discriminatory power; the autosomal STR markers are highly polymorphic, or, in other words, there are many variations in allelic size among individuals (Montelius & Lindblom, 2012). It allows one to assign a unique DNA profile to every individual, which can then be compared to known genetic samples from relatives or reference databases for identification purposes (Nwawuba et al., 2020). In the case of unidentified persons, the DNA is compared to the DNA of biological relatives—a parent, sibling, or child—to confirm familial relationships (Montelius & Lindblom, 2012). Autosomal STR genetic markers obey the pattern of Mendelian inheritance, which provides a forensic expert with the opportunity to trace with high accuracy genetic similarities and differences between the missing person and their relatives. The utilization of various loci of STRs by forensic experts will enable the establishment of an individualized profile, which would provide a very good match or exclusion of identity even in cases where only distant relatives are available for comparison (Montelius & Lindblom, 2012; Shrivastava et al., 2020).

In the event of mass disasters such as natural calamities, aircraft crashes, terrorist attacks, or industrial accidents, the victims suffer extreme trauma or degradation that renders the use of personal identification documents or visual identification out of the question. In this regard, autosomal STR analysis ensures that the victims of mass disasters are identified even in instances where their remains are fragmented or degraded through DNA extraction (Kumar, 2022). Autosomal STR analysis is particularly effective in those cases where other traditional techniques, such as fingerprinting or dental identification, are not applicable or impossible due to the badly decomposed state of the remains. DNA is extracted from a sample of bones, a tooth, or tissue, and STR profiles are generated that can be compared with samples from relatives or from existing DNA databases (Blau et al., 2022; Kumar, 2022). This is what makes autosomal STR analysis an essential tool in disaster victim identification, where victims are identified and returned to their respective families for closure, all from the complexity of extracting DNA, even in complex scenarios.

One of the major challenges with missing persons and disaster victim identification is that the quality and quantity of DNA are available for analysis. In the sampling from missing persons or disaster victims, DNA samples may be degraded by exposure to environmental harsh conditions such as heat, moisture, or time (Vigeland & Egeland, 2021). Autosomal

STR, in this respect, is very valuable since it can obtain genetic profiles from low-quality or partially degraded DNA samples successfully. This is achieved by targeting shorter DNA fragments, which are referred to as mini-STRs, thereby enhancing the recovery of STR analysis successfully in those cases where traditional STR loci may fail. This, then, is an adaptation which allows forensic scientists to retrieve DNA profiles even from challenging samples, like bones or highly decomposed tissues, ensuring that identification can still be achieved in cases where other methods would be ineffective (Montelius & Lindblom, 2012; Vigeland & Egeland, 2021). Autosomal STR analysis stands out among the most reliable methods applied for missing persons and disaster victim identification under difficult conditions in that it cannot handle degraded samples (Montelius & Lindblom, 2012; Vigeland & Egeland, 2021).

2.8.4 DNA databases

Autosomal STR analysis is, in fact, the basis of establishment and implementation for all forensic DNA databases throughout the world (Nwawuba, Momoh, et al., 2020). These databases, which have become quite instrumental in solving criminal cases, locating missing persons, and identifying victims of disasters, depend on being able to compare DNA profiles among enormous numbers of individuals (Nwawuba Stanley & Akpata Chinyere, 2020). For this reason, STR has been widely applied for these purposes, since it has a very high degree of accuracy, individual discrimination, and is easily standardized between jurisdictions. Some key advantages that come with the use of autosomal STR markers in DNA databases are that most of them have been standardized globally; this allows various countries and regions to share and reference DNA profiles with relative ease (Nwawuba et al., 2020).

The STR markers used in forensic databases are selected to be highly polymorphic so that there will be consistent performance within and among diverse populations. The United States CODIS, for example, uses a standardized set of 20 autosomal STR loci for the generation of DNA profiles, whereas a European Standard Set is used in Europe for the same purposes (Vijayaraghavan et al., 2023). This standardization allows forensic scientists to compare DNA profiles generated in one country with those housed within international databases. The ability to cross-match profiles coming from different regions increases the possibility of solving cross-border crimes, identifying international suspects, or identifying missing persons who might have crossed through jurisdictions (Dash et al., 2023; Uguen et al., 2024). Autosome markers are compatible with the existing databases; hence, forensic evidence can be shared with any part of the world and increases the chances of collaboration by law enforcement agencies. DNA databases with autosomal STR profiles assist in the detection of recurrent offenders and aid in solving cold cases (Bodner & Parson, 2020; Link et al., 2023; Uguen et al., 2024).

Whenever there is an increase in the number of individuals in national and international DNA databases, the possibility of finding matches for unsolved cases also increases. Forensic scientists using these databases often re-analyze DNA evidence from past cases by uploading the profiles of the unknown contributors, searching for potential matches within them (Ralf et al., 2024; Uguen et al., 2024). In many instances, the offenders may have been of unknown origin to law enforcement but later identified through their inclusion in a DNA database after a different offense. This has gone on to solve many cold cases, sometimes years and decades after the original incident. Second, the presence of an individual's DNA profile in the database allows law enforcement to detect repeat offenders, preventing further crimes by linking multiple incidents to the same individual. As Ralf et al. (2024) note, this approach greatly expands the utility of DNA databases in forensic investigations; at the same time, it raises important ethical and privacy concerns.

It implies that the more people have genetic profiles stored in the database, the more misuse or unauthorized access to sensitive genetic information will occur. This kind of wide collection and storage of DNA profiles, especially for individuals who have not been convicted of an offense, raises debates on privacy rights, data security, and informed consent (Wyner et al., 2020). In view of these concerns, almost all countries have promulgated strict guidelines and regulations concerning the collection, storage, and usage of DNA profiles in forensic databases. These regulations are geared at ensuring the privacy of individuals while at the same time, the benefits of DNA databases accrue to public safety and justice (Bodner & Parson, 2020; Shrivastava et al., 2020). Ethics will continue to play an important role in building on the extent of DNA databases and their utilization to strike a balance between the needs of law enforcement and the rights of individuals.

2.9 Resolution of cases using autosomal short tandem repeat analysis

2.9.1 Case report I

Many crime scenes yield biological traces such as bloodstains, semen, or even the body of the victim that might become critical evidence. Further analysis of this evidence by forensic scientists establishes the link between the crime and the suspect through matching these biological traces with the suspect by DNA profiling. Where no immediate suspect exists, DNA samples collected from the crime scene can be matched against DNA profiles stored in a National DNA Database. Such matching would yield a "hit" or a match of an individual where the profiles perfectly match the crime scene evidence. This would involve the collection of biological traces from the crime scene as evidence and taking DNA samples from suspects (Chung & Fung, 2013; Jakovski et al., 2017). An example of this process is shown in, where DNA evidence from the crime scene is compared with samples from two suspects. This analysis revealed that suspect 2 had a DNA profile matching the evidence from the crime scene since the repeat sequences of three specific STR loci were from the evidence: D3S1358, vWa, and FGA (Chisum and Turvey, 2011; Nwawuba et al., 2020).

Part of your document could not be imported as it contained an unsupported element. For more information about unsupported elements, please visit the Support Center.

2.9.2 Case report II

A case wherein an elderly woman was found dead in her barn. An investigation was opened to reveal who committed the crime. A cigarette filter and one bottle of Coke were found within the scene of the crime and were taken and sent for DNA analysis. Autopsies of the victim yielded blood and nail debris that were similarly taken for DNA profiling. The forensic analysis required DNA extraction using a Qiagen mini kit, PCR amplification using an Identifier Amp Kit, and DNA typing using a 310 Genetic Analyzer capillary electrophoresis. DNA profiles obtained from the cigarette filter and bottle were possible, but came out a different male unidentified profile. Later on, the unidentified DNA profile was compared with the Macedonian National DNA database in search of a match. Eventually, a hit was created linking the DNA to a convicted prisoner who had previously been charged with drug trafficking and had been on temporary release for the weekend at the time of the crime (Jakovski et al., 2017).

Case study I in the current study highlights the critical role that forensic DNA profiling, in particular the use of STR markers, plays in individualization, identification, mitigation, and giving strong evidence for the resolution of crimes.

TABLE 2.1 DNA profiles from the victim, suspect, and collected items.

Locus	Repeat motif	Victim	Coke bottle	Cigarette filter	Suspect
D8S1179	Compound TCTA/TCTG	14, 14	12, 14	12, 14	12, 14
D21S11	Complex TCTA/TCTG	28, 29	28, 29	28, 29	28, 29
D7S820	Simple GATA	9, 11	8, 8	8, 8	8, 8
CSF1PO	simple AGAT	11, 12	11, 11	11, 11	11, 11
D3S1358	Compound TCTA/TCTG	15, 19	15, 18	15, 18	15, 18
TH01	Simple TCAT	8, 8	9, 9.3	9, 9.3	9, 9.3
D13S317	Simple TATC	12, 14	11, 14	11, 14	11, 14
D16S539	Simple GATA	11, 11	10, 11	10, 11	10, 11
D2S1338	Compound TGCC/TTCC	23, 24	18, 24	18, 24	18, 24
D19S433	Simple AGAA	13, 15	14, 15	14, 15	14, 15
vWA	Compound TCTA/TCTG	17, 17	16, 18	16, 18	16, 18
TPOX	Simple AATG	9, 11	8, 11	8, 11	8, 11
D18S51	Simple AGAA	13, 16	14, 16	14, 16	14, 16
D5S818	Simple AGAT	11, 11	12, 12	12, 12	12, 12
FGA	Compound CTTT/TTCC	21, 22	21, 24	21, 24	21, 24
Amelogenin		XX	XY	XY	XY

Source: Adopted from Jakovski, Z., Ajanovska, R. J., Stankov, A., Poposka, V., Bitoljanu, N., Belakaposka, V. (2017). The power of forensic DNA databases in solving crime cases. Macedonia Forensic Science International: *Genetics Supplement Series,* 6, e275. Available from https://doi.org/10.1016/j.fsigss.2017.09.085, http://www.elsevier.com.

Without DNA profiling's assistance, it would not have been feasible to identify or personalize the suspects based on the case report. As presented in Table 2.1, the suspect's DNA sample obtained through buccal swabbing after the 15 STR loci tests matched perfectly with the DNA retrieved from the various items discovered at the crime scene (coke bottle and cigarette filter).

2.10 Future perspective and conclusion

Autosomal STR has become an indispensable tool in forensic DNA analysis, playing a very important role in human identification, kinship determination, and criminal investigations. High polymorphism, together with low mutation rates and standardization of the loci utilized in forensic databases, results in robust discriminatory power that has set a basis for accurate and reliable forensic practices worldwide. Success notwithstanding, complex DNA mixtures and partial or degraded samples, and issues with cross-jurisdictional comparisons are some concerns raised for further improvement. The future will likely expand the capacity of autosomal STR profiling, using NGS and other rapidly emerging technologies, including improved computational tools to break down mixtures. Future innovations incorporating SNPs, epigenetic markers, and AI-driven algorithms may progress toward increasing the resolution and applicability of STR analysis, especially in cases involving degraded samples and in distant kinship scenarios. However, with these developments, ethical considerations that go along with such strides-for example, with data privacy and increasing uses of forensic genetic genealogy-must also be weighed and set in a balance. As forensic science continues to evolve, autosomal STR analysis will remain at the heart; however, its meaning is bound to shift within a wider frame of combined genomic approaches. Further development and ongoing fine-tuning of the technology of autosomal STR, along with its governance on ethical grounds, will be very important for its continued role in forensic investigations.

References

Abdi, G., Tarighat, M. A., Jain, M., Tendulkar, R., Tendulkar, M., & Barwant, M. (2024). Revolutionizing genomics: Exploring the potential of next-generation sequencing. In *Advances in bioinformatics (2nd ed.)*. Iran: Springer Nature, 1–33. https://link.springer.com/book/10.1007/978-981-99-8401-5.

Blau, S., Ranson, D., & de Boer, H. (2022). Disaster victim identification: Traditional approaches and changing practices. In G. N. Rutty (Ed.). *Essentials of autopsy practice* (pp. 1–10). Cham: Springer. https://doi.org/10.1007/978-3-031-11541-7_6.

Bodner, M., & Parson, W. (2020). The STRidER report on two years of quality control of autosomal STR population datasets. *Genes, 11*(8), 901. https://doi.org/10.3390/genes11080901.

Budowle, B., & Sajantila, A. (2024). Short tandem repeats — How microsatellites became the currency of forensic genetics. *Nature Reviews. Genetics, 25*(7), 450–451. https://doi.org/10.1038/s41576-024-00721-1.

Bukyya, J. L., Avinash Tejasvi, M. L., Avinash, A., Chanchala, H. P., Talwade, P., Afroz, M. M., Pokala, A., Neela, P. K., Shyamilee, T. K., & Srisha, V. (2021). DNA profiling in forensic science: A review. *Global Medical Genetics, 08*(04), 135–143. https://doi.org/10.1055/s-0041-1728689.

Butler, J. M. (2015). The future of forensic DNA analysis. *Philosophical Transactions of the Royal Society B: Biological Sciences, 370*(1674), 20140252. https://doi.org/10.1098/rstb.2014.0252.

Børsting, C., & Morling, N. (2015). Next generation sequencing and its applications in forensic genetics. *Forensic Science International: Genetics, 18*, 78–89. https://doi.org/10.1016/j.fsigen.2015.02.002.

Carracedo, A., Beatriz, S., & Maria, V.L. (2018). Forensic DNA typing technologies: A review. In M. J. Bogusz (Ed.), *Forensic science handbook of analytical separations* (Vol. 6, pp. 955–957).

Butler, J. M. (2022). Recent advances in forensic biology and forensic DNA typing: INTERPOL review 2019–2022. *Forensic Science International: Synergy, 6*, 100311. https://doi.org/10.1016/j.fsisyn.2022.100311.

Chintalaphani, S. R., Pineda, S. S., Deveson, I. W., & Kumar, K. R. (2021). An update on the neurological short tandem repeat expansion disorders and the emergence of long-read sequencing diagnostics. *Acta Neuropathologica Communications, 9*(1). https://doi.org/10.1186/s40478-021-01201-x, https://actaneurocomms.biomedcentral.com/.

Chistiakov, D. A., Hellemans, B., & Volckaert, F. A. M. (2006). Microsatellites and their genomic distribution, evolution, function, and applications: A review with special reference to fish genetics. *Aquaculture (Amsterdam, Netherlands), 255*(1–4), 1–29. https://doi.org/10.1016/j.aquaculture.2005.11.031.

Chisum, W. J., & Turvey, B. E. (2011). Forensic science. In *Crime reconstruction (2nd ed.)*. United States: Elsevier, 3–18. https://www.sciencedirect.com/book/9780123864604.

Cho, S., Shin, E., Park, Y. G., Choi, S. H., Choe, E. K., Bae, J. H., Lee, J. E., & Lee, S. D. (2024). A novel approach of kinship determination based on the physical length of genetically shared regions of chromosomes. *Genes and Genomics, 46*(5), 577–587. https://doi.org/10.1007/s13258-023-01485-4, https://www.springer.com/journal/13258.

Chung, Y. K., & Fung, W. K. (2013). Identifying contributors of two-person DNA mixtures by familial database search. *International Journal of Legal Medicine, 127*(1), 25–33. https://doi.org/10.1007/s00414-012-0666-x.

Connell, J. R., Lea, R. A., Haupt, L. M., & Griffiths, L. R. (2024). Mitochondrial DNA analysis in population isolates: Challenges and implications for human identification. *Current Molecular Biology Reports, 10*(1), 1–8. https://doi.org/10.1007/s40610-023-00155-4.

Dash, H. R., Elkins, K. M., & Al-Snan, N. R. (2023). Role of forensic DNA databases in criminal identification. In *Advancements in forensic DNA analysis*. Singapore: Springer, 1–15. https://doi.org/10.1007/978-981-99-6195-5_9.

Depienne, C., & Mandel, J. L. (2021). 30 years of repeat expansion disorders: What have we learned and what are the remaining challenges? *Cell American Journal of Human Genetics, 108*(5), 764–785. https://doi.org/10.1016/j.ajhg.2021.03.011, http://www.elsevier.com/wps/find/journaldescription.cws_home/713561/description#description.

El-Alfy, S. H., & Abd El-Hafez, A. F. (2012). Paternity testing and forensic DNA typing by multiplex STR analysis using ABI PRISM 310 genetic analyzer. *Journal of Genetic Engineering and Biotechnology, 10*(1), 101–112. https://doi.org/10.1016/j.jgeb.2012.05.001.

Ellegren, H. (2004). Microsatellites: Simple sequences with complex evolution. *Nature Reviews. Genetics, 5*(6), 435–445. https://doi.org/10.1038/nrg1348.

Fan, H., & Chu, J. Y. (2007). A brief review of short tandem repeat mutation. *Genomics, Proteomics and Bioinformatics, 5*(1), 7–14. https://doi.org/10.1016/S1672-0229(07)60009-6, http://www.elsevier.com/wps/find/journaldescription.cws_home/707637/description#description.

Ge, J., & Budowle, B. (2020). How many familial relationship testing results could be wrong? *PLoS Genetics, 16*(8). https://doi.org/10.1371/JOURNAL.PGEN.1008929, https://dx.plos.org/10.1371/journal.pgen.1008929.

Gymrek, M. (2017). A genomic view of short tandem repeats. *Current Opinion in Genetics & Development, 44*, 9–16. https://doi.org/10.1016/j.gde.2017.01.012.

Haddrill, P. R. (2021). Developments in forensic DNA analysis. *Emerging Topics in Life Sciences, 5*(3), 381–393. https://doi.org/10.1042/ETLS20200304, https://portlandpress.com/emergtoplifesci/article/5/3/381/228192/Developments-in-forensic-DNA-analysis.

Hölzl-Müller, P., Bodner, M., Berger, B., & Parson, W. (2021). Exploring STR sequencing for forensic DNA intelligence databasing using the Austrian National DNA Database as an example. *International Journal of Legal Medicine, 135*(6), 2235–2246. https://doi.org/10.1007/s00414-021-02685-x.

Jakovski, Z., Ajanovska, R. J., Stankov, A., Poposka, V., Bitoljanu, N., & Belakaposka, V. (2017). The power of forensic DNA databases in solving crime cases. *Macedonia Forensic Science International: Genetics Supplement Series, 6*, e275. https://doi.org/10.1016/j.fsigss.2017.09.085, http://www.elsevier.com.

Keerti, A., & Ninave, S. (2022). DNA fingerprinting: Use of autosomal short tandem repeats in forensic DNA typing. *Cureus, 14*(10). https://doi.org/10.7759/cureus.30210.

Kling, D., Phillips, C., Kennett, D., & Tillmar, A. (2021). Investigative genetic genealogy: Current methods, knowledge and practice. *Forensic Science International: Genetics, 52*, 102474. https://doi.org/10.1016/j.fsigen.2021.102474.

Kumar, S. (2022). DNA-based human identification in mass-disaster cases. In *Handbook of DNA forensic applications and interpretation*. Saudi Arabia: Springer Nature, 149–175. https://doi.org/10.1007/978-981-19-0043-3.

Link, V., Zavaleta, Y.J.A., Reyes, R.J., Ding, L., Wang, J., Rohlfs, R.V., & Edge, M.D. (2023). Microsatellites used in forensics are located in regions unusually rich in trait-associated variants. bioRxiv, United States. https://www.biorxiv.org, https://doi.org/10.1101/2023.03.07.531629.

Liu, Y., Xu, J., Chen, M. A., Wang, C., & Li, S. (2019). A unified STR profiling system across multiple species with whole genome sequencing data. *BMC Bioinformatics, 20*(Suppl 24), 671. https://doi.org/10.1186/s12859-019-3246-y.

Lynch, M. (2013). Science, truth, and forensic cultures: The exceptional legal status of DNA evidence. *Studies in History and Philosophy of Science Part C: Studies in History and Philosophy of Biological and Biomedical Sciences, 44*(1), 60–70. https://doi.org/10.1016/j.shpsc.2012.09.008.

Mahalinga, R., U., Munuswamy, U., Raghunath, R., Dhanapal, T., & Nithyanandam, M. (2021). Short tandem repeat mutations in paternity analysis. In H. R. Dash, P. Shrivastava, & J. A. Lorente (Eds.). *Handbook of DNA profiling*. Singapore: Springer. https://doi.org/10.1007/978-981-15-9364-2_15-1.

Marrone, M., Tarantino, F., Stellacci, A., Baldassarra, S. L., Cazzato, G., Vinci, F., & Dell'Erba, A. (2021). Forensic analysis and identification processes in mass disasters: Explosion of gunpowder in the fireworks factory. *Molecules (Basel, Switzerland), 27*(1), 244. https://doi.org/10.3390/molecules27010244.

McDonald, C., Taylor, D., & Linacre, A. (2024). PCR in forensic science: A critical review. *Genes, 15*(4), 438. https://doi.org/10.3390/genes15040438.

McKiernan, H. E., & Danielson, P. B. (2017). *Molecular diagnostic applications in forensic science*. Elsevier BV, 371–394. https://doi.org/10.1016/b978-0-12-802971-8.00021-3.

Mengge, W., Chen, H., Luo, L., Huang, Y., Duan, S., Yuan, H., Tang, R., Liu, C., & He, G. (2024). Forensic investigative genetic genealogy: Expanding pedigree tracing and genetic inquiry in the genomic era. *Journal of Genetics and Genomics*. https://doi.org/10.1016/j.jgg.2024.06.016.

Montelius, K., & Lindblom, B. (2012). DNA analysis in disaster victim identification. *Forensic Science, Medicine, and Pathology, 8*(2), 140–147. https://doi.org/10.1007/s12024-011-9276-z.

Nadia, P., Magalhães, M., Conde-Sousa, E., Gomes, C., Pereira, R., Alves, C., Gusmão, L., & Amorim, A. (2013). Assessing paternities with inconclusive STR results: The suitability of bi-allelic markers. *Forensic Science International: Genetics, 7*, 16–21.

Nizami, S. B., Kazmi, S. Z. H., Abid, F., Babar, M. M., Noor, A., Zaidi, Khan, S. U., Hasan, H., Ali, M., & Gul, A. (2018). Omics approaches in forensic biotechnology: Looking for ancestry to offence. In V. K. Gupta, & A. K. Verma (Eds.). *Omics technologies and bio-engineering* (pp. 111–129). Academic Press.

Nwawuba, S. U., Mohammed, K. A., Bukola, A. T., Omusi, P. I., & Ayevbuomwan, D. E. (2020). Forensic DNA profiling: Autosomal short tandem repeat as a prominent marker in crime investigation. *The Malaysian Journal of Medical Sciences, 27*(4), 22–35. https://doi.org/10.21315/mj (SU *et al.*, 2020).27.4.3.

Nwawuba, S. U., Momoh, S. M., Nwokolo, C. C., Ojo, O. G., Ajayi, J. O., Irabor, I. M., & Oboh, B. O. (2020). Key DNA profiling markers for identification: A mini review. *Pharmacy & Pharmacology International Journal, 8*(6), 337–343. https://doi.org/10.15406/ppij.2020.08.00315.

Nwawuba Stanley, U., & Akpata Chinyere, B. N. (2020). Awareness level on the role of forensic DNA database in criminal investigation in Nigeria: A case study of Benin City. *Journal of Forensic Science and Research, 4*(1), 007–014. https://doi.org/10.29328/journal.jfsr.1001019.

Obafunwa, J. O., Ajayi, O., & Okoye, M. I. (2018). Medical evidence and proof of cause of death in Nigerian courts. *Medicine, Science, and the Law, 58*(2), 122–134. https://doi.org/10.1177/0025802418754576.

Omar, I. S., Hafiza, M. Y. N., Zafarina, Z., Norasikin, M. N. S., Azmel, M. I. S., Shafini, M. Y., & NurWaliyuddin, H. Z. A. (2024). Forensic short tandem repeat markers alteration in cancerous tissues: A scoping review. *Egyptian Journal of Forensic Sciences, 14*(1). https://doi.org/10.1186/s41935-024-00387-0, https://www.springer.com/journal/41935.

O'Brien, E., Daeid, N. N., & Black, S. (2015). Science in the court: Pitfalls, challenges and solutions. *Philosophical Transactions of the Royal Society B: Biological Sciences, 370*, 20150062. https://doi.org/10.1098/rstb.2015.0062.

Panneerchelvam, S., & Norazmi, M. N. (2003). Forensic DNA profiling and database. *Journal of Medical Sciences, 10*(2), 20–26.

Prahlow, J. A., Cameron, T., Arendt, A., Cornelis, K., Bontrager, A., Suth, M. S., Black, L., Tobey, R., Pollock, S., Stur, S., Cotter, K., & Gabrielse, J. (2017). DNA testing in homicide investigations. *Medicine, Science, and the Law, 57*(4), 179–191. https://doi.org/10.1177/0025802417721790, https://journals.sagepub.com/home/MSL.

Pyrek, K. M. (2007). *DNA: Convicting the guilty, exonerating the innocent*. Elsevier BV, 291–340. http://doi.org/10.1016/b978-012370861-8/50013-9.

Ralf, A., Zieger, M., & Kayser, M. (2024). Considerations on expanding criminal offender DNA databases with Y-STR profiles. *Journal of Law and the Biosciences, 11*(2), lsae017. https://doi.org/10.1093/jlb/lsae017.

Robertson, J. (2013). Understanding how forensic science may contribute to miscarriages of justice. *Australian Journal of Forensic Sciences, 45*(2), 109–112. https://doi.org/10.1080/00450618.2013.790477.

Satam, H., Joshi, K., Mangrolia, U., Waghoo, S., Zaidi, G., Rawool, S., Thakare, R. P., Banday, S., Mishra, A. K., Das, G., & Malonia, S. K. (2023). Next-generation sequencing technology: Current trends and advancements. *Biology, 12*(7). https://doi.org/10.3390/biology12070997, www.mdpi.com/journal/biology.

Shrivastava, P., Dixit, S., Srivastava, V., Dash, H. R., & Kumawat, R. K. (2020). Autosomal STR typing and case studies. In *Forensic DNA typing: Principles, applications and advancements*. India: Springer Singapore, 257–275. https://link.springer.com/book/10.1007/978-981-15-6655-4.

Shrivastava, P., Kumawat, R. K., & Dixit, S. (2021). Statistical evaluation of paternity using most recent STR-based multiplex system. *Proceedings of the National Academy of Sciences, India Section B: Biological Sciences, 91*(3), 575–580.

Soniya, E. V., & Suresh, K., U. (2022). DNA profiling for mass disaster victim identification. In H. R. Dash, P. Shrivastava, & J. A. Lorente (Eds.). *Handbook of DNA profiling* (pp. 503–518). Singapore: Springer. https://doi.org/10.1007/978-981-16-4318-7_24.

Tanudisastro, H. A., Deveson, I. W., Dashnow, H., & MacArthur, D. G. (2024). Sequencing and characterizing short tandem repeats in the human genome. *Nature Reviews. Genetics, 25*(7), 460–475. https://doi.org/10.1038/s41576-024-00692-3, https://www.nature.com/nrg/.

Thelingwani, R. S., Jonhera, C. A., & Masimirembwa, C. (2024). Analysis of data and common mutations encountered during routine parentage testing in Zimbabwe. *Scientific Reports, 14*, 1385. https://doi.org/10.1038/s41598-024-51987-8.

Uguen, K., Michaud, J. L., & Génin, E. (2024). Short tandem repeats in the era of next-generation sequencing: From historical loci to population databases. *European Journal of Human Genetics, 32*, 1037–1044. https://doi.org/10.1038/s41431-024-01666-z.

Van-Oorschot, R. A., Ballantyne, K. N., & Mitchell, R. J. (2010). Forensic trace DNA: A review. *Investigative Genetics, 1*(1), 14.

Verbiest, M. A., Lundström, O., Xia, F., Baudis, M., Bilgin Sonay, T., & Anisimova, M. (2024). Short tandem repeat mutations regulate gene expression in colorectal cancer. *Scientific Reports, 14*(1). https://doi.org/10.1038/s41598-024-53739-0, https://www.nature.com/srep/.

Verbiest, M., Maksimov, M., Jin, Y., Anisimova, M., Gymrek, M., & Bilgin Sonay, T. (2023). Mutation and selection processes regulating short tandem repeats give rise to genetic and phenotypic diversity across species. *Journal of Evolutionary Biology, 36*(2), 321–336. https://doi.org/10.1111/jeb.14106.

Vermylen, Y. (2010). The role of the forensic expert in criminal procedures according to Belgian Law. *Forensic Science International, 201*(1-3), 8–13. https://doi.org/10.1016/j.forsciint.2010.04.017.

Vigeland, M. D., & Egeland, T. (2021). Joint DNA-based disaster victim identification. *Scientific Reports, 11*(1). https://doi.org/10.1038/s41598-021-93071-5, www.nature.com/srep/index.html.

Vijayaraghavan, P., Batalov, S., Ding, Y., Sanford, E., Kingsmore, S. F., Dimmock, D., Hobbs, C., Bainbridge, M., & Galli, A. (2023). The genomic landscape of short tandem repeats across multiple ancestries. *PLoS One, 18*(1), e0279430. https://doi.org/10.1371/journal.pone.0279430.

Vitsios, D., Dhindsa, R. S., Middleton, L., Gussow, A. B., & Petrovski, S. (2021). Prioritizing non-coding regions based on human genomic constraint and sequence context with deep learning. *Nature Communications, 12*(1). https://doi.org/10.1038/s41467-021-21790-4, http://www.nature.com/ncomms/index.html.

Walsh, S. J. (2023). Forensic science in the criminal justice system: The good, the bad and the academy. *Australian Journal of Forensic Sciences, 55*(3), 285–294. https://doi.org/10.1080/00450618.2023.2200913.

Wet, S. D., Oosthuizen, H., & Visser, J. (2011). DNA profiling and the law in South Africa. *PER: Potchefstroom Electronic Law Journal, 14*(4), 171–207.

Wyner, N., Barash, M., & McNevin, D. (2020). Forensic autosomal short tandem repeats and their potential association with phenotype. *Frontiers in Genetics, 11*, 884. https://doi.org/10.3389/fgene.2020.00884.

Zhang, B. C., Biddanda, A., Gunnarsson, Á. F., Cooper, F., & Palamara, P. F. (2023). Biobank-scale inference of ancestral recombination graphs enables genealogical analysis of complex traits. *Nature Genetics, 55*(5), 768–776. https://doi.org/10.1038/s41588-023-01379-x, https://www.nature.com/ng/.

Zhang, W., & Yu, Y. Y. (2007). Polymorphisms of short tandem repeat of genes and breast cancer susceptibility. *European Journal of Surgical Oncology (EJSO), 33*(5), 529–534. https://doi.org/10.1016/j.ejso.2006.11.027.

Chapter 3

Detection and identification of forensically relevant body fluids

Hirak Ranjan Dash[1,2], **and Abhilash Kumar Das**[2]

[1]*Department of Forensic Science, National Forensic Sciences University, Delhi Campus, New Delhi, India,* [2]*School of Forensic Sciences, Centurion University of Technology and Management, Bhubaneswar, Odisha, India*

3.1 Introduction

Crime is any action committed that constitutes an offense and is punishable by the law of a state. As per the modern definition of crime, there are four basic elements of a typical crime, that is, *actus reus*, *mens rea*, injury, and involvement of human(s) (Chan & Simester, 2011). *Actus reus* is the Latin synonym for "guilty act." Similarly, *mens rea* is the Latin synonym for "guilty mind." However, most of the forensic biological investigation of a crime deals with the last two elements of a crime, that is, establishing the nature of an injury and proving the involvement of a human. In most of the crimes, when physical injury happens to a human body, there will be the release of body fluids. Due to the Locard's exchange principle, such body fluids are exchanged among the victim(s), perpetrator(s), and the crime scene, which can be called as the physical evidence (Fig. 3.1). Bulk of the physical evidences constitute the body fluids including blood, semen, saliva, vaginal fluids, tear, sweat and others. Forensic analysis of such body fluids helps in establishing the victim, suspect, crime scene, as well as the weapon used to a crime.

In a crime scene, body fluids are found either in visible conditions or in the latent form. The conventional body fluid analysis involves their detection, presumptive and confirmatory identification, occasional detection of species origin and blood group, followed by source attribution by DNA profiling techniques (An et al., 2012). Many chemical tests, immunological tests, protein catalytic activity tests, spectroscopic, and microscopic techniques have been developed for body fluid identification (BFI). Such conventional techniques mostly provide presumptive identification of body fluids and can detect only one body fluid at a time. Recently, many genetic-based approaches using either DNA methylation and RNA profiling have been developed for the identification of forensically relevant body fluids (Tian et al., 2020).

Forensically relevant body fluids are found in trace quantities, and one type of body fluid is often mixed with other body fluid types. Such varied natures of the body fluids found at the crime scene make their detection challenging. Butler (2015) explained the advanced forensic DNA protocols to possess the characteristics as described in the Olympic motto, that is, Citius (faster), Altius (higher), and Fortius (stronger). The advanced body fluid detection protocols should also follow similar characteristics. Besides, the body fluid detection is mostly required in the crime scene prior to the collection of the samples for further laboratory analysis. Thus the advanced body fluid detection techniques should also possess the

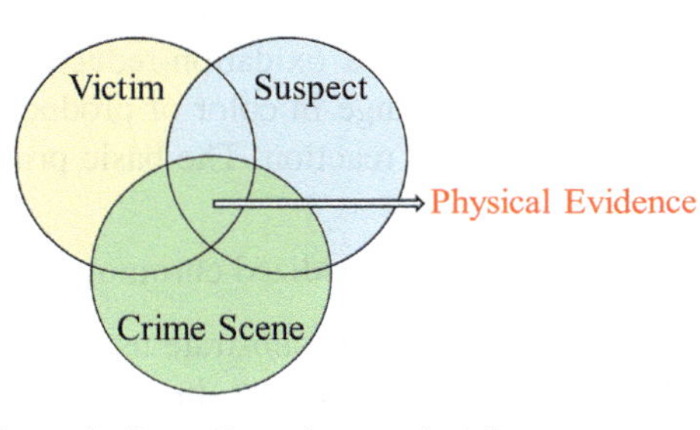

FIGURE 3.1 Origin of physical/biological evidences by Locard's exchange principle.

Investigative and Predictive DNA Testing. DOI: https://doi.org/10.1016/B978-0-443-30128-5.00010-2

characteristics of transportability. In this regard, the point of care tests (PoCTs) are the most suitable techniques for the detection of body fluids in a crime scene.

In many instances, the body fluids are found as trace evidences. A limited quantity of samples is recovered from the crime scene and is available for analysis. Thus the body fluid detection and identification technique should also facilitate nondestructive testing in the crime scene (Vyas et al., 2020). Some of the conventional techniques, such as Luminol test, crystal tests for blood, and microscopic confirmation of spermatozoa, have not been changed drastically since their inception. However, other tests utilizing enzyme-based detection of body fluids such as blood, semen, and saliva have been modified significantly as per the technological advancements (Virkler & Lednev, 2009). The present chapter describes the advanced techniques in the human body fluids detection for forensic interest, and their respective pros and cons have been described in detail.

3.2 Forensically relevant constituents of body fluids

Body fluids can be defined as the various liquids of the body resulting from the different physiological activities that play a pivotal role in maintaining well-being, nutrient transportation, and waste removal from the body. Thus more scientifically, they are the body's physiological fluids. Body fluids can be divided into two types, that is, secretory body fluids and excretory body fluids. Secretory body fluids involve the movement of body fluids within the body, such as blood, lymph, saliva, semen, vaginal fluid. Whereas, the excretory body fluids help in the removal of wastes from the body that including sweat, urine, and feces (Jaiswal & Kumar, 2023). Another way of classifying body fluids is (1) intracellular fluid and (2) extracellular fluid. Intracellular fluids are found within the cells which constitutes approximately 55% of total body water. Various organic substances such as glucose, amino acids, fatty acids, hormones, enzymes, and inorganic substances such as sodium, potassium, calcium, magnesium, chloride, phosphate, and sulfate constitute the intracellular fluids. Similarly, the extracellular body fluids constitute approximately 45% of total body water and include the interstitial fluid (present in between the cells), plasma (present in blood), lymph, synovial fluid, cerebrospinal fluid, pleural, pericardial, and peritoneal fluids. In most of the violent crimes, such body fluids are exchanges among the personnel involved in the crime and play a vital source of evidence for forensic biological analysis.

Body fluids are found often in violent crimes, and blood is most commonly encountered in a wide variety of crimes. Injury of any nature, such as cutting by a sharp weapon, can result in leaving blood evidence in the crime scene. Similarly, semen, vaginal fluids and menstrual blood are most commonly associated with the sexual assault cases. Saliva can also be recovered from the crime associated with oral sexual intercourse. Other body fluids such as urine, sweat, and feces are rarely recovered from the crime scenes (Williams, 2020).

Each body fluid has a unique composition. The presence of specific components in one body fluid compared to another is the fundamental basis of its identification. There are several components that commonly occurred in more than one body fluid. However, the relative abundance of these components varies among different body fluids. For example, large amount of amylase is found in Saliva compared to a smaller amount of amylase in semen and vaginal fluid (Auvdel, 1986). Similarly, the ratio of citrate to lactate varies between semen and vaginal fluid. Vaginal secretion contains a high level of lactic acid and androgenic stimulation resulting in semen production contains a high level of citric acid (Aldunate et al., 2013). Additionally, Urea can be found in urine, semen, and sweat. However, due to the presence of a higher concentration of urea in urine makes it a suitable indicator for urine detection. The forensically relevant components of commonly occurred body fluids are given in Table 3.1.

3.3 Blood and blood stains

3.3.1 Conventional techniques

Most of the presumptive tests of blood and blood stains relied on the presence of hemoglobin, the unique compound present in the blood. Such tests rely on the heme-catalyzed oxidation-reduction reaction. The colorless substrates are catalyzed due to this oxidation reaction, resulting in a change in color or producing chemiluminescence or fluorescence. The peroxidase activity of Heme facilitates this oxidation reaction. The basic principle of this oxidation reduction–based detection of hemoglobin is as follows:

$$H_2O_2 + \text{Chromogen (colorless)} \rightarrow \text{Oxidized chromogen (colored)} + H_2O$$

The principle of these tests remains the same, except for the substrate used. The Kastle–Meyer test, also known as the phenolphthalin assay, was developed by Kastle in 1901. In this test, in the presence of the peroxidase activity of heme, phenolphthalein (colorless compound) is oxidized to form phenolphthalein which appears pink under alkaline conditions.

TABLE 3.1 Forensically relevant components of routinely occurred body fluids in the crime scene.

Blood	Semen	Saliva	Vaginal fluid	Urine	Sweat
• Hemoglobin • Fibrinogen • Erythrocytes • Albumin • Glucose • Immunoglobulins	• Acid phosphatase • Prostate-specific antigen • Spermatozoa • Choline • Spermine • Semenogelin • Flavin • Citric acid • Lactic acid • Fructose • Urea • Ascorbic acid • Immunoglobulins	• Amylase • Lysozyme • Mucin • Buccal epithelial cells • Thiocyanate • Potassium • Bicarbonate • Phosphorous • Glucose • Immunoglobulins	• Acid phosphatase • Lactic acid • Citric acid • Urea • Vaginal peptidase • Glycogenated epithelial cells • Acetic acid • Pyridine • Squalene • Immunoglobulins	• Urea • Creatinine • Uric acid • Choline • Tamm-Horsfall glycoprotein (uromodulin)	• Urea • Lactic acid • Choline • Sodium • Potassium • Immunoglobulins

Similarly, a colorless Leucomalachite Green can be oxidized to form oxidized Malachite green in the presence of heme to develop a blue–green color. Many Benzidine derivatives such as orthotolidine and tetramethylbenzidine have also been explored for their suitability in the redox reaction–based detection of heme compounds. Similarly, other substrates such as luminol (3-aminophthalhydrazide) and fluorescin are routinely used as substrates for the chemiluminescence and fluorescence-based detection of blood (Barni et al., 2007). Recently, spectroscopy-based techniques have gained importance for the nondestructive detection of blood and blood stains. Analysis of the biological fingerprint region of blood in the range of 900–1800 cm^{-1} using attenuated total reflectance Fourier transform infrared spectroscopy (ATR-FTIR) has provided a simple, selective, nondestructive, and rapid approach for the analysis of blood stains (Wei et al., 2024). Similarly, the origin of specific blood Raman peaks at 1658 cm^{-1} (amide I), 1003 cm^{-1} (phenylalanine), 754 cm^{-1} (pyrrole ring) showed a huge potential in utilizing Raman spectroscopy for the detection and identification of blood evidence (Kistenev et al., 2023).

Rapid immunochromatographic (IC) kits, earlier used for the detection of blood in feces for colon cancer screening, are being used for forensic purposes. The IC kits rely on the principle of separation of molecules based on their ability to migrate on solid support by capillary flow, followed by the detection of a target molecule by antigen–antibody reaction (Cuttaia et al., 2024). Most of the blood detection kits target the detection of human hemoglobin (hHb) or glycophorin A (Gly-A) by monoclonal or polyclonal antibodies. The commonly used commercially available IC kits for the detection of human blood and their characteristics are given in Table 3.2 as per the information provided by the kit manufacturers.

3.3.2 Molecular techniques

Messenger RNA (mRNA) is differentially expressed in a tissue-specific manner that can be explored for the differentiation of multiple body fluids collected from a crime scene. Thus human blood-specific transcriptome profile allows its detection by a routine Reverse Transcriptase Polymerase Chain Reaction (RT-PCR) assay. The most widely used target genes for the detection of human blood are *HBA1* (codes for hemoglobin α1), *PBGD* (codes for porphobilinogen deaminase), and SPTB (codes for β-spectrin) (Li, 2015; Zubakov et al., 2009) utilized nine genes, that is, *ALOX5AP*, *AMICA1*, *AQP9*, *ARHGAP26*, *C1QR1*, *C5R1*, *CASP1*, *MNDA*, and *NCF2* for successful detection of older blood stains (13–16 years) by RT-PCR assay. Recently, reverse transcriptase–based recombinase polymerase amplification technique has been explored for the rapid detection of hemoglobin beta mRNA from crude blood lysate without the purification of RNA (Kubo et al., 2022).

Various studies have reported tissue-specific differential DNA methylation for the identification of body fluids. Frumkin et al. (2011) reported the first study on DNA methylation-based detection of blood stains for forensic identification purposes. Park et al. (2014) reported cg06379435 and cg08792630 as CpG sites as DNA methylation markers for the identification of blood samples. Similarly, various other studies have also identified other DNA methylation markers, that is, cg01543184, cg26285698, and cg03363565, for successful identification of blood stains (Lee et al., 2016). Most of the studies relied on the bisulfite sequencing method to study DNA methylation patterns. In this technique, the isolated DNA

TABLE 3.2 Characteristics of commercially available immunochromatographic kits for the detection of human blood.

Kits	Target	Sensitivity (human hemoglobin)	Sensitivity (blood)
ABAcard® Hematrace®	Human Hemoglobin (hHb)	50 ng/mL	0.3 nL
Hexagon OBTI	hHb	100 ng/mL	0.6 nL
SERATEC® HemDirect	hHb	40 ng/mL	0.25 nL
RSID™ blood	Glycophorin A	12.5 ng/mL	75 nL
Bluestar® OBTI	hHb	50 ng/mL	0.3 nL
BLUESTAR® Identi-HEM®	hHb	10 ng/mL	0.06 nL

is subjected to bisulfite treatment, which converts methylated cytosines into uracil. DNA Sequencing reveals the position and pattern of cytosine methylation (Gomaa et al., 2021). However, the recent development of Oxford Nanopore Sequencing allows the detection of CpGs without bisulfite conversion (López-Catalina et al., 2024).

MicroRNAs (miRNAs), a class of noncoding RNAs of 18–24 nucleotides, have also been explored as a suitable marker for the detection of blood evidence. Many studies revealed a tissue-specific expression of miRNAs either by microarray assay or RT-PCR technique of massively parallel sequencing (MPS). Zubakov et al. (2010) identified miR-144 and miR-185 as potential miRNA markers for the detection of blood from 1-year-old samples yielding a few picograms of total RNA. Gradually, with the advent of the MPS technique, many miRNA candidate markers have been identified for the detection of blood samples in a forensic scenario which include miR-451a, miR-144-3p, miR-16, miR-185, miR-126, miR-150, miR-486, miR-203a, miR-943, miR-200b, and miR-142-3p (Rocchi et al., 2020). Though these advanced molecular markers hold huge promise for the forensic identification of blood, they are yet to be adopted in the routine investigation practice of most forensic science laboratories.

3.4 Semen and seminal stains

Forensic-based detection of semen/seminal fluid plays an important role in investigating cases, mostly involving a sexual assault. Seminal fluid is the liquid or semigelatinous cellular suspension containing spermatozoa and the secretions of the accessory male reproductive organs. The presence or absence of seminal fluid in an object provides crucial information regarding the timing and identity of a sexual activity (Yao et al., 2020). The commonly used techniques and their principles have been discussed in detail in the subsequent sections.

3.4.1 Conventional techniques

Most of the conventional techniques used for the detection and identification of seminal fluid involve the presence of different forensically relevant compounds present in it. Some of the forensically relevant compounds found in a seminal fluid include choline, flavin, and lecithin of seminal vesicle origin, acid phosphatase, prostate-specific antigen, and spermine of prostate origin and spermatozoa produced by the testes. The most common presumptive test used for the screening of seminal fluid is the Acid phosphatase test. The acid phosphatase enzyme present in the semen can hydrolyze a variety of phosphate esters, facilitating the removal of the phosphate group. Subsequently, when a stabilized diazonium salt is added to it, an insoluble colored precipitate is formed at the site of acid phosphatase activity (Lewis et al., 2013). The most common substrates used for testing acid phosphatase activity include α-naphthyl phosphate, thymolphthalein monophosphate, phenyl phosphate, 4-nitrophenyl phosphate, and 4-methylumbelliferone phosphate. Crystal tests are also employed as presumptive tests for the detection of the presence of either choline or spermine. When florence reagent (potassium iodide + iodine + water) is added to the test slide containing seminal fluid, rhomboidal shape dark crystals of choline periodide can be observed. Similarly, when a suspected seminal stain is reacted with picric acid, yellow needle-shaped spermine picrate crystals are observed under a microscope (Ndayizeye & Sharma, 2024). However, such tests are presumptive in nature, as the target compound, such as choline, is also found in meat and dairy products.

The most commonly used confirmatory test for the detection of seminal fluid is the finding of spermatozoa under a microscope. The widely used Christmas tree staining technique differentially stains sperm head, neck, and tail. The primary stain nuclear fast red stains the nuclei of spermatozoa in red color in the presence of aluminum ions. The picroindigocarmine (PIC) stains the neck and tail portion green. In comparison to other available staining techniques, that is, hematoxylin–eosin, and alkaline fuchsin, the Christmas tree staining technique has been reported to be more suitable for the detection of spermatozoa, that is, > 35% (Allery et al., 2001). More recently, the fluorescent staining techniques have gained importance, such as the use of SPERM HY-LITER™. Use of such fluorescent dyes coupled with fluorescence microscope, flow cytometry, and computer analysis systems has increased the efficiency of spermatozoa detection in forensic case work samples (Golomingi et al., 2022). However, nonfinding of spermatozoa in conditions such as azoospermia, that is, lack of spermatozoa in the ejaculate, can generate a false negative result in case work samples.

Several studies have demonstrated the use of various protein markers for the confirmatory detection of seminal fluid. Some of the routinely described semen-specific protein markers include semenogelin 1, semenogelin 2, prostatic acid phosphatase, and prostate-specific antigen (Steendam et al., 2013). Although enzyme-linked immunosorbent assay (ELISA) is considered to be most suitable for the detection of a target antigen, the analysis complexity and cost involvement have disregarded this technique for forensic use. The lateral flow immunochromatogenic strips are commonly used for rapid determination of seminal fluid as PoCT.Sato et al. (2004) described a one-step IC assay using monoclonal and polyclonal antibodies against recombinant Sg-II (63 kDa). This testing technique neither showed any cross reactivity with saliva, urine, breast milk, serum or vaginal secretions nor with the semen from horses, dogs, swine and bulls. Similarly, other lateral flow assays based POCT kits include RSID semen test, ABAcard-p30, BLUESTAR Identi-PSA, and the PSA Semiquant test from SERATEC (Bruijns et al., 2023). Though such tests provide rapid and confirmatory detection of seminal fluid, consumption of the samples when available in minute quantities limits their application in forensics. Thus the advanced molecular techniques are currently being employed for the detection and identification of semen.

3.4.2 Molecular techniques

More recently, mRNA-based semen detection has gained importance due to its specificity and sensitivity. Various established mRNA markers for the RT-PCR–based detection of seminal fluid include protamine 1 (*PRM1*), protamine 2 (*PRM2*), and Semenogelin 1 (*SEMG1*). Recently, Kubo et al. (2022) described a single incubation step, that is, 42°C for 20 minutes for reverse transcription-recombinase polymerase amplification of mRNA markers. Haas et al. (2009) reported the usability of mRNA markers such as *PRM1*, *PRM2*, *SEMG1*, and *PSA* for the detection of semen in both azoospermic and normozoospermic men. This RT-PCR–based detection of mRNA markers has also been reported to detect semen samples of minute quantities (approximately 0.1 μL) and in aged stains up to 20 years. Besides forensic BFI, five mRNA markers, that is, *KLK3*, *TGM4*, *SEMG1*, *PRM1*, and *PRM2* showed their potential in distinguishing between normal and infertile semen (Tian et al., 2021).

The first study by Hanson et al. (2009) identified miR-135b and miR-10b as suitable miRNA markers for the detection of human semen. Subsequently, other semen-specific miRNA markers have also been explored, that is, miR-135a, miR-891a, miR-135a-5p, miR-888-5p, miR-891a-5p, miR-374 for efficient identification of semen stains in 1-year-old samples (Rocchi et al., 2020). Other candidate miRNA markers, that is, hsa-miR-943, hsa-miR-135a, hsa-miR-10a, and hsa-miR-507 have also shown differential expression patterns in semen, which have been validated by microarray data analysis as well as RT-PCR data (Zubakov et al., 2010). With the advent of NGS technique, expression pattern of as many as 2588 human mature miRNAs have been characterized in forensic semen stains and nine miRNAs, that is, hsa-miR-891a, hsa-miR-888, hsa-miR-429, hsa-miR-449a, hsa-miR-34b, hsa-miR-2392, hsa-miR-3197, hsa-miR-30a, and hsa-miR-196b showed potentially showed semen-specific nature (Wang et al., 2017).

Due to the limited stability of mRNAs and miRNAs, semen-specific DNA methylation patterns have been explored as an alternative and accurate method for the forensic detection of seminal fluid. Park et al. (2014) suggested the use of *cg23521140* and *cg17610929* DNA methylation markers for the detection of semen samples. Ghai et al. (2020) identified a novel tissue-specific differentially methylated region, that is, *ZNF282,* which displayed hypomethylation in semen-specific samples. Other epigenetic markers, that is, ZC3H12D, DACT1, USP49, cg17610929, cg26763284, cg17621389, cg17610929, cg06379435 cg22407458, and cg05656364 have also been reported for their usability in the detection of human seminal fluid (Park et al., 2014).

3.5 Vaginal fluid

Vaginal fluid is not commonly encountered in routine forensic cases except in sexual cases occasionally. However, limited number of tests are available for testing vaginal fluids due to their varied nature and composition. Due to the complex nature of the female reproductive system, the composition of vaginal fluid varies a lot with the different stages of the mensural cycle (Tindal et al., 2024). Besides, lack of any vaginal fluid specific characteristic proteins makes its identification difficult. Various conventional and molecular methods available for the forensic detection of vaginal fluids is discussed further.

3.5.1 Conventional techniques

Most of the forensic examinations of vaginal fluid is concentrated on the identification of glycogenated epithelial cells. The inner epithelial lining of vagina is comprised of squamous epithelial cells containing abundant glycogen compounds. Many histological staining techniques such as Papanicolaou (Pap) staining and Lugol's staining techniques are routinely used for such tests (Papanicolaou, 1942). Such staining techniques can be considered as presumptive tests as buccal cells of oral cavity also exhibit the similar properties. Immunodiffusion and immunoelectrophoresis techniques have also been explored using antihuman vaginal fluid precipitin. However, these tests are rarely performed in forensic case work samples due to the unavailability of high titer antibodies in the samples (Sakurada et al., 2020).

After the completion of the Human Microbiome Project, it was reported that different bodily fluids harbor characteristics microbial compositions. Vaginal microbial community of healthy females was found to be dominated by different species of *Lactobacillus*, that is, *L. crispatus*, *L. gasseri*, *L. iners*, and *L. jensenii* (Liao et al., 2023). In this regard, Akutsu et al. (2012) developed an RT-PCR–based detection of 16 S ribosomal RNA genes of *L. crispatus*, *L. jensenii* and *Atopobium vaginae* for effective forensic identification of vaginal fluid. Exploitation of vaginal microbiome by NGS analysis in 240 clinical vaginal samples showed 1619 different sequences. Further, microarray analysis coupled with probabilistic approach has been postulated to be the most suitable method for the vaginal microbiome–based detection (Benschop et al., 2012).

Vaginal fluid is often associated with the menstrual blood. Thus it becomes challenging for a forensic biologist to differentiate between peripheral blood and menstrual blood. Many IC techniques have been developed targeting degradation products of fibrinolysis for identification of menstrual blood (Fig. 3.2). During menstruation, fibrinolysis (endogenous degradation of fibrin) of blood clots helps in the easy removal of menstrual blood. D-dimer, the small protein fragment specific for fibrinolysis, has been reported to be present in menstrual blood 200 times higher than that of peripheral blood (Holtkötter et al., 2017). Though D-dimer has been demonstrated to be a suitable marker for the detection of menstrual blood, few IC kits showed a false positive reaction for D-dimer in blood (injury), nasal blood, postmortem blood, and wound crust (Konrad et al., 2023). Thus the molecular markers have been explored for their efficient identification, which has been discussed further.

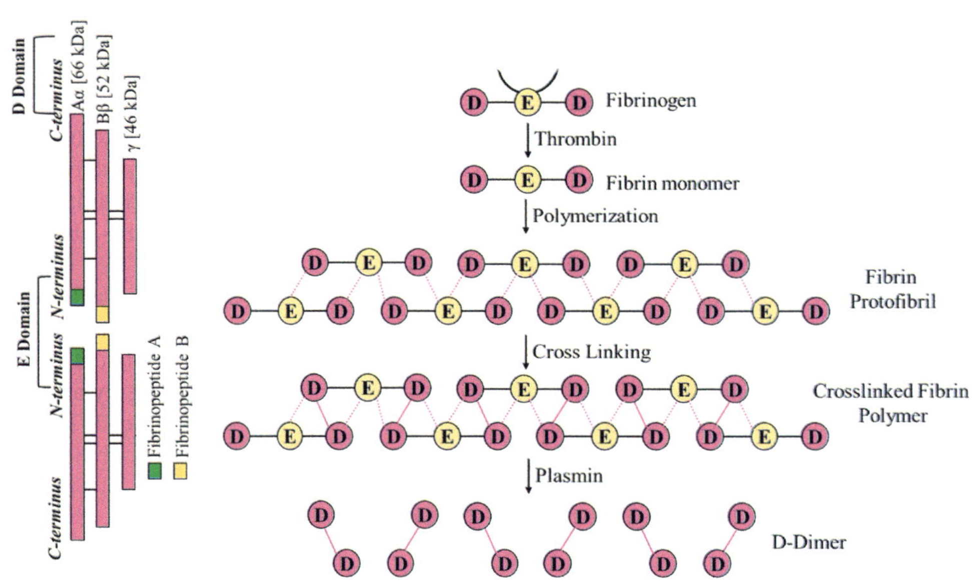

FIGURE 3.2 Process of formation of D-dimer. Fibrinogen is clotted by thrombin to form fibrin monomers. Fibrin monomers polymerize to form fibrin protofibril, which undergo cross-linking to form cross-linked fibrin polymer. Proteolysis by plasmin liberates E-domain from D-dimer.

3.5.2 Molecular techniques

The most widely used molecular basis of detection of vaginal fluid includes the molecular identification of vaginal commensal microbes. *Lactobacillus* sp. including *L. crispatus*, *L. iners*, *L. gasseri*, and *L. jensenii* have been established as the predominant commensals of vagina in women of childbearing age. Thus different multiplex RT-PCR assays have been established targeting genome-specific identification of these commensal microbes from microflora DNA. Giampaoli et al. (2012) developed a multiplex PCR assay for the simultaneous detection of *L. crispatus*, *L. gasseri*, *S. aureus*, *S. salivarius*, *S. mutans*, and *Enterococcus* spp. and successfully validated the protocol for detection of vaginal fluid in fresh as well as stored undergarment samples. Relative quantification of *Lactobacillus* (*L. crispatus*, *L.* gasseri, *L. iners,* and *L. jensenii*) levels compared to human specific primer pair (HSP) as an endogenous control has also been established as a useful technique for the detection of vaginal fluid in forensic samples. ΔCt of the expression level of *Lactobacillus* conserved primers and HSP was also found useful in distinguishing vaginal fluid from other body fluids including saliva, semen and skin surface samples (Doi et al., 2014). However, the microbiome analysis is strongly influenced by the presence of exogenous microflora contaminated from the matrix where vaginal fluid is found. Analysis of 16 S rRNA on the V3-V4 regions suggested the stable occurrence of *Lactobacillus* and *Gardnerella* genera in all types of substrates, even after exposure (Liao et al., 2023). Various primer sets used for the targeted amplification of vaginal commensals have been compiled in Table 3.3.

The simultaneous detection of vaginal mRNAs and *Lactobacilli* has shown its usefulness for rapid and sensitive detection of the vaginal fluid. Jakubowska et al., (2013) developed a hexaplex targeting human beta-defensin 1, mucin 4 (MUC4), menstrual blood marker matrix metalloproteinase 11, house-keeping gene glucose 6-phosphate dehydrogenase (*G6PDH*), and the 16S-23S rRNA of *L. crispatus*, and *L. gasseri/L. johnsonii.* This hexaplex showed potential in detecting the presence of vaginal fluid when mixed with semen and blood in stains of > 2 years old. Similarly, 10 genes covering 39 SNPs have been identified and validated in a NGS panel for the identification of vaginal mucosa (Neis et al., 2024). Multiplexing of various genes targeting *ESR1*, *SERPINB13*, *KLK13*, *CYP2B7P1*, and *MUC4* genes have also been established for the detection of vaginal fluid. However, limited studies are available for vaginal fluid specific detection of miRNA or DNA methylations (Sakurada et al., 2020).

Menstrual blood is occasionally found to be mixed with vaginal fluid. Initial evaluation of tissue-specific expression of various mRNA markers targeting hormone receptors, intermediate filaments, matrix metalloproteinases (MMP), heat shock proteins, cytokines, and growth factors identified MMP to be a suitable marker for the detection of menstrual blood in forensic samples (Bauer & Patzelt, 2002). The expression level of MMP-7 compared to a house-keeping gene GAPDH

TABLE 3.3 Useful primer sets for the targeted amplification of vaginal commensals for the detection of vaginal fluids in forensic biological samples.

Target commensal	Target region	Primer sequences (5′-3′)	Amplicon size (bp)	References
L. crispatus	16S-23S ISR	F: CAGAGCAAGCGGAAGCACA	279	Fleming and Harbison (2010)
		R: CATCTCTGCATTGGGTTCCC		
L. gasseri		F: GAGAAAGCCAAGCGGAAGC	253	
		R: TTGCTTACTTACTGCTCCCCG		
Lactobacillus spp.	Conserved region	F: AGAGGAGAGTGGAACTCCATG	205	Doi et al. (2014)
		R: GTTAGCTGCAGCACTGAGAG		
L. crispatus	EU161625	F: GCACTAACAGCCGAAGAAGG	NA	Giampaoli et al. (2012)
		R: TTCGGATATCTCCGGATCAC		
L. gasseri	DQ445775	F: AGATTGAAGAGCGACCGAGA		
		R: CCTTCATCGGCTTCTAGTGC		
L. crispatus	16 S rRNA	F: CAGCAGTAGGGAATCTTC	145	Lewis and Williams (2022)
		R: CTGGTTGATTACCGTCAA		

was found to be useful in distinguishing menstrual blood from other body fluids, including the peripheral blood (Bauer & Patzelt, 2008). Analysis of three mRNA markers, that is, PAEP, LAPR3, and HOXA10, has also shown their usefulness in distinguishing menstrual blood from venous blood, semen, and saliva. Out of these three mRNA markers, LAPR3 and PAEP showed tissue-specific expression exclusively in menstrual blood, whereas HOXA10 was reported to be expressed in venous blood, menstrual blood, and semen samples (Liang et al., 2018). Recently, a 4-plex system has been developed and validated for investigation of sexual assault cases by targeting a positive control, digestive control, menstrual blood specific marker and vaginal fluid specific marker simultaneously (Hsu et al., 2025). Thus such molecular markers–based identification of vaginal fluid and menstrual blood are highly useful in investigating the sexual assault cases.

3.6 Saliva and saliva stains

Compared to other body fluids, the detection and identification of saliva stains is highly challenging due to their transparent nature. It is one of the most common pieces of evidence found in routine cases such as homicide, sexual assault, suicidal hanging, burglary, and bite marks (Upadhyay et al., 2023). In most of the criminal cases, saliva stains are found to be mixed with other body fluids, making the analysis process more complicated. Some of the important conventional and molecular techniques for both presumptive and confirmatory analysis of saliva stains have been discussed further.

3.6.1 Conventional techniques

Visualization of suspect stains using an alternative light source, such as at 470 nm excitation wavelength, coupled with orange goggles, can enable fluorescence-based detection of saliva stains (Denny et al., 2018). However, the most common technique used for the detection of saliva is the use of the starch-Iodine assay. A dark blue complex is formed when iodine mixes with the amylose, whereas amylopectin results in a reddish-purple color. Human salivary amylase is a major component of saliva, and it has the potential to break down starch to monosaccharides and disaccharides. Such amylase-mediated breakdown of starch prevents the development of color upon addition of iodine, resulting in a presumptive positive result for saliva. Many variants of Amylase detection techniques, such as the use of Phadebas reagent, SALIgAE® kit (Abacus Diagnostics), and SALIgAE spray kit, have been developed over the years for rapid and sensitive detection of salivary amylase (Pang & Cheung, 2008). However, most of these tests are not specific for human salivary amylase (HSA) and show cross reactivity with human pancreatic amylase, leading to the generation of false positive results.

For confirmatory detection of saliva, many IC techniques are also being used. In this regard, the RSID®-Saliva kit provides the definitive HSA identification. This kit utilizes a dual antibody system, with labeled and immobilized anti-HSA antibodies, and an antiglobulin control. When the extracted sample is loaded into the well, the development of a pink line in both test and control zones conclusively identifies the presence of human salivary amylase. RSID®-Saliva kit has the potential to detect HSA in as low as 1 μL of sample within 10 minutes. It also shows high specificity for HSA and does not show any cross-reactivity with other body fluids or saliva of other animal species tested (Old et al., 2009). ELISA is also widely used in both detecting and quantifying HSA employing anti-HSA antibodies. In a forensic setup, antibody–antigen–antibody sandwich ELISA is commonly used, where the target enzyme produces colorimetric or fluorescent signals (Quarino et al., 2005). The intensity of such signals is measured by spectrophotometric techniques and is directly proportional to the amount of bound-HSA. Despite of high specificity and sensitivity of ELISA-based detection of HSA, it has several limitations, including more time-consuming, and requiring multiple incubation and washing steps. Garcia et al. (2018) developed screen-printed carbon electrodes (SPCEs) for amperometric detection of human salivary α-amylase. In this test, salivary α-amylase breaks down starch to a reducing sugar, maltose, in the primary step. Maltose, being a reducing sugar, converts $[Fe(CN)_6]^{3-}$ into $[Fe(CN)_6]^{4-}$ in a second reaction, which is detected by the SPCE. This is a rapid and PoC test for the confirmatory detection of saliva stains in forensic biological samples.

3.6.2 Molecular techniques

Various mRNA, miRNA, and methylation markers have been explored for the confirmatory detection of human saliva. Watanabe et al. (2017) postulated two saliva-specific mRNA markers, that is, Statherin (*STATH*) and Histatin 3 (*HTN3*), for the detection of saliva stains in fresh and 1-year -ld samples. Subsequently, a multiplex system targeting *STATH*, *HTN3*, *PRM2* (Protamin 2), and *SEMG1* (Semenogelin 1) has been developed for the detection of saliva as well as distinguishing it from seminal fluid (Sakurada et al., 2009). Whole genome expression analysis using the Affymetrix U133 plus2 GeneChip showed five stable mRNA markers of saliva origin, that is, *SPRR3*, *SPRR1A*, *KRT4*, *KRT6A*, and *KRT13* (Zubakov et al., 2007). These five markers showed significantly higher expression ($ddC_t > 15$) in saliva compared to semen and other body fluids tested. Both *SPRR1A* and *SPRR3* genes code for cornified envelope precursor proteins. They are predominantly expressed in oral and esophageal epithelia, resulting in terminal differentiation of keratinocytes. Thus these stable saliva-specific mRNA markers can be used as an alternative for rapid and sensitive detection of this body fluid in forensic biological samples.

Due to their stability compared to mRNA, miRNA analysis has been preferred for forensic identification of saliva. Omelia et al. (2013) identified miR-205 for the selective detection of saliva in human samples. Another study revealed higher or moderate expression of miR-200c-3p, miR-203a, and miR-205-5p markers in saliva using Hydrolysis probes based quantitative Real Time Polymerase Chain Reaction (qRT-PCR) assay (Wang et al., 2015). Similarly, DNA methylation markers have also been explored for the detection of saliva in forensic samples. Frumkin et al. (2011) analyzed methylation patterns at 15 loci and identified the value of methylation ratios, such as L91762/L68346, and L76138/L26688 as an indicator for the tissue-specific methylation pattern. Thus molecular approaches of saliva detection not only provide its confirmatory detection, but also increase the sensitivity and specificity of these tests.

3.7 Urine

Urine is one of the most common biological types of evidence found in crime scenes, including sexual assault, homicide, and other case types. The urinary excretory system of the body helps in removing the toxic wastes out of the body, and in this process, the urine formation takes place in the nephrons found in the kidney. Water is the major constituent of urine. Urea is the most abundant waste product found in urine, constituting approximately 8 g/L (Song et al., 2009). Besides, other major constituents of urine include creatinine, uric acid, phosphate, sulfate, chloride, sodium, and potassium ions. A small amount of protein is also present in the urine of healthy individuals. Most of the presumptive and confirmatory tests detect the presence of those compounds which have been discussed further.

Presence of characteristic yellow stains, fluorescence under visible and UV light, and distinctive odor may indicate the presence of urine stains. Determination of uric acid due to its UV absorption at 293 nm is recommended for presumptive detection of urine. Detection of urea and creatinine, respectively, by para-dimethylaminocinnamaldehyde (DMAC) test and Jaffe color test are the routinely used presumptive tests for urine identification. In the DMAC assay, 0.1% of DMAC solution is added to a filter paper containing extracted material, and the presence of urine develops a pink or magenta colored product. DMAC specifically reacts with urea and does not show cross reactivity with other compounds such as creatinine, ammonia, or uric acid (Ong et al., 2012). The development of color within 30 minutes can be considered as a positive test while if the color develops after 30 minutes, it can be considered as a negative result. Though this test is specific to urea, the presence of urea in other body fluids such as saliva, semen, sweat and vaginal fluids can give false positive reactions. Similarly, creatinine in urine can be detected by using the Jaffe color test. When a urine stain is applied with Picric acid, it binds with the creatinine to form creatinine picrate and a bright red color is produced. However, due to safety concerns of the use of Picric acid, the DMAC test is the most preferred technique for urine detection in most of the forensic science laboratories (Ong et al., 2012). A PoC test based on lateral flow assay has also been developed to detect creatinine, that is, Uritrace™ (Abacus Diagnostics, West Hills, CA). Similarly, urea in urine can also be detected by urease assay (Ramirez et al., 2022). Urease breaks down urea to form ammonia and CO_2. Further, ammonia is detected by bromothymol blue to exhibit a blue color.

One of the most commonly used confirmatory tests for urine identification is the detection of Tamm-Horsfall protein (THP). THP is a glycoprotein produced exclusively by the renal epithelial cells inside the Loop of Henle. Though the complete function of THP is not understood, some of the studies have identified THP as a specialized defense molecule of the urinary tract (Weichhart et al., 2005). Radioimmune assay and ELISA have been developed for the detection of THP. However, the time consumption and cost involved in these tests have limited their use in the forensic setup. Thus the IC techniques have been developed using polyclonal rabbit antibody specific for THP for rapid confirmatory detection of urine stains (Akutsu, Watanabe, et al., 2012). RSID™-Urine kit targets THP by IC technique and can detect as low as 0.5 μL of urine within 15 min. Similarly, the detection of five major components of 17-ketosteroid conjugates, that is, androsterone glucuronide, androsterone sulfate, DHEA sulfate, etiocholanolone glucuronide, and etiocholanolone sulfate, is also considered to be a confirmatory test for urine detection. Gas chromatography conjugated with mass spectrometry (GC-MS) techniques have been developed and validated to analyze those ketosteroid conjugates for forensic application (Miyata et al., 2012). However, limited studies are available for the molecular-based detection of urine.

3.8 Sweat

Out of all the body fluids described, sweat is the least commonly encountered body fluid in the crime scene. Whenever it is present, it provides a useful source for DNA as recovered from wrist bands, sweatbands, or hats. However, the forensic detection of sweat is highly complicated as the constituents of sweat mimic other body fluids. The most commonly used presumptive test for sweat includes the identification of relative concentrations of sodium, phosphorus, sulfur, chlorine, potassium, calcium, and other metal traces (Virkler & Lednev, 2009). The occurrence of large chlorine peaks is usually used to distinguish sweat from other body fluids.

Dermcidin has been found as a useful potential biomarker for the detection of human sweat. Dermcidin is a class of human antimicrobial peptides and acts as an epithelial barrier for infection. Dermcidin is encoded by the *DCD* gene and is categorically expressed in eccrine sweat glands. Western blotting using the monoclonal G-81 antibody has been developed for the detection of sweat samples (Minami et al., 2004). Dermcidin detection using qRT-PCR assay and ELISA is widely used in forensic setup (Sakurada et al., 2010). Besides *DCD*, other genes such as orneodesmosin (*CDSN*), late cornified envelope 1 C (*LCE1C*), filaggrin (*FLG*), and desmocollin 1 have also been evaluated and found useful for the detection of sweat or skin characteristics to infer the human origin of the touched contact surfaces (Akutsu et al., 2018).

3.9 Other emerging techniques

Nowadays, the body fluid detection techniques are more confined toward the rapid nondestructive techniques, including vibrational spectroscopy. Techniques such as Raman spectroscopy, FTIR, and nuclear magnetic resonance (NMR) leverage the unique spectral signatures of biological molecules to differentiate between various body fluids. Raman spectroscopy, particularly when coupled with discriminant functional analysis, has demonstrated high accuracy (up to 99%) in identifying saliva, blood, and semen (Muro et al., 2016). Similarly, ATR-FTIR coupled with chemometric methods allows for the estimation of bloodstain age and the sensitive detection of bodily fluids on fabric substrates (Takamura et al., 2018).

NMR spectroscopy, focusing on the relative abundance of metabolite markers, complements these techniques by providing detailed compositional information. Metabolomic studies using ^{1}H NMR, often combined with principal component analysis, have proven effective in differentiating urine and saliva (Lenz et al., 2003). Research has also established the reliability of NMR in assessing bio-specimen storage effects and analytical reproducibility (Barton et al., 2008). Furthermore, advancements integrating NMR with mass spectrometry and filtration methods have enhanced metabolite identification in serum, aiding in disease diagnostics and forensic investigations (Derveaux et al., 2021). The resulting NMR spectra are influenced by experimental conditions and sample preparation, highlighting the need for standardized protocols.

3.10 Conclusion and future directions

BFI to detect the nature of body fluid and source attribution are two important aspects of forensic biology. Over the years, many technological advancements have contributed significantly to the rapid and sensitive detection of body fluids. The basic serology concepts have been aligned with the advanced technologies, such as lateral flow assay and IC techniques, to yield a better result. BFI is primarily useful during crime scene investigation prior to the sample collection. Hence, the rapid, PoCTs are the need of the hour for an effective forensic analysis. In this regard, Butler (2015) proposed a few characteristics of the future forensic DNA analysis techniques. The future technologies should follow the Olympic motto, that is, faster, higher, and stronger. The same features apply to the body fluid detection techniques. In future, we may witness the techniques having faster results, robustness, and stronger admissibility in the court of law.

Forensic biological analysis is always challenged by limited sample quantity. In many instances, traces of body fluids are recovered. If a large quantity of sample is consumed for the presumptive identification of body fluid, the sample may not be available for individual identification by DNA analysis. Thus the nondestructive techniques like vibrational spectroscopy-based body fluid detection have gained importance. Besides, the molecular techniques, such as tissue-specific DNA methylation pattern, analysis are also highly useful. In this case, the same sample can be used for body fluid analysis as well as for individualization tests by DNA profiling.

References

Akutsu, T., Motani, H., Watanabe, K., Iwase, H., & Sakurada, K. (2012). Detection of bacterial 16S ribosomal RNA genes for forensic identification of vaginal fluid. *Legal Medicine, 14*(3), 160–162. https://doi.org/10.1016/j.legalmed.2012.01.005.

Akutsu, T., Watanabe, K., & Sakurada, K. (2012). Specificity, sensitivity, and operability of RSID™-urine for forensic identification of urine: Comparison with ELISA for Tamm-Horsfall protein. *Journal of Forensic Sciences*. https://doi.org/10.1111/j.1556-4029.2012.02174.x.

Akutsu, T., Watanabe, K., Takamura, A., & Sakurada, K. (2018). Evaluation of skin- or sweat-characteristic mRNAs for inferring the human origin of touched contact traces. *Legal Medicine, 33*, 36–41. https://doi.org/10.1016/j.legalmed.2018.05.003.

Aldunate, M., Tyssen, D., Johnson, A., Zakir, T., Sonza, S., Moench, T., Cone, R., & Tachedjian, G. (2013). Vaginal concentrations of lactic acid potently inactivate HIV. *Journal of Antimicrobial Chemotherapy, 68*(9), 2015–2025. https://doi.org/10.1093/jac/dkt156.

Allery, J.-P., Telmon, N., Mieusset, R., Blanc, A., & Rougé, D. (2001). Cytological detection of spermatozoa: Comparison of three staining methods. *Journal of Forensic Sciences, 46*(2), 349–351. https://doi.org/10.1520/jfs14970j.

An, J. H., Shin, K. J., Yang, W. I., & Lee, H. Y. (2012). Body fluid identification in forensics. *BMB Reports, 45*(10), 545–553. https://doi.org/10.5483/BMBRep.2012.45.10.206, http://bmbreports.org/jbmb/pdf.php?data=MTIxMjE3MTVAcGRmX3JhaW50cmFjZV9sZWV5c0AlNUI0NS0xMCU1RDEyMTAyOTIxMTBfJTI4NTQ1LTU1MyUyOUJNQl8xMi0yMDYucGRm.

Auvdel, M. J. (1986). Amylase levels in semen and saliva stains. *Journal of Forensic Sciences, 31*(2), 426–431. https://doi.org/10.1520/jfs12272j.

Barni, F., Lewis, S. W., Berti, A., Miskelly, G. M., & Lago, G. (2007). Forensic application of the luminol reaction as a presumptive test for latent blood detection. *Talanta, 72*(3), 896–913. https://doi.org/10.1016/j.talanta.2006.12.045.

Barton, R. H., Nicholson, J. K., Elliott, P., & Holmes, E. (2008). High-throughput 1H NMR-based metabolic analysis of human serum and urine for large-scale epidemiological studies: Validation study. *International Journal of Epidemiology, 37*(1), i31. https://doi.org/10.1093/ije/dym284.

Bauer, M., & Patzelt, D. (2002). Evaluation of mRNA markers for the identification of menstrual blood. *Journal of Forensic Sciences, 47*(6), 1278–1282. https://doi.org/10.1520/jfs15560j.

Bauer, M., & Patzelt, D. (2008). Identification of menstrual blood by real time RT-PCR: Technical improvements and the practical value of negative test results. *Forensic Science International, 174*(1), 55–59. https://doi.org/10.1016/j.forsciint.2007.03.016.

Benschop, C. C. G., Quaak, F. C. A., Boon, M. E., Sijen, T., & Kuiper, I. (2012). Vaginal microbial flora analysis by next generation sequencing and microarrays: Can microbes indicate vaginal origin in a forensic context? *International Journal of Legal Medicine, 126*(2), 303–310. https://doi.org/10.1007/s00414-011-0660-8.

Bruijns, B., Tiggelaar, R., Knotter, J., & van Dam, A. (2023). Use of lateral flow assays in forensics. *Sensors, 23*(13), 6201. https://doi.org/10.3390/s23136201.

Butler, J. M. (2015). The future of forensic DNA analysis. *Philosophical Transactions of the Royal Society B: Biological Sciences, 370*(1674), 20140252. https://doi.org/10.1098/rstb.2014.0252.

Chan, W., & Simester, A. P. (2011). Four functions of mens rea. *The Cambridge Law Journal, 70*(2), 381–396. https://doi.org/10.1017/s0008197311000547.

Cuttaia, C., Di Stefano, B., Sorçaburu Ciglieri, S., Vetrini, R., Previderè, C., & Fattorini, P. (2024). Immunochromatographic detection of human blood: A forensic review. *Separations, 11*(3), 66. https://doi.org/10.3390/separations11030066.

Denny, S. E., Nazeer, S. S., Sivakumar, T. T., Nair, B. J., & Jayasree, R. S. (2018). Forensic application of fluorescence spectroscopy: An efficient technique to predict the presence of human saliva. *Journal of Luminescence, 203*, 696–701. https://doi.org/10.1016/j.jlumin.2018.07.022.

Derveaux, E., Thomeer, M., Mesotten, L., Reekmans, G., & Adriaensens, P. (2021). Detection of lung cancer via blood plasma and 1H-NMR metabolomics: Validation by a semi-targeted and quantitative approach using a protein binding competitor. *Metabolites, 11*(8), 537. https://doi.org/10.3390/metabo11080537.

Doi, M., Gamo, S., Okiura, T., Nishimukai, H., & Asano, M. (2014). A simple identification method for vaginal secretions using relative quantification of Lactobacillus DNA. *Forensic Science International: Genetics, 12*, 93–99. https://doi.org/10.1016/j.fsigen.2014.05.005.

Fleming, R. I., & Harbison, S. (2010). The use of bacteria for the identification of vaginal secretions. *Forensic Science International: Genetics, 4*(5), 311–315. https://doi.org/10.1016/j.fsigen.2009.11.008.

Frumkin, D., Wasserstrom, A., Budowle, B., & Davidson, A. (2011). DNA methylation-based forensic tissue identification. *Forensic Science International: Genetics, 5*(5), 517–524. https://doi.org/10.1016/j.fsigen.2010.12.001.

Garcia, P. T., Guimarães, L. N., Dias, A. A., Ulhoa, C. J., & Coltro, W. K. T. (2018). Amperometric detection of salivary A-amylase on screen-printed carbon electrodes as a simple and inexpensive alternative for point-of-care testing. *Sensors and Actuators, B: Chemical, 258*, 342–348. https://doi.org/10.1016/j.snb.2017.11.068.

Ghai, M., Naidoo, N., Evans, D. L., & Kader, F. (2020). Identification of novel semen and saliva specific methylation markers and its potential application in forensic analysis. *Forensic Science International: Genetics, 49*, 102392. https://doi.org/10.1016/j.fsigen.2020.102392.

Giampaoli, S., Berti, A., Valeriani, F., Gianfranceschi, G., Piccolella, A., Buggiotti, L., Rapone, C., Valentini, A., Ripani, L., & Romano Spica, V. (2012). Molecular identification of vaginal fluid by microbial signature. *Forensic Science International: Genetics, 6*(5), 559–564. https://doi.org/10.1016/j.fsigen.2012.01.005.

Golomingi, R., Haas, C., Dobay, A., Kottner, S., & Ebert, L. (2022). Sperm hunting on optical microscope slides for forensic analysis with deep convolutional networks – A feasibility study. *Forensic Science International: Genetics, 56*, 102602. https://doi.org/10.1016/j.fsigen.2021.102602.

Gomaa, R., Nader, L., & Jamal, J. (2021). Application of DNA methylation-based markers in identification of mixed body fluid evidences simulating crime scene scenarios. *Egyptian Journal of Forensic Sciences, 11*(12). https://doi.org/10.1186/s41935-021-00226-6.

Haas, C., Muheim, C., Kratzer, A., Bär, W., & Maake, C. (2009). mRNA profiling for the identification of sperm and seminal plasma. *Forensic Science International: Genetics Supplement Series, 2*(1), 534–535. https://doi.org/10.1016/j.fsigss.2009.08.109.

Hanson, E. K., Lubenow, H., & Ballantyne, J. (2009). Identification of forensically relevant body fluids using a panel of differentially expressed microRNAs. *Analytical Biochemistry, 387*, 303–314. https://doi.org/10.1016/j.ab.2009.01.037.

Holtkötter, H., Filho, C. R. D., Schwender, K., Stadler, C., Vennemann, M., Pacheco, A. C., & Roca, G. (2017). Forensic differentiation between peripheral and menstrual blood in cases of alleged sexual assault-validating an immunochromatographic multiplex assay for simultaneous detection of human hemoglobin and D-dimer. *International Journal of Legal Medicine, 132*, 683–690. https://doi.org/10.1007/s00414-017-1719-y.

Hsu, C. T., Tsai, L. C., Liu, K. L., Lin, Y. C., Huang, N. E., Lee, J. C. I., Linacre, A., & Hsieh, H. M. (2025). Identification of menstrual blood and vaginal fluid using a 4-plex MSRE-PCR system applicable to alleged sexual assault cases. *International Journal of Legal Medicine*. https://doi.org/10.1007/s00414-025-03471-9, https://www.springer.com/journal/414.

Jaiswal, N., & Kumar, A. (2023). Human secretory and excretory fluids, molecular constituents, and their biotherapeutic perspective against fungal pathogen *Candida albicans*. *Rendiconti Lincei. Scienze Fisiche e Naturali. 34*(4), 1105–1113. https://doi.org/10.1007/s12210-023-01207-8.

Jakubowska, J., MacIejewska, A., Pawłowski, R., & Bielawski, K. P. (2013). MRNA profiling for vaginal fluid and menstrual blood identification. *Forensic Science International: Genetics, 7*(2), 272–278. https://doi.org/10.1016/j.fsigen.2012.11.005.

Kistenev, Y. V., Borisov, A. V., Samarinova, A. A., Colón-Rodríguez, S., & Lednev, I. K. (2023). A novel Raman spectroscopic method for detecting traces of blood on an interfering substrate. *Nature Research, Russian Federation Scientific Reports, 13*(1). https://doi.org/10.1038/s41598-023-31918-9, https://www.nature.com/srep/.

Konrad, H., Hartung, B., & Poetsch, M. (2023). (Un) Reliable detection of menstrual blood in forensic casework-evaluation of the Seratec® PMB test with mock samples. *International Journal of Legal Medicine, 138*, 781–786. https://doi.org/10.1007/s00414-023-03138-3.

Kubo, S., Niimi, H., & Kitajima, I. (2022). Rapid detection of blood and semen mRNA markers by reverse transcription-recombinase polymerase amplification. *Forensic Science International Genetics, 58*, 102665. https://doi.org/10.1016/j.fsigen.2022.102665.

Lee, H. Y., Lee, S. D., & Shin, K. J. (2016). Forensic DNA methylation profiling from evidence material for investigative leads. *BMB Reports, 49*, 359–369. https://doi.org/10.5483/BMBRep.2016.49.7.070.

Lenz, E. M., Bright, J., Wilson, I. D., Morgan, S. R., & Nash, A. F. P. (2003). A 1H NMR-based metabonomic study of urine and plasma samples obtained from healthy human subjects. *Journal of Pharmaceutical and Biomedical Analysis, 33*(5), 1103–1115. https://doi.org/10.1016/s0731-7085(03)00410-2.

Lewis, C., & Williams, S. J. S. (2022). Design and optimization of a 16S microbial qPCR multiplex for the presumptive identification of feces, saliva, vaginal and menstrual secretions. *Journal of Forensic Sciences, 67*, 1660–1667. https://doi.org/10.1111/1556-4029.15029.

Lewis, J., Baird, A., McAlister, C., Siemieniuk, A., Blackmore, L., McCabe, B., O'Rourke, P., Parekh, R., Watson, E., Wheelhouse, M., & Wilson, N. (2013). Improved detection of semen by use of direct acid phosphatase testing. *Science and Justice, 53*(4), 385–394. https://doi.org/10.1016/j.scijus.2013.04.009.

Li, R. (2015). *Forensic biology, second edition*. United States: Taylor and Francis, 1–534. http://doi.org/10.1201/b18209, http://www.tandfebooks.com/doi/book/10.1201/b18209.

Liang, Q., Sun, H., Wu, X., Ou, X., Gao, G., Jin, Y., & Tong, D. (2018). Development of new mRNA markers for the identification of menstrual blood. *Annals of Clinical and Laboratory Science, 48*(1), 55–62. http://www.annclinlabsci.org/content/48/1/55.full.pdf.

Liao, L., Ye, L., Huang, L., Yao, T., Liang, X., Chen, L., & Shen, M. (2023). Changes in the microbial community after vaginal fluid exposure in different simulated forensic situations. *Forensic Science International, 349*, 111766. https://doi.org/10.1016/j.forsciint.2023.111766.

López-Catalina, A., Costes, V., Peiró-Pastor, R., Kiefer, H., & González-Recio, O. (2024). Oxford nanopore sequencing as an alternative to reduced representation bisulphite sequencing for the identification of CpGs of interest in livestock populations. *Livestock Science, 279*, 105377. https://doi.org/10.1016/j.livsci.2023.105377.

Minami, Y., Uede, K., Sagawa, K., Kimura, A., Tsuji, T., & Furukawa, F. (2004). Immunohistochemical staining of cutaneous tumours with G-81, a monoclonal antibody to dermcidin. *British Journal of Dermatology, 151*(1), 165–169. https://doi.org/10.1111/j.1365-2133.2004.06079.x.

Miyata, T., Takekawa, K., & Ohmori, T. (2012). Analysis of 17-ketosteroids in human urine by gas chromatography-mass spectrometry: Conditioning of acid-hydrolysis of conjugate form and application of extractive-derivatization method. *Japanese Journal of Forensic Science, 17*, 35–43. https://doi.org/10.3408/jafst.17.35.

Muro, C. K., Doty, K. C., de Souza Fernandes, L., & Lednev, I. K. (2016). Forensic body fluid identification and differentiation by Raman spectroscopy. *Forensic Chemistry, 1*, 31–38. https://doi.org/10.1016/j.forc.2016.06.003, http://www.journals.elsevier.com/forensic-chemistry.

Ndayizeye, C., & Sharma, S. (2024). Detectability of semen remained in a condom under various conditions. *Forensic Science International, 361*, 112146. https://doi.org/10.1016/j.forsciint.2024.112146.

Neis, M., Groß, T., Schneider, H., Schneider, P. M., & Courts, C. (2024). Comprehensive body fluid identification and contributor assignment by combining targeted sequencing of mRNA and coding region SNPs. *Forensic Science International: Genetics, 73*, 103125. https://doi.org/10.1016/j.fsigen.2024.103125.

Old, J. B., Schweers, B. A., Boonlayangoor, P. W., & Reich, K. A. (2009). Developmental validation of RSID™-saliva: A lateral flow immunochromatographic strip test for the forensic detection of saliva. *Journal of Forensic Sciences, 54*(4), 866–873. https://doi.org/10.1111/j.1556-4029.2009.01055.x.

Omelia, E. J., Uchimoto, M. L., & Williams, G. (2013). Quantitative PCR analysis of blood- and saliva-specific microRNA markers following solid-phase DNA extraction. *Analytical Biochemistry, 435*, 120–122. https://doi.org/10.1016/j.ab.2012.12.024.

Ong, S. Y., Wain, A., Groombridge, L., & Grimes, E. (2012). Forensic identification of urine using the DMAC test: A method validation study. *Science and Justice, 52*(2), 90–95. https://doi.org/10.1016/j.scijus.2011.07.007.

Pang, B. C. M., & Cheung, B. K. K. (2008). Applicability of two commercially available kits for forensic identification of saliva stains. *Journal of Forensic Sciences, 53*(5), 1117–1122. https://doi.org/10.1111/j.1556-4029.2008.00814.x.

Papanicolaou, G. N. (1942). A new procedure for staining vaginal smears. *Science, 95*(2469), 438–439. https://doi.org/10.1126/science.95.2469.438.

Park, J. L., Kwon, O. H., Kim, J. H., Yoo, H. S., Lee, H. C., Woo, K. M., Kim, S. Y., Lee, S. H., & Kim, Y. S. (2014). Identification of body fluid-specific DNA methylation markers for use in forensic science. *Forensic Science International: Genetics, 13*, 147–153. https://doi.org/10.1016/j.fsigen.2014.07.011, http://www.elsevier.com.

Quarino, L., Dang, Q., Hartmann, J., & Moynihan, N. (2005). An ELISA method for the identification of salivary amylase. *Journal of Forensic Sciences, 50*(4), 873–876. https://doi.org/10.1520/jfs2004417, http://onlinelibrary.wiley.com/journal/10.1111/(ISSN)1556-4029.

Ramirez, J. M. D., Gomez, J., Obernuefemann, C. L. P., Gualberto, N. C., & Walker, J. N. (2022). Semi-quantitative assay to measure urease activity by urinary catheter-associated uropathogens. *Frontiers in Cellular and Infection Microbiology, 12*, 859093. https://doi.org/10.3389/fcimb.2022.859093.

Rocchi, A., Chiti, E., Maiese, A., Turillazzi, E., & Spinetti, I. (2020). MicroRNAs: An update of applications in forensic science. *Diagnostics, 11*, 32. https://doi.org/10.3390/diagnostics11010032.

Sakurada, K., Akutsu, T., Fukushima, H., Watanabe, K., & Yoshino, M. (2010). Detection of dermcidin for sweat identification by real-time RT-PCR and ELISA. *Forensic Science International, 194*(1-3), 80–84. https://doi.org/10.1016/j.forsciint.2009.10.015.

Sakurada, K., Ikegaya, H., Fukushima, H., Akutsu, T., Watanabe, K., & Yoshino, M. (2009). Evaluation of mRNA-based approach for identification of saliva and semen. *Legal Medicine, 11*(3), 125–128. https://doi.org/10.1016/j.legalmed.2008.10.002.

Sakurada, K., Watanabe, K., & Akutsu, T. (2020). Current methods for body fluid identification related to sexual crime: Focusing on saliva, semen, and vaginal fluid. *Diagnostics (Basel), 10*, 693. https://doi.org/10.3390/diagnostics10090693.

Sato, I., Kojima, K., Yamasaki, T., Yoshida, K., Yoshiike, M., Takano, S., Mukai, T., & Iwamoto, T. (2004). Rapid detection of semenogelin by one-step immunochromatographic assay for semen identification. *Journal of Immunological Methods, 287*(1-2), 137–145. https://doi.org/10.1016/j.jim.2004.01.017.

Song, M. F., Li, Y. S., Ootsuyama, Y., Kasai, H., Kawai, K., Ohta, M., Eguchi, Y., Yamato, H., Matsumoto, Y., Yoshida, R., & Ogawa, Y. (2009). Urea, the most abundant component in urine, cross-reacts with a commercial 8-OH-dG ELISA kit and contributes to overestimation of urinary 8-OH-dG. *Free Radical Biology and Medicine, 47*(1), 41–46. https://doi.org/10.1016/j.freeradbiomed.2009.02.017.

Steendam, K. V., Ceuleneer, M. D., Dhaenens, M., Hoofstat, D. V., & Deforce, D. (2013). Mass spectrometry-based proteomics as a tool to identify biological matrices in forensic science. *International Journal of Legal Medicine, 127*, 287–298. https://doi.org/10.1007/s00414-012-0747-x.

Takamura, A., Watanabe, K., Akutsu, T., & Ozawa, T. (2018). Soft and robust identification of body fluid using Fourier transform infrared spectroscopy and chemometric strategies for forensic analysis. *Scientific Reports, 8*(1). https://doi.org/10.1038/s41598-018-26873-9.

Tian, H., Bai, P., Tan, Y., Li, Z., Peng, D., Xiao, X., Zhao, H., Zhou, Y., Liang, W., & Zhang, L. (2020). A new method to detect methylation profiles for forensic body fluid identification combining ARMS-PCR technique and random forest model. *Forensic Science International: Genetics, 49*. https://doi.org/10.1016/j.fsigen.2020.102371, http://www.elsevier.com.

Tian, H., Huang, S., Bai, P., Xiao, X., Peng, D., Zhao, H., Liu, Y., Feng, Q., Liao, M., Li, F., & Liang, W. (2021). The effect of infertile semen on the mRNA-based body fluid identification. *Electrophoresis, 42*(16), 1614–1622. https://doi.org/10.1002/elps.202000238, http://onlinelibrary.wiley.com/journal/10.1002/(ISSN)1522-2683.

Tindal, K., Filby, C. E., Cousins, F. L., Ellery, S. J., Vollenhoven, B., Palmer, K., Gordon, A., Gargett, C. E., & Davies-Tuck, M. (2024). The composition of menstrual fluid, its applications, and recent advances to understand the endometrial environment: A narrative review. *F&S Reviews, 5*(3). https://doi.org/10.1016/j.xfnr.2024.100075, https://www.sciencedirect.com/science/journal/26665719.

Upadhyay, M., Shrivastava, P., Verma, K., & Joshi, B. (2023). Recent advancements in identification and detection of saliva as forensic evidence: A review. *Egyptian Journal of Forensic Sciences, 13*(1). https://doi.org/10.1186/s41935-023-00336-3.

Virkler, K., & Lednev, I. K. (2009). Analysis of body fluids for forensic purposes: From laboratory testing to non-destructive rapid confirmatory identification at a crime scene. *Forensic Science International, 188*(1-3), 1–17. https://doi.org/10.1016/j.forsciint.2009.02.013, www.elsevier.com/locate/forsciint.

Vyas, B., Halámková, L., & Lednev, I. K. (2020). A universal test for the forensic identification of all main body fluids including urine. *Forensic Chemistry, 20*, 100247. https://doi.org/10.1016/j.forc.2020.100247.

Wang, Z., Zhang, J., Wei, W., Zhou, D., Luo, H., Chen, X., & Hou, Y. (2015). Identification of saliva using microRNA biomarkers for forensic purpose. *Journal of Forensic Sciences, 60*(3), 702–706. https://doi.org/10.1111/1556-4029.12730China, http://onlinelibrary.wiley.com/journal/10.1111/(ISSN)1556-4029.

Wang, Z., Zhao, X., & Hou, Y. (2017). Exploring of microRNA markers for semen stains using massively parallel sequencing. *Forensic Science International: Genetics Supplement Series, 6*, e107–e109. https://doi.org/10.1016/j.fsigss.2017.09.039.

Watanabe, K., Akutsu, T., Takamura, A., & Sakurada, K. (2017). Practical evaluation of an RNA-based saliva identification method. *Science and Justice, 57*(6), 404–408. https://doi.org/10.1016/j.scijus.2017.07.001, http://www.scienceandjusticejournal.com/contact.

Wei, C. T., You, J. L., Weng, S. K., Jian, S. Y., Lee, J. C. L., & Chiang, T. L. (2024). Enhancing forensic investigations: Identifying bloodstains on various substrates through ATR-FTIR spectroscopy combined with machine learning algorithms. *Spectrochimica Acta - Part A: Molecular and Biomolecular Spectroscopy, 308*. https://doi.org/10.1016/j.saa.2023.123755, https://www.sciencedirect.com/science/journal/13861425.

Weichhart, T., Zlabinger, G. J., & Säemann, M. D. (2005). The multiple functions of Tamm-Horsfall protein in human health and disease: A mystery clears up. *Wiener Klinische Wochenschrift, 117*, 316–322. https://doi.org/10.1007/s00508-005-0353-8.

Williams, G. A. (2020). Body fluid identification: A case for more research and innovation. *Forensic Science International: Reports, 2*, 100096. https://doi.org/10.1016/j.fsir.2020.100096.

Yao, T., Han, X., Guan, T., Wang, Z., Zhang, S., Liu, C., Liu, C., & Chen, L. (2020). Effect of indoor environmental exposure on seminal microbiota and its application in body fluid identification. *Forensic Science International, 314*, 110417. https://doi.org/10.1016/j.forsciint.2020.110417.

Zubakov, D., Boersma, A. W. M., Choi, Y., Van Kuijk, P. F., Wiemer, E. A. C., & Kayser, M. (2010). MicroRNA markers for forensic body fluid identification obtained from microarray screening and quantitative RT-PCR confirmation. *International Journal of Legal Medicine, 124*(3), 217–226. https://doi.org/10.1007/s00414-009-0402-3.

Zubakov, D., Hanekamp, E., Kokshoorn, M., van IJcken, W., & Kayser, M. (2007). Stable RNA markers for identification of blood and saliva stains revealed from whole genome expression analysis of time-wise degraded samples. *Legal Medicine, 122*, 135–142. https://doi.org/10.1007/s00414-007-0182-6.

Zubakov, D., Kokshoorn, M., Kloosterman, A., & Kayser, M. (2009). New markers for old stains: Stable mRNA markers for blood and saliva identification from up to 16-year-old stains. *International Journal of Legal Medicine, 123*(1), 71–74. https://doi.org/10.1007/s00414-008-0249-z.

Chapter 4

Analysis of mitochondrial DNA and its aid in forensic investigation

Christian Haarkötter, María Isabel Medina-Lozano, Francisco Santisteban-Álvarez, Xiomara Gálvez, Diana C. Vinueza-Espinosa, María Saiz, Daniel Díaz-Ruiz, Juan Carlos Álvarez, and José Antonio Lorente

Faculty of Medicine, Laboratory of Genetic Identification & Human Rights (LABIGEN-UGR), Department of Legal Medicine, University of Granada, Granada, Spain

4.1 Introduction

Most of the efforts on human identification have focused on nuclear DNA and short tandem repeats (STRs) markers. However, mitochondrial DNA (mtDNA) analysis has opened new possibilities for identification due to its small size, maternal inheritance, and the number of copies per cell. In this chapter, the characteristics, applications, and analysis procedures regarding the most recent advances in next-generation sequencing (NGS) will be discussed.

4.2 Mitochondrial DNA structure

Unlike other mammalian organelles, mitochondria contain a small amount of their DNA, known as mtDNA. This encodes several proteins essential for cellular respiration. Each mitochondrion can contain one or more copies of mtDNA, which are located in the mitochondrial matrix (Mishra & Chan, 2014). The mtDNA is distributed throughout the mitochondrial matrix because it is crucial for maintaining mitochondrial function.

mtDNA is packaged by several proteins. The mtDNA-protein complex is called the nucleoid. This packaging is necessary due to space limitations, as the length of the mtDNA in mammals is approximately 5 μm, while the width of the mitochondrion is about 0.5 μm (Gustafsson et al., 2016). The major nucleoid protein is mitochondrial transcription factor A (mtTFA), which acts as an mtDNA transcription factor and controls the distribution and organization of the nucleoid (Maniura-Weber, 2004).

However, mtDNA is a histone-free, double-stranded, closed-loop DNA of approximately 16,569 base pairs (Fig. 4.1). The two strands that form it are referred to as the heavy (H) and light (L) strands, named after their different densities due to their varying compositions of purine bases (adenine and guanine) (Amorim et al., 2019). In human cells, mtDNA encodes 37 genes: 11 messenger RNAs (mRNA) translated into 13 proteins, two ribosomal RNAs (rRNA, 12S, and 16S), and 22 tRNAs (Gustafsson et al., 2016). The oxidative phosphorylation system (OXPHOS) has a dual genetic origin; some subunits are encoded by nuclear genes, translated on cytosolic ribosomes, and imported into mitochondria, while others are encoded by mtDNA and translated on mitochondrial ribosomes. Although only 13 subunits are encoded by mtDNA, they are essential because the OXPHOS system collapses in the absence of mtDNA expression. These 13 polypeptides are the core subunits of the OXPHOS complexes I, III, IV, and V (Gustafsson et al., 2016).

Another feature of the mammalian mtDNA genome is the absence of introns, with only one long noncoding region known as the control region. The control region contains a promoter responsible for the transcription of each of the mtDNA strands (Gustafsson et al., 2016). This region is also called the displacement loop (D-loop). The D-loop contains three regions of high sequence variability (Fig. 4.1) compared to the rest of the genome in the population: hypervariable 1 (HV-I), hypervariable 2 (HV-II), and hypervariable 3 (HV-III) (Brandstätter et al., 2004). This hypervariability results from the lack of mtDNA repair mechanisms and the low fidelity of mtDNA polymerase, leading to a significantly higher mutation rate in the mitochondrial genome compared to the nuclear genome. Most of the sequence variation between individuals is found in the HV-I (positions 16,024–16,365) and HV-II (positions 73–340) regions (Greenberg et al., 1983).

Investigative and Predictive DNA Testing. DOI: https://doi.org/10.1016/B978-0-443-30128-5.00007-2

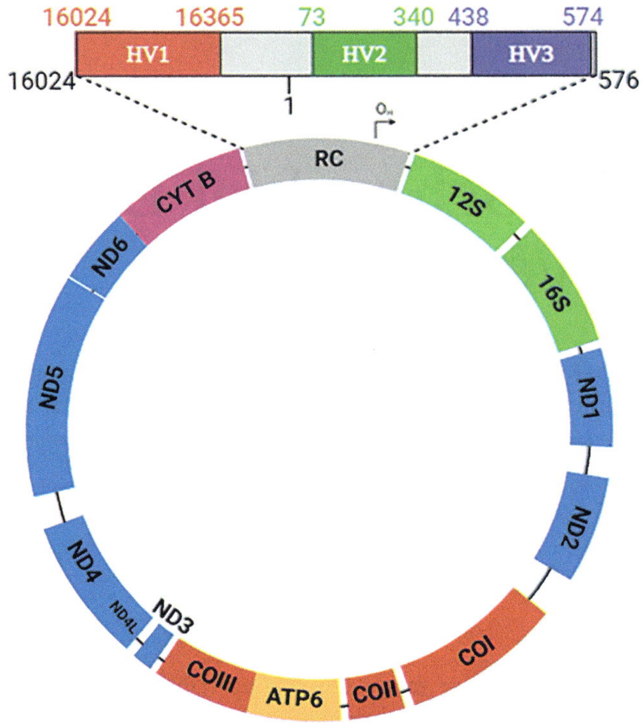

FIGURE 4.1 The human mitochondrial DNA (mtDNA) consists of an outer heavy (H) strand and an inner light (L) strand, both forming a circular DNA molecule. The mtDNA encodes 37 genes, including 2 ribosomal RNAs and 22 transfer RNAs essential for protein synthesis and the respiratory chain. The hypervariable regions (HV1, HV2, and HV3) within the control or D-loop region are highlighted.
Source: Created with Biorender.

The HV-III region (positions 438–574) contains additional polymorphic positions useful in samples with indistinguishable HV-I/HV-II (Lutz et al., 2000). The high interindividual variability makes the analysis of HV regions very useful in forensic evidence.

4.3 Characteristics of mitochondrial DNA

mtDNA has become a valuable tool in forensic investigations. This type of DNA has enabled numerous identifications of criminal or uniparental relatives, and victims of mass disasters and/or accidents. We will discuss the characteristics that make this genome so important in the forensic field.

4.3.1 Maternal inheritance

mtDNA is transmitted uniparentally, from mother to offspring, in a vertical, non-Mendelian pattern. Most mitochondria in the male gamete are in the tail, as a substantial amount is required for motility. Upon fertilization, the tail is detached, leaving the sperm head with only a minimal number of mitochondria. In contrast, the oocyte contains many mitochondria, as it requires significant energy to develop a new organism. The mechanism behind the elimination of paternal mtDNA remains enigmatic. Previous studies have indicated that the mitochondria entering the oocyte during fertilization are eliminated, except for mutations, through processes known as ubiquitination and mitophagy (Sutovsky et al., 2000).

In addition, subsequent studies utilizing digital droplet polymerase chain reaction (ddPCR) have demonstrated that throughout spermatogenesis, culminating in the formation of mature spermatozoa, there is a degradation of mtDNA, remaining as explained before in the sperm tail. This process results in spermatozoa that are devoid of mtDNA and incapable of maintaining or replicating this genome. Additionally, the studies indicate that male gametes lack mtTFA, the nucleoid protein responsible for maintaining, protecting, and transcribing mtDNA (Lee et al., 2023).

In various ways, there are both indications and certainties that mtDNA is uniparentally inherited from mother to child, except in cases where mutations occur. For instance, a study involving a 28-year-old man with mitochondrial myopathy revealed that the mutation originated from paternal mtDNA, which constituted 90% of the patient's muscle mtDNA (Schwartz & Vissing, 2002).

4.3.2 High copy number

In terms of copy number, mtDNA is notably abundant (Ballard, 2016). Each mitochondrion contains approximately 1–15 copies of the mitochondrial genome, with an average of 4.6 copies (Satoh, 1991). Considering the presence of multiple mitochondria, depending on the cell type and location, this results in an average of hundreds to thousands (220 ± 6.2–1720 ± 162 molecules) of mtDNA copies per cell (Robin & Wong, 1988), as opposed to the two copies of nuclear DNA, one from each chromosome. Due to the high number of copies in the human body and its resistance to extreme environmental conditions, mtDNA is easier to obtain when nuclear DNA is highly degraded. This characteristic underscores the importance of mtDNA in forensic fields.

4.3.3 Mutation rate

mtDNA exhibits a high mutation rate within populations due to the rapid evolution of this genome. This phenomenon can be attributed to two types of mutations: spontaneous or point mutations, and damage-induced mutations resulting from large genomic rearrangements.

Spontaneous mutations primarily arise from the inherent infidelity of the polymerase, which can insert an incorrect nucleotide opposite a chemically modified normal one, leading to base substitutions. Conversely, damage-induced mutations are predominantly caused by the loss of large DNA segments through processes such as deletion, inversion, duplication, insertion, or translocation. Among these, deletions have been the most extensively studied (Khrapko & Turnbull, 2014). These deletions can occur at low levels, potentially having minimal functional impact, but they can accumulate or clone, resulting in mitochondrial dysfunction. Large deletions, such as the "common deletion" of 4977 base pairs, can lead to severe mitochondrial diseases (Tuppen et al., 2023).

Several methods exist for estimating the mutation rate of mtDNA, including phylogenetic and pedigree-based approaches. Phylogenetic methods rely on phylogenetic trees of different haplotypes and their rates of genetic divergence, enabling the study of the deep history of a population based on mutations found with notable frequency. In contrast, pedigree-based methods are more suitable for examining recent history, as they estimate mutations that have occurred recently and have not yet become fixed in the population (Connell et al., 2022).

Understanding the mutation rate in a certain population is crucial for comprehending the molecular clock, which posits that molecular sequences evolve at a constant rate. Consequently, the number of accumulated mutations is proportional to the time of evolution, allowing for the establishment of time scales for dating evolutionary events and determining the age of different lineages (Soares et al., 2009). In most cases, mutations in this genome are of heteroplasmic origin.

4.3.4 Heteroplasmy

It is common for mtDNA to be homoplasmic; that is, when inherited from the mother, it produces an individual with a single haplotype of that genome. However, with the development of sequencing methods, heteroplasmic individuals with more than one haplotype were observed (Stewart & Chinnery, 2015). In the medical field, it is crucial to determine the level of heteroplasmy in an individual, as exceeding the critical threshold can trigger chronic metabolic deterioration and cause various mitochondrial diseases (Pereira et al., 2021). Heteroplasmy can be caused by several processes (Polovina et al., 2020):

1 **De novo somatic mutations**: These mutations generate a variety of haplotypes that differ from each other and from the maternally inherited haplotype. The frequency of these variants in individuals is influenced by the interaction between mutations and genetic drift, as natural selection does not act in somatic tissues, although it has been observed to act in the germline (Lieber et al., 2019).
2 **Paternal mtDNA leakage**: This occurs when mtDNA from the sperm enters the ovule during fertilization. Although mechanisms exist to prevent this, it can occasionally happen, as previously mentioned.
3 **Maternal transmission**: This occurs through heteroplasmic eggs. When the maternal germline or its ancestors have acquired heteroplasmy due to mutations or the incorporation of paternal mtDNA, the mtDNA of the egg is already heteroplasmic. Once heteroplasmy is present in the egg, it is transmitted to the offspring in the usual manner.

4.3.5 Molecular damage

mtDNA is more susceptible than nuclear DNA to molecular damage, being one of the reasons for this phenomenon the absence of a complex chromatin organization in mtDNA, since the presence of histones organization in DNA strands makes it less susceptible to molecular damage compared to naked DNA (Yakes & Van Houten, 1997). Heteroplasmy events are valuable in forensic analysis as they enhance the strength of a potential match and even enable the distinction between maternal relatives. However, it has been observed that molecular damage and a low quantity of mtDNA templates for analysis are interrelated, impacting the quality of the results obtained. This is because molecular damage and heteroplasmy can be confused (Holland et al., 2021).

4.4 Applications of mitochondrial DNA

mtDNA has several unique properties that make it valuable for genetic analysis. First, it is inherited exclusively from the mother, making it a powerful tool for tracing maternal lineages through generations. Second, because it is found in multiple copies in each mitochondrion (and each cell contains numerous mitochondria), it is relatively easier to extract and analyze even under degrading conditions, such as in ancient or poorly preserved remains. Also, mtDNA is exposed to a higher mutation rate than nuclear DNA, making it a useful marker for tracking evolutionary changes over shorter periods.

Therefore, mtDNA has broad applications in various disciplines such as anthropology, evolutionary biology, forensic genetics, medicine, and ecology.

4.4.1 Mitochondrial DNA in evolution and phylogeny studies

1. Evolutionary history and phylogeny. mtDNA (mtDNA) has played a central role in evolutionary and phylogenetic studies because its maternal inheritance and lack of recombination make it a very stable genetic marker. Phylogeny, the study of the evolutionary relationships between species, has made extensive use of mtDNA because of its ability to trace lineages and build phylogenetic trees showing how species have evolved over time. One of the major achievements in the use of mtDNA in phylogenetic studies has been the construction of the human evolutionary tree. The pioneering studies by Cann, Stoneking, and Wilson in 1987 used mtDNA to propose the "mitochondrial Eve" hypothesis, which identified an African woman who lived around 200,000 years ago as the maternal ancestor of all modern humans (Cann et al., 1987). This study revolutionized molecular anthropology by providing genetic evidence for the "Out of Africa" hypothesis, which suggests that modern humans originated in Africa and then migrated to other parts of the world. Phylogenetic studies using mtDNA have not been restricted to humans. In studies of primates, for example, mtDNA has been crucial in determining the evolutionary relationships between species. A recent study on the phylogeny of great apes showed that the bonobo (*Pan paniscus*) and the common chimpanzee (*Pan troglodytes*) are more closely related to each other than to humans, confirming previous morphological studies (Prüfer et al., 2012).

2. Estimating the time of divergence between species. The use of molecular clocks based on mtDNA has allowed scientists to estimate when two species shared a common ancestor. This approach uses the rate of mutation accumulation in mtDNA to calculate the time of divergence between species. However, the accuracy of these molecular clocks depends on several factors, including the species-specific mutation rate and calibration with well-dated fossil events. In studies of marine mammals, mtDNA has been used to estimate that dolphins and whales diverged from a common ancestor between 34 and 45 million years ago, coinciding with major geological changes that led to the emergence of these cetacean (McGowen et al., 2020).

3. mtDNA in biogeographic studies. Biogeography, the study of the distribution of species over time and space, has also benefited greatly from mtDNA analysis. For instance, ancient mtDNA analysis of pre-Columbian skeletons indicates how the initial peopling of the Americas took place, and its comparison to modern mtDNA distribution suggests the impact of European colonization in the substantial loss of these mtDNA lineages (Llamas et al., 2016).

4.4.2 Forensic applications of mitochondrial DNA

1. Use in human identification: mtDNA is particularly valuable in forensic genetics when nuclear DNA samples are highly degraded or insufficient for analysis. Because mtDNA has many more copies per cell than nuclear DNA, it is more likely to survive under adverse conditions, making it useful for analyzing ancient human remains or highly degraded biological samples. This makes it an important tool for identifying individuals in cases of unidentified human remains or mass disasters. It is useful to refer to mtDNA in biological investigations as maternally inherited because mtDNA is found in the mitochondria, and these are inherited maternally. This makes it useful in cases involving distant relatives with maternal inheritance. mtDNA analysis was crucial in the identification of victims of the September 11, 2001 attacks in New York. Due to the severe degradation of samples, mtDNA was an important tool in identifying many victims from bone and tissue remains by comparing mtDNA to living maternal relatives (Biesecker et al., 2005).

2. mtDNA in criminal investigations: Mitochondrial DNA has also been successfully used in numerous criminal cases, particularly in situations where nuclear DNA was unavailable or insufficient for analysis. Unlike nuclear DNA, mtDNA can provide information from small samples, such as rootless hairs, which are a common source of evidence at crime scenes. In a famous case in the UK, mtDNA was used to identify the perpetrator of a murder in the 1990s. A hair found at the crime scene was used to link a suspect, leading to his conviction (Parsons & Coble, 2001).

3. Identification of victims of mass disasters and conflicts: The use of mtDNA has been crucial in the identification of victims of war and mass disasters, where human remains are often degraded. In the case of victims of the war in the former Yugoslavia, mtDNA analysis enabled the identification of thousands of missing persons, providing emotional and legal closure for their families. Specifically, mtDNA was used to identify the victims of the 1995 Srebrenica massacre (Holland et al., 1993).

4.4.3 Mitochondrial DNA in disease studies

1. mtDNA mutations and their role in disease: mtDNA mutations have been associated with several human diseases, particularly those affecting tissues with high energy demands, such as muscle and the nervous system. Because mtDNA is inherited maternally and does not undergo recombination, mutations accumulate over time, and some of these mutations can have pathological effects. One of the best-known mitochondrial diseases is Leber's Hereditary Optic Neuropathy, which causes sudden loss of vision, particularly in young adults. This disease is caused by point mutations in the mtDNA that reduce the cell's ability to produce energy (Mackey et al., 1996). Other disorders associated with mtDNA mutations include mitochondrial myopathies, progressive external ophthalmoplegia, and Kearns-Sayre syndrome.

2. mtDNA and age-related diseases: Mitochondrial dysfunction has also been linked to several age-related diseases such as Alzheimer's, Parkinson's, and cardiovascular disease. As people age, the number of mtDNA mutations increases, contributing to a decline in mitochondrial function and the development of degenerative diseases. In Alzheimer's disease, for example, studies have found an increased frequency of certain mtDNA mutations in patients, suggesting that mitochondrial dysfunction plays a role in the onset and progression of the disease (Coskun et al., 2010).

3. Therapies targeting mtDNA: As mitochondrial diseases are often caused by mutations in mtDNA, therapies aimed at correcting these mutations are being investigated. One promising approach is mitochondrial replacement therapy, in which defective mitochondria are replaced with healthy ones to prevent the transmission of mitochondrial diseases from mother to child. This technique, also known as 'three-parent baby' technology, has been tested in clinical trials and offers hope to families with a history of mitochondrial disease (Craven et al., 2011).

4.4.4 Mitochondrial DNA and cancer

1. mtDNA mutations in cancer: Recent studies have shown that mutations in mtDNA can contribute to the development of cancer. Mutations in mtDNA can affect cellular energy production and apoptosis, both critical processes in cancer progression. Because mtDNA has a higher mutation rate than nuclear DNA, these mutations accumulate more rapidly, potentially leading to tumourigenesis. In breast and prostate cancer, for example, specific mutations in mtDNA have been identified that can promote tumour growth by altering mitochondrial function and the production of reactive oxygen species, which can damage cellular components and promote genetic instability (Petros et al., 2005).

2. Mitochondrial metabolism in cancer: Mitochondria play a key role in cellular metabolism, and cancer cells often display altered mitochondrial metabolism to support rapid growth and proliferation. This metabolic reprogramming, known as the "Warburg effect,“ involves a shift from oxidative phosphorylation to glycolysis, even in the presence of oxygen. This shift in energy production allows cancer cells to grow rapidly but also makes them more dependent on their mitochondria for survival. Recent research has focused on targeting mitochondrial metabolism as a potential therapeutic strategy in cancer treatment. By inhibiting mitochondrial respiration or disrupting the mitochondria's ability to produce energy, it may be possible to selectively kill cancer cells without harming normal cells (Vyas et al., 2016).

3. Mitochondria as a therapeutic target: Given the importance of mitochondria in cancer cell survival and proliferation, targeting mtDNA and function is being explored as a novel approach to cancer treatment. One strategy is to use drugs that selectively target mitochondrial metabolism or induce mitochondrial dysfunction in cancer cells. These therapies may be particularly effective in cancers with mtDNA mutations or metabolic reprogramming. One example is the use of drugs such as metformin, which inhibits mitochondrial complex I and reduces the production of ATP in cancer cells. This has shown promise in preclinical studies for the treatment of certain cancers (Wheaton et al., 2014).

4.4.5 mtDNA in animal conservation and biodiversity

1. Use in genetic diversity monitoring: In conservation biology, mtDNA is widely used to assess genetic diversity within and between populations of endangered species. By analyzing mtDNA, scientists can determine levels of inbreeding, gene flow, and population structure, which are essential for developing conservation strategies. For example, in studies of the genetic diversity of the giant panda, mtDNA analysis has shown that the population is divided into several distinct groups, with important implications for its conservation (Zhan et al., 2006).

2. Tracking species migration and population history: mtDNA is also used to track species migrations and understand the historical population dynamics of endangered animals. By comparing mtDNA sequences from different populations, researchers can reconstruct migration routes and identify population bottlenecks or expansions that occurred in the past. For example, mtDNA has been used to study the migration patterns of sea turtles, providing valuable information for conservation efforts to protect their nesting sites (Bowen et al., 1997).

3. mtDNA and hybridisation studies: In some cases, mtDNA is used to study hybridization between species or subspecies, particularly in animals where hybridization is a threat to the genetic integrity of the species. An example is the study of hybridization between the European wildcat and domestic cat, where mtDNA analysis has helped to identify hybrid individuals and assess the extent of genetic introgression (Randi et al., 2001).

4.5 Advantages and disadvantages of mitochondrial DNA analysis

Human mtDNA analysis is a valuable tool in forensic science for the identification of human remains or missing persons cases. Often, this analysis is performed on complex samples such as rootless hairs, bones, blood, and saliva. mtDNA analysis can serve as a complementary tool to facilitate identification (Carracedo et al., 2000).

The following are the advantages and disadvantages of using mtDNA in forensic science for identification and case solving.

4.5.1 Advantages of mitochondrial DNA

(1) Maternal inheritance: mtDNA, unlike genomic DNA, is inherited exclusively through the maternal pathway, allowing the study of maternal lineages as it is relatively easy to obtain mtDNA from a maternal relative (Butler, 2005). Therefore, the mtDNA sequence is not unique to a particular individual; however, it is shared by a group of individuals belonging to the same maternal lineage, transmitted from mother to child, as shown in Fig. 4.2.

(2) Resistance to degradation: One of the main advantages of mtDNA analysis is its remarkable resistance to degradation. Because of its location in the mitochondria, which are more abundant in cells than the nucleus, mtDNA can be recovered from samples that have undergone significant degradation, such as old bones, teeth, or remains in adverse conditions. mtDNA is particularly well preserved in these samples due to the encapsulation of the DNA in the tissue between layers of keratin in the case of hair and hydroxyapatite in the case of bone (Linch et al., 2001). This recovery capability is critical in cases where nuclear DNA is unavailable.

(3) Missing persons identification: mtDNA is very useful in situations where samples are particularly degraded, such as the human remains of missing persons. There are two copies of autosomal genetic markers per cell, whereas there are

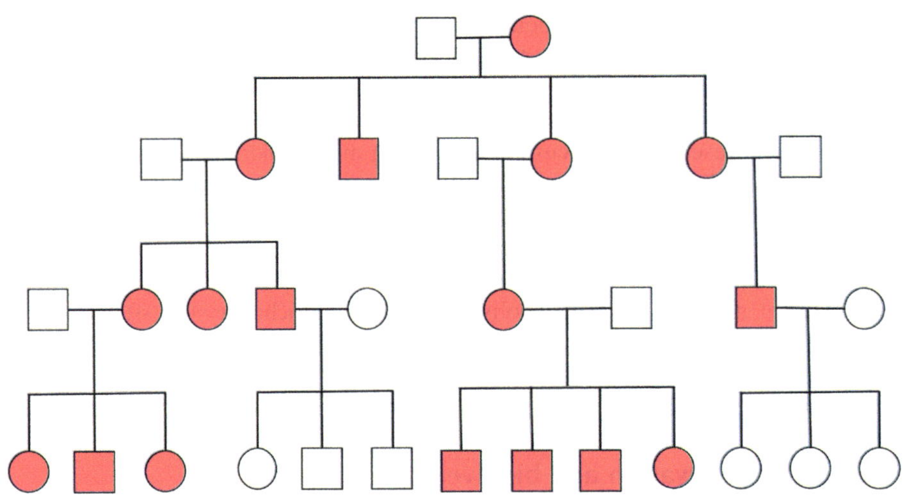

FIGURE 4.2 Maternal inheritance of mtDNA through four generations. In pink, male (square) and female (circle) individuals who share the same mtDNA.

thousands of mtDNA molecules. For this reason, if samples from close relatives are available and STR typing does not provide information, there is still the possibility of obtaining results by typing mtDNA (Alvarez-Cubero et al., 2012).

(4) Ubiquity in forensic samples: mtDNA can be obtained from a variety of biological sources such as hair, bone, teeth, saliva, or body fluids. This is particularly useful in situations where samples are limited or difficult to obtain. The availability of different types of samples broadens the possibilities for analysis in forensic investigations (Daeid et al., 2021).

4.5.2 Disadvantages of mitochondrial DNA

(1) Heteroplasmy: The coexistence of multiple mtDNA variants in a single cell (heteroplasmy) can complicate the interpretation of genetic results. This can lead to confusion in identification as an mtDNA profile may not be unique to an individual, especially if multiple variants are present (Elson et al., 2001).

(2) Lower discriminatory power: Compared to nuclear DNA, mtDNA has a lower discriminatory power because it is inherited matrilineally, and the genetic variability of its sequence is lower than that of the nuclear genome. The profiles obtained can be shared by several individuals of the same family or lineage, which results in a much more limited discriminatory power. This can lead to confusion when trying to establish the family relationship between different remains and an individual (Parsons & Coble, 2001). For this reason, mtDNA analysis is considered a complementary method rather than the primary focus of analysis.

(3) Criticism of the revised Cambridge sequence reference (rCRS): The rCRS has been the subject of several criticisms in the field of genetics and molecular biology. One of the main criticisms is that although the rCRS is used as a standard for mtDNA sequencing, it may not reflect the genetic variability of all human populations. This can lead to misinterpretation in genealogical and evolutionary studies. In addition, some researchers have pointed out that rCRS is based on a single individual, which may not be representative of global genetic diversity. This has led to the need to consider other reference sequences that may better capture variability between different populations (Bandelt et al., 2014). Another important criticism is that the rCRS does not include certain mutations that may be relevant to the identification of specific lineages, which may limit its usefulness in kinship studies and in understanding the evolutionary history of populations (Torroni et al., 1998).

(4) Cost and resources: mtDNA analysis can be expensive and requires specialized resources for highly degraded samples. Conventional techniques for obtaining genetic information by Sanger require more than 8 amplification reactions and a large number of sequences per Sanger, in addition to a large amount of DNA extract. However, with new technologies such as NGS, more genetic information can be obtained in less time due to its high robustness and sensitivity (Vinueza-Espinosa et al., 2023).

4.6 Mitochondrial DNA analysis procedure

In ancient and forensic DNA, mtDNA analysis is sometimes the only solution to obtain genetic information when nuclear DNA is highly degraded. From a technical point of view, the first step in obtaining genetic information, including mtDNA, is DNA extraction. Forensic mtDNA analysis uses materials with little initial DNA, including teeth, hair, and bones. From these materials, DNA must be carefully extracted and often purified from PCR inhibitors, which may be co-extracted. Nowadays, numerous studies explore methods for DNA extraction from highly degraded samples, which enable greater DNA recovery. These methods are applied to the retrieval of fragmented DNA with significant molecular damage (Haarkötter et al., 2023; Loreille et al., 2007). In fact, these new methods and strategies have enabled laboratories to obtain results using STRs (Haarkötter et al., 2024) or SNPs (Emery et al., 2024) from skeletal remains, where a few years ago, only mtDNA results were possible.

The next step in mtDNA analysis is DNA quantification. To quantify the amount of mtDNA, laboratories often conduct a nuclear DNA quantification assay and estimate the mtDNA amount by assuming a fixed ratio between nuclear and mtDNA. In addition, more recent approaches involving real-time PCR allow for the direct characterisation of the number of mtDNA molecules in a cell (Xavier et al., 2023).

4.6.1 Mitochondrial DNA sequencing methodologies

Once the DNA template is obtained, two methods for analyzing mtDNA could be used, the conventional technique of Sanger sequencing (SS) and/or next generation sequencing (NGS). While SS is still the most commonly used method, NGS is more sensitive and has produced many successful results. However, NGS remains costly, especially when processing a small number of samples.

In 1977, Sanger presented the first DNA sequencing technology (Sanger et al., 1977), also called the chain termination method and now known as first-generation sequencing. The integration of ddNTPs into newly synthesized DNA strands results in the termination of the elongation process, providing information about the specific nucleotide found at each position in the sequence (Fig. 4.3: step 3). Sanger sequencing can generate reads from 25 to 1200 nucleotides.

If genetic identification is the goal, at least two regions of the control region must be analysed: HV1 (16024–16365) and HV2 (73–340). The DNA input should be amplified using primers flanking these regions (Fig. 4.3: step 1). PCR amplification of mtDNA is usually done with 34 to 38 cycles. Protocols for highly degraded DNA specimens even call for 42 cycles (Gabriel et al., 2001). In addition, small amplicons are used to improve the success of amplification; in the same way, in the case of STRs, the use of smaller-sized PCR products improves the recovery of information from the original DNA template. Mini-Mito primer sets have been developed to amplify smaller portions of HV1, HV2, and HV3 (Fig. 4.4) (Eichmann & Parson, 2008; Gabriel et al., 2001).

The use of overlapping mini amplicons is sometimes referred to as the 'ancient DNA' approach and can recover abundant DNA from a sample that would otherwise not yield results using a standard protocol. This approach has been used successfully to recover information from Neanderthal bones thousands of years old (Krings et al., 1997). The Mitominis strategy was refined to develop the commercial PowerSeq® CRM Nested System kit (Promega), which enables the amplification and sequencing of the three hypervariable regions in a single multiplex reaction using NGS. In a study, this kit was successfully tested on highly degraded skeletal remains, achieving over 85% recovery of mtDNA information. Additionally, the NGS kit reduces the risk of contamination due to minimal sample handling throughout the entire technical process (Vinueza-Espinosa et al., 2023).

After amplifying the hypervariable regions, the amplification of DNA fragments can be verified using agarose gel electrophoresis or a Bioanalyzer. The latter allows for the analysis of thousands of DNA fragments from a minimal extract volume (1–5 µlL, based on size (pb) and intensity (RFUs), through graphical interpretation (Fig. 4.3: step 2). The PCR products must be purified to eliminate primers that were not consumed during the reaction. ExoSAP-IT (ThermoFisher) or MSB Spin PCRapace (Stratec) are recommended for PCR product cleanup. Finally, both strands (forward and reverse) of each amplicon should be sequenced. Sequencing can be performed using the Big Dye Terminator kit (Thermo Fisher Scientific), which enables the generation of high-quality sequences through a genetic analyzer based on capillary electrophoresis (conventional technique) (Fig. 4.3: step 3). The cleanup of unconsumed ddNTPs from the sequencing reaction must also be performed. The BigDye XTerminator Kit (Thermo Fisher Scientific) and Optima™ DTR are some recommendations that can be used for this purpose.

4.6.2 Data review, editing, and interpretation of mtDNA results

Human mtDNA was first sequenced in 1981 in the laboratory of Frederick Sanger in Cambridge, England, by Anderson. The Anderson sequence, also referred to as the Cambridge reference sequence (CRS), was created in 1981. In 1999, the original placental material used by Anderson and his co-workers to generate the CRS was re-sequenced, resulting in a rCRS. The latter is now the accepted standard for comparison and is available in GenBank under the NCBI reference sequence NC_012920.1. BioEdit, along with Chromas or SeqScape, are some of the recommended software that can be used to align the sequences obtained, the results of our assays, to rCRS and to obtain the differences or polymorphisms in contrast to rCRS (Butler, 2012) (Fig. 4.5).

The chromatograms resulting from Sanger sequencing must be checked to verify the quality of the sequence. Various types of chromatograms can be produced. Figure IV shows some examples of sequence quality. Good quality sequencing data is characterized by well-defined nucleotides with no errors in the base calling (Fig. 4.6A). Acceptable quality sequencing data is indicated by low base intensity but with no errors in base calling (Fig. 4.6B). Poor quality sequences are identified by poorly defined peaks, which may occur when the template concentration is too low (Fig. 4.6C), when there is excessive noise due to an overabundance of DNA, poor primers, incorrect primers added to the template, or blocked capillaries in the sequencer (Fig. 4.6D). A gradual loss of signal at larger sizes is typically found in raw data from mtDNA sequences derived from highly degraded samples (Fig. 4.6E). This loss of signal is also frequent in scenarios with an excess of DNA, as well as homopolymeric regions such as the one found in HV1; in both cases, there is a meaningful loss of signal in raw data.

The sequencing process does not always produce a clear and unambiguous set of data for every single base. Certain regions, like the C-stretches, can be difficult to interpret and may even be excluded from the final analysis. DNA sequencing is carried out in both forward and reverse directions, allowing for the comparison of complementary strands as a quality control measure. This approach enables the possibility of obtaining sequences from both strands; for instance, in the case of a C-stretch, the same strand can be sequenced twice in separate reactions. The objective is to achieve at least

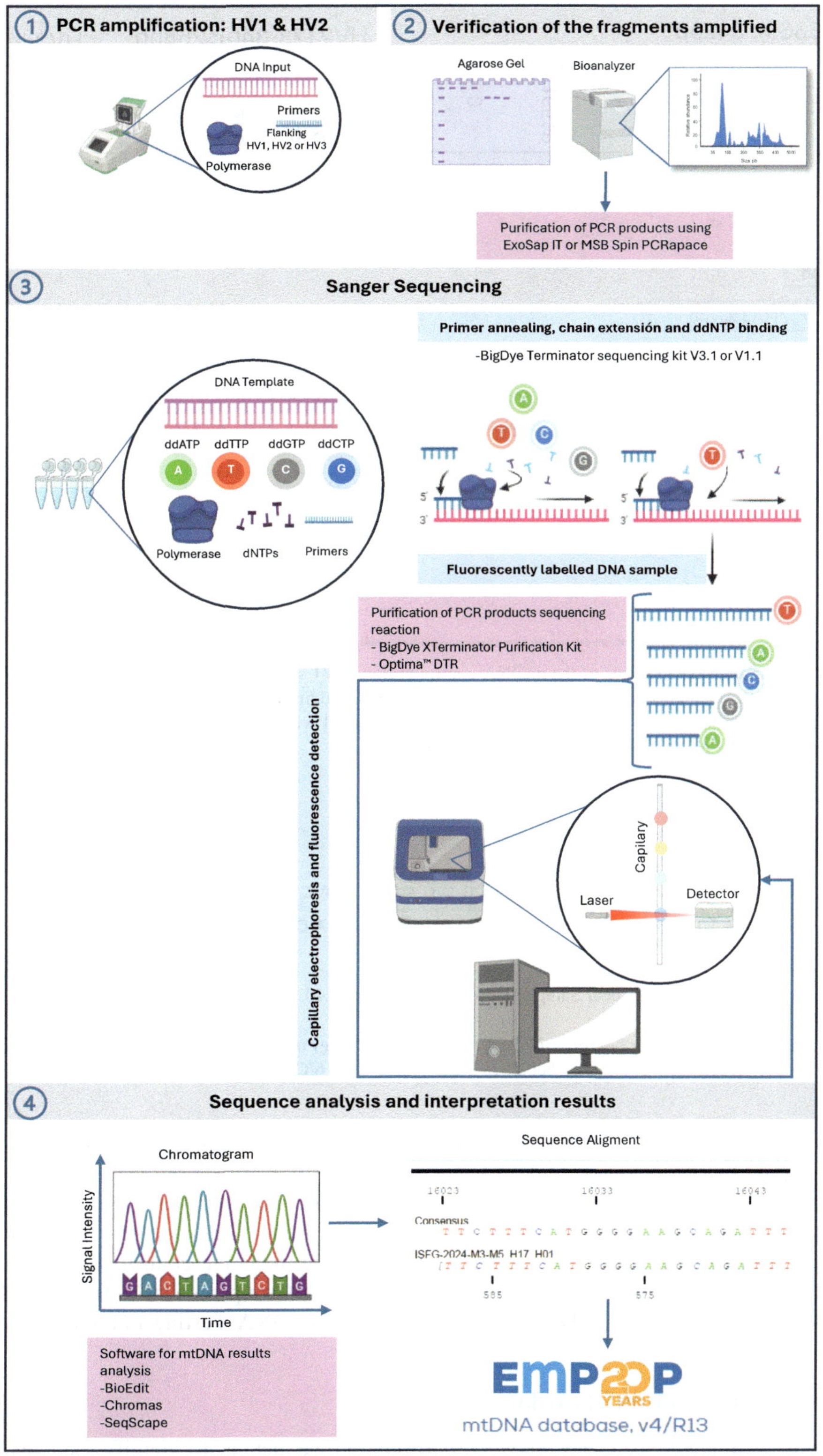

FIGURE 4.3 Workflow of mtDNA analysis from DNA extract by Sanger Sequencing (SS). **Step 1:** PCR amplification using primers flanking at least the HV1 and HV2 regions. **Step 2:** Verification of the mtDNA PCR products by agarose gel or using a bioanalyzer technique. The DNA fragments must be purified to remove primer dimers. **Step 3:** Sequencing of purified PCR products by SS. **Step 4:** Analysis of sequencing and interpretation results. *mtDNA*, mitochondrial DNA; *HV*: hypervariable.

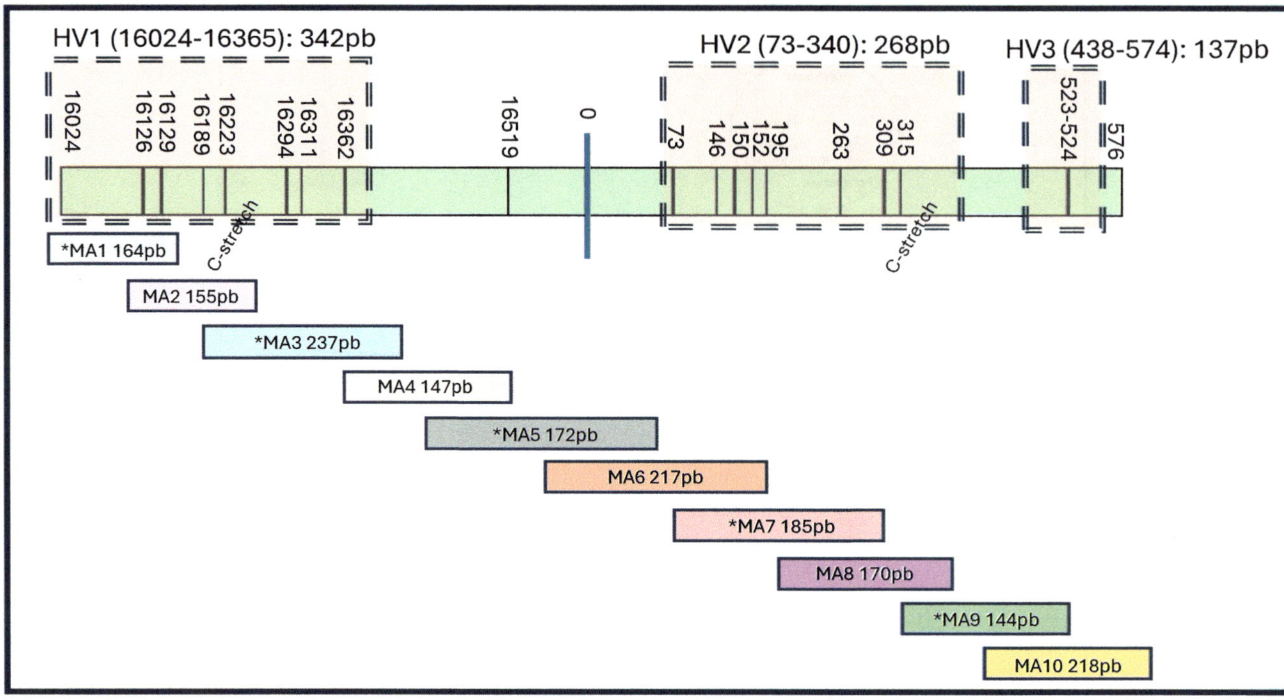

FIGURE 4.4 The three hypervariable (HV) regions of the mtDNA control region. HV1 spans nucleotide positions 16024–16365 (342 bp), HV2 spans positions 73–340 (268 bp), and HV3, which is rarely analysed in forensic testing. Additionally, a schematic representation and sequencing strategy, "mitominis," uses miniamplicons to analyse the mtDNA control region. Bars marked with an asterisk (*) represent DNA fragments co-amplified in multiplex 1, while the remaining mini amplicons are amplified in multiplex 2.
Adapted from (Eichmann & Parson, 2008).

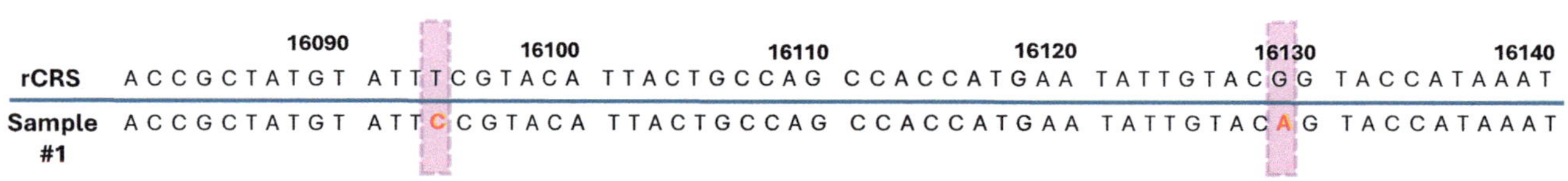

FIGURE 4.5 Sequence alignments for a hypothetical Sample#1 positions 16071–16140. The polymorphisms or differences that must be reported for the sample#1 are 16093 C, 16129 A.
Adapted fromButler, J.M. (2012). *Advanced topics in forensic DNA typing: Methodology advanced topics in forensic dna typing: methodology*. Elsevier Inc., United States. http://www.sciencedirect.com/science/book/978012374513210.1016/C2011-0-04189-3.

double coverage, ensuring that every nucleotide is evaluated either by sequencing both the top and bottom strands or by sequencing the same strand twice.

Positive and negative controls must also be processed alongside the samples at each step of the procedure. Contamination is assessed using reagent blanks and negative controls. Reagent blanks are used to monitor contamination from the extraction phase through to the final sequence analysis, while negative controls track contamination from the amplification stage to the final sequence analysis (Scientific Working Group on DNA Analysis Methods. SWGDAM Interpretation Guidelines for Mitochondrial DNA Analysis by Forensic DNA Testing Laboratories, 2024).

4.6.3 Reporting statistics and mtDNA database

The genetic comparison of mtDNA sequences between two or more sequences to identify matches by maternal lineage typically results in three outcomes: exclusion, inconclusiveness, or failure to exclude. Table 4.1 provides some recommendations for mtDNA interpretation. In cases where there is a genetic match (NOT EXCLUSION), it is necessary to demonstrate that the match is not a random coincidence. In current practice, the rarity of a particular mtDNA type among unrelated individuals is determined by counting how many times a specific haplotype (sequence) appears in an mtDNA database. The most widely used mtDNA database in forensic genetics is EMPOP (https://empop.online/).

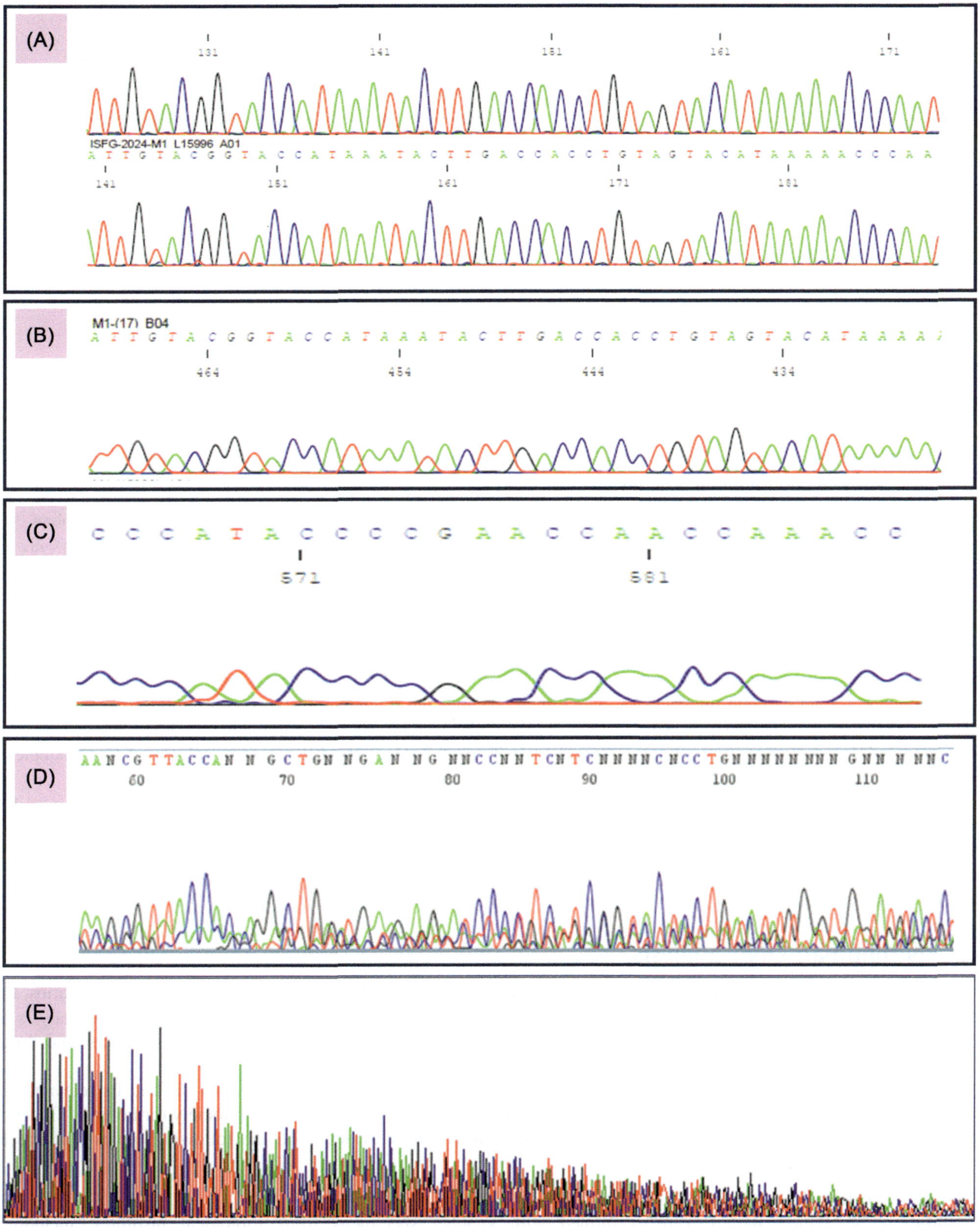

FIGURE 4.6 Chromatogram comparison of the quality of the mtDNA sequences obtained by Sanger Sequencing. (A) Good quality sequence. (B) Acceptable quality sequence. (C) Poor quality sequence, low intensity of the peaks. (D) Poor quality sequence with a much noise, not well defined basecalling. (E) A gradual loss of signal at larger sizes from degraded samples.

TABLE 4.1 Example mtDNA sequence and interpretation for known and question sample pairs.

Sample ID	Sequences Results	Interpretation	Recommendation
Sample #1	TATTGTACGGG	CANNOT EXCLUDE	If the sequences from the questioned and known samples being compared share a common base at each position, the sample cannot be excluded as originating from the same individual or maternal lineage.
Sample #2	TATTGTACGGG		
Sample #1	TATTG**C**ACGGG	INCONCLUSIVE	The sequences are identical at every position except one, with no indication of heteroplasmy or base mixtures
Sample #2	TATTGTACGGG		
Sample #1	TATTG**C**AC**A**GG	EXCLUSION	If there are two or more nucleotide differences between the questioned and known samples, the sample can be excluded as originating from the same individual or maternal lineage
Sample #2	TATTGTACGGG		

Source: Adapted from Advanced topics in forensic DNA typing: Methodology book, Chapter 14.

TABLE 4.2 Reporting format with differences from the rCRS: Sample #1 and Sample #2 may be genetically related, as they share the same polymorphisms in the analyzed region.

Sample ID	Analysis range	Polymorphisms
1	16071–16140	16093 C,16129 C
2	16071–16140	16093 C,16129 C

Polymorphisms are entered into the database, which then provides the frequency of the specific mtDNA sequence. This value is then used to calculate a likelihood ratio (LR), which is reported in identification findings. The manner of expressing mtDNA sequence comparisons is described in Table 4.2. Once the LR has been calculated (for example, a hypothetical LR=10,000), the following conclusion should be reported:

"The mtDNA (mtDNA) sequence obtained from the questioned sample (Sample 1) matches the sequence from the reference sample (Sample 2) at all analyzed positions. The calculated Likelihood Ratio (LR) for this match is 10,000, indicating that it is 10,000 times more likely that this match would be observed if both samples originate from the same maternal lineage rather than from unrelated individuals. Based on this value, the samples cannot be excluded as sharing a common maternal origin."

4.7 Nonhuman DNA analysis

Since DNA is the basis of life, and there are many contexts in which living beings are involved in forensic scenarios, the applications of nonhuman DNA analysis have increased in recent years. These applications include DNA analysis of wildlife, insects, geolocation, linking suspects to crime scenes, counterfeit food or living-derived products, and drugs (Dash et al., 2023).

In the following lines, we are going to outline the potential uses of nonhuman DNA analysis and its relevance in forensic cases. Regarding the actual application of this technology in court, the International Society for Forensic Genetics has outlined several recommendations for the use of nonhuman DNA evidence on the following topics (Linacre & Gusmão, Hecht, et al., 2011):

- **Sample collection**. Use of integrity and traceability procedures.
- **Validation**. Validation studies were performed with voucher specimens from zoological institutes, including: intraspecies and interspecies studies, sensitivity, specificity, reproducibility, and mixed samples.
- **Methodology**. Justification of the choice of locus/loci used in species identification and the use of previously referenced nucleotide sequence and primer map. Primer's specificity and reproducibility should be ensured. If STRs are tested,

tetrameric STRs are preferred. The allelic ladder should contain sequenced alleles and be suitable for most samples. If mutations are encountered, the mutation rate should be assessed if possible.
- **Reporting**. Use of relevant population and forensic parameters, application of kinship factors, likelihood ratio approach.
- **Accreditation**. Laboratories performing this procedure should consider their accreditation to the ISO 17025 standard.

In general terms, nonhuman DNA analysis is performed by amplifying specific regions with universal primers, followed by sequencing the amplicons. The obtained sequence is then matched with reference sequences using a Basic Local Alignment Search Tool engine (Jun et al., 2011).

4.7.1 Animals

For animals, mitochondrial regions with low mutation rates are chosen, such as cytochrome B (*cytB*), cytochrome c oxidase subunit I (*COI*), or NADH dehydrogenase subunits one and two (*ND1*, *ND2*). Among these, *COI* was chosen for the molecular taxonomic assignment of all animals. However, it shows low species discrimination in many related taxa, which may suggest the study of other regions (Meiklejohn et al., 2021).

Animal DNA testing has several applications in forensic genetics. Wildlife forensic science addresses two main issues: the identification of a particular species and the identification of a member of that species, all within the framework of protecting animals at both national and international legislation (Linacre & Tobe, 2011). Other applications include animal DNA detection methods to uncover food fraud (Karabasanavar et al., 2017), use as a trace evidence in crime scenes (Lyons et al., 2014), and the identification of forensically relevant cadaveric arthropods (Samerjai et al., 2020).

4.7.2 Plants

Forensic botany has proven to be useful in several criminal cases, including establishing a link between criminal, victim, and crime scene, determining the cause of death, determining the place of death, or estimating the postmortem interval (Oliveira et al., 2023). In contrast to animal species identification, finding an adequate region for DNA barcoding in plants has been challenging; the Consortium for the Barcode of Life Plant Working Group has proposed two coding regions from the chloroplast genome: *rbcL* (ribulose 1,5-biphosphate carboxylase) and *matK* (maturase K), which successfully discriminated 72% of the species they tested (Hollingsworth et al., 2009).

Pollen grains have been widely analyzed in forensic cases due to their combination of dispersal mechanisms, abundance, and resistance to mechanical and chemical resistance, allowing pollen grains to survive for up to 21 days in the intestinal tract (Alotaibi et al., 2020). While traditional methods of pollen analysis included microscopy and spectroscopy, DNA metabarcoding has emerged as a reliable method for pollen identification (Prudnikow et al., 2023).

Plants DNA analysis has demonstrated its usefulness not only in the characterization of plant species but also in the individualization of the plant which may allow prosecutors to connect different illicit plant growers, as has been the case with *Cannabis sativa*(Coyle et al., 2003). Other uses of plant DNA analysis include detecting counterfeit products. For example, the embedded pollen DNA in honey can be analyzed to determine its origin (Khansaritoreh et al., 2020).

4.7.3 Microorganisms

Currently, the 16S ribosome DNA (rDNA) is the most studied region for bacterial identification because it is easily isolated, ubiquitous, and highly conserved (Janda, 2021). However, despite having potential applications, microbiome forensics still needs to address several requirements, such as standardization, the creation of a reliable database, and evaluation in many more samples (Nagasawa et al., 2013).

Forensic microbiology started with the fear of a possible bioterrorist attack after the 9/11 attacks. Thanks to the development of sequencing technology, its applications are developing rapidly. Currently, there are three areas of study in forensic science: identification, based on an individual's characteristic microbiome; post-mortem interval estimation, studying the microorganisms involved in body decay; and geolocation possible with the specific microbiota of a specific area (Haarkötter et al., 2021).

4.8 Next generation sequencing

Molecular biology has witnessed a rapid evolution in recent times, with notable advancements particularly in the area of sequencing. Many of these developments have contributed to the advent of NGS. A key feature common to all NGS techniques is the massive parallel sequencing strategy, whereby thousands of reads are generated simultaneously (Alvarez-Cubero et al., 2017).

The first commercial application of NGS was in 2005 (Margulies et al., 2005). The technique was initially referred to as "massively parallel sequencing" due to its ability to facilitate the simultaneous sequencing of numerous DNA strands, in contrast to the single-strand sequencing approach employed in Sanger sequencing. The advent of rapid genotyping advancements in recent years has led to a transition in mtDNA analyses from the traditional Sanger sequencing to the NGS methods, which offer enhanced benefits such as significantly higher throughputs, reduced analysis time, and opportunities for standardization and automation (Bruijns et al., 2018).

4.8.1 Analysis and sequencing platforms

Forensic laboratories may efficiently and cost-effectively produce mitogenome data using NGS techniques. Commercial NGS kits offer enrichment approaches that use small amplicons to target the mtDNA genome, which are suitable for a range of sample qualities. Whole genome sequencing of mtDNA based on the MPS platform has the potential to obtain the complete genetic information of mtDNA in a single experiment, which could significantly enhance the forensic application of mtDNA. During the data analysis process, the mtDNA whole genome sequence is obtained by the alignment and merging of overlapping sequencing reads (Barbaro, 2023).

All NGS platforms necessitate a sample preparation process based on the same procedure. mtDNA typing is dependent on the generation of PCR libraries from previously amplified products of mtDNA. The library is typically prepared in two stages (Fig. 4.7): firstly, the amplification of specific regions; and secondly, library preparation (addition and indexing of adapters) (Barbaro, 2023).

The present sequencing of mtDNA technologies is based on the sequencing by synthesis method, which represents a further evolution of the established Sanger sequencing process. This involves repeated cycles of synthesis, whereby additional nucleotides are incorporated into the growing chain. The addition of each of the four nucleotides is performed iteratively, ensuring that only a single dNTP is responsible for the observed signal. In the absence of the next nucleotide in the reaction, elongation is prevented (Goodwin et al., 2016). The current sequencing by synthesis methodologies differ from the original Sanger sequencing method in that they rely on the generation of shorter reads with high sequence coverage, resulting in millions to billions of short DNA sequence reads (50–300 nucleotides) (Slatko et al., 2018).

Prior to sequencing, libraries have to go through a process of template, enrichment, and capture that will depend on the chemistry and the commercial kit used.

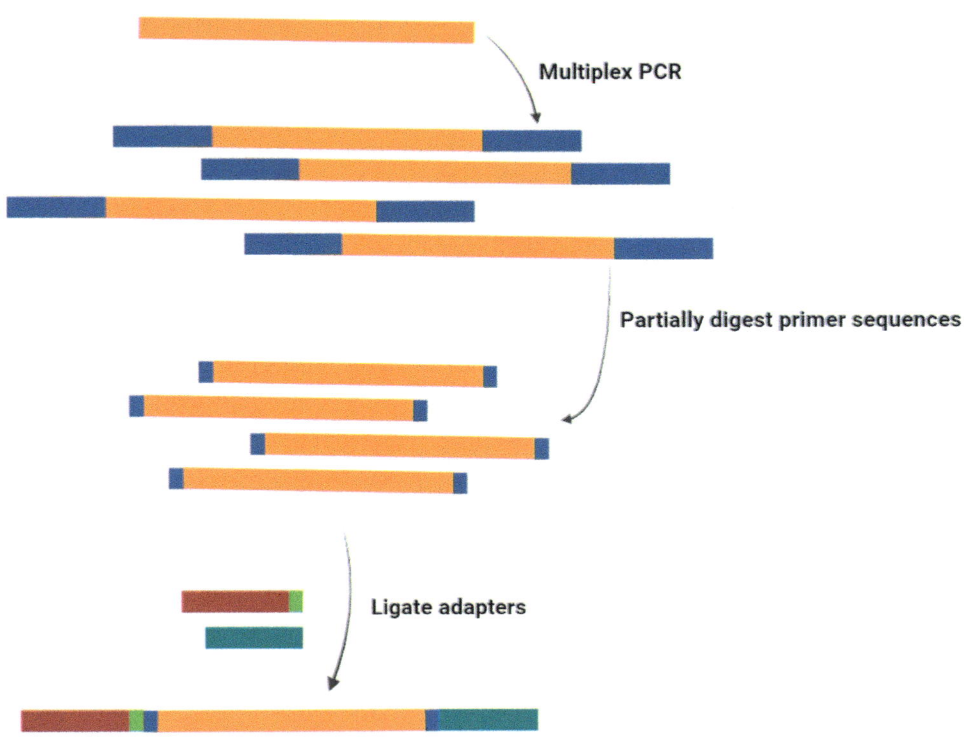

FIGURE 4.7 Generation of PCR libraries.

In emulsion PCR, DNA templates are subjected to fragmentation and ligation to adapter sequences. They are then captured within micelles utilizing a bead that has been coated with complementary adapters, primers, deoxynucleotides, and DNA polymerase. The polymerase chain reaction is performed within the micelle, with each bead becoming coated with thousands of copies of the target DNA sequence (Barbaro, 2023).

In solid-phase bridge amplification, DNA templates are ligated to adapters and subsequently bind with primers that have been immobilized onto a solid support surface (Fig. 4.8). The free extremity interacts with other primers, thereby creating a bridge structure. Using PCR, replicate strands are produced from the immobilized primers whilst unbound DNA is removed. The new double-stranded DNA is partially denatured, allowing the free end of the original template to bind with another primer sequence. The generation of additional free templates is facilitated by the use of reverse primers in strand displacement. Each of these free templates can bind to a new primer (Barbaro, 2023).

In 2010, Applied Biosystems initiated the commercialization of the Ion Torrent, a nonoptical sequencing platform comprising integrated circuits based on the utilization of a semiconductor ion (Fig. 4.9B). This method is based on the iterative addition of each of the four dNTPs to a sequencing reaction, with the objective of ensuring that only one nucleotide is responsible for the signal. The absence of the subsequent nucleotide prevents the sequencing reaction from continuing. During DNA synthesis, when a novel dNTP is incorporated, a proton (H+) is discharged, and a slight alteration in pH is detected by a sensor (a field-effect transistor). This enables the identification of the bases (A, C, G, and T) that are incorporated into the chain. The chemical signals are converted into digital information (Barbaro, 2023).

The system developed by Illumina employs terminator molecules that are analogous to those utilized in Sanger sequencing. dNTPs are labeled with a base-specific, cleavable fluorophore and blocked by a 30-O-azidomethyl group. In the sequencing process, all four dNTPs are added. When a single dNTP is incorporated, it blocks chain elongation. The emitted color allows the identification of which nucleotide was incorporated in each cluster (Fig. 4.9). Subsequently, the

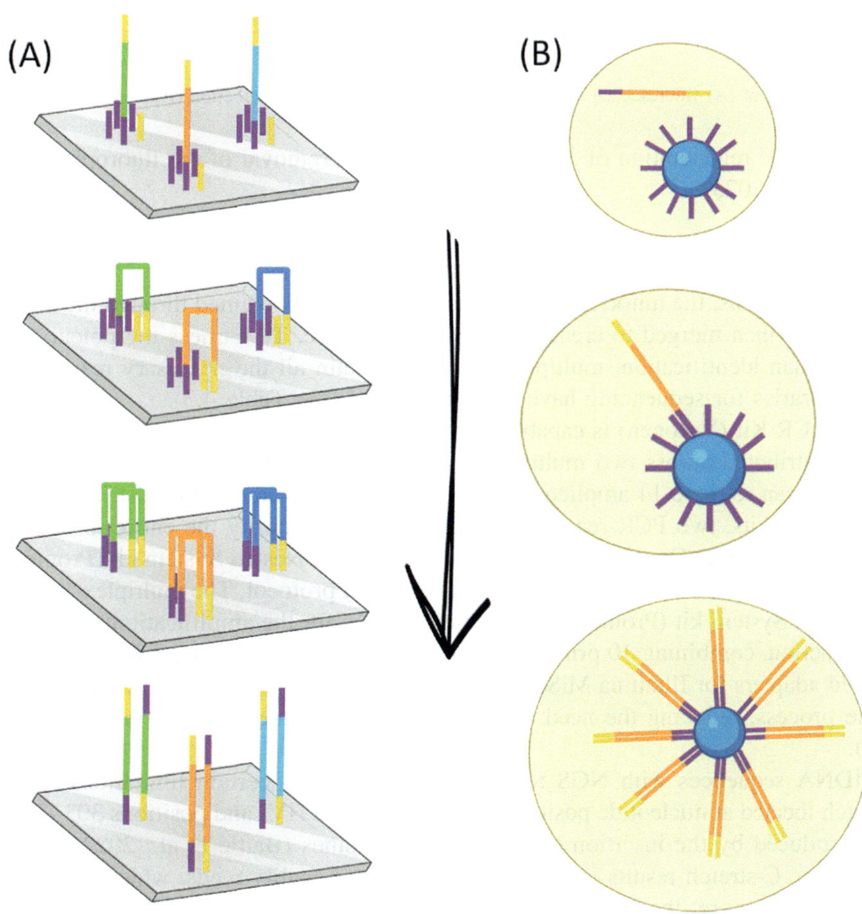

FIGURE 4.8 Enrichement and template techniques of (A) bridge amplification (Illumina) (B) emulsion PCR (Applied Biosystems).

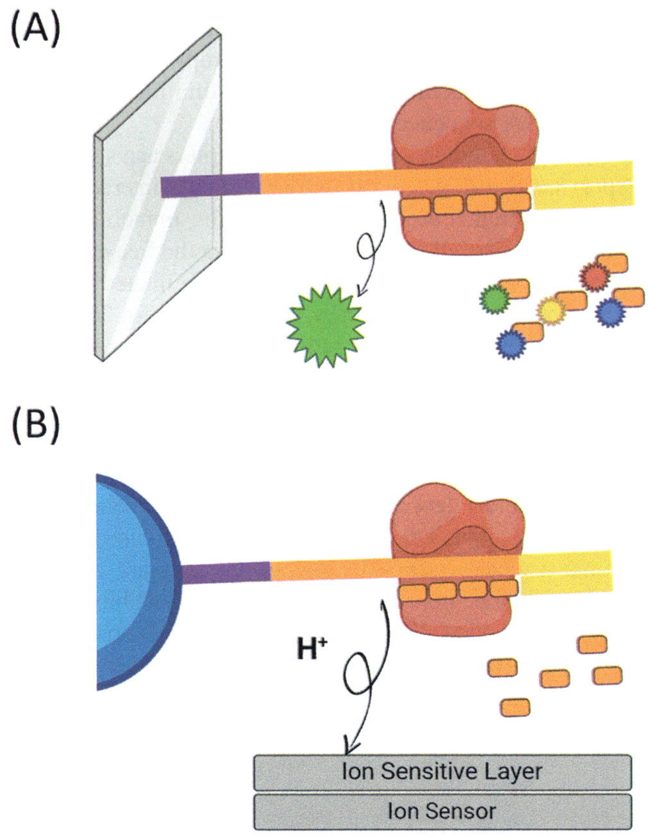

FIGURE 4.9 Sequencing techniques of (A) fluorescence detection (Illumina) and (B) proton detection (Applied Biosystems).

dye is cleaved, resulting in the regeneration of 30-OH. Following the removal of the fluorophore and blocking group, a new cycle can start (Barbaro, 2023).

4.8.2 Commercial kits

During the process of data analysis, the mitochondrial genome sequence is obtained through the alignment of overlapping sequencing reads, which are then merged to create a complete sequence. To harness the potential of massively parallel sequencing (MPS) for human identification, multiplex kits that contain all the necessary reagents for targeting specific markers and preparing libraries for sequencing have been developed (see Table 4.3).

The ForenSeq mtDNA-CR Kit (Verogen) is capable of retrieving 18 overlapping amplicons of 60–150 bp through the use of 18 primer pairs distributed across two multiplexes (England & Harbison, 2020). The Precision ID mtDNA-CR Panel Kit (Thermo Fisher) can retrieve 14 amplicons of 163 bp using 14 primer pairs distributed across two multiplexes (Barrio et al., 2020). In both kits, two PCR reactions are conducted to amplify the entire mtDNA control, followed by adapter ligation and sample indexing. On the other hand, PowerSeq CRM Nested System kit (Promega Corporation) offers the advantage of reducing the workflow through a nested amplification protocol. The multiplex-PCR strategy employed by the PowerSeq CRM Nested System kit (Promega Corporation) allows for the amplification of the entire mtDNA-CR in a single multiplex PCR reaction, combining 10 primer pairs to generate 10 overlapping short amplicons (147–237 bp). The incorporation of indexed adapters for Illumina MiSeq (Illumina, Inc., San Diego, CA, USA) in a single PCR reaction step further streamlines the process, reducing the need for amplicon manipulation and the risk of contamination (Vinueza-Espinosa et al., 2023).

The analysis of mtDNA sequences with NGS techniques is subject to certain limitations. In particular, regions of homopolymeric C stretch located at nucleotide positions 16164–16193 in HVI and positions 303–315 in HVII are prone to present heteroplasmy produced by the insertion or deletion of cytosines (Battle et al., 2022; Forster et al., 2010). The presence of homopolymeric C-stretch results in the generation of low-quality reads, which reduces the number of reads that pass quality filters. Furthermore, the alignment of these regions is typically poor (Gutierrez et al., 2022; Seo et al., 2015; Sturk-Andreaggi et al., 2020)

TABLE 4.3 Commercial kits for the analysis of mtDNA by next-generation sequencing

	Sequencing platform	Panel	Primer set	Amplicon number	Average amplicon size (bp)	DNA input (pg)	Template	Sequencing detection	Manufacturer
Control Region	Ion S5	Precision ID mtDNA Control Region Panel	2	14	153	30	Emulsion PCR	Proton	Thermo Fisher Scientific
	MiSeq FGx	ForenSeq mt DNA Control Region Kit	2	18	118	100	Solid-phase bridge amplification	Fluorescence	Verogen Qiagen
	MiSeq FGx	PowerSeq® CRM Nested System Custom	1	10	144–237	50	Solid-phase bridge amplification	Fluorescence	Promega
Whole mt genome	Ion S5	Precision ID mtDNA Whole Genome Panel	2	162	163	30	Emulsion PCR	Proton	Thermo Fisher Scientific
	MiSeq FGx	ForenSeq mtDNA Whole Genome Kit	2	245	131	100	Solid-phase bridge amplification	Fluorescence	Verogen Qiagen

4.8.3 Applications

NGS is increasingly being applied to the research of mitochondrial genome (Craven et al., 2010; Schlieben & Prokisch, 2023; Yang et al., 2022), including the genetic structure of various populations (Sturk-Andreaggi et al., 2023), maternal genealogy (Peng et al., 2023), and mtDNA variants (Shukla et al., 2023). The application of this technology has enabled significant advancements in whole mitochondrial genome research, including a notable enhancement in the sensitivity of detecting heterogeneity and the ability to detect point heteroplasmy at a frequency as low as 0.15% (Guo et al., 2013).

The capacity to analyze the entire mitochondrial genome markedly enhances the discriminatory power in comparison with control region analysis alone, enabling the most comprehensive resolution of matrilineal geographic ancestry. A study involving 44 samples from 4- generations, 3-generations, and 2-generations pedigrees observed a stable transmission of homoplasmic variants among maternal pedigrees and variable differences in heteroplasmic variants between maternal relatives (Peng et al., 2023).

Developmental validation of these types of MPS-based kits has confirmed their utility for human identification purposes. Sensitivity studies on the ForenSeq mtDNA control region kit (Verogen/Qiagen) showed successful sequencing readings when working with DNA input as low as 1 pg, reaching coverages from 89% to 100%. Comparisons of control region coverage and STR profile success reveal that, even in compromised bone samples; nearly complete control region coverage can be achieved when low-quality STR profiles are obtained (Gutierrez et al., 2022). Concordance studies between different types of samples from the same individual, shown that sequences recovered from rootless hairs are concordant with buccal reference samples, but heteroplasmy at 11.4% detected in one hair may be undetected in the buccal reference. Furthermore, concordance studies between Sanger sequencing and NGS techniques, that heteroplasmy between 10% and 20% can be underdetected with Sanger sequencing. When different NGS chemistries are compared, PowerSeq CRM Nested System and ForenSeq mtDNA Control Region kit, concordance between forensic samples is observed for both kits (Gutierrez et al., 2022).

Validation studies on the Precision ID mtDNA Whole Genome Panel reveal that more than 90% of the mtGenome reaches the 20X read depth threshold when genomic DNA input is as low as 2.3 pg (Cihlar et al., 2020).

Sensitivity studies on the ForenSeq mtDNA Whole Genome kit (Verogen/Qiagen) showed successful sequencing readings when working with DNA input as low as 2.5 pg. Furthermore, analysis of different types of forensic samples such as buccal swabs, blood stain samples, unrooted hairs, bones, teeth, and ancient bone samples analyzed with ForenSeq mtDNA Whole Genome (Verogen/Qiagen) demonstrated that the sequencing coverage of fresh samples like buccal swabs is relatively complete as soon as for bone and aged bloodstain samples. However, sequencing results from teeth or unrooted hair were very poor (Liu et al., 2023).

In addition, all these kits have been successfully used in highly degraded skeletal remains (Templeton et al., 2013). A study based on 52 age-old skeletal remains using the Precision ID mtDNA Control Region Panel achieved fifty full and two partial CR haplotypes. However, data from Sanger sequencing yielded 28 full and 24 partial HVI sequences for the same bone samples (Ta et al., 2021). Another study implemented three different protocols, based on PCR modification conditions, to determine mtDNA-CR sequences of 41 individuals from different time periods with the PowerSeq® CRM Nested System Custom kit. The results show that it is possible to recover more than 85% of the mtDNA-CR from ancient human skeletal remains when the PCR manufacturer's recommendations are modified with 35 cycles, and longer denaturation and extension steps (Vinueza-Espinosa et al., 2023).

4.8.4 Future endeavors

Recently, the study of mtDNA through next-generation technologies has gone into the next step.

In contrast to the current technologies that are based on the amplification of short-length fragments, new developments constructed on long fragment amplification may reduce the interference of homologous ngDNA segments and reduce the co-amplification of nuclear mitochondrial sequences (NUMTs) and may lessen the influence of the mtDNA complex structure and produce more balanced sequencing data. A new MPS platform, the DNA nanoball sequencing platform, has been recently launched by Roche based on DNA single-strand circularization and DNA nanoball making techniques, far from the present one available in the analysis of mtDNA (Chen et al., 2024). Validation studies on this technology have been released to evaluate its forensic efficiency. However, none of them evaluate theirefficiency on real ancient or degraded samples or DNA input lower than 31.25 pg.

Oxford Nanopore Technology, based on long reads;works by identifying DNA bases by measuring the changes in electrical conductivity generated as DNA strands pass through a biological pore, so the DNA molecule is directly read. To date, this technology has been tested to detect microdeletions on the mtDNA genome associated with several mitochondrial diseases (Frascarelli et al., 2023).

References

Alotaibi, S. S., Sayed, S. M., Alosaimi, M., Alharthi, R., Banjar, A., Abdulqader, N., & Alhamed, R. (2020). Pollen molecular biology: Applications in the forensic palynology and future prospects: A review. *Saudi Journal of Biological Sciences, 27*(5), 1185–1190. https://doi.org/10.1016/j.sjbs.2020.02.019, http://www.ksu.edu.sa/sites/Colleges/CollegeofScinces/BotanyDepartment/SJBS/Default.aspx.

Alvarez-Cubero, M. J., Saiz, M., Martinez-Gonzalez, L. J., Alvarez, J. C., Eisenberg, A. J., Budowle, B., & Lorente, J. A. (2012). Genetic identification of missing persons: DNA analysis of human remains and compromised samples. *Pathobiology, 79*(5), 228–238. https://doi.org/10.1159/000334982.

Alvarez-Cubero, M. J., Saiz, M., Martínez-García, B., Sayalero, S. M., Entrala, C., Lorente, J. A., & Martinez-Gonzalez, L. J. (2017). Next generation sequencing: An application in forensic sciences? *Annals of Human Biology, 44*(7), 581–592. https://doi.org/10.1080/03014460.2017.1375155.

Amorim, A., Fernandes, T., & Taveira, N. (2019). Mitochondrial DNA in human identification: A review. *PeerJ, 7*, e7314. https://doi.org/10.7717/peerj.7314.

Ballard, D. (2016). Analysis of mitochondrial control region using sanger sequencing. *Methods in Molecular Biology, 1420*, 143–155. https://doi.org/10.1007/978-1-4939-3597-0_12, http://www.springer.com/series/7651.

Bandelt, H. J., Kloss-Brandstätter, A., Richards, M. B., Yao, Y. G., & Logan, I. (2014). The case for the continuing use of the revised Cambridge reference sequence (rCRS) and the standardization of notation in human mitochondrial DNA studies. *Journal of Human Genetics, 59*(2), 66–77. https://doi.org/10.1038/jhg.2013.120.

Barbaro, A. (2023). Overview of NGS platforms and technological advancements for forensic applications. In *Next generation sequencing (NGS) technology in DNA analysis*. Italy: Elsevier, 35–58. https://www.sciencedirect.com/book/9780323991445, https://doi.org/10.1016/B978-0-323-99144-5.00019-6.

Barrio, P. A., García, Ó., Phillips, C., Prieto, L., Gusmão, L., Fernández, C., Casals, F., Freitas, J. M., González-Albo, M.del C., Martín, P., Mosquera, A., Navarro-Vera, I., Paredes, M., Pérez, J. A., Pinzón, A., Rasal, R., Ruiz-Ramírez, J., Trindade, B. R., & Alonso, A. (2020). The first GHEP-ISFG collaborative exercise on forensic applications of massively parallel sequencing. *Forensic Science International: Genetics, 49*. https://doi.org/10.1016/j.fsigen.2020.102391 102391.

Battle, S. L., Puiu, D., Verlouw, J., Broer, L., Boerwinkle, E., Taylor, K. D., Rotter, J. I., Rich, S. S., Grove, M. L., Pankratz, N., Fetterman, J. L., Liu, C., & Arking, D. E. (2022). A bioinformatics pipeline for estimating mitochondrial DNA copy number and heteroplasmy levels from whole genome sequencing data. *NAR Genomics and Bioinformatics, 4*(2). https://doi.org/10.1093/nargab/lqac034, https://academic.oup.com/nargab.

Biesecker, L. G., Bailey-Wilson, J. E., Ballantyne, J., Baum, H., Bieber, F. R., Brenner, C., Budowle, B., Butler, J. M., Carmody, G., Conneally, P. M., Duceman, B., Eisenberg, A., Forman, L., Kidd, K. K., Leclair, B., Niezgoda, S., Parsons, T. J., Pugh, E., Shaler, R., ... Walsh, A. (2005). DNA identifications after the 9/11 World Trade Center attack. *Science, 310*(5751), 1122–1123. https://doi.org/10.1126/science.1116608.

Bowen, B. W., Clark, A. M., Abreu-Grobois, F. A., Chaves, A., Reichart, H. A., & Ferl, R. J. (1997). Global phylogeography of the ridley sea turtles (Lepidochelys spp.) as inferred from mitochondrial DNA sequences. *Genetica, 101*(3), 179–189. https://doi.org/10.1023/A:1018382415005, www.wkap.nl/journalhome.htm/0016-6707.

Brandstätter, A., Niederstätter, H., & Parson, W. (2004). Monitoring the inheritance of heteroplasmy by computer-assisted detection of mixed basecalls in the entire human mitochondrial DNA control region. *International Journal of Legal Medicine, 118*(1), 47–54. https://doi.org/10.1007/s00414-003-0418-z.

Bruijns, B., Tiggelaar, R., & Gardeniers, H. (2018). Massively parallel sequencing techniques for forensics: A review. *Electrophoresis, 39*(21), 2642–2654. https://doi.org/10.1002/elps.201800082.

Butler, J. M. (2005). *Forensic DNA typing*. Elsevier.

Butler, J. M. (2012). *Advanced topics in forensic DNA typing: Methodology*. United States: Elsevier Inc. http://www.sciencedirect.com/science/book/9780123745132, 10.1016/C2011-0-04189-3.

Cann, R. L., Stoneking, M., & Wilson, A. C. (1987). Mitochondrial DNA and human evolution. *Nature, 325*(6099), 31–36. https://doi.org/10.1038/325031a0.

Carracedo, A., Bär, W., Lincoln, P., Mayr, W., Morling, N., Olaisen, B., Schneider, P., Budowle, B., Brinkmann, B., Gill, P., Holland, M., Tully, G., & Wilson, M. (2000). DNA Commission of the International Society for Forensic Genetics: Guidelines for mitochondrial DNA typing. *Forensic Science International, 110*(2), 79–85. https://doi.org/10.1016/s0379-0738(00)00161-4.

Chen, M., Chen, C., Li, N., Su, Y., Cui, W., Huang, Y., Cai, M., & Zhu, B. (2024). Forensic efficiency evaluation of a mtDNA whole genome sequencing system constructed with long fragment amplification strategy on DNA nanoball sequencing platform. *Forensic Science International: Genetics, 73*, 103126. https://doi.org/10.1016/j.fsigen.2024.103126.

Cihlar, J. C., Amory, C., Lagacé, R., Roth, C., Parson, W., & Budowle, B. (2020). Developmental Validation of a MPS Workflow with a PCR-based short amplicon whole mitochondrial genome panel. *Genes, 11*(11), 1345. https://doi.org/10.3390/genes11111345.

Connell, J. R., Benton, M. C., Lea, R. A., Sutherland, H. G., Chaseling, J., Haupt, L. M., Wright, K. M., & Griffiths, L. R. (2022). Pedigree derived mutation rate across the entire mitochondrial genome of the Norfolk Island population. *Scientific Reports, 12*(1). https://doi.org/10.1038/s41598-022-10530-3, www.nature.com/srep/index.html.

Coskun, P. E., Wyrembak, J., Derbereva, O., Melkonian, G., Doran, E., Lott, I. T., Head, E., Cotman, C. W., & Wallace, D. C. (2010). Systemic mitochondrial dysfunction and the etiology of Alzheimer's disease and down syndrome dementia. *Journal of Alzheimer's Disease, 20*(2), S293. https://doi.org/10.3233/JAD-2010-100351, www.iospress.nl/journal/journal-of-alzheimers-disease/.

Coyle, H. M., Palmbach, T., Juliano, N., Ladd, C., & Lee, H. C. (2003). An overview of DNA methods for the identification and individualization of marijuana. *Croatian Medical Journal, 44*(3), 315–321.

Craven, L., Tuppen, H. A., Greggains, G. D., Harbottle, S. J., Murphy, J. L., Cree, L. M., Murdoch, A. P., Chinnery, P. F., Taylor, R. W., Lightowlers, R. N., Herbert, M., & Turnbull, D. M. (2010). Pronuclear transfer in human embryos to prevent transmission of mitochondrial DNA disease. *Nature, 465*(7294), 82–85. https://doi.org/10.1038/nature08958.

Craven, L., Elson, J. L., Irving, L., Tuppen, H. A., Lister, L. M., Greggains, G. D., Byerley, S., Murdoch, A. P., Herbert, M., & Turnbull, D. (2011). Mitochondrial DNA disease: New options for prevention. *Human Molecular Genetics, 20*(2), 168–174. https://doi.org/10.1093/hmg/ddr373.

Daeid, N. N., Hackman, L., & Court, D. S. (2021). Mitochondrial DNA in forensic use. *Emerging Topics in Life Sciences, 5*(3), 415–426. https://doi.org/10.1042/etls20210204.

Dash, H. R., Elkins, K. M., & Al-Snan, N. R. (2023). *Advancements in non-human forensic DNA analysis.* Springer Science and Business Media LLC, 79–89. 10.1007/978-981-99-6195-5_6.

Eichmann, C., & Parson, W. (2008). Mitominis': multiplex PCR analysis of reduced size amplicons for compound sequence analysis of the entire mtDNA control region in highly degraded samples. *International Journal of Legal Medicine, 122*(5), 385–388. https://doi.org/10.1007/s00414-008-0227-5.

Elson, J. L., Andrews, R. M., Chinnery, P. F., Lightowlers, R. N., Turnbull, D. M., & Howell, N. (2001). Analysis of European mtDNAs for recombination. *The American Journal of Human Genetics, 68*(1), 145–153. https://doi.org/10.1086/316938.

Emery, M. V., Bolhofner, K., Spake, L., Ghafoor, S., Versoza, C. J., Rawls, E. M., Winingear, S., Buikstra, J. E., Loreille, O., Fulginiti, L. C., & Stone, A. C. (2024). Targeted enrichment of whole-genome SNPs from highly burned skeletal remains. *Journal of Forensic Sciences, 69*(5), 1558–1577. https://doi.org/10.1111/1556-4029.15482, http://onlinelibrary.wiley.com/journal/10.1111/(ISSN)1556-4029.

England, R., & Harbison, S. (2020). A review of the method and validation of the MiSeq FGx™ forensic genomics solution. *WIREs Forensic Science, 2*(1). https://doi.org/10.1002/wfs2.1351.

Forster, L., Forster, P., Gurney, S. M. R., Spencer, M., Huang, C., Röhl, A., & Brinkmann, B. (2010). Evaluating length heteroplasmy in the human mitochondrial DNA control region. *International Journal of Legal Medicine, 124*(2), 133–142. https://doi.org/10.1007/s00414-009-0385-0.

Frascarelli, C., Zanetti, N., Nasca, A., Izzo, R., Lamperti, C., Lamantea, E., Legati, A., & Ghezzi, D. (2023). Nanopore long-read next-generation sequencing for detection of mitochondrial DNA large-scale deletions. *Frontiers in Genetics, 14.* https://doi.org/10.3389/fgene.2023.1089956.

Gabriel, M. N., Huffine, E. F., Ryan, J. H., Holland, M. M., & Parsons, T. J. (2001). Improved MtDNA sequence analysis of forensic remains using a "mini-primer set" amplification strategy. *Journal of Forensic Sciences, 46*(2), 247–253. https://doi.org/10.1520/jfs14957j.

Goodwin, S., McPherson, J. D., & McCombie, W. R. (2016). Coming of age: Ten years of next-generation sequencing technologies. *Nature Reviews Genetics, 17*(6), 333–351. https://doi.org/10.1038/nrg.2016.49, http://www.nature.com/reviews/genetics.

Greenberg, B. D., Newbold, J. E., & Sugino, A. (1983). Intraspecific nucleotide sequence variability surrounding the origin of replication in human mitochondrial DNA. *Gene, 21*(1-2), 33–49. https://doi.org/10.1016/0378-1119(83)90145-2.

Guo, Y., Li, C. I., Sheng, Q., Winther, J. F., Cai, Q., Boice, J. D., & Shyr, Y. (2013). Very low-level heteroplasmy mtDNA variations are inherited in humans. *Journal of Genetics and Genomics, 40*(12), 607–615. https://doi.org/10.1016/j.jgg.2013.10.003.

Gustafsson, C. M., Falkenberg, M., & Larsson, N. G. (2016). Maintenance and expression of mammalian mitochondrial DNA. *Annual Review of Biochemistry, 85*, 133–160. https://doi.org/10.1146/annurev-biochem-060815-014402, http://arjournals.annualreviews.org/loi/biochem.

Gutierrez, R., Roman, M. G., Harrel, M., Hughes, S., LaRue, B., & Houston, R. (2022). Assessment of the ForenSeq mtDNA control region kit and comparison of orthogonal technologies. *Forensic Science International: Genetics, 59*, 102721. https://doi.org/10.1016/j.fsigen.2022.102721.

Haarkötter, C., Saiz, M., Gálvez, X., Medina-Lozano, M. I., Álvarez, J. C., & Lorente, J. A. (2021). Usefulness of microbiome for forensic geolocation: A review. *Life, 11*(12), 1322. https://doi.org/10.3390/LIFE11121322.

Haarkötter, C., Vinueza-Espinosa, D. C., Gálvez, X., Saiz, M., Medina-Lozano, M. I., Lorente, J. A., & Álvarez, J. C. (2023). A comparison between petrous bone and tooth, femur and tibia DNA analysis from degraded skeletal remains. *Electrophoresis, 44*(19-20), 1559–1568. https://doi.org/10.1002/elps.202300097, http://onlinelibrary.wiley.com/journal/10.1002/(ISSN)1522-2683.

Haarkötter, C., Isabel Medina-Lozano, M., Vinueza-Espinosa, D. C., Saiz, M., Gálvez, X., Carlos Álvarez, J., & Antonio Lorente, J. (2024). Evaluating the efficacy of three Y-STRs commercial kits in degraded skeletal remains. *Science & Justice, 64*(5), 543–548. https://doi.org/10.1016/j.scijus.2024.07.007.

Holland, C. A., McElhoe, J. A., Gaston-Sanchez, S., & Holland, M. M. (2021). Damage patterns observed in mtDNA control region MPS data for a range of template concentrations and when using different amplification approaches. *International Journal of Legal Medicine, 135*(1), 91–106. https://doi.org/10.1007/s00414-020-02410-0, link.springer.de/link/service/journals/00414/index.htm.

Holland, M. M., Fisher, D. L., Mitchell, L. G., Rodriquez, W. C., Canik, J. J., Merril, C. R., & Weedn, V. W. (1993). Mitochondrial DNA sequence analysis of human skeletal remains: Identification of remains from the Vietnam War. *Journal of Forensic Sciences, 38*(3), 542–553. https://doi.org/10.1520/jfs13439j.

Hollingsworth, P. M., Forrest, L. L., Spouge, J. L., Hajibabaei, M., Ratnasingham, S., Bank, M.van der, Chase, M. W., Cowan, R. S., Erickson, D. L., Fazekas, A. J., Graham, S. W., James, K. E., Kim, K. J., John Kress, W., Schneider, H., van AlphenStahl, J., Barrett, S. C. H., van den Berg, C., Bogarin, D., ... Little, D. P. (2009). A DNA barcode for land plants. *Proceedings of the National Academy of Sciences of the United States of America, 106*(31), 12794–12797. https://doi.org/10.1073/pnas.0905845106, http://www.pnas.org/content/106/31/12794.full.pdf United Kingdom.

Janda, JM. (2021). Taxonomic Classification of Bacteria. In L. H. Green, & E. Goldman (Eds.). *Practical Handbook of Microbiology* (pp. 161–167). Boca Raton: CRC Press.

Jun, J., Han, S. H., Jeong, T. J., Park, H. C., Lee, B., & Kwak, M. (2011). Wildlife forensics using mitochondrial DNA sequences: Species identification based on hairs collected in the field and confiscated tanned Felidae leathers. *Genetics Society of Korea, South Korea Genes and Genomics, 33*(6), 721–726. https://doi.org/10.1007/s13258-011-0080-7, http://www.springerlink.com/content/1976-9571.

Karabasanavar, N., Girish, P. S., Kumar, D., & Singh, S. P. (2017). Detection of beef adulteration by mitochondrial D-loop based species-specific polymerase chain reaction. *International Journal of Food Properties, 20*, 1–8. https://doi.org/10.1080/10942912.2017.1369103.

Khansaritoreh, E., Salmaki, Y., Ramezani, E., Akbari Azirani, T., Keller, A., Neumann, K., Alizadeh, K., Zarre, S., Beckh, G., & Behling, H. (2020). Employing DNA metabarcoding to determine the geographical origin of honey. *Heliyon, 6*(11), e05596. https://doi.org/10.1016/j.heliyon.2020.e05596.

Khrapko, K., & Turnbull, D. (2014). Mitochondrial DNA mutations in aging. *Progress in Molecular Biology and Translational Science, 127*, 29–62. https://doi.org/10.1016/B978-0-12-394625-6.00002-7, http://www.elsevier.com/books/book-series/progress-in-molecular-biology-and-translational-science#.

Krings, M., Stone, A., Schmitz, R. W., Krainitzki, H., Stoneking, M., & Pääbo, S. (1997). Neandertal DNA sequences and the origin of modern humans. *Cell, 90*(1), 19–30. https://doi.org/10.1016/S0092-8674(00)80310-4, https://www.sciencedirect.com/journal/cell.

Lee, W., Zamudio-Ochoa, A., Buchel, G., Podlesniy, P., Marti Gutierrez, N., Puigròs, M., Calderon, A., Tang, H.-Y., Li, L., Mikhalchenko, A., Koski, A., Trullas, R., Mitalipov, S., & Temiakov, D. (2023). Molecular basis for maternal inheritance of human mitochondrial DNA. *Nature Genetics, 55*(10), 1632–1639. https://doi.org/10.1038/s41588-023-01505-9.

Lieber, T., Jeedigunta, S. P., Palozzi, J. M., Lehmann, R., & Hurd, T. R. (2019). Mitochondrial fragmentation drives selective removal of deleterious mtDNA in the germline. *Nature, 570*(7761), 380–384. https://doi.org/10.1038/s41586-019-1213-4, http://www.nature.com/nature/index.html.

Linacre, A., & Tobe, S. S. (2011). An overview to the investigative approach to species testing in wildlife forensic science. *Investigative Genetics, 2*(1). https://doi.org/10.1186/2041-2223-2-2Australia, http://www.investigativegenetics.com/content/2/1/2.

Linacre, A., Gusmão, L., Hecht, W., Hellmann, A. P., Mayr, W. R., Parson, W., Prinz, M., Schneider, P. M., & Morling, N. (2011). ISFG: Recommendations regarding the use of non-human (animal) DNA in forensic genetic investigations. *Forensic Science International: Genetics, 5*(5), 501–505. https://doi.org/10.1016/j.fsigen.2010.10.017.

Linch, C. A., Whiting, D. A., & Holland, M. M. (2001). Human hair histogenesis for the mitochondrial DNA forensic scientist. *Journal of Forensic Sciences, 46*(4), 844–853. https://doi.org/10.1520/jfs15056j.

Liu, G., Zheng, Y., Wu, Q., Feng, T., Xia, Y., Chen, D., Ren, L., Bai, X., Li, Q., Chen, D., Lv, M., Liao, M., Liang, W., Zhang, L., & Qu, S. (2023). Assessment of ForenSeq mtDNA whole genome kit for forensic application. *International Journal of Legal Medicine, 137*(6), 1693–1703. https://doi.org/10.1007/s00414-023-03084-0.

Llamas, B., Fehren-Schmitz, L., Valverde, G., Soubrier, J., Mallick, S., Rohland, N., Nordenfelt, S., Valdiosera, C., Richards, S. M., Rohrlach, A., Romero, M. I. B., Espinoza, I. F., Cagigao, E. T., Jiménez, L. W., Makowski, K., Reyna, I. S. L., Lory, J. M., Torrez, J. A. B., Rivera, M. A., Burger, R. L., ... Haak, W. (2016). Ancient mitochondrial DNA provides high-resolution time scale of the peopling of the Americas. *Science Advances, 2*(4). https://doi.org/10.1126/sciadv.1501385.

Loreille, O. M., Diegoli, T. M., Irwin, J. A., Coble, M. D., & Parsons, T. J. (2007). High efficiency DNA extraction from bone by total demineralization. *Forensic Science International: Genetics, 1*(2), 191–195. https://doi.org/10.1016/j.fsigen.2007.02.006.

Lutz, S., Wittig, H., Weisser, H.-J., Heizmann, J., Junge, A., Dimo-Simonin, N., Parson, W., Edelmann, J., Anslinger, K., Jung, S., & Augustin, C. (2000). Is it possible to differentiate mtDNA by means of HVIII in samples that cannot be distinguished by sequencing the HVI and HVII regions? *Forensic Science International, 113*(1-3), 97–101. https://doi.org/10.1016/s0379-0738(00)00222-x.

Lyons, L. A., Grahn, R. A., Kun, T. J., Netzel, L. R., Wictum, E. E., & Halverson, J. L. (2014). Acceptance of domestic cat mitochondrial DNA in a criminal proceeding. *Forensic Science International: Genetics, 13*, 61–67. https://doi.org/10.1016/j.fsigen.2014.07.007, http://www.elsevier.com.

Mackey, D. A., Oostra, R. J., Rosenberg, T., Nikoskelainen, E., Bronte-Stewart, J., Poulton, J., Harding, A. E., Govan, G., Bolhuis, P. A., Norby, S., Bleeker-Wagemakers, E. M., Savontaus, M. L., Chan, C., & Howell, N. (1996). Primary pathogenic mtDNA mutations in multigeneration pedigrees with Leber hereditary optic neuropathy [5]. *American Journal of Human Genetics, 59*(2), 481–485.

Maniura-Weber, K. (2004). Transient overexpression of mitochondrial transcription factor A (TFAM) is sufficient to stimulate mitochondrial DNA transcription, but not sufficient to increase mtDNA copy number in cultured cells. *Nucleic Acids Research, 32*(20), 6015–6027. https://doi.org/10.1093/nar/gkh921.

Margulies, M., Egholm, M., Altman, W. E., Attiya, S., Bader, J. S., Bemben, L. A., Berka, J., Braverman, M. S., Chen, Y.-J., Chen, Z., Dewell, S. B., Du, L., Fierro, J. M., Gomes, X. V., Godwin, B. C., He, W., Helgesen, S., Ho, C. H., Irzyk, G. P., ... Rothberg, J. M. (2005). Genome sequencing in microfabricated high-density picolitre reactors. *Nature, 437*(7057), 376–380. https://doi.org/10.1038/nature03959.

McGowen, M. R., Tsagkogeorga, G., Álvarez-Carretero, S., Dos Reis, M., Struebig, M., Deaville, R., Jepson, P. D., Jarman, S., Polanowski, A., Morin, P. A., & Rossiter, S. J. (2020). Phylogenomic resolution of the cetacean tree of life using target sequence capture. *Systematic Biology, 69*(3), 479–501. https://doi.org/10.1093/sysbio/syz068, http://sysbio.oxfordjournals.org/.

Meiklejohn, K. A., Burnham-Curtis, M. K., Straughan, D. J., Giles, J., & Moore, M. K. (2021). Current methods, future directions and considerations of DNA-based taxonomic identification in wildlife forensics. *Forensic Science International: Animals and Environments, 1*, 100030. https://doi.org/10.1016/j.fsiae.2021.100030.

Mishra, P., & Chan, D. C. (2014). Mitochondrial dynamics and inheritance during cell division, development and disease. *Nature Reviews Molecular Cell Biology, 15*(10), 634–646. https://doi.org/10.1038/nrm3877, http://www.nature.com/molcellbio.

Nagasawa, S., Motani-Saitoh, H., Inoue, H., & Iwase, H. (2013). Geographic diversity of *Helicobacter pylori* in cadavers: Forensic estimation of geographical origin. *Forensic Science International, 229*(1-3), 7–12. https://doi.org/10.1016/j.forsciint.2013.02.028.

Oliveira, M., Azevedo, L., Ballard, D., Branicki, W., & Amorim, A. (2023). Using plants in forensics: State-of-the-art and prospects. *Plant Science, 336*, 111860. https://doi.org/10.1016/j.plantsci.2023.111860.

Parsons, T. J., & Coble, M. D. (2001). Increasing the forensic discrimination of mitochondrial DNA testing through analysis of the entire mitochondrial DNA genome. *Croatian medical journal, 42*(3), 304–309.

Peng, D., Geng, J., Yang, J., Liu, J., Wang, N., Wu, R., & Sun, H. (2023). Whole mitochondrial genome detection and analysis of two- to four-generation maternal pedigrees using a new massively parallel sequencing panel. *Genes, 14*(4), 912. https://doi.org/10.3390/genes14040912.

Pereira, C. V., Gitschlag, B. L., & Patel, M. R. (2021). Cellular mechanisms of mtDNA heteroplasmy dynamics. *Critical Reviews in Biochemistry and Molecular Biology, 56*(5), 510–525. https://doi.org/10.1080/10409238.2021.1934812, http://www.tandfonline.com/loi/ibmg20.

Petros, J. A., Baumann, A. K., Ruiz-Pesini, E., Amin, M. B., Sun, C. Q., Hall, J., Lim, S. D., Issa, M. M., Flanders, W. D., Hosseini, S. H., Marshall, F. F., & Wallace, D. C. (2005). MtDNA mutations increase tumorigenicity in prostate cancer. *Proceedings of the National Academy of Sciences of the United States of America, 102*(3), 719–724. https://doi.org/10.1073/pnas.0408894102.

Polovina, E. S., Parakatselaki, M. E., & Ladoukakis, E. D. (2020). Paternal leakage of mitochondrial DNA and maternal inheritance of heteroplasmy in Drosophila hybrids. *Nature Research, Greece Scientific Reports, 10*(1). https://doi.org/10.1038/s41598-020-59194-x, www.nature.com/srep/index.html.

Prudnikow, L., Pannicke, B., & Wünschiers, R. (2023). A primer on pollen assignment by nanopore-based DNA sequencing. *Frontiers in Ecology and Evolution, 11*. https://doi.org/10.3389/fevo.2023.1112929.

Prüfer, K., Munch, K., Hellmann, I., Akagi, K., Miller, J. R., Walenz, B., Koren, S., Sutton, G., Kodira, C., Winer, R., Knight, J. R., Mullikin, J. C., Meader, S. J., Ponting, C. P., Lunter, G., Higashino, S., Hobolth, A., Dutheil, J., Karakoç, E., ... Pääbo, S. (2012). The bonobo genome compared with the chimpanzee and human genomes. *Nature, 486*(7404), 527–531. https://doi.org/10.1038/nature11128.

Randi, E., Pierpaoli, M., Beaumont, M., Ragni, B., & Sforzi, A. (2001). Genetic identification of wild and domestic cats (felis silvestris) and their hybrids using bayesian clustering methods. *Molecular Biology and Evolution, 18*(9), 1679–1693. https://doi.org/10.1093/oxfordjournals.molbev.a003956.

Robin, E. D., & Wong, R. (1988). Mitochondrial DNA molecules and virtual number of mitochondria per cell in mammalian cells. *Journal of Cellular Physiology, 136*(3), 507–513. https://doi.org/10.1002/jcp.1041360316.

Satoh, M. (1991). Organization of multiple nucleoids and DNA molecules in mitochondria of a human cell. *Experimental Cell Research, 196*(1), 137–140. https://doi.org/10.1016/0014-4827(91)90467-9.

Samerjai, C., Sukontason, K. L., Sontigun, N., Sukontason, K., Klong-klaew, T., Chareonviriyaphap, T., Kurahashi, H., Klimpel, S., Kochmann, J., Saeung, A., Somboon, P., & Wannasan, A. (2020). Mitochondrial DNA-based identification of forensically important flesh flies (Diptera: Sarcophagidae) in Thailand. *Insects, 11*(1), 2. https://doi.org/10.3390/insects11010002.

Sanger, F., Nicklen, S., & Coulson, A. R. (1977). DNA sequencing with chain-terminating inhibitors. *Proceedings of the National Academy of Sciences, 74*(12), 5463–5467. https://doi.org/10.1073/pnas.74.12.5463.

Schlieben, L. D., & Prokisch, H. (2023). Genetics of mitochondrial diseases: Current approaches for the molecular diagnosis. *Handbook of Clinical Neurology, 194*, 141–165. https://doi.org/10.1016/B978-0-12-821751-1.00011-7, http://www.sciencedirect.com/science/journal/00729752.

Schwartz, M., & Vissing, J. (2002). Paternal Inheritance of mitochondrial DNA. *New England Journal of Medicine, 347*(8), 576–580. https://doi.org/10.1056/nejmoa020350.

Scientific Working Group on DNA Analysis Methods (2024). SWGDAM Interpretation Guidelines for Mitochondrial DNA Analysis by Forensic DNA Testing Laboratories. https://www.swgdam.org/_files/ugd/4344b0_9784320e1188491fbdd0dbf040ca69ae

Seo, S. B., Zeng, X., King, J. L., Larue, B. L., Assidi, M., Al-Qahtani, M. H., Sajantila, A., & Budowle, B. (2015). Underlying data for sequencing the mitochondrial genome with the massively parallel sequencing platform Ion Torrent™ PGM™. *BMC Genomics, 16*(1). https://doi.org/10.1186/1471-2164-16-S1-S4, http://www.biomedcentral.com/bmcgenomics.

Shukla, P., Mukherjee, S., Patil, A., & Joshi, B. (2023). Molecular characterization of variants in mitochondrial DNA encoded genes using next generation sequencing analysis and mitochondrial dysfunction in women with PCOS. *Gene, 855*, 147126. https://doi.org/10.1016/j.gene.2022.147126.

Slatko, B. E., Gardner, A. F., & Ausubel, F. M. (2018). Overview of next-generation sequencing technologies. *Current Protocols in Molecular Biology, 122*(1). https://doi.org/10.1002/cpmb.59, http://onlinelibrary.wiley.com/book/10.1002/0471142727.

Soares, P., Ermini, L., Thomson, N., Mormina, M., Rito, T., Röhl, A., Salas, A., Oppenheimer, S., Macaulay, V., & Richards, M. B. (2009). Correcting for purifying selection: An improved human mitochondrial molecular clock. *The American Journal of Human Genetics, 84*(6), 740–759. https://doi.org/10.1016/j.ajhg.2009.05.001.

Stewart, J. B., & Chinnery, P. F. (2015). The dynamics of mitochondrial DNA heteroplasmy: Implications for human health and disease. *Nature Reviews Genetics, 16*(9), 530–542. https://doi.org/10.1038/nrg3966, http://www.nature.com/reviews/genetics.

Sturk-Andreaggi, K., Parson, W., Allen, M., & Marshall, C. (2020). Impact of the sequencing method on the detection and interpretation of mitochondrial DNA length heteroplasmy. *Forensic Science International: Genetics, 44*, 102205. https://doi.org/10.1016/j.fsigen.2019.102205.

Sturk-Andreaggi, K., Bodner, M., Ring, J. D., Ameur, A., Gyllensten, U., Parson, W., Marshall, C., & Allen, M. (2023). Complete mitochondrial DNA genome variation in the Swedish population. *Genes, 14*(11), 1989. https://doi.org/10.3390/genes14111989.

Sutovsky, P., Moreno, R. D., Ramalho-Santos, J., Dominko, T., Simerly, C., & Schatten, G. (2000). Ubiquitinated sperm mitochondria, selective proteolysis, and the regulation of mitochondrial inheritance in mammalian embryos. *Biology of Reproduction, 63*(2), 582–590. https://doi.org/10.1095/biolreprod63.2.582, https://academic.oup.com/biolreprod/issue.

Ta, M. T. A., Nguyen, N. N., Tran, D. M., Nguyen, T. H., Vu, T. A., Le, D. T., Le, P. T., Do, T. T. H., Hoang, H., & Chu, H. H. (2021). Massively parallel sequencing of human skeletal remains in Vietnam using the precision ID mtDNA control region panel on the Ion S5™ system. *International Journal of Legal Medicine, 135*(6), 2285–2294. https://doi.org/10.1007/s00414-021-02649-1, link.springer.de/link/service/journals/00414/index.htm.

Templeton, J. E. L., Brotherton, P. M., Llamas, B., Soubrier, J., Haak, W., Cooper, A., & Austin, J. J. (2013). DNA capture and next-generation sequencing can recover whole mitochondrial genomes from highly degraded samples for human identification. *Investigative Genetics, 4*(1). https://doi.org/10.1186/2041-2223-4-26Australia, http://www.investigativegenetics.com/content/4/1/26.

Torroni, A., Bandelt, H. J., D'Urbano, L., Lahermo, P., Moral, P., Sellitto, D., Rengo, C., Forster, P., Savontaus, M. L., Bonné-Tamir, B., & Scozzari, R. (1998). mtDNA analysis reveals a major late paleolithic population expansion from southwestern to northeastern Europe. *American Journal of Human Genetics, 62*(5), 1137–1152. https://doi.org/10.1086/301822, http://www.elsevier.com/wps/find/journaldescription.cws_home/713561/description#description.

Tuppen, H. A. L., Reeve, A. K., & Vincent, A. E. (2023). Single cell analysis of mitochondrial DNA deletions. *Methods in Molecular Biology, 2615*, 443–463. https://doi.org/10.1007/978-1-0716-2922-2_29, https://www.springer.com/series/7651.

Vinueza-Espinosa, D. C., Cuesta-Aguirre, D. R., Malgosa, A., & Santos, C. (2023). Mitochondrial DNA control region typing from highly degraded skeletal remains by single-multiplex next-generation sequencing. *Electrophoresis, 44*(17-18), 1423–1434. https://doi.org/10.1002/elps.202200052, http://onlinelibrary.wiley.com/journal/10.1002/(ISSN)1522-2683.

Vyas, S., Zaganjor, E., & Haigis, M. C. (2016). Mitochondria and cancer. *Cell, 166*(3), 555–566. https://doi.org/10.1016/j.cell.2016.07.002, https://www.sciencedirect.com/journal/cell.

Wheaton, W. W., Weinberg, S. E., Hamanaka, R. B., Soberanes, S., Sullivan, L. B., Anso, E., Glasauer, A., Dufour, E., Mutlu, G. M., Budigner, G. R. S., & Chandel, N. S. (2014). Metformin inhibits mitochondrial complex I of cancer cells to reduce tumorigenesis. *eLife, 2014*(3). https://doi.org/10.7554/eLife.02242, http://elifesciences.org/content/elife/early/2014/05/07/eLife.02444.full.pdf.

Xavier, C., Sutter, C., Amory, C., Niederstätter, H., & Parson, W. (2023). NuMY—A qPCR assay simultaneously targeting human autosomal, Y-chromosomal, and mitochondrial DNA. *Genes, 4*(8), 1645. https://doi.org/10.3390/genes14081645.

Yakes, F. M., & Van Houten, B. (1997). Mitochondrial DNA damage is more extensive and persists longer than nuclear DNA damage in human cells following oxidative stress. *Proceedings of the National Academy of Sciences of the United States of America, 94*(2), 514–519. https://doi.org/10.1073/pnas.94.2.514.

Yang, Z., Slone, J., & Huang, T. (2022). Next-generation sequencing to characterize mitochondrial genomic DNA heteroplasmy. *Current Protocols, 2*(5). https://doi.org/10.1002/cpz1.412.

Zhan, X., Li, M., Zhang, Z., Goossens, B., Chen, Y., Wang, H., Bruford, M. W., & Wei, F. (2006). Molecular censusing doubles giant panda population estimate in a key nature reserve. *Current Biology, 16*(12), R451. https://doi.org/10.1016/j.cub.2006.05.042.

Chapter 5

Law and practice of forensic DNA profiling in India: Perspectives and challenges

Shreya Rastogi[1,2,3], and Maria Divya Sahayaselvan[3,4,5]

[1]*National Law University, Delhi, India,* [2]*LLM New York University School of Law, New York, NY, United States,* [3]*The Square Cirlce Clinic, NALSAR University of Law, Hyderabad, Telangana, India,* [4]*Biotechnology, St. Joseph's College, Bengaluru, Karnataka, India,* [5]*Forensic Science, University of Strathclyde, Glasgow, Scotland*

5.1 Introduction

Among the various disciplines within forensic science, DNA profiling holds a unique and indispensable position due to its ability to identify and distinguish between individuals. The earliest reported use of DNA profiling in a forensic investigation in India was in a disputed paternity case in 1989.[1] Using the restriction fragment length polymorphism technique, DNA fragments were visualized on an autoradiograph. The DNA bands of the putative father were compared with the DNA profiles of the biological mother and the child. Based largely on the evidence of the prosecution witness, a senior scientist from Centre for Cellular and Molecular Biology, Hyderabad who conducted the DNA examination, the Kerala High Court held that the DNA report conclusively proved the paternity. Since this case, significant advancements have been made in the field of DNA profiling, accompanied by a growing body of research on the methodological limitations and inherent challenges in its forensic application. Even in India, the practice of forensic DNA profiling has advanced significantly, and its use has dramatically increased in criminal investigations and prosecutions, particularly in serious criminal cases involving sexual offenses and murder. Recognizing its transformative potential, recent years have seen focused efforts to integrate DNA analysis more comprehensively into India's legal and judicial framework.

A significant milestone in this evolution was the 2005 amendment to the erstwhile Code of Criminal Procedure, 1973 (CrPC),[2] which introduced provisions for collecting biological evidence during medical examinations of accused individuals and victims of sexual assault. This development significantly enhanced the importance of DNA profiling as courts increasingly emphasized its necessity in ensuring robust prosecution. Subsequent legislative developments have also sought to expand the use of DNA profiling. Notably, these include the introduction of the DNA Technology (Use and Application) Regulation Bill, 2019 (DNA Bill)[3] and the Criminal Procedure (Identification) Act, 2022 (CPIA).[4] In July 2023, three major criminal laws, that is, the Indian Penal Code, 1860 (IPC); the Code of Criminal Procedure,

1. Kunhiraman v. Manoj, Kerala High Court, (1991) 2 KLT 190. Srivastava, A., Harshey, A., Das, T. et al. (2022). Impact of DNA evidence in criminal justice system: Indian legislative perspectives. *Egyptian Journal of Forensic Science,* 12, 51). https://doi.org/10.1186/s41935-022-00309-y.

2. The Code of Criminal Procedure (Amendment) Act (2005). https://www.mha.gov.in/sites/default/files/2022-09/TheCCP%28Amendment%29Act%2C2005%5B1%5D.pdf.

3. This Bill was the first attempt to create a regulatory framework for DNA laboratories, which is currently absent for any kind of forensic laboratory. The bill was referred to the Parliamentary Standing Committee, whose report was submitted in February 2021. The Committee recommended changes to a range of aspects covered within the Bill. However, the Bill was withdrawn in July 2023. Text of the Bill and the Parliamentary Standing Committee Report can be found at PRS Legislative Research, The DNA Technology (Use and Application) Regulation Bill, 2019. https://prsindia.org/billtrack/the-dna-technology-use-and-application-regulation-bill-2019.

4. The Criminal Procedure (Identification) Act (2022) replaced the Identification of Prisoners Act, 1920, and introduced powers for the collection of biological, physical, and behavioral measurements from convicted persons, arrestees, those under preventive detention, and those who have been ordered to give security for reasons covered under sections 107-110 CrPC or any other person who has been ordered to give their measurements by a court, as it may be expeditious for the investigation. The text of the CPIA can be found here https://www.mha.gov.in/sites/default/files/2022-11/CriminalPro_14112022%5B1%5D.pdf.

Investigative and Predictive DNA Testing. DOI: https://doi.org/10.1016/B978-0-443-30128-5.00016-3

1973 (CrPC) and the Indian Evidence Act, 1872 (IEA) were replaced with the Bharatiya Nyaya Sanhita 2023 (BNS), Bharatiya Nagarik Suraksha Sanhita 2023 (BNSS) and Bharatiya Sakshya Adhiniyam 2023 (BSA) respectively. While a vast majority of the provisions across these three statutes were the same as the erstwhile laws, in the context of forensic evidence, emphasis was laid on proper collection of evidence by mandating the presence of a forensic expert at the crime scene for offenses punishable with more than 7 years. Besides these legislative changes, the Central and state governments have also launched initiatives towards enhancing the forensic infrastructure and upgrading FSLs.[5]

Despite these developments, the law and practice of forensic DNA profiling in India faces significant challenges. DNA analysis is still lacking in many cases as it is either not collected during investigation, or there is improper collection due to inadequate training of investigating officers. Forensic science laboratories (FSLs) also struggle with operational bottlenecks, which lead to long delays in the analysis of evidence and may impact the reliability of results, given the risks of improper storage conditions and contamination within the laboratories. Moreover, the absence of a standardized regulatory framework for the administration and functioning of FSLs has led to wide disparities in practices for forensic analysis. There are no clear standards establishing the quality benchmarks that FSLs must adhere to, nor mechanisms to evaluate the scientific validity of forensic methods or the reliability of expert procedures. While the Directorate of Forensic Science Services (DFSS) oversees Central FSLs, it lacks direct control over the State and Regional FSLs, which constitute the bulk of the forensic laboratories and handle the majority of forensic examinations.[6] Without any statutory authority, the DFSS cannot uniformly enforce quality standards across laboratories. As a result, DNA divisions across different laboratories may follow inconsistent and scientifically unreliable procedures for analysis.

Beyond laboratory inefficiencies, concerns also arise in the judicial appreciation of DNA evidence. Courts often admit forensic reports, including DNA profiling reports, without rigorous scrutiny. There is an absence of clear legal standards for assessing the admissibility and evidentiary weight of forensic evidence. Section 39 BSA (erstwhile section 45 IEA) allows courts to rely on the opinion of "persons specially skilled," known as experts, in matters related to science. However, the provision itself does not have a definitive criteria for determining what constitutes as "science" or "scientific," who qualifies as an "expert," and how the validity and reliability of forensic findings will be assessed. Consequently, judges who often lack scientific expertise are left to determine the credibility of forensic evidence based on subjective perceptions. Adding to these challenges, legal provisions such as Section 329 BNSS (erstwhile Section 293 CrPC) exempts government scientific experts at a specific designation from testifying in court unless summoned by the court, further limiting the ability of the defense to challenge forensic conclusions.

This chapter examines the systemic issues surrounding forensic DNA profiling in India, both in laboratory practices and in judicial interpretation. It argues that the practice of forensic DNA profiling within forensic science laboratories, as well as the appreciation of DNA evidence by courts, suffers from several fundamental issues, making such evidence scientifically unreliable. Part II outlines the current landscape of evidentiary and forensic practices, detailing the existing legal framework and the administrative structure governing FSLs. While Part III delves into specific issues with the practice at different stages, including sample collection, laboratory analysis, Part IV covers the issues with the judicial scrutiny of DNA evidence. Using case studies from decided cases, Part V illustrates the issues and challenges in the practice of DNA profiling in India. Finally, Part VI proposes comprehensive reforms, advocating for statutory regulation of forensic laboratories, adherence to scientific protocols, and the establishment of clear legal standards for evaluating forensic evidence. By addressing these issues, this chapter seeks to enhance the credibility and reliability of forensic DNA profiling in India, ensuring its effective role in the pursuit of justice.

5.2 The current landscape

5.2.1 Legislative framework

India's legal framework that governs criminal proceedings and trials is primarily anchored in the Bharatiya Nyaya Sanhita 2023 (BNS), Bharatiya Nagarik Suraksha Sanhita 2023 (BNSS), and the Bharatiya Sakshya Adhiniyam 2023 (BSA)

5. Under the Nirbhaya Fund Scheme, the procurement of forensic kits for sexual assault cases for police stations and medical institutions, and the establishment and upgradation of DNA analysis and cyber forensic divisions were conducted. https://pib.gov.in/PressReleasePage.aspx?PRID=1601842.

6. Project 39A, Forensic Science India Report (2023), p. 269 [FSIR]. https://www.project39a.com/fsir-report.

which came into force from July 1, 2024.[7] As per the repeals clause in the BNSS and BSA,[8] any existing investigation, trial, inquiry or appeal will continue as per the procedure under the erstwhile CrPC and IEA.

5.2.1.1 Collection of forensic evidence

The framework governing the collection and use of forensic evidence is primarily rooted in the BNSS, supplemented by judicial precedents. It grants investigative authorities specific powers to collect, examine, and utilize forensic evidence in criminal investigations, while aiming to uphold procedural fairness and accountability. Section 51 and 52 BNSS (Sections 53 and 53A in erstwhile CrPC) govern the medical examination of the accused and collection of biological evidence for DNA profiling when there are reasonable grounds to believe that such an examination could yield evidence related to the commission of an offense. Similarly, Section 184 BNSS (s. 164A CrPC) focuses on the medical examination of rape victims, mandating that it be conducted within 24 hours of reporting the offense, with the victim's consent or that of an authorized representative. The introduction of these provisions significantly enhanced the importance of DNA profiling in criminal investigations and prosecutions. In Krishan Kumar Malik v. State of Haryana[9] the Supreme Court emphasized that with the insertion of Section 53A CrPC, it became 'necessary' for the prosecution to conduct DNA tests to substantiate its case against the accused. Building on this precedent, the Supreme Court in Chotkau v. State of Uttar Pradesh[10] ruled that the failure to produce DNA evidence, despite no resistance or obstacle from the accused or any other party, creates a critical gap in the prosecution's case. The court further held that such a failure raises serious doubts about the prosecution's narrative.

Apart from retaining provisions from CrPC regarding evidence collection from the accused and victim, the BNSS also introduces new provisions to expand the collection and use of forensic evidence in the investigation and adjudication of criminal cases.[11] However, these changes are likely to present newer challenges and would require strengthening of the existing forensic science system for meaningful implementation. Section 176 BNSS mandates the collection of forensic evidence by forensic experts at crime scenes for offenses punishable with imprisonment for seven years or more, aiming to standardize evidence collection by qualified individuals across states. However, implementation of this provision would require significant investment to increase the availability of qualified government and private forensic practitioners and mechanisms to assess their competence.[12] Additionally, Section 349 BNSS expands the Magistrate's power to order the collection of forensic samples. Unlike Section 311A CrPC, which limited collection to signatures and handwriting samples from arrested individuals, Section 349 includes fingerprints and voice samples. However, the provision lacks a requirement to establish the person's connection to the offense or the relevance of the sample to the investigation.[13]

5.2.1.2 Examination of expert evidence

In the Indian legal framework, the examination of expert evidence in any criminal or civil proceedings is governed by Section 45 IEA/Section 39 BSA. Similar to other common law jurisdictions, these provisions carve out an exception to the general rule against opinion evidence, allowing courts to rely on expert evidence in areas where specialized knowledge is necessary.

While Section 45 IEA specifies certain areas of expertise (foreign law, science, art, handwriting and finger impressions), Section 39 BSA broadens this scope by permitting expert opinions in "any other field." However, neither provision lays down clear standards for determining the admissibility of expert evidence or assessing its probative value. This legislative

7. The Bharatiya Nyaya Sanhita, 2023. https://www.indiacode.nic.in/bitstream/123456789/20062/1/a2023-45.pdf; The Bharatiya Nagarik Suraksha Sanhita, 2023. https://www.indiacode.nic.in/bitstream/123456789/20099/3/aa2023-46.pdf; The Bharatiya Sakshya Adhiniyam, 2023. https://www.indiacode.nic.in/bitstream/123456789/20063/1/a2023-47.pdf.

8. Sections 531 BNSS and 170 BSA.

9. Criminal Appeal No. 1252 of 2011, Supreme Court, dated 04.07.2011, para 44.

10. Criminal Appeal No. 361362 of 2018, Supreme Court, dated 28.09.2022, para 81. In this case, the court concluded that the prosecution's omission to subject the appellant to a medical examination was detrimental and ultimately fatal to its case.

11. Statement of Objects and Reasons, BNSS.

12. FSIR, p. 112.

13. The P39A Criminal Law Blog, Criminal Law Bills 2023 Decoded #17: Forensic Evidence (2023). https://p39ablog.com/2023/11/criminal-law-bills-2023-decoded-17-forensic-evidence/.

gap has prompted courts to develop their own standards for evaluating expert opinions.[14] Yet, the resulting jurisprudence remains fragmented, offering no consistent framework to determine what qualifies as "science", or who qualifies as an expert or how should scientific evidence be examined by a court.

In the absence of clear legal guidelines, judges often assess the strength of expert opinions based on their subjective perception of the reliability of the forensic discipline.[15] Courts tend to rely on scientific textbooks and literature without conducting meaningful fact-finding on the current state of scientific research. This approach fosters arbitrariness, with different courts adopting varying methods to evaluate forensic evidence.[16] The depth of scrutiny also fluctuates between disciplines, with some fields receiving more rigorous evaluation than others. A more troubling consequence of this lack of uniform standards is the courts' reliance on "junk science"—forensic techniques and disciplines whose scientific validity remains unproven or has been discredited by empirical studies.[17]

5.2.1.3 *Exemption of scientific experts from testifying*

In India, the reliance on forensic expert opinion in judicial proceedings is primarily governed by Section 293 CrPC, 1973 which traces its origins to Section 370 CrPC, 1861. Section 293 CrPC permits reports from designated "Government Scientific Experts" to be admitted as evidence without requiring their examination as witnesses in court. It also grants discretion to the court to summon the expert as a witnesses, however does not provide any guidance on the manner in which such discretion may be exercised. The scope of the experts covered under this exemption has been further expanded under Section 329 BNSS. While retaining the cateogries of government scientific experts as listed in section 293 CrPC, it allows exemption of any other scientific expert that may be specified or certified by the state and central government. This exemption allowing admission of forensic reports without the testimony of experts hinders the judicial scrutiny of their forensic reports. This exemption is justified on grounds of administrative ease for the experts, who are few in number and cannot be conveniently spared from thier work within the laboratory for testifying in every criminal case where their reports are being submitted.[18,19] Further, the presumption that the expertise and credibility of these specific experts covered within this provision, eliminates the need for their cross-examination creates an arbitrary distinction between forensic examiners within the same discipline, solely based on their designation.[20]

While Section 329(2) BNSS (293(2) CrPC) empowers courts to summon experts when necessary, this largely depends on a party making a specific application with reasons to justify the request. In practice, forensic reports are often admitted without cross-examination, limiting the court's ability to assess their scientific reliability. Section 330 BNSS imposes further constraints by limiting the summoning of experts to instances where the report's genuineness is challenged before the recording of the evidence begins in the trial. This mirrors Section 294 CrPC but imposes stricter limitations,[21]

---

14. The Supreme Court has consistently held that expert evidence is advisory, and that courts must evaluate the forensic report and form its independent judgement on the strength of its conclusions (State of Himachal Pradesh v. Jai Lal, Criminal Appeal No. 530 of 1997, Supreme Court, dated 13.09.1999, para 18 [Jai Lal]; Ramesh Chandra Agrawal v. Regency Hospital Ltd., Civil Appeal No. 5991 of 2002, Supreme Court, dated 11.09.2009, para 20 [Ramesh Chandra Agrawal]; Pattu Rajan v. State of Tamil Nadu, Criminal Appeal Nos. 680-81 of 2009, Supreme Court, dated 29.3.2019, para 51 [Pattu Rajan].

15. FSIR, p. 244.

16. The lack of clear standards for evaluating forensic evidence has led to inconsistencies in how courts interpret and apply the same type of evidence, resulting in varying decisions. For instance, in Santosh Kumar Singh v. State through CBI, Criminal Appeal No. 87 of 2007, Supreme Court, dated 06.10.2010, para 71 [Santosh Kumar], the Supreme Court described DNA profiling as an "exact science." Similarly, in In Reference v. Vinod alias Rahul Chouhta, Criminal Reference No. 1 of 2018, High Court of Madhya Pradesh, dated 08.08.2018 [Vinod Chouhta], the court considered a positive DNA test result to be a decisive piece of evidence, ruling that the expert's testimony was unnecessary for cross-examination. In contrast, in Prakash Nishad @ Kewat v. State of Maharashtra, Criminal Appeal Nos. 1637-1637 of 2023, Supreme Court, dated 19.05.2023 [Kewat], the Supreme Court observed that while advancements in science and technology have improved the accuracy of DNA evidence, it is not yet infallible. The Court criticized the lower courts for failing to thoroughly scrutinize the evidence, pointing to several irregularities that led to the chain of custody being broken.

17. In Mukesh & Anr. v. State (NCT of Delhi), Criminal Appeal Nos. 607-608 of 2017, Supreme Court, dated 5.5.2017, the Supreme Court at para 237-241 cited literature on bitemark analysis to conclude that it allows for the identification of persons through comparison with suspected such injuries. However, the scientific research on bitemark analysis unequivocally points to the lack of a scientific foundation for such analysis. See National Institute of Standards and Technology (NIST), Bitemark Analysis: A NIST Scientific Foundation Review (2023). https://nvlpubs.nist.gov/nistpubs/ir/2023/NIST.IR.8352.pdf.

18. FSIR, pg 152.

19. 41st Report of the Law Commission of India on revisions to the Code of Criminal Procedure 1898, Chapter XLI, Pg 325.

20. FSIR, p. 253.

21. 248th Report of the Parliamentary Standing Committee on Home Affairs, Rajya Sabha, on the Bharatiya Sakshya Bill, 2023, p. 22 of Dissent. https://prsindia.org/files/bills_acts/bills_parliament/2023/SC_Report_Bharatiya_Sakshya_Bill_2023.pdf.

assuming that expert testimony is needed only to verify the report's authenticity rather than to assess its scientific reliability.

A more serious issue arises when forensic reports are proved solely through investigating officers who lack scientific expertise rather than the experts who prepared them. These officers cannot be examined on any aspects related to the scientific validity of methodology used, the manner of examination and adherence to quality control measures. This also precludes inquiry into the expert's qualifications. The lack of legal clarity on who qualifies as an expert exacerbates this issue. Section 45 IEA/Section 39 BSA only considers "special skills" as a criteria for determining the expert without specifying qualifications, training standards, or experience. Although courts have expanded the interpretation to include necessary knowledge and experience,[22] the standards remain unclear.

Despite these concerns, courts routinely presume government forensic examiners to be qualified and exempt them from testifying in court. This unchecked reliance on potentially unqualified experts undermines the adversarial process and heightens the risk of flawed forensic evidence being used to secure convictions.

5.2.2 Developments to the framework

Significant changes have been introduced to the legal framework governing forensic science in India, particularly concerning DNA evidence. Two major legislative developments in this regard are the introduction of the DNA Technology (Use and Application) Regulation Bill, 2019 (DNA Bill) and the CPIA. These changes aimed to enhance the collection, analysis, and utilization of DNA and other forensic data for investigative and judicial purposes, but they both face significant challenges in their implementation due to several practical and constitutional concerns.

5.2.3 DNA Technology (Use and Application) Regulation Bill, 2019

The DNA Bill sought to establish a robust mechanism for the identification of specific categories of individuals, including offenders, suspects, undertrials, missing persons, and unidentified deceased persons, through DNA analysis. Central to its framework was the creation of DNA databanks and a system for the accreditation of laboratories authorized to perform DNA analysis. Accredited laboratories were required to share DNA data obtained through casework with the databanks, ensuring centralized access to the DNA profiles stored in different indices. The bill was introduced in the Lok Sabha in July 2019 and was referred to the Parliamentary Standing Committee on Science and Technology in February 2021, recommending several changes to the draft. A major critique of the Bill was the broad powers to collect bodily substances from arrested individuals and the subsequent databasing of DNA profiles of suspected persons and undertrials.[23] This broad mandate raised practical challenges, including the difficulty of analyzing a vast number of samples, the prohibitive costs of maintaining such a database, and potential ethical and privacy concerns.[24] Moreover, while the Bill stipulated that only accredited laboratories would be allowed to perform DNA profiling and thier data would be uploaded onto the DNA database, the widespread issues with implementation of best practices for DNA collection, storage, and analysis undermines this intent. The Bill also failed to address key operational specifics, such as how the data will be processed and protocols for the deletion of data. These gaps raised questions about the feasibility and reliability of the proposed framework. The Bill was withdrawn altogether in July 2023 as the wide powers of collection and databasing that were covered under the Bill were enacted as part of another law, the CPIA which was passed by both houses of the Parliament.

5.2.4 Criminal Procedure (Identification) Act, 2022

The CPIA broadened the scope of forensic evidence collection by authorizing the collection of "measurements" or samples from individuals, including arrestees, detainees, and convicts, to aid criminal investigations and the prevention of crimes.[25] It grants a wide range of police and prison officers the power to collect such samples, and allow states to designate an agency to collect, preserve, store, and share them. Under this Act, the National Crime Records Bureau (NCRB) is tasked with maintaining a centralized database of these records.

---

22. Jai Lal, para 17.

23. 340th Report of the Parliamentary Standing Committee on Science and Technology, Rajya Sabha, on the DNA Technology (Use and Application) Regulation Bill, 2019 (2021), p. 11. https://prsindia.org/files/bills_acts/bills_parliament/2019/DNA%20Report.pdf.

24. NLSIR Online, Critique on the DNA Technology Bill, 2019: Preventing Bodily Incursion and Parliamentary Excursion (2023). https://www.nlsir.com/post/critique-on-the-dna-technology-bill-2019-preventing-bodily-incursion-and-parliamentary-excursion.

25. The Criminal Procedure (Identification) Act, 2022. https://www.indiacode.nic.in/bitstream/123456789/19029/1/a2022-11_.pdf.

While the Act aims to improve criminal investigations by expanding the collection and databasing of samples collected from individuals, it has been criticized on several grounds.[26] Contemporary scientific research suggests that many of the "measurements" included under the Act may not effectively serve its stated purpose of crime prevention and investigation. Thus, the broad nature of biological, physical and behavioural "measurements" that can be collected, without the need to link their collection to the prevention, investigation, or prosecution of an offense, shows the lack of a determining principle and raises concerns of arbitrariness under Article 14 of the Indian Constitution. Moreover, the Act lacks an information-sharing framework, which raises concerns about the collection of measurements and potential breaches of confidentiality. Section 4(3) of the Act empowers State governments and Union Territories to notify any agency to collect, preserve, and share sensitive personal data without specifying restrictions on the nature of such agencies. This opens the possibility of delegating these functions to private, unregulated entities, which could compromise individuals' privacy and the confidentiality of biological samples or measurement records.

Additionally, the Act raises serious concerns regarding the the right to privacy due to a lack of rational connection between the expansive powers of collection of personal data under the Act and the legitimate state goal of improving crime detection and prevention. The absence of scientific evidence to substantiate the effectiveness of such measures further weakens the justification for these rights-infringing provisions. A public interest litigation[27] is currently pending before the Delhi High Court challenging the constitutionality of the CPIA, seeking judicial review of the provisions. The petitioner contended that compelling individuals to provide "measurements" violates the standard of "substantive due process" required to restrict personal libertyand is in clear violation of Article 21 of the Indian Constitution.

5.2.5 The Forensic Science System

In India, the practice of forensic science is still largely limited to the public sector, with private laboratories engaging in only a few forensic disciplines. Government FSLs in India function within a three-tiered administrative structure comprising Central, State, and Regional Forensic Science Laboratories. Currently, there are 119 functional laboratories in the country.[28] As of September 2025, only 21 laboratories are accredited,[29] since accreditation is not mandatory in order to process case samples.

The DFSS under the federal Ministry of Home Affairs (MHA) presently oversees the forensic science system in India. According to its charter,[30] DFSS is responsible for capacity building at the central level and providing technical and financial support to state laboratories. Its wide mandate includes promoting research and development (R&D) in forensic science, promoting quality management in forensic testing by arranging technical and financial support for the development of quality standards and uniform standard operating procedures (SOPs), and formulating plans for capacity-building in forensics. Therefore, it publishes best practices and model SOPs for forensic practice, but these are in the form of guidelines. DFSS cannot exercise monitoring powers to ensure compliance with these standards by FSLs, as it does not have any statutory powers; therefore, it acts merely as an administrative body rather than a regulator.

The bulk of the FSLs in India come under the control of different State home departments which administer the State Forensic Science Laboratories (SFSLs) and Regional Forensic Science Laboratories (RFSLs) within that state, either directly or through their police departments.[31] RFSLs are under the administrative and financial control of the SFSLs. Therefore, their heavy dependance on SFSLs for budgeting, receipt of funds and expenditure can lead to bottlenecks and cause delays in casework.[32] Additionally, nearly half of the state and regional laboratories are under the financial and administrative control of the police department, which severely affects their functioning and independence. The forensic medicine department, on the other hand, is separately managed by the Ministry of Health and Family Welfare through the Directorate General of Health Services.[33]

26. Project 39A, Research Brief: An Analysis of the Criminal Procedure (Identification) Act, 2022. https://www.project39a.com/identification-act.
27. Harshit Goel v. Union of India through Home Secretary & Anr., W.P. (CRL) 869/2022, Delhi High Court.
28. FSIR, p. 53. As per the DFSS website, there are 7 Central FSLs, 32 State FSLs and 80 Regional FSLs.
29. In India, forensic laboratories are accredited by the National Accreditation Board for Testing and Calibration Laboratories (NABL).
30. MHA, Resolution No. 25020/50/2010-PM. II dated 29.11.2010, Gazette of India, Part I, Section I, dated 18.12.2010, p. 1871.
31. FSIR, p. 40–41.
32. FSIR, p. 60, 84–85.
33. FSIR, p. 42.

5.3 Issues with the practice of forensic DNA analysis

5.3.1 Lack of quality standards

DNA technology is continuously evolving, leading to highly sensitive methods of forensic analysis. However, despite these advancements, they are prone to errors[34] like any other scientific process. While errors cannot be entirely eliminated as they are inevitable, steps can be taken to minimize their impact. Establishing a robust quality management system in DNA testing laboratories and ensuring its strict adherence is one such effort, particularly given the critical role of DNA evidence in the criminal justice system.

5.3.1.1 Absence of standard operating procedures for collection and storage

DNA evidence is highly susceptible to contamination and degradation; therefore, following standardized procedures is essential at every stage—collection, packaging, transport, storage, and analysis. The reliability of DNA evidence does not solely depend on the quality management at the laboratory, but also on maintaining the sample's integrity from the moment of collection.

A major challenge in India is the lack of a uniform SOP governing the collection, handling, and analysis of DNA evidence. While the DFSS has issued guidelines for investigating officers[35] and medical professionals[36] on the collection, storage, and transportation of biological samples, compliance with these guidelines is not mandatory. This has led to significant inconsistencies in forensic practices across states and districts, raising concerns about the reliability of forensic evidence. A critical issue arising from the absence of SOPs is that evidence collection at crime scenes is primarily conducted by investigating officers, many of whom lack adequate training in forensic procedures.[37] Limited awareness of the forensic significance of various types of evidence often leads to mishandling, contamination, and degradation of crucial samples. While some FSLs have initiated training programs for investigators, these efforts remain sporadic and insufficient to address systemic deficiencies.[38] Although the introduction of Section 176 BNSS, which mandates the presence of forensic experts at crime scenes, is a step towards standardizing evidence collection, its implementation remains a challenge. Without a clear and enforceable SOP, inconsistencies in forensic practices will continue to undermine the reliability of DNA evidence.

Another pressing concern is the improper storage of forensic samples. In India, DNA evidence is often not sent directly to FSLs but is instead stored for prolonged periods in police or court evidence storage rooms, which frequently lack appropriate storage conditions. Even after reaching forensic laboratories, samples may remain unexamined for extended periods unless an urgent request is made for analysis. Such delays and improper handling further compromise the quality and admissibility of DNA evidence in legal proceedings.[39] Recently, in view of the glaring chain of custody issues, the Supreme Court in Kattavelli @ Devakar has issued detailed directions to be followed in all cases where DNA evidence is involved. These directions mark a significant judicial intervention to strengthen procedural safeguards and reinforce the need to preserve the integrity of forensic evidence at every stage.[40]

---

34. Morgan, J. (2023) Wrongful convictions and claims of false or misleading forensic evidence, *Journal of Forensic Sciences, 68* (3). https://doi.org/10.1111/1556-4029.15233. In this paper, the author talks about how DNA analysis is prone to analyst error. He gives an example of the Houston Police Department Laboratory to emphasize the importance of rigorous and enforced standards.

35. DFSS, Guidelines for Collection, Storage, and Transportation of Crime Scene Biological Samples for Investigating Officers. http://dfs.nic.in/pdfs/IO%20-Forensic%20evidence-Guidelines%20for%20%20IO.pdf.

36. DFSS, Guidelines for Forensic Medical Examination in Sexual Assault Cases (2018). http://dfs.nic.in/pdfs/MO-Forensic%20examiniation-%20Guidelines%20%20for%20MO.pdf.

37. Sahajpal, V., Bhandari, D. (2024). DNA profiling in India: Addressing issues of sample preservation, databasing, marker selection, & statistical approaches, *Science & Justice*, *64*(4), 389-396. https://doi.org/10.1016/j.scijus.2024.05.003.

38. FSIR, p. 148.

39. In Kewat, the Supreme Court in para 61 observed that the samples were sent for analysis after an "unexplained delay," raising concerns about the "possibility of contamination" and the resulting potential for diminished evidentiary value, which could not be "reasonably ruled out."

40. In Kattavelli @ Devakar v. State of Tamil Nadu, Criminal Appeal No. 1672 of 2019, Supreme Court, dated 15.07.2025, the Supreme Court acknowledged the absence of standard procedures for evidence collection, noting its impact on investigations, and accordingly issued directions to standardize the process. It mandated that collection must be done in the presence of independent witnesses, and clear information regarding the case and articles recorded at that time. The investigating officer must ensure that biological evidence reaches the forensic science laboratory within 48 hours. If delays occur, the evidence must be preserved in appropriate conditions, and the reasons for delay should be clearly documented in the case diary. Additionally, the Court directed that once DNA samples are stored pending trial, appeal, or other legal proceedings, no package shall be opened, altered, or resealed without express authorization from the Trial Court. From the point of collection to the final adjudication (whether conviction or acquittal), a Chain of Custody Register must be maintained. Every movement of the evidence must be recorded, with counter-signatures and stated reasons for each transfer. This register must be made part of the Trial Court record. Failure to maintain or produce it will require the investigating officer to explain the lapse.

5.3.1.2 Lack of working procedure manuals for DNA profiling

Forensic DNA profiling is a multistep process, with each stage involving different kits, instruments, and testing methodologies. According to best practices,[41] forensic laboratories should conduct internal validation of all DNA analysis methods before applying them to casework.[42] The results of these validation studies should be documented and used to create a working procedure manual (WPM)[43] that outlines laboratory-specific parameters, interpretation guidelines, and troubleshooting procedures.

Internal validation differs from developmental validation, which is conducted by manufacturers before a kit or method is introduced. While developmental validation establishes the scientific basis of a method, manufacturers explicitly recommend that forensic laboratories perform internal validation to account for variations in local conditions such as temperature, humidity, and available resources. However, in India, most forensic science laboratories do not conduct internal validation,[44] and rely solely on manufacturer-provided validation data. This approach leads to inconsistencies in testing procedures, even within the same institution, resulting in a lack of standardization across forensic laboratories.

Many forensic laboratories also do not develop their own WPMs and instead rely on publicly available resources,[45] such as the WPMs published by the DFSS.[46] However, this manual serves only as general guidance, as the data it provides is not tailored to the specific conditions of individual forensic laboratories. Even laboratories that have developed their own WPMs often lack mechanisms for periodic review and updates. Without regularly updated WPMs, forensic laboratories may fail to incorporate advancements in DNA analysis techniques or adapt to evolving forensic standards.[47] This lack of standardization can significantly undermine the reliability and admissibility of DNA evidence in court.

5.3.1.3 Gaps in maintaining chain of custody

A significant risk associated with DNA evidence is contamination, which has become a growing concern with the increased sensitivity of modern forensic techniques. The capability of the newer methodologies to generate a DNA profile even with small amounts of the genetic material means it can also detect DNA that was left behind through casual contact or actions like sneezing. Handling of DNA evidence, therefore, demands extra care as the reliability of the evidence is dependent on preserving its integrity throughout the investigation and its examination in the laboratory. Ensuring a secure chain of custody from collection till the end of DNA analysis is critical for maintaining the evidence's admissibility and reliability in court.[48] The chain of custody serves as a documented record tracing the movement and handling of items collected for forensic examination,[49] without which the evidence risks exclusion from trial or diminished credibility.

To enhance traceability, forensic laboratories across the world have gradually transitioned from manual record-keeping to electronic systems like the Laboratory Information Management Systems (LIMS). LIMS is a software-based tool that "collects, creates, and stores all data related to forensic examinations within a crime laboratory." In addition to storing the case information, it documents the chain of custody of the exhibit from its entry through analysis and ultimately its

41. Organization of Scientific Area Committees for Forensic Science (OSAC), Standard for Internal Validation of Forensic DNA Analysis Methods, pg 5. https://www.nist.gov/document/standard-internal-validation-forensic-dna-analysis-methods; Scientific Working Group on DNA Analysis Methods (SWGDAM), Validation Guidelines for DNA Analysis Methods (2016) at pg 9-11, discuss the different aspects of the DNA profiling process that should be tested in validation studies. https://www.swgdam.org/_files/ugd/4344b0_813b241e8944497e99b9c45b163b76bd.pdf; Forensic Science Regulator Guidance, Validation (2020). FSR-G-201, Issue 2, p. 16. https://assets.publishing.service.gov.uk/media/5f6b1a3de90e077ca292204f/201_-_FSR-G-201_Validation_Guidance_Issue_2.pdf.

42. Forensic Science Regulator, Draft Code of Practice 2025 (Version 2), p. 61. https://assets.publishing.service.gov.uk/media/67dabc1b1a60f7964302841b/E03313596+-+CoP+Forensic+Science+Regulator+2025_Elay.pdf. [FSR Code of Practice].

43. NIST, Forensic DNA Interpretation and Human Factors: Improving Practice Through a Systems Approach (2024), p. 28 [NIST Human Factors Report]. https://nvlpubs.nist.gov/nistpubs/ir/2024/NIST.IR.8503.pdf.

44. FSIR, p. 212–213.

45. FSIR, p. 207.

46. DFSS, WPM Forensic DNA Testing (2019). http://dfs.nic.in/pdfs/PP%20PDFs/DNA%20manual%20final%20aug%202019%20(1)-merged%20(1)-protected.pdf.

47. FSIR, p. 195.

48. In Kewat para 62, the Supreme Court emphasised the importance of chain of custody clarifying that it implies "right from the time of taking of the sample, to the time its role in the investigation and processes subsequent, is complete, each person handling said piece of evidence must duly be acknowledged in in the documentation, so as to ensure that the integrity is uncompromised."

49. National Institute of Justice (NIJ), Chain of Custody (Law 101: Legal Guide for the Forensic Expert). https://nij.ojp.gov/nij-hosted-online-training-courses/law-101-legal-guide-forensic-expert/pretrial/pretrial-motions/chain-custody.

transfer back to the investigating authorities.[50] In contrast, in India, chain of custody documentation remains predominantly handwritten, with laboratories recording varying details.[51] This inconsistency is further exacerbated by frequent cancellations and overwriting, raising concerns about the sample's integrity and reliability. Moreover, while many countries use tamper-resistant packaging to seal the evidence samples, investigators in India still use wax seals, which can detach easily. This compromises the integrity of forensic exhibits, as it becomes unclear whether they remained securely sealed throughout the process.

Further, judicial scrutiny in India often focuses solely on the transfer of forensic exhibits from the crime scene to the laboratory, neglecting their handling within the laboratory. This oversight is concerning, as reviewing internal chain of custody records is crucial for ensuring transparency in forensic processes and verifying that evidence integrity was preserved throughout the analysis.

5.3.1.4 *Anticontamination measures are not regularly undertaken*

Even the slightest amount of contamination can compromise forensic evidence, leading to inaccurate conclusions. Therefore, beyond maintaining a clear chain of custody, forensic laboratories must implement stringent anticontamination measures to prevent such occurrences effectively.[52] Some of the important measures to prevent contamination include:

Segregation of Work Space: A well-designed forensic DNA laboratory is essential for minimizing contamination, as designated areas are required for specific processes. Evidence and reference samples should be examined in separate areas,[53] since reference samples contain higher concentrations of DNA than evidence samples. Likewise, pre-amplification and post-amplification procedures must be conducted in physically distinct spaces, as pre-amplification samples contain low DNA concentrations, whereas post-amplification samples contain significantly higher amounts.[54] To further minimize contamination risks, sample processing should follow a unidirectional workflow, moving systematically from one stage to the next to prevent mix-ups and cross-transfer. However, many forensic laboratories in India struggle with limited space, making it difficult to implement these critical separation measures.[55]

Dedicated Equipment: Laboratories should assign separate equipment for pre- and post-amplification processes. Using the same instruments for both stages increases the likelihood of DNA transfer, compromising the integrity of forensic analysis. Dedicated workstations, pipettes, and tools for each stage of DNA profiling are essential to maintaining sample purity. Unfortunately, Indian forensic laboratories often lack sufficient equipment, forcing analysts to use shared instruments, which increases the risk of contamination.[56]

Restricted Access to Authorized Personnel: Limiting access to forensic workspaces is crucial in preventing unknown DNA from entering the laboratory environment.[57] Only authorized personnel should be permitted in DNA analysis areas, and access should be strictly monitored.[58] By restricting entry, laboratories can better control contamination risks and ensure that only trained professionals handle forensic samples. However, many forensic laboratories in India lack critical safeguards, such as restricting entry to evidence storage areas to necessary personnel, implementing camera surveillance, and maintaining a digitised logbook. Moreover, laboratories do not maintain an internal chain of custody to document the movement and handling of exhibits from the time of receipt until the completion of examination.[59] This lack of oversight not only raises concerns about contamination but also increases the risk of tampering, further compromising the integrity of forensic evidence.

---

50. Forensic Technology Center of Excellence-National Institute of Justice (FTCoE-NIJ), A Landscape Study of LIMS for Forensic Crime Laboratories (2020), pg 5. https://forensiccoe.org/private/62a8d0a8f071c; FTCoE,LIMS in Forensic Science Service Provider Laboratories: Current State and Next Generation (2024), pg 7. https://forensiccoe.org/private/6697e14783371.

51. FSIR, p.154.

52. FSR Code of Practice, pp. 57–60.

53. SWGDAM, Contamination Prevention and Detection Guidelines for Forensic DNA Laboratories (2017), pp. 4-7. [SWGDAM Contamination Prevention and Detection Guidelines]. https://www.swgdam.org/_files/ugd/4344b0_c4d4dbba84f1400a98eaa2e48f2bf291.pdf.

54. Kampmann M, Tfelt Hansen J, Børsting C. (2024), Cleaning protocols in forensic genetic laboratories, *International Journal of Legal Medicine, 138*(2): 365-376. https://doi.org/10.1007/s00414-024-03232-0.

55. FSIR, pp. 179-180.

56. FSIR, p. 182, 230.

57. SWGDAM, Quality Assurance Standards for Forensic DNA Testing Laboratories (2020), p. 21. [SWGDAM Quality Assurance Standards]. https://www.swgdam.org/_files/ugd/4344b0_d73afdd0007c4ed6a0e7e2ffbd6c4eb8.pdf.

58. FSR Code of Practice, p. 55.

59. FSIR, p. 184.

Staff Elimination Database: A staff DNA elimination database[60] helps forensic laboratories identify and rule out contamination from laboratory personnel. By maintaining records of all authorized individuals' DNA profiles, analysts can compare unexpected DNA traces found in samples against known personnel, ensuring contamination is detected and accounted for. Despite its importance, Indian forensic laboratories currently lack such a database,[61] making it difficult to distinguish between genuine evidence and inadvertent contamination from lab personnel.

Personal Protective Equipment (PPE): Proper use of PPE is a fundamental safeguard against DNA contamination.[62] Laboratory personnel should wear disposable PPE gowns, gloves, masks, and hairnets to prevent their DNA from coming into contact with forensic samples. PPE must be changed regularly, particularly after handling surfaces or different samples, to reduce contamination risks. However, many forensic laboratories operate without adequate safety protocols. Since most are not accredited, they are not subject to routine safety inspections, including checks for compliance with health and safety standards. Also, there are no standardized contamination control guidelines, including the use of essential items like PPE gowns, and hairnets, when protective gear should be replaced.[63]

Use of Controls: Positive and negative controls must be incorporated at every stage of DNA profiling to ensure the reliability of forensic analysis.[64] Positive controls verify that the DNA profiling process and the equipments are functioning correctly, while negative controls detect cross-contamination by ensuring no unintended DNA is present. However, forensic laboratories in India either fail to use these controls or neglect to document their use, raising concerns about the integrity of forensic evidence.

Beyond the laboratory, judicial oversight on contamination prevention has been inconsistent. Courts often rely uncritically on the opinions of forensic experts without scrutinizing whether contamination risks were adequately mitigated and whether the contamination prevention safeguards were properly followed. While the Supreme Court has emphasised the need for adherence with quality control and quality assurance standards and contamination prevention measures,[65] the consequences of non-compliance have not been stated clearly. This ambiguity has led to inconsistent application of quality standards across cases, further weakening forensic evidence reliability.

5.3.1.5 No scrutiny of expert proficiency

The competence and proficiency of forensic experts are critical for ensuring that valid and reliable forensic examinations are undertaken. Experts are required to possess a strong foundation in the relevant scientific and technical fields, along with proficiency in report writing and delivering expert testimony.[66] Given the continuous advancements in forensic technology and methodology, ongoing education and skill development are essential to stay updated and maintain professional standards.

However, delays in recruitment of scientific personnel in FSLs and rising pressures of pending casework, result in analysts being rotated between laboratories or divisions, to roles outside their areas of qualifications and expertise. Further, there is a lack of dedicated traning and assessment periods and analysts often receive an "on-the-job training", the duration and quality of which vary widely across FSLs.[67]Additionally, analysts rarely participate in regular proficiency testing, further impacting the reliability of forensic examinations.

---

60. Basset P, Castella V. Positive impact of DNA contamination minimization procedures taken within the laboratory. Forensic Sci Int Genet. 2019 Jan; 38:232–235. https://doi.org/10.1016/j.fsigen.2018.11.013. This study highlights the need for an elimination database to avoid wasting resources on following up on profiles that arise from the examination; FSR Code of Practice, p. 56.

61. FSIR, p. 230.

62. SWGDAM Contamination Prevention and Detection Guidelines, pp. 8–14.

63. FSIR, p. 183, 186, 230.

64. SWGDAM Quality Assurance Standards, pg 26; SWGDAM Contamination Prevention and Detection Guidelines, p. 15.

65. Manoj & Ors. v. State of Madhya Pradesh, Criminal Appeal Nos. 248-250 of 2015, Supreme Court, dated 20.05.2022, para 141 [Manoj & Ors.]. The Supreme Court highlighted the need to ensure quality in the testing and eliminate the possibility of contamination of evidence; it also held that being an opinion, the probative value of such evidence has to vary from case to case.; Anil v. State of Maharashtra, Criminal Appeal Nos. 1419-20 of 2012, Supreme Court, dated 20.02.2014, para 18 [Anil]. The Supreme Court held that while examining DNA evidence, quality control measures must be ensured for the analysis of the samples as well as the techniques.

66. SWGDAM Quality Assurance Standards, pg 12. Butler, J.M. Summarized Information from Publicly Accessible Validation Studies, Proficiency Testing Results, and Interlaboratory Comparison Data, Supplemental Document to DNA Mixture Interpretation: A NIST Scientific Foundation Review NIST IR 8351sup2 (2024), p. 31. https://doi.org/10.6028/NIST.IR.8351sup2.

67. FSIR, p. 113.

Another significant concern is the absence of a comprehensive technical review mechanism in forensic practice.[68] In many cases, reports are finalized and submitted without undergoing an independent review process. While some DNA reports include a supervisor's signature alongside that of the analyst, they often fail to indicate whether the supervisor conducted a thorough review of the analysis or independently arrived at the same conclusions.

Moreover, forensic experts do not provide details about their qualifications and experience in their reports,[69] making it difficult for the courts to adequately assess their expertise. Courts have traditionally assumed that government forensic experts possess the necessary qualifications, as there is no legal mandate requiring scrutiny of their credentials. This represents a significant gap in the legal framework. However, recent shifts in judicial attitudes emphasize the need for forensic experts to testify as witnesses and be subject to cross-examination.[70] This allows the defense to challenge incriminating evidence and provides courts with a crucial opportunity to evaluate the competence and proficiency of forensic experts, ensuring a more rigorous and fair evidentiary process.

5.3.2 Reliability concerns with DNA profiling process

DNA evidence is widely relied upon by courts because, unlike many other forensic disciplines that often lack a scientific basis, DNA profiling is grounded in decades of life sciences research. However, its accuracy and reliability depend on proper analysis. Without adherence to rigorous protocols, its credibility and effectiveness may be compromised.

5.3.2.1 Absence of reliable reference sample

The purpose of DNA profiling in a criminal investigation is to identify the source of the unknown DNA profile obtained from an evidence sample recovered from a crime scene, the victim or the accused. This can only be achieved if a reliable reference sample is available from a suspect, the victim or potential contributor is available for comparison. Once the DNA profile of the evidence sample is generated, it is compared with the reference sample of a known individual to determine whether that individual is included or excluded as a potential contributor. This comparison plays a pivotal role in forensic investigations, as it helps establish a link between an individual and the crime. Without a reliable reference sample, the accuracy and validity of DNA analysis can be significantly compromised.[71]

Despite its critical importance, the collection and use of reliable reference samples in forensic investigations remain inconsistent and often inadequate in India. In many cases, reference samples are not collected, or are incorrectly substituted with the evidence samples which are assumed to have the DNA profile of the known contributor. Without a reliable reference profile, the unknown DNA profile obtained evidence sample cannot be interpreted and compared. However, judicial oversight of this issue has also been inconsistent. While some courts fail to rigorously scrutinize the reliability of reference samples presented in cases,[72] others have taken a more proactive approach in questioning their validity.[73] The lack of stringent judicial review in many instances has allowed flawed forensic practices to persist, further jeopardizing the integrity of DNA evidence.

5.3.2.2 Incorrect interpretation of DNA mixtures

When a DNA sample contains more than two alleles at a locus, it indicates the presence of a mixture from two or more contributors.[74] Unlike single-source DNA samples, where genotype identification is straightforward, interpreting mixed

68. FSIR, pg 150-151. The report reveals that a review mechanism was operational in only two FSLs. In several laboratories, there was only one reporting scientist assigned per division, which made it impossible for another examiner to independently verify the findings. Furthermore, one FSL disclosed that technical reviews were conducted solely in cases where a "negative result" was obtained, highlighting a significant gap in quality assurance practices.

69. Forensic Science Regulator Guidance, Expert Report Guidance, Issue 4 (2020), p. 9. https://assets.publishing.service.gov.uk/media/5ea6fa0586650c0314ac74f0/Statement_and_Report_Guidance_Appendix_Issue_4.pdf.

70. In Anokhilal v. State of Madhya Pradesh, Criminal Reference No. 6 of 2022, Madhya Pradesh High Court judgment dated 11.09.2023, para 14, Irfan Bhayu v. State of Madhya Pradesh, Crimina Appeal No. 220 of 2022, Supreme Court judgment dated 16.01.2025 para 29-30.

71. NIST Human Factors Report, p. 26.

72. In Rahul & Ors., the female fraction of DNA from a vaginal and anal swab was used as a reference sample for comparison with evidence samples. However, the court failed to recognize that such a sample does not constitute a reliable reference sample.

73. In State of Madhya Pradesh v. Anokhilal (Sessions Case No. 53/2013 judgment dated 19.03.2024), a stain from the clothing of the deceased was used as the reference sample, despite the availability of the deceased's body for direct sample collection. The trial court acknowledged this as a significant error, highlighting the importance of adhering to proper protocols for collecting reference samples.

74. In forensic DNA profiling, the composition of DNA is examined at specific locations referred to as genetic markers or loci (singular locus). These loci have short repetitive sequences, known as Short Tandem Repeats (STRs), which are examined. While the same STR appears in all individuals, the number of repetitions may vary across individuals. The variable number of repeats is referred to as alleles.

DNA samples is far more complex.[75] Instead of simply identifying a single genotype, analysts must determine the possible genotype combinations at each locus. A common but incorrect approach to mixture interpretation involves comparing the alleles present in the mixed sample with those in a reference sample to assess whether a suspect's DNA could be included or excluded. While such comparisons are sometimes unavoidable (e.g., indistinguishable mixture), this method is scientifically inadequate.[76] Instead, peak height ratios should be calculated to determine the correct genotype combination.[77] Peak height ratios provide a quantitative measure of DNA contribution from different individuals, allowing forensic analysts to assess whether the allelic peaks are shared between contributors and whether the individual genotypes can be reliably distinguished. It is also important to recognize that the difficulty of interpretation varies across loci, and in cases where contributors have similar DNA quantities, the mixture may be indistinguishable.

DNA mixture interpretation can be divided into two primary approaches—binary method and probabilistic genotyping (PGS).[78] Binary methods use fixed thresholds to interpret the data and work under the assumption of whether a specific genotype is present or absent. However, this method has significant limitations as it cannot account for allelic drop-out in case of low-template or degraded DNA samples, or allelic drop-in due to high sensitivity of methods used.[79] In contrast, PGS represents a significant advancement, utilizing a broader range of data to evaluate possible genotypes and calculate probabilities after accounting for possibities for allelic drop-out, allelic drop-in, stutter products. Recognizing its advantages, forensic laboratories worldwide have increasingly been adopting PGS to address the complexities of interpreting mixed DNA profiles.[80]

Despite these advancements, forensic laboratories in India continue to rely on the outdated binary method for interpreting mixed DNA samples. This concern is further exacerbated since mixed DNA samples are often interpreted without calculating peak height ratios to determine the possible genotypes. In such cases, DNA analysts solely depend on checking whether the individual alleles (and not the allele pairs or genotypes) found in the evidence sample are consistent with the suspected contributor's reference DNA profile, which is scientifically inaccurate.[81]

5.3.2.3 *Lack of statistical evaluation*

After the genotypes detected in the questioned sample are compared with the genotypes in the evidence sample, and found to be the same, then it is considered to be an inclusion (match). In order to assess the significance of this inclusion, it is necessary to conduct statistical analysis.[82] Depending on the use of binary or probabilistic genotyping approach, random match probability (RMP), combined probability of inclusion (CPI) or likelihood ratios (binary/discrete/continuous) may be calculated to represent the statistical significance of the DNA inclusion.[83] Relying solely on the presence or absence of alleles is inadequate for drawing scientifically valid conclusions.

In India, the integration of statistical analysis into standard DNA reporting practices remains largely absent.[84] Many forensic laboratories do not routinely perform any statistical evaluations, and even when they do, these assessments are often excluded from the reports submitted to courts. This omission creates a misleading perception in courts, where a DNA match may be viewed as definitive proof rather than as evidence with inherent probabilistic considerations. Further, courts across India also routinely admit and rely on DNA reports without any statistical calculations. Recognizing this issue, the Supreme Court, in Manoj & Ors,[85] noted the importance of statistical analysis in forensic DNA interpretation. Despite this clear guidance, the forensic laboratories have been slow to adopt statistical evalution as part of DNA profiling.

---

75. NIST, DNA Mixture Interpretation: A NIST Scientific Foundation Review (2024), p. 36. https://doi.org/10.6028/NIST.IR.8351 [NIST DNA Mixture Interpretation].

76. SWGDAM, Interpretation Guidelines for Autosomal STR Typing by Forensic DNA Testing Laboratories (2017), p. 20 [SWGDAM Interpretation Guidelines].

77. FSIR (Annexure I), p. 299.

78. SWGDAM Quality Assurance Standards, pg 36; NIST DNA Mixture Interpretation, p. 44.

79. NIST, DNA Mixture Interpretation: A NIST Scientific Foundation Review, p. 46. https://doi.org/10.6028/NIST.IR.8351.

80. NIST Human Factors Report, pg 70. Butler, J.M. History of DNA Mixture Interpretation. Supplemental Document to DNA Mixture Interpretation: A NIST Scientific Foundation Review, NIST IR 8351sup1 (2024), p. 10. https://doi.org/10.6028/NIST.IR.8351sup1.

81. Butler, John. Advanced Topics in Forensic DNA Typing: Interpretation, Academic Press, (2015) p.16, 87; SWGDAM Interpretation Guidelines p. 37–38.

82. SWGDAM Interpretation Guidelines, p. 36.

83. NIST, DNA Mixture Interpretation: A NIST Scientific Foundation Review, p. 45–48.

84. FSIR, p. 227, 232.

85. Manoj & Ors, para 148–150.

This lack of scientific rigour not only weakens the reliability of DNA evidence but also raises broader concerns about the standards applied to forensic evidence in the judicial process.

5.3.2.4 Poor reporting practices in DNA analysis

The law permits courts to rely on expert opinions in specialized fields where judicial knowledge may be insufficient. In forensic DNA analysis, the expert can only aid the fact-finder in determining the source of the biological material. It cannot address how or when the DNA was deposited on the said item.[86] It is also essential to recognize that forensic DNA profiling is probabilistic in nature, it examines only a small portion of the genome, making it statistically possible for two unrelated individuals to share the same DNA profile by coincidence. Therefore, the mere presence of a person's DNA at a crime scene does not automatically establish guilt.

Despite this, DNA reports in India frequently use misleading terms such as "match" or "identical" without providing statistical context. These terms can create misconceptions about the certainty of the findings, leading to erroneous inferences about the source of the DNA.[87] Further, forensic reports must be clear, comprehensive, and technically accurate. It should contain sufficient detail to stand independently as a reliable scientific document.[88] However, DNA reports in India often fall short of these standards. Typically, they include only a brief summary outlining the items received, the methods used, and the expert's final opinion. Critical details such as underlying assumptions, limitations, and an allelic table (data from EPGs) are frequently omitted. Beyond the reports themselves, essential supporting records such as analytical worksheets, instrument logbooks, internal chain of custody records, case receipt details, and EPGs are rarely submitted.[89] Further, the electronic raw data files for all the samples and the controls, on the basis of the which the EPGs are preapred, are not submitted. These materials provide critical insights into the forensic process, including any deviations from standard protocols and the manner in which the samples were handled. This lack of transparency weakens the credibility of forensic evidence as it fails to communicate to end-users the nuances and constraints of the findings, leading to potential misinterpretation of the evidence.

Indian courts have started recognizing these deficiencies in expert evidence. Recently, placing reliance on Rahul and Ors. v. State (NCT of Delhi),[90] the High Court of Maharashtra in two cases[91] held that "when the expert gives no real data in support of his evidence, such evidence even though admissible may be excluded from consideration, as it affords no assistance in arriving at the correct value." The Court stated that such evidence, without its underlying basis, fails to provide adequate assistance to prove the guilt of the accused/applicant. The High Court remanded both the cases back to the Trial Court, instructing them to summon the necessary documentation for a proper evaluation.

5.3.3 Bias in forensic DNA profiling

Cognitive bias refers to "class of effects by which an individual's pre-existing beliefs, expectations, motives, and situational context may influence their collection, perception, or interpretation of information, or thier resulting judgmnts, decisions and confidence."[92] Forensic DNA profiling, despite its scientific foundation, is susceptible to cognitive biases that can influence analysis and interpretation.[93] While sophisticated software assists in generating DNA profiles from

---

86. NIST Human Factors Report, p. 24.

87. NIST Human Factors Report, p. 89, 98.

88. SWGDAM Quality Assurance Standards, pp. 31-32; NIST Human Factors Report, p. 30.

89. In Anokhilal, the same Trial Court that had previously pronounced the death sentence twice eventually acquitted the accused during a third trial. This reversal occurred after the court acknowledged a critical error in interpreting the DNA evidence. Specifically, the court noted that it was incorrect to infer that the presence of male DNA in the intimate samples of the deceased belonged to Anokhilal, especially when the DNA findings explicitly excluded him as the source. During the proceedings, the forensic expert admitted that Anokhilal's DNA was not found in the samples, yet the report was ambiguously written to state only that "male DNA was found." This lack of clarity in reporting led to the erroneous interpretation by the courts that the DNA implicated Anokhilal.

90. In Rahul & Ors., the Supreme Court explicitly addressed the unique evidentiary value of DNA reports, directing trial and appellate courts to examine "the underlying basis of the findings in DNA reports" and "whether the techniques were reliably applied by the expert" for the expert opinion to be considered.

91. Bharatkumar Dhaniram Kori v. State of Maharashtra, Confirmation Case No. 2 of 2022, High Court of Bombay, dated 29.07.2024 para 16; Sanjay Baban Katkar v. State of Maharashtra, Criminal Appeal No. 1313 of 2023, High Court of Bombay, dated 29.07.2024 para 18.

92. Spellman BA, Eldridge H, Bieber P., Challenges to reasoning in forensic science decisions. Forensic Science International Synergy, 2022; 4:100200. https://doi.org/10.1016/j.fsisyn.2021.100200.

93. NIST Human Factors Report, p. 12.

evidence and reference samples, it is the analysts who ultimately review, verify the information provided by the software, and interpret the results. This reliance on human judgment introduces subjectivity, as different examiners may interpret the same DNA data differently based on their expertise, training, or preexisting beliefs.

Two major forms of bias affect DNA analysis—confirmation bias and contextual bias.[94] Confirmation bias occurs when an examiner's pre-existing beliefs and expectations influence their interpretation of results in a manner that conform with those beliefs.[95] For instance, if an analyst examines a suspect's reference profile before analyzing the evidence sample, they may unconsciously expect to see the same alleles in both, shaping their conclusions. Contextual bias arises when external information which would be task-irrelevant, such as details about the crime, confession by the suspect, or other investigative findings, affects the analyst's judgment.[96] This risk is heightened when forensic experts also participate in crime scene investigations or interact closely with police and witnesses.

In India, contextual bias is a significant concern due to the structural relationship between forensic laboratories and law enforcement agencies. A majority of SFSLs and RFSLs function under police supervision, creating an environment where forensic analysts may face external pressure to align their findings with investigative narratives.[97] Moreover, police officers often include unnecessary case details in forwarding letters when submitting exhibits for DNA analysis. While intended to provide background information, these details can inadvertently bias forensic examiners by exposing them to information unrelated to the technical aspects of DNA profiling.

To mitigate contextual bias, one of the widely recommended best practice is Linear Sequential Unmasking (LSU),[98] a method that ensures analysts receive only relevant information at appropriate stages of the forensic process. LSU prevents exposure to extraneous details until after objective assessments are made. However, Indian forensic laboratories have yet to adopt LSU, largely due to the absence of a structured case management system.[99] Consequently, analysts receive full case details upfront, increasing the risk of subjective influence. Besides, when reference samples (such as a suspect's DNA) are submitted alongside evidence samples, forensic examiners tend to analyze them together. This practice significantly heightens the likelihood of confirmation bias, compromising the objectivity of the analysis.

Bias also extends to judicial decision-making regarding DNA evidence. Judges often perceive DNA profiling as infallible, assuming it to be 100% accurate and conclusive proof of identity, without fully understanding its probabilistic nature or the risks of contamination, leading to uncritical acceptance of forensic reports.[100] This over-reliance can result in misinterpretation of evidence, especially when reports lack detailed statistical analysis. The issue is further exacerbated by legal provisions such as Section 293 CrPC / Section 329 BNSS, which exempts forensic experts from appearing in court to testify. As a result, defense counsel and judges have limited opportunities to question or clarify forensic findings, reducing the likelihood of identifying potential errors or biases.

5.4 Issues with the judicial scrutiny of DNA evidence

While courts in India increasingly recognize the significance of DNA evidence in criminal investigations, particularly its ability to establish identity and resolve complex cases, as stated previously, the legal framework for assessing its reliability remains inadequate. The absence of clear guidelines for evaluating expert opinions has forced courts to develop their own standards. However, since judges typically lack forensic science training, they rely on textbooks, outdated literature, or broad legal principles to interpret DNA findings, many of which are scientifically outdated, inaccurate, or lack rigorous validation.[101] This reliance on flawed sources has led to the development of legal standards that do not align with current

94. Dror I. Cognitive and Human Factors in Expert Decision Making: Six Fallacies and the Eight Sources of Bias, Analytical Chemistry (2020) 92 (12): p. 7998. https://pubs.acs.org/doi/10.1021/acs.analchem.0c00704.

95. NIST Human Factors Report, p. 12.

96. NIST Human Factors Report, p. xvi.

97. FSIR, p. 81.

98. Quigley-McBride, A., Dror, I.E., Roy, T., Garrett, B.L., Kukucka, J. A practical tool for information management in forensic decisions: Using Linear Sequential Unmasking-Expanded (LSU-E) in casework, Forensic Science International: Synergy (2022), Volume 4. https://doi.org/10.1016/j.fsisyn.2022.100216; NIST Human Factors Report, p. 33–35.

99. FSIR, p. 150.

100. Santosh Kumar, para 71; Vinod Chouhta, para 40.

101. In Santosh Kumar, the Trial Court rejected the DNA evidence by relying on certain textbooks as there was no signs of rape on the body. However, the High Court reversed the finding by the Trial Court by observing that it would be dangerous for the court to discard the evidence of an expert witness by referring to certain texts and books without putting those texts to the expert and taking his opinion thereon.

forensic science, resulting in inconsistencies in judicial evaluation and varied approaches to assessing DNA evidence across different courts.

Besides, while courts have a legal duty to assess the accuracy and reliability of the DNA analysis, they are often unable to conduct an independent review due to the lack of access to underlying data and supporting materials. As noted earlier, DNA reports submitted in court are typically limited to brief summaries that exclude critical components such as statistical analysis, electropherogram, and procedural documentation. Without these details, judges cannot effectively assess whether proper scientific protocols were followed or whether the results are valid. However, a shift is beginning to emerge, with some courts recognizing the need for deeper examination of forensic reports and expert testimony.

Moreover, unlike in other jurisdictions where the admissibility of forensic evidence is determined before it is presented to the fact-finder,[102] Indian courts often decide admissibility at the stage of final arguments, after judges have already been exposed to the entire body of evidence. This increases the risk of confirmation bias, particularly when dealing with highly persuasive evidence like DNA.[103] In India, the same judge that conducts the trial also determines both the admissibility and weight of expert evidence. While parties can object to the submission of forensic reports, such objections are typically addressed much later in the trial, when judges may have already formed opinions about the case. This procedural structure further undermines impartial scrutiny of DNA evidence.

A related concern is the judiciary's reliance on forensic reports without summoning the expert who prepared them. As previously noted, courts frequently accept DNA reports without questioning the qualifications, training, or methodology of government forensic experts, assuming their competence without verification. This practice prevents meaningful cross-examination, making it impossible to determine whether the DNA analysis was conducted correctly. Recent judicial precedents have begun addressing these concerns. In Irfan Bhayu Mevati v. State of Madhya Pradesh,[104] the Supreme Court set aside the conviction after the DNA report central to the prosecution's case was exhibited by a police officer instead of the forensic expert. The Court emphasized that without expert testimony, the DNA evidence could not be meaningfully evaluated, violating the accused's right to a fair trial. Similarly, recent High Court judgments[105] have echoed this concern, noting that forensic reports typically present only final conclusions without explaining the underlying scientific methodology or data. However, such rulings remain inconsistent and isolated.

Ultimately, while courts have taken initial steps toward improving their scrutiny of DNA evidence, significant gaps remain. Without clear legal standards, procedural safeguards, and independent review mechanisms, the judicial reliance on forensic DNA evidence continues to be vulnerable to misinterpretation and an overstatement of its probative value.

5.5 Case study

This compilation presents case studies from the author's pro bono litigation efforts,[106] involving the defense of individuals sentenced to death. As part of the legal proceedings during the confirmation of the death sentence, underlying forensic documentation was sought either through applications under Section 91 CrPC[107] or requests under the Right to Information Act, 2005.[108]

---

102. Daubert v. Merrell Dow Pharmaceuticals, Inc., 509 U.S. 579 (1993). In this landmark decision, the United States Supreme Court established that federal district court judges must serve as 'gatekeepers' to prevent jurors from being influenced by unreliable expert testimony. This role requires judges to evaluate not only the expert's qualifications but also whether their testimony is grounded in a reliable scientific methodology and applied consistently to the facts of the case. Recognizing that such determinations cannot be effectively made during the urgency of a trial, judges typically conduct a separate pre-trial hearing outside the jury's presence to assess the admissibility of the proposed evidence.

103. FSIR, p. 247.

104. Criminal Appeal Nos. 220 of 2022, Supreme Court, Order dated 16.01.2025, para 29-30 [Irfan Bhayu].

105. Baburao Ukandu Sangerao v. State of Maharashtra, Criminal Confirmation Case No. 1 of 2021, High Court of Bombay (Aurangabad Bench), dated 15.09.2023; Anokhilal v. State of Madhya Pradesh, Criminal Reference No. 6 of 2022, High Court of Madhya Pradesh, dated 11.09.2023; Bharatkumar Dhaniram Kori v. State of Maharashtra, Confirmation Case No. 2 of 2022, High Court of Bombay, dated 29.07.2024; Sanjay Baban Katkar v. State of Maharashtra, Criminal Appeal No. 1313 of 2023, High Court of Bombay, dated 29.07.2024.

106. The authors are members of The Square Circle Clinic (Clinic), a criminal justice and legal aid initiative at the NALSAR University of Law, Hyderabad. One of the main areas of work for the Clinic, is to provide pro bono legal representation to prisoners sentenced to death. As part of this work, the authors are involved in examining the forensic evidence in these cases and raising appropriate challenges to the unreliable forensic evidence.

107. Section 91, The Code of Criminal Procedure. https://www.indiacode.nic.in/show-data?actid=AC_CEN_5_23_000010_197402_1517807320555&sectionId=22468&sectionno=91&orderno=101.

108. The Right to Information Act, 2005, The Gazette of India, Part II-Section I. https://rti.gov.in/rti-act.pdf.

The author gained access to critical underlying records, including worksheets, EPGs, equipment logbooks,[109] chain of custody documents, receipts, registers, and was also able to cross-examine the forensic examiners who conducted the DNA analysis. These case studies provide a rare insight into the responses elicited during cross-examinations, highlighting systemic deficiencies in forensic practices, procedural lapses, and the absence of standardized protocols within Indian forensic laboratories. By documenting these findings, the case studies aim to demonstrate the gaps in forensic procedures, the risks of wrongful convictions, and the urgent need for improved scientific rigor in forensic evidence analysis.

5.5.1 Baburao Ukandu Sangerao v. State of Maharashtra[110]

It is a case of rape and murder of a minor girl aged 4–5 years old. The evidence presented against the accused included the fact that the victim was last seen with him, he was found naked in the bushes near the victim's body, and a DNA report linking him to the crime. At the trial, the DNA expert did not testify, invoking Section 293 CrPC, and the court accepted the DNA report without further examination.

The trial court relied on the DNA report to conclude that none other than the accused had committed the rape and murder. Consequently, the court concluded that the evidence proved the accused kidnaped, raped, and committed sodomy on the deceased victim. Given that the crime fell under the "rarest of rare"[111] category, the accused was sentenced to death. During the confirmation of the death sentence at the High Court, an application was filed seeking access to the underlying documentation of the DNA report and permission to cross-examine the DNA expert. The High Court directed the Trial Court to record additional evidence by examining the DNA expert. Subsequently, worksheets, equipment logbooks, kit procedure manuals, and electropherograms were obtained for analysis.

Upon a detailed review of the laboratory documentation and cross-examination of the DNA expert, significant errors and deficiencies in the forensic analysis were identified, including:

Contamination of the victim's sample: The control bloodstain sample prepared from the victim was mixed with an evidence sample and was considered as the reference sample of the deceased. The DNA expert admitted to mixing the sample while explaining that he observed multiple alleles at only two loci and, in the absence of a "Y" allele on the amelogenin marker, assumed the sample was of female origin.

Issues with the EPG: Numerous unmarked peaks were observed in the EPG. The expert acknowledged during the cross-examination that the GeneMapper IDX software might have assigned allele numbers to these peaks, which he later deleted based on his personal judgment and experience. Additionally, instances were identified where alleles were marked despite the absence of corresponding peaks in the electropherogram, and the expert admitted to marking alleles in one or more instances where no visible peak was present.

Incorrect identification of the accused as a contributor: In one sample reported to contain the accused's DNA, analysis revealed that 13 out of 24 loci in the EPG did not match the accused's DNA. Despite this, the expert had erroneously reported that the accused contributed to the sample.

Improper handling of reference and evidence samples: As per the equipment logbooks, it was observed that the victim's reference sample and the evidence samples from both the accused and the victim were processed together during the DNA extraction step. The expert disclosed during his examination that samples from other unrelated cases were also processed alongside the current case samples. The expert further acknowledged that no negative controls or extraction blanks were run during each batch of extraction.

Procedural lapses: The expert admitted to failing to perform DNA quantitation on the extracted evidence samples. He also did not calculate the peak height ratio to determine which alleles constituted genotypes during interpretation. In addition, the DNA report lacked statistical analysis, which the expert admitted was not practiced in his laboratory.

Lack of qualification and training: The expert held an MSc Chemistry degree and had no formal training in DNA examination. He had previously worked in the biology division and admitted that his only training in DNA profiling occurred on the job. He had not undergone any proficiency testing as well.

Given these serious procedural and analytical lapses, the High Court dismissed the DNA evidence as unreliable and "not free from doubt." However, the High Court upheld the accused's guilt based on other circumstantial evidence and commuted the death sentence to life imprisonment.

---

109. Documented record maintained for each major piece of laboratory equipment. It logs every instance the equipment is used, who used it, what was analyzed, and when the analysis occurred.

110. Criminal Confirmation Case No. 1 of 2021, High Court of Bombay (Aurangabad Bench), dated 15.09.2023.

111. Bachan Singh v. State of Punjab, Criminal Appeal No. 273 of 1979, Supreme Court of India, dated 09.05.1980.

5.5.2 Anokhilal v. State of Madhya Pradesh[112]

In this case, the accused was sentenced to death by the Trial Court in 2013 for the rape and murder of a minor, based primarily on DNA evidence. During the trial, the DNA expert invoked an exemption under Section 293 CrPC and did not testify in court. The High Court upheld the death sentence. However, the Supreme Court, upon review in 2019, identified several procedural shortcomings, including the counsel representing the accused had been appointed on the very day of the trial and was required to defend the accused during the framing of charges with no sufficient opportunity to prepare the case. Also, the DNA report, which played a critical role in the conviction, was received at the fag end of the trial, leaving the defense counsel unprepared to examine or challenge it. As a result, the Supreme Court remanded the case for fresh consideration.

In the second trial in 2021, the trial court issued summons to the DNA expert. However, after the summons were unanswered for four months, they were canceled and the trial court proceed to admit the report under section 293 CrPC without the expert's evidence. Despite this, the trial court convicted the accused once again and sentenced him to death. In the second confirmation hearing in 2023, the High Court set aside the conviction and remanded the case back to the trial court for the third time. The High Court ordered that laboratory documentation be made accessible and that forensic experts be summoned for examination. Upon scrutiny of the laboratory documentation and cross-examination of the DNA expert, several critical issues were identified:

Incorrect reference sample of the deceased: No blood or other biological sample from the deceased was collected as a reference. Instead, among two bloodstains found on the deceased's clothing, one was determined to be of female origin and was used as the reference DNA profile of the deceased. However, the GeneMapper software marked the amelogenin marker for this bloodstain as "XY," suggesting male origin. Despite this discrepancy, the expert reported the DNA as female in the DNA analysis report and treated it as the reference profile of the deceased.

Incorrect Interpretation: For the vaginal and anal swab samples collected from the deceased, the GeneMapper software marked only one peak, "X." However, a handwritten "Y" was observed on the EPG. During cross-examination, the expert claimed that based on his experience and scientific understanding, a detectable "Y" was present. However, he failed to substantiate this claim due to the absence of raw data.

Incorrect reporting: The DNA report combined the results of the deceased's vaginal and anal swab samples into a single column, with the expert explaining that both profiles were identical. However, analysis of the respective EPGs revealed discrepancies at two marker locations, where the samples displayed different alleles. These differences were not reported in the DNA report.

Exclusion of the accused: Analysis of the vaginal and anal swab samples indicated the presence of male DNA, but the DNA did not match that of the accused. The accused's DNA was excluded as a contributor to the samples. Despite this, the DNA expert failed to explicitly report that the male DNA in the samples did not belong to the accused.

In addition to the scientific errors, there were significant lapses in the chain of custody of the samples, both while in police custody and at the RFSL, before being transferred to the SFSL for DNA analysis. Given that the proper interpreation of the DNA evidence showed the presence of another male individual's DNA in the vaginal and anal samples, and the weak nature of the other circumstantial evidence, the same trial court that had sentenced the accused to death twice ultimately acquitted him during the third retrial. The court also took note of the other substantial shortcomings of the DNA examination conducted in this case.

5.5.3 Jitendra @ Jeetu, Babu @ Ketan, Sanni @ Devendra v. State of Madhya Pradesh

This case pertains to the homicide of a minor girl, whose death was attributed to asphyxiation by muddy fluid. Medical examination also revealed head injuries and evidence of sexual assault. The prosecution was unable to conclusively establish the last seen theory or substantiate the circumstantial evidence. As a result, DNA evidence became the crucial link connecting the three accused to the crime. DNA from the deceased's vaginal smear and undergarments reportedly matched the profiles of the three accused. The trial court sentenced all three accused to death without summoning the DNA expert for testimony.

Upon confirmation, the High Court upheld the convictions, stating that the DNA report led to a "definite and inescapable conclusion" that the accused were the sole perpetrators. The High Court did not independently assess the reliability or accuracy of the DNA evidence, dismissing concerns about sealing and sample integrity as overly technical and instead relying entirely on the DNA report. The three accused appealed the High Court's decision, but the Supreme

112. Criminal Reference No. 6 of 2022, Madhya Pradesh High Court, Order dated 11.09.2023.

Court dismissed their petitions without providing reasons for upholding the conviction or sentence. Subsequently, one of the co-accused, Jitendra, filed a review petition. While his sentence was commuted, the Court did not conduct a detailed evaluation of the DNA evidence or the merits of the conviction. While review petitions for the other two accused were still pending, a Right to Information (RTI) application was filed with SFSL Sagar, requesting the laboratory documentation related to the DNA examination in this case. Reviewing the documentation revealed the following scientific errors in the DNA profiling process and interpretation.

Nonexamination of the deceased's reference profile: The FSL received the blood samples of the three accused, the deceased's vaginal smear slides, and her undergarments for analysis. However, no reference DNA sample from the deceased was collected during the post-mortem examination. Both the vaginal smear and the undergarments contained mixed DNA profiles. Without the deceased's reference DNA profile, it is impossible to distinguish between alleles originating from the deceased and those belonging to external contributors.

Incorrect inclusion of Jitendra: In the vaginal swab of the deceased, Jitendra's alleles were absent in eight marker locations. According to standard forensic practice, if an individual's genotype differs at even a single locus, the individual is excluded as a contributor to the DNA profile. However, the DNA report concluded Jitendra as an inclusion in the DNA obtained from the deceased's vaginal swab.

Inconsistencies between the vaginal swab and smear slide: The vaginal smear slide is prepared by smearing the vaginal swab onto a glass slide, and both are expected to yield identical DNA profiles since they originate from the same source. However, in this case, there were significant differences in the DNA profiles of the swab and the slide. At eight marker locations, discrepancies were observed between the alleles in the vaginal swab and those in the smear slide. Despite these inconsistencies, no explanation was provided in the DNA report or laboratory documentation to account for such differences.

Discrepancies between the EPG and DNA report: The DNA report is a representation of the EPG, therefore, it should reflect the same data from the EPG; however, in this case, several inconsistencies were observed between them. In the DNA profile of the vaginal swab, in seven marker locations, alleles present in the EPG were omitted from the DNA report. Similarly, in the DNA profile of the deceased's undergarments, in six marker locations, alleles reported in the DNA report were absent from the EPG.

Missing information and quality control issues: The DNA report and the accompanying laboratory documentation lacked crucial details, including information on DNA quantification, statistical analysis, and the use of controls. This omission is particularly significant in this case, as visible anomalous peaks were detected in at least five marker locations across the samples. However, there was no indication in the documentation that these samples were reanalyzed to ensure accuracy.

Despite the apparent issues with the interpretation and reporting of DNA evidence, the Supreme Court dismissed concerns, citing the absence of errors in the chain of custody. Eventually, the review petitions of the remaining two accused resulted in the commutation of their death sentences by the Supreme Court.

These case studies illustrate a troubling reality—the systemic issues within both forensic laboratories and the judicial process in India. Unlike other countries that mandate strict adherence to established standards for forensic evidence[113] to be deemed admissible, India lacks clear scientific and legal standards for such evidence. This absence of standardized protocols leads to significant inconsistencies in laboratory practices, including how forensic analyses are conducted and how results are interpreted, both across different laboratories and within the same laboratory.

One of the most pressing concerns in forensic examination in India is the lack of transparency, particularly due to the absence of comprehensive documentation in laboratory reports. A critical gap is the failure to preserve electronic raw data generated by the GeneMapper software during DNA analysis. This raw data, stored within the software and exportable for backup, is essential for ensuring the integrity and reproducibility of forensic findings. However, in India, forensic laboratories often do not maintain backups of this data. Frequently, systems become non-functional, and more concerningly, laboratory examiners are often unfamiliar with the process of exporting this data. The Court and the defense are asked to rely on EPGs instead of the raw data, but EPGs represent only the final analysis and do not reveal any edits such as additions, deletions, or changes to allele calls made during interpretation. Given the continuing reliance on binary methods of DNA mixture interpretation, the lack of electronic raw data gains more significance. This raises serious

113. TFSC, Final Audit Report For Austin Police Department Forensic Services Division DNA Section (2016), pg 12-25. https://static.texastribune.org/media/documents/ APD_Audit_Final_report_071116.pdf. Following a site inspection by the Texas Forensic Science Commission (TFSC), an investigation into the DNA section of the Austin Police Department's Forensic Science Division was conducted. It highlighted issues regarding standards followed during DNA mixture interpretation, inadequate validation studies, subjectivity in DNA mixture interpretation, contamination events, and issues related to leadership and training. Consequently, the DNA division was closed due to non-conformity with ISO standards.

concerns about the examiners' training and competence, and whether the analyses are conducted correctly, given their apparent limited understanding of the software's capabilities.

This lack of transparency is further compounded by the judiciary's routine acceptance of DNA reports without rigorous scrutiny, allowing critical gaps in the forensic process to go unexamined. As demonstrated in the case studies above, essential procedures such as DNA quantitation, statistical analysis, use of controls are frequently omitted, yet courts continue to rely on these incomplete reports. While DNA evidence can be a powerful tool in the pursuit of justice, these case studies reveal a deeper problem: the failure of the justice system to critically evaluate the reliability of forensic evidence. The widespread belief in the unfailing accuracy of DNA evidence, combined with inadequate laboratory standards and judicial oversight, poses a significant risk of miscarriages of justice. Addressing these issues requires systemic reforms, including the establishment of uniform forensic standards, ensuring comprehensive documentation of forensic analyses, and promoting judicial awareness about the nuances and limitations of forensic evidence. Without these measures, the likelihood of wrongful convictions or unjust outcomes due to flawed forensic evidence remains unacceptably high.

5.6 Path towards reform

5.6.1 Audit of DNA divisions

Not all FSLs house every forensic division; their setup depends on the types of cases they typically oversee. As a result, many FSLs lack a dedicated DNA division. In some states, only a single FSL may have the capability to conduct DNA analysis, placing the burden of the entire state's DNA examinations on that one laboratory. But the demand for DNA analysis has been growing rapidly, driven by the 2005 amendments to the CrPC and recent legislative developments. Meeting this demand will require not only upgrading existing DNA laboratories but also establishing new ones. However, this expansion must be informed by a comprehensive understanding of the current landscape, including the challenges faced by existing DNA divisions and the prerequisites for setting up new ones. An audit would provide valuable insights into the strengths and weaknesses of existing laboratories, revealing areas for improvement and guiding the establishment of new facilities.

While relatively few reports have been published over the past two decades, they provide some insight into the forensic landscape in India, particularly highlighting critical concerns such as insufficient financial, personnel, and infrastructural resources in forensic laboratories.[114] Unfortunately, the recommendations outlined in these reports have largely gone unimplemented. The Forensic Science India Report by Project 39A at National Law University Delhi attempted to audit the forensic practices, including DNA divisions in all FSLs, but its scope did not encompass all laboratories with a DNA division.[115] Beyond this report, there is a significant lack of comprehensive data on the functioning and challenges faced by DNA divisions across the country.

5.6.2 Development of laboratory-specific working procedure manual

Forensic laboratories must develop and implement their own WPM based on validation studies that account for each laboratory's unique infrastructure, available technology, and operational conditions. The WPM should provide clear guidance on analytical testing, reporting procedures, corrective actions, precautionary measures, and safety instructions to maintain the integrity of forensic examinations. Implementing a laboratory-specific WPM is crucial for promoting consistency in DNA analysis, minimizing intra-laboratory variations, and ensuring adherence to scientifically validated protocols.

Furthermore, WPMs must be integrated into broader quality assurance frameworks that promote continuous improvements in forensic methodology, sample handling, and casework management. To maintain scientific credibility, these manuals must be periodically reviewed and updated in response to technological advancements and evolving laboratory conditions. By implementing and adhering to a structured WPM, laboratories can uphold high standards of quality, minimize errors, and produce scientifically robust DNA evidence that withstands legal scrutiny.[116]

---

114. National Human Rights Commission, State of the Art of Forensic Science: For Better Criminal Justice (1999), MHA, Perspective Plan on Indian Forensics (2010), and Report on Scientific Performance Audit of DFSS HQ and its CFSLs (2011).

115. FSIR, p. 223.

116. NIST Human Factors Report, p. 28.

5.6.3 Continuous training for analyst

The shortage of scientific staff in FSLs has led to frequent rotation of available personnel between divisions, often with minimal on-the-job training. This is particularly concerning for DNA divisions, given the technical complexity of DNA analysis. Analysts must possess not only foundational expertise but also the ability to adapt to unforeseen challenges during the analysis process. For instance, if an analysis yields uninterpretable DNA results, the analyst should be able to identify the underlying cause, assess potential issues with the procedure, and implement alternative methods to obtain the result. Without proper training, analysts may simply report the DNA as uninterpretable, rendering the sample unusable and potentially compromising critical evidence in a case.

Even for analysts with formal academic training in DNA-related fields, continuous education and skill development are essential.[117] Advanced training programmes should address both routine procedures and rare, complex scenarios, equipping analysts to handle deviations from standard protocols effectively. Regularly updated training also ensures that analysts remain informed about the latest technologies, methodologies, and best practices in DNA forensics.[118]

5.6.4 Statistical evaluation in DNA reporting

DNA reports must incorporate statistical analysis to provide a meaningful interpretation of the forensic results.[119] Given that forensic DNA analysis examines only a small portion of an individual's genetic profile, there is a risk of coincidental matches with unrelated individuals. Statistical evaluation quantifying the likelihood of a match occurring within the general population provides crucial context, enabling recipients of the DNA report, including judges and other legal professionals, to understand how rare or common the DNA profile is. This will ensure that the rarity of the given DNA profile is accurately assessed, preventing misinterpretation and undue reliance on DNA evidence.

A simple declaration of a 'match' based solely on identical alleles is insufficient and can lead to courts overestimating the probative value of DNA findings.[120] Standardizing the use of statistical evaluation in DNA reporting will enhance transparency, minimize the risk of wrongful convictions, and align forensic practices with internationally accepted scientific standards.

5.6.5 Need for regulation of forensic practice

The DFSS was established by the MHA with the mandate to enhance and promote forensic science practices across the country. Its primary role includes formulating policies, plans, and legislation to ensure quality, capacity, and capacity-building for forensic services. The DFSS is also tasked with fostering best practices by promoting quality management, developing scientific standards, and creating uniform protocols for forensic procedures, which should be widely disseminated to ensure consistency. In addition, it connects with academia and R&D initiatives to advance forensic science in India.[121]

However, the effectiveness of the DFSS is limited by its lack of direct control over the SFSLs and RFSLs, which handle the majority of forensic casework in the country. Reiterating the point explained above, the DFSS cannot mandate compliance with its guidelines or enforce adherence to scientific or quality standards due to the absence of legislative authority. These SFSLs and RFSLs often function under the administrative control of the police, who may lack the technical knowledge and understanding of the operational and resource requirements of forensic laboratories. This unregulated structure has resulted in significant inconsistencies and inefficiencies in forensic practices. Procedures and protocols not only vary widely between laboratories but can also differ within the same laboratory, creating a fragmented system. Laboratories often follow disparate manuals and methodologies, which undermines the credibility and reliability of forensic evidence.

To overcome this limitation, a robust regulatory framework is urgently needed to standardize forensic practices across all laboratories in the country.[122] This framework should include uniform rules, procedures, and quality standards

---

117. NIST Human Factors Report, p. 241.

118. AAFS Standards Board, Standard for Forensic DNA Analysis Training Programs (2019). https://www.aafs.org/sites/default/files/media/documents/022_Std_e1.pdf.

119. SWGDAM Interpretation Guidelines, p. 36.

120. SWGDAM Interpretation Guidelines, p. 36.

121. Directorate of Forensic Science Services, DFSS Brochure (2023). http://dfs.nic.in/pdfs/Final%20brochure%2024%20Jan%202023.pdf.

122. FSIR, p. 269.

applicable to CFSLs, SFSLs, and RFSLs alike. Furthermore, it must include mechanisms for regular audits and compliance checks to ensure that laboratories adhere to established guidelines. If laboratories fail to comply, clear and enforceable actions must be prescribed to hold them accountable. Regulating forensic science practice through legislative authority and centralized oversight is critical for improving the quality, consistency, and credibility of forensic services. Without such regulation, it will remain challenging to address the systemic issues plaguing forensic laboratories and to achieve the desired level of reliability in forensic science, which is essential for ensuring justice.

5.7 Conclusion

The increasing reliance on DNA evidence in the judicial process highlights its ability to provide objective, scientifically grounded inputs on critical questions of fact. However, the realization of this vision faces significant hurdles stemming from a dual-faceted problem: systemic shortcomings in the forensic science system and critical gaps within the legal framework. Despite the push to integrate DNA evidence more broadly, these foundational issues undermine its reliability, validity, and utility in criminal justice proceedings.

The forensic system, as it stands, is plagued by under-resourced laboratories, inconsistent practices, and a lack of standardized protocols, which compromise the quality and accuracy of DNA analysis. Laboratories face chronic understaffing, inadequate training, and a dire absence of regulatory oversight, further complicating the production of reliable forensic reports. Moreover, the legal framework exhibits an incomplete understanding of the complexities of DNA evidence. Judges and lawyers often lack the technical expertise to scrutinize forensic findings adequately, leading to blind reliance on DNA evidence or misinterpretation of its scientific nuances. The absence of explicit legal standards for evaluating the admissibility and weight of DNA evidence only exacerbates this problem, resulting in inconsistent rulings and potentially unjust outcomes.

Addressing these challenges requires a comprehensive, systemic overhaul. On the forensic side, this means investing in infrastructure, establishing a central regulatory authority, mandating quality assurance practices, and fostering continuous training for forensic analysts. For the legal system, it involves educating legal professionals on the principles of forensic science, creating clear standards for the admissibility and evaluation of expert evidence, and ensuring mechanisms for independent review of forensic findings. Fostering collaboration between forensic scientists and the judiciary can bridge knowledge gaps and promote scientifically informed decision-making. Only through such efforts can DNA evidence truly serve as a cornerstone of a fair and reliable judicial process.

Part II

Module II: Technological advancements in forensic DNA analysis

Chapter 6

Understanding Rapid DNA and its application

Tugba Unsal Sapan
Faculty of Engineering and Natural Science, Department of Forensic Sciences, Institute of Addiction and Forensic Sciences, Üsküdar University, Istanbul, Turkey

6.1 Introduction

DNA analysis has found application in forensic sciences, the determination of paternity, and criminal investigations. In paternity investigations, biological samples taken from the mother, suspected father, and child are analyzed to determine whether the individual can be the father of the child. In criminal investigations, biological fluids and tissues (blood, semen, saliva, hair, bone, etc.) collected from crime scenes and their stains and residues are identified by short tandem repeat (STR) loci (Chan, 1992; Edwards et al., 1992; Weber et al., 2002).

Several commercial kits have been developed for the easy acquisition of STR profiles, which are used worldwide. Even with these ready-made kits, the process of obtaining profiles in routine studies is long and tedious. However, the time to reach the perpetrator is very important. Catching the perpetrator as soon as possible can prevent a second victimization in that process (Larue et al., 2014).

Rapid DNA technologies, which have been developed in addition to the devices used routinely for this purpose, are fully automated DNA analyzers that allow us to obtain STR profiles. These are complex systems that combine routine laboratory steps such as isolation, DNA quantification, polymerase chain reaction, capillary electrophoresis, and data analysis to generate DNA profiles in a short time. They reduce the number of analysts, reducing the risk of contamination, and yet they are compatible with CODIS in as little as 75–90 minutes (Cornelis, 2019; Hennessy et al., 2014; Holland & Wendt, 2015; Salceda et al., 2017).

Rapid DNA technologies, which are known to work with reference samples and are still under development, have been used since the 2010s. It has also been stated that there may be difficulties in solving cases in complex paternity and paternity cases studied in forensic sciences that require meticulousness and expertise in the evaluation of the results of Rapid DNA technologies, which can also report with the software it is connected to without the need for an expert (Hennessy et al., 2014; Holland & Wendt, 2015; Salceda et al., 2017).

In this section, all areas of use of Rapid DNA technologies in Forensic Sciences, their working principles, and the conveniences or difficulties brought by their use are examined.

6.2 Historical development of forensic genetic studies

Developments in molecular techniques have found a parallel place in forensic science studies. In 1900, with the definition of the ABO blood grouping system, individuals were placed in different groups according to their blood types, and the first step was taken in the development of forensic hemogenetics. In 1915, Leone Lattes published a book describing the use of ABO typing to solve a paternity case. Thereafter, variations in erythrocyte enzymes, serum proteins, hemoglobin, and leukocyte antigens were exploited. In the 1960s and 1970s, advances in molecular biology, including Sanger sequencing and Southern blotting, opened up new possibilities for forensic science.

With the advances in molecular biology techniques that began in the 1980s, it is now possible to examine regions of the genome. These developments enabled forensic sciences to work with polymorphic regions at the DNA level. In 1984, the

Investigative and Predictive DNA Testing. DOI: https://doi.org/10.1016/B978-0-443-30128-5.00009-6

DNA fingerprinting method, which is unique to the individual, was developed with DNA polymorphic regions, and its discrimination power was increased to high rates in interpersonal identification (Jeffreys & Wilson, 1985).

Thus, a new era began in the field of forensic sciences. Soon, these polymorphic regions (DNA fingerprints), which are unique to the individual, found application in forensic sciences, paternity determination, and criminal investigations. In paternity studies, biological samples taken from the mother, suspected father, and child are analyzed, and in criminal investigations, biological fluids and tissues (blood, semen, saliva, hair, bone, etc.) collected from the crime scene, as well as their stains and residues, are identified and used as evidence in courts (Chan, 1992; Edwards et al., 1992; Weber et al., 2002).

Each nucleated human cell has two complete copies of the genome. The genome is the "haploid genetic complement of a living organism" and contains approximately 3,200,000,000 base pairs (bp) of information organized into 23 chromosomes in humans. There are polymorphic genes on the genome, one from the mother and one from the father. These polymorphic regions on DNA have provided a new opportunity to analyze forensic cases (Jeffreys & Wilson, 1985; Jeffreys et al., 1995). Two types of polymorphisms are commonly used in forensic genetics;

- Single nucleotide polymorphisms(SNP);
- Two important categories of repeating sequences are widely used: mini-satellites, also called variable number of tandem repeats (VNTRs); and microsatellites, also called short repeats (STRs).

VNTRs have repeat sequences ranging from 6 to 100 bp in our genome. These regions are represented thousands of times in some alleles; variation in the number of repeats results in alleles ranging in size from 500 bp to 30 kb. VNTRs were the first polymorphisms used in DNA profiling and have been successfully used in forensic case studies for several years. However, the use of VNTRs was limited by the type of sample that could be successfully analyzed, as large amounts of high molecular weight DNA were required, and there are limitations in their interpretation (Jeffreys & Wilson, 1985; Jeffreys et al., 1995). Polymerase chain reaction (PCR), the technology that can amplify specific regions of DNA, has demonstrated that DNA profiles can be generated from only a few cells. It also shortened the time needed to create a DNA profile, could be used with degraded DNA, and allowed almost any polymorphism in the genome to be analyzed. This was soon followed by PCR analysis of Short Tandem Repeats (STRs), which are currently the most widely used genetic markers in forensic science (Saiki et al., 1985; Stoneking et al., 1991).

6.2.1 DNA analysis with conventional systems

Obtaining DNA profiles in conventional systems consists of 5 basic steps. These are DNA isolation, DNA quantification, PCR (amplification), electrophoresis, and data analysis.

6.2.1.1 DNA extraction

It involves isolating total cellular DNA suitable for forensic DNA analysis. DNA extraction has two main purposes: the first is to extract enough DNA from a sample to perform DNA profiling. The second is to extract DNA that is pure enough for subsequent downstream analysis. There are many methods available to extract DNA, that is., the organic method, magnetic-beads-based method, or column-based method. Which method to use depends on several factors, including sample type and quantity (Goodwin et al., 2007).

6.2.1.2 DNA quantification

Once the DNA has been isolated, accurate measurement of the DNA is important for subsequent analysis. Adding the right amount of DNA to a PCR will ensure the best quality results in the shortest time. Adding too much or not enough DNA will generate a non-informative DNA profile.

6.2.1.3 Polymerase chain reaction

PCR exploits the enzymatic processes of DNA replication. During each cell cycle, the entire DNA content of a cell is copied. This replication of DNA is performed outside the cell in vitro to replicate specific regions of DNA. A PCR has the components: Mold DNA, at least two primers, Taq polymerase, magnesium chloride, deoxynucleotide triphosphates, and DNA polymerase (Goodwin et al., 2007).

6.2.1.4 Capillary electrophoresis

After DNA has been amplified using PCR, the length of the products is precisely measured. Applied Biosystems provides the most widely used capillary electrophoresis systems, all of which are capable of multicolor detection. Samples are transferred into capillaries using electrokinetic injection, a voltage is applied, and charged molecules, including amplified DNA fragments and internal lane size standards, move into the capillary. After injection, a constant voltage is applied to the capillary, and the PCR products move towards the positively charged anode, passing through the polymer that fills the capillary and acts as a screening matrix. Urea and 2-pyrrolidinone in the gel polymer and a temperature of 60°C help to prevent the formation of any secondary structures during electrophoresis (Rosenblum et al., 1997). During the electrophoresis period, an argon ion laser shines through a small glass. PCR products labeled with fluorescent dyes emit fluorescence that is detected by the instrument camera charge coupled device (CCD) and then recorded by the acquisition software (Butler, 2001).

6.2.1.5 Analysis and interpretation of results

The spectra of the dyes used to label PCR products give peaks according to the dye color. After data collection, GeneScan or GeneMapper ID software removes the spectral overlap in the profile and calculates the sizes of the amplified DNA fragments, giving the DNA profile (Goodwin et al., 2007).

6.3 Rapid DNA technology

Rapid DNA technologies, which are complex systems that combine routine laboratory steps such as DNA extraction, PCR, and electrophoresis, which we call conventional systems, have entered the DNA analysis market with IntegenX (RapidHIT ID), LGC (ParaDNA), and NetBio (DNAscan).

6.3.1 The ParaDNA intelligence system

ParaDNA is a system developed by Laboratory of the Government Chemist (LGC, UK), based on HyBeacon technology that allows for the screening and preliminary analysis of potential evidence or the identification of individuals within 75 minutes. The ParaDNA system is designed to be used by non-specialists (e.g., police). The ParaDNA system consists of two parts:

1. A unit designed for DNA amplification and fluorescent dye-based electrophoresis, including software that controls the entire amplification-detection process.
2. A consumable box consisting of a sample collector and a ready-to-use PCR reaction platen containing the reagents required for amplification and DNA profiling.

Cellular material is collected from an evidence sample with the disposable ParaDNA sample collector, and the cells obtained with the sample collector are transferred to reaction plates in the consumable box (Fig. 6.1). The ParaDNA system generates DNA profiles including five STR loci (D3S1358, D16S539, D8S1179, D18S51, and TH01) and the amelogenin locus using HyBeacon technology (Blackman et al., 2015; Dawnay et al., 2014; Li et al., 2019).

Although the DNA profiles generated can be used to screen suspects, they are less discriminative than other Rapid DNA systems available on the market. ParaDNA was acquired by Life Technologies Corporation in 2013. However, ParaDNA is currently discontinued (Cornelis, 2019).

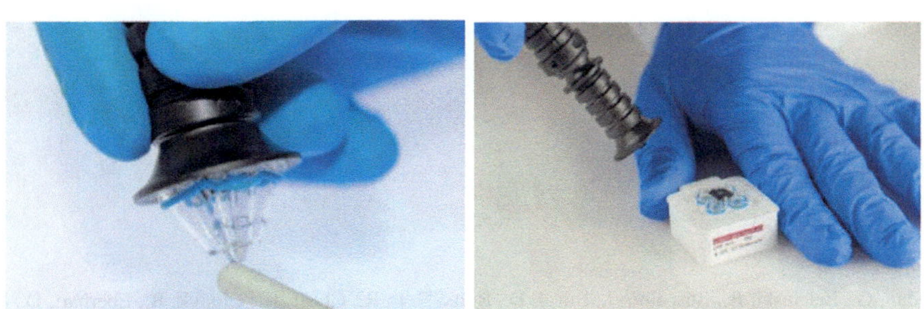

FIGURE 6.1 Working principle of the ParaDNA system.
From Doole, S. (2017). *Forensic Science International: Genetics Supplement Series, 6*, e492–e493. https://doi.org/10.1016/j.fsigss.2017.09.185.

6.3.2 DNAscan/ANDE Rapid DNA Analysis platform

A consortium of US government agencies, including the Federal Bureau of Investigation (FBI), the Department of Defense, and the Department of Homeland Security, addressed the need for Rapid DNA technology by establishing the Accelerated Nuclear DNA Equipment (ANDE) program. The DNAscan/ANDE Rapid DNA Analysis system was developed in 2011 as a result of these meetings. ANDE systems were developed and tested by NetBio before becoming commercially available in their current format. In 2013, Tan et al. reported a fully integrated system for the automated generation of DNA profiles (Tan et al., 2013).

This Rapid DNA analysis platform consists of three components. A swab, a cassette, and the benchtop DNAscan system are swab-driven, with each swab loaded into a disposable cassette containing all the lyophilized reagents necessary to perform the complete analysis. The cassettes have a capacity of five swabs. The cassette is loaded into the DNAscan device. DNA extraction is performed based on guanidine-based lysis and silica binding. After PCR amplification, a profile is generated with CE. STR profiles consisting of more than 16 STR loci can be easily generated following the quality standards described in the Rapid DNA law, ensuring full compatibility with CODIS DNA databases (Carney et al., 2019; Cornelis, 2019; Moreno et al., 2017) (Fig. 6.2).

6.3.3 RapidHIT system

Another Rapid DNA technology was developed in 2012 under the name RapidHIT 200 (eight cartridges) and was introduced by IntegenX Inc. RapidHIT ID (1 cartridge) systems were also developed in the process of RapidHIT systems. The RapidHIT 200 System was developed to run 8 samples simultaneously to generate a DNA profile in approximately 2 hours. For this instrument, two cartridges were developed, a cartridge containing PowerPlex 16 System (Promega Corporation, Madison, WI) loci and a cartridge containing GlobalFiler Express PCR Amplification Kit (Thermo Fischer Scientific) loci (Date-Chong et al., 2016) (Fig. 6.3).

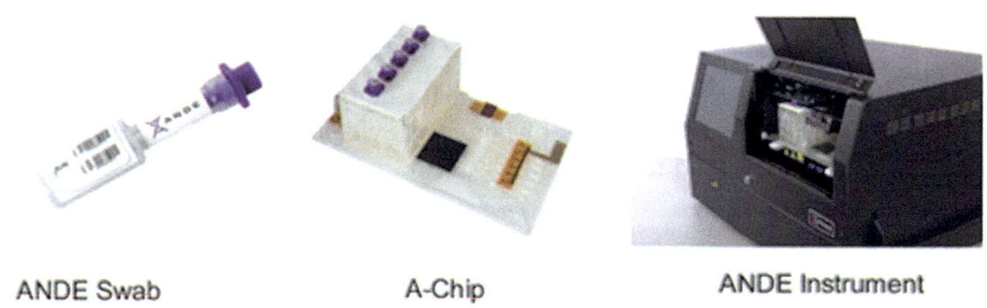

FIGURE 6.2 DNAscan/ANDE Rapid DNA Analysis system.

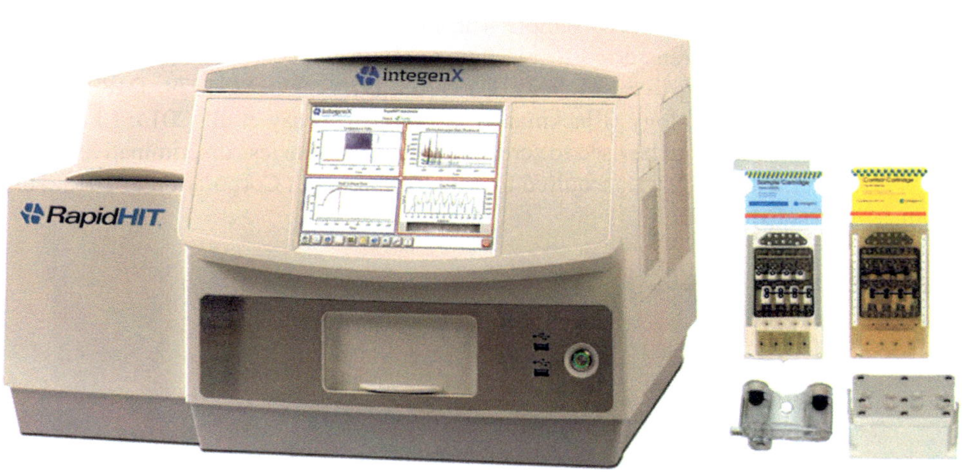

FIGURE 6.3 RapidHIT 200 system.
From Jovanovich, S., Bogdan, G., Belcinski, R., Buscaino, J., Burgi, D., Butts, E. L. R., Chear, K., Ciopyk, B., Eberhart, D., El-Sissi, O., Franklin, H., Gangano, S., Gass, J., Harris, D., Hennessy, L., Kindwall, A., King, D., Klevenberg, J., Li, Y., … Williams, S. (2015). Developmental validation of a fully integrated sample-to-profile rapid human identification system for processing single-source reference buccal samples. *Forensic Science International: Genetics, 16*, 181–194. https://doi.org/10.1016/j.fsigen.2014.12.004.

These systems have been developed and updated under the name RapidHIT ID system. The RapidHIT ID system is a fully automated DNA profiling system for STR-based identification. It generates CODIS-compliant DNA profiles in less than 90 minutes for presumed single-source DNA samples. RapidHIT ID is optimized for expert-free use in decentralized environments.

6.3.4 RapidHIT ID System

In the RapidHIT ID system, direct PCR (amplification) without DNA extraction, capillary electrophoresis, and analysis of data are automated within the system. Once an STR profile is generated, RapidHIT ID transfers the data to a central computer running RapidLINK software (IntegenX) for processing, including manual profile review, where automated results can be reported. The system also has a built-in fingerprint reader and camera for access and automated inspection tracking, and authentication. The RapidHIT ID System generates CODIS-compliant DNA profiles in less than 90 minutes after less than one minute of application. The system uses a single-use cartridge for each sample tested. For the capillary electrophoresis module, there is a primary cartridge chamber with a 150-use consumable (Buscaino et al., 2018; Hennessy et al., 2014; Salceda et al., 2017) (Fig. 6.4).

Rapid DNA technology allows the study of reference samples (buccal swabs) using the reference samples kit, while the crime scene investigation samples cartridge allows the analysis of swabs taken from the surface, the filter part of the cigarette butt or bone samples by first cleaning the outer surfaces, removing soft tissue, scraping and placing them in the well in powder form (Application note: Bone sample processing on the RapidHIT ID System with RapidINTEL cartridges, 2020)

6.3.4.1 Kits used in RapidHIT ID technology

Three types of special cartridges are available for RapidHIT ID:

- RapidHIT ID ACE GlobalFiler Express sample cartridge: developed for buccal swab samples.
- RapidHIT ID ACE NGM SElect Express sample cartridge: developed for buccal swab samples.
- RapidHIT INTEL sample cartridge: developed for blood and other crime scene samples (Figs. 6.5;6.6;Fig. 6.7) (Murakami et al., 2020; Thermo Fisher Scientific, 2022).

6.3.4.2 Working steps of the RapidHIT ID System

Working steps with the RapidHIT ID System

- Sample collection;
- RapidHIT ID System sample insertion;
- RapidHIT ID System consists of the following steps.

With the RapidHIT ID System, DNA extraction, quantification, PCR (amplification), Electrophoresis, and reporting of results with the RapidHIT ID System v1.1.2 software are performed automatically.

For sample placement in the RapidHIT device, swabs taken from the mouth, swabs taken from the surface for crime scene samples, or bone powders scraped with a grinder for bone samples are placed in the well in the card system of RapidHIT technology, and the device is started by inserting the cartridge into the system. The biological sample in the swab compartment is first lysed. After DNA lysis, the lysate is transferred from the swab chamber to the PCR chamber in

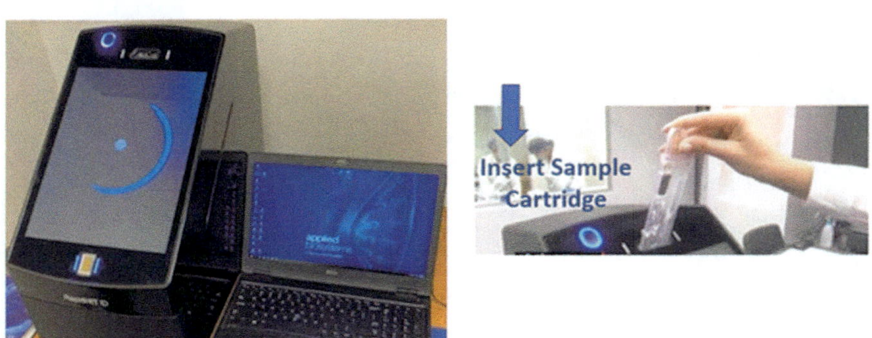

FIGURE 6.4 RapidHIT ID system.

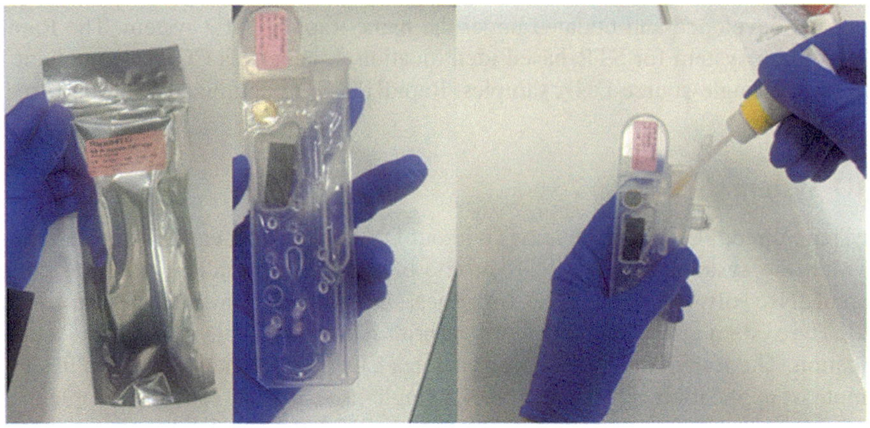

FIGURE 6.5 RapidHIT ID ACE GlobalFiler Express sample cartridge (buccal swabs).

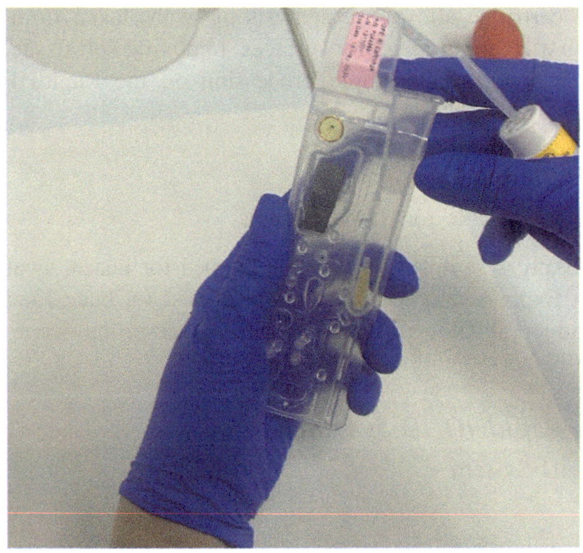

FIGURE 6.6 RapidHIT INTEL sample cartridge.

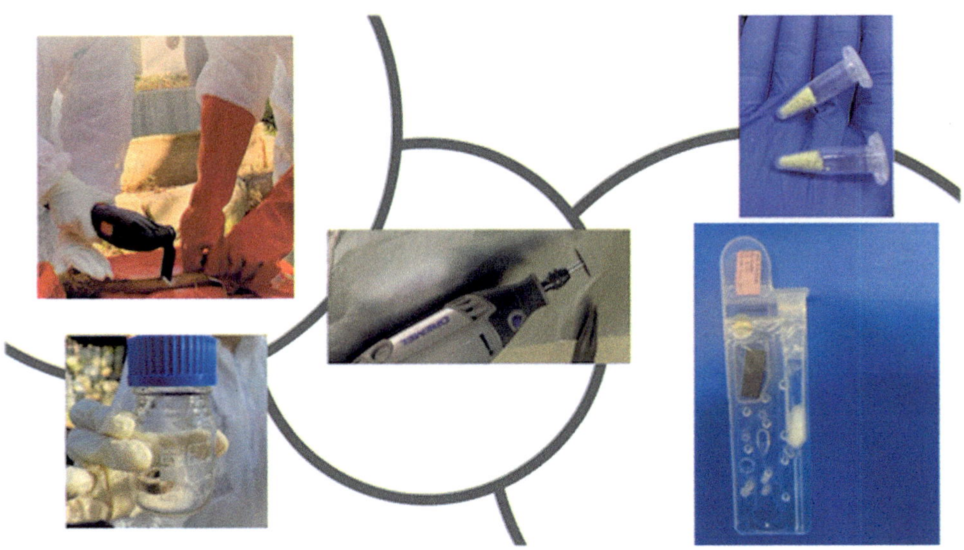

FIGURE 6.7 RapidHIT INTEL sample cartridge (for bone samples).

the sample cartridge. The PCR chamber contains a 1.2 mm diameter paper disk and captures the lysate. Primer and master mixes are then simultaneously pumped into the PCR chamber. Thermal cycling occurs inside the PCR chamber. It is pumped from the PCR chamber. The resulting mixture is then transported through a heat denaturation zone (95°C) and into the primary cartridge, where it is injected into electrophoresis capillaries. DNA fragments are separated and optically detected. RapidHIT ID processes the data and generates a DNA profile (Fig. 6.8) (Hennessy et al., 2014; Salceda et al., 2017).

1. Sample cartridge system;
2. Primary cartridge system (capillary electrophoresis module).

6.3.4.3 RapidHIT ID software system and reporting of results

The system includes RapidLINK software, a Genemapper ID-X-like software. This software allows the analysis and control of STR profiles generated by rapid DNA systems located in the laboratory or remotely. The software has four functions.

1. **RapidLINK Match App**
   It identifies a match between two DNA profiles and reports a likelihood ratio that the match did not occur due to random chance from a member of the population with the same DNA profile as the suspect.
2. **RapidLINK Familial App**

   Familial associations of DNA profiles

- Parent-child
- Grandmother-grandfather-grandson
- Identifies the aunt/uncle-nephew connection.

It can provide indirect access to the perpetrator from family relationships in unsolved murders, and identification of disaster victims and missing persons.

3. **RapidLINK Kinship App**
   Identifies the familial relationship between suspected parents and children by revealing the DNA relationship between them using LR and PI calculations.

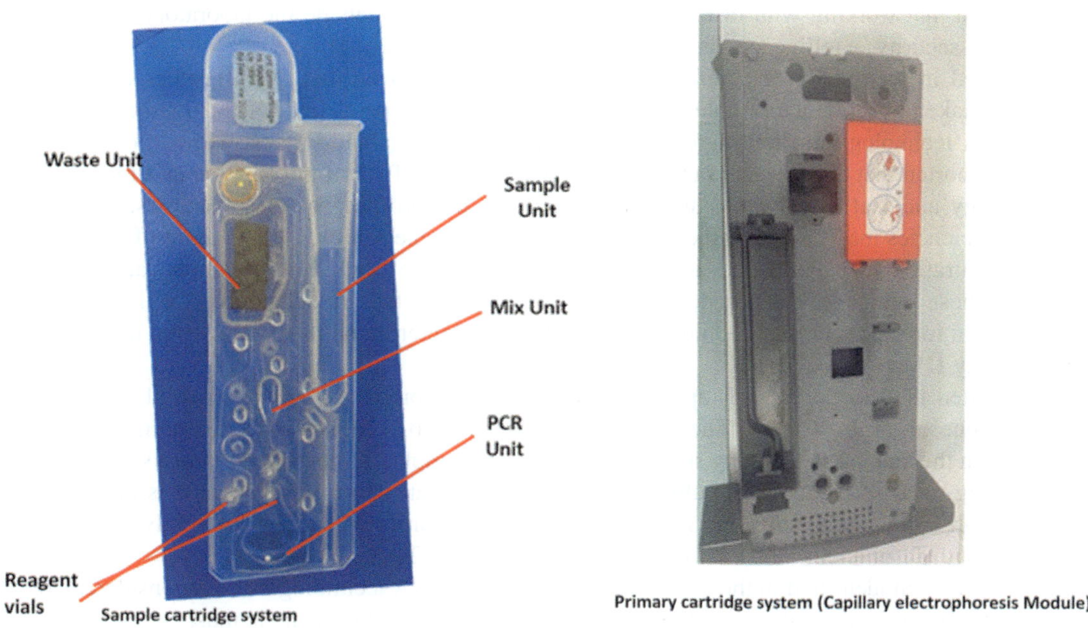

FIGURE 6.8 Internal mechanism of the RapidHIT ID System sample and primary cartridge system.

4. RapidLINK Staff Elimination Database

It provides a contamination alert from DNAs stored in the system against possible contamination of personnel or other persons (26)

6.3.5 Advantages and disadvantages of Rapid DNA technologies

Several commercial kits have been developed for the easy acquisition of STR profiles, which are used worldwide. Even with these ready-made kits, the process of obtaining profiles in routine studies is long and tedious. However, the time to reach the perpetrator is very important. Catching the perpetrator as soon as possible can prevent a second victimization in that process (Larue et al., 2014).

Rapid DNA technologies, developed in addition to the devices used routinely for this purpose, are fully automated DNA analyzers that allow us to obtain STR profiles. These are sophisticated systems that combine routine laboratory steps such as isolation, DNA quantification, polymerase chain reaction, capillary electrophoresis, and data analysis to generate DNA profiles in a short time (Cornelis, 2019; Hennessy et al., 2014; Holland & Wendt, 2015; Salceda et al., 2017).

ParaDNA is designed to screen and pre-analyze potential evidence or identify individuals within 75 minutes and can be used by non-specialists (e.g., police). However, it only covers 5 loci and is less discriminative than other Rapid DNA systems available on the market.

The DNAscan device allows direct analysis of samples in 90 minutes. However, with a net weight of over 50 kg, it is not considered a portable device. It is considered more suitable for use in a laboratory setting to produce results quickly without the need for human intervention, or during disasters where normal infrastructure is destroyed (Cornelis, 2019).

Compact and easy to use, the RapidHIT ID System provides a more innovative rapid DNA platform for generating forensic DNA profiles than other systems. With cartridges developed for reference and crime scene samples, STR DNA profiles can be generated in just 1 minute by breaking the swab into the cartridge and inserting the cartridge into the device after 90 minutes. Applied Biosystems RapidLINK software enables integration with CODIS for DNA matching, determination of family and kinship relationships, as well as personnel matches. In addition to the advantages of the RapidHIT system, which is still in production and use, such as ease, speed, and no need for user control, the disadvantages that have emerged and may emerge in the studies have been reported.

Larue et al. (2014) reported that for unsuccessful results obtained with rapid DNA, the swab can be removed from the cartridge and re-analyzed in this system or by conventional methods. However, they reported that this issue should be supported by additional studies, and in this case, the samples studied will not only cause a loss of time but also raise suspicion. They also stated that what is important in forensic sciences is to analyze the physical evidence obtained in the most accurate and shortest time (Larue et al., 2014).

Hennesey et al. While drawing attention to the use of DNA analysis outside the laboratory and the limitations of their research, they reported that working with fast DNA technology inside and outside the laboratory will contribute to law enforcement, national security, intelligence, and defense (Hennessy et al., 2014).

Hares et al. reported in their study that rapid DNA technology has proven that DNA profiles can be obtained from buccal swab samples taken as reference samples without the presence of experts, and for this reason, studies to prove the success of rapid DNA technology in profiling crime scene samples have gained interest. However, based on the results they obtained, they reported that crime scene samples could not provide sufficient information as a result of working with rapid DNA technology and that there were partial deficiencies in the profiling of these samples. They stated that the reasons for these deficiencies may be due to the differences in crime scene samples, the amount of crime scene samples, and the DNA concentration to be obtained accordingly, the age of the crime scene sample, and the environmental conditions to which these samples are exposed. In addition, they reported that mixed samples, which are very difficult to analyze and interpret in forensic sciences, should still be analyzed and interpreted by experts, even if they are studied with rapid DNA technology (Hares et al., 2020)

Kitayama et al. tested the success rate of rapid DNA technology by preparing different sample groups. One of these sample groups was buccal swab samples. The situation they wanted to compare is to determine the success rates of profiling depending on the amount of DNA by studying buccal swab samples, which are reference samples, with fast DNA technology. According to their results, they published their research that the profiling success is high as a result of the analysis of buccal swab samples with sufficient DNA amount, while the profiling success is slightly lower as a result of the analysis of samples with insufficient DNA amount (Kitayama et al., 2020).

Since problems have been identified in the analysis results of rapid DNA technologies, it is considered appropriate that the use of rapid DNA technologies alone in the paternity analysis is not sufficient; it must be supported by applying conventional DNA analysis steps and only reported by experts. In addition, while extracting products and PCR products of the same sample can be restudied in DNA analysis performed by conventional methods, this is not possible for the moment in rapid DNA technology. Because the biological sample taken is directly placed in the sample cartridge and is spent. In cases where the analysis needs to be repeated, the questionable reusability of the samples studied with rapid DNA technology is seen as another disadvantage of rapid DNA technology (Ünsal Sapan et al., 2023).

6.4 Conclusion

Reducing the time required for DNA analysis in forensic genetics can provide significant benefits both in determining ancestry relationships for legal proceedings and in criminal cases. The rapid identification of individuals plays a crucial role in solving crimes and apprehending perpetrators. Furthermore, in situations such as mass disasters, the rapid identification of victims is crucial.

Rapid DNA technologies, introduced to speed up DNA analysis, have been developed to streamline the process in a single step. This not only minimizes the risk of human error and contamination but also provides a significant time advantage by delivering 24 STR locus profiles in as little as 90 minutes.

However, despite the fact that they were launched to be able to provide results without an expert analyst due to the shortage of profiles, there is a general opinion that results should not be reported without an expert, and that the results should be confirmed with conventional systems when necessary.

References

Application note: Bone sample processing on the RapidHIT ID System with RapidINTEL cartridges. Thermo Fisher Scientific Inc, Thermo Fisher Scientific Inc. (2020).

Blackman, S., Dawnay, N., Ball, G., Stafford-Allen, B., Tribble, N., Rendell, P., Neary, K., Hanson, E. K., Ballantyne, J., Kallifatidis, B., Mendel, J., Mills, D. E. K., & Wells, S. (2015). Developmental validation of the ParaDNA® intelligence system - A novel approach to DNA profiling. *Forensic Science International: Genetics, 17*, 137–148. https://doi.org/10.1016/j.fsigen.2015.04.009, http://www.elsevier.com.

Buscaino, J., Barican, A., Farrales, L., Goldman, B., Klevenberg, J., Kuhn, M., Lin, F., Nguyen, P., Salceda, S., Schueren, R., Smith, C., Troup, C., Tsou, D., Vangbo, M., & King, D. (2018). Evaluation of a rapid DNA process with the RapidHIT® ID system using a specialized cartridge for extracted and quantified human DNA. *Forensic Science International: Genetics, 34*, 116–127. https://doi.org/10.1016/j.fsigen.2018.02.010, http://www.elsevier.com.

Butler, J. (2001). *Forensic DNA typing: biology and technology behind STR markers*. Academic Press.

Carney, C., Whitney, S., Vaidyanathan, J., Persick, R., Noel, F., Vallone, P. M., Romsos, E. L., Tan, E., Grover, R., Witkowski, R. S., French, J. L., & Selden, R. F. (2019). Developmental validation of the ANDE™ rapid DNA system with FlexPlex™ assay for arrestee and reference buccal swab processing and database searching. *Forensic Science International: Genetics, 40*, 120–130. https://doi.org/10.1016/j.fsigen.2019.02.016, http://www.elsevier.com.

Chan, L. (1992). Advances in molecular biology with applications in clinical medicine. *Klinisches Labor, 38*(1), 2–4.

Cornelis, S., Forensic Lab-On-A-Chip DNA analysis (Doctorate thesis). (2019).

Date-Chong, M., Hudlow, W. R., & Buoncristiani, M. R. (2016). Evaluation of the RapidHIT™ 200 and RapidHIT GlobalFiler® Express kit for fully automated STR genotyping. *Forensic Science International: Genetics, 23*, 1–8. https://doi.org/10.1016/j.fsigen.2016.03.001, http://www.elsevier.com.

Dawnay, N., Ahmed, R., & Naif, S. (2014). The ParaDNA® screening system — A case study in bringing forensic R&D to market. *Science & Justice, 54*(6), 481–486. https://doi.org/10.1016/j.scijus.2014.08.002.

Edwards, A., Hammond, H. A., Jin, L., Caskey, C. T., & Chakraborty, R. (1992). Genetic variation at five trimeric and tetrameric tandem repeat loci in four human population groups. *Genomics, 12*(2), 241–253. https://doi.org/10.1016/0888-7543(92)90371-X.

Goodwin, W., Linacre, A., & Hadi, S. (2007). *An introduction to forensic genetics*. John Wiley & Sons.

Hares, D. R., Kneppers, A., Onorato, A. J., & Kahn, S. (2020). Rapid DNA for crime scene use: Enhancements and data needed to consider use on forensic evidence for State and National DNA Databasing – An agreed position statement by ENFSI, SWGDAM and the Rapid DNA Crime Scene Technology Advancement Task Group. *Forensic Science International: Genetics, 48*. https://doi.org/10.1016/j.fsigen.2020.102349, http://www.elsevier.com.

Hennessy, L. K., Mehendale, N., Chear, K., Jovanovich, S., Williams, S., Park, C., & Gangano, S. (2014). Developmental validation of the GlobalFiler® express kit, a 24-marker STR assay, on the RapidHIT® System. *Forensic Science International: Genetics, 13*, 247–258. https://doi.org/10.1016/j.fsigen.2014.08.011, http://www.elsevier.com.

Holland, M., & Wendt, F. (2015). Evaluation of the RapidHIT™ 200, an automated human identification system for STR analysis of single source samples. *Forensic Science International: Genetics, 14*, 76–85. https://doi.org/10.1016/j.fsigen.2014.08.010.

Jeffreys, A. J., & Wilson, V. (1985). Hypervariable regions in human DNA. *Genetical Research, 45*, 213 213.

Jeffreys, A. J., Allen, M. J., Armour, J. A. L., Collick, A., Dubrova, Y., Fretwell, N., Guram, T., Jobling, M., May, C. A., Neil, D. L., & Neumann, R. (1995). Mutation processes at human minisatellites. *Electrophoresis, 16*(1), 1577–1585. https://doi.org/10.1002/elps.11501601261.

Kitayama, T., Fukagawa, T., Watahiki, H., Mita, Y., Fujii, K., Unuma, K., Sakurada, K., Uemura, K., Sekiguchi, K., & Mizuno, N. (2020). Evaluation of Rapid DNA system for buccal swab and disaster victim identification samples. *Legal Medicine, 46*, 101713. https://doi.org/10.1016/j.legalmed.2020.101713.

Larue, B. L., Moore, A., King, J. L., Marshall, P. L., & Budowle, B. (2014). An evaluation of the RapidHIT® system for reliably genotyping reference samples. *Forensic Science International: Genetics, 13*, 104–111. https://doi.org/10.1016/j.fsigen.2014.06.012, http://www.elsevier.com.

Li, M., Tao, R., Zhou, W., Li, Y., Meng, M., Zhang, Y., Yu, L., Chen, L., Bian, Y., & Li, C. (2019). Validation studies of the ParaDNA® intelligence system with artificial evidence items. *Forensic Science Research, 6*(1), 84–91.

Moreno, L. I., Brown, A. L., & Callaghan, T. F. (2017). Internal validation of the DNAscan/ANDE™ Rapid DNA Analysis™ platform and its associated PowerPlex® 16 high content DNA biochip cassette for use as an expert system with reference buccal swabs. *Forensic Science International: Genetics, 29*, 100–108. https://doi.org/10.1016/j.fsigen.2017.03.022, http://www.elsevier.com.

Murakami, C., Irie, W., Sasaki, C., Nakamaru, N., Sakamoto, M., Nagato, J., & Satoh, F. (2020). Individual identification using the RapidHIT™ ID system for forensic samples. *Legal Medicine, 47*, 101776. https://doi.org/10.1016/j.legalmed.2020.101776.

Rosenblum, B. B., Oaks, F., Menchen, S., & Johnson, B. (1997). Improved single-strand DNA sizing accuracy in capillary electrophoresis. *Nucleic Acids Research, 25*(19), 3925–3929. https://doi.org/10.1093/nar/25.19.3925.

Saiki, R. K., Scharf, S., Faloona, F., Mullis, K. B., Horn, G. T., Erlich, H. A., & Arnheim, N. (1985). Enzymatic amplification of β-globin genomic sequences and restriction site analysis for diagnosis of sickle cell anemia. *Science (New York, N.Y.), 230*(4732), 1350–1354. https://doi.org/10.1126/science.2999980.

Salceda, S., Barican, A., Buscaino, J., Goldman, B., Klevenberg, J., Kuhn, M., Lehto, D., Lin, F., Nguyen, P., Park, C., Pearson, F., Pittaro, R., Salodkar, S., Schueren, R., Smith, C., Troup, C., Tsou, D., Vangbo, M., Wunderle, J., & King, D. (2017). Validation of a rapid DNA process with the RapidHIT ® ID system using GlobalFiler ® Express chemistry, a platform optimized for decentralized testing environments. *Forensic Science International: Genetics, 28*, 21–34. https://doi.org/10.1016/j.fsigen.2017.01.005.

Stoneking, M., Hedgecock, D., Higuchi, R. G., Vigilant, L., & Erlich, H. A. (1991). Population variation of human mtDNA control region sequences detected by enzymatic amplification and sequence-specific oligonucleotide probes. *American Journal of Human Genetics, 48*(2), 370–382. http://www.elsevier.com/wps/find/journaldescription.cws_home/713561/description#description.

Tan, E., Turingan, R. S., Hogan, C., Vasantgadkar, S., Palombo, L., Schumm, J. W., & Selden, R. F. (2013). Fully integrated, fully automated generation of short tandem repeat profiles. *Investigative Genetics, 4*(1). https://doi.org/10.1186/2041-2223-4-16, http://www.investigativegenetics.com/content/4/1/16.

Sample Collection Field Reference Guide For the Thermo Fisher Scientific Applied Biosystems™ RapidHIT™ ID System. Thermo Fischer Scientific.. (2024). https://assets.thermofisher.com/TFS-Assets%2FGSD%2FReference-Materials%2Frhid-field-reference-nfstc-reference-guide.pdf.

Thermo Fisher Scientific *RapidLINK™ software v1.3.3 user guide*. https://assets.thermofisher.com/TFS-Assets/LSG/manuals/MAN0025746-RapidHITID-v1.3.3-ROW-UG.pdf.

Ünsal Sapan, T., Yağmur Kartal, N., Meherremli, Ş., Erdem Obut, M., Yilancioğlu, K., & Atasoy, S. (2023). Success rate of rapid DNA technology in kinship analysis for forensic purpose. *Forensic Science, Medicine and Pathology*. https://doi.org/10.1007/s12024-023-00735-x.

Weber, J. L., David, D., Heil, J., Fan, Y., Zhao, C., & Marth, G. (2002). Human diallelic insertion/deletion polymorphisms. *American Journal of Human Genetics, 71*(4), 854–862. https://doi.org/10.1086/342727, http://www.elsevier.com/wps/find/journaldescription.cws_home/713561/description#description.

Chapter 7

Rapidly mutating Y-short tandem repeats in forensic DNA analysis

Arash Alipour Tabrizi
Legal Medicine Research Center, Legal Medicine Organization, Tehran, Iran

7.1 Human Y-chromosome specifications and markers

Humans with X and Y-chromosomes within the cell's nucleus are phenotypically male. The Y-chromosome, typically found in males, triggers the cascade of male development, is essential for male fertility, and is inherited in its entirety from father to son (Court, 2021; Diegoli, 2015).

Molecular, biological, and cytogenetic studies on the human Y-chromosome give us an insight into the presence of the biggest nonrecombining block in the human genome, along with many structural variants within the Y-chromosome, which might be deletions, duplications, and inversions. The Y-chromosome's complex structure, including long palindromes and tandem repeats, makes it difficult to properly assemble. It is the latest human chromosome to be entirely sequenced (Adnan et al., 2021; Li et al., 2020; Rhie et al., 2023).

Furthermore, a mere 5% of the human Y-chromosome consists of three pseudo-autosomal regions (PARs), namely PAR1 (Yp), PAR2 (Yq) and PAR3. The PARs are regions of chromatin that include genes PAR1 (2.6 Mb) and PAR2 (0.32 Mb). These regions exhibit a dense clustering of genes and experience genetic recombination with comparable segments of the X chromosome during the process of meiosis.

Approximately, 2% of the general population has PAR3 (2.3 Mb) in length, which was first elaborated in 2013. Surprisingly, PAR3 and some parts of the X-chromosome (Xq21.3) share more than 98% sequence homology (Court, 2021; Dash et al., n.d.).

Hence, the remaining 95% consists of nonrecombining sections of the Y-chromosome (NRY) and 78 functional protein-coding sequences that encode 27 distinct proteins. The male-specific region (MSY) encodes the developmental factors of the testes. It is composed of repetitive sequences that are physically linked together and do not undergo recombination. This region serves as a lineage marker as it contains a single haplotype that is passed down directly along the paternal lineage.

While the Y-chromosome can tolerate significant deletions without causing major health issues, the loss of the Y-chromosome (LOY) can usually be detected in the blood samples of males, particularly in healthy males over the age of 16. LOY can have an impact on the outcomes of forensic investigations, so it is advisable to take into account the potential effects of LOY on forensic analyses.

NRY is extensively utilized in forensic genetics and population genetic research because of its male inheritance pattern. Due to its vulnerability to genetic drift caused by a small effective population size and cultural practices that involve mating within the same patrilineal line, the NRY can be effectively used to study genetic differentiation and population structure across different geographic regions. It can provide valuable insights into human population history by considering demographic factors. Additionally, the NRY can be utilized in forensic applications such as paternity testing, kinship analysis, and familial searching to determine male lineage and establish genealogical relationships (Ballantyne et al., 2010; Court, 2021; Dash et al., n.d.; Roewer et al., 2020).

As a result of the MSY, Y-haplotypes are predominantly inherited without significant changes from one generation to the next, except for the occasional alterations produced by mutations. The predominant forensic markers utilized on the Y-chromosome are short tandem repeats (STRs) and single-nucleotide polymorphisms (SNPs). Given their low mutation

Investigative and Predictive DNA Testing. DOI: https://doi.org/10.1016/B978-0-443-30128-5.00011-4

rate of 10^{-9} substitutions per site per year, distinctive differences in human evolution are frequently caused by SNPs and several small insertions/deletions (indels) (Solé-Morata et al., 2017).

For the past 25 years, STR analysis has been extensively utilized as the most reliable method for individualizing and identifying caseworks in forensic genetics. In addition to autosomal STRs, the male human genome also has several hundred Y-STRs that are situated on the MSY region of the Y-chromosome. Forensic investigators now have access to more advanced and modern tools thanks to commercial Y-STR kits that include more markers (Ralf et al., 2020; Xiong et al., 2022).

The X and Y chromosomes involve homologous genes that encode the protein amelogenin, which is essential for the demineralization of tooth enamel and is used as a sex marker in forensic DNA kits. Commercial autosomal STR typing kits have been updated to include Y-STR markers, such as DYS391, which is located on the long arm of the Y-chromosome. These markers are included due to deletion and point mutation events that result in the Null-Amelogenin phenomenon. Additionally, one insertion–deletion locus (rs2032678 locus) is also included in the updated kits.

The significance of the Y-chromosome in forensic genetics has been undervalued in comparison to autosomal markers (Court, 2021).

Y-STRs are repetitive regions characterized by a repetition motif of 2–13 base pairs, with a recurrence ranging from 7 to 40. The variability in the number of short tandem repetitions among individuals enhances the efficacy of these markers for human identification. These polymorphisms are typically created through the process of strand-slippage replication, with DNA polymerase slippage most frequently resulting in the generation of one repeat unit difference.

The advantages of Y-STRs are as follows:

- A low quantity of DNA is sufficient, making it acceptable for a degraded sample.
- Results that meet the criteria include samples with a high proportion of feminine DNA and a low proportion of male DNA.

The results are promptly obtained within a time frame of 1–2 working days (Jankauskiene et al., 2017).

Lutz Roewer identified the first human Y-STR 32 years ago, which was promptly utilized in a forensic context (Ralf et al., 2023).

The minimal haplotype (MHT) set consists of nine Y-loci, including DYS19, DYS389I, DYS389II, DYS390, DYS391, DYS392, DYS393, and DYS385a, b. These Y-STRs were initially suggested by Kayser et al. for their use in forensic science and paternity testing. The selection of these Y-STRs was in view of the general criteria proposed by international working groups such as ISFG[1] and SWGDAM[2]. The aim was to enhance the overall discrimination capacity (DC) (Puch-Solis et al., 2024; Soldati et al., 2022).

The primary factor contributing to the variability in the MSY is the accumulation of novel mutations, especially SNPs and STRs (D'Atanasio et al., 2019).

In contrast to Y-chromosomal SNPs (Y-SNPs), which have a very low mutation rate of approximately 2×10^{-8} per base per generation, Y-STRs usually represent individual haplotypes and have substantially greater mutation rates (around 10^{-3} per locus per generation on average). Furthermore, Y-STR mutation rates exhibit significant diversity, spanning from approximately 10^{-2} to 10^{-4} per locus per generation (Boattini et al., 2019).

The population's demographic background significantly affects the diversity of a marker or haplotype. Y-STRs are quite effective in one group of people, but their usefulness is restricted when applied to populations with distinct demographic attributes (Ullah et al., 2022).

In general, highly polymorphic loci are typically used for forensic purposes. However, when it comes to paternity testing, markers with a high mutation rate need to be selected with caution. Hence, it is imperative to estimate reliable mutation rates to effectively utilize Y-STRs and ensure exact interpretation of genetic profiles.

For optimal results in forensic cases requiring Y-loci analysis, a panel of Y-STRs should consist of markers with a gene diversity (GD) value of at least 0.70. This ensures a high level of accuracy and precision in individual identification, complementing the findings obtained from autosomal STRs. A multiplex set's discriminating capacity (DC) has been improved by gradually increasing the number of Y-STRs, as not all markers in the kits achieve the ideal GD value (Eskandarion et al., 2024; Soldati et al., 2022).

Several commercially available Y-STR multiplex systems have been developed as reliable tools for resolving various forensic cases (as seen in Table 7.1). Available kits like STRtyper-27Y (Health Gene Technologies, China), the

---

1. International Society for Forensic Genetics.

2. Scientific Working Group on DNA Analysis Methods.

TABLE 7.1 Y-chromosomal short tandem repeats (Y-STRs) and rapidly mutating Y-chromosomal short tandem repeats (RM Y-STRs) included in commercially available multiplex systems.

Kit/loci/repeated motif	Minimal haplotype	AmpFlSTR Yfiler	PowerPlex Y23 System	AmpFlSTR Yfiler Plus	AmpFlSTR Yfiler Platinum
Tetra (TAGA)	DYS19	DYS19	DYS19	DYS19	DYS19
Tetra (TCTG)(TCTA)	DYS389I	DYS389I	DYS389I	DYS389I	DYS389I
Tetra (TCTG)(TCTA)	DYS389II	DYS389II	DYS389II	DYS389II	DYS389II
Tetra (TCTA)(TCTG)	DYS390	DYS390	DYS390	DYS390	DYS390
Tetra (TCTA)	DYS391	DYS391	DYS391	DYS391	DYS391
Tetra (GAAA)	DYS385a/b	DYS385 a/b	DYS385 a/b	DYS385 a/b	DYS385 a/b
Three (TAT)	DYS392	DYS392	DYS392	DYS392	DYS392
Tetra (AGAT)	DYS393	DYS393	DYS393	DYS393	DYS393
Tetra (TCTR) compound		DYS437	DYS437	DYS437	DYS437
Penta (TTTTC)		DYS438	DYS438	DYS438	DYS438
Tetra (AGAT)		DYS439	DYS439	DYS439	DYS439
Tetra (AGAT)		DYS456	DYS456	DYS456	DYS456
Tetra (GAAA)		DYS458	DYS458	DYS458	DYS458
Hexa (AGAGAT)		DYS448	DYS448	DYS448	DYS448
Tetra (TSTA) compound		DYS635	DYS635	DYS635	DYS635
Tetra (TAGA)		Y-GATA-H4	Y-GATA-H4	Y-GATA-H4	Y-GATA-H4
Three (CTT)			DYS481	DYS481	DYS481
Tetra (ATCT)			DYS533	DYS533	DYS533
Tetra (GATA)			DYS549		DYS549
RM Y-STR—tetra (TTTC) simple			DYS570	DYS570	DYS570
RM Y-STR—etra (AAAG) simple			DYS576	DYS576	DYS576
Penta (CTTTT)			DYS643		DYS643
RM Y-STR—tetra (AAAG) complex				DYS627	DYS627
RM Y-STR—tetra (AAAG) complex				DYS518	DYS518
RM Y-STR—tetra (TTTC) complex				DYS449	DYS449
RM Y-STR—tetra (RAAG) complex and multicopy				DYF387S1a/b	DYF387S1a/b
Tetra (ATAG)				DYS460	DYS460
Penta AAAC/AAAT					DYS593
Tetra TTTC					DYS557
Tetra GATA					DYS522
Tetra TAGA					DYS444
Tetra GGAA					DYS527a/b
Hexa [GAA][GGAGAA]					DYS596
Penta GTTTT					DYS645
Penta TAAWA					DYS447

PowerPlex Y23 System (Promega, USA), and the Yfiler Plus PCR Amplification Kit (Thermo Fisher Scientific, USA) usually contain 16–27 markers. Forensic genetics analyses and population studies have both confirmed the usefulness of all of these methods (Albastaki et al., 2022).

The Yfiler Plus PCR Amplification kit, manufactured by Thermo Fisher Scientific, USA, utilizes six-dye technology to amplify 27 Y-STR loci, which include seven rapidly mutating markers (RM Y-STRs) and 11 mini-Y-STRs(Jankauskiene et al., 2017).

Thermo Fisher Scientific has released a new Y-STR kit known as the Yfiler Platinum PCR Amplification Kit, which contains 11 new Y-STR loci than its predecessor, the Yfiler Plus kit. The kit amplifies 41 Y-chromosome markers, including 38 Y-STRs and 3 insertion/deletion polymorphisms (rs771783753, rs759551978, rs199815934). This system's discriminatory capability was greatly increased by amplifying a large number of loci, beyond other kits on the market (Albastaki et al., 2022; Lee et al., 2023).

7.2 Application of Y-chromosomal short tandem repeats in forensic DNA typing

For more than three decades, the uniparental nature of Y-STRs has consistently demonstrated its effectiveness in distinguishing male DNA from mixed stains in sexual assault cases, without interference from the DNA of female victims. This has made it a valuable tool in forensic genetics. In addition, if a convicted felon's Y-STR profile does not match the crime scene sample, he will be exonerated. It has facilitated the tracking of paternal ancestry and the examination of population admixture history, in cases where the analysis of autosomal DNA is not effective. Additionally, they have a significant impact in cases involving missing individuals, identifying victims of disasters, and assessing patrilineal relationships in complex kinship research. If an autosomal DNA match is not found, the Y-STR haplotype can still be utilized for familial searches in forensic cases because it is possessed by both close and distant male relatives. Moreover, they are also especially important tools for population inference and inferring ancient human migration trajectories and timing (Court, 2021; Fan et al., 2021; Li et al., 2020; Wang et al., 2022; Ye et al., 2022).

Beginning as early as 12 weeks of gestation, Y-STR analysis has shown its usefulness for noninvasively fetal sex determination and fetal male paternity analysis using maternal plasma DNA (Diegoli, 2015).

Y-STRs have also been used to look at male ancestry. Clusters of data representing Y-haplotypes are a nonrandom distribution caused by the linear transmission pattern of haplotypes. These groups are associated with the population's ethno-linguistic composition, specifically through paternal lineages. Even though there was archeological evidence that pointed to a patrilineal relationship, a male individual was excluded from the old Königsfelder paternal lineage using four-marker Y-STR haplotypes.

In recent times, Y-STR haplotypes have been associated with surnames in countries where descent is traced through the male line. They are utilized to assess the degree to which male lineages have a shared origin. In forensic casework, there have been suggestions that the Y-haplotype in question could potentially predict the surname of a suspect and offer an investigative lead in cases where autosomal typing has failed to do so (Diegoli, 2015; Roewer et al., 2020).

Determining the male contributor with autosomal STR profiling is particularly challenging because of the uneven proportions of male and female DNA in mixtures. This is especially true when multiple males have contributed to the DNA mixture obtained from vaginal or rectal swabs. Even after attempting to separate male sperm cells from female epithelial cells through differential lysis, there are often insufficient male cells to generate a useful profile using this method. In cases where differential extraction is unable to differentiate between male and female cells, the best way to collect genetic data is to target Y-STRs. These instances could involve situations where the offender is a male who does not have sperm in their semen (azoospermia), cases where a female has been touched or penetrated by someone and nonsperm DNA has been left behind in small amounts, or instances where sexual assault stains have degraded over time. Physical evidence can only be recovered in situations of sexual assault or rape if the crime is reported promptly after it happens. However, the time frame for recovering such evidence has been extended by utilizing Y-STR profiling. Additionally, Y-STRs can be beneficial in cases where there is a mix of male and female DNA, particularly in situations where the male offender's saliva or epithelial cells are relevant. This can include examining bite marks in sexual assault or collecting fingernail scrapings from the victim afterwards.

Depending on the total amount of male DNA present, currently available Y-STR multiplexes can detect male DNA in male: female mixtures at ratios as low as 1:24,000. Furthermore, with only 62.5 pg. of male DNA, these multiplexes may produce complete Y-haplotypes. The analysis of sexual assault mixes involving several donors has been made possible by recent advancements in producing Y-STR profiles from single sperm cells collected using laser microdissection (Court, 2021; Diegoli, 2015; Puch-Solis et al., 2024; Ralf et al., 2020).

When the father's reference sample is not available, Y-STRs can help resolve paternity scenarios and explore paternal lineage. Whenever this is the case, it is possible to establish relatedness by comparing the results to those of a paternal male relative. Y-STRs are especially valuable for excluding paternity, making them a valuable tool in resolving cases about missing persons or bodies in mass graves. Under such conditions, it is advantageous to examine a substantial number of Y-STRs while refraining from analyzing the Y-STRs that undergo mutations at a faster rate (Court, 2021; Diegoli, 2015; Gopinath et al., 2016; Javed et al., 2022).

In 2021, Bini et al. utilized forensic Y-STR haplotypes and archaeogenetics methodologies in conjunction with historical, archaeological, and anthropological data to examine population patterns and relationships in Roccapelago, a tiny community situated in the Northern Central Apennines. A total of nearly 400 partially mummified individuals, dating from the late 16th to the 18th century, were found in the church's crypt. The Yfiler Plus's capacity to differentiate between closely related males is particularly relevant in cases of mass disasters where multiple victims from the same family are buried together. This is important due to the potential for false exclusions caused by mutations when comparing the DNA of deceased individuals to that of their living relatives in kinship testing.

The molecular research demonstrated the utility of the Y-chromosome in identifying historically significant remains and uncovering patterns of kinship within societies (Bini et al., 2021).

Y-STRs polymorphisms that evolve rapidly, with an average mutation rate of around 10^{-3} per locus per generation, have been found to be valuable for assessing paternal ties within short to medium timeframes. These polymorphisms allow for the identification of common paternal ancestry among male lineages (Ballantyne et al., 2014).

Due to the limited amount of DNA in forensic DNA samples, the enhanced sensitivity of Yfiler Plus, which is at least 10–20 times more sensitive than Yfiler, guarantees the generation of the highest possible allele information for both inclusionary and exclusionary objectives. The enhanced sensitivity may further facilitate the identification of minor contributors in mixed male samples and the generation of profiles for additional case scenarios where Yfiler has not demonstrated efficacy, such as digital rapes. The inclusion of extra loci in Yfiler Plus may lead to a higher haplotype diversity (HD), which leads to lower match probabilities when compared to Yfiler haplotypes. For exclusionary purposes, the increased number of haplotype data generated by the additional loci will be useful, especially when dealing with low-level or partial profiles. The presence of large background levels (1000-fold) of female DNA slightly inhibits the intensity of the male profile obtained with Yfiler. The Yfiler Plus does not exhibit the same phenomenon, making it more capable of detecting small quantities of male DNA in a mixed sample with much higher yields of female DNA. The Yfiler Plus demonstrates exceptional performance even when dealing with nonprobative casework materials that are impacted by inhibitors and DNA degradation. Yfiler Plus is more effective than Yfiler in determining the accurate number of contributors in a mixed profile due to the presence of more loci. This is because the likelihood of having alleles that are not shared among individuals is higher (Henry et al., 2015).

Y-STRs are also used in population genetics and genetic anthropological research for a number of purposes, with an emphasis on unrelated males. Y-STRs can offer insights into the geographical and chronological origins and dispersal of Y-SNP-derived haplogroups. The utilization of Y-STRs with varying mutation rates enables the examination of events that took place at different temporal scales.

In 2024, Ralf et al. developed the Male Pedigree Toolbox (MPT), a software tool designed to assist with Y-STR investigation, including forensics. This software enables automated analysis of Y-STR data in pedigree and population datasets, specifically focusing on patrilineal genealogical relationships. The MPT utilizes Y-STR pedigree data to predict mutation rates and male relative differentiation rates. It can help identify ancestral haplotypes in a pedigree and visualize genetic diversity (GD) across all branches of a family tree. Additionally, it can use machine learning to provide probabilistic classifications, aiding in the determination or verification of the pedigree's structure and the degree of relatedness among males, even in cases when individuals are closely related and possess substantially similar haplotypes. The MPT is well-suited for intricate pedigrees, particularly those with a large number of individuals, and is capable of conducting these studies rapidly and methodically. Forensic cases and genealogy benefit greatly from the program's prediction-based features when unclear information about a crime scene stain's donor or the pedigree structure is available. It should be noted that the software is not without limitations. For example, the system lacks the capability to handle samples that have incomplete data. Hence, in partial Y-haplotypes, it is necessary to eliminate either the particular sample itself or the loci with missing data before proceeding with the analysis. Two comparable Y-STR haplotypes may be observed in forensic casework. If one haplotype is inherited from a known individual but the other's identity is unknown, this comparison could indicate further investigation. If Y-STRs with high mutation rates match, it suggests a potential close relationship between the two males.

Various techniques have been suggested to estimate the time since the most recent common ancestor (tMRCA) using Y-STR haplotypes. A unique approach employing simulation and machine learning has been suggested to ascertain the link between the donors of two comparable Y-STR haplotypes. The method's precision and accuracy have been properly proven in previous studies. MPT holds significant promise for widespread use in the domains of forensic and genealogical genetics, as well as in the integration of both for patrilineal investigative genetic genealogy (Ralf et al., 2024).

The ISFG strongly recommends preserving any material for Y-STR typing when a mixture of male and female DNA is expected. An instance where Y-STR testing is required is when a vaginal swab is analyzed to detect the presence of seminal fluid, but no sperm cells are found (Roewer et al., 2020).

The use of Y-STRs in routine case analysis is limited mainly because they cannot differentiate between an individual and their lineage, and efforts to complete a separate Y-haplotype analysis might result in valuable case material.

7.3 How to interpret Y-chromosomal short tandem repeats

Since their discovery in 1990, Y-STRs have been employed to generate unique haplotypes of different paternal lineages (Wang et al., 2022).

Estimating haplotype frequencies necessitates a reference database that accurately reflects the population substructures, as well as a statistical model that examines the target haplotype in relation to the dominant subpopulation clusters. The United States and Germany have implemented national norms that adhere to these concepts. Variations in the rules might be observed in the selection of the subpopulations. Germany routinely suggests the YHRD-embedded Western European metapopulation, even though the SWGDAM guidelines advocate utilizing the US database containing subpopulations (Roewer et al., 2020).

In 2024, Puch-Solis et al. suggested guidelines for utilizing a mutation model in the evaluation of Y-STRs. This study focuses on a crucial step necessary for integrating a mutation model into a Y-STR interpretation approach. This integration aims to enhance the utilization of Y-STRs in criminal investigations. We chose a mutation model that takes into account not only the mutation rate but also considers multistep mutations and the preservation of the allele range. This allows us to better represent the observed mutation data. To enhance the accuracy of mutation models that accurately represent the mutation patterns of Y-STRs, researchers should not only report the quantity of detected mutations in a familial study, but also include information on the number of steps involved in each mutation and, ideally, the size of both mutant and nonmutated alleles. Considering allele length is crucial in the development of mutation models that incorporate mutation rate as a variable dependent on allele length (Puch-Solis et al., 2024).

Based on this understanding, specific instructions were developed for the use of Y-STR markers in circumstances when there is a discrepancy in the profiles. Considering the generally observed mutation rates of around 10^{-3} and the occurrence of two mutations at different Y-STR markers of a father and his son's Y-haplotype, it is recommended to use a threshold of three or more differences to exclude relatedness between two profiles. It's conceivable for related individuals separated by only one generation to have one or two differences as a effect of random chance (Diegoli, 2015).

The assumption of independence for markers in the NRY lacks biological explanation and is considered erroneous. Finding the frequency of complete Y-haplotypes in the absence of factorization is necessary for calculating match probability. A possible solution could apply the "count estimate" of the haplotype frequency in a reference database to calculate the match probability. To do this, divide the one plus total number of haplotype observations in the database by the total size of the database plus one (Caliebe et al., 2015).

Roewer provided a summary of Y-STR interpretation methods, emphasizing the drawbacks of counting methods for being excessively conservative and the unsuitability of the discrete Laplace (DL) method for loci with multiple copies and profiles that have null or intermediate alleles, which goes against the stepwise mutation model (Puch-Solis et al., 2024).

To accurately interpret Y-chromosomal data in a real case, it is necessary to have knowledge of the level of paternal relatedness within the suspect population and their familial connections (Roewer et al., 2020).

7.4 Mutations on Y-chromosome

Generally, Y-STRs have a similar possibility of mutating as autosomal STRs. However, they are specifically chosen based on considerable diversity measurements. The increased frequency of mutations between dads and sons is presumably in response to the enormous number of cell divisions during spermatogenesis. As a result, Y-STR haplotypes will be more variable than single autosomal STRs, and to effectively estimate frequencies, Y-STR haplotype reference databases must be significantly larger than autosomal STR allele reference databases.

Mutations exhibit nonrandom distribution across the chromosome, with certain areas being more susceptible to mutational events than others, which undergo more frequent mutations on average (Court, 2021).

The unique GD of the MSY has been primarily influenced by the absence of crossing over, haploidy, and male-specificity. In the absence of crossing over, it is generally observed that alleles at various polymorphic loci on MSY are typically inherited together, unless there are recurring mutations (Iacovacci et al., 2017).

Evolutionary conclusions made within timeframes of tens to hundreds of generations heavily rely on the precision of the Y-STR mutation rate estimations that are typically employed. Moreover, in forensic scenarios where Y-STRs are used for purposes like paternity testing, particularly in situations involving male children and deceased putative fathers, it is crucial to have a precise understanding of the variability of the Y-STRs being used to establish reliable paternity probability. Such knowledge is crucial in genealogical research that seeks to determine the connection between potentially closely or distantly related males (Ballantyne et al., 2010).

Additional forms of mutations include alterations in the flanking area, where the length of the repeat varies without affecting the sequence. For example, the electropherogram showed changes in mobility due to mutations upstream of DYS481, which resulted in improper off-ladder allele reporting. This suggests that any alteration in the area surrounding this locus could affect the secondary structure, resulting in a change in mobility. There are also mutations in the primer binding site, as well as duplications and deletions of large sections within a region. Furthermore, single-nucleotide insertions and deletions can occur (Court, 2021; Puch-Solis et al., 2024).

Multiple branches of human genetics, such as evolutionary genetics and population history, rely on Y-chromosome DNA variants to fill in the gaps in our understanding of male genetic structure. The aims of the investigation involve the modeling of regional and worldwide human evolution, the mapping of global migratory trends, and the continuous monitoring of cultural and demographic aspects such as polygyny, endogamy, patrilocality, and extrapair paternity. Furthermore, Y-chromosome DNA analysis is an essential tool for genetic genealogy, community genetic projects, and personalized family tree investigations. Besides, it is used for legal and investigative purposes to determine paternal genetic history and determine male lineages (Albastaki et al., 2022).

7.5 Factors influenced mutation

One of the initial factors that was found to affect mutation rates of microsatellites was the length of the repeat, which showed a significant inverse correlation. The impact of repeat length on the mutability of Y-STRs was seen, although this link was somewhat unclear due to variations in the numbers of each repeat class, particularly due to the abundance of tetranucleotide Y-STRs in the dataset. It has long been believed that the probability of a slippage mutation is mainly affected by the number of repeats in a specific allele. Kayser et al. investigated the impact of adjacent repetitions on the variability of Y-STR loci. They discovered a stronger correlation with the length of the longest homogeneous sequence. We were able to demonstrate that faulty repeat stretches also have a significant impact on the variability of a certain Y-STR. The maximum estimate of the effect size (78%) underscores the parameter's strong impact on Y-STR mutability, and similar effects are expected for STRs in general.

When analyzing the total number of repeats inside a single Y-STR region, there is a clear positive association between motif complexity and mutation rate. Although the Y-STRs did not demonstrate a significant impact of the sequence motif on the mutation rate, motifs with asymmetries between purine and pyrimidine bases (such as AAAG, AGAT, and AAGG) showed increased variability and diversity. This suggests that these motifs may contribute to the increase of repeat lengths. By thoroughly examining the characteristics of these loci that undergo rapid mutations, such as having short repeat lengths, a great number of repeats, and high complexity, it may be possible to identify autosomal RM STRs. These can be used in conjunction with the RM Y-STRs for various applications in the future. As a result, it appears that point mutations play a critical role in maintaining repeat lengths throughout extended periods of evolution, as indicated by the great complexity of many STRs. However, it is worth noting that point mutations occur at a far slower pace (by at least 10 times or more) compared to the slippage bias mechanism. Thus, it is acceptable to assume that the exceptional mutational properties of the RM Y-STRs, as discussed for Europeans, are probably applicable to other groups as well. However, further research on mutation rates in different groups will be necessary in the future. In 2010, researchers looked at a larger set of 186 tri-, tetra-, penta-, and hexanucleotide Y-STR markers and calculated the mutation rates. An analysis was conducted using a specific group of 13 markers that have higher mutation rates compared to the usual Y-STRs used in forensic investigations (i.e., greater than 10^{-3}). The results showed that there was not a significant difference between the number of repeat losses and gains (ratio of 1.16:1), but there was a significant difference between the number of single-repeat changes and multirepeat changes (ratio of 25.23:1). The variability of Y-STR loci is mostly regulated by the overall quantity of repetitions. Nevertheless, Y-STR mutability is influenced by other factors, including the complexity of repeats, the length of the repeating motif in base pairs, and the age of the father (Ballantyne et al., 2010).

7.6 Classification of Y-chromosomal short tandem repeats

7.6.1 Simple versus multicopy

A simple Y-STR is composed of a singular, distinctive motif that repeats. Since the normal human karyotype contains only one Y-chromosome, the majority of Y-STRs exhibit a single-copy trait. Conversely, certain Y-STRs generate multiple alleles because of the presence of multiple copies of their unique sequences. Traditionally, DYS numbers are allocated to markers that exist in just one copy, while DYF numbers are employed for markers that have many copies.

For example, DYS385 is a locus that is always duplicated and consists of two linked sub-loci with a GAAA as a repeated motif in dissimilar positions. The two alleles, referred to as "a" for the smaller one and "b" for the larger one, often have different sizes. Employing only one set of primers amplifies both alleles and treats them as a single haplotype.

Y-STR variants in multicopy Y-STR loci, like DYS385 and DYF387S1, are more prevalent than those observed in commonly utilized autosomal STRs (Court, 2021).

Y-STR loci that contain multiple copies (DYF403S1 has four copies, DYF399S1, DYF404S1, DYF387S1, DYS385, DYS527, and DYS459 each hold two copies) are particularly linked to palindromic sequences in the Y-chromosome's q11.23 band (Baeta et al., 2018).

To enhance discrimination power, it is currently favored to incorporate new genetic markers that possess a wide range of alleles. Moreover, multicopy Y-STRs exhibit considerable promise in offering a substantial degree of GD given to the excessive number of copies found in the Y-chromosome. Developing a new multiplex that combines informative markers could prove beneficial in addressing the challenges associated with familial searching in practical applications. According to published genetic data, all of the multicopy loci we utilized exhibited GD values higher than 0.7, with the exception of DYF383S1 (average GD value: 0.601) and DYS389I/II (average GD values: 0.582, 0.714). Among the multicopy loci, seven had GD values exceeding 0.8, while two (DYS464, DYS385) surpassed 0.9 (Shang et al., 2021).

The DYF403S1 has a complex sequence structure and consists of four copies. Among the four copies, three are classified as subtype a. When these three copies were amplified in the multiplex system, their fragment sizes overlapped during capillary electrophoresis (CE) (Lee et al., 2017).

In addition to several benefits of utilizing multicopy Y-STRs, certain drawbacks arise in some forensic scenarios. When faced with a mixture of evidence, multicopy Y-STRs are not the preferred markers for accurately determining the number of male perpetrators.

Interpreting the allele dosages of equal-length alleles posed difficulties when dealing with poor quality or low quantities of DNA in forensic samples. Furthermore, when determining genealogical ties, it is advantageous to exclusively utilize data from single-copy markers, because the data displayed by multicopy markers is often not traceable to their ancestral condition (Rogalla et al., 2015).

7.6.2 Slow mutating versus rapidly mutating

The Y-STRs frequently employed in forensic investigations and casework analysis exhibit slow to moderate mutation rates. Consequently, these collections of markers are incapable of distinguishing individuals belonging to the same patrilineage (Alghafri et al., 2018).

Holandric inheritance and the nonrecombining nature of the Y-chromosome result in the transmission of an identical and unchanged Y-STR haplotype from father to son, with the exception of rare mutational occurrences. Due to the greater number of multiplication events in male gametes compared to female egg cells, it is more likely to find mutational events in father–son pairs than in mother–son pairs (Boattini et al., 2019; Court, 2021).

On the basis of their mutation rates, Y-STRs can be classified into three distinct groups: those that mutate slowly (slowly mutation [SM], lower than 1.0×10^{-4}), those that mutate moderately (moderately mutating [MM], ranging from 1.0×10^{-4} to 1.0×10^{-2}), and those that mutate rapidly (rapidly mutating [RM], greater than 1.0×10^{-2}).

According to an alternative categorization, Y-STRs can be put into four groups based on the rates at which they undergo mutations:

SM with a mutation rate (μ) less than 0.002, MM with a mutation rate between 0.002 and 0.004, fast-mutating (FM) variant has a mutation rate (μ) between 0.004 and 0.010, and RM with an estimated rate of $0.010 \leq \mu$ (Xiong et al., 2022).

In 2021, Ralf and colleagues presented RMplex, a novel multiplex test that effectively examines 30 Y-STRs with elevated mutation rates. This assay integrates the techniques of CE and massively parallel sequencing (MPS). The test includes all 26 currently identified RM Y-STRs as well as 4 FM Y-STRs and focuses on a total of 44 genetic locations that are distinct to males. The examination of these 30 Y-STRs with the recently created multiplex assays effectively separated about 60% of the father–son pairs, more than 80% of the brother pairs, and over 75% of the grandfather–grandson pairs (Ralf et al., 2021).

Enhancing the differentiation between closely related males using RMplex analysis was the focus of a study carried out by Neuhuber et al. (2022). The earlier recommendation for classifying the 49 Y-STRs into four groups was reevaluated using the updated consensus mutation rates.

They identified one Y-STR (DYS1013) as an FM Y-STR out of the 24 RM Y-STRs. DYS403S1b, DYS626, and DYS570, previously categorized as FM Y-STRs, were identified as RM Y-STRs, while DYS389II was classified as an MM Y-STR. Out of the 13 previously classified as MMY-STRs, DYS460 and DYS389I are now classified as FM-Y-STRs. The three remaining Y-STRs, namely DYS448, DYS392, and DYS438, were categorized as (SM) Y-STRs. According to this updated classification, RMplex consists of 24 RM Y-STRs, which account for 80% of the total, and 6 FM Y-STRs, which make up 20%. On the other hand, Yfiler Plus comprises 5 RM Y-STRs (20%), 4 FM Y-STRs (16%), 13 MM Y-STRs (52%), and 3 SM Y-STRs (12%).

Yu Na Oh and colleagues in 2014 conducted a study on haplotype and mutation analysis of recently proposed Y-STRs in Korean father–son pairs. The findings revealed that Koreans exhibited significantly lower mutation rates for the DYF387S1, DYS518, and DYS570 loci compared to other groups, where these loci have mutation rates exceeding 1%.

The mutation rates at most genetic locations did not show significant differences compared to rates observed in other groups. However, the mutation rates for DYF387S1, DYS518, and DYS570 were notably lower in the Korean population compared to other populations (Oh et al., 2015).

The observed variations in mutation rates could result in the reclassification of certain Y-STRs in different populations. For instance, based on European data, Y-STRs such as DYF387S1, DYS403S1b, and DYS570 would be categorized as RM Y-STRs; however, Asian data indicates that they would be classified as FM Y-STRs. Additional data from many major populations is required to have a more comprehensive understanding of the impact of population on Y-STR mutation rates. The utilization of Yfiler Plus resulted in a father–son distinction rate of 13.4%. The rate of differentiation is more than double the prior rate achieved for the predecessor Yfiler kit. Significantly, RMplex demonstrated superior performance compared to Yfiler Plus, with an estimated rate of discrimination between fathers and sons of 41.9%. Out of the total 209 father–son pairs analyzed using RMplex, 75.1% were found to have just one mutated Y-STR marker, 20.6% had mutations at two markers, and 4.3% had mutations at three markers. This finding may be influenced by the impacts of sample size, as the more Y-STRs analyzed, particularly in the case of RM Y-STRs, the bigger population sample size required to achieve accurate diversity estimates (Neuhuber et al., 2022).

Currently, commonly used commercially available Y-STR kits predominantly consist of MM and RM Y-STRs. The SM Y-STRs exhibit a lower likelihood of undergoing mutations within male relatives belonging to the same pedigree. This characteristic renders them appropriate markers for conducting forensic studies involving pedigrees. When conducting familial searching in DNA databanks, it is more advantageous to choose a set of Y-STR loci with low to medium mutation rates.

The SureID PathFinder Plus Kit enables the simultaneous amplification of 41 loci using a configuration of six dyes. The set of loci includes all 17 loci from the Yfiler kit, as well as an additional 24 Y-STR loci with documented mutation rates. The loci DYS557, DYS593, DYS596, and DYS645 exhibit a high degree of polymorphism. On the other hand, DYS449, DYS518, DYS549, DYS570, DYS576, DYS627, DYS643, DYF387S1a/b, and DYF404S1 are Y-STR loci that undergo rapid mutation. Lastly, DYS388, DYS444, DYS447, DYS460, DYS481, DYS522, DYS527a/b, and DYS533 are loci with a low to medium mutation rate. Therefore, this kit has the potential to serve as an innovative application kit for distinguishing between related males by incorporating loci with high mutation rates, as well as for conducting familial searches by utilizing Y-STR loci with low to medium mutation rates. Furthermore, the incorporation of three Y-InDels—rs199815934, rs771783753, and rs759551978—which demonstrate a minimal likelihood of undergoing mutations, may potentially contribute to the process of familial searching (Fan et al., 2021; Zeng et al., 2022).

When conducting evolutionary investigations, Y-STRs can be employed in addition to SNP markers; nevertheless, a large number of markers is required to draw more precise conclusions (Hadi et al., 2022).

Rogalla et al. (2015) conducted a study on a multiplex assay that amplified 13 single-copy Y-STRs with rapid and moderate mutation rates. The results showed that the markers DYS518 and DYS458 had the highest mutation rates (Rogalla et al., 2015).

Regarding STR mutations and regardless of the genetic transmission mode, it is widely accepted that:

1. Mutations that only change one repeat, known as "single-step mutations," are more common than those that change multiple repeats at once.
2. Mutation is more likely to occur in longer alleles compared to shorter ones.

3. Longer alleles tend to experience repeat losses, while shorter alleles are more prone to gains rather than losses.
4. Advanced paternal age increases the likelihood of allele mutations.
5. Paternal mutations are more common than maternal ones in autosomal and X-linked transmission mechanisms.

In 2024, Antao-Sousa et al. performed research on the modeling of Y-STR mutation. They focused on analyzing Y-chromosomal transmission by determining fragment lengths. The findings indicated:

- The likelihood of single-repeat gain or loss mutations was 23.2 times higher than that of mutations involving a larger number of repeats.
- The longer alleles exhibited a mutation rate that was approximately 2.55 times higher than that of the shorter alleles.
- Longer alleles have a 1.7 times higher likelihood of losing repetitions compared to gaining them.
- Shorter alleles exhibited a symmetrical pattern, with an occurrence of repeat gains that was twice as frequent as reductions.

The average mutation rate varied between 0.0005 (DYS438 and DYS643) and 0.0170 (DYS547). Notably, the ratio of single-step/multistep mutations differed considerably between markers, with DYS439 having a ratio of 1.25 (representing equal frequency) and DYS449 showing a ratio of 98. The structure of the recurring motif corresponds with this variation.

The results provide evidence for a direct relationship between the rate of mutations and the age of the father. Specifically, the age group of 51–60 had the highest mutation rate (Antão-Sousa et al., 2024).

The incorporation of novel (SM) Y-STRs into the existing Y-STR panels could be beneficial in forensic casework, specifically in exclusion situations where there are minor inconsistencies. Similarly, the minimal rate of mutation in SM Y-STRs can aid in elucidating the genetic connections between Y-chromosome lineages (Baeta et al., 2018).

In 2022, Ying Zeng and colleagues developed a single multiplex typing system that consists of 45 Y-STR markers. This system is designed for familial searching and database construction, and it includes a Y-STR marker that undergoes rapid mutation. The Y45 kit outperformed the Yfiler Plus kit, which had been widely used before, in terms of male identification capabilities due to the higher HD and DC (Zeng et al., 2022).

Mutation rates of 42 Y-STRs in the Chinese Han population residing in Eastern China were studied in 2018. Wu et al. showed mutation rate variations at RM Y-STR loci, with lower rates observed at DYS552 ($<2.84 \times 10^{-3}$) and DYS630 ($<4.86 \times 10^{-3}$). Their findings suggest that these mutation rates may be influenced by the position of the locus and the complexity of the repetitive and surrounding regions of the Y-chromosome. Single-copy markers (DYS627, DYS570, DYS449, and DYS576) positioned on the Y-chromosome short arm (Yp) have mutation rates exceeding 10^{-2}. These markers exhibit greater mutation rates compared to multicopy loci, including DYF387S1a/b, DYS385a/b, DYS459a/b, and DYS527a/b. As mentioned earlier, alleles that were longer tended to see a reduction in the number of repetitions, while shorter alleles showed a significant and more frequent increase in the number of repeats (Wu et al., 2018).

7.7 Strategies to estimate rapidly mutating Y-chromosomal short tandem repeats mutation rates

There are multiple established methods for accurately determining the mutation rate of Y-STRs, each with its own benefits and limitations. There are three primary methods for analyzing Y-STR data that are most suitable for accurately estimating the mutation rates of Y-STRs in a population:

- Examination of father/son pairs.
- Examination of genetically linked individuals from deep-rooted pedigrees.
- An examination of individual or small pools of sperm cells.

Until now, the most commonly used approach in most published studies for estimating Y-STR mutation rates and differentiating male relatives has been the examination of father/son pairs. Autosomal STR typing was conducted on samples to approve father–son pairings to identify any discrepancies between each father and his son by Y-STR analysis. Consequently, the examination of father–son pairs fails to include the possible influence of numerous time-scale-dependent variables on mutation rates, including an extension of intergenerational time or substantial alterations in lifestyles. Furthermore, to obtain an accurate calculation of mutation rates, it is necessary to have access to a sufficiently large number of father–son pairs. Alternatively, the estimation of mutation rates for Y-STRs can be done using either a population-based technique or by analyzing deep-rooted male pedigrees. These methods rely on the high-resolution Y-SNP-based phylogeny and have the advantage of requiring a smaller number of male samples.

Nevertheless, when dealing with RM Y-STRs that have mutation rates exceeding 10^{-2}, a smaller number of father–son pairs is sufficient to establish a reliable mutation rate. A minimum of 10 RM Y-STR mutations might be detectable by analyzing 1000 father–son pairs.

Performing a deep-rooted pedigree analysis enabling the study of different male relationships and a great deal of meiotic divisions. This method has been shown to be more efficient in achieving a sufficient number of meiosis events required for precise calculation of mutation rates in a limited male sample. Deep-rooted pedigree studies are preferable to father–son pair studies in discriminating between male relatives since they incorporate both closely and distantly related males. However, it is unable to determine the direction (forward or backward) of mutations and the exact number of mutations observed in a deep-rooted pedigree. This creates ambiguities in assuming the ancestral Y-STR haplotype, as there is a potential risk of concealed backward mutations for RM Y-STRs. Therefore, a deep-rooted examination is the most suitable approach for studying the mutation rates of SM Y-STR and MM Y-STR (Čokić et al., 2019; Ralf et al., 2020, 2023, 2024).

Moreover, assumptions on the number of generations between the tested persons and tMRCA are crucial for population-based methodologies and, to a lesser extent, deep-rooted pedigree studies. These assumptions can result in erroneous assessments of the mutation rates.

Both of these methodologies have the drawback of potentially having parallel mutations, hidden mutations, and multistep mutations. These factors can all contribute to a greater probability of error in the estimation of mutation rates. Thus, when it comes to RM Y-STRs, it is more advantageous to directly observe father–son pairs to establish mutation rates, especially if a significant number of pairs are available. Moreover, conducting direct observations on father–son pairs enables the characterization of the expansion or reduction of repeated sequences, as well as the distinction between single-step and multistep changes. A locus-specific approach to mutability prediction score was created, where four molecular factors have an impact on the mutability of Y-STRs, as mentioned before.

- The number of consecutive repetitions is the primary factor that increases the probability of mutations of Y-STRs.
- The number of repetitions extends within a sequence.
- Being a marker with one copy or multiple copies.
- Tetranucleotide repeats, which refer to the size of the repeat motif, have been observed to be more prone to mutation compared to other motifs.

The rate at which the Y-STR marker mutates is calculated by dividing the total number of detected mutations by the number of father–son pairs examined. The mutation rate is expressed as the number of mutations per marker per meiosis.

The most predominant and effective determinant of Y-STR mutability is the number of Y-STR repeats (length). Long alleles of Y-STRs have a higher likelihood of developing mutations at these loci in comparison to short alleles of Y-STRs. The mutation rate of a Y-STR differs between populations as a result of distinct Y-SNP haplogroups (Ralf et al., 2020).

Natural germline mutations can be used to differentiate Y-STR haplotypes between a father and his male children. Typically, the loci that have the highest degree of variation and mutation rates are the most useful markers for forensic casework. However, it is crucial to have accurate estimations of mutation rates to effectively use Y-STRs and correctly interpret the results. Incorrect utilization of RM Y-STRs can result in a mistaken exclusion of biological paternity (Oh et al., 2015).

In 2016, Adnan et al. performed an investigation on the Pakistani population, which found that the multicopy nature of DYS399S1 allows for potential mutations. As anticipated, the previous investigation found that DYS399S1 had the highest mutation rate, measuring 7.7×10^{-2}, which can be attributed to the cumulative impact of numerous copies. DYS612 is an additional robust RM Y-STR marker used to distinguish between male relatives. Unlike other markers, DYS612 is a single-copy Y-STR with a significantly lower mutation rate of 1.4×10^{-2}. Consequently, obtaining additional RM Y-STR data from male relatives is a valuable method to improve the accuracy of distinguishing between individuals, particularly when dealing with males who are not closely related. The rate of differentiation between fathers and sons has shown a consistent decrease from the original study conducted by Ballantyne et al. up to now. Initially, the rate was 70% based on 20 pairs, which decreased to 49% based on 39 pairs, and further decreased to 27% with a significant increase in the number of father–son pairs to 2327. Utilization of RM Y-STRs instead of Yfiler resulted in a threefold increase in the differentiation rate of all related men, from 9.6% to 35.3% (Adnan et al., 2016).

A comparison study conducted by Arwin Ralf et al. in 2023 analyzed 9379 pairs of males utilizing 49 Y-STRs. The study compared mutation rate estimates derived from father–son pairs to the mutation rates based on pedigrees. The two methods of assessing mutation rates yielded significant variations in six Y-STRs: DYF1000, DYF403S1a, DYS612, DYS1013, DYS442, and DYS448. The pedigree-based mutation rate estimates for DYS1000, DYF403S1, and DYS612 were found to be much higher than the rates based on father–son pairs. This study demonstrates that utilizing pedigrees is a highly effective method to get an idea of how often mutations occur and how much males differ from their relatives in Y-STRs, even in Y-STRs with high mutation rates. This study illustrates that employing pedigrees is a remarkably efficient approach to ascertaining the frequency of mutations and the extent of male differentiation in Y-STRs, even in Y-STRs with elevated mutation rates. The results suggest that using Y-STR data to predict the level of patrilineal consanguinity is feasible. Additionally, the inclusion of RM Y-STRs in RMplex yielded more accurate predictions than Y-STRs with lower mutation rates. The rates of discrimination among male relatives based on the analysis of 30 RMplex Y-STRs were found to be 43%, 84%, 96%, 99%, and 100% for relatives separated by 1, 4, 6, 9, and 12 meioses, respectively. These rates were much higher than those obtained using 25 standard Y-STRs. The utilization of machine learning algorithms to predict the level of patrilineal consanguinity showed high accuracy and reasonable precision when employing RM Y-STRs (Ralf et al., 2023).

Estimating the time to the most recent common ancestor (tMRCA) is of utmost importance in genealogical research. To precisely estimate tMRCA among paternal lineages, it is necessary to determine all Y-STR variants and accurately evaluate Y-STR mutation rates. The CE system may fail to detect variations in sequences of two Y-STR alleles that have the same length (isometric). Sequence analysis methods are useful for identifying identical alleles with repeat pattern arrangements and SNPs in the flanking region or within the STR repeat unit. This helps to improve the discrimination of different alleles and avoid underestimating tMRCA. Next-generation sequencing has become popular in forensic science for classifying Y-STR variants between alleles with the same size but distinct sequences. These techniques are particularly advantageous in population genetics and forensic investigations. It is recommended to analyze both SM Y-STRs and MM Y-STRs to increase the chances of identifying distant relatives of a suspect in a forensic case. On the other hand, FM Y-STRs and RM Y-STRs are used to differentiate even among close relatives in forensic familial searching (Claerhout et al., 2019; Kasu et al., 2019).

In 2019, Kasu and colleagues conducted a study on the novel variation of Y-chromosome STR sequences at the following loci: DYS710, DYS518, DYS385, DYS644, DYS612, DYS626, DYS504, DYS481, DYS447, and DYS449. The variation caused by size homoplasy was clearly more informative than the presence of SNPs for identifying isometric alleles. Out of the 10-Y-STR panel, the loci DYS710, DYS449, and DYS447 accounted for 76% of the total size homoplasy events. These loci have complicated sequence structures and nucleotide compositions, which make them more useful for distinguishing among shared male haplotypes (Kasu et al., 2019).

Evaluating the mutation rate of Y-STRs plays an essential role in various interdisciplinary fields, such as forensic genetics, genetic genealogy, and population genetic investigations. When calculating Y-STR mutation rates, it is important to account for small differences across Y haplogroups, not only the averages. This evaluation can offer insights into the paternal relatedness among males and provide a close estimate of tMRCA. In 2019, Boattini et al. conducted a study to determine the mutation rates of Y-STRs and tMRCA using deep-rooted Italian pedigrees. The results have confirmed that the Y-STR mutation rate assay, when used in conjunction with the genealogy approach, can provide accurate estimates of tMRCA. This is true even when there is a limited number of male samples available. The assay takes into account the potential of multistep mutations and reverse mutations, which may not be significant within the time frame typically used in forensic casework and genetic genealogy. Furthermore, both the deep-rooted and father–son analysis methodological frameworks offer highly accurate and comparable estimations for the Y-STR mutation rate, with the exception of inconsistencies seen in RM Y-STRs (Boattini et al., 2019).

7.8 Rapidly mutating Y-chromosomal short tandem repeats

The present application of NRY in forensics is substantially limited by the inability to draw individual-level judgments, which is typically necessary in forensic investigations. It is possible to hypothesize that if Y-STRs with significantly higher mutation rates than the ones currently applied were accessible, it might be feasible to distinguish male relatives at the individual level. This would resolve the present issue with Y-chromosome applications in forensics. A more precise approach to assessing Y-STR mutation rates involves directly observing the transmission between fathers and sons, provided that a significant number of genetic transfers (meiosis) be examined by evaluating quite a lot of father–son pairs. Recently, it has been proposed that the sequence motif has a significant role in the variation of mutation rates among loci, with certain motifs showing a higher rate of mutation (Ballantyne et al., 2010).

Ballantyne et al. introduced the idea of RM Y-STR in 2010. Rapidly mutating Y-STRs are Y-STRs with mutation rates in the order of 10^{-2}, as commonly defined. RM Y-STRs experience a minimum of 1 mutation every 100 generations per locus (Otagiri et al., 2022; Wang et al., 2022).

Relying only on Y-STR in practical forensic situations appears insufficient. Even if a sample and a suspect have a perfect Y-STR match, the contributor cannot be distinguished from his ancestral paternal descendants. In response to this issue, scientists methodically conducted research and identified novel Y-STRs, known as RM Y-STRs, which exhibit significant mutation rates. RM Y-STRs possess a higher discriminatory capability, enabling the potential distinction between closely related male individuals. Over the past decade, numerous studies have been conducted to amplify RM Y-STRs in multiplex systems, taking advantage of their significant discriminatory ability. Both the PowerPlex Y23 and Yfiler Plus commercial kits now feature a number of these markers in addition to the classic Y-STR markers typically used in forensic laboratories (Albastaki et al., 2022; Diegoli, 2015).

In 2010, a comprehensive empirical study was conducted to analyze the mutation rate of Y-STRs. This study introduced a set of 14 RM Y-STRs, which included DYF387S1, DYF399S1, DYF403S1a/b, DYF404S1, DYS449, DYS518, DYS526a/b, DYS547, DYS570, DYS576, DYS612, DYS626, and DYS627. These Y-STRs have proven to be a highly effective tool for differentiating between male relatives who share the same paternal lineage. In the past 10 years, several investigations conducted on various groups of males, whether they were connected or not, have shown that these 14 RM Y-STRs are useful in the field of forensic genetics. An empirical investigation, using 1616 DNA-confirmed father–son pairs, has recently discovered 12 new RM Y-STRs. Later on, a comprehensive panel called RMplex, comprised of 26 RM Y-STRs, was introduced in the field of Y-chromosome analysis. The presence of these new markers, which have a high frequency of mutations, offers a more effective means of differentiating males who are linked through their paternal lines (Wang et al., 2022).

Ballantyne et al. conducted a study on a significant number of Y-STRs in father–son pairs with high mutation rates. Their objective was to enhance the differentiation power among male relatives who have the same Y-haplotype. Thirteen RM Y-STRs (DYF387S1, DYF399S1, DYF403S1, DYF404S1, DYS449, DYS518, DYS526, DYS547, DYS570, DYS576, DYS612, DYS626, and DYS627) are chosen to distinguish among males residing in various continents. The Yfiler Plus kit now incorporates seven of these RM loci. This collection of RM Y-STR markers enhances the ability to differentiate between closely related individuals by identifying at least one observed mutation. In 56.3% of cases, brothers are identified by each other, whereas Yfiler only identifies them in 10% of cases (Court, 2021).

In 2012, Ballantyne once again presented an in-depth analysis of 13 RM Y-STRs that displayed greater capacity to distinguish between male lineages in 51 populations worldwide. When comparing RM Y-STRs to Yfiler, the average male relative differentiation is 4.4 times better, which is a statistically significant advantage. The RM Y-STR panel demonstrates a high level of discriminatory power in distinguishing closely related men. It is capable of distinguishing over 50% of father–son pairs, as well as 60% of brothers, using RM Y-STRs. In contrast, the Yfiler panel only achieves a distinction rate of 7.7% for father–son pairs and 8% for brothers. As a result, RM Y-STRs present a great opportunity to solve most of the problems that forensic genetics has encountered with Y-chromosome analysis (Ballantyne et al., 2012).

In 2020, Ralf et al. conducted a study on the discovery and characterization of novel Y-STR markers that exhibit rapid mutation. The results revealed that the combination of all 25 RM Y-STRs significantly enhanced the ability to differentiate between male relatives in all pairs. Out of the father–son pairs, 44% showed differentiation via one round of meiosis. For brothers and grandfather–grandson pairs, the percentage was 69% after two rounds of meiosis. Uncle–nephew pairs showed differentiation in 83% of cases after three rounds of meiosis. Finally, cousins showed differentiation in 90% of cases after four rounds of meiosis. In all cases, differentiation was indicated by at least one mutation. As previously declared, there is a direct correlation between the father's age and the chance of Y-STR mutations. These findings emphasize the importance of considering the age distribution of fathers when studying Y-STR mutability using father–son pairs, and it is crucial to take this factor into account when interpreting mutation outcomes. Sequence motifs may vary significantly in line with Y-STR mutation rates. One possible explanation is that specific repeat motifs (such as AAG, AGG, AAAG, and AAGG) cause the creation of secondary structures like triplex DNA. But the question remains if this has a direct effect on mutability or if it mostly changes the direction of mutations, leading to the extension of repeat stretches and greater mutability. Further dedicated studies are needed to fully understand this phenomenon (Ralf et al., 2020).

The study conducted by Ballantyne et al. in 2014 provided valuable insights into several elements of rapidly mutating Y-STRs (RM Y-STRs), through a comprehensive collaborative analysis including 52 distinct populations worldwide. Across all tested populations, there were comparable high levels of HD at the population level. The 13 RM Y-STR set consistently yielded more accurate estimates compared to the 17-loci Yfiler set in all diversity measures used. The use of RMY-STRs increased the global HD from 0.99995 to 0.999997, surpassing the diversity achieved with Yfiler. Consequently, the inclusion of regional (metapopulation) reference databases for estimating haplotype frequencies in forensic and other applications is significantly reduced when using RM Y-STRs compared to conventional Y-STRs included in the Yfiler kit. However, these data also show that even in populations with a significant number of undisclosed paternal links, like endogamous groups, the RM Y-STR set is still effective in distinguishing a significant proportion of men. The first estimate of the differentiation between fathers and sons was 70% based on a sample size of only 20 pairs.

However, using a larger sample size of 2378 pairs, the differentiation rate has now been determined to be 26.9% (with a 95% confidence interval of 25.1%–28.7%). Prior suggestions propose that having a minimum of three mutations at any Y-STR haplotype, particularly those included within the 17 Yfiler, is enough to rule out the paternity. Within the existing dataset, it was observed that 3.8% of the DNA-confirmed father–son pairs had three or four RM Y-STR mutations, or even three mutations have been detected occasionally among 17 Yfiler loci. Therefore, it is recommended to raise the threshold for exclusion when utilizing the RM Y-STRs. Instead of implementing a fixed limit, a dynamic threshold will be used; and this threshold will be decided by answering two primary questions: how many Y-STRs were examined and what data exists about their mutation rates in the population. This inspection must be applied to conventional Y-STR sets like Yfiler, which seems more suitable for paternity and kinship testing compared to RM Y-STRs due to their moderate mutation rates. It is essential to accurately determine mutation rates for the specific population to effectively account for any discrepancies that may arise from mutational occurrences. Hence, just having knowledge of the average mutation rates of Y-STR markers, which fall within the range of approximately 2–4 $\times 10^{-3}$, is not sufficient from a statistical standpoint (Ballantyne et al., 2014; Diegoli, 2015).

The lower extent of significant pairwise FST comparisons observed with the RM Y-STRs as well as the decreased within-population variation, provides further evidence that RM Y-STRs are less effective in detecting genetic population substructure compared to Yfiler-included Y-STRs. The worldwide data employing RM Y-STRs shows very little population substructure, so it appears match probability estimates may not need to be adjusted for population substructure. Additionally, there may be a reduced need to know the source population of the sample donor beforehand to accurately estimate haplotype frequencies. Hence, it is more reasonable to provide a probabilistic evaluation of the likelihood of seeing a certain number of mutations in a particular meiotic transfer, considering the number of meiotic transfers and the locus-specific mutation rates of the Y-STRs utilized. The Y-STRs used in the evaluation and the underlying mutation rates are major factors in the recommendation for this type of evaluation; RM Y-STRs are anticipated to have substantially higher estimations (Ballantyne et al., 2012).

While RM Y-STRs improve discrimination between unrelated people, they also improve it between males of the same paternal lineage, which could allow for the differentiation of first-degree relatives (Rapone et al., 2016).

The recently introduced 13 RM Y-STRs exhibit a mutation median rate that is 6.5 times higher than that of traditional Y-STR markers. Their capacity to distinguish between persons is improved by this greater mutation rate, which solves problems with conventional Y-STRs and makes them useful in a variety of forensic investigations. A DC of 0.983 was indicated by the GD value of 0.9997. The GD value of RM Y-STR markers was particularly high in Indian populations. The locus DYF399S1 has the greatest GD value of 0.99219 out of all of these markers. The DYF403S1a came in second with 0.98181, the DYF404S1 in third with 0.93009, and the DYF387S1 in fourth place with 0.92511 (Shrivastava et al., 2019).

Earlier investigations have demonstrated variations in mutation rates among males from different populations for the initial set of 13 RM Y-STRs. Claerhout et al. conducted a study on European males to investigate the impact of Y-SNP haplogroups on Y-STR mutation rates. Unlike previous studies, they included Y-SNP haplogroup data. The findings revealed that certain Y-SNP haplogroups, specifically haplogroups I and J, exhibited lower average mutation rates compared to haplogroup R1b. Shorter alleles in males with haplogroups I and J could point towards lower Y-STR mutation rates for some Y-STRs. On the other hand, in certain groups, the passing down of longer Y-STR alleles from male ancestors can lead to an increased likelihood of mutations occurring in these Y-STRs in future generations. An analysis of allele frequencies in the DYS712 Y-STR gene showed that the Japanese population had a greater number of variations in their alleles compared to the European population. The extended Y-STR alleles at DYS712 were particularly prevalent in the samples of Japanese males who belong to a specific subset of haplogroup O1b2 (O-P49), which is entirely missing in Europeans. Thus, the Y-STR mutation rates used for interpreting RMplex casework among Europeans may be problematic when applied to suspects with non-European paternal heritage. RMplex demonstrates a much greater differentiation rate (52%) for non-European father–son pairs, specifically Japanese, compared to the current state-of-the-art commercial Y-STR kit Yfiler Plus (14%) (Otagiri et al., 2022).

Otagiri et al. conducted a study in 2021 to investigate the mutation rates of 25 Y-STRs in the Japanese population. Their findings provided more evidence that population and haplogroup variations may explain why European and Japanese populations exhibit such varied mutation rates. From the seven RMY-STRs that are part of the Yfiler Plus, only DYS576 and DYS458 show a mutation rate greater than 1.0×10^{-2} per locus/generation. Out of the 22 mutations, 21 occurred in a single step, while one mutation in DYS392 switched three steps away. Thus, it is not guaranteed that the mutation rates of the RM Y-STRs provided by Yfiler Plus are higher compared to similar markers in the Japanese population. However, they are recommended for use in determining paternal relationships rather than distinguishing patrilineal relatives in the Japanese population (Otagiri et al., 2021).

A 2019 study conducted by Iyavoo examined the mutation rates of 13 rapidly mutating Y-STRs in a Dravidian lineage. The study found that various populations exhibited varying rates of RM Y-STR mutations. The study highlighted the need to consider the mutation rates in different populations when including the RM Y-STRs. This consideration can enhance the ability to distinguish between males that belong to the same paternal lineage (Iyavoo et al., 2019).

Adnan et al. utilized 13 RM Y STRs that had been found earlier to distinguish males who were separated by different numbers of generations. They were able to differentiate 27% of males separated by one generation, 46% of males separated by two generations, 54% of males separated by three generations, and 62% of males separated by four generations. Ralf et al. discovered a set of 12 RM Y STRs that are complementary and can distinguish between males who are separated by one, two, or three generations with an accuracy of 27%, 47%, and 61%, respectively (Ay et al., 2018).

A study on 18 deep-rooted pedigrees from the Sindhi endogamous population of Pakistan by Rakha et al. in 2018 focused on investigating 13 RM YSTRs. The DYF399S1, DYS449, DYS518DYS547, and DYF403S1b2 loci exhibited higher levels of mutation compared to other RM Y-STRs. Additionally, the panel of RM Y-STR markers showed a significant increase in discrimination power (29% greater) when compared to the Y-STR markers often used in commercial kits such as the AmpFlSTR Yfiler kit (Rakha et al., 2018).

Franz and colleagues used RMplex in 2022 to identify the father's lineage in 499 father–son pairs from Europe, which accounted for 42% of the male relatives. On the other hand, out of 530 pairs, Yfiler Plus managed a mere 14% differentiation rate for male paternal relatives. Using RMplex in conjunction with Yfiler Plus outperformed using RMplex alone in distinguishing 48% of male paternal relations. Similarly, a total differentiation rate of 66% in 97 endogamous pedigrees was achieved for male pairs with paternal relation, who were separated by one to seven meiosis. This greatly exceeds the differentiation rates achievable with commercially available multiplexes, such as PowerPlex Y23. Since DYS570 and DYS576 have mutation rates of more than 1%, they are categorized as Y-STR loci with rapid mutations. (Javed et al., 2022; Neuhuber et al., 2022)

In 2022, Albastaki and colleagues successfully created a new multiplex kit called 26 RM-Yplex that can be used to determine a person's full Y-STR data. This kit is able to provide accurate results even when only a very small amount of DNA (0.125 pg.) is available, such as in a mixture of male and female DNA, and has been tested and showed to produce complete profiles at male to female ratios ranging from 1:400 to 1:3200 (Albastaki et al., 2022).

To assess the efficacy of Yfiler Plus, in African populations characterized by significant levels of endogamy, Della Rocca et al. conducted a study in 2019. The Yfiler Plus kit has a low discrimination power of 0.90, which is attributed to endogamy and ethnic fragmentation rather than close familial relationships. This emphasizes the need to include more RM Y-STRs in commercial kits to enhance the routine discrimination power in populations with high levels of endogamy (Della Rocca et al., 2019).

The Yfiler Plus and 82 Y-SNPs were found to have a rather low DC in African populations with high ethnic fragmentation and linguistic variety in the central Sahel by Della Rocca et al. (2020). This was particularly evident in rural areas, where the average DC was 0.83. One possible explanation for the lower DC in rural areas is the fact that different ethnic groups tend to live in relative isolation from one another. In contrast, urban populations had a higher DC, with an average of 0.96. Moreover, the genetic differences between populations residing in nearby villages are far greater than those observed inside the countries of Cameroon and Chad. Hence, it is crucial to systematically choose a suitable reference population database for the statistical assessment of Y-haplotype matches, particularly in criminal investigations that occur in rural regions characterized by fragmented ethnic groups or influenced by endogamy that originates from social structure (Della Rocca et al., 2020).

In 2019, D'Atanasio et al. conducted a study to examine the genetic characteristics of the growing population in northern Africa. They used rapidly mutating Y-STRs to analyze the genetic makeup. The results revealed that northern Africa has complex genetic features, highlighting the need to study multiple populations from various countries for forensic analysis. Additionally, the study found that RM Y-STRs were particularly useful in distinguishing individuals from groups with low genetic variability within their population due to recent and rapid expansion (D'Atanasio et al., 2019).

The ability of RM Y-STRs to determine genetic variations in the population of Qatar was assessed and compared to those for the global population. Almohammed et al. found that the set of 13 RM Y-STR markers exhibited impressive capability to identify unique genetic patterns in the Qatari population, with a high level of GD and reliability. They were able to completely determine the haplotypes (genetic profiles) of individuals in the national Qatari population and the broader Middle Eastern group using RM Y-STRs. Marker DYF399S1 in the Qatar population displayed particularly high GD (0.992). Additionally, markers DYS626, DYS526a, and DYS627 exhibited high levels of GD across multiple populations (Almohammed & Hadi, 2019).

In another analysis of rapidly mutating Y-STRs in the Pakistani population in 2019, Farhatullah et al. calculated the HD to be 0.999925 and found that the GD values for all the loci exceeded 0.5. The loci with the highest GD values were DYF399S1, DYF403S1a, and DYF404S1, with values of 0.99419, 0.98252, and 0.93061, respectively. These findings again indicate that the selected 13 rapidly mutating Y-STRs have a far more powerful discriminatory power in Pakistani populations (Farhatullah et al., 2019).

Referring to Mitchell et al., a literature review and Y-STR meta-analysis were performed in 2024 to enhance the approach for identifying historical military remains. The findings revealed that even when only a partial DNA haplotype is obtained from degraded remains, there is still a risk of encountering low Y-STR discrimination power. However, it was concluded that RM Y-STRs are not suitable for historical military identifications due to their significantly higher mutation rate, which is 10 times greater than other Y-STRs. However, applying precise data on mutation rates and variations in other genetic markers is crucial to determine which ones could be useful in resolving historical cases (Mitchell et al., 2024).

In 2019, Wang et al. conducted a study on the RM Y-STRs in the Chinese Yi community. Of all the loci, the DYF399S1 exhibited the highest mutation rate, with DYF403S1a coming in second. In general, RM Y-STRs were able to distinguish father–son pairs with a success rate of 19%, which was similar to that of the Chinese Han population. In comparison to prior studies on father–son couples, the differentiation rate in the Chinese population was marginally lower than that observed in European and Pakistani populations (26.9% and 24.3%, respectively) (Wang et al., 2019).

In 2022, Wang et al. performed a study to examine the rate of mutation and paternal differentiation based on 30 highly mutating Y-STRs in Chinese father–son pairs, comparing them to the rates observed in European reference populations. The results revealed significant variations in the mutation rates of DYS570, DYS399S1, DYS547, DYS612, and DYF403S1b between the two populations, with differences spanning many times. Remarkably, 10 RM Y-STR loci had very moderate mutation rates, ranging from 10^{-3} to 10^{-2} per meiosis, while 2 FM Y-STR loci displayed high mutation rates above 10^{-2} per meiosis in the Chinese population. A lesser percentage of Chinese paternity pairs, approximately 40.06%, were identified as different using the RMplex method compared to European populations. A two-step increase in the copy number was detected at DYF1001. A higher copy number is associated with a greater probability of mutation at the particular locus. DYF387S1 showed a significantly higher mutation rate in the Chinese population than in the Austrian population. As the number of RM Y-STRs used increased, the rate of discrimination between fathers and sons improved dramatically in both groups. However, significant variations were identified, indicating differences in mutagenesis between the populations from different continents (Wang et al., 2022) (Fig. 7.1).

7.9 Application of rapidly mutating Y-chromosomal short tandem repeats

Autosomal STRs are frequently employed as a standard to trace the origin of a biological specimen from a crime scene, as contrasted to utilizing Y-chromosome STR markers. Autosomal STRs are highly informative genetic markers that can be used to generate a unique profile for a person. Y-STRs, when used as lineage markers, have the ability to identify a cluster of male relatives rather than a specific individual. Hence, if there is a perfect match between a piece of evidence and a suspect, it does not rule out the possibility that his father, brothers, sons, and other male relatives could also be potential sources, as they all possess the same Y-haplotype (unless there has been a mutation) (Diegoli, 2015).

Andersen and Balding conducted a simulation analysis on three genetic marker systems: Yfiler (consisting of 17 loci without RMYSTs), PowerPlex Y23 (consisting of 23 loci with two RM Y-STRs), and Yfiler Plus (consisting of 27 loci with seven RM Y-STRs). The simulation findings revealed that 95% of all complete Y-STR matches were separated by less than 50–100 meioses, taking into account the number of loci and the population's history. Hence, prior to conducting match probability calculations for Y-STR analysis, it is imperative to consider the level of pairwise relatedness among the suspect population. An effective method to address potential issues linked to relatedness and improve the reliability of Y-STR matches is to raise the mutation rate of the Y-STR panel to a minimum of 50% or higher by incorporating several RM Y-STRs (Caliebe & Krawczak, 2018).

In 2010, Ballantyne et al. explored two situations regarding the application of RM Y-STRs in forensic genetics. In cases involving several perpetrators, such as rape cases involving male relatives, traditional Y-STR profiling may not be sufficient to distinguish between suspects. Furthermore, in forensic cases when the investigator aims to eliminate individuals who are not the perpetrators and traditional Y-STR profiling is not a practical method, an alternative approach is required. RM Y-STRs have the potential to transition forensic Y-STR analysis from focusing on male lineage differentiation to specializing on individual male identification in the future. Consequently, the end goal of Y-STR analysis is expected to obtain complete haplotype resolution. The Middle East population showed the greatest notable improvement in DC, increasing from 84% with Yfiler to 100% with RM Y-STRs. The extent of DC improvement varied

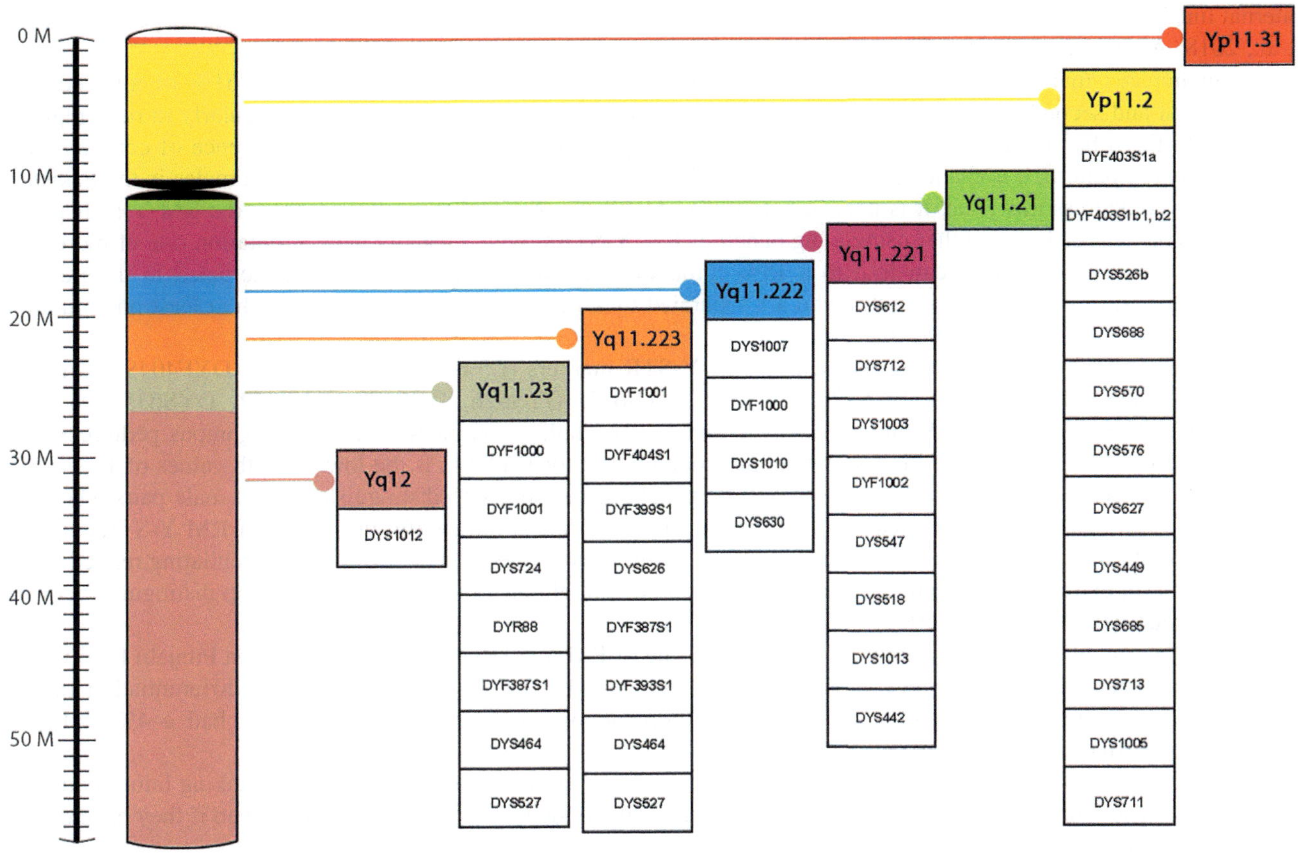

FIGURE 7.1 Relative positions of rapidly mutating Y-chromosomal short tandem repeats (RM Y-STRs) on the human Y-chromosome. Available at: https://ybrowse.org/gb2/gbrowse/chrY/.

across geographic parts of the world. HD improved from 0.9994 with Yfiler to 0.99996 with RM Y-STRs on a global scale (Ballantyne et al., 2010, 2012).

As well, a higher diversity of RM Y-STRs can effectively distinguish unrelated men, and they are useful in cases when differentiation between both close and distant male relatives is needed. RM Y-STRs are particularly advantageous in cases when autosomal STRs are uninformative, unclear, or mixed (male/female), notably in situations involving several male perpetrators who are related or in sexual assault cases. If a commonly used Y-STR kit like Yfiler does not exclude a person, it is recommended to utilize the RM Y-STR set. This will help decrease the likelihood of accidental matches in cases involving male relatives who have committed crimes, such as family rapes. Furthermore, utilizing RM Y-STRs could potentially exclude paternal male relatives from further examination in certain cases, thereby effectively countering claims such as the brother's defense in a court setting (Ay et al., 2018; Ballantyne et al., 2012; Henry et al., 2015).

To identify victims of mass disasters (DVI), such as plane crashes or mass military graves from past war conflicts, where there are multiple related victims or groups of related male soldiers, investigators require a robust tool to distinguish between these individuals. RM Y-STRs, being highly discriminative markers, have the potential to enhance the interpretability of partial profiles, which are often encountered in mass disaster cases. Furthermore, RM Y-STRs could prove highly useful in mass disasters where the standard Y-STR analysis cannot be used to reunite the remains belonging to related male victims. The exceptional ability to resolve cases where Y-haplotypes cannot be distinguished from conventional Y-STRs makes them highly effective tools in mass disaster identification.

For instance, RM-Yplex, which is composed of 13 RM Y-STRs, which include 9 single-copy loci and 4 multiple-copy Y-STRs was used to examine bones and tooth samples from 55 skeletons found in a degraded mass grave in Slovenia dating back to the Second World War. Remarkably, 56% of the samples yielded complete profiles. The RM Y-STRs that are most frequently dropped out are DYS403S1b, DYS547, DYS627, and DYS526b, which are the longest loci. Furthermore, unsolved cold cases of sexual assault involving deceased suspects who are genetically related through the paternal line can be reexamined by utilizing RM Y-STR panels on the skeletal or dental remains of the suspects (Alghafri et al., 2018; Ambers et al., 2018; Niederstätter et al., 2016).

Consanguineous marriages are prevalent in the Middle East and popular in Pakistan, with 61% of all weddings including first and second cousins. This trend is more pronounced towards the paternal side, particularly in rural regions, compared to the urban population (http://www.pbscensus.gov.pk). Because of the frequent occurrence of consanguinity and the observed pattern, it is challenging to genetically differentiate between related and unrelated males in the Pakistani community. To address this issue, it is necessary to utilize highly discriminative markers such as the RM Y-STR panel. Researchers have already found that using a big bunch of RM Y-STRs (which usually have a mutation rate of more than 1% per locus every generation) can lead to a 15% cumulative mutation rate. This rate is anticipated to differentiate between over half of closely related males who are separated by no more than 20 meioses through at least one mutation (Pickrahn et al., 2016; Ullah et al., 2022).

In 2022, Faqeeha Javed and colleagues used 19 RM Y-STR markers (DYF399S1, DYF387S1, DYF404S1, DYS526 (b), DYS449, DYS518, DYS547, DYS570, DYS576, DYS612, DYS626, DYS627, DYS442, DYS635, DYS464, DYS389 (II), DYF393S1, DYS385, and DYS549) to analyze 162 father–son pairs from 97 endogamous pedigrees. The mutation rate for DYF399S1 is the highest, while the mutation rate for DYS442 is the lowest, with values of 1.79×10^{-1} and 6.17×10^{-3}, respectively. This specific RM Y-STR panel has the ability to distinguish 66% of male pairs who have paternal ancestral ties and are separated by one to seven meiosis. Hence, the careful utilization of RM Y-STRs and the establishment of population-specific mutation rates should effectively address the difficulty in differentiating related males from Pakistan. In extremely endogamous societies, this approach might also improve the capacity to distinguish between male close relatives (Javed et al., 2022).

In 2022, Nazir et al. conducted a study on the mutation rate of RM Y-STRs. The study focused on Punjabi family trees from Pakistan spanning multiple generations and found that the 13 RM Y-STRs panel had a differentiation rate of 26%–85% in consanguineous Pakistani populations. In comparison, conventional Y-STRs only had a 4% ability to differentiate paternal lineages (Nazir et al., 2022).

Utilizing RM Y-STRs has the potential to be advantageous in the field of genetic genealogy. By taking into account the mutation rates and numerous mutant markers, it becomes feasible to identify closely related men even if they have distant paternal ties (Ralf et al., 2020).

Among certain populations, such as Chinese residents of natural villages, it is common for people to maintain the same surname for many generations. However, using traditional SM Y-STRs and MM Y-STRs may not be sufficient to accurately differentiate between thousands of patrilineal noncontributors of a crime scene specimen and the male suspect. Ye et al. (2022) have demonstrated that in patrilocal populations with small effective population sizes, a limited number of 13 RM Y-STR haplotypes out of the 50 analyzed have proven to be crucial in reducing the need for collecting samples from potential suspects in forensic investigations. These haplotypes have significantly aided in narrowing down the pool of individuals under investigation (Li et al., 2020; Ye et al., 2022).

The genetic relatedness between individuals with the same surname can be verified by comparing their Y-chromosome haplotypes, specifically by examining the correlation between the Y-SNP sub-haplogroup and the existence of maximum seven mutant Y-STRs. The latter is derived from Walsh's formula, which suggests that out of the 46 studied Y-STRs, it is exceedingly improbable to discover more than seven mutations among individuals with the same surname who are separated by less than 30 generations (Čokić et al., 2019).

Irish surnames are passed down from fathers and are believed to have origins in Norse culture. Examination of SM Y-STRs and RM Y-STRs indicated the presence of a little amount of Scandinavian heritage among the so-called Norse surname group. This suggests that Norse colonial settlements may have mostly comprised Indigenous Irish individuals rather than Scandinavians (McEvoy et al., 2006).

As previously mentioned, while accurately estimating the probability of a Y-STR haplotype match, make sure to take into account the potential presence of population structure. This will ensure the use of appropriately comprehensive haplotype reference databases. RM Y-STRs have been demonstrated to effectively reduce signals of population structure due to their elevated mutation rates. Thus, the utilization of RM Y-STRs has the potential to mitigate the issue of population substructure and reduce the need for subsequent adjustments (Lacerenza et al., 2017) (Table 7.2).

7.10 The impact of rapidly mutating Y-chromosomal short tandem repeats on forensic genetic databanks

The Y-STR Haplotype Reference Database (YHRD), accessible at: https://yhrd.org, is the most extensive and commonly utilized database in the fields of forensic genetics and population genetics. The Y-haplotype database consists of about

TABLE 7.2 Rapidly mutating Y-short tandem repeatss implemented in multiplex amplification assays.

Loci/motif length	RMplex	26 RM-Yplex	Microreader RM-Y ID system
DYS612 (three)	Yellow	Green	Green
DYF403S1a (tetra—multicopy)	Red	Green	Green
DYS712 (tetra)	Blue	Green	NA
DYS1007 (tetra)	Red	Green	NA
DYS1003 (tetra)	Green	Green	NA
DYF1000 (three—multicopy)	Green	Purple	NA
DYF1001 (tetra—multicopy)	Yellow	Yellow	NA
DYS1012 (tetra)	Green	Yellow	NA
DYS1010 (tetra)	Green	Yellow	NA
DYS724 (tetra—multicopy)	Green	Blue	NA
DYF403S1b1, b2 (tetra—multicopy)	Red	Blue	Blue
DYS526b (tetra)	Blue	Purple	Green
DYS688 (three)	NA	Purple	NA
DYF1002 (tetra—multicopy)	Yellow	Purple	NA
DYR88 (tetra—multicopy)	Red	Red	NA
DYS547 (tetra)	Green	Red	Red
DYF404S1 (tetra—multicopy)	Red	Red	Red
DYF399S1 (tetra—multicopy)	Green	Yellow	Yellow
DYS570 (tetra)	Blue	Blue	NA
DYS576 (tetra)	Yellow	Purple	NA
DYS626 (tetra)	Blue	Yellow	Yellow
DYS627 (tetra)	Blue	Blue	Yellow
DYS518 (tetra)	Green	Green	Green
DYS449 (tetra)	Green	Red	Red
DYF387S1a/b (tetra—multicopy)	Red	Red	Red
DYS685 (tetra)	NA	Yellow	NA
DYF393S1 (tetra)	Blue	NA	NA
DYS713 (tetra)	Blue	NA	Yellow
DYS1005 (tetra)	Green	NA	NA
DYS1013 (tetra)	Red	NA	NA
DYS711 (three)	Blue	NA	NA
DYS442 (Tetra)	Red	NA	NA
DYS630 (tetra)	NA	NA	Blue
DYS464 (tetra—multicopy)	NA	NA	Blue
DYS527a/b (tetra)	NA	NA	Yellow

350,000 Y-STR MHTs, which include over 106,000 Yfiler plus haplotypes. These haplotypes incorporate RM Y-STR markers from 329 different populations and 61 countries across the globe. The Yfiler Plus panel was expanded to include DYS549 and DYS643, which are originally found in PowerPlex Y23, to achieve a Ymax set consisting of 29 Y-STRs. Based on the assessment of mutation rates of Y-STRs in release 69 (August 2024), it has been determined that DYS627, DYS576, and DYS518 are the most prone to mutation, with rates of 1.4×10^{-2}, 1.19×10^{-2}, and 1.08×10^{-2} mutations per locus per generation, respectively. The Y-STR markers DYS438, DYS392, and DYS643 exhibit the lowest mutation rates among the 29 markers studied, with rates of 3.95×10^{-4}, 4.45×10^{-4}, and 9.98×10^{-4} mutations per locus per generation, respectively.

In 2014, China began developing a national Y-STR database to aid in the solving of cold cases. Over the course of 10 years, numerous forensic cases were successfully solved by utilizing 16 to 25 Y-STR markers. Currently, the YHRD database has the biggest collection of Y-STR data from China, consisting of 130,000 Yfiler haplotypes and around 67,000 Yfiler Plus haplotypes. (Ralf et al., 2020; Wang et al., 2017).

With the expansion of a database containing conventional Y-STR haplotypes (MHTs), the likelihood of a factitious match between unknown male perpetrators and their paternal lineages has significantly increased. This is due to the lower mutation rates of SM Y-STRs and MM Y-STRs, as well as the lower DC estimated from these systems. As a result, several national databases choose to include RM Y-STRs as additional markers in their Y-STR database panels (Zhou et al., 2022).

Additionally, calculating Y-STR match probability appears to be straightforward because of the minimal population substructure observed with RM Y-STRs in our comprehensive global RM Y-STR database, in contrast to the traditional Y-STR haplotypes. There are two primary reasons that might be addressed in relation to this matter:

Utilizing RM Y-STRs may eliminate the need for population substructure correction; therefore, prior to estimating haplotype frequency, it is not essential to ascertain the source population to which the suspect belongs. Thus, it would be more reliable to use a worldwide RM Y-STR haplotype database instead of using multiple subpopulations (metapopulation) to accurately estimate frequencies for each real casework.

Furthermore, in nonsuspect forensic cases, where Y-STRs can tell us a lot about the donor's ethnicity, high-resolution RM Y-STRs could aid in the complete individualization of even closely related suspects. This level of identification cannot be achieved even with thorough analysis of male relatives spanning five generations using 47 Y-STR markers. By looking at a decrease in the number of individuals who possess the same Y-STR data as the suspect, investigators were able to eliminate innocent individuals as potential sources of the trace evidence. This decrease was attributed to the probability of mutation events occurring (Robino et al., 2015; Roewer et al., 2020).

In 2016, Rapone et al. conducted a study on 203 unrelated males, encompassing various geographical groups (North, Center, South, and Sardinia) in Italy. Y-STR diversity investigations, together with Y-SNP-based studies, provide evidence that the Sardinians exhibit significant genetic differentiation from the other three Italian groups. Therefore, it is necessary to consider them as a distinct subpopulation in Y-STR haplotype database (Rapone et al., 2016).

In 2017, Lacerenza et al. implemented a study on the Sardinian population. They found that even when using extended Y-STR haplotypes, the specific genetic characteristics of Sardinians living in Italy were not adequately addressed by including RM Y-STRs. Therefore, they recommend incorporating Sardinian-specific data into the Italian Y-haplotype reference databases used for Y-STR match calculations (Lacerenza et al., 2017).

Caliebe and Krawczak (2018) examined the match probabilities of Y-STR haplotypes derived from reference populations that do not accurately represent the actual suspect population. The main factor that can affect the probability of Y-STR haplotypes matching is the degree of pairwise relatedness within the suspect population. This factor is not clearly defined and does not match accurately to the reference database population used to calculate the frequency of haplotypes obtained from crime scene samples. They discussed that the match probability is influenced by the relatedness exclusively through the total mutation rate (μ) of the Y-STR panel. As the overall rate of genetic mutation increased, the impact of genetic relatedness in the suspect population became insignificant. Consequently, using highly mutating RM Y-STR markers for trace donor identification reduces the likelihood of finding a Y-chromosomal match, even among the suspect's close relatives (Caliebe & Krawczak, 2018).

The integration of rapidly mutating Y-STRs and the adoption of kits containing a range of RM Y-STRs for the existing Y-STR databases could yield the aforementioned benefits. However, the estimated time and cost needed to upgrade the current Y-STR haplotype databases are the primary obstacles hindering the expansion of these databases. Thus, the immediate modification of databases with the hundred thousand Y-STR profiles is not feasible.

Investigative DNA databases have reported an increase in requests for familial searches in an effort to link criminals to their possible relatives. This requires considering both SM Y-STRs, which are essential for detecting probable familial connections among male individuals, and RM markers. Training and guidelines from authorized organizations are necessary to interpret RM Y-STR results accurately and identify mutation events that occur between haplotypes. In forensic casework, it is necessary for new statistical approaches to accurately calculate the weight of an RM Y-STR haplotype match. A recommended technique that may be useful to real forensic casework is to use locus-specific mutation rates and the number of meiotic transfers to formulate a probability assessment of the possibility of discovering a certain number of mutations during a typical meiotic transfer (Ballantyne et al., 2012, 2014; Henry et al., 2015).

7.11 Disadvantages of rapidly mutating Y-chromosomal short tandem repeats

Y-STR typing has many applications outside forensic investigations, including genealogical fields and the study of human evolution and migratory events through paternal lineages. The concept of employing Y-STRs as a molecular clock, which allows for the reconstruction of haplotype genealogies throughout history, relies on the basic idea that the average mutation rates in Y-STR haplotypes remain relatively constant over time. As a default, mutations are supposed to happen randomly, independent of particular haplogroups, populations, or periods. Therefore, as RM Y-STRs have much higher mutation rates, they could accelerate the molecular clock estimations and were not assumed to be the best candidates for phylogenetic studies. On the contrary, SM Y-STR variations have been suited to estimate tMRCA, to enhance the resolution, reduce complexities of phylogenetic relationships, and explore information about their phylogeny due to the lesser possibility of a random convergence in SM Y-STRs. Additionally, published results suggested the potential of combining SM Y-STR and RM Y-STR systems to address different evolutionary time windows (Baeta et al., 2018; Bosch et al., 1999).

DNA analysis via autosomal STRs is always prioritized to generate an eligible suspect's DNA profile for database search. If the autosomal could not provide an interpretable profile or results are inconclusive for the male component, Y-STR analysis is the choice as the second priority, which highlights the minor male portion for successful amplification. As an individual matching the crime scene profile, the RM Y-STR panel could be applied to exclude possible close male relatives. Although the application of RM Y-STRs is a powerful method for individualization of crime scene male profiles, the quantity of DNA that can be extracted from the crime scene specimen may be the main limitation of this strategy (Roewer et al., 2020).

The use of RM Y-STRs in paternity testing or mass catastrophe victim identification, through comparing Y-haplotypes among paternal bonds, can result in incorrect exclusions due to their high mutation rate, which can lead to the incorrect exclusion of actual paternity. The existing threshold of three or more differences between two Y-STR haplotypes within a single germline transmission, which is used to interpret paternity applications, has to be modified based on the mutation rates particular to each genetic locus. Furthermore, certain RM Y-STRs, such as DYF403S1b, may possess a large number of repeats, which is crucial for their increased mutation rates. However, their long sizes of the PCR amplicons may have adverse impacts on the quality of Y-haplotypes derived from severely degraded DNA, which is commonly encountered in cases of mass disaster human remains and poorly preserved or vulnerable crime scene samples (Ambers et al., 2018; Baeta et al., 2018; Ballantyne et al., 2012; Čokić et al., 2019; Henry et al., 2015; Zgonjanin et al., 2017).

Out of the 13 Y-STRs that mutate rapidly, there are four markers (DYF387S1, DYF399S1, DYF403S1, and DYF404S1) that have multiple alleles. These markers have the highest ability to differentiate between individuals with similar Y-STR profiles. However, interpreting these markers in mixtures of DNA from multiple male contributors is challenging and is usually avoided to simplify the analysis. Nevertheless, they can be beneficial when analyzing DNA from a single source (Alghafri, 2015).

As previously mentioned, RM Y-STRs have the ability to transform Y-STR haplotypes from being used simply for identifying paternal lineages to becoming a highly effective tool for personal individualization. Therefore, it is advised that any potential privacy concerns in the study setting and the publication of RM Y-STR individual haplotypes should be carefully considered and restricted to reports on newly discovered mutations or population-based haplotype statistics (Ralf et al., 2021).

7.12 Advances in rapidly mutating Y-chromosomal short tandem repeats typing

Due to the increasing number of crime scene items and sexual assault-related samples, forensic genetic laboratories have faced a surplus of pending cases. To address this issue, the rapid screening of multiple suspects within a limited time frame has emerged as an acceptable method to enhance the efficiency and timeliness of forensic investigations. In mass disaster tragedies, it is crucial to examine and check the close relatives of deceased individuals, as this can provide significant benefits.

Due to the fast polymerase enzymes and high-speed thermocyclers, which can complete 28 cycles in under 16 minutes, forensic scientists can improve PCR procedures for forensic samples. By using smaller PCR amplicons and microfluidic systems for electrophoresis instead of current CE instruments, the overall processing time of forensic samples can be reduced. For instance, four RM Y-STRs (DYS526a/b, DYS576, DYS626, and DYS570) with amplicon sizes ranging from 137 to 402 bp were amplified utilizing the Philisa rapid thermocycler (Streck) and a specifically built polymerase called

Z-Taq (Takara) in under 17 minutes with high accuracy. Electrophoresis of PCR products is performed using microfluidic chips with separation channels measuring 1.5 centimeters in length. Applying such a separation step requires less than 80 seconds. This approach has the capability to reduce the analysis time to under 25 minutes per Y-haplotype following sample extraction. This method has proven beneficial for screening sexual assault samples to exclude or presumptively identify male suspects (Gibson-Daw et al., 2017).

The Microreader RM-Y ID System is comprised of 17 RM Y-STRs, including 11 single-copy loci and six multicopy RM Y-STRs. These Y-STRs are labeled with five different fluorescent dyes: DYS630, DYS464, DYF403S1b1, DYF403S1b2, DYS612, DYS518, DYF403S1a, DYS526, DYF399S1, DYS626, DYS627, DYS527, DYS713, DYF404S1, DYF387S1, DYS449, and DYS547. Among complex Y-STRs, DYF399S1 had the greatest mutation rate per locus per generation at 8.8×10^{-2} (95% CI 5.60×10^{-2} to 1.30×10^{-1}).

The Yfiler Platinum PCR Amplification Kit and the Microreader RM-Y ID System were employed to analyze fifty father–son pairs. The Yfiler Platinum PCR Amplification Kit was able to differentiate 27 pairs of Y-haplotypes among the 50 analyzed random father–son pairs. This number was subsequently increased to 45 pair haplotypes for the Microreader RM-Y ID System, indicating the ability to differentiate even among close male relatives of the same paternal lineage (Zhou, Song, et al., 2022).

The current approach for analyzing Y-STRs relies on CE systems, which offer greater speed and cost-effectiveness compared to MPS.

Scientifically, the MPS analysis of Y-STRs at least benefited from three fields in comparison to CE systems: Initially, there was the possibility of analyzing over 100 Y-STRs simultaneously with MPS, as opposed to nearly 30 markers with CE systems. Secondly, MPS reveals variation in the sequence of the alleles rather than just its length. Consequently, two alleles with the same size but distinct sequences (isometric variants) could be distinguished via MPS, which are not differentiable with CE analysis. The third advantage of the MPS approach is the smaller size of MPS amplicons compared to those implemented in CE. This feature makes the MPS method particularly valuable for analyzing problematic samples, such as degraded DNA. Thus, it is recommended that the analysis of current SM Y-STR and MM Y-STR loci using MPS assays may be far more beneficial than the commonly used CE systems for distinguishing between male relatives, despite the higher cost and longer running time of the MPS procedures compared to CE systems. In 2022, Bredemeyer et al. conducted a comparison between the MPS assay of Y-STRs and CE in 64 father–son pairs to determine the cost and benefits of MPS in Y-STR analysis. The findings revealed that mutations are most frequently found in tetranucleotide repeat motifs and are mainly limited to the longest uninterrupted repeat sequence (LUS) of a compound or complex Y-STR marker. The MPS analysis reveals a similar number of mutational events (13 mutations out of 64 father–son pairs) compared to the CE analysis. These mutations are primarily length-based variants. This similarity may be attributed to the lower mutation rates of the traditional Y-STRs. Thus, the MPS analysis failed to distinguish more father–son pairs compared to CE using the PowerPlex Y23 System. Therefore, forensic MPS techniques based on traditional MM Y-STRs and SM YTRs may not provide significant benefits for the individualization of males.

Nevertheless, after investigating the dataset using RM Y-STRs, CE discovered eight more supplementary mutations. Therefore, for distinguishing between father–son pairs, it is more efficient to use a comprehensive analysis of additional RM Y-STRs using CE rather than focusing only on the MPS for the commonly used Y-STRs. On top of that, Huszar et al. (2018) suggested the use of MPS for RM Y-STRs is a potentially effective approach to enhance discriminatory power, especially because the results indicate an improvement in allele diversity (Bredemeyer et al., 2022; Huszar et al., 2018).

A subsequent investigation conducted by Moonand et al. in 2022 employed MPS analysis to investigate the Y-STR sequence variation. The study reported the detection of 195 previously undisclosed alleles by MPS, which were not detectable using the CE technique. The MPS analysis results in a 3.69-fold increase of DYS518 alleles, which is the highest increase observed among the 31 Y-STRs, primarily due to changes in the repeat region. Remarkably, allele 39 of DYS518 shows that it can be differentiated into 7 distinct iso-alleles. In DYF399S1, SNPs located in the flanking region were also responsible for the majority of the observed allele increases. Male individualization could be advanced with the use of discriminative MPS assays for RM Y-STRs because of their diverse sequences (Moon et al., 2022).

7.13 Conclusion

The introduction of RM Y-STRs as a supplement to the commonly used slowly and moderately mutation Y-STRs could bring forensic genetics even closer to the goal of male individualization through Y-chromosome marker analysis. More research is needed to gather comprehensive data on mutation rates at these new genomic sites within different worldwide populations with diverse genetic backgrounds. Furthermore, RM Y-STR investigations on other forensically relevant body parts, such as hairs, in a large population of male individuals should be considered to explore the in vivo effects of somatic mutability on RM Y-STRs.

References

Adnan, A., Rakha, A., Nazir, S., Alghafri, R., Hassan, Q., Wang, C. C., & Lu, J. (2021). Forensic features and genetic legacy of the Baloch population of Pakistan and the Hazara population across Durand line revealed by Y-chromosomal STRs. *International Journal of Legal Medicine, 135*(5), 1777–1784. https://doi.org/10.1007/s00414-021-02591-2.

Adnan, A., Ralf, A., Rakha, A., Kousouri, N., & Kayser, M. (2016). Improving empirical evidence on differentiating closely related men with RM Y-STRs: A comprehensive pedigree study from Pakistan. *Forensic Science International: Genetics, 25*, 45–51. https://doi.org/10.1016/j.fsigen.2016.07.005.

Albastaki, A., Naji, M., Lootah, R., Almheiri, R., Almulla, H., Alreyami, A., Almarri, I., Yu, Z., Zhang, Y., Shu, C., & Alghafri, R. (2022). Design and development of novel single multiplex system incorporating 26 rapidly mutating Y-STRs; 26 RM Yplex. *Electrophoresis, 43*(18–19), 1911–1919. https://doi.org/10.1002/elps.202200099.

Alghafri, R. (2015). An evaluation of rapidly mutating Y-STR multi-allelic markers. *Forensic Science International: Genetics Supplement Series, 5*, e647–e649. https://doi.org/10.1016/j.fsigss.2015.10.011.

Alghafri, R., Zupanič Pajnič, I., Zupanc, T., Balažic, J., & Shrivastava, P. (2018). Rapidly mutating Y-STR analyses of compromised forensic samples. *International Journal of Legal Medicine, 132*(2), 397–403. https://doi.org/10.1007/s00414-017-1600-z.

Almohammed, E., & Hadi, S. (2019). A study of genetic analysis using novel rapidly mutating Y-STR multiplex for Qatari population. *Forensic Science International: Genetics Supplement Series, 7*(1), 875–877. https://doi.org/10.1016/j.fsigss.2019.11.008.

Ambers, A., Votrubova, J., Vanek, D., Sajantila, A., & Budowle, B. (2018). Improved Y-STR typing for disaster victim identification, missing persons investigations, and historical human skeletal remains. *International Journal of Legal Medicine, 132*(6), 1545–1553. https://doi.org/10.1007/s00414-018-1794-8.

Antão-Sousa, S., Gusmão, L., Modesti, N. M., Feliziani, S., Faustino, M., Marcucci, V., Sarapura, C., Ribeiro, J., Carvalho, E., Pereira, V., Tomas, C., de Pancorbo, M. M., Baeta, M., Alghafri, R., Almheiri, R., Builes, J. J., Gouveia, N., Burgos, G., Pontes, M.de L., ... Pinto, N. (2024). Microsatellites' mutation modeling through the analysis of the Y-chromosomal transmission: Results of a GHEP-ISFG collaborative study. *Forensic Science International: Genetics, 69*. https://doi.org/10.1016/j.fsigen.2023.102999.

Ay, M., Serin, A., Sevay, H., Gurkan, C., & Canan, H. (2018). Genetic characterisation of 13 rapidly mutating Y-STR loci in 100 father and son pairs from South and East Turkey. *Annals of Human Biology, 45*(6–8), 506–515. https://doi.org/10.1080/03014460.2018.1559353.

Baeta, M., Núñez, C., Villaescusa, P., Ortueta, U., Ibarbia, N., Herrera, R. J., Blazquez-Caeiro, J. L., Builes, J. J., Jiménez-Moreno, S., Martínez-Jarreta, B., & de Pancorbo, M. M. (2018). Assessment of a subset of slowly mutating Y-STRs for forensic and evolutionary studies. *Forensic Science International: Genetics, 34*, e7–e12. https://doi.org/10.1016/j.fsigen.2018.03.008.

Ballantyne, K. N., Goedbloed, M., Fang, R., Schaap, O., Lao, O., Wollstein, A., Choi, Y., Van Duijn, K., Vermeulen, M., Brauer, S., Decorte, R., Poetsch, M., Von Wurmb-Schwark, N., De Knijff, P., Labuda, D., Vézina, H., Knoblauch, H., Lessig, R., Roewer, L., ... Kayser, M. (2010). Mutability of Y-chromosomal microsatellites: Rates, characteristics, molecular bases, and forensic implications. *American Journal of Human Genetics, 87*(3), 341–353. https://doi.org/10.1016/j.ajhg.2010.08.006.

Ballantyne, K. N., Keerl, V., Wollstein, A., Choi, Y., Zuniga, S. B., Ralf, A., Vermeulen, M., De Knijff, P., & Kayser, M. (2012). A new future of forensic Y-chromosome analysis: Rapidly mutating Y-STRs for differentiating male relatives and paternal lineages. *Forensic Science International: Genetics, 6*(2), 208–218. https://doi.org/10.1016/j.fsigen.2011.04.017.

Ballantyne, K. N., Ralf, A., Aboukhalid, R., Achakzai, N. M., Anjos, M. J., Ayub, Q., Balažic, J., Ballantyne, J., Ballard, D. J., Berger, B., Bobillo, C., Bouabdellah, M., Burri, H., Capal, T., Caratti, S., Cárdenas, J., Cartault, F., Carvalho, E. F., Carvalho, M., ... Kayser, M. (2014). Toward male individualization with rapidly mutating Y-chromosomal short tandem repeats. *Human Mutation, 35*(8), 1021–1032. https://doi.org/10.1002/humu.22599.

Boattini, A., Sarno, S., Mazzarisi, A. M., Viroli, C., De Fanti, S., Bini, C., Larmuseau, M. H. D., Pelotti, S., & Luiselli, D. (2019). Estimating Y-Str mutation rates and tmrca through deep-rooting Italian pedigrees. *Scientific Reports, 9*(1). https://doi.org/10.1038/s41598-019-45398-3.

Bini, C., Cilli, E., Sarno, S., Traversari, M., Fontani, F., Boattini, A., Pelotti, S., & Luiselli, D. (2021). Twenty-seven Y-chromosome short tandem repeats analysis of Italian mummies of the 16th and 18th centuries: An interdisciplinary research. *Frontiers in Genetics, 12*. https://doi.org/10.3389/fgene.2021.720640.

Bredemeyer, S., Roewer, L., & Willuweit, S. (2022). Next generation sequencing of Y-STRs in father-son pairs and comparison with traditional capillary electrophoresis. *Forensic Sciences Research, 7*(3), 484–489. https://doi.org/10.1080/20961790.2021.1898078.

Bosch, E., Calafell, F., Santos, F. R., Pérez-Lezaun, A., Comas, D., Benchemsi, N., Tyler-Smith, C., & Bertranpetit, J. (1999). Variation in short tandem repeats is deeply structured by genetic background on the human Y chromosome. *Am. J. Hum. Genet, 65*.

Caliebe, A., & Krawczak, M. (2018). Match probabilities for Y-chromosomal profiles: A paradigm shift. In *Forensic Science International: Genetics, Vol. 37*. Elsevier Ireland Ltd, 200–203. https://doi.org/10.1016/j.fsigen.2018.08.009.

Caliebe, A., Jochens, A., Willuweit, S., Roewer, L., & Krawczak, M. (2015). No shortcut solution to the problem of Y-STR match probability calculation. *Forensic Science International: Genetics, 15*, 69–75. https://doi.org/10.1016/j.fsigen.2014.10.016.

Claerhout, S., Van der Haegen, M., Vangeel, L., Larmuseau, M. H. D., & Decorte, R. (2019). A game of hide and seq: Identification of parallel Y-STR evolution in deep-rooting pedigrees. *European Journal of Human Genetics, 27*(4), 637–646. https://doi.org/10.1038/s41431-018-0312-2.

Čokić, V. P., Kecmanović, M., Zgonjanin Bosić, D., Jakovski, Z., Veljković, A., Katić, S., Keckarević Marković, M., & Keckarević, D. (2019). A comprehensive mutation study in wide deep-rooted R1b Serbian pedigree: mutation rates and male relative differentiation capacity of 36 Y-STR markers. *Forensic Science International: Genetics, 41*, 137–144. https://doi.org/10.1016/j.fsigen.2019.04.007.

Court, D. S. (2021). The y chromosome and its use in forensic dna analysis. In *Emerging topics in life sciences, Vol. 5*. Portland Press Ltd., 427–441. https://doi.org/10.1042/ETLS20200339, Issue 3.

D'Atanasio, E., Iacovacci, G., Pistillo, R., Bonito, M., Dugoujon, J. M., Moral, P., El-Chennawi, F., Melhaoui, M., Baali, A., Cherkaoui, M., Sellitto, D., Trombetta, B., Berti, A., & Cruciani, F. (2019). Rapidly mutating Y-STRs in rapidly expanding populations: Discrimination power of the Yfiler Plus multiplex in northern Africa. *Forensic Science International: Genetics, 38*, 185–194. https://doi.org/10.1016/j.fsigen.2018.11.002.

Dash, H.R., Shrivastava, P., & Lorente, J.A. (n.d.). *Handbook of DNA profiling*.

Della Rocca, C., Alladio, E., Barni, F., Cannone, F., D'Atanasio, E., Trombetta, B., Berti, A., & Cruciani, F. (2019). Low discrimination power of the Yfiler™ plus PCR amplification kit in African populations. Do we need more RM Y-STRs? *Forensic Science International: Genetics Supplement Series, 7*(1), 671–673. https://doi.org/10.1016/j.fsigss.2019.10.133.

Della Rocca, C., Cannone, F., D'Atanasio, E., Bonito, M., Anagnostou, P., Russo, G., Barni, F., Alladio, E., Destro-Bisol, G., Trombetta, B., Berti, A., & Cruciani, F. (2020). Ethnic fragmentation and degree of urbanization strongly affect the discrimination power of Y-STR haplotypes in central Sahel. *Forensic Science International: Genetics, 49*. https://doi.org/10.1016/j.fsigen.2020.102374.

Diegoli, T. M. (2015). Forensic typing of short tandem repeat markers on the X and Y chromosomes. *Forensic Science International: Genetics, 18*, 140–151. https://doi.org/10.1016/j.fsigen.2015.03.013.

Eskandarion, M. R., Tabrizi, A. A., Shirkoohi, R., Raoofian, R., Naji, M., Pazhoomand, R., Salari, H., Samadirad, B., Sabouri, A., Zohour, M. M., Namazi, H., Farhadi, P., Baratieh, Z., Sayyari, M., Dadgarmoghaddam, M., Safdarian, E., Nikbakht, A., Golshan, F., Baybordi, F., ... Mohebbi, R. (2024). Haplotype diversity of 17 Y-STR in the Iranian population. *BMC Genomics, 25*(1). https://doi.org/10.1186/s12864-024-10217-1.

Fan, G., Pan, L., Tang, P., Zhou, Y., Liu, M., & Luo, X. (2021). Technical note: developmental validation of a novel 41-plex Y-STR system for the direct amplification of reference samples. *International Journal of Legal Medicine, 135*(2), 409–419. https://doi.org/10.1007/s00414-020-02326-9.

Farhatullah, M., Absar, M., Rauf, S., Tracy, M., Almulla, H., Almheiri, R., Adam, A., Alghafri, R., & Khan, M. R. (2019). Evaluation of rapidly mutating Y-STRs in Pakistani population. *Forensic Science International: Genetics Supplement Series, 7*(1), 245–247. https://doi.org/10.1016/j.fsigss.2019.09.094.

Gibson-Daw, G., Albani, P., Gassmann, M., & McCord, B. (2017). Rapid microfluidic analysis of a Y-STR multiplex for screening of forensic samples. *Analytical and Bioanalytical Chemistry, 409*(4), 939–947. https://doi.org/10.1007/s00216-016-9950-9.

Gopinath, S., Zhong, C., Nguyen, V., Ge, J., Lagacé, R. E., Short, M. L., & Mulero, J. J. (2016). Developmental validation of the Yfiler® Plus PCR Amplification Kit: An enhanced Y-STR multiplex for casework and database applications. *Forensic Science International: Genetics, 24*, 164–175. https://doi.org/10.1016/j.fsigen.2016.07.006.

Hadi, S., Yao, J., & Adnan, A. (2022). Editorial: Role of Y chromosome in molecular anthropology, forensics, and genetic genealogy. *Frontiers in Genetics, 13*. https://doi.org/10.3389/fgene.2022.863455.

Henry, J., Simon, C., & Linacre, A. (2015). The benefits and limitations of expanded Y-chromosome short tandem repeat (Y-STR) loci. *Forensic Science International: Genetics Supplement Series, 5*, e28–e30. https://doi.org/10.1016/j.fsigss.2015.09.012.

Huszar, T. I., Jobling, M. A., & Wetton, J. H. (2018). A phylogenetic framework facilitates Y-STR variant discovery and classification via massively parallel sequencing. *Forensic Science International: Genetics, 35*, 97–106. https://doi.org/10.1016/j.fsigen.2018.03.012.

Iacovacci, G., D'Atanasio, E., Marini, O., Coppa, A., Sellitto, D., Trombetta, B., Berti, A., & Cruciani, F. (2017). Forensic data and microvariant sequence characterization of 27 Y-STR loci analyzed in four Eastern African countries. *Forensic Science International: Genetics, 27*, 123–131. https://doi.org/10.1016/j.fsigen.2016.12.015.

Iyavoo, S., Alghafri, R., Almheiri, R., & Haizel, T. (2019). Evaluation of 13 rapidly mutating Y-STRs on a Dravidian pedigree. *Forensic Science International: Genetics Supplement Series, 7*(1), 216–217. https://doi.org/10.1016/j.fsigss.2019.09.083.

Jankauskiene, J., Kukiene, J., Ivanova, V., & Aleknaviciute, G. (2017). Population data and forensic genetic evaluation with the Yfiler™ Plus PCR Amplification kit in the Lithuanian population. *Forensic Science International: Genetics Supplement Series, 6*, e606–e607. https://doi.org/10.1016/j.fsigss.2017.10.009.

Javed, F., Shafique, M., McNevin, D., Javed, M. U., Shehzadi, A., & Shahid, A. A. (2022). Empirical Evidence on Enhanced Mutation Rates of 19 RM-YSTRs for Differentiating Paternal Lineages. *Genes, 13*(6). https://doi.org/10.3390/genes13060946.

Kasu, M., Fredericks, J., Fraser, M., Labuschagne, C., Lesaoana, M., & D'Amato, M. E. (2019). Novel Y-chromosome short tandem repeat sequence variation for loci DYS710, DYS518, DYS385, DYS644, DYS612, DYS626, DYS504, DYS481, DYS447 and DYS449. *International Journal of Legal Medicine, 133*(6), 1681–1689. https://doi.org/10.1007/s00414-019-02056-7.

Lacerenza, D., Aneli, S., Di Gaetano, C., Critelli, R., Piazza, A., Matullo, G., Culigioni, C., Robledo, R., Robino, C., & Calò, C. (2017). Investigation of extended Y chromosome STR haplotypes in Sardinia. *Forensic Science International: Genetics, 27*, 172–174. https://doi.org/10.1016/j.fsigen.2016.12.009.

Lee, E. Y., Lee, H. Y., Kwon, S. Y., Oh, Y. N., Yang, W. I., & Shin, K. J. (2017). A multiplex PCR system for 13 RM Y-STRs with separate amplification of two different repeat motif structures in DYF403S1a. *Forensic Science International: Genetics, 26*, 85–90. https://doi.org/10.1016/j.fsigen.2016.10.019.

Lee, S., McAnany, P. K., Chang, C. W., Norona, W., Short, M. L., Mulero, J. J., & Zhong, C. (2023). Developmental validation of applied biosystems YFiler platinum casework PCR amplification kit. *Scientific Reports, 13*(1). https://doi.org/10.1038/s41598-023-41788-w.

Li, M., Zhou, W., Zhang, Y., Huang, L., Wang, X., Wu, J., Meng, M., Wang, H., Li, C., & Bian, Y. (2020). Development and validation of a novel 29-plex Y-STR typing system for forensic application. *Forensic Science International: Genetics, 44*. https://doi.org/10.1016/j.fsigen.2019.102169.

McEvoy, B., Brady, C., Moore, L. T., & Bradley, D. G. (2006). The scale and nature of Viking settlement in Ireland from Y-chromosome admixture analysis. *European Journal of Human Genetics, 14*(12), 1288–1294. https://doi.org/10.1038/sj.ejhg.5201709.

Mitchell, M. R., Chaseling, J., Jones, L., White, T., Bernie, A., Haupt, L. M., Griffiths, L. R., & Wright, K. M. (2024). Improving the strategy to identify historical military remains: a literature review and Y-STR meta-analysis. *Forensic Sciences Research, 9*(1). https://doi.org/10.1093/fsr/owad050.

Moon, M. H., Hong, S. R., & Shin, K. J. (2022). Sequence variations of 31 Y-chromosomal short tandem repeats including 9 rapidly mutating Y-chromosomal short tandem repeats analyzed by massively parallel sequencing in three U.S. population groups and Korean population. *Journal of Korean Medical Science, 37*(6). https://doi.org/10.3346/JKMS.2022.37.E40.

Nazir, S., Adnan, A., Rehman, R. A., Al-Qahtani, W. S., Alsaleh, A. B., Al-Harthi, H. S., Safhi, F. A., Almheiri, R., Lootah, R., Alreyami, A., Almarri, I., Wang, C. C., Rakha, A., & Hadi, S. (2022). Mutation rate analysis of RM Y-STRs in deep-rooted multi-generational Punjabi pedigrees from Pakistan. *Genes, 13*(8). https://doi.org/10.3390/genes13081403.

Neuhuber, F., Dunkelmann, B., Grießner, I., Helm, K., Kayser, M., & Ralf, A. (2022). Improving the differentiation of closely related males by RMplex analysis of 30 Y-STRs with high mutation rates. *Forensic Science International: Genetics, 58*. https://doi.org/10.1016/j.fsigen.2022.102682.

Niederstätter, H., Berger, B., Kayser, M., & Parson, W. (2016). Differences in urbanization degree and consequences on the diversity of conventional vs. rapidly mutating Y-STRs in five municipalities from a small region of the Tyrolean Alps in Austria. *Forensic Science International: Genetics, 24*, 180–193. https://doi.org/10.1016/j.fsigen.2016.07.009.

Oh, Y. N., Lee, H. Y., Lee, E. Y., Kim, E. H., Yang, W. I., & Shin, K. J. (2015). Haplotype and mutation analysis for newly suggested Y-STRs in Korean father-son pairs. *Forensic Science International: Genetics, 15*, 64–68. https://doi.org/10.1016/j.fsigen.2014.09.023.

Otagiri, T., Sato, N., Asamura, H., Parvanova, E., Kayser, M., & Ralf, A. (2022). RMplex reveals population differences in RM Y-STR mutation rates and provides improved father-son differentiation in Japanese. *Forensic Science International: Genetics, 61*. https://doi.org/10.1016/j.fsigen.2022.102766.

Otagiri, T., Sato, N., Shiozaki, T., Harayama, Y., Hayashi, T., Kobayashi, K., & Asamura, H. (2021). Mutation analysis for 25 Y-STR markers in Japanese population. *Legal Medicine, 50*. https://doi.org/10.1016/j.legalmed.2021.101860.

Pickrahn, I., Müller, E., Zahrer, W., Dunkelmann, B., Cemper-Kiesslich, J., Kreindl, G., & Neuhuber, F. (2016). Yfiler® Plus amplification kit validation and calculation of forensic parameters for two Austrian populations. *Forensic Science International: Genetics, 21*, 90–94. https://doi.org/10.1016/j.fsigen.2015.12.014.

Puch-Solis, R., Pope, S., & Tully, G. (2024). Considerations on the application of a mutation model for Y-STR interpretation. *Science and Justice, 64*(2), 180–192. https://doi.org/10.1016/j.scijus.2024.01.001.

Rakha, A., Oh, Y. N., Lee, H. Y., Hussain, S., Waryah, A. M., Adnan, A., & Shin, K. J. (2018). Discriminating power of rapidly mutating Y-STRs in deep rooted endogamous pedigrees from Sindhi population of Pakistan. *Legal Medicine, 34*, 17–20. https://doi.org/10.1016/j.legalmed.2018.08.001.

Ralf, A., Lubach, D., Kousouri, N., Winkler, C., Schulz, I., Roewer, L., Purps, J., Lessig, R., Krajewski, P., Ploski, R., Dobosz, T., Henke, L., Henke, J., Larmuseau, M. H. D., & Kayser, M. (2020). Identification and characterization of novel rapidly mutating Y-chromosomal short tandem repeat markers. *Human Mutation, 41*(9), 1680–1696. https://doi.org/10.1002/humu.24068.

Ralf, A., Montiel González, D., Zandstra, D., van Wersch, B., Kousouri, N., de Knijff, P., Adnan, A., Claerhout, S., Ghanbari, M., Larmuseau, M. H. D., & Kayser, M. (2023). Large-scale pedigree analysis highlights rapidly mutating Y-chromosomal short tandem repeats for differentiating patrilineal relatives and predicting their degrees of consanguinity. *Human Genetics, 142*(1), 145–160. https://doi.org/10.1007/s00439-022-02493-2.

Ralf, A., van Wersch, B., Montiel González, D., & Kayser, M. (2024). Male pedigree toolbox: A versatile software for Y-STR data analyses. *Genes, 15*(2). https://doi.org/10.3390/genes15020227.

Ralf, A., Zandstra, D., Weiler, N., van Ijcken, W. F. J., Sijen, T., & Kayser, M. (2021). RMplex: An efficient method for analyzing 30 Y-STRs with high mutation rates. *Forensic Science International: Genetics, 55*. https://doi.org/10.1016/j.fsigen.2021.102595.

Rapone, C., D'Atanasio, E., Agostino, A., Mariano, M., Papaluca, M. T., Cruciani, F., & Berti, A. (2016). Forensic genetic value of a 27 Y-STR loci multiplex (Yfiler® Plus kit) in an Italian population sample. *Forensic Science International: Genetics, 21*, e1–e5. https://doi.org/10.1016/j.fsigen.2015.11.006.

Rhie, A., Nurk, S., Cechova, M., Hoyt, S. J., Taylor, D. J., Altemose, N., Hook, P. W., Koren, S., Rautiainen, M., Alexandrov, I. A., Allen, J., Asri, M., Bzikadze, A. V., Chen, N. C., Chin, C. S., Diekhans, M., Flicek, P., Formenti, G., Fungtammasan, A., ... Phillippy, A. M. (2023). The complete sequence of a human Y chromosome. *Nature, 621*(7978), 344–354. https://doi.org/10.1038/s41586-023-06457-y.

Robino, C., Ralf, A., Pasino, S., De Marchi, M. R., Ballantyne, K. N., Barbaro, A., Bini, C., Carnevali, E., Casarino, L., Di Gaetano, C., Fabbri, M., Ferri, G., Giardina, E., Gonzalez, A., Matullo, G., Nutini, A. L., Onofri, V., Piccinini, A., Piglionica, M., ... Kayser, M. (2015). Development of an Italian RM Y-STR haplotype database: Results of the 2013 GEFI collaborative exercise. *Forensic Science International: Genetics, 15*, 56–63. https://doi.org/10.1016/j.fsigen.2014.10.008.

Roewer, L., Andersen, M. M., Ballantyne, J., Butler, J. M., Caliebe, A., Corach, D., D'Amato, M. E., Gusmão, L., Hou, Y., de Knijff, P., Parson, W., Prinz, M., Schneider, P. M., Taylor, D., Vennemann, M., & Willuweit, S. (2020). DNA commission of the International Society of Forensic Genetics (ISFG): Recommendations on the interpretation of Y-STR results in forensic analysis. *Forensic Science International: Genetics, 48*. https://doi.org/10.1016/j.fsigen.2020.102308.

Rogalla, U., Woźniak, M., Swobodziński, J., Derenko, M., Malyarchuk, B. A., Dambueva, I., Koziński, M., Kubica, J., & Grzybowski, T. (2015). A novel multiplex assay amplifying 13 Y-STRs characterized by rapid and moderate mutation rate. *Forensic Science International: Genetics, 15*, 49–55. https://doi.org/10.1016/j.fsigen.2014.11.004.

Shang, L., Ding, G., Mo, X., Sun, J., Sun, H., Yu, Z., & Li, W. (2021). A novel multiplex of 12 multicopy Y-STRs for forensic application. *Journal of Forensic Sciences, 66*(5), 1901–1907. https://doi.org/10.1111/1556-4029.14774.

Shrivastava, P., Alghafri, R., Almheiri, R., Mishra, A., Alfalasi, H., Khoory, S., Damji, R., Kishan Kumawat, R., Kumar Choudhary, S., Dixit, S., Kaitholia, K., Ranjan Dash, H., Ben Trivedi, V., & Shrivastava, D. (2019). Investigation on rapidly mutating Y-STRs multiplex in Indian population: A pilot study. *Forensic Science International: Genetics Supplement Series, 7*(1), 805–806. https://doi.org/10.1016/j.fsigss.2019.10.184.

Soldati, G., Turrina, S., & De Leo, D. (2022). Forensic assessment on the application of a virtual pool of 30 Y-STRs. *Forensic Science International: Genetics Supplement Series, 8*, 317–320. https://doi.org/10.1016/j.fsigss.2022.10.075.

Solé-Morata, N., García-Fernández, C., Urasin, V., Bekada, A., Fadhlaoui-Zid, K., Zalloua, P., Comas, D., & Calafell, F. (2017). Whole Y-chromosome sequences reveal an extremely recent origin of the most common North African paternal lineage E-M183 (M81). *Scientific Reports, 7*(1). https://doi.org/10.1038/s41598-017-16271-y.

Ullah, M. F., Ali, A., Majeed, A., Ijaz, A., Albastaki, A., Khan, M. I., Khan, M. R., Alghamdi, H. A., & Alshahrani, M. A. (2022). Population and mutational assessment of novel repeats in 13RM Y-STRs in unrelated males born in Gilgit, Pakistan. *Saudi Journal of Biological Sciences, 29*(5), 3177–3183. https://doi.org/10.1016/j.sjbs.2022.01.055.

Wang, F., Song, F., Wang, X., Song, M., Zhou, Y., Liu, J., Wang, Z., & Hou, Y. (2022). Mutation analysis for newly suggested 30 Y-STR loci with high mutation rates in Chinese father-son pairs. *Scientific Reports, 12*(1). https://doi.org/10.1038/s41598-022-20014-z.

Wang, M., Wang, Z., Zhang, Y., He, G., Liu, J., & Hou, Y. (2017). Forensic characteristics and phylogenetic analysis of two Han populations from the southern coastal regions of China using 27 Y-STR loci. *Forensic Science International: Genetics, 31*, e17–e23. https://doi.org/10.1016/j.fsigen.2017.10.009.

Wang, Q., Jin, B., An, G., Zhong, Q., Chen, M., Luo, X., Li, Z., Jiang, Y., Liang, W., & Zhang, L. (2019). Rapidly mutating Y-STRs study in Chinese Yi population. *International Journal of Legal Medicine, 133*(1), 45–50. https://doi.org/10.1007/s00414-018-1894-5.

Wu, W., Ren, W., Hao, H., Nan, H., He, X., Liu, Q., & Lu, D. (2018). Mutation rates at 42 Y chromosomal short tandem repeats in Chinese Han population in Eastern China. *International Journal of Legal Medicine, 132*(5), 1317–1319. https://doi.org/10.1007/s00414-018-1784-x.

Xiong, C., Yang, C., Wu, W., Zeng, Y., Lin, T., Chen, L., Liu, H., Liu, C., Du, W., Wang, M., Sun, H., & Liu, C. (2022). Development and validation of a multiplex typing system with 32 Y-STRs for forensic application. *Forensic Science International, 339*. https://doi.org/10.1016/j.forsciint.2022.111409.

Ye, Y., An, Y., Yang, Y., Wu, H., Zheng, Y., & Liao, L. (2022). Assessment of the forensic application of 50 Y-STR markers in a large pedigree. *Forensic Sciences Research, 7*(2), 207–210. https://doi.org/10.1080/20961790.2020.1802827.

Zeng, Y., Chen, L., Wang, M., Yang, C., Liu, H., Xiao, C., Liu, C. H., Li, Y., Xu, Q., Du, W., & Liu, C. (2022). The validation of a single multiplex typing system with 45 Y-STR markers for familial searching and database construction. *Frontiers in Genetics, 13*. https://doi.org/10.3389/fgene.2022.842004.

Zgonjanin, D., Alghafri, R., Almheiri, R., Antov, M., Toljic, D., Vukovic, R., & Petkovic, S. (2017). Mutation rate at 13 rapidly mutating Y-STR loci in the population of Serbia. *Forensic Science International: Genetics Supplement Series, 6*, e377–e379. https://doi.org/10.1016/j.fsigss.2017.09.171.

Zhou, Y., Cui, W., Wu, B., & Zhu, B. (2022). Development and validation of a new multiplex Y-STR panel designed to increase the power of discrimination. *Electrophoresis, 43*(18–19), 1899–1910. https://doi.org/10.1002/elps.202100313.

Zhou, Y., Song, F., Dai, H., Wang, S., Zhang, K., Wei, X., Wang, X., & Luo, H. (2022). Developmental validation of the Microreader™ RM-Y ID System: A new rapidly mutating Y-STR 17-plex system for forensic application. *International Journal of Legal Medicine, 136*(2), 501–512. https://doi.org/10.1007/s00414-021-02632-w.

Chapter 8

The use of biallelic genetic markers in forensic DNA analysis

Anna Barbaro
President Worldwide Association of Women Forensic Experts (WAWFE), Reggio Calabria, Italy

8.1 Introduction

Forensic DNA analysis has revolutionized the field of criminal justice, providing a powerful tool for identifying individuals, establishing biological relationships, and solving crimes. New technologies are continually developed and validated to enhance the ability of laboratories to recover DNA results with greater sensitivity and informativeness (Butler et al., 2004). Traditionally, forensic DNA analysis has relied heavily on short tandem repeats (STRs), which are highly polymorphic and provide a high degree of discrimination between individuals (Goodwin et al., 2011). However, STRs are not without limitations, particularly when dealing with degraded or low-quality DNA samples (Jobling & Gill, 2004).

In recent years, biallelic genetic markers, especially single-nucleotide polymorphisms (SNPs), have emerged as valuable alternatives and complements to STRs in forensic applications (Sobrino et al., 2005). Biallelic markers are characterized by the presence of two alleles at a specific locus, which makes them simpler in terms of mutation patterns and evolutionary stability compared to multi-allelic markers like STRs (Phillips, 2012). SNPs are the most common type of biallelic markers and occur frequently throughout the genome, making them abundant and informative for forensic purposes (Shriver et al., 1997). The application of biallelic markers in forensic DNA analysis offers several advantages. Their lower mutation rate and higher stability make them particularly useful for analyzing old, degraded, or highly damaged samples where STRs might fail (Butler et al., 2007). Moreover, the shorter amplicon lengths required for SNP analysis enhance their performance with fragmented DNA (Gill et al., 2004). Additionally, the dense distribution of SNPs across the genome provides high-resolution data, which is beneficial for individual identification and kinship analysis (Sanchez et al., 2006).

Beyond individual identification, certain SNPs are informative for biogeographical ancestry and phenotypic prediction, offering valuable leads in criminal investigations where no direct suspect is available (Kayser & Schneider, 2009). These capabilities highlight the multifaceted utility of biallelic markers in forensic science.

Technological advancements, such as NGS and high-throughput genotyping platforms, have further facilitated the adoption of SNPs in forensic laboratories (Kidd et al., 2014). These technologies enable the simultaneous analysis of large panels of SNPs, enhancing the efficiency and comprehensiveness of forensic investigations (Hofreiter et al., 2015).

Despite their advantages, the use of biallelic markers in forensic DNA analysis also presents challenges. Population stratification, ethical considerations, and the need for extensive reference databases are critical factors that must be addressed to ensure accurate and responsible use (Buckleton et al., 2011).

8.2 Single-nucleotide polymorphisms

A SNP is a single base change in a DNA sequence, occurring when one nucleotide (A, T, C, or G) is replaced by another. SNPs represent the most common form of natural genetic variation in the human genome, accounting for approximately 90% of this variation and serving as a major genetic source of phenotypic variability (Collins et al., 1998). It is estimated that around 7 million SNPs occur with a minimum allele frequency (MAF) of 5% in the genome, with an additional 7 million occurring with a MAF of 1% (Sachidanandam et al., 2001).

Investigative and Predictive DNA Testing. DOI: https://doi.org/10.1016/B978-0-443-30128-5.00004-7

The distribution of SNPs varies across human chromosomes, with the Y and X chromosomes exhibiting less genetic variation than autosomes due to their lower chromosome numbers (Jobling & Tyler-Smith, 2003). The most common type of SNP involves alleles A and G on one strand, and T and C on the complementary strand. Although theoretically, a SNP could have three or four alleles, most common SNPs are biallelic, meaning they have only two alleles: the major and minor alleles, with the minor allele being less frequent at a locus (Brookes, 1999).

The distribution of SNPs varies between human populations; an allele common in one geographical or ethnic group may be less frequent or rare in another (Jorde et al., 2000). This variation is further reduced by linkage disequilibrium between closely linked SNP markers, which creates haplotype blocks separated by recombination hotspots (Gabriel et al., 2002). SNPs arise initially from rare spontaneous mutations and reach appreciable population frequencies through genetic drift and selection. The spontaneous mutation rate is low (approximately 10^{-8} per base pair per generation), making subsequent mutations at the same locus unlikely (Nachman & Crowell, 2000).

SNPs can be found within coding or noncoding regions of genes, or in intergenic regions between genes, in both nuclear and mtDNA (Venter et al., 2001). Since only about 3%–5% of the DNA sequence codes for proteins, most SNPs are located outside of coding sequences (International Human Genome Sequencing Consortium, 2001). Due to the degeneracy of the genetic code, a SNP within a coding sequence does not always change the amino acid sequence of the protein, often resulting in a silent mutation (synonymous SNP) (Koonin & Galperin, 2003). Conversely, nonsynonymous SNPs produce variations in polypeptide sequences: missense changes result in different amino acids, while nonsense changes lead to premature stop codons (Krawczak et al., 2000). SNPs in noncoding regions can affect gene splicing, transcription factor binding, or the sequence of noncoding RNA.

Variations in DNA sequences can predispose individuals to diseases or influence their responses to diseases, environmental factors (such as bacteria, viruses, toxins, and chemicals), drugs, and therapies. They can also affect individual phenotypes, although these phenotypic effects depend on the presence of multiple polymorphisms and their positions within a given region (Wang et al., 1998). In recent years, significant research efforts from both public (Human Genome Project - https://www.genome.gov/human-genome-project) and private (SNP Consortium - http://snp.cshl.org) sectors have generated high-density SNP maps, providing a framework for further research studies and forensic applications (Collins et al., 2003).

SNPs offer many advantages, such as their abundance in the genome, low mutation rates (10^-8), reduced amplicon sizes, high-throughput genotyping, and potential for phenotypic trait prediction and ancestry studies (Kwok et al., 1996). However, they also have disadvantages, including low polymorphism and discrimination power, the need to analyze a large number of SNPs to achieve high discrimination power, and difficulties in detecting body fluid mixtures due to their generally biallelic nature (Brumfield et al., 2003).

8.3 Features of biallelic markers

The main advantages of SNPs are summaried as follows (Sobrino & Carracedo, 2005):

1. *High Resolution:* SNPs provide higher resolution due to their abundance and distribution throughout the genome. This high density can help differentiate between individuals, especially in complex forensic cases (Jobling and Gill, 2004).
2. *Stability and Low Mutation Rate*: Biallelic markers like SNPs have lower mutation rates compared to STRs (approximately 10^{-8} vs 10^{-3}) ensuring more stable and reliable results over generations, which is crucial in complex paternity cases (Butler, 2006).
3. *Efficiency in Degraded Samples*: SNPs are advantageous in analyzing highly degraded DNA samples because, as single-base markers, they can be amplified with shorter polymerase chain reaction (PCR) products (around 50–90 bp) (Wang et al., 2015).
4. *Forensic DNA Phenotyping (FDP)*: It involves predicting an individual's externally observable traits, such as appearance, biogeographic ancestry, and age, using DNA from crime scene samples. This technique offers valuable investigative leads for identifying unknown suspects when traditional forensic STR profiling is insufficient. Certain SNPs are key in determining these characteristics, aiding in investigative leads (Kayser et al., 2023; Kayser & de Knijff, 2011).

Main disadvantages include:

1. *Limited Number of Alleles:* Typically, two alleles for each SNP locus.
2. *Difficulties in Mixture Interpretation:* The limited number of alleles restricts or even prevents reliable interpretation of mixtures due to the possibility that the multiple contributors have overlapping alleles (Butler et al., 2007).

3. *Requirement for Large Numbers of SNPs:* Due to the low level of polymorphisms, more SNPs are required to obtain the equivalent discrimination power or random match probabilities than STRs. Generally, 40–60 SNPs have been suggested in order to approximate the power of 13–15 STR loci. Obviously assays with a larger number of loci are more sensitive to the quantity and integrity of the DNA input, which is an issue when amplifying limited DNA samples (Pontes et al., 2015).

8.4 Applications in forensic DNA analysis

8.4.1 Individual identification

STR analysis has been the cornerstone of forensic identification due to its high discrimination power and established protocols. However, SNPs provide additional benefits that complement STRs. SNPs are more abundant in the human genome and have lower mutation rates, making them stable genetic markers. Due to the small size (a single base variation), they can be analyzed from highly degraded DNA samples where STRs might fail, thereby increasing the chances of obtaining a usable genetic profile from compromised forensic evidence (Butler, 2006).

However, the low polymorphism of individual SNPs implies that a high number of markers must be analyzed to achieve the same discrimination power obtained by the traditional analysis of 10–15 autosomal STRs. This means that although each SNP provides less individual discriminatory power, the combined analysis of many SNPs can still yield highly reliable identification results.

In fact, a panel of 52 SNPs has been developed specifically for human identification (Sanchez et al., 2006). Recently, a novel multiplex assay of SNP-STR markers for forensic purposes has been developed (Wei et al., 2018).

Biallelic markers can complement STR analysis for confirming identities in criminal cases, mass disasters, and missing persons (Gill et al., 2001).

In criminal cases, the integration of SNP analysis with STR profiling can provide a more comprehensive genetic analysis. SNPs can be particularly useful in situations where the DNA sample is degraded or the quantity is limited, which often challenges traditional STR analysis. By analyzing both STRs and SNPs, forensic scientists can achieve higher confidence in individual identification, reducing the risk of false positives or negatives (Gill et al., 2001).

SNP analysis could successfully confirm identities and provide additional genetic information that is not available through STR analysis alone.

SNPs also play a crucial role in identifying missing persons, and especially in mass disaster scenarios, such as plane crashes, natural disasters, or terrorist attacks, when the victim's identification can be exceptionally challenging due to the severe degradation of biological samples. SNPs are particularly advantageous in these contexts because of their ability to produce reliable results from degraded DNA. SNP panels specifically designed for forensic applications can quickly and accurately identify individuals, even when only minute or damaged DNA samples are available (Pakstis et al., 2012).

A notable application of SNPs in mass disaster victim identification is the aftermath of the 2004 Indian Ocean tsunami. The deployment of SNP genotyping allowed for the identification of victims when traditional methods were inadequate, demonstrating the critical role of SNPs in forensic investigations under extreme conditions.

8.4.2 Kinship testing

Due to their inheritance pattern, biallelic markers are useful in paternity and kinship analysis, providing robust statistical support (Butler, 2011). In the case of missing persons, SNP-based kinship analysis can be employed to establish familial relationships, thereby aiding in the identification of individuals. This method has been used effectively in various countries to match unidentified remains with families of missing persons, providing closure to many cases (Butler, 2006). Incorporating SNPs alongside autosomal STR analysis in paternity/kinship caseworks allows for obtaining paternity index and probability values that are equivalent to or higher than those achieved with more STR loci, but at reduced costs.

8.4.3 Biogeographical ancestry

Certain SNPs can provide insights into an individual's ancestry, which can be useful in narrowing down suspect pools (Shriver et al., 2003).

Lineage SNPs located on the mitochondrial DNA (mtDNA) genome and the Y chromosome have proven highly informative for evolutionary research and kinship analysis due to their lack of recombination and low mutation rates. Y-chromosomal haplogroups assigned from male-specific Y-chromosomal SNPs (Y-SNPs) allow paternal lineage identification than paternal bio-geographic ancestry inference, both being relevant in forensic genetics. These characteristics make lineage SNPs especially valuable in cases where reference and evidence samples are separated by multiple generations. This could enhance the ability to identify missing persons when only a limited number of family members are available for comparison.

In cases where no suspect is identified from an evidentiary sample's STR profile, even after a database search, or where information about the perpetrator is unreliable, such as from eyewitness accounts, knowing the biogeographic ancestry of an individual can provide valuable investigative leads. While forensic STR loci are powerful for identity confirmation, they are less effective for determining an individual's genetic biogeographical ancestry due to significant allele-sharing among populations. Ancestry informative markers (AIMs) are SNPs distributed across the human genome that occur at varying frequencies among different populations and can indicate the ancestral origin of a sample or person (Phillips et al., 2007).

There are three types of SNPs considered as AIMs (or AISNPs): Y-chromosome SNPs (YSNPs), mitochondrial SNPs (mt-SNPs), and autosomal SNPs (aSNPs). The first two define paternal and maternal haplogroups, and they have been historically used for evolution studies because of their low recombination rates, their nonrandom geographical distribution, and their well-known global frequency distribution. Interestingly, Y-SNPs show a better genetic differentiation with geographical distance than mtDNA or autosomal SNPs due to patrilocality (Terrado-Ortuño & May, 2024).

The main issue when inferring ancestry using nonautosomal markers is that although they are highly accurate when recent ancestors were from the same region, they only represent half of the lineage, and they can lead to misinterpretation of complex origins. Therefore, autosomal SNPs are proposed in combination with parental SNPs to infer admixed ancestries (Cheung et al., 2018; Sobrino et al., 2005).

8.4.4 Externally visible characteristics prediction

Predictive SNPs can infer physical traits such as eye color, hair color, and skin pigmentation, aiding investigative efforts by generating an identikit of a perpetrator based on biological evidence left at a crime scene (Walsh et al., 2014). DNA-based phenotypic prediction is particularly useful in cases where there are no suspects, or the suspect's DNA does not match the evidence or any profiles in a DNA database (Pośpiech et al., 2022).

Predictive DNA analysis of externally visible characteristics (EVCs) is playing an increasingly important role in modern forensic and anthropological investigations. This approach offers a valuable means of identifying unknown individuals from DNA samples that standard forensic DNA analysis cannot resolve. The forensic genetics community has embraced the concept of generating EVC information from DNA, and ongoing research is focused on developing reliable methods and systems for DNA-based predictions (Chaitanya et al., 2018; Liu et al., 2009, 2014, 2016).

Notably, whole genome analyses have enhanced our understanding of the genetics of hair color, allowing predictions beyond the previously well-understood red hair color associated with variations in the MC1R gene (Grimes et al., 2001). Several methods have been proposed for hair color prediction using DNA variants.

In 2010, Walsh developed the IrisPlex system, a highly sensitive multiplex genotyping assay consisting of six SNPs, which can predict blue and brown eye color from DNA samples with over 90% precision (Walsh et al., 2011). Subsequently, in 2013, the HIrisPlex system was developed (Walsh et al., 2013) to simultaneously predict eye and hair color by analyzing 24 SNPs in a single multiplex assay. By 2018, the HIrisPlex-S system was released, which extended predictions to include eye, hair, and skin color. This system consists of two SNaPshot-based multiplex assays targeting a total of 41 SNPs: 17 for skin color prediction and the previous 24 from HIrisPlex for eye and hair color, 19 of which also contribute to skin color prediction (Chaitanya et al., 2018).

The HIrisPlex prediction system is the most validated and widely used method for pigmentation prediction in both forensic and anthropological contexts (Chaitanya et al., 2013; King et al., 2014). A mathematical algorithm is used for color prediction, achieving very high accuracy for categories such as red hair and blue or brown eyes. Further studies aim to improve the prediction accuracy of intermediate eye colors and age-related hair color changes (e.g., darkening from blond to brown/black).

Other interesting quantitative traits are height and weight. Predicting adult height from DNA has important implications in forensic DNA phenotyping. In 2014 it has been introduced a prediction model consisting of 180 height-associated SNPs since the SNPs account for 40% of all variation in height for individuals of European ancestry, and 10%–20% of variation for people of non-European ancestry (Liu et al., 2014; Yengo et al., 2022).

Weight is highly polygenic and greatly influenced by environmental factors such as social class, income, education, family size, housing, urban locations, etc. (Kayser, 2013). Although it has been immensely studied in clinical research, few studies are directed to its incorporation into FDP. On the other hand, an individual's weight is usually measured using the body mass index (BMI), and it is said to have a heritability around 60 to 70%, despite knowing that epigenetic factors have more influence than genetic ones (Cho et al., 2021).

The most ambitious and challenging phenotype to incorporate is facial morphology. The human face represents a multidimensional set of correlated, mostly symmetric, complex phenotypes with high heritability. This is illustrated by a large degree of facial similarity between monozygotic twins and, albeit less so, between other relatives (Djordjevic et al., 2016).

Although environmental factors have a small effect and the genetic component is strong, it is highly polygenic, with limited understanding of the underlying genetic mechanisms (Claes et al., 2014). The FaceBase Consortium (https://www.facebase.org/) and the International Visible Trait Genetics (VisiGen) Consortium have identified numerous markers associated with human morphology. Researchers typically start by focusing on individual facial features before expanding to whole-face predictions (Stephan et al., 2019; VisiGen, 2019). Some of the specific traits studied are eye morphology (eyelid fold), epicanthal index, palpebral fissure distance, and inclination. Other traits of interest under investigation include hair morphology and texture (straight, wavy, curly), hair thickness, male-pattern baldness (MPB), handedness or relative hand skill, eyebrows, and beard (Marcińska et al., 2015; Phillips et al., 2006; Pośpiech et al., 2015; Pośpiech et al., 2018; Terrado-Ortuño & May, 2024; Tozzo et al., 2021).

8.5 SNPs genotyping methods

In clinical research and forensics, choosing the appropriate SNP genotyping method depends on factors such as the number of SNPs to be analyzed, the required throughput, available resources, and the specific application. The development of these advanced techniques has significantly enhanced our ability to study genetic variation and its implications for health and disease (Kwok, 2001).

8.5.1 SNaPshot

The commonly used method for analyzing SNPs is the SNaPshot® reaction, a primer extension-based technique that enables rapid and robust multiplexing of up to 10 SNPs. This method involves the single-base extension of an unlabeled oligonucleotide at the 3’ end of the base immediately adjacent to the SNP site. Dideoxynucleotides, each labeled with different fluorescent dyes, are incorporated during this extension process (Kricka, 2002). Unlabeled, noncomplementary oligonucleotide sequences of varying lengths serve as mobility modifiers, ensuring that the extended SNaPshot primers differ by size and color. These modified primers can then be separated and detected via capillary electrophoresis (CE). using automated DNA sequencers, facilitating high-throughput SNP analysis (Syvänen, 2001).

The SNaPshot® method offers several advantages, including high sensitivity, specificity, and the ability to multiplex, making it suitable for various applications such as genotyping, genetic linkage analysis, and the detection of mutations associated with diseases (Butler, 2005). The method allows for efficient and accurate genotyping of large panels of SNPs, facilitating comprehensive forensic analyses (Kline et al., 2005).

8.5.2 DNA microarray

This technique relies on the preferential binding of complementary single-stranded nucleic acid sequences. It involves the immobilization of single-stranded DNA (ssDNA) probes onto a transducer solid surface, which then hybridize with their complementary target DNA sequences. Detection and quantification of DNA hybridization (probe-target) are achieved through labeled targets (fluorophores, silver, chemiluminescence). This labeling enables the detection of thousands of probes within the microarray. DNA microarrays consist of ordered arrays of spots containing probe DNA molecules, which are used to measure the quantity of target nucleic acid molecules (Bumgarner, 2013; Rahmati, 2016).

DNA microarrays are used for various applications, including the detection of SNPs and the genotyping or sequencing of mutant genomes. They are useful for their specificity, speed, reusability, continuity, selectivity, and sensitivity (Sassolas et al., 2008). However, the method has lower sensitivity compared to PCR, and it is also less susceptible to sample cross-contamination (Ballin et al., 2009). This reduced sensitivity can be a limitation in certain applications, but the overall utility of DNA microarrays in specific detection systems remains significant. DNA microarray assays have been demonstrated as a potential feature in forensic genetic genealogy (Urtiaga et al., 2022).

8.5.3 TaqMan assays

TaqMan assay is a highly specific and sensitive method, but is limited in multiplexing capability compared to SNaPshot®. TaqMan® assays are renowned for their robustness and ease of automation, although their high cost can be prohibitive for large-scale studies. Developed by Thermo Fisher Scientific, the TaqMan™ SNP genotyping assay emerged as a state-of-the-art technology alongside Roche's LightCycler®. This assay utilizes the 5′ exonuclease activity of Taq DNA polymerase during PCR reactions, integrating allele-specific oligonucleotides (ASO), PCR, and fluorescence resonance energy transfer (FRET) for precise genetic analysis.

The TaqMan™ assay employs a single probe per reaction, a significant simplification compared to the LightCycler®'s requirement for two probes. In this assay, the probe is labeled with a fluorophore at the 5′ end and a quencher at the 3′ end. The close proximity of the fluorophore to the quencher prevents fluorescence emission when the probe is intact. During the PCR process, the probe anneals to the target DNA during the annealing step. As Taq DNA polymerase extends the primer during the extension step, it encounters the probe and cleaves it from the 5′ end due to its exonuclease activity. This cleavage separates the fluorophore from the quencher, resulting in the emission of fluorescence. The fluorescence intensity correlates with the amount of amplified DNA, indicating the presence of the target DNA (Livak et al., 1995).

Originally, the TaqMan™ assay utilized two probes, each complementary to different polymorphic sites or alleles, with distinct fluorophores attached. This setup enabled the FRET emission to correspond with the specific sequence of the target DNA, allowing for accurate discrimination between alleles (Livak, 1999; Livak et al., 1995). Recently, advancements have allowed the TaqMan™ assay to be adapted for multiplex reactions. These adaptations enable the use of various combinations of fluorophores and quenchers, allowing simultaneous detection of multiple targets within a single reaction. This multiplexing capability significantly enhances the throughput and efficiency of genetic analysis, making it suitable for complex and large-scale studies (Wong & Medrano, 2005) (Tables 8.1 and 8.2).

While both TaqMan™ and LightCycler® assays leverage the principle of FRET and ASO for SNP genotyping, the TaqMan™ assay's requirement for only one probe per reaction streamlines the process and reduces reagent costs. Additionally, the flexibility in adapting to multiplex formats further positions the TaqMan™ assay as a versatile tool in genomic research and clinical diagnostics (Wong & Medrano, 2005).

8.5.4 Next-generation sequencing

In recent decades, innovative technologies have been developed and applied across various fields of biological research (Bruijns, 2018). One of the most transformative of these technologies is NGS, which was introduced for commercial use in 2005. Initially referred to as "massively parallel sequencing or MPS" NGS enables the simultaneous sequencing of numerous DNA strands, as opposed to the single-strand approach used in Sanger sequencing.

One of the key strengths of NGS is its ability to perform millions to billions of sequencing reactions simultaneously across multiple samples, allowing for the analysis of various genomic traits in a single run at reduced costs.

Additionally, NGS requires lower sample inputs, offers higher accuracy, and can detect variants at lower allele frequencies compared to Sanger sequencing (Metzker, 2010).

It's commonly employed to determine the DNA or RNA sequence of entire genomes or targeted regions, enabling a wide array of applications in different areas such as disease diagnosis, agrigenomics forensics (Yaran et al., 2014).

The high-throughput capacity of NGS has significantly advanced the detection of SNPs in forensic science, enabling the targeting of SNP sets that are not routinely accessible with traditional CE-based methods. Unlike traditional methods such as SNaPshot or array systems, NGS allows the simultaneous analysis of a larger number of markers from even low quantities of DNA. This capability is particularly valuable in forensic applications where sample amounts are often limited.

NGS technology facilitates the targeting of microhaplotypes, which have shown considerable promise in forensics for both identification and ancestry purposes (Pakstis et al., 2012). Microhaplotypes, which consist of closely linked SNPs, provide high-resolution data that can enhance the discriminatory power of forensic analyses.

The advent of NGS has led to the discovery of millions of SNPs, greatly expanding the genetic markers available for forensic investigations, enabling high-throughput analysis of multiple markers simultaneously (van der Gaag et al., 2014).

Additionally, NGS offers advantages such as higher accuracy and the ability to detect variants at lower allele frequencies compared to traditional methods. This increased sensitivity is crucial for forensic cases involving degraded or mixed DNA samples, where minor allele detection can provide critical evidence.

Several custom panels of large SNP assays have been developed, allowing for comprehensive genetic profiling in forensic contexts. For instance some Authors, Zhang et al. (2017), Wang et al. (2019) have developed extensive SNP panels tailored for forensic use, demonstrating the practical application of NGS in creating robust and informative genetic assays.

TABLE 8.1 Main commercial multiplexes for single-nucleotide polymorphisms (SNPs) analysis.

Name	Markers	Company	Application
ForenSeq DNA Signature kit	DPMA: 27 global autosomal STRs, 24 Y-STRs, 7 X-STRs, 94 identity SNPs DPMB: 27 global autosomal STRs, 24 Y-STRs, 7 X-STRs, 94 identity SNPs, 22 phenotypic SNPs, 56 biogeographical ancestry SNPs	Verogen	Identity biogeographical ancestry (BGA)
ForenSeq Imagen Kit	DPME: 22 hair SNPs, 6 eye color SNPs, 36 skin color SNPs and 14 Y-SNPs (total of 55 SNPs) DPMF (Geo): 22 hair SNPs, 6 eye color SNPs, 36 skin color SNPs, 56 BGA SNPs and 14 Y-SNPs (total of 111 SNPs)	Verogen	Biogeographical ancestry (BGA) External visbile characteristics (EVC)
ForenSeq Kintelligence Kit	10,230 SNP	Verogen	Forensic genetic genealogy (FGG),
Precision ID Ancestry Panel	165 SNPs	Applied Biosystems TermoFisher Scientific	Biogeographical ancestry (BGA)
Precision ID Identity Panel	34 Y-Clade SNPs and 90 autosomal SNPs	Applied Biosystems TermoFisher Scientific	Identity
Ion AmpliSeq VISAGE-Basic Tool Research Panel	153 SNPs	Applied Biosystems TermoFisher Scientific	Biogeographical ancestry (BGA) External visible characteristics (EVC)
Ion AmpliSeq DNA Phenotyping Panel	23 SNPs + 1 INDEL	Applied Biosystems TermoFisher Scientific	External Visible Characteristics (EVC)
Ion AmpliSeq PhenoTrivium Panel	200 autosomal SNPs and 120 Y chromosomal SNPs	Applied Biosystems TermoFisher Scientific	Biogeographical ancestry (BGA) External visible characteristics (EVC)
Ion AmpliSeq HID Y-SNP Research Panel	859 Y-SNPs	Applied Biosystems TermoFisher Scientific	Paternal lineage identification and bio-geographic ancestry (BGA)

TABLE 8.2 Main advantages and disadvantages of forensic biallelic markers.

Advantages	Disadvantages
Low mutation rate	Low discrimination power
High chance of success in degraded DNA samples	Lower sensitivity than STRs
High multiplex capacity	Difficulty in the analysis of mixed samples
Fast and fully automated is possible	Few forensic validated commercial kit
Information about ancestry and phenotype	

The VISAGE Consortium created an appearance and ancestry single assay, named the VISAGE Basic Tool (BT) that consists of a total of 153 AISNPs for origin inference (most of them part of the EUROFORGEN Global AIM-MPs ancestry panel), two SNPs from Kidd's panel, 11 SNPs from the Precision ID ancestry panel and the 41 SNPs from the HIrisPlex-S panel for pigmentation inference (Xavier et al., 2020).

Few commercial multiplexes are available, such as, for example, the ForenSeq DNA Signature Prep Kit (Verogen) that allows to analyze 27 global autosomal STRs, 24 Y-STRs, 7 X-STRs, 94 identity SNPs, 22 phenotypic SNPs (for eye and hair color), 56 biogeographical ancestry SNPs (European, American, African, and East Asian).

8.6 Challenges and considerations

8.6.1 Population stratification

Variability in allele frequencies across different populations can significantly complicate statistical interpretations in forensic DNA analysis. This phenomenon, known as population stratification, occurs because allele frequencies vary between populations due to factors such as genetic drift, selection, mutation, and historical migration patterns (Kayser, 2015). For example, genetic drift can lead to random changes in allele frequencies in small populations, while natural selection can increase the frequency of advantageous alleles. Additionally, mutations introduce new genetic variations, and historical migrations cause gene flow between populations, further contributing to genetic diversity (Tishkoff & Verrelli, 2003).

The complications arising from population stratification are particularly pronounced in mixed or admixed populations, where individuals have ancestry from multiple distinct genetic backgrounds. In such cases, allele frequencies may not follow a uniform distribution, leading to biases in forensic analysis. These biases can manifest as inflated match probabilities or erroneous exclusion of potential suspects, thus undermining the reliability of forensic evidence (Buckleton et al., 2011).

To mitigate these issues, it is essential to develop and utilize population-specific databases that accurately reflect the genetic diversity of the populations being studied. These databases should include comprehensive allele frequency data for various loci across different populations. Additionally, advanced statistical methods, such as structured association tests and principal component analysis (PCA), can be employed to correct for population stratification. Structured association tests can control for population stratification by incorporating population structure into the analysis, thereby reducing false-positive rates. PCA, on the other hand, identifies axes of genetic variation, which can be used to adjust for population stratification in forensic analyses.

Moreover, the incorporation of AIMs can further enhance the precision of forensic DNA analysis. AIMs are genetic markers that exhibit substantial frequency differences between populations from different geographical regions. By analyzing these markers, forensic scientists can infer the biogeographical ancestry of an individual, which aids in the correct interpretation of genetic evidence in the context of population stratification (Shriver et al., 2003).

Population-specific databases and the use of AIMs not only improve the accuracy of forensic analyses but also facilitate the development of probabilistic genotyping approaches. Probabilistic genotyping leverages statistical models to interpret complex DNA mixtures and degraded samples, incorporating population genetic data to estimate the likelihood ratios of different genetic profiles. This approach enhances the robustness of forensic conclusions, especially in challenging cases involving low-template DNA or mixed samples.

In summary, addressing population stratification in forensic DNA analysis requires a multifaceted approach that includes the creation and utilization of detailed population-specific databases, the application of advanced statistical methods, and the integration of AIMs. These strategies collectively improve the accuracy and reliability of forensic evidence, reducing the risk of erroneous conclusions and ensuring that justice is served effectively.

8.6.2 Ethical and legal implications

The use of genetic markers in forensic science, particularly for phenotypic prediction and ancestry inference, raises significant ethical and privacy concerns (Morling et al., 2002). Phenotypic prediction involves deducing physical traits such as hair color, eye color, and skin pigmentation from DNA, while ancestry inference aims to identify the biogeographical origin of an individual. These capabilities can provide valuable leads in criminal investigations, especially when traditional methods fail to yield suspects. However, they also pose risks of misuse and discrimination. For instance, genetic data might be used to reinforce racial stereotypes or disproportionately target certain ethnic groups, exacerbating existing social inequalities (Kayser & Schneider, 2009).

Phenotypic prediction, although useful in narrowing down potential suspects, can lead to profiling based on physical characteristics, which may not always be accurate or fair. The predictive accuracy of these markers can vary, and there is a risk of stigmatizing individuals who share similar traits with a suspect profile. Additionally, ancestry inference, while aiding in the identification of an individual's geographical origins, can inadvertently highlight population-specific genetic traits, leading to potential misuse in areas such as employment, insurance, or law enforcement practices (Phillips, 2012).

To address these concerns, robust ethical guidelines and legal frameworks must be established and adhered to, ensuring that genetic data is used responsibly, and that individuals' privacy rights are protected (Williams & Johnson, 2004). These guidelines should include clear policies on the scope and limitations of genetic data use in forensic contexts. Informed consent is paramount; individuals should be fully aware of how their genetic information will be used and the potential implications. Transparency in the use of genetic information, including clear communication about the limitations and potential errors in phenotypic prediction and ancestry inference, is essential to maintain public trust (Jobling and Gill, 2004).

Strict data security measures are also critical components of ethical forensic practice. Genetic information must be stored and processed in a manner that safeguards against unauthorized access and breaches. Encryption, secure databases, and regular audits are necessary to protect sensitive genetic data from misuse or leaks. Additionally, ethical oversight committees should be established to review and monitor the use of genetic markers in forensic investigations, ensuring compliance with ethical standards and legal requirements (Knoppers et al., 2012).

In summary, while the use of genetic markers for phenotypic prediction and ancestry inference holds great promise for forensic science, it is accompanied by significant ethical and privacy challenges. Addressing these challenges requires a comprehensive approach involving robust ethical guidelines, informed consent, transparency, and stringent data security measures to ensure that the benefits of these technologies are realized without compromising individual rights or societal equity.

8.7 Insertion/deletion polymorphisms

Insertion/deletion INDELs are the second most common type of polymorphism in the human genome, following SNPs (Carvalho et al., 2011; Mullaney et al., 2010) estimating that they represent 15.6% of the polymorphisms discovered. Indels involve the insertion or deletion of one or more DNA bases and occur in both coding and noncoding regions, in autosomal and sexual chromosomes. In coding regions, frameshift indels, which are not multiples of three base pairs, can disrupt the reading frame, leading to altered protein products with potential impacts on gene expression and disease (Mills et al., 2011; Mullaney et al., 2010).

INDELs are increasingly used in forensic science as supplementary markers to STR analysis due to their distinct advantages. They are biallelic, widely distributed throughout the genome, and they exhibit lower mutation rates compared to STRs, high heterozygosity rate and small amplicon lengths (60–200 bp). These characteristics make Indels valuable for analyzing highly degraded DNA or very small DNA quantities (Cortellini et al., 2018; Weber-Lehmann et al., 2014) and enhance the resolution of kinship cases without the high mutation rates associated with STRs (Pereira et al., 2009). Furthermore, as they do not produce stutter artifacts or microvariants, Indels are a promising option for resolving complex mixture interpretation.

InDels are well-suited for multiplexing, and their amplified products can be easily separated using the CE platforms commonly available in forensic genetic laboratories worldwide. Due to these advantages, InDels have gained significant attention from forensic researchers over the years as a complement to STRs. They are useful for identification, biogeographic ancestry analysis, evolutionary studies, and determining kinship relationships.

In 2019 was developed a novel multiplex panel named the AGCU InDel 50 Kit that simultaneously amplify 47 advanced A-InDels, Amelogenin, and 2 Y-InDels (Chen, 2019).

In recent years, several Authors (Fan et al., 2015; Qu et al., 2020) described a new kind of DNA markers named Multi-InDel as characterized by two or more tightly linked InDel loci within a short length of physical position, usually 200–300 nucleotides. Multi-InDels have excellent applications in ancestry inference and forensics (Huang et al., 2014; Li et al., 2021; Sun et al., 2019).

Only few Commercial kits are available for Indels analysis such as the Investigator DIPplex for the analysis of 30 INDELs (Qiagen) or the InDelPlex kit (GENOMICA) introduced in 2017 that enable the multiplex analysis of 38 Indels, providing robust tools for forensic investigations (Cortellini et al., 2018). Despite needing many INDELs to match the discrimination power of STRs, their length polymorphism allows for simple allele length measurement, adding to their utility in forensic casework (Li et al., 2021).

Indels are useful in various forensic applications, including identification, biogeographic ancestry, genealogical research, evolutionary studies, and kinship analysis. Their high heterozygosity rate and the ability to perform multiplex PCR studies with small amplicon lengths (60–200 bp) further enhance their applicability in challenging forensic scenarios (Mullaney et al., 2010; Pereira et al., 2009; Weber-Lehmann et al., 2014).

In addition to SNPs, INDELs can also be effectively analyzed using simple PCR amplification followed by CE separation (PCR-to-CE) or through NGS techniques. The PCR-to-CE approach is widely used in forensic laboratories due to its efficiency and compatibility with existing equipment, allowing for the precise separation and detection of amplified products. On the other hand, NGS offers a more advanced alternative, providing higher resolution and the ability to analyze multiple genetic markers simultaneously, which is particularly valuable for complex forensic cases. These methodologies enhance the versatility of INDELs in forensic applications, making them a robust tool for genetic analysis.

References

Ballin, N. Z., Vogensen, F. K., & Karlsson, A. H. (2009). Species determination – Can we detect and quantify meat adulteration? *Meat Science, 83*(2), 165–174.

Brookes, A. J. (1999). The essence of SNPs. *Gene, 234*(2), 177–186.

Brumfield, R. T., Beerli, P., Nickerson, D. A., & Edwards, S. V. (2003). The utility of single nucleotide polymorphisms in inferences of population history. *Trends in Ecology & Evolution, 18*(5), 249–256.

Buckleton, J., Bright, J., & Taylor, D. (2011). *Forensic DNA Evidence interpretation*. CRC Press.

Butler, J. M., Buel, E., Crivellente, F., & McCord, B. R. (2004). Forensic DNA typing by capillary electrophoresis using the ABI Prism 310 and 3100 genetic analyzers for STR analysis. *Electrophoresis, 25*(10-11), 1397–1412.

Bumgarner, R. (2013). DNA microarrays: Types, Applications and their future. *Curr Protoc Mol Biol, 22*. https://doi.org/10.1002/0471142727.mb2201s101.

Butler, J. M. (2005). Forensic DNA typing: Biology. In *Technology, and genetics of STR markers*. Elsevier ISBN: 9780121479527.

Butler, J. M. (2006). Genetics and genomics of core short tandem repeat loci used in human identity testing. *Journal of Forensic Sciences, 51*(2), 253–265.

Butler, J. M., Coble, M. D., & Vallone, P. M. (2007). STRs vs. SNPs: Thoughts on the future of forensic DNA testing. *Forensic Sci Med Pathol, 3*, 200–205.

Butler, J. M. (2011). *Advanced topics in forensic DNA typing: Methodology*. Elsevier ISBN: 9780123878236.

Carvalho, A., Laura Cainé, L., Carvalho, R., Maria, F., & Pinheiro, M. F. (2011). Application of indels (investigator DIPplex) in mixture samples. *Forensic Science International: Genetics Supplement Series, 3*(1), e351–e352.

Chaitanya, L., Breslin, K., Zuñiga, S., Wirken, L., Pośpiech, E., Kukla-Bartoszek, M., Sijen, T., Knijff, P., Liu, F., Branicki, W., Kayser, M., & Walsh, S. (2018). The HIrisPlex-S system for eye, hair and skin colour prediction from DNA: Introduction and forensic developmental validation. *Forensic Science International: Genetics, 35*, 123–135.

Chaitanya, I., Pajnič, I. Z., Walsh, S., Balažic, J., Zupanc, T., Kayser, M., Liu, J., et al. (2013). Development and validation of a 47-locus INDEL panel for forensic application. *Forensic Science International: Genetics, 7*(4), 488–496.

Chen, L. (2019). Developmental validation of a novel six-dye typing system with 47 A-InDels and 2 Y-InDels. *Forensic Science International: Genetics, 40*. https://doi.org/10.1016/j.fsigen.2019.02.009.

Cheung, E. Y. Y., Gahan, M. E., & McNevin, D. (2018). Predictive DNA analysis for biogeographical ancestry. *Australian Journal of Forensic Sciences, 50*(6), 651–658.

Cho, S., Lee, E. H., Kim, H., Lee, J. M., So, M. H., Ahn, J. J., et al. (2021). Validation of BMI genetic risk score and DNA methylation in a Korean population. *International Journal of Legal Medicine, 135*(4), 1201–1212.

Claes, P., Hill, H., & Shriver, M. D. (2014). Toward DNA-based facial composites: Preliminary results and validation. *Forensic Science International: Genetics, 13*, 208–216.

Collins, F. S., Brooks, L. D., & Chakravarti, A. (1998). A DNA polymorphism discovery resource for research on human genetic variation. *Genome Research, 8*(12), 1229–1231.

Collins, F. S., Morgan, M., & Patrinos, A. (2003). The Human Genome Project: lessons from large-scale biology. *Science (New York, N.Y.), 300*(5617), 286–290.

Cortellini, V., et al. (2018). Forensic performance of a novel 38-plex assay for INDELs genotyping. *Forensic Science International: Genetics Supplement Series, 6*, e543–e545.

Djordjevic, J., Zhurov, A. I., Richmond, S., & Visigen Consortium (2016). Genetic and environmental contributions to facial morphological variation: A 3D Population-Based twin study. *PLoS One, 11*, e0162250.

Fan, G., Ye, Y., Luo, H., & Hou, Y. (2015). Use of multi-InDels as novel markers to analyze 13 X-chromosome haplotype loci for forensic purposes. *Electrophoresis, 36*, 2931–2938.

Gabriel, S. B., Schaffner, S. F., Nguyen, H., Moore, et al. (2002). The structure of haplotype blocks in the human genome. *Science, 296*(5576), 2225–2229.

Gill, P., et al. (2001). The evolution of DNA databases—Recommendations for new European STR loci. *Forensic Science International, 123*(1), 123–129.

Gill, P., Fereday, L., Morling, N., & Schneider, P. M. (2004). The evolution of DNA databases—Recommendations for new European STR loci. *Forensic Science International, 156*(2-3), 242–244.

Goodwin, W., Linacre, A., & Hadi, S. (2011). *An introduction to forensic genetics*. John Wiley & Sons.

Grimes, E. A., Noake, P. J., Dixon, L., & Urquhart, A. (2001). Sequence polymorphism in the human melanocortin 1 receptor gene as an indicator of the red hair phenotype. *Forensic Science International, 122*(2-3), 124–129.

Hofreiter, M., Paijmans, J. L. A., Goodchild, H., Speller, C. F., Barlow, A., Fortes, G. G., Thomas, J, A., Ludwig, A., & Collins, M, J, (2015). The future of ancient DNA: Technical advances and conceptual shifts. *Bioessays, 37*(3). https://doi.org/10.1002/bies.201400160.

Huang, J., Luo, H., Wei, W., & Hou, Y. (2014). A novel method for the analysis of 20 multi-Indel polymorphisms and its forensic application. *Electrophoresis, 35*, 487–493.

International Human Genome Sequencing Consortium. (2001). Initial sequencing and analysis of the human genome. *Nature, 409*(6822), 860–921.

International Visible Trait Genetics (VisiGen) Consortium. (2019). Novel genetic loci affecting facial shape variation in humans. *eLife, 8* Article e49898.

Jobling, M. A., & Tyler-Smith, C. (2003). The human Y chromosome: An evolutionary marker comes of age. *Nature Reviews. Genetics, 4*(8), 598–612.

Jobling, M. A., & Gill, P. (2004). Encoded evidence: DNA in forensic analysis. *Nature Reviews. Genetics, 5*(10), 739–751.

Kayser, M., Branicki, W., Parson, W., & Phillips, C. (2023). Recent advances in forensic DNA phenotyping of appearance, ancestry and age. *Forensic Science International: Genetics, 65*, 102870.

Kayser, M., & de Knijff, P. (2011). Improving human forensics through advances in genetics, genomics and molecular biology. *Nature Reviews. Genetics, 12*(3), 179–192.

Kayser, M., & Schneider, P. M. (2009). DNA-based prediction of human externally visible characteristics in forensics: Motivations, scientific challenges, and ethical considerations. *Forensic Science International: Genetics, 3*(3), 154–161.

Kayser, M. (2013). Forensic DNA phenotyping: DNA testing for externally visible characteristics. *Encyclopedia of Forensic Sciences: Second Edition*, 369–374.

Kayser, M. (2015). Forensic use of Y-chromosome DNA: A general overview. *Human Genetics, 136*(5), 621–635.

Kidd, J. M., Gravel, S., Byrnes, J., Moreno-Estrada, A., Musharoff, S., et al. (2014). Population genetic inference from personal genome data: Impact of ancestry and admixture on human genomic variation. *The American Journal of Human Genetics, 95*(4), 740–749.

King, T., Fortes, G., Balaresque, P., et al. (2014). Identification of the remains of King Richard III. *Nat Commun, 5*, 5631.

Kline, M. C., et al. (2005). SNP typing by amplicon length discrimination: Analysis of forensic SNPs with microarray technology. *Journal of Forensic Sciences, 50*(2), 377–385.

Knoppers, B. M., Dove, E. S., Litton, J. E., & Nietfeld, J. J. (2012). Questioning the limits of genomic privacy. *Am J Hum Genet, 91*(3). https://doi.org/10.1016/j.ajhg.2012.05.027.

Koonin, E. V., & Galperin, M. Y. (2003). *Sequence - evolution - function: computational approaches in comparative genomics*. Kluwer Academic.

Krawczak, M., Ball, E. V., & Cooper, D. N. (2000). Neighboring-nucleotide effects on the rates of germ-line single-base-pair substitution in human genes. *American Journal of Human Genetics, 63*(2), 474–488.

Kricka, L. J. (2002). Clinical and biochemical applications of luciferases and luciferins. *Analytical and Bioanalytical Chemistry, 373*(8), 1–6.

Kwok, P. Y., Carlson, C., Yager, T. D., Ankener, W., Nickerson, D. A., & Ward, R. (1996). High-throughput genotyping of single nucleotide polymorphisms by competitive PCR and fluorescence resonance energy transfer. *Genomics, 23*(1), 138–144.

Kwok, P. Y. (2001). Methods for genotyping single nucleotide polymorphisms. *Annual Review of Genomics and Human Genetics, 2*(1), 235–258.

Liu, F., Hendriks, A. E., Ralf, A., Boot, A. M., Benyi, E., Savendahl, L., Oostra, B. A., Duijn, C., van Hofman, A., Rivadeneira, F., Uitterlinden, A. G., Drop, S. L., & Kayser, M. (2014). Common DNA variants predict tall stature Europeans. *Human Genetics, 133*, 587–597.

Liu, F., van Duijn, K., Vingerling, J. R., Hofman, A., Uitterlinden, A. G., Janssens, A. C., & Kayser, M. (2009). Eye color and the prediction of complex phenotypes from genotypes. *Current Biology, 19*(5), 192–193.

Liu, F., Hamer, M. A., Heilmann, S., Herold, C., Moebus, S., Hofmann, A., et al. (2016). Genetic prediction of male pattern baldness. *European Journal of Human Genetics: EJHG, 24*, 895–902.

Li, J., Lin, L., Jiang, B., Wang, C., Zeye, M. M. J., Wen, D., et al. (2021). An 18 Multi-InDels panel for analysis of highly degraded forensic biological samples. *Electrophoresis, 42*(9-10), 1143–1152.

Livak, K. J. (1999). Allelic discrimination using fluorogenic probes and the 5′ nuclease assay. *Genetic Analysis: Biomolecular Engineering, 14*(5-6), 143–149.

Livak, K. J., Flood, S. J., Marmaro, J., Giusti, W., & Deetz, K. (1995). Oligonucleotides with fluorescent dyes at opposite ends provide a quenched probe system useful for detecting PCR product and nucleic acid hybridization. *PCR Methods and Applications, 4*(6), 357–362.

Marcińska, M., Pośpiech, E., Abidi, S., Andersen, J. D., van den Berge, M., Carracedo, et al. (2015). Evaluation of DNA variants associated with androgenetic alopecia and their potential to predict male pattern baldness. *PLoS ONE, 10*, e0127852.

Metzker, M. L. (2010). Sequencing technologies—The next generation. *Nature Reviews Genetics, 11*(1), 31–46.

Mills, R. E., et al. (2011). Mapping copy number variation by population-scale genome sequencing. *Nature, 470*(7332), 59–65.

Morling, N., et al. (2002). Recommendations on the use of Y-STRs in forensic analysis. *Forensic Science International, 157*(2-3), 131–136.

Mullaney, J. M., Mills, R. E., Pittard, W. S., & Devine, S. E. (2010). Small insertions and deletions (INDELs) in human genomes. *Human Molecular Genetics, 19*(R2), R131–R136.

Nachman, M. W., & Crowell, S. L. (2000). Estimate of the mutation rate per nucleotide in humans. *Genetics, 156*(1), 297–304.

Pakstis, A. J., et al. (2012). SNPs for a universal individual identification panel. *Human Genetics, 131*(4), 485–497.

Pereira, R., et al. (2009). A new multiplex for human identification using insertion/deletion polymorphisms. *Electrophoresis, 30*(21), 3682–3690.

Phillips, C. (2012). Forensic genetic analysis of bio-geographical ancestry. *Forensic Science International: Genetics, 6*(4), 314–325.

Phillips, C., Barbaro, A., Lareu, M.v, Salas, A., & Carracedo, A. (2006). Initial study of candidate genes on chromosome two for relative hand skill. *International Congress Series/Excerpta Medica, 1288*, 798–800.

Phillips, C., Salas, A., Sánchez, J. J., Fondevila, M., Gómez-Tato, A., Alvarez-Dios, J., Calaza, M., de Cal, M. C., Ballard, D., Lareu, M. V., & Carracedo, A. (2007). Inferring ancestral origin using a single multiplex assay of ancestry-informative marker SNPs. *Forensic Science International: Genetics, 1*(3). https://doi.org/10.1016/j.fsigen.2007.06.008.

Pontes, M. L., Fondevila, M., Laréu, M. V., & Medeiros, R. (2015). SNP markers as additional information to resolve complex kinship cases. *Transfusion Medicine and Hemotherapy: Offizielles Organ der Deutschen Gesellschaft fur Transfusionsmedizin und Immunhamatologie, 42*(6), 385–388.

Pośpiech, E., Karłowska-Pik, J., Marcińska, M., Abidi, S., Andersen, J. D., Berge, M. V. D., et al. (2015). Evaluation of the predictive capacity of DNA variants associated with straight hair in Europeans. *Forensic Science International: Genetics, 19*, 280–288.

Pośpiech, E., Chen, Y., Kukla-Bartoszek, M., Breslin, k, et al. (2018). Towards broadening forensic DNA Phenotyping beyond pigmentation: Improving the prediction of head hair shape from DNA. *Forensic Science International-Genetics Volume, 37*, 241–251.

Pośpiech, E., Teisseyre, P., Mielniczuk, J., & Branicki, W. (2022). Predicting physical appearance from DNA data: Towards genomic solutions. *Genes, 13*(1), 121.

Qu, S., Lv, M., Xue, J., Zhu, J., Wang, L., Jian, H., et al. (2020). Multi-Indel: A microhaplotype marker can be typed using capillary electrophoresis platforms. *Frontiers in Genetics, 11*, 567082.

Rahmati, S., et al. (2016). *Application of DNA microarray technology in the identification of meat species.*

Sachidanandam, R., Weissman, D., Schmidt, S. C., Kakol, J. M., Stein, L. D., et al. (2001). A map of human genome sequence variation containing 1.42 million single nucleotide polymorphisms. *Nature, 409*(6822), 928–933.

Sanchez, J. J., Phillips, C., Børsting, C., Balogh, K., Bogus, M., Fondevila, M., et al. (2006). A multiplex assay with 52 single nucleotide polymorphisms for human identification. *Electrophoresis, 27*(9), 1713–1724.

Sassolas, A., et al. (2008). DNA biosensors and microarrays. *Chemical Reviews, 108*(1), 109–139.

Shriver, M. D., et al. (2003). Skin pigmentation, biogeographical ancestry and admixture mapping. *Human Genetics, 112*(4), 387–399.

Shriver, M. D., Jin, L., Boerwinkle, E., Deka, R., Ferrell, R. E., & Chakraborty, R. (1997). A novel measure of genetic distance for highly polymorphic tandem repeat loci. *Molecular Biology and Evolution, 14*(9), 985–993.

Syvänen, A. C. (2001). Accessing genetic variation: genotyping single nucleotide polymorphisms. *Nature Reviews Genetics, 2*(12), 930–942.

Sobrino, B., Brion, M., & Carracedo, A. (2005). SNPs in forensic genetics: A review on SNP typing methodologies. *Forensic Science International, 154*(2-3), 181–194.

Sobrino, B., & Carracedo, A. (2005). SNP typing in forensic genetics: A review. *Methods in Molecular Biology, 297*, 107–126.

Stephan, C. N., Caple, J. M., Guyomarc, P., Claes, P., Guyomarc'h, P., & Claes, P. (2019). An overview of the latest developments in facial imaging. *Forensic Sciences Research, 4*(1), 10–28.

Sun, K., Yun, L., Zhang, C., Shao, C., Gao, T., Zhao, Z., et al. (2019). Evaluation of 12 Multi-InDel markers for forensic ancestry prediction in Asian populations. *Forensic Science International: Genetics, 43*, 102155.

Terrado-Ortuño, N., & May, P. (2024). Forensic DNA phenotyping: a review on SNP panels, genotyping techniques, and prediction models. *Forensic Sci Res, 10*(1). https://doi.org/10.1093/fsr/owae013.

Terrado-Ortuño, N., & May, P. (2024). Forensic DNA phenotyping: A review on SNP panels, genotyping techniques, and prediction models. *Forensic Sciences Research, 2024*, owae013.

Tishkoff, S. A., & Verrelli, B. C. (2003). Patterns of human genetic diversity: implications for human evolutionary history and disease. *Annu Rev Genomics Hum Genet, 4*. https://doi.org/10.1146/annurev.genom.4.070802.110226.

Tozzo, P., Politi, C., Delicati, A., Gabbin, A., & Caenazzo, L. (2021). External visible characteristics prediction through SNPs analysis in the forensic setting: A review. *Frontiers in Bioscience, 26*(10), 828–850.

Urtiaga, G. O., Domingues, W. B., Komninou, E. R., Amanda, W. S., Martins, A. W. S., et al. (2022). DNA microarray for forensic intelligence purposes: High-density SNP profiles obtained directly from casework-like samples with and without a DNA purification step. *Forensic Science International, 332*, 111181.

van der Gaag, K. J., et al. (2014). Massively parallel sequencing of short tandem repeats—Population data and mixture analysis results for the PowerSeq™ system. *Forensic Science International: Genetics, 12*, 77–89.

Venter, J. C., Adams, M. D., Myers, E. W., Li, P. W., Mural, R. J., Sutton, G. G., et al. (2001). The sequence of the human genome. *Science, 291*(5507), 1304–1351.

Walsh, S., Liu, F., et al. (2011). IrisPlex: A sensitive DNA tool for accurate prediction of blue and brown eye colour in the absence of ancestry information. *Forensic Science International: Genetics, 5*(3), 170–180.

Walsh, S., Liu, F., Wollstein, A., Kovatsi, L., Ralf, A., Kosiniak-Kamysz, A., Branicki, W., & Kayser, M. (2013). The HIrisPlex system for simultaneous prediction of hair and eye colour from DNA. *Forensic Science International: Genetics,* (1), 98–115.

Walsh, S., et al. (2014). Developmental validation of the IrisPlex system: Determination of blue and brown iris colour for forensic intelligence. *Forensic Science International: Genetics, 9*, 150–160.

Wang, D. G., Fan, J. B., Siao, C. J., Berno, A., Young, P., et al. (1998). Large-scale identification, mapping, and genotyping of single-nucleotide polymorphisms in the human genome. *Science, 280*(5366), 1077–1082.

Wang, Z., et al. (2015). Forensic SNP genotyping with SNaPshot: A comprehensive review. *Electrophoresis, 36*(1), 80–96.

Wang, H., et al. (2019). High-resolution forensic SNP panel for individual identification and ancestry inference. *Forensic Science International: Genetics, 42*, 92–99.

Weber-Lehmann, J., et al. (2014). STR typing of forensic stain samples: An approach to achieve high sensitivity and specificity. *Forensic Science International: Genetics, 12*, 72–79.

Wei, T., Liao, F., Wang, Y., Pan, C., Xiao, C., & Huang, D. (2018). A novel multiplex assay of SNP-STR markers for forensic purpose. *PLoS One, 13*(7), e0200700.

Williams, R., & Johnson, P. (2004). Wonderment and dread': representations of DNA in ethical disputes about forensic DNA databases. *New Genet Soc, 23*(2). https://doi.org/10.1080/1463677042000237035.

Wong, M. L., & Medrano, J. F. (2005). Real-time PCR for mRNA quantitation. *Biotechniques, 39*(1), 75–85.

Xavier, C., de la Puente, M., Mosquera-Miguel, A., Freire-Aradas, A., Kalamara, V., Vidaki, A., et al. (2020). Development and validation of the VISAGE AmpliSeq basic tool to predict appearance and ancestry from DNA. *Forensic Science International: Genetics, 48*, 102336.

Yaran, Y., Xie, B., & Yan, J. (2014). Application of next-generation sequencing technology in forensic science. *Genomics Proteomics Bioinformatics, 12*(5). https://doi.org/10.1016/j.gpb.2014.09.001.

Yengo, L., et al. (2022). A saturated map of common genetic variants associated with human height from 5.4 million individuals of diverse ancestries. *Nature, 610*(7933), 704–712.

Zhang, S., et al. (2017). Development of a SNP assay for identification and ancestry inference in Chinese populations. *Forensic Science International: Genetics, 28*, 1–7.

Chapter 9

Epigenetics and its application in forensic analysis

Graham A Williams
Centre for Biomedicine, Hull York Medical School, University of Hull, Hull, United Kingdom

9.1 Introduction

Fragment analysis of short tandem repeats (STRs) is used worldwide to link suspects to stains recovered from crime scenes and items (Bukyya et al., 2021). This process, commonly known as DNA profiling, was first used in 1985 to secure the conviction of Colin Pitchfork, using variable number tandem repeats. Since then, the process has substantially improved and developed into a highly specific, sensitive, and stable analytical process. These improvements have been technical and interpretational (such as the increased need for mixture calculations). However, given all this improvement, the basic principle of DNA profiling remains the same. A DNA profile from the crime stain and a DNA profile from a suspect are compared, and a statistical match probability is applied for a possible match. This process is considered evident as this match and statistics can be reported in a court of law.

However, there is a significant flaw in that if there is no suspect, then there is nothing to compare a DNA profile with. Efforts then turn to using that DNA profile to try and identify a suspect. Then, once a suspect has been identified, a direct evidential comparison can be conducted. This process of identifying the suspect is the Investigative process. There are several ways to gain further information from the genetic material, including gender analysis and estimation of geographical origin.

Should standard STR analysis be unfruitful, then it can be necessary to conduct alternative, more specialist analytical techniques. Several distinct types of information can be obtained from genetic or epigenetic analysis, such as mRNA, miRNA, and DNA methylation. This chapter will focus on miRNA and DNA methylation strategies in forensic analysis; however, it will also discuss mRNA to provide context.

9.2 Body fluid identification

Once a DNA profile is obtained, this can be used to ascertain which individual can be attributed to the stain. However, it is often necessary to link a DNA profile with a body fluid, to show that the DNA profile in a sexual assault case came from the semen detected, or that the DNA profile of the victim found on the suspect's clothing in a physical assault came from the blood. Therefore, despite the increasing sensitivity and specificity of DNA profiling, it is still necessary to be able to identify the body fluid (Harbison & Fleming, 2016; Sakurada et al., 2020). DNA profiles can be considered inadmissible in a court of law if it cannot be said which body fluid the DNA came from. These body fluids include blood, semen, saliva, and vaginal secretions. While conventional methods such as microscopic examination, chemical tests, and spectroscopic techniques have long been employed for body fluid identification, they are often limited by sensitivity, specificity, and susceptibility to environmental degradation. In recent years, genetic approaches to body fluid identification (BFID), such as mRNA, microRNA (miRNA), and DNA methylation profiling has emerged as a more robust alternative (Lee et al., 2012; Roeder & Haas, 2013; Wang et al., 2013).

MicroRNAs are small noncoding RNA molecules, typically 20–24 nucleotides in length, which regulate gene expression and play significant roles in biological processes (Ha & Kim, 2014; Hanson et al., 2009; He & Hannon, 2004; Pritchard et al., 2012; Zubakov et al., 2010). Their exceptional stability and specificity under diverse environmental conditions make them ideal candidates for forensic investigations. Unlike other RNA molecules, miRNAs are resistant to

Investigative and Predictive DNA Testing. DOI: https://doi.org/10.1016/B978-0-443-30128-5.00008-4

enzymatic degradation, allowing for their detection in trace or degraded forensic samples. Therefore, MicroRNA expression patterns can be specific to different body fluids, enabling their discrimination with high accuracy (Liu et al., 2021; Wang et al., 2013; Weber et al., 2010; Zubakov et al., 2010).

Early studies in the field employed microarray sequencing to explore miRNA expression, progressively expanding the range of detectable markers. Subsequent studies utilized next-generation sequencing (NGS) and real-time quantitative PCR to enhance the precision and speed of miRNA profiling. Regardless of how the markers are characterized, a range of miRNA markers has been identified for BFID purposes as shown in Table 9.1 (Hanson et al., 2009; Wang et al., 2013; Weber et al., 2010; Zubakov et al., 2010).

Several key studies have shaped the field of miRNA-based body fluid identification. For example, initial work by Hanson et al. in (2009) and Zubakov et al. (2010) established foundational miRNA panels for distinguishing body fluids. Subsequent research by Weber et al. (2010) expanded miRNA profiling to include 12 body fluids, providing a comprehensive reference for forensic investigations. Further innovations, such as capillary electrophoresis (CE) and the development of probabilistic discriminant models, have refined miRNA analysis techniques, enabling their application to complex forensic scenarios involving mixed or environmentally compromised samples (Pritchard et al., 2012; van der Meer & Williams, 2015; Williams, Uchimoto, Coult, World, & Beasley, 2013).

The use of miRNA in forensic science is not without challenges. The overlap in miRNA expression profiles between certain body fluids and the influence of pathological conditions on miRNA regulation necessitates careful marker selection and interpretation. However, some of the advantages of using miRNA profiling have demonstrated consistent reliability, even in cases involving aged or laundered samples (Pritchard et al., 2012; Williams, Uchimoto, Coult, World, Beasley, & Avenell, 2013).

9.3 mRNA profiling

Whilst it is not an epigenetic messenger RNA (mRNA) profiling is an area that has received much attention in genetic-based BFID, and it is currently used in casework in a few countries, namely The Netherlands and New Zealand. Therefore, it is useful to provide some information in this area for context.

The specificity of mRNA expression in different cell types and tissues makes it an ideal marker for distinguishing body fluids. Unlike other biomolecular approaches, mRNA profiling is compatible with CE platforms commonly used in forensic laboratories. This compatibility allows for coanalysis of body fluid identification and STR profiling (Alvarez et al., 2004; Bowden et al., 2011; Tran, 2020).

Common mRNA multiplexes have shown efficacy in analyzing complex mixtures, such as those found in sexual assault cases, where multiple body fluids from different individuals may be present. The kit's robust design ensures reliable

TABLE 9.1 Body fluid–specific microRNA markers.

microRNA marker	Body fluid	Function/specificity
miR-451	Blood	Highly expressed in erythrocytes, specific to blood.
miR-16	Blood	Abundant in blood plasma, useful for blood identification.
miR-205	Saliva	Specific to epithelial cells in saliva.
miR-124a	Brain tissue (proxy for saliva)	Associated with brain tissue but detectable in saliva due to neural origin.
miR-135b	Saliva	Linked to epithelial differentiation, detected in saliva.
miR-891a	Semen	Highly specific to sperm cells, useful in semen identification.
miR-372	Semen	Expressed in sperm cells, related to fertility processes.
miR-518c	Semen	Associated with trophoblast cells, found in semen.
miR-412	Vaginal fluid	Linked to vaginal epithelial cells, expressed in vaginal fluid.
miR-1246	Vaginal fluid	Highly expressed in vaginal epithelial cells.

Source: Modified from Hanson et al. (2009), Wang et al. (2013), Weber et al. (2010), and Zubakov et al. (2010).

results even in challenging conditions, such as degraded or environmentally exposed samples (Lin et al., 2015; Lin et al., 2016; Sirker et al., 2016).

However, cross-reactivity among markers in closely related body fluids, such as saliva and vaginal fluids, underscores the need for continued refinement of marker panels. Additionally, the influence of pathological conditions on mRNA expression warrants further investigation to ensure that mRNA is usable regardless of inherent conditions.

Finally, DNA methylation refers to the "switched" status of genes, whereby blood-specific markers are expressed would have the "blood genes" switched on. Thus characterizing the epigenetic status of certain markers can help indicate the body fluid type present in the sample. The main techniques used for DNA analysis in forensic science are SNaPShot analysis and Massively Parallel Sequencing, see Table 9.2 (An et al., 2013; Fu et al., 2015; Lee et al., 2012; Park et al., 2014).

One of the advantages of using DNA methylation is the reduction of separate workflows. Unlike mRNA and miRNA analysis, the same single extraction technique can be used, and then an aliquot undergoes STR analysis, and another undergoes DNA methylation analysis. This also allows for a stronger association between the STR profile and the body fluid. Finally, there is the perception that DNA methylation is more stable than using mRNA or miRNA, as the DNA molecule is much more stable than the RNA molecule. However, it is worth noting that DNA methylation refers to the secondary structure of the DNA molecule, and it is unclear how stable these are. Additionally, the structure and function of miRNA are degradation resistant. Thus miRNA may be more stable than DNA methylation analysis (Li et al., 2021).

9.4 Temporal predictive genomics—the challenge of terminology

One area of predictive genomics is the estimation of an individual's age. However, there are two different areas, and the authors do not always clarify this. Age can be estimated for an individual and a body fluid stain. When discussing age estimation, this is in general referring to individual age estimation. However, the term can also relate to stain age estimation, reflecting the infancy of this research subdiscipline. Currently, the sub–discipline of stain age estimation seems to refer to itself as such, but other terms can also be used, such as time since deposition (TSD) and about semen specifically, time since intercourse. There can be some confusion among the researchers using terms such as stain age estimation to estimate individual age from stains, rather than the age of the stain itself. Thus it is suggested that, the word "age" should relate to the age of the individual and that "time" should be used against the stain—TSD and refers to the period between a body fluid being shed during a crime and the time of its subsequent analysis in a laboratory.

TABLE 9.2 List of body fluid–specific DNA methylation markers.

DNA methylation marker	Body fluid	Function/specificity
ELOVL2	Blood	Encodes a fatty acid elongase, shows age-specific methylation in blood.
ASPA	Blood	Linked to nervous system development, age-associated methylation in blood.
ITGA2B	Blood	Encodes integrin alpha 2b, a platelet-specific gene with methylation changes in blood.
PDE4C	Blood	Phosphodiesterase 4 C, a marker for age estimation in blood.
FHL2	Blood	FHL2 is involved in transcription regulation, methylation increases with age in blood.
EDARADD	Saliva	Encodes EDAR-associated death domain, methylation changes observed in saliva.
KLF14	Saliva	Known as a master regulator of obesity, methylation changes in saliva with age.
C1orf132	Semen	C1orf132, linked to age-dependent methylation in semen.
NPTX2	Vaginal fluid	NPTX2, related to sy.0naptic plasticity, methylation variations in vaginal fluid.
TRIM59	Vaginal fluid	TRIM59, linked to immune regulation, methylation specific to vaginal fluid.

Source: Modified from An et al. (2013), Fu et al. (2015), Lee et al. (2012), and Park et al. (2014).

9.5 Stain age estimation or time since deposition

The subdiscipline of TSD is the prediction or estimation of the time between the deposition of a body fluid stain and its subsequent recovery (Anderson et al., 2005; Asaghiar & Williams, 2020; Bremmer et al., 2012; Doty et al., 2017; Manasatienkij & Nimnual, 2021). The purpose of this is to provide a timeline of a criminal incident. However, the value of this question is to allow for the distinction between alternative hypotheses, such as the allegation and the defense. For example, if a bloodstain is obtained from a crime scene that matches that of a suspect, then a defense could be that the suspect was at that location previously and had another injury, for example, a nosebleed. Thus this provides defense for the presence of his DNA at the crime scene. The ability to be able to estimate the time since the blood was deposited would help address this issue (Anderson et al., 2005; Asaghiar & Williams, 2020; Bremmer et al., 2012; Doty et al., 2017; Manasatienkij & Nimnual, 2021).

Currently, this capability is not in use, but as a research field, this has gained considerable traction in recent years. Much of the research in this area relates to bloodstains; however, applying this research to semen stains would be of great value in sexual assaults, particularly between acquaintances who have had a previously consensual sexual relationship, where it may be necessary to differentiate between recent sexual intercourse and previous sexual intercourse (within a range of up to 7 days).

There are two genetically-based approaches to this: DNA methylation and RNA degradation (mRNA and miRNA).

Manasatienkij et al. and many others have used a common technique of looking at the degradation profile of mRNA markers, with some looking at blood-specific markers, such as HBB or housekeeping genes (GAPDH, B2M, and SPTB) via qPCR. This looked at a period of 150 days (or approximately 5 months) and developed a predictive correlation equation. The authors reported a margin of error of 23 days; therefore within a 150-day period, they could predict the TSD give or take 23 days (Manasatienkij & Nimnual, 2021).

Others have looked at different markers, for example, looking at hypoxia-related markers. The study by Asaghiar and Williams (2020) explores the use of hypoxia-sensitive RNA markers, namely vascular endothelial growth factor A and hypoxia-inducible factor 1 alpha, for predicting the age of body fluid stains—blood, saliva, and semen—over a period of four weeks. Hypoxia-sensitive markers present a potential solution, as changes in oxygen concentration following deposition could influence RNA expression and degradation rates. By quantifying these changes, it may be possible to infer the time since deposition (Asaghiar & Williams, 2020).

Efforts in this area have looked at a wide range of different time ranges, with the first studies by Anderson et al covering a period of months, with further efforts endeavoring to target shorter time frames, to a matter of weeks (Anderson et al., 2005; Asaghiar & Williams, 2020). The study by Asaghiar et al is one of the shortest time frames explored (matter of days and weeks), which should be more relevant to most criminal investigations. Other nongenetic techniques, such as Colorimetric analysis, have explored shorter time frames (Lins & Blazek, 1982).

A major consideration in any temporal investigation is the margin of error or time range (i.e., 3 days old give or take 1 day, giving a range of 2–4 days since deposition). Many other works have reduced the time ranges, and many prediction models have been formulated, generating a shorter time frame. There is also work looking at reducing these ranges, with the model giving shorter time frames.

9.6 Individual age estimation

As well as externally visible characteristics, it is useful to gain an estimation of the individual's age, which can not only provide intelligence in terms of identifying potential suspects but also from a civil perspective, concerning differentiating between adults and children. For example, in asylum seeker/immigration cases, an adult may claim to be a child to gain a more favorable outcome. There have been a range of different approaches to this, with a more established method using odontology. However, this technique is imprecise. A technique was developed where Vidaki, López, et al. (2017) investigated DNA methylation-based forensic age prediction using artificial neural networks (ANNs) and NGS (Vidaki, Ballard, et al., 2017). The study aimed to develop accurate age prediction models based on epigenetic changes at age-related CpG sites in whole blood and other biological samples. Researchers analyzed data from 1156 whole blood samples across diverse age groups (2–90 years) using Illumina methylation platforms and identified 16 CpG sites strongly correlated with chronological age. By leveraging ANNs, the study achieved a prediction accuracy of $R^2 = 0.96$, with a mean absolute error (MAE) of 3.3 years for the training set and 4.4 years for blind testing. Validation in saliva samples yielded similar accuracy, indicating the model's versatility for nonblood tissues. Additionally, the study evaluated an NGS-based approach for quantifying methylation at these sites, achieving a lower accuracy (MAE = 7.5 years).

Yamagishi et al. used machine learning to target the DNA methylation state of the ELOVL2 gene. The principle behind this is that as the stain is exposed to the environment, it will cause the methylation of a range of genetic markers. Castagnola et al. explored other markers and showed a similar approach (Castagnola et al., 2024; Yamagishi et al., 2024).

Yamagitshi et al.'s approach was to explore the models for interpreting the methylation data and derive a prediction model. They used an ANN. This model provided an MAE of 5.3 years and a root mean square error of 7.3 years. This study, unlike others, has a larger dataset (n = 549), which provides high confidence in its ability to predict the age of a bloodstain, give or take 5.3 years (Yamagishi et al., 2024).

Like stain age estimation, the driving force behind this research is to achieve more precision, so that accurate time estimates can be provided with a minimal range or margin for error. Both stain age prediction and individual age prediction have similar challenges in that it is not the technique, but how it is interpreted. All the studies covered include a prediction model, with some including Machine Learning. Currently, there has been no consistent approach in predictive models.

9.7 Monozygotic twins

Monozygotic (or identical) twins share the same DNA profile. Therefore, if there is one offender and both twins are potential suspects, then the DNA profile cannot be used as evidence, as it is unable to differentiate between these two individuals. There have been a few cases where a conventional DNA-based examination could not be conducted due to this limitation. However, epigenetics can be used, as it is environmentally influenced. Therefore, if one of the twins were to have a different behavior or a different habit, then this may potentially cause variation in the epigenome. Thus characterizing the epigenome might allow for the discrimination of the monozygotic twin via a comparison between the crime scene stain and each of the twins (Stewart et al., 2015; Vidaki et al., 2017).

Numerous studies have identified two main approaches to characterize MZ twins' epigenomes.

One of the first approaches is whole genome sequencing (WGS) to identify any postsomal mutations. If the twins have different single-nucleotide variants, this can be detected via WGS. However, at least three samples will be needed, including the twins and the crime scene sample, and therefore, this can be expensive (considering the limited budget of police forces) and would only be used in the most serious cases (Hwa et al., 2024; Vidaki et al., 2017).

The other approach is epigenetic and focuses entirely on DNA methylation. Within DNA methylation, the main techniques used have been WGS following bisulfite treatment, which has the same challenges as WGS for mutation analysis. A more common technique is the use of pyrosequencing, which can provide a quantitative DNA methylation analysis at targeted loci. There is also methylation-specific PCR, which again targets specific loci on the epigenome. All these processes have a bisulfite treatment stage, through which unmethylated CpG islands are converted. It is these differences that are picked up (Vidaki, López, et al., 2017; Xu et al., 2015).

Environmental change of the epigenome is typically through the methylation state of CpG islands. Therefore, if a CpG island is unmethylated, then it will be converted via the bisulfite treatment step, whereas if a CpG island is methylated, then it will not be converted, and therefore the sequence of the genome itself changes as a result. It is the detection of these changes that allows for the differentiation of monozygotic twins, due to differences in their sample genome following bisulfite treatment.

Much of the processes target specific regions and quantify them accurately. However, a study by Stewart et al took more of a blanket approach and looked to ascertain whether there was a difference or not, rather than what that difference is. This technique used a combination of bisulfite treatment and high-resolution melting curve analysis (HRMA). HRMA quantifies the number of hydrogen bonds in the sample (AT having two and CG having three), and therefore, if the sequence changes, the number of hydrogen bonds changes. HRMA can detect a single base pair difference. It is just not capable of saying where this difference is. This is a substantially cheaper approach to use (Stewart et al., 2015).

Given these techniques, regardless of the capability of the tests, there will be some fundamental principles to appreciate. One is that the twins' epigenome will continue to change over time. Therefore, if there is some time between the crime scene stain being deposited and the reference DNA sample from the twins being recovered, then the epigenome may vary to the extent that neither twin's epigenome would match that of the crime scene sample.

It is also unknown how stable a postdeposition stain's epigenome is; therefore, it is possible that the epigenome within the stain could alter over time before it is recovered and analyzed, thus rendering it unsuitable for comparison.

One observation made by the researchers is that age is a key factor in that the older the twins, the more diverse their epigenomes are likely to be, depending on lifestyle and environmental factors. However, it is not possible to state at which age this occurs—again due to environmental factors—for example, if one twin started smoking and the other did not, then this could affect the genome, but it depends on what age this happens.

Finally, a greater understanding of the dynamics in what factors make the biggest changes, and how stable or permanent these changes are.

9.8 Medical implications

Whilst there is potential to glean additional information about a potential offender from crime scene stains left behind, there is little understood about the implications of medical disorders upon the expression of analyzed markers. For example, HBB is a marker of interest for the identification of blood (Lynch & Fleming, 2021); however, the expression changes in those with leukemia (Fluhr et al., 2017). Thus there is a significant risk of medical information being obtained from crime scene stains, which may lead to abuse. However, it is worth noting that genetic disorders such as Down's syndrome can be observed in STR profiles (such as a trisomy being present on a D21 locus), and this has not led to any known incidences of abuse, given that such forensic tests are not validated for diagnostic purposes (Fluhr et al., 2017). If such tests are expanded in capability and scope, then the opportunities for abuse increase.

Little work has been done that focuses on understanding the influence of various diseases on the expression of forensic biomarkers, particularly those present in semen and vaginal fluids. For example, forensic biomarkers represent a broad spectrum of molecules integral to forensic science. MicroRNAs, for instance, are noncoding RNAs that regulate gene expression and are remarkably stable in bodily fluids, making them valuable indicators of conditions such as cancer and HIV (Li et al., 2004; Pilakka-Kanthikeel et al., 2011; Rupaimoole & Slack, 2017; Tao & Freudenheim, 2010; Triboulet et al., 2007). Similarly, mRNAs serve as tissue-specific markers, enabling the identification of body fluids such as semen and vaginal secretions. DNA methylation, an epigenetic modification, plays a crucial role in distinguishing between body fluids and detecting diseases. Other biomarkers, including prostate-specific antigen and Y-chromosome DNA, have proven effective for semen detection, though their specificity can vary.

By exploring the effects of diseases on these forensic biomarkers, assessing the accuracy of biomarkers in body fluids such as semen and vaginal fluids, and identifying factors that contribute to variations in biomarker expression, a good understanding can be gained.

Myriad studies have revealed significant insights into the role of forensic biomarkers in detecting diseases. DNA methylation markers, for example, were found to be elevated in endometrial and prostate cancers (Li et al., 2004; Tao & Freudenheim, 2010). Moreover, miRNAs in semen and vaginal fluids displayed altered expressions in cancer patients, facilitating early detection and diagnosis. Infectious diseases, such as HIV and human papillomavirus, were associated with substantial changes in biomarker expression (Pilakka-Kanthikeel et al., 2011; Triboulet et al., 2007). These changes often reflect immune responses, with proinflammatory cytokines and specific miRNA profiles indicating viral replication and systemic inflammation. In the context of male reproductive health, conditions like prostatitis or varicocele were linked to increased sperm DNA fragmentation and elevated levels of specific miRNAs, such as miR-34c-5p. These variations provide critical information for forensic analyses, particularly when distinguishing between fertile and infertile individuals (Chen et al., 2018).

The variability of biomarker expression across populations is influenced by numerous factors, including lifestyle, health conditions, and genetic predispositions. For instance, treatments for diseases such as antiretroviral therapy in HIV were shown to alter biomarker efficacy (Pilakka-Kanthikeel et al., 2011).

Several challenges and limitations affect the utility of forensic biomarkers. Tumor heterogeneity, for example, leads to inconsistencies in cancer biomarker expression, complicating detection. Additionally, environmental and genetic factors, along with coinfections, significantly impact biomarker levels. Therefore, the analysis of forensic biomarkers may be impacted by the disease state of that individual. For example, blood-specific markers are downregulated in those with leukemia, thus potentially impacting the efficacy of such tests. However, a darker side is that an individual may be identified as having a particular disease, and then people are being profiled because of that disease, thus entering a significant ethical quagmire.

There is significant capability in obtaining information from the epigenome that can support the investigative process, both from utilizing STR profiles effectively in conjunction with well-regulated databases and characterizing a wide range of different markers alongside effective interpretation processes, such as predictive models.

References

Alvarez, M., Juusola, J., & Ballantyne, J. (2004). An mRNA and DNA co-isolation method for forensic casework samples. *Analytical Biochemistry, 335*(2), 289–298.

An, J. H., Choi, A., Shin, K., Yang, W. I., & Lee, H. Y. (2013). DNA methylation-specific multiplex assays for body fluid identification. *International Journal of Legal Medicine, 127*, 35–43.

Anderson, S., Howard, B., Hobbs, G. R., & Bishop, C. P. (2005). A method for determining the age of a bloodstain. *Forensic Science International, 148*(1), 37–45.

Asaghiar, F., & Williams, G. A. (2020). Evaluating the use of hypoxia sensitive markers for body fluid stain age prediction. *Science & Justice, 60*(6), 547–554.

Bowden, A., Fleming, R., & Harbison, S. (2011). A method for DNA and RNA co-extraction for use on forensic samples using the Promega DNA IQ™ system. *Forensic Science International: Genetics, 5*(1), 64–68.

Bremmer, R. H., De Bruin, K. G., Van Gemert, M. J., Van Leeuwen, T. G., & Aalders, M. C. (2012). Forensic quest for age determination of bloodstains. *Forensic Science International, 216*(1-3), 1–11.

Bukyya, J. L., Tejasvi, M. L. A., Avinash, A., Chanchala, H. P., Talwade, P., Afroz, M. M., Pokala, A., Neela, P. K., Shyamilee, T. K., & Srisha, V. (2021). DNA profiling in forensic science: A review. *Global Medical Genetics, 8*(04), 135–143.

Castagnola, M. J., Medina-Paz, F., & Zapico, S. C. (2024). Uncovering forensic evidence: A path to age estimation through DNA methylation. *International Journal of Molecular Sciences, 25*(9), 4917.

Chen, Y., Chen, S., Zhang, J., Wang, Y., Jia, Z., Zhang, X., Han, X., Guo, X., Sun, X., & Shao, C. (2018). Expression profile of microRNAs in expressed prostatic secretion of healthy men and patients with IIIA chronic prostatitis/chronic pelvic pain syndrome. *Oncotarget, 9*(15), 12186.

Doty, K. C., Muro, C. K., & Lednev, I. K. (2017). Predicting the time of the crime: Bloodstain aging estimation for up to two years. *Forensic Chemistry, 5*, 1–7.

Fluhr, S., Krombholz, C. F., Meier, A., Epting, T., Mücke, O., Plass, C., Niemeyer, C. M., & Flotho, C. (2017). Epigenetic dysregulation of the erythropoietic transcription factor KLF1 and the β-like globin locus in juvenile myelomonocytic leukemia. *Epigenetics: Official Journal of the DNA Methylation Society, 12*(8), 715–723.

Fu, X. D., Wu, J., Wang, J., Huang, Y., Hou, Y. P., & Yan, J. (2015). Identification of body fluid using tissue-specific DNA methylation markers. *Forensic Science International: Genetics Supplement Series, 5*, e151–e153.

Ha, M., & Kim, V. N. (2014). Regulation of microRNA biogenesis. *Nature Reviews Molecular Cell Biology, 15*(8), 509–524.

Hanson, E. K., Lubenow, H., & Ballantyne, J. (2009). Identification of forensically relevant body fluids using a panel of differentially expressed microRNAs. *Analytical Biochemistry, 387*(2), 303–314.

Harbison, S. A., & Fleming, R. I. (2016). Forensic body fluid identification: State of the art. *Research and Reports in Forensic Medical Science*, 11–23.

He, L., & Hannon, G. J. (2004). MicroRNAs: Small RNAs with a big role in gene regulation. *Nature Reviews Genetics, 5*(7), 522–531.

Hwa, H., Lin, C., Yu, Y., Linacre, A., & Lee, J. C. (2024). DNA identification of monozygotic twins. *Forensic Science International: Genetics, 69*, 102998.

Lee, H. Y., Park, M. J., Choi, A., An, J. H., Yang, W. I., & Shin, K. (2012). Potential forensic application of DNA methylation profiling to body fluid identification. *International Journal of Legal Medicine, 126*, 55–62.

Li, L., Okino, S. T., & Dahiya, R. (2004). DNA methylation in prostate cancer. *Biochimica Et Biophysica Acta (BBA)-Reviews on Cancer, 1704*(2), 87–102.

Li, Z., Chen, D., Wang, Q., Tian, H., Tan, M., Peng, D., Tan, Y., Zhu, J., Liang, W., & Zhang, L. (2021). mRNA and microRNA stability validation of blood samples under different environmental conditions. *Forensic Science International: Genetics, 55*, 102567.

Lin, M., Albani, P. P., & Fleming, R. (2016). Degraded RNA transcript stable regions (StaRs) as targets for enhanced forensic RNA body fluid identification. *Forensic Science International: Genetics, 20*, 61–70.

Lin, M., Jones, D. F., & Fleming, R. (2015). Transcriptomic analysis of degraded forensic body fluids. *Forensic Science International: Genetics, 17*, 35–42.

Lins, G., & Blazek, V. (1982). The use of remission analysis for direct colorimetric determination of age of blood stains. *Zeitschrift Für Rechtsmedizin, 88*, 13–22.

Liu, Y., He, H., Xiao, Z., Ji, A., Ye, J., Sun, Q., & Cao, Y. (2021). A systematic analysis of miRNA markers and classification algorithms for forensic body fluid identification. *Briefings in Bioinformatics, 22*(4), bbaa324.

Lynch, C., & Fleming, R. (2021). RNA-based approaches for body fluid identification in forensic science. *Wiley Interdisciplinary Reviews: Forensic Science, 3*(4), e1407.

Manasatienkij, C., & Nimnual, A. (2021). Forensic blood stain aging using reverse transcription real-time PCR. *Forensic Science International: Reports, 3*, 100205.

Park, J., Kwon, O., Kim, J. H., Yoo, H., Lee, H., Woo, K., Kim, S., Lee, S., & Kim, Y. S. (2014). Identification of body fluid-specific DNA methylation markers for use in forensic science. *Forensic Science International: Genetics, 13*, 147–153.

Pilakka-Kanthikeel, S., Saiyed, Z. M., Napuri, J., & Nair, M. P. (2011). MicroRNA: Implications in HIV, a brief overview. *Journal of Neurovirology, 17*, 416–423.

Pritchard, C. C., Cheng, H. H., & Tewari, M. (2012). MicroRNA profiling: Approaches and considerations. *Nature Reviews Genetics, 13*(5), 358–369.

Roeder, A. D., & Haas, C. (2013). mRNA profiling using a minimum of five mRNA markers per body fluid and a novel scoring method for body fluid identification. *International Journal of Legal Medicine, 127*, 707–721.

Rupaimoole, R., & Slack, F. J. (2017). MicroRNA therapeutics: Towards a new era for the management of cancer and other diseases. *Nature Reviews Drug Discovery, 16*(3), 203–222.

Sakurada, K., Watanabe, K., & Akutsu, T. (2020). Current methods for body fluid identification related to sexual crime: Focusing on saliva, semen, and vaginal fluid. *Diagnostics, 10*(9), 693.

Sirker, M., Schneider, P. M., & Gomes, I. (2016). A 17-month time course study of human RNA and DNA degradation in body fluids under dry and humid environmental conditions. *International Journal of Legal Medicine, 130*, 1431–1438.

Stewart, L., Evans, N., Bexon, K. J., van der Meer, D. J., & Williams, G. A. (2015). Differentiating between monozygotic twins through DNA methylation-specific high-resolution melt curve analysis. *Analytical Biochemistry, 476*, 36–39.

Tao, M. H., & Freudenheim, J. L. (2010). DNA methylation in endometrial cancer. *Epigenetics: Official Journal of the DNA Methylation Society, 5*(6), 491–498.

Tran, A. (2020). Co-extraction methods of DNA and total RNA for body fluid identification: A review.

Triboulet, R., Mari, B., Lin, Y., Chable-Bessia, C., Bennasser, Y., Lebrigand, K., Cardinaud, B., Maurin, T., Barbry, P., & Baillat, V. (2007). Suppression of microRNA-silencing pathway by HIV-1 during virus replication. *Science, 315*(5818), 1579–1582.

van der Meer, D. J., & Williams, G. A. (2015). Performing body fluid identification with microRNAs using capillary electrophoresis. *Forensic Science International: Genetics Supplement Series, 5*, e592–e594.

Vidaki, A., Ballard, D., Aliferi, A., Miller, T. H., Barron, L. P., & Court, D. S. (2017). DNA methylation-based forensic age prediction using artificial neural networks and next generation sequencing. *Forensic Science International: Genetics, 28*, 225–236.

Vidaki, A., López, C. D., Carnero-Montoro, E., Ralf, A., Ward, K., Spector, T., Bell, J. T., & Kayser, M. (2017). Epigenetic discrimination of identical twins from blood under the forensic scenario. *Forensic Science International: Genetics, 31*, 67–80.

Wang, Z., Zhang, J., Luo, H., Ye, Y., Yan, J., & Hou, Y. (2013). Screening and confirmation of microRNA markers for forensic body fluid identification. *Forensic Science International: Genetics, 7*(1), 116–123.

Weber, J. A., Baxter, D. H., Zhang, S., Huang, D. Y., How Huang, K., Jen Lee, M., Galas, D. J., & Wang, K. (2010). The microRNA spectrum in 12 body fluids. *Clinical Chemistry, 56*(11), 1733–1741.

Williams, G., Uchimoto, M. L., Coult, N., World, D., & Beasley, E. (2013). Body fluid mixtures: Resolution using forensic microRNA analysis. *Forensic Science International: Genetics Supplement Series, 4*(1), e292–e293.

Williams, G., Uchimoto, M. L., Coult, N., World, D., Beasley, E., & Avenell, P. (2013). Characterisation of body fluid-specific microRNA markers by capillary electrophoresis. *Forensic Science International: Genetics Supplement Series, 4*(1), e274–e275.

Xu, J., Fu, G., Yan, L., Craig, J. M., Zhang, X., Fu, L., Ma, C., Li, S., & Cong, B. (2015). LINE-1 DNA methylation: A potential forensic marker for discriminating monozygotic twins. *Forensic Science International: Genetics, 19*, 136–145.

Yamagishi, T., Sakurai, W., Watanabe, K., Toyomane, K., & Akutsu, T. (2024). Development and comparison of forensic interval age prediction models by statistical and machine learning methods based on the methylation rates of ELOVL2 in blood DNA. *Forensic Science International: Genetics, 69*, 103004.

Zubakov, D., Boersma, A. W., Choi, Y., van Kuijk, P. F., Wiemer, E. A., & Kayser, M. (2010). MicroRNA markers for forensic body fluid identification obtained from microarray screening and quantitative RT-PCR confirmation. *International Journal of Legal Medicine, 124*, 217–226.

Chapter 10

Microbial DNA in forensic science: a tool for predicting postmortem interval

Joe Adserias-Garriga
Department of Applied Forensic Sciences, Mercyhurst University, Erie, PA, United States

10.1 Introduction

Postmortem Interval (PMI) is the time since the occurrence of a person's death until the finding of the remains. Estimating PMI represents a key aspect of criminal investigations (Carter et al., 2007; Zapico & Adserias-Garriga, 2022).

This estimation is of utmost importance, as it provides valuable information for the criminal investigation, assisting in the identification of the deceased by comparing the estimated time of death with known times of disappearance or last contact. Additionally, PMI represents a key component of the reconstruction of the events around the death, by narrowing the time frame of the death and providing invaluable insights to establish a timeline of events. Due to its significance in forensic investigations, several PMI estimation methods have been developed based on the changes throughout the cadaver breakdown that may include the analysis of early postmortem body changes and the decomposition stages, assessment of the taphonomical factors, and the analysis of biochemical and histological postmortem processes.

One of the most limiting and therefore challenging aspects of PMI estimation methods are the high number of factors affecting the decomposition process and, consequently, PMI estimation, including temperature, humidity, moisture and water exposure, scavengers and insect activity, and the presence of wounds among others (Adserias-Garriga et al., 2017a; Gunn & Pitt, 2012; Paczkowski & Schütz, 2011; Zapico & Adserias-Garriga, 2022).

10.1.1 Methods for postmortem interval estimation

Due to the relevance of PMI estimation in forensic casework, several methods have been developed throughout time, each with limitations and relevance. Classical approaches of PMI are based on physical and chemical changes that take place in the body after death. These changes may be divided into early and late postmortem stagea (Megyesi et al., 2005). The early postmortem changes include: Algor Mortis, which is the cooling of the body after death, wherein the body temperature approaches the ambient temperature. The hypothalamus in the body, which regulates body temperature, stops working after death. The rule of thumb is that the body cools at approximately 1°C/hour. However, there are various factors influencing this rate, such as ambient temperature, body size, and even if the body is covered with clothes or other material. A common method to approximate PMI used body cooling includes Henssge's nomogram, in which rectal temperature, average environmental temperature, body weight, and clothing are considered (Henssge, 1988). This method works very well in indoor scenes, where the changes in temperature are usually not as great and variable. The error for the nomogram method has been reported as high as 7 hours on a 20-hour scale (Wilk et al., 2020).

Rigor Mortis, which consists of the stiffening of the muscles that the body experiences after death. Rigor mortis occurs due to the depletion of adenosine triphosphate, which is needed to relax the muscles. Time and temperature impact the setting and duration of rigor mortis. Livor Mortis, which refers to lividity's appearance, is the discoloration of the skin and internal organs due to gravitational settling of blood in dependent parts of the body. In appearance, it is a kind of reddish-purple discoloration. Gravity force affects the process of its appearance; the position of the body may, therefore, be assessed with the observation of the location and stage of the lividities on the body.

These early postmortem changes (algor mortis, rigor mortis, and livor mortis) may overlap and do not occur in sharp time frames. They are most useful in estimating the PMI at indoor death scenes and within the first 24 to 72 hours

Investigative and Predictive DNA Testing. DOI: https://doi.org/10.1016/B978-0-443-30128-5.00005-9

following death. Beyond this period, other methods must be employed for PMI estimation. After the early postmortem changes, the body undergoes further changes associated with decomposition, which include the following stages. Bloating, which is characterized by body structures distention due to the production of gases in the body, mostly hydrogen sulfide, caused by high bacterial activities, mostly within the anaerobic environment of the intestine (Dent et al., 2004). Active Decay is a stage characterized by very rapid mass loss through insect and bacterial activities and the eventual rupture of the abdominal cavity (Dent et al., 2004). Advanced decay, which takes place when insect activity slackens because most of the soft tissue has been ingested and liquefied (Dent et al., 2004). Skeletonization is the final stage of decomposition, wherein only bones, dry skin, and hair remain (Dent et al., 2004). These later postmortem stages occur in a predictable fashion; however, their length can vary significantly because of environmental conditions and the state of the remains. In an attempt to achieve higher accuracy in the PMI estimations, other methods have been developed beyond the observation of the remains' appearance.

In that sense, Forensic Entomology focuses on the insect's colonization and activity on the body for the estimation of PMI. An estimate of the PMI, or more accurately, the postcolonization interval, is carried out through the identification and analysis of the insect families and species found on the corpse. In the estimation of PMI beyond 72 hours, Forensic entomology is one of the methods normally used. Moreover, the environment where the body is decomposing plays a key role due to insect access to the body. Taphonomy provides valuable information to estimate PMI based on changes to the remains that have occurred because of the environment in which decomposition took place. However, it requires very specific knowledge of the environment, and it is especially useful in outdoor death scenes.

The latter stages of decomposition, as well as methods relying on insect activity and taphonomical changes, can be applied beyond 72 hours, nevertheless, these methodologies aren't as accurate as those relying on the early changes, which can provide estimates of the time since death within hours. After the initial 72 hours, the accuracy of these methods diminishes. Traditional methods to estimate PMI are based on a summation of physical and biological changes occurring in the body postmortem. Although these methods can provide a general estimate of the time since death, they are limited by several factors that may affect their accuracy. As such, there is a need for the development of more precise and reliable methods for PMI estimation. Methods based on thanatochemistry, thanatobiology, proteomics, and microbial analysis have been developed with the aim of higher the objectivity and the accuracy of time-since-death estimates. The main problem of most of the methods developed is that they estimate the PMI within the first hours after death, and very few methods can be used for longer PMIs. Additionally, the longer the PMI a method can estimate, the lower the accuracy.

Thanatochemistry is the study of postmortem chemical changes that occur in a decomposing body. It primarily involves the quantification of byproducts produced during organic decay processes Zapico & Adserias-Garriga, 2022). This method considers the action of factors, such as temperature and humidity, that can influence reaction rates and byproducts produced. Basically, thanatochemistry analyzes the chemical processes of decomposition, such as autolysis, the breakdown of cells by hydrolytic enzymes, which release carbohydrates, proteins, minerals, and fats that bacteria then utilize in putrefaction.

Thanatobiology is the analysis of postmortem DNA and RNA degradation as well as the analysis of cell death signaling pathways and proteins (Zapico & Adserias-Garriga, 2022). Basically, DNA and RNA breakdown through the time elapsed after the death events correlated to the PMI. Although research is showing promising results, it has only shown usefulness in the early postmortem period, and further research is required to explore its applicability in casework.

Luminol chemiluminescence, which reacts to the presence of blood, has also been explored as a PMI indicator (Ermida et al., 2024). Since the amount of hemoglobin in skeletonized remains decreases over time, the chemiluminescence reaction is expected to decrease with increased PMI (Creamer & Buck, 2009; Ermida et al., 2017). An important limitation of this technique is that the results are impacted by taphonomic conditions, and further research is needed in that sense.

Proteomics can also be applied to estimate PMI, analyzing protein degradation and posttranslational modification. This measurement of changes over time in terms of protein expression and amino acid sequences has shown correlations with PMI (Procopio et al., 2018). Specific proteins have been found to decrease in abundance with increasing PMI, while others show an increase with the PMI. Even though the application of proteomics in PMI estimation is still relatively new, and its applicability is still under research, studies have so far shown the potential of proteomics in estimating PMI, even for longer periods of time.

In the last years, the changes in microbial communities throughout the decomposition have been researched as a potential means of PMI estimation. This chapter explores these changes and their usefulness in estimating time since death.

10.2 Antemortem and postmortem microbial succession

The human microbiome refers to the collection of microorganisms that live in and on the human body, and it is composed of commensal, symbiotic, and pathogenic microorganisms (Human Microbiome Project Consortium, 2012). During life, the microbial community is closely regulated through the environment, immune function, and physiological processes in the host organism. Microbial community composition varies immensely between sites, with each habitat supporting a specific group of microbes attuned to site-specific requirements. For example, microbial communities in gut microbiota include microbes such as *Firmicutes* and *Bacteroidetes*, responsible for processes such as digestion and suppression of immune function (Human Microbiome Project Consortium, 2012). Microbial communities in skin microbiota, in contrast, include a complex mixture of facultative and aerobic microbes such as *Staphylococcus*, *Propionibacterium*, and *Corynebacterium*, responsible for barrier function and suppression of pathogens (Grice & Segre, 2011).

The antemortem microbial community is in constant flux, affected by disease, age, and diet. For example, long-term diseases such as obesity and diabetes are characterized by microbial imbalance, or dysbiosis, and can modulate the abundance of specific microbes. In addition, the use of antibiotics can cause a loss in microbial diversity, providing a substrate for colonization for facultative pathogens (Dethlefsen et al., 2008). Following death, microbial processes for homeostasis no longer occur, and rapid and extreme variation in microbial community occurs. The basic changes in the microbiome after death: overgrowth of endogenous microbes; colonization of new microenvironments within the body; and exogenous microorganisms colonization. Postmortem endogenous microbe overgrowth occurs because, in early decomposition, immune defenses break down, allowing an unregulated overgrowth of endogenous microorganisms. As decomposition advances, microorganisms from the environment surrounding the body start to colonize the decomposing tissues. For example, epidermis is colonized soon by such saprophytic soil microbes such as *Acinetobacter* and *Pseudomonas* and fungi such as *Aspergillus* and *Candida*. The microbiome changes following death respond to the shutdown of physiologic processes in the host, especially the loss of immune function, which permits the overgrowth of facultative microbes. Differences between ante and postmortem microbial communities have significant implications in the overall process of decomposition, and they are key for understanding the cadaver breakdown. There are remarkable distinguishing features between antemortem and postmortem microbial communities.

The antemortem microbial community has high diversity and site-specificity, with each site having its function-related diversity in microbes present in it. In contrast, postmortem communities have less diversity and uniformity of taxa in all parts of the body (Pechal et al., 2018). The metabolic activity and function of the microbial community change drastically as well. During life, microbial metabolism is closely coordinated with host physiology, with microbes contributing to processes such as the absorption of nutrients and immune function. After death, microbial metabolism shifts towards tissue degradation, with production of gases, acids, and decomposition compounds ((Javan et al., 2016).

Additionally, antemortem microbial communities have relatively invariant composition over time, with slow shifts in response to fluctuations in host physiology. In contrast, postmortem communities exhibit rapid and predictable successions, with specific taxa dominating at specific decomposition phases (Metcalf et al., 2016). The high environmental influence over the microbial community is another distinctive signature of the postmortem microbiome. Whereas antemortem communities are most sensitive to factors derived from the host, postmortem communities respond most to factors in the environment, such as temperature, humidity, and oxygen availability. Environmental factors drive microbial succession rate and direction and generate decomposition signatures in alternative environments (Carter, 2005).

These differences between antemortem and postmortem microbial communities involve important implications for forensics investigations, such as their potential to provide important information regarding PMI. However, in the present times, it proves challenging to standardize these techniques and methods, in particular, controlling factors impacting microbial succession.

10.2.1 Postmortem microbial succession

Postmortem microbial succession refers to the changes in microbial communities that occur in and on a body after death. The changes after the death of any organism lead to the formation of new niches and variation in conditions; hence, changes in the microbial communities occur. Autolysis is the initial stage of decomposition by which the cells in the organism begin to break down by their own enzymatic action; the products of this degradation are nutrient-rich cellular constituents that microbes metabolize. Putrefaction refers to the decomposition of organic material by microorganisms, resulting in the breakdown of tissues and gas production (Can et al., 2014; Hyde et al., 2013).

The first microbial communities that benefit from the autolysis byproducts and initiate the putrefaction process are indigenous microbes that are colonizing certain areas of the human body. About 10-100 trillion microbes have been estimated in a human body, many of which populate the gut, being this habitat one of the most populated ones in the human body, the next major habitat would be the oral cavity (Adserias-Garriga et al., 2017b; Ursell et al., 2012). Upon death, the immune system no longer functions, internal temperatures shift, and physiological barriers break down; the composition of local microbiomes shifts and translocates to other areas of the body (Alan & Sarah, 2012; Burcham et al., 2019).

So the body's indigenous microorganisms are the ones present in the very first stages of decomposition, since putrefaction is initiated by indigenous representatives in different body areas. As decomposition advances, microbial populations shift according to the new microenvironments created as conditions change, where one of the most significant factors (if not the most) is oxygen availability. So that, as decomposition progresses, the microbial community shifts from aerobic to anaerobic, due to the depletion of oxygen within the body and the accumulation of gases like carbon dioxide, hydrogen sulfide, and volatile organic compounds (Adserias-Garriga et al., 2017a). And there is a transference between the body and the substrate of microbial organisms.

So that, the changes in the microbial communities can be summarized these general changes: initiation of the putrefaction by indigenous microorganisms, shifts in the microbial populations, and presence of opportunistic microorganisms, and interchange between the body and substrate microbiome representative. However, different body areas will show slight differences in their microbial community composition, mostly in the early stages of decomposition. Therefore, a complex microbial succession occurs as different groups of microorganisms colonize the body sequentially throughout the decomposition process, and each group of microbes plays a specific role in this process.

10.2.2 Postmortem microbial succession in the gut

The gut microbiome is the most populated microbial community in the human body, playing a critical role in digestion, immune modulation, and overall health (Human Microbiome Project Consortium, 2012). After death, the gut microbiome undergoes rapid and dramatic changes, driven by the breakdown of host regulatory mechanisms, which allows endogenous gut bacteria, such as *Enterobacteriaceae* and *Clostridium*, to proliferate without the limiting factors that used to function during life. Anaerobic conditions develop due to the lack of oxygen, favoring the growth of facultative and obligate anaerobes (Metcalf et al., 2016). In the bloat stage, microbial fermentation of carbohydrates and proteins produces gases (e.g., methane, hydrogen sulfide), leading to abdominal bloating. *Clostridium* species, particularly *C. perfringens*, dominate the gut microbiome during this stage. In later stages of decay, the gut microbiome becomes less diverse as nutrient availability declines, and environmental microbes from the surrounding soil begin to colonize the gut, introducing taxa such as *Pseudomonas* and *Acinetobacter* (Pechal et al., 2018) (Table 10.1).

10.2.3 Postmortem microbial succession in the oral cavity

Research evidence shows that the oral cavity is a valuable and informative area for assessing postmortem microbial succession due to its unique characteristics and the role it plays in decomposition (Adserias-Garriga et al., 2017a). The oral cavity contains a high bacterial load, comprising about 26% of total bacterial cells within the human microbiome. The large population offers considerable material for analysis; thus, changes in microbial communities over time can be detected (Human Microbiome Project Consortium, 2012). Additionally, the oral cavity presents a high microbial diversity, hosting one of the most diverse microbiomes in the human body, apart from the gastrointestinal tract (Adserias-Garriga et al., 2017b), and at the same time the oral cavity possesses a well-defined indigenous microbiome which allows the identification and comparison between the antemortem and postmortem microbial communities. Moreover, accessibility to the oral cavity surfaces permits an easy and noninvasive sampling. All these attributes make the oral cavity a highly valuable source for estimating the PMI in forensic investigations. Throughout the decomposition process, some microbial groups decrease, while others increase their abundance, reflecting the changing conditions and resource availability within the microenvironment they are colonizing.

In the fresh stage of decomposition, basically the indigenous microbiome exhibits an overgrowth, so the microbial groups found in the early stages of decomposition will consist of the microbial representatives of the oral cavity, including *Lactobacillaceae, Staphylococcaceae, Gemellaceae, Carnobacteriaceae, Aerococcaceae, Veillonellaceae, Streptococcaceae, Campylobacteraceae, Micrococcaceae, Bifidobacteriaceae, Actinomycetaceae*, and *Corynebacteriaceae* (Adserias-Garriga et al., 2017b). These are replaced by other groups as decomposition advances, and new niches are created as the conditions change throughout the decomposition process. As decomposition progresses, an initial decrease in Firmicutes and Actinobacteria relative abundances is observed; at the outset of the bloat stage, Tenericutes show a transient appearance that won't last for a long time (Adserias-Garriga et al., 2017b). This "flash" appearance of Tenericutes is caused by the specific nutritional and metabolic requirements of this group and the specific environmental physicochemical conditions that take place in that stage. This makes Tenericutes a good indicator of the bloat stage of the decomposition process. Passing the bloat stage and entering the active decay, Firmicutes experience a remarkable increase. However, the Firmicutes community present in the early stages, and the Firmicutes community colonizing in the later stages are drastically different; Early Firmicutes community is represented by oral microbiome

TABLE 10.1 Summary of the bacterial succession throughout the different stages of decomposition in the different body sites.

Body site	Stage of decomposition	Dominant microbial taxa	Diagnostic utility
Gut	Fresh	*Enterobacteriaceae, Clostridium*	Early indicators of decomposition due to lack of oxygen, proliferation of anaerobes
	Bloat	*Clostridiumperfringens*	Production of gases (methane, hydrogen sulfide) causing bloating
	Active decay	Decline in gut diversity; Introduction of *Pseudomonas* and *Acinetobacter*	Transition to environmental colonization; loss of endogenous microbiota
	Advanced decay	Soil-dwelling taxa such as *Pseudomonas, Acinetobacter*	Integration of gut microbiome with surrounding soil microbial communities
Oral cavity	Fresh	*Lactobacillaceae, Staphylococcaceae, Gemellaceae, Carnobacteriaceae, Veillonellaceae, Streptococcaceae*	Indigenous oral microbiome overgrowth
	Bloat	Transient appearance of *Tenericutes*	Indicator of bloat stage due to specific environmental conditions
	Active decay	Increase in *Firmicutes*, including soil-indigenous *Clostridiales* and *Bacillaceae*	Shift from oral to environmental microbial colonization
	Advanced decay	Soil-dwelling microbes dominate	Final microbial integration with surrounding soil
Skin	Fresh	*Staphylococcus, Corynebacterium, Propionibacterium*	Persistence of host-associated skin taxa
	Bloat	Malassezia (transient), introduction to *Aspergillus* and *Candida*	Breakdown of skin barriers, fungal growth
	Active decay	Saprophytic fungi (*Aspergillus, Penicillium*), bacterial dominance by *Pseudomonas* and *Acinetobacter*	Utilization of organic compounds from decomposition tissues
	Advanced decay	*Cladosporium, Fusarium, Alternaria*	Soil fungal colonization, integration with environmental microbiota
Internal organs	Fresh	*Enterobacteriaceae, Clostridium*	Initial colonization via blood stream and lymphatic system
	Bloat	*Clostridium* species dominate	Accelerated tissue breakdown
	Active decay	*Pseudomonas, Acinetobacter*	Colonization by environmental microbes
	Advanced decay	Indistinguishable from soil microbes	Complete microbial integration with surrounding soil

phylotypes, while late Firmicutes community is represented mainly by soil-indigenous opportunistic microorganism, including *Clostridiales* and *Bacillaceae* (Adserias-Garriga et al., 2017b) (Table 10.1).

10.2.4 Postmortem microbial succession in the skin

The skin microbiome is highly diverse and varies across different regions of the body. After death, the skin microbiome undergoes rapid changes as environmental microbes colonize the remains. In early postmortem periods, the skin microbiome is initially dominated by host-associated taxa such as *Staphylococcus*, *Corynebacterium*, and *Propionibacterium*. As decomposition progresses, these taxa are gradually replaced by environmental microbes, including fungal representatives. Among those are *Malassezia*, a genus of lipophilic yeasts commonly found on human skin, which may persist briefly after death but declines as decomposition progresses. As the body is exposed to the environment, airborne and soil-dwelling fungi begin to colonize the skin, such as *Aspergillus* and *Candida* species. The bloat stage is characterized by the breakdown of skin barriers and the release of nutrients from decomposing tissues, creating favorable conditions for fungal growth. Saprophytic fungi, such as *Aspergillus* and *Penicillium*, become more abundant, utilizing organic compounds released during this stage of decomposition. The skin bacterial population becomes dominated by certain groups, such as *Pseudomonas* and *Acinetobacter*, which thrive in moist environments (Hawksworth & Wiltshire, 2015). As decomposition progresses, the skin becomes increasingly colonized by soil-dwelling fungi, such as *Cladosporium*, *Fusarium*, and *Alternaria*. These fungi thrive in moist, nutrient-rich environments and play a key role in breaking down complex organic compounds. In the latter stages of decomposition, the skin microbiome is largely composed of soil-dwelling microbes, reflecting the integration of the remains into the surrounding environment (Pechal et al., 2018) and the skin microbiome becomes indistinguishable from that of the surrounding soil, reflecting the integration of the remains into the ecosystem (Table 10.1).

10.2.5 Microbial Succession in Internal organs

Internal organs become colonized by microbes after death as the body's barriers break down. The liver and spleen are among the first internal organs to be colonized by gut bacteria, which spread via the bloodstream and lymphatic system (Javan et al., 2019). *Enterobacteriaceae* and *Clostridium* are commonly found in these organs during the early stages of decomposition. The bloat stage represents an acceleration of microbial colonization of internal organs, in which *Clostridium* species play a key role in tissue breakdown. The heart and lungs are colonized by environmental microbes, particularly *Pseudomonas* and *Acinetobacter* (Metcalf et al., 2016). In the latter stages of decomposition, Internal organs are largely degraded, and their microbiomes become indistinguishable from those of the surrounding soil (Table 10.1).

10.2.6 Factors affecting postmortem microbial succession

As previously mentioned, decomposition is a highly complex process where a wide range of factors impact the composition and development of microbial communities within the body. Extrinsic or environmental factors, including temperature, moisture levels, oxygen availability, and pH levels, are key in defining microbial communities. Intrinsic factors, defined by physical attributes and the condition of the body, may also impact microbial communities during decomposition, such as the presence of injuries, illnesses, or medical treatments. In addition, the presence of insects, rodents, and other scavengers can disrupt existing microbial communities.

Intrinsic factors, or those host-derived characteristics that influence microbial colonization, include antemortem health status, physiological conditions, and individual-specific traits. Antemortem health status may impact the composition of the microbial community. Chronic conditions such as diabetes or obesity can alter the composition of the antemortem microbiome, creating a microbial signature that will persist after death (at least in the very early PMI). It has been established that individuals with diabetes often exhibit an overgrowth of *Enterobacteriaceae* in the gut, which can accelerate postmortem colonization of this group of bacteria in the surrounding tissues (Javan et al., 2019). Active infections at the time of death can also introduce pathogenic microbes that will impact the postmortem microbiome. *Staphylococcus aureus* or *Clostridium difficile* infections can lead to localized microbial blooms during decomposition due to active infections at the time of death. Additionally, medical treatment, such as recent antibiotic treatment before death, can suppress certain microbial taxa, reducing diversity and delaying the proliferation of decomposer communities. This can result in atypical succession patterns, complicating PMI estimation (Pechal et al., 2018). The host's physiology itself can also impact the microbial community composition. Age-related changes in the microbiome have been identified in terms of diversity, dominating taxa, and overall functional roles. Pediatric microbiomes are less diverse initially but increase in diversity with age; adult microbiomes are the most diverse and stable throughout the lifespan (in healthy conditions); and geriatric microbiomes exhibit a reduced diversity (Bäckhed et al., 2015; Biesbroek et al., 2014; Claesson et al., 2012; Crielaard et al., 2011; Hollister et al., 2015; (Metcalf et al., 2016) ; Shibagaki et al., 2017). In addition, hormonal differences between males and females can affect the microbial community structure. For example, estrogen has been shown to modulate the abundance of *Lactobacillus* in the gut, which may influence postmortem microbial dynamics (Javan et al., 2019). Moreover, body composition is another body factor that can affect the microbial community composition, since adipose tissue, which is rich in lipids, can serve as a nutrient source for lipolytic bacteria such as *Pseudomonas* and *Acinetobacter*, accelerating decomposition in obese individuals (Carter, 2005).

Extrinsic factors, or the environmental conditions and other external influences that impact microbial colonization after death, include temperature, oxygen availability, humidity, substrate, and geographic location. Temperature is one of the most critical factors affecting microbial activity. Higher temperatures accelerate microbial metabolism, leading to faster decomposition and more rapid succession of microbial communities. Conversely, cold temperatures can slow microbial activity, preserving remains for extended periods (Metcalf et al., 2016; Adserias-Garriga et al., 2017a). Oxygen Availability is a key factor determining the microbial community changes throughout the decomposition. Aerobic microbes dominate in the presence of oxygen, but as decomposition progresses, and oxygen levels decrease in several body parts, anaerobic taxa become predominant (Adserias-Garriga et al., 2017a; 2017b). This is of particular importance in buried remains or sealed environments (Javan et al., 2019). Moisture levels establish the availability of water, which is a key resource for microbial growth. Therefore, high humidity promotes the proliferation of hydrophilic bacteria and fungi, while dry conditions favor desiccation-resistant taxa such as *Bacillus* and *Aspergillus* .

The substrate where the body decomposes affects the microorganisms that will colonize the body in the later stages of decomposition. From all substrates, soil has been extensively studied. The composition of soil microbiomes varies widely depending on factors such as pH, nutrient content, and organic matter, among others. Remains deposited in nutrient-rich soils may exhibit faster decomposition due to the presence of decomposer microbes, while acidic or nutrient-poor soils can slow microbial activity (Carter, 2005). In addition, urban environments, which are often characterized by higher levels of pollution and human activity, can introduce unique microbial taxa to decomposing remains. In contrast, rural settings may harbor soil-specific microbes that influence colonization patterns (Pechal et al., 2018). Another extrinsic factor that has been described as affecting the decomposition microbial succession is the presence of insects, rodents, carnivores, and other scavengers, which introduce exogenous microbes and create physical openings that facilitate microbial invasion. It has been established that insect larvae can accelerate tissue breakdown and alter microbial succession timelines .

Despite the number of factors affecting the microbial succession after death, several studies have demonstrated that community composition changes occur in a somewhat predictable fashion through the course of decomposition, which allows scientists to exploit this as a potential tool to estimate PMI (Adserias-Garriga et al., 2017a; 2017b; Burcham et al., 2019; Hyde et al., 2013; Javan et al., 2017).

10.3 Analyzing postmortem microbial succession through microbial DNA

The study of the postmortem microbiome involves a multidisciplinary approach combining microbial ecology and forensic science, and utilizing molecular techniques, bioinformatics tools, and controlled experiments to analyze microbial communities in decomposing remains. The microbiome analysis involves sample collection, microbial DNA extraction and sequencing, and the application of bioinformatics and data analysis. Sampling can involve swabs and tissue samples or soil samples, depending on the approach. Swabs and tissue samples can be collected from various body sites to capture microbial communities; For fresh or early stages of decay, skin, muscle, and organs may be sampled, while in later stages of decay, bones and teeth only can be sampled, as microbial DNA can persist in these tissues for much longer periods. On the other hand, soil surrounding or beneath the remains can be sampled to assess the body-environment microbial communities exchange (Adserias-Garriga et al., 2017a).

Microbial DNA is extracted from the samples collected using commercial kits or protocols designed for complex matrices. Microbial extraction from soil samples should be conducted using specific protocols (or kits) since soil humic acids inhibit the posterior PCR (Adserias-Garriga et al., 2017a). Sequencing techniques include 16S rRNA Gene Sequencing, which targets the hypervariable regions of the 16S rRNA gene to identify bacterial taxa (Caporaso et al., 2010); 16S rRNA gene sequencing is the most frequently applied method for bacterial identification . The 16S rRNA gene contains hypervariable regions that can serve as biomarkers for taxonomic classification and phylogenetic analysis of the microbiome of death. High-throughput sequencing of the 16S rRNA gene can investigate the development and evolution of microbiota involved in the different stages of decomposition. Amplicons of the 16S rRNA gene are sequenced to classify microbial taxa associated with cadaver organs. In the case of analyzing fungal groups, internal transcribed spacer (ITS) sequencing will be performed, which targets the ITS region to identify fungal taxa (Blaalid et al., 2013).

Additionally, shotgun metagenomics can be applied, sequencing all DNA in a sample, providing information on both taxonomic composition and functional potential (Quince et al., 2017). This is a tool for identifying microorganisms in a sample that involves sequencing all extracted DNA segments without targeting PCR amplification of a specific region. It also allows mapping of the reads to certain metabolic functions, detailing the involved potential metabolism within that environment; hence, this technique provides both identification and function information about microorganisms. Moreover, it gets rid of PCR-based approaches, such as 16S rRNA sequencing, due to amplification bias (Adserias-Garriga et al., 2024).

The application of bioinformatics and data analysis requires that the raw sequencing data be processed to remove low-quality reads, adapters, and contaminants using tools like QIIME2, Mothur, or DADA2. Then, processed reads are aligned against reference databases (e.g., SILVA, Greengenes for bacteria; UNITE for fungi) to identify microbial taxa. Once the different taxa are identified, different analyses can be applied to assess the community structure, such as the calculation of alpha and beta diversities. Additionally, functional analysis can be conducted on the metagenomics sequencing products to predict microbial metabolic pathways and functional genes using tools like HUMAnN2 or PICRUSt2.

10.4 Conclusions

PMI estimation is a critical aspect of criminal investigations, helping to identify the deceased and reconstruct events around the death. Due to the importance of PMI information, research efforts have been devoted in the lasts years to improve the accuracy of the more classical approaches to estimate time since death, and especially, to develop methods applicable to remains with long PMIs. In that sense, more quantitative and objective methods are emerging, such as the analysis of the postmortem microbial succession. Even though there is scientific evidence that the analysis of the microbiome can be a valuable tool for PMI estimation, further research is needed to refine and standardize the methodologies, identify robust biomarkers, and likely, integrate it with other forensic techniques to estimate PMI. So that, although not applicable to casework yet, the microbiome composition analysis has great potential since it can provide valuable information to estimate PMI.

References

Adserias-Garriga, J., Hernández, M., Quijada, N. M., Rodríguez Lázaro, D., Steadman, D., & Garcia-Gil, J. (2017a). Daily thanatomicrobiome changes in soil as an approach of postmortem interval estimation: An ecological perspective. *Forensic Science International, 278*, 388–395. https://doi.org/10.1016/j.forsciint.2017.07.017 Epub2017Aug7.PMID:28818754.

Adserias-Garriga, J., Quijada, N. M., Hernandez, M., Rodriguez Lazaro, D., Steadman, D., & Garcia-Gil, L. J. (2017b). Dynamics of the oral microbiota as a tool to estimate time since death. *Mol. Oral Microbiol, 32*(6), 511e516. https://doi.org/10.1111/omi.12191.(b).

Alan, G., & Sarah, J. P. (2012). Microbes as forensic indicators. *Tropical Biomedicine, 29*(3), 311e330. https://www.ncbi.nlm.nih.gov/pubmed/23018494.

Biesbroek, G., Tsivtsivadze, E., Sanders, E. A., Montijn, R., Veenhoven, R. H., Keijser, B. J., & Bogaert, D. (2014). Early respiratory microbiota composition determines bacterial succession patterns and respiratory health in children. *American Journal of Respiratory and Critical Care Medicine, 190*(11), 1283–1292.

Blaalid, R., Kumar, S., Nilsson, R. H., Abarenkov, K., Kirk, P. M., & Kauserud, H. (2013). ITS1 versus ITS2 as DNA metabarcodes for fungi. *Molecular Ecological Resources, 13*(2), 218–224. https://doi.org/10.1111/1755-0998.12065 Epub2013Jan25.PMID:23350562.

Burcham, Z. M., Pechal, J. L., Schmidt, C. J., Bose, J. L., Rosch, J. W., Benbow, M. E., & Jordan, H. R. (2019). Bacterial community succession, transmigration, and differential gene transcription in a controlled vertebrate decomposition model. *Frontiers in Microbiology, 10*, 745. https://doi.org/10.3389/fmicb.2019.00745.

Bäckhed, F., Roswall, J., Peng, Y., Feng, Q., Jia, H., Kovatcheva-Datchary, P., Li, Y., Xia, Y., Xie, H., Zhong, H., Khan, M. T., Zhang, J., Li, J., Xiao, L., Al-Aama, J., Zhang, D., Lee, Y. S., Kotowska, D., Colding, C., ... Wang, J. (2015). Dynamics and stabilization of the human gut microbiome during the first year of life. *Cell Host & Microbe, 17*(5), 690–703.

Can, I., Javan, G. T., Pozhitkov, A. E., & Noble, P. A. (2014). Distinctive than atomicrobiome signatures found in the blood and internal organs of humans. *Journal of Microbiological Methods, 106*, 1e7. https://doi.org/10.1016/j.mimet.2014.07.026.

Caporaso, J. G., Kuczynski, J., Stombaugh, J., Bittinger, K., Bushman, F. D., Costello, E. K., Fierer, N., Peña, A. G., Goodrich, J. K., Gordon, J. I., Huttley, G. A., Kelley, S. T., Knights, D., Koenig, J. E., Ley, R. E., Lozupone, C. A., McDonald, D., Muegge, B. D., Pirrung, M., ... Knight, R. (2010). QIIME allows analysis of high-throughput community sequencing data. *Nature Methods, 7*(5), 335–336.

Carter, D.O. (2005). Forensic taphonomy: processes associated with cadaver decomposition in soil. Ph.D. thesis, James Cook University.

Carter, D. O., Yellowlees, D., & Tibbett, M. (2007). Cadaver decomposition in terrestrial ecosystems. *Die Naturwissenschaften, 94*(1), 12–24. https://doi.org/10.1007/s00114-006-0159-1.

Claesson, M. J., Jeffery, I. B., Conde, S., Power, S. E., O'Connor, E. M., Cusack, S., Harris, H. M., Coakley, M., Lakshminarayanan, B., O'Sullivan, O., Fitzgerald, G. F., Deane, J., O'Connor, M., Harnedy, N., O'Connor, K., O'Mahony, D., van Sinderen, D., Wallace, M., Brennan, L., ... O'Toole, P. W. (2012). Gut microbiota composition correlates with diet and health in the elderly. *Nature, 488*(7410), 178–184.

Creamer, J., & Buck, A. (2009). The assaying of haemoglobin using luminol chemiluminescence and its application to the dating of human skeletal remains. *Luminescence, 24*(5), 311–316. https://doi.org/10.1002/bio.1110.

Crielaard, W., Zaura, E., Schuller, A. A., Huse, S. M., Montijn, R. C., & Keijser, B. J. (2011). Exploring the oral microbiota of children at various developmental stages of their dentition in relation to their oral health. *BMC Medical Genomics, 4*(1), 22.

Dent, B., Forbes, S. L., & Stuart, B. (2004). Review of human decomposition processes in soil. *Environmental Earth Science, 45*(4), 576e585. https://doi.org/10.1007/s00254-003-0913-z.

Dethlefsen, L., Huse, S., Sogin, M. L., & Relman, D. A. (2008). The pervasive effects of an antibiotic on the human gut microbiota, as revealed by deep 16S rRNA sequencing. *PLoS Biol. 6*(11), e280. https://doi.org/10.1371/journal.pbio.0060280 PMID: 19018661; PMCID: PMC2586385.

Ermida, C., Navega, D., & Cunha, E. (2017). Luminol chemiluminescence: contribution to postmortem interval determination of skeletonized remains in Portuguese forensic context. *International Journal of Legal Medicine, 131*(4), 1149–1153. https://doi.org/10.1007/s00414-017-1547-0.

Ermida, C., Cunha, E., & Ferreira, M. T. (2024). Luminol and the postmortem interval estimation - Influence of taphonomic factors. *International Journal of Legal Medicine, 138*(3), 1109–1116. https://doi.org/10.1007/s00414-023-03132-9 Epub2023Nov24. PMID:37996553; PMCID:PMC11003912.

Grice, E. A., & Segre, J. A. (2011). The skin microbiome. *Nature Reviews. Microbiology, 9*(4), 244–253.

Gunn, A., & Pitt, S. A. (2012). Microbes as forensic indicators. *Tropical Biomedicine, 29*(3), 311–330 PMID:23018494.

Hawksworth, D. L., & Wiltshire, P. E. J. (2015). Forensic mycology: the use of fungi in criminal investigations. *Forensic Science International, 249*, 1–6.

Henssge, C (1988). Death time estimation in case work. I. The rectal temperature time of death nomogram. *Forensic Sci. Int. 38*, 209–236.

Hollister, E. B., Riehle, K., Luna, R. A., Weidler, E. M., Rubio-Gonzales, M., Mistretta, T. A., Raza, S., Doddapaneni, H. V., Metcalf, G. A., Muzny, D. M., Gibbs, R. A., Petrosino, J. F., Shulman, R. J., & Versalovic, J. (2015). Structure and function of the healthy pre-adolescent pediatric gut microbiome. *Microbiome, 3*, 36. https://doi.org/10.1186/s40168-015-0101-x PMID:26306392;PMCID:PMC4550057.

Hyde, E. R., Haarmann, D. P., Lynne, A. M., Bucheli, S. R., & Petrosino, J. F. (2013). The living dead: Bacterial community structure of a cadaver at the onset and end of the bloat stage of decomposition. *PLoS ONE, 8*(10), e77733. https://doi.org/10.1371/journal.pone.0077733.

Human Microbiome Project Consortium. (2012). Structure, function and diversity of the healthy human microbiome. *Nature, 486*, 207–214.

Javan, G. T., Finley, S. J., Abidin, Z., & Mulle, J. G. (2016). The Thanatomicrobiome: A Missing Piece of the Microbial Puzzle of Death. *Front. Microbiol. 7*, 225.

Javan, G. T., Finley, S. J., Smith, T., Miller, J., & Wilkinson, J. E. (2017). Cadaver thanatomicrobiome signatures: The ubiquitous nature of *Clostridium* species in human decomposition. *Frontiers in Microbiology, 8*, 2096. https://doi.org/10.3389/fmicb.2017.02096 PMID:29163394; PMCID:PMC5670113.

Megyesi, M. S., Nawrocki, S. P., & Haskell, N. H. (2005). Using accumulated degree-days to estimate the postmortem interval from decomposed human remains. *Journal of Forensic Sciences, 50*(3), 618e626. https://www.ncbi.nlm.nih.gov/pubmed/15932096.

Metcalf, J. L, Xu, Z. Z, Weiss, S, Lax, S, Van Treuren, W, Hyde, E. R., Song, S. J, Amir, A, Larsen, P., & Sangwan, N. (2016). Microbial community assembly and metabolic function during mammalian corpse decomposition. *Science, 351*, 158–162.

Paczkowski, S., & Schütz, S. (2011). Post-mortem volatiles of vertebrate tissue. *Applied Microbiology and Biotechnology, 91*(4), 917–935. https://doi.org/10.1007/s00253-011-3417-x Epub2011Jul1.PMID:21720824;PMCID:PMC3145088.

Pechal, J. L., Crippen, T. L., Benbow, M. E., Tarone, A. M., Dowd, S., & Tomberlin, J. K. (2018). The potential use of bacterial community succession in forensics as described by high throughput metagenomic sequencing. *International Journal of Legal Medicine, 132*(2), 449–462.

Procopio, N., Williams, A., Chamberlain, A. T., & Buckley, M. (2018). Forensic proteomics for the evaluation of the post-mortem decay in bones. *Journal of Proteonomics, 177*, 21e30. https://doi.org/10.1016/j.jprot.2018.01.016.

Quince, C., Walker, A. W., Simpson, J. T., Loman, N. J., & Segata, N. (2017). Shotgun metagenomics, from sampling to analysis. *Nature Biotechnology, 35*(9), 833–844. https://doi.org/10.1038/nbt.3935, 10.1038/nbt1217-1211b PMID:28898207.

Shibagaki, N., Suda, W., Clavaud, C., Bastien, P., Takayasu, L., Iioka, E., Kurokawa, R., Yamashita, N., Hattori, Y., Shindo, C., Breton, L., & Hattori, M. (2017). Aging-related changes in the diversity of women's skin microbiomes associated with oral bacteria. *Scientific Reports, 7*(1), 10567. https://doi.org/10.1038/s41598-017-10834-9 PMID:28874721;PMCID:PMC5585242.

Ursell, L. K., Metcalf, J. L., Parfrey, L. W., & Knight, R. (2012). Defining the human microbiome. *Nutrition Reviews, 70*(Suppl. 1), S38eS44. https://doi.org/10.1111/j.1753-4887.2012.00493.x.

Wilk, L. S., Hoveling, R. J. M., Edelman, G. J., Hardy, H. J. J., van Schouwen, S., van Venrooij, H., & Aalders, M. C. G. (2020). Reconstructing the time since death using noninvasive thermometry and numerical analysis. *Sci. Adv. 6*, eaba4243. https://doi.org/10.1126/sciadv.aba4243.

Zapico, S. C., & Adserias-Garriga, J. (2022). Postmortem interval estimation: New approaches by the analysis of human tissues and microbial communities' changes. *Forensic Science, 2*, 163–174. https://doi.org/10.3390/forensicsci2010013.

Chapter 11

Probabilistic genotyping techniques and their applications

Kelly M. Elkins
Chemistry and Forensic Science Department, Towson University, Towson, MD, United States

11.1 Introduction

Body fluid stains can originate from a single source or be mixed in a recovered forensic evidence sample and have two or more contributors. It is common in sexual assault cases to identify male and female contributors in a sample. However, mixed samples can result from multiple female and male contributors as well. An important question in forensic DNA analysis is determining if the suspect contributed DNA to the sample. Other questions include determining the number of contributors to mixed samples in mass disaster, gang rape, and other cases. These are challenging questions that become even more difficult in cases of degradation, low quantities of contributed DNA, polymerase chain reaction (PCR) inhibition, and allele dropout. In many cases, the contributor ratio is such that the profiles cannot be deconvoluted using traditional methods. To gather the data for the contributor determination, the sample must be screened, sampled, and the DNA must be extracted, quantified, amplified at the targets, and electrophoresed or sequenced to generate the profile as for any DNA identification analysis.

In 1985, Peter Gill and collaborators demonstrated a method for differential extraction of mixed stains that included vaginal epithelial cells and sperm cells on swabs and bloodstains with semen on cotton cloth (Gill et al., 1985). Using the differential extraction method, the epithelial cells in the evidence sample could be preferentially lysed, leaving the sperm cells intact for lysis in a subsequent step. Differential extraction yields two samples each, which may originate from a single male contributor or two contributors, or more per fraction.

In cases where there is a person of interest and a body fluid sample can be collected from that person, a single-source profile can be generated and compared to the profiles generated from the epithelial and sperm cell fractions. A suspect profile may be excluded or a random match probability (RMP) may be computed. In other cases, if there is no person of interest at the time, the investigation proceeds without a suspect profile or profiles.

Further, in cases in which differential extraction or the extraction method used yields a mixed source sample, a mixed profile is generated using the DNA typing process. This may occur if two or more persons contribute sperm or epithelial cells to a sample that coelute in the respective elution steps. Deconvolution of two-person mixtures was described from peak height and peak area data in 1998 (Clayton et al., 1998). Once the mixed profile is deconvoluted, each contributor's profile can be analyzed manually using the same statistics used for single source samples such as RMP, combined probability of inclusion, combined probability of exclusion, and likelihood ratios (LRs) or using software tools (Butler, 2005). However, in many cases, there are more than two contributors, or the contributor ratio is such that the profiles cannot be deconvoluted using this method. The challenge of interpreting complex case profiles has led to the development of software tools that employ mathematical probability and statistical approaches to mixture deconvolution; these are probabilistic genotyping software (PGS) tools. PGS employs advanced statistical models and calculates an LR statistic that a individual of interest contributed to a DNA sample. PGS tools were developed by mathematicians and forensic scientists over a decade ago and are now used by organizations and forensic laboratories to solve cases throughout the world. They can improve the accuracy of analyses in complex cases with multiple contributors and reduce analyst subjectivity. Additionally, PGS can ingest partial profiles and data previously considered too complex for analysis.

Investigative and Predictive DNA Testing. DOI: https://doi.org/10.1016/B978-0-443-30128-5.00022-9

11.2 Probabilistic genotyping tools

As for DNA extraction, quantitation, and typing, there are multiple PGS tools available for mixed profile analysis (Table 11.1). PGS tools accept data input, model DNA profiles of potential contributors to the mixture, calculate LRs under different hypotheses, deconvolute the mixture, compute statistics, and generate reports (Fig. 11.1). PGS tools model biological data, typically raw electropherograms from capillary electrophoresis, but also inputted alleles, using statistical methods. PGS software can quickly evaluate scenarios with the assigned number of contributors, and if their profiles are present and overlapping in the profiles. PGS tools compute LRs of the probability of a peak contributing to the profile versus not contributing based upon peak heights, stutter, and allele drop in/dropout. Laboratories set LR numerical cutoffs for inclusion and exclusion based upon their internal validation and laboratory guidelines. PGS theory was described by Gill et al., and an early tool LoComationN was published by Gill et al. (2007) shortly after. Additional software tools followed (Table 11.1).

PGS tools use binary, discrete or semicontinuous, and continuous models (Coble & Bright, 2019). Allele marker calls are discrete values. Binary models use assigned values of 1 or 0, whether the peak is observed or not, to make yes or no decisions of whether a genotype is associated with a contributor. Discrete models or semicontinuous models consider alleles that are present and absent, and the weights of probabilities of drop in/dropout in the profile, but not peak height, in their calculations. A drawback to this approach is that the data collected in the experiment is not fully used. However, considering only allelic information makes the computation much faster than other approaches. In comparison, continuous models incorporate allele calls, peak height, heterozygote peak balance, and stutter ratios (Gill et al., 2021).

The two leading PGS systems used in the United States are TrueAllele (Perlin et al., 2011) and STRmix™ (Greenspoon et al., 2024; Taylor et al., 2013). Both are commercial products. Other PGS software includes Lab Retriever, forensim (Haned, 2011), Forensic Statistical Tool (Mitchell et al., 2012), LR Mix (Gill & Haned, 2013), BP Sentry, EuroForMix (Bleka et al., 2016), CaseSolver (Bleka et al., 2019), and DNAStatistX. LR Mix consists of open source code written in R and included in forensim (Gill & Haned, 2013). TrueAllele is a program created by Mark Perlin and Cybergenetics in 1995 (Perlin et al., 1995), patented in 1996, and commercialized. It was used to analyze the human remains recovered from the New York World Trade Center beginning in 2001 (Cybergenetics, 2023, https://www.cybgen.com/information/newsroom/2023/) and the first court case using probabilistic genotyping in 2009. STRmix™ and FaSTR (Lin et al., 2021) are products produced by New Zealand's Institute of Environmental Science and Research. STRmix™ has been used in

TABLE 11.1 Probabilistic genotyping interpretation software tools.

Probabilistic genotyping software tool	References
LoComationN	Gill et al. (2007)
TrueAllele	Perlin et al. (2011)
Forensim	Haned (2011)
Forensic Statistical Tool	Mitchell et al. (2012)
BP Sentry	bpsentry.com
LR Mix	Gill and Haned (2013)
STRmix™	Taylor et al. (2013)
EuroForMix	Bleka et al. (2016)
CaseSolver	Bleka et al. (2019)

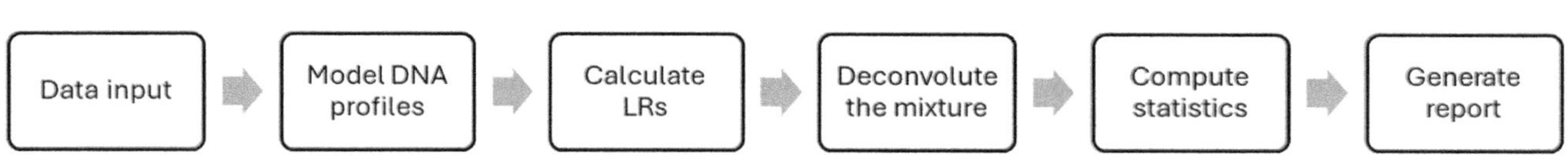

FIGURE 11.1 Process of implementing probabilistic genotyping software tools. *LR*, Likelihood ratio.

forensic casework since 2012. Both the TrueAllele and STRmix™ tools employ Markov chain Monte Carlo sampling algorithms to test statistical probability distributions and predict reasonable genotypes.

TrueAllele uses a Bayesian (Bayes' rule) (Bayes, 1763) probability model. In TrueAllele, analytical and stochastic thresholds and calibration are used to model the data, and dropout is not considered. Short tandem repeat (STR) peak heights, area under the peaks, and base pair length are considered. Background noise, PCR variation, and mixture weight are computed using probability calculations. PCR stutter and efficiency values are inputted from research on these issues. DNA degradation, coancestry, paternity, and relatedness are considered. The contributor and genotype at each locus are input as well as a mixture weight. The output is a visualization of the LR of separate genotype probabilities for the evidence to have been deposited by the suspect, as compared to coincidence.

STRmix™ also employs a Bayesian approach and searches the genotype information in a DNA profile against the database of alleles at the loci within model parameters. It has been used in forensic casework since 2012 (Gill et al., 2021). The software uses input loci and alleles, expected stutter ratios, and peak heights to compare peak balance for various potential outcomes. Drop out and drop in are both considered, as are the effects of changing the analytical threshold. The algorithms can test many more scenarios than a human in an exponentially shorter time from a few minutes to an hour, improving efficiency in solving cases. Analysts must specify how many contributors contributed to the sample based on the electropherogram; STRmix™ makes calculations based on this and the other input values. As humans inherit two copies of each STR allele from their biological parents, a locus with six peaks in the electropherogram indicates at least three contributors to the sample. STRmix™ evaluates the possible scenarios and outputs an LR for each based on the data.

In FaSTR, the raw .hid file profile from the evidence sample, such as a firearm, is uploaded, and the user indicates the kit used to generate the data. The user assigns the number of contributors or the software can estimate the number of contributors using an algorithm. Upon selection, the analysis is moved to STRmix™ for deconvolution analysis. The report summarizes the template relative fluorescence unit and mixture proportion for each number of contributor possibilities and log LR, as well as where the allele and stutter variances lie compared to normal distributions. If the report yields profiles, they can be used for database (e.g., Combined DNA Index System, CODIS) interrogation. The suspect's profile can be compared to the possible solutions outputted along the Bell curve, and an LR can be computed. If there is no suspect, the profile can be analyzed using the database LR (DBLR) product to search for common contributors in DNA mixtures and determine the possible genotypes of the contributors.

11.3 Validation studies of probabilistic genotype software tools

New tools, including PGS tools, must undergo developmental and internal validation prior to introduction into the forensic DNA analysis workflow. In developmental validation, accuracy studies are performed to determine if the calculations of the PGS systems are correct, and LRs are correctly computed using mixtures with equal and imbalanced mixture proportions where the contributor profiles are known. Other developmental validation studies include those of sensitivity, specificity, and precision (ANSI/ASB Standard 018, 2020). Precision studies of the same data and statistical parameters are conducted to evaluate the range of LRs outputted as no two runs are expected to yield identical values; the values should fall in the range of expected values. Limitations of the system should be uncovered and disseminated. Accuracy studies cannot be conducted if the contributor profiles are unknown. Each set of protocols and parameters used for casework must be validated in internal validations as well as studies of accuracy, sensitivity, specificity, and precision. The Scientific Working Group on DNA Analysis Methods (SWGDAM) PGS validation was disseminated in 2015 (Scientific Working Group on DNA Analysis Methods, 2015).

Reviews of validations and comparative studies have been published extensively and can be found elsewhere (Haned et al., 2016) is one). A few are mentioned here. International validations include the European Network of Forensic Science Institutes Best Practice Manual PGS validation in 2017 and United Kingdom Forensic Science Regulator Mixture Interpretation and response to President's Council of Advisors on Science and Technology PGS validation in 2018 (Bright et al., 2018; FSR-G-222). STRmix™ performed very well for up to three-person mixtures in which each contributor was present at a minimum 20% level; analysts correctly interpreted three-person mixtures 98% of the time. 2825 three to six contributor mixtures were analyzed in the study (Bright et al., 2018). A subsequent STRmix™ interlab study with 174 participants was conducted in 2019 using two complex mixtures from the PROVEDit set (Bright et al., 2019). The variation was within a factor of 10 but considered minor as it was well outside the threshold (Bright et al., 2019).

Recently, John Butler et al. (2024) have published a DNA Mixture Interpretation: A National Institute of Standards and Technology (NIST) Scientific Foundational Review (NISTIR 8351). In this report, the team documents and describes the scientific principles and theory of currently used methods and practices, catalogs the literature that describes the state of the field, including capabilities and limitations, and assesses the tests' reliability using publicly available data.

11.4 Considerations for use

The software must be validated by the lab prior to use for casework. Analysts must adhere to their lab's standard operating procedures (SOP) for use and implementation. Training, supervision, and peer review are essential for all analysts to ensure accurate and reliable results. The software must be maintained, including essential updates.

11.5 Artificial specific intelligence

PGS is considered artificial specific intelligence as it is not strictly machine learning but does employ large datasets and statistical models to make and model predictions. However, in the proprietary models, humans are unable to view each decision the software makes.

11.5.1 Case study

Probabilistic genotyping has been used to aid in solving cold cases and overturning wrongful convictions. In a case in Texas in 2010, DNA was recovered from a sample taken under the victim's fingernails. The DNA profile that was generated indicated the presence of the victim's DNA, as well as an unknown male contributor. Lydell Grant was suspected by an eyewitness of committing the crime. When the analyst compared his profile to that of the unknown male contributor, she found several loci in which one allele could not be excluded and several other loci where none of the suspect's alleles were present in her report. In sum, 26 alleles were not present in the victim's or Lydell's profiles. The analyst stated that "no conclusions will be made regarding Lydell Grant as a possible contributor to this mixture." Based on eyewitness testimony, Lydell Grant was convicted of murder. In years that followed, with the help of the Texas Innocence Project and legal and forensic students and faculty, TrueAllele was used to analyze the mixture. A contributor STR profile was generated and uploaded to CODIS. The hit led investigators to an entirely different suspect and led to his conviction. Lydell was exonerated and released from prison (Alexander Knoops, 2021; Ambers & Ware, 2021; Shapley, 2020).

11.6 Advantages of probabilistic genotyping

Discussions with crime lab directors have indicated that PGS is revolutionizing DNA mixture analysis and aiding their labs in solving many more cases. PGS has been demonstrated to detect contamination in collection, processing, or extraction from investigators or lab staff in quality assurance applications (Taylor et al., 2019). An advantage of PGS is that the software computes outcomes without knowledge of defendant profiles. However, the analyst still must analyze the data, call peaks, and consider the LRs when writing the report.

PGS applications have been accepted by courts worldwide. US labs began to adopt PGS approximately 10 years ago, and more than 50 US labs now use PGS tools. Locally, in Maryland, for example, the Baltimore City Police Department has adopted TrueAllele, and the Maryland State Police Forensic Science Division has adopted STRmix™. The software computes LRs for the contributor propositions and possibilities. TrueAllele has been used in over 1000 US cases and almost all of the US states. A recent study showed that STRmix™ and TrueAllele arrived at the same result 91% of the time (Greenspoon et al., 2024).

11.7 Drawbacks of probabilistic genotyping

There are drawbacks to PGS tools. Firstly, PGS products are commercial, and the code is private and unpublished. Analysts and researchers are not being able to see inside the proprietary "black-box" and directly compare testing of lab evidence data with multiple PGS systems.

This is a narrow example of a wider issue in science surrounding data sharing. Even though authors publish papers based on their data, frequently they do not share the complete datasets or make them available in a usable format (Elkins, 2021). A discussion of this issue is ongoing in the journal *Forensic Science International: Genetics* (Guide for Authors). The submission guidelines for authors state,

> *"This journal encourages and enables you to share data that supports your research publication where appropriate, and enables you to interlink the data with your published articles. Research data refers to the results of observations or experimentation that validate research findings, which may also include software, code, models, algorithms, protocols, methods and other useful materials related to the project." (https://www.sciencedirect.com/journal/forensic-science-international-genetics/publish/guide-for-authors)*

Until code and data are required to be deposited as a condition of publication, many authors will not willingly share (My students conducted a data sharing study in 2023 (Wallace & Elkins, 2024) in which 38% of authors queried shared the requested, published data. Although this is a significant improvement over publications of the same from 2009, in which 10% of authors shared the requested data, the problem persists.)

To address the black-box and shared data set issues, the United States National Institute of Standards and Technology has produced a product, Forensic DNA Research Grade Test Material 10235, with intact and degraded mixtures with explicit informed consent for public sharing. Another drawback in the application of PGS, as well as other emerging technologies in forensics, is that the technologies need to be validated by the labs with the types of samples they are being used to apply to casework prior to use in casework. For example, PGS software must be validated for the number of contributors in question (SWGDAM guidelines). Another issue is that the LRs calculated using different PGS systems may be different when the same data is inputted. There is not a lot of data on this at this time because lab data is not publicly available, leading to only limited comparative studies of common datasets.

Furthermore, analysis of mixed profile electropherograms generated from crime scene samples is challenging. Addressing the challenges is part of the reason for PGS tools, but there is still a human element to testing and interpretation. Analysts use SWGDAM and ASB guidelines to guide their decision-making to distinguish artifacts and true peaks. However, variations arise from differences in human decision-making and parameters fed to the programs, as well as differences in statistical models used by the programs themselves. In complicated mixtures with multiple contributors, the number of stutter peaks increases with the stutter possible for each peak, leading to increased overlap of true peaks and stutter. STR alleles can be detected at low levels, and some alleles may be undetected or drop below the threshold ("dropout"). PCR DNA polymerase artifacts, including stutter and incomplete adenylation as well as dye pull-up, can lead to low-level peaks that must be differentiated from the alleles contributed by the contributors. In a study, 79% of analysts estimates of the number of contributors were correct (Hicklin et al., 2023). This is relevant because programs such as STRmix™ require the analyst to indicate the number of contributors.

Analysts must take care to avoid overestimating the number of contributors, as demonstrated in previous studies. However, due to peak overlap, with six alleles called at a locus but four (or more) contributors and the analyst must also take care to consider the higher contributor ratio. In studies, analysts almost always underestimated the number of contributors in 2–3 contributor mixtures (Hicklin et al., 2023). Hicklin et al. (2023) showed that most incorrect determinations of the contributors were overestimates. The NOCit software tool by SoftGenetics can aid in making the number of contributor determinations for STRmix™, MaSTR, and EuroForMix. Labs report using STRmix™ for five or less contributors, and most have validated the tool for four or less. Nonetheless, the computed LR values varied even though the reported analysis was correct in terms of exclusions and inclusions (Bright et al., 2019).

11.8 Ongoing research and future work

Additional research and studies are still needed. For example, more comparison studies are needed. A recent study showed that STRmix™ and TrueAllele deviate with low template and degraded contributor samples (Greenspoon et al., 2024). Additionally, reporting formats from different labs need to be standardized so that juries are given the interpretation in the same format for a given result. In a recent study of perceptions of forensic scientists on PGS tools, several themes emerged (Aalbers et al., 2023). One was that "transparency is key" (Aalbers et al., 2023).

Much progress has been made. In the past year, labs have demonstrated the capability of single-cell analysis systems using multiple myeloma (MM), laser capture microdissection, and DEPArray (Huffman & Ballantyne, 2023). Determining DNA profiles of single cells eliminates the need for mixture deconvolution and PGS systems. A separate issue arises: is the presence of a single cell relevant to a case when several studies have shown secondary and tertiary transfer of cells from individuals who were never present at a scene? DNA testing demonstrates only that a particular profile is present and not when the source of the DNA was deposited.

PGS results from various programs have been shown to be comparable in various comparison studies. Probabilistic genotyping has embedded itself into the forensic toolkit. The challenges that remain include determining assumptions, choosing the best tool, and ensuring that the tools lead to the same results and interpretations for a profile or dataset.

References

Aalbers, S. E., Khan, A. T., & Weir, B. S. (2023). Perceptions of forensic scientists on statistical models, sequence data, and ethical implications for DNA evidence evaluations: A qualitative assessment. *Forensic Science International Synergy, 6*, 100335. https://doi.org/10.1016/j.fsisyn.2023.100335.

Alexander Knoops, G.-J. (2021). *Probabilistic genotyping: A possible new legal avenue to prevent and redress miscarriages of justice*. IntechOpen. https://doi.org/10.5772/intechopen.98247.

Ambers, A., & Ware, M. (2021). *The wrongful conviction of Lydell Grant for murder in Texas*.

ANSI/ASB Standard 018. (2020). *Standard for validation of probabilistic genotyping systems, American Academy of Forensic Sciences*. https://www.aafs.org/sites/default/files/media/documents/018_Std_e1.pdf. Accessed August 15, 2024.

Bayes, T. R. (1763). An essay towards solving a problem in the doctrine of chances. *Philosophical Transactions, 53*, 370–418.

Bleka, Ø., Storvik, G., & Gill, P. (2016). EuroForMix: An open source software based on a continuous model to evaluate STR DNA profiles from a mixture of contributors with artefacts. *Forensic Science International: Genetics, 21*, 35–44. https://doi.org/10.1016/j.fsigen.2015.11.008.

Bleka, Ø., Prieto, L., & Gill, P. (2019). CaseSolver: An investigative open source expert system based on EuroForMix. *Forensic Science International: Genetics, 41*, 83–92. https://doi.org/10.1016/j.fsigen.2019.03.024.

Bright, J. A., Cheng, K., Kerr, Z., McGovern, C., Kelly, H., Moretti, T. R., Smith, M. A., Bieber, F. R., Budowle, B., Coble, M. D., Alghafri, R., Allen, P. S., Barber, A., Beamer, V., Buettner, C., Russell, M., Gehrig, C., Hicks, T., Charak, J., ... Buckleton, J. (2019). STRmix™ collaborative exercise on DNA mixture interpretation. *Forensic Science International: Genetics, 40*, 1–8. https://doi.org/10.1016/j.fsigen.2019.01.006, http://www.elsevier.com.

Bright, J. A., Richards, R., Kruijver, M., Kelly, H., McGovern, C., Magee, A., McWhorter, A., Ciecko, A., Peck, B., Baumgartner, C., Buettner, C., McWilliams, S., McKenna, C., Gallacher, C., Mallinder, B., Wright, D., Johnson, D., Catella, D., Lien, E., ... Buckleton, J. S. (2018). Internal validation of STRmix™–A multi laboratory response to PCAST. *Forensic Science International: Genetics, 34*, 11–24. https://doi.org/10.1016/j.fsigen.2018.01.003, http://www.elsevier.com.

Butler, J. M. (2005). *Forensic DNA Typing* (2nd ed). Elsevier Academic Press.

Butler, J. B., Iyer, H., Press, R., Taylor, M. K., Vallone, P. M., & Willis, S. (2024). DNA mixture interpretation: A national institute of standards and technology (NIST) scientific foundational review (NIST IR 8351). https://www.nist.gov/spo/forensic-science-program/dna-mixture-interpretation-nist-scientific-foundation-review. Accessed August 1, 2025.

Clayton, T. M., Whitaker, J. P., Sparkes, R., & Gill, P. (1998). Analysis and interpretation of mixed forensic stains using DNA STR profiling. *Forensic Science International, 91*(1), 55–70. https://doi.org/10.1016/s0379-0738(97)00175-8.

Coble, M. D., & Bright, J. A. (2019). Probabilistic genotyping software: An overview. *Forensic Science International: Genetics, 38*, 219–224. https://doi.org/10.1016/j.fsigen.2018.11.009.

DNA Mixture Interpretation (FSR-G-222) Forensic Science Regulator Guidance. 31 July. Forensic Science Regulator, (3), https://assets.publishing.service.gov.uk/media/5f607d8be90e076cdbd9227b/G222_DNA_Mixture_Interp__Issue3.pdf. Accessed August 15, 2024.

Elkins, K.M. (2021) Ethics of data sharing. In ACS Symposium Series. Schelble, S.M. and Elkins, K.M. Eds. International Ethics in Chemistry: Developing Common Values across Cultures. Washington, DC: ACS Books. pp 113–124.

Gill, P., & Haned, H. (2013). A new methodological framework to interpret complex DNA profiles using likelihood ratios. *Forensic Science International: Genetics, 7*(2), 251–263. https://doi.org/10.1016/j.fsigen.2012.11.002.

Gill, P., Benschop, C., Buckleton, J., Bleka, Ø., & Taylor, D. (2021). A review of probabilistic genotyping systems: EuroForMix, DNAStatistX and STRmix™. *Genes, 12*(10), 1559. https://doi.org/10.3390/genes12101559.

Gill, P., Jeffreys, A. J., & Werrett, D. J. (1985). Forensic application of DNA 'fingerprints' *Nature, 318*(6046), 577–579. https://doi.org/10.1038/318577a0.

Gill, P., Kirkham, A., & Curran, J. (2007). LoComatioN: A software tool for the analysis of low copy number DNA profiles. *Forensic Science International, 166*(2-3), 128–138. https://doi.org/10.1016/j.forsciint.2006.04.016.

Greenspoon, S. A., Schiermeier-Wood, L., & Jenkins, B. C. (2024). A tale of two PG systems: A comparison of the two most widely used continuous probabilistic genotyping systems in the United States. *Journal of Forensic Sciences*. https://doi.org/10.1111/1556-4029.15571.

Haned, H. (2011). Forensim: An open-source initiative for the evaluation of statistical methods in forensic genetics. *Forensic Science International: Genetics, 5*(4), 265–268. https://doi.org/10.1016/j.fsigen.2010.03.017.

Haned, H., Gill, P., Lohmueller, K., Inman, K., & Rudin, N. (2016). Validation of probabilistic genotyping software for use in forensic DNA casework: Definitions and illustrations. *Science & Justice, 56*(2), 104–108. https://doi.org/10.1016/j.scijus.2015.11.007.

Hicklin, R. A., Richetelli, N., Emerick, B. L., Bever, R. A., & Davoren, J. M. (2023). Variation in assessments of suitability and number of contributors for DNA mixtures. *Forensic Science International: Genetics, 65*. https://doi.org/10.1016/j.fsigen.2023.102892, http://www.elsevier.com.

Huffman, K., & Ballantyne, J. (2023). Single cell genomics applications in forensic science: Current state and future directions. *iScience, 26*(11), 107961. https://doi.org/10.1016/j.isci.2023.107961.

Mitchell, A. A., Tamariz, J., O'Connell, K., Ducasse, N., Budimlija, Z., Prinz, M., & Caragine, T. (2012). Validation of a DNA mixture statistics tool incorporating allelic drop-out and drop-in. *Forensic Science International: Genetics, 6*.

Perlin, M. W., Lancia, G., & Ng, S.-K. (1995). Toward fully automated genotyping: Genotyping microsatellite markers by deconvolution. *American Journal of Human Genetics, 57*(5), 1199–1210.

Perlin, M. W., Legler, M. M., Spencer, C. E., Smith, J. L., Allan, W. P., Belrose, J. L., & Duceman, B. W. (2011). Validating trueallele ® DNA mixture interpretation. *Journal of Forensic Sciences, 56*(6), 1430–1447. https://doi.org/10.1111/j.1556-4029.2011.01859.x.

Scientific Working Group on DNA Analysis Methods. (2015). *Guidelines for the Validation of Probabilistic Genotyping Systems*. https://www.swgdam.org/_files/ugd/4344b0_22776006b67c4a32a5ffc04fe3b56515.pdf. Accessed August 15, 2024.

Shapley J. (2020). Texas man close to exoneration after computer algorithm leads to new suspect. NBCNews. https://www.nbcnews.com/news/us-news/texas-man-close-exoneration-after-computer-algorithm-leads-new-suspect-n1124946. Accessed August 1, 2025.

Taylor, D., Bright, J. A., & Buckleton, J. (2013). The interpretation of single source and mixed DNA profiles. *Forensic Science International: Genetics, 7*(5), 516–528. https://doi.org/10.1016/j.fsigen.2013.05.011.

Taylor, D., Rowe, E., Kruijver, M., Abarno, D., Bright, J. A., & Buckleton, J. (2019). Inter-sample contamination detection using mixture deconvolution comparison. *Forensic Science International: Genetics, 40*, 160–167. https://doi.org/10.1016/j.fsigen.2019.02.021, http://www.elsevier.com.

Wallace, R. M., & Elkins, K. M. (2024). A study of data sharing by authors published in scientific journals. *Journal of Forensic Science Education, 6*(1).

Further reading

Lin, M.-H., Lee, S.-I., Zhang, X., Russell, L., Kelly, H., Cheng, K., Cooper, S., Wivell, R., Kerr, Z., Morawitz, J., & Bright, J.-A. (2021). Developmental validation of FaSTR™ DNA: Software for the analysis of forensic DNA profiles. *Forensic Science International: Reports, 3*. https://doi.org/10.1016/j.fsir.2021.100217.

Chapter 12

Applications of genetics in archaeological investigations: exploring human migration and genetic evolution in Indian subcontinent

Bhavna Ahlawat[1], Aparna Dwivedi[1,2], Snigdha Konar[1,2], Shristi Gupta[1,2], and Niraj Rai[1,2]

[1]*Birbal Sahni Institute of Palaeosciences, Lucknow, Uttar Pradesh, India,* [2]*Academy of Scientific and Innovative Research (AcSIR), Ghaziabad, Uttar Pradesh, India*

12.1 Introduction

Since 60–40 KYA, modern humans began settling across Eurasia and other regions of the world. The most profound changes have been witnessed in the last 10,000 years, driven by practices such as agriculture, culture, and spoken languages (Narasimhan et al., 2019). The archaeological records of Southeast Asia indicate archaic hominins arriving at least 1.6 million years ago, with anatomically modern humans settling by 50,000 yBP. The most significant shift in human culture is marked by the spread of agriculture and changing climatic conditions (Lipson et al., 2018).

South Asia exhibits high genetic differentiation and extensive population structuring due to its evolutionary antiquity and the prevalence of endogamy among different population groups. It has served as a critical route for population dispersal and migration. Many Indian and South Asian societies consist of tribal groups and urban dwellers, divided and ranked into castes, with these groups remaining highly endogamous in contemporary times (Majumder, 2010; Reich et al., 2009). India, as a major part of South Asia, has experienced significant influences on its genetic structure through its social history, cultural practices, migration, and agricultural movements (Majumder, 2010; Narasimhan et al., 2019).

Contemporary South Asians are predominantly intramarrying ethnic populations with a diverse range of cultural and linguistic groups. These language groups are geographically distributed within India and South Asia and largely do not overlap. The northern region is inhabited by Indo-European speakers, the southern by Dravidian speakers, the northeastern by Tibeto-Burman speakers, and numerous Austro-Asiatic speakers, mainly tribal groups residing in remote areas across the continent. It is estimated that India contains more than 400 tribal groups and over 4000 caste groups. Due to strict social norms and the intramarrying nature of these populations, gene flow between tribal populations or between tribal and caste populations is minimal.

Genomic and anthropological evidence suggests that tribal populations are indigenous to India, while historical and anthropological studies show that the caste system in India has not always been homogeneous, with varying admixture from immigrants from Central and Western Asia via the northwestern corridor (Majumder, 2010; Narasimhan et al., 2019).

This chapter aims to study the dispersal of modern humans in South Asia over the last 45,000 years, utilizing archaeological evidence and ancient genomic data. Our understanding of how modern humans spread into South Asia has significantly advanced due to the successful sequencing of ancient human DNA. The chapter emphasizes discoveries that have enhanced our knowledge of the initial migration of modern humans into South Asia, subsequent human movements and interactions, and the rapid dispersal of humans across Asia within the last 10,000 years.

Investigative and Predictive DNA Testing. DOI: https://doi.org/10.1016/B978-0-443-30128-5.00006-0

12.2 Archaeological and genetic evidence supporting dispersal of modern humans in South Asia

12.2.1 Archaeological evidence

India has a wealth of archaeological sites that represent a wide range of cultures, dating from the Lower Paleolithic to the present day. Numerous hominin fossils and stone tool assemblages from the Paleolithic period, along with absolute dates from sites like Attirampakkam, Hathnora I and II, Netankheri I and II, Jwalapuram (JWP 9), Mehtakheri, Bhimbetka, Isampur, Tikoda, Bori Morgaon, and Nagarjuna Konda, provide significant evidence supporting the out-of-Africa theory (Derricourt, 2005; Joshi, 1964; Korisettar & Petraglia, 1998; Mishra, 2000; Pappu et al., 2003; Possehl & Ghosh, 1991; Potts & Teague, 2010; Sankhyan, 1997; Sonakia, 1984).

The transition from the Lower to the Middle Paleolithic shows behavioral modernity in stone tool assemblages, including specialized projectile points, microliths made of fine rock cores, and the use of bone for objects, art forms, and standardization (Bednarik, 1997; Gaillard & Mishra, 2001; Mishra, 2006). Sites such as Bhimbetka in the Narmada Valley, Chirki near Nevasa (Maharashtra), Pipiya from Orsang Valley (Gujarat), and others in Rajasthan and the Garo Hills support the modernity of these stone assemblages (Bain, 2015; Mishra, 1982, 2006). However, the Indian subcontinent has not yielded any human remains from the Middle Paleolithic period (Schug et al., 2016). The shift in stone and bone tool assemblages marks the presence of Upper Paleolithic cultures in India (Neumayer, 1993). Evidence from the Kurnool caves, dating back to 17,390 B.P., supports the transition from Middle to Upper Paleolithic. Fire activity has been found at both the Kurnool caves and the Muchchatla Chintamanu Gavi cave complex (Bednarik, 2016; Misra, 1977; Neumayer, 1993; Paddayya et al., 2000), though no human skeletal remains have been found (Mishra, 2006).

The Mesolithic period marks the transition from Paleolithic hunting and gathering to the settled Neolithic civilization, characterized by microliths. Evidence of rock art and human activities has been found at sites like Sarai Nahar Rai, Mahadaha, Damdama, Baghai Khor, Bagor Bhimbetka, and others (Bednarik, 1997; Lorblanchet, 1992; Mathpal, 1978; Misra, 1971, 1973; Reade, 1979).

The emergence of Neolithic cultures and the onset of sedentary lifestyles in the Indian subcontinent can be linked to developments in East and Southwest Asia. During the Neolithic period, climate, environment, and population growth influenced the development of new stone tool technologies and organized ceramics, transforming human cultures (Bednarik, 1997; Dales & Allchin, 1964; Jarrige, 2001). Sites like Mehrgarh, known for its early evidence of urbanization and cotton production, and other Neolithic sites like Koldihwa, Paiyampalli, Hallur, and Burzahom offer strong evidence of early human civilization in South Asia (Misra & Bellwood, 1985; Misra, 1977; Mohammad, 2023).

During the Chalcolithic period, communities along the Indus River saw the emergence of the Indus Civilization (2600–1900 Before Common Era [BCE]), marked by advancements in stone technology and the discovery of metals like copper and bronze (Lyn Wadley et al., 1983). Evidence of trade between Egypt and Mesopotamia, along with significant material culture, has been discovered at locations such as Harappa, Mohenjo Daro, Farmana, Lothal, Kalibangan, and Dholavira. The earliest evidence of a plowed field was found at Kalibangan, Rajasthan (Mamatamayee, 2017; Possehl & Ghosh, 1991).

Human skeletal remains have been collected from various Indus Valley sites, including Rakhigarhi, Lothal, Harappa, and Dholavira (Table 12.1). The Bronze Age is observed through cultures such as Praghas, Rangpur, Ahar-Banas, and Jorwe (Bisht & Asthana, 1977; Bisht, 1978; Jacobson, 1975). Large skeletal assemblages have been found at Daimabad, Nevasa, and Inamgaon. By the early 13th century BCE, large-scale iron smelting was occurring in India (Dhavalikar, 1988; Dhavalikar, Sankalia, et al., 1988; Lal, 1997).

Archaeological evidence from Brahmagiri shows continuous habitation from the Upper Neolithic through the Iron Age and into the Early Historic period. Chalcolithic sites in northern regions have yielded iron objects and significant deposits of iron slag and ore. Painted Grey Ware (PGW), dating from 1100 to 500/400 BCE, succeeded by Northern Black Polished Ware (NBPW), is found in the upper Ganga plains and beyond, marking trade and migration (Feldman et al., 2019). The "South Indian megalithic burial complex" during the Iron Age includes most megalithic constructions on the Indian subcontinent, with sites like Agiripalli, Adichchannallur, and many others providing evidence of human skeletons (Schug et al., 2016) (Table 12.1).

The Early Historic era, beginning around the reign of Ashoka (3rd century BCE), marks the end of South India's prehistoric period and the rise of transnational trade, the Brahmi script, and urban growth. Sites such as Adi Badri, Harsh-ka-Tila, Nagarjuna Konda, and others have yielded human remains, providing crucial insight into early historic human populations (Ambekar, 2021; Bhavna Ahlawat et al., 2024; Bhavna Ahlawat et al., 2023) (Table 12.1).

TABLE 12.1 Comprehensive categorization of different time periods with cultures and their archaeological sites.

Cultural period	Stone tool assemblages	Human skeleton remains	Material culture	Sites
Paleolithic	Acheulian stone tools, Sohanian stone tools, Mousterian (Levalloisian) stone tool, Projectile points, Flake tool, blades, Scrapers, Bone tools	Hominin fossil reported	Ostrich eggshell beads, sculpted bone item, engraved chert core, rock arts	Attirampakkam (Pappu et al., 2003), Hathnora I and II (Sonakia, 1984), Netankheri I and II (Sankhyan, 1997), Jwalapuram (JWP 9) (Petraglia et al., 2007), Mehtakheri (Ghosh & Mishra, 1993), Bhimbetka, Isampur, Tikoda, Bori (Korisettar et al., 1989), Morgaon (Mishra et al., 2009), Nagarjuna Konda, Muchchatla Chintamanu Gavi, Chirki, Pipiya (Locality V); Budha Pushkar, Singi Talav, Garo hills.
Mesolithic	Microliths	Human Skeleton remains unearthed	Rock arts	Sarai Nahar Rai, Mahadaha, Damdama, Baghai Khor, Bagor Bhimbetka, Gormati-ni-Khan Dhansura, Langhnaj, Lekhahiya, Loteshwar
Neolithic	New stone tool technology along with the preceding tools	Human skeletons recovered	Pottery, pit dwellings, mud–brick buildings, ecofacts, evidence of beginning of domestication and agriculture	Koldihwa, Paiyampalli, Hallur, Burzahom, Sanganakallu, Chopani Mando, Chirand, Baijapalle, Gandlur, Hallur, Hullikallu, Ieej, Tekkalakota, burzahom, the Cachar Hills, Garo Hills, Naga Hills, Daojali Hading, Salbalgiri, Sarutaru, Marakdola, Lahuradewa.
Chalcolithic/ Bronze Age	Copper and bronze tools along with the previous stone tool industry	Human skeletons recovered	Pottery, agricultural tools, terracotta models, plowed field, fire pits, houses, drainage system, granaries, ecofacts, game piece, weights, compass, drainage system, seals and tablets.	Harappa, Mohenjo Daro, Farmana, Lothal, Kalibangan, Dholavira, Rakhigarhi, Nagwada, Ran Daliyo, Rupar, Pirak, Allahdino, Ahar, Gufkral, Praghas, Rangpur, Ahar-Banas, Kayatha, Malwa, Jorwe, Savalda, Late Harappan, Daimabad, Nevasa, and Inamgaon
Iron Age and Megalithic	Iron tools	Human skeletal remains recovered, urn and extended burials	Megaliths, ashmound, pottery, iron artifacts, ornaments, and ecofacts	Abhaipur, Agara, Orathur, Amilkoni, Agara Orathur, Bhagwanpura, Harsh-ka-Tila, Jainal-Naula, Agiripalli, Adichchannallur, Amirthamangalam, Banimilia Bahera, Bhagimohari, Cheramangad, Chichali, Dhamma (Linga), Dhanora, Halingali, Hazurnagar.
Early historic and medieval period	Modern tool assemblages	Recovered	Forts, ramparts, towers, inscriptions, sculptures, monasteries, man-made caves, paintings, pottery, pillars, coins, seals, copper hoards.	Adi Badri, Balupur, Harsh-ka-Tila (Thanesar), Indragarh, Jotsoma, Kausambi, Khairwada, Khamaran, Kumhar tekri, Leshmi, Maidam 2, Nagarjuna Konda, Oakleu Loukol, Pandu Rajar Dhibi.

12.2.2 Genetic evidence

Present-day South Asian populations show a high level of genetic admixture. Recent studies involving ancient DNA (aDNA) have revealed the presence of a distinct and deeply diverged ancestral lineage known as the Ancient Ancestral South Indian (AASI) lineage (Kerdoncuff, 2024; Reich et al., 2009). This AASI lineage covers a small percentage of both ancient and present-day South Asia. It serves as a reference population and is most accurately represented by contemporary Onge individuals from the Andamanese Islands (Thangaraj et al., 2005). However, in-depth genetic analyses have revealed that the divergence between the AASI lineage and the ancestry of modern Onge is both ancient and considerable (Thangaraj & Rai, 2019).

The AASI lineage is found at low levels in nearly all existing Indian populations, except in southern Indian populations where this lineage is prevalent, emphasizing its significant historical impact on the genetic composition of South Asian populations (Reich et al., 2009). The discovery and characterization of the AASI lineage through aDNA studies offer important insights into the complex population history of South Asia, highlighting the contributions of deeply divergent ancestral populations to the genetic diversity observed in the region today. Further research into the origins and interactions of the AASI lineage will continue to enhance our understanding of human migration and admixture in South Asia. The major ancestry sources of South Asia are further classified as: (1) ancient Iranian farmers, (2) ancient Eurasian Steppe pastoralists, and (3) indigenous South Asians (AASIs) (Kerdoncuff, 2024).

The differentiation and admixture patterns in Asia between 45,000 and 10,000 years ago are of significant interest, particularly those associated with ancient South Asian (AASI) and Australasian (AA) populations. Although no Upper Paleolithic individuals sampled so far have been directly linked to these lineages, analysis of current populations from South Asia and Australasia suggests the presence of complex genetic structures and evidence of admixture. Genetic studies on ancient individuals from the Upper Paleolithic period (45,000–10,000 years ago) in Asia have primarily focused on specimens excavated from regions like China, Mongolia, and Russia, largely due to the better preservation of genetic material in colder and drier environments (Kerdoncuff, 2024; Narasimhan et al., 2018; Narasimhan et al., 2019).

aDNA data associated with the East Asian (ESEA) lineage during the Upper Paleolithic have helped elucidate genetic structures and admixture events within ESEA populations, providing insights into the population dynamics and migrations in ancient Asia during this critical period. However, direct genetic evidence from Upper Paleolithic individuals associated with specific lineages like AASI and AA remains largely unexplored, while ongoing research utilizing aDNA continues to uncover genetic diversity, population movements, and interactions that shaped prehistoric populations in Asia (Majumder, 2010; Narasimhan et al., 2018; Narasimhan et al., 2019; Reich et al., 2009; Thangaraj et al., 2005).

During the Neolithic and Copper Ages, individuals from hunter-gatherer backgrounds in northern Eurasia mixed with people from different southern regions, creating distinct genetic profiles known as clines. These clines, such as the Southwest Asian Cline and the Indus Periphery Cline, emerged from the blending of populations from southwest Asia and the Indus region with those from northern Eurasia. This interaction reflects complex patterns of migration, genetic exchange, and cultural influence during the transition to early agricultural societies. The formation of these clines illustrates the diverse history of human movement and interaction in ancient times (Narasimhan et al., 2018; Narasimhan et al., 2019; Shinde et al., 2019; Thangaraj & Rai, 2019).

Anatolian farmers played a pivotal role in the Neolithic revolution, as various studies reveal a decreasing pattern of Anatolian farmer-related admixture from west to east during the Copper and Bronze Ages. This decline suggests movement and mixing of people with local groups they encountered (Shinde et al., 2019), leading to the Southwest Asian Cline. In the easternmost part of the Southwest Asian Cline, individuals from the 3rd millennium BCE show the smallest proportions of Anatolian farmer-related admixture, combined with admixture related to West Siberian Hunter-Gatherers (WSHG), suggesting an impact of North Eurasian-related ancestry in Turan before the migration of Yamnaya Steppe pastoralists into the region (Shinde et al., 2019).

The Yamnaya Steppe pastoralists were not the source of this North Eurasian-related ancestry, as they had more ancestry related to Eastern European Hunter-Gatherers and carried specific mitochondrial and Y chromosome haplogroups not found in Iran and Turan during this period. The ancestry of the Yamnaya Steppe pastoralists eventually spread southward into Central Asia by the second millennium BCE. The admixture between Iranian farmer-related and Ancient Ancestral South Indian (AASI), Andamanese Hunter Gatherer (AHG) ancestry indicates these groups were in contact long before the mature Indus Valley Civilization (IVC) (approximately 2600–1900 BCE) (Narasimhan et al., 2019; Shinde et al., 2019).

Starting around 3000 BCE, the source of Yamnaya Steppe pastoralist ancestry in Central and South Asia came from a secondary migration event, rather than the initial eastward expansion. Studies indicate that the most likely source of the Steppe ancestry found in South Asia is from Middle to Late Bronze Age Steppe groups, which intermixed with the

Bactria-Margiana Archaeological Complex (BMAC) in Kazakhstan. This admixture likely occurred in the first half of the second millennium BCE, when Steppe ancestry was introduced into local South Asian populations (Narasimhan et al., 2019; Shinde et al., 2019).

Present-day South Asian groups are highly admixed, evidence of this admixture can be seen across India, from the mainland to the islands. Mainland Indian populations, along with Andamanese, Papuans, and Aboriginal Australians, are hypothesised to carry genomic contributions from an extinct hominin population, although the exact source remains unclear. The distinctive morphology of the Andamanese population suggests strong adaptive selection (Mondal et al., 2016; Thangaraj et al., 2005).

The genetic affinity of Himalayan and adjoining populations (HAAPs) and populations of Northeast India reveals novel genetic components, with a mixture of East and South Asian ancestries. HAAP shares a high genetic affinity with East Asian populations, particularly from north of the Himalayas, and has significant levels of Ust'-Ishim and Denisovan ancestry (Tamang et al., 2018). Northeast Indian populations show an ancestral origin from East Asia, followed by admixture with South Asians within the last 1000 years. The Ahom dynasty from Thailand in the 12th century also contributed to this genetic amalgamation in Northeast India, with modern Ahom populations displaying substantial admixture with local Tibeto-Burman populations (Silva et al., 2017; Kumar, Rajpal, et al., 2024).

In Ladakh, genetic research aligns with archaeological evidence, revealing diverse migrations and settlements. Ladakh populations show genetic similarities with Tibetans, Sherpas, and Northwest Indian groups like Gujjars and Kashmiri Muslims, influenced by historical events like the Mughal invasion. The Changpa community of Ladakh, isolated by geography and Buddhist practices, shows a strong founder effect compared to the Brokpa and Monero groups, with genetic markers indicating high-altitude adaptation (Francfort et al., 1990; Kumar et al., 2024a; Lo Bue & Bray, 2014).

Lakshadweep, a cluster of islands in the Arabian Sea, demonstrates distinct paternal and maternal ancestry patterns. Paternal ancestry inclines towards populations from the Maldives, North India, and Pakistan, while maternal ancestry shows a strong founder effect and gene flow primarily from Southern Indian states. Despite being a maritime crossroads, the Lakshadweep population exhibits a strong founder effect and diverse ancestries (Kumar et al., 2024b; Mustak et al., 2019).

Indian Jews also show genetic evidence of Middle Eastern affinity and local South Asian populations. Although genetically diverse, some groups show closer relations due to geographic proximity. Roman Catholics in India, part of the larger Konkani-speaking Gaud Saraswat community, show South Asian and Indo-European ancestry, with genetic evidence of Jewish components in some regions. The paternal ancestry of Roman Catholics points towards North Indian or Indo-European origins, while their maternal ancestry shows South Asian roots (Chaubey et al., 2016; Kumar et al., 2021; Rai et al., 2012; Sachin Kumar et al., 2024).

In conclusion, significant population movements and admixture events have shaped the genetic landscape of South Asia throughout both historical and prehistorical periods.

12.3 Three ancestry that succeeded each other in time in South Asia

12.3.1 Pre-2000 BCE Indus Periphery

The majority of the ancestors of present-day South Asians can be traced back to people fitting on the Indus Periphery Cline. Formal modeling shows that this contribution from Indus Periphery Cline individuals to later South Asians, rather than westward gene flow introducing unique South Asian ancestry onto the Iranian plateau, explains the significant shared ancestry observed between present-day South Asians and early Holocene Iranians. This shows significant genetic continuity and influence of populations associated with the Indus Periphery Cline on the ancestry of modern South Asians. It suggests that interactions and demographic movements within the broader South Asian region during ancient times played a crucial role in shaping the genetic makeup of present-day populations in the region (Kerdoncuff, 2024; Reich et al., 2009).

12.3.2 The 2000 BCE to 3000 BCE

The Steppe ancestry represents a genetic continuum resulting from admixture between two distinct sources during the Bronze Age. One end of this cline corresponds to individuals along the Indus Periphery, indicating a genetic affinity with populations from that region. The other end of the Steppe is characterized by a mixture comprising approximately 41% Central Steppe Middle to Late Bronze Age (MLBA) ancestry and 59% from a subgroup of the Indus Periphery with relatively high Iranian farmer-related ancestry (Kerdoncuff, 2024; Reich et al., 2009).

This analysis suggests that the Steppe ancestry emerged as a result of complex population interactions and migrations during the Bronze Age. The presence of Central Steppe Middle to Late Bronze Age (Central Steppe MLBA) ancestry, combined with Iranian farmer-related ancestry within specific subgroups of the Indus Periphery Cline, underscores the dynamic nature of genetic admixture and cultural exchanges that shaped the genetic landscape of Eurasia during this period. The Steppe Cline reflects the intricate genetic history of populations in these regions, highlighting the contributions of multiple ancestries to the ancestry of contemporary populations (Narasimhan et al., 2019; Reich et al., 2009).

The source of the Western Steppe (Early to Middle Bronze Age) EMBA ancestry that ultimately spread into Central Asia and South Asia was not from the initial eastward expansion but instead from a secondary expansion. This secondary expansion involved a group with approximately 67% Western Steppe EMBA ancestry and around 33% ancestry from a point along the European Cline.

This indicates that the migration and spread of Western Steppe ancestry into Central Asia and South Asia involved a population that had a significant genetic contribution from the Western Steppe region, along with a notable proportion of ancestry associated with populations along the European Cline. This secondary expansion likely played a crucial role in shaping the genetic composition of populations in these regions during ancient times, contributing to the genetic diversity observed in contemporary populations (Kerdoncuff, 2024; Narasimhan et al., 2019).

12.3.3 Modern Indian Cline

Present-day South Asian populations can be modeled using various sources, including Central Steppe MLBA (Middle to Late Bronze Age), representing populations from the Central Steppe during this period, characterized by specific Steppe-related ancestry. The second model supports Indus Periphery Cline individuals, referring to those from the broader region around the ancient IVC, exhibiting genetic affinity with populations along the Indus Periphery Cline. Lastly, the AHG (AASI) or a subgroup of Indus Periphery Cline individuals with relatively high AHG-related ancestry reflects genetic contributions from ancient populations associated with South Asia (Narasimhan et al., 2018; Narasimhan et al., 2019; Reich et al., 2009; Thangaraj & Rai, 2019).

By combining these populations as sources, we can better understand the complex admixture patterns that have shaped the genetic makeup of present-day South Asians. These models highlight significant contributions of Steppe-related, Indus Periphery Cline, and ancestral South Indian ancestry in shaping South Asia's genetic diversity today. The presence of AHG-related ancestry underscores deep historical connections and interactions between ancient populations across the Indian subcontinent (Reich et al., 2009).

12.4 Methodology, challenges, and precautions

An archaeogenetics study requires a multidisciplinary approach involving archaeologists, historians, anthropologists, geneticists, bioinformaticians, and others. Working with ancient bone samples demands adherence to best practices, considering the sensitivities of local cultural and religious communities. Once samples are acquired, contamination must be strictly limited.

Since ancient samples are rare and limited in quantity, nondestructive techniques should be prioritized before moving to aDNA analysis, which is destructive and requires minimal samples (Fulton, 2012). aDNA work is conducted in a clean lab to minimize contamination risks. The lab is separate from the postamplification area and equipped with air handling units, HEPA filters, and a positive air pressure system. All surfaces are treated with UV light, wiped with 10% bleach (v/v), followed by 75% ethanol. Researchers wear full PPE, including a full-body suit, face masks, shoe covers, headgear, and gloves. Rooms for drilling, DNA extraction, and DNA library preparation are separate to reduce cross-contamination. All work is conducted in laminar flow hoods and biosafety cabinets, cleaned before and after each use (Fulton, 2012).

The petrous bone and teeth dentine (at the root) are prioritized for aDNA extraction due to their preservation potential (Pinhasi et al., 2019). Upon entering the clean lab, samples are UV treated, properly labeled, and opened only in the fume hood to avoid contamination. The samples are cleaned to remove dust and further decontaminated using bleach and alcohol (Korlević et al., 2015). Surface layers are removed using a Dremel disk and sand-blaster to expose the clean inner part, from which powder is obtained (Pinhasi et al., 2015).

For DNA extraction, the powdered sample is incubated with lysis buffer containing 0.5 M EDTA and proteinase K (Dabney & Meyer, 2019; Dabney et al., 2013). DNA is then separated using silica, with the binding buffer containing Guanidine-HCl, 2-propanol, and acetate (Rohland & Hofreiter, 2007). DNA binds to silica at low pH, and after desalting with ethanol, it is dissolved in a low-salt buffer. A modified method using silica-based spin columns can improve results but is more costly (Dabney et al., 2013).

The next step is DNA library preparation, which can vary depending on the project requirements. aDNA often has C-to-T misincorporations that can bias results. To reduce this, the DNA can be treated with Uracil-DNA-glycosylase (UDG) and Endonuclease-VIII to remove uracil caused by cytosine deamination. However, this process shortens the DNA and can affect damage pattern detection, which is essential for aDNA authentication (Briggs et al., 2010).

A DNA library contains overlapping DNA fragments from a single organism. Double-stranded libraries involve repairing DNA ends and ligating double-stranded adapters, while single-stranded libraries use single-stranded adapters. Both types of libraries are indexed to identify them after sequencing. Single-stranded libraries are advantageous in aDNA studies as they increase the percentage of endogenous DNA (Kircher et al., 2012; Meyer & Kircher, 2010).

DNA libraries are amplified by PCR and quality-checked. Tropical samples often yield low endogenous DNA, which requires a high depth of coverage, making sequencing expensive. To increase endogenous DNA percentages, target enrichment can be used, where biotinylated probes hybridize with specific genomic regions, followed by magnetic isolation (Devault et al., 2014). However, this method may introduce biases toward the targeted SNPs, requiring correction of the probes (Günther & Nettelblad, 2019; Lachance & Tishkoff, 2013).

Illumina sequencing, popular for its cost-effectiveness and high data output, uses sequencing by synthesis technology. A newer technology from Ultima Genomics reduces costs further by replacing the flow cell with an open substrate (Almogy et al., 2022). After sequencing, the data is processed, analyzed, and interpreted based on the project's objectives. Demultiplexing identifies and separates sequences based on their indexes, and reads are trimmed to remove low-quality or adapter sequences. These reads are then aligned to a reference genome using BWA or Bowtie2 (Langmead & Salzberg, 2012; Li & Durbin, 2010).

Postmortem DNA damage is analyzed to confirm aDNA authenticity, with tools like mapDamage2 quantifying damage patterns, particularly in non-USER-treated or partially-USER-treated libraries. The aligned sequences are used for downstream population genomic analyses (Jónsson et al., 2013).

12.5 The importance of tropical samples in past human settlement studies

aDNA is prone to damage from heat, and tropical samples face additional challenges due to postmortem DNA decay caused by diagenetic and taphonomic processes (Allentoft et al., 2012; Briggs et al., 2007; Dabney et al., 2013; Hofreiter et al., 2001). However, advancements in sequencing technology, such as target enrichment, have enabled sequencing of complete genomes from Middle Pleistocene samples (Gansauge & Meyer, 2014; Meyer et al., 2016; Seguin-Orlando et al., 2013). Despite this progress, tropical samples remain difficult to sequence, creating a geographical bias in paleogenomics studies. DNA decay from heat and time limits recovery from tropical environments (Adler et al., 2011; Allentoft et al., 2012; Hofreiter et al., 2015; Kaestle & Smith, 2001; Kistler et al., 2017; Lindahl, 1993).

Successful aDNA studies from tropical regions, such as the Caribbean and Southeast Asia, have proven valuable despite these challenges (Damgaard et al., 2015; Fernández et al., 2014; Gutiérrez-García et al., 2014; Jina, 1996; Kehlmaier, 1846; Mendisco, 1660; Schroeder et al., 2015; Shinde et al., 2019). The study of tropical samples is essential for understanding ancient human settlements, as these regions hold much of the world's biodiversity and historical human populations (Brown et al., 2015; Buzas, 2002).

12.6 Processing of tropical samples

UDG treatment is helpful in reducing error rates from cytosine deamination in aDNA. However, UDG-treated libraries cannot be used for damage pattern analysis, so a non-UDG-treated library must be prepared alongside to rule out contamination. Enrichment methods, though effective, can be biased toward certain sequences. Preparing multiple libraries from the same or different DNA extractions and merging the data can improve sequence yield and SNP detection (Nieves-Colón et al., 2018; Shinde et al., 2019) (Fig. 12.1).

12.7 Shreds of evidence from the ancient human remains from varied locations to explain the peopling of South Asia

12.7.1 Rakhigarhi, Haryana

Genetic findings from IVC site Rakhigarhi, suggest parallels between South Asia and Europe regarding the introduction of farming. Farming expansions in both regions may have involved large-scale population movements,

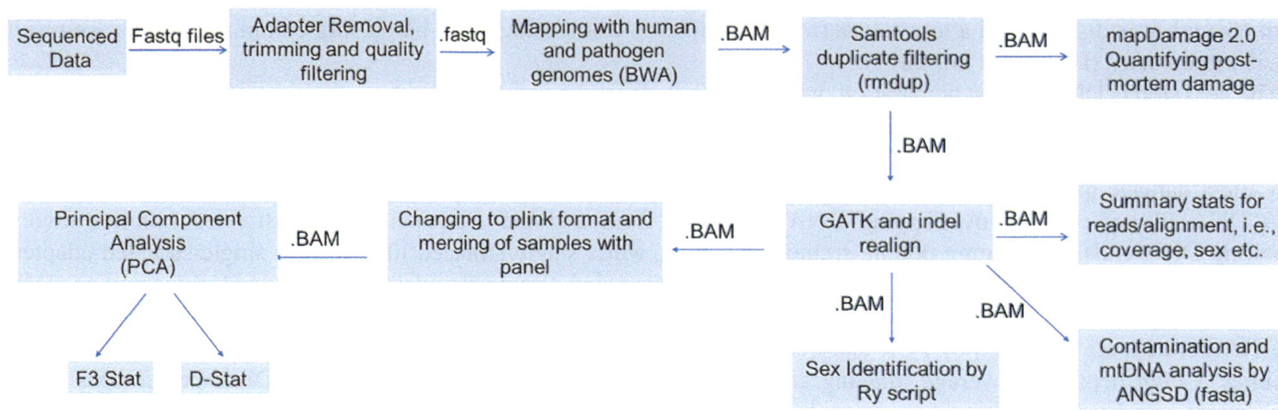

FIGURE 12.1 Showing the flowchart of bioinformatic genomic data analysis.

as seen in Europe after 6500 BCE with the expansion of Western Anatolian farmers. The ancient South Asian farmers of the IVC had minimal genetic ancestry related to Anatolian or Iranian farmers, arguing against the migration from Anatolia through Iran associated with farming spread. Instead, findings support the theory of Indo-European language spread from Eastern Europe via Central Asia around the first half of the 2nd millennium BCE, a route supported by genetic evidence linking Steppe pastoralist ancestry in South Asia with Bronze Age Eastern Europe (Shinde et al., 2019) (Fig. 12.2, Table 12.2).

12.7.2 Pattanam, Kerala

The genetic evidence from the Pattanam archaeological site shows the presence of culturally, religiously, and ethnically diverse groups. Through cultural and commercial encounters, the area was enriched in the diverse genetic makeup and clearly connected the indigenous groups to other populations such as Greco-Romans, the Egyptians, and the West Asians. The shred of evidence retrieved from the mtDNA analysis distinctly shows the diverse demographic settlement included both local populations and the influx of foreign populations, particularly from the Middle East and Europe. Pattanam played a significant role in transoceanic exchanges between 100 BCE and 300 CE (Common Era). Maritime trade and cultural interactions involving ancient Mediterranean, West Asian, Red Sea, African, and Asian cultures have been identified and influence the genetic makeup of the region and contemporary populations of the state (Bhavna Ahlawat et al., 2023) (Fig. 12.2, Table 12.2).

12.7.3 Sanjan, Gujarat

The ancient site of Sanjan depicts the possible presence of a large city that existed between the 8th and 13th centuries, from where hundreds of skeletal remains were excavated. The Parsi populace is established by the presence of many well-established structures that proves the existence of Zoroastrian culture. Parsis are limited to a small community in India with only a handful of 60,000 individuals. Parsis are known to practice Zoroastrian religion, native to Iran, who fled the Muslim harassment in their homeland and landed on the western coast of India around 750 CE. The archaeological and genetic evidence suggests the Parsi occupation of Sanjan city during the 8th to the 13th centuries (Chaubey et al., 2017; Kumar et al., 2023; Thangaraj & Rai, 2019) (Fig. 12.2, Table 12.2).

12.7.4 Roopkund Lake, Chamoli, Uttrakhand

The unforeseen nature of skeletal remains found at Roopkund Lake shows the genetically, temporally, and isotopically heterogeneous composition of the groups. Presently, Roopkund is a religious pilgrimage route; however, no such reliable evidence shows the presence of pilgrimage ritual until the late 19th century. The nearby temples dating back to the 8th and 10th centuries suggest a potential origin of the religious route around the same time. Genetic analysis of the skeletal remains suggests that skeletal remains from Roopkund Lake belong to two different groups of individuals traveling there at different time periods, belonging to present-day Aouth Asians and unrelated men and women, born in the eastern Mediterranean during the period of the Ottoman Empire. The diet of these individuals is identified to be terrestrial.

FIGURE 12.2 Archeological sites with potential ancient DNA work representing published and unpublished work from Indian subcontinent. Map lines delineate study areas and do not necessarily depict accepted national boundaries.

TABLE 12.2 Comprehensive categorization of different archaeological sites with description of ancient DNA analysis from excavated human remains.

Location	Time period	Genetic ancestry
Rakhigarhi, Haryana	Iranian-related lineage contributing to the IVC Cline split earlier than Belt Cave at 10,000 BCE and Ganj Dareh at 8000 BCE	Migration from Eastern Europe via Central Asia around the first half of the 2nd millennium BCE, low genetic ancestry related to Anatolian or Iranian farmers, arguing against the migration from Anatolia through Iran associated with farming spread.
Pattanam, Kerala	Human remains belonging between 100 BCE to 300 CE.	Genetic ancestry from the Middle East and Europe.
Sanjan, Gujarat	Human remains belonging to 750 CE	Middle Eastern genetic ancestry.
Roopkund Lake, Chamoli, Uttrakhand	Two different groups of human remains belonging to approximately 800 CE and approximately 1800 CE separately	Genetic ancestry from three groups, present-day South Asians, present-day West Eurasians, and East Asian-related ancestry.
Vadnagar, Gujarat	Human remains belonging between 12th and 18th centuries CE.	Genetic ancestry from local ancestral components, West Eurasia, and Central Asia, particularly Tajikistan.
St. Augustine Ruins, Vela, Goa	Human remains belonging Queen Ketevan of Georgia—1624 to 1627	Genetic ancestry from Georgia and adjoining regions.
Ajnala, Amritsar, Punjab	Human remains belonging to Indian soldiers—1857	Genetic ancestry from populations of the Gangetic plain and the eastern regions of South Asia.

The potential hypothesis suggests that people traveled to Roopkund Lake and its associated route for pilgrimage and eventually died in the Himalayas. However, what is interesting to note is that Hindu practices were relatively rare in the eastern Mediterranean during the Ottoman Empire (Harney et al., 2019) (Fig. 12.2, Table 12.2).

12.7.5 Vadnagar, Gujarat

The ancient township of Vadnagar reveals a history of cultural diversity and exchange spanning from the 2nd century BCE to the present day. This continuously inhabited site has witnessed a complex sequence of cultural periods, resulting in a rich and diverse settlement over time. While previous archaeological studies highlighted the cultural heritage of Vadnagar, genetic studies to identify its ancestry were lacking until now. The complete mitogenomes of medieval individuals from the Vadnagar archaeological site explore the cosmopolitan nature of the present-day population and investigate patterns of migration and cultural exchange through trade and religious practices. Study revealed a heterogeneous medieval population in Vadnagar, characterized by a blend of local ancestral components and central Asian genetic ancestry. Specifically, the presence of a Central Asian component associated with mitochondrial haplotype U2e was identified, which is not shared with any individuals from India but shows connections to individuals from the Bronze Age in Tajikistan and earlier coalescence periods. It can be concluded that Vadnagar in western India during the medieval period was culturally and genetically rich, reflecting both local indigenous elements and influences from western Eurasia (Bhavna Ahlawat et al., 2024) (Fig. 12.2).

12.7.6 St. Augustine ruins, Vela, Goa

Augustinian friars carried remains of the queen to Goa and kept in a chapel in the St. Augustine convent. Genetic analysis on the Human remains excavated showed U1b based on specific mutations in the mtDNA coding regions, which show high to moderate frequency in the Georgia and adjoining regions. The research aligns with historical records, such as Silva Rego's descriptions of tombstones, which suggest associations between the bone remains and Friar Guilherme de Sto. Agostinho and Friar Hyeronimo da Cruz. The overall genetic and archaeological evidence supports the hypothesis that represents the remains of Queen Ketevan of Georgia. Since the remains were brought to India by Augustinian friars, the study supports the migration from different regions to India during the medieval times (Niraj Rai et al., 2014) (Fig. 12.2, Table 12.2).

12.7.7 Ajnala, Amritsar, Punjab

Unidentified human skeletons found in an abandoned well at Ajnala, Punjab, were suspected to be Indian soldiers involved in the 1857 mutiny. Genetic analysis shows the presence of West Eurasian-specific haplogroups HV0e, U3, and U7, albeit in smaller proportions. Comparative analysis indicated that the genetic profile of these sepoy samples more closely resembles populations from the Gangetic plain rather than local Punjabi populations, suggesting a genetic affinity with populations from the eastern regions of South Asia. These findings challenge the assumption of local ancestry and suggest a genetic link with populations from eastern India, shedding light on the identity and origins of these historical figures (Sehrawat et al., 2022) (Fig. 12.2, Table 12.2).

12.8 Discussion and conclusion

The application of genetics in archaeological investigation has revolutionized our understanding of ancient human migration, genetic evolution, and population dynamics. In the context of the Indian subcontinent, aDNA analysis has provided insights into the region's deep, complex population history, revealing ancient connections between South Asia and other regions, and shedding light on the genetic evolution of its populations over millennia.

South Asia, particularly India, has long been a critical region for studying ancient human migrations due to its strategic geographic location at the crossroads of major migratory routes between Africa, the Middle East, Central Asia, and Southeast Asia. Through aDNA studies, archaeologists have been able to trace key migratory events and population movements that shaped the genetic structure of contemporary South Asians.

For example, genetic data supports the theory of an early wave of migration associated with the *Out of Africa* event, which brought anatomically modern humans into South Asia as early as 60–50,000 years ago. Later migrations from regions like Iran and Central Asia during the Neolithic period (approximately 10,000 years ago) introduced agricultural practices and significantly influenced the genetic makeup of South Asia. The genetic signatures of ancient Iranian farmers, Steppe pastoralists, and indigenous South Asian hunter-gatherers (AASI) are detectable in modern populations across the region, reflecting centuries of admixture and population interactions (Narasimhan et al., 2018; Narasimhan et al., 2019; Reich et al., 2009; Thangaraj & Rai, 2019).

One of the most significant applications of genetics in archaeology has been in the study of the IVC, one of the world's earliest urban civilizations. The genetic analysis of ancient individuals from the Indus Periphery Cline has provided clues about the population structure of the IVC and its interaction with neighboring regions. Unlike other early civilizations that show a strong influence from Anatolian and Near Eastern agriculturalists, genetic studies suggest that IVC populations lacked significant Anatolian farmer-related ancestry, pointing to the eastward spread of agriculture from the Iranian plateau rather than large-scale westward migrations (Narasimhan et al., 2019; Thangaraj & Rai, 2019).

This revelation has shifted our understanding of the origins of agriculture in South Asia, indicating that farming practices were likely adopted locally with limited external genetic influence. Furthermore, ongoing aDNA research is expected to clarify the role of the IVC in facilitating interactions between the Steppe pastoralists and South Asian populations, potentially contributing to the spread of Indo-European languages into the region.

A major focus of genetic research in South Asia has been the impact of Steppe pastoralist migrations during the Bronze Age (approximately 2000–1500 BCE). These migrations brought significant genetic contributions from the Eurasian Steppe into northern India, particularly among populations associated with the early Vedic culture. This has been linked to the spread of Indo-European languages across the region. However, the exact timing and route of these migrations remain topics of active investigation, as aDNA from regions like Central Asia, Turan, and the BMAC continues to reveal new complexities in these interactions (Shinde et al., 2019).

The introduction of Steppe-related ancestry into the Indian subcontinent represents one of the most critical genetic events in shaping modern South Asian populations. It highlights the long-standing role of migration in the development of cultural and linguistic diversity in the region.

While migration and admixture events have had a significant impact, the indigenous populations of South Asia, particularly in southern India and among tribal groups, retain high levels of AASI ancestry. This lineage reflects the deep evolutionary history of South Asia, with genetic contributions predating the migrations of Iranian farmers and Steppe pastoralists. In fact, genetic evidence shows that these indigenous populations have undergone significant natural selection in response to local environmental conditions.

For instance, genetic studies have identified signals of positive selection for traits such as skin pigmentation, lactose tolerance, and immunity-related genes. These genetic adaptations reflect the evolutionary pressures faced by ancient populations in South Asia, including changes in diet, climate, and pathogen exposure (Metspalu et al., 2018). The study of these adaptive traits through aDNA has implications not only for understanding human evolution but also for exploring how ancient populations responded to their environments and how these responses continue to influence modern populations.

Despite the groundbreaking discoveries, aDNA research in South Asia faces several challenges. The region's hot and humid climate accelerates the degradation of DNA, making it difficult to recover high-quality genetic material from ancient remains. This has created a geographical bias in aDNA studies, as temperate regions with better DNA preservation conditions, such as Europe and Siberia, have been overrepresented in the global genetic dataset.

Additionally, South Asia's complex social structure, characterized by strict endogamy and caste systems, complicates the study of genetic admixture and population dynamics. Endogamous practices have led to high levels of genetic differentiation among groups, creating a mosaic of genetic diversity that is difficult to disentangle without sufficient ancient samples.

The cultural and religious significance of human remains in South Asia also poses ethical challenges in conducting aDNA research. Collaborative efforts with local communities, anthropologists, and archaeologists are essential to navigate these issues and ensure that scientific investigations are conducted with respect for cultural heritage (Silva et al., 2017).

Despite the challenges, the future of aDNA research in South Asia holds immense potential. As technological advances in sequencing continue to improve the recovery and analysis of degraded DNA, more ancient genomes from the region will likely become available, providing a clearer picture of South Asia's genetic evolution. Understanding the genetic ancestry of isolated and underrepresented populations, such as the indigenous tribes of the Andaman Islands, the Himalayan communities, and the populations of the northeastern hill regions, will fill critical gaps in our knowledge of human migration and adaptation.

Further investigation into the genetic structure of ancient South Asian societies will also have important implications for understanding the spread of agricultural practices, language distributions, and social hierarchies. Unraveling these historical processes is not only essential for reconstructing the past but also for informing studies on modern population health, disease susceptibility, and genetic disorders, which are influenced by the unique genetic history of South Asia (Basu et al., 2016; Metspalu et al., 2018; Reich et al., 2009).

References

Adler, C. J., Haak, W., Donlon, D., & Cooper, A. (2011). Survival and recovery of DNA from ancient teeth and bones. *Journal of Archaeological Science, 38*(5), 956–964. https://doi.org/10.1016/j.jas.2010.11.010.

Allentoft, M. E., Collins, M., Harker, D., Haile, J., Oskam, C. L., Hale, M. L., Campos, P. F., Samaniego, J. A., Gilbert, T. P. M., Willerslev, E., Zhang, G., Scofield, R. P., Holdaway, R. N., & Bunce, M. (2012). The half-life of DNA in bone: Measuring decay kinetics in 158 dated fossils. *Royal Society, Australia Proceedings of the Royal Society B: Biological Sciences, 279*(1748), 4724–4733. https://doi.org/10.1098/rspb.2012.1745, http://rspb.royalsocietypublishing.org/content/279/1748/4724.full.pdf+html.

Ambekar, A.S. (2021). Recently reported Sikotara Mata Temple at Vadnagar and Diva Dandi from Vadgam3Khambhat.

Almogy, G., Pratt, M., Oberstrass, F., Lee, L., Mazur, D., Beckett, N., Yamaguchi, K., Mayshar, Y., Nawy, T., Hall, L., Mola, J., Regev, A., & Lipson, D. (2022). Cost-efficient whole genome-sequencing using novel mostly natural sequencing-by-synthesis chemistry and open fluidics platform. bioRxiv. https://doi.org/10.1101/2022.05.29.493900.

Bain, W. K. (2015). A preliminary study on the Stone age artifacts of NetanKhedi, Sehore district of Central Narmada valley in Madhya Pradesh. *Humankind, 11*, 31–49.

Basu, A., Sarkar-Roy, N., & Majumder, P. P. (2016). Genomic reconstruction of the history of extant populations of India reveals five distinct ancestral components and a complex structure. *Proceedings of the National Academy of Sciences of the United States of America, 113*(6), 1594–1599. https://doi.org/10.1073/pnas.1513197113, http://www.pnas.org/content/113/6/1594.full.pdf.

Bednarik, R. (1997). The role of Pleistocene beads in documenting hominid cognition. *Rock Art Research, 14*, 27–40.

Bednarik, R. G. (1997). The earliest evidence of ocean navigation. *International Journal of Nautical Archaeology, 26*(3), 183–191. https://doi.org/10.1111/j.1095-9270.1997.tb01331.x.

Bednarik, R. G. (2016). Pleistocene stone pendant from Western Australia. *Australian Archaeology, 45*(1), 32–34. https://doi.org/10.1080/03122417.1997.11681591.

Bhavna Ahlawat, L. K., Abhijit Ambekar, J. S. S., & Yadubir Singh Rawat, N. R. (2024). Ancient mitogenomes suggest complex maternal history of one of the oldest settlements of western India. *Mitochondrion, 76*, 101871. https://doi.org/10.1016/j.mito.2024.101871.

Bhavna Ahlawat, L. K., Parayil John Cherian, J. S. S., & Niraj Rai, K. T. (2023). Deciphering the West Eurasian genetic footprints in ancient South India. *Genes, 14*(5), 963. https://doi.org/10.3390/genes14050963.

Bisht, R., & Asthana, S. (1977). Banawali and some other recently excavated Harappan sites in India. *South Asian Archaeology,* (1), 223–240.

Bisht, R. S. (1978). Banawali, A new Harappan site in Haryana. *Man and Environment, 2*, 86–88.

Briggs, A. W., Stenzel, U., Johnson, P. L. F., Green, R. E., Kelso, J., Prüfer, K., Meyer, M., Krause, J., Ronan, M. T., Lachmann, M., & Pääbo, S. (2007). Patterns of damage in genomic DNA sequences from a Neandertal. *Proceedings of the National Academy of Sciences, 104*(37), 14616–14621. https://doi.org/10.1073/pnas.0704665104.

Briggs, A. W., Stenzel, U., Meyer, M., Krause, J., Kircher, M., & Pääbo, S. (2010). Removal of deaminated cytosines and detection of in vivo methylation in ancient DNA. *Nucleic Acids Research, 38*(6), e87. https://doi.org/10.1093/nar/gkp1163.

Brown, T. A., Cappellini, E., Kistler, L., Lister, D. L., Oliveira, H. R., Wales, N., & Schlumbaum, A. (2015). Recent advances in ancient DNA research and their implications for archaeobotany. *Vegetation History and Archaeobotany, 24*(1), 207–214. https://doi.org/10.1007/s00334-014-0489-4, http://link.springer-ny.com/link/service/journals/00334/index.htm.

Buzas, M. A. (2002). Foraminiferal densities over five years in the Indian River Lagoon, Florida: A model of pulsating patches. *The Journal of Foraminiferal Research, 32*(1), 68–92. https://doi.org/10.2113/0320068.

Chaubey, G., Singh, M., Rai, N., Kariappa, M., Singh, K., Singh, A., Pratap Singh, D., Tamang, R., Selvi Rani, D., Reddy, A. G., Kumar Singh, V., Singh, L., & Thangaraj, K. (2016). Genetic affinities of the Jewish populations of India. *Scientific Reports, 6*(1). https://doi.org/10.1038/srep19166.

Chaubey, G., Tamang, R., Pennarun, E., Dubey, P., Rai, N., Upadhyay, R. K., Meena, R. P., Patel, J. R., van Driem, G., Thangaraj, K., Metspalu, M., & Villems, R. (2017). Reconstructing the population history of the largest tribe of India: The Dravidian speaking Gond. *European Journal of Human Genetics, 25*(4), 493–498. https://doi.org/10.1038/ejhg.2016.198.

Dabney, J., Knapp, M., Glocke, I., Gansauge, M. T., Weihmann, A., Nickel, B., Valdiosera, C., García, N., Pääbo, S., Arsuaga, J. L., & Meyer, M. (2013). Complete mitochondrial genome sequence of a Middle Pleistocene cave bear reconstructed from ultrashort DNA fragments. *Proceedings of the National Academy of Sciences of the United States of America, 110*(39), 15758–15763. https://doi.org/10.1073/pnas.1314445110Germany, http://www.pnas.org/content/110/39/15758.full.pdf+html.

Dabney, J., & Meyer, M. (2019). Extraction of highly degraded DNA from ancient bones and teeth. *Methods in Molecular Biology, 1963*, 25–29. https://doi.org/10.1007/978-1-4939-9176-1_4, https://www.springer.com/series/7651.

Dales, G. F., & Allchin, F. R. (1964). Neolithic cattle-keepers of South India: A study of the Deccan Ashmounds. *Journal of the American Oriental Society, 84*(1), 93. https://doi.org/10.2307/597078.

Damgaard, P. B., Margaryan, A., Schroeder, H., Orlando, L., Willerslev, E., & Allentoft, M. E. (2015). Improving access to endogenous DNA in ancient bones and teeth. *Nature Publishing Group, Denmark Scientific Reports, 5*. https://doi.org/10.1038/srep11184, http://www.nature.com/srep/index.html.

Derricourt, R. (2005). Getting "Out of Africa": Sea crossings, land crossings and culture in the Hominin Migrations. *Journal of World Prehistory, 19*(2), 119–132. https://doi.org/10.1007/s10963-006-9002-z.

Devault, A. M., Golding, G. B., Waglechner, N., Enk, J. M., Kuch, M., Tien, J. H., Shi, M., Fisman, D. N., Dhody, A. N., Forrest, S., Bos, K. I., Earn, D. J. D., Holmes, E. C., & Poinar, H. N. (2014). Second-pandemic strain of Vibrio cholerae from the Philadelphia cholera outbreak of 1849. *New England Journal of Medicine, 370*(4), 334–340. https://doi.org/10.1056/NEJMoa1308663, http://www.nejm.org/doi/pdf/10.1056/NEJMoa1308663.

Dhavalikar, M. K. (1988). *The first farmers of the Deccan.*

Dhavalikar, M. K., Sankalia, H. D., & Ansari, Z. D. (1988). *Excavations at Inamgaon, 1.*

Feldman, M., Master, D. M., Bianco, R. A., Burri, M., Stockhammer, P. W., Mittnik, A., Aja, A. J., Jeong, C., & Krause, J. (2019). Ancient DNA sheds light on the genetic origins of early Iron Age Philistines. *Science Advances, 5*(7). https://doi.org/10.1126/sciadv.aax0061.

Fernández, E., Pérez-Pérez, A., Gamba, C., Prats, E., Cuesta, P., Anfruns, J., Molist, M., Arroyo-Pardo, E., Turbón, D., & Williams, S. M. (2014). Ancient DNA analysis of 8000 B.C. Near Eastern farmers supports an early neolithic pioneer maritime colonization of Mainland Europe through Cyprus and the Aegean islands. *PLoS Genetics, 10*(6), e1004401. https://doi.org/10.1371/journal.pgen.1004401.

Francfort, H.-P., Klodzinski, D., & Mascle, G. (1990). Pétroglyphes archaïques du Ladakh et du Zanskar. *Arts asiatiques, 45*(1), 5–27. https://doi.org/10.3406/arasi.1990.1275.

Fulton, T. L. (2012). Setting Up an ancient DNA laboratory. *Methods in Molecular Biology, 840*, 1–11. https://doi.org/10.1007/978-1-61779-516-9_1.

Gaillard, C.,& Mishra, S. (2001). The Lower Palaeolithic in South Asia. Origine des peuplements et chronologie des cultures paléolithiques dans le Sud-Est Asiatique. Semenanjung-Artcom, 73–91.

Gansauge, M. T., & Meyer, M. (2014). Selective enrichment of damaged DNA molecules for ancient genome sequencing. *Genome Research, 24*(9), 1543–1549. https://doi.org/10.1101/gr.174201.114, http://genome.cshlp.org/content/24/9/1543.full.pdf+html.

Gutiérrez-García, T. A., Vázquez-Domínguez, E., Arroyo-Cabrales, J., Kuch, M., Enk, J., King, C., & Poinar, H. N. (2014). Ancient DNA and the tropics: A rodent's tale. *Biology Letters, 10*(6). https://doi.org/10.1098/rsbl.2014.0224.

Ghosh, A., & Mishra, S. (1993). Recent archaeological discoveries in India. *Man and Environment, 18*(2), 129–137.

Günther, T., & Nettelblad, C. (2019). Anna Di Rienzo, The presence and impact of reference bias on population genomic studies of prehistoric human populations. *PLOS Genetics, 15*(7), e1008302. https://doi.org/10.1371/journal.pgen.1008302.

Harney, É., Nayak, A., Patterson, N., Joglekar, P., Mushrif-Tripathy, V., Mallick, S., Rohland, N., Sedig, J., Adamski, N., Bernardos, R., Broomandkhoshbacht, N., Culleton, B. J., Ferry, M., Harper, T. K., Michel, M., Oppenheimer, J., Stewardson, K., Zhang, Z., Harashawaradhana, Bartwal, M. S., ... Rai, N. (2019). Ancient DNA from the skeletons of Roopkund Lake reveals Mediterranean migrants in India. *Nature Communications, 10*(1). https://doi.org/10.1038/s41467-019-11357-9.

Hofreiter, M., Jaenicke, V., Serre, D., Von Haeseler, A., & Pääbo, S. (2001). DNA sequences from multiple amplifications reveal artifacts induced by cytosine deamination in ancient DNA. *Nucleic Acids Research, 29*(23), 4793–4799. https://doi.org/10.1093/nar/29.23.4793, https://academic.oup.com/nar/issue.

Hofreiter, M., Paijmans, J. L. A., Goodchild, H., Speller, C. F., Barlow, A., Fortes, G. G., Thomas, J. A., Ludwig, A., & Collins, M. J. (2015). The future of ancient DNA: Technical advances and conceptual shifts. *BioEssays, 37*(3), 284–293. https://doi.org/10.1002/bies.201400160.

Jacobson, J. (1975). Early Stone Age habitation sites in eastern Malwa. *Proceedings of the American Philosophical Society, 119*(4), 280–297.

Jarrige, J.-F. (2001). *Mehrgarh Neolithic: The updated sequence, 1*, 128–141.

Jina, P. S. (1996). *Ladakh: The land and the people.* Indus Publishing.

Joshi, R. (1964). *Acheulian succession in central India.*

Jónsson, H., Ginolhac, A., Schubert, M., Johnson, P. L. F., & Orlando, L. (2013). mapDamage2.0: Fast approximate Bayesian estimates of ancient DNA damage parameters. *Bioinformatics, 29*(13), 1682–1684. https://doi.org/10.1093/bioinformatics/btt193.

Kaestle, F. A., & Smith, D. G. (2001). Ancient mitochondrial DNA evidence for prehistoric population movement: The numic expansion. *American Journal of Physical Anthropology, 115*(1), 1–12. https://doi.org/10.1002/ajpa.1051.

Kehlmaier, C. (1846). *Tropical ancient DNA reveals relationships of the extinct Bahamian giant tortoise Chelonoidis alburyorum, 284.*

Kerdoncuff, E. (2024). 000 years of evolutionary history of India: Insights from ~2,700 whole genome sequences. *bioRxiv, 50.*

Kircher, M., Sawyer, S., & Meyer, M. (2012). Double indexing overcomes inaccuracies in multiplex sequencing on the Illumina platform. *Nucleic Acids Research, 40*(1), e3. https://doi.org/10.1093/nar/gkr771.

Kistler, L., Ware, R., Smith, O., Collins, M., & Allaby, R. G. (2017). A new model for ancient DNA decay based on paleogenomic meta-analysis. *Nucleic Acids Research, 45*(11), 6310–6320. https://doi.org/10.1093/nar/gkx361.

Korisettar, R., & Petraglia, M. D. (1998). The archaeology of the Lower Palaeolithic: Background and overview. *Early Human Behaviour in Global Context.*

Korisettar, R., Rajaguru, S. N., & Mishra, S. (1989). Quaternary geology and archaeology of the Upper Krishna Basin. *Man and Environment, 13*, 87–116.

Korlević, P., Gerber, T., Gansauge, M. T., Hajdinjak, M., Nagel, S., Aximu-Petri, A., & Meyer, M. (2015). Reducing Microbial and Human Contamination in DNA Extractions from Ancient Bones and Teeth. BioTechniques, 59(2), 87–93. https://doi.org/10.2144/000114320.

Kumar, L., Ahlawat, B., Kumar, S., Mushrif-Tripathy, V., & Rai, N. (2023). Maternal ancestry of first Parsi settlers of India using ancient mitogenome. *Mitochondrion, 71*, 104–111. https://doi.org/10.1016/j.mito.2023.06.004.

Kumar, L., Farias, K., Prakash, S., Mishra, A., Mustak, M. S., Rai, N., & Thangaraj, K. (2021). Dissecting the genetic history of the Roman Catholic populations of West Coast India. *Human Genetics, 140*(10), 1487–1498. https://doi.org/10.1007/s00439-021-02346-4.

Kumar, L., Rajpal, R., Ahlawat, B., Sehrawat, J. S., Spalzin, S., Fonia, R. S., Thangaraj, K., & Rai, N. (2024a). The maternal genetic origin and diversity of the extant populations of the Ladakh region in India. *Mitochondrion, 75*, 101828. https://doi.org/10.1016/j.mito.2023.101828.

Kumar, S., Singh, P. P., Pasupuleti, N., Shendre, S. S., Sequeira, J. J., Babu, I., Mustak, M. S., Rai, N., & Chaubey, G. (2024b). Genetic evidence for a single founding population of the Lakshadweep Islands. *Molecular Genetics and Genomics, 299*(1). https://doi.org/10.1007/s00438-024-02110-z.

Lachance, J., & Tishkoff, S. A. (2013). SNP ascertainment bias in population genetic analyses: Why it is important, and how to correct it. *BioEssays, 35*(9), 780–786. https://doi.org/10.1002/bies.201300014.

Lal, B. B. J. (1997). *The earliest civilization of South Asia (rise, maturity and decline).*

Langmead, B., & Salzberg, S. L. (2012). Fast gapped-read alignment with Bowtie 2. *Nature Methods, 9*(4), 357–359. https://doi.org/10.1038/nmeth.1923, https://www.nature.com/nmeth/.

Lindahl, T. (1993). Instability and decay of the primary structure of DNA. *Nature, 362*(6422), 709–715. https://doi.org/10.1038/362709a0.

Lipson, M., Cheronet, O., Mallick, S., Rohland, N., Oxenham, M., Pietrusewsky, M., Pryce, T. O., Willis, A., Matsumura, H., Buckley, H., Domett, K., Nguyen, G. H., Trinh, H. H., Kyaw, A. A., Win, T. T., Pradier, B., Broomandkhoshbacht, N., Candilio, F., Changmai, P., ... Reich, D. (2018). Ancient genomes document multiple waves of migration in Southeast Asian prehistory. *Science, 361*(6397), 92–95. https://doi.org/10.1126/science.aat3188.

Li, H., & Durbin, R. (2010). Fast and accurate long-read alignment with Burrows-Wheeler transform. *Bioinformatics, 26*(5), 589–595. https://doi.org/10.1093/bioinformatics/btp698.

Lo Bue, E., & Bray, J. (2014). *Art and Architecture in Ladakh: Cross-cultural Transmissions in the Himalayas and Karakoram.* BRILL. http://doi.org/10.1163/9789004271807.

Lorblanchet, M. (1992). *Rock art in the Old World.*

Lyn Wadley, B., Allchin, R., & Allchin (1983). The rise of civilization in India and Pakistan. *Archaeological Bulletin, 38*(137), 41. https://doi.org/10.2307/3888217.

Majumder, P. P. (2010). The human genetic history of South Asia. *Current Biology, 20*(4), R184. https://doi.org/10.1016/j.cub.2009.11.053.

Mamatamayee, C. (2017). The Pastoralists of Kalibangan. *Journal of Human Ecology, 3*(4), 271–278. https://doi.org/10.1080/09709274.1992.11907936.

Mathpal, Y. (1978). *Prehistoric rock paintings of Bhimbetka central India.*

Mendisco, F. (1660). *Where are the Caribs? Ancient DNA from ceramic period human remains in the Lesser Antilles, 370.*

Metspalu, M., Mondal, M., & Chaubey, G. (2018). The genetic makings of South Asia. *Estonia Current Opinion in Genetics and Development, 53*, 128–133. https://doi.org/10.1016/j.gde.2018.09.003, http://www.elsevier.com/locate/gde.

Meyer, M., Arsuaga, J.-L., de Filippo, C., Nagel, S., Aximu-Petri, A., Nickel, B., Martínez, I., Gracia, A., de Castro, J. M. B., Carbonell, E., Viola, B., Kelso, J., Prüfer, K., & Pääbo, S. (2016). Nuclear DNA sequences from the Middle Pleistocene Sima de los Huesos hominins. *Nature, 531*(7595), 504–507. https://doi.org/10.1038/nature17405.

Meyer, M., & Kircher, M. (2010). Illumina sequencing library preparation for highly multiplexed target capture and sequencing. *Cold Spring Harbor Protocols, 2010*(6). https://doi.org/10.1101/pdb.prot5448.

Mishra, S. (2000). Recent research trends in South Asian archaeology 121 137 Excavations at the early Acheulian site of Morgaon.

Mishra, S. (1982). On the effects of basalt weathering on the distribution of lower Paleolithic sites in the Deccan. *Bulletin of the Deccan College Research Institute, 41*, 107–115.

Mishra, S. (2006). The Indian Lower Palaeolithic. *Bulletin of the Deccan College Research Institute, 66*, 47–94.

Mishra, S., Rajaguru, S. N., & Korisettar, R. (2009). Palaeoenvironment and hominin dispersals in South Asia during the Quaternary. *Quaternary International, 199*(1–2), 1–15.

Misra, V. (1971). Two late Mesolithic settlements in Rajasthan: A brief review of investigations. *Poona University Journal, 35*, 59–77.

Misra, V. D. (1977). *Some aspects of Indian archaeology*.

Misra, V. N. (1973). Bagor – A late mesolithic settlement in north-west India. *World Archaeology, 5*(1), 92–110. https://doi.org/10.1080/00438243.1973.9979556.

Misra, & Bellwood (1985). *Recent advances in Indo-Pacific Prehistory: Proceedings of the international symposium held at Poona*. BRILL. http://doi.org/10.1163/9789004644472.

Mohammad, S. (2023). Neolithic settlements in Kashmir Valley: An explanation of Gufkral and Urzohom. *The Mythic Society Quarterly Journal, 114*(1). https://doi.org/10.58844/gizh7976.

Mondal, M., Casals, F., Xu, T., Dall'Olio, G. M., Pybus, M., Netea, M. G., Comas, D., Laayouni, H., Li, Q., Majumder, P. P., & Bertranpetit, J. (2016). Genomic analysis of Andamanese provides insights into ancient human migration into Asia and adaptation. *Nature Genetics, 48*(9), 1066–1070. https://doi.org/10.1038/ng.3621.

Mustak, M. S., Rai, N., Naveen, M. R., Prakash, S., Carlus, S. J., Pasupuleti, N., Srivastava, A., Singh, P. P., Babu, I., Dubey, P. K., Chaubey, G., & Thangaraj, K. (2019). The peopling of Lakshadweep Archipelago. *Scientific Reports, 9*(1). https://doi.org/10.1038/s41598-019-43384-3, http://www.nature.com/srep/index.html.

Narasimhan, V. M., Patterson, N., Moorjani, P., Lazaridis, I., Lipson, M., Mallick, S., Rohland, N., Bernardos, R., Kim, A. M., Nakatsuka, N., Olalde, I., Coppa, A., Mallory, J., Moiseyev, V., Monge, J., Olivieri, L. M., Adamski, N., Broomandkhoshbacht, N., Candilio, F., ... Reich, D. (2018). The genomic formation of South and Central Asia. *bioRxiv*. https://doi.org/10.1101/292581, https://www.biorxiv.org.

Narasimhan, V. M., Patterson, N., Moorjani, P., Rohland, N., Bernardos, R., Mallick, S., Lazaridis, I., Nakatsuka, N., Olalde, I., Lipson, M., Kim, A. M., Olivieri, L. M., Coppa, A., Vidale, M., Mallory, J., Moiseyev, V., Kitov, E., Monge, J., Adamski, N., ... Reich, D. (2019). The formation of human populations in South and Central Asia. *Science, 365*(6457). https://doi.org/10.1126/science.aat7487, https://science.sciencemag.org/content/365/6457/eaat7487/tab-pdf.

Neumayer, E. (1993). *Lines on stone. The prehistoric rock art of India*.

Nieves-Colón, M. A., Ozga, A. T., Pestle, W. J., Cucina, A., Tiesler, V., Stanton, T. W., & Stone, A. C. (2018). Comparison of two ancient DNA extraction protocols for skeletal remains from tropical environments. *Journal of Physical Anthropology, 166*(4), 824–836. https://doi.org/10.1002/ajpa.23472, http://onlinelibrary.wiley.com/journal/10.1002/(ISSN)1096-8644.

Niraj Rai, N. T., Manvendra Singh, G. C., Aditya Nath Jha, L. S., & Thangaraj, K. (2014). Relic excavated in western India is probably of Georgian Queen Ketevan. *Mitochondrion, 14*(1), 1–6. https://doi.org/10.1016/j.mito.2013.12.002.

Paddayya, K., Jhaldiyal, R., & Petraglia, M. D. (2000). Excavation of an Acheulian workshop at Isampur, Karnataka (India). *Antiquity, 74*(286), 751–752. https://doi.org/10.1017/s0003598x00060269.

Pappu, S., Gunnell, Y., Taieb, M., Brugal, J.-P., & Touchard, Y. (2003). Excavations at the Palaeolithic site of Attirampakkam, South India: Preliminary findings. *Current Anthropology, 44*(4), 591–598. https://doi.org/10.1086/377652.

Petraglia, M. D., Haslam, M., Fuller, D. Q., Boivin, N., Clarkson, C., & Korisettar, R. (2007). Middle Paleolithic assemblages from the Indian subcontinent before and after the Toba super-eruption. *Science, 317*(5839), 114–116. https://doi.org/10.1126/science.1141564.

Pinhasi, R., Fernandes, D. M., Sirak, K., & Cheronet, O. (2019). Isolating the human cochlea to generate bone powder for ancient DNA analysis. *Nature Protocols, 14*(4), 1194–1205. https://doi.org/10.1038/s41596-019-0137-7.

Pinhasi, R., Fernandes, D., Sirak, K., Novak, M., Connell, S., Alpaslan-Roodenberg, S., Gerritsen, F., Moiseyev, V., Gromov, A., Raczky, P., ... Petraglia, M. D. (2015). Optimal Ancient DNA yields from the inner ear part of the human petrous bone. *PLOS One, 10*(6), e0129102. https://doi.org/10.1371/journal.pone.0129102.

Possehl, G. L., & Ghosh, A. (1991). An encyclopaedia of Indian archaeology. *Journal of the American Oriental Society, 111*(1), 207. https://doi.org/10.2307/603838.

Potts, R., & Teague, R. (2010). Behavioral and environmental background to 'out-of-Africa I' and the arrival of homo erectus in East Asia. In *Vertebrate Paleobiology and Paleoanthropology*. United States: Springer, 67–85. http://doi.org/10.1007/978-90-481-9036-2_5, http://www.springer.com/series/6978.

Rai, N., Chaubey, G., Tamang, R., Pathak, A. K., Singh, V. K., Karmin, M., Singh, M., Rani, D. S., Anugula, S., Yadav, B. K., Singh, A., Srinivasagan, R., Yadav, A., Kashyap, M., Narvariya, S., Reddy, A. G., van Driem, G., Underhill, P. A., Villems, R., ... Yao, Y.-G. (2012). The phylogeography of Y-chromosome haplogroup H1a1a-M82 reveals the likely Indian origin of the European Romani populations. *PLoS ONE, 7*(11), e48477. https://doi.org/10.1371/journal.pone.0048477.

Reade, J. (1979). Early etched beads and the Indus-Mesopotamia trade. British Museum, Dept. of West Asiatic Antiquities, British Museum.

Reich, D., Thangaraj, K., Patterson, N., Price, A. L., & Singh, L. (2009). Reconstructing Indian population history. *Nature, 461*(7263), 489–494. https://doi.org/10.1038/nature08365.

Rohland, N., & Hofreiter, M. (2007). Comparison and optimization of ancient DNA extraction. *BioTechniques, 42*(3), 343–352. https://doi.org/10.2144/000112383.

Sachin Kumar, P. P. S., Nagarjuna Pasupuleti, V. M. T., Milan Kumar, C., Gyaneshwer, C., & Niraj, R. (2024). The genetic admixture and assimilation of Ahom: A historic migrant from Thailand to India. *Human Molecular Genetics, 33*(11), 1015–1019. https://doi.org/10.1093/hmg/ddae054.

Sankhyan, A. R. (1997). Fossil clavicle of a Middle Pleistocene hominid from the Central Narmada Valley, India. *Journal of Human Evolution, 32*(1), 3–16. https://doi.org/10.1006/jhev.1996.0117.

Schroeder, H., Ávila-Arcos, M. C., Malaspinas, A.-S., Poznik, G. D., Sandoval-Velasco, M., Carpenter, M. L., Moreno-Mayar, J. V., Sikora, M., Johnson, P. L. F., Allentoft, M. E., Samaniego, J. A., Haviser, J. B., Dee, M. W., Stafford, T. W., Ludovic Orlando, A. S., Bustamante, E. W. C. D., & Gilbert, M. P. (2015). Genome-wide ancestry of 17th-century enslaved Africans from the Caribbean. *Proceedings of the National Academy of Sciences, 112*(12), 3669–3673. https://doi.org/10.1073/pnas.1421784112.

Schug, G. R., Walimbe, S. R., & Lahiri, S. (2016). Where are they now? The human skeletal remains from India. In *Asia in the Past*. Wiley, 496–533. http://doi.org/10.1002/9781119055280.ch32.

Seguin-Orlando, A., Schubert, M., Clary, J., Stagegaard, J., Alberdi, M. T., Prado, J. L., Prieto, A., Willerslev, E., Orlando, L., & Hoheisel, J. D. (2013). Ligation bias in Illumina next-generation DNA libraries: Implications for sequencing ancient genomes. *PLoS ONE, 8*(10), e78575. https://doi.org/10.1371/journal.pone.0078575.

Sehrawat, J. S., Agrawal, S., Sankhyan, D., Singh, M., Kumar, S., Prakash, S., Rajpal, R., Chaubey, G., Thangaraj, K., & Rai, N. (2022). Pinpointing the geographic origin of 165-year-old human skeletal remains found in Punjab, India: Evidence from mitochondrial DNA and stable isotope analysis. *Frontiers in Genetics, 13*. https://doi.org/10.3389/fgene.2022.813934.

Shinde, V., Narasimhan, V. M., Rohland, N., Mallick, S., Mah, M., Lipson, M., Nakatsuka, N., Adamski, N., Broomandkhoshbacht, N., Ferry, M., Lawson, A. M., Michel, M., Oppenheimer, J., Stewardson, K., Jadhav, N., Kim, Y. J., Chatterjee, M., Munshi, A., Panyam, A., ... Reich, D. (2019). An ancient Harappan genome lacks ancestry from Steppe Pastoralists or Iranian farmers. *Cell, 179*(3), 729. https://doi.org/10.1016/j.cell.2019.08.048.

Silva, M., Oliveira, M., Vieira, D., Brandão, A., Rito, T., Pereira, J. B., Fraser, R. M., Hudson, B., Gandini, F., Edwards, C., Pala, M., Koch, J., Wilson, J. F., Pereira, L., Richards, M. B., & Soares, P. (2017). A genetic chronology for the Indian Subcontinent points to heavily sex-biased dispersals. *BMC evolutionary biology, 17*(1), 88. https://doi.org/10.1186/s12862-017-0936-9.

Sonakia, A. J. (1984). *The skull cap of early man and associated mammalian fauna from Narmada Valley alluvium, Hoshangabad area, 113*, 159–172 I. R.O.T.G.S.O.

Tamang, R., Chaubey, G., Nandan, A., Govindaraj, P., Singh, V. K., Rai, N., Mallick, C. B., Sharma, V., Sharma, V. K., Shah, A. M., Lalremruata, A., Reddy, A. G., Rani, D. S., Doviah, P., Negi, N., Hadid, Y., Pande, V., Vishnupriya, S., van Driem, G., ... Thangaraj, K. (2018). Reconstructing the demographic history of the Himalayan and adjoining populations. *Human Genetics, 137*(2), 129–139. https://doi.org/10.1007/s00439-018-1867-2.

Thangaraj, K., Chaubey, G., Kivisild, T., Reddy, A. G., Singh, V. K., Rasalkar, A. A., & Singh, L. (2005). Reconstructing the origin of Andaman Islanders. *Science, 308*(5724), 996. https://doi.org/10.1126/science.1109987.

Thangaraj, K., & Rai, N. (2019). Peopling of India: Ancient DNA perspectives. *Journal of Biosciences, 44*(3). https://doi.org/10.1007/s12038-019-9874-6, http://www.ias.ac.in/jbiosci/.

Chapter 13

Digital polymerase chain reaction: a revolution in forensic DNA analysis

Licínio Manco[1,2], and Helena Correia Dias[1,3,4]

[1]Research Centre for Anthropology and Health (CIAS), University of Coimbra, Coimbra, Portugal, [2]Department of Life Sciences, University of Coimbra, Coimbra, Portugal, [3]Instituto Nacional de Medicina Legal e Ciências Forenses (INMLCF), I.P., Coimbra, Portugal, [4]Laboratory of Forensic Anthropology (LFA), Centre for Functional Ecology (CFE), University of Coimbra, Coimbra, Portugal

13.1 The digital polymerase chain reaction technology

The polymerase chain reaction (PCR) is an in vitro technique developed by Kary Mullis in 1985 to amplify DNA (Saiki et al., 1985). This process uses two DNA oligonucleotides, known as primers, which hybridize to each strand of the target DNA. By adding DNA polymerase to the reaction, millions of copies of a specific segment of DNA between the primers can be generated from a small amount of starting material. The PCR process typically involves a series of temperature cycles that are repeated 30–40 times. Each cycle consists of three main steps: denaturation of the DNA duplexes, hybridization of the primers that flank the target sequence, and elongation of the primers by the DNA polymerase enzyme. With each cycle, the number of target DNA molecules doubles, leading to exponential amplification. Eventually, the process reaches a saturation point where the amplification signal tapers off and ceases to grow as the PCR reagents (primers and nucleotides) are consumed and the Polymerase enzyme activity is inhibited by the large quantity of amplified double-stranded DNA (Quan et al., 2018).

The PCR technique has advanced through three significant stages since the discovery of the high-temperature resistant polymerase known as Taq polymerase (Saiki et al., 1988). With technological advancements, conventional PCR thermal cyclers have evolved into real-time fluorescence PCR thermal cyclers, also known as quantitative PCR (qPCR) machines, and further developed into digital PCR platforms. In traditional PCR, the amplified products with conventional PCR thermal cyclers are analyzed at the end of the reaction using gel electrophoresis, providing an endpoint measurement. Real-time qPCR builds upon this by measuring the amount of PCR product after each amplification cycle using a fluorescent readout, which allows for real-time monitoring of the PCR process. A typical amplification plot in real-time PCR displays a sigmoidal-shaped curve (on a linear scale) consisting of a baseline phase, followed by an exponential phase that finally reaches the nonexponential plateau phase. The exponential phase is the most efficient stage of amplification, where the amount of PCR product doubles with each cycle, assuming a complete amplification efficiency. Real-time PCR can be utilized for both qualitative and quantitative analysis of nucleic acids. In specific applications, such as allelic discrimination, qualitative detection is adequate. Real-time quantitative PCR enables the relative quantification of nucleic acid targets by using the cycle number at which enough amplified product accumulates to yield a detectable fluorescence signal in the exponential phase of PCR, called the cycle of quantification (Cq) or threshold cycle (Ct) (Quan et al., 2018). The reaction will have a high or a low Ct inversely related to the number of copies of the target template present at the start of the amplification reaction. This relationship forms the basis for the quantitative aspect of real-time PCR. The absolute amount of the target sequence in a qPCR reaction is determined concerning a standard curve generated from a sample of known quantity (Quan et al., 2018).

Digital polymerase chain reaction (dPCR) is a significant advancement over traditional PCR methods, allowing for the direct quantification of target nucleic acids in a sample without the need for calibration. dPCR uses the same primers and probes as qPCR but achieves a higher level of sensitivity and precision. In dPCR, the sample is divided into many independent PCR subreactions, ensuring that each partition contains either a few or no target sequences. Each partition undergoes thermocycling separately, allowing for independent PCR amplification. After the amplification process, the

Investigative and Predictive DNA Testing. DOI: https://doi.org/10.1016/B978-0-443-30128-5.00017-5

partitions are analyzed for fluorescence to determine the concentration of the target sequence by considering the number of positive partitions relative to the total number of partitions. The quantification in dPCR relies on the random distribution of molecules in multiple partitions, which follows a Poisson distribution. This means that the distribution of target molecules across the partitions occurs independently and at a consistent rate. Poisson statistics enable the calculation of the initial number of targets based on the number of positive and negative partitions. As a result, dPCR inherently provides absolute quantification. Unlike qPCR, dPCR does not depend on calibration curves for sample quantification, thus avoiding issues related to variations in reaction efficiencies. The proportion of PCR-positive partitions is sufficient to determine the target sequence concentration without the need for calibration (Quan et al., 2018; Xu et al., 2023).

Recent advances in microfluidic techniques have significantly contributed to the ongoing revolution in digital quantification by introducing efficient partitioning methods. Two partitioning principles are currently utilized in digital PCR instruments: droplet-based and chip-based (Quan et al., 2018). BioRad's droplet digital PCR (ddPCR) systems exemplify the droplet-based method, while the Applied Biosystems QuantStudio Absolute Q Digital PCR System (ThermoFisher Scientific) can represent the chip-based approach. The BioRad's ddPCR System automates the complete workflow of ddPCR technology, which includes droplet generation, thermal cycling, droplet reading, and data analysis. The QX200 ddPCR systems can read up to two channels in fluorescein (FAM) and hexachloro-fluorescein spectra from approximately 20,000 oil droplets. The QuantStudio Absolute Q Digital PCR System (ThermoFisher Scientific) employs a microfluidic array plate that contains 16 dPCR reaction units, each consisting of 20,480 fixed array microchambers. Additionally, a new digital real-time PCR instrument, known as LOAA (OPTOLANE Technologies), has recently been developed to integrate partitioning, thermocycling, and imaging into a single dPCR instrument.

The ddPCR technology has been widely applied in the last decade due to its advantages of high throughput, high sensitivity, low reagent consumption, low cross-contamination, and absolute quantification ability. With ddPCR, instead of reaction chambers, reagents are dispersed into nanoliter-sized droplets, and each droplet represents a PCR reaction. After end-point PCR, each droplet is individually analyzed in a two-color optical detection system. The method for performing is based on water-oil emulsion droplet technology. A sample is fractionated to about 20,000 droplets, and PCR amplification of the template molecules in each droplet uses reagents and workflows similar to those used for most standard TaqMan probe-based assays. Regarding the laboratory processes, in the sample preparation, the DNA sample is mixed with PCR reagents, including primers and fluorescent probes specific to the target of interest. A microfluidic device divides the reaction mixture into thousands of monodisperse droplets. Each droplet contains one, a few, or no copies of the target DNA. Droplets are subjected to PCR amplification in thermal cycles (denaturation, annealing, and extension). During amplification, fluorescent probes bind to target DNA and emit a fluorescent signal. After amplification, the drops are passed through a fluorescence detector that counts the number of fluorescent (positive) and nonfluorescent (negative) drops. The proportion of positive partitions is used to calculate the concentration of the target sequence using the Poisson distribution to calculate the absolute amount of target DNA in the original sample, without the need for a standard curve (Quan et al., 2018; Xu et al., 2023).

The ddPCR technique has been widely applied in basic research, clinical diagnostics, and environmental testing for the detection of rare mutations, copy number variations, gene rearrangements, chimerism analysis, fetal sex determination, or pathogen detection (D'Aversa et al., 2018; Okano et al., 2018; Olmedillas-López et al., 2022; Santurtún et al., 2017; Xu et al., 2023; Zhang et al., 2022). For disease diagnosis, ddPCR is widely applied in cancer diagnosis, through the detection of genetic mutations, tumor biomarkers, and circulating tumor DNA (ctDNA) in blood samples. Moreover, ddPCR is employed in liquid biopsies to monitor tumor burden and treatment response in cancer patients by analyzing ctDNA in the blood. It can also be used for single-cell analysis or to assess cell-free DNA (cfDNA) in plasma. Additionally, ddPCR is a highly effective technique used to detect and quantify pathogens, such as bacteria, viruses, and fungi, in both clinical and environmental samples, including water and soil. It is also utilized for pathogen detection in food, which is essential for ensuring food safety and preventing foodborne disease outbreaks (D'Aversa et al., 2018; Okano et al., 2018; Olmedillas-López et al., 2022; Santurtún et al., 2017; Xu et al., 2023; Zhang et al., 2022). These applications demonstrate the versatility and importance of ddPCR in various fields, including biomedical research, clinical diagnostics, and environmental monitoring.

13.2 Droplet digital polymerase chain reaction in forensic science

ddPCR has emerged as an essential tool in forensic research for analyzing casework samples, particularly those containing low quantities or highly degraded DNA. The accurate analysis of trace samples can be critical in criminal investigations. Recently, ddPCR has been applied in various nucleic acid detection methods relevant to forensic investigations.

In a study conducted by Fu et al. (2025), a ddPCR system was developed to evaluate mixed samples containing both male and female DNA. The goal of the study was to detect trace amounts of target DNA in highly unbalanced mixtures and to determine the mixing ratios. This is extremely important for forensic criminal cases because small amounts of DNA from a perpetrator can be obscured by larger amounts of DNA from the victim during crime-scene investigations. The researchers specifically quantified the nonrecombining regions of the X and Y chromosomes in the DNA samples of both males and females using duplex ddPCR. The system allowed for the absolute quantification of low-abundance trace samples across various mixing ratios. In another recent study by Bae et al. (2025), ddPCR technology was used to highlight the significance of human and animal-related biological evidence at crime scenes. The authors developed a method to simultaneously identify and quantify human, dog, and cat DNA in mixed samples. The study selected three markers: HLA-DRA as a human-specific marker, OR6D7 as a dog-specific marker, and FLAI-K as a cat-specific marker. In mixture tests, each DNA sample was accurately detected, even in samples containing trace amounts of all three DNA types, demonstrating the effectiveness of ddPCR for the accurate identification and quantification of different DNA samples simultaneously.

For forensic casework applications, DNA analysis is a crucial and unique source of identification, being a primary method that permits the identification of individuals from small evidence that contains only low quantities or degraded DNA. However, in forensic contexts, the origin of the evidence is also very important for the establishment of a link with the crime scene. In some cases, forensic RNA analysis can address some of these limitations (Bauer, 2007). For instance, mRNA profiling can identify body fluids and be an important supplement to DNA analysis when understanding the type and origin of evidence is essential. Moreover, RNA analysis can offer insights into diseases and mechanisms that lead to death, potentially becoming a valuable tool for diagnosing the cause of death in forensic pathology research. Other possible applications include determining the age of damages and injuries, as well as the postmortem interval, which is also a relevant issue for forensic cases. Therefore, molecular identification of body fluids through the analysis of cell-specific mRNA expression is already emerging as a new technique to complement DNA analysis in forensic investigations (Bauer, 2007). Currently, most follow-up experiments involving RNA necessitate reverse transcription (RT) procedures. Research indicates that the RT step can vary and significantly impact subsequent data analysis, especially in the context of forensic trace samples (Gao et al., 2024). Some examples of RNA analysis based on ddPCR can address some of these limitations. A recent study by Lee et al. (2024) developed a ddPCR assay for identifying body fluids, specifically saliva-semen mixtures. The researchers used body fluid-specific mRNA markers HTN3 and PRM1, which are unique to saliva and semen, respectively, as well as the housekeeping gene marker GAPDH. They employed a two-step RT ddPCR technique with a QX600 ddPCR System from Bio-Rad Laboratories. The study demonstrated that this method had enough accuracy and sensitivity to detect as little as 1.5 copies per microliter of complementary DNA. In this study, saliva-semen mixtures with ratios ranging from 1:50 to 50:1 were successfully identified. Therefore, RT ddPCR proves to be valuable in forensic contexts, particularly for analyzing mixed samples such as those found in sexual assault cases, where low DNA quantities often lead to ambiguous results with traditional RT qPCR (RT-qPCR). The authors also suggested that additional markers could enable the identification of other body fluids, such as vaginal fluid and menstrual blood (Lee et al., 2024).

In a second study, Dass and Ghai (2024) reported on the identification of biological fluids based on their microbial profiles. The focus of this research was on using the microbiome content of vaginal fluid and saliva for differentiation. One objective was to employ ddPCR as a novel method for quantifying bacterial species, both individually and in mixtures of these two body fluids, using a panel of 16S rRNA bacterial markers. Four specific bacterial markers were selected for analysis, and all four species were successfully detected in both mixtures and aged samples. The research highlighted the sensitivity of ddPCR for bacterial quantification in both pure and mixed body fluids, which could assist in crime scene reconstruction and provide valuable investigative leads (Dass & Ghai, 2024).

Based on the aforementioned studies, it is clear that ddPCR is crucial in forensic research today for both DNA and RNA analyses. Due to its ability for absolute quantification, ddPCR can detect even samples with minimal nucleic acid content, significantly enhancing detection efficiency. Also, it highlights geneticists' concern about investigating new methodologies and approaches to overcoming pre-existing problems.

13.3 Application of droplet digital polymerase chain reaction to DNA methylation analysis

DNA methylation (DNAm) is an epigenetic regulatory mark that affects phenotype without a change in genotype (Moore et al., 2013). In the mammalian genome, DNAm involves the addition of a methyl group to the C5 position of cytosine, resulting in the formation of 5-methylcytosine (5mC). Typically, methylation occurs at dinucleotide CpG sites across the genome (5'-CpG-3'), which are sequences of cytosine (C) followed by guanine (G) linked by a phosphate at the DNA molecule. It is now well recognized that DNAm, in concert with other regulators, is a major epigenetic factor influencing gene activities. The DNAm process plays a crucial role in regulating gene expression by either attracting proteins that promote gene repression or preventing transcription factors from binding to DNA. Differentiated cells develop a stable and unique DNAm pattern that regulates gene expression.

Recently, DNAm has become an important marker in forensic genetics that can provide valuable insights for solving crimes. By identifying body fluids based on tissue-specific DNAm patterns, investigators can better understand the activities related to the evidence collected in crime scenes. Variations in DNAm can indicate environmental exposures, such as cigarette smoking and alcohol consumption, which may help to predict the forensically relevant lifestyle of an unknown trace donor (Lee et al., 2016). Additionally, utilizing DNAm for age estimation is a new active field in dPCR applications for forensic science that can help to narrow down the number of potential suspects, particularly when the DNA profile obtained from the evidence does not match any known individuals (Correia Dias et al., 2022).

During development, the pattern of DNAm across the genome undergoes significant changes due to a dynamic interplay of both de novo methylation and demethylation processes. A study by whole-genome bisulfite sequencing (WGBS) and CpG-site DNAm microarrays in newborn and centenarian genomes using blood showed that the centenarian DNA had a lower DNAm content and a reduced correlation in the methylation status of neighboring CpGs in comparison with the more homogeneously methylated newborn DNA (Heyn et al., 2012). The biological significance of this relative lack of neighbor site methylation concordance in the centenarian versus the newborn might be related to a deficient targeting by the enzymes that add (DNA methyltransferases) or remove (tet methylcytosine dioxygenases) methyl-groups (Heyn et al., 2012). As we age, the levels of DNAm change within the human genome. Overall, there is a global decrease in methylation, while specific sites may experience an increase. Most CpG sites throughout the genome tend to lose methylation, whereas CpG islands, which are typically found in gene promoters, generally gain methylation (Heyn et al., 2012).

Multiple technologies have been developed for the detection, analysis, and quantification of DNAm levels. These can be categorized into three main subgroups based on the pretreatment of DNA before analysis: (1) enzyme digestion, which utilizes restriction endonucleases; (2) affinity enrichment, involving antibodies or methyl-binding proteins that are specific for methylated CpG sites; and (3) sodium bisulfite conversion (Freire-Aradas et al., 2017). Sodium bisulfite conversion is considered the gold standard method for DNAm analysis, as most current methods rely on it as the initial step for identifying methylated CpG sites. This chemical modification facilitates the differentiation between methylated cytosine (5mC) and nonmethylated cytosine. When DNA is treated with sodium bisulfite, nonmethylated cytosine is converted to uracil, while methylated cytosine remains unchanged. During PCR amplification, uracil is amplified as thymine (T). As a result, the final sequence reflects 5mC as cytosine (C) and nonmethylated cytosine as thymine (T) (Yang et al., 2006). After sodium bisulfite conversion and PCR amplification, the quantitative measurement of methylation at a specific CpG site can be determined through various methods (Freire-Aradas et al., 2020). These methods include the EpiTYPER assay/Matrix-Assisted Laser Desorption/Ionization Time of Flight (MALDI-TOF) mass spectrometry (Suchiman et al., 2015), DNAm microarrays (Lin et al., 2013), SNaPshot (Jung et al., 2019), and different sequencing techniques, including pyrosequencing (the most widely used method) (Tost & Gut, 2007), Sanger sequencing (Jiang et al., 2010), and massively parallel sequencing (MPS) (Naue et al., 2018). The proportion of DNAm at a specific CpG site, as measured by one of these methods, is the methylation β-value. This value is determined by calculating the ratio of methylated (C) to unmethylated (T) signals. The fraction is typically calculated using the formula: M/(M+U), where M represents the signal for methylated molecules and U represents the signal for unmethylated molecules. DNAm is measured on a scale of 0–1 (or 0%–100%): a value of 0 indicates a complete absence of methylation, while a value of 1 indicates full methylation at the examined CpG site (Freire-Aradas et al., 2017).

The development of a fluorescence-based qPCR (TaqMan) assay for measuring DNAm levels has significantly improved the detection and quantification of methylated alleles using real-time qPCR. In this method, primers and/or fluorescent probes are designed to target bisulfite-converted DNA sequences. Quantitative data on methylated alleles are obtained through real-time PCR, with corrections made for DNA loading and normalization to a methylated DNA standard (Danenberg et al., 2000). Following this procedure, the ddPCR technology has been employed in TaqMan probe-based assays using methylation-specific primers to quantify both methylated and unmethylated DNA after bisulfite conversion (Pharo et al., 2018; Yu et al., 2015). To our knowledge, the first ddPCR study utilizing this approach was conducted in the context of colorectal cancer (Yu et al., 2015). This study accurately quantified methylated NTRK3 and methylated EVL alleles in bisulfite-converted DNA samples extracted from colorectal cancer tissue. The probe sequence designed for the methylated CpGs was synthesized with a FAM reporter. The relative amount of DNAm in each sample was compared to a C-LESS-C1 assay, which amplifies the total amount of DNA in a duplex PCR reaction. The C-LESS probe was synthesized with a VIC reporter (Yu et al., 2015, 2018). In another study, Hayashi et al. (2015) utilized ddPCR to target the methylation of the Paired box 5 (PAX5) gene promoter in bisulfite-treated DNA from head and neck squamous cell carcinomas. This method demonstrated increased sensitivity compared to traditional quantitative methylation-specific real-time PCR (Hayashi et al., 2015). Additionally, a circulating cfDNA-based system for detecting primary breast cancer using a methylation array for 12 epigenetic biomarkers was tested with ddPCR technology (Uehiro et al., 2016). Quantitative Methylation-Specific ddPCR assays have also been applied in the study of autoimmune diseases (Husseiny et al., 2020) and genomic imprinting (Mitake et al., 2021). Following these studies, ddPCR was introduced as a novel tool for predicting epigenetic age based on DNAm measurement (Han et al., 2020; Shi et al., 2018).

13.4 DNA methylation age estimation with droplet digital polymerase chain reaction

In forensic science, positive identification is a systematic process that confirms the identity of an individual, such as a victim, perpetrator, missing person, or human remains (Ubelaker et al., 2018). This process is always comparative, meaning that it requires specific data for comparison to establish identification successfully. Positive identification can be achieved through primary identification methods, including DNA analysis (forensic genetics), fingerprint analysis (dactyloscopy), and odontology (forensic dentistry) (Ubelaker et al., 2018). In forensic genetics, the process of positive identification involves comparing the genetic profile obtained from reference samples to the genetic profile obtained from forensic evidence. This forensic evidence can come from various sources of DNA, such as blood, buccal swabs, teeth, bones, hair, clothing, semen, or body parts, depending on the specifics of the case. The most common method for DNA profiling today is based on an individual's genetic profile established by autosomal polymorphic DNA loci known as short tandem repeats (STRs). Advances in DNA technology have allowed the extraction of useful DNA profiles even from severely damaged evidence. The STR analysis using the multiplex PCR method was introduced in the early 1990s and has since become the primary method for personal identification in criminal investigations of suspects or victims, paternity testing, and identification of missing persons in mass disasters (Parson et al., 2016). In many forensic contexts, it is also necessary to investigate other genetic markers, such as the Y-STRs, for the paternal lineage and could be essential in cases of sexual assault, mitochondrial DNA (mtDNA), which helps in accessing the maternal lineage, and X-STRs, which can be useful in kinship investigations in cases of complex relationships (Stasi et al., 2023). Over the years, the sensitivity and discrimination capability of genetic analysis have significantly improved due to the development of various commercial kits that incorporate different numbers of STRs (Wyner et al., 2020). Standard panels, consisting of 15–27 STR loci, are now routinely used in DNA profiling.

Additionally, Single Nucleotide Polymorphism (SNP) typing serves as a valuable tool in forensic genetics, allowing for the analysis of even smaller and more degraded DNA samples. The genetic profile information obtained from SNPs can provide new investigative information and help determine human identities. This is especially relevant for various applications, including: (1) identity-informative SNPs; (2) lineage-informative SNPs; (3) ancestry-informative SNPs; (4) phenotype-informative SNPs; and (5) pharmacogenetic SNPs (Novroski & Cihlar, 2022). Examples include the extensively investigated SNPs predictive for eye, hair, skin color, and biogeographical ancestry, which have become the most accurate and widely used in Forensic DNA Phenotyping (Schneider et al., 2019). These externally visible characteristics can help the identification process and guide the entire criminal investigation.

Sometimes, DNA profiles obtained in forensic contexts by STR and SNP analysis can fail to identify individuals when there is no available suspect or when the DNA evidence does not match any suspect or person in a forensic DNA database (Moore et al., 2013). In such cases, estimating an individual's age is a crucial step in the forensic identification process, as it helps narrow down the pool of potential suspects. Age estimation can be essential in forensic science for identifying both deceased and living individuals. For deceased persons, particularly those with skeletal remains, age estimation can lead to exclusion from suspect lists. For living individuals, age estimation has a vital role in resolving immigration cases, determining the status of minors, and establishing criminal responsibility, among other scenarios (Correia Dias et al., 2022). Beyond determining an individual's age, it can also predict age-related phenotypic traits, such as hair color or baldness.

Initial approaches for forensic age determination were originally based on the examination of bones and teeth with several different anthropological methods (Baccino & Schmitt, 2006). In addition to skeletal methods, molecular approaches for age estimation have increasingly been reported based on the gradual alterations to biomolecules that occur during the life cycle of an individual as a result of the aging process. Five main groups of molecular alteration have been assessed through DNA or protein-based methodologies: (1) mtDNA deletions; (2) shortening of telomeres; (3) advanced glycation end-products; (4) aspartic acid racemization; and (5) signal-joint T-cell receptor excision circles (Freire-Aradas et al., 2017). Those methodologies comprise the molecular framework for forensic age estimation from its origins up to 2010. Since then, significant advances have been achieved in developing new DNA tests to infer individual age from a biological trace. The change of emphasis that enabled these advances was a shift in focus from characteristics of DNA and proteins toward epigenetic changes to DNA, exploiting increased knowledge of human epigenetic patterns from clinical studies. This shift has opened the door to the identification and development of age-correlated markers that have provided robust age prediction models (APMs) for forensic use (Freire-Aradas et al., 2017).

It is well established that aging cells undergo a variety of accumulated epigenetic changes driven by both intrinsic and extrinsic processes. Epigenetic drift refers to the changes in DNAm patterns in an individual's epigenome over time (Declerck & Vanden Berghe, 2018). It can be influenced by genetics and environmental exposure, such as smoking, alcohol consumption, exposure to toxins, eating habits, and physical activity. Research suggests that the impact of these stochastic and environmental influences on DNAm levels increases with age (Declerck & Vanden Berghe, 2018). Overall, alterations in DNAm patterns have been recognized as biomarkers of aging, which form the basis for "epigenetic clocks." This suggests that epigenetic drift is not entirely random. The concept of the epigenetic clock is a directional phenomenon resulting from nonrandom events, and it involves specific regions of the genome where DNAm levels remain strongly associated with age (Freire-Aradas et al., 2017). Consequently, in recent years, there have been multiple efforts to utilize DNAm signatures as predictors of chronological age. These changes in DNAm patterns can be consistent across individuals and, remarkably, a small number of these sites can be employed to create highly accurate age estimators (Horvath, 2013). These specific age-related DNAm changes hold considerable promise for applications in forensic science, particularly given the accessibility of samples like saliva and blood. As a consequence, the study of DNAm patterns has emerged as one of the most important areas in aging research these days (Castagnola et al., 2024).

As DNAm has emerged as a highly accurate age estimator, evaluating DNAm changes in age-associated genes has been explored as the most promising epigenetic modification for age prediction for human forensic contexts (Castagnola et al., 2024; Freire-Aradas et al., 2017; Horvath, 2013). Most new DNAm data analysis relies on bisulfite-based methods that provide accurate quantitative data and single-base-pair resolution. This process begins with a bisulfite-conversion step, followed by DNA amplification using target-specific primers and a methylation detection method. The use of WGBS and array-based DNAm analysis enabled comprehensive scanning of DNAm patterns, leading to the significant discovery of potential CpG age-related markers (Freire-Aradas et al., 2017). Notably, regardless of the specific human ethical groups and body fluids involved, these models converge on the CpG islands of the ELOVL2, FHL2, KLF14, MIR29B2, PDE4C, and TRIM59 genes (Freire-Aradas et al., 2017). This suggests that these regions may contain DNAm changes that are largely consistent with the progression of age. Many APMs have been developed so far. These models are either tissue-specific (i.e., developed for one specific tissue type) or multitissue APMs developed for a combined training set of different tissues (Freire-Aradas et al., 2020). Most tissue-specific DNAm-based APMs were developed using blood from healthy individuals, showing a mean absolute deviation (MAD) between predicted and chronological ages between 3 and 10 years. On the other hand, relevant forensic tissues, such as buccal swabs, teeth, bones, semen, urine, and hair, have been only weakly investigated, and the development of multitissue APMs is still an area of research with few publications (Correia Dias et al., 2022; Freire-Aradas et al., 2017).

Some of the most accurate "epigenetic clocks" developed so far rely on a large number of markers (Horvath, 2013) and/or require time-consuming or expensive techniques, such as pyrosequencing, mass spectrometry, or MPS (Castagnola et al., 2024; Freire-Aradas et al., 2020). Other ways of evaluating DNAm, like SNaPshot or Sanger sequencing, are semiquantitative and depend on measuring fluorescence using dyes with different signal intensities, which can reduce accuracy and may not be ideal for precise DNAm quantification (Castagnola et al., 2024; Freire-Aradas et al., 2020). Furthermore, differences in DNAm evaluation can occur due to the specificity of each method (Freire-Aradas et al., 2020) or even within the same method but using different equipment with varied software settings (So et al., 2024). As a result, the pros and cons of these DNAm evaluation methods do not currently allow for the selection of a preferred methodology for implementation in age estimation.

As previously mentioned, ddPCR technology has been introduced as a tool for evaluating DNAm patterns. Due to its precise and accurate capabilities for detecting methylation, several studies have also explored the use of bisulfite ddPCR for estimating DNAm age (Dias & Manco, 2024; Han et al., 2020; Lee et al., 2022; Manco & Dias, 2022; Shi et al., 2018; Zhou et al., 2024). To analyze DNAm patterns, ddPCR has been conducted using two approaches. In the first approach, two fluorescence-labeled probes are used: one to assess the methylation status of the gene of interest and another for a reference gene that is not affected by CpG methylation (Yu et al., 2018). The reference gene serves as a control for the total amount of amplifiable DNA in the ddPCR assay. Using this method, our group (Manco & Dias, 2022) developed a duplex ddPCR assay targeting simultaneously five CpGs in a selected region of the ELOVL2 gene and using the reference gene C-LESSC1, which amplifies regardless of its methylation status. A sample set of 56 blood samples from healthy individuals, aged 1–94 years old, was submitted to bisulfite conversion, followed by ddPCR analysis. The results showed a strong correlation between DNAm captured from the five CpG sites of the *ELOVL2* gene and chronological age ($R = 0.892$; $P = 2.84 \times 10^{-20}$), explaining 79.2% of age variation. The mean absolute deviation (MAD) between predicted and chronological ages was 10.07 years (Table 13.1).

TABLE 13.1 Review of studies using the droplet digital polymerase chain reaction (ddPCR) technology in DNA methylation evaluation for age estimation purposes.

Study	Tissue/N	Age range	CpGs or genes analyzed	Findings[a]	Accuracy of age prediction model (MAD/MAE)[b] (Spearman's rho)[c]	Statistics
Manco and Dias (2022)	Blood/56 (training set)	1–94 years old	ELOVL2 (using the reference gene C-LESSC1)	R^2 = 0.892 (DNAm simultaneously captured from five CpG sites)	MAD 10.07 years; rho = 0.904	Simple linear regression
Shi et al. (2018)	Blood/124 (training set) (78 boys and 46 girls)	6–15 years old	DDO, PRPH2, DHX8, ITGA2 B, and unknown (cg 22398226)	R = 0.36 (ITGA2 B), 0.47 (DDO), 0.47 (unknown), 0.57 (DHX8), 0.64 (PRPH2)	Combination model incorporating skeletal age (SA) and dental age (DA) with the methylation levels of age-associated CpG sites	
Han et al. (2020)	Blood/40 (training set); 40 (validation set)	21–72 years old	Reliable assays for CCDC102B, COL1A1, MEIS1-AS3, FHL2, PDE4C, ASPA, IGSF11; (hampered assays for ITGA2B and ELOVL2)	Multivariable model using 7 CpGs R^2 = 0.94 (in training set); R^2 = 0.89 (in validation set)	MAE 1.9 years (in training set); 2.9 years (in validation set)	Multiple linear regression
Lee et al. (2022)	Saliva/76	23 to 58 years old	6 CpG markers SLC12A5 (cg07547549), KLF14 (cg14361627), TSSK6 (cg08928145), CNGA3 (cg19671120), SST (cg00481951), and TBR1 (cg12757011)	Four methylation markers in genes SST, KLF14, TSSK6, and SLC12A5 with R^2 = 0.6614, 0.6089, 0.2725, and 0.3594, respectively.	MAD 3.3 years (RMSE of 4.19); rho = 0.922	Multiple linear regression
Dias and Manco (2024)	Blood/58 (training set)	1–93 years old	ELOVL2 (5 CpG sites simultaneously), FHL2, and PDE4C	R = 0.970 (DNAm captured on ELOVL2 [squared], FHL2, and PDE4C)	MAD 4.657 years (RMSE 6.044), rho = 0.976	Multiple linear regression
Zhou et al. (2024)	Blood/327	20–100 years old	Multiplex system (ASPA, IGSF1, COL1A1, and MEIS1-AS3) (four CpG sites; eight fluorescent clusters)	R^2 = 0.4878 (ASPA), 0.4832 (IGSF1), 0.6881 (COL1A1), and 0.6475 (MEIS1-AS3)	MAE 4.722 years, (testing set); MAE 2.0036 years (training set)	Random forest regression

[a] *Pearson correlation between DNAm and chronological age.*
[b] *MAD (mean absolute deviation) or MAE (mean absolute error) between predicted and chronological ages.*
[c] *Spearman correlation between predicted and chronological ages.*

However, most studies using ddPCR for DNAm age have utilized a duplex assay targeting methylated and unmethylated DNA in the target gene. This method employs a probe-based PCR technique with two fluorescence-labeled probes that target the CpG of interest: one probe binds to the methylated allele, and the other to the unmethylated allele. In each droplet, ideally containing only one bisulfite-converted DNA molecule, PCR amplification occurs, and one of the two fluorescence-labeled probes binds depending on the allele in the droplet. After amplification, the fluorescence signal is read for each droplet, and the ratio between the two fluorescent probes can be calculated to reveal the percentage of methylation of the analyzed CpG. As this technology facilitates parallel PCR reactions in numerous micro-droplets, it is expected to reduce PCR bias for methylated/nonmethylated strands that may occur in conventional PCR.

In a study conducted in 2018 following this approach, Shi et al. (2018) examined the DNAm profiling from whole blood samples of 124 randomly selected Chinese children (78 boys and 46 girls) aged between 6 and 15 years. They used the ddPCR technique to analyze CpG sites located at the genes DDO, PRPH2, DHX8, ITGA2 B, and one unknown gene (cg 22398226), all of which were identified as age-associated genes using the Illumina HumanMethylation450 BeadChip. The authors found a correlation between the methylation levels of CpG sites and chronological age, ranging from –0.36 to –0.64 ($P < .001$). The data were then used to develop an age estimation model in combination with skeletal and dental ages, resulting in the highest accuracy of age estimation for both boys (MAE = 0.47 years, $R^2 = 0.886$) and girls (MAE = 0.33 years, $R^2 = 0.941$) (Table 13.1).

Using this same approach of methylated/unmethylated primer/probe mix, Han et al. (2020) established in 2020 reliable ddPCR assays for the age-associated genes CCDC102B, COL1A1, MEIS1-AS3, FHL2, PDE4C, ASPA, and IGSF11. The ddPCR results showed a clear correlation with chronological age for the tested CpGs, even slightly higher than those observed by pyrosequencing. The multivariable model based on ddPCR measurement of these seven CpGs provided relatively reliable age predictions in an independent validation set of 40 blood samples ($R^2 = 0.89$), with a median error of only 2.9 years (Table 13.1). This indicates the high accuracy of age predictions obtained with ddPCR. The assay yielded negative results for ITGA2B and ELOVL2 genes, justified by neighboring CpG sites.

In 2022, Lee et al. (2022) used the same duplex ddPCR approach to analyze four age-related CpG sites for DNAm-based age prediction in saliva samples of 76 individuals. In this study, the set of experiments was composed of four single sets of ddPCR reactions targeting each of the four CpGs, cg00481951 (SST), cg14361627 (KLF14), cg08928145 (TSSK6), and cg07547549 (SLC12A5). The DNAm ratio at these four CpG sites exhibited a strong correlation with chronological age ($R^2 = 0.6614, 0.6089, 0.2725,$ and 0.3594), similar to the results obtained by the SNaPshot method at the same CpG markers in a study by (Hong et al., 2017) ($R^2 = 0.4791, 0.6347, 0.4341,$ and 0.5486). The methylation values of the four CpGs and the chronological ages of the 76 individuals allowed the authors to establish a strong correlation (Spearman's rho = 0.922) between predicted and chronological ages with a MAD of 3.3 years (see Table 13.1).

In a study published in 2024, our group (Dias & Manco, 2024) examined CpGs in the three age-correlated genes *ELOVL2, FHL2*, and *PDE4C* using blood from living individuals. We used the duplex ddPCR approach targeting methylated and unmethylated DNA sequences after bisulfite conversion. Simple linear regression revealed a strong correlation between DNAm levels and chronological age for FHL2 (R = 0.948) and PDE4C (R = 0.819), addressing only one CpG for each gene. The ELOVL2 gene, evaluating five CpG sites simultaneously, revealed a strong age correlation (R = 0.887) in simple linear regression, and a very strong age correlation (R = 0.926) when using quadratic regression analysis. The multivariable APM with the methylation information captured on *ELOVL2, FHL2*, and *PDE4C* genes, revealed a robust age correlation (R = 0.970), explaining 93.7% of age variance, displaying a MAD between chronological and predicted ages of 4.657 years (RMSE = 6.044) (Table 13.1).

All these ddPCR assays using DNAm analysis for age estimation purposes were established with the QX100/QX200 Droplet Digital PCR Systems (Bio-Rad, CA, United States), where each ddPCR reaction targets the CpG sites individually for each gene. However, we must consider that although the accuracy of the ddPCR single CpG assay has been extensively validated in all these studies, the detection of only one CpG site by ddPCR in a single test proves to be costly and labor-intensive. Most recently, in 2024, to address these limitations, Zhou et al. (2024) implemented a novel high-throughput multiplex ddPCR assay allowing for the simultaneous detection of four CpGs (eight targets), supported by the novel developed ddPCR instrument Sniper DQ24 Digital PCR System (Sniper, Suzhou, Jiangsu, China). This ddPCR assay, incorporating four distinct fluorophores (FAM, Cy5, ROX, and VIC) (one CpG site corresponding to two fluorescent clusters), enabled the absolute quantification of CpG sites in genes ASPA, IGSF11, MEIS1-AS3, and COL1A1. As in multiplexed dPCR assays, different target sequences can be coded with different fluorescent colors, but also with different intensities. Every single assay optimized a combination of primer and probe concentrations for different sets of integrated assays, with two probes (for methylated and nonmethylated sequences) targeting two different CpG sites in generated water-in-oil nanodroplets (Zhou et al., 2024). In blood samples of Chinese unrelated donors, aged 20–100 years old, the authors obtained correlation coefficient squared values of 0.4878 (ASPA), 0.4832 (IGSF1), 0.6881 (COL1A1), and 0.6475

(MEIS1-AS3), with the testing set exhibiting a mean absolute error of 4.5923 years (Table 13.1), indicating the robustness and accuracy of the APM obtained in this alternative multiplexing strategy. Therefore, the results demonstrate the utility of multiplex ddPCR in DNAm detection, making it a valuable preliminary step for developing novel multiplex ddPCR approaches in age estimation through DNAm.

13.5 Conclusion

The field of forensic genetics is continuously evolving, marked by the development of new methodologies and laboratory techniques aimed at enhancing the efficiency, accuracy, and speed of DNA analysis. One notable advancement is ddPCR, an advanced third-generation PCR technology that allows for the absolute quantification of nucleic acids without the need for a standard curve. In this method, a sample is divided into thousands of tiny microdroplets, each containing one or a few copies of DNA molecules. These microdroplets are then subjected to PCR amplification and analyzed individually. This innovative ddPCR methodology has become an essential tool in forensic research, effectively used for analyzing both DNA and RNA samples in various forensic contexts. Recently, the ddPCR technology has been explored for the evaluation of DNAm. As a key indicator of aging, DNAm has been extensively examined in recent years for developing age estimation models using various technologies. Several studies have examined the use of bisulfite ddPCR technology for assessing DNAm to predict age, yielding accurate results. A significant advancement in ddPCR would be the expansion of single duplex assays to multiplex systems, which can differentiate targets based on varying fluorescence amplitudes or by utilizing different probe concentrations. This development would greatly enhance the ddPCR technique, providing improvements in cost-effectiveness, efficiency, and labor requirements.

References

Baccino, E., & Schmitt, A. (2006). Determination of adult age at death in the forensic context. In A. Schmitt, E. Cunga, & J. Pinheiro (Eds.). *Forensic anthropology and medicine: Complementary sciences from recovery to cause of death.* Totiwa, NJ: Humana Press.

Bae, H. J., Jeong, K. S., Kim, J. E., Hwang, E. M., & Yoo, S. Y. (2025). Beyond the mix: Advancements in simultaneous detection and quantification of human, dog, and cat DNA. *Journal of Forensic Sciences, 70*(1), 38–48. https://doi.org/10.1111/1556-4029.15644.

Bauer, M. (2007). RNA in forensic science. *Forensic Science International. Genetics, 1*(1), 69–74. https://doi.org/10.1016/j.fsigen.2006.11.002.

Castagnola, M. J., Medina-Paz, F., & Zapico, S. C. (2024). Uncovering forensic evidence: A path to age estimation through DNA methylation. *International Journal of Molecular Sciences, 25*(9), 4917. https://doi.org/10.3390/ijms25094917.

Correia Dias, H., Cunha, E., Corte Real, F., & Manco, L. (2022). Challenges and (un)certainties for DNAm age estimation in future. *Forensic Sciences, 2*(3), 601–614. https://doi.org/10.3390/forensicsci2030044.

D'Aversa, E., Breveglieri, G., Pellegatti, P., et al. (2018). Non-invasive fetal sex diagnosis in plasma of early weeks pregnants using droplet digital PCR. *Molecular Medicine, 24*, 14. https://doi.org/10.1186/s10020-018-0016-7.

Dass, M., & Ghai, M. (2024). Development of a multiplex PCR assay and quantification of microbial markers by ddPCR for identification of saliva and vaginal fluid. *Forensic Science International, 362*, 112147. https://doi.org/10.1016/j.forsciint.2024.112147.

Declerck, K., & Vanden Berghe, W. (2018). Back to the future: Epigenetic clock plasticity towards healthy aging. *Mechanisms of Ageing and Development, 174*, 18–29. https://doi.org/10.1016/j.mad.2018.01.002.

Dias, H. C., & Manco, L. (2024). Predicting age from blood by droplet digital PCR using a set of three DNA methylation markers. *Forensic Science International, 356*, 111950. https://doi.org/10.1016/j.forsciint.2024.111950.

Danenberg, E. C. A., Kawakami, K. D., Saltz, K., Blake, L. B., Shibata, C., Danenberg, D., & Laird, P. V. (2000). P. W. MethyLight: A high-throughput assay to measure DNA methylation. *Nucleic Acids Research, 28*, E32.

Freire-Aradas, A., Phillips, C., & Lareu, M. V. (2017). Forensic individual age estimation with DNA: From initial approaches to methylation tests. *Forensic Science Review, 29*(2), 121–144.

Freire-Aradas, A., Pośpiech, E., Aliferi, A., Girón-Santamaría, L., Mosquera-Miguel, A., Pisarek, A., Ambroa-Conde, A., Phillips, C., Casares de Cal, M. A., Gómez-Tato, A., Spólnicka, M., Woźniak, A., Álvarez-Dios, J., Ballard, D., Court, D. S., Branicki, W., Carracedo, Á., & Lareu, M. V. (2020). A comparison of forensic age prediction models using data from four DNA methylation technologies. *Frontiers in Genetics, 11*, 932.

Fu, X., Li, X., Wang, Y., Xie, M., Wen, D., Tang, X., Wang, C., Jia, H., Liu, Y., Chen, S., Wang, Y., Zha, L., & Li, J. (2025). Discovery unbalanced DNA mixtures and evaluation mixing ratio via a droplet digital PCR method. *International Journal of Legal Medicine, 139*(1), 67–76. https://doi.org/10.1007/s00414-024-03306-z.

Gao, N., Huo, Y., Yu, D., Cheng, F., Wang, T., Zhang, X., Zhang, L., Hu, W., Li, J., Yuan, P., Liu, J., Wang, Y., & Yan, J. (2024). Evaluation of reverse transcription yield of RNA standards and forensic samples based on droplet digital PCR. *Biochemical and Biophysical Research Communications, 711*, 149909. https://doi.org/10.1016/j.bbrc.2024.149909.

Han, Y., Franzen, J., Stiehl, T., Gobs, M., Kuo, C. C., Nikolić, M., Hapala, J., Koop, B. E., Strathmann, K., Ritz-Timme, S., & Wagner, W. (2020). New targeted approaches for epigenetic age predictions. *BMC Biology, 18*(1), 71.

Hayashi, M., Guerrero-Preston, R., Sidransky, D., & Koch, W. M. (2015). Paired box 5 methylation detection by droplet digital PCR for ultra-sensitive deep surgical margins analysis of head and neck squamous cell carcinoma. *Cancer Prevention Research (Philadelphia, Pa.), 8*(11), 1017–1026. https://doi.org/10.1158/1940-6207.CAPR-15-0180.

Heyn, H., Li, N., Ferreira, H. J., Moran, S., Pisano, D. G., Gomez, A., Diez, J., Sanchez-Mut, J. V., Setien, F., Carmona, F. J., Puca, A. A., Sayols, S., Pujana, M. A., Serra-Musach, J., Iglesias-Platas, I., Formiga, F., Fernandez, A. F., Fraga, M. F., Heath, S. C., Valencia, A., ... Esteller, M. (2012). Distinct DNA methylomes of newborns and centenarians. *Proceedings of the National Academy of Sciences of the United States of America, 109*(26), 10522–10527. https://doi.org/10.1073/pnas.1120658109.

Hong, S. R., Jung, S. E., Lee, E. H., Shin, K. J., Yang, W. I., & Lee, H. Y. (2017). DNA methylation-based age prediction from saliva: High age predictability by combination of 7 CpG markers. *Forensic science international. Genetics, 29*, 118–125. https://doi.org/10.1016/j.fsigen.2017.04.006.

Horvath, S. (2013). DNA methylation age of human tissues and cell types. *Genome Biology, 14*(10), R115.

Husseiny, M. I., Fahmy, A., Du, W., Gu, A., Garcia, P., Ferreri, K., & Kandeel, F. (2020). Development of quantitative methylation-specific droplet digital PCR (ddMSP) for assessment of natural tregs. *Frontiers in Genetics, 11*, 300. https://doi.org/10.3389/fgene.2020.00300.

Jiang, M., Zhang, Y., Fei, J., Chang, X., Fan, W., Qian, X., Zhang, T., & Lu, D. (2010). Rapid quantification of DNA methylation by measuring relative peak heights in direct bisulfite-PCR sequencing traces. *Laboratory Investigation; A Journal of Technical Methods and Pathology, 90*(2), 282–290.

Jung, S. E., Lim, S. M., Hong, S. R., Lee, E. H., Shin, K. J., & Lee, H. Y. (2019). DNA methylation of the ELOVL2, FHL2, KLF14, C1orf132/MIR29B2C, and TRIM59 genes for age prediction from blood, saliva, and buccal swab samples. *Forensic Science International. Genetics, 38*, 1–8.

Lee, H. Y., Lee, S. D., & Shin, K. J. (2016). Forensic DNA methylation profiling from evidence material for investigative leads. *BMB Reports, 49*(7), 359–369.

Lee, H. M., Hee Hwang, J., Min Seong, K., Jin Ahn, J., Jun Kim, S., Yong Hwang, S., & Lim, S. K. (2022). Application of droplet digital PCR method for DNA methylation-based age prediction from saliva. *Legal medicine (Tokyo, Japan), 54*, 101992.

Lee, S. E., Koh, U. N., Fakhr, R., & Lim, S. K. (2024). Development of two step reverse transcription droplet digital PCR (RT-ddPCR) for simultaneous identification of saliva and semen. *Legal Medicine (Tokyo, Japan), 71*, 102536. https://doi.org/10.1016/j.legalmed.2024.102536.

Lin, Q., Wagner, W., & Zenke, M. (2013). Analysis of genome-wide DNA methylation profiles by BeadChip technology. *Methods in Molecular Biology (Clifton, N.J.), 1049*, 21–33.

Manco, L., & Dias, H. C. (2022). DNA methylation analysis of ELOVL2 gene using droplet digital PCR for age estimation purposes. *Forensic Science International, 333*, 111206.

Mitake, M., Hirano, S., & Kishino, T. (2021). Imprinting analysis by droplet digital PCR coupled with locked nucleic acid TaqMan probes. *Epigenetics: Official Journal of the DNA Methylation Society, 16*(7), 729–740. https://doi.org/10.1080/15592294.2020.1823160.

Moore, L. D., Le, T., & Fan, G. (2013). DNA methylation and its basic function. *Neuropsychopharmacology: Official Publication of the American College of Neuropsychopharmacology, 38*(1), 23–38. https://doi.org/10.1038/npp.2012.112.

Naue, J., Sänger, T., Hoefsloot, H. C. J., Lutz-Bonengel, S., Kloosterman, A. D., & Verschure, P. J. (2018). Proof of concept study of age-dependent DNA methylation markers across different tissues by massive parallel sequencing. *Forensic Science International. Genetics, 36*, 152–159.

Novroski, N. M. M., & Cihlar, J. C. (2022). Evolution of single-nucleotide polymorphism use in forensic genetics. *WIREs Forensic Science, 4*(6), e1459. https://doi.org/10.1002/wfs2.1459.

Okano, T., Tsujita, Y., Kanegane, H., et al. (2018). Droplet digital PCR-based chimerism analysis for primary immunodeficiency diseases. *Journal of Clinical Immunology, 38*, 300–306. https://doi.org/10.1007/s10875-018-0497-8.

Olmedillas-López, S., Olivera-Salazar, R., García-Arranz, M., et al. (2022). Current and emerging applications of droplet digital PCR in oncology: An updated review. *Molecular Diagnosis & Therapy, 26*, 61–87. https://doi.org/10.1007/s40291-021-00562-2.

Parson, W., Ballard, D., Budowle, B., Butler, J. M., Gettings, K. B., Gill, P., Gusmão, L., Hares, D. R., Irwin, J. A., King, J. L., Knijff, P., Morling, N., Prinz, M., Schneider, P. M., Neste, C. V., Willuweit, S., & Phillips, C. (2016). Massively parallel sequencing of forensic STRs: Considerations of the DNA commission of the International Society for Forensic Genetics (ISFG) on minimal nomenclature requirements. *Forensic Science International. Genetics, 22*, 54–63. https://doi.org/10.1016/j.fsigen.2016.01.009.

Pharo, H. D., Andresen, K., Berg, K. C. G., Lothe, R. A., Jeanmougin, M., & Lind, G. E. (2018). A robust internal control for high-precision DNA methylation analyses by droplet digital PCR. *Clinical Epigenetics, 10*, 24. https://doi.org/10.1186/s13148-018-0456-5.

Quan, P. L., Sauzade, M., & Brouzes, E. (2018). dPCR: A technology review. *Sensors (Basel, Switzerland), 18*(4), 1271. https://doi.org/10.3390/s18041271.

Saiki, R. K., Gelfand, D. H., Stoffel, S., Scharf, S. J., Higuchi, R., Horn, G. T., Mullis, K. B., & Erlich, H. A. (1988). Primer-directed enzymatic amplification of DNA with a thermostable DNA polymerase. *Science (New York, N.Y.), 239*(4839), 487–491. https://doi.org/10.1126/science.2448875.

Saiki, R. K., Scharf, S., Faloona, F., Mullis, K. B., Horn, G. T., Erlich, H. A., & Arnheim, N. (1985). Enzymatic amplification of beta-globin genomic sequences and restriction site analysis for diagnosis of sickle cell anemia. *Science (New York, N.Y.), 230*(4732), 1350–1354. https://doi.org/10.1126/science.2999980.

Santurtún, A., Riancho, J. A., Arozamena, J., et al. (2017). Indel analysis by droplet digital PCR: A sensitive method for DNA mixture detection and chimerism analysis. *International Journal of Legal Medicine, 131*, 67–72. https://doi.org/10.1007/s00414-016-1422-4.

Schneider, P. M., Prainsack, B., & Kayser, M. (2019). The use of forensic DNA phenotyping in predicting appearance and biogeographic ancestry. *Deutsches Arzteblatt International, 116*(51–52), 873–880. https://doi.org/10.3238/arztebl.2019.0873.

Shi, L., Jiang, F., Ouyang, F., Zhang, J., Wang, Z., & Shen, X. (2018). DNA methylation markers in combination with skeletal and dental ages to improve age estimation in children. *Forensic Science International. Genetics, 33*, 1–9. https://doi.org/10.1016/j.fsigen.2017.11.005.

So, M. H., Lee, J. E., & Lee, H. Y. (2024). Strategies to deal with genetic analyzer-specific DNA methylation measurements. In *Electrophoresis*. Advance online publication10.1002/elps.202300185 https://doi.org/10.1002/elps.202300185.

Stasi, A., Mir, T. U. G., Pellegrino, A., Wani, A. K., & Shukla, S. (2023). Forty years of research and development on forensic genetics: A bibliometric analysis. *Forensic Science International. Genetics, 63*, 102826. https://doi.org/10.1016/j.fsigen.2023.102826.

Suchiman, H. E. D., Slieker, R. C., Kremer, D., Slagboom, P. E., Heijmans, B. T., & Tobi, E. W. (2015). Design, measurement and processing of region-specific DNA methylation assays: The mass spectrometry-based method EpiTYPER. *Frontiers in Genetics, 6*, 1.

Tost, J., & Gut, I. G. (2007). DNA methylation analysis by pyrosequencing. *Nature Protocols, 2*, 2265–2275.

Ubelaker, D. H., Shamlou, A., & Kunkle, A. (2018). Contributions of forensic anthropology to positive scientific identification: A critical review. *Forensic Sciences Research, 4*(1), 45–50. https://doi.org/10.1080/20961790.2018.1523704.

Uehiro, N., Sato, F., Pu, F., Tanaka, S., Kawashima, M., Kawaguchi, K., Sugimoto, M., Saji, S., & Toi, M. (2016). Circulating cell-free DNA-based epigenetic assay can detect early breast cancer. *Breast Cancer Research: BCR, 18*(1), 129. https://doi.org/10.1186/s13058-016-0788-z.

Wyner, N., Barash, M., & McNevin, D. (2020). Forensic autosomal short tandem repeats and their potential association with phenotype. *Frontiers in Genetics, 11*, 884. https://doi.org/10.3389/fgene.2020.00884.

Xu, D., Zhang, W., Li, H., Li, N., & Lin, J.-M. (2023). Advances in droplet digital polymerase chain reaction on microfluidic chips. *Lab on a Chip, 23*, 1258–1278.

Yang, I., Park, I. Y., Jang, S. M., Shi, L. H., Ku, H. K., & Park, S. R. (2006). Rapid quantification of DNA methylation through dNMP analysis following bisulfite-PCR. *Nucleic Acids Research, 34*(8), e61. https://doi.org/10.1093/nar/gkl257.

Yu, M., Carter, K. T., Makar, K. W., Vickers, K., Ulrich, C. M., Schoen, R. E., Brenner, D., Markowitz, S. D., & Grady, W. M. (2015). MethyLight droplet digital PCR for detection and absolute quantification of infrequently methylated alleles. *Epigenetics: Official Journal of the DNA Methylation Society, 10*(9), 803–809. https://doi.org/10.1080/15592294.2015.1068490.

Yu, M., Heinzerling, T. J., & Grady, W. M. (2018). DNA methylation analysis using droplet digital PCR. *Methods in Molecular Biology (Clifton, N.J.), 1768*, 363–383. https://doi.org/10.1007/978-1-4939-7778-9_21.

Zhang, L., Parvin, R., Fan, Q., & Ye, F. (2022). Emerging digital PCR technology in precision medicine. *Biosensors & Bioelectronics, 211*, 114344. https://doi.org/10.1016/j.bios.2022.114344.

Zhou, Y., Wang, Y., Song, M., Jiang, L., Sun, C., Wang, S., Yao, H., Wang, Z., Wang, X., Liu, C., Luo, H., & Song, F. (2024). A high-throughput droplet digital PCR system aiming eight DNA methylation targets for age prediction. *Journal of Pharmaceutical and Biomedical Analysis, 240*, 115943.

Chapter 14

Role of nanotechnology in forensic DNA analysis

Kamayani Vajpayee, Vidhi Paida, and Ritesh K. Shukla
Biological and Life Sciences, School of Arts and Sciences, Ahmedabad University, Central Campus, Navrangpura, Ahmedabad, Gujarat, India

14.1 Introduction

Contemporary criminal investigations depend mainly on forensic DNA profiling to establish individualization (Alketbi, 2023; Vajpayee et al., 2020). Since its inception in 1984–85, significant progress has been achieved in improving the sensitivity, precision, and effectiveness of DNA profiling techniques (Dash et al., 2023; Haddrill, 2021). The introduction of *nanotechnology* in forensic DNA analysis has become one of the most exciting advancements, holistically transforming forensic genetics (Paikrao et al., 2022; Pandya & Shukla, 2018; Patel et al., 2018; Prasad et al., 2016; Tambo & Ablateye, 2020).

Nanotechnology, which takes advantage of distinctive characteristics exhibited by materials at the nanoscale, has long been known to revolutionize several scientific domains like information technology, medicine, agriculture, and healthcare (Fulekar, 2010; Luis et al., 2019; Mihranyan et al., 2012; Nasrollahzadeh et al., 2019; Nie et al., 2007; Shukla et al., 2021; Silva, 2004; Vázquez-Núñez et al., 2020). However, it is now being used at a preliminary stage in forensic genetics, where it brings sophisticated sample preparation, amplification, and analytical methods (Fathi-karkan et al., 2024). Forensic experts can now use nanomaterials and nanodevices, exploiting nanoscale phenomena to overcome prevalent issues in DNA analysis—complex sample matrices, damaged DNA, and limited sample quantity (Vajpayee, 2023).

This amalgamation of nanotechnology into forensic DNA analysis is thought to significantly increase the sensitivity, specificity, and reliability of DNA profiling techniques (see Fig. 14.1). DNA may be successfully extracted from small quantities of complex biological evidence, including degraded samples (Nadar et al., 2021; Singh & Samal, 2021). Nanoscale devices, like nanopore sequencers and nanofluidic platforms, provide fast, high-throughput DNA analysis with low sample requirements to satisfy the need for rapid forensic investigations (Bazyar, 2023; Bruijns et al., 2016; Shivakumar, 2024), Figs. 14.1 and 14.2.

Moreover, nanotechnology has revolutionized DNA analysis instrumentation and techniques, significantly improving data precision and productivity (Adams & Barbante, 2013; Mukhtar et al., 2021). Several nano-enabled DNA analysis techniques work significantly better than conventional ones since nanomaterials tend to enhance the detection signals (Wu et al., 2014). Gold nanoparticles (AuNPs), for instance, improve fluorescence inbound probes, making it possible to detect even low-concentration DNA targets with extreme sensitivity (Craciun et al., 2024). A microfluidic system has also vastly decreased reaction time and the required sample volume. DNA testing is now simpler and more efficient owing to these portable devices, which have also reduced the number of reagents used and hastened the entire analysis procedure (Hopwood et al., 2010).

In addition, nanotechnology facilitates the development of advanced biosensors and imaging techniques for precise DNA identification and characterization. By enabling comprehensive genetic profiling, ancestry determination, and the identification of single-nucleotide polymorphisms (SNPs) associated with phenotypic traits, these nanotechnological developments not only ramped up forensic DNA analysis processes but also improved the scope of forensic investigations (Yáñez-Sedeño et al., 2014, 2019).

The standardization of procedures, the validation of innovative techniques, and ethical and legal concerns are some of the obstacles and constraints preventing the application of nanotechnology in forensic DNA analysis (Costanzo et al., 2023). To ensure that nanotechnology is used in forensic research ethically and responsibly, these concerns must be overcome.

Investigative and Predictive DNA Testing. DOI: https://doi.org/10.1016/B978-0-443-30128-5.00023-0

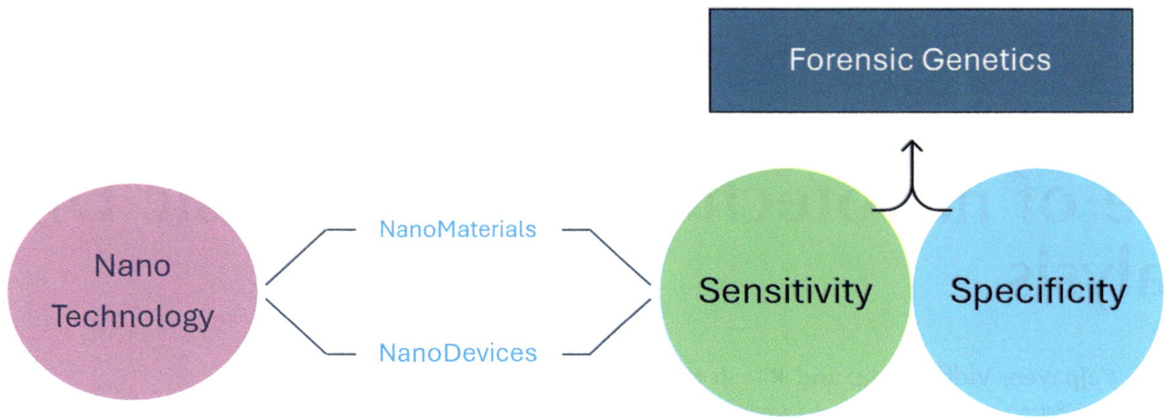

FIGURE 14.1 Role of nanotechnology in forensic genetics, which enhances the sensitivity and specificity of DNA analysis.

In this regard, the present chapter provides an extensive discussion of the application of nanotechnology in forensic DNA analysis, covering key concepts, current developments, standard procedures, challenges, and potential applications. By carefully analyzing nanotechnological advancements and their consequences for forensic science, it aims to promote forensically sound practices and explain the possible implications of nanotechnology in forensic DNA analysis.

14.2 Fundamentals of nanotechnology

The critical component of nanotechnology is *nanoparticles*. These particles are defined by their nano-scale dimensions. Metals (such as gold and silver), metal oxides (such as titanium dioxide and zinc oxide), nonmetals [carbon nanotubes (CNTs), graphene oxide (GO) nanoparticles], and organic counterparts (such as liposomes and dendrimers) are among the wide variety of nanoparticle compositions (Sajid, 2022). The complex interaction between composition, shape, and size determines the distinct properties and functions of these nanoparticles. These nanoparticles are classified based on their size and shape. This classification includes 1d structures, such as nanotubes; 2D nanosheets; 3D architectures, such as nanospheres and nanocapsules; and 0D structures, such as quantum dots (QDs) (Rizwan et al., 2021; Sannino, 2021).

Distinct categories, delineated by spatial limitations at the nanoscale, provide customized characteristics, expanding the range of potential uses in various scientific and technological fields. The full potential of nanotechnology in multiple fields, such as electronics, medicine, energy, and materials science, depends on an understanding of these principles (Ali Mazari et al., 2021; Prakash Sharma et al., 2018). Nanoparticles are highly versatile, with applications in environmental science (Ali Mazari et al., 2021; Kusiak et al., 2022; Saif et al., 2016), electronics (Payal & Pandey, 2022), and medicine (Malik et al., 2023). Their unique features (Khan et al., 2019) originate from phenomena that happen at the nanoscale are as follows (see Fig. 14.2):

Size-dependent properties: The physical, chemical, and biological properties of materials change significantly when they are reduced to nanoscale dimensions. For example, because of their high surface area-to-volume ratio and quantum confinement effects, nanoparticles may show enhanced electrical conductivity, optical properties, and reactivity (Salatin et al., 2015).

Surface effects: Since a significant fraction of atoms at the nanoscale are found at or close to the surface of nanoparticles or nanostructures, surface interactions are a significant factor in determining how nanomaterials behave. Surface modifications, like functionalization with organic or inorganic molecules, can modify the surface of nanomaterials to customize their functionality and properties for particular uses (Salatin et al., 2015).

Quantum effects: When materials approach nanometer dimensions, quantum effects become more prominent, influencing their electronic, optical, and magnetic properties. Quantum mechanics controls the behavior of particles at the nanoscale, giving rise to phenomena like quantum confinement, tunneling, and the distinct optical properties of QDs (Zhu et al., 2016).

Self-assembly: Nanoparticles can arrange themselves into intricate architectures with precise control over their composition, size, and morphology. Self-assembly processes are harnessed to fabricate nanomaterials and nanostructures

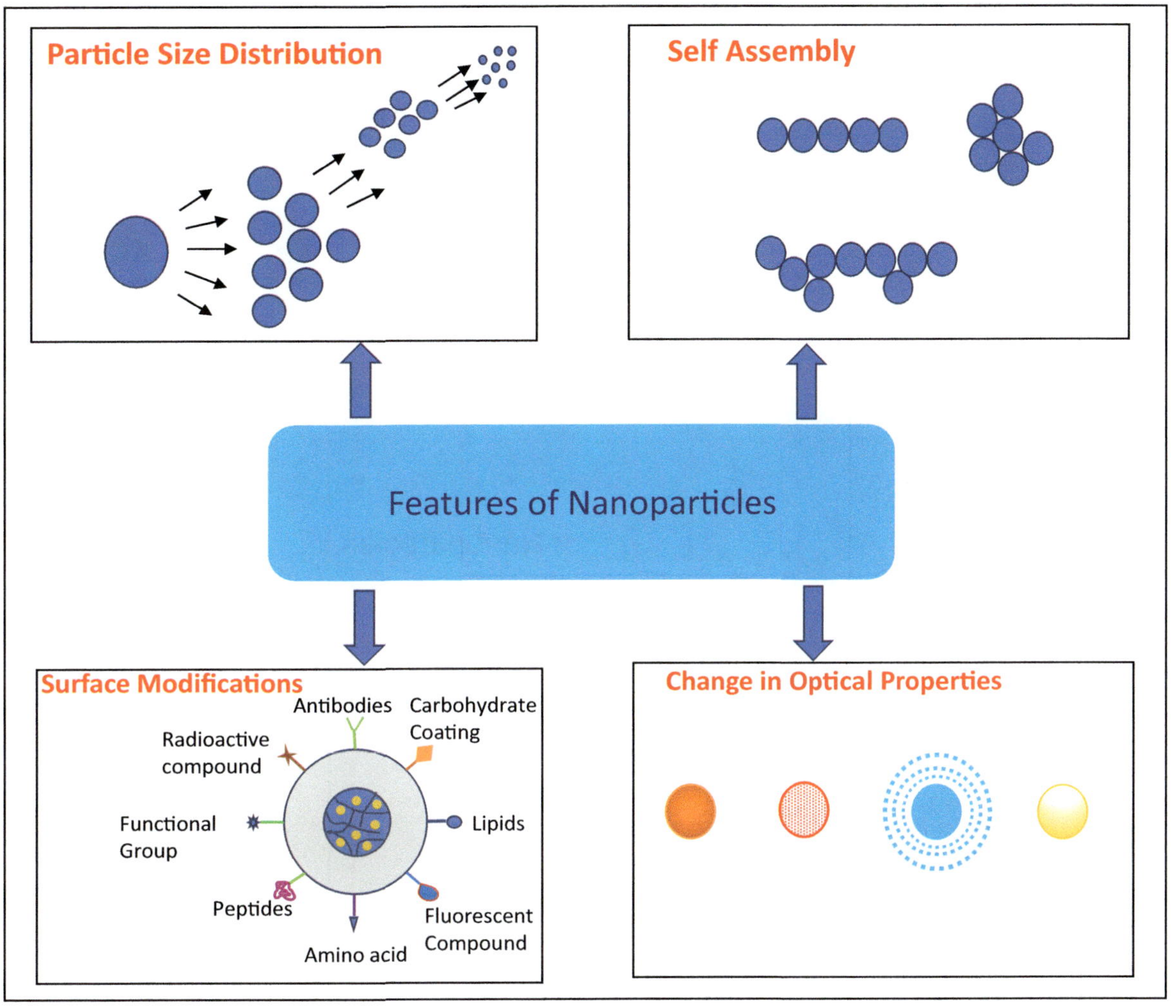

FIGURE 14.2 Features of nanoparticles.

with tailored properties for various applications. Nanotechnology takes advantage of certain materials' ability to self-assemble into ordered structures at Fig. 14.3 nanoscale (Böker et al., 2007; Genix & Oberdisse, 2018; Luo et al., 2015).

Novel fabrication techniques: A wide variety of fabrication techniques designed to work with materials at the nanoscale are included in nanotechnology. These technologies include bottom-up processes like chemical synthesis, where nanomaterials are constructed atom by atom or molecule by molecule, and top-down ways like lithography, where bulk materials are chiseled down to nanoscale levels. The production of various nanomaterials and nanostructures with particular characteristics and functions is made possible by the distinct benefits and drawbacks of each fabrication process (Bhatia, 2016; Habibi et al., 2020; Li et al., 2019; Thomas et al., 2015) (see Fig. 14.3).

These characteristics highlight their critical functions in drug delivery, imaging, and diagnosis functions (Baetke et al., 2015 Nagaraju et al., 2021; Singh and Sharma, 2023). Their functionalization increases the precision and effectiveness of therapeutic approaches by enabling tuned interventions. Nanoparticles are advantageous because of their numerous applications, enhanced reactivity, and adjustable characteristics that allow exact material qualities. However, disadvantages include potential toxicity, environmental impacts, and ethical issues with their use in biological systems. (Kiio & Park, 2021; Shukla et al., 2021). A thorough understanding and regulation are necessary to balance utilizing the benefits of nanoparticles and minimizing any potential risks.

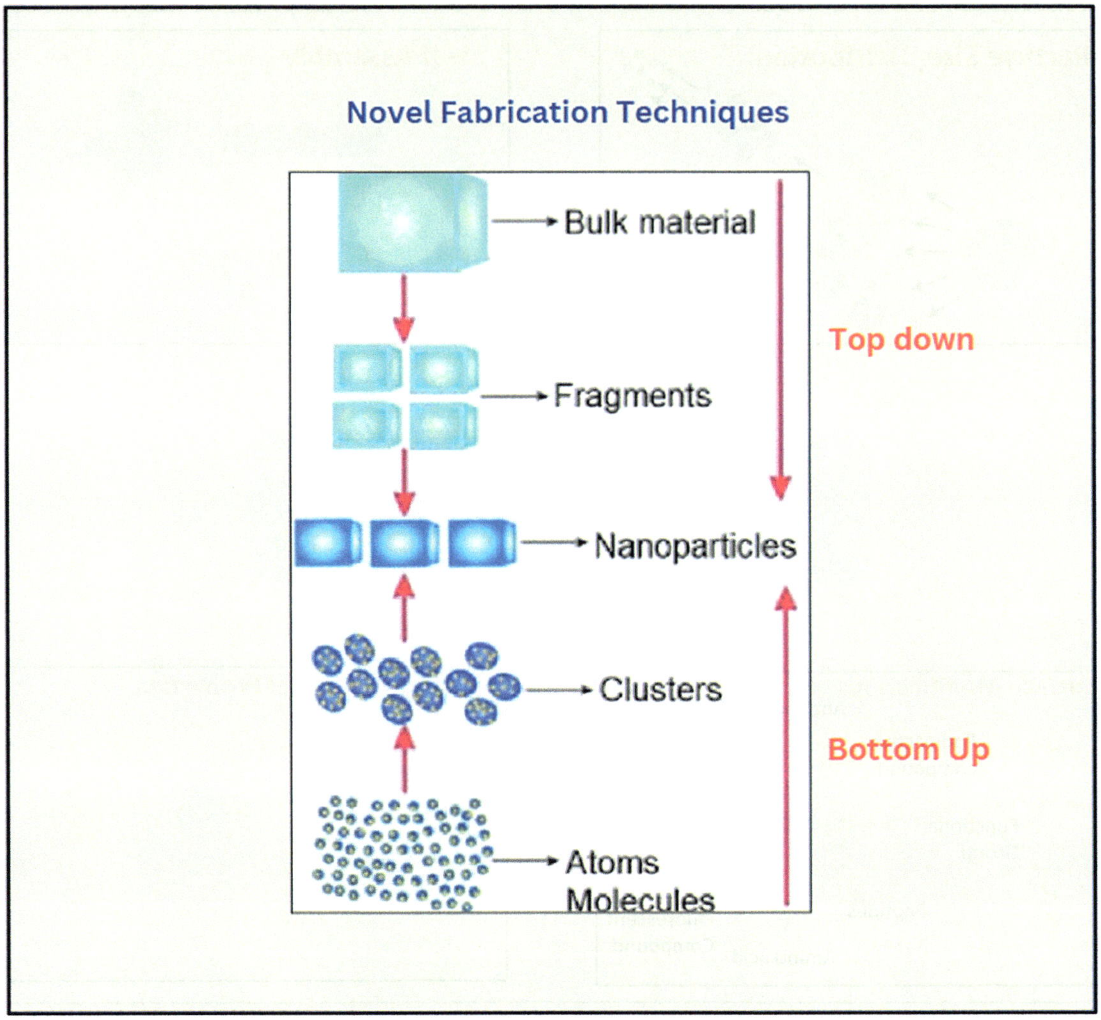

FIGURE 14.3 Fabrication techniques used for synthesizing nanoparticles.

14.3 Key nanomaterials and their properties relevant to DNA analysis

Nanomaterials make a difference in DNA analysis because of their diverse and unique characteristics, enhancing the sensitivity, specificity, and efficiency of detection methods (Mohanraj & Chen, 2006). These materials exhibit behaviors at the nanoscale that differ significantly from those of their bulk counterparts, such as increased surface area, quantum effects, and unique optical and magnetic properties (Shukla et al., 2021). These characteristics make them ideal for various technological applications of DNA and molecular biology (see Table 14.1).

One of the most commonly used nanomaterials in DNA analysis is AuNPs (Watthanapanpituck et al., 2014). Their exceptional surface area allows them to bind an extensive range of biomolecules, including DNA probes, significantly increasing detection sensitivity. Additionally, the electrons on the surface of AuNPs vibrate with incident light due to surface plasmon resonance (SPR), producing a solid scattering or absorption effect (Fong & Yung, 2013; Ghosh & Pal, 2007). This applies to spectroscopic and colorimetric methods of DNA detection. AuNPs can also be easily functionalized with DNA strands to achieve targeted hybridization with complementary sequences (Samanta & Medintz, 2016). These nanoparticles also stabilize DNA and lessen the effects of PCR inhibitors, which increases the efficiency of PCR (Kaur & Sharma, 2022; Vajpayee, 2023).

Silver nanoparticles (AgNPs) are helpful for optical signal-based DNA detection devices since they have plasmonic solid properties like gold. However, the antibacterial properties of AgNPs are particularly well-known. (Tufail & Liaqat, 2022),

TABLE 14.1 Table enlisting the key nanomaterials and their properties essential to Forensic DNA Profiling.

S. No.	Nanomaterial type	Properties essential to DNA analysis	References
1	Gold nanoparticles	Strong Plasmonic properties, thermal conductivity, SPR property, and easy functionalization help in DNA detection systems and facilitate PCR amplification.	Vajpayee (2023); Fong and Yung (2013); Ghosh and Pal (2007); Kaur and Sharma (2022)
2	Silver nanoparticles	strong plasmonic properties; Surface-Enhanced Raman Scattering; Antibacterial properties aid in optical signal-based DNA detection devices, nano-PCR	Tufail and Liaqat (2022); Marin et al. (2015)
3	Magnetic nNanoparticles	Superparamagnetic property helps in DNA isolation.	Nadar et al., 2021; Tang (2020); El-Deen and Hussain (2023)
4	Carbon nanotubes	Electrical conductivity and mechanical strength make them ideal for application in electrochemical DNA biosensors.	Tambo and Ablateye (2020); Díez-Pascual et al. (2022)
5	Graphene oxide	Fluorescence-quenching capabilities and high electrical conductivity make it a valuable component for developing electrochemical biosensors.	Liu et al. (2013); Zhang et al. (2017); Fan Li et al. (2013)
6	Silica nanoparticles	Biocompatibility, which can be used in hybridization detection	Paikrao et al., 2022; Pascual et al. (2015); Rawtani et al. (2019).
7	Nanodiamonds	Fluorescence and biocompatibility aid in labeling and imaging	Mochalin et al. (2020); Wu and Weil (2017)
8.	Quantum dots	biocompatibility and fluorescence aid in optical sensing of DNA detection, nano-PCR	Barker and Xiao (2005); Ellis et al. (2020); Fuming et al. (2015); Ma et al. (2009)

which could assist in preventing contamination during DNA analysis. They are also used in fluorescence-based DNA detection methods, which enhance fluorescent signals and make detecting even small amounts of DNA easier. Additionally, AgNPS play a crucial role in surface-enhanced Raman scattering (SERS), a technique that amplifies the Raman signal of DNA and facilitates the detection of specific sequences or structural changes in the DNA molecule (Marin et al., 2015).

Likewise, attention has been focused on the potential of *magnetic nanoparticles* (MNPs) for DNA separation and purification. Since these nanoparticles are superparamagnetic—they only become magnetic when subjected to an external magnetic field—DNA may be readily isolated from intricate biological components. Functionalized MNPs may capture target DNA sequences, and a magnetic field can separate the attached DNA to increase the yield and purity of DNA for use in forensic or diagnostic applications (El-Deen & Hussain, 2023; Nadar et al., 2021; Tang, 2020).

On the other hand, *silica nanoparticles* (SiNPs) have been commended for their biocompatibility and simplicity of surface modification. They are widely used in DNA analysis to transport and encapsulate DNA into cells—a critical step in the study of genetic modification. SiNPs can also be used in hybridization detection, where target DNA binding is measured using optical or electrochemical signals following the immobilization of DNA probes on their surface (Paikrao et al., 2022; Pascual et al., 2015; Rawtani et al., 2019).

Another exciting family of nanomaterials is CNTs, which are well-known for their remarkable electrical conductivity and mechanical strength. These features make them ideal for application in electrochemical DNA biosensors since DNA hybridization can be detected through changes in electrical impulses. Their vast surface area enables efficient immobilization of DNA probes, improving the experiment's sensitivity. CNTs have been employed in DNA delivery in addition to sensing, serving as carriers to transfer DNA into cells. (Díez-Pascual et al., 2022; Tambo & Ablateye, 2020).

Graphene and GO nanoparticles are becoming progressively more significant in DNA analysis. Although GO's fluorescence-quenching capabilities are used in DNA assays, graphene's high electrical conductivity makes it a valuable component for developing electrochemical biosensors. Fluorescence-based DNA hybridization enables sensitive detection since the quenching effect is reversed when labeled DNA binds to target areas. (Liu et al., 2013; Zhang et al., 2017; Fan Li et al., 2013).

Last but not least, nanodiamonds offer a fresh approach to DNA analysis, particularly for labeling and imaging. Their fluorescence and biocompatibility allow them to track proteins or DNA without causing damage to biological systems. Functionalized nanodiamonds can also be used in DNA biosensors, which offer a susceptible and accurate platform for detecting DNA interactions (Mochalin et al., 2020; Wu & Weil, 2017).

14.4 Nano-based techniques for DNA isolation

DNA profiling technique has been transformed by incorporating nanomaterials into forensic DNA extraction and isolation, especially for complicated and degraded forensic evidence (Fathi-karkan et al., 2024; Kawatra et al., 2018). Nanoparticles improve each stage of DNA isolation and purification—from lysis and binding to washing and elution—by reacting specifically with DNA, making the process more efficient (see Table 14.2). Compared to conventional techniques, this selective interaction results in a better yield and purity of DNA.

Cell lysis is the first step in any DNA isolation process, during which nanoparticles facilitate DNA release and cell disintegration (Islam et al., 2017). It has been demonstrated that some metal oxide nanoparticles, especially those coated in *silica*, increase the effectiveness of lysis buffers, resulting in more successful DNA extraction from challenging forensic samples such as dried blood stains, hair, and bone (He et al., 2007). When working with limited or deteriorated biological samples, this improved lysis step gives forensic investigators access to high-molecular-weight DNA, which is essential for downstream analysis.

After lysis, nanoparticles bound the isolated DNA explicitly, making it simple to separate from other cellular components (Lodha et al., 2017; Min et al., 2014). In this stage, MNPs are especially useful because they may be modified to have surfaces with strong electrostatic DNA interactions, such as silica or DNA-binding ligands. These DNA-bound nanoparticles may be efficiently separated from the solution by applying an external magnetic field, which reduces contamination and DNA loss (Saiyed et al., 2008). Furthermore, *carbon-based nanoparticles* such as *GO* and *CNTs provide significant binding affinities for DNA*, which aid in the high recovery rates of DNA isolation from intricate forensic matrices (Eivazzadeh-Keihan et al., 2022; Premkumar & Geckeler, 2011).

A critical stage in DNA extraction is removing inhibitors, which often stay with forensic evidence and impede later procedures like PCR. Nanoparticles bind and neutralize common PCR inhibitors such as melanin, humic chemicals, or metallic ions to facilitate this purification process (Vajpayee, 2023). Specifically, silica-coated nanoparticles have shown a high capacity for DNA binding while enabling efficient removal of impurities. This washing phase is achievable by nanoparticles to prepare inhibitor-free DNA for successful PCR and other forensic assays (Pan et al., 2012; Shen et al., 2009).

Following washing and purification, DNA needs to be separated from the nanoparticles in a way that can be examined forensically. Nanoparticles that release DNA in response to particular stimuli, including pH or ionic strength changes, can be used to optimize this stage (Ganta et al., 2008). Because of their fluorescence or selective binding capabilities, AuNPs and QDs can even be employed to improve DNA recovery by allowing for the exact extraction of high-quality DNA (Ellis et al., 2020; Sedighi & Krull, 2016). Targeted binding and recovery of particular DNA fragments are made possible by AuNPs with thiol-modified probes (Liu & Lämmerhofer, 2019). This is especially useful in forensic casework where the sample numbers are constrained and great precision is required.

TABLE 14.2 List of nanomaterials currently utilized in forensic DNA sample preparation methodology.

S. No	Steps in sample preparation	Nanomaterials currently being utilized	References
1.	DNA isolation/ cell lysis	Magnetic nanoparticles silica carbon-based nanoparticles, such as graphene oxide and carbon nanotubes,	Saiyed et al. (2008); Eivazzadeh-Keihan et al., 2022; Premkumar and Geckeler (2011)
2.	DNA purification	Gold nanoparticles, silver nanoparticles, zinc oxide nanoparticles, titanium oxide nanoparticles	Dun Pan et al. (2012); Shen et al. (2009)
3.	Separation of nanoparticle-bound DNA	Gold nanoparticles, quantum dots	Ellis et al. (2020); Sedighi and Krull (2016)

14.5 Nanotechnology in DNA amplification

Nanotechnology has made significant strides in DNA amplification, particularly in increasing the effectiveness of the PCR, a crucial tool in molecular biology (Pan et al., 2012). From forensic science to clinical diagnostics, the capacity of PCR to amplify small amounts of DNA to detectable levels is essential (Mullis, 1990). However, low template concentrations, the presence of inhibitors, and nonspecific amplification can all restrict the effectiveness of PCR and jeopardize the results (Vajpayee, 2023). Nanoparticles have been incorporated into the PCR reaction mix to circumvent these issues (Ali et al., 2018). The use of nanotechnology in DNA amplification is examined in the current section, with an emphasis on how nanomaterials can improve PCR sensitivity, specificity, and efficiency (see Fig. 14.4).

Implications on PCR specificity and sensitivity: In forensic DNA analysis, nanomaterials provide groundbreaking advances in modifying and improving the efficiency, specificity, sensitivity, and inhibitory control of PCR. Increasing *efficiency* is one of the primary impacts of nanotechnology on PCR, mainly when template DNA is limited or inhibitors are present. Nanomaterials such as CNTs and AuNPs are crucial for enhancing PCR performance, given that they enhance heat dispersion and elevate enzyme activity. AuNPs, known for their remarkable thermal conductivity, improve the thermal dynamics of the PCR reaction mixture (Yuce et al., 2014; Shen & Zhang, 2013; Sang et al., 2017). The three thermal phases employed in PCR—denaturation, annealing, and extension—need to be carefully regulated to get the optimum results. (McDonald et al., 2024).

AuNPs enhance DNA amplification by maintaining the stability of the DNA polymerase enzyme at elevated temperatures, particularly in the presence of inhibitors. Additionally, they guarantee uniform DNA melting during denaturation and improve primer binding during annealing by distributing heat throughout the reaction mixture (Karunanathie et al., 2022). Even with minimal template DNA, CNTs' enormous surface area helps stabilize DNA, improve polymerase-template interaction, and maintain ionic balance, increasing PCR efficiency (Cui et al., 2004; Yang et al., 2022).

Nanotechnology also dramatically improves PCR's sensitivity and specificity, which is essential for precisely amplifying only the target DNA sequence (Pelt-Verkuil & Witt, 2019). Primers or probes can be used to functionalize nanomaterials such as QDs and SiNPs, increasing selectivity by focusing on the desired DNA sequence (Pelt-Verkuil & Witt, 2019). For example, QDs ensure selective binding due to their fluorescent properties, while SiNPs provide a firm surface for primer attachment, reducing nonspecific amplification like primer dimers (Ma et al., 2009; Zhang & Wang, 2012). In multiplex PCR, where multiple targets are amplified at once, AuNPs further stabilize primer-template interactions, which are particularly crucial (Li, 2006).

Furthermore, nanotechnology has improved sensitivity by allowing the detection of trace amounts of DNA frequently collected in forensic casework. MNPs make extracting DNA from challenging samples easier, enabling highly sensitive amplification (Nadar et al., 2021; Tang, 2020; Bai et al., 2015; Graham et al., 2005), while AgNPs improve the interface between polymerase and DNA while enhancing fluorescence signals (Tufail & Liaqat, 2022; Wan et al., 2009; Wang et al., 2007).

Implications on PCR inhibition: The inhibition of DNA amplification by environmental contaminants, hemoglobin, melanin, etc., is a significant obstacle in PCR, especially in forensic and diagnostic settings. These inhibitors could negatively impact DNA polymerase activity, leading to inadequate or inaccurate amplification results. Nanomaterials have shown promise in overcoming this challenge. For instance, AuNPs can neutralize or adsorb inhibitor molecules, protecting

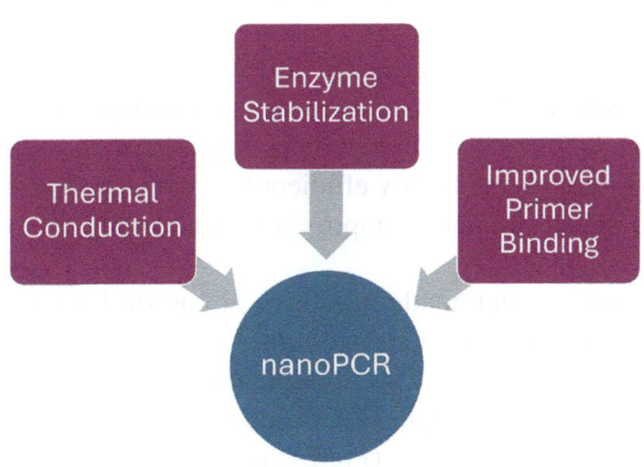

FIGURE 14.4 Implications of nano-associated polymerase chain reaction (PCR).

the polymerase enzyme and boosting the overall effectiveness of PCR (Huang et al., 2008; Li et al., 2016; Li et al., 2005; Li, 2005). Similarly, *GO* has demonstrated the capacity to adsorb these inhibitors, making successful amplification possible even in situations where other factors would generally be limiting (Gabriel et al., 2018; Jeong et al., 2015; Jia et al., 2012; Khaliq, 2012; Wang, & et al., 2017).

Nanotechnology has not only improved the traditional PCR but also stimulated the development of innovative amplification techniques like nano-PCR (Ali et al., 2018). This method goes beyond what conventional PCR can accomplish regarding efficiency, specificity, and sensitivity by directly integrating nanomaterials such as AuNPs (Lou & Zhang, 2013), MNPs (Aysan et al., 2017), and CNTs (Cui et al., 2004) into the amplification process (Pan et al., 2012; Sang et al., 2017). Nano-PCR allows quicker and more precise amplification cycles by stabilizing enzymes and improving heat conductivity. Thus, nano-PCR allows portable diagnostic tools to deliver accurate and timely results even in environments with limited resources or remote locations (Shen et al., 2009; Yang et al., 2022; Sang et al., 2017; Wang et al., 2015; Xue et al., 2021).

14.6 Nanomaterials for DNA sequencing

In order to understand next-level, more intricate genetic information for individualization, forensic geneticists often use DNA sequencing methods like next-generation sequencing (NGS). Modern sequencing technology depends on nanomaterials owing to their unique physiochemical properties. This includes developments in sample processing, sensing methods, and direct sequencing procedures.

Nanoenabled DNA sequencing methods: Nanopore sequencing is one of the most promising nanomaterial-driven DNA sequencing technologies. However, there are other similar nanomaterial-based techniques as well. The following are some crucial methods in which nanotechnology is essential:

Nanopore sequencing: Nanopore sequencing uses a membrane-embedded nanoscale pore. When a DNA strand passes through this pore, the movement of the nucleotide bases disrupts the ionic current that is passing through the pore. Due to the distinct differences in this ionic current produced by the nucleotides A, T, C, and G, the sequence can be read in real time. Both biological (such as αhemolysin) and solid-state (like graphene or silicon nitride) nanopores are commonly used. Nanomaterials like graphene, silicon-based nanostructures, and other 2D materials are used to create ultra-thin, sensitive membranes for nanopores. These materials can carefully control pore size and improve the signal-to-noise ratio, enabling high-throughput and accurate sequencing (Branton et al., 2008; Leggett & Clark, 2017; Wang et al., 2015).

The precision of sequencing can be increased by using *graphene* to detect minute changes in electrical current when DNA passes through the pore. Forensic geneticists frequently use nanopore sequencing for analyzing short tandem repeat (STR) markers (Ren et al., 2021), SNPs (Cornelis et al., 2017), mitochondrial DNA (mtDNA) analysis (Hall et al., 2020), Familial Identification, Microbial DNA analysis in case of geolocation and bioterrorism, wildlife forensics, InDels, and DNA methylation markers (de Bruin et al., 2025; Tytgat & Van Nieuwerburgh, 2023).

Sequencing based on QDs: Semiconductor nanocrystals or the QDs can be made to emit specific wavelengths of light upon stimulation. They can be attached to nucleotide bases so that an optical readout of DNA sequences can be obtained by using the light released to detect the incorporation of a particular nucleotide throughout sequencing. Due to their small size and strong fluorescence, enabling high sensitivity and multiplexing capabilities, QDs make good reporters in sequencing platforms, producing different hues for each type of nucleotide, increasing the efficiency of the sequencing process (Zhang et al., 2005; Rastgoo et al., 2018).

MNP-based sequencing: MNPs can be used to alter and purify DNA molecules, often in combination with other sequencing methods. Sometimes, magnetic fields are used to direct nanoparticles functionalized with specific DNA probes to target molecules in order to enhance DNA collection and detection. These nanoparticles reduce the duration of sample preparation and increase throughput by facilitating highly efficient DNA strand separation. Furthermore, the application of magnetic fields ensures a more controlled environment, improving the accuracy of sequencing steps (Tang, 2020; Graham et al., 2005; Li et al., 2013).

Nanowire-Based Sequencing: Nanowires, often made of silicon or carbon-based materials, can be used to detect DNA hybridization processes. When a complementary DNA strand binds itself to a probe attached to the nanowire, the electrical conductivity of the nanowire changes significantly. This modification reveals the DNA sequence. Nanowires produce incredibly sensitive electrical impulses that enable rapid DNA sequence detection. They can detect even trace amounts of the target DNA due to their high surface-to-volume ratio (Premkumar & Geckeler, 2011; Namdari et al., 2016; Myndrul & Iatsunskyi, 2019).

14.7 Nanotechnology-assisted DNA analysis instrumentation

Nanotechnology has transformed DNA analysis tools and methods, resulting in significant improvements in productivity and data accuracy (Ren et al., 2021; Cornelis et al., 2017). Since nanomaterials improve detection signals, they considerably enhance a number of DNA analysis techniques (Hall et al., 2020). For example, AuNPs enhance fluorescence in linked probes, enabling extremely sensitive detection of even low-abundance DNA targets (de Bruin et al., 2025).

Additionally, microfluidic systems utilizing nanoscale components have greatly reduced reaction times and sample volume requirements. These portable devices have made DNA testing easier and more effective, reducing the need for reagents and speeding up the entire analysis process (Tytgat & Van Nieuwerburgh, 2023). Furthermore, nanoscale components enable the simultaneous detection of several DNA targets in a single test, which is essential for applications including genetic profiling (DeLuca et al., 2020). This multiplexing ability is made possible by real-time parallel processing of DNA sequences using QDss and nanopore arrays (Tambo & Ablateye, 2020; Zareef et al., 2024; Elingarami et al., 2013). Furthermore, by allowing researchers to more accurately gather and examine complex datasets, the integration of digital technologies and nanotechnology has increased the depth of data analysis and contributed to a better knowledge of genetic material and its applications (Elingarami et al., 2013; Ullah et al., 2024).

Biosensors, a key example of nanoscale components in DNA analysis, have advanced significantly as a result of nanotechnology (Pandit, 2016). Together with biological recognition elements like proteins and nucleic acids, these devices use transducers enhanced by nanomaterials, which improve sensitivity, selectivity, and reaction times (Teles & Fonseca, 2008). For instance, observable signals are produced when AuNPs hybridize with target DNA (Watthanapanpituck et al., 2014). Similarly, due to their electrical conductivity, *CNTs* have also been utilized to enhance the electrochemical responses in field-effect transistor-based biosensors (Alabsi et al., 2020).

Microfluidic devices are enhanced by nanotechnology by increasing the binding efficiency of DNA probes via nanostructured surfaces (Wang et al., 2021). Nanochannel-based devices enable rapid separation of DNA fragments, simplifying DNA analysis and sequencing. The incorporation of nanosensors onto these chips has greatly enhanced the simultaneous readout of multiple DNA targets and enabled real-time monitoring of biochemical activities in microfluidic systems (de Roo et al., 2023).

Nanoparticles are now essential to signal amplification because they improve detection capabilities above conventional methods. For instance, *MNPs'* exceptional ability to separate and concentrate DNA from complicated materials makes them highly valuable in clinical diagnostics (Berensmeier, 2006). *AgNPs* further increase detection sensitivity by employing SERS, which enables DNA identification at incredibly low quantities (Gao et al., 2013). Meanwhile, nanofluidic devices improve DNA manipulation in limited places by offering precise control over fluids and molecules. Nanofluidic transistors enable the quick and accurate detection of DNA hybridization events by regulating the ionic current (Tegenfeldt et al., 2004).

By extending the capabilities of traditional DNA analysis, these developments in nanotechnology have created new research opportunities in the field of forensic science. As nanotechnology advances, it may be able to enhance DNA analysis significantly, making genetic tools more sophisticated, practical, and effective for future applications.

14.8 Application of nanotechnology in forensic DNA profiling

Nanotechnology can find applicability in several fields of forensic genetics and provide an advanced, more robust approach to solving complex cases (see Fig. 14.5) (Ullah et al., 2024). *Human and nonhuman STR typing* is one such field where nanotechnology improves its ability to generate genetic profiles from limited or degraded samples. The nanoparticles, such as silica, silver, or gold, selectively bind to the DNA markers, which otherwise might not be detectable due to degradation in a sample, thus making profiling sound and robust. Even in scenarios where evidence is contaminated by inhibitor molecules such as humics and melanin, nanotechnology provides scope for positive identifications based on enhanced precision of such DNA profiles (Paikrao et al., 2022; de Bruin et al., 2025; Lodha et al., 2016; Chauhan et al., 2017), Fig. 14.5

Likewise, in case of touch DNA analysis, the DNA can be bonded to and enhanced by nanomaterials due to the very small sample size, increasing the likelihood of producing a reliable genetic profile. This boosts confidence levels in identifying suspects together with things or locations because it would only be possible to link suspects in extremely complicated situations involving only trace evidence through intricate forensic analyses (Kaur & Sharma, 2022; Peng et al., 2019).

Also, nanotechnology can be helpful while studying *mtDNA analysis*. Being more abundant in cells and being retrievable even from highly degraded samples, mtDNA is especially useful in cases where nuclear DNA availability is

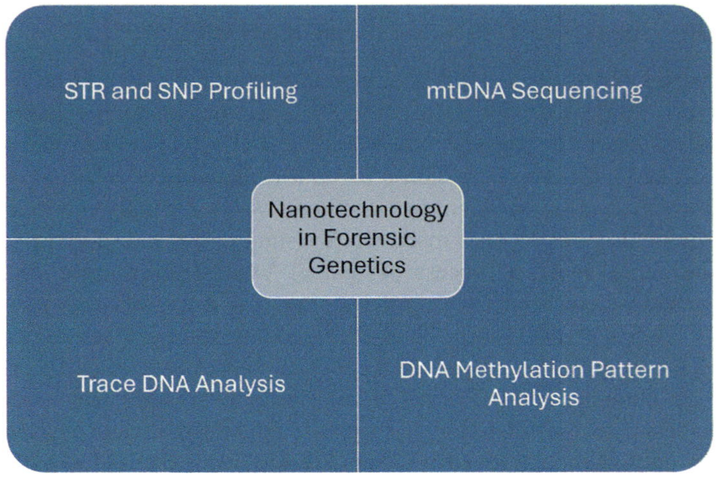

FIGURE 14.5 Domains of forensic genetics where nanotechnology can be applied for improved results.

less. By using nanotechnology, mtDNA can be selectively targeted with the least possibility of contamination to improve the maternal lineages study in ancient DNA research and the identification of missing persons and victims of disasters. Scientists can isolate and sequence the mtDNA with a higher degree of success when nanoparticles are employed if the conventional nuclear DNA profiling proves abortive (Kaur & Sharma, 2022; Tytgat & Van Nieuwerburgh, 2023; Wang et al., 2022).

Another area where nanotechnology plays a very significant role is *DNA phenotyping* (FDP), or predicting physical characteristics from genetic material. For unidentified suspects, FDP is particularly helpful in generating investigation leads. With assays based on nanotechnology, the specificity of the phenotypic markers identified to be associated with skin tone, eye color, and hair color would improve, giving a physical profile that would help identify a victim or narrow down a suspect list. Scientists, via nanotechnology, could attain helpful information from complex evidence through their ability to enhance sensitivity in the detection of genetic markers, particularly from deteriorated or contaminated samples (Tambo & Ablateye, 2020; Mukhtar et al., 2021).

Forensic epigenetics, or, for that matter, DNA methylation analysis, is advanced with nanotechnology-based applications. Nanotechnology-based sensors can detect minute changes in chemical levels; hence, they are more sensitive than conventional methods and impart an added layer of information in forensic practice. This enables forensic experts to create a more thorough profile of people that goes beyond static genetic data, giving investigations that need knowledge of a person's life history and environmental background additional depth (Kaur & Sharma, 2022; Niha, 2018).

14.9 Challenges and future perspectives

Encouragingly, nanotechnology is applied in the analysis of forensic DNA. However, there are limitations and challenges that must be adequately confronted to realize its full potential. Major concerns regarding nanoparticle-based methods are *scalability and reproducibility* (Paliwal et al., 2014; Bellan et al., 2011). Uniform production of nanoparticles with appropriate size, surface charge, and functionalization is necessary for precise DNA extraction and analysis. However, it remains very difficult to standardize the fabrication of nanoparticles, as slight changes can easily result in differences in DNA binding and recovery (Somorjai & Park, 2008). This is further complicated by the issue of *standardization* among different types of samples, such as contaminated, low-quantity, or degraded samples, since the nature of evidence in forensic cases is highly diversified (Heavey, 2022).

Since novel technologies need to be thoroughly verified to meet the standards of forensic laboratories, *regulatory compliance* for their applications is another barrier to the widespread use of nanoparticles in forensic casework (Ullah et al., 2024; Murphy, 2007). The potential *toxicity and environmental impact* of the nanoparticles are additional significant disadvantages (Sharma et al., 2009; Shukla et al., 2021, 2011a, 2011b, 2013, 2014; Vallabani et al., 2011). Even though carbon-based nanomaterials, including GO and MNPs, have appeared to be very promising for DNA extraction, long-term effects on human health and safety in the environment remain an issue. This possibility of *bioaccumulation* and possibly hazardous result may restrict the application of some of these nanoparticles, particularly in the handling and disposal of forensic materials, which should only be performed by personnel outside the laboratory (Mo et al., 2016; Rim et al., 2013).

Furthermore, some of these chemical reactions between some types of nanoparticles and DNA are likely to eventually result in *DNA degradation*, especially when specific environmental conditions are concerned (Rim et al., 2013). This possibility presents a challenge for evidence preservation because DNA samples need to remain stable and uncontaminated over long periods of time, Fig. 14.6

Future advances in the synthesis of nanoparticles and functionalization techniques may help overcome these limitations and improve the appropriateness of nanotechnology for forensic uses. New research in the area of *green synthesis techniques*, which create nanoparticles from biological agents or environmentally benign chemicals, may eliminate such environmental concerns, besides being more sustainable in the method to employ nanoparticles (Parveen et al., 2016). In addition, *biocompatible nanoparticles* may lessen toxicity risks associated with using the nanoparticles in forensic applications (Sathish Sundar et al., 2016). Other research interests also include *machine learning-driven design and microfluidic synthesis* to increase the reproducibility of nanoparticle properties and the possibility of getting a more consistent DNA separation across different sample types and laboratories (Liu et al., 2022; Fong et al., 2021).

Beyond the current challenges, the future aspects of nanotechnology can be seen in further revolutionizing the forensic analysis of DNA. A very promising approach is to *design multifunctional nanoparticles* capable of performing different tasks, such as binding to DNA, lysis of cells, and the removal of contaminants (de Dios & Díaz-García, 2010). This adaptability would significantly reduce the processing time and resources required to gain forensic DNA. One development in this area is the use of *nanosensors* to allow real-time on-site forensic DNA analysis. These sensors could provide quick investigative data without needing much lab processing because they would allow fast scanning and preliminary analysis on-site at the crime scene based on concepts like electrochemical detection or fluorescence (Abbasi et al., 2023). The *incorporation of AI and machine learning* in the study of nanoparticles facilitates the predictive prediction of interactions of nanoparticles with pollutants and DNA, thus potentially maximizing the forensic applications of DNA. These models could make it possible to alter the properties of nanoparticles for specific sample types or forensic situations, thereby improving DNA recovery and purity in challenging samples (Wang et al., 2017). QDs and other advanced nanomaterials may eventually find applications in high-resolution imaging and detection in forensic DNA analysis to provide molecular-level detail unavailable at present (Costanzo et al., 2023; Moore, 2015; Bhatia, 2022). Thus, the new developments hold promising answers to the problems of reproducibility, environmental impact, and regulatory compliance of nanotechnology in forensic DNA analysis. The accuracy, efficiency, and portability of forensic DNA analysis may be significantly improved in the future by developing safer, more dependable, and multifunctional nanoparticle systems. As these developments continue, the emerging field of nanotechnology is poised to drive a significant transformation in forensic science.

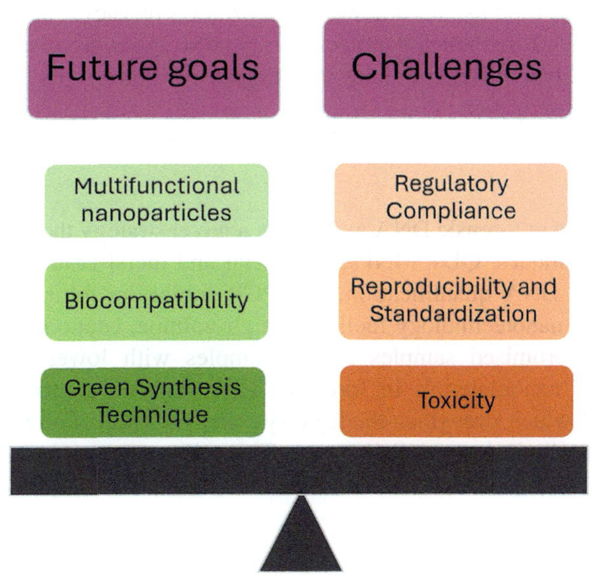

FIGURE 14.6 The challenges with adapting nanotechnology in forensic genetics and the future expectations from the technology to overcome these challenges.

14.10 Ethical and legal concerns

The application of nanotechnology in forensic DNA analysis raises numerous ethical concerns, particularly those related to *privacy, consent, and data protection.*

Data security and privacy are important ethical factors. To protect genetic data from being misused or unauthorized access, stringent guidelines regarding proper data storage are required. The continual growth of DNA databases that contain genetic information about individuals increases the chances of data breaches, resulting in the misuse or public disclosure of private genetic information.

Given the rate at which forensic nanotechnology is evolving, existing legal frameworks typically are unable to keep pace with it. A legislative gap exists as several current legislations do not translate the use of nanotechnology in forensic DNA analysis, which could result in inconsistent interpretation and application of genetic data. To balance scientific credibility, forensic science typically follows strict evidentiary criteria, and these standards need to be amended to take into consideration the special facets and challenges presented by nanotechnology-based methods. To guarantee that results are similar across labs and forensic jurisdictions, for example, improved validation and repeatability requirements in nanoparticle-assisted DNA analysis must be approved, along with standardized protocols.

By preventing any possible challenges to evidence that could affect case outcomes, these modifications would help guarantee that DNA evidence obtained using nanotechnology is accepted and consistently understood in courtrooms. The ethics and admissibility of evidence based on nanoparticles are complex since national regulations governing the handling of forensic DNA data vary greatly. The use of forensic genetic data is strictly prohibited by law in many jurisdictions, requiring strong guidelines for data protection, informed permission, and data retention restrictions. However, other jurisdictions may not have such laws, which could result in differences in the use and perception of forensic nanotechnology. The ethical and legal control of forensic DNA analysis utilizing nanotechnology might be standardized by including these methods and creating international rules.

This would stop the abuse of genetic data and further ensure consistency across jurisdictions. Concerns regarding the ecological impact of nanoparticles used in forensic techniques raise additional ethical and legal issues. Many nanoparticles, especially those containing metals or synthetic substances, can be harmful to the environment and human health if not treated appropriately, even while nanotechnology greatly improves forensic capabilities. Since the current legislation does not adequately address these challenges, guidelines for secure consumption and disposal of nanomaterials in forensic contexts must be developed. Methods for disposing of nanoparticles and reducing workplace exposure will be crucial to protecting forensic experts and reducing environmental contamination.

This will bring advances in forensic technology into compliance with environmental and ethical norms. Future legislative initiatives must appropriately handle these moral and legal issues while striking a balance between people's fundamental rights and the potential of nanotechnology in forensic science. As forensic science continues to adopt nanotechnology, it will be essential to establish a framework that protects privacy, guarantees data security, and maintains strict standards for evidence acceptance. Although nanotechnology holds great promise for the future, its ethical and legal incorporation into forensic DNA analysis will necessitate an anticipatory strategy that focuses on transparency, confidence among the public, and responsible innovation.

14.11 Conclusion

In summary, the use of nanomaterials in forensic DNA analysis is a novel strategy that improves the efficiency, sensitivity, and specificity of conventional techniques. QDs, MNPs, and metal and metal oxide nanoparticles all exhibit significant potential for sample identification, DNA quantification, and PCR inhibition mitigation. In addition to improving the accuracy of forensic investigations, nanotechnology facilitates high-volume, economical, and noninvasive testing, which helps with DNA analysis in compromised samples and/or samples with lower concentrations of DNA. As these developments proceed, the use of nanomaterials in forensic DNA analysis contributes to its improved adaptability, dependability, and reliability, all of which strengthen the legal system as a whole. Forensic applications of nanotechnology require continued research and interdisciplinary collaboration. This will guarantee both advancements in technology and adherence to forensic standards.

References

Abbasi, A. S., Rehman, R. A., Anwar, A., & Mehak, U. A. (2023). International activities in nano-forensics. In *Modeling and simulation of functional nanomaterials for forensic investigation*. Pakistan: IGI Global, 290–310. https://doi.org/10.4018/978-1-6684-8325-1.ch015.

Adams, F. C., & Barbante, C. (2013). Nanoscience, nanotechnology and spectrometry. *Spectrochimica Acta - Part B Atomic Spectroscopy, 86*, 3–13. https://doi.org/10.1016/j.sab.2013.04.008.

Alabsi, S. S., Ahmed, A. Y., Dennis, J. O., Khir, M. H. M., & Algamili, A. S. (2020). A review of carbon nanotubes field effect-based biosensors. *IEEE Access, 8*, 69509–69521. https://doi.org/10.1109/access.2020.2987204.

Ali, Z., Jin, G., Hu, Z., Wang, Z., Khan, M. A., Dai, J., & Tang, Y. (2018). A review on NanoPCR: History, mechanism and applications. *Journal of Nanoscience and Nanotechnology, 18*(12), 8029–8046. https://doi.org/10.1166/jnn.2018.16390.

Alketbi, S. K. (2023). The role of DNA in forensic science: A comprehensive review. *International Journal of Science and Research Archive, 02*, 814–829.

Aysan, A. B., Knejzlík, Z., Ulbrich, P., Šoltys, M., Zadražil, A., & Štěpánek, F. (2017). Effect of surface functionalisation on the interaction of iron oxide nanoparticles with polymerase chain reaction. *Colloids and Surfaces B: Biointerfaces, 153*, 69–76. https://doi.org/10.1016/j.colsurfb.2017.02.005, http://www.elsevier.com/locate/colsurfb.

Baetke, S. C., Lammers, T., & Kiessling, F. (2015). Applications of nanoparticles for diagnosis and therapy of cancer. *The British Journal of Radiology, 88*(1054), 20150207. https://doi.org/10.1259/bjr.20150207.

Bai, Y., Cui, Y., Paoli, G. C., Shi, C., Wang, D., & Shi, X. (2015). Nanoparticles Affect PCR primarily via surface interactions with PCR components: Using amino-modified silica-coated magnetic nanoparticles as a main model. *ACS Applied Materials & Interfaces, 7*(24), 13142–13153. https://doi.org/10.1021/am508842v.

Barker, Y., & Xiao (2005). Quantum dot-based quantitative cellular imaging for biomedicine. *Cancer Biomarkers, 1*, 233–239.

Bazyar, H. (2023). On the application of microfluidic-based technologies in forensics: A review. *Sensors (Basel, Switzerland), 23*(13), 5856. https://doi.org/10.3390/s23135856.

Bellan, L. M., Wu, D., & Langer, R. S. (2011). Current trends in nanobiosensor technology. *Wiley Interdisciplinary Reviews: Nanomedicine and Nanobiotechnology, 3*(3), 229–246. https://doi.org/10.1002/wnan.136.

Berensmeier, S. (2006). Magnetic particles for the separation and purification of nucleic acids. *Applied Microbiology and Biotechnology, 73*(3), 495–504. https://doi.org/10.1007/s00253-006-0675-0.

Bhatia, S. (2016). Nanoparticles types, classification, characterization, fabrication methods and drug delivery applications. In *Natural polymer drug delivery systems*. Cham: Springer Nature, 33–93. https://doi.org/10.1007/978-3-319-41129-3_2.

Bhatia, T. (2022). Novel nanomaterials in forensic investigations: A review. *Materials Today: Proceedings, 50*, 1071–1079.

Böker, A., He, J., Emrick, T., & Russell, T. P. (2007). Self-assembly of nanoparticles at interfaces. *Soft Matter, 3*(10), 1231–1248. https://doi.org/10.1039/b706609k.

Branton, D., Deamer, D. W., Marziali, A., Bayley, H., Benner, S. A., Butler, T., Di Ventra, M., Garaj, S., Hibbs, A., Huang, X., Jovanovich, S. B., Krstic, P. S., Lindsay, S., Ling, X. S., Mastrangelo, C. H., Meller, A., Oliver, J. S., Pershin, Y. V., Ramsey, J. M., ... Schloss, J. A. (2008). The potential and challenges of nanopore sequencing. *Nature Biotechnology, 26*(10), 1146–1153. https://doi.org/10.1038/nbt.1495.

Bruijns, B., Van Asten, A., Tiggelaar, R., & Gardeniers, H. (2016). Microfluidic devices for forensic DNA analysis: A review. *Biosensors, 6*(3), 41. https://doi.org/10.3390/bios6030041.

de Bruin, D. D. S. H., Haagmans, M. A., van der Gaag, K. J., Hoogenboom, J., Weiler., N. E. C., Tesi, N., Salazar, A., Zhang, Y., Holstege, H., Reinders, M., M'charek, A. A., Sijen, T., & Henneman, P. (2025). Exploring nanopore direct sequencing performance of forensic STRs, SNPs, InDels, and DNA methylation markers in a single assay. *Forensic Science International: Genetics, 74*, 1031154. https://doi.org/10.1016/j.fsigen.2024.103154.

Chauhan, V., Singh, V., & Tiwari, A. (2017). Applications of nanotechnology in forensic investigation. *International Journal of Life-Sciences Scientific Research, 3*(3). https://doi.org/10.21276/ijlssr.2017.3.3.13.

Cornelis, S., Gansemans, Y., Deleye, L., Deforce, D., & Van Nieuwerburgh, F. (2017). Forensic SNP genotyping using nanopore MinION sequencing. *Scientific Reports, 7*(1). https://doi.org/10.1038/srep41759.

Costanzo, H., Gooch, J., & Frascione, N. (2023). Nanomaterials for optical biosensors in forensic analysis. *Talanta, 253*, 123945. https://doi.org/10.1016/j.talanta.2022.123945.

Craciun, A.-M., Astilean, S., Focsan, M., & Lamy de la Chapelle, M. (2024). Gold nanoparticles conjugated with fluorophore-labeled DNA: Overview of sensing and imaging applications. *TrAC Trends in Analytical Chemistry, 180*, 117913. https://doi.org/10.1016/j.trac.2024.117913.

Cui, D., Tian, F., Kong, Y., Titushikin, I., & Gao, H. (2004). Effects of single-walled carbon nanotubes on the polymerase chain reaction. *Nanotechnology, 15*(1), 154–157. https://doi.org/10.1088/0957-4484/15/1/030.

Dash, H. R., Elkins, K. M., & Al-Snan, N. R. (2023). *Advancements in forensic DNA analysis advancements in forensic DNA analysis*. India: Springer Nature, 1–156. https://doi.org/10.1007/978-981-99-6195-5.

DeLuca, M., Shi, Z., Castro, C. E., & Arya, G. (2020). Dynamic DNA nanotechnology: Toward functional nanoscale devices. *Nanoscale Horizons, 5*(2), 182–201. https://doi.org/10.1039/c9nh00529c.

Díez-Pascual, A. M., Cruz, D. L., & Redondo, A. L. (2022). Advanced carbon-based polymeric nanocomposites for forensic analysis. *Polymers, 14*(17). https://doi.org/10.3390/polym14173598.

de Dios, A. S., & Díaz-García, M. E. (2010). Multifunctional nanoparticles: Analytical prospects. *Analytica Chimica Acta, 666*(1-2), 1–22. https://doi.org/10.1016/j.aca.2010.03.038.

Eivazzadeh-Keihan, R., Bahojb Noruzi, E., Chidar, E., Jafari, M., Davoodi, F., Kashtiaray, A., Ghafori Gorab, M., Masoud Hashemi, S., Javanshir, S., Ahangari Cohan, R., Maleki, A., & Mahdavi, M. (2022). Applications of carbon-based conductive nanomaterials in biosensors. *Chemical Engineering Journal, 442*, 136183. https://doi.org/10.1016/j.cej.2022.136183.

El-Deen, A. K., & Hussain, C. M. (2023). Magnetic analytical extractions of forensic samples: Latest developments and future perspectives. *Trends in Environmental Analytical Chemistry, 39*, e00209. https://doi.org/10.1016/j.teac.2023.e00209.

Elingarami, S., Li, X., & He, N. (2013). Applications of nanotechnology, next generation sequencing and microarrays in biomedical research. *Journal of Nanoscience and Nanotechnology, 13*(7), 4539–4551. https://doi.org/10.1166/jnn.2013.7522.

Ellis, G. A., Dean, S. N., Walper, S. A., & Medintz, I. L. (2020). Quantum dots and gold nanoparticles as scaffolds for enzymatic enhancement: Recent advances and the influence of nanoparticle size. *Catalysts, 10*(1), 83. https://doi.org/10.3390/catal10010083.

Fathi-karkan, S., Easwaran, Easwaran C., Kharaba, Z., Rahdar, A., & Pandey, S. (2024). Unlocking mysteries: The cutting-edge fusion of nanotechnology and forensic science. *BioNanoScience, 14*(3), 3572–3598. https://doi.org/10.1007/s12668-024-01542-6.

Fong, A. Y., Pellouchoud, L., Davidson, M., Walroth, R. C., Church, C., Tcareva, E., Wu, L., Peterson, K., Meredig, B., & Tassone, C. J. (2021). Utilization of machine learning to accelerate colloidal synthesis and discovery. *Journal of Chemical Physics, 154*(22). https://doi.org/10.1063/5.0047385.

Fong, K. E., & Yung, L. Y. L. (2013). Localized surface plasmon resonance: A unique property of plasmonic nanoparticles for nucleic acid detection. *Nanoscale, 5*(24), 12043–12071. https://doi.org/10.1039/c3nr02257a.

Fulekar, M. (2010). *Nanotechnology: Importance and applications*. IK International Pvt Ltd.

Fuming, S., Yang, Y., Hexiang, Z., Meirong, M., & Zhizhou, Z. (2015). CdTe quantum dots accelerate the speed of Pfu-based polymerase chain reaction. *Journal of Experimental Nanoscience, 10*(6), 476–482. https://doi.org/10.1080/17458080.2013.843208, http://www.tandf.co.uk/journals/titles/17458080.asp.

Gabriel, S., Rasheed, A. K., Siddiqui, R., Appaturi, J. N., Fen, L. B., & Khan, N. A. (2018). Development of nanoparticle-assisted PCR assay in the rapid detection of brain-eating amoebae. *Parasitology Research, 117*(6), 1801–1811. https://doi.org/10.1007/s00436-018-5864-0.

Ganta, S., Devalapally, H., Shahiwala, A., & Amiji, M. (2008). A review of stimuli-responsive nanocarriers for drug and gene delivery. *Journal of Controlled Release, 126*(3), 187–204. https://doi.org/10.1016/j.jconrel.2007.12.017.

Gao, F., Lei, J., & Ju, H. (2013). Label-free surface-enhanced raman spectroscopy for sensitive DNA detection by DNA-mediated silver nanoparticle growth. *Analytical Chemistry, 85*(24), 11788–11793. https://doi.org/10.1021/ac4032109.

Genix, A. C., & Oberdisse, J. (2018). Nanoparticle self-assembly: From interactions in suspension to polymer nanocomposites. *Soft Matter, 14*(25), 5161–5179. https://doi.org/10.1039/c8sm00430g, http://pubs.rsc.org/en/journals/journal/sm.

Ghosh, S. K., & Pal, T. (2007). Interparticle coupling effect on the surface plasmon resonance of gold nanoparticles: From theory to applications. *Chemical Reviews, 107*(11), 4797–4862. https://doi.org/10.1021/cr0680282.

Graham, D. L., Ferreira, H. A., Feliciano, N., Freitas, P. P., Clarke, L. A., & Amaral, M. D. (2005). Magnetic field-assisted DNA hybridisation and simultaneous detection using micron-sized spin-valve sensors and magnetic nanoparticles. *Sensors and Actuators B: Chemical, 107*(2), 936–944. https://doi.org/10.1016/j.snb.2004.12.071.

Habibi, N., Quevedo, D. F., Gregory, J. V., & Lahann, J. (2020). Emerging methods in therapeutics using multifunctional nanoparticles. *WIREs Nanomedicine and Nanobiotechnology, 12*(4). https://doi.org/10.1002/wnan.1625.

Haddrill, P. R. (2021). Developments in forensic DNA analysis. *Emerging Topics in Life Sciences, 5*(3), 381–393. https://doi.org/10.1042/ETLS20200304, https://portlandpress.com/emergtoplifesci/article/5/3/381/228192/Developments-in-forensic-DNA-analysis.

Hall, C. L., Zascavage, R. R., Sedlazeck, F. J., & Planz, J. V. (2020). Potential applications of nanopore sequencing for forensic analysis. *Forensic Science Review, 32*(1), 23–54. http://www.forensicsciencereview.com.

He, X., Huo, H., Wang, K., Tan, W., Gong, P., & Ge, J. (2007). Plasmid DNA isolation using amino-silica coated magnetic nanoparticles (ASMNPs). *Talanta, 73*(4), 764–769. https://doi.org/10.1016/j.talanta.2007.04.056.

Heavey, A. L. (2022). Toward a common language for quality issues in forensic science. *Wiley Interdisciplinary Reviews: Forensic Science, 4*.

Hopwood, A. J., Hurth, C., Yang, J., Cai, Z., Moran, N., Lee-Edghill, J. G., Nordquist, A., Lenigk, R., Estes, M. D., Haley, J. P., McAlister, C. R., Chen, X., Brooks, C., Smith, S., Elliott, K., Koumi, P., Zenhausern, F., & Tully, G. (2010). Integrated microfluidic system for rapid forensic DNA analysis: Sample collection to DNA profile. *Analytical Chemistry, 82*(16), 6991–6999. https://doi.org/10.1021/ac101355r.

Huang, S.-H., Yang, T.-C., Tsai, M.-H., Tsai, I.-S., Lu, H.-C., Chuang, P.-H., Wan, L., Lin, Y.-J., Lai, C.-H., & Lin, C.-W. (2008). Gold nanoparticle-based RT-PCR and real-time quantitative RT-PCR assays for detection of Japanese encephalitis virus. *Nanotechnology, 19*(40), 405101. https://doi.org/10.1088/0957-4484/19/40/405101.

Islam, M. S., Aryasomayajula, A., & Selvaganapathy, P. R. (2017). A review on macroscale and microscale cell lysis methods. *Micromachines, 8*(3). https://doi.org/10.3390/mi8030083.

Jeong, H. Y., Baek, S. H., Chang, S. J., Yang, M., Lee, S. J., Lee, K. G., & Park, T. J. (2015). A hybrid composite of gold and graphene oxide as a PCR enhancer. *RSC Advances, 5*(113), 93117–93121. https://doi.org/10.1039/c5ra12932j, http://pubs.rsc.org/en/journals/journalissues.

Jia, J., Sun, L., Hu, N., Huang, G., & Weng, J. (2012). Graphene enhances the specificity of the polymerase chain reaction. *Small (Weinheim an der Bergstrasse, Germany), 8*(13), 2011–2015. https://doi.org/10.1002/smll.201200139.

Karunanathie, H., Kee, P. S., Ng, S. F., Kennedy, M. A., & Chua, E. W. (2022). PCR enhancers: Types, mechanisms, and applications in long-range PCR. *Biochimie, 197*, 130–143. https://doi.org/10.1016/j.biochi.2022.02.009.

Kaur, L., & Sharma, S. G. (2022). Forensic DNA analysis: A powerful investigative tool. In *Crime scene management within forensic science: Forensic techniques for criminal investigations*. India: Springer Nature, 1–40. https://doi.org/10.1007/978-981-16-6683-4_1.

Kawatra, R., Dhanda, S., Singh, J., Kumar, A., & Kumar, R. (2018). Role of nanotechnology in advancement, automation and miniaturization of DNA fingerprinting. *Annals of Agri Bio Research, 23*(1), 11–20. http://agribiop.com/category/annals-of-agri-bio-research/.

Khaliq, A. (2012). Enhancing the efficiency of polymerase chain reaction using graphene nanoflakes. *Nanotechnology, 23*(45).

Khan, I., Saeed, K., & Khan, I. (2019). Nanoparticles: Properties, applications and toxicities. *Arabian Journal of Chemistry, 12*(7), 908–931. https://doi.org/10.1016/j.arabjc.2017.05.011, http://colleges.ksu.edu.sa/Arabic%20Colleges/CollegeOfScience/ChemicalDept/AJC/default.aspx (ScienceDirecthttp://www.sciencedirect.com/science/journal/18785352).

Kiio, T. M., & Park, S. (2021). Physical properties of nanoparticles do matter. *Journal of Pharmaceutical Investigation, 51*(1), 35–51. https://doi.org/10.1007/s40005-020-00504-w, http://www.springer.com/biomed/journal/40005.

Kusiak, M., Oleszczuk, P., & Jośko, I. (2022). Cross-examination of engineered nanomaterials in crop production: Application and related implications. *Journal of Hazardous Materials, 424*. https://doi.org/10.1016/j.jhazmat.2021.127374, http://www.elsevier.com/locate/jhazmat.

Leggett, R. M., & Clark, M. D. (2017). A world of opportunities with nanopore sequencing. *Journal of Experimental Botany, 68*(20), 5419–5429. https://doi.org/10.1093/jxb/erx289, http://jxb.oxfordjournals.org/.

Li, A., Zhou, B., Alves, C. S., Xu, B., Guo, R., Shi, X., & Cao, X. (2016). Mechanistic studies of enhanced PCR using PEGylated PEI-entrapped gold nanoparticles. *ACS Applied Materials & Interfaces, 8*(39), 25808–25817. https://doi.org/10.1021/acsami.6b09310.

Li, F., Pei, H., Wang, L., Lu, J., Gao, J., Jiang, B., Zhao, X., & Fan, C. (2013). Nanomaterial-based fluorescent DNA analysis: A comparative study of the quenching effects of graphene oxide, carbon nanotubes, and gold nanoparticles. *Advanced Functional Materials, 23*(33), 4140–4148. https://doi.org/10.1002/adfm.201203816.

Li, H., Huang, J., Lv, J., An, H., Zhang, X., Zhang, Z., Fan, C., & Hu, J. (2005). Nanoparticle PCR: Nanogold-assisted PCR with enhanced specificity. *Angewandte Chemie International Edition, 44*(32), 5100–5103. https://doi.org/10.1002/anie.200500403.

Li, M. (2005). Enhancing the efficiency of a PCR using gold nanoparticles. *Nucleic Acids Research, 33*(21), e184. https://doi.org/10.1093/nar/gni183.

Li, M. (2006). Enhancing the efficiency of a polymerase chain reaction using gold nanoparticles. *Nucleic Acids Research, 34*(1), 396. https://doi.org/10.1093/nar/gkj457 –396.

Li, N., Shang, Y., Han, Z., Wang, T., Wang, Z.-G., & Ding, B. (2019). Fabrication of metal nanostructures on DNA templates. *ACS Applied Materials & Interfaces, 11*(15), 13835–13852. https://doi.org/10.1021/acsami.8b16194.

Li, S., Liu, H., Jia, Y., Mou, X., Deng, Y., Lin, L., Liu, B., & He, N. (2013). An automatic high-throughput single nucleotide polymorphism genotyping approach based on universal tagged arrays and magnetic nanoparticles. *Journal of Biomedical Nanotechnology, 9*(4), 689–698. https://doi.org/10.1166/jbn.2013.1568.

Liu, B., Sun, Z., Zhang, X., & Liu, J. (2013). Mechanisms of DNA sensing on graphene oxide. *Analytical Chemistry, 85*(16), 7987–7993. https://doi.org/10.1021/ac401845p.

Liu, S., & Lämmerhofer, M. (2019). Functionalized gold nanoparticles for sample preparation: A review. *Electrophoresis, 40*(18-19), 2438–2461. https://doi.org/10.1002/elps.201900111, http://onlinelibrary.wiley.com/journal/10.1002/.

Liu, Y., Li, S., & Liu, Y. (2022). Machine Learning-driven multiobjective optimization: An opportunity of microfluidic platforms applied in cancer research. *Cells, 11*(5).

Lodha, A., Pandya, A., & Shukla, R. (2016). Nanotechnology: An applied and robust approach for forensic investigation. *Forensic Res Criminol Int J, 2*(1).

Lodha, A., Ansari, N., Shah, S., Rao, M. V., & Menon, S. K. (2017). Isolation of PCR ready-human DNA using copper nanoparticles from skeletal remains. *Forensic Science International, 270*, 146–152. https://doi.org/10.1016/j.forsciint.2016.12.003.

Lou, X., & Zhang, Y. (2013). Mechanism studies on nanoPCR and applications of gold nanoparticles in genetic analysis. *ACS Applied Materials and Interfaces, 5*(13), 6276–6284. https://doi.org/10.1021/am4013209.

Luis, A. I. S., Campos, E. V. R., de Oliveira, J. L., & Fraceto, L. F. (2019). Trends in aquaculture sciences: From now to use of nanotechnology for disease control. *Reviews in Aquaculture, 11*(1), 119–132. https://doi.org/10.1111/raq.12229, http://onlinelibrary.wiley.com/journal/10.1111/.

Luo, D., Yan, C., & Wang, T. (2015). Interparticle forces underlying nanoparticle self-assemblies. *Small, 11*(45), 5984–6008. https://doi.org/10.1002/smll.201501783, http://onlinelibrary.wiley.com/journal/10.1002/.

Ma, L., He, S., Huang, J., Cao, L., Yang, F., & Li, L. (2009). Maximizing specificity and yield of PCR by the quantum dot itself rather than property of the quantum dot surface. *Biochimie, 91*(8), 969–973. https://doi.org/10.1016/j.biochi.2009.04.020.

Malik, S., Muhammad, K., & Waheed, Y. (2023). Emerging applications of nanotechnology in healthcare and medicine. *Molecules, 28*(18). https://doi.org/10.3390/molecules28186624, http://www.mdpi.com/journal/molecules.

Marin, S., Vlasceanu, G., Tiplea, R., Bucur, I., Lemnaru, M., Marin, M., & Grumezescu, A. (2015). Applications and toxicity of silver nanoparticles: A recent review. *Current Topics in Medicinal Chemistry, 15*(16), 1596–1604. https://doi.org/10.2174/1568026615666150414142209.

Ali Mazari, S., Ali, E., Abro, R., Ahmed Khan, F. S., Ahmed, I., Ahmed, M., Nizamuddin, S., Siddiqui, T. H., Hossain, N., Mubarak, N. M., & Shah, A. (2021). Nanomaterials: Applications, waste-handling, environmental toxicities, and future challenges – A review. *Journal of Environmental Chemical Engineering, 9*(2), 105028. https://doi.org/10.1016/j.jece.2021.105028.

McDonald, C., Taylor, D., & Linacre, A. (2024). PCR in forensic science: A critical review. *Genes, 15*(4).

Mihranyan, A., Ferraz, N., & Strømme, M. (2012). Current status and future prospects of nanotechnology in cosmetics. *Progress in Materials Science, 57*(5), 875–910. https://doi.org/10.1016/j.pmatsci.2011.10.001.

Min, J. H., Woo, M. K., Yoon, H. Y., Jang, J. W., Wu, J. H., Lim, C. S., & Kim, Y. K. (2014). Isolation of DNA using magnetic nanoparticles coated with dimercaptosuccinic acid. *Analytical Biochemistry, 447*(1), 114–118. https://doi.org/10.1016/j.ab.2013.11.018, http://www.elsevier.com/inca/publications/store/6/2/2/7/8/1/index.htt.

Mo, Y., Gu, A., Tollerud, D. J., & Zhang, Q. (2016). *Nanoparticle toxicity and environmental impact oxidative stress and biomaterials*. United States: Elsevier, 117–143. https://doi.org/10.1016/B978-0-12-803269-5.00005-X, http://www.sciencedirect.com/science/book/9780128032695.

Mochalin, V., Shenderova, O., Ho, D., & Gogotsi, Y. (2020). The properties and applications of nanodiamonds. In *Nano-enabled medical applications*. London: Informa, 313–350. https://doi.org/10.1201/9780429399039-11.

Mohanraj, V., & Chen, Y. (2006). Nanoparticles-A review. *Tropical Journal of Pharmaceutical Research, 5*(1), 561–573. https://doi.org/10.4314/tjpr.v5i1.14634.

Moore, J. R. (2015). *Expanding the use of RNA in forensic science: A quantum dot molecular beacon approach*.

Mukhtar, M., Sargazi, S., Barani, M., Madry, H., Rahdar, A., & Cucchiarini, M. (2021). Application of nanotechnology for sensitive detection of low-abundance single-nucleotide variations in genomic DNA: A review. *Nanomaterials, 11*(6), 1384. https://doi.org/10.3390/nano11061384.

Mullis, K. B. (1990). The unusual origin of the polymerase chain reaction. *Scientific American, 262*(4), 56–65. https://doi.org/10.1038/scientificamerican0490-56.

Murphy, E. (2007). The new forensics: Criminal justice, false certainty, and the second generation of scientific evidence. *California Law Review, 95*(3), 721–797.

Myndrul, V., & Iatsunskyi, I. (2019). Nanosilicon-based composites for (Bio) sensing applications: Current status, advantages, and perspectives. *Materials, 12*(18).

Nadar, S. S., Kelkar, R. K., Pise, P. V., Patil, N. P., Patil, S. P., Chaubal-Durve, N. S., Bhange, V. P., Tiwari, M. S., & Patil, P. D. (2021). The untapped potential of magnetic nanoparticles for forensic investigations: A comprehensive review. *Talanta, 230*, 122297. https://doi.org/10.1016/j.talanta.2021.122297.

Nagaraju, G. P., Srivani, G., Dariya, B., Chalikonda, G., Farran, B., Behera, S. K., Alam, A., & Kamal, M. A. (2021). Nanoparticles guided drug delivery and imaging in gastric cancer. *Seminars in Cancer Biology, 69*, 69–76. https://doi.org/10.1016/j.semcancer.2020.01.006, http://www.elsevier.com/inca/publications/store/6/2/2/9/4/3/index.htt.

Namdari, P., Daraee, H., & Eatemadi, A. (2016). Recent advances in silicon nanowire biosensors: Synthesis methods, properties, and applications. *Nanoscale Research Letters, 11*(1). https://doi.org/10.1186/s11671-016-1618-z, http://www.springer.com/materials/nanotechnology/journal/11671.

Nasrollahzadeh, M., Sajadi, S. M., Sajjadi, M., & Issaabadi, Z. (2019). Applications of nanotechnology in daily life. *Interface Science and Technology, 28*, 113–143. https://doi.org/10.1016/B978-0-12-813586-0.00004-3, http://www.elsevier.com/wps/find/bookdescription.cws_home/BS_3710/description#description.

Nie, S., Xing, Y., Kim, G. J., & Simons, J. W. (2007). Nanotechnology applications in cancer. *Annual Review of Biomedical Engineering, 9*(1), 257–288. https://doi.org/10.1146/annurev.bioeng.9.060906.152025.

Niha, A. (2018). Forensic nanotechnology in forensic genetics. *Peer Review Journal of Forensic & Genetics Science, 1*(2). https://doi.org/10.32474/PRJFGS.2018.01.000107.

Paikrao, H. M., Tajane, D. S., Patil, A. S., & Dipale, A. D. (2022). Applications of nanotechnology in forensic science. In *Nanotechnology in the life sciences*. India: Springer Science and Business Media B.V, 257–276. http://springer.com/series/15921, https://doi.org/10.1007/978-3-030-82918-6_11.

Paliwal, R., Babu, R. J., & Palakurthi, S. (2014). Nanomedicine scale-up technologies: Feasibilities and challenges. *AAPS PharmSciTech, 15*(6), 1527–1534. https://doi.org/10.1208/s12249-014-0177-9.

Pan, D., Mi, L., Huang, Q., Hu, J., & Fan, C. (2012). Genetic analysis with nanoPCR. *Integrative Biology (United Kingdom), 4*(10), 1155–1163. https://doi.org/10.1039/c2ib20076g.

Pandit, S. (2016). Nanotechnology-based biosensors and its application. *The Pharma Innovation*.

Pandya, A., & Shukla, R. K. (2018). New perspective of nanotechnology: role in preventive forensic. *Egyptian Journal of Forensic Sciences, 8*(1). https://doi.org/10.1186/s41935-018-0088-0, https://link.springer.com/journal/41935/7/1/page/1.

Parveen, K., Banse, V., & Ledwani, L. (2016). Green synthesis of nanoparticles: Their advantages and disadvantages. *AIP Conference Proceedings, 1724*. https://doi.org/10.1063/1.4945168, http://scitation.aip.org/content/aip/proceeding/aipcp.

Pascual, L., Baroja, I., Aznar, E., Sancenón, F., Marcos, M. D., Murguía, J. R., Amorós, P., Rurack, K., & Martínez-Máñez, R. (2015). Oligonucleotide-capped mesoporous silica nanoparticles as DNA-responsive dye delivery systems for genomic DNA detection. *Chemical Communications, 51*(8), 1414–1416. https://doi.org/10.1039/C4CC08306G.

Patel, P., Kansara, K., Singh, R., Shukla, R. K., Singh, S., Dhawan, A., & Kumar, A. (2018). Cellular internalization and antioxidant activity of cerium oxide nanoparticles in human monocytic leukemia cells. *International Journal of Nanomedicine, 13*, 39–41. https://doi.org/10.2147/IJN.S124996, https://www.dovepress.com/cellular-internalization-and-antioxidant-activity-of-cerium-oxide-nano-peer-reviewed-fulltext-article-IJN.

Payal, P., & Pandey (2022). Role of nanotechnology in electronics: A review of recent developments and patents. *Recent Patents on Nanotechnology, 16*(1), 45–66. https://doi.org/10.2174/1872210515666210120114504.

Pelt-Verkuil, R., & Witt (2019). Principles of PCR. *Molecular Diagnostics: Part 1: Technical Backgrounds and Quality Aspects,* 131–215.

Peng, D., Huang, M., Xiao, Y., Zhang, Y., Lei, L., & Zhu, J. (2019). Highly-selective recognition of latent fingermarks by La-sensitized Ce nanocomposites via electrostatic binding. *Chemical Communications, 55*(71), 10579–10582. https://doi.org/10.1039/c9cc04257a.

Prakash Sharma, V., Sharma, U., Chattopadhyay, M., & Shukla, V. N. (2018). Advanced applications of nanomaterials: A review. *Materials Today: Proceedings, 5*(2), 6376–6380. https://doi.org/10.1016/j.matpr.2017.12.248.

Prasad, V., Lukose, S., & Prasad, L. (2016). Emerging forensic applications of nanotechnology. *International Journal of Engineering and Allied Sciences, 2*, 1–8.

Premkumar, T., & Geckeler, K. E. (2011). DNA and carbon-based nanomaterials. In *Materials Science of DNA*77–120. https://doi.org/10.1201/b11290-5.

Rastgoo, M., Tabatabaei, S. M., & Fathipour, M. (2018). A first-principles study on DNA sequencing using graphene quantum dot. *European Physical Journal B, 91*(6). https://doi.org/10.1140/epjb/e2018-80666-y, http://link.springer.de/link/service/journals/10051/index.htm.

Rawtani, D., Tharmavaram, M., Pandey, G., & Hussain, C. M. (2019). Functionalized nanomaterial for forensic sample analysis. *TrAC Trends in Analytical Chemistry, 120*, 115661. https://doi.org/10.1016/j.trac.2019.115661.

Ren, Z. L., Zhang, J. R., Zhang, X. M., Liu, X., Lin, Y. F., Bai, H., Wang, M. C., Cheng, F., Liu, J. D., Li, P., Kong, L., Bo, X. C., Wang, S. Q., Ni, M., & Yan, J. W. (2021). Forensic nanopore sequencing of STRs and SNPs using Verogen's ForenSeq DNA Signature Prep Kit and MinION. *International Journal of Legal Medicine, 135*(5), 1685–1693. https://doi.org/10.1007/s00414-021-02604-0, link.springer.de/link/service/journals/00414/index.htm.

Rim, K. T., Song, S. W., & Kim, H. Y. (2013). Oxidative DNA damage from nanoparticle exposure and its application to workers' health: A literature review. *Safety and Health at Work, 4*(4), 177–186. https://doi.org/10.1016/j.shaw.2013.07.006.

Rizwan, M., Shoukat, A., Ayub, A., Razzaq, B., & Tahir, M. B. (2021). Chapter 3 - Types and classification of nanomaterials. In *Nanomaterials: Synthesis, characterization, hazards and safety. A volume in Micro and nano technologies*. Pakistan: Elsevier, 31–54. https://doi.org/10.1016/B978-0-12-823823-3.00001-X.

de Roo, R., Mapes, A., van Cooten, M., van Hooff, B., Kneppers, S., Kokshoorn, B., Valkenburg, T., & de Poot, C. (2023). Introducing a rapid DNA analysis procedure for crime scene samples outside of the laboratory—A field experiment. *Sensors (Basel, Switzerland), 23*(8), 4153. https://doi.org/10.3390/s23084153.

Saif, S., Tahir, A., & Chen, Y. (2016). Green synthesis of iron nanoparticles and their environmental applications and implications. *Nanomaterials, 6*(11), 209. https://doi.org/10.3390/nano6110209.

Saiyed, Z. M., Ramchand, C. N., & Telang, S. D. (2008). Isolation of genomic DNA using magnetic nanoparticles as a solid-phase support. *Journal of physics. Condensed matter, 20*(20), 204153. https://doi.org/10.1088/0953-8984/20/20/204153.

Sajid, M. (2022). Nanomaterials: Types, properties, recent advances, and toxicity concerns. *Current Opinion in Environmental Science & Health, 25*, 100319. https://doi.org/10.1016/j.coesh.2021.100319.

Salatin, S., Maleki Dizaj, S., & Yari Khosroushahi, A. (2015). Effect of the surface modification, size, and shape on cellular uptake of nanoparticles. *Cell Biology International, 39*(8), 881–890. https://doi.org/10.1002/cbin.10459, http://onlinelibrary.wiley.com/journal/10.1002/.

Samanta, A., & Medintz, I. L. (2016). Nanoparticles and DNA-a powerful and growing functional combination in bionanotechnology. *Nanoscale, 8*(17), 9037–9095. https://doi.org/10.1039/c5nr08465b, http://www.rsc.org/publishing/journals/NR/Index.asp.

Sang, F. M., Li, X., & Liu, J. (2017). Development of nano-polymerase chain reaction and its application. *Chinese Journal of Analytical Chemistry, 45*(11), 1745–1753. https://doi.org/10.1016/S1872-2040(17)61051-X, https://www.journals.elsevier.com/chinese-journal-of-analytical-chemistry.

Sannino, D. (2021). Types and classification of nanomaterials. In *Nanotechnology: Trends and future applications*. Italy: Springer Singapore, 15–38. https://link.springer.com/book/10.1007/978-981-15-9437-3, https://doi.org/10.1007/978-981-15-9437-3_2.

Sathish Sundar, D., Gover Antoniraj, M., Senthil Kumar, C., S. Mohapatra, S., N. Houreld, N., & Ruckmani, K. (2016). Recent trends of biocompatible and biodegradable nanoparticles in drug delivery: A review. *Current Medicinal Chemistry, 23*(32), 3730–3751. https://doi.org/10.2174/0929867323666160607103854.

Sedighi, A., & Krull, U. J. (2016). Rapid immobilization of oligonucleotides at high density on semiconductor quantum dots and gold nanoparticles. *Langmuir, 32*(50), 13500–13509. https://doi.org/10.1021/acs.langmuir.6b03840, http://pubs.acs.org/journal/langd5.

Sharma, V., Shukla, R. K., Saxena, N., Parmar, D., Das, M., & Dhawan, A. (2009). DNA damaging potential of zinc oxide nanoparticles in human epidermal cells. *Toxicology Letters, 185*(3), 211–218. https://doi.org/10.1016/j.toxlet.2009.01.008.

Shen, C., & Zhang, Z. (2013). An overview of nanoparticle-assisted polymerase chain reaction technology. In *Bio-nanotechnology: A revolution in food, biomedical and health sciences*. Australia: Blackwell Publishing Ltd, 97–106. https://doi.org/10.1002/9781118451915.ch5, http://onlinelibrary.wiley.com/book/10.1002/9781118451915.

Shen, C., Yang, W., Ji, Q., Maki, H., Dong, A., & Zhang, Z. (2009). NanoPCR observation: different levels of DNA replication fidelity in nanoparticle-enhanced polymerase chain reactions. *Nanotechnology, 20*(45), 455103. https://doi.org/10.1088/0957-4484/20/45/455103.

Shivakumar, N., Dr (2024). Recent Advances in biological nanodevices and biosensors: Insights into applications and technological innovations. *Malaysian NANO-An International Journal, 4*(1), 86–101. https://doi.org/10.22452/mnij.vol4no1.6.

Shukla, R. K., Badiye, A., Vajpayee, K., & Kapoor, N. (2021). Genotoxic potential of nanoparticles: Structural and functional modifications in DNA. *Frontiers in Genetics, 12*. https://doi.org/10.3389/fgene.2021.728250, https://www.frontiersin.org/journals/genetics#.

Shukla, R. K., Sharma, V., Pandey, A. K., Singh, S., Sultana, S., & Dhawan, A. (2011a). ROS-mediated genotoxicity induced by titanium dioxide nanoparticles in human epidermal cells. *Toxicology In Vitro, 25*(1), 231–241. https://doi.org/10.1016/j.tiv.2010.11.008.

Shukla, R. K., Kumar, A., Gurbani, D., Pandey, A. K., Singh, S., & Dhawan, A. (2013). TiO_2 nanoparticles induce oxidative DNA damage and apoptosis in human liver cells. *Nanotoxicology, 7*(1), 48–60. https://doi.org/10.3109/17435390.2011.629747.

Shukla, R. K., Kumar, A., Vallabani, N. V. S., Pandey, A. K., & Dhawan, A. (2014). Titanium dioxide nanoparticle-induced oxidative stress triggers DNA damage and hepatic injury in mice. *Nanomedicine: Nanotechnology, Biology, and Medicine, 9*(9), 1423–1434. https://doi.org/10.2217/nnm.13.100, http://www.futuremedicine.com/loi/nnm.

Shukla, R. K., Kumar, A., Pandey, A. K., Singh, S. S., & Dhawan, A. (2011b). Titanium dioxide nanoparticles induce oxidative stress-mediated apoptosis in human keratinocyte cells. *Journal of Biomedical Nanotechnology, 7*(1), 100–101. https://doi.org/10.1166/jbn.2011.1221India http://docserver.ingentaconnect.com/deliver/connect/asp/15507033/v7n1/s55.pdf?expires=1329722431&id=67298637&titleid=11665&accname=Elsevier&checksum=D5310DB2DF2A00810C92E2846CA02EAE.

Silva, G. A. (2004). Introduction to nanotechnology and its applications to medicine. *Surgical Neurology, 61*(3), 216–220. https://doi.org/10.1016/j.surneu.2003.09.036, http://www.elsevier.com/locate/surneu.

Singh, S., & Sharma, H. (2023). Emerging applications of nanotechnology in drug delivery and medical imaging: Review. *Current Radiopharmaceuticals, 16*(4), 269–283. https://doi.org/10.2174/1874471016666230621120453, http://www.eurekaselect.com/672/journal/current-radiopharmaceuticals.

Singh, S., & Samal, N. (2021). Nanotechnology: A powerful tool in forensic science for solving criminal cases. *Arab Journal of Forensic Sciences and Forensic Medicine, 3*(2), 273–296. https://doi.org/10.26735/yddj4516.

Somorjai, G. A., & Park, J. Y. (2008). Colloid science of metal nanoparticle catalysts in 2D and 3D structures. Challenges of nucleation, growth, composition, particle shape, size control and their influence on activity and selectivity. *Topics in Catalysis, 49*(3-4), 126–135. https://doi.org/10.1007/s11244-008-9077-0.

Tambo, F., & Ablateye, D. N. O. (2020). A review on the role of emerging revolutionary nanotechnology in forensic investigations. *Journal of Applied and Natural Science, 12*(4), 582–591. https://doi.org/10.31018/jans.v12i4.2415, https://journals.ansfoundation.org/index.php/jans/article/view/2415/1937.

Tang, C. (2020). Application of magnetic nanoparticles in nucleic acid detection. *Journal of Nanobiotechnology, 18*, 1–19.

Tegenfeldt, J. O., Prinz, C., Cao, H., Huang, R. L., Austin, R. H., Chou, S. Y., Cox, E. C., & Sturm, J. C. (2004). Micro- and nanofluidics for DNA analysis. *Analytical and Bioanalytical Chemistry, 378*(7), 1678–1692. https://doi.org/10.1007/s00216-004-2526-0.

Teles, F., & Fonseca, L. (2008). Trends in DNA biosensors. *Talanta, 77*(2), 606–623. https://doi.org/10.1016/j.talanta.2008.07.024.

Thomas, S. C., Harshita, Mishra, P. K., & Talegaonkar, S. (2015). Ceramic nanoparticles: Fabrication methods and applications in drug delivery. *Current Pharmaceutical Design, 21*(42), 6165–6188. https://doi.org/10.2174/1381612821666151027153246, http://www.ingentaconnect.com/content/ben/cpd.

Tufail, M. S., & Liaqat, I. (2022). Silver nanoparticles and their applications-A comprehensive review. *Pure and Applied Biology,* (1), 315–330.

Tytgat, O., & Van Nieuwerburgh, F. (2023). Applications of nanopore sequencing for forensic analysis. In *Next generation sequencing (NGS) technology in DNA analysis*. Belgium: Elsevier, 85–98. https://www.sciencedirect.com/book/9780323991445, https://doi.org/10.1016/B978-0-323-99144-5.00027-5.

Ullah, M. F., Khan, Y., Khan, M. I., Abdullaeva, B. S., & Waqas, M. (2024). Exploring nanotechnology in forensic investigations: Techniques, innovations, and future prospects. *Sensing and Bio-Sensing Research, 45*, 100674. https://doi.org/10.1016/j.sbsr.2024.100674.

Vajpayee, K. (2023). PCR inhibitors and facilitators-Their role in forensic DNA analysis. *Forensic Science International*.

Vajpayee, K., Sagar, D. C., & Dash, H. R. (2020). Forensic DNA typing: Inception, methodology, and technical advancements. In *Forensic DNA Typing: Principles, Applications and Advancements*. India: Springer Singapore, 3–26. https://link.springer.com/book/10.1007/978-981-15-6655-4, https://doi.org/10.1007/978-981-15-6655-4_1.

Vallabani, N. V. S., Mittal, S., Shukla, R. K., Pandey, A. K., Dhakate, S. R., Pasricha, R., & Dhawan, A. (2011). Toxicity of graphene in normal human lung cells (BEAS-2B). *Journal of Biomedical Nanotechnology, 7*(1), 106–107. https://doi.org/10.1166/jbn.2011.1224India, http://docserver.ingentaconnect.com/deliver/connect/asp/15507033/v7n1/s58.pdf?expires=1329722472&id=67298644&titleid=11665&accname=Elsevier&checksum=431E6109B89401EF008A5E486D230F96.

Vázquez-Núñez, E., Molina-Guerrero, C. E., Peña-Castro, J. M., Fernández-Luqueño, F., & de la Rosa-Álvarez, M. G. (2020). Use of nanotechnology for the bioremediation of contaminants: A review. *Processes, 8*(7), 826. https://doi.org/10.3390/pr8070826.

Wan, W., Yeow, J. T. W., & Van Dyke, M. I. (2009). 12 2009/12 2009 Effect of silver and titanium dioxide nanoparticles on PCR efficiency.In 9th IEEE Conference on nanotechnology,, IEEE NANO 2009, Canada, pp. 458–461 Canada Effect of silver and titanium dioxide nanoparticles on PCR efficiency..

Wang, D. X., Wang, J., Wang, Y. X., Du, Y. C., Huang, Y., Tang, A. N., Cui, Y. X., & Kong, D. M. (2021). DNA nanostructure-based nucleic acid probes: Construction and biological applications. *Chemical Science, 12*(22), 7602–7622. https://doi.org/10.1039/d1sc00587a, http://pubs.rsc.org/en/journals/journal/sc.

Wang, J., Cheng, Y., Zhang, M., Zhao, H., Lin, P., Yi, L., Tong, M., & Cheng, S. (2015). Development of a nanoparticle-assisted PCR (nanoPCR) assay for detection of mink enteritis virus (MEV) and genetic characterization of the NS1 gene in four Chinese MEV strains. *BMC Veterinary Research, 11*(1). https://doi.org/10.1186/s12917-014-0312-6.

Wang, p, et al. (2017). *Nonlinear Analysis: Hybrid Systems,* 1–10.

Wang, Q., Li, J., & Cao, X. (2007). Silver nanoparticles enhance the specificity of repeated long PCR amplification. *Tianjin Keji Daxue Xuebao, 22*.

Wang, W., Sedykh, A., Sun, H., Zhao, L., Russo, D. P., Zhou, H., Yan, B., & Zhu, H. (2017). Predicting nano–bio interactions by integrating nanoparticle Libraries and Quantitative Nanostructure Activity Relationship Modeling. *ACS Nano, 11*(12), 12641–12649. https://doi.org/10.1021/acsnano.7b07093.

Wang, Y., Yang, Q., & Wang, Z. (2015). The evolution of nanopore sequencing. *Frontiers in Genetics, 5*.

Wang, Z., Qin, L., Liu, J., Jiang, L., Zou, X., Chen, X., Song, F., Dai, H., & Hou, Y. (2022). Forensic nanopore sequencing of microhaplotype markers using QitanTech's QNome. *Forensic Science International: Genetics, 57*, 102657. https://doi.org/10.1016/j.fsigen.2021.102657.

Watthanapanpituck, K., Kiatpathomchai, W., Chu, E., & Panvisavas, N. (2014). Identification of human DNA in forensic evidence by loop-mediated isothermal amplification combined with a colorimetric gold nanoparticle hybridization probe. *International Journal of Legal Medicine, 128*(6), 923–931. https://doi.org/10.1007/s00414-014-1018-9.

Wu, L., Xiong, E., Zhang, X., Zhang, X., & Chen, J. (2014). Nanomaterials as signal amplification elements in DNA-based electrochemical sensing. *Nano Today, 9*(2), 197–211. https://doi.org/10.1016/j.nantod.2014.04.002.

Wu, Y., & Weil, T. (2017). Nanodiamonds for biological applications. *De Gruyter, Physical Sciences Reviews, 2*(6). https://doi.org/10.1515/psr-2016-0104, http://www.degruyter.com/view/j/psr.

Xue, Z., You, M., Peng, P., Tong, H., He, W., Li, A., Mao, P., Xu, T., Xu, F., & Yao, C. (2021). Taqman-MGB nanoPCR for highly specific detection of single-base mutations. *International journal of nanomedicine, 16*, 3695–3705. https://doi.org/10.2147/ijn.s310254.

Yáñez-Sedeño, P., Agüí, L., Villalonga, R., & Pingarrón, J. M. (2014). Biosensors in forensic analysis. A review. *Analytica Chimica Acta, 823*, 1–19. https://doi.org/10.1016/j.aca.2014.03.011.

Yáñez-Sedeño, P., Agüí, L., Campuzano, S., & Pingarrón, J. (2019). What electrochemical biosensors can do for forensic science? Unique features and applications. *Biosensors, 9*(4), 127. https://doi.org/10.3390/bios9040127.

Yang, Z., Shen, B., Yue, L., Miao, Y., Hu, Y., & Ouyang, R. (2022). Application of nanomaterials to enhance polymerase chain reaction. *Molecules, 27*(24), 8854. https://doi.org/10.3390/molecules27248854.

Yuce, M., Kurt, H., Mokkapati, V. R. S. S., & Budak, H. (2014). Employment of nanomaterials in polymerase chain reaction: Insight into the impacts and putative operating mechanisms of nano-additives in PCR. *RSC Advances, 4*(69), 36800–36814. https://doi.org/10.1039/c4ra06144f, http://pubs.rsc.org/en/journals/journalissues.

Zareef, I., Riaz, A., Salahuddin, Fatima, M., Anwar, A., & Munawar, A. (2024). *Unveiling the power of nanomaterials in the area of forensics*. Springer Science and Business Media LLC, 29–57. https://doi.org/10.1007/978-3-031-57843-4_2.

Zhang, C. Y., Yeh, H. C., Kuroki, M. T., & Wang, T. H. (2005). Single-quantum-dot-based DNA nanosensor. *Nature Materials, 4*(11), 826–831. https://doi.org/10.1038/nmat1508, http://www.nature.com/nmat/.

Zhang, H., Zhang, H., Aldalbahi, A., Zuo, X., Fan, C., & Mi, X. (2017). Fluorescent biosensors enabled by graphene and graphene oxide. *Biosensors and Bioelectronics, 89*, 96–106. https://doi.org/10.1016/j.bios.2016.07.030.

Zhang, Y., & Wang, T. H. (2012). Quantum dot enabled molecular sensing and diagnostics. *Theranostics, 2*(7), 631–654. https://doi.org/10.7150/thno.4308, http://www.thno.org/v02p0631.pdf.

Zhu, et al. (2016). *Computer Networks,* 1–14.

Chapter 15

Advancements in wildlife forensics

Sweety Sharma, and Bhawna Bangwal

Department of Forensic Biology and Biotechnology, Delhi Campus, National Forensic Sciences University, Ministry of Home Affairs, Government of India, Delhi, India

15.1 Introduction

Wildlife encompasses the plants and animals flourishing in the wild without any introduction or intervention by humans; it also includes the minuscule, often overlooked microorganisms that are found in the wild. These unvoiced plants and animals dwelling in the wild have always been prone to hunting, poaching, stealing, etc. by human beings for a plethora of reasons like the thrill, entertainment, illicit trade, financial gain, keeping souvenirs, etc., making a large number of species to become endangered while dragging several species to extinction. The wild flora and fauna are vulnerable to such crimes because nature has provided them with some features or substances that are highly demanded by the people and amount to great money, for example, elephant - tusks; tiger, lion - claws, skin; snake - skin, venom; crocodile - skin; tortoise - shell; musk deer - musk scent; and many more (Gouda et al., 2020). A key difference in human and alleged wildlife crimes is that there can often be no "victim" to deliver evidence for the investigation (Johnson et al., 2014).

Wildlife crime has emerged as one of the biggest international crimes as the laws enforced in wildlife conservation are not very strict, accompanied by less harsh penalties, but the gross profit is comparable with other big international crimes like narcotics trading, laws and punishments associated with the same are very rigorous. Wildlife crimes give rise to a myriad of harmful repercussions, including the peril of species extinction, loss of biodiversity, disturbances in the food chain, and setbacks to a nation's economic well-being (Ogden et al., 2009).

Deoxyribonucleic acid (DNA) and ribonucleic acid (RNA) are two main genetic molecules found in living beings. DNA, being double-stranded, has an edge over RNA by having more stability and resistance to degradation. Moreover, DNA can be extracted from most of the tissues imaginable with the pros and cons of each site like it is present in good amounts in soft parenchymal tissues but this DNA is very prone to degradation, and DNA present in hard tissues like bones and teeth although present in inconspicuous quantities has the longest life. DNA analysis is primarily used when other morphologically identifiable features of the wildlife sample have been obliterated by cutting, cooking, processing, etc. to an extent that little to no conclusion can be drawn about the species of the given sample such as in the case of burnt bones or skin, deteriorated body parts, snake venom, cooked meat, medicinal preparations, ornaments made up of tusks, horns, claws, etc. (Ellis, 2013; Gupta et al., 2005; Mainka & Mills, 1995; Vipin et al., 2016).

Different types of samples are analyzed in wildlife forensic investigations, such as animal body as a whole in both dead and living conditions. Sometimes the traces of evidence, such as body fluids or processed body parts of animals, can also involve DNA. Various items, such as exoskeletons such as nails, horns, skeletal remains, skin, and body parts of animals (genitals, limbs, head, and internal organs), are also encountered as sample in wildlife cases (Gondhali & Mishra, 2022).

DNA has been used in human identification for a very long time, and the fundamentals of the same can be applied to wildlife conservation as DNA possesses several properties and factors that can answer the four major questions of conservation genetics, namely, "what is the species of the questioned material?" that is, species identification, "where did the questioned material come from?" that is, identification of geographic location, "who did the questioned material come from?" that is, individual identification and "was the questioned material bred in captivity?" that is, deciphering whether the specimen has been stolen from the wild or was legally bred in captivity (Ogden et al., 2009).

With the advent of the polymerase chain reaction (PCR) technique, different DNA extraction methods, short tandem repeat (STR) and mitochondrial markers, single-nucleotide polymorphism (SNP) typing, DNA sequencing, next-generation sequencing (NGS), forensically informative nucleotide sequencing (FINS), on-site species identification tools,

Investigative and Predictive DNA Testing. DOI: https://doi.org/10.1016/B978-0-443-30128-5.00034-5

etc. several advancements and developments have been made to increase the effectiveness and reduce the cost at the same time, aiding in conducting the species determination from the DNA extracted from the anticipated sample with ease and a diminished financial outlay.

15.2 Advancement in DNA techniques

Wildlife DNA forensics has flourished as a fusion of conservation genetics and the real-world application of forensic genetics. This chapter explores the latest cutting-edge innovations and advancements in wildlife DNA techniques, specifically in extraction, amplification, and sequencing.

15.2.1 Advancement in DNA extraction methods

In wildlife forensics, advanced kits for DNA extraction play a significant role in extracting high-quality DNA from various challenging samples such as degraded bones, tissues, feathers, and hairs. Specific kits are designed to efficiently extract the DNA from these types of challenging wildlife samples. To choose the appropriate DNA extraction method based on the applications and nature of the sample. Table 15.1 organizes the different DNA extraction kits used for the wildlife samples with their descriptions, advantages, and applications.

A study was conducted by Carvajal-Agudelo et al. (2021) to compare the efficiency of various DNA extraction techniques employed in wildlife cases, and it was found that the commercial technologies/methods like Isolate II Genomic DNA (Bioline®), Genomic DNA Isolation Kit (Norgen®), PureLink™ Genomic DNA Mini Kit (Invitrogen®), and DNeasy Blood and Tissue Kit (Qiagen®) produced lower amounts of DNA, but with highest purity and minimum variations because of the standardized procedures based on silica columns. Due to their low variation, these commercial kits can be used in automatization lines to reduce time consumption, increase efficiency, obtain reproducible results, and obtain high output, that is, attending more samples in the same amount of time. An exception to the qualities of the commercially available kits is Wizard® Genomic DNA Purification Kit by Promega® which yielded a higher amount of DNA but with higher variation and poor quality as it doesn't use silica-based columns and involves the use of solutions instead of the former, which eventually contributes to such divergence from the other commercially available counterparts. The same study showed that the noncommercial methods like UltraPure™ Phenol: Chloroform: Isoamyl Alcohol (25:24:1, v/v) and guanidine thiocyanate or guanidium thiocyanate reagent yielded the highest amounts of DNA along with the highest variation and low efficiency. However, these noncommercial kits are inexpensive, feasible, and easy to prepare.

15.2.2 Advancement in the amplification process of DNA

15.2.2.1 Polymerase chain reaction techniques

PCR is the process that helps amplify extracted DNA for sequencing and better analysis. Since the evidence present at the crime scene is present in trace amounts, therefore the DNA extracted from the same is also present in very small volumes which limits the possibility of performing several tests to increase the accuracy of results, here comes the role of PCR which increases the amounts of extracted DNA several folds and allows to perform multiple tests, reanalysis if the test goes wrong, easy sequencing using various methods, etc. (Li, 2011).

Conventional PCR (cPCR) involves the extraction of DNA to isolate it in its pure form, free of the cellular content that encapsulates it. It is present in most laboratories worldwide and hence is involved in more cases of species identification (Ojaveer et al., 2014). Ziegler et al. (2007) conducted a study to diagnose a parasite named *Cryptosporidium* spp in the fecal samples obtained from dairy cattle and wild rodents using PCR, aiming at four gene targets: the 70-kD heat shock protein gene, the *Cryptosporidium* oocyst wall protein gene, the 18S small-subunit rRNA gene, and the thrombospondin-related adhesive proteinC2 (TRAP-C2) gene. The study established the advantages of PCR for having lower limits of detection (LoD) and being able to distinguish between species of the parasite, which was not possible by other tests involving floatation method and ELISA. The competency of differentiating the species of the parasite helped the researchers to better understand which species are host-specific and which can go for multiple hosts. The PCR showed higher sensitivity and specificity than the floatation method and ELISA in the detection of *Cryptosporidium* spp. in samples collected from cattle, but its sensitivity and specificity appeared to decrease when samples from wild rodents were analyzed. This study suggested that an amalgamation of conventional and molecular diagnostic techniques, such as PCR, should be used to detect this parasite in the samples (Barbaro et al., 2004; Martin & Linacre, 2020; Ottens et al., 2013; Van Oorschot et al., 2003).

TABLE 15.1 Different DNA extraction kits used for the wildlife samples and their descriptions, advantages, and applications.

Method	Description	Advantage	Applications	References
FTA® card	Paper is impregnated with chemicals that lyse cells and denature proteins, releasing DNA trapped in the paper for amplification and analysis.	- Ease of use for various sample types (wet/dried blood, clots, tissue, saliva, and buccal swabs) - Simple processing: punch, wash, and amplify - Allows multiple sample collection with boundaries - Prevents bacterial growth and deals with free radicals/pollutants - Good shelf life: up to 4 years at room temperature - Easy transportation	Amplification and analysis	Smith and Burgoyne (2004)
ISOLATE II genomic DNA kit	Combines proteinase K lysis with silica-membrane purification for DNA extraction from various samples.	DNA recovery in ~20 min	qPCR	
		Efficient lysis and column binding even from small samples	End-point PCR	Bioline I Meridian Bioscience (n.d.)
		High-purity DNA suitable for downstream processing	Southern, dot, and slot blotting genotyping	
		Suitable for human, nonhuman, and microbial samples	Restriction digestion	
			Next-generation sequencing	
			Bisulfite conversion/ methylation analysis	
Genomic DNA isolation kit	Uses a rapid spin-column protocol for DNA extraction from various samples.	Procedure ~80 min including incubation	- qPCR - PCR	Norgen Biotek Corp. (n.d.)
		Versatile: animal tissues, cells, body fluids, viruses, swabs	End-point PCR	
		High-purity DNA suitable for downstream processing	Southern, dot, and slot blotting genotyping	
		Excellent yield: high yield from small samples	Restriction digestion	
			Next-generation sequencing	
PureLink™ genomic DNA mini Kit	Uses silica-based, microcentrifuge, spin-column technology for DNA extraction from various samples.	Uses silica-based, microcentrifuge, spin-column technology for DNA extraction from various samples	Genotyping	Thermo Fisher Scientific (n.d.)
		High-quality DNA: minimal contamination, good for downstream processing	Real-time qPCR	
		- High performance: optimized buffers and spin-column design for good DNA extraction	Southern blotting	
			Sequencing	
			PCR	

(*Continued*)

TABLE 15.1 (Continued)

Method	Description	Advantage	Applications	References
Wizard® genomic DNA purification Kit	Solution-based system for DNA extraction from blood, plant and animal tissue, cell culture, yeast, and bacteria.	- Fast: process completion in ~60 min	Genotyping	Promega Corporation (n.d.)
		Versatility: suitable for human, nonhuman, and microbial samples	Real-time qPCR	
			Southern blotting	
			Sequencing	
			PCR	
DNeasy blood and tissue kit	Combines silica-based columns with microspin tubes for DNA extraction from various samples.	Versatility: suitable for human, nonhuman, and microbial sample	Life sciences	QIAGEN (n.d.)
		Expeditious: process time ~20–60 min	Veterinary science	
		- High performance: optimized buffers and spin-column design for good	Genotyping	
		High-quality DNA: pure DNA free from protein and salt residues	Forensics	
		High-throughput technology: 96-well format available for large sample testing		
		Sustainable: reduced use of plastic and cardboard		
UltraPure™ phenol:chloroform alcohol	Mixture of phenol, chloroform, and isoamyl alcohol for nucleic acid extraction.	- Efficient separation - Produces high-purity DNA/RNA - Isoamyl alcohol reduces foaming and stabilizes the extraction process.	- Nucleic acid extraction	Thermo Fisher Scientific (n.d.)
Guanidine thiocyanate	Guanidium thiocyanate-phenol-chloroform extraction method for RNA and DNA extraction.	- Strongly denatures proteins and inactivates RNases/DNases - Yields high-quality RNA	RNA extraction	Thermo Fisher Scientific (India) (n.d.)

PCR, Polymerase chain reaction; *qPCR*, quantitative polymerase chain reaction.

15.2.2.2 Real-time quantitative polymerase chain reaction

Real-time quantitative PCR (qPCR) is a modification of PCR that works in real time, that is, we can monitor the amplification process of desired DNA fragments while they are in the middle of the process. This technique combines amplification and quantitation of target DNA and hence is more beneficial and strategic than cPCR. This high-throughput and sensitive technique is often used in investigating cases involving wildlife.

Xia et al. (2018) conducted a study to compare the performance of conventional PCR with real-time qPCR in detecting the presence of a highly invasive mollusc species namely the "golden mussel" (*Limnoperna fortune)* using environmental DNA (eDNA) obtained from laboratory aquaria and natural irrigation channels. The study revealed a higher sensitivity of qPCR (LoD = 1×10^{-7} ng/μL) than cPCR (LoD = 1×10^{-6} ng/μL), the former also attained a higher detection rate in both laboratory (100%) and field (68.6%) samples than the latter, which only secured 87.9% detection rate in laboratory and 47.1% in field samples. Conte et al. (2019) also developed a qPCR assay to identify and quantify mitochondrial DNA (mtDNA) from the buccal swabs of African (- *Loxodonta africana*) and Asian (- *Elephas maximus*) elephants. This assay is useful for forensic purposes to get an idea of whether the sample in question, such as ivory, ornament, weapon, etc., has originated from an Asian or African elephant, and can help link the sample with the living elephant or its carcass. Keatley et al. (2020) created a rapid multiplex qPCR that can quickly, efficiently (more than 90%), sensitively (detection limit of five copies of DNA per PCR cycle), and specifically (without any cross-contamination) distinguish between the DNA from six species of big cats namely tiger (*Panthera tigris*), Cheetah (*Acinonyx jubatus*), Leopard (*Panthera pardus*), Lion (*Panthera leo*), Snow leopard (*Panthera uncia*), and Jaguar (*Panthera onca*) from wildlife products such as skin, claws, fur, teeth, etc.

In conclusion, the qPCR proved to be more advantageous than cPCR owing to the methods of detection opted for by it and the provision of real-time results. qPCR detects the PCR products by measuring the fluorescence in each PCR plate during the exponential stage of the PCR process. In contrast, cPCR utilizes UV ray visualization of ethidium bromide-stained gel, measuring the products at the plateau stage of PCR (Smith & Osborn, 2009). qPCR also prevents the degradation of products at high reaction cycles because the reagents get exhausted.

15.2.2.3 Direct polymerase chain reaction

This kind of PCR technique has been used for many years for sample amplification from various biological materials; however, its use in forensics has emerged very recently (Linacre et al., 2010). In 1993, Panaccio et al. (1993) presented the first description of direct PCR amplification from various tissues, including liver, brain, testes, ovaries, heart, kidney, skin, muscles, and skeletal. In the reported study, it was said that this technique has applications in forensic science, origin determination of meat samples, and genetic analysis of biopsy samples. This method omits the step of DNA extraction and directly puts the sample, collected by any means, into the amplification process without performing any treatment beforehand. The purpose of this technique is to enhance the availability of DNA templates for further amplification by excluding the steps of extraction. However, it is recorded that a major loss that is up to 76%–83% can occur during the step of extraction (Linacre et al., 2010; Van Oorschot et al., 2003). The DNA trapped in the FTA card can be amplified directly without any prior purification. Direct PCR makes use of universal primers and sometimes, alternate buffers, which can cause a problem if the laboratory workflow is bound to follow the accredited procedures and where the introduction of a new scientific procedure can be a lengthy task (Martin & Linacre, 2020). The results of direct PCR are found to be very similar to those obtained by following proper DNA isolation before amplification (Kitpipit et al., 2013). A study conducted by Kitpipit et al. (2013), reported that they successfully amplified commonly found wildlife samples such as bloodstain, muscle tissues, hair, skin, bone, ear, urine, and feces of common victim species like Tiger, White Rhino, and Asian elephant using direct PCR and were able to match the samples with their corresponding species. In the cited study, a three-step process was used: (1) amplification of DNA using 16S rRNA and cytochrome b universal primers, (2) FINS, and (3) matching the sequence using the Basic Local Alignment Search Tool (BLAST) database. The workflow of Digital and conventional PCR is given in Fig. 15.1. Table 15.2 lists the advantages and disadvantages of the Direct PCR method.

15.2.2.4 Digital polymerase chain reaction

It is a third-generation PCR technique that is used for absolute detection of very small traces of target DNA from the bulk through portioning reaction. In digital PCR (dPCR), the extracted DNA is portioned or spread randomly in a large number of compartments such that no or one or a limited molecules are present in each compartment. When a compartment contains more than one molecule, the competition between them is still reduced to a great extent, and is corrected using Poisson's statistics. The absorbance from each compartment is observed, each compartment acts as a micro reactor, and the reaction with one target molecule is given the value "1" known as a "positive reaction," while the compartment having no target molecule will give a "negative reaction" and will be assigned a value of "0" (Mao et al., 2019). The fluorescence

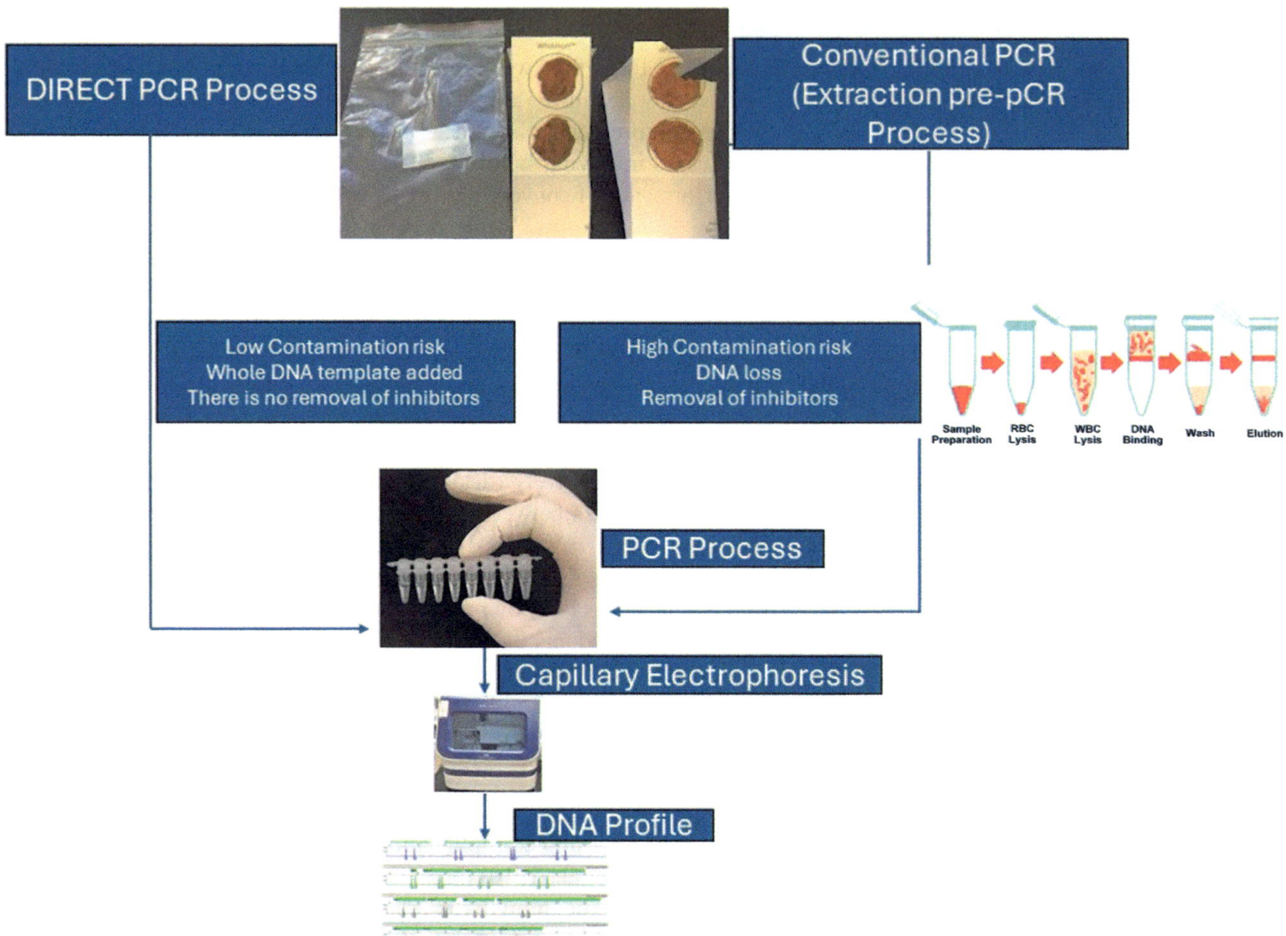

FIGURE 15.1 Workflow of conventional and direct-polymerase chain reaction (PCR).

is a result of the separation of fluorophore (fluorescent dye) from the quencher dye. During the amplification process, the fluorescent signature of every compartment is produced which is captured by an imaging device. This technique is very sensitive that a single DNA molecule is capable of producing its florescent signature which can be taken into calculation. The number of compartments showing bright fluorescence equal to the target DNA molecule. Depending upon the type of compartment, dPCR is of two types; chip dPCR and droplet dPCR water in oil droplets. This dPCR requires a minuscule sample volume, making it efficient for forensic applications where sample availability is limited which is particularly important in wildlife forensics where sample availability is often limited. It can also work effectively with degraded nucleic acids, which is often a challenge in wildlife forensic samples, and demonstrates a high tolerance to PCR inhibitors, which are commonly present in biological materials. Its high sensitivity allows for the detection of trace amounts of DNA, with sensitivity levels reaching as low as 0.1%, and in some cases, 0.001%. This is particularly advantageous for trace DNA identification. The method produces highly readable data, ensuring accurate analysis and interpretation. Additionally, it allows for absolute quantification, as highlighted by Quan et al. (2018), providing precise measurements of DNA quantity. The low error rate associated with this technology further enhances its reliability. Moreover, it can be combined withNGS, expanding its applicability and providing comprehensive genetic information and species identification. However, the disadvantage of dPCR is that it is costly and has a narrow dynamic range. The workflow of dPCR is illustrated in Fig. 15.2.

15.2.3 Advancement in DNA sequencing techniques

DNA nucleotide sequencing involves sequencing the nucleotides particularly the nitrogenous bases (A, T, G, and C) of a particular DNA region and matching the produced DNA sequence with a reference database to decipher the degree of

TABLE 15.2 Advantages and disadvantages of the direct-PCR technique.

Advantages	Disadvantages
Rapid, robust, and reliable technique.	Direct PCR skips the process of "inhibitor removal," leading to potential increased inhibition in the PCR. It should be used with samples of known origin assumed to be free of inhibitors (Martin & Linacre, 2020).
Increases sensitivity of amplification (Dinis et al., 2017; Martin et al., 2018; Templeton et al., 2013, 2015).	DNA quantification is mandatory in many forensic science laboratories due to accreditation guidelines, preventing the use of direct PCR until guidelines are revised.
	As DNA quantification is mandatory in many working forensic science laboratories (FSLs) due to accreditation and adherence to guidelines, direct PCR can't be employed in such labs until the guidelines are revised (Federal Bureau of Investigation, 2011)
Applicable to various samples: DNA trapped in FTA cards such as blood or saliva (Barbaro et al., 2008; Park et al., 2008), tissues (Sorensen et al., 2016), hair (Ottens et al., 2013; Templeton et al., 2013; Handt & Linacre, 2015), fingernail clippings (Ottens et al., 2015; Tie & Uchigasaki, 2014), fibers or fabrics (Blackie et al., 2016; Brito et al., 2015; Linacre et al., 2010; Ottens et al., 2013).	Direct PCR often consumes the entire sample, leaving no material for secondary analysis required for accreditation or admissibility (National Association of Testing Authorities NATA, 2020)
Optimal for "touch DNA" containing trace amounts of DNA, conserving all available DNA for amplification, producing quality results (Martin & Linacre, 2020).	High sensitivity can detect and amplify 'non-self-DNA' present on evidence, complicating mixture analysis (Buckingham et al., 2016; Goray et al., 2016; Graham & Rutty, 2008; Magee et al., 2018; van den Berge et al., 2016).
Can be used for amplifying DNA traces obtained by adhesive and semiadhesive tape lifting (Martin & Linacre, 2020).	

PCR, Polymerase chain reaction.

closeness. Usually, around 500 bases of DNA nucleotides are sequenced to provide a species-specific sequence for identification. DNA sequencing also provides data for the development of genetic markers like SNP, microsatellites, insertions, deletions, etc. Many free DNA sequence analysis tools and softwares are available, such as Alfresco, ALP, Arioc, AutoCSA, BaMM, etc. (Bioinformatics Home, n.d.).

15.2.3.1 Sanger sequencing

Sanger sequencing also known as chain termination method. It is a first-generation DNA sequencing technique developed by Sanger and his colleagues in 1977. In wildlife forensics, species differentiation can be carried out based on the DNA barcodes. It is useful for the identification of species, which is crucial in wildlife protection law, moreover where the morphological identification is quite challenging. By using Sanger sequencing, the genetic profiles of individual animals can be generated, and it is crucial to link specific individuals to a crime scene (Meiklejohn et al., 2021). Sanger sequencing uses a radiolabelled dideoxynucleotide (ddNTP) that lacks a hydroxyl group (OH) at the 3' position of the deoxyribonucleotide molecule. Examples of ddNTP are Adenine (ddATP), thymine (ddTTP), cytosine (ddCTP), and guanine (ddGTP). These ddNTPs are present in very low concentrations in the working solution (also containing polymerase enzyme, primers, and dNTPs) and stop the reaction of DNA polymerase due to the absence of the OH group at the third position. In the classical method, four tubes were used, each having one type of ddNTP which resulted in the production of DNA fragments of variable lengths due to the random attachment of the specific ddNTP on the fragment and breaking of fragments at varying lengths. These obtained DNA fragments were then subjected to gel electrophoresis to get the possible position of the nucleotide in the target DNA. When the electrophoresis of the DNA fragments from the rest three test tubes is also done, the sequence of the target DNA can be deciphered (Butler, 2005; Li, 2008) (Fig. 15.3).

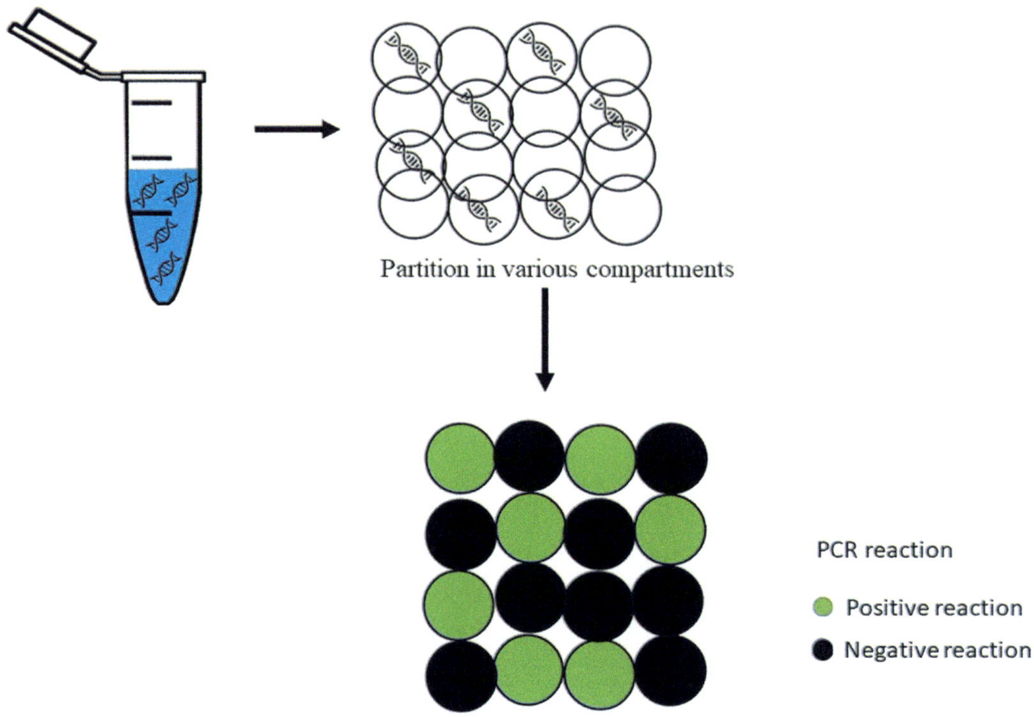

FIGURE 15.2 Workflow of digital polymerase chain reaction (PCR).

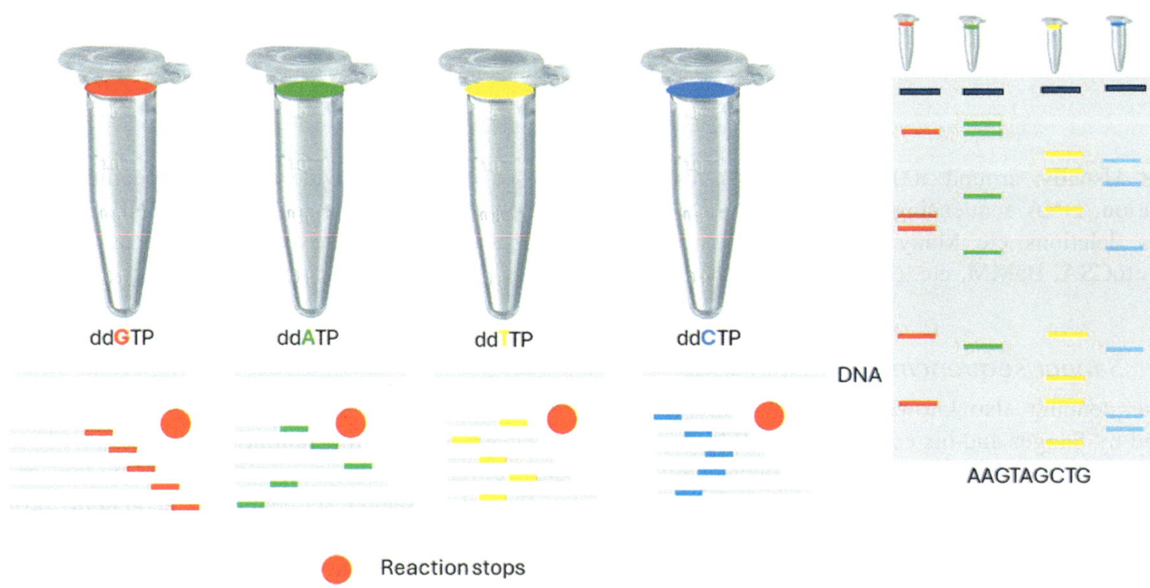

FIGURE 15.3 Working principle of classical Sanger sequencing (Saini et al., 2023).

15.2.3.2 Modern Sanger sequencing

It is the same as the classic Sanger sequencing, the only difference lies in the use of different colored fluorescent ddNTPs instead of radiolabeled ddNTPs. This technique also reduces the workload and time consumed by cutting down the test tubes from four to one, as the entire reaction can take place in a single test tube because each ddNTP will give a different fluorescence. The type of electrophoresis used is capillary electrophoresis, and the fluorescence is observed by exciting the DNA fragments with a laser source (Sigma-Aldrich, n.d.).

15.2.3.3 Next-generation sequencing

The advanced DNA sequencing method of DNA profiling is known as the Next Generation Sequencing (NGS), also known as the high-throughput or mass parallel sequencing.This technique allows for the rapid sequencing of large quantities of DNA with greater efficiency (Johnson et al., 2014). However, despite its efficiency, NGS is yet not widely used for creating molecular signatures of wildlife species due to the high overall costs associated with - sequencing reagents, sample-processing,- library preparation, and data storage (Gouda et al., 2020). DNA profiles obtained from NGS can be used to check the authenticity of wildlife products as many wildlife products available in the local markets and online sites are suspected to be fraudulent. Studies conducted by Rajpoot et al. (2018) and Sharma et al. (2019) were able show that the genitals of Indian monitor lizards (possession and trade of which are prohibited under the Indian Wildlife (Protection) Act, 1972) were being cut, dried, and sold as "Hatha Jodi," the root of an herbaceous plant called *Martynia annua* (tiger's claw) used in Hindu religious practices, as both have a strong morphological resemblance and are difficult to distinguish by naked and unskilled eye. Cottenet et al. (2020) gauged the performance of NGS in the analysis of meat samples from European and Asian local markets to establish adulteration, meat mixing, and mislabelling; the NGS was successfully able to identify and detect meat species in mixtures up to 1% (w/w). Melville et al. (2017) employed a genome complexity reduction strategy involving restriction enzymes, combined with next-generation sequencing [Diversity Arrays Technology sequencing(DArTseq)] to comprehend how the scale patterns of the agamid lizard (*Ctenophorus caudicinctus*) contribute to its evolutionary and geographical progression. DArTseq, is a novel NGS technique and appears promising for phylogenetic investigations. The application of NGS for solving wildlife crime cases is still very restrained due to its cost, but it holds immense opportunities for wildlife cases in the future because of its sensitivity and high-throughput approach.

15.2.3.4 Forensically informative nucleotide sequencing

It is a rapid, reproducible, and reliable technique that is used for species identification for conservation and food safety purposes in extreme cases involving highly processed biological samples. It is particularly used in wildlife cases. FINS works by combining already established procedures to decipher the closest related species. It involves four stages as shown in Fig. 15.4, first: DNA extraction from the biological sample using appropriate methods depending upon the sample type and condition, second: feeding the extracted DNA to PCR to amplify the desired DNA segments, third: producing the nucleotide sequencing of the amplified DNA, and fourth: subjecting the obtained nucleotide sequence to a phylogenetic analysis to determine the closely related species to the given sample (Bartlett & Davidson, 1992).

Thakur et al. (2013) applied FINS to the extracted DNA from shed-off feathers and found nine polymorphic sites on the Cytochrome b gene at positions 15011, 15145, 15160, 15181, 15199, 15257, 15259, 15271, and 15330 which can be used to discriminate between *Pavo cristatus* (Indian peafowl)and *Pavo muticus* imperator (Indonesian peafowl). Further Rajpoot et al. (2017) extracted DNA from fourteen skin samples collected from dead Indian Varanus species individuals, applied FINS to the two mitochondrial genes (Cyt b and 16S rRNA) and made phylogenetic trees to show that there are two major clades (a group of organisms that includes a single ancestor and all of its descendants) present in Indian Varanidae viz. Clade A consists of the Yellow monitor and Desert monitor lizard, along with subclade A containing Bengal monitor lizard, and Clade B represented by Water monitor lizard. Later in 2018, Bahuguna (2018) extracted DNA from the taxidermy skin samples and performed FINS of Cyt b (1100 bp) and observed seven haplotypes (a group of genes within an organism that was inherited together from a single parent); 816 conserved sites, 284 variable sites, 84 parsimony informative sites (containing at least two types of nucleotides and at least two of them occur with a minimum frequency of two), and 200 singleton sites (containing at least two types of nucleotides with at most one occurring many times); and 12S rRNA (377 bp) and found five haplotypes, 312 conserved sites, 59 variable sites, 17 parsimony informative sites, and 41 singleton sites.

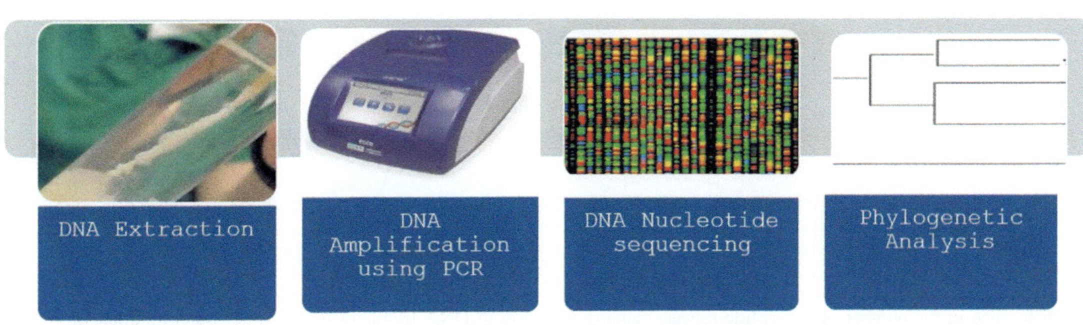

FIGURE 15.4 The steps of forensically informative nucleotide sequencing technique. *PCR*, Polymerase chain reaction.

15.2.4 Advancement in DNA markers for wildlife forensics

15.2.4.1 Single nucleotide polymorphism

SNP refers to the change in a single nucleotide or base pair of the DNA sequence, which acts as an identifying marker and helps in species identification as well as individualization if the questioned and the reference sample belong to a common origin. SNP doesn't require long fragments of high-quality DNA, but this advantage brings along a limitation of its own that only a small amount of information is available for comparison, unlike whole genome sequencing. Natesh et al. (2017) found and amplified a total of 339 SNPs in short fragments of 40 bp of tiger (*P. tigris*) for its conservation purposes.

15.2.4.2 Microsatellites

Also known as STRs are the repeated nucleotide sequences of very small length that are found in the noncoding region of DNA and are employed in DNA fingerprinting for identification purposes. Microsatellites can be divided into three categories based on the number of repeating units, viz. class I (> 20 bp), class II (11–20 bp), and class III (< 11 bp) (Temnykh et al., 2001; Varshney et al., 2002). Differences in the length of microsatellites cause a difference in the DNA fragments obtained after PCRs that can be resolved under electrophoresis and visualized by using fluorescent dyes. STRs are commonly used in individual identification as these polymorphic sequences are unique to an individual and their close relatives, but are extremely rare to be found in highly unrelated individuals (Lumen Learning, n.d.).

If two samples exhibit the same STR pattern, then they are more likely to have originated from a common source. Population identification becomes evident when there is a need to investigate whether an individual has been lawfully killed from a location where its hunting is legal, or illegally poached from a protected or conserved area. It should be noted that the gene pool of two or more highly isolated populations will have noticeable differences when sequenced (Fig. 15.5).

White et al. (2012) used STR markers for the identification of very rare and expensive black cockatoos (*Calyptorhynchus* spp.) endemic to Australia that were illegally procured or killed. They obtained the DNA profiles from the sample and compared the particular loci with the available cockatoo database to identify the sample in question.

15.2.4.3 Mitochondrial DNA markers

mtDNA is a small, circular, double-stranded structure, and about 100 s of copies are present in each cell. It is only 16.5 kb in size and encodes for only 37 genes, that is, 13 protein-coding genes, two rRNAs, and 22 tRNAs (Mitra et al., 2018). mtDNA is highly utilized in wildlife forensics because of its ability to resist degradation and destruction due to its own rigid membrane which is high in protein content (so it can be extracted from extremely degraded, aged, or even burnt samples where nuclear DNA (nuDNA) is almost impossible to find and extract) (Anderson et al., 1981), highly conserved nature as it only recombines with its identical copies in the cell, lack of proofreading during mtDNA replication so there are more chances for mutation to occur as compared to nuDNA which will allow for more unique sequences among individuals for comparison (Gray, 1989), and exhibition of maternal lineage as it is only acquired from the mother and not from the father (unlike the nuDNA) (Clayton et al., 2004; Gondhali & Mishra, 2022; Rastogi et al., 2007).

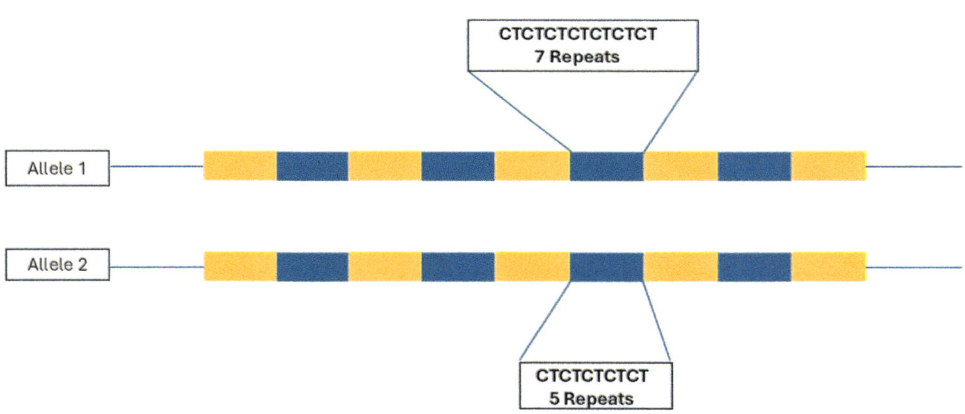

FIGURE 15.5 Difference in the number of nucleotide repeats causing variation in DNA fragment length (microsatellite alleles).

The mitochondrial markers used for species identification are cytochrome b (Cyt b) gene, cytochrome c oxidase subunit I (COI) gene, 12S and 16S rRNA segment and control region (D loop) in animals; *rbcL* and *matK* (plastid genes) in plants (Gondhali & Mishra, 2022; Johnson et al., 2014).

MtDNA markers have several advantages like, they can be used for species identification when the morphometric analysis fails to do so and can help in studying the phylogenetic relationships between species (Mitra et al., 2018). On the other hand, the disadvantages of mtDNA markers are that they cannot be used in case of mixed, complex, or contaminated DNA samples. If a legally protected hybrid species is produced by a nonprotected mother, then its mtDNA will have the same sequence as that of the nonprotected mother because of the maternal lineage of mtDNA, which can cause problems in the jurisdiction (Zhang & Hewitt, 1996). Heteroplasmy also poses problems when using mtDNA markers for identification (Breton et al., 2007; Crochet et al., 2003).

15.2.4.4 Nuclear DNA markers

The nuDNA is used for individual identification of the animal. Individual identification is done by STR or microsatellite profiling, which is unique to every individual, therefore, if there is a high percentage of similarity between the STR profiles of the unknown and the reference sample, then there is a high possibility that the two have originated from a common source. The most commonly employed nuDNA markers are STRs, random amplified polymorphic DNA, and amplified fragment length polymorphism (Mitra et al., 2018). The existing STR markers used for individualization purposes do not hold much forensic value, despite the vast availability of sequence data. This is basically due to the prevalence of dinucleotide STR markers in the literature, which have a subdued forensic value due to the production of stutter products (Singh et al., 2004).

15.2.5 DNA metabarcoding

DNA barcoding involves sequencing short DNA sequences (barcodes) from standardized regions and comparing the obtained sequence with a reference database for species identification (Staats et al., 2016). DNA metabarcoding is an advanced form of DNA barcoding, and it involves NGS for the simultaneous detection of multiple species in a complex mixture sample in a relatively shorter amount of time. It uses universal PCR primers that can mass-amplify one or many taxonomically informative sequences. DNA metabarcoding is very important in cases where there is a highly degraded or processed biological sample, such as in traditional medicines and potions, a sample suspected to contain DNA from various individuals, a cooked sample, etc. DNA metabarcoding finds its huge application in food safety, including food mislabeling, adulteration, substitution, etc. (Fajardo et al., 2010; Hanner et al., 2011; Wong & Hanner, 2008). The steps involved in DNA metabarcoding are given in Fig. 15.6.

For animals, the standard barcode is mitochondrial cytochrome *c* oxidase I (COI or COX1, CO1) (Hebert et al., 2003) due to its use from earlier times to study relationships of closely related species, as it has a high level of interspecific variation (Trontelj et al., 2005). It has a superior discrimination power for the identification of various species as given in Table 15.3.

Some other mitochondrial genes that can be used as barcodes are cytochrome b, 12S, and 16S rRNA. These barcodes can be used when the received samples are highly degraded, as mtDNA has an innate resistance to withstand degradation and time. Cytochrome b (*cytb*) can be used for the identification of mammals and hunted game meats (D'Amato et al., 2013; Galimberti et al., 2013; Kitano et al., 2007). Kitano et al. (2007) developed 12 S and 16 S rRNA mini-barcodes for

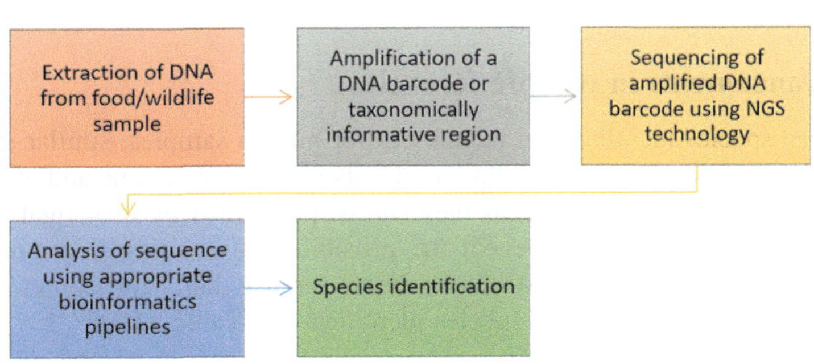

FIGURE 15.6 Various steps involved in DNA metabarcoding. *NGS*, Next-generation sequencing.

TABLE 15.3 Discrimination power for the identification of various species using the *COI* gene.

Taxonomic group	Identification success rate	References
Spiders	100%	Barrett and Hebert (2005)
Birds	98%–100%	Hebert et al. (2004)
Butterflies	97.9%	Hajibabaei et al. (2006)
Fish	93%–98%	Ward et al. (2005)
Reptiles	72.7%–100%	Nagy et al. (2012)

the identification of a large number of vertebrates. Likewise, Karlsson and Holmlund (2007) used 16S rRNA mini-barcodes for the identification of 28 different mammals, including domestic and game animals.

In plants, the mitochondrial genes are not of much use because the plant mitochondrial genome evolved too slowly to be used in DNA barcoding (Palmer & Herbon, 1988). Therefore, two plastid sequences viz. *rbc*L (ribulose bisphosphate carboxylase) having 599 bp and *mat*K (Maturase K) having 841 bp are used as "core" DNA barcodes in plants. The two most widely used supplementary loci are the nuclear ribosomal ITS (nrITS) (Chen et al., 2010) and plastid intergenic spacer *psb*A-*trn*H region (Kress et al., 2005). A better alternative to nrITS is a subset of the ribosomal cassette (nrITS2), which provides greater amplification and sequencing success compared to the entire nrITS region and has a successful identification rate of 92.7% at the species level (Chen et al., 2010). Coghlan et al. (2012) demonstrated that DNA metabarcoding can be successfully applied to traditional Chinese medicines, which contain a complex mixture of various plant and animal ingredients and can be found in any form, like capsules, crystals, powders, tablets, and even herbal tea.

15.2.5.1 Environmental DNA metabarcoding

eDNA consists of fragments of DNA shed into the environment by various organisms. These DNA fragments can be extracted from environmental samples such as soil or water (Freeland, 2017). Characterizing eDNA can indicate the presence or absence of species at specific sites. Using eDNA eliminates the need to identify and locate individual organisms, providing crucial insights into quantifying biodiversity (Freeland, 2017). In a study conducted by Xia et al. (2018), eDNA from laboratory aquaria and field channel water samples was detected to assess the presence or absence of a highly invasive mollusc species named the "golden mussel" (*L. fortune*). The eDNA metabarcoding technique is a truly noninvasive method for discovering and identifying endangered and rare species in diverse ecosystems, such as aquatic environments. Genetic traces left by animals in the environment can be easily retrieved using this method. This technique has significant advantages, as it does not damage the habitats of the species studied or cause harm to any threatened species or ecosystems during sampling. This advanced method is used worldwide to study the biodiversity of aquatic environments, yet no one in India is currently focusing on this new technology in forensic science. However, it has the potential to become a next-generation tool in aquatic ecosystem conservation and biodiversity research (Sahu et al., 2023). To develop eDNA metabarcoding markers, the amplification region must be effective, specific, and relatively short, as eDNA degrades quickly (Freeland, 2017; Othman et al., 2021; Ruppert et al., 2019).

15.2.6 Field-based advancements in wildlife forensics

With the advent of field-based species identification techniques for human samples, similar steps have been followed to bring a revolution in the field of wildlife forensics. Field-based tests not only save time and effort but also reduce the need for highly professional experts to come to the field, as they can be performed by investigative officials with little to no knowledge about wildlife forensics. They also prevent the possibility of sending a genuine sample for unnecessary analysis by experts using robust bench-mounted instruments in the forensic lab. Morrison et al. (2018) enumerates many tools and technologies that can be used for on-site species identification (Table 15.4).

TABLE 15.4 Advanced tools and technologies that can be used for on-site species identification.

Method	Description	Reference(s)
Recombinase polymerase amplification	A highly sensitive and specific isothermal method of DNA amplification that can take place in a single tube, without needing a thermal cycler.	James and MacDonald (2015)
TwistFlow™ Red Snapper	A lateral flow detection kit that is specific to Red snapper (*Lutjanus campechanus*) DNA giving either positive or negative results in less than 20 min, and only requires a one-step biochemical reaction before performing the assay, which can be easily performed by any nonexpert.	(SeafoodSource, 2013)
Real-time nucleic acid sequence-based amplification	Developed for RNA amplification, it has similar efficiency to benchtop RNA purification instruments but with reduced sensitivity. Takes about 80 min and allows detection of sequences using a fluorometer, which doubles as a heating device for optimal temperatures.	Ulrich et al. (2015)
MinION™	An on-site DNA/RNA sequencing device by Oxford Nanopore Technologies. Lightweight, portable, smaller than a mobile phone, and connects to a laptop via USB 3.0. Suitable for remote locations, but requires expert knowledge. Studies show an error rate as high as 38%, reducible to as low as 5% with improved data analysis.	Jain et al. (2015); Kilianski et al. (2015); Laver et al. (2015); Mikheyev and Tin (2014)

15.2.7 Genetic databases for wildlife forensics

Genetic Databases for Wildlife Forensics are indispensable tools for the identification of species. These repositories encompass with millions of sequences (Gondhali & Mishra, 2022).

15.2.7.1 GenBank

GenBank (https://www.ncbi.nlm.nih.gov/genbank/) is an open-source DNA database started in the year of 1992; built and distributed by National Center for Biotechnology Information, a division of National Library of Medicine in the US. This database is enriched by the information provided by authors, scientists, research scholars, etc. (Benson et al., 2009). Each GenBank entry contains a characteristics table that highlights biologically relevant sections, a brief description of the sequence, the scientific name and taxonomy of the source organism, and bibliographic references. These segments include repetition regions, transcription units, coding sequences, and the protein translations corresponding to them, and mutation or modification sites. This publicly available database contains nucleotide sequences for more than 300,000 organisms. This stored nucleotide sequence treasure can be used to compare with the sequence obtained from the questioned sample to decipher its species, genus, or other taxa, which is very helpful in jurisdiction cases related to wildlife crimes. It also incorporated a sequence search and alignment tool, which is known as the Basic Local Alignment Search Tool (BLAST). While most of the data has open access, some authors might patent their findings or keep their work on request-based access to avoid any tampering or misuse of their work (Gondhali & Mishra, 2022; National Center for Biotechnology Information, 1982).

15.2.7.2 Barcode of Life Data Systems

It is also a publicly accessible DNA sequence database that can be used for academic as well as legal purposes. It is a web-based platform having a very user-friendly Graphical User Interface, the homepage of which shows four major icons: vPublic Data Portal: here the searches can be made in terms of geography, taxonomy, depository, etc., and these can be summarized, plotted on high-resolution maps, and downloaded. Barcode Index Number Database: used when taxonomic data is lacking or unverified. It is used to look for sequences that closely resemble the given sequence. Publications: holds a collection of published barcodes and the papers that have used the barcode results. Primer Database: it contains a registry of primers used for the generation of barcodes. This catalog is maintained by the users of Barcode of Life Data Systems (BOLD Systems, n.d.) - (https://v3.boldsystems.org/index.php/databases).

15.2.7.3 ForCyt DNA database of wildlife species

This is comparatively a newer, publicly available database project developed by (Ahlers et al., 2017)- (http://www.forcyt.org/). Unlike other DNA databases that use a single gene sequence, it utilizes the sequence of the entire mtDNA genome to increase the robustness of wildlife applications. The use of mtDNA sequence enables the comparison process to take place in any lab, even where different mtDNA genes are used for the identification of species. Once sequenced, the library for mtDNA data is prepared and is published only when it is meticulously reviewed and strictly scrutinized to maintain the accuracy and quality of data going online. This available information can be utilized by authors, research scholars, legal officials, etc.

15.3 Challenges in wildlife DNA forensics

15.3.1 Lack of reference database

A large number of species that are victims of wildlife crimes are either endangered or on the verge of extinction, due to which it is becoming exceedingly difficult to obtain any kind of sample from them. It creates a big hurdle in the functioning of wildlife forensics (Kanthaswamy, 2024).

15.3.2 Poor accessibility

Many species of flora and fauna are protected by various legislation of CITES, as a result of which their transport or trade is highly monitored and controlled, creating a need for citing special requests from the authorities to procure the sample for creating a reference database (Kanthaswamy, 2024).

15.3.3 Zoonoses/zoonotic diseases

Public health concerns over the occurrence and communication of various types of zoonotic diseases, which are escalating globally. These are the diseases that spread from nonhuman organisms to humans and can range from mild diseases to fatal ones. Wildlife trade is being identified as one of the main drivers to spread these diseases. Zoonoses can be broadly classified into Bacterial zoonoses (caused by bacteria), viral zoonoses (caused by viruses), mycotic/ fungal zoonoses (caused by fungi), and parasitic zoonoses (caused by parasites). If proper care and instructions are not followed, the person handling the sample might catch a zoonotic disease. The following are some examples of zoonoses (Table 15.5) (Jones et al., 2008; National Center for Biotechnology Information [NCBI], n.d.; Smart et al., 2021).

15.3.4 Poor handling

Many times, wildlife samples are not taken seriously like human samples, and their handling, packaging, and transportation show signs of carelessness, as a result of which the sample received at the laboratory is not present in its best form. This can lead to contamination of DNA in the sample, initiation of degradation, growth of microbes, etc., which produce a negative effect on sample identification by acting as a hindrance or inhibitor in several protocols (Kanthaswamy, 2024).

15.3.5 Highly processed samples

It is well known that most of the samples received in wildlife crime cases are either chopped, cooked, crafted, or prepared something so processed that it looks nothing like its original form. Here comes the difficult part, since the morphometric features of the sample are obliterated, the officials find it complex to decipher whether the sample in question has originated from a protected species or is just a genuine sample (unnecessary analysis of which might waste the time and resources of an equipped laboratory), this is where field-based assays come into play (Gupta et al., 2005).

15.3.6 Lack of awareness

Wildlife forensics is still a niche field that requires a lot more research and exploration. Many countries still lack highly equipped and wildlife-focused laboratories. Moreover, many scientists of conventional FSLs don't take a wildlife case as

TABLE 15.5 Example of zoonoses.

Disease	Animal host
Bacterial zoonoses	
Anthrax	Cattle, horses, sheep, pigs, dogs, bison, elk, white-tailed deer, goats, and mink
Tuberculosis	Cattle, sheep, swine, deer, wild boars, camels, and bison
Bubonic plague	Rock squirrels, wood rats, ground squirrels, prairie dogs, mice, voles, chipmunks, and rabbits
Tularemia	Rabbits, squirrels, muskrats, deer, sheep, bull snakes, wild rodents, beavers, cats, and dogs
Pasteurellosis	Poultry, pigs, cattle, buffaloes, sheep, goats, deer, cats, dogs, and antelope
Clostridioides difficile infection	Cattle, horses, and birds
Viral zoonoses	
Rabies	Cattle, horses, cats, dogs, bats, monkeys, wolves, skunks, rabbits, and coyotes
Avian influenza	Ducks, chickens, turkeys, dogs, cats, pigs, whales, horses, seals, and wild birds
Ebola virus disease (Ebola hemorrhagic fever)	Monkeys, gorillas, chimpanzees, apes, and wild antelopes
AIDS	Monkeys and chimpanzees
West Nile fever	Horses, birds, and reptiles
Hantavirus infection (hantavirus pulmonary syndrome)	Deer mice, cotton rats, rice rats, white-footed mice, shrews, and moles
Parasitic zoonoses	
Trichinellosis	Pigs, dogs, cats, rats, and other wild species
Visceral larva migrans	Birds, emus, cats, chinchillas, porcupines, prairie dogs, rabbits, weasels, woodchucks, and wood rats
Cryptococcosis	Dogs, cattle, horses, sheep, goats, birds, and wild animals
Fungal/mycotic zoonoses	
Blastomycosis	Mostly dogs, cats, and less common in horses, ferrets, deer, wolves, African lionizers, bottle-nosed dolphins, and sea lionizers
Cryptococcosis	Cats, dogs, cattle, horses, sheep, goats, birds, and wild animals
Sporotrichosis	Dogs, cats, horses, cows, camels, dolphins, goats, mules, birds, pigs, rats, and armadillos

they find it beyond their expertise and knowledge, along with this they have a heavy workload of human samples to be analyzed and reported for jurisdiction, and accepting a wildlife case will only add more to the burden.

15.4 Conclusion

Wildlife DNA forensics plays an important role in combating wildlife crimes by aiding in species identification, geographic origin determination, individual identification, and distinguishing between specimens bred in captivity or stolen from the wild. The integration of DNA analysis techniques in conservation genetics has led to significant advancements in the field, offering a powerful tool for wildlife conservation efforts. In forensic analysis, PCR methods like direct PCR and conventional PCR are useful instruments for amplifying DNA material. Direct PCR amplifies the sample directly without the need for DNA extraction and purification, which is a step in conventional PCR. The benefits of direct PCR include heightened sensitivity, quick and accurate findings, and the capacity to amplify different kinds of samples. In wildlife forensics, advanced DNA markers like SNP, Microsatellite, mtDNA, nuDNA, and DNA metabarcoding are used for an assortment of purposes, including population identification, species identification, individualization, and phylogenetic relationship analysis. With the advent of instruments like real-time nucleic acid sequence-based amplification, TwistFlow™ Red Snapper, recombinase polymerase amplification, and MinION™, field-based assays for wildlife forensics have made tremendous strides. By enabling on-site species identification, with little to no taxonomic knowledge, these instruments diminish the need for forensic lab analysis of genuine samples and cut down on the number of specialized experts on site. Genetic databases such as GenBank, BOLD, and ForCyt play a crucial role in wildlife forensics by providing a vast collection of nucleotide sequences that can be used to identify species and aid in jurisdiction cases related to wildlife crimes.

References

Ahlers, N., Creecy, J., Frankham, G., Johnson, R. N., Kotze, A., Linacre, A., McEwing, R., Mwale, M., Rovie-Ryan, J. J., & Sitam, F. (2017). ForCyt'DNA database of wildlife species. *Forensic Science International: Genetics Supplement Series, 6*, 466.

Anderson, S., Bankier, A. T., Barrell, B. G., De Bruijn, M. H. L., Coulson, A. R., Drouin, J., Eperon, I. C., Nierlich, D. P., Roe, B. A., Sanger, F., Schreier, P. H., Smith, A. J. H., Staden, R., & Young, I. G. (1981). Sequence and organization of the human mitochondrial genome. *Nature, 290*(5806), 457–465. https://doi.org/10.1038/290457a0.

Bahuguna, A. (2018). Forensically informative nucleotide sequencing (FINS) for species and subspecies of genus Prionailurus (Mammalia: Carnivora: Felidae) through mitochondrial genes (12SrRNA and cytochrome b) by using old taxidermy samples. *Mitochondrial DNA Part B, 3*(2), 615–619. https://doi.org/10.1080/23802359.2018.1462115.

Barbaro, A., Staiti, N., Cormaci, P., & Saravo, L. (2004). DNA profiling by different extraction methods. *International Congress Series, 1261*(C), 562–564. https://doi.org/10.1016/s0531-5131(03)01647-9.

Barbaro, A., Cormaci, P., Teatino, A., & Barbaro, A. (2008). Use of "AnyDirect PCR buffer" for PCR amplification of washed bloodstains: A case report. *Forensic Science International: Genetics Supplement Series, 1*(1), 11–12. https://doi.org/10.1016/j.fsigss.2007.10.201.

Barrett, R. D. H., & Hebert, P. D. N. (2005). Identifying spiders through DNA barcodes. *Canadian Journal of Zoology, 83*(3), 481–491. https://doi.org/10.1139/z05-024.

Bartlett, S. E., & Davidson, W. S. (1992). FINS (forensically informative nucleotide sequencing): A procedure for identifying the animal origin of biological specimens. *BioTechniques, 12*(3), 408–411.

Benson, D. A., Karsch-Mizrachi, I., Lipman, D. J., Ostell, J., & Sayers, E. W. (2009). GenBank. *Nucleic Acids Research, 38*(1), D46. https://doi.org/10.1093/nar/gkp1024.

Bioinformatics Home. (n.d.). "67 Free DNA Sequence Analysis Tools – Software and Resources," Bioinformatics Home. [Online]. https://bioinformaticshome.com/tools/DNA-sequence-analysis/DNA-sequence-analysis.html#gsc.tab=0. [Accessed: Aug. 7, 2025].

Bioline | Meridian Bioscience. (n.d.). ISOLATE II Genomic DNA Kit. https://www.bioline.com/isolate-ii-genomic-dna-kit.html.

Blackie, R., Taylor, D., & Linacre, A. (2016). DNA profiles from clothing fibers using direct PCR. *Forensic Science, Medicine, and Pathology, 12*(3), 331–335. https://doi.org/10.1007/s12024-016-9784-y, http://www.springerlink.com/content/120561/.

BOLD Systems, (n.d.). https://v3.boldsystems.org/index.php/databases.

Breton, S., Beaupré, H. D., Stewart, D. T., Hoeh, W. R., & Blier, P. U. (2007). The unusual system of doubly uniparental inheritance of mtDNA: Isn't one enough? *Trends in Genetics, 23*(9), 465–474. https://doi.org/10.1016/j.tig.2007.05.011.

Brito, P., Gouveia, N., Bogas, V., Serra, A., Balsa, F., Andrade, L., São-Bento, M., Cunha, P., Lopes, V., & Anjos, M. J. (2015). Direct amplification of casework samples in fabrics using GlobalFiler® PCR amplification kit. *Forensic Science International: Genetics Supplement Series, 5*, e487. https://doi.org/10.1016/j.fsigss.2015.09.193.

Buckingham, A. K., Harvey, M. L., & van Oorschot, R. A. H. (2016). The origin of unknown source DNA from touched objects. *Forensic Science International: Genetics, 25*, 26–33. https://doi.org/10.1016/j.fsigen.2016.07.015, http://www.elsevier.com.

Butler, J. M. (2005). *Forensic DNA typing: Biology, technology, and genetics of STR markers*. Elsevier.

Carvajal-Agudelo, J. D., Trujillo-Betancur, M. P., Velásquez-Guarín, D., Ramírez-Chaves, H. E., Pérez-Cárdenas, J. E., & Rivera-Páez, F. A. (2021). Field blood preservation and DNA extraction from wild mammals: Methods and key factors for biodiversity studies. *Universidad de Ciencias Aplicadas y Ambientales U.D.C.A, Colombia Revista U.D.C.A Actualidad and Divulgacion Cientifica, 24*(1). https://doi.org/10.31910/rudca.v24.n1.2021.1766, https://revistas.udca.edu.co/index.php/ruadc/article/view/1766/2135.

Chen, S., Yao, H., Han, J., Liu, C., Song, J., Shi, L., Zhu, Y., Ma, X., Gao, T., Pang, X., Luo, K., Li, Y., Li, X., Jia, X., Lin, Y., Leon, C., & Gilbert, M. T. P. (2010). Validation of the ITS2 region as a novel DNA barcode for identifying medicinal plant species. *PLoS ONE, 5*(1), e8613. https://doi.org/10.1371/journal.pone.0008613.

Clayton, T. M., Guest, J. L., Urquhart, A. J., & Gill, P. D. (2004). A basis for anomalous band patterns encountered during DNA STR profiling. *Journal of Forensic Sciences, 49*(6). https://doi.org/10.1520/jfs2003145.

Coghlan, M. L., Haile, J., Houston, J., Murray, D. C., White, N. E., Moolhuijzen, P., Bellgard, M. I., & Bunce, M. (2012). Deep sequencing of plant and animal DNA contained within traditional Chinese medicines reveals legality issues and health safety concerns. *PLoS Genetics, 8*(4). http://www.plosgenetics.org/article/fetchObjectAttachment.action?uri=info%3Adoi%2F10.1371%2Fjournal.pgen.1002657&representation=PDF https://doi.org/10.1371/journal.pgen.1002657Australia, http://www.plosgenetics.org/article/fetchObjectAttachment.action?uri=info%3Adoi%2F10.1371%2Fjournal.pgen.1002657&representation=PDF.

Conte, J., Potoczniak, M. J., Mower, C., & Tobe, S. S. (2019). ELEquant: A developmental framework and validation of forensic and conservation real-time PCR assays. *Molecular Biology Reports, 46*(2), 2093–2100. https://doi.org/10.1007/s11033-019-04660-7.

Cottenet, G., Blancpain, C., Chuah, P. F., & Cavin (2020). Evaluation and application of a next generation sequencing approach for meat species identification. *Food Control, 110*.

Crochet, P. A., Chen, J. Z., Pons, J. M., Lebreton, J. D., Hebert, P. D. N., & Bonhomme, F. (2003). Genetic differentiation at nuclear and mitochondrial loci among large white-headed gulls: Sex-biased interspecific gene flow? *Society for the Study of Evolution, France Evolution, 57*(12), 2865–2878. https://doi.org/10.1111/j.0014-3820.2003.tb01527.x, http://www.blackwell-synergy.com/toc/evo/61/11.

D'Amato, M. E., Alechine, E., Cloete, K. W., Davison, S., & Corach, D. (2013). Where is the game? Wild meat products authentication in South Africa: A case study. *Investigative Genetics, 4*(1). https://doi.org/10.1186/2041-2223-4-6, http://www.investigativegenetics.com/content/4/1/6 South Africa.

Dinis, V., Bento, A. M., Teixeira, A. L., Gouveia, N., Bogas, V., Porto, M. J., & Corte-Real, F. (2017). Comparative study between a direct DNA quantification methodology and the standardized methodology in the forensic workflow. *Forensic Science International: Genetics Supplement Series, 6*, e216. https://doi.org/10.1016/j.fsigss.2017.09.076.

Ellis, R. (2013). *Tiger bone & rhino horn: The destruction of wildlife for traditional Chinese medicine*. Island Press.

Fajardo, V., González, I., Rojas, M., García, T., & Martín, R. (2010). A review of current PCR-based methodologies for the authentication of meats from game animal species. *Trends in Food Science & Technology, 21*(8), 408–421. https://doi.org/10.1016/j.tifs.2010.06.002.

Federal Bureau of Investigation. (2011). Quality Assurance Standards for Forensic DNA Testing Laboratories. FBI. https://ucr.fbi.gov/lab/biometric-analysis/codis/quality-assurance-standards-for-forensic-dna-testing-laboratories.

Freeland, J. R. (2017). The importance of molecular markers and primer design when characterizing biodiversity from environmental DNA. *Canadian Science Publishing, Canada Genome, 60*(4), 358–374. https://doi.org/10.1139/gen-2016-0100, http://pubs.nrc-cnrc.gc.ca.

Galimberti, A., De Mattia, F., Losa, A., Bruni, I., Federici, S., Casiraghi, M., Martellos, S., & Labra, M. (2013). DNA barcoding as a new tool for food traceability. *Food Research International, 50*(1), 55–63. https://doi.org/10.1016/j.foodres.2012.09.036.

Gondhali, U., & Mishra, A. (2022). Wildlife DNA profiling and its forensic relevance. In *Handbook of DNA profiling*. India: Springer, 823–838. https://link.springer.com/referencework/10.1007/978-981-15-9364-2, https://doi.org/10.1007/978-981-16-4318-7_41.

Goray, M., Fowler, S., Szkuta, B., & van Oorschot, R. A. H. (2016). Shedder status—An analysis of self and non-self DNA in multiple handprints deposited by the same individuals over time. *Forensic Science International: Genetics, 23*, 190–196. https://doi.org/10.1016/j.fsigen.2016.05.005.

Gouda, S., Kerry, R. G., Das, A., & Chauhan, N. S. (2020). Wildlife forensics: A boon for species identification and conservation implications. *Forensic Science International, 317*. https://doi.org/10.1016/j.forsciint.2020.110530, www.elsevier.com/locate/forsciint.

Graham, E. A. M., & Rutty, G. N. (2008). Investigation into "normal" background DNA on adult necks: Implications for DNA profiling of manual strangulation victims. *Journal of Forensic Sciences, 53*(5), 1074–1082. https://doi.org/10.1111/j.1556-4029.2008.00800.x.

Gray, M. W. (1989). Origin and evolution of mitochondrial DNA. *Annual Review of Cell Biology, 5*, 25–50. https://doi.org/10.1146/annurev.cb.05.110189.000325.

Gupta, S. K., Verma, S. K., & Singh, L. (2005). Molecular insight into a wildlife crime: The case of a peafowl slaughter. *Forensic Science International, 154*(2-3), 214–217. https://doi.org/10.1016/j.forsciint.2004.12.010, www.elsevier.com/locate/forsciint.

Hajibabaei, M., Janzen, D. H., Burns, J. M., Hallwachs, W., & Hebert, P. D. N. (2006). DNA barcodes distinguish species of tropical Lepidoptera. *Proceedings of the National Academy of Sciences of the United States of America, 103*(4), 968–971. https://doi.org/10.1073/pnas.0510466103.

Handt, O., & Linacre, A. (2015). The end of bad hair days. *Forensic Science International: Genetics Supplement Series, 5*, e146. https://doi.org/10.1016/j.fsigss.2015.09.059, http://www.elsevier.com.

Hanner, R., Becker, S., Ivanova, N. V., & Steinke, D. (2011). FISH-BOL and seafood identification: Geographically dispersed case studies reveal systemic market substitution across Canada. *Mitochondrial DNA, 22*(1), 106–122. https://doi.org/10.3109/19401736.2011.588217.

Hebert, P. D. N., Cywinska, A., Ball, S. L., & DeWaard, J. R. (2003). Biological identifications through DNA barcodes. *Proceedings of the Royal Society B: Biological Sciences, 270*(1512), 313–321. https://doi.org/10.1098/rspb.2002.2218.

Hebert, P. D. N., Stoeckle, M. Y., Zemlak, T. S., & Francis, C. M. (2004). Identification of birds through DNA barcodes. *PLoS Biology, 2*(10). https://doi.org/10.1371/journal.pbio.0020312Canada, http://www.plosbiology.org/plosonline/?request=get-document&doi=10.1371%2Fjournal.pbio.0020312.

Jain, M., Fiddes, I. T., Miga, K. H., Olsen, H. E., Paten, B., & Akeson, M. (2015). Improved data analysis for the MinION nanopore sequencer. *Nature Methods, 12*(4), 351–356. https://doi.org/10.1038/nmeth.3290, http://www.nature.com/nmeth/.

James, A., & MacDonald, J. (2015). Recombinase polymerase amplification: Emergence as a critical molecular technology for rapid, low-resource diagnostics. *Expert Review of Molecular Diagnostics, 15*(11), 1475–1489. https://doi.org/10.1586/14737159.2015.1090877.

Johnson, R. N., Wilson-Wilde, L., & Linacre, A. (2014). Current and future directions of DNA in wildlife forensic science. *Forensic Science International: Genetics, 10*(1), 1–11. https://doi.org/10.1016/j.fsigen.2013.12.007, http://www.elsevier.com.

Jones, K. E., Patel, N. G., Levy, M. A., Storeygard, A., Balk, D., Gittleman, J. L., & Daszak, P. (2008). Global trends in emerging infectious diseases. *Nature, 451*(7181), 990–993. https://doi.org/10.1038/nature06536, http://www.nature.com/nature/index.html.

Kanthaswamy, S. (2024). Review: Wildlife forensic genetics—Biological evidence, DNA markers, analytical approaches, and challenges. *Animal Genetics, 55*(2), 177–192. https://doi.org/10.1111/age.13390, http://onlinelibrary.wiley.com/journal/10.1111/(ISSN)1365-2052.

Karlsson, A. O., & Holmlund, G. (2007). Identification of mammal species using species-specific DNA pyrosequencing. *Forensic Science International, 173*(1), 16–20. https://doi.org/10.1016/j.forsciint.2007.01.019, www.elsevier.com/locate/forsciint.

Keatley, S., Botero, A., Fosu-Nyarko, J., Pallant, L., Northover, A., & Thompson, R. C. A. (2020). Species-level identification of trypanosomes infecting Australian wildlife by high-resolution melting - real time quantitative polymerase chain reaction (HRM-qPCR). *International Journal for Parasitology: Parasites and Wildlife, 13*, 261–268. https://doi.org/10.1016/j.ijppaw.2020.11.003.

Kilianski, A., Haas, J. L., Corriveau, E. J., Liem, A. T., Willis, K. L., Kadavy, D. R., Rosenzweig, C. N., & Minot, S. S. (2015). Bacterial and viral identification and differentiation by amplicon sequencing on the MinION nanopore sequencer. *GigaScience, 4*(1). https://doi.org/10.1186/s13742-015-0051-z, http://www.gigasciencejournal.com/content/4/1/12.

Kitano, T., Umetsu, K., Tian, W., & Osawa, M. (2007). Two universal primer sets for species identification among vertebrates. *International Journal of Legal Medicine, 121*(5), 423–427. https://doi.org/10.1007/s00414-006-0113-y.

Kitpipit, T., Thanakiatkrai, P., & Chotigeat, W. (2013). Direct PCR-FINS: Wildlife species identification without DNA extraction. *Forensic Science International: Genetics Supplement Series, 4*(1), e364. https://doi.org/10.1016/j.fsigss.2013.10.185.

Kress, W. J., Wurdack, K. J., Zimmer, E. A., Weigt, L. A., & Janzen, D. H. (2005). Use of DNA barcodes to identify flowering plants. *Proceedings of the National Academy of Sciences of the United States of America, 102*(23), 8369–8374. https://doi.org/10.1073/pnas.0503123102.

Laver, T., Harrison, J., O'Neill, P. A., Moore, K., Farbos, A., Paszkiewicz, K., & Studholme, D. J. (2015). Assessing the performance of the Oxford nanopore technologies MinION. *Biomolecular Detection and Quantification, 3*, 1–8. https://doi.org/10.1016/j.bdq.2015.02.001.

Li, R. (2008). *Forensic biology: Identification and DNA analysis of biological evidence*. Taylor and Francis, 1–414. https://doi.org/10.4324/9781420043440, https://www.routledge.com/Forensic-Biology-Identification-and-DNA-Analysis-of-Biological-Evidence/author/p/book/9781420043440.

Li, R. (2011). *Forensic biology*. United States of America: CRC Press.

Linacre, A., Pekarek, V., Swaran, Y. C., & Tobe, S. S. (2010). Generation of DNA profiles from fabrics without DNA extraction. *Forensic Science International: Genetics, 4*(2), 137–141. https://doi.org/10.1016/j.fsigen.2009.07.006.

Lumen Learning. (n.d.). Reading: Genetic information used for identification. Western Michigan University Biology 1. August 7, 2025, from https://courses.lumenlearning.com/wm-biology1/chapter/reading-genetic-information-used-for-identification.

Magee, A. M., Breathnach, M., Doak, S., Thornton, F., Noone, C., & McKenna, L. G. (2018). Wearer and non-wearer DNA on the collars and cuffs of upper garments of worn clothing. *Forensic Science International: Genetics, 34*, 152–161. https://doi.org/10.1016/j.fsigen.2018.02.011, http://www.elsevier.com.

Mainka, S. A., & Mills, J. A. (1995). Wildlife and traditional Chinese medicine: Supply and demand for wildlife species. *Journal of Zoo and Wildlife Medicine*, 193–200.

Mao, X., Liu, C., Tong, H., Chen, Y., & Liu, K. (2019). Principles of digital PCR and its applications in current obstetrical and gynecological diseases. *American Journal of Translational Research, 11*(12), 7209–7222. http://www.ajtr.org/files/ajtr0101419.pdf.

Martin, B., & Linacre, A. (2020). Direct PCR: A review of use and limitations. *Science and Justice, 60*(4), 303–310. https://doi.org/10.1016/j.scijus.2020.04.003, http://www.scienceandjusticejournal.com/contact.

Martin, B., Blackie, R., Taylor, D., & Linacre, A. (2018). DNA profiles generated from a range of touched sample types. *Forensic Science International: Genetics, 36*, 13–19. https://doi.org/10.1016/j.fsigen.2018.06.002, http://www.elsevier.com.

Meiklejohn, K. A., Burnham-Curtis, M. K., Straughan, D. J., Giles, J., & Moore, M. K. (2021). Current methods, future directions and considerations of DNA-based taxonomic identification in wildlife forensics. *Forensic Science International: Animals and Environments, 1*. https://doi.org/10.1016/j.fsiae.2021.100030, https://www.journals.elsevier.com/forensic-science-international-animals-and-environments.

Melville, J., Haines, M. L., Boysen, K., Hodkinson, L., Kilian, A., Smith Date, K. L., Potvin, D. A., & Parris, K. M. (2017). Identifying hybridization and admixture using SNPs: Application of the DArTseq platforminphylogeographic research on vertebrates. *Royal Society Open Science, 4*(7). https://doi.org/10.1098/rsos.161061, http://rsos.royalsocietypublishing.org/content/royopensci/4/7/161061.full.pdf.

Mikheyev, A. S., & Tin, M. M. Y. (2014). A first look at the Oxford Nanopore MinION sequencer. *Molecular Ecology Resources, 14*(6), 1097–1102. https://doi.org/10.1111/1755-0998.12324Japan, http://onlinelibrary.wiley.com/journal/10.1111/(ISSN)1755-0998.

Mitra, I., Roy, S., & Haque, I. (2018). Application of molecular markers in wildlife DNA forensic investigations. *Journal of Forensic Science and Medicine, 4*(3), 156–160. https://doi.org/10.4103/jfsm.jfsm_23_18, http://www.jfsmonline.com/aboutus.asp.

Morrison, J., Watts, G., Hobbs, G., & Dawnay, N. (2018). Field-based detection of biological samples for forensic analysis: Established techniques, novel tools, and future innovations. *Forensic Science International, 285*, 147–160. https://doi.org/10.1016/j.forsciint.2018.02.002, www.elsevier.com/locate/forsciint.

Nagy, Z. T., Sonet, G., Glaw, F., Vences, M., & Chave, J. (2012). First large-scale DNA barcoding assessment of reptiles in the biodiversity hotspot of Madagascar, based on newly designed COI primers. *PLoS ONE, 7*(3), e34506. https://doi.org/10.1371/journal.pone.0034506.

Natesh, M., Taylor, R., Prost, S., Hadly, E., Petrov, D., & Ramakrishnan, U. (2017). SNP genotyping to monitor wild tigers for conservation. *Canadian Journal of Biotechnology, 1*(Special Issue), 19. https://doi.org/10.24870/cjb.2017-a7.

National Association of Testing Authorities (NATA). 2020. Specific Accreditation Criteria: ISO/IEC 17025 Application Document, Life Sciences – Appendix (SAC-07). Issued November 2020; Effective February 2021.In text citation: (Federal Bureau of Investigation, 1998)Full Reference: Federal Bureau of Investigation, "Quality Assurance Standards for Forensic DNA Testing Laboratories" and "Quality Assurance Standards for Convicted Offender DNA Databasing Laboratories," Forensic Science Communications, 2(3).

National Center for Biotechnology Information (NCBI) (1982). Bethesda (MD): National Library of Medicine (US), National Center for Biotechnology Information; [1982] – [cited 2025 Aug 7]. https://www.ncbi.nlm.nih.gov/genbank/.

Norgen Biotek Corp. Genomic DNA Isolation Kit. https://norgenbiotek.com/product/genomic-dna-isolation-kit.

Ogden, R., Dawnay, N., & McEwing, R. (2009). Wildlife DNA forensics - Bridging the gap between conservation genetics and law enforcement. *Endangered Species Research, 9*(3), 179–195. https://doi.org/10.3354/esr00144, http://www.int-res.com/articles/esr2009/9/n009p179.pdf.

Ojaveer, H., Galil, B. S., Minchin, D., Olenin, S., Amorim, A., Canning-Clode, J., Chainho, P., Copp, G. H., Gollasch, S., Jelmert, A., Lehtiniemi, M., McKenzie, C., Mikuš, J., Miossec, L., Occhipinti-Ambrogi, A., Pećarević, M., Pederson, J., Quilez-Badia, G., Wijsman, J. W. M., & Zenetos, A. (2014). Ten recommendations for advancing the assessment and management of non-indigenous species in marine ecosystems. *Marine Policy, 44*, 160–165. https://doi.org/10.1016/j.marpol.2013.08.019.

Othman, N., Haris, H., Fatin, Z., Najmuddin, M. F., Sariyati, N. H., Md-Zain, B. M., & Abdul-Latiff, M. A. B. (2021). A review on environmental DNA (eDNA) metabarcoding markers for wildlife monitoring research. In *IOP Conference Series: Earth and Environmental Science*. Malaysia: IOP Publishing Ltd. 1 https://doi.org/10.1088/1755-1315/736/1/012054, https://iopscience.iop.org/journal/1755-1315.

Ottens, R., Taylor, D., Abarno, D., & Linacre, A. (2013). Successful direct amplification of nuclear markers from a single hair follicle. *Humana Press Inc., Australia Forensic Science, Medicine, and Pathology, 9*(2), 238–243. https://doi.org/10.1007/s12024-012-9402-6, http://www.springerlink.com/content/120561/.

Ottens, R., Taylor, D., & Linacre, A. (2015). DNA profiles from fingernails using direct PCR. *Forensic Science, Medicine, and Pathology, 11*(1), 99–103. https://doi.org/10.1007/s12024-014-9626-8, http://www.springerlink.com/content/120561/.

Ottens, R., Templeton, J., Paradiso, V., Taylor, D., Abarno, D., & Linacre, A. (2013). Application of direct PCR in forensic casework. *Forensic Science International: Genetics Supplement Series, 4*(1), e47. https://doi.org/10.1016/j.fsigss.2013.10.024.

Palmer, J. D., & Herbon, L. A. (1988). Plant mitochondrial DNA evolved rapidly in structure, but slowly in sequence. *Journal of Molecular Evolution, 28*(1-2), 87–97. https://doi.org/10.1007/BF02143500.

Panaccio, M., Georgesz, M., Hollywell, C., & Lew, A. (1993). Direct PCR from solid tissues without DNA extraction. *Nucleic Acids Research, 21*(19), 4656. https://doi.org/10.1093/nar/21.19.4656.

Park, S. J., Kim, J. Y., Yang, Y. G., & Lee, S. H. (2008). Direct STR amplification from whole blood and blood- or saliva-spotted FTA® without DNA purification. *Journal of Forensic Sciences, 53*(2), 335–341. https://doi.org/10.1111/j.1556-4029.2008.00666.x.

Promega Corporation. n.d. Wizard® Genomic DNA Purification Kit. Catalog No. A1120. https://www.promega.in/products/nucleic-acid-extraction/genomic-dna/wizard-genomic-dna-purification-kit/catNum=A1120&accordion0=0#specifications.

QIAGEN. n.d. DNeasy Blood & Tissue Kit. Catalog No. 69504. https://www.qiagen.com/us/products/discovery-and-translational-research/dna-rna-purification/dna-purification/genomic-dna/dneasy-blood-and-tissue-kit.

Quan, P.-L., Sauzade, M., & Brouzes, E. (2018). dPCR: A technology review. *Sensors, 18*(4), 1271. https://doi.org/10.3390/s18041271.

Rajpoot, A., Kumar, V. P., Bahuguna, A., & Kumar, D. (2017). Forensically informative nucleotide sequencing (FINS) for the first time authentication of Indian Varanus species: implication in wildlife forensics and conservation. *Mitochondrial DNA Part A, 28*(6), 892–900. https://doi.org/10.1080/24701394.2016.1202943.

Rajpoot, A., Kumar, V. P., Bahuguna, A., Singh, T., Joshi, S., & Kumar, D. (2018). Wildlife forensics in battle against veneration frauds in Uttarakhand, India: identification of protected Indian monitor lizard in items available in the local market under the name of Hatha Jodi. *Mitochondrial DNA Part B, 3*(2), 925–932. https://doi.org/10.1080/23802359.2018.1501284.

Rastogi, G., Dharne, M. S., Walujkar, S., Kumar, A., Patole, M. S., & Shouche, Y. S. (2007). Species identification and authentication of tissues of animal origin using mitochondrial and nuclear markers. *Meat Science, 76*(4), 666–674. https://doi.org/10.1016/j.meatsci.2007.02.006.

Ruppert, K. M., Kline, R. J., & Rahman, M. S. (2019). Past, present, and future perspectives of environmental DNA (eDNA) metabarcoding: A systematic review in methods, monitoring, and applications of global eDNA. *Global Ecology and Conservation, 17*, e00547. https://doi.org/10.1016/j.gecco.2019.e00547.

Sahu, A., Kumar, N., Pal Singh, C., & Singh, M. (2023). Environmental DNA (eDNA): Powerful technique for biodiversity conservation. *Journal for Nature Conservation, 71*, 126325. https://doi.org/10.1016/j.jnc.2022.126325.

Saini, M., Gaurav, Bm, H., Kumar, J., Sanu, K., Kumar, H., Pradesh, K., & Vishvafidyalaya (2023). Popular article sequencing techniques. *The science world Magazine, 3*(9), 2378–2393.

SeafoodSource. (2013). Red snapper DNA ID kit in development. https://www.seafoodsource.com/news/supply-trade/red-snapper-dna-id-kit-in-development.

Sharma, C. P., Kumar, A., Vipin, Sharma, V., Singh, B., Kumar, G. C., & Gupta, S. K. (2019). Online selling of wildlife part with spurious name: A serious challenge for wildlife crime enforcement. *International Journal of Legal Medicine, 133*(1), 65–69. https://doi.org/10.1007/s00414-018-1795-7, link.springer.de/link/service/journals/00414/index.htm.

Sigma-Aldrich. (n.d.). Sanger Sequencing. Sigma-Aldrich. [08.08.2025], from Sigma-Aldrich Technical Documents. https://www.sigmaaldrich.com/IN/en/technical-documents/protocol/genomics/sequencing/sanger-sequencing.

Singh, A., Gaur, A., Shailaja, K., Bala, B. S., & Singh, L. (2004). A novel microsatellite (STR) marker for forensic identification of big cats in India. *Forensic Science International, 141*(2-3), 143–147. https://doi.org/10.1016/j.forsciint.2004.01.015, www.elsevier.com/locate/forsciint.

Smart, U., Cihlar, J. C., & Budowle, B. (2021). International Wildlife Trafficking: A perspective on the challenges and potential forensic genetics solutions. *Forensic Science International: Genetics, 54*, 102551. https://doi.org/10.1016/j.fsigen.2021.102551.

Smith, C. J., & Osborn, A. M. (2009). Advantages and limitations of quantitative PCR (Q-PCR)-based approaches in microbial ecology. *FEMS Microbiology Ecology, 67*(1), 6–20. https://doi.org/10.1111/j.1574-6941.2008.00629.x.

Smith, L. M., & Burgoyne, L. A. (2004). Collecting, archiving and processing DNA from wildlife samples using FTA® databasing paper. *BMC Ecology, 4*. https://doi.org/10.1186/1472-6785-4-4, http://www.biomedcentral.com/1472-6785/4/4.

Sorensen, A., Berry, C., Bruce, D., Gahan, M. E., Hughes-Stamm, S., & McNevin, D. (2016). Direct-to-PCR tissue preservation for DNA profiling. *International Journal of Legal Medicine, 130*(3), 607–613. https://doi.org/10.1007/s00414-015-1286-z, link.springer.de/link/service/journals/00414/index.htm.

Staats, M., Arulandhu, A. J., Gravendeel, B., Holst-Jensen, A., Scholtens, I., Peelen, T., Prins, T. W., & Kok, E. (2016). Advances in DNA metabarcoding for food and wildlife forensic species identification. *Analytical and Bioanalytical Chemistry, 408*(17), 4615–4630. https://doi.org/10.1007/s00216-016-9595-8, link.springer.de/link/service/journals/00216/index.htm.

Temnykh, S., DeClerck, G., Lukashova, A., Lipovich, L., Cartinhour, S., & McCouch, S. (2001). Computational and experimental analysis of microsatellites in rice (*Oryza sativa* L.): Frequency, length variation, transposon associations, and genetic marker potential. *Genome Research, 11*(8), 1441–1452. https://doi.org/10.1101/gr.184001.

Templeton, J., Ottens, R., Paradiso, V., Handt, O., Taylor, D., & Linacre, A. (2013). Genetic profiling from challenging samples: Direct PCR of touch DNA. *Forensic Science International: Genetics Supplement Series, 4*(1), e224. https://doi.org/10.1016/j.fsigss.2013.10.115.

Templeton, J. E. L., Taylor, D., Handt, O., Skuza, P., & Linacre, A. (2015). Direct PCR improves the recovery of DNA from various substrates. *Journal of Forensic Sciences, 60*(6), 1558–1562. https://doi.org/10.1111/1556-4029.12843.

Thakur, M., Singh, S. K., Shukla, M., Sharma, L. K., & Mohan, N. (2013). Identification of galliformes through forensically informative nucleotide sequencing (FINS) and its implication in wildlife forensics. *Journal of Forensic Research Science, 3*.

Thermo Fisher Scientific. (n.d.). Invitrogen™ PureLink™ Genomic DNA Mini Kit. Catalog No. K182001. https://www.thermofisher.com/order/catalog/product/K182001.

Thermo Fisher Scientific. (n.d.). Invitrogen™ UltraPure™ Phenol:Chloroform:Isoamyl Alcohol (25:24:1, v/v). Catalog No. 15593031. https://www.thermofisher.com/order/catalog/product/15593031.

Thermo Fisher Scientific (India). (n.d.). Guanidium Thiocyanate Lysis Buffers. https://www.thermofisher.in/chemicals/en/chemical-categories/life-science/guanidium-thiocyanate-lysis-buffers.html.

Tie, J., & Uchigasaki, S. (2014). Detection of short tandem repeat polymorphisms from human nails using direct polymerase chain reaction method. *Electrophoresis, 35*(21-22), 3188–3192. https://doi.org/10.1002/elps.201400061, http://onlinelibrary.wiley.com/journal/10.1002/(ISSN)1522-2683.

Trontelj, P., MacHino, Y., & Sket, B. (2005). Phylogenetic and phylogeographic relationships in the crayfish genus *Austropotamobius* inferred from mitochondrial COI gene sequences. *Molecular Phylogenetics and Evolution, 34*(1), 212–226. https://doi.org/10.1016/j.ympev.2004.09.010.

Ulrich, R. M., John, D. E., Barton, G. W., Hendrick, G. S., Fries, D. P., & Paul, J. H. (2015). A handheld sensor assay for the identification of grouper as a safeguard against seafood mislabeling fraud. *Food Control, 53*, 81–90. https://doi.org/10.1016/j.foodcont.2015.01.022.

van den Berge, M., Ozcanhan, G., Zijlstra, S., Lindenbergh, A., & Sijen, T. (2016). Prevalence of human cell material: DNA and RNA profiling of public and private objects and after activity scenarios. *Forensic Science International: Genetics, 21*, 81–89. https://doi.org/10.1016/j.fsigen.2015.12.012.

Van Oorschot, R. A. H., Phelan, D. G., Furlong, S., Scarfo, G. M., Holding, N. L., & Cummins, M. J. (2003). Are you collecting all the available DNA from touched objects? *International Congress Series, 1239*(C), 803–807. https://doi.org/10.1016/S0531-5131(02)00498-3.

Varshney, R. K., Thiel, T., Langridge, N. S. P., & Graner, A. (2002). Cellular and molecular biology letters 2 A. In *Silico analysis on frequency and distribution of microsatellites in ESTs of some cereal species*. Germany: BioMed Central Ltd, 537–546. http://www.springerlink.com/content/120185/7.

Vipin, Sharma, V., Sharma, C. P., Kumar, V. P., & Goyal, S. P. (2016). Pioneer identification of fake tiger claws using morphometric and DNA-based analysis in wildlife forensics in India. *Forensic Science International, 266*, 226–233. https://doi.org/10.1016/j.forsciint.2016.05.024, www.elsevier.com/locate/forsciint.

Ward, R. D., Zemlak, T. S., Innes, B. H., Last, P. R., & Hebert, P. D. N. (2005). DNA barcoding Australia's fish species. *Philosophical Transactions of the Royal Society B: Biological Sciences, 360*(1462), 1847–1857. https://doi.org/10.1098/rstb.2005.1716, http://rstb.royalsocietypublishing.org/.

White, N. E., Dawson, R., Coghlan, M. L., Tridico, S. R., Mawson, P. R., Haile, J., & Bunce, M. (2012). Application of STR markers in wildlife forensic casework involving Australian black-cockatoos (*Calyptorhynchus* spp.). *Forensic Science International: Genetics, 6*(5), 664–670. https://doi.org/10.1016/j.fsigen.2011.10.003.

Wong, E. H. K., & Hanner, R. H. (2008). DNA barcoding detects market substitution in North American seafood. *Food Research International, 41*(8), 828–837. https://doi.org/10.1016/j.foodres.2008.07.005.

Xia, Z., Johansson, M. L., Gao, Y., Zhang, L., Haffner, G. D., MacIsaac, H. J., & Zhan, A. (2018). Conventional versus real-time quantitative PCR for rare species detection. *Ecology and Evolution, 8*(23), 11799–11807. https://doi.org/10.1002/ece3.4636, http://onlinelibrary.wiley.com/journal/10.1002/(ISSN)2045-7758.

Zhang, D. X., & Hewitt, G. M. (1996). Nuclear integrations: Challenges for mitochondrial DNA markers. *Trends in Ecology and Evolution, 11*(6), 247–251. https://doi.org/10.1016/0169-5347(96)10031-8, https://www.sciencedirect.com/journal/trends-in-ecology-and-evolution.

Ziegler, P. E., Santucci, F., Lindergard, G., Nydam, D. V., Wade, S. E., Schaaf, S. L., Chang, Y. F., & Mohammed, H. O. (2007). Evaluation of polymerase chain reaction diagnosis of Cryptosporidium spp in dairy cattle and wildlife. *Veterinary Therapeutics, 8*(2), 148–159.

Chapter 16

Usefulness of microsatellite markers in forensic botany

Nurdan Sandikci
Department of Forensic Sciences, Faculty of Engineering and Natural Sciences, Üsküdar University, İstanbul, Türkiye

16.1 Overview of forensic botany

Plants are fundamental to life on Earth, playing a pivotal role in maintaining ecological balance and supporting countless life forms. Through the process of photosynthesis, plants convert solar energy into essential molecules, releasing oxygen and forming the basis of food chains. Beyond their ecological significance, plants are indispensable to human industries, finding applications in pharmaceuticals, cosmetics, food production, textiles, and even construction. Their contribution extends to esthetic and environmental aspects, as they provide habitats for various species and enhance the quality of life through urban greenery and reforestation efforts.

Given their ubiquitous presence and diverse uses, plants frequently become incidental participants in forensic investigations. Forensic botany, the scientific application of botanical evidence in legal contexts, focuses on analyzing plant materials to address legal questions. This discipline bridges the fields of biology, ecology, and criminal justice, utilizing botanical expertise to unravel complex cases where plants or plant-based materials are linked to criminal activity.

Botanical evidence can serve multiple functions in forensic investigations. It may establish connections between a suspect, a crime scene, or an object by identifying plant materials transferred during the commission of a crime. For example, soil or pollen adhering to clothing or vehicles can indicate the geographical origin of a suspect's movements. Similarly, seeds, leaves, or wood fragments found at a crime scene may help pinpoint the location of the crime or validate alibis. In cases involving narcotics, forensic botany aids in tracing the origin of illicit drugs derived from plants, such as opium, marijuana, or cocaine, by analyzing their botanical and genetic characteristics.

The admissibility of botanical evidence in court depends on its scientific rigor and interpretive accuracy. Courts require that such evidence undergo thorough analysis to meet evidentiary standards. A multidisciplinary approach is often necessary, combining botany with other forensic sciences like DNA analysis or chemical testing, to establish robust links between evidence and the crime.

Plant evidence encompasses a variety of forms, including physical structures, chemical compounds, and genetic material. Physical evidence may involve plant fragments like leaves, seeds, stems, or pollen grains. Pollen, in particular, holds significant forensic value due to its resilience and species-specific characteristics, enabling investigators to infer the geographical and seasonal context of a crime. Chemical compounds, such as alkaloids and terpenes, can also provide crucial information. For instance, the unique chemical profiles of certain plants may identify their origin, serving as a "chemical fingerprint" in criminal cases (Hall & Byrd, 2012; Miller Coyle, 2004).

In many forensic cases, traditional morphological analyses of plant material are complemented or replaced by genetic studies, which offer higher precision. Morphology-based identification relies on observable traits such as leaf shape, flower structure, and seed characteristics. While effective in many cases, this method has limitations when dealing with degraded, fragmented, or microscopic samples. For example, leaf fragments or single grains of pollen may lack the distinct features needed for accurate identification.

Genetic analysis, on the other hand, delves into the molecular level to extract and interpret DNA from plant materials. This approach allows investigators to achieve objectives such as species identification, determining the population of origin, or tracing individual plants involved in a case. Techniques such as DNA barcoding, which employ specific regions

Investigative and Predictive DNA Testing. DOI: https://doi.org/10.1016/B978-0-443-30128-5.00002-3

of the genome as identifiers, have become indispensable in forensic botany. Commonly analyzed loci include chloroplast and nuclear genes like rbcL, matK, ITS, and trnH-psbA, which provide reliable markers for plant identification

Forensic botany also intersects with ecology, as the interactions of plants with their environment can yield valuable clues. For example, understanding a plant's ecological role, such as its relationship with specific soil types or pollinators, can help narrow down the possible locations of a crime. Additionally, alkaloid content—chemical compounds often unique to specific plant families—can further aid in determining a plant's identity and potential origin.

The application of forensic botany has proven transformative in several high-profile cases. For instance, in the early 1990s, forensic botanists used pollen analysis to link a suspect to a burial site, a groundbreaking application that demonstrated the discipline's potential. Similarly, genetic analysis has been employed to trace the origin of illegal timber, combat wildlife trafficking, and verify the authenticity of herbal medicines.

In conclusion, forensic botany represents a powerful tool in criminal investigations, leveraging the diverse attributes of plants to provide critical insights. From physical characteristics to genetic markers, the discipline offers a range of methods to analyze plant evidence and establish connections between suspects, crime scenes, and objects. As advancements in technology continue to refine the field, forensic botany's role in the justice system is poised to grow, offering new ways to uncover the truth and solve complex cases.

16.2 The role of genetic markers in forensic botany: identification, origin determination, and genotyping

In forensic botany, there are instances where the morphological or anatomical analysis of plant samples is inconclusive due to the absence of distinguishing features or the presence of only small or damaged fragments. In such cases, genetic analysis becomes a crucial tool, allowing for the identification of plant species based on DNA sequences. By sequencing specific segments of DNA from different samples, it is possible to compare their sequences and determine genetic similarities or differences, providing a reliable means of identification when traditional methods fail. One such method is DNA barcoding, a rapidly growing field that uses specific regions of the genome to identify unknown species. By comparing the DNA sequence of an unknown sample with a reference database, species identification can be achieved with high accuracy. For plants, common genetic loci such as rbcL, matK, the trnH-psbA spacer, and internal transcribed spacer (ITS) regions are frequently used in such analyses. Combining multiple loci increases the likelihood of successful species identification, particularly when only a small number of samples are available for analysis (Dunning & Savolainen, 2010)

The typical workflow for genetic analysis begins with DNA extraction from plant tissues, which may include leaves, stems, or fruits. This is followed by the amplification of the targeted loci through polymerase chain reaction (PCR), after which the amplified regions are sequenced using appropriate techniques. The resulting DNA sequences are then compared against large public databases, such as the GenBank database maintained by the National Center for Biotechnology Information (NCBI), which houses a vast and growing collection of reference sequences (National Center for Biotechnology Information, 2025).

The determination of the origin, population, or general region of a plant sample is often achieved through a combination of sequencing-based methods, with genetic marker-based approaches playing a central role. Phylogenetic analysis, which reconstructs the evolutionary history of samples from a wide range of species, is a key method in this context. By analyzing DNA sequences, researchers can identify patterns that distinguish samples from different geographic regions. In cases where more detailed information is required, population-level studies using genetic markers such as microsatellites can provide finer resolution for origin determination (Manel et al., 2005).

In forensic botany, genetic marker data are employed to statistically include or exclude a sample of unknown origin from reference populations. However, the effectiveness of these studies is constrained by the absence of a comprehensive, standardized database akin to the Combined DNA Index System (CODIS), which is used in human identification. To conduct a forensic investigation that aims to match plant species to evidence, it is often necessary to build a reference database specifically for the case. This process involves collecting and genotyping numerous samples from the areas surrounding the crime scene where the plant was found. By genotyping both nearby and distant samples, researchers can estimate the probability of a match and thereby strengthen the forensic evidence (Hall & Byrd, 2012).

Once the decision is made to incorporate genetic data into a forensic investigation, and a suitable laboratory has been identified to conduct the analyses, the next step involves selecting an appropriate genetic marker for the study. This decision is influenced by several factors, including the laboratory's expertise, the specific objectives of the analysis, and budgetary constraints. In recent years, microsatellites have emerged as the preferred genetic markers for a wide range of analyses due to their high polymorphism and resolution capabilities. However, in certain cases, alternative markers such as randomly amplified polymorphic DNA (RAPD) and intersimple sequence repeats (SSRs) may also be utilized. These markers, while less common than microsatellites, can offer valuable insights depending on the specific requirements of the study and the nature of the plant sample being analyzed. After the marker selection, it is critical to ensure that the chosen methodology can successfully genotype samples that are comparable to the case sample(s). Fresh leaves from the same species are typically the most suitable material for optimizing genotyping protocols. At this stage, however, the protocols must be rigorously tested on samples that closely resemble the actual forensic sample to ensure accurate and reliable results.

To increase the probability of successful genotyping in forensic botany, researchers frequently employ the use of witness samples, which are distinct from case samples, intending to optimize protocols. This preliminary step helps refine the process and ensures the selected marker sufficiently captures the genetic diversity needed to test the hypothesis. Once the protocols are established and the suitability of the chosen markers is confirmed, the genotyping of forensic evidence can proceed using standard analytical procedures. When comparing plant evidence from a crime scene to a specific plant, maintaining the cleanliness of the laboratory is paramount. Cross-contamination between samples can compromise the results, and thus, strict contamination controls are necessary throughout the process. To further ensure the accuracy and reproducibility of genotyping, two to three independent subsamples should be collected and analyzed when sufficient evidence is available. Even if only a single DNA extraction is feasible, the sample should still undergo multiple rounds of processing to validate the results.

Since the PCR method used in genotyping is sensitive to secondary plant compounds or fragmented DNA, there are instances where DNA fragments may fail to amplify. In such cases, performing repeat PCR runs serves as a preventive measure to mitigate the effects of nonamplifiable fragments and ensure the reliability of the data.

16.3 Microsatellite markers: an overview

Microsatellites, also referred to as short tandem repeats (STRs) or SSRs, are highly versatile and polymorphic genetic markers found throughout the eukaryotic genome. They consist of short sequences of DNA, typically 1–6 base pairs in length, repeated in a tandem pattern. These regions are notable for their high variability, which arises from the addition or deletion of repeat units, often through replication slippage or errors in DNA replication. This variability makes microsatellites invaluable in distinguishing individuals within a population, species, or even family, as each may exhibit unique patterns of repeat numbers at specific loci.

Microsatellite loci are not strictly defined by the number of repeat units they contain, and there is some debate within the scientific community about the threshold required to classify a sequence as a microsatellite or a minisatellite. While some researchers argue that a minimum of three, four, or even five repeat units is necessary, others suggest that even two repeat units can suffice for distinguishing genetic differences, particularly in specific forensic or ecological contexts (Treangen & Salzberg, 2008). This ambiguity underscores the diverse applications of microsatellites, as their utility often depends on the context in which they are analyzed.

Microsatellites are multiallelic, meaning they exhibit a high degree of variability among individuals within a species. This characteristic is especially valuable in genetic studies where resolving fine-scale differences is crucial. Due to their high mutation rate, which is attributed to the inherent instability of repeat regions, microsatellites serve as sensitive markers for studying genetic diversity, population structure, and evolutionary processes. Additionally, their codominant inheritance pattern allows researchers to distinguish between homozygous and heterozygous states at a given locus, providing more detailed genetic information than dominant markers.

The application of microsatellites is widespread in various scientific fields, including human forensics, animal genetics, plant sciences, and evolutionary biology. Their ability to be amplified through PCR makes them particularly attractive for genetic studies. PCR-based amplification is not only efficient but also highly sensitive, allowing the analysis of even small or degraded DNA samples. This capability has made microsatellite markers the gold standard for DNA profiling in human forensic investigations, where they are used to establish identity, familial relationships, and connections to crime scenes.

The first description of a microsatellite sequence was reported in 1984 by Weller et al., who identified the $(GGAT)_{165}$ repeat in the human hemoglobin gene. This discovery marked the beginning of a new era in genetic research. Subsequently, the term "microsatellite" was adopted to describe similar sequences in the genome of the yeast *Saccharomyces cerevisiae*. Over time, the utility of microsatellites expanded beyond humans and yeast, as geneticists began using them to explore population structures, evolutionary patterns, and entire genome sequences in various species. For instance, microsatellites have been employed to study the genetic diversity and breeding history of plants such as wheat (*Triticum aestivum*) and pine (*Pinus taeda*), which are agriculturally and ecologically significant (Dib et al., 1996).

One of the primary steps in utilizing microsatellites involves the amplification of target loci using PCR. This process requires primers specifically designed for each locus to ensure accurate and efficient amplification. Once amplified, the PCR products are analyzed to determine the size of the microsatellite alleles, which corresponds to the number of repeat units present. This information is then compared against known data, often stored in databases such as the NCBI STRbase. The STRbase, maintained by the NCBI, is a valuable resource that provides detailed information on microsatellite markers, including their sequences, chromosomal locations, and applications in forensic science (National Institute of Standards and Technology, n.d.).

Microsatellites are not only important for individual identification but also serve as critical tools in studies of population genetics and evolutionary biology. Their high mutation rate allows researchers to detect subtle genetic changes over short timescales, making them ideal for studying recent evolutionary events or tracking gene flow within and between populations. Additionally, their ubiquity across the genome ensures that microsatellites can be found in coding, noncoding, and regulatory regions, providing a comprehensive view of genetic variation.

Beyond basic research, the practical applications of microsatellites have had far-reaching implications. In forensics, they are used to link suspects to crime scenes, identify human remains, and establish biological relationships in paternity cases. In conservation biology, microsatellites help monitor genetic diversity in endangered species, guide breeding programs, and assess the impact of habitat fragmentation on genetic connectivity. In agriculture, they are used to develop molecular markers for crop improvement, enabling the identification of desirable traits and the management of genetic resources.

Despite their many advantages, the use of microsatellites is not without challenges. Their high mutation rates, while beneficial for detecting genetic variation, can complicate the interpretation of evolutionary relationships. Additionally, the development of species-specific primers is often required, which can be time-consuming and costly. Nonetheless, advancements in genomic technologies, such as next-generation sequencing (NGS), have facilitated the discovery of new microsatellite loci and streamlined their application in various scientific disciplines.

In summary, microsatellites are versatile and powerful genetic markers that have revolutionized our ability to study genetic variation and solve complex problems in forensics, ecology, and agriculture. Their high polymorphism, codominant inheritance, and adaptability to PCR-based analysis make them indispensable tools in modern genetics. As technological advancements continue to enhance their utility, microsatellites are poised to remain a cornerstone of genetic research and its applications for years to come.

16.4 Role of microsatellite markers in forensic botany

The identification and analysis of plant material as evidence in forensic investigations play a critical role in connecting suspects to crime scenes, determining the origin of illicit substances, and verifying the authenticity of botanical products. Depending on the nature of the crime, plant-based evidence can provide valuable insights, often serving as the missing link in solving complex cases. Microsatellite markers have emerged as a cornerstone of forensic botany, offering precision and reliability in genetic analyses that complement traditional morphological and histological methods.

Historically, the identification of botanical evidence relied heavily on morphological and histological analyses. These approaches involve examining the physical structures of plant material, such as leaves, seeds, and fruits, to identify species or origins (Fig. 16.1). While effective in many scenarios, morphological identification has notable limitations. For example, it requires specialized knowledge of plant taxonomy and often depends on intact or relatively complete samples. When evidence is degraded, fragmented, or microscopic, morphological methods may not yield reliable results (Abdul-Muneer, 2014).

To overcome these limitations, the integration of DNA-based techniques has revolutionized forensic botany. The PCR and advancements in genetic databases like GenBank have enabled the precise identification of plant species, even when only trace amounts of material are available (National Center for Biotechnology Information, 2025). DNA-based approaches can achieve high accuracy, irrespective of the condition of the sample, providing a powerful alternative to traditional methods.

One of the most widely used DNA-based methods is DNA barcoding, which relies on the analysis of species-specific sequences at one or more genetic loci. Microsatellites and variable number tandem repeats found in the chloroplast or nuclear genome are frequently used for this purpose. DNA barcoding involves matching unknown plant material to a reference database of known sequences. However, this process requires robust databases that include a diverse array of species, populations, and genetic variations to ensure accurate matching.

Microsatellites are especially effective in forensic applications due to their high levels of polymorphism. These short, repetitive DNA sequences differ in the number of repeat units between individuals, making them highly discriminative for

FIGURE 16.1 Stem, leaves, and fruits of *Ephedra* sp.

both interspecific (between species) and intraspecific (within species) analyses. Forensic investigators use this variability to identify species, trace the origin of plant material, and differentiate between individual plants, even within a single population.

An important feature of microsatellites is their codominant inheritance pattern, which follows Mendelian principles. This means that both alleles at a locus are expressed, allowing forensic scientists to identify heterozygous and homozygous states. This property is particularly useful when analyzing plant material that may have undergone hybridization or cross-pollination, as it enables precise genetic profiling.

Microsatellite markers have transformed forensic botany, offering unparalleled precision and reliability in the identification of plant material. Their high polymorphism, codominant inheritance, and adaptability to PCR-based techniques make them indispensable tools for investigating crimes involving plant evidence. As forensic science continues to evolve, the integration of microsatellites into forensic workflows will undoubtedly enhance the accuracy and scope of botanical investigations (Coyle et al., 2003; Gigliano, 2001).

16.4.1 Species identification

The identification of plant species from forensic evidence plays a crucial role in criminal investigations, as trace botanical materials can establish connections between individuals, objects, and crime scenes. Various forms of plant-derived evidence, such as leaf fragments, seeds, fruit pieces, wood splinters, and pollen grains, may be inadvertently transferred onto clothing, vehicles, or personal belongings. In some cases, entire plants or their distinctive parts can serve as key evidence in linking a suspect to a particular location or providing insights into the circumstances surrounding a crime. The presence of specific plant materials on victims or suspects can be indicative of their movement, proximity to certain environments, or interaction with a crime scene.

In forensic investigations, DNA-based methodologies have significantly enhanced species identification by providing a more precise and objective means of classifying plant materials. Traditional morphological and histological analyses, while still valuable, are often limited when working with degraded or fragmented samples. Molecular techniques, such as DNA barcoding and genetic marker analysis, have revolutionized forensic botany by allowing for species identification based on genetic sequences rather than relying solely on physical characteristics. These methods have become particularly important in cases where forensic evidence consists of microscopic or highly processed plant materials that lack identifiable morphological traits.

Several genetic loci have been extensively studied and established as reliable markers for species identification in plants. In evolutionary and taxonomic studies, nuclear genes such as 18S, ITS1, and ITS2 (ITS regions) have been widely used due to their high interspecific variability and their ability to distinguish closely related species. These regions provide valuable phylogenetic information, making them useful for forensic applications where plant classification is required at the species or genus level. Similarly, chloroplast genes play a fundamental role in forensic plant identification, given their uniparental inheritance and relatively slow mutation rates, which contribute to their stability across generations.

Among the chloroplast genes commonly employed in forensic botany, ribulose-1,5-bisphosphate carboxylase/oxygenase large subunit (rbcL) is one of the most widely utilized genetic markers. The rbcL gene, which encodes a key enzyme in the photosynthetic pathway, is approximately 1400 base pairs in length and exhibits a moderate level of sequence divergence, making it suitable for distinguishing plant species at both the genus and family levels. Additionally, other chloroplast genes such as matK (maturase K) and psbA-trnH (a noncoding intergenic spacer region) have demonstrated considerable success in both intraspecific and interspecific plant identification. The matK gene, in particular, has been recognized for its high substitution rate and ability to provide species-level resolution, whereas the psbA-trnH intergenic spacer region is highly variable and has been widely used for differentiating closely related taxa.

The increasing availability of reference genetic databases, such as GenBank and the Barcode of Life Data Systems (BOLD) (*Barcode of Life Data System (BOLD)*, 2025) has further facilitated the identification of plant species in forensic investigations. These databases contain vast collections of curated DNA sequences from diverse plant taxa, allowing forensic scientists to compare unknown samples against known reference sequences for accurate species identification. The establishment of these publicly accessible resources has enhanced the reliability of forensic plant identification, enabling the standardization of genetic markers used in forensic casework.

By integrating DNA-based techniques with traditional botanical analyses, forensic scientists can achieve a higher degree of accuracy and reproducibility in plant species identification. As forensic botany continues to evolve, advancements in molecular biology, NGS, and bioinformatics are expected to further refine species identification methodologies, thereby strengthening the role of plant evidence in legal proceedings.

16.4.1.1 Case studies

DNA from a plant sample collected from a suspect's cargo box was used as evidence for the first time in history in a 1992 murder trial in the United States. A polymorphic 10-base pair DNA region of seed capsules of the Palo Verde tree (*Cercidium* sp.) was amplified and identified by Random Amplification of polymorphic DNA (RAPD) analysis, and the seed capsules were found to match a reference sample taken from a tree near the site of the body (*Plant CSI: Solvingcrimes with plants*, 2023).

With the developing technology and its widespread use, the use of microsatellites in DNA analysis of plant evidence in forensic cases has become widespread.

Microsatellites played a critical role in a murder case involving the bodies of an unborn baby and a pregnant woman found next to three large sand oak trees (*Quercus geminata*) in a shallow cemetery in Florida. Three leaf samples found in the suspect's car were identified by an expert botanist, two as *Q. geminata* and the third as another oak species. Using four different microsatellite loci developed for Quercus *macrocarpa* and Quercus *petraea*, Craft et al. analyzed samples of 24 different *Q. geminata* trees collected within 10 km of the crime scene and reported that these loci were intraspecific and interspecific polymorphic. Following these results, leaf samples collected from the suspect's car were analyzed, and although not all loci were amplified, the amplified loci were excluded, indicating that they did not match the trees at the burial site. This study is an example of the use of microsatellites in forensic botanical applications, helping to preserve innocence, which is one of the goals of forensic science (Craft et al., 2007).

16.4.2 Determining geographic origin

The identification of botanical evidence through DNA analysis has proven to be an invaluable tool in forensic and environmental sciences, allowing not only for species identification but also for determining the geographical origin of plant material. Establishing the provenance of botanical evidence is a crucial aspect of forensic investigations, as it provides critical insights that can support law enforcement efforts, validate legal claims, and contribute to environmental conservation. The ability to trace plant material back to its geographic source relies on the genetic variation present within and between plant populations, which is shaped by evolutionary processes, environmental factors, and patterns of human-mediated dispersal.

Determining the geographical origin of plant evidence is essential for several key forensic and ecological applications:

- *Linking Suspects to Crime Scenes*: Botanical evidence can serve as a powerful tool in criminal investigations, particularly when plant material recovered from a suspect's clothing, vehicle, or personal belongings can be linked to a specific location. Seeds, wood fragments, pollen, or leaf material may have been inadvertently transferred during the commission of a crime, providing forensic experts with a means to associate individuals with a particular environment. This is especially valuable in cases where a suspect's alibi contradicts the geographical distribution of the plant evidence. The ability to genetically determine the origin of such evidence is particularly useful when investigating crimes that span multiple locations, such as kidnapings, illegal waste dumping, or clandestine drug operations.
- *Combating Illegal Trade and Trafficking*: DNA-based geographic tracing methods play a vital role in identifying the provenance of plant species that are subject to illegal trade. The illicit harvesting and sale of endangered or protected plant species, including rare medicinal plants, hardwood timber, and plants with high economic value, pose significant threats to biodiversity and environmental sustainability. Forensic analysis of genetic markers allows authorities to determine whether confiscated plant material originates from restricted or protected areas, thereby facilitating the enforcement of conservation laws and international treaties, such as the Convention on International Trade in Endangered Species of Wild Fauna and Flora (CITES) (*Convention on International Trade in Endangered Species of Wild Fauna and Flora*, n.d.). One particularly relevant example is the tracing of Ephedra species, which serve as a precursor for the illicit synthesis of methamphetamine. By analyzing the genetic signatures of Ephedra samples, forensic experts can identify cultivation regions and disrupt illegal supply chains that contribute to global drug trafficking networks.
- *Verification of Legal Claims in Trade and Commerce*: In commercial industries that rely on plant-derived products, such as timber, herbal medicines, and food products, verifying the geographical authenticity of materials is essential for regulatory compliance, fraud prevention, and environmental sustainability. The illegal logging of rare hardwood species, for instance, has significant economic and ecological consequences, and forensic botanical methods have been used to determine whether timber products are sourced from legal or illegally harvested forests. Similarly, the authenticity of herbal medicines and dietary supplements can be confirmed through genetic analysis to ensure that the raw materials originate from approved sources and meet required safety standards. By employing DNA-based techniques, regulatory agencies can detect fraudulent labeling, identify cases of species substitution, and enforce trade regulations to protect consumers and natural resources.
- *Supporting Biodiversity and Environmental Protection*: Geographic origin determination plays a crucial role in monitoring invasive plant species, preventing habitat destruction, and managing biodiversity conservation efforts. The introduction of nonnative plant species into new environments can have devastating ecological consequences, as invasive species often outcompete native flora and disrupt local ecosystems. By analyzing the genetic composition of plant populations, forensic botanists can track the movement and spread of invasive species, thereby informing management strategies to mitigate their impact. Additionally, genetic techniques are used to assess the extent of illegal deforestation and land conversion activities, particularly in regions where protected forests and wildlife reserves are under threat from human encroachment.

Determining the dominance of a plant species in a population can be useful in obtaining information about its geographical location. The origin of a plant can be determined by phylogenetic analysis, which involves the reconstruction of the evolutionary history of samples through the examination of DNA sequences. This approach allows for the investigation of a wide range of species. Microsatellites with high intraspecific and interspecific polymorphism are frequently employed in population-level studies to ascertain the population of origin.

The integration of molecular genetic techniques with traditional botanical expertise has significantly enhanced the accuracy and reliability of geographic tracing in forensic botany. As forensic science continues to evolve, the development of comprehensive reference databases, improvements in DNA sequencing technologies, and advancements in bioinformatics are expected to further refine geographic origin determination methods. The analysis of genetic marker data allows for the probabilistic assignment of an unknown sample to a known population, or the statistical exclusion of an unknown sample as originating from a given population (Manel et al., 2005; Petit & Vendramin, 2007). This will enable forensic experts to provide more precise and scientifically defensible conclusions in criminal investigations, environmental monitoring, and regulatory enforcement.

16.4.2.1 Case studies

In the context of drug smuggling investigations, forensic scientists have employed the use of plant DNA analysis to ascertain the provenance of narcotic plants, including the Cannabis sativa species, commonly known as marijuana.

Mexican authorities seized a considerable quantity of marijuana and were thus obliged to ascertain its geographical source in order to dismantle the criminal networks responsible for its production and distribution. The DNA markers from the marijuana plants were compared to reference samples collected from disparate regions of Mexico. The use of microsatellite markers enabled the identification of specific cultivation sites, thereby revealing the regions where the cannabis had been illicitly cultivated. This approach enabled law enforcement agencies to delineate the geographical areas of cultivation and to identify the specific drug cartels responsible for the illicit trafficking operations (Howard et al., 2008).

The palm oil industry has been the subject of intense scrutiny due to allegations of illegal deforestation, particularly in Indonesia and Malaysia. Forensic botany has been employed to ascertain the provenance of palm oil products, with a view to identifying illegal plantations established on previously forested land. Investigators utilized genetic markers, including microsatellite DNA, to trace the provenance of the palm oil back to its original plantation. By comparing the DNA from palm oil products to reference samples from legally certified plantations and illegally deforested areas, it was possible for the relevant authorities to determine the source of the oil. The evidence was instrumental in the prosecution of companies that were contributing to illegal deforestation. Furthermore, it provided a scientific foundation for the implementation of sustainable sourcing practices within the palm oil industry (Hayati et al., 2004).

16.5 Advantages and limitations of microsatellites in forensic botany

Microsatellite markers, also referred to as STRs, have become an indispensable tool in forensic botany due to their high discriminatory power, reproducibility, and applicability across a wide range of plant species. These molecular markers consist of short, tandemly repeated DNA sequences, typically 1–6 base pairs in length, which are widely distributed throughout the genome. The key advantage of microsatellites lies in their high level of polymorphism, which allows for precise differentiation between species, populations, and even individual plants. This feature is particularly valuable in forensic investigations where plant evidence is used to establish links between suspects, crime scenes, and objects. High polymorphism of microsatellites, which results from variations in the number of repeat units present at a given locus makes them highly effective for forensic applications, as it increases the probability of distinguishing between closely related individuals or plant populations. Unlike other genetic markers that may exhibit limited variation, microsatellites provide a high-resolution tool for genetic fingerprinting, making them ideal for forensic casework that requires precise species or individual identification (Butler, 2009).

Another key advantage of microsatellites is their codominant inheritance, meaning that both alleles at a given locus are expressed in a heterozygous individual. This inheritance pattern allows for the determination of both maternal and paternal contributions to a genotype, providing more informative data than dominant markers such as amplified fragment length polymorphisms or RAPD. The ability to distinguish between heterozygous and homozygous genotypes enhances the statistical power of forensic analyses, as it enables more accurate comparisons between samples collected from different locations or individuals (Hall & Byrd, 2012).

Microsatellites are also highly amenable to PCR-based amplification, which makes them particularly useful in forensic applications where the amount of available DNA is often limited. PCR amplification allows forensic scientists to generate reliable genetic profiles from small or degraded samples, which is crucial in cases where plant evidence has been exposed to environmental factors such as heat, moisture, or microbial degradation. The ability to amplify microsatellite loci from trace amounts of DNA increases the likelihood of obtaining usable genetic data even when the sample quality is suboptimal.

Additionally, microsatellite markers are highly versatile and can be applied across various forensic contexts, including species identification, geographic origin determination, and individual plant discrimination. Their use extends beyond criminal investigations to include applications in environmental forensics, conservation biology, and regulatory compliance. For example, microsatellites have been used to trace the illegal trade of endangered plant species, authenticate the provenance of timber and herbal medicines, and monitor biodiversity loss due to habitat destruction (Hall & Byrd, 2012).

Despite their numerous advantages, microsatellites are not without limitations. One of the primary challenges associated with their use is their high mutation rate, which can lead to repeat slippage during DNA replication. While this characteristic contributes to their high polymorphism, it can also introduce complexities in forensic analysis, particularly when attempting to establish long-term genetic relationships between populations. Additionally, high mutation rates can sometimes result in somatic mutations that may cause discrepancies between closely related samples, complicating the interpretation of forensic evidence.

Another limitation is that microsatellite markers often require species-specific primer development, which can be a time-consuming and resource-intensive process. Unlike universal genetic markers such as ribosomal DNA (rDNA) sequences or chloroplast DNA barcodes, microsatellite loci must be individually characterized for each species before they can be used in forensic investigations. This requirement poses a challenge for forensic cases involving plant species that have not been extensively studied, as the lack of reference databases can limit the utility of microsatellite markers in species identification.

Furthermore, microsatellites can be susceptible to allelic dropout, particularly in cases where DNA degradation has occurred. Allelic dropout refers to the preferential amplification of one allele over another at a heterozygous locus, leading to false homozygosity in the resulting genetic profile. This phenomenon is especially problematic in forensic botany, where plant samples collected from crime scenes may have been exposed to harsh environmental conditions that degrade DNA integrity. To mitigate the impact of allelic dropout, forensic scientists often use multiple replicates of PCR amplification and cross-validate results with additional genetic markers.

In summary, microsatellite markers offer a powerful and highly discriminatory tool for forensic botany, providing high-resolution genetic profiles that can be used for species identification, population assignment, and individual plant differentiation. Their high polymorphism, codominant inheritance, and compatibility with PCR-based techniques make them invaluable in forensic casework. However, challenges such as high mutation rates, species-specific primer requirements, and the risk of allelic dropout must be carefully managed to ensure the reliability and accuracy of microsatellite-based forensic analyses. As forensic technology continues to advance, ongoing research into the development of standardized microsatellite marker panels and the integration of NGS approaches is expected to further enhance the utility of these markers in forensic investigations (Goldstein & Schlötterer, 2023) (Figs. 16.2 and 16.3).

16.6 Future directions and developments in microsatellite analysis for forensic botany

The continued advancement of microsatellite marker technology is expected to significantly impact forensic botany, expanding its applications and improving its accuracy in species identification, population differentiation, and geographic origin determination. As forensic genetics progresses, new developments in microsatellite analysis are shaping the field by enhancing marker discovery, improving analytical methodologies, increasing accessibility to reference databases, and integrating complementary molecular techniques. These advancements collectively contribute to the establishment of more robust, reliable, and universally accepted forensic botanical protocols.

16.6.1 Expansion of microsatellite marker discovery through genomic technologies

A major development in forensic botany is the discovery of novel microsatellite loci through advancements in genomic sequencing technologies. Traditionally, microsatellite markers were identified through labor-intensive methods such as clone-based sequencing and primer development, which limited the number of available loci. However, recent breakthroughs in NGS technologies have enabled the rapid identification of new microsatellite loci across a wide range of plant species. This expansion enhances marker diversity, allowing for finer resolution in species identification and increasing the discriminatory power of microsatellite-based forensic investigations (Goldstein & Schlötterer, 2023). By leveraging high-throughput sequencing (HTS), researchers can now screen entire genomes for polymorphic microsatellite regions, thereby accelerating the discovery of markers that are more informative and applicable across various forensic scenarios.

The improved resolution provided by these new microsatellite markers is particularly beneficial in distinguishing closely related plant species, which is often a challenge in forensic casework. Many forensic investigations require the identification of plant material at the species or even individual level, particularly in cases involving rare, protected, or illicitly traded plant species. The availability of more diverse and highly polymorphic microsatellite loci increases the likelihood of successful differentiation between species and populations, making forensic analysis more precise and legally defensible.

16.6.2 Advancements in high-throughput sequencing and microsatellite genotyping

The adoption of HTS technologies has revolutionized microsatellite analysis by enabling the simultaneous screening of thousands of microsatellite markers across multiple samples. Unlike traditional methods that require labor-intensive, locus-by-locus screening, HTS allows for large-scale genotyping, thereby significantly improving the efficiency and resolution of forensic botanical studies (Goncalves et al., 2025). This technological shift has not only expanded the available marker repertoire but also increased the ability to analyze degraded or limited DNA samples, which are commonly encountered in forensic investigations.

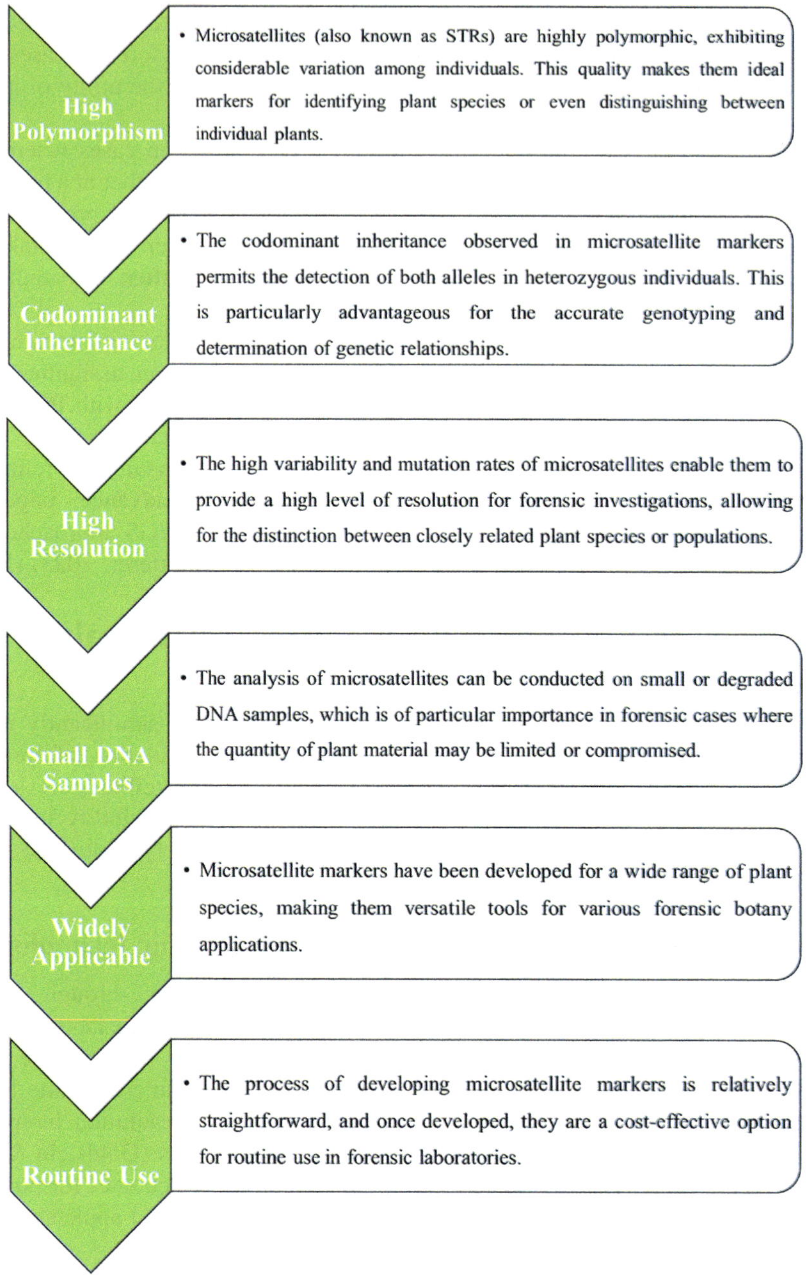

FIGURE 16.2 Advantages of microsatellites in forensic botany.

Furthermore, HTS-based microsatellite analysis minimizes the occurrence of artifacts such as allelic dropout and stutter peaks, which can sometimes complicate genotyping results in conventional PCR-based microsatellite studies. By integrating computational bioinformatics tools, forensic scientists can now process large datasets with greater accuracy and identify genotypic variations at an unprecedented scale. These improvements enhance the reliability of forensic botany, ensuring that microsatellite markers continue to be a powerful tool for species identification and evidence interpretation in legal cases.

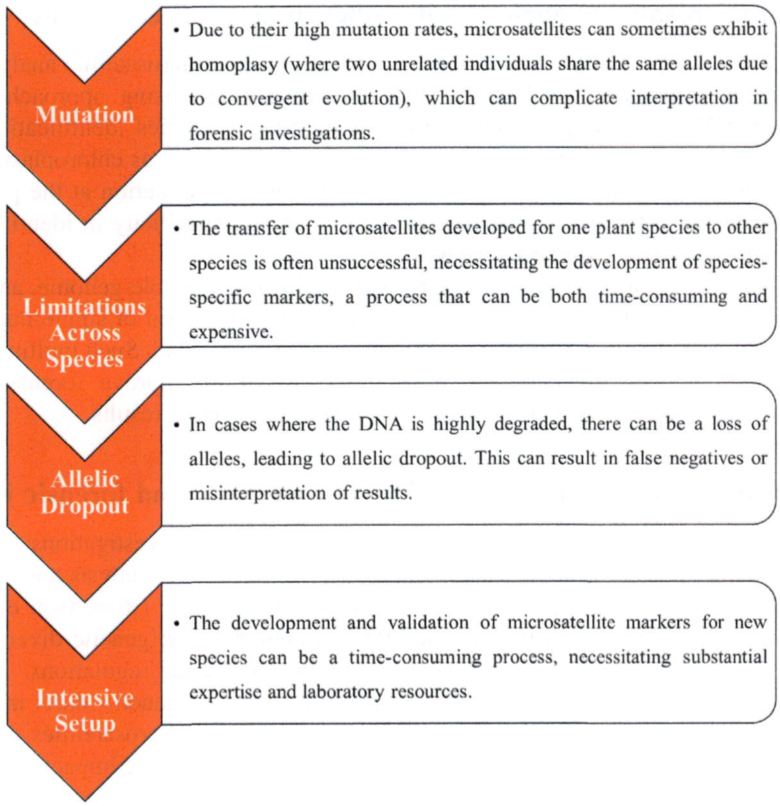

FIGURE 16.3 Limitations of microsatellites in forensic botany.

16.6.3 Development of comprehensive reference databases

The establishment of extensive microsatellite reference databases is another critical advancement in forensic botany. These databases serve as repositories for allele frequency data, genetic profiles of plant species, and population-level variations, providing forensic scientists with a valuable resource for comparative analysis. Existing databases such as GenBank and the BOLD have already contributed to forensic investigations by offering access to curated genetic sequences. However, the expansion of microsatellite-specific repositories, such as the one developed by Schuelke, represents a significant step toward standardizing forensic microsatellite analysis (Schuelke, 2000).

The availability of well-structured, species-specific microsatellite databases enhances the accuracy of forensic investigations by allowing for more precise matching of unknown plant samples to reference populations. Moreover, the development of geographically annotated genetic databases could further refine forensic methodologies by enabling scientists to determine the likely geographic origin of a given plant sample based on microsatellite variation patterns. Such resources are particularly useful in cases involving illegal logging, drug trafficking, and invasive species monitoring, where identifying the provenance of plant material is essential for legal and environmental enforcement.

16.6.4 Standardization of microsatellite analysis protocols

For forensic botany to gain broader acceptance in the legal system, the standardization of microsatellite analysis protocols is essential. Inconsistencies in DNA extraction, amplification conditions, and genotyping methodologies can lead to discrepancies in forensic casework, potentially affecting the admissibility of botanical evidence in court. Recent efforts have focused on the development of internationally recognized forensic botany guidelines to ensure consistency and reproducibility across different laboratories. Doyle proposed a comprehensive methodology for DNA extraction, microsatellite amplification, and genotyping, emphasizing the need for uniform procedures in forensic investigations. The establishment of standardized forensic protocols will facilitate the cross-validation of results, strengthen the reliability of forensic botanical evidence, and promote its widespread use in law enforcement (Mawarni et al., 2021).

16.6.5 Integration of microsatellite analysis with other molecular techniques

To maximize the accuracy and comprehensiveness of forensic investigations, microsatellite analysis is increasingly being integrated with complementary molecular techniques. One particularly promising approach is the combination of microsatellite markers with DNA barcoding, which allows for multilocus species identification and genetic profiling (Meudt & Clarke, 2007). DNA barcoding relies on standardized genetic loci, such as chloroplast genes (*rbcL* and *matK*), to classify plant species, while microsatellites provide higher-resolution differentiation at the population and individual levels. By utilizing both techniques, forensic scientists can achieve greater specificity in identifying plant evidence and provide a more robust genetic profile for forensic casework.

Additionally, the integration of microsatellite analysis with NGS-based whole-genome analysis holds significant promise for forensic applications. Whole-genome sequencing enables the detection of single-nucleotide polymorphisms, structural variations, and other genetic markers that complement microsatellite data. Such multimarker approaches offer a more comprehensive view of genetic diversity and have the potential to improve forensic discrimination power, particularly in cases where microsatellite data alone may not provide conclusive results.

16.6.6 Expansion of microsatellite applications in environmental and forensic investigations

The versatility of microsatellite markers extends beyond traditional forensic investigations to include environmental forensics, ecological monitoring, and wildlife crime investigations. Microsatellite analysis has been increasingly applied to cases involving illegal logging, plant poaching, and invasive species tracking (Robertson et al., 2021). In addition, forensic botanists are exploring the potential of microsatellites in monitoring the genetic diversity of endangered plant species, aiding conservation efforts, and ensuring compliance with environmental regulations.

In the context of climate change and habitat destruction, the ability to track genetic shifts in plant populations using microsatellite markers provides valuable data for biodiversity conservation. Microsatellites are particularly useful in identifying genetic bottlenecks, assessing population connectivity, and determining the impact of human activities on plant species distributions. These applications highlight the expanding role of forensic botany in addressing global environmental challenges and reinforcing conservation policies.

16.6.7 Ethical and legal considerations in forensic microsatellite analysis

As microsatellite analysis becomes increasingly integrated into forensic science, it is essential to address the ethical and legal implications associated with its use. The collection, storage, and analysis of genetic data must be conducted with strict adherence to data protection laws and ethical guidelines to prevent misuse. Concerns regarding genetic privacy, bioprospecting, and unauthorized use of plant DNA must be carefully managed to maintain public trust in forensic botany. The implementation of clear regulatory frameworks and ethical oversight mechanisms will be crucial in ensuring that microsatellite-based forensic investigations uphold the principles of justice and scientific integrity.

16.7 Conclusion

Microsatellite markers have emerged as an essential tool in forensic botany, providing an unparalleled level of precision in plant species identification, individual differentiation, and geographic origin determination. Their high polymorphism and codominant inheritance make them particularly valuable in forensic investigations, where traditional botanical methods such as morphological, anatomical, or histological analyses may be insufficient, particularly in cases where plant material is highly fragmented, degraded, or lacks distinctive physical characteristics. The ability of microsatellite markers to detect subtle genetic variations among individuals within a species or population significantly enhances the discriminatory power of forensic investigations, facilitating the establishment of links between crime scenes, suspects, and objects associated with criminal activity. As a result, the integration of microsatellite markers into forensic workflows has substantially strengthened the evidentiary value of botanical material, increasing its credibility in legal contexts and reinforcing the role of forensic botany as a recognized subdiscipline within forensic science.

Despite the numerous advantages offered by microsatellite markers, several technical and methodological challenges remain that must be addressed to fully optimize their application in forensic casework. One of the primary limitations is their high mutation rate, which, while beneficial in distinguishing between closely related individuals, can sometimes introduce complexities in interpreting genetic relationships and population dynamics. Additionally, allelic dropout, particularly in degraded samples, can lead to false homozygosity, potentially compromising the reliability of genotyping results. This issue is particularly relevant in forensic investigations, where DNA extracted from plant evidence is often subjected to environmental stressors such as heat, moisture, and microbial degradation.

Another critical limitation is the requirement for species-specific primer development, which can be both time-consuming and resource-intensive. Unlike universal genetic markers such as rDNA or chloroplast DNA barcodes, microsatellite loci must be characterized individually for each species before they can be used in forensic applications. This necessity poses a significant challenge, particularly when dealing with forensic cases involving rare or underresearched plant species, where limited genetic reference data may be available. The lack of standardized microsatellite reference databases, analogous to human forensic DNA databases such as CODIS, further restricts the widespread implementation of microsatellite analysis in forensic botany. Without comprehensive population-level allele frequency data for various plant species, the statistical robustness of forensic conclusions drawn from microsatellite genotyping remains constrained.

Looking ahead, ongoing advancements in genomic technologies, NGS, and high-throughput microsatellite genotyping hold great promise for overcoming these limitations. The increasing availability of whole-genome sequencing data is facilitating the discovery of new microsatellite loci, enhancing marker diversity, and improving species identification at finer taxonomic levels. Additionally, improvements in DNA extraction and amplification techniques for degraded samples are expected to mitigate challenges associated with allelic dropout, increasing the reliability of forensic plant genotyping. The development of global microsatellite reference databases, containing population-level genetic profiles of plant species commonly encountered in forensic investigations, would further strengthen the applicability of microsatellite markers in forensic botany by enabling more precise statistical analysis of plant evidence.

In conclusion, microsatellite markers have revolutionized forensic botany, providing a highly discriminatory and scientifically robust approach to plant species identification, individual differentiation, and geographic origin determination. Their ability to generate high-resolution genetic profiles has significantly expanded the evidentiary potential of botanical material in forensic investigations, offering crucial insights into criminal cases, illegal trade, and environmental law enforcement. As forensic botany continues to evolve, the integration of molecular genetic techniques, bioinformatics, and traditional botanical expertise will be essential in further refining methodologies and expanding the scope of forensic plant analysis. With continued advancements in genetic technologies, forensic botany is poised to become an even more integral component of modern forensic science, contributing not only to criminal investigations but also to biodiversity conservation, ecological monitoring, and the enforcement of environmental regulations.

References

Abdul-Muneer, P. M. (2014). Application of microsatellite markers in conservation genetics and fisheries management: Recent advances in population structure analysis and conservation strategies. *Genetics Research International*(1).

Barcode of Life Data System (BOLD). (2025). https://www.ibol.org/phase1/bold/

Butler, J. M. (ed.). (2009). *Fundamentals of forensic DNA typing*. Academic Press.

Convention on International Trade in Endangered Species of Wild Fauna and Flora. https://cites.org/eng/disc/text.php..

Coyle, H. M., Shutler, G., Abrams, S., Hanniman, J., Neylon, S., Ladd, C., Palmbach, T., & Lee, H. C. (2003). A simple DNA extraction method for marijuana samples used in amplified fragment length polymorphism (AFLP) analysis. *Journal of Forensic Sciences, 48*(2), 343–347. https://doi.org/10.1520/jfs2001207, http://onlinelibrary.wiley.com/journal/10.1111/(ISSN)1556-4029.

Craft, K. J., Owens, J. D., & Ashley, M. V. (2007). Application of plant DNA markers in forensic botany: Genetic comparison of *Quercus* evidence leaves to crime scene trees using microsatellites. *Forensic Science International, 165*(1), 64–70. https://doi.org/10.1016/j.forsciint.2006.03.002, https://www.elsevier.com/locate/forsciint.

Dib, C., Fauré, S., Fizames, C., Samson, D., Drouot, N., Vignal, A., Millasseau, P., Marc, S., Hazan, J., Seboun, E., Lathrop, M., Gyapay, G., Morissette, J., & Weissenbach, J. (1996). A comprehensive genetic map of the human genome based on 5,264 microsatellites. *Nature, 380*(6570), 152–154. https://doi.org/10.1038/380152a0.

Dunning, L. T., & Savolainen, V. (2010). Broad-scale amplification of matK for DNA barcoding plants, a technical note. *Botanical Journal of the Linnean Society, 164*(1), 1–9. https://doi.org/10.1111/j.1095-8339.2010.01071.x.

Gigliano, G. S. (2001). *Cannabis sativa* L.: Botanical problems and molecular approaches in forensic investigations. *Forensic Science Review, 13*(1), 1–17.

Goldstein, D. B., & Schlötterer, C. (eds.). (2023). *Microsatellites: Evolution and applications.* Oxford University Press.

Goncalves, A. L., García, M. V., Chancerel, E., Lepais, O., & Heuertz, M. (2025). High-throughput sequence-based microsatellite genotyping for the non-model Neotropical tree species *Anadenanthera colubrina* (Leguminosae). *Plant Ecology and Evolution, 158*(1), 43–52.

Hall, D. W., & Byrd, J. H. (2012). *Forensic botany: A practical guide*. United States: John Wiley and Sons. http://onlinelibrary.wiley.com/book/10.1002/9781119945734.

Hayati, Wickneswari, Maizura, & Rajanaidu (2004). Genetic diversity of oil palm (*Elaeis guineensis* Jacq.) germplasm collections from Africa: Implications for improvement and conservation of genetic resources. *Theoretical and Applied Genetics, 108*(7), 1274–1284. https://doi.org/10.1007/s00122-003-1545-0.

Howard, C., Gilmore, S., Robertson, J., & Peakall, R. (2008). Application of new DNA markers for forensic examination of *Cannabis sativa* seizures – Developmental validation of protocols and a genetic database. National Drug Law Enforcement Research Fund (NDLERF).

Manel, S., Gaggiotti, O. E., & Waples, R. S. (2005). Assignment methods: Matching biological questions with appropriate techniques. *Trends in Ecology and Evolution, 20*(3), 136–142. https://doi.org/10.1016/j.tree.2004.12.004, https://www.elsevier.com/inca/publications/store/3/0/3/3/9.

Mawarni, S. N., Khairunnisa, D., Larasati, I., Rizqo, N., Erfianti, T., Kusumaningrum, H. P., Ferniah, R. S., & Pujiyanto, S. (2021). Application of the Doyle and Doyle method for DNA isolation from pomelo yellow orange (*Citrus maxima* Merr), lime orange (*C. limon*), and sunkist orange (*C. sinensis*). *Journal of Physics: Conference Series.*

Meudt, H. M., & Clarke, A. C. (2007). Almost forgotten or latest practice? AFLP applications, analyses and advances. *Trends in Plant Science, 12*(3), 106–117. https://doi.org/10.1016/j.tplants.2007.02.001.

National Center for Biotechnology Information. (2025). https://www.ncbi.nlm.nih.gov/.

National Institute of Standards and Technology. Short tandem repeat DNA internet database (STRBase). https://strbase-archive.nist.gov/.

Miller Coyle, H. (2004). *Forensic botany principles and applications to criminal casework.* CRC Press.

Petit, R. J., & Vendramin, G. G. (2007). Plant phylogeography based on organelle genes: An introduction. In *Phylogeography of Southern European Refugia: Evolutionary perspectives on the origins and conservation of european biodiversity*. France: Springer Netherlands, 23–97. https://doi.org/10.1007/1-4020-4904-8_2, http://www.springerlink.com/openurl.asp?genre=book&isbn=978-1-4020-4903-3.

Plant CSI: Solving crimes with plants. (2023). https://mountaintimes.info/2023/03/15/plant-csi-solving-crimes-with-plants/

Robertson, J. M., Damaso, N., & Meiklejohn, K. A. (2021). DNA-based analysis of plant material in forensic investigations. In H. R. Dash, P. Shrivastava, & J. A. Lorente (Eds.). *Handbook of DNA profiling*. Springer.

Schuelke (2000). An economic method for the fluorescent labeling of PCR fragments. *Nature Biotechnology, 18*(2), 233–234. https://doi.org/10.1038/72708.

Treangen, T. J., & Salzberg, S. L. (2008). Repetitive DNA and next-generation sequencing: Computational challenges and solutions. *Nature Reviews Genetics, 13*(1), 36–46.

Chapter 17

DNA barcoding as a useful tool in the identification of endangered plants

Gagan Kumar Panigrahi[1], Rutupurna Das[1], and Kunja Bihari Satapathy[2]
[1]*Department of Zoology, School of Applied Sciences, Centurion University of Technology and Management, Bhubaneswar, Odisha, India,* [2]*School of Applied Sciences, Centurion University of Technology and Management, Bhubaneswar, Odisha, India*

17.1 Introduction

Integration of advanced technologies in operations must take place rapidly to enhance the identification of species and exceed the boundaries of traditional approaches. Molecular techniques, especially DNA barcoding integrated with morphological identification, have been necessary tools in meeting demand in a scientifically sophisticated manner (Ahmed et al., 2022). Biological diversity is the difference between various animals, plants, fungi, and prokaryotes that live in terrestrial, marine, and freshwater ecosystems and form ecological complexes. Biodiversity encompasses individual species, different species, and ecosystems. Biodiversity plays a fundamental role in maintaining and conserving ecological balance (Al-Shuhaib & Hashim, 2023). So far, only a small number of the estimated 9 million species have been described (Ankola et al., 2021). To assess the extinction rates of species, scientists have only managed to document about 10%–15% of global biodiversity (Antil et al., 2023). Significantly, around 86%–91% or 7.2 million species in the world are yet to be described due to several limiting factors such as reduced finance to support taxonomy, lesser trained taxonomists, and more importantly, the unavailability of efficient ways of species identification. For gaining a deep understanding of this largely unexplored biodiversity, reliable identification methods, a well-trained taxonomy workforce, and adequate funding are important (Aylagas et al., 2016). The traditional taxonomic methods have always involved a reliance on morphological characters that are labor-intensive, dependent upon the availability of specialized specialists, and often misleading when cryptic species or phenotypic plasticity are involved. The conventional taxonomists have also declined globally (Baena-Bejarano et al., 2023; Panigrahi et al., 2021; Patro et al., 2024). Therefore, the identification of several microorganisms and invertebrates may sometimes require only DNA-based molecular methods, especially when the live cultures or physical specimens are unavailable (Franco-Duarte et al., 2019). Several advantages are offered by molecular techniques compared to the traditional ones. They offer standardized tools for direct comparison purposes among varied users. They do not require specific taxonomic expertise and can be used for the analysis of environmental samples with high diversity (Patro et al., 2023). DNA barcoding has come to be a new molecular marker for species identification over the last decade, whereby one or more standardized short genetic markers taken from the DNA of an organism are used to classify it within a particular species. This approach is very useful in cases where conventional morphological characterizations are not possible or are not feasible (Antil et al., 2023). DNA barcoding is an approach wherein the DNA extract from an unknown species is matched against a database of reference sequences compiled by Paul Hebert and colleagues. The creation of a public reference database has greatly enhanced accurate taxonomy identification for a vast variety of species. This application has been discovered in various scientific fields, including conservation biology and evolutionary studies. DNA barcoding is applied in identifying and tracking patented organisms in the field of biotechnology science (Balaban et al., 2019). It helps determine the taxonomic identity of damaged organisms or fragments found in products, food, or stomach contents. Its application ranges from the food industries to forensics and combating illicit trade and hunting of endangered species (Odah, 2024). Species would first need to be taxonomically described for their inclusion in the Barcode of Life Data Systems (BOLDs), as the Consortium for the Barcode of Life aims at the goals of describing all species (Bergmann et al., 2013). The practice is known to enable researchers to actively cope with the analytical, technical, and fundamental challenges associated with the implementation of DNA barcoding. DNA barcoding enables researchers to use molecular

Investigative and Predictive DNA Testing. DOI: https://doi.org/10.1016/B978-0-443-30128-5.00020-5

techniques and public reference databases for accurate taxonomic identification, conservation efforts, and challenges across multiple fields.

17.2 Plant identification foremost part of forensic sciences

Plants are essential to most aspects of human life, and provide fundamental nutrition and medicinal resources. However, the existing poisonous plants are threat to humans and animals. Contact or intake with such poisonous plants can lead to illness or even death. Toxic properties within such plants are associated with several poisonous metabolites such as alkaloids, glycosides, toxalbumins, saponins, and proteins, which cause some of them to cause serious health issues with slight exposure (Bezeng et al., 2017; Bingpeng et al., 2018; Blazej et al., 2018). Forensic investigations revealed that more than 80% of the poisoning cases were accidental, while the rest were from intentional poisoning. Cases of poisoning reported with evil intent, psychotropic use, and drug-facilitated sexual assault included those resulting from plants *like Cannabis sativa*and *Erythrina velutina*. Ornamental household plants, including Dieffenbachia, Euphorbia, Epipremnum, and Toxicodendron, have been associated with severe acute intoxications and deaths, mostly in children 1–18 years old, with the majority of cases and deaths occurring in those aged 1–6 years. International reports also report emergency cases consisting of *Datura stramonium*, *Nerium oleander*, and *Ricinus communis*. The consumption of plant seeds as a suicide attempt is more commonly identified in Asian regions, involving *Abrus precatorius, Cascabelathevetia, Cleistanthus-collinus,* and *Cerberaodollam*. The emergencies other than direct poisonings that such a mixing of toxic plant material with medicinal herbs can bring about are the instances where some medicinal plants are inherently toxic in nature (Bohmann et al., 2020; Botha et al., 2023). Such harmful effects of herbal remedies are often seen when such medicines are wrongly identified during collection and preparation. It is reported that the toxic parts of plants are found in the global trade of natural health products and dietary supplements, thus bringing about public health concerns to consumers. Factors include the general prevalence of poisonous plants, the coincidence of toxic species with useful varieties, their pleasing appearance and similarity to edible crops, and a general lack of information on the potential dangers (Behera et al., 2024; Cahyaningsih et al., 2022; Carlin et al., 2020). For all these poisoning instances, it is therefore necessary that, in forensic toxicology, species identification needs to be very rapid, especially for the more common toxic plants. Poisoning worldwide has been on the increase, forcing the establishment of poison control centers around the globe in order to prevent poisoning as well as treat poisoning cases. Treatment of the poisoned persons may involve specific antidotes on an emergency basis. Since there are different groups of poisonous plants containing different toxic metabolites, identification of the specific plant species involved is critical for successful treatment (Cawthorn et al., 2012; Das & Panigrahi, 2024). However, visual examination of vomit or excreta samples may be difficult as the same are often found to be small and fragmented, devoid of morphological features that may be identified. DNA barcoding can reliably identify ingested plants from such samples since it does not rely on morphological characteristics. DNA barcoding enables taxonomic identification of species using small pieces of tissue from any part of a plant and is increasingly used in forensic toxicology, research on avian diets, raw drug authentication, and monitoring the trade in timber. It is unlikely that contamination by human or microbial DNA will affect the analysis, because the used markers are plant-specific (Cermakova et al., 2023; Majumder & Panigrahi, 2024a, 2024b). The cost of DNA sequencing has decreased appreciably, thus it has become available for assured species identification in many countries. Forensic identification of plants based on DNA barcoding may improve the establishment of standardized procedures and protocols as recommended by the US National Research Council of the National Academy of Sciences. A reference DNA barcode library of 100 poisonous plants was established, and the major phytochemicals in these species were mapped onto a phylogenetic tree to clearly understand the difference between plants and the distribution across taxa (Chan et al., 2020; Chan et al., 2022; Panigrahi & Satapathy, 2020a, 2020b, 2020c;).

17.3 General idea of the DNA barcoding process

DNA barcoding is a molecular-based technique aimed at species identification and cataloging using the amplification and sequencing of standardized, short DNA regions. It is an extremely useful tool for identifying species, where there is a need to identify the organism not easily by its morphology alone (Friel et al., 2021). By simply comparing sequences from the organism with those available in the databases, scientists can quickly and reliably identify species, which is always important for biodiversity studies, conservation, and safety in the food chain. DNA barcoding involves different significant steps (Fig. 17.1).

Organisms are collected from their natural habitats, including animals, plants, fungi, and environmental samples such as soil and water. Appropriate collection and preservation methods, freezers, alcohol, etc., ensure that DNA does not degrade

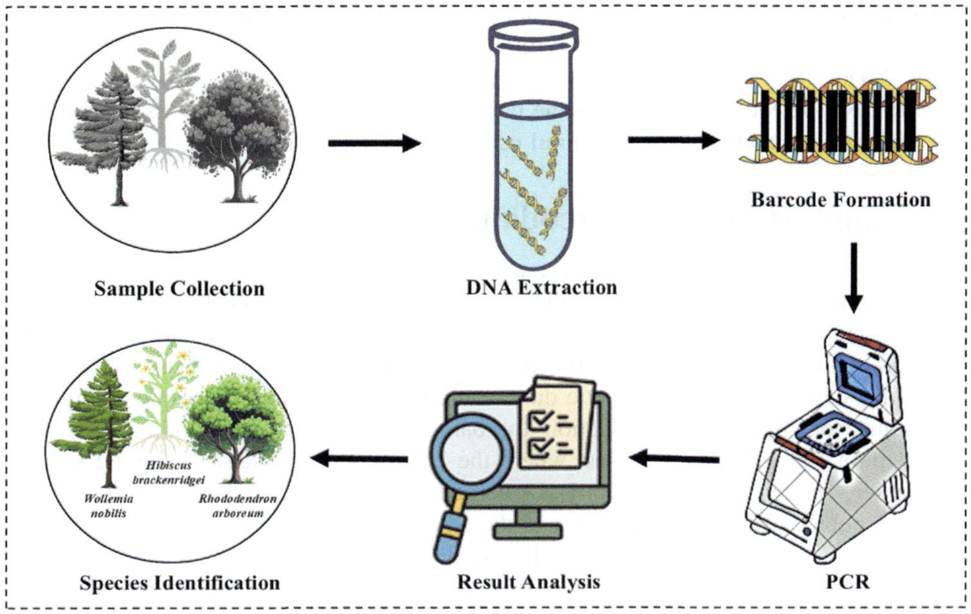

FIGURE 17.1 Steps involved in species identification by using DNA barcoding.

at analysis time. To expose the cell walls to allow the release of DNA, cells need to be broken open. This can be done through either physical or chemical methods. Physical methods include grinding, while chemical methods involve the use of detergents and enzymes (Cochrane et al., 2016; Sahoo et al., 2024; Singh et al., 2024). The isolated DNA is subsequently purified to remove contaminants like proteins and lipids that may interfere with subsequent analysis. Regions of DNA are chosen as barcodes. For animals, the most commonly used was cytochrome c oxidase subunit I, a mitochondrial gene. In plants, the most commonly used genes were rbcL and matK. These gene regions should be sufficiently variable to distinguish species but sufficiently conserved to allow reliable amplification (De Groot et al., 2011). Polymerase chain reaction is the amplification of the DNA region of interest, creating millions of copies of the target sequence. This is a cyclic process involving denaturation, annealing, and extension steps (Fig. 17.1). It will involve optimization of conditions such as temperature and concentration reagents for high efficiency and specificity. The DNA is then sequenced to establish the nucleotides in a particular order (Fig. 17.1). Sanger sequencing and next-generation sequencing (NGS) are just some of the sequencing technologies involved. Sequences obtained are controlled for quality before use (DeSalle, 2006; Sahoo & Satapathy, 2021). Species identification is made possible by comparison of sequences to a reference database, like BOLD or GenBank. Bioinformatics tools are used to assess similarity among sequences and reconstruct phylogenies to show evolutionary relationships (Dohna & Kochzius, 2016; Panigrahi & Satapathy, 2021; Sahoo, Satapathy, & Panigrahi et al., 2023; Sahoo, Satapathy, et al., 2023; Sahoo, Sethi, et al., 2023). The new species discovered can be cross-checked with sequences in an existing database. This increases the accuracy of identification. Joining morphological measurements with genetic material may offer even further precision for species identification.

17.4 Importance of DNA barcoding

DNA barcoding provides fast and accurate species identification. It is especially important for species that are cryptic or morphologically similar. Such accuracy is required for accurate biodiversity estimations. DNA barcoding is an important instrument used in monitoring biodiversity and ecosystem integrity (Erckie et al., 2022). As species composition is monitored over time, research via DNA barcoding becomes feasible for the study of species richness. DNA barcoding identifies endangered species, hence enabling conservation experts to structure management efforts targeted at specific threatened or endangered species (Farooq et al., 2020; Ferri et al., 2015).

Food safety and fraud detection: It is becoming increasingly applied for species verification of food products, thus avoiding erroneous labeling, and this technique ensures authenticity, therefore guaranteeing consumer safety and sustainability (Fernandes et al., 2021) (Fig. 17.1).

Ecological research: DNA barcoding enhances the capabilities of ecological studies, enabling taxonomers to identify which species of interest are engaged in ecological activities like predator-prey relationships, and the body composition of any community (Fig. 17.1).

Public engagement: DNA barcoding will involve the public in citizen science endeavors in biodiversity conservation through increased awareness and appreciation of natural ecosystems (Ford et al., 2009).

17.5 Historical context of DNA barcoding

Advances in molecular biology, genetics, and bioinformatics gave rise to DNA barcoding as a pivotal tool for species identification and biodiversity assessment, largely applied in botany. This account deals with the key figures, milestones, and significant studies shaping this field. Molecular systematics was its intellectual ancestor and made use of DNA sequences to understand evolutionary relationships (Forrest et al., 2019). Actual pioneering work by the likes of Walter Fitch and David M. Hillis demonstrated how genetic data would come to clarify taxonomic ambiguities. Mitochondrial DNA usage also became something of a foundation for barcoding studies within animal phylogenetics. This was especially influential within animal species studies, where the cytochrome c oxidase subunit I gene became the most commonly adopted marker. The formal definition of DNA barcoding is attributed to Paul Hebert, a Canadian biologist. In a ground-breaking paper published in the Proceedings of the National Academy of Sciences, Hebert and his colleagues proposed applying a standardized gene region, COI, for the identification of species within animals. Hebert's ambition was to reach further than animals alone, as there was a requirement to evolve a universal barcoding system within forms of life, not only in animals but also in plants. Following Hebert's first report, the focus became exploring good plant barcodes. Researchers discovered that the chloroplast genome may be barcodable. The most chosen markers, like rbcL and matK, have been named as great candidates for plant DNA barcoding (Fu et al., 2019, 2022; Panigrahi et al., 2024). It was in 2004 that the BOLD was inaugurated to minimize the storage and distribution of DNA barcoding data. BOLD lets researchers construct, compare, and interpret DNA sequences, as well as fosters international collaboration in biodiversity research. Botanical DNA barcodes have marked a significant advance in outlining methods of identification and taxonomy of plant species. The works of such visionaries as Paul Hebert, among other collaborative efforts, have been crucial in putting forward DNA barcoding as one of the most important tools in assessing biodiversity and in active conservation and ecological research (Garcia et al., 2024). Applications of the field will likely continue to evolve towards furthering our understanding of plant diversity and the critical roles played by plants in ecosystems all around the world. DNA barcoding has significantly advanced our ability to resolve taxonomic ambiguities and classify poorly understood plant groups.

17.6 Barcoding in taxonomy

Cryptic Species: Many plant species share similar morphologies that make them hard to tell apart. DNA barcoding identifies cryptic species with a similar appearance but shows some genetic divergence by characterizing a set of genetic markers that should be identified. This is important for accurate assessments of biodiversity and hence correct conservation activities. Traditional morphology has, on more than one occasion, led to the misidentification of specimens, particularly within taxonomically complex groups. DNA barcoding is, however, an evidence-based, unbiased tool that minimizes the risk of incorrect identification (Geiger et al., 2021). DNA barcodes aid in the resolution of evolutionary relationships among plant species. Using genetic data in building a phylogenetic tree, researchers can gain insight into how different species are related, thus allowing the classification of poorly understood groups. Barcode-generated genotypic information can lead to taxonomic revisions and could clean up outdated or inappropriate names (Fig. 17.1). This is particularly helpful with taxa that have been notoriously problematic to place in morphological spaces, such as those with developmental plasticity or convergent evolution. DNA barcoding can be used to rapidly assess plant diversity in many ecosystems (Guo et al., 2023). Even more importantly, the areas of high biodiversity where most species have yet to be described or are characterized only poorly are biodiversity hotspots. Rapid identification of species then transforms recording and monitoring changes of biodiversity into easy work. DNA barcoding takes a step forward in the assessment of genetic diversity within and between populations because of its ability to correctly identify species. This is the foundation for effective planning and management for the conservation of species.

17.7 The science behind DNA barcoding

The core of plant DNA barcoding is dependent on some specific genetic markers selected based on evidence for their efficiency in distinguishing species. Among the most widely used markers are rbcL, matK, and ITS.

Ribulose bisphosphate carboxylase/oxygenase large subunit (rbcL)**:** The rbcL gene encodes for the protein, which forms a constituent component of the ribulose bisphosphate carboxylase/oxygenase enzyme. This enzyme is crucial in the Calvin cycle, where, through its enzymic activity, it catalyzes the fixation of carbon dioxide.rbcL is not conserved as much across so many plant species that facilitates amplification of the gene from diverse taxa. At the same time, it contains enough variability among species to be properly differentiated. It is widely used in phylogenetic work, and a very large amount of sequences are present in the databases, thus, it can easily be relied upon for barcoding. The rbcL marker is applied in almost all taxonomical studies from angiosperms and gymnosperms to other plant groups. It is applicable to a wide range of development.

Maturase K (*matK*): *matK* is a protein-coding gene that serves in chloroplast pre-mRNA for splicing introns. It is of vital importance to make a functional chloroplast. In comparison with rbcL, matK does have a higher rate of nucleotide substitution; therefore, it is quite efficient in distinguishing species that are relatively close relatives. In use with rbcL, the matK gene brings greater resolution to plant species identification; more significantly, where rbcL alone is insufficient in delivering differentiation within difficult groups. Similar to rbcL, matK has a good deal of data in genetic databases that aid the comparative analysis of the species efficiently (Fig. 17.2).

17.8 Internal transcribed spacer

The internal transcribed spacer (ITS) region is a DNA region that is not coded but lies between rRNA genes. It resides within the nuclear genome and is being transcribed to form rRNA, but is not being included in the rRNA product. The ITS region is very variable between species, thus highly useful in the identification of closely related taxa, also in resolving species complexes. ITS is present in all eukaryotic organisms; thus, its use can be extended over a wide variety of plant groups, such as ferns, angiosperms, and gymnosperms. The ITS region has been the subject of several studies and is abundant in genetic databases for use by researchers.

17.9 DNA barcoding link with bioinformatics

In DNA barcoding, bioinformatics tools and databases play a critical role in the comparison of sequences and the recognition of species. These resources facilitate further analysis of genetic data for taxonomic identification purposes, thereby contributing to an understanding of biodiversity (Kress & Erickson, 2008). Some of the key databases and software tools used routinely in practice are described below.

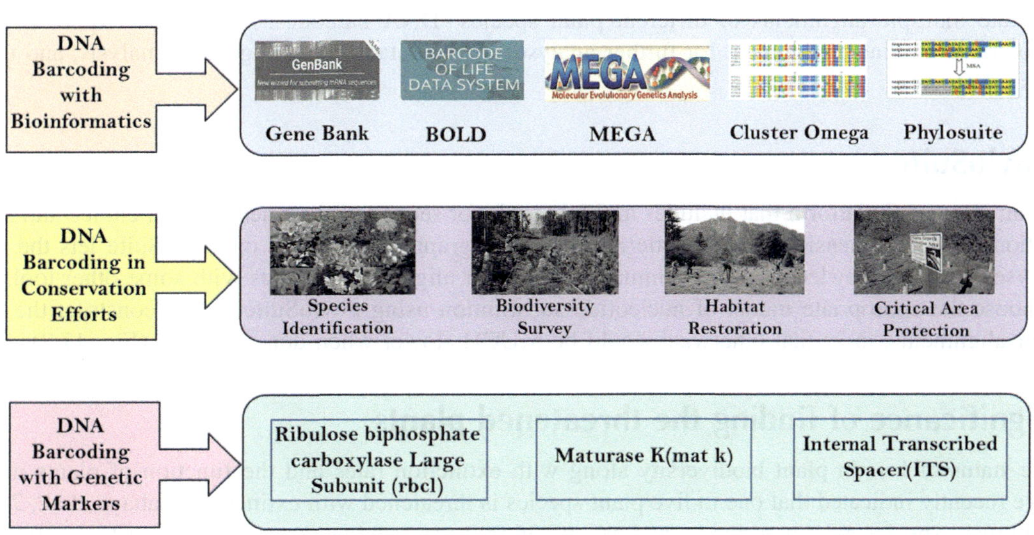

FIGURE 17.2 DNA barcoding in association with different bioinformatics tools, conservation effortsand specific genetic markers.

17.10 GenBank

GenBank is the general-purpose database maintained by the National Center for Biotechnology Information (NCBI). GenBank contains a large collection of nucleotide sequences obtained from a variety of organisms. Research scientists can submit their DNA sequenced data to GenBank so that the data becomes accessible to the research scientist community. Users can do BLAST searches to match their own sequences with the GenBank database with a view to finding possible matches and species (Meiklejohn et al., 2019). After DNA barcoding, one could input the sequence obtained into the BLAST tool in search of similar sequences. On making a match in a known species, it could establish or confirm the identity of the sample (Fig. 17.2).

17.11 Barcode of Life Data System

BOLD is a DNA barcoding database specifically designed to store DNA barcoding data. It contains associated metadata, including species descriptions, geographically and ecologically specific information. Authors of a DNA barcode paper may enter their DNA barcode data in BOLD. BOLD has several identification tools; one of which is "Identification Engine," "Identification Engine," which compares query sequences against its database to propose possible species matches. The DNA sequence could be uploaded to BOLD by a botanist to identify a plant species. The system could then analyze the sequence, producing possible matches with their corresponding confidence score to identify in the shortest time possible (Fig. 17.2).

17.12 Molecular Evolutionary Genetics Analysis

Molecular Evolutionary Genetics Analysis (MEGA) is software for molecular evolutionary data analysis. Alignment of sequences, construction of phylogenetic trees, among other types of evolutionary analyses are available. Sequences can be aligned through ClustalW and Muscle algorithms in MEGA. After the sequences have been aligned, an individual will be able to build his or her phylogenetic trees based on maximum likelihood or neighbor-joining. Once the sequences are obtained from BOLD, MEGA will be utilized in order to align the sequences and then construct a phylogenetic tree which will represent the relationship of evolution between species of interest (Fig. 17.2).

17.13 Clustal Omega

Clustal Omega is one of the most frequently used tools in multiple sequence alignment, with a high potential for large-scale datasets and good-quality alignments. The users only need to input DNA sequences in the FASTA format, and then the results of the multiple sequence alignment from Clustal Omega could be further analyzed (Fig. 17.2). A researcher wants to generate multiple alignments of different plant species' DNA barcodes. She would require running Clustal Omega first to create a reliable alignment for further analyses of her data, like phylogenetic analysis, and the conserved regions across all species.

17.14 PhyloSuite

PhyloSuite is an integrated platform that includes multiple tools for sequence alignment, model choice and selection, and tree construction for comprehensive phylogenetic analysis. The graphical interface of PhyloSuite lets the user perform complex analyses without knowledge of the command line. After aligning sequences with some other tool, a researcher might then choose the appropriate model of nucleotide substitution using PhyloSuite and/or construct the phylogenetic tree from their alignments, tasks that otherwise would be MUCH slower when done by hand (Fig. 17.2).

17.15 Significance of finding the threatened plants

Document the status of loss in plant biodiversity along with extinction rates and the function of plants in ecosystems. Estimates have recently indicated that one in five plant species is threatened with extinction (Lahaye et al., 2008). That is, nearly 400,000 plant species exist worldwide, and around 36,000 are listed as threatened by the International Union for Conservation of Nature. Studies indicate that the current rate of extinction for plants is 100 to 1000 folds higher than the natural background rate and is mainly due to human activities. The high rates of extinctions occur in the tropical regions, especially the Amazon rainforest and Southeast Asia, due to deforestation and habitat destruction (Larridon et al., 2020).

The Mediterranean Basin is defined as a biodiversity hotspot. A lot of endemic plant species have been experiencing increased decline rates, and many of the species face threats from urbanization and climate change (Lestari et al., 2018).

17.16 Habitat destruction

"The main causes of habitat destruction are urbanization, agriculture, and logging. The Food and Agriculture Organization has estimated that every year around 10 million hectares of forest are lost, which in turn reduces the plant diversity."

17.17 Climate change

Climate change is altering habitats and many plant species are shifting their ranges. Already, the threat of extinction based on climatic conditions changes may affect 30% of all plant species by 2050.

17.18 Invasive species

Non-native plants may outcompete native species for space and resources, leading to local biodiversity declines. In this region, invasive species account for about 42% of plant extinctions since 1750.

17.19 Overexploitation

Most plant populations face unsustainable harvesting for timber, medicinal plants, and ornamental species. It is estimated that 2000 plant species are used commercially, and most of these are harvested unsustainably. For most forms of life to be alive, there is a need to have oxygen. This oxygen is produced through photosynthesis by plants (Head et al., 2014). Plants take in carbon dioxide, thus slowing down the effects of global warming. Plants form the base component of food webs. They are primary producers, supporting herbivores and, in turn, higher trophic levels. Loss of plant diversity can cause the collapse of entire ecosystems and lead to the loss of associated animal species that rely on them for food and shelter (Hartvig et al., 2015).

17.20 Cultural and economic significance

Most human societies depend on plant diversity for food, medicine, and raw materials. Traditionally, 80% of the population utilizes biodiversity in the form of traditional medicine, mostly plant-based (Hebert et al., 2003).

17.21 Conservation initiatives

International undertakings, including the Convention on Biological Diversity (CBD), focus on halting the loss of biodiversity by 2020 and improving the sustainable management of natural resources. Protected Areas have increased in number worldwide, with over 15% of terrestrial regions and 7% of marine regions being declared protected, although enforcement and management are yet challenges (Hebert et al., 2025).

17.22 Restoration projects

Different ecological restoration projects are going on all around the globe to restore damaged ecosystems. The United Nations Decade on Ecosystem Restoration 2021–30 aims at restoring 350 million hectares of degraded land (Hollingsworth et al., 2016).

17.23 Public engagement

This is one of the more popular applications of citizen science projects and educational initiatives to increase public awareness regarding plant biodiversity and get communities actively involved in conservation. Loss of plant diversity is, however, emerging as a significant global issue with important implications at the ecological and human well-being levels (Hoveka et al., 2022). High rates of extinction and habitat degradation are leading to a new urgency in conservation and restoration efforts. Crucial lessons can be learned by understanding how plants function in the ecosystem to help develop sound strategies for safeguarding and preserving botanical diversity for generations to come (Hoveka et al., 2020).

Ecosystem Services: State the different types of ecosystem services provided by plants, such as carbon sequestration, provision of habitat, and medicinal use. Plants are essential components of ecosystems when it comes to a wide range of life-supporting services for living organisms (Lobo et al., 2013). Ecosystem services are, in turn, divided into four categories-provisioning, regulating, supporting, and cultural services (Huang et al., 2024). Let's consider some of the major ecosystem services that plants provide below:

17.24 Carbon sequestration

The process of photosynthesis involves the conversion of carbon dioxide absorbed from the atmosphere into organic matter. Such processes help minimize the concentration of greenhouse gases and, hence, reduce the impacts of climate change. Forests, grasslands, and wetlands act as the best carbon sequestration agents. Carbon sequestration via forest formation alone is estimated to reach about 2.6 billion metric tons per year. Carbon gets fixed into trunks, branches, and leaves of plant biomass and in the soil in the form of organic matter—a long-term carbon storage solution (Ki et al., 2009). Plants serve as habitats for different organisms such as animals, insects, fungi, and microorganisms. Varying plant communities correspond to the supporting diversity in animal populations, hence creating ecosystems with high biodiversity. For example, thousands of species of birds and mammals live in forests; many insects are found in forests, and amphibians and aquatic life in wetlands (Knot et al., 2020). Healthy plant communities help ecosystems be resilient to disturbances, such as climate change, disease, and invasive species.

17.25 Water regulation and quality

The vascular plants use the water, absorb it, and transpire it; thereby moderating local and regional climates. Through their roots, the plants hold soil, preventing erosion and run-off, thereby limiting sedimentation in water bodies through holding the river banks and shorelines intact (Kraaij et al., 2018). Wetland plants filter off pollutants in water; thus, ensuring that clean water quality is maintained, which leads to clean drinking water for the biotic community and human communities alike.

17.26 Medicinal applications

Most cultures rely on plants for traditional medicine. Practically 80% of the world's population uses plant-based drugs as the basis for their medical health care. Most modern drugs have originated from plant-based compounds. Aspirin is derived from willow bark, and quinine, used to treat malaria, comes from the cinchona tree (Krishnamurthy & Francis, 2012). Vast is the share of studies in plants for novel medicinal compounds, and still, there are many potential drugs that are yet to be discovered.

17.27 Food production

Plants are a basic diet for human nutrition, with fruits, vegetables, grains, and nuts providing much-needed vitamins, minerals, and energy. Different types of crop species increase food security as several crops become available. The diversity acts as a cushion against pest and disease attacks and climate variability. Most of the plants act as feed for animals; hence, they are part of the agricultural systems and also support the economies of different countries worldwide. Plants have significant cultural and spiritual values in many societies. They often form part of cultures, rituals, and local customs (Loman et al., 2012).

The existence of green space, parks, and gardens in a city can greatly ornament the urban environment. This extends to creating recreational spaces and enhancing the psychological well-being of human beings. The beauty of plants gives rise to landscapes and affects design in urban spaces to inform human enjoyment and connection with nature (Lou et al., 2010). Plants offer a greater number of ecosystem services, the most critical significance for maintaining ecological balance and supporting human life. From carbon sequestration and habitat provision to medicinal uses and food production, the benefits of plants are diverse and so far-reaching (Lobl et al., 2023). Much effort is required in the recognition and protection of these services to foster biodiversity, sustain ecosystems, and ensure a healthy environment for future generations. As threats like climate change, habitat loss, and pollution heighten, the need to save plant diversity increases for the preservation of such invaluable ecosystem services.

17.28 Case studies of successful applications

A number of detailed case studies of specific endangered plant species identified with success by DNA barcoding are presented, including their geographic locations and species descriptions, as well as the significance of the findings. DNA barcoding has proven to be a very useful tool for the identification and conservation of endangered plant species. Several detailed case studies focusing on some of the species that have been the beneficiaries of DNA barcoding are also presented (Meiklejohn et al., 2019).

Hibiscus brackenridgei Habitat: Native range: Hawaii; Known only from the islands of Oahu, Maui, and the island of Hawaii. This shrubby perennial has massive, showy yellow flowers with a maroon center. Coastal to lowland habitats. Different studies have been conducted to gain knowledge about various species' genetic diversity. DNA barcoding is used in genetic research and involves the extraction of DNA samples from the selected species (Poovitha et al., 2016). This information is quite significant to conservation endeavors, as it shows the potential of increase in resistance to environmental changes as genetic variation is sustained (Fig. 17.1). The detection has been applied in the course of restoration to maintain diverse genetic stock, which plays a crucial role in the survival of the species.

Rhododendron arboreum: This plant is an eastern Himalayan species. It is reported to occur in Nepal, Bhutan, and northern India. *Rhododendron arboreum* is a broad-leaved evergreen shrub or small tree, reaching a maximum height of 15 m. It has great, leathery leaves and bright red to pink flowers that bloom in clusters. It clarified through DNA barcoding the taxonomic status of *Rhododendron arboreum* so it can now be better separated from the closely related species (Ren et al., 2023). This clarifies conservation efforts because one would know more focused areas that may need habitat protection and restoration activities (Fig. 17.1). Moreover, it also highlights the effect of climate change on its distribution, so advanced conservation is considered.

Wollemia nobilis: This conifer species is rare. It was discovered in 1994 and its appearance is very characteristic: tall, straight trunks; and unique, fern-like foliage. It has sometimes been called the "living fossil," since its lineage is ancient [38]. DNA barcoding has also confirmed the genetic identity of this species and its relations with other species of conifers. The genetic data have had an impact on conservation management activities, especially on propagation and reintroduction efforts (Fig. 17.1). These are important since the Wollemi Pine is easily threatened by factors like habitat destruction and climate change.

Euphorbia purpurea: This is a Mediterranean plant, endemic to southern France and southern parts of Italy. *E. purpurea* has purple-colored stems and small, greenish-yellow flowers. Its growth takes place on rocky or disturbed sites. DNA barcoding is used to identify the *E. purpurea* accurately and help differentiate it from other Euphorbia species. Such clarification bears relevant implications for conservation, for instance, protection of specific habitats of this species from the danger of urban development and agricultural expansion (Peirson et al., 2014).

17.29 DNA barcoding challenges

Identifying and conserving plant species using various methods, including DNA barcoding, poses several challenges. It is worth looking into the issues that would emerge in terms of hybrid species, incomplete genetic databases, or difficulties in addressing highly related species (Menegon et al., 2017).

Hybrid species: Identification Problems Hybrid species, which are a hybrid of two different species, often have characteristics from both parent species that make their identification hard. Morphologically, they appear to be a combination of their parent species, which is always confusing (Mishra et al., 2017). DNA barcoding often fails to classify hybrids, especially when the hybrids contain genetic material from one of the parent species (Fig. 17.3). This may, therefore, lead to ambiguous results that complicate conservation efforts. The fact of the hybrid's presence may become a contradictory attempt on the part of the conservators, as it is not clear whether to conserve the hybrid or one of its parents. Hybridization could also threaten the genetic integrity of the native species.

Low-grade genetic databases: DNA barcoding requires reference samples in large genetic databases to find out the species correctly. Inadequate databases may result in misidentifications, particularly for less-known or endemic species. Some regions have limited genetic sampling, which can skew our understanding of species diversity and distribution. Without adequate representation, conservation policies may overlook critical areas or species (Munjal et al., 2019). Taxonomic classifications are often revised as new genetic information comes in. Incomplete databases do not necessarily represent the changed taxonomic system, and hence, may contain old or wrong conservation priorities.

Closely related species: Many related species are morphologically similar or indistinguishable, which makes identification very difficult. The genetic differences may appear with DNA barcoding, but morphological traits do not clarify well in these cases. This means that the region selected for DNA barcoding may be too poor in resolution to

FIGURE 17.3 Challenges in DNA barcoding during different steps.

distinguish between closely related species. Practically, this means that one cannot produce the correct biodiversity assessments and therefore, will be producing ineffective conservation strategies. Successful conservation requires the ability to resolve evolutionary relationships among closely related species (Munjal et al., 2019). It is very difficult to formulate management plans that allow for every species' special needs without a clear resolution of these relationships by genetic data (Fig. 17.3).

Field challenges: Practical issues encountered in the field, for example, degradation of samples, and what is required for documentation. In practice, there are several fieldwork problems researchers in plant conservation and identification have to deal with, especially regarding sample degradation and proper documentation. Here's an overview of those challenges:

Sample degradation: Deterioration of samples is very frequently encountered, especially when samples are collected from fields as they are exposed to moisture, temperature fluctuations, UV light, etc. Both the morphological characters and the integrity of genetic material may be affected. Time Constraints Samples degrade when long periods are taken in processing samples. DNA may deteriorate if samples are not stored or analyzed within a short time (Malakasi et al., 2019). The foreign DNA might contaminate the specimens from the environment or even from other plant species, and also from the researcher's handling. This complicates genetic analysis and leads to false outcomes (Fig. 17.3).

Need for proper documentation: The information documented in field records may be inconsistent and/or inadequate, leaving holes where critical pieces of information about the location, habitat, and conditions under which samples were taken are missing. In the course of the study, samples may get misplaced and wrongly labeled. Proper labeling, therefore, means data for each sample should correspond correctly (Mamathaba et al., 2022). Researchers or institutions might use different methods in the documentation and storage of data, which may make it hard to integrate and compare findings from various studies.

Solutions: A suitable collection technique. All the techniques and materials should be followed by researchers to avoid degradation. For example, drying plant tissues using silica gel or preservation in appropriate buffers preserves DNA. Formulating raw processing or preservation procedures for samples collected can be a useful method of minimizing the risks of degradation (Mallott et al., 2018). Standardized documentation protocols may be adopted in developing them to ensure uniformity in data collection and record-keeping. This can include standardized formats for field notes, sample labels, and metadata. The awareness of the best documentation and handling of samples does make the collected data of good quality and trustworthy (Mani et al., 2021). Technology can be utilized to facilitate the continuation of documentation efforts by using mobile applications or databases. In that case, the entry of real-time data is feasible, and the correct entry of data with samples is assured.

Ethical issues: Look at the ethical issues from genetic research on endangered species, such as consent, ownership, and implications on biodiversity. Genetic research in endangered species raises major ethical issues, starting from the questions of consent and ownership, through effects on biodiversity (Mankga et al., 2013). These are some of the aspects elaborated below (Fig. 17.3).

Consent: Several endangered species are from ecosystems traditionally controlled by indigenous peoples and local communities. Carrying out any genetic research without their consent would be a disregard for their rights and an infringement of their right to be custodians of land. Ethical research requires obtaining informed consent from affected communities (Manzanilla et al., 2022). This means clearly explaining the research's purpose, potential risks, benefits, and how the findings will be used. Involving local and indigenous peoples in the research process instills confidence and ensures that there is cooperation, so their knowledge becomes respected. Collaboration can improve the appropriateness and success of conservation measures.

Ownership: Samples of genetic material from endangered species may be taken for research or commercial application and raising questions of ownership. "Who owns the genetic material, and who receives the benefits? Benefits must flow from the commercialization of genetic resources derived from their ecosystems. Thus, benefits must be equitably shared with local communities and countries. Ethical frameworks need to define ownership and seal deals in a way that communities contributing to research are equitably compensated and recognized (Margulies et al., 2023). The nature of cooperation in these relationships could facilitate more effective conservational relationships.

Impacts on biodiversity: Involvements in the manipulation of genetics, including CRISPR, have ethical implications for meddling in the genetic background of endangered species. But interventions of this nature, though beneficial in conservation, carry unintended ecological consequences. Genetic enhancement to increase the survival potential may decrease the genetic diversity responsible for species resilience. In this regard, ethical considerations need to address the cumulative long-term impacts on the affected populations and ecosystems (Maurin et al., 2014). A cautious approach of precaution must be taken by genetic research so that the natural evolution of endangered species and their integrity are not jeopardized. Ethics should dictate short-term solutions for the long-term preservation of biodiversity and ecosystem health.

17.30 Integrating DNA barcoding with conservation efforts

Overemphasizing genetics research could overshadow the other important conservation measures that including habitat protection and restoration. Such a narrow focus could lead to neglect of the more holistic ecological contexts relevant to species endangerment. Ethical research dictates a holistic approach, integrating genetic insights into traditional conservation methods. Thus, efforts to save endangered species will be complete and effective since they are directed toward the roots of their decline (Meyer & Paulay, 2005). DNA barcoding is a great method for discovering species using short sequences of DNA, and it represents an important component in informing targeted conservation actions. Here is how DNA barcoding can really help to direct efforts such as habitat restoration and protection of critical areas:

Species identification: Identification of Endangered Species: DNA barcoding enables the proper identification of endangered or threatened species that might easily be overlooked through traditional morphological methods (Meier et al., 2006). Their

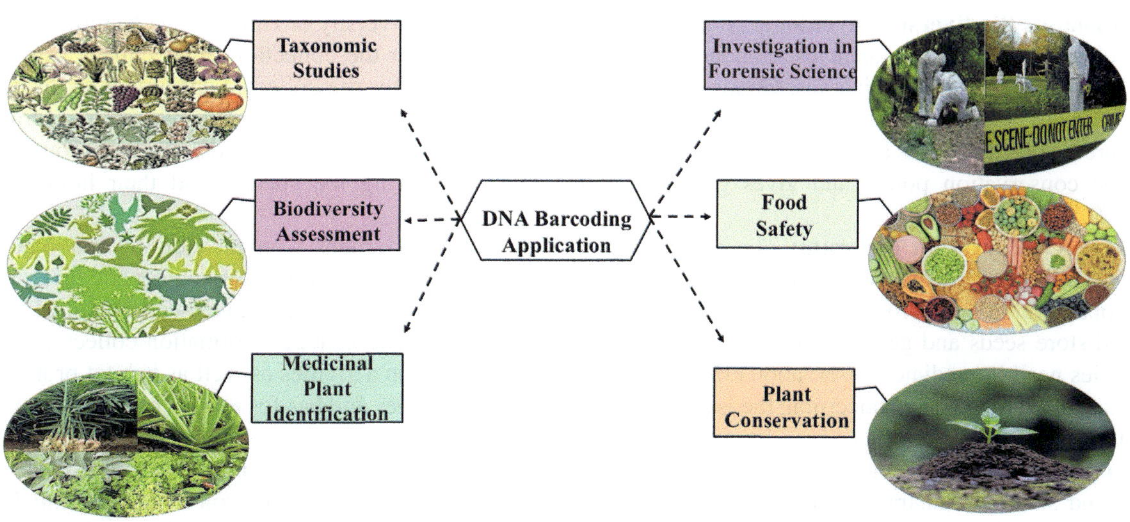

FIGURE 17.4 Application of DNA barcoding in various fields.

recognition supports setting priority conservation actions. It can reveal cryptic species, or those morphologically the same but genetically different, so that conservationists can know the actual biodiversity level of an area (Fig. 17.4).

Biodiversity surveys: DNA barcoding records the species in a particular habitat, which, over time, would provide baseline data to guide future conservation practices and monitor changes through time (McDonnell et al., 2021). The use of DNA barcoding assessments consistently facilitates the tracking of ecosystem health by studying shifts in species composition, which might reflect environmental stress or degradation (Fig. 17.4).

Focused habitat restoration: Keystone species identification: DNA barcoding can identify keystone or indicator species that are crucial in maintaining the health of an ecosystem. Restoration can concentrate on habitat enhancement for habitats whose supporting species require such habitat enhancement. The specific flora and fauna will provide practitioners with information about the requirements that such particular endangered species require with regard to habitat, thereby preparing them better for restoration (Murphy et al., 2020).

Critical areas protection: DNA barcoding is useful in the identification of biodiversity hotspots, which refer to areas with a high concentration of endangered or endemic species. That is necessary for the conservation of general biodiversity. DNA barcoding data would help in the establishment of new protected areas through evidence that areas with unique or vulnerable species are granted the required conservation attention.

Invasive species management: DNA barcoding can rapidly identify invasive species threatening the native flora and fauna. Early identification means timely management and control measures to protect native ecosystems (Fig. 17.4). Knowing which native species are being outcompeted by invasives can guide restoration efforts in reintroducing or bolstering native populations (Nakazato & Jinbo, 2022).

Community engagement and education: DNA barcoding programs can engage local communities in biodiversity monitoring, promoting the awareness and stewardship of ecosystems in the locality. The scientific evidence that is delivered by DNA barcoding can inform policy-makers so that their conservation strategy will have strong evidence.

17.31 Collaborative projects

One of the best means of conservation is collaborative efforts between NGOs, universities, and governments. Some of the most remarkable projects and what they do and achieve are as follows:

Amazon Conservation Team (ACT): ACT partners with local indigenous communities, the university, and government agencies. ACT works together with indigenous peoples to create precise maps of their traditional territories, merging ecological and cultural knowledge. The scientists use DNA barcoding to classify species of plants and animals, which aids in increasing the reliability of biodiversity assessments of the Amazon rainforest (Niskanen et al., 2023). The outcome of the project has led to the establishment of conservation areas through mapped indigenous areas. This ensures biodiversity conservation. The success story assists the indigenous people to secure their legal rights towards land, ensuring them better and increased participation in biodiversity conservation.

Hawaiian Islands Biodiversity Inventory: A collaborative effort by the University of Hawaii, The Nature Conservancy, and multiple state and federal agencies. Conduct extensive field surveys for the collection of specimens from different types of habitats dispersed over islands. Obtain a DNA barcoding analysis for the collected specimens to determine the species involved, which includes many endemics in Hawaii. The project has led to the discovery of several species hitherto unrecognized; it gives a glimpse into the extent of biodiversity on the Hawaiian Islands. The data from the project informs the conservation policy and gives guidance to the effort of saving the species and their home (Niemann et al., 2022).

Global Tree Conservation Program: The partnership is with Botanic Gardens Conservation International, several universities, and governmental environmental agencies. DNA barcoding is employed in assessing genetic diversity in tree species globally, although it focuses intensely on endangered species. The project collaborates with botanical gardens to harvest and store seeds and genetic material from threatened tree species. The genetic information collected determines which species need immediate conservation efforts, hence guiding restoration activities as well as habitat protection. The program has raised awareness concerning the issues of the conservation of trees and has fostered public participation in the conservation process.

European Union's LIFE Program: Projects under the LIFE Fund are mainly the contribution of NGOs, research institutes, and state administrations in the European Union member states. Most of the projects implemented with the assistance of the European Union's LIFE program are composed of an integrated approach that involves habitat restoration, species monitoring, and community participation. DNA barcoding promotes the monitoring of the population and health of species, particularly in protected environments (Ogier et al., 2019). Many endangered species have benefited from the program, with evidence shown in the rising count of populations and improving habitats.

The acquired knowledge is reflected in the formulation of biodiversity policies and conservation measures in EU regions.

Wildlife Conservation Society (WCS) and Local Governments in Madagascar: While operating in Madagascar, the WCS works with local governments to effectively enforce conservation efforts. WCS, local governments, universities, and offices in Madagascar. WCS conducts ecological monitoring using both conventional methods and DNA barcoding which helps to identify biodiversity in key habitats. The project involves local communities in conservation work while promoting responsible behaviors that result in habitat conservation. Through the installation and expansion of various protected sites, the project has protected many habitats for a certain number of endemic species. By providing conservation goals and enhancing the local economy, these projects develop sustainable livelihoods (Pezzini et al., 2023).

These collaborative efforts show clearly how partnerships among NGOs, universities, and governments caninfluence specific conservation outcomes. Different skills and resources are combined in these endeavors that lead to the effective understanding of biodiversity challenges and improvement of the role of communities locally, thereby continuing sustainable practices with benefits toward people and the environment.

DNA barcoding research influenced the policies and regulations in conserving the species: The results of DNA barcoding studies have greatly influenced conservation policies and regulations through robust data, providing information for making decisions. Here are some of the ways through which DNA barcoding has impacted conservation:

Species identification and protection: DNA barcoding facilitates the accurate identification of endangered species covered under the law. For example, accurate identification at a high taxonomic level has been pivotal for the enforcement of various laws, such as the US Endangered Species Act. The discovery of new species through DNA barcoding may result in adding such species to the protected lists, thereby increasing the number of conserved species (Phillips et al., 2024).

Monitoring and compliance: Regulatory agencies have adopted DNA barcoding in monitoring programs to track compliance with conservation legislation. For instance, DNA barcoding can be applied for the species-related legality verification of confiscated products. DNA barcoding has helped the Convention on International Trade in Endangered Species of Wild Fauna and Flora to better control the traded species and has stronger control against illegal wildlife trafficking (Pirie et al., 2016).

Biodiversity assessments and planning: DNA barcoding provides critical information for evaluating biodiversity, which guides conservation planners in focusing on high species-richness and endemism regions for prioritization. DNA barcoding data inform the design of protected areas and reserves. Information on species richness and emerging threats from DNA barcoding studies aids adaptive management of conserved areas through monitoring shifts in species composition and emerging threats so that policy interventions are made in time.

Habitat restoration and land use planning: DNA barcoding can identify species that are of special interest to restoration work. This means any conservation actions made are directed towards maintaining or restoring the biodiversity of certain areas. This directedness impacts land use planning and habitat management policies. Policymakers can, through determination of the ecological functions of the various species, integrate ecosystem services into the confines of conservation regulations and policies that support conserving them (Pauthenier & Faulon, 2014).

Community engagement and education: DNA barcoding can facilitate community-based conservation through the education of local communities about their biodiversity. Policies that involve the community have usually been more effective as well as sustainable. Results from DNA barcoding studies are very useful for creating awareness among members of the public concerning biodiversity issues, which affects opinion and contributes to conservation policies.

International cooperation: DNA barcoding is part of global biodiversity initiatives, such as the Global Genome Biodiversity Network, which enables most nations to be able to cooperate on efforts and data sharing. Such cooperation would eventually help build harmonized regulations across borders. Results from DNA barcoding research can be utilized in tailoring training programs for conservation practitioners in most countries to strengthen local capacities in implementing effective conservation policies through genetic information (Peirson et al., 2014). DNA barcoding has become a powerful tool in shaping the policies and regulations that define conservation. This scientific approach further heightens the effectiveness of conservation strategies while enhancing compliance with laws, facilitating decision-making processes, and ultimately contributing toward more sustainable resource management and biodiversity conservation (Phillips et al., 2022).

17.32 Citizen science associated with DNA barcoding

Citizen science is one opportunity to make biodiversity research more robust through collective action, increase public awareness of the need for protection, and accelerate community involvement in conservation. This would be a quest into the possibility of involving the public in this endeavor: platforms and tools that may serve to achieve such an involvement as follows:

Citizen science initiatives: Citizen science refers to projects that allow the public to participate in scientific research. The citizen may be involved in sampling, surveys, to the analysis of DNA barcoding data. This network platform allows users to document biodiversity by uploading photographs of plants and animals. Data obtained from this process can identify species, and some projects involve DNA barcoding for confirmation (Porter & Hajibabaei, 2018). The Great Sunflower Project: Here, participants grow sunflowers and provide information on visitations made by pollinators. With DNA barcoding, one can even tag and identify the species of pollinator involved, which would further help in the scientific data for conservational studies.
Workshops and training programs: One can train the public with workshops on DNA barcoding and how to use it. Further, empowering local communities to take part in biodiversity monitoring increases the worth of this method. Hands-on Workshops: Running workshops that guide trainees to collect samples and learn the basic DNA principles of barcoding. These can be achieved through hands-on sessions held in local parks or nature reserves. Incorporate DNA barcoding into school curricula, thereby engaging students in practical scientific research and training on biodiversity and conservation.
Online platforms perpetual for data sharing and analysis: Web platforms would help in the gathering, dissemination, and processing of data, making DNA barcoding projects easier for people to engage with.

GenBank and BOLDs: These databases give the user access to genetic data as well as the opportunity to upload their personal data. Public input to these databases can better make them, hence helping in the identification of species.

DNA Subway: A learning resource that provides equipment for students and members of the community to conduct DNA barcoding activities from sampling to data analysis; this facilitates experiential learning (Ren et al., 2023).

Mobile applications in monitoring biodiversity: Mobile applications can empower users to collect and submit samples or records easily, thus promoting public participation in DNA barcoding.

Examples: Seek by iNaturalist: The app encourages users to explore the biodiversity around them and receive immediate identification help. Their observations can be added through linking projects to DNA barcoding initiatives. PlantNet: The app makes it easy to identify species of plants through photos. Users may contribute to citizen science projects that could eventually involve further DNA barcoding for verification (Rach et al., 2008).

Social media and online campaigns: Social media can provide a means of contacting the widest possible audience and raising awareness concerning DNA barcoding and its application to conservation issues. Use the facilities of Instagram, Twitter, and Facebook to disseminate information on DNA barcoding activities currently in train, like features on discoveries and competitions, and involve the public at large in those activities. Contest ideas like community challenges (for instance, "Find and Identify" campaigns) can ensure public participation is not just active but also encouraged through recognition or even reward contributions.

Community-based conservation projects: Community involvement in DNA barcoding conservation can ultimately reinforce local stewardship of biodiversity. Local collaborations with NGOs, schools, and community groups for conducting DNA barcoding studies focusing on local issues for conservation. This makes more difference since the community becomes more sold on and called upon. Feedback loops from research conducted to communities on their contributions and findings, thereby enhancing ownership and pride in the sense of belonging to conservation efforts. It is certainly going to bring a much-heightened dimension to biodiversity research and conservation (Rivero et al., 2006). By utilizing citizen science programs, education-related activities, mobile applications, and social media, researchers can engage with the community and raise awareness while they set the ball rolling for people to work towards remarkable goals in the midst of biodiversity conservation efforts. It strengthens the scientific process and encourages a closer bond between the public and nature while striving for biodiversity conservation.

17.33 Organizations and initiatives

Enumerate some organizations that are leaders in DNA barcoding research and contributions towards conservation, along with descriptions of their respective missions. Here is a list of organizations and leaders in DNA barcoding research and conservation, along with a description of their missions and contributions:

International Barcode of Life Project (IBOL): IBOL plans to develop a comprehensive DNA barcode reference library for the world's biodiversity as a support in species identification and conservation efforts. IBOL works collaboratively with researchers worldwide to standardize methods in DNA barcoding, develop genetic databases, and promote public awareness of biodiversity (Rowley et al., 2007). Their contributions benefit conservation planning and monitoring through the provision of critical genetic information.

BOLDs: BOLD is an online facility that hosts a repository for DNA barcode sequences and their associated metadata to catalyze research into biodiversity. BOLD contributes as the research reservoir, enabling the identification of species, data sharing, and collaboration. It encourages the application of DNA barcoding in conservation, ecology, and taxonomy through its toolkit and access to a constantly developing database of barcoded species.

The Nature Conservancy (TNC): TNC is working to protect the lands and waters on which all life depends. Using science to attack challenging environmental problems. TNC uses DNA barcoding to serve the purposes of conservation planning, tracking biodiversity, and monitoring ecosystem health. Most projects with TNC have to do with the implementation of molecular methods in ascertaining species distributions and supporting habitat conservation programs.

Smithsonian Institution: The mission of the Smithsonian is the increase and diffusion of knowledge, with particular emphasis on natural history, biodiversity, and conservation. Various research centers in the Smithsonian Institute contribute to DNA barcoding projects, which aid in documenting biodiversity, monitoring species threatened by extinction, and conservation. The Smithsonian Conservation Biology Institute employs genetic technology to strengthen wildlife preservation (Ruhsam et al., 2015).

Royal Botanic Gardens, Kew: Kew aims to inspirie and provide science-based plant conservation for the entire world to make plants and fungi understood, valued, and conserved. Kew executes broad DNA barcoding research to document plant species, increase biodiversity information, and help in global conservation activities. Their study contributes to feeding into restoration projects and aiding in formulating policies for plant conservation.

World Wildlife Fund (WWF): WWF's conservation efforts are dedicated to nature and to addressing the greatest threats to the diversity of life on Earth. WWF applies DNA barcoding as part of conservation efforts to monitor wildlife populations and assess diversity as it fights the illegal wildlife trade. Its work improves the management of species and informs conservation policy.

University of Guelph: The University of Guelph serves as a global leader in agricultural and environmental research and aims to foster a better quality of life through leadership in the biosciences. This university hosts the Centre for Biodiversity Genomics, specializing in DNA barcoding and its practical utility in conservation, ecology, and biodiversity studies. Their works include protocol development and promotion of the science of genetic identification.

Australian National Wildlife Collection (ANWC): ANWC deals with the collection, preservation, and research of Australia's unique fauna and flora, promoting conservation and research. ANWC contributes to the identification of species and understanding of biodiversity through DNA barcoding in Australia. Through these research contributions, endangered species and habitats can be conserved properly.

National Park Service (NPS): Preserve the ecological and historical integrity of national parks but allow public access. NPS uses DNA barcoding to monitor biodiversity in national parks, the health of an ecosystem, and to make decisions on management (Roy et al., 2010). Research can be utilized in the monitoring of species abundance trends as well as informing conservation decisions.

NatureServe: NatureServe intends to provide scientific information regarding the status of species and ecosystems' conservation. NatureServe partners with other organizations to integrate DNA barcoding information into biodiversity assessments. All of this work contributes to conservation planning by including information on species distribution and their needs for conservation action.

17.34 Future directions and innovations

Among the new DNA sequencing technologies are single-molecule sequencing and other cutting-edge techniques that promise, in large part, to greatly assist in improving DNA barcoding. These technologies can improve the accuracy, efficiency, and scale of species identification and biodiversity assessment. Here is a closer look at these technologies and some of their potential impacts:

Single-molecule sequencing: Single-molecule sequencing reads out individual DNA molecules directly, with no amplification by PCR, compared to traditional sequencing methods. Technologically, this includes PacBio's Single Molecule, Real-Time sequencing and Oxford Nanopore sequencing. Much greater accuracy and resolution: Single-molecule sequencing produces reads that are much longer than several other systems; they can resolve complex regions of

DNA that are generally difficult with reads of shorter length (Schweikle et al., 2023). This may ultimately lead to much more accurate species identification, especially where species are closely related or cryptic. The single-molecule sequencing does not require the process of PCR amplification, therefore diminishing the chances of bias that is introduced during amplification, thus it means a realistic representation of the contents of genetic material within a sample (Suaste-Dzul et al., 2019).

Metabarcoding: Metabarcoding integrates DNA barcoding and high-throughput sequencing applications. This helps scientists analyze several species in one go from samples obtained from the environment (such as soil and water). Metabarcoding provides a fast approach to generating highly integrated species profiles throughout an environment, thereby illuminating the entire community structure and state of the environment. It may be best to observe over time with this method the changes in biodiversity by detecting changes in species composition that may have resulted from environmental stressors, invasive species, or other effects of climate change.

Portable sequencing technologies: The Oxford Nanopore MinION has brought portability to DNA sequencing technologies, thus, allowing DNA sequencing in the field. DNA barcoding from the field researchers can make for immediate identification of species and quick decisions on conservation actions (Satam et al., 2023). Portable sequencers will provide researchers with access to DNA barcoding even in remote locations; hence, studies that would have otherwise been impossible due to distances into biodiversity-rich and under-sampled areas.

Better bioinformatics tools: As sequencing technologies advance, so do the bioinformatics tools targeted toward the analysis and interpretation of large datasets produced by the advanced sequencing methods. Advanced algorithms and machine learning techniques can help process complex data to yield more accurate identifications and further insights into species relationships. With better bioinformatics, it is possible to integrate genetic data with more ecological, geographical, and environmental data for understanding biodiversity (Weigand et al., 2019).

Synthetic biology applications: Synthetic biology and advances in gene synthesis may make it possible to design personal bar codes or markers for certain specific conservation needs. Synthetic DNA barcodes should be developed for specific taxa or areas to upgrade the level of precision in addition to enhancing detection sensitivity in species identification applications (Shah et al., 2007). Specific species should be tracked effectively for conservation programs, especially for endangered or threatened species, with custom barcoding tools. More specifically, they will increase accuracy and efficiency with data collection in real-time, helping biodiversity assessments improve and leading to more effectively determined conservation strategies. The use of these new technologies allows for a closer look at the biodiversity of the world and better protection of the same.

17.35 Conclusion

Generally, DNA is extracted and followed by the amplification of the barcode region, and then DNA sequencing is performed to lead to species identification using databases as comparisons. While the use of COI for animals was recognized as a universal marker, the employment of the other markers could be recommended for the relevant groups in the sense of using ITS for plants and 16S rRNA for bacteria, while Cytb could be used for fishes. DNA barcoding technologies such as NGS, metabarcoding, and miniaturized devices have recently been achieved, thus unlocking the field for the high-throughput analysis of complex environmental samples and on-site species identification (Stevenson et al., 2023). Nonetheless, DNA barcoding is not problem-free. Issues like incomplete reference databases, DNA degradation, the presence of pseudogenes and NUMTs, and complicated evolutionary scenarios such as hybridization and introgression cause troubles with accurate species identification. Besides these, there should also be standard protocols that emphasize ethics and quality control of such collaborations. In reality, the inclusion of multilocus barcoding techniques, expansion of reference databases, and the designing of improved bioinformatics tools and algorithms is making DNA barcoding increasingly reliable and useful. It is an area where there are still studies and technological improvements, promising to deal with the existing challenges and further refine this powerful tool for biodiversity assessment and species identification in various ecosystems.

References

Ahmed, S., Ibrahim, M., Nantasenamat, C., Nisar, M. F., Malik, A. A., Waheed, R., Ahmed, M. Z., Ojha, S. C., & Alam, M. K. (2022). Pragmatic applications and universality of DNA barcoding for substantial organisms at species level: a review to explore a way forward. *BioMed Research International, 2022*(1), 1846485.

Al-Shuhaib, M. B. S., & Hashim, H. O. (2023). Mastering DNA chromatogram analysis in Sanger sequencing for reliable clinical analysis. *Journal of Genetic Engineering and Biotechnology, 21*(1), 115.

Ankola, K., Mahadevegowda, L., Melichar, T., & Boregowda, M. H. (2021). DNA barcoding: Nucleotide signature for identification and authentication of livestock. *Advances in animal genomics* (pp. 299–308). Academic Press.

Antil, S., Abraham, J. S., Sripoorna, S., Maurya, S., Dagar, J., Makhija, S., Bhagat, P., Gupta, R., Sood, U., Lal, R., & Toteja, R. (2023). DNA barcoding, an effective tool for species identification: a review. *Molecular biology reports, 50*(1), 761–775.

Aylagas, E., Borja, A., Irigoien, X., & Rodriguez-Ezpeleta, N. (2016). Benchmarking DNA metabarcoding for biodiversity-based monitoring and assessment. *Frontiers in Marine Science, 3*, 96.

Baena-Bejarano, N., Reina, C., Martínez-Revelo, D. E., Medina, C. A., Tovar, E., Uribe-Soto, S., Neita-Moreno, J. C., & Gonzalez, M. A. (2023). Taxonomic identification accuracy from BOLD and GenBank databases using over a thousand insect DNA barcodes from Colombia. *Plos one, 18*(4), e0277379.

Balaban, M., Moshiri, N., Mai, U., Jia, X., & Mirarab, S. (2019). TreeCluster: Clustering biological sequences using phylogenetic trees. *PloS one, 14*(8), e0221068.

Behera, A., Panigrahi, G. K., & Sahoo, A. (2024). Nonsense-mediated mRNA decay in human health and diseases: Current understanding, regulatory mechanisms and future perspectives. *Molecular Biotechnology,* 1–15.

Bergmann, T., Rach, J., Damm, S., DeSalle, R., Schierwater, B., & Hadrys, H. (2013). The potential of distance-based thresholds and character-based DNA barcoding for defining problematic taxonomic entities by CO 1 and ND 1. *Molecular ecology resources, 13*(6), 1069–1081.

Bezeng, B. S., Davies, T. J., Daru, B. H., Kabongo, R. M., Maurin, O., Yessoufou, K., van der Bank, H., & Van der Bank, M. (2017). Ten years of barcoding at the African Centre for DNA Barcoding. *Genome, 60*(7), 629–638.

Bingpeng, X., Heshan, L., Zhilan, Z., Chunguang, W., Yanguo, W., & Jianjun, W. (2018). DNA barcoding for identification of fish species in the Taiwan Strait. *PloS one, 13*(6), e0198109.

Blazej, P., Wnetrzak, M., Mackiewicz, D., & Mackiewicz, P. (2018). Optimization of the standard genetic code according to three codon positions using an evolutionary algorithm. *PloS one, 13*(8), e0201715.

Bohmann, K., Mirarab, S., Bafna, V., & Gilbert, M. T. P. (2020). Beyond DNA barcoding: the unrealized potential of genome skim data in sample identification. *Mol. Ecol. 29*(14), 2521–2534. https://doi.org/10.1111/mec.15507.

Botha, D., du Plessis, M., Siebert, F., & Barnard, S. (2023). Introducing an rbc L and a trn L reference library to aid in the metabarcoding analysis of foraged plants from two semi-arid eastern South African savanna bioregions. *Plos one, 18*(5), e0286144.

Cermakova, E., Lencova, S., Mukherjee, S., Horka, P., Vobruba, S., Demnerova, K. & Zdenkova, K., 2023. Identification of Fish Species and Targeted Genetic Modifications Based on DNA Analysis: State of the Art. Foods (Basel, Switzerland), 12 (1): 228.

Chan, A. H. E., Chaisiri, K., Morand, S., Saralamba, N. & Thaenkham, U., 2020. Evaluation and utility of mitochondrial ribosomal genes for molecular systematics of parasitic nematodes. Parasites & vectors, 13 (1): 364.

Chan, A. H. E., Saralamba, N., Saralamba, S., Ruangsittichai, J. & Thaenkham, U., 2022. The potential use of mitochondrial ribosomal genes (12S and 16S) in DNA barcoding and phylogenetic analysis of trematodes. BMC genomics, 23 (1): 104.

Cahyaningsih, R., Compton, L. J., Rahayu, S., Magos Brehm, & Maxted, N. (2022). DNA barcoding medicinal plant species from Indonesia. *Plants, 11*(10), 1375.

Carlin, E., Teren, G., & Ganswindt, A. (2020). Non-invasive assessment of body condition and stress-related fecal glucocorticoid metabolite concentrations in African elephants (*Loxodonta africana*) roaming in fynbos vegetation. *Animals, 10*(5), 814.

Cawthorn, D. M., Steinman, H. A., & Witthuhn, R. C. (2012). Evaluation of the 16S and 12S rRNA genes as universal markers for the identification of commercial fish species in South Africa. *Gene, 491*(1), 40–48.

Cochrane, S. K., Andersen, J. H., Berg, T., Blanchet, H., Borja, A., Carstensen, J., Elliott, M., Hummel, H., Niquil, N., & Renaud, P. E. (2016). What is marine biodiversity? Towards common concepts and their implications for assessing biodiversity status. *Frontiers in Marine Science, 3*, 248.

Das, R., & Panigrahi, G. K. (2024). Messenger RNA surveillance: Current understanding, regulatory mechanisms, and future implications. *Molecular Biotechnology*. https://doi.org/10.1007/s12033-024-01062-4, https://www.springer.com/journal/12033.

De Groot, G. A., During, H. J., Maas, J. W., Schneider, H., Vogel, J. C., & Erkens, R. H. (2011). Use of rbcL and trnL-F as a two-locus DNA barcode for identification of NW-European ferns: an ecological perspective. *PLoS one, 6*(1), e16371.

DeSalle, R. O. B. (2006). Species discovery versus species identification in DNA barcoding efforts: response to Rubinoff. *Conservation Biology, 20*(5), 1545–1547.

Dohna, T. A., & Kochzius, M. (2016). Obstacles to molecular species identification in sea anemones (*Hexacorallia: Actiniaria*) with COI, a COI intron, and ITS II. *Marine Biodiversity, 46*(1), 291–297.

Erckie, L., Adedoja, O., Geerts, S., van Wyk, & Boatwright, J. S. (2022). Impacts of an invasive alien Proteaceae on native plant species richness and vegetation structure. *South African Journal of Botany, 144*, 332–338.

Farooq, Q., Shakir, M., Ejaz, F., Zafar, T., Durrani, K., & Ullah, A. (2020). Role of DNA barcoding in plant biodiversity conservation. *Scholars International Journal of Biochemistry, 3*(3), 48–52.

Fernandes, T. J., Amaral, J. S., & Mafra, I. (2021). DNA barcode markers applied to seafood authentication: An updated review. *Critical reviews in food science and nutrition, 61*(22), 3904–3935.

Ferri, G., Corradini, B., Ferrari, F., Santunione, A. L., Palazzoli, F., & Alu, M. (2015). Forensic botany II, DNA barcode for land plants: Which markers after the international agreement? *Forensic Science International: Genetics, 15*, 131–136.

Ford, C. S., Ayres, K. L., Toomey, N., Haider, N., VAN ALPHEN, STAHL, Kelly, L. J., Wikstrom, N., Hollingsworth, P. M., Duff, R. J., Hoot, S. B., & Cowan, R. S. (2009). Selection of candidate coding DNA barcoding regions for use on land plants. *Botanical Journal of the Linnean Society, 159*(1), 1–11.

Forrest, L. L., Hart, M. L., Hughes, M., Wilson, H. P., Chung, K. F., Tseng, Y. H., & Kidner, C. A. (2019). The limits of Hyb-Seq for herbarium specimens: impact of preservation techniques. *Frontiers in ecology and evolution, 7*, 439.

Friel, J., Bombarely, A., Fornell, C. D., Luque, F., & Fernandez-Ocana, A. M. (2021). Comparative analysis of genotyping by sequencing and whole-genome sequencing methods in diversity studies of Olea europaea L. *Plants, 10*(11), 2514.

Franco-Duarte, R., Cernakova, L., Kadam, S., S. Kaushik, K., Salehi, B., Bevilacqua, A., Corbo, M.R., Antolak, H., Dybka-Stępień, K., Leszczewicz, M. and Relison Tintino, S., 2019. Advances in chemical and biological methods to identify microorganisms—from past to present. *Microorganisms, 7*(5), p.130.

Fu, C. N., Mo, Z. Q., Yang, J. B., Cai, J., Ye, L. J., Zou, J. Y., Qin, H. T., Zheng, W., Hollingsworth, P. M., Li, D. Z., & Gao, L. M. (2022). Testing genome skimming for species discrimination in the large and taxonomically difficult genus Rhododendron. *Molecular Ecology Resources, 22*(1), 404–414.

Fu, C. N., Wu, C. S., Ye, L. J., Mo, Z. Q., Liu, J., Chang, Y. W., Li, D. Z., Chaw, S. M., & Gao, L. M. (2019). Prevalence of isomeric plastomes and effectiveness of plastome super-barcodes in yews (Taxus) worldwide. *Scientific Reports, 9*(1), 2773.

Garcia, S., Kovarik, A., Maiwald, S., Mann, L., Schmidt, N., Pascual-Díaz, J. P., Vitales, D., Weber, B., & Heitkam, T. (2024). The dynamic interplay between ribosomal DNA and transposable elements: a perspective from genomics and cytogenetics. *Molecular Biology and Evolution, 41*(3), msae025.

Geiger, M., Koblmüller, S., Assandri, G., Chovanec, A., Ekrem, T., Fischer, I., Galimberti, A., Grabowski, M., Haring, E., Hausmann, A., & Hendrich, L. (2021). Coverage and quality of DNA barcode references for Central and Northern European Odonata. *PeerJ, 9*, e11192.

Guo, C., Luo, Y., Gao, L. M., Yi, T. S., Li, H. T., Yang, J. B., & Li, D. Z. (2023). Phylogenomics and the flowering plant tree of life. *Journal of Integrative Plant Biology, 65*(2), 299–323.

Hartvig, I., Czako, M., Kjaer, E. D., Nielsen, L. R., & Theilade, I. (2015). The use of DNA barcoding in identification and conservation of rosewood (Dalbergia spp.). *PLoS One, 10*(9), e0138231.

Head, S. R., Komori, H. K., LaMere, S. A., Whisenant, T., Van Nieuwerburgh, F., Salomon, D. R., & Ordoukhanian, P. (2014). Library construction for next-generation sequencing: overviews and challenges. *Biotechniques, 56*(2), 61–77.

Hebert, P. D., Cywinska, A., Ball, S. L., & DeWaard, J. R. (2003). Biological identifications through DNA barcodes. *Proceedings of the Royal Society of London. Series B: Biological Sciences, 270*(1512), 313–321.

Hebert, P. D., Floyd, R., Jafarpour, S., & Prosser, S. W. (2025). Barcode 100k specimens: in a single nanopore run. *Molecular Ecology Resources, 25*(1), e14028.

Hollingsworth, P. M., Li, D. Z., van der Bank, M., & Twyford, A. D. (2016). Telling plant species apart with DNA: from barcodes to genomes. *Philosophical Transactions of the Royal Society B: Biological Sciences, 371*(1702), 20150338.

Hoveka, L. N., van Der Bank, M., Bezeng, B. S., & Davies, T. J. (2020). Identifying biodiversity knowledge gaps for conserving South Africa's endemic flora. *Biodiversity and Conservation, 29*(9), 2803–2819.

Hoveka, L. N., van der Bank, M., & Davies, T. J. (2022). Winners and losers in a changing climate: how will protected areas conserve red list species under climate change? *Diversity and Distributions, 28*(4), 782–792.

Huang, Y., Wang, M., Liu, C., & He, G. (2024). Comprehensive landscape of non-CODIS STRs in global populations provides new insights into challenging DNA profiles. *Forensic Science International. Genetics, 70*, 103010. https://doi.org/10.1016/j.fsigen.2024.103010.

Ki, J. S., Zhang, R., Zhang, W., Huang, Y. L., & Qian, P. Y. (2009). Analysis of RNA polymerase beta subunit (rpoB) gene sequences for the discriminative power of marine Vibrio species. *Microbial ecology, 58*(4), 679–691.

Knot, I. E., Zouganelis, G. D., Weedall, G. D., Wich, S. A., & Rae, R. (2020). DNA barcoding of nematodes using the MinION. *Frontiers in Ecology and Evolution, 8*, 100.

Kraaij, T., Baard, J. A., Arndt, J., Vhengani, L., & Van Wilgen, B. W. (2018). An assessment of climate, weather, and fuel factors influencing a large, destructive wildfire in the Knysna region, South Africa. *Fire Ecology, 14*(2), 1–12.

Kress, W. J., & Erickson, D. L. (2008). DNA barcodes: genes, genomics, and bioinformatics. *Proceedings of the National Academy of Sciences, 105*(8), 2761–2762.

Krishnamurthy, P., & Francis, R. A. (2012). A critical review on the utility of DNA barcoding in biodiversity conservation. *Biodiversity and conservation, 21*(8), 1901–1919.

Lahaye, R., Van der Bank, M., Bogarin, D., Warner, J., Pupulin, F., Gigot, G., Maurin, O., Duthoit, S., Barraclough, T. G., & Savolainen, V. (2008). DNA barcoding the floras of biodiversity hotspots. *Proceedings of the National Academy of Sciences, 105*(8), 2923–2928.

Larridon, I., Villaverde, T., Zuntini, A. R., Pokorny, L., Brewer, G. E., Epitawalage, N., Fairlie, I., Hahn, M., Kim, J., Maguilla, E., & Maurin, O. (2020). Tackling rapid radiations with targeted sequencing. *Frontiers in plant science, 10*, 1655.

Lestari, D. A., Purbowati, E., Sutopo, S., & Kurnianto, E. (2018). Amino acid sequence based on Cytochrome b gene in Kejobong goat and its genetic relationships among several local goats in Asia. *Veterinary World, 11*(8), 1196.

Lobl, Ivan, Bernhard Klausnitzer, Matthias Hartmann, and Frank-Thorsten Krell. 2023. "The silent extinction of species and taxonomists—An appeal to science policymakers and legislators." *Diversity* 15, no. 10 (2023): 1053.

Lobo, J., Costa, P. M., Teixeira, M. A., Ferreira, M. S., Costa, M. H., & Costa, F. O. (2013). Enhanced primers for amplification of DNA barcodes from a broad range of marine metazoans. *BMC ecology, 13*(1), 34.

Loman, N. J., Misra, R. V., Dallman, T. J., Constantinidou, C., Gharbia, S. E., Wain, J., & Pallen, M. J. (2012). Performance comparison of benchtop high-throughput sequencing platforms. *Nature biotechnology, 30*(5), 434–439.

Lou, S. K., Wong, K. L., Li, M., But, P. P. H., Tsui, S. K. W., & Shaw, P. C. (2010). An integrated web medicinal materials DNA database: MMDBD (Medicinal Materials DNA Barcode Database). *BMC genomics, 11*(1), 402.

Majumder, S., & Panigrahi, G. K. (2024b). Advancements in contemporary pharmacological innovation: Mechanistic insights and emerging trends in drug discovery and development. *Intelligent Pharmacy*. https://doi.org/10.1016/j.ipha.2024.10.001.

Majumder, S., & Panigrahi, G. K. (2024a). Molecular docking and ADME evaluation of plant-based bioactive molecules targeting nonsense-mediated mRNA decay pathway factors to modulate tumorigenesis Plant. *Science Today, 11*(4), 1148–1155. https://doi.org/10.14719/pst.3737.

Malakasi, P., Bellot, S., Dee, R., & Grace, O. M. (2019). Museomics clarifies the classification of Aloidendron (Asphodelaceae), the iconic African tree aloes. *Frontiers in Plant Science, 10*, 1227.

Mallott, E. K., Garber, P. A., & Malhi, R. S. (2018). trnL outperforms rbcL as a DNA metabarcoding marker when compared with the observed plant component of the diet of wild white-faced capuchins (*Cebus capucinus*, Primates). *PloS one, 13*(6), e0199556.

Mamathaba, M. P., Yessoufou, K., & Moteetee, A. (2022). What does it take to further our knowledge of plant diversity in the megadiverse South Africa? *Diversity, 14*(9), 748.

Mani, S., Osborne, C. P., & Cleaver, F. (2021). Land degradation in South Africa: Justice and climate change in tension. *People and. Nature, 3*(5), 978–989.

Mankga, L. T., Yessoufou, K., Moteetee, A. M., Daru, B. H., & van der Bank, M. (2013). Efficacy of the core DNA barcodes in identifying processed and poorly conserved plant materials commonly used in South African traditional medicine. *ZooKeys, (365),*, 215.

Manzanilla, V., Teixidor-Toneu, I., Martin, G. J., Hollingsworth, P. M., de Boer, & Kool, A. (2022). Using target capture to address conservation challenges: Population-level tracking of a globally-traded herbal medicine. *Molecular Ecology Resources, 22*(1), 212–224.

Margulies, J. D., Moorman, F. R., Goettsch, B., Axmacher, J. C., & Hinsley, A. (2023). Prevalence and perspectives of illegal trade in cacti and succulent plants in the collector community. *Conservation Biology, 37*(3), e14030.

Maurin, O., Davies, T. J., Burrows, J. E., Daru, B. H., Yessoufou, K., Muasya, A. M., Van der Bank, & Bond, W. J. (2014). Savanna fire and the origins of the 'underground forests' of A frica. *New Phytologist, 204*(1), 201–214.

McDonnell, A. J., Baker, W. J., Dodsworth, S., Forest, F., Graham, S. W., Johnson, M. G., Pokorny, L., Tate, J., Wicke, S., & Wickett, N. J. (2021). Exploring Angiosperms353: Developing and applying a universal toolkit for flowering plant phylogenomics. *Applications in Plant Sciences, 9*(7), 10–1002.

Meier, R., Shiyang, K., Vaidya, G., & Ng, P. K. (2006). DNA barcoding and taxonomy in Diptera: a tale of high intraspecific variability and low identification success. *Systematic biology, 55*(5), 715–728.

Meiklejohn, K. A., Damaso, N., & Robertson, J. M. (2019). Assessment of BOLD and GenBank–Their accuracy and reliability for the identification of biological materials. *PloS one, 14*(6), e0217084.

Menegon, M., Cantaloni, C., Rodriguez-Prieto, A., Centomo, C., Abdelfattah, A., Rossato, M., Bernardi, M., Xumerle, L., Loader, S., & Delledonne, M. (2017). On site DNA barcoding by nanopore sequencing. *PLoS One, 12*(10), e0184741.

Meyer, C. P., & Paulay, G. (2005). DNA barcoding: error rates based on comprehensive sampling. *PLoS biology, 3*(12), e422.

Mishra, P., Kumar, A., Nagireddy, A., Shukla, A. K., & Sundaresan, V. (2017). Evaluation of single and multilocus DNA barcodes towards species delineation in complex tree genus Terminalia. *PloS one, 12*(8), e0182836.

Munjal, G., Hanmandlu, M., and Srivastava, S., 2019. Phylogenetics algorithms and applications. In *Ambient Communications and Computer Systems: RACCCS-2018* (pp. 187-194). Singapore: Springer Singapore.

Murphy, B., Forest, F., Barraclough, T., Rosindell, J., Bellot, S., Cowan, R., Golos, M., Jebb, M., & Cheek, M. (2020). A phylogenomic analysis of Nepenthes (Nepenthaceae). *Molecular Phylogenetics and Evolution, 144*, 106668.

Nakazato, T., & Jinbo, U. (2022). Cross-sectional use of barcode of life data system and GenBank as DNA barcoding databases for the advancement of museomics. *Frontiers in Ecology and Evolution, 10*, 966605.

Niemann, H. J., Bezeng, B. S., Orton, R. D., Kabongo, R. M., Pilusa, M., & Van der Bank, M. (2022). Using a DNA barcoding approach to facilitate biosecurity: Identifying invasive alien macrophytes traded within the South African aquarium and pond plant industry. *South African Journal of Botany, 144*, 364–376.

Niskanen, T., Lücking, R., Dahlberg, A., Gaya, E., Suz, L. M., Mikryukov, V., Liimatainen, K., Druzhinina, I., Westrip, J. R., Mueller, G. M., & Martins-Cunha, K. (2023). Pushing the frontiers of biodiversity research: Unveiling the global diversity, distribution, and conservation of fungi. *Annual review of Environment and resources, 48*(1), 149–176.

Odah, M. (2024). The double helix of justice: The crucial role of DNA in advancing criminal investigations.

Ogier, J. C., Pagès, S., Galan, M., Barret, M., & Gaudriault, S. (2019). rpoB, a promising marker for analyzing the diversity of bacterial communities by amplicon sequencing. *BMC microbiology, 19*(1), 171.

Panigrahi, G. K., & Satapathy, K. B. (2020c). Sacrificed surveillance process favours plant defense: A review. *Plant Archives, 20*(1), 2551–2559.

Panigrahi, G. K., & Satapathy, K. B. (2020a). *Arabidopsis* DCP5, a decapping complex protein interacts with ubiquitin-5 in the processing bodies. *Plant Archives, 20*(1), 2243–2247.

Panigrahi, G. K., & Satapathy, K. B. (2020b). Formation of *Arabidopsis* poly(A)-specific ribonuclease associated processing bodies in response to pathogenic infection. *Plant Archives, 20*(2), 4907–4912.

Panigrahi, G. K., & Satapathy, K. B. (2021). Pseudomonas syringae pv. Syringae Infection Orchestrates the fate of the Arabidopsis J domain containing cochaperone and decapping protein factor 5. *Physiological and Molecular Plant Pathology, 113*. https://doi.org/10.1016/j.pmpp.2020.101598, http://www.elsevier.com/inca/publications/store/6/2/2/9/3/2/index.htt.

Panigrahi, G. K., Sahoo, S. K., Sahoo, A., Behera, S., Sahu, S. R., Dash, A., & Satapathy, K. B. (2021). Bioactive molecules from plants: A prospective approach to combat SARS-Cov-2. *Advances in Traditional Medicine.* 1–14.

Panigrahi, G. K., Sahoo, A., & Satapathy, K. B. (2024). The processing body component varicose plays a multiplayer role towards stress management in *Arabidopsis*. *Plant Physiology Reports,* 1–10.

Patro, A. K., Sahoo, A., Satapathy, K. B., & Panigrahi, G. K. (2024). Nonsense-mediated mRNA decay: Physiological significance, mechanistic insights and future implications. *Pathology, Research and Practice, 264*, 155677.

Patro, I., Sahoo, A., Nayak, B. R., Das, R., Majumder, S., & Panigrahi, G. K. (2023). Nonsense-mediated mRNA decay: Mechanistic insights and physiological significance. *Molecular Biotechnology,* 1–15.

Peirson, J. A., Riina, R., Mayfield, M. H., Ferguson, C. J., Urbatsch, L. E., & Berry, P. E. (2014). Phylogenetics and taxonomy of the New World leafy spurges, Euphorbia section Tithymalus (Euphorbiaceae). *Botanical Journal of the Linnean Society, 175*(2), 191–228.

Pezzini, F. F., Ferrari, G., Forrest, L. L., Hart, M. L., Nishii, K., & Kidner, C. A. (2023). Target capture and genome skimming for plant diversity studies. *Applications in Plant Sciences, 11*(4), e11537.

Phillips, J. D., Gillis, D. J., & Hanner, R. H. (2022). Lack of statistical rigor in DNA barcoding likely invalidates the presence of a true species' barcode gap. *Frontiers in ecology and evolution, 10*, 859099.

Phillips, J. D., Griswold, C. K., Young, R. G., Hubert, N., & Hanner, R. H. (2024). A measure of the DNA barcode gap for applied and basic research. *DNA Barcoding: Methods and Protocols* (pp. 375–390). Springer.

Pirie, M. D., Oliver, E. G. H., Mugrabi de Kuppler, A., Gehrke, B., Le Maitre, N. C., Kandziora, M., & Bellstedt, D. U. (2016). The biodiversity hotspot as evolutionary hot-bed: spectacular radiation of Erica in the Cape Floristic Region. *BMC Evolutionary Biology, 16*(1), 190.

Poovitha, S., Stalin, N., Balaji, R., & Parani, M. (2016). Multi-locus DNA barcoding identifies matK as a suitable marker for species identification in Hibiscus L. *Genome, 59*(12), 1150–1156.

Porter, T. M., & Hajibabaei, M. (2018). Over 2.5 million COI sequences in GenBank and growing. *PloS one, 13*(9), e0200177.

Pauthenier, C., & Faulon, J. L. (2014). PrecisePrimer: an easy-to-use web server for designing PCR primers for DNA library cloning and DNA shuffling. *Nucleic acids research, 42*(W1), W205–W209.

Rach, J., DeSalle, R., Sarkar, I. N., Schierwater, B., & Hadrys, H. (2008). Character-based DNA barcoding allows discrimination of genera, species and populations in Odonata. *Proceedings of the Royal Society B: Biological Sciences, 275*(1632), 237–247.

Ren, C. Q., Zhang, D. Q., Liu, X. Y., & Zhang, J. Q. (2023). Genomic data provide a robust phylogeny backbone for Rhodiola L.(Crassulaceae) and reveal extensive reticulate evolution during its rapid radiation. *Molecular Phylogenetics and Evolution, 186*, 107863.

Rivero, E. R., Neves, A. C., Silva-Valenzuela, M. G., Sousa, S. O., & Nunes, F. D. (2006). Simple salting-out method for DNA extraction from formalin-fixed, paraffin-embedded tissues. *Pathology-Research and Practice, 202*(7), 523–529.

Rowley, D. L., Coddington, J. A., Gates, M. W., Norrbom, A. L., Ochoa, R. A., Vandenberg, N. J., & Greenstone, M. H. (2007). Vouchering DNA-barcoded specimens: test of a nondestructive extraction protocol for terrestrial arthropods. *Molecular Ecology Notes, 7*(6), 915–924.

Roy, L., Dowling, A. P., Chauve, C. M., & Buronfosse, T. (2010). Diversity of phylogenetic information according to the locus and the taxonomic level: an example from a parasitic mesostigmatid mite genus. *International Journal of Molecular Sciences, 11*(4), 1704–1734.

Ruhsam, M., Rai, H. S., Mathews, S., Ross, T. G., Graham, S. W., Raubeson, L. A., Mei, W., Thomas, P. I., Gardner, M. F., Ennos, R. A., & Hollingsworth, P. M. (2015). Does complete plastid genome sequencing improve species discrimination and phylogenetic resolution in Araucaria? *Molecular Ecology Resources, 15*(5), 1067–1078.

Sahoo, A., & Satapathy, K. B. (2021). Differential expression of Arabidopsis EJC core proteins under short-day and long-day growth conditions. *Plant Science Today, 8*(4), 815–819. https://doi.org/10.14719/PST.2021.8.4.1214, https://horizonepublishing.com/journals/index.php/PST/article/view/1214.

Sahoo, A., Satapathy, K. B., & Panigrahi, G. K. (2023). Ectopic expression of disease resistance protein promotes resistance against pathogen infection and drought stress in Arabidopsis. *Physiological and Molecular Plant Pathology, 124*(101949), 1–7.

Sahoo, A., Satapathy, K. B., & Panigrahi, G. K. (2024). Security check: Plant immunity under temperature surveillance. *Journal of Plant Biochemistry and Biotechnology,* 1–4.

Sahoo, A., Satapathy, K. B., Sahoo, S. K., & Panigrahi, G. K. (2023). Microbased biorefinery for gold nanoparticle production: Recent advancements, applications and future aspects. *Preparative Biochemistry & Biotechnology, 53*(6), 579–590.

Sahoo, A., Sethi, J., Satapathy, K. B., Sahoo, S. K., & Panigrahi, G. K. (2023). Nanotechnology for precision and sustainable agriculture: Recent advances, challenges and future implications. *Nanotechnology for Environmental Engineering, 8*(3), 775–787.

Satam, H., Joshi, K., Mangrolia, U., Waghoo, S., Zaidi, G., Rawool, S., Thakare, R. P., Banday, S., Mishra, A. K., Das, G., & Malonia, S. K. (2023). Next-generation sequencing technology: current trends and advancements. *Biology, 12*(7), 997.

Schweikle, S., Häser, A., Wetters, S., Raisin, M., Greiner, M., Rigbers, K., Fischer, U., Pietsch, K., Suntz, M., & Nick, P. (2023). DNA barcoding as new diagnostic tool to lethal plant poisoning in herbivorous mammals. *Plos one, 18*(11), e0292275.

Shah, M. M., Iihara, H., Noda, M., Song, S. X., Nhung, P. H., Ohkusu, K., Kawamura, Y., & Ezaki, T. (2007). dnaJ gene sequence-based assay for species identification and phylogenetic grouping in the genus Staphylococcus. *International Journal of Systematic and Evolutionary Microbiology, 57*(1), 25–30.

Singh, S. P., Panigrahi, G. K., & Sahoo, A. (2024). Optimizing nanoparticles use for growth promotion in zebrafish: Insights on concentration-dependent effects. *Nanotechnology for Environmental Engineering, 9*(4), 757–767.

Stevenson, D. W., Ramakrishnan, S., de Santis Alves, C., Coelho, L. A., Kramer, M., Goodwin, S., Ramos, O. M., Eshel, G., Sondervan, V. M., Frangos, S., & Zumajo-Cardona, C. (2023). The genome of the Wollemi pine, a critically endangered "living fossil" unchanged since the Cretaceous, reveals extensive ancient transposon activity. *bioRxiv*.

Suaste-Dzul, A. P., Rodríguez-Vélez, J. M., Rodríguez-Vélez, B., Arredondo-Bernal, H. C., & Gallou, A. (2019). Non-destructive DNA extraction methods for entomophagous insects with emphasis on biological control. *Genome, 62*(4), 287–293.

Weigand, H., Beermann, A. J., Čiampor, F., Costa, F. O., Csabai, Z., Duarte, S., Geiger, M. F., Grabowski, M., Rimet, F., Rulik, B., & Strand, M. (2019). DNA barcode reference libraries for the monitoring of aquatic biota in Europe: Gap-analysis and recommendations for future work. *Science of the Total Environment, 678*, 499–524.

Chapter 18

Prenatal paternity testing and forensic applications of noninvasive prenatal paternity testing technology

Andrei Semikhodskii

Genecode Diagnostics LLC, Kyiv, Ukraine

18.1 Introduction

Rape is one of the most serious crimes against a person, leaving victims with deep physical, emotional, and psychological trauma. The problem of sexual violence is becoming increasingly relevant, as reflected in the growing number of reported cases worldwide. According to official data, sexual offenses, including rape, remain one of the most common and underreported forms of violence.

In recent decades, there has been a worldwide trend toward an increase in the number of reported rapes and other sexual crimes. According to the UK's Office of National Statistics (ONS), the police recorded 193,566 sexual offenses in England and Wales in the year ending March 2022, the highest level on record. This figure is 31% higher than the year ending March 2021, and 19% higher than the year ending March 2020 (ONS, 2023). In Canada, the rate of police-reported sexual assault increased 38% between 2017 and 2022 (Conroy, 2017). Similar trends are reported in the European Union (Eurostat, 2024).

In many cases, rape results in an unwanted pregnancy of the victim. It is estimated that rape-related pregnancy is 5.0% per rape among victims of reproductive age (aged 12–45 years) with 50% of the pregnant victims opting for an abortion (Holmes et al., 1996).

Pregnancy resulting from rape remains an ongoing and critical public health concern. Besides the acute psychological trauma and physical effect on the victim who chose to terminate the pregnancy (Wallin Lundell et al., 2013), it is often not the child of the rapist but the child of the woman's husband or partner, which may be destroyed as a result of the abortion, leading to even more traumatic experience and emotional stress. Early identification of such an unwanted pregnancy and determining the paternity of the fetus will not only lead to early detection of the crime but also benefit the victim.

Forensic DNA analysis is one of the most powerful tools in the arsenal of modern law enforcement agencies for solving rape cases. Thanks to scientific advances in genetics, investigative bodies significantly changed the approach to investigating rape cases. They can now identify criminals with high precision, even if a significant amount of time has passed since the crime was committed.

The use of DNA analysis in rape investigations plays a critical role in identifying the perpetrator. This is especially true when the victim becomes pregnant following the assault, but for some reason, biological samples could not be acquired from the victim or the crime scene. In such situations, DNA analysis provides a unique opportunity for obtaining crucial evidence tracing the perpetrator to the victim long after the incident by analyzing a DNA sample of the fetus while the victim is pregnant or of the newborn baby *postpartum*. For more than 30 years, prenatal DNA analysis has been used for determining parentage both for forensic applications and for personal information of future parents, initially using invasive procedures (Waye, 1993) and recently by a noninvasive approach (Jiang et al., 2016). This chapter will examine these methods, discussing the latter methodology in more detail.

Investigative and Predictive DNA Testing. DOI: https://doi.org/10.1016/B978-0-443-30128-5.00001-1

18.2 Principles of paternity analysis

Parentage analysis is based on a simple presumption that in the absence of mutation, a child must receive one allele matching each parent at every locus tested (Tracey, 2001). This allele is called an "obligate allele." When for all the molecular markers tested, the alleged father (AF) matches the child (C) at all obligate alleles, the paternity cannot be excluded and the weight of evidence in favor of the presumed parentage is calculated, taking into account allelic frequencies of the markers analyzed. This weight of evidence is known as the combined parentage index (CPI). When a genetic inconsistency (a mismatch between the genotypes of AF and C) is observed, it may be an indication of paternity exclusion; however, the possibility of the mismatch being caused by mutation must always be considered (Gjertson et al., 2007). If the CPI value of > 10,000 is reached, the genetic evidence is thought to be in favor of parentage, and the mismatched alleles (if any) are presumed to be mutations (Li et al., 2017). Otherwise, analysis of additional markers is required. Mismatches of one or two alleles are commonly observed in parentage testing. Taking mutation events into account, the CPI when two mutations are suspected is low, bordering on single digits. When three inconsistencies are present, the CPI is usually close to 0, and this is deemed to be sufficient to issue a paternity exclusion result (Gjertson et al., 2007).

18.2.1 DNA markers used for paternity testing

18.2.1.1 Simple tandem repeat markers

Microsatellites or simple tandem repeat (STR) sequences are the genetic markers of choice for postnatal and invasive prenatal parentage testing. STRs are DNA sequences made up of a unit that can vary in length from 2 to 7 bp, repeated many times. Although thousands of STRs are found in the human genome, only a small subset of them is used for DNA identification and parentage analysis purposes (Butler, 2006). Forensic STR multiplexes commonly employ tetranucleotide STRs, which improve the efficiency of polymerase chaine reaction (PCR) amplification and minimizes spurious DNA peaks, although pentanucleotide STRs, like Penta E and Penta D, are also included in some STR marker panels. There are simple and complex STR loci. In STRs, the repeat units are constant throughout the repeat region. In contrast, complex STRs vary in their motif composition (tri, tetra, and pentanucleotide repeats) and may contain sequence polymorphisms within individual repeat units (Tilanus, 2006).

Modern STR analysis for paternity determination or human identification purposes consists of generating PCR amplicons of 100–500 bp using one of several commercially available STR multiplexes, like PowerPlex® Fusion (Promega, United States) or VeriFiler™ Plus (ThermoFisher, United States), followed by separation of the products of amplification by capillary electrophoresis on an Applied Biosystems™ 3500 Genetic Analyzer (ThermoFisher, United States) or a similar high-throughput capillary electrophoresis (CE) instrument.

Despite being the main markers for human identification and relatedness testing, STRs suffer from several problems complicating their use. The relatively large size of PCR amplicons (100–500 bp) presents an issue when dealing with degraded DNA. Although this can be mitigated by using reduced-size amplicon mini-STRs (amplicon size, 60–200 bp), it is still a challenge when fixed histological preparations or samples from decomposed bodies are presented for analysis. As for noninvasive prenatal paternity testing (NIPPT), STR analysis has very limited use due to the nature of circulated fetal DNA in the mother's bloodstream (discussed below).

Other disadvantages of STR markers include increased mutation rate (Tracey, 2001), which often causes a mismatch between the alleged parent and the child and limited information content per locus (Wenk, 2004) since the need for 15–20 markers or more to achieve convincing levels of CPI.

18.2.1.2 Single-nucleotide polymorphism markers

Single-nucleotide polymorphisms (SNPs) are the most common type of genetic variation among humans. An SNP represents a difference in a single nucleotide between alleles. Most SNPs used for human identification purposes are biallelic, but recently, triallelic SNPs have been shown to be forensically promising (Phillips et al., 2020). Compared with STRs, SNPs perform better when analyzing degraded samples (Kayser & Knijff de, 2011), have a significantly higher number of available loci (Børsting & Morling, 2015), and a significantly lower mutation rate (Schwark et al., 2012). One major disadvantage of SNPs in parentage testing is low discriminative power, which has to be offset by increasing the number of typed SNPs. Because SNPs can be analyzed in highly fragmented DNA samples, they are the marker of choice for NIPPT.

18.2.1.3 Insertion–deletion polymorphism markers

Insertion–deletion polymorphisms (InDels) are DNA length variation markers characterized by the presence or absence (insertion or deletion) of a certain DNA sequence (typically from 1 to 50 bp). They are diallelic, with great length variation between alleles, have a lower mutation rate compared to STRs, and are widely distributed throughout the genome, constituting about 20% of human genetic variation (Santos et al., 2010). Like SNPs, InDels have low DP (Cortellini et al., 2020) and are mostly used as supplementary markers in STR or SNP panels.

18.2.1.4 Deletion–insertion polymorphism-simple tandem repeat markers

Deletion–insertion polymorphism + STR (DIP-STR) markers are deletion–insertion polymorphisms linked to or located within STR (Castella et al., 2013). These markers are biallelic with differences between the "short" and "long" alleles of 3–15 bp. The compound structure of DIP-STRs makes them very promising for identity testing, mixture deconvolution in cases of degraded or low-quality DNA, or for cell-free DNA analysis (Oldoni et al., 2015).

18.2.1.5 Microhaplotype markers

Microhaplotypes (MHs) are a novel type of molecular marker with a broad range of uses in forensics, parentage and kinship testing, missing person identification, mixture deconvolution, and medical genetics applications. A MH locus is a short (less than 300 bp) segment of DNA, characterized by the presence of two or more closely linked SNPs that have at least three allelic combinations (i.e., "haplotypes") (Oldoni et al., 2019). Because of the short distance between the SNPs, each MH is thought to be inherited as a single block that is being passed over from generation to generation (Kidd et al., 2014). Being multi-SNP markers, MHs can provide, on a per locus basis, a higher DP than individual SNPs. A big advantage of MHs over STRs is that the absence of repetitive regions removes the problem of stutters caused by DNA polymerase slippage during amplification (Staadig & Tillmar, 2021). Due to the low mutation rate of SNPs, MHs can complement standard STRs in parentage and kinship tests, especially in deficient family pedigrees (Oldoni et al., 2019) or used on their own (Staadig & Tillmar, 2021).

18.2.2 Genotyping technology

18.2.2.1 Capillary electrophoresis

Capillary electrophoresis is currently a standard method for STR and InDel analysis and can also be used for SNP genotyping. STR and InDel alleles are identified by the length of a PCR-amplified fragment and are called relative to known-sized allelic ladders, with a number value indicating the number of complete repeats observed. However, a major disadvantage of this approach is that it disregards any sequence-based variation that may exist between alleles of the same size. Another significant drawback of the CE genotyping technology is its limited multiplex capability due to the small number of dye channels available for simultaneous analysis (Grandell et al., 2016). The introduction of next-generation sequencing (NGS) within the framework of parentage testing provides solutions for the above limitations.

18.2.2.2 Next-generation sequencing

NGS was developed in the 1990s and became commercially available in 2005. The term "next generation" implies the next step in the development of DNA sequencing technology from the first generation (Sanger sequencing). NGS methods can be grouped into two major categories: sequencing by hybridization, using DNA microarrays, and sequencing by synthesis (SBS)—pyrosequencing, sequencing by ligation, nanopore semiconductor-based technology (Ion Torrent™, ThermoFisher, United States), and bridge amplification technology developed by Illumina (United States) (Bruijns et al., 2018; Slatko et al., 2018). All these technologies share satisfactory accuracy, high throughput, and a lower cost per base compared to Sanger sequencing (Liu et al., 2012). SBS is mainly used in medical genetics for diagnostic applications such as identifying disease-related SNPs or for molecular karyotyping (Slatko et al., 2018). NGS based on SBS has both medical and forensic applications. This is the type of NGS used for paternity and kinship testing (Semikhodskii, 2024).

All the SBS technologies rely on high sequence coverage achieved by massive parallel sequencing (MPS), where millions of reads from a DNA fragment produced at the same time are detected on the same chip or flow cell in parallel (hence the name) (Saiz et al., 2020). Such a high number of sequencing runs is necessary to compensate for the elevated error rate of NGS relative to Sanger sequencing. After MPS, an accurate sequence of a DNA fragment is obtained as a consensus sequence of many thousands of reads of the same fragment (Slatko et al., 2018).

The MPS approach has several principal advantages over CE, with important implications for paternity analysis. One of the crucial features of MPS is a high level of multiplexing, both in the number of molecular markers and in the number of samples. Because MPS does not require size separation between amplicons, it overcomes the multiplexing issue of current CE methods, allowing typing a significantly larger number of DNA markers (not only STRs but also SNPs, InDels, MH, as well as mtDNA) simultaneously in a single assay (Daniel et al., 2015). In addition, many samples can be multiplexed in a single sequencing run by using barcoding, which further increases the throughput (Grandell et al., 2016). Also, because NGS requires much smaller amplicons than the CE-based STR analysis, it allows obtaining genetic information from badly degraded genomic DNA. This is especially important for NIPPT, where analyzed fetal DNA is of very small size.

In length-based CE genotyping, only the size of the allele is taken into consideration. In contrast, MPS could detect not only fragment length polymorphisms but also obtain detailed sequencing information about each allele analyzed (Gettings et al., 2015). This allows the discovery of previously unknown alleles and mutational events and the investigation of isoalleles and their parental origin (Rockenbauer et al., 2014), thus helping to resolve many issues affecting parentage and kinship analysis pertinent to CE.

A number of NGS platforms and kits for the simultaneous analysis of forensically important DNA markers have been developed during the past 15 years (Li et al., 2021; Saiz et al., 2020). Currently, Illumina MiSeq/Verogen MiSeq FGx (Illumina/Verogen, United States) and Ion GeneStudio S5 systems (ThermoFisher, United States) are the mostly used instruments for forensic identity and relatedness testing. Recently, MGI developed novel NGS sequencers DNBSEQ-G50 and MGISEQ-2000 (MGI Tech, Co., Ltd, China), which promise to be interesting alternatives to these two systems. The most popular commercial NGS multiplexes for forensic and relatedness purposes and the platforms for their analysis are given in Table 18.1.

18.3 Prenatal paternity testing

As has been mentioned, in some situations, it is important to be able to perform a paternity test while the child is still *in utero*. If it is suspected that the pregnancy may have resulted from a sexual assault or a proscribed relationship, confirmation of the identity of the biological father allows the pregnant woman to make an informed decision as to the termination or continuation of the pregnancy. Prenatal paternity tests can be invasive, and noninvasive depending on the type of child's biological sample used for analysis. Testing products of conception after the pregnancy was terminated can also be classified as a prenatal analysis.

18.3.1 Prenatal paternity testing using termination products

One of the approaches for prenatal paternity testing is the analysis of conception products after artificial or spontaneous pregnancy termination using STR markers (Karger et al., 2001). The combined tissue obtained as a result of pregnancy termination, in many instances, lacks clear separation into maternal and fetal parts, making physical dissection challenging or impossible. Genotyping DNA from pregnancy termination products often produces a mixed profile containing alleles from both the mother and the fetus, complicating the interpretation of the results (Jusic et al., 2023; Semikhodskii, 2007). In such cases, special software for mixture deconvolution, like LRmix Studio, is required for data analysis (Jusic et al., 2023).

18.3.2 Invasive prenatal paternity testing

Invasive prenatal paternity testing of an unborn child relies on amniocentesis, performed at gestational week (GW) 15–24 (Kondili & Miniati, 2011; Nata et al., 1993) or chorionic villus sampling (CVS), performed at GW 8–12 (Lobbiani et al., 1991) to obtain a fetal DNA sample. These procedures are stressful, associated with various pregnancy-related complications, pose a serious risk of intrauterine infection, damage to the fetus, and have a chance of miscarriage of about 1% (Akolekar et al., 2015). Invasive prenatal paternity testing is usually performed by STR genotyping coupled with CE.

18.3.3 Noninvasive prenatal paternity testing

In the early 2010s, a new, noninvasive methodology of paternity testing was developed. This approach, called NIPPT, is based on the analysis of circulating cell-free fetal DNA (ccffDNA) in maternal blood.

18.3.3.1 Circulating cell-free fetal DNA

In 1997, Lo et al. (1997) reported the presence of ccffDNA in the blood of pregnant women. This type of DNA primarily originates from trophoblasts undergoing apoptosis (Jackson, 2003). Additional sources of the ccffDNA include the

TABLE 18.1 Overview of commercially available next-generation sequencing (NGS) kits for forensic identification, parentage, and kinship testing.

Kit name	Manufacturer	Instrument	Autosomal simple tandem repeat (STR)	Y-STR	X-STR	SNP	mtDNA	MH
PowerSeq® 46GY[a]	Promega, United States	MiSeq®	22 + Amel	23	-	-	-	-
PowerSeq® CRM Nested System	Promega, United States	MiSeq®	-	-	-	-	+[b]	-
ForenSeq DNA Signature Kit	Verogen, United States	MiSeq ® FGx™	27 + Amel	24	7	172[c]	-	-
ForenSeq MainstAY Kit	Verogen, United States	MiSeq ® FGx™	27 + Amel	25	-	-	-	-
ForenSeq mtDNA Whole Genome Kit	Verogen, United States	MiSeq ® FGx™	-	-	-	-	+[d]	-
ForenSeq mtDNA Control Region Kit	Verogen, United States	MiSeq ® FGx™	-	-	-	-	+[b]	-
ForenSeq Kintelligence Kit	Verogen, United States	MiSeq ® FGx™	-	-	-	10,230	-	-
Precision ID GlobalFiler™ NGS STR Panel v2[e]	ThermoFisher, United States	HID Ion S5™ HID Ion GeneStudio™S5 System	31 + Amel	3	-	-	-	-
MGIEasy Signature Identification Library Prep Kit	MGI Tech Co., China	MGISEQ-2000	51 + Amel	48	27	227[f]	+[g]	-
MGIEasy mtDNA Whole Genome Amplification Kit[c]	MGI Tech Co., China	DNBSEQ-G400RS	-	-	-	-	+	-
MGIEasy Pa-SNPs Genotyping Kit[h]	MGI Tech Co., China	DNBSEQ-G50 DNBSEQ-G400	-	-	-	2010	-	-
QIAseq Investigator Missing Persons SNP panel[i]	Qiagen, United States	MiSeq®Ion S5 System	-	-	-	1289	-	45
QIAseq Investigator ID SNP panel[j]	Qiagen, United States	MiSeq® Ion S5 System	-	-	-	139	-	-
QIAseq Human Mitochondria panel[d]	Qiagen, United States	MiSeq® Ion S5 System	-	-	-	-	+	-

(Continued)

TABLE 18.1 (Continued)

Kit name	Manufacturer	Instrument	Autosomal simple tandem repeat (STR)	Y-STR	X-STR	SNP	mtDNA	MH
QIAseq Investigator Microhaplotype panel	Qiagen, United States	MiSeq® Ion S5 System	-	-	-	-	-	+
QIAseq Investigator Human Mitochondria Control Region panel[b]	Qiagen, United States	MiSeq® Ion S5 System	-	-	-	-	+	-

[a] The loci included are US Combined DNA Index System (CODIS) core and European Standard Set loci, Amelogenin, and the Y-STR loci included in the PowerPlex® Y23 System.
[b] mtDNA control region.
[c] 94 identity-informative SNPs, 56 ancestry-informative SNPs, and 22 phenotypic-informative SNPs.
[d] Full (16,569 bp) mitochondrial genome.
[e] Includes 21 CODIS STR markers, nine additional multiallelic STR markers, and 4 sex determination markers.
[f] 145 identity-informative SNPs, 53 ancestry-informative SNPs, and 29 phenotypic-informative SNPs.
[g] Seven amplicons covering the whole hypervariable region of mtDNA.
[h] The kit is specifically designed for use in NIPPT, paternity testing, and individual identification.
[i] The loci include 1200 triallelic kinship SNPs, 34 X-chromosome SNPs, and 55 ancestry-informative SNPs. Furthermore, the panel includes a set of 46 MHs.
[j] The panel is specifically designed to provide additional discrimination power to parentage and kinship tests.

transference of fetal DNA through the placenta and constant placental expansion throughout pregnancy (Sekizawa et al., 2003). The relative amount of the ccffDNA varies significantly, ranging from as low as 1% to as high as 40% (Muzzey, 2018), generally falling between 10% and 20% during the later stages of pregnancy (Lun et al., 2008). The ccffDNA first becomes detectable from GW 4 to 5 (Illanes et al., 2007), and typically disappears within two days after birth (Lo et al., 1999). By GW 9–10, the ccffDNA levels often exceed 3%–4% (Christiansen et al., 2019), which is sufficient for various noninvasive tests, such as aneuploidy screening (Renga, 2018), monogenic disorder detection (Chang et al., 2021; Chen et al., 2021), blood group genotyping (Alshehri & Jackson, 2021), and paternity testing (Chang et al., 2019).

The ccffDNA consists of short fragments, usually less than 160 base pairs (bp) in length, approximately corresponding to the length of the DNA wound around a nucleosomal core unit (Luger et al., 1997). The placenta-derived cell-free DNA fragments are shorter in length (143 bp) than maternally derived DNA fragments (166 bp) (Renga, 2018). Although circulating cell-free DNA in the maternal bloodstream is highly fragmented, the entire fetal genome could still be determined from maternal plasma (Kitzman et al., 2012).

18.3.3.2 Types of DNA markers used for noninvasive prenatal paternity testing

The small size of ccffDNA fragments bears implications for the choice of genetic markers used for NIPPT. While STRs are the preferred marker system for invasive prenatal and postnatal paternity testing, they are less effective for NIPPT. Due to the very short length of ccffDNA fragments, STR alleles longer than 160 bp are extremely rarely amplified. Initial efforts to use STRs in the AmpFLSTR Identifiler kit (ThermoFisher, United States) for NIPPT yielded limited results, detecting only a few informative, paternally inherited alleles (Wagner et al., 2009). Similarly, the ForenSeq DNA Signature Prep Kit (Illumina, United States) produced suboptimal outcomes (Shen et al., 2021). Additionally, STR amplification often introduces PCR artifacts that can be misinterpreted as paternally inherited fetal alleles (Christiansen et al., 2019; Shen et al., 2021). The heavy dilution of ccffDNA by maternal cell-free DNA adds further technical challenges. Because of this, genetic markers that use shorter amplicons and less prone to PCR artifacts, such as SNPs (Chang et al., 2019; Zhang et al., 2018) and MHs (Bai et al., 2020; Ou & Qu, 2020; Wang et al., 2020), provide more informative results for such a highly fragmented mixed DNA. Analysis of a great number of these markers by MPS allows reliable and robust noninvasive paternity testing of an unborn child.

The main challenge in developing an SNP-based NIPPT is the choice of fetal SNPs and their accurate genotyping in the presence of large amounts of maternal DNA. Although millions of SNPs have been identified in the human genome, SNPs selected for NIPPT must be high frequency (Chang et al., 2019; Jiang et al., 2016) and not be associated with medical conditions. Detecting fetal DNA relies on choosing SNPs where the mother is homozygous, and the fetus is heterozygous. These are referred to as "effective SNPs."

Accurate estimation of fetal DNA levels in maternal plasma is essential for correctly identifying and assessing paternally inherited alleles. Since only effective SNPs are used to calculate the weight of evidence, approximately half of the analyzed SNP markers may be uninformative in any given study (Jiang et al., 2016; Yang et al., 2017). This influences the required number of SNPs in the panel to achieve the desired confidence level in the results.

The GW of maternal blood collection is crucial for achieving reliable NIPPT results. Christiansen et al. (2019) examined the likelihood of detecting paternally inherited alleles in maternal blood using the Precision ID Identity Panel (ThermoFisher, United States). They found that at GW 7, only a moderate number of paternally inherited fetal SNP alleles could be identified; by GW 12, these alleles were detectable in most of the pregnant women; and by GW 20, paternally inherited fetal SNP alleles were observed in all participants.

Various studies support the feasibility of detecting fetal-specific alleles in maternal blood for paternity determination. Jiang et al. (2016) developed an NIPPT based on MPS of maternal plasma DNA using a custom panel containing 5000 to 8000 high-frequency SNPs. The results from sequencing-based NIPPT were fully consistent with invasive prenatal paternity tests using the AmpFLSTR Identifiler (ThermoFisher, United States) STR multiplex, correctly determining paternity in all 17 cases tested.

Similarly, Chang and colleagues (2019) developed a custom panel of 5457 SNPs to evaluate paternity in 349 cases involving pregnancies between GW 6 and GW 35, along with nine negative controls from nonpregnant women with prior fertility experience. Each mother-fetus pair was compared against all males in a blinded approach, accurately identifying biological fathers from unrelated males in all cases. In negative controls, no fetal SNPs from past pregnancies were detected. A subset of 50 SNPs from the panel yielded a combined probability of exclusion of 0.9999 in full trio paternity testing.

In another study aimed at refining NIPPT, Tam et al. (2020) employed MPS to interrogate a 356-SNP panel, with samples collected between GW 7 and GW 20. After analysis, 108–174 SNPs were classified as effective and included in paternity calculations across 15 family trios, resulting in paternity probabilities exceeding 99.9999%. All paternity results were confirmed by CE-based STR analysis using either amniotic fluid collected at GW 16–19 or buccal swabs collected *postpartum*. The high specificity of this approach was further validated by its ability to distinguish biological fathers from their siblings and by a clear separation between the CPIs calculated for biological fathers versus those for 60 unrelated men.

Promising approaches for NIPPT using DIP-STR (Damour et al., 2023) and mini-STR markers (Song, Xiao, Zhou, Yu, Wang, Shao, Duan, et al; Song, Xiao, Zhou, Yu, Wang, Shao, Liang, 2022) have also recently been reported. These types of markers can complement SNP panels, increasing the reliability of results, especially in samples with low levels of ccffDNA.

18.3.3.3 Bioinformatic approaches to the analysis of noninvasive prenatal paternity testing data

NIPPT using MPS demands advanced bioinformatics to handle large datasets from genotyping. Currently, there is no consensus in the literature regarding the optimal biostatistical approach to ensure accurate paternity results. One perspective approach advocates for using the probability of exclusion (PE) or paternity index (PI) to estimate NIPPT accuracy (Drábek & Cereda, 2014; Jiang et al., 2016). Another view, however, argues that traditional PI-based methods are insufficient for reliable paternity determination. Standard paternity testing methods rely on pure child DNA, but the ccffDNA in maternal plasma is mixed with a significantly larger proportion of maternal DNA, making traditional methods challenging for NIPPT analysis (Ryan et al., 2014). To address this, special bioinformatic algorithms that account for fetal DNA fraction have been proposed, allowing for accurate detection of fetal-specific alleles and enhancing the diagnostic power of NIPPT (Ryan et al., 2013).

18.3.3.4 Commercial availability of noninvasive prenatal paternity testing marker panels

Natera Inc. (United States) was the first company to successfully commercialize NIPPT service for public use, utilizing over 300,000 SNPs on Illumina's Human-Cyto12-SNP BeadChip microarray (Ryan et al., 2013). Among 36,400 tests with an unrelated male as the alleged father, 99.95% correctly excluded paternity, with only 0.05% yielding indeterminate results. Another US company, DNA Diagnostic Center, developed Certainty™ NIPPT, based on MPS analysis of 2688 SNPs (DNA Diagnostic Center, 2024), with maternal blood collection starting from GW 10. This is currently the only NIPPT approved by the Association for the Advancement of Blood and Biotherapies (former American Association of Blood Banks). Additionally, several other genetic testing companies, like BGI (China), provide NIPPT as a service.

Unlike the many commercially available STR multiplexes for CE-based genotyping, only one MPS-based panel, the MGIEasy Pa-SNPs Genotyping Kit (MGI Tech Co., China) (Li et al., 2021), was specifically designed and commercially available for NIPPT at the time of writing. Given the growing number of paternity tests worldwide, it is likely that more MPS panels tailored for NIPPT will become available soon.

18.4 Conclusions

DNA analysis plays a crucial role in investigating rape cases by aiding in the identification of the perpetrator. Prenatal paternity testing offers vital evidence by linking the perpetrator to the victim through analysis of a DNA sample from products of conception after pregnancy termination or from the fetus while the victim is pregnant. Noninvasive testing approaches enable this process without posing significant risks to the health of the mother or fetus.

Invasive methods, such as amniocentesis and CVS, allow for accurate genetic analysis by directly sampling fetal tissue, making these methods highly reliable. However, they carry a certain risk of complications, including miscarriage, infection, and other maternal and fetal risks. Despite their effectiveness, the invasive nature of these tests has driven researchers and clinicians to seek safer alternatives, leading to the rapid development of NIPPT.

NIPPT has transformed the field of prenatal testing by allowing examination to be performed using a sample of blood drawn from the mother. This method is based on analyzing fetal DNA circulating in the maternal bloodstream, thus eliminating the physical risks associated with invasive sampling. NIPPT offers remarkable potential for providing early, accurate, and safe paternity results, which is crucial for the well-being of the mother and child. Further development of MPS methodology, marker panels, and specific bioinformatics algorithms will enhance the reliability and precision of NIPPT, making it a preferred option for forensic prenatal paternity testing.

Despite recent advances, current NIPPT technology is not without issues. Accurate detection of paternal alleles depends on several factors, including the gestational age at which the sample is collected, the concentration of fetal DNA in the maternal bloodstream, and the technical limitations of existing marker panels. There are still challenges due to significant amounts of maternal DNA in the plasma, and increasing the accuracy of the test will require further refinement in technology and methodology. Additionally, as NIPPT is still a developing technology that is not widely commercialized, high testing costs limit its adoption in forensic casework.

As NIPPT continues to evolve, addressing these challenges will be critical for the future of prenatal paternity testing. Ongoing improvements in DNA sequencing accuracy, reduction in testing costs, and enhanced bioinformatics approaches promise to make noninvasive testing more widely accessible and reliable. It is also essential to develop proper regulatory standards to ensure the testing accuracy, reliability, and ethical application in both clinical and legal settings. Prenatal paternity testing will likely see further development in noninvasive approaches, leading to even greater reliability and expanded use. Continued investment in research, technology, and ethical frameworks will ensure that NIPPT becomes an accessible and powerful tool for forensic and legal applications.

References

Akolekar, R., Beta, J., Picciarelli, G., Ogilvie, C., & D'Antonio, F. (2015). Procedure-related risk of miscarriage following amniocentesis and chorionic villus sampling: A systematic review and meta-analysis. *Ultrasound in Obstetrics & Gynecology, 45*(1), 16–26. https://doi.org/10.1002/uog.14636.

Alshehri, A. A., & Jackson, D. E. (2021). NonInvasive prenatal fetal blood group genotype and its application in the management of hemolytic disease of fetus and newborn: Systematic review and meta-analysis. *Transfusion Medicine Reviews, 35*(2), 85–94. https://doi.org/10.1016/j.tmrv.2021.02.001.

Bai, Z., Zhao, H., Lin, S., Huang, L., He, Z., Wang, H., & Ou, X. (2020). Evaluation of a microhaplotype-based noninvasive prenatal test in twin gestations: Determination of paternity, zygosity, and fetal fraction. *Genes, 12*(1), 26. https://doi.org/10.3390/genes12010026.

Bruijns, B., Tiggelaar, R., & Gardeniers, H. (2018). Massively parallel sequencing techniques for forensics: A review. *Electrophoresis, 39*(21), 2642–2654. https://doi.org/10.1002/elps.201800082.

Butler, J. M. (2006). Genetics and genomics of core short tandem repeat loci used in human identity testing. *Journal of Forensic Sciences, 51*(2), 253–265. https://doi.org/10.1111/j.1556-4029.2006.00046.x.

Børsting, C., & Morling, N. (2015). Next generation sequencing and its applications in forensic genetics. *Forensic Science International: Genetics, 18*, 78–89. https://doi.org/10.1016/j.fsigen.2015.02.002.

Castella, V., Gervaix, J., & Hall, D. (2013). DIP – STR: Highly sensitive markers for the analysis of unbalanced genomic mixtures. *Human Mutation, 34*(4), 644–654. https://doi.org/10.1002/humu.22280.

Chang, L., Yu, H., Miao, X., Zhang, J., & Li, S. (2019). Development and comprehensive evaluation of a noninvasive prenatal paternity testing method through a scaled trial. *Forensic Sci Int Genet. 43*, 102158. https://doi.org/10.1016/j.fsigen.2019.102158.

Chang, L., Zhu, X., Li, R., Wu, H., Chen, W., Chen, J., Liu, H., Li, S., & Liu, P. (2021). A novel method for noninvasive diagnosis of monogenic diseases from circulating fetal cells. *Prenatal Diagnosis, 41*(4), 400–408. https://doi.org/10.1002/pd.5796.

Chen, C., Li, R., Sun, J., Zhu, Y., Jiang, L., Li, J., Fu, F., Wan, J., Guo, F., An, X., Wang, Y., Fan, L., Sun, Y., Guo, X., Zhao, S., Wang, W., Zeng, F., Yang, Y., Ni, P., ... Liao, C. (2021). Noninvasive prenatal testing of α-thalassemia and β-thalassemia through population-based parental haplotyping. *Genome Medicine, 13*(1). https://doi.org/10.1186/s13073-021-00836-8.

Christiansen, S. L., Jakobsen, B., Børsting, C., Udengaard, H., Buchard, A., Kampmann, M. L., Grøndahl, M. L., & Morling, N. (2019). Non-invasive prenatal paternity testing using a standard forensic genetic massively parallel sequencing assay for amplification of human identification SNPs. *International Journal of Legal Medicine, 133*(5), 1361–1368. https://doi.org/10.1007/s00414-019-02106-0, link.springer.de/link/service/journals/00414/index.htm.

Conroy, S. (2017). Recent trends in police-reported clearance status of sexual assault and other violent crime in Canada. *Statistics Canada. Juristat.*

Cortellini, V., Brescia, G., Carnevali, E., Cerri, N., Correa, H. S. D., Nespeca, P., Severini, S., Tommolini, F., Tozzo, P., Verzeletti, A., & Caenazzo, L. (2020). Genetic data and comparative study of 38 autosomal InDel markers in three Italian population groups. *Forensic Science International: Genetics, 44*, 102170. https://doi.org/10.1016/j.fsigen.2019.102170.

Damour, G., Baumer, K., Legardeur, H., & Hall, D. (2023). Early noninvasive prenatal paternity testing by targeted fetal DNA analysis. *Scientific Reports, 13*(1). https://doi.org/10.1038/s41598-023-39367-0.

Daniel, R., Santos, C., Phillips, C., Fondevila, M., van Oorschot, R. A. H., Carracedo, Á., Lareu, M. V., & McNevin, D. (2015). A SNaPshot of next generation sequencing for forensic SNP analysis. *Forensic Science International: Genetics, 14*, 50–60. https://doi.org/10.1016/j.fsigen.2014.08.013.

Drábek, J., & Cereda, G. (2014). Interpreting noninvasive prenatal paternity tests. *Genetics in Medicine, 16*(10), 793–794. https://doi.org/10.1038/gim.2014.100.

DNA Diagnostic Center (2024). https://dnacenter.com/blog/dna-diagnostics-center-ddc-worlds-largest-provider-private-dna-paternity-tests-announces-companys-exclusive-license-first-non-invasive-prenatal-paternity-test-using-snp-microa/ (Accessed 21 November 2024).

Eurostat (2024). Intentional homicide and sexual offences by legal status and sex of the person involved. doi: 10.2908/CRIM_HOM_SOFF.

Gettings, K. B., Aponte, R. A., Vallone, P. M., & Butler, J. M. (2015). STR allele sequence variation: Current knowledge and future issues. *Forensic Science International: Genetics, 18*, 118–130. https://doi.org/10.1016/j.fsigen.2015.06.005, http://www.elsevier.com.

Gjertson, D. W., Brenner, C. H., Baur, M. P., Carracedo, A., Guidet, F., Luque, J. A., Lessig, R., Mayr, W. R., Pascali, V. L., Prinz, M., Schneider, P. M., & Morling, N. (2007). ISFG: Recommendations on biostatistics in paternity testing. *Forensic Science International: Genetics, 1*(3-4), 223–231. https://doi.org/10.1016/j.fsigen.2007.06.006.

Grandell, I., Samara, R., & Tillmar, A. O. (2016). A SNP panel for identity and kinship testing using massive parallel sequencing. *International Journal of Legal Medicine, 130*(4), 905–914. https://doi.org/10.1007/s00414-016-1341-4, link.springer.de/link/service/journals/00414/index.htm.

Holmes, M. M., Resnick, H. S., Kilpatrick, D. G., & Best, C. L. (1996). Rape-related pregnancy: Estimates and descriptive characteristics from a national sample of women. *American Journal of Obstetrics and Gynecology, 175*(2), 320–324. https://doi.org/10.1016/s0002-9378(96)70141-2.

Jackson, L. (2003). Fetal cells and DNA in maternal blood. *Prenatal Diagnosis, 23*(10), 837–846. https://doi.org/10.1002/pd.705.

Jiang, H., Xie, Y., Li, X., Ge, H., Deng, Y., Mu, H., Feng, X., Yin, L., Du, Z., Chen, F., & He, N. (2016). Noninvasive prenatal paternity testing (NIPAT) through maternal plasma DNA sequencing: A pilot study. *PLoS One, 11*(9), e0159385. https://doi.org/10.1371/journal.pone.0159385.

Jusic, B., Pilav, A., Dzehverovic, M., & Cakar, J. (2023). Analysis of aborted fetal material using autosomal STR markers in forensic cases of sexual assault. *Journal of Forensic and Legal Medicine, 94*, 102468. https://doi.org/10.1016/j.jflm.2022.102468.

Karger, B., Rand, S. P., & Duchesne, A. (2001). DNA analysis of abortion material assisted by histology screening. *The American Journal of Forensic Medicine and Pathology, 22*(4), 397–399. https://doi.org/10.1097/00000433-200112000-00013.

Kayser, M., & Knijff de, P. (2011). Improving human forensics through advances in genetics, genomics and molecular biology. *Nature Reviews. Genetics, 12*, 179–192. https://doi.org/10.1038/nrg2952.

Kidd, K. K., Pakstis, A. J., Speed, W. C., Lagacé, R., Chang, J., Wootton, S., Haigh, E., & Kidd, J. R. (2014). Current sequencing technology makes microhaplotypes a powerful new type of genetic marker for forensics. *Forensic Science International: Genetics, 12*, 215–224. https://doi.org/10.1016/j.fsigen.2014.06.014.

Kitzman, J. O., Snyder, M. W., Ventura, M., Lewis, A. P., Qiu, R., Simmons, L. E., Gammill, H. S., Rubens, C. E., Santillan, D. A., Murray, J. C., Tabor, H. K., Bamshad, M. J., Eichler, E. E., & Shendure, J. (2012). Noninvasive whole-genome sequencing of a human fetus. *Science Translational Medicine, 4*(137). https://doi.org/10.1126/scitranslmed.3004323.

Kondili, A., & Miniati, P. (2011). Prenatal samples used as DNA evidence in rape cases. *Forensic Science International: Genetics Supplement Series, 3*, e548–e549. https://doi.org/10.1016/j.fsigss.2011.10.016.

Li, H., Zhao, X., Ma, K., Cao, Y., Zhou, H., Ping, Y., Shao, C., Xie, J., & Liu, W. (2017). Applying massively parallel sequencing to paternity testing on the Ion Torrent Personal Genome Machine. *Forensic Science International: Genetics, 31*, 155–159. https://doi.org/10.1016/j.fsigen.2017.09.007.

Li, R., Shen, X., Chen, H., Peng, D., Wu, R., & Sun, H. (2021). Developmental validation of the MGIEasy signature identification library prep kit, an all-in-one multiplex system for forensic applications. *International Journal of Legal Medicine, 135*(3), 739–753. https://doi.org/10.1007/s00414-021-02507-0.

Liu, L., Li, Y., Li, S., Hu, N., He, Y., Pong, R., Lin, D., Lu, L., & Law, M. (2012). Comparison of next-generation sequencing systems. *Journal of Biomedicine & Biotechnology, 2012*, 251364. https://doi.org/10.1155/2012/251364.

llanes, S., Denbow, M., Kailasam, C., Finning, K., & Soothill, P. W. (2007). Early detection of cell-free fetal DNA in maternal plasma. *Early Human Development, 83*(9), 563–566. https://doi.org/10.1016/j.earlhumdev.2006.11.001.

Lo, Y. M., Corbetta, N., Chamberlain, P. F., Rai, V., Sargent, I. L., Redman, C. W., & Wainscoat, J. S. (1997). Presence of fetal DNA in maternal plasma and serum. *Lancet, 350*(9076), 485–487. https://doi.org/10.1016/S0140-6736(97)02174-0.

Lo, Y. M., Zhang, J., Leung, T. N., Lau, T. K., Chang, A. M., & Hjelm, N. M. (1999). Rapid clearance of fetal DNA from maternal plasma. *American Journal of Human Genetics, 64*(1), 218–224. https://doi.org/10.1086/302205.

Lobbiani, A., Nocco, A., Vedrietti, P., Brambati, B., & Colucci, G. (1991). Prenatal paternity testing by DNA analysis. *Prenatal Diagnosis, 11*(5), 343–346. https://doi.org/10.1002/pd.1970110512.

Luger, K., Mäder, A. W., Richmond, R. K., Sargent, D. F., & Richmond, T. J. (1997). Crystal structure of the nucleosome core particle at 2.8 A resolution. *Nature, 389*(6648), 251–260. https://doi.org/10.1038/38444.

Lun, F. M., Chiu, R. W., Chan, K. C., Leung, T. Y., Lau, T. K., & Lo, Y. M. (2008). Microfluidics digital PCR reveals a higher than expected fraction of fetal DNA in maternal plasma. *Clinical Chemistry, 54*(10), 1664–1672. https://doi.org/10.1373/clinchem.2008.111385.

Muzzey, D. (2018). The technology and bioinformatics of cell-free DNA-based NIPT. In L. Page-Christiaens, & H.-G. Klein (Eds.). *Noninvasive prenatal testing (NIPT). Applied genomics in prenatal screening and diagnosis*. Academic Press.

Nata, M., Nagae, M., Aoki, Y., Sagisaka, K., & Uehara, S. (1993). Prenatal paternity testing with DNA analyses. *International Journal of Legal Medicine, 106*(3), 160–162. https://doi.org/10.1007/BF01225240.

Office for National Statistics (ONS). (2023). Sexual offences prevalence and trends, England and Wales: Year ending March 2022. https://www.ons.gov.uk/peoplepopulationandcommunity/crimeandjustice/articles/sexualoffencesprevalenceandtrendsenglandandwales/yearendingmarch2022 (Accessed 21 October 2024).

Oldoni, F., Castella, V., & Hall, D. (2015). A novel set of DIP-STR markers for improved analysis of challenging DNA mixtures. *Forensic Science International: Genetics, 19*, 156–164. https://doi.org/10.1016/j.fsigen.2015.07.012.

Oldoni, F., Kidd, K. K., & Podini, D. (2019). Microhaplotypes in forensic genetics. *Forensic Science International: Genetics, 38*, 54–69. https://doi.org/10.1016/j.fsigen.2018.09.009.

Ou, X., & Qu, N. (2020). Noninvasive prenatal paternity testing by target sequencing microhaps. *Forensic Science International: Genetics, 48*, 102338. https://doi.org/10.1016/j.fsigen.2020.102338.

Phillips, C., Amigo, J., Tillmar, A. O., Peck, M. A., de la Puente, M., Ruiz-Ramírez, J., Bittner, F., Idrizbegović, Š., Wang, Y., Parsons, T. J., & Lareu, M. V. (2020). A compilation of tri-allelic SNPs from 1000 genomes and use of the most polymorphic loci for a large-scale human identification panel. *Forensic Science International: Genetics, 46*, 102232. https://doi.org/10.1016/j.fsigen.2020.102232.

Renga, B. (2018). Non invasive prenatal diagnosis of fetal aneuploidy using cell free fetal DNA. *European Journal of Obstetrics, Gynecology, and Reproductive Biology, 225*, 5–8. https://doi.org/10.1016/j.ejogrb.2018.03.033.

Rockenbauer, E., Hansen, S., Mikkelsen, M., Børsting, C., & Morling, N. (2014). Characterization of mutations and sequence variants in the D21S11 locus by next generation sequencing. *Forensic Science International: Genetics, 8*, 68–72.

Ryan, A., Baner, J., Demko, Z., Hill, M., Sigurjonsson, S., Baird, M. L., & Rabinowitz, M. (2013). Informatics-based, highly accurate, noninvasive prenatal paternity testing. *Genetics in Medicine: Official Journal of the American College of Medical Genetics, 15*(6), 473–477. https://doi.org/10.1038/gim.2012.155.

Ryan, A., Demko, Z., Sigurjonsson, S., & Rabinowitz, M. (2014). Response to Drábek and Cereda. *Genetics in Medicine: Official Journal of the American College of Medical Genetics, 16*(10), 794. https://doi.org/10.1038/gim.2014.99.

Saiz, M., Martínez-Gonzalez, L. J., Lorente, J. A., & Alvarez-Cubero, M. J. (2020). Applications of next-generation sequencing in forensic field. In P. Shrivastava, H. R. Dash, J. A. Lorente, & J. Imam (Eds.). *Forensic DNA typing: Principles, applications and advancements*Singapore: Springer10.1007/978-981-15-6655-4_23.

Santos, N. P., Ribeiro-Rodrigues, E. M., Ribeiro-Dos-Santos, A. K., Pereira, R., Gusmão, L., Amorim, A., Guerreiro, J. F., Zago, M. A., Matte, C., Hutz, M. H., & Santos, S. E. (2010). Assessing individual interethnic admixture and population substructure using a 48-insertion-deletion (INSEL) ancestry-informative marker (AIM) panel. *Human Mutation, 31*(2), 184–190. https://doi.org/10.1002/humu.21159.

Schwark, T., Meyer, P., Harder, M., Modrow, J. H., & von Wurmb-Schwark, N. (2012). The SNPforID assay as a supplementary method in kinship and trace analysis. *Transfusion Medicine and Hemotherapy, 39*(3), 187–193. https://doi.org/10.1159/000338855.

Sekizawa, A., Yokokawa, K., Sugito, Y., Iwasaki, M., Yukimoto, Y., Ichizuka, K., Saito, H., & Okai, T. (2003). Evaluation of bidirectional transfer of plasma DNA through placenta. *Human Genetics, 113*(4), 307–310. https://doi.org/10.1007/s00439-003-0987-4.

Semikhodskii, A. (2007). *Dealing with DNA evidence: A legal guide*. London: Routledge Cavendish.

Semikhodskii, A. (2024). Application of NGS technology for parentage testing and relatedness analysis. In H. R. Dash, K. M. Elkins, & N. R. Al-Snan (Eds.). *Next generation sequencing (NGS) technology in DNA analysis* (pp. 219–241). Academic Press. https://doi.org/10.1016/B978-0-323-99144-5.00008-1.

Shen, X., Li, R., Li, H., Gao, Y., Chen, H., Qu, N., Peng, D., Wu, R., & Sun, H. (2021). Noninvasive prenatal paternity testing with a combination of well-established SNP and STR markers using massively parallel sequencing. *Genes (Basel), 12*(3), 454. https://doi.org/10.3390/genes12030454.

Slatko, B. E., Gardner, A. F., & Ausubel, F. M. (2018). Overview of next-generation sequencing technologies. *Current Protocols in Molecular Biology/, 122*(1), e59. https://doi.org/10.1002/cpmb.59.

Song, W., Xiao, N., Zhou, S., Yu, W., Wang, N., Shao, L., Duan, Y., Chen, M., Pan, L., Xia, Y., Zhang, L., & Liu, M. (2022). Non-invasive prenatal paternity testing by analysis of Y-chromosome mini-STR haplotype using next-generation sequencing. *PLoS One, 17*(4). https://doi.org/10.1371/journal.pone.0266332 e0266332.

Song, W., Xiao, N., Zhou, S., Yu, W., Wang, N., Shao, L., & Liang, X. (2022). Non-invasive prenatal paternity testing using mini-STR-based next-generation sequencing: a pilot study. *Journal of Laboratory Medicine, 46*(5), 337–344. https://doi.org/10.1515/labmed-2021-0191.

Staadig, A., & Tillmar, A. (2021). Evaluation of microhaplotypes in forensic kinship analysis from a Swedish population perspective. *International Journal of Legal Medicine, 135*(4), 1151–1160. https://doi.org/10.1007/s00414-021-02509-y.

Tam, J. C. W., Chan, Y. M., Tsang, S. Y., Yau, C. I., Yeung, S. Y., Au, K. K., & Chow, C. K. (2020). Noninvasive prenatal paternity testing by means of SNP-based targeted sequencing. *Prenatal Diagnosis, 40*(4), 497–506. https://doi.org/10.1002/pd.5595.

Tilanus, M. G. (2006). Short tandem repeat markers in diagnostics: What's in a repeat? *Leukemia, 20*(8), 1353–1355. https://doi.org/10.1038/sj.leu.2404273.

Tracey, M. (2001). Short tandem repeat-based identification of individuals and parents. *Croatian Medical Journal, J42*(3), 233–238.

Wagner, J., Dzijan, S., Marjanović, D., & Lauc, G. (2009). Non-invasive prenatal paternity testing from maternal blood. *International Journal of Legal Medicine, 123*(1), 75–79. https://doi.org/10.1007/s00414-008-0292-9.

Wallin Lundell, I., Sundström, P. I., Frans, O., Helström, L., Högberg, U., Moby, L., Nyberg, S., Sydsjö, G., Georgsson Öhman, S., Östlund, I., & Skoog Svanberg, A. (2013). The prevalence of posttraumatic stress among women requesting induced abortion. *The European Journal of Contraception & Reproductive Health Care, 18*(6), 480–488. https://doi.org/10.3109/13625187.2013.828030.

Wang, J. Y. T., Whittle, M. R., Puga, R. D., Yambartsev, A., Fujita, A., & Nakaya, H. I. (2020). Noninvasive prenatal paternity determination using microhaplotypes: A pilot study. *BMC Medical Genomics, 13*(1), 157. https://doi.org/10.1186/s12920-020-00806-w.

Waye, J. S. (1993). Prenatal paternity testing following sexual assault: A novel application of forensic DNA typing. *Canadian Society of Forensic Science Journal, 26*(2), 81–85. https://doi.org/10.1080/00085030.1993.10757461.

Wenk, R. E. (2004). Testing for parentage and kinship. *Current Opinion in Hematology, 11*(5), 357–361. https://doi.org/10.1097/01.moh.0000137914.80855.8a.

Yang, D., Liang, H., Gao, Y., Lin, S., He, Z., Gao, J., Sun, H., Li, Q., Ma, X., & Ou, X. (2017). Noninvasive fetal genotyping of paternally inherited alleles using targeted massively parallel sequencing in parentage testing cases. *Transfusion, 57*(6), 1505–1514. https://doi.org/10.1111/trf.14077.

Zhang, S., Han, S., Zhang, M., & Wang, Y. (2018). Non-invasive prenatal paternity testing using cell-free fetal DNA from maternal plasma: DNA isolation and genetic marker studies. *Legal Medicine, 32*, 98–103. https://doi.org/10.1016/j.legalmed.2018.03.009.

Chapter 19

Secondary transfer and its involvement in forensic investigation

Francesco Sessa
Department of Medical, Surgical and Advanced Technologies "G.F. Ingrassia," University of Catania, Catania, Italy

19.1 Introduction

Forensic science, in its pursuit of uncovering the truth, relies heavily on the analysis of trace evidence that can link a suspect, victim, or crime scene. Even the slightest, seemingly minor clues can be quite valuable in the field of criminal investigations (Chakravarty & Pandya, 2022; Sijen & Harbison, 2021). These minute evidence found at crime scenes can provide important information about the perpetrator's identity and the chronology of events, eventually assisting in the pursuit of justice for the victims (Butler, 2023). Indeed, in the last decades, the evolution of DNA profiling techniques has revolutionized forensic science, allowing investigators to generate comprehensive DNA profiles from minute biological samples (Haddrill, 2021). The development of advanced amplification and detection technologies, such as mini-short tandem repeats (STRs), next-generation sequencing, and low-copy number analysis, has further refined DNA profiling (Jordan & Mills, 2021). These techniques allow for the extraction and amplification of DNA from as few as 10 cells (Kanokwongnuwut et al., 2021). Modern DNA profiling techniques offer new hope in solving cold cases, even decades after the crime was committed. With the ability to generate complete DNA profiles from just a few cells, investigators can now revisit old evidence that may have been insufficient for earlier technologies. Moreover, these techniques may be useful as well as exonerating innocents (Caglià et al., 2011; Dash et al., 2022). This can be particularly useful for cases where physical evidence like clothing, weapons, or surfaces at a crime scene was retained.

However, the enhanced sensitivity of these techniques also presents challenges. One of the key concerns is the issue of DNA transfer, which occurs when DNA is moved from one person or object to another (Sessa et al., 2023). Primary transfer refers to the direct transfer of DNA when someone physically contacts a surface or object, such as a victim's skin or clothing. Secondary transfer, however, occurs indirectly, such as when someone touches an object, and their DNA is subsequently transferred to another individual or surface without direct contact (Buckingham et al., 2016; Thornbury et al., 2021; Van Oorschot et al., 2019). This raises the possibility of detecting DNA that belongs to an innocent individual who may not have been involved in the crime but left trace DNA that was transferred (Morgan, 2023).

This chapter explores the intricacies of secondary transfer, its mechanisms, and how forensic scientists assess and interpret this phenomenon in real-world investigations. We will also examine its limitations, potential for misinterpretation, and recent advancements that enhance the precision of forensic analysis.

19.2 Understanding transfer in forensics

The basic principle governing forensic transfer stems from Locard's Exchange Principle, which states that "every contact leaves a trace." When two objects or individuals come into contact, there is a transfer of material between them (Trejos et al., 2020). This transfer could involve fibers, hair, soil, skin cells, bodily fluids, or any number of small particles that could serve as evidence in a criminal case.

As previously described, transfer can be classified into two types:

- **Primary Transfer**: This occurs when material is transferred directly from one source (e.g., a person or object) to another during contact or with an indirect deposition. Direct deposition transfer happens when someone touches an

Investigative and Predictive DNA Testing. DOI: https://doi.org/10.1016/B978-0-443-30128-5.00021-7

object or surface, depositing DNA through skin cells or contact, such as leaving fingerprints. However, indirect deposition transfer involves the deposition of body fluids, such as saliva, sweat, or blood, without direct contact. For example, a person can leave their DNA on a surface by sneezing (saliva) or sweating (sweat droplets) onto an object. In both cases, the DNA comes directly from the individual, but the mechanism of transfer differs. It's crucial to understand these distinctions when interpreting forensic evidence, as both types of primary transfer can place DNA at a scene, even without direct physical touch. This nuanced understanding helps differentiate how biological material is deposited, providing clarity in forensic investigations (Van Oorschot et al., 2019).

- **Secondary Transfer**: Secondary transfer is always indirect. It occurs when DNA is transferred to an object or person via an intermediary, without direct contact from the person whose DNA is detected. For example, Person A shakes hands with Person B, leaving their DNA on Person B's hand. If Person B then touches a surface, Person A's DNA might be found on that surface, even though Person A never touched it directly. Secondary transfer can continue through multiple steps, leading to DNA being detected in places the person whose DNA it is has never visited or touched (Van Oorschot et al., 2019).

While primary transfer is often considered straightforward, secondary transfer adds layers of complexity, as it may create misleading links between individuals or locations not directly involved in the criminal activity (Romero-García et al., 2019).

In general, the transfer is influenced by key variables that impact the interpretation of DNA evidence in forensic investigations: transfer, persistence, prevalence, and recovery (van Szkuta et al., 2019; Van Oorschot et al., 2019).

DNA transfer refers to how DNA moves from one surface to another and may be influenced by the nature of contact, the amount of DNA shed, and the type of surface involved. For example, DNA is more likely to transfer effectively to smooth surfaces like glass than to porous materials like fabric (Jansson et al., 2022; Sessa et al., 2023).

Persistence refers to how long DNA remains on a surface after it is deposited. Environmental conditions, such as temperature, humidity, and exposure to sunlight, can degrade DNA over time. DNA tends to persist longer in dry, cool environments and may degrade more quickly when exposed to heat, moisture, or chemicals (Bhoyar et al., 2024; Poetsch et al., 2022).

Prevalence concerns how frequently DNA is found in a particular location or on an object. In high-traffic areas or shared items, multiple individuals' DNA may be prevalent, making it more challenging to determine the source or relevance of any one DNA sample (Jansson et al., 2022; Oldoni et al., 2016; Sessa et al., 2022).

Recovery is the ability to collect and analyze DNA from a surface. This depends on the tools and techniques used, such as swabbing or taping, and the surface type. Efficient recovery methods are crucial for collecting trace amounts of DNA, especially when dealing with minimal samples (Kuffel et al., 2024; Tozzo et al., 2022; Währer et al., 2023).

Together, these variables significantly influence the interpretation of DNA evidence and its probative value in forensic investigations.

19.3 The mechanisms of secondary transfer

As previously described, secondary transfer refers to the indirect transfer of DNA, where a person's genetic material is found on an object or surface they never directly touched. This process introduces complexities in forensic science, as it can place a person's DNA at a crime scene or on an object without them being physically present. Understanding secondary transfer is crucial for accurately interpreting DNA evidence, allowing to evaluate "the weight of DNA evidence" in the courtroom.

Based on recent publications, several factors affect the likelihood and extent of secondary DNA transfer:

- **Amount of DNA Shed**: Based on a recent publication, individuals may be subdivided into three categories. "High shedders" release a significant amount of DNA, making it easier for their genetic material to be transferred during contact with objects or people. "Low shedders" release minimal DNA, making it harder to detect their genetic material, even after contact. "Medium shedders" fall between, releasing moderate amounts of DNA under typical conditions (Goray et al., 2016).
- **Nature of the Surface**: Smooth, non-porous surfaces (e.g., glass, metal) tend to retain DNA more effectively than porous materials like fabric or wood. Additionally, the texture of the intermediate surface can influence how much DNA is picked up and transferred (Daly et al., 2012; Korzik et al., 2023).
- **Type of Contact**: The amount and type of contact are important factors. A brief, light touch might not transfer as much DNA as prolonged or repeated contact. For example, a handshake lasting several seconds may transfer more DNA than a quick brush of the hand (McColl et al., 2017; Szkuta et al., 2018).

- **Time and Environmental Conditions**: DNA degrades over time, especially when exposed to heat, moisture, or sunlight. The timing between the original DNA deposit and the secondary transfer can influence how much DNA is transferred and how likely it is to persist at the secondary location (Bhoyar et al., 2024; Poetsch et al., 2022).
- **Presence of Biological Fluids**: DNA transfer is facilitated by fluids like sweat, saliva, or blood. If DNA is transferred through such fluids, it can adhere more effectively to surfaces and remain there for longer periods (Van Oorschot et al., 2021).
- **Previous Activities of Individuals**: These activities, such as washing hands or engaging in tasks before touching objects, can significantly influence secondary DNA transfer. Washing hands reduces the amount of DNA available for transfer by removing skin cells, sweat, and other biological material, thus lowering the likelihood of DNA being passed on indirectly. Conversely, not washing hands or handling multiple objects can increase DNA buildup on a person's hands, making secondary transfer more likely. Activities like eating, touching one's face, or interacting with surfaces shortly before an event may increase DNA transfer due to the accumulation of biological material, impacting forensic interpretations (Goray & van Oorschot, 2015; Goray et al., 2024).

Summarizing the dynamic of the events (Sessa et al., 2023), a secondary transfer DNA may occur in the following situations (see Fig. 19.1):

- **Person-to-Person Transfer**: A common scenario involves two individuals shaking hands. Person A leaves DNA on Person B's hand through primary transfer. When Person B touches an object, such as a knife or a glass, they may transfer Person A's DNA to that item. If the object is later found at a crime scene, Person A's DNA may be detected, even though they were never there. This transfer could occur among multiple intermediaries, increasing the complexity of interpreting forensic evidence (A).
- **Object-to-Person Transfer**: Handling shared items, such as a pen or keyboard, that pass between several individuals. If Person A uses the pen and leaves DNA on it, and Person B later handles the same pen, Person A's DNA may be transferred to Person B's hands or another surface Person B touches. This could lead to Person A's DNA being found on objects they never came into direct contact with (B).

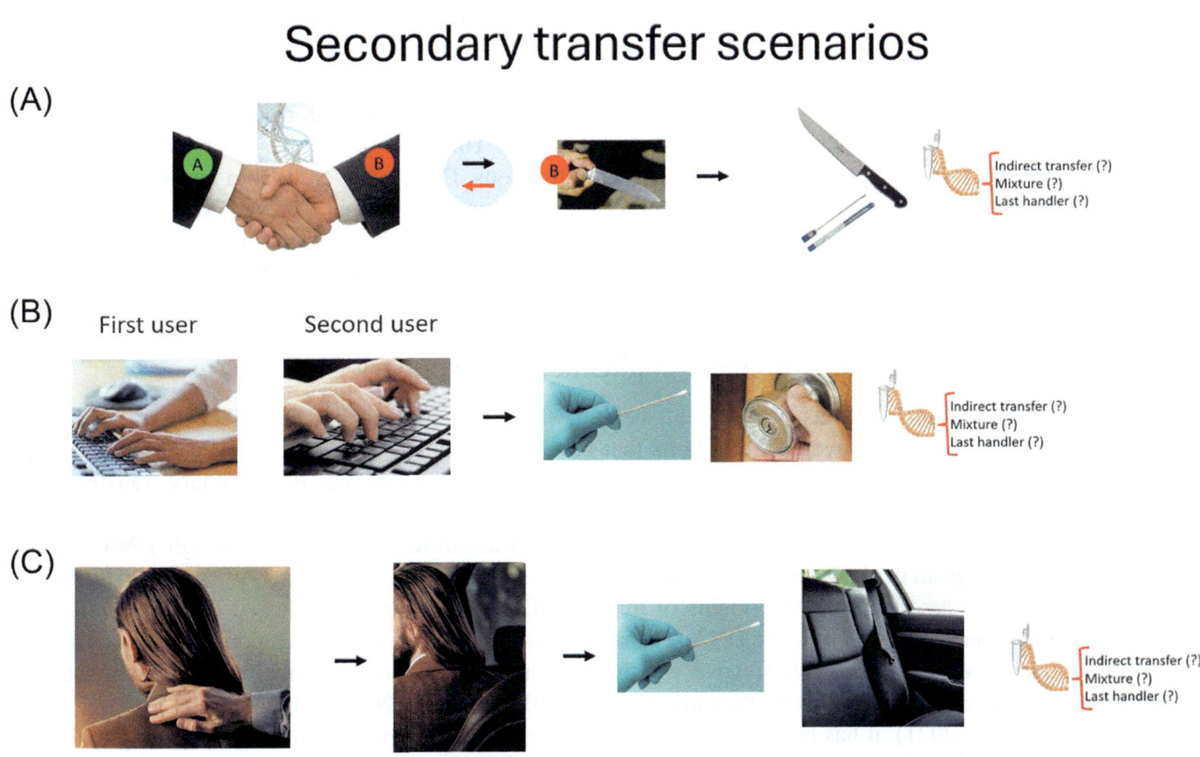

FIGURE 19.1 (A) *Person-to-person transfer*: After handshaking, person A transfers his DNA onto person B's hand through primary contact. When Person B touches a knife, he may transfer Person A's DNA onto that item. (B) *Object-to-person transfer*: A keyboard is handled by two subjects. If the first user leaves DNA on it, subsequently the second user touching the same item could transfer Person A's DNA on other surfaces, such as an handle. (C) *Surface-to-surface transfer*: Person A touches Person B's jacket; Person B then sits on a car seat, Person A's DNA may be indirectly transferred to this surface, even though Person A never came into direct contact with it.

- **Surface-to-Surface Transfer**: For instance, if Person A touches Person B's clothing, their DNA may be transferred to the fabric. If Person B then sits on a chair or bed, some of Person A's DNA could be transferred to those surfaces, even though Person A never touched them directly (C).

19.4 The role of secondary transfer in forensic investigations

Secondary transfer evidence has the potential to be both highly informative and highly misleading in forensic investigations. It is crucial for forensic scientists to distinguish between primary and secondary transfer, as misinterpreting the origin of trace evidence can lead to wrongful convictions or misdirection of the investigation.

19.4.1 Fibers, hair, and environmental factors

Fibers and hair are among the most common forms of trace evidence encountered in forensic investigations, often involved in both primary and secondary transfer events. Due to their lightweight nature and the ease with which they adhere to clothing, skin, or various surfaces, fibers, and hair can be easily transferred between individuals, objects, or environments. This capability can complicate the forensic interpretation of trace evidence, as it raises the risk of attributing an erroneous connection between individuals or crime scenes.

In 1987, Gaudette and Tessarolo described the possibility of secondary transfer of human scalp hair, where individuals carry hair from previous interactions and subsequently transfer it to other locations, objects, or individuals. Based on their findings, the transfer is influenced by factors like the type of surface, hair characteristics, and environmental conditions (Gaudette & Tessarolo, 1987). These results were confirmed by Dachs et al. (2003) who clarified other important aspects. Considering that the hair could be used as a DNA source, forensic investigators must remain cautious when evaluating hair evidence, particularly given the persistence of scalp hair on clothing, which resembles the behavior of transferred fibers.

Recent research highlights the importance of understanding the mechanisms underpinning fiber transfer to ensure robust forensic evaluations. A study investigating the impact of washing activities on fiber release into wastewater and transfer from donor garments to receiver swatches found that significant fiber release occurs during wash cycles. The results of this study revealed that garments subjected to repetitive washes showed a progressive decline in fiber transfer, suggesting that investigators should carefully consider this variable when interpreting fiber evidence in forensic casework (Galais et al., 2024).

In recent years, environmental DNA has also emerged as a novel source of forensic information, with research showing that humans constantly shed DNA into their surrounding environments. This DNA can either remain suspended in the air or settle on surfaces in the form of indoor dust. In forensic investigations, DNA recovered from air and dust can provide valuable insights, particularly when no visible traces of DNA are present (Fantinato et al., 2023). Further evidence suggests that the mere presence of an individual in a room is sufficient to facilitate direct or indirect DNA transfer to surfaces and surrounding environments: combining surface DNA with dust and air samples collected from the same environment improved the understanding of DNA transfer events in indoor settings (Fantinato et al., 2024).

19.4.2 Touch DNA and touch microbiome: implication in secondary transfer

Touch DNA refers to the minute amounts of DNA left behind when a person touches an object or surface. It is often composed of skin cells and can be recovered from items that have been handled or briefly brushed against. While touch DNA has become an invaluable tool in forensic investigations, it is highly susceptible to secondary transfer due to the small quantity of DNA and its ease of movement.

A recent study demonstrated that a mere two seconds of contact was sufficient to release enough DNA to obtain an interpretable profile (Buckingham et al., 2016; Comte et al., 2019; Jansson et al., 2022; Pfeifer & Wiegand, 2017; Sessa et al., 2019). Recently, it has been described that the origins of this trace DNA may include nucleated epithelial cells, anucleate corneocytes, fragmentary nuclei, or even cell-free DNA. The sensitivity of modern forensic techniques, as well as the methods used for collecting touch DNA, can improve the recovery of this evidence even in scenarios involving indirect transfer, raising critical questions about the reliability of touch DNA in courtrooms (Burrill et al., 2019; Schulte et al., 2023).

Recently (Procopio et al., 2021), it has been highlighted the utility of "touch microbiome" studies as a parallel forensic tool to touch DNA. Researchers found that unique donor-specific microbial taxa could persist on surfaces like glass slides for up to 30 days, offering potential for personal identification in cases where touch DNA fails. Recent findings (Procopio et al., 2024) suggest that skin microbiomes transferred to clothing could serve as additional forensic markers. This type of investigation could be considered complementary to touch DNA evidence, especially in cases where genetic material is too degraded or insufficient for profiling.

Nevertheless, although microbiome signatures could serve as an additional source of forensic intelligence, expanding the potential for identification beyond conventional genetic markers, the problem related to the indirect transfer should be carefully evaluated.

19.5 Challenges in investigating secondary transfer

Secondary transfer introduces unique challenges, and investigators must be wary of the following issues:

19.5.1 Risk of misinterpretation

The key risk with secondary transfer is misinterpretation: modern forensic techniques, while incredibly sensitive, can sometimes detect trace evidence that is not relevant to the crime in question. Low-copy DNA analysis can amplify DNA profiles from just a few cells (Morgan, 2023). If forensic scientists or investigators assume that trace evidence found on a suspect or at a crime scene is the result of primary transfer, they may incorrectly link a suspect to a crime, despite the evidence originating from an innocent secondary source. Research has shown that secondary transfer DNA can produce interpretable profiles even without direct contact. For instance, a study demonstrated that after just two minutes of hand-to-hand contact, secondary transfer DNA was detected in 85% of the samples tested, and in some cases, the secondary contributor was the major or only DNA profile recovered (Cale et al., 2016). Moreover, as previously described, research into DNA degradation has shown that environmental conditions, including temperature and humidity, can also impact the quality and amount of DNA recovered, further complicating forensic investigations (Bhoyar et al., 2024).

In this way, it should be important to evaluate the possibility of a secondary transfer case by case to establish the "weight of the evidence."

19.5.2 Ethical concerns in secondary transfer

The growing use of forensic DNA analysis, especially in cases involving secondary transfer, raises significant ethical concerns. The collection and analysis of DNA, even from indirect or secondary transfer, can have far-reaching implications for individual privacy. DNA contains deeply personal and sensitive information, not only about a person's identity but also their genetic traits and health. Misuse of this data could lead to unintended invasions of privacy or even discrimination based on genetic information. The expansion of DNA databases and familial DNA searching further amplifies these concerns, as individuals who have never been involved in criminal activity may find their genetic material linked to investigations through distant relatives (García, 2021; Guillen et al., 2000; Ogbogu & Ahmed, 2022; Oosthuizen & Howes, 2022).

Another ethical consideration involves the issue of informed consent when collecting DNA. While suspects and convicted individuals may be legally required to provide DNA samples, there are concerns over collecting DNA from individuals indirectly, such as from public places or secondary transfer scenarios (Machado & Silva, 2009). This could lead to unintended consequences, such as DNA from individuals not involved in criminal activities being stored and analyzed without their knowledge or consent.

Moreover, in cases where DNA is highly degraded, as may occur in certain environmental conditions, ethical questions arise about the reliability and use of such evidence in court. Degraded DNA can be difficult to interpret and may produce ambiguous results, which could be exploited to either wrongfully convict or exonerate a suspect (Bhoyar et al., 2024). Forensic scientists have an ethical obligation to communicate the limitations of such evidence clearly to avoid misleading interpretations.

Finally, the use of DNA databases also raises ethical concerns about bias and discrimination. Studies have indicated that certain populations, particularly minority groups, may be overrepresented in forensic DNA databases due to disparities in law enforcement practices. This can perpetuate systemic inequalities in the criminal justice system, as individuals from these groups may face a higher likelihood of wrongful accusation or conviction simply due to their overrepresentation in such databases (Oosthuizen & Howes, 2022).

19.6 Advances in understanding secondary transfer

Recent advancements in forensic science are helping to address the complexities of secondary transfer. For instance, researchers are developing more sophisticated models of how trace evidence behaves in different environments. Additionally, new laboratory techniques are being refined to better differentiate between primary and secondary transfer.

19.6.1 Statistical models

Statistical models are increasingly being employed to quantify the likelihood of secondary transfer in forensic investigations. These models help forensic scientists assess whether trace evidence, such as DNA or fibers, is more likely the result of primary or secondary transfer. Factors such as the type of material, environmental conditions, and duration of contact between surfaces are incorporated into these models to provide a probabilistic understanding of how the evidence may have been transferred.

In the context of secondary transfer, forensic scientists are increasingly moving from the question of "Whose DNA is this?" to "How did this DNA get here?" This shift in focus acknowledges that DNA profiling technology has become so sensitive that it can detect trace material transferred easily through everyday interactions. As a result, the evaluation of DNA evidence now often requires models that go beyond identifying the source and addressing the activities that may have led to its transfer (Biedermann et al., 2016).

This evolution in forensic science has created challenges for experts who must assist courts in understanding the complexity of DNA evidence in cases involving secondary transfer. While some forensic scientists remain cautious about venturing beyond source identification due to a lack of data on transfer mechanisms and persistence, others advocate for integrating activity-based propositions into their evaluations (McKenna, n.d.). This shift is essential to prevent the misrepresentation of DNA results in court, particularly in scenarios where the probative strength of evidence is influenced by how it may have been transferred.

Bayesian statistical models have gained traction in this area due to their capacity to deal with the inherent uncertainties in forensic evidence. The Bayesian approach allows forensic experts to assess the likelihood of different transfer scenarios by considering not only the type of trace evidence but also the environmental and situational context. For example, Bayesian analysis can help determine the probability that a piece of DNA evidence was transferred secondarily through casual, indirect contact, rather than through direct involvement in the crime. This kind of probabilistic reasoning has become essential in helping juries and judges understand the relative significance of trace evidence in criminal trials (Buckleton et al., 2014). For example, a recent study introduced a Bayesian model for estimating the number of dust samples required to identify all occupants of a location. This study showcases the potential of air and dust as novel sources of DNA evidence in forensic investigations. By applying a Bayesian approach to the analysis of dust samples, forensic scientists can now consider previously overlooked sources of trace evidence, further broadening the scope of secondary transfer investigations (Fantinato et al., 2023).

In cases where no clear propositions exist, Bayesian models enable forensic experts to act as investigators, developing explanations to account for observed evidence. This dual role—investigator and evaluator—highlights the flexibility of Bayesian models in dealing with uncertainties and ensuring that forensic testimony is grounded in a probabilistic framework that reflects the real-world complexities of evidence transfer.

19.6.2 Improvements in trace evidence analysis

Advancements in forensic technology, such as advanced microscopy and chemical analysis, are enabling more precise examination of trace evidence. These innovations are critical in determining whether a transfer occurred directly or indirectly and have significantly improved forensic investigations.

For instance, a recent study explored the long-term stability of DNA deposited on materials like fabric, steel, and rubber over a period of nine months. The study revealed that while fabric samples maintained informative STR profiles even after nine months in different environments, steel and rubber only retained useful DNA data for shorter periods, especially under external conditions. This research highlights how environmental factors can influence DNA persistence, aiding in the forensic analysis of touch DNA evidence (Kaesler et al., 2023).

In addition to DNA analysis, the role of RNA molecules, particularly mRNA and miRNA, has gained attention. These molecules provide valuable information about the type of body fluids involved and the time since deposition (TSD). A recent study developed a method using reverse transcription quantitative PCR to estimate the age of biological stains, specifically saliva and semen. The degradation behavior of RNA markers over time exhibited distinct patterns, allowing forensic scientists to predict the TSD of body fluids, which can be crucial in narrowing down the timeline of a crime (Alshehhi & Haddrill, 2019).

Furthermore, the application of Raman spectroscopy paired with chemometrics has improved the analysis of bloodstains, one of the most encountered forms of trace evidence. In a recent proof-of-concept study, bloodstains aged in extreme thermal environments were analyzed to predict their TSD. This technique demonstrated that bloodstains degraded more quickly under high temperatures, such as those inside a vehicle exposed to direct sunlight, compared to those in ambient conditions. Such tools allow investigators to make informed predictions about when a bloodstain was deposited, which can be instrumental in crime scene investigations (Barber et al., 2024).

Finally, forensic scientists are exploring innovative methods to estimate the TSD of trace biological samples, such as touch DNA, using imaging flow cytometry. This approach, combined with advanced machine learning (ML) models like gradient boosting machines and generalized linear mixed model least absolute shrinkage and selection operator (LASSO), has shown high accuracy in predicting the age of biological evidence. These models provide reliable predictions of TSD with a high degree of accuracy, which can be crucial for determining the relevance of biological evidence in forensic casework (Gentry et al., 2023).

These improvements in trace evidence analysis could be very useful in order to interpret the origin, age, and transfer mechanisms of biological materials encountered at crime scenes.

19.6.3 Artificial intelligence and the future of forensic investigation

The integration of artificial intelligence (AI) in forensic science is set to revolutionize the analysis of secondary transfer and trace evidence. AI, particularly through ML algorithms, enhances the capacity to process large, complex forensic datasets, significantly improving both the accuracy and efficiency of evaluations. AI allows forensic scientists to model and simulate transfer events with remarkable precision by accounting for variables such as surface type, temperature, and humidity. For example, AI can predict how trace evidence like fibers or DNA behaves under different environmental conditions, providing deeper insights into whether evidence was likely transferred directly or secondarily (Sessa et al., 2024).

AI's applications in forensic genetics are particularly transformative. It has already been effectively used in haplogroup analysis to expedite and enhance the classification of DNA samples, as well as in analyzing STR profiles, minimizing the risk of misinterpretation. By distinguishing between primary and secondary contributors in complex DNA mixture profiles, AI helps forensic experts handle challenging evidence while reducing errors (Sessa et al., 2024; Veldhuis et al., 2022).

Furthermore, AI plays a key role in evaluating DNA mixtures. In cases where DNA from multiple individuals is mixed, AI-driven algorithms use probabilistic models to deconvolute the data, helping scientists interpret even low-quality or complex DNA samples that would have been difficult to analyze with traditional methods. Explainable AI techniques further enhance this process by offering not just accurate predictions but also transparent reasoning, allowing forensic experts to understand why specific conclusions were reached, improving trust in the results (Veldhuis et al., 2022).

The future of AI in forensic investigations looks promising. As AI technologies continue to evolve, they are expected to become standard tools in solving crimes, offering more reliable and precise forensic analyses. This will ultimately reduce human error and bias, ensuring that justice is served with greater accuracy and efficiency.

19.7 Case studies: secondary transfer in action

In this section, three published cases were analyzed, illustrating how secondary transfer has influenced the course of investigations in real courtroom cases, occasionally leading to wrongful accusations due to misinterpreted evidence (Table 19.1).

19.7.1 Case study 1: the Stephen Lawrence case

Stephen Lawrence, a Black teenager, was murdered in a racially motivated attack in 1993 in South East London. The initial police investigation failed to secure convictions due to insufficient evidence. However, in 2011, forensic scientists discovered tiny fibers and bloodstains linking two of the suspects, Gary Dobson and David Norris, to Lawrence's clothing. The prosecution used this newly discovered evidence to suggest that the suspects had direct contact with Lawrence during the attack.

During the trial, the defense argued that the fibers and blood could have been transferred secondarily. The defense highlighted that evidence bags containing Lawrence's clothing had come into contact with items related to the suspects during the original investigation. This raised the possibility of cross-contamination, which could explain the presence of Lawrence's DNA and fibers on the suspects' clothing. Despite these concerns, Dobson and Norris were convicted based on the forensic evidence, although questions remain about the possibility of secondary transfer influencing the case (Buchanan, 2006).

TABLE 19.1 Report of main information for each selected case.

Case and year	Reference	Summary of the case	Possible secondary transfer event
Stephen Lawrence Case (UK, 1993)	Buchanan (2006)	Stephen Lawrence was murdered in a racially motivated attack in 1993. In 2011, fibers and DNA linked suspects to Lawrence's clothing, leading to convictions.	Defense argued that DNA and fibers may have been transferred via cross-contamination in evidence bags during police investigations.
Amanda Knox Trial (Italy, 2007)	Gill (2016)	Amanda Knox and Raffaele Sollecito were accused of the murder of Meredith Kercher. DNA evidence played a crucial role, though later overturned on appeal.	Knox's DNA could have been secondarily transferred onto the kitchen knife through casual use. Bra clasp DNA possibly contaminated after being left at the crime scene for a month.
Lukis Anderson Case (USA, 2012)	McCrane and Mulligan (2023)	Lukis Anderson was wrongfully charged with murder after his DNA was found on the victim's body. His alibi showed he was hospitalized at the time of the crime.	Paramedics who treated Anderson earlier may have transferred his DNA to the crime scene while responding to the murder.

This case underscores the importance of carefully controlling the chain of custody for evidence and the risk of unintentional secondary transfer through cross-contamination. The controversy surrounding the forensic findings illustrates how even microscopic fibers or DNA can raise doubts about the integrity of evidence in high-profile cases.

19.7.2 Case study 2: the Amanda Knox trial

The Amanda Knox trial exemplifies the risks of wrongful conviction when forensic evidence is mishandled or misinterpreted. The case emphasizes the need for rigorous forensic protocols, careful interpretation of low-template DNA evidence, and judicial skepticism of forensic testimony that may be flawed.

The 2007 murder of Meredith Kercher, a British exchange student, in Perugia, Italy, became one of the most scrutinized legal cases of the 21st century. It was not only the horrific nature of the crime but also the involvement of Amanda Knox, an American exchange student, and her Italian boyfriend, Raffaele Sollecito, that turned the case into a global media spectacle. Knox and Sollecito were eventually convicted, exonerated, re-convicted, and finally acquitted in 2015 after nearly 8 years of legal battles. Another individual, Rudy Guede, was separately convicted of the crime.

On November 1, 2007, Meredith Kercher was found brutally murdered in the apartment she shared with Amanda Knox in Perugia. The crime scene presented a chaotic picture—Kercher's throat had been slashed, and her room showed signs of a struggle. Almost immediately, Knox and Sollecito were implicated in the crime based on circumstantial evidence, odd behavior (such as Knox's perceived lack of grief), and supposed inconsistencies in their alibis.

The Italian police arrested the couple within days, alongside Rudy Guede, a local man with a criminal record. Guede's DNA was found on multiple items in the murder room, including Kercher's body and clothing, which pointed clearly to his involvement in the murder. Despite this, the focus of the investigation shifted heavily toward Knox and Sollecito, largely due to circumstantial behavioral evidence and DNA profiles that, while far less conclusive, were presented as crucial.

The prosecution's case against Knox and Sollecito rested primarily on forensic DNA evidence, which at first glance seemed damning but was later shown to be fundamentally flawed. The first key piece of evidence was a kitchen knife found in a drawer at Sollecito's apartment. The knife was alleged to be the murder weapon, as it had Knox's DNA on the handle and a trace amount of Kercher's DNA on the blade. However, there was no blood on the knife, and forensic experts pointed out that the small quantity of DNA found on the blade could have been the result of contamination or secondary transfer. Knox, who lived with Kercher and had access to common kitchen utensils, could have used the knife for everyday tasks unrelated to the murder.

A second critical piece of evidence was a DNA profile attributed to Raffaele Sollecito found on a bra clasp in Kercher's bedroom. The bra clasp had been forcibly removed from Kercher during the attack, but it was only collected by investigators 46 days after the murder. The clasp had been left at the crime scene during that period, leading to concerns about contamination. Forensic experts argued that the clasp, having been exposed to various environmental factors, could have easily picked up DNA from individuals who were not involved in the murder.

The contrast between the evidence against Rudy Guede and the evidence against Knox and Sollecito is stark. Guede's DNA was found throughout the murder room and on Kercher's body, whereas no DNA belonging to Knox or Sollecito was discovered in the murder room itself. The lack of DNA evidence linking Knox to the crime scene should have been a significant red flag in the prosecution's case. Yet, the focus remained on the knife and bra clasp, despite the strong possibility that the DNA evidence was either unrelated to the murder or the result of contamination.

The case saw an extraordinary series of trials, retrials, and legal appeals. In 2009, Knox and Sollecito were convicted of Kercher's murder based on the DNA evidence and the prosecution's narrative, which portrayed Knox as a jealous roommate who killed Kercher in a violent sexual game gone wrong. However, from the beginning, forensic experts criticized the way evidence had been collected and handled, noting the absence of strict contamination protocols and questioning the reliability of low-template DNA samples.

In 2011, an appellate court overturned their convictions, citing serious doubts about the integrity of the forensic evidence. The court noted that the DNA on the knife and bra clasp could not be trusted due to potential contamination, and forensic experts pointed out the flaws in the prosecution's interpretation of DNA evidence. However, the legal saga was far from over. In 2013, Italy's Supreme Court vacated the acquittals, and Knox and Sollecito were once again convicted in a 2014 retrial.

It wasn't until March 2015 that the Italian Supreme Court, in its final ruling, definitively exonerated Knox and Sollecito. The court criticized the investigation, the collection of evidence, and the prosecution's theories, stating that there was no credible evidence linking the pair to the murder. The final judgment, known as the Marasca-Bruno motivation report, highlighted the illogicality of much of the prosecution's case and offered guidance on how courts should handle complex forensic evidence in the future. It also recognized the potential for miscarriages of justice when forensic science is misapplied or misunderstood (Gill, 2016).

One of the major lessons from the Knox trial is the importance of proper evidence handling and forensic protocols. In the initial investigation, police failed to follow basic contamination controls, leaving crucial pieces of evidence, such as the bra clasp, exposed to environmental contamination for weeks. The defense argued that Knox's DNA on the kitchen knife could have been the result of secondary transfer, unrelated to the murder. The Marasca-Bruno report emphasized this point, warning against concluding low-template or mixed DNA samples without considering the possibility of contamination or innocent transfer.

The judicial process in the case also raised questions about how courts handle complex forensic evidence. The Marasca-Bruno report pointed out that judges, who often lack scientific training, may struggle to evaluate complex forensic testimony. In the Knox trial, the judges' reliance on the prosecution's forensic experts—who presented DNA evidence as conclusive proof of guilt—led to wrongful convictions. The case serves as a reminder of the need for critical peer review in forensic science and for judicial caution when dealing with scientific evidence that can be misinterpreted.

19.7.3 Case study 3: the case of Lukis Anderson (USA, 2012)

In 2012, Lukis Anderson, a homeless man in California, was charged with the murder of Raveesh Kumra, a wealthy tech executive. Anderson's DNA was found on Kumra's body, leading police to arrest him. However, Anderson had a solid alibi—he was hospitalized for alcohol intoxication at the time of the murder, making it physically impossible for him to have been involved.

The explanation for the presence of Anderson's DNA came through secondary transfer. Paramedics who had treated Anderson earlier in the day had also been involved in responding to Kumra's murder scene. It is believed that Anderson's DNA was transferred from the paramedics' equipment or clothing to Kumra during their interactions. This led to a wrongful charge based on DNA evidence that was secondarily transferred through the actions of first responders.

This case illustrates the importance of understanding secondary transfer in the context of DNA analysis. It also raises questions about the standards for evidence handling in emergency and medical services, and how forensic science can account for such scenarios to prevent wrongful accusations (McCrane & Mulligan, 2023).

19.8 Conclusion

Secondary transfer has emerged as a complex yet critical issue in forensic science, particularly in an era of heightened sensitivity in evidence detection. The ability to detect trace amounts of DNA that have been indirectly transferred has expanded the reach of forensic analysis, enabling investigators to revisit cold cases and identify perpetrators from minuscule samples. However, this increased sensitivity also poses significant challenges. The potential for secondary transfer raises questions about the accuracy of evidence interpretation, as it is possible for DNA or other traces to be found at crime scenes without the person in question having been present. Misinterpretation of such evidence has led to wrongful accusations and convictions in several high-profile cases.

The risk of misinterpretation is particularly high when forensic scientists and investigators fail to account for the indirect nature of some transfers. This can lead to errors in linking individuals to crime scenes, as illustrated by cases like the Amanda Knox trial, where misinterpretation of low-template DNA and secondary transfer played a significant role in wrongful convictions.

Environmental factors, such as temperature, humidity, and the texture of surfaces, further complicate the interpretation of trace evidence.

Recent advancements in forensic science are helping to address these challenges. Bayesian statistical models, for instance, provide a structured, probabilistic approach to evaluating the likelihood of secondary transfer. These models consider multiple factors, such as the type of evidence, environmental conditions, and the nature of contact, offering forensic scientists a more nuanced framework for assessing trace evidence. By incorporating these models into forensic testimony, courts can better understand the complexities of transfer scenarios, reducing the risk of wrongful convictions based on misinterpreted evidence.

In addition to statistical modeling, AI and ML are revolutionizing forensic investigations. AI algorithms can analyze large datasets, improving the efficiency and accuracy of evaluating secondary transfer scenarios. These tools can model how evidence behaves under different conditions, offering deeper insights into the likelihood of primary or secondary transfer. AI's capacity to deconvolute complex DNA mixtures also aids in distinguishing between true and false donors, enhancing the objectivity of forensic analysis.

References

Alshehhi, S., & Haddrill, P. R. (2019). Estimating time since deposition using quantification of RNA degradation in body fluid-specific markers. *Forensic Science International, 298*, 58–63. https://doi.org/10.1016/j.forsciint.2019.02.046, www.elsevier.com/locate/forsciint.

Barber, A. P., Weber, A. R., & Lednev, I. K. (2024). Raman spectroscopy to determine the time since deposition of heated bloodstains. *Forensic Chemistry, 37*. https://doi.org/10.1016/j.forc.2024.100549, https://www.sciencedirect.com/science/journal/24681709.

Bhoyar, L., Mehar, P., & Chavali, K. (2024). An overview of DNA degradation and its implications in forensic caseworks. *Egyptian Journal of Forensic Sciences, 14*(1). https://doi.org/10.1186/s41935-024-00389-y.

Biedermann, A., Champod, C., Jackson, G., Gill, P., Taylor, D., Butler, J., Morling, N., Hicks, T., Vuille, J., & Taroni, F. (2016). Evaluation of forensic DNA traces when propositions of interest relate to activities: Analysis and discussion of recurrent concerns. *Frontiers in Genetics, 7*. https://doi.org/10.3389/fgene.2016.00215.

Buchanan, J. (2006). Assessing the impact of the Stephen Lawrence Inquiry. *Probation Journal, 53*(2), 173–176. https://doi.org/10.1177/026455050605300209.

Buckingham, A. K., Harvey, M. L., & van Oorschot, R. A. H. (2016). The origin of unknown source DNA from touched objects. *Forensic Science International: Genetics, 25*, 26–33. https://doi.org/10.1016/j.fsigen.2016.07.015, http://www.elsevier.com.

Buckleton, J., Bright, J. A., Taylor, D., Evett, I., Hicks, T., Jackson, G., & Curran, J. M. (2014). Helping formulate propositions in forensic DNA analysis. *Science and Justice, 54*(4), 258–261. https://doi.org/10.1016/j.scijus.2014.02.007, http://www.scienceandjusticejournal.com/contact.

Burrill, J., Daniel, B., & Frascione, N. (2019). A review of trace "Touch DNA" deposits: Variability factors and an exploration of cellular composition. *Forensic Science International: Genetics, 39*, 8–18. https://doi.org/10.1016/j.fsigen.2018.11.019.

Butler, J. M. (2023). Recent advances in forensic biology and forensic DNA typing: INTERPOL review 2019–2022. *Forensic Science International: Synergy, 6*. https://doi.org/10.1016/j.fsisyn.2022.100311, https://www.journals.elsevier.com/forensic-science-international-synergy.

Caglià, A., Stefanoni, P., & La Rosa, A. (2011). Cold cases: New technologies for DNA analysis allow the reopening and solution of unsolved cases. *Forensic Science International: Genetics Supplement Series, 3*(1), e230. https://doi.org/10.1016/j.fsigss.2011.09.001.

Cale, C. M., Earll, M. E., Latham, K. E., & Bush, G. L. (2016). Could secondary DNA transfer falsely place someone at the scene of a crime? *Journal of Forensic Sciences, 61*(1), 196–203. https://doi.org/10.1111/1556-4029.12894.

Chakravarty, M., & Pandya, P. (2022). Introduction to forensic DNA typing and current trends. In *Handbook of DNA Profiling*. India: Springer, 29–49. https://link.springer.com/referencework/10.1007/978-981-15-9364-2.

Comte, J., Baechler, S., Gervaix, J., Lock, E., Milon, M.-P., Delémont, O., & Castella, V. (2019). Touch DNA collection – Performance of four different swabs. *Forensic Science International: Genetics, 43*, 102113. https://doi.org/10.1016/j.fsigen.2019.06.014.

Dachs, J., McNaught, I. J., & Robertson, J. (2003). The persistence of human scalp hair on clothing fabrics. *Forensic Science International, 138*(1-3), 27–36. https://doi.org/10.1016/j.forsciint.2003.07.014.

Daly, D. J., Murphy, C., & McDermott, S. D. (2012). The transfer of touch DNA from hands to glass, fabric and wood. *Forensic Science International: Genetics, 6*(1), 41–46. https://doi.org/10.1016/j.fsigen.2010.12.016.

Dash, H. R., Vajpayee, K., & Agarwal, R. (2022). A glimpse of famous cases in history solved by DNA typing. In *Handbook of DNA Profiling*. India: Springer, 357–373. http://doi.org/10.1007/978-981-16-4318-7_14, https://link.springer.com/referencework/10.1007/978-981-15-9364-2.

Fantinato, C., Fonneløp, A. E., Bleka, Ø., Vigeland, M. D., & Gill, P. (2023). The invisible witness: Air and dust as DNA evidence of human occupancy in indoor premises. *Scientific Reports, 13*(1). https://doi.org/10.1038/s41598-023-46151-7, https://www.nature.com/srep/.

Fantinato, C., Gill, P., & Fonneløp, A. E. (2024). Investigative use of human environmental DNA in forensic genetics. *Forensic Science International: Genetics, 70*. https://doi.org/10.1016/j.fsigen.2024.103021, https://www.sciencedirect.com/science/journal/18724973.

Galais, V., Gannicliffe, C., Dugard, P., Wilson, S., Daéid, N. N., & Ménard, H. (2024). Exploring the influence of washing activities on the transfer and persistence of fibres in forensic science. *Forensic Science International, 361*, 112078. https://doi.org/10.1016/j.forsciint.2024.112078.

García, Ó. (2021). Forensic genealogy: Social, ethical, legal and scientific implications. *Spanish Journal of Legal Medicine, 47*(3), 112–119. https://doi.org/10.1016/j.remle.2020.06.002.

Gaudette, B. D., & Tessarolo, A. A. (1987). Secondary transfer of human scalp hair. *Journal of Forensic Sciences, 32*(5), 1241–1253. https://doi.org/10.1520/jfs11175j.

Gentry, A. E., Ingram, S., Philpott, M. K., Archer, K. J., & Ehrhardt, C. J. (2023). Preliminary assessment of three quantitative approaches for estimating time-since deposition from autofluorescence and morphological profiles of cell populations from forensic biological samples. *PLoS One, 18*(10). https://doi.org/10.1371/journal.pone.0292789, https://journals.plos.org/plosone/article?id=10.1371/journal.pone.0292789.

Gill, P. (2016). Analysis and implications of the miscarriages of justice of Amanda Knox and Raffaele Sollecito. *Forensic Science International: Genetics, 23*, 9–18. https://doi.org/10.1016/j.fsigen.2016.02.015.

Goray, M., & van Oorschot, R. A. H. (2015). The complexities of DNA transfer during a social setting. *Legal Medicine, 17*(2), 82–91. https://doi.org/10.1016/j.legalmed.2014.10.003, www.elsevier.com/locate/legalmed.

Goray, M., Fowler, S., Szkuta, B., & van Oorschot, R. A. H. (2016). Shedder status—An analysis of self and non-self DNA in multiple handprints deposited by the same individuals over time. *Forensic Science International: Genetics, 23*, 190–196. https://doi.org/10.1016/j.fsigen.2016.05.005.

Goray, M., Taylor, D., Bibbo, E., Fantinato, C., Fonneløp, A. E., & Gill, P. (2024). Oorschot, emerging use of air EDNA and its application to forensic investigations – A review. *Electrophoresis*.

Guillen, M., Lareu, M. V., Pestoni, C., Salas, A., & Carracedo, A. (2000). Ethical-legal problems of DNA databases in criminal investigation. *Journal of Medical Ethics, 26*(4), 266–271. https://doi.org/10.1136/jme.26.4.266.

Haddrill, P. R. (2021). Developments in forensic DNA analysis. *Emerging Topics in Life Sciences, 5*(3), 381–393. https://doi.org/10.1042/ETLS20200304, https://portlandpress.com/emergtoplifesci/article/5/3/381/228192/Developments-in-forensic-DNA-analysis.

Jansson, L., Swensson, M., Gifvars, E., Hedell, R., Forsberg, C., Ansell, R., & Hedman, J. (2022). Individual shedder status and the origin of touch DNA. *Forensic Science International: Genetics, 56*, 102626. https://doi.org/10.1016/j.fsigen.2021.102626.

Jordan, D., & Mills, D. E. (2021). Past, present, and future of DNA typing for analyzing human and non-human forensic samples. *Frontiers in Ecology and Evolution, 9*. https://doi.org/10.3389/fevo.2021.646130.

Kaesler, T., Kirkbride, K. P., & Linacre, A. (2023). Persistence of touch DNA on commonly encountered substrates in different storage conditions. *Forensic Science International*. https://doi.org/10.1016/J.FORSCIINT.2023.111728.

Kanokwongnuwut, P., Martin, B., Taylor, D., Kirkbride, K. P., & Linacre, A. (2021). How many cells are required for successful DNA profiling? *Forensic Science International: Genetics, 51*. https://doi.org/10.1016/j.fsigen.2020.102453, http://www.elsevier.com.

Korzik, M. L., De Alcaraz-Fossoul, J., Adamowicz, M. S., & San Pietro, D. (2023). Preliminary study: DNA transfer and persistence on non-porous surfaces submerged in spring water. *States Genes, 14*(5). https://doi.org/10.3390/genes14051045, http://www.mdpi.com/journal/genes/.

Kuffel, A., Nic Daeid, N., & Gray, A. (2024). Comparison of swabbing and cutting-out DNA collection methods from cotton, paper, and cardboard surfaces. *Forensic Science International: Synergy, 8*, 100453. https://doi.org/10.1016/j.fsisyn.2023.100453.

Machado, H., & Silva, S. (2009). Informed consent in forensic DNA databases: Volunteering, constructions of risk and identity categorization. *BioSocieties, 4*(4), 335–348. https://doi.org/10.1017/s1745855209990329.

McColl, D. L., Harvey, M. L., & van Oorschot, R. A. H. (2017). DNA transfer by different parts of a hand. *Forensic Science International: Genetics Supplement Series, 6*, e29. https://doi.org/10.1016/j.fsigss.2017.09.014, http://www.elsevier.com.

McCrane, S. M., & Mulligan, C. J. (2023). An innovative transfer DNA experimental design and qPCR assay: Protocol and pilot study. *Journal of Forensic Sciences, 68*(3), 990–1000. https://doi.org/10.1111/1556-4029.15243, http://onlinelibrary.wiley.com/journal/10.1111/(ISSN)1556-4029.

McKenna, L. (n.d.). Understanding DNA results within the case context: Importance of the alternative proposition.

Morgan, J. (2023). Wrongful convictions and claims of false or misleading forensic evidence. *Journal of Forensic Sciences, 68*(3), 908–961. https://doi.org/10.1111/1556-4029.15233.

Ogbogu, U., & Ahmed, N. (2022). Legal, and social implications (ELSI) research: Methods and approaches. *Curr Protoc*.

Oldoni, F., Castella, V., & Hall, D. (2016). Shedding light on the relative DNA contribution of two persons handling the same object. *Forensic Science International: Genetics, 24*, 148–157. https://doi.org/10.1016/j.fsigen.2016.07.002.

Oosthuizen, T., & Howes, L. M. (2022). The development of forensic DNA analysis: New debates on the issue of fundamental human rights. *Forensic Science International: Genetics, 56*. https://doi.org/10.1016/j.fsigen.2021.102606, http://www.elsevier.com.

Pfeifer, C. M., & Wiegand, P. (2017). Persistence of touch DNA on burglary-related tools. *International Journal of Legal Medicine, 131*(4), 941–953. https://doi.org/10.1007/s00414-017-1551-4, http://link.springer.de/link/service/journals/00414/index.htm.

Poetsch, M., Markwerth, P., Konrad, H., Bajanowski, T., & Helmus, J. (2022). About the influence of environmental factors on the persistence of DNA—A long-term study. *International Journal of Legal Medicine, 136*(3), 687–693. https://doi.org/10.1007/s00414-022-02800-6.

Procopio, N., Lovisolo, F., Sguazzi, G., Ghignone, S., Voyron, S., Migliario, M., Renò, F., Sellitto, F., D'Angiolella, G., Tozzo, P., Caenazzo, L., & Gino, S. (2021). Touch microbiome as a potential tool for forensic investigation: A pilot study. *Journal of Forensic and Legal Medicine, 82*. https://doi.org/10.1016/j.jflm.2021.102223, http://www.elsevier.com.

Procopio, N., Sguazzi, G., Eriksson, E. V., Ogbanga, N., McKell, F. C., Newton, E. P., Magni, P. A., Bonicelli, A., & Gino, S. (2024). Transferability of human and environmental microbiome on clothes as a tool for forensic investigations. *Genes, 15*(3). https://doi.org/10.3390/genes15030375, http://www.mdpi.com/journal/genes/.

Romero-García, C., Rosell-Herrera, R., Revilla, C. J., Baeza-Richer, C., Gomes, C., Palomo-Díez, S., Arroyo-Pardo, E., & López-Parra, A. M. (2019). Effect of the activity in secondary transfer of DNA profiles. *Forensic Science International: Genetics Supplement Series, 7*(1), 578–579. https://doi.org/10.1016/j.fsigss.2019.10.096.

Schulte, J., Rittiner, N., Seiberle, I., Kron, S., & Schulz, I. (2023). Collecting touch DNA from glass surfaces using different sampling solutions and volumes: Immediate and storage effects on genetic STR analysis. *Journal of Forensic Sciences, 68*(4), 1133–1147. https://doi.org/10.1111/1556-4029.15305.

Sessa, F., Salerno, M., & Pomara, C. (2022). The interpretation of mixed DNA samples historical: Perspective and current developments. In *Handbook of DNA Profiling*. Italy: Springer, 997–1017. https://link.springer.com/referencework/10.1007/978-981-15-9364-2.

Sessa, F., Salerno, M., Bertozzi, G., Messina, G., Ricci, P., Ledda, C., Rapisarda, V., Cantatore, S., Turillazzi, E., & Pomara, C. (2019). Touch DNA: impact of handling time on touch deposit and evaluation of different recovery techniques: An experimental study. *Scientific Reports, 9*(1). https://doi.org/10.1038/s41598-019-46051-9.

Sessa, F., Pomara, C., Esposito, M., Grassi, P., Cocimano, G., & Salerno, M. (2023). Indirect DNA transfer and forensic implications: A literature review. *Genes, 14*(12), 2153. https://doi.org/10.3390/genes14122153.

Sessa, F., Esposito, M., Cocimano, G., Sablone, S., Karaboue, M. A. A., Chisari, M., Albano, D. G., & Salerno, M. (2024). Artificial intelligence and forensic genetics: Current applications and future perspectives. *Applied Sciences, 14*(5), 2113. https://doi.org/10.3390/app14052113.

Sijen, T., & Harbison, S. A. (2021). On the identification of body fluids and tissues: A crucial link in the investigation and solution of crime. *Genes, 12*(11), 1728. https://doi.org/10.3390/genes12111728.

Szkuta, B., Ballantyne, K. N., Kokshoorn, B., & van Oorschot, R. A. H. (2018). Transfer and persistence of non-self DNA on hands over time: Using empirical data to evaluate DNA evidence given activity level propositions. *Forensic Science International: Genetics, 33*, 84–97. https://doi.org/10.1016/j.fsigen.2017.11.017, http://www.elsevier.com.

Szkuta, B., Ansell, R., Boiso, L., Connolly, E., Kloosterman, A. D., Kokshoorn, B., McKenna, L. G., Steensma, K., & van Oorschot, R. A. H. (2019). Assessment of the transfer, persistence, prevalence and recovery of DNA traces from clothing: An inter-laboratory study on worn upper garments. *Forensic Science International: Genetics, 42*, 56–68. https://doi.org/10.1016/j.fsigen.2019.06.011, http://www.elsevier.com.

Thornbury, D., Goray, M., & van Oorschot, R. A. H. (2021). Transfer of DNA without contact from used clothing, pillowcases and towels by shaking agitation. *Science and Justice, 61*(6), 797–805. https://doi.org/10.1016/j.scijus.2021.10.005, http://www.scienceandjusticejournal.com/contact.

Tozzo, P., Mazzobel, E., Marcante, B., Delicati, A., & Caenazzo, L. (2022). Touch DNA sampling methods: Efficacy evaluation and systematic review. *International Journal of Molecular Sciences, 23*(24), 15541. https://doi.org/10.3390/ijms232415541.

Trejos, T., Koch, S., & Mehltretter, A. (2020). Scientific foundations and current state of trace evidence—A review. *Forensic Chemistry, 18*, 100223. https://doi.org/10.1016/j.forc.2020.100223.

Van Oorschot, R. A. H., Szkuta, B., Meakin, G. E., Kokshoorn, B., & Goray, M. (2019). DNA transfer in forensic science: A review. *Forensic Science International: Genetics, 38*, 140–166. https://doi.org/10.1016/j.fsigen.2018.10.014, http://www.elsevier.com.

Van Oorschot, R. A. H., Meakin, G. E., Kokshoorn, B., Goray, M., & Szkuta, B. (2021). DNA transfer in forensic science: Recent progress towards meeting challenges. *Genes, 12*(11). https://doi.org/10.3390/genes12111766, https://www.mdpi.com/2073-4425/12/11/1766/pdf.

Veldhuis, M. S., Ariëns, S., Ypma, R. J. F., Abeel, T., & Benschop, C. C. G. (2022). Explainable artificial intelligence in forensics: Realistic explanations for number of contributor predictions of DNA profiles. *Forensic Science International: Genetics, 56*. https://doi.org/10.1016/j.fsigen.2021.102632, http://www.elsevier.com.

Währer, J., Kehm, S., Allen, M., Brauer, L., Eidam, O., Seiberle, I., Kron, S., Scheurer, E., & Schulz, I. (2023). The DNA-buster: The evaluation of an alternative DNA recovery approach. *Forensic Science International: Genetics, 64*, 102830. https://doi.org/10.1016/j.fsigen.2023.102830.

Part III

Module III: Predictive forensic DNA analysis

Chapter 20

Molecular autopsy for analysis of sudden unexplained death cases

Amnuay Kleenayoon[1], and Viroj Wiwanitkit[2]
[1]*Samraong, Cambodia,* [2]*Research Center, Chandigarh University, Chandigarh, Punjab, India*

20.1 Introduction

Due to the limits of traditional autopsy techniques in diagnosing the cause of death, cases of sudden unexplained death (SUD) pose a considerable problem for forensic pathologists (Hudzik et al., 2018) (Fig. 20.1). Therefore more sophisticated molecular-based methods must be used in these cases. Examining genetic, metabolic, and epigenetic components is part of the molecular autopsy, which has become a useful method for analyzing SUD cases (Castiglione et al., 2021; Tester & Ackerman, 2006). An overview of the molecular autopsy approach is given in this chapter, along with an analysis of its potential benefits and challenges in determining the cause of death in cases of SUD. It also goes into several molecular techniques used in the research, such as next-generation sequencing (NGS), gene expression profiling, and DNA methylation analysis. In addition, the need for genetic counseling and family screening in the context of SUD cases is also discussed. Molecular autopsy's interdisciplinary approach has the potential to provide important new information on the cause of unexpected fatalities (Castiglione et al., 2021; Tester & Ackerman, 2006).

The sudden and unexpected death of a person is one of the biggest challenges facing forensic pathology. Conventional autopsy methods frequently fall short of providing a conclusive answer, leaving investigators and families grieving and impeding the creation of preventative measures (Alqarawi et al., 2022). The application of molecular-based methods, also referred to as "molecular autopsy," has become more popular recently as a means of addressing this problem (Castiglione et al., 2021; Tester & Ackerman, 2006). A thorough method known as "molecular autopsy" examines genetic, biochemical, and epigenetic elements in an attempt to identify the underlying cause of unexpected deaths. Researchers and forensic pathologists want to find the answers that are concealed from view by exploring the molecular mechanisms and interactions within a person's genetic composition (Castiglione et al., 2021; Tester & Ackerman, 2006).

In diagnosing the cause of death, especially in cases of SUD, this chapter explores the possible advantages and difficulties of the molecular autopsy technique. It provides an overview of the different approaches and procedures used in studies involving molecular autopsies, with a particular emphasis on three major approaches: NGS, gene expression profiling, and DNA methylation analysis. In molecular autopsy investigations, DNA methylation analysis is essential because it looks at the chemical changes made to DNA molecules. Certain diseases or disorders are linked to specific DNA methylation patterns that can shed light on the cause of mortality (Campuzano & Sarquella-Brugada, 2023; Castiglione et al., 2021; Heathfield et al., 2018; Yamamoto et al., 2024). Through the comparison of methylation profiles between persons who died suddenly and unexplainedly and control subjects, researchers can detect genetic abnormalities or modifications that may have had a role in the fatality. In molecular autopsy, gene expression profiling is yet another crucial instrument (Campuzano & Sarquella-Brugada, 2023; Castiglione et al., 2021; Heathfield et al., 2018; Yamamoto et al., 2024). Through the examination of the gene expression levels in a subject's tissues, scientists might spot any dysregulation or anomalies that might have contributed to the unexpected death (Campuzano & Sarquella-Brugada, 2023; Castiglione et al., 2021; Heathfield et al., 2018; Yamamoto et al., 2024). The identification of any abnormalities and the provision of important hints on the cause of death can be facilitated by comparing the gene expression patterns of affected individuals with those of healthy controls.

Investigating molecular autopsies involves the innovative method of NGS. With the use of NGS, scientists may quickly sequence vast volumes of DNA or RNA samples to find genetic variants and mutations that might be the cause of

Investigative and Predictive DNA Testing. DOI: https://doi.org/10.1016/B978-0-443-30128-5.00012-6

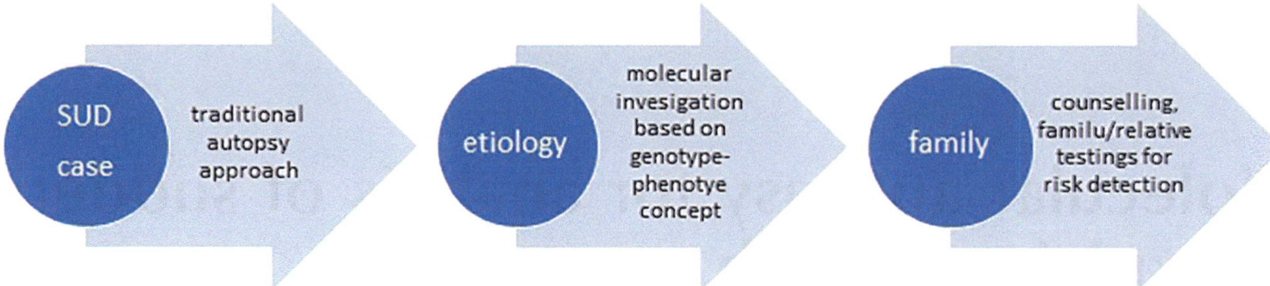

FIGURE 20.1 Conceptual steps for molecular autopsy in sudden unexplained death (SUD).

unexpectedly high death rates (Campuzano & Sarquella-Brugada, 2023; Castiglione et al., 2021; Heathfield et al., 2018; Yamamoto et al., 2024). Full genome analysis and targeted gene panels associated with genetic disorders or cardiovascular diseases provide extensive insights into the genetic causes that may have contributed to the death (Methner et al., 2016). This chapter covers the molecular methods used in molecular autopsy investigations as well as the significance of genetic counseling and family screening. Family screening can identify potential hereditary variables associated with SUD cases by analyzing medical and genetic histories among affected families. On the other hand, genetic counseling is essential in helping affected families comprehend and take care of inherited illnesses.

This chapter emphasizes the multidisciplinary aspect of the molecular autopsy approach and the cooperation of genetic counselors, clinicians, forensic pathologists, and geneticists (Michaud et al., 2009). The potential for a thorough investigation and a deeper knowledge of the reasons for SUDs is greatly increased by bringing together experts from other professions. In summary, a promising method for identifying the cause of unexpected deaths is the molecular autopsy. Researchers and forensic pathologists can reveal the hidden variables driving these terrible events by using advanced molecular tools, such as gene expression profiling, DNA methylation analysis, and NGS, to examine genetic, metabolic, and epigenetic aspects. In addition, family screening and genetic counseling are essential for identifying the underlying genetic susceptibility and providing support to impacted families (Campuzano & Sarquella-Brugada, 2023; Castiglione et al., 2021; Heathfield et al., 2018; Yamamoto et al., 2024). The molecular autopsy approach's multidisciplinary character offers great promise for solving the puzzles surrounding unexpected fatalities and eventually advancing the creation of prophylactic measures.

20.2 Molecular autopsy

Molecular autopsy has become a potent weapon in forensic science in recent years. Finding genetic changes or anomalies that may have aided in the person's death, provides a fresh method for examining the cause of SUD (Campuzano & Sarquella-Brugada, 2023; Castiglione et al., 2021; Heathfield et al., 2018; Yamamoto et al., 2024). Our understanding of genetic heart illnesses, metabolic problems, and other conditions that might cause sudden death has greatly advanced as a result of this ground-breaking technique (Campuzano & Sarquella-Brugada, 2023; Castiglione et al., 2021; Heathfield et al., 2018; Yamamoto et al., 2024). We shall give a thorough overview of molecular autopsy in this essay, covering its roots in history, methods, clinical uses, and limitations. The necessity to determine the causes of sudden infant death syndrome (SIDS) and sudden cardiac death (SCD) in young people who did not exhibit any obvious abnormalities during traditional postmortem examination gave rise to the idea of molecular autopsy. It became more and more clear that the underlying genetic or molecular defects causing these terrible events could not be found using conventional autopsy techniques. Consequently, scientists and medical professionals began investigating the possibility of using molecular methods to bridge this diagnostic divide (Table 20.1).

Analysis of DNA from postmortem tissue samples, such as blood, heart tissue, or saliva, is the main process of a molecular autopsy. Different molecular methods may be used, based on the probable reason for death. One way to check for particular genetic variants linked to hereditary cardiac illnesses like long QT syndrome (a condition involving prolonged QT interval on an ECG) or hypertrophic cardiomyopathy is through targeted gene sequencing or gene panel testing. More thorough methods, such as whole-exome sequencing (WES) and genome-wide association studies, can find genetic variants or mutations previously unidentified as being connected to sudden death (Alzahrani et al., 2023; Klassen et al., 2012; Reid et al., 2017). The method of investigating inexplicable deaths has been completely transformed by the incorporation of molecular autopsy into standard forensic investigations. Through the precise identification of the genetic defects causing sudden death, medical professionals can reliably diagnose related family members who may be susceptible to the same fate. Molecular autopsy has also made it possible to identify and help at-risk patients early on, which may have

TABLE 20.1 Molecular autopsy data.

Data	Details
1. DNA sequencing data	This kind of data examines a person's DNA sequence to find genetic abnormalities or variations linked to particular illnesses or ailments.
2. Gene expression data	The levels of gene activity in a tissue or sample are measured using this type of data, which offers insights into the use and regulation of genes in a particular setting.
3. Epigenetic data	Studying DNA alterations that can affect gene expression without altering the sequence is known as epigenetic data, and it provides information about how lifestyle and environment affect gene activity.
4. Protein expression data	This type of data provides important insights into the functional output of genes by quantifying the amounts of proteins generated in a tissue or sample by genes.
5. Metabolomics data	The measurement and quantification of tiny molecules (metabolites) in a sample is known as metabolomics data, and it is used to gain insight into the metabolic pathways and activities that are occurring in a biological system.
6. Histopathology data	This sort of data comprises microscopic tissue examinations that help with illness diagnosis and classification by identifying aberrant tissue patterns and characterizing cellular morphologies.
7. Radiological imaging data	Analyzing radiological data entails finding structural or anatomical anomalies that could be linked to a patient's illness from medical imaging studies like MRIs and CT scans.
8. Clinical phenotype data	To help establish links between molecular abnormalities and observable symptoms, clinical phenotype data collects information about a patient's physical attributes, medical history, and other pertinent clinical factors.
9. Pathogen detection data	In order to help understand the processes behind infectious diseases, this sort of data focuses on detecting the presence of pathogens, such as bacteria or viruses, within a sample.
10. Pharmacogenomics data	Personalized medical techniques and drug therapy optimization are made possible by pharmacogenomics data, which investigates the relationship between an individual's genetic composition and their reaction to pharmaceuticals.

prevented further deaths. The findings may also help in the creation of tailored treatments or genetic counseling for hereditary disease sufferers and their families (Tester & Ackerman, 2012).

Molecular autopsy has limitations, even if it provides important information on the genetic causes of unexpected death. Interpreting the importance of genetic variants found using these techniques is one of the major challenges. It can be difficult to differentiate between harmful and benign mutations, and not all discovered mutations will directly cause sudden death. Moreover, the lack of genetic anomalies does not rule out the possibility that the sudden death event was caused by the existence of additional environmental variables or illnesses (Conrad & Barker, 2010). It is important to carefully consider the ethical issues raised by the use of molecular autopsies in forensic medicine. Because postmortem DNA testing is a sensitive operation and the bereaved often feel emotional anguish, it can be difficult to obtain informed consent from families (Hudzik et al., 2018; Semsarian et al., 2015). It is imperative to protect genetic data and maintain privacy to avoid unforeseen outcomes like stigmatization or discrimination based on genetic predispositions.

A significant change in the way that unexpected, unexplained deaths are investigated is represented by molecular autopsy. It allows for early intervention and preventative measures and offers important insights into the underlying causes through the examination of genetic and molecular abnormalities (Campuzano & Sarquella-Brugada, 2023; Castiglione et al., 2021; Heathfield et al., 2018; Yamamoto et al., 2024). Its therapeutic uses help enhance genetic counseling and personalized treatment while providing affected families with hope. However, to guarantee this technique's appropriate application in forensic medicine, it will be imperative to tackle its ethical implications and overcome its limits (Hudzik et al., 2018; Semsarian et al., 2015).

20.3 Background of sudden unexplained death

The term "sudden unexplained death" refers to an individual's unexpected and sudden death that is still unsolved even after a comprehensive inquiry. For years, this occurrence has baffled researchers, medical professionals, and families, leaving them in search of answers and resolutions. SUD frequently affects people who appear healthy and have never had a medical history (Alqarawi et al., 2022; Bagnall et al., 2014; Hudzik et al., 2018; Steinberg et al., 2016). It can hit without warning and impact people of various ages, even adults. It is quite difficult for medical personnel to ascertain the reason for death and give families appropriate answers when there is no obvious cause or underlying disease (Alqarawi et al., 2022; Bagnall et al., 2014; Hudzik et al., 2018; Steinberg et al., 2016).

The occurrence of sudden deaths without apparent reason has been documented in medical literature for several decades, marking the beginning of the history of SUD. However, SUD did not receive much attention or be acknowledged as a separate medical condition until the latter half of the 20th century. Medical researchers have been more interested in learning about the underlying mechanisms and potential risk factors connected with SUD as a result of the abrupt and unexpected character of these deaths (Alqarawi et al., 2022; Bagnall et al., 2014; Hudzik et al., 2018; Steinberg et al., 2016). When doctors first started looking into SUD cases, they mostly concentrated on the possibility that heart problems were the reason. Undiagnosed structural heart problems, such as cardiac arrhythmias and hereditary diseases like long QT syndrome, have been implicated in several sudden fatalities. Following the discovery of these underlying heart problems, many screening and diagnostic procedures were put in place to find people who were at risk and maybe avert deadly consequences (Ottaviani & Ramos, 2023; Steinberg et al., 2016).

Although cardiac anomalies are frequently observed in cases of SUD, it is important to understand that not all occurrences of SUD are caused by these diseases. Traditional cardiac theories of SUD are no longer sufficient to fully comprehend the condition. Numerous factors, including genetic susceptibilities, environmental triggers, and even certain drugs, have been shown by researchers to contribute to SUD (Alqarawi et al., 2022; Bagnall et al., 2014; Hudzik et al., 2018; Steinberg et al., 2016). These discoveries have expanded the field of study for SUD and encouraged researchers to look at risk factors and complex causes. A further challenge to the SUD inquiry is the absence of established pathology and autopsy protocols (Alqarawi et al., 2022; Bagnall et al., 2014; Hudzik et al., 2018; Steinberg et al., 2016). Sometimes, the lack of visible structural abnormalities and limitations in current procedures prevents autopsies from being able to determine the cause of death. The absence of standardization impedes the discovery of common etiological components by making it challenging to compare and analyze SUD cases across various demographics and countries (Alqarawi et al., 2022; Bagnall et al., 2014; Hudzik et al., 2018; Steinberg et al., 2016).

The psychological impact on SUD-affected families is profound (Risgaard, 2016). One experiences intense grief and annoyance when a loved one passes away unexpectedly and mysteriously. Families frequently have a difficult time accepting their loss and finding closure. Feelings of injustice and powerlessness are sometimes brought on by ignorance regarding SUD. The psychological effects of SUD on families have led to the development of support groups and counseling programs aimed at assisting them in coping with the bereavement process (Risgaard, 2016). Over time, there has been a noticeable increase in the efforts made to address the issues related to SUD. Scientists are still looking into the underlying causes of SUD as well as possible risk factors. In order to get a thorough understanding of SUD, collaborative efforts seek to incorporate data from several sources, including genetic, environmental, and clinical aspects. In order to improve data collection and analysis, policies are being designed to standardize autopsy processes and improve surveillance systems (Alqarawi et al., 2022; Bagnall et al., 2014; Hudzik et al., 2018; Steinberg et al., 2016).

In summary, unexpected deaths that happen suddenly continue to be a terrible and puzzling occurrence. It takes the lives of people of all ages without any prior notice or discernible reason. Over time, the study of SUD has changed, focusing on a wider spectrum of causes while initially concentrating on cardiac abnormalities in the early stages (Alqarawi et al., 2022; Bagnall et al., 2014; Hudzik et al., 2018; Steinberg et al., 2016). The development of a comprehensive understanding of the etiology of SUD has been impeded by the absence of conventional autopsy protocols and pathologic evaluation. However, combined efforts from researchers, doctors, and support systems are getting us closer to solving the puzzles underlying SUD and offering comfort to the families impacted by this sad occurrence (Risgaard, 2016).

20.4 Importance of molecular autopsy in sudden unexplained death cases

Molecular autopsy, sometimes referred to as postmortem genetic testing, is the process of analyzing a person's genetic composition after their passing. In order to determine any inherited or acquired diseases that may have contributed to their sudden unexplained death (SUD), a thorough analysis of genes and the mutations linked with them is conducted (Farrugia et al., 2015; Khalikov et al., 2021; Neubauer et al., 2021, 2022; Semsarian & Hamilton, 2012). This novel method of comprehending the fundamental reasons behind unexpected deaths has numerous significant ramifications (Farrugia et al., 2015; Khalikov et al., 2021; Neubauer et al., 2021, 2022; Semsarian & Hamilton, 2012). First off, molecular autopsy can accurately determine the cause of death in situations where more conventional autopsy techniques might be unable to reveal underlying genetic problems. Molecular autopsies provide a more accurate cause of death by analyzing a person's

DNA and mutations to find any genetic variations or abnormalities that may have contributed to their demise (Farrugia et al., 2015; Khalikov et al., 2021; Neubauer et al., 2021, 2022; Semsarian & Hamilton, 2012).

Secondly, the risk to surviving family members can be assessed by molecular autopsy. Through the identification of genetic abnormalities linked to the unexpected death, this method can determine whether other family members are susceptible to the same illness. In order to lower the likelihood of future unexpected deaths within the family, preventive steps can be taken with the help of this information, which is essential for genetic counseling. Additionally, measures for prevention and treatment are informed by molecular autopsy (Farrugia et al., 2015; Khalikov et al., 2021; Neubauer et al., 2021, 2022; Semsarian & Hamilton, 2012). When a genetic mutation linked to SUD is found, it can help determine the best course of treatment for other family members who might also have the mutation. To lower the chance of unexpected death, family members can be examined for the same mutation and given the necessary medical care if it is connected to a cardiac problem, for instance (Turkgenc et al., 2023).

Molecular autopsy is useful in clinical settings, but it also has a big impact on forensic inquiries. Forensic investigators may find this method useful in situations where the cause of death is first unknown. Finding genetic anomalies that might have contributed to the death can assist in determining the reason and improve comprehension of the events leading up to the unexpected death. Furthermore, the results of molecular autopsies add to our understanding of the genetic underpinnings of SUDs (Farrugia et al., 2015; Khalikov et al., 2021; Neubauer et al., 2021, 2022; Semsarian & Hamilton, 2012). Researchers can better understand the underlying mechanisms, create targeted medicines, and identify and manage individuals at risk by gathering data on genetic abnormalities linked to unexpected fatalities. This advances the field and facilitates the creation of evidence-based plans for averting unexpected deaths (Farrugia et al., 2015; Khalikov et al., 2021; Neubauer et al., 2021, 2022; Semsarian & Hamilton, 2012).

To sum up, molecular autopsies are an essential tool for figuring out what causes sudden, unexpected deaths, pinpointing genetic risk factors, and developing preventative and therapeutic plans (Farrugia et al., 2015; Khalikov et al., 2021; Neubauer et al., 2021, 2022; Semsarian & Hamilton, 2012). The process of examining a person's genetic composition and detecting any anomalies or mutations offers important insights that influence specific instances as well as further scientific understanding in this area.

20.5 Methodologies and techniques used in molecular autopsy

A scientific method known as "molecular autopsy" is used to look into situations of sudden, unexpected death and reveal underlying genetic causes and contributory variables. Pathologists and researchers can uncover potential genetic mutations and anomalies that may have contributed to an individual's demise by using this unique approach, which integrates a variety of procedures and techniques. The approaches and procedures that are most frequently utilized in molecular autopsy will be covered in the paragraphs that follow. Scientists can identify mutations affecting the structure and function of proteins associated with a variety of illnesses, such as cardiomyopathies and arrhythmias, using WES (Markwerth et al., 2021; Michaud et al., 2020; Orlíčková et al., 2022).

One of the main techniques in molecular autopsy is WES (Markwerth et al., 2021; Michaud et al., 2020; Orlíčková et al., 2022). This technique is used to sequence the coding sections of genes found in the human genome. By concentrating on particular genomic regions that are more likely to have pathogenic variations, WES is a useful technique for detecting genetic alterations that may have contributed to the sudden death event (Markwerth et al., 2021; Michaud et al., 2020; Orlíčková et al., 2022). Targeted gene sequencing is another important technique used in molecular autopsy investigations, in addition to WES. Through targeted gene sequencing, pathologists can efficiently and accurately examine a small number of genes that are known to be connected to particular disorders (Markwerth et al., 2021; Michaud et al., 2020; Orlíčková et al., 2022). This method can provide important information about potentially lethal genetic variants while saving time and money. Targeted gene sequencing finds particular mutations relevant to the sudden death case under inquiry by examining individual genes or panels of genes.

Furthermore, genetic variant databases and bioinformatics technologies are frequently used in molecular autopsy (Sun et al., 2023). Databases of genetic variants gather data on known benign and harmful genetic variations. Pathologists and scientists can find possibly clinically significant differences by using these databases as a reference (Ginsberg & Mirnics, 2006; Sun et al., 2023). The enormous amount of genetic data produced by sequencing methods requires analysis, and bioinformatics tools are essential for this process. These technologies significantly facilitate the molecular autopsy process by helping with variant annotation, interpretation, and identification of putative disease-causing variations. Comparative genomic hybridization (CGH) arrays are another method that is commonly used in molecular autopsy (CGH arrays). DNA copy number changes, such as deletions, duplications, and amplifications, can be found using CGH arrays (Ginsberg & Mirnics, 2006; Sun et al., 2023). Large-scale genetic changes, such as chromosomal rearrangements or structural differences, that may have contributed to abrupt death can be found using this high-resolution method. These results offer vital details regarding possible syndromes or ailments that might have aided in the person's death (Ginsberg & Mirnics, 2006; Sun et al., 2023).

Finally, functional analyses and validation methods are crucial parts of the molecular autopsy research. In vitro experiments are used in functional research to determine how genetic differences affect the regulation and function of proteins (Ginsberg & Mirnics, 2006). To assess the functional effects of discovered mutations, these investigations may use cellular models, protein expression tests, or electrophysiological studies (Ginsberg & Mirnics, 2006). Sanger sequencing and polymerase chain reaction (PCR) are two validation procedures that are commonly used to rule out possible technical artifacts and validate the presence of particular mutations (Ginsberg & Mirnics, 2006). In conclusion, a variety of approaches and techniques are employed by the discipline of molecular autopsy to examine cases of sudden unexpected death (Table 20.2). Important steps in the molecular autopsy process include WES, targeted gene sequencing, CGH arrays, genetic variation databases, functional studies, and validation methods. By using these techniques, pathologists and researchers can learn important things about the genetic foundation of sudden death occurrences, which might help us understand the underlying causes and possibly develop preventive measures in the future.

20.6 Key findings and discoveries through molecular autopsy

An innovative method in forensic science called "molecular autopsy" has transformed the way that sudden and unexplained deaths are investigated. This cutting-edge method combines conventional autopsy techniques with genetic analysis to identify underlying genetic abnormalities and offer vital information about the cause of death (Campuzano & Sarquella-Brugada, 2023; Castiglione et al., 2021; Heathfield et al., 2018; Yamamoto et al., 2024). By means of comprehensive investigation and testing, researchers have achieved important breakthroughs with molecular autopsy, unlocking the secrets of hitherto unexplained fatalities. Finding genetic alterations linked to a range of cardiovascular disorders is one of the main discoveries made by molecular autopsy (Campuzano & Sarquella-Brugada, 2023; Castiglione et al., 2021; Heathfield et al., 2018; Yamamoto et al., 2024). These types of mutations are very important for SCD, especially in the young. Through careful examination of the genetic makeup of cadavers, scientists have been able to pinpoint several gene variations that

TABLE 20.2 Comparative analysis of molecular autopsy findings.

Analysis	Details
1. Genetic analysis	Identifying possible mutations that cause disease by comparing the genetic variants of the dead person with a reference database.
2. Histopathological analysis	Analyzing tissue samples to look for unusual cellular patterns or structures in order to identify the cause of death.
3. Toxicological analysis	Examining biological samples to check for the presence of poisons, medicines, or other materials that might have played a role in the death.
4. Metabolic analysis	Evaluating enzymatic activity and metabolic pathways to find any underlying metabolic dysfunctions or diseases.
5. Biochemical analysis	Measuring the concentrations of particular molecules, including hormones or enzymes, in order to spot abnormalities or imbalances that might be connected to the cause of death.
6. Electron microscopy	Examining tissue samples at the cellular or subcellular level for structural anomalies using high-resolution imaging methods.
7. Cardiac analysis	Assessing the heart's composition and operation using methods such as echocardiography or electrocardiography to ascertain whether any cardiac defects were a factor in the death.
8. Radiological analysis	Analyzing imaging data, such as CT or X-rays, to find any injuries or anatomical anomalies that might have led to the death.
9. Microbiological analysis	Checking different samples for the presence of infectious agents or diseases that might have contributed to the death.
10. Immunohistochemical analysis	Evaluating particular protein expression patterns in tissue samples in order to find abnormal protein signaling or indicators connected to particular illnesses or ailments.

predispose people to irregular cardiac rhythms and anatomical abnormalities. Thanks to this knowledge, focused medicines and interventions that may be able to avert these potentially fatal illnesses have been developed.

The identification of genetic markers for hereditary arrhythmia syndromes, such as long QT syndrome and Brugada syndrome, is another significant finding in Molecular Autopsy research (Semsarian & Ingles, 2016; Spears & Gollob, 2015). Abnormal electrical activity in the heart is a hallmark of many disorders, frequently leading to abrupt cardiac arrest. Scientists have discovered particular gene mutations that can greatly raise a person's likelihood of experiencing these illnesses through genetic analysis. With the use of this knowledge, genetic testing protocols for at-risk individuals and their families have been put into place, allowing for early detection and preventive interventions (Semsarian & Ingles, 2016; Spears & Gollob, 2015). Furthermore, significant information regarding SIDS patients has been found using molecular autopsy. Through the analysis of the genetic material of newborns who have tragically died without apparent explanation, scientists have found mutations in genes associated with brain development, heart function, and respiratory regulation (Semsarian & Ingles, 2016; Spears & Gollob, 2015). These findings have opened up new paths for future research and possible therapies by shedding light on the possible underlying genetic variables that contribute to SIDS (Semsarian & Ingles, 2016; Spears & Gollob, 2015).

Furthermore, molecular autopsy is also being used for illnesses other than cardiovascular ones. Researchers have discovered genetic anomalies linked to hereditary metabolic problems, neurological issues, and even certain types of cancer thanks to this novel technique (Bonilla et al., 2018; Kumar et al., 2021). Through the deciphering of these fundamental genetic pathways, scientists aim to create more individualized and focused treatment plans for those afflicted with these disorders (Bonilla et al., 2018; Kumar et al., 2021). To sum up, Molecular Autopsy has yielded a multitude of significant results and insights that have significantly enhanced our comprehension of unexpected and unexplained fatalities. Through studying the genetic composition of departed people, researchers have discovered important gene abnormalities underlying a number of cardiovascular diseases, hereditary arrhythmia syndromes, SIDS, and other ailments. These findings have not only made it easier to identify the cause of mortality, but they have also created new avenues for focused therapies, genetic testing, and prevention (Bonilla et al., 2018; Kumar et al., 2021; Semsarian & Ingles, 2016). Molecular Autopsy holds the potential to solve additional riddles and save countless lives as research and development proceed.

20.7 Challenges and limitations in molecular autopsy

Within forensic pathology, the fast-developing area of "molecular autopsy" uses molecular tools and genetic testing to look into sudden unexpected deaths. Although molecular autopsy has enormous potential to shed light on the underlying genetic disorders that lead to these cases, researchers and doctors must overcome a number of obstacles and constraints before it can be widely used. First off, getting a high-quality DNA sample from the deceased is one of the biggest obstacles in molecular autopsy. Since autopsies are usually conducted after a considerable amount of time has passed since the death, DNA deterioration becomes a prevalent problem (Campuzano & Sarquella-Brugada, 2023; Castiglione et al., 2021; Heathfield et al., 2018; Yamamoto et al., 2024). A trustworthy genetic profile might be difficult to produce because of variables like postmortem time, corpse decomposition, and environmental circumstances that can all affect the DNA sample's integrity (Campuzano & Sarquella-Brugada, 2023; Castiglione et al., 2021; Heathfield et al., 2018; Yamamoto et al., 2024). Therefore, in order to increase DNA extraction and guarantee correct analysis, strict protocols and specific techniques are needed.

Finding and interpreting genetic variations that might be causing unexpected death presents a big problem in molecular autopsy. It is difficult to differentiate between benign and harmful genetic variants due to the enormous complexity of the human genome, which is made up of millions of genetic variations (Campuzano & Sarquella-Brugada, 2023; Castiglione et al., 2021; Heathfield et al., 2018; Yamamoto et al., 2024). The existence of uncommon and unique genetic variants that have not been thoroughly described in the scientific literature exacerbates this even more. Furthermore, it can be very difficult to discern between a genetic variant that is the only cause of death and an accidental discovery (Martínez-Barrios et al., 2023). This requires a thorough understanding of the disease mechanisms and a rigorous evaluation of the clinical association (Martínez-Barrios et al., 2023). Moreover, one of the main obstacles to the widespread use of molecular autopsies is their expense. The affordability of molecular procedures and genetic testing, in addition to the need for specialist knowledge for data interpretation, can severely restrict the applicability of this strategy. Because of this, molecular autopsy is often saved for situations in which a traditional autopsy is unable to determine the cause of death and there is a possibility that a genetic mutation may be present. The lack of funding and resource availability in many forensic labs and medical systems keeps molecular autopsy from being widely used as a standard diagnostic technique.

An additional difficulty is the ethical questions raised by molecular autopsy. As genetic testing becomes increasingly common, several ethical concerns surface, such as consent, privacy, and how to handle accidental results. Getting the deceased person's or their family members' informed consent might be difficult because sensitive decisions about genetic testing and data exchange need to be made. Moreover, the possibility of obtaining genetic data that could be pertinent to the family members can pose ethical problems concerning sharing such results (Campuzano & Sarquella-Brugada, 2023; Castiglione et al., 2021; Heathfield et al., 2018; Yamamoto et al., 2024). To guarantee the proper application of molecular autopsy in a way that respects the autonomy and confidentiality of those concerned, these ethical conundrums must be carefully considered.

To sum up, molecular autopsy has a great deal of potential to clarify the genetic components causing unexpected, untimely deaths. However, for it to be implemented successfully, several obstacles and constraints must be removed. Obtaining high-quality DNA samples, identifying genetic variants, financial constraints, and ethical considerations are some of these hurdles. To advance the science of molecular autopsy and maximize its potential to offer vital insights into the underlying genetic causes of sudden unexpected deaths, it will be imperative to address these challenges (Campuzano & Sarquella-Brugada, 2023; Castiglione et al., 2021; Heathfield et al., 2018; Tashjian et al., 2019; Yamamoto et al., 2024).

20.8 Ethical considerations in conducting molecular autopsy

The rapidly developing discipline of molecular autopsy, often called postmortem molecular genetic testing, looks for the genetic cause of unexpected inexplicable death in people, particularly in young people. This method entails using DNA samples from the deceased to perform a variety of genetic tests, the main objective of which is to find any genetic mutations or variations that may have contributed to the person's unexpected death. Although molecular autopsy has significant potential to improve our understanding of genetic predispositions to SCD or other unexplained fatalities, there are several ethical issues that should be carefully considered before doing this procedure. First and foremost, informed permission is essential for molecular autopsy (Shade et al., 2019; Weil, 2023). Getting the deceased person's informed consent is logistically challenging because this operation is done after death. To ensure that the intentions of the deceased are honored, procedures and guidelines pertaining to informed consent for molecular autopsy must be established (Shade et al., 2019; Weil, 2023). This could entail getting approval from a person's heirs or legal guardians, depending on the person's stated wishes before passing away. The family needs to be given clear and honest information about the testing's possible advantages, drawbacks, and ramifications so they may make an educated decision.

Genetic information privacy and secrecy are additional ethical considerations. Important genetic information is uncovered by molecular autopsies, which can help determine future health risks for surviving family members or offspring in addition to providing insight into the cause of death. Strict guidelines and procedures must be established for the handling, sharing, and preservation of genetic material gleaned via molecular autopsies (Shade et al., 2019; Weil, 2023). Protecting the deceased person's privacy and averting any genetic prejudice may require anonymizing and de-identifying the genetic data. Careful ethical consideration is needed to strike a balance between protecting the family's right to privacy and the significance of sharing pertinent genetic information with them (Shade et al., 2019; Weil, 2023). Additionally, there may be ethical issues with the use of molecular autopsy in specific cultural and religious contexts. There are many distinct customs and beliefs among various cultures and religions about postmortem care, including genetic testing (Shade et al., 2019; Weil, 2023). When considering the application of molecular autopsy, these practices and beliefs must be observed. Molecular autopsy can be made to conform to cultural norms and values by using culturally sensitive communication techniques, including religious or cultural leaders in conversations, and offering substitute genetic testing choices, when applicable (Shade et al., 2019; Weil, 2023).

Fair access to molecular autopsy is an additional ethical factor. A molecular autopsy's sophisticated technologies and genetic testing can be costly and out of reach for some groups of people or those with low incomes. This gives rise to worries over possible disparities in the availability of precise diagnosis, treatment, and preventative measures (Riepert, 2010; Shade et al., 2019; Weil, 2023). It should be a priority to guarantee that molecular autopsy is accessible, affordable, and available to everyone, irrespective of their geographic location or socioeconomic standing. To solve these issues and advance fair access to molecular autopsy, cooperation between researchers, legislators, funding organizations, and healthcare practitioners is necessary (Riepert, 2010; Shade et al., 2019; Weil, 2023). Finally, it is necessary to think about the ethical ramifications of unintentional discoveries made during molecular autopsy. Unexpected genetic discoveries that are unrelated to the cause of death but could have a substantial impact on the health of surviving family members or future generations are referred to as incidental findings. When disclosing such incidental results, guidelines should be developed that take into account the family's desires, the availability of successful interventions, and the possible advantages and dangers (Basso et al., 2017). For the field of molecular autopsy, there is still an ethical difficulty in striking a balance between the need to deliver valuable information for medical management and the possible psychological impact on persons receiving unintentional discoveries (Basso et al., 2017). In conclusion, a number of ethical issues need to be taken into account, even if molecular autopsy has great potential to reveal the genetic origins of unexpected, inexplicable deaths.

These factors include getting informed permission, protecting genetic information's privacy and confidentiality, honoring cultural and religious values, fostering equitable access, and handling incidental findings. In order to ensure that molecular autopsy is conducted responsibly and ethically, ethical guidelines and frameworks must be established. This will help to ensure that the deceased person's autonomy, privacy, and cultural values are respected and that the benefits to their surviving family members and future generations are maximized.

20.9 Future directions and advancements in molecular autopsy

Molecular autopsy is a new and exciting method in forensic pathology that has the potential to completely change how sudden and unexplained deaths are handled. Through the application of cutting-edge molecular methods like PCR and NGS, researchers have been able to identify the underlying genetic variables causing these deaths (Campuzano & Sarquella-Brugada, 2023; Castiglione et al., 2021; Heathfield et al., 2018; Müllauer, 2021; Yamamoto et al., 2024). This helps identify potential inheritable disorders that may affect surviving family members in addition to offering insightful information about the cause of death. Looking ahead, molecular autopsy has several important future paths and breakthroughs that have the potential to greatly improve its diagnostic skills. Extending the use of genetic testing beyond the traditional panel-based method is one of the main areas of focus. Panel tests have limitations in their capacity to uncover novel or uncommon mutations, even though they are useful in identifying known genetic variants linked to particular illnesses. WES and whole-genome sequencing are being investigated as more thorough methods to find both known and novel genetic variations to get around this restriction.

Furthermore, it is anticipated that the advancement of molecular autopsy would be greatly aided by the incorporation of machine-learning techniques and bioinformatics tools. Accurately identifying probable disease-causing genetic variations requires quick processing, analysis, and interpretation of the massive amount of data provided by sequencing technology. Researchers hope to anticipate the pathogenicity of genetic mutations and automate variant detection by creating advanced machine-learning models and bioinformatics pipelines. In addition to saving time, this integration will increase the precision and effectiveness of molecular autopsy (Buja et al., 2019). The application of multi-omics techniques is a key avenue for future development in molecular autopsy. A more thorough knowledge of the molecular pathways behind sudden death can be obtained by including additional omics data, such as transcriptomics, proteomics, and metabolomics, in addition to the usual focus on DNA genomic analysis (Buja et al., 2019). By combining numerous omics datasets, scientists will be able to find changes in protein levels, gene expression, and metabolic pathways in addition to genetic variants, providing a more comprehensive understanding of the disease processes.

Furthermore, there is a lot of potential for increasing the diagnostic accuracy of molecular autopsy through the creation of new biomarkers. These days, the majority of molecular autopsy studies concentrate on finding genetic variations linked to arrhythmias, channelopathies, and cardiovascular illnesses. On the other hand, adding circulating tumor DNA, microRNAs, and epigenetic markers to the repertoire of biomarkers can improve the identification of other hereditary illnesses, such as cancer and neurological diseases. The diagnosis of disorders that were previously undetectable by traditional genetic testing alone may be aided by these developing biomarkers (Giard & Van Den, 1999; Horowitz & Naritoku, 2007). Moreover, the broad use of molecular autopsy depends on its incorporation into standard medicolegal inquiries. Establishing standard operating protocols and recommendations for the application of molecular autopsy in postmortem examinations requires collaborations between forensic pathologists, geneticists, bioinformaticians, and other pertinent stakeholders (Giard & Van Den, 1999; Horowitz & Naritoku, 2007). This entails specifying the proper test indications, standardizing lab procedures, protecting the privacy and security of data, and educating the forensic industry and legal community.

In conclusion, there are a number of fascinating new paths and developments in the field of molecular autopsy that bode well for the field's future. To further enhance molecular autopsy diagnostic skills, it is imperative to expand the scope of genetic testing, integrate bioinformatics and machine learning, incorporate multi-omics techniques, discover novel biomarkers, and ensure wider application. Molecular autopsy has the potential to be a vital tool in forensic pathology as technology advances, offering priceless insights into the genetic reasons driving unexpected and inexplicable deaths.

20.10 Conclusions

To sum up, forensic pathologists face a great deal of difficulty when dealing with SUD cases because conventional autopsy techniques frequently leave the cause of death unclear. However, there is now renewed hope for figuring out the underlying reasons for these unfortunate events, thanks to the development of molecular autopsy as a useful tool for SUD case investigations. An outline of the molecular autopsy technique has been given in this chapter, along with an emphasis on its possible advantages and difficulties in determining the cause of death in SUD cases. The capacity of molecular autopsy to examine genetic, metabolic, and epigenetic elements—aspects that conventional autopsies are unable to completely capture—is one of its primary benefits. Through the application of sophisticated molecular methods like gene expression profiling, NGS, and DNA methylation analysis, scientists can identify genetic variants, mutations, or epigenetic alterations that were previously unidentified and may be involved in SUD.

By analyzing an individual's genetic and metabolic composition in great detail, these molecular approaches enable researchers to find possible genetic markers or biomarkers linked to SUD. This information can aid in the diagnosis of the cause of death in particular situations, as well as advance our understanding of the causes of unexpected, unexplained deaths in general. Moreover, molecular autopsy can be used to find possible therapeutic targets for the treatment and prevention of SUD. Nevertheless, there are certain unique difficulties associated with the molecular autopsy method. First and foremost, specific knowledge and resources are needed for the investigation of genetic, biochemical, and epigenetic variables. Consequently, not all forensic laboratories will be able to fully adopt this strategy, particularly in environments with limited resources. Furthermore, interpreting the data produced by molecular techniques calls for an interdisciplinary approach combining specialists in pathology, genetics, bioinformatics, and other disciplines.

The ethical issues surrounding genetic analysis and informing families of results present another difficulty. It is unclear how to address and educate the deceased person's family members about any potential familial or hereditary problems that may be revealed by genetic information gleaned by molecular autopsy. As a result, family screening and genetic counseling are essential to the thorough assessment of SUD cases. To sum up, the discipline of molecular autopsy has a lot of potential for determining the cause of death in unexpected deaths. This method provides a more thorough understanding of the underlying etiology of SUD by combining genetic, biochemical, and epigenetic analysis. However, more study, standardization of procedures, and cooperation between specialists in other fields are required to fully realize the promise of molecular autopsy.

Going ahead, initiatives should be taken to improve molecular autopsy accessibility and viability for forensic labs across the globe. This can be accomplished through developing policies and procedures for molecular autopsies, imparting molecular methods training to researchers and forensic pathologists, and enhancing the accessibility and cost-effectiveness of molecular diagnostic instruments. In addition, comprehensive genetic counseling and family screening should be the primary focus of thoughtful consideration of the ethical concerns of genetic analysis and information disclosure. This will make it possible to guarantee that people and their families get the assistance and direction they require in order to comprehend and take care of any potential genetic disorders.

To sum up, the molecular autopsy method has the potential to completely change how sudden, inexplicable deaths are investigated. We may learn vital information about the reasons behind unexpected, inexplicable deaths by figuring out the genetic, physiological, and epigenetic elements that underlie these sad occurrences.

References

Alqarawi, W., Alsaeed, A., Alghamdi, M., Hersi, A., Kashour, T., & Aljerian, K. (2022). Medical autopsy for sudden unexplained death in Saudi Arabia: A call to action. *Journal of the Saudi Heart Association, 34*(3), 142–147. https://doi.org/10.37616/2212-5043.1310.

Alzahrani, S. A., Alswaimil, N. F., Alammari, A. M., Al Saeed, W. H., & Menezes, R. G. (2023). Postmortem genetic testing in sudden unexpected death: A narrative review. *Cureus*. https://doi.org/10.7759/cureus.33728.

Bagnall, R. D., Das K, J., Duflou, J., & Semsarian, C. (2014). Exome analysis-based molecular autopsy in cases of sudden unexplained death in the young. *Heart Rhythm: The Official Journal of the Heart Rhythm Society, 11*(4), 655–662. https://doi.org/10.1016/j.hrthm.2014.01.017, http://www.elsevier.com/inca/publications/store/7/0/2/3/3/3/index.htt.

Basso, C., Aguilera, B., Banner, J., Cohle, S., d'Amati, G., de Gouveia, R. H., di Gioia, C., Fabre, A., Gallagher, P. J., Leone, O., Lucena, J., Mitrofanova, L., Molina, P., Parsons, S., Rizzo, S., Sheppard, M. N., Mier, M. P. S., Kim Suvarna, S., Thiene, G., ... Michaud, K. (2017). Guidelines for autopsy investigation of sudden cardiac death: 2017 update from the Association for European Cardiovascular Pathology. *Virchows Archiv, 471*(6), 691–705. https://doi.org/10.1007/s00428-017-2221-0, https://link.springer.de/link/service/journals/00428/index.htm.

Bonilla, J. C., Parra-Medina, R., Chaves, J. J., Campuzano, O., Sarquella-Brugada, G., Brugada, R., & Brugada, J. (2018). Molecular autopsy in sudden cardiac death. *Archivos de Cardiologia de Mexico, 88*(4), 306–312. https://doi.org/10.1016/j.acmx.2018.06.001, http://www.elsevier.es/es-revista-archivos-cardiologia-mexico-293-pdf-S1405994018300697.

Buja, L. M., Barth, R. F., Krueger, G. R., Brodsky, S. V., & Hunter, R. L. (2019). The importance of the autopsy in medicine: Perspectives of pathology colleagues. *Academic Pathology, 6*. https://doi.org/10.1177/2374289519834041.

Campuzano, O., & Sarquella-Brugada, G. (2023). Autopsy in sudden cardiac death. *Global Cardiology Science and Practice, 2023*(1), 2023.

Castiglione, V., Modena, M., Aimo, A., Chiti, E., Botto, N., Vittorini, S., Guidi, B., Vergaro, G., Barison, A., Rossi, A., Passino, C., Giannoni, A., Di Paolo, M., & Emdin, M. (2021). Molecular autopsy of sudden cardiac death in the genomics era. *Diagnostics, 11*(8), 1378. https://doi.org/10.3390/diagnostics11081378.

Conrad, P., & Barker, K. K. (2010). The social construction of illness: Key insights and policy implications. *Journal of Health and Social Behavior, 51*(1), S67. https://doi.org/10.1177/0022146510383495, http://hsb.sagepub.com/archive/.

Farrugia, A., Keyser, C., Hollard, C., Raul, J. S., Muller, J., & Ludes, B. (2015). Targeted next generation sequencing application in cardiac channelopathies: Analysis of a cohort of autopsy-negative sudden unexplained deaths. *Forensic Science International, 254*, 5–11. https://doi.org/10.1016/j.forsciint.2015.06.023, https://www.elsevier.com/locate/forsciint.

Giard, R. W. M., & Van Den, J. G. (1999). Tweel, De waarheid na de dood. *Nederlands Tijdschrift Voor Geneeskunde, 143*(47), 2345–2347.

Ginsberg, S. D., & Mirnics, K. (2006). *Functional genomic methodologies* (Vol. 815). Elsevier BV, 15–40. https://doi.org/10.1016/s0079-6123(06)58002-1.

Heathfield, L., Martin, L., & Ramesar, R. (2018). A systematic review of molecular autopsy studies in sudden infant death cases. *Journal of Pediatric Genetics, 07*(04), 143–149. https://doi.org/10.1055/s-0038-1668079.

Horowitz, R. E., & Naritoku, W. Y. (2007). The autopsy as a performance measure and teaching tool. *Human Pathology, 38*(5), 688. https://doi.org/10.1016/j.humpath.2007.01.001, http://www.elsevier.com/inca/publications/store/6/2/3/1/3/9/index.htt.

Hudzik, B., Hudzik, M., Lekston, A., & Gasior, M. (2018). Sudden unexplained cardiac deaths in young adults: A call for multidisciplinary approach. *Acta Cardiologica, 73*(1), 7–12. https://doi.org/10.1080/00015385.2017.1351234, http://www.tandfonline.com/loi/tacd20.

Khalikov, A. A., Kuznetsov, K. O., Iskuzhina, L. R., & Khalikova, L. V. (2021). Forensic aspects of sudden autopsy-negative cardiac death. *Federation Sudebno-Meditsinskaya Ekspertiza, 64*(3), 59–63. https://doi.org/10.17116/sudmed20216403159, mediasphera.ru/journal/sudebno-meditsinskaya-ekspertiza.

Klassen, T. L., Von Rden, E. L., Drabek, J., Noebels, J. L., & Goldman, A. M. (2012). Comparative analytical utility of DNA derived from alternative human specimens for molecular autopsy and diagnostics. *Journal of Molecular Diagnostics, 14*(5), 451–457. https://doi.org/10.1016/j.jmoldx.2012.04.005.

Kumar, A., Avishay, D. M., Jones, C. R., Shaikh, J. D., Kaur, R., Aljadah, M., Kichloo, A., Shiwalkar, N., & Keshavamurthy, S. (2021). Sudden cardiac death: Epidemiology, pathogenesis and management. *Reviews in Cardiovascular Medicine, 22*(1), 147–158. https://doi.org/10.31083/J.RCM.2021.01.207, https://rcm.imrpress.com/EN/10.31083/j.rcm.2021.01.207.

Markwerth, P., Bajanowski, T., Tzimas, I., & Dettmeyer, R. (2021). Sudden cardiac death—Update. *International Journal of Legal Medicine, 135*(2), 483–495. https://doi.org/10.1007/s00414-020-02481-z, https://link.springer.de/link/service/journals/00414/index.htm.

Martínez-Barrios, E., Grassi, S., Brión, M., Toro, R., Cesar, S., Cruzalegui, J., Coll, M., Alcalde, M., Brugada, R., Greco, A., Ortega-Sánchez, M. L., Barberia, E., Oliva, A., Sarquella-Brugada, G., & Campuzano, O. (2023). Molecular autopsy: Twenty years of post-mortem diagnosis in sudden cardiac death. *Frontiers in Medicine, 10*. https://doi.org/10.3389/fmed.2023.1118585, https://journal.frontiersin.org/journal/medicine.

Methner, D. N. R., Scherer, S. E., Welch, K., Walkiewicz, M., Eng, C. M., Belmont, J. W., Powell, M. C., Korchina, V., Doddapaneni, H. V., Muzny, D. M., Gibbs, R. A., Wolf, D. A., Sanchez, L. A., & Kahn, R. (2016). Postmortem genetic screening for the identification, verification, and reporting of genetic variants contributing to the sudden death of the young. *Genome Research, 26*(9), 1170–1177. https://doi.org/10.1101/gr.195800.115, http://genome.cshlp.org/content/26/9/1170.full.pdf+html.

Michaud, K., Basso, C., d'Amati, G., Giordano, C., Kholová, I., Preston, S. D., Rizzo, S., Sabatasso, S., Sheppard, M. N., Vink, A., & van der Wal, A. C. (2020). Diagnosis of myocardial infarction at autopsy: AECVP reappraisal in the light of the current clinical classification. *Virchows Archiv, 476*(2), 179–194. https://doi.org/10.1007/s00428-019-02662-1, https://link.springer.de/link/service/journals/00428/index.htm.

Michaud, K., Fellmann, F., Abriel, H., Beckmann, J. S., Mangin, P., & Elger, B. S. (2009). Molecular autopsy in sudden cardiac death and its implication for families: Discussion of the practical, legal and ethical aspects of the multidisciplinary collaboration. *Swiss Medical Weekly, 139*(49-50), 712–718. http://www.smw.ch/docs/PdfContent/smw-12837.pdf.

Müllauer, L. (2021). Molecular pathology of cancer: The past, the present, and the future. *Journal of Personalized Medicine, 11*(7), 676. https://doi.org/10.3390/jpm11070676.

Neubauer, J., Kissel, C. K., Bolliger, Barbon, D., Thali, M. J., Kloiber, D., Bode, P. K., Kovacs, B., Graf, U., Maspoli, Berger, W., Saguner, A. M., & Haas, C. (2022). Benefits and outcomes of a new multidisciplinary approach for the management and financing of sudden unexplained death cases in a forensic setting in Switzerland. *Forensic Science International.*

Neubauer, J., Wang, S., Russo, G., & Haas, C. (2021). Re-evaluation of single nucleotide variants and identification of structural variants in a cohort of 45 sudden unexplained death cases. *International Journal of Legal Medicine, 135*(4), 1341–1349. https://doi.org/10.1007/s00414-021-02580-5, https://link.springer.de/link/service/journals/00414/index.htm.

Orlíčková, J., Zeman, M., Vojtíšek, T., & Slabý, O. (2022). The importance of molecular autopsy in forensic medicine. *Casopis Lekaru Ceskych, 161*(5), 207–211. http://www.prolekare.cz/casopis-lekaru-ceskych.

Ottaviani, G., & Ramos, S. G. (2023). Autopsy for medical diagnostics: Finding the cause of sudden unexpected death through investigation of the cardiac conduction system by serial sections. *Diagnostics, 13*(11), 1919. https://doi.org/10.3390/diagnostics13111919.

Reid, K. M., Maistry, S., Ramesar, R., & Heathfield, L. J. (2017). A review of the optimisation of the use of formalin fixed paraffin embedded tissue for molecular analysis in a forensic post-mortem setting. *Forensic Science International, 280*, 181–187. https://doi.org/10.1016/j.forsciint.2017.09.020, https://www.elsevier.com/locate/forsciint.

Riepert, T. (2010). Zur Rechtsmedizin und ihren Berührungspunkten mit der Pathologie. *Der Pathologe, 31*(4), 248–255. https://doi.org/10.1007/s00292-010-1284-3.

Risgaard, B. (2016). Sudden cardiac death: A nationwide cohort study among the young. *Danish Medical Journal, 63*(12). http://www.danmedj.dk/portal/pls/portal/!PORTAL.wwpob_page.show?_docname=11379505.PDF&pTitle=Sudden%20cardiac%20death:%20A%20nationwide%20cohort%20study%20among%20the%20young.

Semsarian, C., & Hamilton, R. M. (2012). Key role of the molecular autopsy in sudden unexpected death. *Heart Rhythm: the Official Journal of the Heart Rhythm Society, 9*(1), 145–150. https://doi.org/10.1016/j.hrthm.2011.07.034, http://www.elsevier.com/inca/publications/store/7/0/2/3/3/3/index.htt.

Semsarian, C., & Ingles, J. (2016). Molecular autopsy in victims of inherited arrhythmias. *Journal of Arrhythmia, 32*(5), 359–365. https://doi.org/10.1016/j.joa.2015.09.010, http://www.elsevier.com/wps/find/journaldescription.cws_home/727758/description#description.

Semsarian, C., Ingles, J., & Wilde, A. A. M. (2015). Sudden cardiac death in the young: The molecular autopsy and a practical approach to surviving relatives. *European Heart Journal, 36*(21), 1290–1296. https://doi.org/10.1093/eurheartj/ehv063, http://eurheartj.oxfordjournals.org/.

Shade, J., Coon, H., & Docherty, A. R. (2019). Ethical implications of using biobanks and population databases for genetic suicide research. *American Journal of Medical Genetics, Part B: Neuropsychiatric Genetics, 180*(8), 601–608. https://doi.org/10.1002/ajmg.b.32718, http://onlinelibrary.wiley.com/journal/10.1002/(ISSN)1552-485X.

Spears, D. A., & Gollob, M. H. (2015). Genetics of inherited primary arrhythmia disorders. *Application of Clinical Genetics, 8*, 215–233. https://doi.org/10.2147/TACG.S55762, https://www.dovepress.com/getfile.php?fileID=27097.

Steinberg, C., Laksman, Z. W. M., & Krahn, A. D. (2016). Sudden cardiac death: A reappraisal. *Trends in Cardiovascular Medicine, 26*(8), 709–719. https://doi.org/10.1016/j.tcm.2016.05.006, https://www.elsevier.com/locate/tcm.

Sun, Y. X., Gong, X. J., Hao, X. L., Tian, Y. X., Chen, Y. M., Zhang, B., & Yan, C. X. (2023). Screening of genes co-associated with sudden infant death syndrome and infectious sudden death in infancy and bioinformatics analysis of their regulatory networks. *Journal of Forensic Medicine, 39*(5), 433–440. https://doi.org/10.12116/j.issn.1004-5619.2022.420803, http://www.fyxzz.cn/EN/1004-5619/home.shtml.

Tashjian, R. S., Williams, R. R., Vinters, H. V., & Yong, W. H. (2019). Autopsy biobanking: Biospecimen procurement, integrity, storage, and utilization. *Methods in Molecular Biology, 1897*, 77–87. https://doi.org/10.1007/978-1-4939-8935-5_8, http://www.springer.com/series/7651.

Tester, D. J., & Ackerman, M. J. (2012). The molecular autopsy: Should the evaluation continue after the funeral? *Pediatric Cardiology, 33*(3), 461–470. https://doi.org/10.1007/s00246-012-0160-8.

Tester, D. J., & Ackerman, M. J. (2006). The role of molecular autopsy in unexplained sudden cardiac death. *Current Opinion in Cardiology, 21*(3), 166–172. https://doi.org/10.1097/01.hco.0000221576.33501.83.

Turkgenc, B., Baydar, C. L., Deniz, I., Akcay, A., Ergoren, M. C., Sag, S. O., Yakicier, M. C., & Temel, S. G. (2023). From death to life/back to the future: Detailed premorbid clinical and family history can save lives and address the final diagnosis in sudden unexplained deaths with negative autopsy. *Applied Immunohistochemistry and Molecular Morphology, 31*(10), 690–696. https://doi.org/10.1097/PAI.0000000000001163, http://www.appliedimmunohist.com.

Weil, C. J. (2023). Ethical, legal, and policy issues surrounding biospecimen research conducted or supported in the USA. *Biopreservation and Biobanking, 21*(1), 14–22. https://doi.org/10.1089/bio.2021.0094, http://www.liebertonline.com/cpt.

Yamamoto, T., Emoto, Y., Murase, T., Umehara, T., Miura, A., Nishiguchi, M., Ikematsu, K., & Nishio, H. (2024). Molecular autopsy for sudden death in Japan. *Journal of Toxicologic Pathology, 37*(1), 1–10. https://doi.org/10.1293/tox.2023-0080, https://www.jstage.jst.go.jp/article/tox/37/1/37_2023-0080/_pdf.

Chapter 21

Artificial intelligence and machine learning in forensic DNA analysis

Sarah J.M. Alley, and Kelly M. Elkins
Chemistry and Forensic Science Department, Towson University, Towson, MD, United States

21.1 Introduction

In recent years, there has been an increasing interest in the use of artificial intelligence (AI) and machine learning (ML) across a wide range of disciplines ranging from marketing to medicine due to its ability to quickly analyze large volumes of data while recognizing patterns and making predictions (Collins et al., 2021; Xu et al., 2021). In this chapter, we define AI and ML and review the applications of these tools in forensic science and DNA analysis. Although the use of AI is still in its infancy in forensic science, it shows great potential in advancing various facets of forensic science, including DNA analysis, forensic serology (body fluid identification), forensic medicine, and the analysis of digital evidence.

21.2 What is artificial intelligence?

The term "artificial intelligence" was coined by John McCarthy at Dartmouth College in 1956 as "the science of engineering and making intelligent machines" (Collins et al., 2021). Despite the lack of a single definition which is widely accepted by the scientific community, today's AI involves the use of computer systems and algorithms which can "generate novel plans to achieve goals," including the transformation of complex data into conclusions or predictions (Collins et al., 2021; Jarrett & Choo, 2021).

Early researchers focused on the use of AI to solve mathematical problems and, spurred by their success, were perhaps overly optimistic regarding how quickly human-level intelligence in machines could be a reality. Major issues remained and could not be overcome quickly. As a result of underperforming AI programs, much of the AI funding was cut until the 1980s, when universities began exploring the potential of AI to make decisions based on a set of rules, known as "expert systems." Major shortcomings of this method led to a second low point in the development of AI until 2006, when Geoffrey Hinton and his team made a breakthrough in building neural networks and once again sparked major interest in the field of AI (Xu et al., 2021).

Conventional methods of building models consider relationships between all elements of a system to accurately generate predictions. Data is collected on the natural world or engineered systems and modeled using traditional statistical methods. Coefficients can be used to account for the influence of unknown factors which cause the model to deviate from expected values. In expert systems using AI and ML, models can be built using large amounts of non-systematic data. For some datasets, relationships between variables may be unknown, and the amount of data and potential variables may be large. ML methods and modern computational resources are especially useful for the analysis and emergence of trends of large datasets in which relationships and patterns are unknown (Barash et al., 2024).

AI has the potential to automate and streamline data analysis in forensic science. The analysis of large amounts of data manually is time-consuming, challenging, and prone to error (Barash et al., 2024). The use of AI offers a method to process large amounts of text and image data in a short amount of time, reducing human subjectivity while improving accuracy, reproducibility, and computational power (Barash et al., 2024; Galante et al., 2022). Broadly, uses of AI in forensic science include DNA analysis, forensic serology (body fluid identification), forensic medicine, and the analysis of digital evidence (Fig. 21.1).

Investigative and Predictive DNA Testing. DOI: https://doi.org/10.1016/B978-0-443-30128-5.00015-1

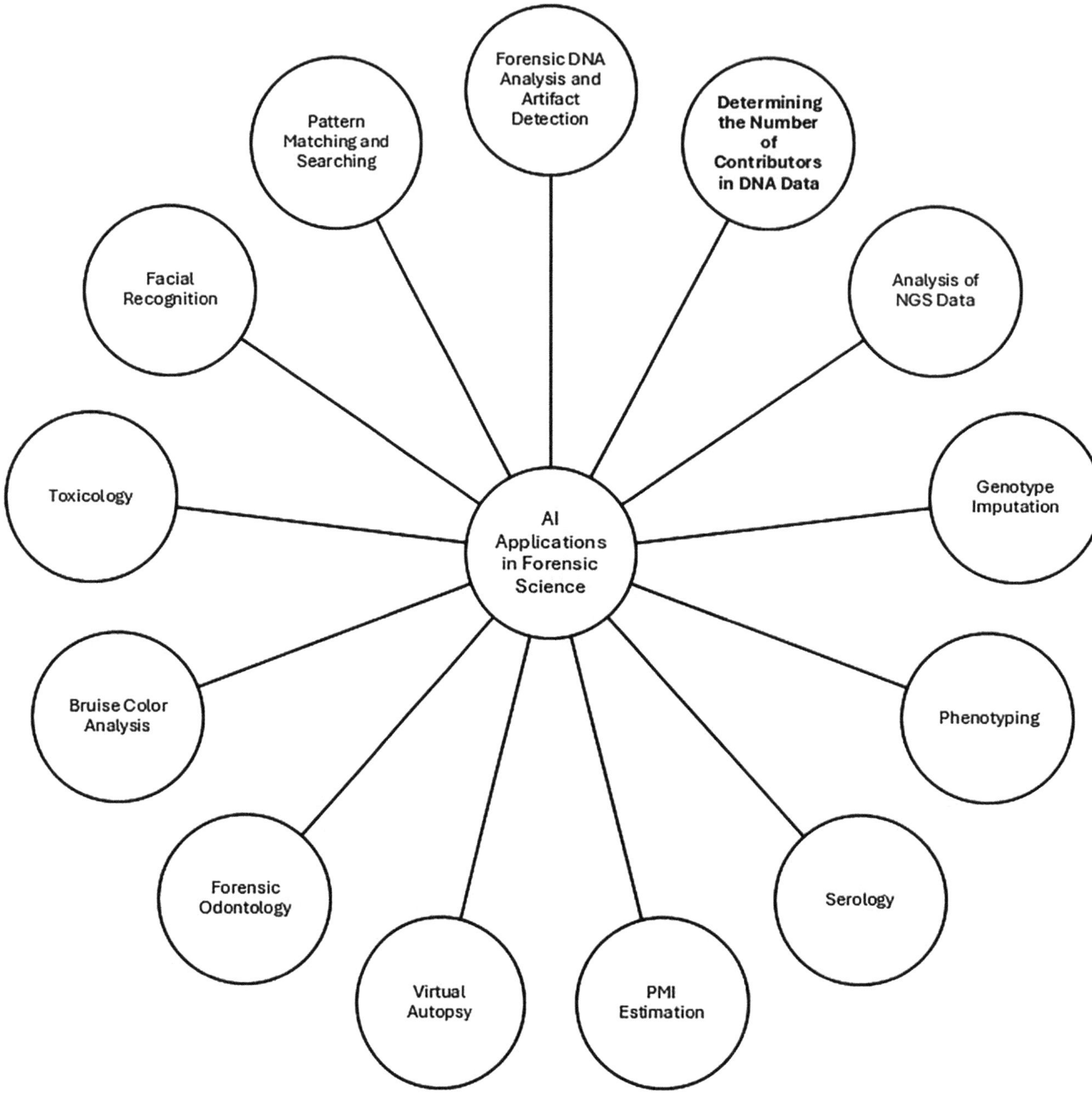

FIGURE 21.1 Applications of artificial intelligence in forensic analysis.

21.3 Understanding artificial intelligence: machine learning versus deep learning

AI is an umbrella term that includes ML and deep learning. ML is a tool that can be fed large amounts of data and use it to make predictions or conclusions (Jarrett & Choo, 2021). ML can be either supervised or unsupervised; the difference is whether the input data is linked to an outcome. For example, media streaming services generate recommendations in an unsupervised fashion when they are based on what the viewer previously watched, but the algorithm does not receive data on whether the viewer liked or disliked the previously watched content. Supervised recommendations, on the other hand, have an input from the viewers indicating whether they liked or disliked content. This data can then be used by the algorithm to predict more accurately content the viewer will like (Jarrett & Choo, 2021). Unsupervised principal component analysis is an example of ML. Deep learning is a subset of ML that relies on the use of artificial neural networks (ANNs) to understand complex data and verify what it has learned (Jarrett & Choo, 2021).

21.4 Artificial intelligence in forensic DNA analysis

Since its inception in the 1980s using restriction fragment length polymorphism analyses, the analysis of genetic material deposited at crime scenes has seen remarkable progress. Rapid technological developments in the late 1990s and early 2000s, namely the adoption of short tandem repeat (STR) analyses but also sequencing, have established the individualizing power of DNA analyses as a compelling form of evidence and a critical facet of forensic science (Jordan & Mills, 2021).

21.5 Artificial intelligence in forensic DNA analysis: electropherogram analysis and artifact detection

Accurate electropherogram analysis is critical for the proper interpretation of STR profiles. In addition to identifying fluorescence peaks from electropherograms as STR amplicons (true alleles), it is important to identify peaks that are not true alleles, including stutter peaks as a result of strand slippage of the *Taq* DNA polymerase during extension and pull-up as a result of software incorrectly assigning fluorescence from an oversaturated peak to another color channel. AI approaches to artifact detection can use all available information to make calls as opposed to using fixed threshold values, thereby reducing situations where artifacts are incorrectly labeled as true allele peaks or true allele peaks are discarded (Barash et al., 2024).

A number of studies have demonstrated the potential of AI to automate artifact detection in electropherograms. Marciano et al. (2018) developed a method of artifact removal that utilized four computational models to extract a final DNA profile from each locus with an overall accuracy of 97.2% at identifying true allele peaks. Adelman et al. (2019) developed a method of detecting pull-ups using a symbolic regression model that can correct peak heights when pull-ups are present for true alleles. Lin et al. (2021) developed the FaSTR DNA analysis software, which uses ANN models and was found to be 99.7% accurate in detecting true allele peaks.

21.6 Artificial intelligence in forensic DNA analysis: electropherogram analysis and determining the number of contributors

Manual interpretation and deconvolution of electropherograms from multiple contributors can prove extremely challenging, especially if one individual only contributed a very small amount to the mixture or if the DNA has been degraded to varying degrees. True allele peaks may be confused with artifacts, and a number of contributor estimates can be difficult. In 2015, Swaminathan et al. (2015) developed an ML approach called NOCIt, which calculates the a posteriori probability (APP) of the number of contributors. It is based on single-source calibration data consisting of known genotypes and considers a wide range of parameters, including peak height, degradation, stutter, noise, dropout, and other parameters to infer the APP for a sample of unknown contributor numbers. Probabilistic genotyping software, such as STRMix and TrueAllele, incorporates factors such as expected allele frequencies and stutter peak heights at a given locus to deconvolute mixtures. AI-based methods have also been applied to identify a number of contributors from next-generation sequencing (NGS) data (Benschop et al., 2019; Crysup et al., 2023).

21.7 Artificial intelligence in forensic DNA analysis: next-generation sequencing data

Length-based analyses, such as STR analysis, are typically used for individual identification in forensic science rather than analyzing the DNA sequence itself. However, the introduction of NGS presents new possibilities for sequencing-based approaches in forensics. Compared to traditional Sanger sequencing, which can only process one nucleotide sequence at a time, NGS has the capability to sequence billions of nucleic acid molecules simultaneously and rapidly in a massively parallel fashion. This type of sequencing produces immense quantities of data, which require more sophisticated bioinformatics systems to develop DNA profiles than is necessary for the interpretation of traditional capillary electrophoresis data, STR electropherograms. AI-based approaches for NGS data analysis offer methods to wrangle these large volumes of data into a more meaningful form. Liu et al. (2020) developed Fragsifier, an ML tool that can extract STR sequences of interest within raw NGS data based on a random forest (RF) modeling approach to locate flanking sequences and the repeats themselves. Yang et al. (2022) developed a method of analyzing NGS data of single-nucleotide polymorphisms (SNPs) to aid in the interpretation of mixtures.

21.8 Artificial intelligence in forensic DNA analysis: genotype imputation

Investigative genetic genealogy (IGG) is a method used to generate investigative leads by searching for relatives of samples of unknown provenance and sufficient quality that do not yield a match when their STR profile is run through Combined DNA Index System (CODIS). CODIS contains STR profiles of known offenders, biological evidence found at crime scenes, missing persons, and unidentified remains (Butler, 2005). IGG relies on SNPs, or differences of a single nucleotide in otherwise conserved DNA sequences, and is used by genealogy companies to generate family trees (Wickenheiser, 2019).

Database algorithms used by GEDMatch compare the SNP profile of a sample with an unknown donor to the SNP profiles of individuals who have provided their DNA to direct-to-consumer genetic testing platforms and voluntarily shared their genetic information with law enforcement. These are individuals who have generally placed their genetic information in a public database for the purpose of locating a missing relative. The algorithm searches for matching strings and characterizes relationships based on different lengths and sequences of shared DNA. From this, family trees can be generated. Notably, IGG was used to solve the cold case of the Golden State Killer in 2018 and to generate investigative leads in the 2022 University of Idaho killings (Fixler, 2024; Wickenheiser, 2019).

While STR analyses are based on how many repeats are present at a given locus and a profile for a given individual can be generated based on the length of specific DNA regions detected via capillary electrophoresis, SNP analyses are based on single base differences found throughout the human genome; therefore, a sequencing-based approach is necessary, often whole-genome sequencing (Tillmar et al., 2021). Microarray is a useful technique for generating genotypes for SNP analyses; however, it performs best with a large input of DNA. Evidentiary DNA samples collected from crime scenes often contain limited amounts of DNA and are often degraded. As a result, there are often missing SNPs in an individual's genotype (Tillmar et al., 2021). Genotype imputation is a critical tool in gaining useful information from these challenging samples for investigative leads by inferring what base is present when an SNP is missing in the sequence based on which bases are typically present upstream and downstream (Tillmar et al., 2021).

The main principle behind genotype imputation is that two individuals, who may not even necessarily be related, may share segments of DNA from a common ancestor with genetic variants (such as SNPs), which are conserved through generations. Shared segments can be identified by comparing genetic variants in the sample with variants from a reference panel, and based on the reference, the software has the capacity to predict the missing SNP and generate leads that can be verified by other methods (Tillmar et al., 2021). Although genotype imputation was successfully used by the Swedish Police in 2020 to solve the double murder of a woman and a young boy using Beagle 5.1 software to enhance the profile of an unknown individual found in a mixture with the victim's DNA on the murder weapon, AI is still in its infancy for performing these types of forensic analyses (Tillmar et al., 2021).

The use of AI for genotype imputation of SNPs in high-throughput sequencing data has largely been contained in genome-wide association studies to determine interactions between genes. SNP imputation is well-suited to predictive models to impute missing data, as most SNPs are dependent on one another. In 2008, Sun and Karida (2008) developed the fastPHASE algorithm, which used neural networks to impute missing genotype data with 97% accuracy. In another paper, Wang et al. (2012) presented a method for imputing missing SNPs using nearest-neighbor and weighted nearest-neighbor approaches.

21.9 Artificial intelligence in forensic DNA analysis: suspect phenotyping

Forensic DNA phenotyping (FDP) uses genetic data to predict externally visible characteristics of the donor (e.g., hair color, skin color, and eye color). These characteristics are a valuable tool for generating investigative leads for suspects, victims, or unidentified remains from trace tissue or partial human remains. Traditionally, statistical methods based on multinomial logistic regression were used to develop these predictions and serve as the basis for the IrisPlex (eye color), HIrisPlex (hair and eye color), and HIrisPlex-S (hair, eye, and skin color) systems (Katsara et al., 2021). Katsara et al. (2021) examined the effectiveness of three different ML approaches, ANNs, support vector machines, and RF, and found that ML methods do not currently offer an advantage over traditional multinomial regression models for predicting appearance.

21.10 Artificial intelligence in forensic serology

Despite the powerful evidentiary value of DNA alone to link an individual to the scene of a crime, forensic serology, or the identification of body fluids, provides additional strength to DNA evidence by associating a given individual with a particular fluid type. This information provides context regarding the circumstances under which the individual's DNA was deposited at a crime scene and can be used to corroborate witness accounts of events (Sijen & Harbison, 2021). Traditional methods for body fluid identification include chemical and antibody-based tests; however, the presence of body fluid-specific mRNA biomarkers is emerging as a method for confirming fluid types (Iacob et al., 2019). Iacob et al. (2019) used NGS of targeted mRNA biomarkers, specifically to develop a machine learning-based method of predicting the origin of both single-source samples and mixtures.

Another emerging method for the identification of body fluids is spectral methods, such as Raman spectroscopy and attenuated total reflectance (ATR) Fourier transform infrared (FTIR) spectroscopy, which produce characteristic profiles for a given substance based on how its chemical bonds interact with light (Elkins, 2011; Virkler & Lednev, 2008). The interpretation of spectral data visually is challenging, and often identifying fine-scale variation with the human eye is impossible (Khandasammy et al., 2018). AI-based approaches, including chemometrics, or the use of statistics and computer modeling to interpret spectral data, are critical to reveal the full extent of information that can be obtained from spectral data. Chemometric approaches have the ability not only to identify body fluids but to reveal fine-scale variations between samples of the same fluid type to reveal a wealth of information, including species and donor race (McLaughlin et al., 2014; Mistek et al., 2016).

21.11 Artificial intelligence in forensic medicine: microbial community data to determine postmortem interval

In addition to applications in forensic DNA analysis and serology, AI has been used to estimate postmortem interval (PMI). Microbial communities sampled at different time points after death can be sequenced and used to build predictive models to estimate PMI and minimum PMI. Determination of PMI is an essential part of forensic investigations involving suspicious and unattended deaths (Oliveira & Amorim, 2018). An accurate estimate of PMI can verify witness accounts and suspect alibis, provide validation for developing leads, and clarify the circumstances under which a person died (Belk et al., 2018; Oliveira & Amorim, 2018). Traditional methods to determine PMI involve the use of different techniques depending on the decomposition stage of the body, including *algor mortis*, the drop in body temperature that occurs after death, physicochemical changes known as *rigor mortis*, definable degenerative changes in the skin, changes in biochemical markers present in tissues and body fluids, and insect and scavenger activity (Belk et al., 2018; Metcalf, 2019; Roy et al., 2021). Many of these techniques are based on observation, resulting in subjective estimates that may vary from one practitioner to another. Even if performed with the highest degree of expertise, the estimation of PMI may be a wide time interval that may not provide the level of detail or certainty needed to validate witness accounts or develop leads (Díez López et al., 2022).

Diverse communities of microorganisms drive the decomposition of mammalian remains and undergo successional changes throughout the phases of decomposition, which are relatively consistent across environments and hosts (Metcalf, 2019). During life, microbes play a critical role in human health by mediating processes such as digestion and immune responses (Lutz et al., 2020). After death, microbial communities experience dramatic chemical and biological changes in the physical environment of the host (Hu et al., 2021). The immune system ceases to function, and within hours, gut microbes will migrate throughout the body via the circulatory and lymphatic systems, proliferate in other organs, and start the decomposition process (Oliveira & Amorim, 2018). The early stages of decomposition are characterized by a rapid shift from aerobic to anaerobic microbial communities as oxygen supplies in the body are depleted (Liu et al., 2023). The lack of oxygen and cessation of the circulatory system cause subsequent tissue autolysis, freeing up nutrients to fuel microbial activity and drive the decomposition process (Javan et al., 2019). Microbial fermentation activity results in gas production, causing the bloat stage of decomposition. The body may rupture due to gas pressure, which causes a shift of the microbial community back toward domination by aerobic species due to the influx of atmospheric oxygen. After rupture, the microbial load of the corpse is increased as microbes present flood in from the external environment (Dash & Das, 2020).

The advent of NGS has made such analyses of microbial communities possible. Before the introduction of NGS, most knowledge of the human microbiome was acquired from culture-based approaches, which, in addition to being time-consuming and labor-intensive, fail to capture the diversity of entire communities due to their inherent bias toward organisms that grow well in culture (Roy et al., 2021). NGS has revolutionized the field of microbial forensics by allowing for the rapid sequencing, identification, and analysis of whole microbial communities, including unknown and low-abundance microbes (Roy et al., 2021; Zhang et al., 2019). Traditional sequencing methods, such as Sanger sequencing, can only process one nucleotide sequence at a time, whereas NGS can sequence billions of nucleic acid fragments simultaneously and independently (Lee, 2019). Rapidly declining costs of NGS have allowed for the cost-effective analysis of large numbers of samples (Kuiper, 2016).

ML and modeling are critical in analyzing microbial community data obtained from NGS (Metcalf, 2019). Computer programs trained to identify patterns are commonly applied to microbiome datasets because of their large size and complexity. Additional considerations, such as environmental parameters, can be incorporated into the models to improve their predictive capacity (Metcalf, 2019).

Belk et al. (2018) evaluated criteria that contribute to robust RF models used to determine PMI. Shifts in no single microbial species are informative enough to estimate PMI; therefore, a comprehensive community approach is needed. ML models can be trained to learn relationships between microbial community composition and decomposition time points and use these criteria to estimate PMI. The authors tested different sample types, gene markers, and taxonomic levels to determine which criteria produced the most accurate RF models. Gene markers, including the 16S ribosomal RNA (rRNA) and 18S rRNA genes and the internal transcribed spacer region, were used to analyze skin and grave soil samples; the phyla that produced the most accurate models were reported. One disadvantage of RF modeling is poor predictive capacity at very early or late PMIs; this can be addressed by providing the model with additional datasets. RF models have the greatest resistance against overfitting and are well-suited to be applied to other datasets besides the one on which they were trained.

Zhang et al. (2019) evaluated the abilities of three ML methods (RF, gradient boosting, and neural network) to predict PMI, location of death, and manner of death. The models were trained on datasets of microbial community sequences derived from 16S rRNA sequencing and information regarding the sample and donor, including anatomical sampling location, sex, race, age, body mass, location, and manner of death to estimate PMI. The gradient boosting approach (Xgboost) produced the most accurate predictions (74.56%–87.6% accuracy) but was prone to overfitting. Overfitting occurs when a model performs very well for one dataset but fails to generate reliable predictions for other datasets because its training is too specific to the features of the original dataset. RF and neural network approaches performed similarly (73.6%–86.3% and 70.7%–83% accuracy, respectively), with RF modeling yielding the most stable predictions across anatomical sampling locations. The range in accuracy depends on whether the model was asked to determine PMI, manner of death, or location of death.

In summary, ML algorithms are invaluable to analyze postmortem microbial communities; however, general models that can generate predictions about microbiome data from any given specimen have not yet been developed. The predictive frameworks are currently only able to generate predictions using a dataset used in a specific study (Díez López et al., 2022).

21.12 Artificial intelligence in forensic medicine: virtual autopsy

AI has the potential to assist forensic pathologists in determining the cause of death by analyzing computer tomography and magnetic resonance imaging scans of victims and comparing these images to a database to locate injuries, inflammation, and provide information on the trajectories of bullets and weapons. Increasing the capabilities of virtual autopsies through image analysis by AI provides an approach for increasing autopsy acceptance within cultures that oppose autopsies due to religious or ethical reasons. Additionally, AI-assisted virtual autopsies have the potential to detect the cause of death in most cases while reducing the costs and time associated with a conventional autopsy (O'Sullivan et al., 2017). Additional benefits include reduced risk of disease due to not cutting (Al Matrouk & Dashti, 2025) as well as virtual staining and facial recognition tools.

21.13 Artificial intelligence in forensic medicine: forensic odontology

Forensic odontology utilizes structures in the oral cavity to identify individuals, which is a useful technique, particularly in identifying victims of large-scale accidents or disasters with remains damaged beyond recognition (Khanagar et al., 2021). Typically, forensic odontology analyzes teeth and jawbones, which not only possess patterns unique to a given individual but may also reveal chronological age and biological sex and indicate personal information such as diet, eating patterns, and grinding habits based on wear patterns (Khanagar et al., 2021).

Many approaches exist to odontological analysis, including biochemical and radiographical approaches. Biochemical approaches involving evaluating molecular changes to proteins and tissues associated with the aging process are less subjective than radiological analyses; however, the methods have not been well-standardized (Kurniawan et al., 2024). Radiological approaches include Kvaal's technique, developed in 1995, which estimates age based on the quantity of dental pulp, and the Van Heerdan and Harris and Nortje techniques, which use root growth of the third molars as an indicator of adult age (Kurniawan et al., 2024). Age-related changes to the third molars are key in providing age estimates of adults, which are often more challenging to age as all permanent teeth have erupted (Kurniawan et al., 2024). Age estimates from radiographs are time-consuming and prone to bias, as they are formed by experts relying on years of experience to guide their determination (Kurniawan et al., 2024).

AI is being evaluated to expedite such analyses, allowing them to be performed quickly and with reduced bias. Matsuda et al. (2020) investigated the capabilities of six different convolutional neural network architectures in identifying individuals from orthopantomographs. The authors found that each architecture demonstrated a detection accuracy of 80% or greater, with the VGG16 model exhibiting the highest accuracy of 100% when the model was pretrained and fine-tuned.

High-quality training data is integral to producing accurate estimates from AI models, and fine-tuning models is complex and time-consuming. Unfortunately, models developed based on data from a specific population often do not deliver reliable predictions when presented with data from a population not included in the training data due to variations in dental development and tooth morphology (Kurniawan et al., 2024).

21.14 Artificial intelligence in forensic medicine: bruise color analysis

Age determination of bruises is of particular importance in cases involving domestic violence and violence against women (Tirado & Mauricio, 2021). As bruises heal, they undergo predictable shifts in color due to the degradation of hemoglobin (Georgieva et al., 2005). Bruises are initially red or blueish purple and may turn green due to the presence of biliverdin, brown due to the presence of bilirubin, and yellow due to the accumulation of hemosiderin, before finally fading completely in 10–14 days (Georgieva et al., 2005). However, due to variability in bruise appearance and capability to heal, estimates provided by medical specialists may have low accuracy. AI may provide a method that can more subjectively determine bruise age from images based on physical appearance (Tirado & Mauricio, 2021). Tirado and Mauricio (2021) utilized deep convolutional neural networks to date images of bruises of varying ages (0–2 days and 17–30 days). Although the precision of the model decreased with increasing age of the bruise, the MnasNet classification model had a precision and sensitivity of 97% and 99.5% specificity.

21.15 Artificial intelligence in forensic medicine: toxicological analysis

During autopsy, various specimens, including blood, vitreous fluid from the eyes, urine, bile, stomach contents, and organ tissues, are collected from the body for toxicological analyses. The presence of alcohol, drugs, or poisons can help determine the cause of death and whether criminal activity is a factor (Toxicology, n.d.). Human errors have been reported in the use of spectrophotometers for toxicological analyses, and AI-based approaches can increase accuracy while reducing costs (Helma, 2004). The use of AI may help automate the identification (both qualitatively and quantitatively) of toxic substances, drugs, and metabolites. As of 2020, there are over 160 million substances in the Chemical Abstract Service database that could be used by algorithms for comparison (Gasteiger, 2020).

21.16 Artificial intelligence in the analysis of digital evidence: surveillance footage and facial recognition

AI has also aided in the analysis of digital evidence in forensic investigations. Investigators are often faced with vast amounts of surveillance footage to sift through to locate events related to a crime or search for the whereabouts of a particular individual. Analyzing hours upon hours of surveillance footage is mentally fatiguing and highly prone to human error, and often this footage must be processed in a short amount of time so any leads can be acted upon quickly. AI can assist in matching faces and detecting complex events such as accidents and crimes, which can help reduce the time required by analysts to locate time points of interest (Rigano, 2019).

21.17 Artificial intelligence in the analysis of digital evidence: pattern matching and searching

Evidence such as fingerprints, shoe prints, handwriting, and firearms is considered pattern evidence and is analyzed through pattern matching by comparing the evidence to a known sample from a person of interest. Traditionally, training and experience are used to evaluate pattern evidence, and there is no defined way to quantitate what constitutes "similar enough" to claim a match between an evidence item and a reference. A 2009 NAS report found pattern evidence disciplines to be "rife with subjectivity and lacking in scientific validity" (Carriquiry et al., 2019). In 2017, Hare et al. (2017) proposed using supervised learning algorithms to produce a similarity score for bullets, a quantitative metric of how similar striations on two bullets are.

Prusinowski et al. (2024) reported on a data analysis algorithm using data obtained from physical fit examinations of duct tape and textiles. A physical fit examination is a technique used in the analysis of trace evidence that conveys a high degree of association between two items; however, these analyses are time-consuming, and bias introduced by the analyst may affect the consistency of results. Compared with human analysis, the algorithm had an accuracy above 90% and an improved rate of true positives (matches). However, false positives were observed in high-quality scissor-cut duct tape and textiles, so the authors advise that the algorithm be used in tandem with human analyses.

In sexual assault cases, the presence of semen on swabs recovered from the victim is an essential piece of evidence. The most widely accepted technique for the identification of semen is the visualization of sperm cells under the microscope; however, manually searching for sperm on a microscope slide is often time-consuming, and finding sperm can be very difficult. Only a few sperm cells may be present, and the investigation may hinge on not missing them. The use of neural networks can help automate these searches by scanning microscope images for sperm (Golomingi et al., 2022). Golomingi et al. (Golomingi et al., 2022) trained two convolutional neural networks, a "dominant technology in the field of image classification and pattern recognition," using two different network architectures and achieved an accuracy of over 90%.

21.18 Considerations for future use in forensic science

Despite the powerful ability of AI to quickly recognize patterns in complex datasets and predict future events, several challenges must first be overcome before this novel technology may be routinely used in analyses presented in court (Barash et al., 2024; Katsara et al., 2021). Scientific evidence in most US states is bound by the Daubert Standard, which requires potential sources of error to be described in scientific evidence presented to a court (Zhang et al., 2019). Unfortunately, AI algorithms internalize the processes leading to the final output, making them inaccessible to the user; therefore, the "thought process" of the program is unknown (Galante et al., 2022). In fields such as advertising or marketing, this lack of explanation is typically not an issue, but full transparency, as well as standardization of methods, are major considerations in forensic science, especially when people may face incarceration as a result of the analysis (Barash et al., 2024). Explainable AI is an emerging field of research to develop AI systems that can provide the user with an explanation for how the system arrived at its conclusions (Barash et al., 2024). Nonetheless, AI tools are being used for investigative leads for cases in which the suspect's DNA can be directly tested to verify the hypothesis derived from using the AI tool.

Another major issue with many AI models is overfitting. The model may identify more variables than are necessary for making predictions in the training data. As a result, the model performs very well on the training data but fails to generate reliable predictions on new data due to training that is too specific to the features of the original dataset (Zhang et al., 2019). Additionally, human experts and an upfront investment of time and labor are required to develop and train algorithms for a specific application. Human users must understand and recognize that these methods are not infallible and monitor them carefully to ensure they are producing reliable and accurate results (Barash et al., 2024).

The forensic community is largely unaware of the capabilities of AI, and ML professionals are not aware of the specific requirements needed for forensic analyses. Improved collaboration between the forensic science and computer science fields is necessary to develop programs that are better-suited to the unique demands of forensic science and to educate practitioners (Barash et al., 2024).

References

Adelman, J. D., Zhao, A., Eberst, D. S., & Marciano, M. A. (2019). Automated detection and removal of capillary electrophoresis artifacts due to spectral overlap. *Electrophoresis, 40*(14), 1753–1761. https://doi.org/10.1002/elps.201900060.

Al Matrouk, A., & Dashti, M. (2025). Virtual Autopsy: An Innovative Approach to Determining Cause and Circumstances of Death. In H. R. Dash, K. M. Elkins, & N. R. Al Snan (Eds.). *Advances in Forensic Biology and Genetics* (pp. 231–237). Singapore: Springer.

Barash, M., McNevin, D., Fedorenko, V., & Giverts, P. (2024). Machine learning applications in forensic DNA profiling: A critical review. *Forensic Science International: Genetics, 69*. https://doi.org/10.1016/j.fsigen.2023.102994.

Belk, A., Xu, Z. Z., Carter, D. O., Lynne, A., Bucheli, S., Knight, R., & Metcalf, J. L. (2018). Microbiome data accurately predicts the postmortem interval using random forest regression models. *Genes, 9*(2). https://doi.org/10.3390/genes9020104.

Benschop, C. C. G., van der Linden, J., Hoogenboom, J., Ypma, R., & Haned, H. (2019). Automated estimation of the number of contributors in autosomal short tandem repeat profiles using a machine learning approach. *Forensic Science International: Genetics, 43*. https://doi.org/10.1016/j.fsigen.2019.102150.

Butler, J. M. (2005). *Forensic DNA typing* (2nd ed.). Elsevier Academic Press.

Carriquiry, A., Hofmann, H., Tai, X. H., & VanderPlas, S. (2019). Machine learning in forensic applications. *Significance, 16*(2), 29–35. https://doi.org/10.1111/j.1740-9713.2019.01252.x.

Collins, C., Dennehy, D., Conboy, K., & Mikalef, P. (2021). Artificial intelligence in information systems research: A systematic literature review and research agenda. *International Journal of Information Management, 60*. https://doi.org/10.1016/j.ijinfomgt.2021.102383.

Crysup, B., Mandape, S., King, J. L., Muenzler, M., Kapema, K. B., & Woerner, A. E. (2023). Using unique molecular identifiers to improve allele calling in low-template mixtures. *Forensic Science International: Genetics, 63*. https://doi.org/10.1016/j.fsigen.2022.102807.

Dash, H. R., & Das, S. (2020). Thanatomicrobiome and epinecrotic community signatures for estimation of post-mortem time interval in human cadaver. *Applied Microbiology and Biotechnology*. https://doi.org/10.1007/s00253-020-10922-3.

Díez López, C., Vidaki, A., & Kayser, M. (2022). Integrating the human microbiome in the forensic toolkit: Current bottlenecks and future solutions. *Forensic Science International: Genetics, 56*. https://doi.org/10.1016/j.fsigen.2021.102627.

Elkins, K. M. (2011). Rapid presumptive "fingerprinting" of body fluids and materials by ATR FT-IR spectroscopy. *Journal of Forensic Sciences, 56*(6), 1580–1587. https://doi.org/10.1111/j.1556-4029.2011.01870.x.

Fixler, K. (2024, February 25). How a DNA technique to pin Bryan Kohberger as Idaho murder suspect could shape case law. *The Columbian*. https://www.columbian.com/news/2024/feb/25/how-a-dna-technique-to-pin-bryan-kohberger-as-idaho-murder-suspect-could-shape-case-law/.

Galante, N., Cotroneo, R., Furci, D., Lodetti, G., & Casali, M. B. (2022). Applications of artificial intelligence in forensic sciences: Current potential benefits, limitations and perspectives. *In International Journal of Legal Medicine*. https://doi.org/10.1007/s00414-022-02928-5 Springer Science and Business Media Deutschland GmbH.

Gasteiger, J. (2020). Chemistry in times of artificial intelligence. *ChemPhysChem: A European Journal of Chemical Physics and Physical Chemistry, 21*(20), 2233–2242. https://doi.org/10.1002/cphc.202000518.

Georgieva, L., Dimitrova, T., & Stoyanov, I. (2005). Computer-aided system for the bruise color's recognition, Bulgarian Chapter: Sofia, Bulgaria, pp. 1–6. [Google Scholar].

Golomingi, R., Haas, C., Dobay, A., Kottner, S., & Ebert, L. (2022). Sperm hunting on optical microscope slides for forensic analysis with deep convolutional networks - a feasibility study. *Forensic Science International: Genetics, 56*, 102602. https://doi.org/10.1016/j.fsigen.2021.102602.

Hare, E., Hofmann, H., & Carriquiry, A. (2017). Algorithmic approaches to match degraded land impressions. *Law, Probability and Risk, 16*(4), 203–221.

Helma, C. (2004). Data mining and knowledge discovery in predictive toxicology. *SAR and QSAR in Environmental Research, 15*(5–6), 367–383. https://doi.org/10.1080/10629360412331297407.

Hu, L., Xing, Y., Jiang, P., Gan, L., Zhao, F., Peng, W., Li, W., Tong, Y., & Deng, S. (2021). Predicting the postmortem interval using human intestinal microbiome data and random forest algorithm. *Science and Justice, 61*(5), 516–527. https://doi.org/10.1016/j.scijus.2021.06.006.

Iacob, D., Fürst, A., & Hadrys, T. (2019). A machine learning model to predict the origin of forensically relevant body fluids. *Forensic Science International: Genetics Supplement Series, 7*(1), 392–394. https://doi.org/10.1016/j.fsigss.2019.10.025.

Jarrett, A., & Choo, K. R. (2021). The impact of automation and artificial intelligence on digital forensics. *WIREs Forensic Science, 3*(6). https://doi.org/10.1002/wfs2.1418.

Javan, G. T., Finley, S. J., Tuomisto, S., Hall, A., Benbow, M. E., & Mills, D. E. (2019). An interdisciplinary review of the thanatomicrobiome in human decomposition. *Forensic Science, Medicine, and Pathology, 15*(1), 75–83. https://doi.org/10.1007/s12024-018-0061-0.

Jordan, D., & Mills, D. E. (2021). Past, present, and future of DNA typing for analyzing human and non-human forensic samples. *Frontiers in Ecology and Evolution, 9*. https://doi.org/10.3389/fevo.2021.646130.

Katsara, M. A., Branicki, W., Walsh, S., Kayser, M., & Nothnagel, M. (2021). Evaluation of supervised machine-learning methods for predicting appearance traits from DNA. *Forensic Science International: Genetics, 53*. https://doi.org/10.1016/j.fsigen.2021.102507.

Khanagar, S. B., Vishwanathaiah, S., Naik, S., A Al-Kheraif, A., Devang Divakar, D., Sarode, S. C., Bhandi, S., & Patil, S. (2021). Application and performance of artificial intelligence technology in forensic odontology—A systematic review. *Legal Medicine (Tokyo, Japan), 48*, 101826. https://doi.org/10.1016/j.legalmed.2020.101826.

Khandasammy, S. R., Fikiet, M. A., Mistek, E., Ahmed, Y., Halámková, L., Bueno, J., & Lednev, I. K. (2018). Bloodstains, paintings, and drugs: Raman spectroscopy applications in forensic science. *Forensic Chemistry, 8*, 111–133. https://doi.org/10.1016/j.forc.2018.02.002.

Kuiper, I. (2016). Microbial forensics: Next-generation sequencing as catalyst. *EMBO Reports, 17*(8), 1085–1087. https://doi.org/10.15252/embr.201642794.

Kurniawan, A., Novianti, A., Lestari, F. A., & Ramaniasari, S. M. (2024). Integrating artificial intelligence and adult dental age estimation in forensic identification: A literature review. *World Journal of Advanced Research and Reviews, 21*(2), 1374–1379. https://doi.org/10.30574/wjarr.2024.21.2.0605.

Lee, R. (2019). Metagenomic next generation sequencing: How does it work and is it coming to your clinical microbiology lab? *American Society for Microbiology*. https://asm.org/articles/2019/november/metagenomic-next-generation-sequencing-how-does-it#:~:text=To%20characterize%20a%20bacterial%20genome,single%20genome%20using%20computational%20approaches.

Lin, M. H., Lee, S. I., Zhang, X., Russell, L., Kelly, H., Cheng, K., Cooper, S., Wivell, R., Kerr, Z., Morawitz, J., & Bright, J. A. (2021). Developmental validation of FaSTRTM DNA: Software for the analysis of forensic DNA profiles. *Forensic Science International: Reports, 3*. https://doi.org/10.1016/j.fsir.2021.100217.

Liu, Y. Y., Welch, D., England, R., Stacey, J., & Harbison, S. A. (2020). Forensic STR allele extraction using a machine learning paradigm. *Forensic Science International: Genetics, 44*. https://doi.org/10.1016/j.fsigen.2019.102194.

Liu, R., Zhang, K., Li, H., Sun, Q., Wei, X., Li, H., Zhang, S., Fan, S., & Wang, Z. (2023). Dissecting the microbial community structure of internal organs during the early postmortem period in a murine corpse model. *BMC Microbiology, 23*(1). https://doi.org/10.1186/s12866-023-02786-0.

Lutz, H., Vangelatos, A., Gottel, N., Osculati, A., Visona, S., Finley, S. J., Gilbert, J. A., & Javan, G. T. (2020). Effects of extended postmortem interval on microbial communities in organs of the human cadaver. *Frontiers in Microbiology, 11*. https://doi.org/10.3389/fmicb.2020.569630.

Marciano, M. A., Williamson, V. R., & Adelman, J. D. (2018). A hybrid approach to increase the informedness of CE-based data using locus-specific thresholding and machine learning. *Forensic Science International: Genetics, 35*, 26–37. https://doi.org/10.1016/j.fsigen.2018.03.017.

Matsuda, S., Miyamoto, T., Yoshimura, H., & Hasegawa, T. (2020). Personal identification with orthopantomography using simple convolutional neural networks: A preliminary study. *Scientific reports, 10*(1), 13559. https://doi.org/10.1038/s41598-020-70474-4.

McLaughlin, G., Doty, K. C., & Lednev, I. K. (2014). Raman spectroscopy of blood for species identification. *Analytical Chemistry, 86*(23), 11628–11633. https://doi.org/10.1021/ac5026368.

Metcalf, J. L. (2019). Estimating the postmortem interval using microbes: Knowledge gaps and a path to technology adoption. *Forensic Science International: Genetics, 38*, 211–218. https://doi.org/10.1016/j.fsigen.2018.11.004.

Mistek, E., Halámková, L., Doty, K. C., Muro, C. K., & Lednev, I. K. (2016). Race differentiation by Raman spectroscopy of a bloodstain for forensic purposes. *Analytical Chemistry, 88*(15), 7453–7456. https://doi.org/10.1021/acs.analchem.6b01173.

Oliveira, M., & Amorim, A. (2018). Microbial forensics: new breakthroughs and future prospects. *Applied Microbiology and Biotechnology, 102*(24), 10377–10391. https://doi.org/10.1007/s00253-018-9414-6.

O'Sullivan, S., Holzinger, A., Zatloukal, K., Saldiva, P., Sajid, M. I., & Wichmann, D. (2017). Machine learning enhanced virtual autopsy. *Autopsy & Case Reports, 7*(4), 3–7. https://doi.org/10.4322/acr.2017.037.

Prusinowski, M., Tavadze, P., Andrews, Z., Lang, L., Pulivendhan, D., Neumann, C., Romero, A. H., & Trejos, T. (2024). Experimental results on data analysis algorithms for extracting and interpreting edge feature data for duct tape and textile physical fit examinations. *Journal of Forensic Sciences, 69*(2), 498–514. https://doi.org/10.1111/1556-4029.15449.

Rigano, C. (2019). Using Artificial Intelligence to Address Criminal Justice Needs. *NIJ Journal, 280*. https://www.nij.gov/journals/280/Pages/using-artificialintelligence-to-address-criminal-justice-needs.aspx.

Roy, D., Tomo, S., Purohit, P., & Setia, P. (2021). Microbiome in death and beyond: Current vistas and future trends. *Frontiers in Ecology and Evolution, 9*. https://doi.org/10.3389/fevo.2021.630397.

Sijen, T., & Harbison, S. (2021). On the identification of body fluids and tissues: A crucial link in the investigation and solution of crime. *Genes, 12*(11), 1728. https://doi.org/10.3390/genes12111728.

Sun, Y. V., & Karida, S. L. R. (2008). Imputing missing genotypic data of single-nucleotide polymorphisms using neural networks. *European Journal of Human Genetics, 16*(4), 487–495. https://doi.org/10.1038/sj.ejhg.5201988.

Swaminathan, H., Grgicak, C. M., Medard, M., & Lun, D. S. (2015). NOCIt: A computational method to infer the number of contributors to DNA samples analyzed by STR genotyping. *Forensic Science International: Genetics, 16*, 172–180. https://doi.org/10.1016/j.fsigen.2014.11.010.

Tillmar, A., Fagerholm, S. A., Staaf, J., Sjölund, P., & Ansell, R. (2021). Getting the conclusive lead with investigative genetic genealogy—A successful case study of a 16 year old double murder in Sweden. *Forensic Science International: Genetics, 53*. https://doi.org/10.1016/j.fsigen.2021.102525.

Tirado, J., & Mauricio, D. (2021). Bruise dating using deep learning. *Journal of Forensic Sciences, 66*(1), 336–346. https://doi.org/10.1111/1556-4029.14578.

Toxicology. (n.d.). Mississippi Department of Public Safety. https://www.dps.ms.gov/forensic-laboratories/Crime-Lab/toxicology#:~:text=Biological%20specimens%20including%2C%20blood%2C%20vitreous,the%20Toxicology%20Section%20for%20analysis.

Virkler, K., & Lednev, I. K. (2008). Raman spectroscopy offers great potential for the nondestructive confirmatory identification of body fluids. *Forensic Science International, 181*(1–3), e1–e5. https://doi.org/10.1016/j.forsciint.2008.08.004.

Wang, Y., Cai, Z., Stothard, P., Moore, S., Goebel, R., Wang, L., & Lin, G. (2012). Fast accurate missing SNP genotype local imputation. *BMC Research Notes, 5*. http://www.biomedcentral.com/1756-0500/5/404.

Wickenheiser, R. A. (2019). Forensic genealogy, bioethics and the Golden State Killer case. *Forensic Science International: Synergy, 1*, 114–125. https://doi.org/10.1016/j.fsisyn.2019.07.003.

Xu, Y., Liu, X., Cao, X., Huang, C., Liu, E., Qian, S., Liu, X., Wu, Y., Dong, F., Qiu, C. W., Qiu, J., Hua, K., Su, W., Wu, J., Xu, H., Han, Y., Fu, C., Yin, Z., Liu, M., ... Zhang, J. (2021). Artificial intelligence: A powerful paradigm for scientific research. *Innovation, 2*(4). https://doi.org/10.1016/j.xinn.2021.100179.

Yang, T. W., Li, Y. H., Chou, C. F., Lai, F. P., Chien, Y. H., Yin, H. I., Lee, T. T., & Hwa, H. L. (2022). DNA mixture interpretation using linear regression and neural networks on massively parallel sequencing data of single nucleotide polymorphisms. *Australian Journal of Forensic Sciences, 54*(2), 150–162. https://doi.org/10.1080/00450618.2020.1807050.

Zhang, Y., Pechal, J. L., Schmidt, C. J., Jordan, H. R., Wang, W. W., Eric Benbow, M., Sze, S. H., & Tarone, A. M. (2019). Machine learning performance in a microbial molecular autopsy context: A cross-sectional postmortem human population study. *PLoS One, 14*(4). https://doi.org/10.1371/journal.pone.0213829.

Chapter 22

Prediction of Y-chromosome lineages from *Y-SNPs* and *Y-STRs* for generating investigative leads

Claudia Marina Schellin-Becker[1,2], Vítor Matheus Soares Moraes[3], Danielle Malheiros[4], and Celso Teixeira Mendes-Junior[5]

[1]*Programa de Pós-Graduação em Genética, Universidade Federal do Paraná, Curitiba, PR, Brazil,* [2]*Laboratório de Genética Molecular Forense da Polícia Científica do Paraná, Curitiba, PR, Brazil,* [3]*Departamento de Genética, Faculdade de Medicina de Ribeirão Preto, Universidade de São Paulo, Ribeirão Preto, SP, Brazil,* [4]*Departamento de Genética, Laboratório de Genética Molecular Humana, Universidade Federal do Paraná, Curitiba, PR, Brazil,* [5]*Departamento de Química, Laboratório de Pesquisas Forenses e Genômicas, Faculdade de Filosofia, Ciências e Letras de Ribeirão Preto, Universidade de São Paulo, Ribeirão Preto, SP, Brazil*

22.1 Y-chromosome analysis in forensic genetics

In forensic genetics, routine analysis aims to identify persons by comparing traces and suspects based on a preestablished set of markers (Syndercombe Court, 2021). In the current forensic routine, short tandem repeats (STRs) are the markers of choice (Garcia et al., 2022; Kayser, 2015). They consist of a type of length polymorphism nucleotides organized into short sequences that form blocks of variable sizes (usually two to seven base pairs) and which are consecutively repeated a variable number of times (Moran et al., 2004). Forensically relevant STR markers include STRs located on autosomal and sex (Y and X) chromosomes, which have distinct inheritance patterns (Garcia et al., 2022; Jobling, 2022).

To identify the origin of an STR profile recovered from evidence in a crime scene, a suspect must exist for profiles to be compared. When a suspect is lacking, databases supplied with genetic profiles obtained from convicted criminals and traces may provide a match (Jobling, 2022). However, a case may not always be resolved based on autosomal STR markers alone, especially in the absence of a suspect or matching database hits. While autosomal STR profiling limits searches to a male perpetrator's close relatives, lineage markers like the Y-chromosome and mitochondrial DNA (mtDNA), inherited without changes across generations, can respectively help paternal and maternal lineages to be traced (Butler, 2023; Garcia et al., 2022; Kayser, 2015).

Y-chromosome analysis is a highly effective means of tracing patrilineal ancestry: it connects a man to all his paternal relatives, so it is a powerful investigative tool (Fitzpatrick, 2022; Garcia et al., 2022). Y-chromosome analysis has proven particularly valuable in the examples described as follows.

In the Netherlands in 1999, investigators working on the Vaatstra case initially failed to identify a suspect through autosomal STR markers when they were looking into a young woman's sexual assault and murder. This caused the local population to suspect foreign refugees, which sparked tension and conflicts among residents, refugees, and law enforcement authorities. The staff at the Forensic DNA Research Laboratory of the Leiden University Medical Center was called to help, and they used Y-chromosome STR markers to investigate the perpetrator's paternal biogeographical origin. They discovered the perpetrator was of Western European descent, which shifted suspicion to local individuals. Nevertheless, the case remained unsolved for fourteen years until, in 2012, the police initiated a campaign to collect buccal swabs from over 6600 local volunteers, which ultimately led to the murderer's identification through familial DNA tracing (Kayser, 2017). Similarly, in the Pamela Cahanes case, the murder of a young woman in the USA remained unsolved for 34 years. Initially, the perpetrator was believed to be of European descent. However, in 2015, a Y-STR analysis of the DNA recovered from the crime scene revealed the perpetrator was of African ancestry. This shifted the

Investigative and Predictive DNA Testing. DOI: https://doi.org/10.1016/B978-0-443-30128-5.00013-8

investigators' focus to individuals of African American ancestry, resulting in the identification of a suspect who was present in the same location at the same time as the victim (Fitzpatrick, 2022).

As in the cases described above, after investigators failed to obtain a lead by using autosomal STRs, they employed Y-STRs to identify the suspect's paternal lineage, which shifted the focus of the investigation. In this context, forensic DNA phenotyping, which helps to predict visible external characteristics like skin pigmentation, hair, or eye color, or factors altering some of these characteristics, such as chronological age, can play a major role along with biogeographical ancestry (BGA) inference (Canales Serrano, 2020; Gentile et al., 2019; Hamczyk et al., 2020; Heidegger et al., 2022; Kayser et al., 2023; Kayser, 2017; Montesanto et al., 2020; Porras-Hurtado et al., 2013; Woźniak et al., 2021).

Here, we will focus on the biogeographical inference of paternal haplogroups inferred from Y-chromosome single-nucleotide polymorphisms (SNPs) and Y-STRs, given that the latter have also proven successful in inferring paternal BGA (de Knijff, 2000; Gusmão et al., 2003; Kayser, 2017; Nebel et al., 2001; Syndercombe Court, 2021).

22.2 Biogeographical ancestry in the forensic context

BGA represents the geographical region of origin of an individual's ancestors. Therefore, inferring BGA implies disregarding concepts such as ethnicity or manifestations of culture or tradition (Schneider et al., 2019). Within this context, an individual's BGA can be inferred by understanding genetic diversity among different population groups and the global distribution of these groups (Syndercombe Court, 2021).

The genetic diversity observed in the allelic, haplotypic, and genotypic frequencies of distinct population groups arises from events related to evolutionary processes, including mutations, selection, genetic isolation, migrations throughout history, and genetic drift (Schneider et al., 2019; Syndercombe Court, 2021; Underhill & Kivisild, 2007), as well as progressively increasing populations (Syndercombe Court, 2021). Thus, genetic markers may exhibit different frequencies depending on the geographical region—specific alleles may be common in some regions and rare in others, which makes them highly informative about their likely origin (Schneider et al., 2019). The way polymorphisms not affected by natural selection are distributed has reflected the history of modern humans since populations dispersed from Africa over thousands of years (Phillips, 2015; Syndercombe Court, 2021). Therefore, on the basis of genetic variation within and among populations, phylogenetic analyses can help to organize information about distant ancestors and their hierarchical descent order (Underhill & Kivisild, 2007).

Inferring ancestry is complex, which becomes apparent when one assesses how the geographical distance and geophysical barriers affect the free movement of individuals, random mating, and hence genetic data. For a long time, such barriers defined how mating would occur and strongly influenced how the human population was structured. After mass movements, though, genetic material was progressively exchanged between populations, and such movements were facilitated by successively removing the cultural, social, and geographical barriers originally limiting demographic movements (Phillips, 2015; Underhill & Kivisild, 2007). Given the recent human evolutionary history, BGA analyses are expected to reveal how contributions from various origins prevail in individuals (Phillips, 2015), and genetic data provide a reliable source to infer BGA (Kayser et al., 2023; Schneider et al., 2019). Because genetic diversity is geographically structured (Jobling, 2012), the quality of BGA inference depends on how informative the employed genetic markers are and on prior knowledge of their frequencies in the populations being used as reference (Schneider et al., 2019).

Forensic ancestry analysis relies on genetic differences between global population groups, often defined by continents, but this approach overlooks the greater genetic diversity among individuals within the same population (Phillips & de la Puente, 2021). In forensic contexts, ancestry information plays a crucial role in identifying missing persons, triaging disaster victims, verifying the accuracy of genetic databases, and enhancing familial search systems (Phillips & de la Puente, 2021).

In this scenario, ancestry informative markers (AIMs) are usually the markers of choice as they present genetic variations differing between populations, which makes them valuable to test forensic ancestry. AIMs commonly include autosomal SNPs and insertion-deletion polymorphisms (INDELS), and their allele frequency patterns differ among populations. These markers are particularly useful in forensic tests because they are small, so they are less affected by DNA degradation than usual STRs (Phillips & de la Puente, 2021). Additionally, AIMs represented by SNPs constitute well-established panels that enable BGA origin to be inferred at a continental or subcontinental level (Europe, Sub-Saharan Africa, East Asia, South Asia, Oceania, and the Americas) (Schneider et al., 2019).

AIMs include autosomal, mtDNA, and Y-chromosome markers. They are transmitted by different mechanisms across generations, which allows different strategies to be adopted to understand biogeographical origin. For example, autosomal markers simultaneously reflect where both parents and all their ancestors originated. Moreover, these markers may recombine, so that offspring will carry only half of their respective parental markers. In the case of Y-chromosome markers, they are only transmitted from father to son, which in turn allows only the biogeographical origin of the paternal lineage to be inferred. According to their inheritance pattern, Y-chromosome markers are essentially transmitted to subsequent generations unchanged, so their original lineages remain strongly correlated with continental regions (Fleckhaus & Schneider, 2020; Phillips, 2015).

Uniparental markers like Y-STRs and mtDNA are used to infer ancestry in forensic cases, mainly when male suspects and hair are analyzed. These markers are geographically stratified, but accurately determining ancestry depends on extensive global databases. Although uniparental markers can add valuable details, their single-marker nature may sometimes lead to ancestry being inaccurately represented, especially in recently admixed individuals. Integrating Y-STR and mtDNA data with autosomal variation provides a more detailed understanding of an individual's ancestry (Phillips & de la Puente, 2021).

While Y-chromosome markers are strongly correlated with ancestry, they may not represent an individual's entire ancestry precisely because their uniparental characteristic is related to the paternal lineage, which may culminate in errors when BGA origin is interpreted. This bias is particularly impactful when admixed populations are considered, as these populations have shared ancestries due to their respective demographic histories (Phillips, 2015). Furthermore, uniparental markers in individuals originating from highly admixed populations, such as admixed Americans, are not informative enough when it comes to correlating these markers with phenotypes (Pena & Bortolini, 2004).

Although inferring BGA has limitations, and even though BGA can hardly be defined accurately and is often limited to continental-level ancestry, this topic of research is gradually progressing toward accurately detecting subcontinental ancestry (Kayser et al., 2023; Syndercombe Court, 2021).

22.3 Y-chromosome haplotypes and haplogroups

The human Y-chromosome presents two pseudoautosomal regions (PARs), located on the extremities of its short and long arms (Syndercombe Court, 2021). However, approximately 95% of the length of the Y-chromosome (60 Mb) represents the nonrecombining region (NRY) and consists mainly of repetitive elements (Ali & Hasnain, 2002; Quintana-Murci et al., 2001), inversions, and palindromic sequences (Syndercombe Court, 2021). The NRY also contains the SRY gene, essential for the testis to start developing, and this region is crucial for several other genes involved in the development and maintenance of male germ cells (Quintana-Murci et al., 2001). The NRY, considered the largest nonrecombining block of the human genome, is transmitted intact to male individuals across generations in a haploid state (de Knijff, 2000; Quintana-Murci et al., 2001; Underhill & Kivisild, 2007).

Because recombination is absent in the NRY, and due to the Y-chromosome paternal lineage transmission, any mutation in this region remains in a unique Y-chromosome background. Besides that, genetic drift influences the uniparental inheritance of the Y-chromosome more than it influences autosomes, so more pronounced genetic differences between geographical regions can be elicited simply by chance. Therefore, Y-chromosome markers are extremely informative when it comes to investigating population history and human origins (Babić Jordamović et al., 2021; Kayser, 2017).

Hundreds of polymorphic markers have been identified throughout the NRY (Chen et al., 2021; de Knijff, 2000; Quintana-Murci et al., 2001), including (1) biallelic markers with low mutation rates (in the order of 10^{-9}), which are represented by substitutions (Y-SNPs) and insertion-deletions (INDELS) and constitute unique mutation events (UMEs) in human evolution; (2) moderately fast-evolving microsatellites (Y-STRs) with average mutation rate (in the order of 10^{-3}) (Jannuzzi et al., 2020; Syndercombe Court, 2021); and (3) rapidly evolving minisatellites.

From now on, we will use the term "haplogroup" when we refer to specific Y-SNP alleles composing a given human Y-chromosome and the term "haplotype" to indicate the set of Y-STR alleles associated with a given haplogroup. These terms are detailed below.

22.3.1 Haplogroups

SNPs, polymorphic single nucleotide differences in DNA sequences, can be observed within and among populations (Doo & Kim, 2015). A Y-chromosome haplogroup refers to a lineage connecting male individuals to the same ancestral individual in whom a specific mutational mark originated, which indicates a descent link. With each subsequent mutational event, new connections can be made to form a new branch of this lineage (Chen et al., 2021; de Knijff, 2000; Quintana-Murci et al., 2001; Van Oven et al., 2014). This branch can be either another haplogroup or a haplotype.

Low mutation rates (around 10^{-9}) allow Y-SNPs to represent UMEs in human evolutionary history, so they are excellent for understanding the origin of human populations and migratory routes (Jannuzzi et al., 2020; Syndercombe Court, 2021). These UMEs have a significant impact because they define phylogenetically preserved and deeply rooted haplogroups across broad geographical distributions (Kayser, 2017). Haplogroups and their spatial distributions are closely related, which provides the means to trace migrations and demographic processes that have shaped current populations (Babić Jordamović et al., 2021) since the modern human diaspora (Karafet et al., 2008). Thanks to the numerous polymorphisms resulting from mutations, the NRY has been widely used as a molecular marker in research fields such as evolution, genetic genealogy, forensics, and population history, among others (Karafet et al., 2008; Syndercombe Court, 2021; Van Oven et al., 2014).

The extensive data obtained from Y-SNPs have enabled well-established haplogroup phylogenetic trees to be created and continuously improved (de Knijff, 2022; Hallast et al., 2021; Jannuzzi et al., 2020; Karafet et al., 2008; Syndercombe Court, 2021; Van Oven et al., 2014). In these trees, a standardized nomenclature is assigned to the haplogroups (International Society of Genetic Genealogy, 2019; Schlecht et al., 2008; Van Oven et al., 2014). Eighteen major clades are identified by capital letters (from A to R) and represented by prominent nodes. From these nodes, internal nodes represent lineages, and hierarchical subclades within each major haplogroup are defined by using an alphanumeric system (e.g., R1, R1a, and so on) (Karafet et al., 2008; Syndercombe Court, 2021; Van Oven et al., 2014). The oldest nodes are haplogroups A and B, essentially restricted to Africa, which supports the African ancestry of the human population (Fig. 22.1) (Karafet et al., 2008; Syndercombe Court, 2021; Van Oven et al., 2014). An alternative system including the terminal mutation defining a haplogroup (e.g., E-M81) has also been established (Karafet et al., 2008). Lineages without a defining derived character have been placed at interior nodes of the tree and termed "paragroups" (marked by *) due to their potential paraphyletic nature (Karafet et al., 2008). As novel Y-SNPs are discovered, data need to be reinterpreted, and deeper roots of phylogenetic trees may be described. Continuing identification of new Y-SNPs has added considerable complexity to the already mapped phylogenetic trees. While some Y-SNPs contribute to new expansions of the trees, others may prove less informative because they generate very rare lineages. Therefore, focus has been placed only on the most informative Y-SNP (Van Oven et al., 2014), which should be assigned as the terminal mutation defining a haplogroup (Karafet et al., 2008). A minimalized version of the human Y-chromosome phylogenetic tree (freely available online at http://www.phylotree.org/Y) has been adopted ever since (Van Oven et al., 2014).

The currently accepted versions of the human Y-chromosome phylogenetic tree are documented on the website of the International Society of Genetic Genealogy (ISOGG). This public resource site is continuously updated as advancements in technology allow new relevant variants to be discovered. This resource offers the possibility of elucidating the geographical origin of humankind since the most recent common patrilineal ancestor of modern humans (Underhill et al., 2015; Van Oven et al., 2014).

22.3.2 Haplotypes

We have already provided the definition of STRs at the beginning of this chapter, and here we will focus on the STRs located in the NRY. Y-STR haplotypic blocks exhibit high variability and discriminatory power, which allows Y-chromosomes to be differentiated (de Knijff, 2000; Gusmão et al., 2017; Syndercombe Court, 2021). Therefore, these blocks are routinely employed in forensic investigations such as paternity testing, kinship analyses, and familial searching. Their applicability is particularly important in biological samples obtained from cases of sexual violence, where there is the contribution from a male aggressor and a female victim (Kayser, 2017).

Among the mutational models proposed for the dynamics of STR mutations, the stepwise mutation model offers the most accepted representation underlying the diversity in these markers (Fan & Chu, 2007; Kimura & Ohta, 1978). According to this model, a strand slips while DNA is replicated, causing the template and newly synthesized strands to be incorrectly paired, resulting in one or more repeat units being inserted or omitted (Fan & Chu, 2007; Myers, 2007). Because mutations are not frequent (10^{-3}), and given that most mutations in STRs involve losing or gaining a single repeat unit (Kimura & Ohta, 1978), the resulting haplotypes differ slightly from the original one, so they are closely related and can cluster (Gusmão et al., 2003; Nebel et al., 2001).

A haplogroup consists of Y-chromosomes sharing the same UME (e.g., same Y-SNPs) (de Knijff, 2000), but it may contain many distinct Y-STR haplotypes. This happens because Y-SNPs and Y-STRs present different mutation rates, so that a network covering various closely related haplotypes can be identified within a haplogroup (de Knijff, 2000). Thus, Y-STR haplotypes can be applied to infer haplogroups (Jannuzzi et al., 2020).

22.4 Global population ancestry components

For decades, multiple studies have been conducted to unravel the processes through which all the regions of the globe were peopled, to determine where the populations were originally located, and to estimate the likely routes through which the populations migrated. By analyzing genomic data obtained from modern and ancient populations, the genealogical relationships among humans were determined, migration routes were elucidated, and other insights were gained (Nielsen et al., 2017).

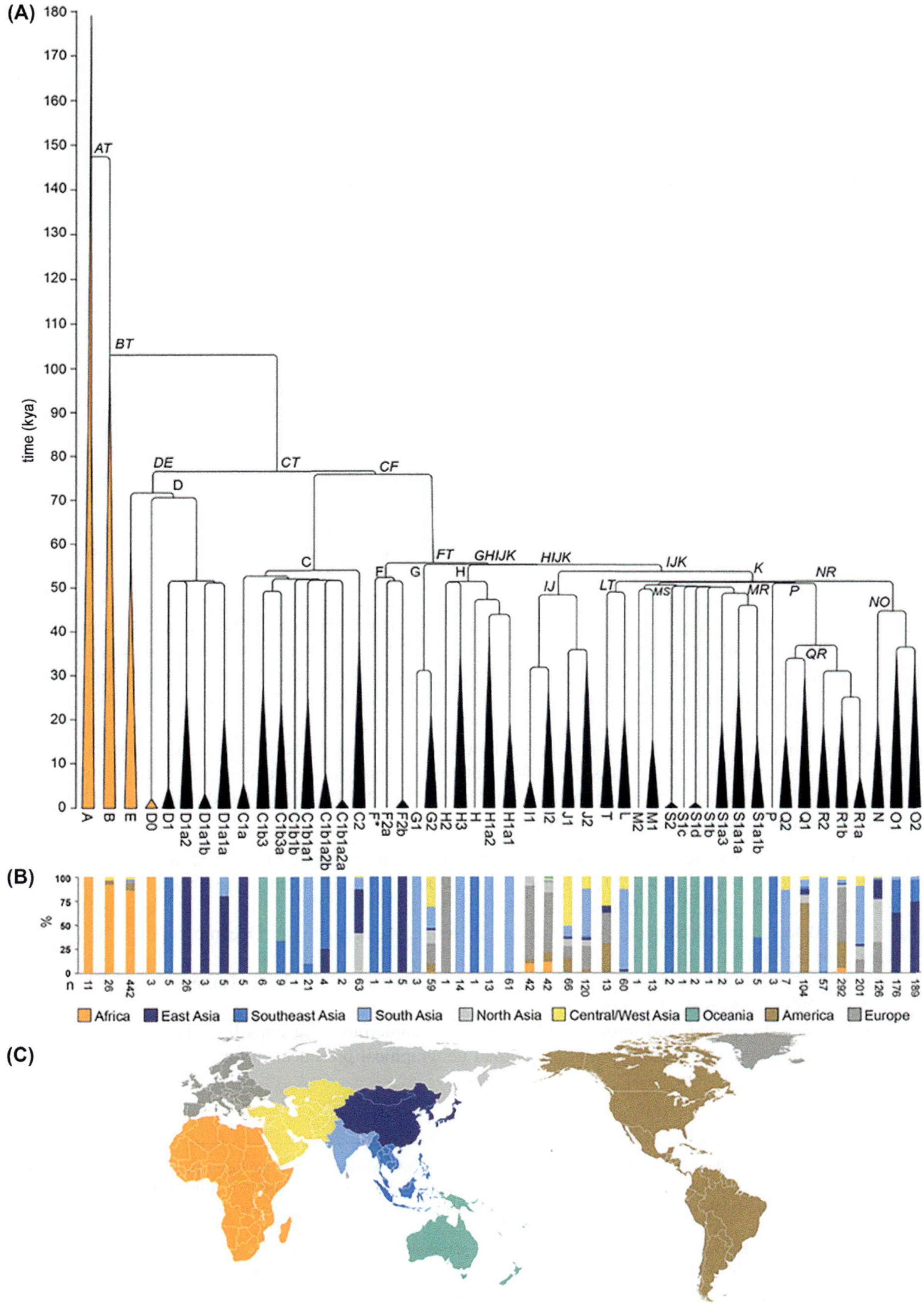
(A)
time (kya)
180
170
160
150
140
130
120
110
100
90
80
70
60
50
40
30
20
10
0
AT
BT
DE
CT
CF
D
C
F
G
FT
GHIJK
HIJK
H
IJ
IJK
K
NR
LT
MS
MR
P
NO
QR
A B E D0 D1 D1a2 D1a1b D1a1a C1a C1b3 C1b3a C1b1b C1b1a1 C1b1a2b C1b1a2a C2 F* F2a F2b G1 G2 H2 H3 H H1a2 H1a1 I1 I2 J1 J2 T L M2 M1 S2 S1c S1d S1b S1a3 S1a1a S1a1b P Q2 Q1 R2 R1b R1a N O1 O2
(B)
%
100
75
50
25
0
n
11 26 442 3 5 26 3 5 5 6 9 1 21 4 2 63 1 1 5 3 59 1 14 1 13 61 42 42 66 120 13 60 1 13 2 1 2 1 2 3 5 3 7 104 57 292 201 126 176 189
Africa
East Asia
Southeast Asia
South Asia
North Asia
Central/West Asia
Oceania
America
Europe
(C)

To improve understanding of how biogeographical origins are inferred in a global context, here we offer a brief overview of the key aspects of human evolutionary history, along with the global distribution of Y-chromosome haplogroups as mapped in the early 2000s (Gusmão et al., 2008; Karafet et al., 2008). However, one must bear in mind that subsequent analyses have raised alternative interpretations regarding the early expansions outside Africa (Hallast et al., 2021).

Genetic data support that the deepest divisions among human populations occurred in sub-Saharan Africa (Llamas et al., 2020; Nielsen et al., 2017), with genes possibly flowing between regions in various parts of the continent (Nielsen et al., 2017). Based on evidence, it has been suggested that anatomically modern humans rapidly dispersed out of Africa, which impacted the genetic variation of non-African populations (Hallast et al., 2021; Henn et al., 2012; Kayser, 2017; Nielsen et al., 2017). Additionally, all non-Africans share approximately 2% Neanderthal ancestry due to interbreeding occurring shortly after Africa was left, and this ancestry is particularly pronounced in East Asian populations (Hallast et al., 2021; Llamas et al., 2020; Nielsen et al., 2017).

Modern European populations likely originated from at least three genetic components entering Europe at different times, which enhanced genetic diversity. Asia and Oceania colonization was also complex, and early human genomes from Asia connected variably to Eurasians, Native Americans, East Asians, and Aboriginal Australians. In Oceania, population divergence resulted from a single founding event, which led Aboriginal Australians and Papuans to be separated. South Pacific populations show a mix of Polynesian, Melanesian, and East Asian ancestries, with Polynesian contact with the Americas being a matter of debate. It is believed that the earliest Americans descended from a mix of Mal'ta-related and unknown East Asian groups entering North America at the end of the Last Glacial Maximum. Migration of the Inuit from the Arctic and their relation to early Arctic inhabitants remain a topic of discussion (Llamas et al., 2020; Nielsen et al., 2017).

After global settlement and isolation for thousands of years, increased movement of people across regions led to admixed populations. These populations arose from genes flowing between distinct groups, often at continental borders, and through large-scale migrations driven by trade, conquest, and slavery (Medina-Muñoz et al., 2023; Phillips, 2015; Suarez-Pajes et al., 2021). Latin Americans are the largest group of admixed populations worldwide (da Cruz et al., 2022; Verdu et al., 2014), and Latin America saw an influx of European, Middle Eastern, Japanese, and Chinese migrants, as well as indigenous people and sub-Saharan Africans (da Cruz et al., 2022). Different colonization practices and origins led to distinct demographic trajectories and shaped the genetic diversity of each Latin American population. This diversity resulted from global human expansion and local admixture events (Elgart et al., 2022; Kayser et al., 2023; Verdu et al., 2014) and can be observed in the Y-chromosome haplogroup frequencies.

The human Y-chromosome phylogenetic tree depicts how each haplogroup is distributed and shows the geographical region where they predominantly occur. This tree reflects the evolutionary history and supplies information about BGA and the dispersion of populations across continents (Van Oven et al., 2014). The global distribution of Y-chromosome haplogroups has been mapped as briefly outlined below.

- The oldest branches of the Y-chromosome phylogeny indicate that sub-Saharan Africa concentrates haplogroups A and B (Gusmão et al., 2008; Hallast et al., 2021; Karafet et al., 2008), some E clades as described below (Gusmão et al., 2008; Hallast et al., 2021), and D0 (Hallast et al., 2021).
- Lineage C, not detected in sub-Saharan Africa, occurs in Asian, Australian, and Oceanic populations (Hallast et al., 2021; Karafet et al., 2008).
- Haplogroup D occurs mainly in East Asia, Southeast Asia (Gusmão et al., 2008; Hallast et al., 2021), and Japan (Gusmão et al., 2008; Karafet et al., 2008).
- Clade E is the most diversified, and three basal branches and numerous subgroups have been identified: E1 and E2 originated in Northeast Africa, while E3 exhibits a broad geographical presence. In turn, subgroup E3a is widespread

FIGURE 22.1 Human Y-chromosome phylogeny and haplogroup distribution. (A) Y-chromosome phylogenetic tree based on 1204 worldwide samples. "kya" means "thousand years ago." (B) Absolute frequencies of specific haplogroups, based on 2319 worldwide samples, and geographical distribution of haplogroups indicated by different colors. (C) Geographical color legend. Map lines delineate study areas and do not necessarily depict accepted national boundaries.
Reproduced from Hallast, P., Agdzhoyan, A., Balanovsky, O., Xue, Y., & Tyler-Smith, C. (2021). A Southeast Asian origin for present-day non-African human Y chromosomes. Human Genetics, 140(2), 299–307. https://doi.org/10.1007/s00439-020-02204-9, under the terms of the Creative Commons Attribution 4.0 International License (http://creativecommons.org/licenses/by/4.0/).

across Africa and among African-Americans, whereas E3b occurs in Western Europe, North Africa, and the Near East (Gusmão et al., 2008).

- Some clades from haplogroup F occur in Asia, and both paragroup F* and haplogroup H are present in the Indian subcontinent, with haplogroup H occurring predominantly there (Gusmão et al., 2008; Karafet et al., 2008).
- The Caucasus region, the Mediterranean areas, and the Middle East present high frequencies of haplogroup G, which is restricted to these regions (Gusmão et al., 2008; Karafet et al., 2008).
- Haplogroups I and R are the most frequent among Europeans (Karafet et al., 2008).
- Haplogroup J seems to have been dispersed by the westward movement of people from the Middle East to North Africa, Europe, Central Asia, Pakistan, and India, where it occurs in high frequencies (Karafet et al., 2008).
- Low frequencies of Haplogroup K exist in dispersed geographic regions worldwide, including India/Pakistan, Oceania, Indonesia, and Australia (Karafet et al., 2008).
- Haplogroup L occurs mainly in the Indian continent, as well as in the Middle East and, very occasionally, in Europe, particularly in Mediterranean countries (Gusmão et al., 2008; Sahakyan et al., 2021).
- The highest frequencies of haplogroup M are restricted to Oceania and Eastern Indonesia (Karafet et al., 2008).
- Haplogroup N is primarily distributed in Northern Eurasia and is almost absent in other regions (Karafet et al., 2008).
- Lineage O represents most chromosomes in East Asia and Oceania (Karafet et al., 2008).
- The rarely detected P haplogroup consists of the Q and R lineages. Low frequencies of this clade occur in the Caucasus and India (Gusmão et al., 2008; Karafet et al., 2008).
- Haplogroup Q is widely distributed in North Eurasia. It occurs at low frequencies in Europe, East Asia, and the Middle East. One of its subclades, group Q3, is almost exclusively associated with Native Americans (Gusmão et al., 2008; Karafet et al., 2008).
- Haplogroup R is mainly represented by two lineages: R1a (currently found in Central and Western Asia, India, and in Slavic populations of Eastern Europe) and R1b (the most common Y haplogroup in Europe) (Gusmão et al., 2008; Karafet et al., 2008).

Along with environmental exposures, genetic ancestry, influenced by allele frequencies and genetic architectures of different ancestral origins, contributes to distinct phenotypic distributions among populations (Elgart et al., 2022). Externally visible characteristics often reflect BGA, and there are notable differences such as lighter features in Europeans, darker features in Africans, and intermediate traits in indigenous populations of Asia and the Americas (Schneider et al., 2019). Nevertheless, these correlations are not absolute and are significantly influenced by admixture dynamics (Bradbury et al., 2019; Schneider et al., 2019).

Using uniparental genetic markers, particularly markers on the Y-chromosome, to infer ancestry only provides information about paternal lineage. However, these markers may not fully capture the complexity of genetic admixture in populations in Latin America, where asymmetric mating among ancestral groups has been documented. Therefore, caution must be taken when correlating BGA with phenotypes of individuals in ancestrally admixed populations, especially when uniparental genetic markers are employed (Bradbury et al., 2019; Kayser et al., 2023; Pena et al., 2011).

22.5 Inferring haplogroups from Y-SNPs and Y-STR haplotypes

Because both kinds of Y-chromosome polymorphic markers, SNPs and STRs, can form clusters on the basis of similarity (haplogroups and haplotypes), they can be employed to infer an individual's BGA (Jannuzzi et al., 2020; Kayser, 2017). Therefore, we will present some approaches and tools employed to infer BGA from each of these sets of markers.

ISOGG lists various tools to predict Y-chromosome haplogroups based on Y-SNPs or Y-STRs, but it does not endorse any. We will focus on some of the most commonly used ones.

22.5.1 Proposed approaches for inferring haplogroups from Y-SNPs

Predicting haplogroups from Y-SNPs is a well-established procedure. Below, we list some of the tools cited by ISOGG as of September 2024.

- Y-Haplo (Poznik, 2016): developed for large datasets and is reliable even in complex situations (missing data and genotyping errors). Available for noncommercial use through 23andMe (International Society of Genetic Genealogy, 2019). An intuitive manual is available at https://github.com/23andMe/yhaplo/blob/master/yhaplo_manual.pdf.
- Y-LineageTracker (2021): free software to combine Y-SNP and Y-STR data obtained from NGS (Next Generation Sequencing). Available at https://github.com/Shuhua-Group/Y-LineageTrackerwithaneasy-to-followmanual.

- YSEQ Clade finder (2020): considers the presence (positive) and absence (negative) of SNPs and relies on the YFull tree. Supports manual entry of a list of SNPs and raw data uploads from genetic testing companies (23andMe, AncestryDNA, and MyHeritage) and processes VCF (Variant Call File) data obtained from NGS. Available at https://github.com/hprovyn/clade-finder.
- Chris Morley's Y-SNP subclade predictor: optimized for Geno 2.0 data (a DNA test provided by National Geographic for deep ancestry, including autosomal, X, and uniparental markers) and transferred to FamilyTreeDNA (FTDNA) and the Full Genomes demo data. Available at ytree.morleydna.com.
- Y-leaf: performs haplogroup predictions from NGS data (BAM or FASTQ files) (Ralf et al., 2018).
- Y-SNP Haplogroup Hierarchy Finder: constructs a hierarchical tree of haplogroups from Y-SNP data by using whole-genome sequencing, providing reports on mutations, haplogroups, and their relationships (Tseng et al., 2022). Available at http://forensic.mc.ntu.edu.tw:9000/DNAToolWeb/ YHGSearch.

22.5.2 Proposed approaches for inferring haplogroups from Y-STRs

While Y-SNPs directly and accurately determine an individual's haplogroup, inferring haplogroups from Y-STR analysis is associated with a higher level of uncertainty. However, Y-STR analysis is conventionally employed and automatically obtained for all cases requiring Y-chromosome analysis for human identification. Therefore, despite the greater complexity and uncertainties associated with Y-STR analysis, inferring haplogroups from Y-STRs allows available data to be immediately used without additional costs or workload. Proposed approaches for inferring haplogroups from a set of Y-STRs include (Table 22.1) (Athey, 2005, 2006):

- **Measuring the Genetic Distance**
  In ancestry inference using genetic distance, the queried haplotype is compared to haplotypes in a database. Genetic distance is calculated as the sum of the differences in repeat values across all markers. For example, if one marker has 10 repeats and another has 11, the difference for that marker is 1. If a haplotype matching the queried one occurs with two or fewer differences, the queried haplotype is assigned to the same haplogroup as the reference haplotype. If no such matching haplotype is found, no haplogroup estimate is provided.
  Although this approach has a prediction success rate of 80%, in some cases no prediction is made even when certain haplogroups can be excluded or when the haplotype likely belongs to a less common haplogroup. Even though the system could theoretically indicate the most likely haplogroups, the tool may fail to provide a definitive result and leave the user without any prediction.
- **Allele Frequency Adjustment**
  An "adjustment" based on allele frequencies for each haplogroup and on the compatibility between the queried haplotype and the pattern of allele frequencies of each STR for each haplogroup has been elaborated to provide an index of how well the person's Y-STR haplotype fits into different ancestral groups. This adjustment has been implemented in a publicly available program since 2004 (http://www.hprg.com/hapest5/), which returns a suitability index for the inserted haplogroup.
  Although the method applying this "adjustment" based on allele frequencies has successfully indicated the true haplogroup, it does not actually offer a prediction score or probability that a haplotype belongs to a certain haplogroup (Athey, 2005).
- **Bayesian Approach**
  The probability that a Y-STR haplotype belongs to a specific haplogroup can be better obtained through a Bayesian approach based on allele frequencies. Applying the Bayesian theory in this context enables one to calculate the probability of a specific haplotype being associated with one of the various possible haplogroups. This probability is based on different Y-STR markers and considers the probabilities of obtaining each test result (i.e., haplogroup assignment) given the haplotype and prior probabilities (a priori) of each haplogroup. With the Bayesian approach, more markers generally improve the probability (provided that the added markers present real discriminatory power). Typically, with 10–20 markers, the probability of one of the haplogroups can be raised to over 99% (Athey, 2006).
  This approach naturally requires extensive data on haplotypes and allele frequencies within each haplogroup (Athey, 2005). Y-STR data for major haplogroups are abundant, but they are often insufficient for minor and non-European haplogroups, especially for less frequently evaluated markers (Athey, 2006).
- **Machine Learning (ML)**

Haplogroups can be accurately assigned from Y-STR data by using ML algorithms, but accuracy depends on the number and type of Y-STRs. Preliminary results suggested that some Y-STRs offer different levels of informativeness for haplogroups, which is likely due to mutation patterns (Schlecht et al., 2008).

TABLE 22.1 Comparison of haplogroup inference methods that use Y-short tandem repeats (STRs).

	Measuring genetic distance	Allele frequency adjustment	Bayesian approach	Machine learning
Application principles	Database comparison: Compares haplotype to database, finds similar matches	Adjusts analysis on the basis of the allele frequencies of Y-STRs in each haplogroup	Calculates probability by using Bayesian theory	Uses algorithms to identify patterns and to predict haplogroups
Criteria for prediction	Few differences (typically two) between haplotypes	Best fit of a haplotype to the expected allele pattern	Considers marker results and prior probabilities	Considers multiple features and patterns, often with complex algorithms
Advantages	Simple and quick	Provides the compatibility index	High accuracy with more markers, can reach > 99% probability	Can achieve high accuracy (> 97%) with appropriate algorithms and training data
Limitations	No prediction is obtained in the absence of a close match	Does not offer a prediction score or probability	Requires extensive allele frequency data	Performance depends on training data size, algorithm choice, and Y-STR panel composition
Example	Finds matching haplotype in a database	Matches one specific Y-STR marker within a particular haplogroup	Uses multiple Y-STR markers	Uses techniques such as support vector machine (SVM) and random forest

Building on this, Bouakaze et al. (2020) addressed biases in predicting haplogroups (Jannuzzi et al., 2020) and the underexplored potential of Y-STR mutation rates by developing an ML program ("PredYMaLe") and a designed composition of Y-STRs with varying mutation rates, selected to maximize the accuracy of haplogroup prediction while meeting forensic requirements for discrimination ("CombYplex"). "PredYMaLe" employs techniques such as support vector machines and random forest to achieve an average accuracy of 97%. The effectiveness of these methods depends on the composition and number of the STR panel, with the size of the training dataset also being a determining factor.

Although the cited ML programs proved effective and adaptable for various STR panels, neither is listed among ISOGG references as of September 2024, which suggests their limited adoption in the genetic genealogy community.

We will detail two free access software programs, Whit Athey's Haplogroup Predictor (HAPEST5) (Athey, 2005, 2006) and the NevGen Haplogroup Predictor (Cetkovic Gentula & Nevski, 2015), cited by ISOGG as of September 2024. In addition, these programs are mentioned (Bouakaze et al., 2020; Jannuzzi et al., 2020; Petrejčíková et al., 2014) for their effectiveness in classifying haplogroups by using Y-STR haplotypes, but they are still considered less efficient than the gold standard method (Y-SNP- based inference) (Babić Jordamović et al., 2021; Emmerova et al., 2017; Jannuzzi et al., 2020; Petrejčíková et al., 2014; Zieger & Utz, 2019). Both programs combine a Bayesian-Allele-Frequency Approach to determine a final fitting coefficient. One of the differences between NevGen's and Athey's predictors is that the former uses the correlation (interdependence) between the values of different STRs to calculate the probabilities of subclades (https://www.nevgen.org/AboutNevGen.html).

We have excluded other Y-STR haplogroup prediction tools cited by ISOGG as of September 2024 from this discussion due to limitations such as outdated or unsupported technology and compatibility issues, which affect their practicality and accessibility.

22.5.2.1 NevGen Y-DNA Haplogroup Predictor

According to the program's website, the NevGen Y-DNA Haplogroup Predictor is a tool designed to determine the likely haplogroup or even its specific subbranch. By analyzing Y-STR data, the predictor provides a fitting score and an associated probability to indicate how well the Y-STR haplotype matches with one or more haplogroups on average. The software can be easily accessed at http://www.nevgen.org/AboutNevGen.html.

22.5.2.1.1 Individual samples

Each sample can be individually analyzed by using the online methodology available at the website Nevgen.org (Fig. 22.2).

On the software's online tool interface, you can choose the methodology (Predictor Model) and the number and order of Y-STR markers, namely FTDNA order (111 Y-STRs), 17 markers of the AmpFLSTR™ Yfiler™ PCR Amplification Kit (Thermo Fisher Scientific), or 23 markers available on the PowerPlex® Y23 System (Promega).

As for the general analysis, you can enter the person's haplotype in two ways (Table 22.2): by typing text in the white box placed above the "Calculate" button (following the order of markers determined by the lower box) or directly selecting the values for each of the determined Y-STRs from drop-down menus.

You should report the four "missing values" or "not considered for the analysis" values as "$" in the text format or "0" in the drop-down menu. Therefore, for the text input, the following line presents how you can type out the haplotype in the white box: 19 13 17 30 15 10 25 $ $ 10 14 18 21 25 13 11 14 13 15 $ $ 16 12.

The output is provided in terms of tables and figures as shown in Figs. 22.3–22.5. The pie chart depicts the probability values of a given Y-STR haplotype belonging to each retrieved Y-chromosome haplogroup. In the current example, the Y-chromosome of this individual of Japanese origin has 100% probability of belonging to haplogroup D1b1 (Fig. 22.3), as expected. Fig. 22.4 shows the distribution of possible haplogroup assignments and indicates the Fitness Scores. Fig. 22.5 shows if the Y-STR alleles of the tested individual fit those of the average individuals belonging to the same inferred haplogroup. Black vertical lines indicate alleles within the expected values for a given Y-STR, while the red vertical lines indicate markers deviating from the standard.

22.5.2.1.2 Batch analysis

If you need to infer more than one haplotype at the same time, doing it one by one can be time-consuming. In such cases, you can run a batch analysis by downloading the software from the website https://www.nevgen.org/NevGenGenealogyTools_v1_1.html by clicking the hyperlink "NevGen Genealogy Tools v1.1.". You do not need to install the program; however, your computer may display compatibility and security warnings when you run the program

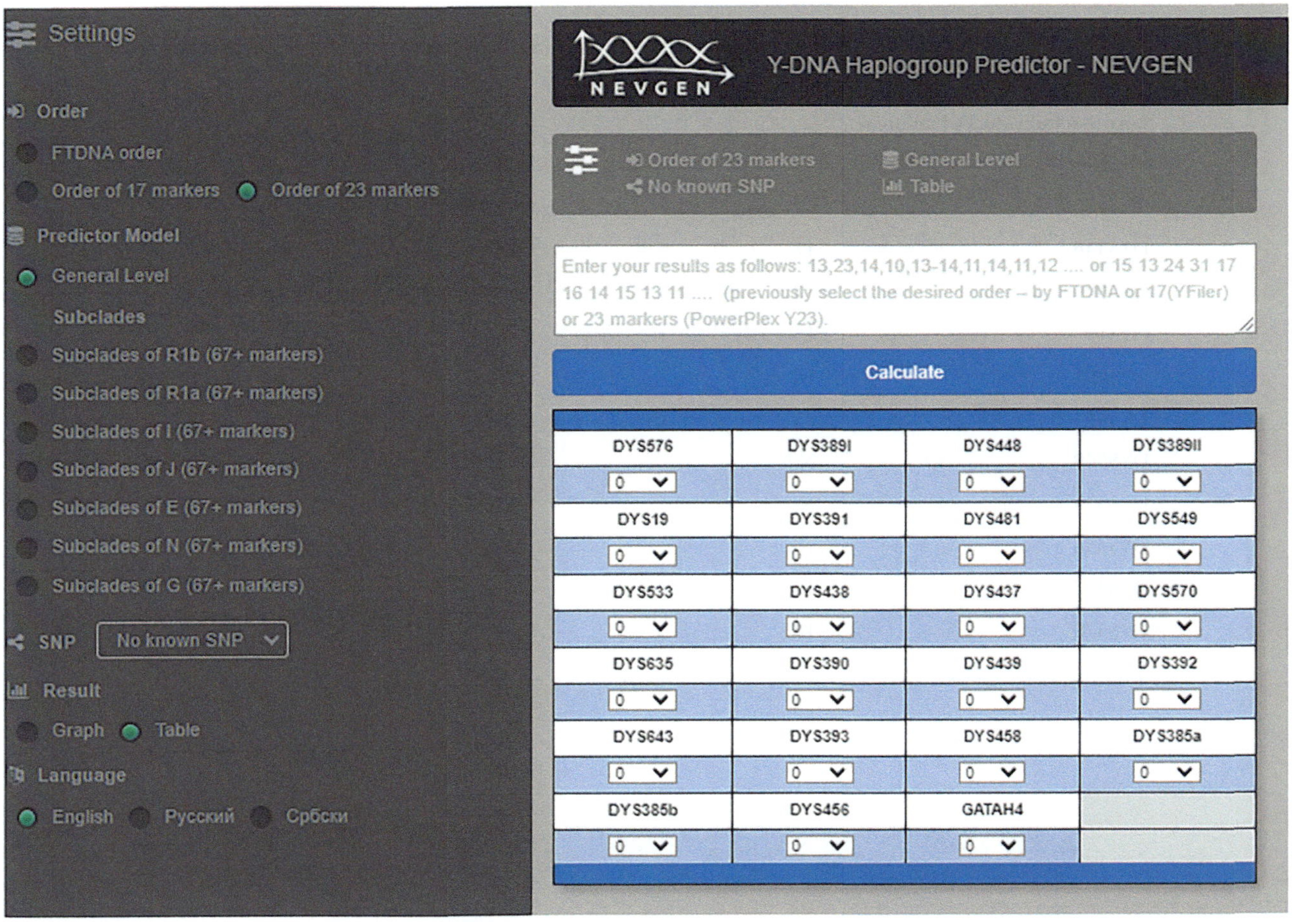

FIGURE 22.2 NevGen Y-DNA Haplogroup Predictor homepage. Screenshot showing the graphical interface of the NevGen Haplogroup Predictor tool.
Screenshot taken from https://www.nevgen.org/.

for the first time because Windows does not distribute it. After the program is executed, you should take the following steps:

Step 1: Select the predictor function in the software (Fig. 22.6).

Step 2: Select the prediction method; in this example, general analysis was selected (Fig. 22.7).

Step 3: Select the forensic set of Y-STRs. Selection ranges from the 17 Y-STR markers available in YFiler™ to the 23 markers available in the PowerPlex Y23® Kit. For this approach, as demonstrated in Fig. 22.7, the program does not provide the option to analyze all the 111 markers available on FTDNA. However, you can assign the 17 or 23 markers in the specified order, as shown in Fig. 22.8. You should systematically order the Y-STRs according to the table displayed on the homepage where the respective kit is selected (Fig. 22.8).

Step 4: Construct a list of haplotypes. Each row will represent the Y-chromosome of a single person, with the alleles separated by single spaces. At the end of each row, there is an optional semicolon ';' so that you can write any notes at the end of the row. Click the "Batch calculate" button to infer haplogroups from Y-STR data. Below is an example following the same Japanese individual used previously, according to the available markers corresponding to the previous test: 19 13 17 30 15 10 25 $ $ 10 14 18 21 25 13 11 14 13 15 $ $ 16 12; HGDP747.

To analyze several individuals simultaneously, just list them one immediately after the other, as indicated below. (Note: to evaluate the inference results, we have included the actual haplogroup determined by Y-SNPs after the HGDP code for each individual): 19 13 18 30 13 10 23 $ $ 11 14 19 22 24 11 14 9 13 16 $ $ 15 12; HGDP837 Q 18 13 20 30 13 10 20 $ $ 11 14 17 22 24 12 15 8 13 17 $ $ 15 11; HGDP1015 Q 18 13 20 28 15 10 22 $ $ 10 15 19 22 23 13 11 11 12 18 $ $ 14 11; HGDP665 I 19 13 20 28 17 10 21 $ $ 10 15 19 22 23 11 11 11 13 17 $ $ 14 13; HGDP670 I 19 13 17 30 15 10 25 $ $ 10 14 18 21 25 13 11 14 13 15 $ $ 16 12; HGDP747 D 19 14 18 29 15 10 20 $ $ 13 14 18 20 22 13 13 12 13 20 $ $ 15 12;

TABLE 22.2 Allelic values for each short tandem repeat (STR) of the PowerPlex® Y23 system associated with a Japanese individual, present in HGDP.

Japanese individual's haplotype			
DYS576	19	DYS635	21
DYS389I1	13	DYS390	25
DYS448	17	DYS439	13
DYS389I2	30	DYS392	11
DYS19	15	DYS643	14
DYS391	10	DYS393	13
DYS481	25	DYS458	15
DYS549	N/A	DYS385a	N/A
DYS533	N/A	DYS385b	N/A
DYS438	10	DYS456	16
DYS437	14	GATA_H4	12
DYS570	18		

Note: The value reported for DYS389I2 is the combined value of the two copies.

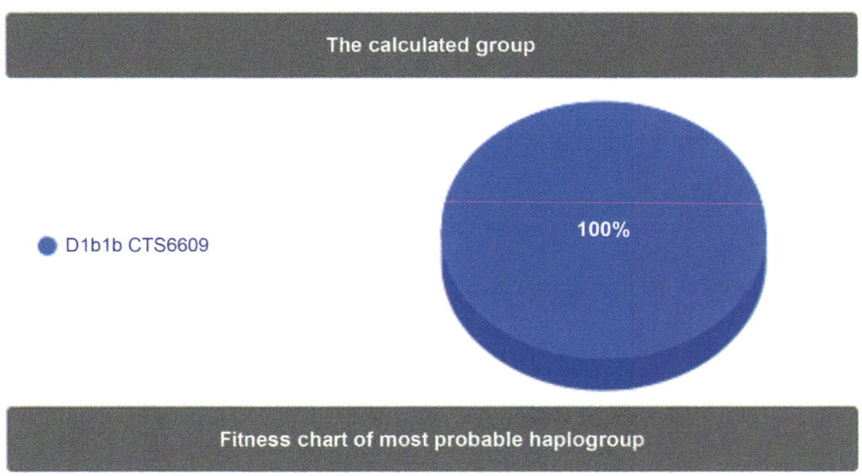

FIGURE 22.3 Probability of the provided haplotype being associated with the retrieved haplogroups listed on the left side of the image. Screenshot showing the graphical interface of the NevGen Haplogroup Predictor tool.
Screenshot taken from https://www.nevgen.org/.

HGDP748 O 17 11 20 27 14 10 22 $ $ 11 15 18 21 23 12 14 12 12 19 $ $ 15 11; HGDP749 O 20 14 19 31 16 10 25 $ $ 10 14 15 21 25 12 11 12 13 15 $ $ 15 11; HGDP752 D 19 12 19 30 15 10 21 $ $ 10 16 $ 19 23 12 12 11 12 18 $ $ 15 12; HGDP753 O 19 13 21 30 16 10 24 $ $ 11 14 18 21 21 12 11 13 15 18 $ $ 16 11; HGDP927 E 15 13 21 31 15 10 27 $ $ 10 14 20 21 21 12 11 14 13 18 $ $ 13 12; HGDP930 E 19 13 20 31 15 9 25 $ $ 11 14 19 23 21 10 11 14 14 18 $ $ 16 11; HGDP932 E.

The output will then be displayed in the graphical interface of the software (Fig. 22.9). The histograms previously depicted in Fig. 22.5 will be displayed on the right side, allowing you to click on any haplogroup you wish to view, but the image may appear incomplete on some computers.

Graphical view of probability and fitness

	Haplogroup	Probability	Fitness	Fitness 2
1	**D1b1b CTS6609**	**100**	**48.27**	**1.04**
2	O1b1 F2320	0	26.41	0.68
3	E1b1b M123* (xM34)	0	26.81	0.52
4	E1b1b V1515	0	20.1	0.57
5	E1b1b V22	0	17.91	0.47
6	E1b1b > V12	0	16.61	0.48
7	E1b1b M123>M34> M84	0	17.02	0.42
8	E1b1b M123> PF4428	0	16.58	0.37
9	E1b1b M123>M34> Z841	0	14.9	0.38
10	E1b1b > V13	0	14	0.31
11	B2b M112	0	12.87	0.32
12	D1b1a M125	0	12.49	0.22
13	I2a1 Isles	0	11.87	0.29
14	I2a2a M223 (for 67+ markers, try level for I-s)	0	10.58	0.29
15	O2a2 F525	0	9.79	0.31
16	E1a M132	0	9.2	0.27
17	R1a M198 (for 67+ markers, try level for R1a-s, 120+ subclades)	0	9.35	0.23
18	J1a2a1a2 P58	0	7.46	0.17
19	J2a1 Z7700	0	7.73	0.21
20	D1b2a CTS220	0	14.49	0.28
21	E1b1b M123> BY45422	0	14.01	0.22
22	C1a2 V20	0	11.49	0.29
23	C1 F3393	0	11.08	0.21
24	E1b1b > V1133	0	10.98	0.24
25	E1b1a V38> M329	0	10.59	0.23

FIGURE 22.4 Inferred haplogroups displayed in a decreasing probability order. Screenshot showing the graphical interface of the NevGen Haplogroup Predictor tool.
Screenshot taken from https://www.nevgen.org/.

The results may exhibit certain variations of the following pattern: 1. ——————— 19 13 18 30 13 10 23 $ $ 11 14 19 22 24 11 14 9 13 16 $ $ 15 12; HGDP837 Q Probability of unsupported subclade: 3.52% Probability=96.20% Fitness=31.42 [0.92] Q M346> >Z780 Probability=0.16% Fitness=23.55 [0.49] Q M346> >M3>Y4276 Probability=0.12% Fitness=22.93 [0.73] Q M346> >M3>M902 Probability=0.00% Fitness=20.40 [0.41] Q M346> >Y4800>Z5902 Probability=0.00% Fitness=15.56 [0.37] R1b (for 67+ markers, try level for R1b-s, 300+ subclades) Probability=0.00% Fitness=17.08 [0.39] T> >CTS11451>PF7455 Probability=0.00% Fitness=13.47 [0.37] E1b1b>V12 Probability=0.00% Fitness=14.65 [0.31] T> >CTS11451>Y4119 Probability=0.00% Fitness=15.97 [0.39] R1b V88 Probability=0.00% Fitness=14.03 [0.31] Q M346> >Y4800>F835>L932 Probability=0.00% Fitness=13.03 [0.39] O2a2 F525 Probability=0.00% Fitness=11.33 [0.22] J1a> >P58> >L858> >Z640>BY74.

In the structure displayed above, the first line indicates the number (in the same order as the input data) of the individual being analyzed (in this case, the first individual from the batch analysis). The second line shows the individual's query

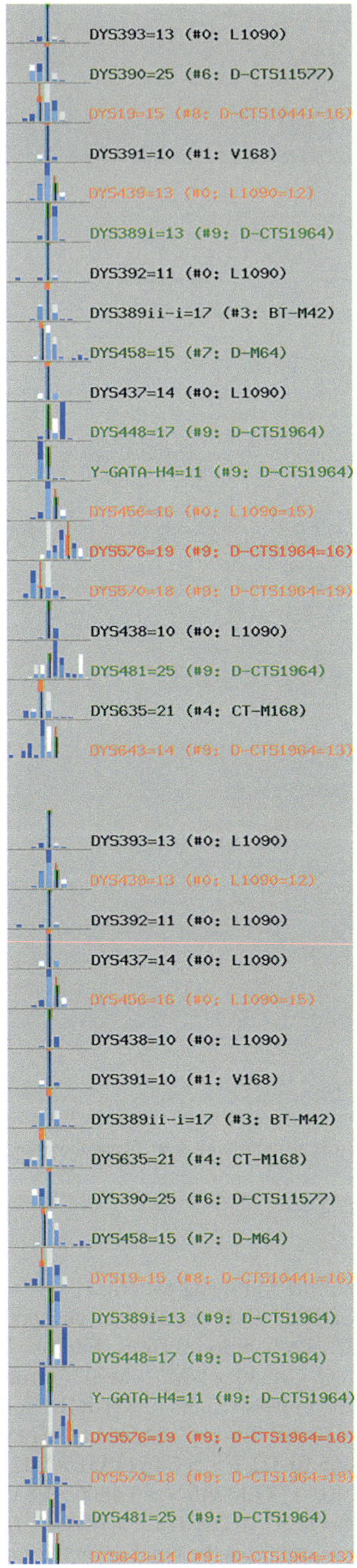

FIGURE 22.5 Histograms displaying allele frequency distributions for each Y-STR marker considered for inferring haplogroups. Black vertical lines indicate alleles within the expected values for a given Y-STR, while red vertical lines indicate markers deviating from the standard. Screenshot showing the graphical interface of the NevGen Haplogroup Predictor tool.
Screenshot taken from https://www.nevgen.org/.

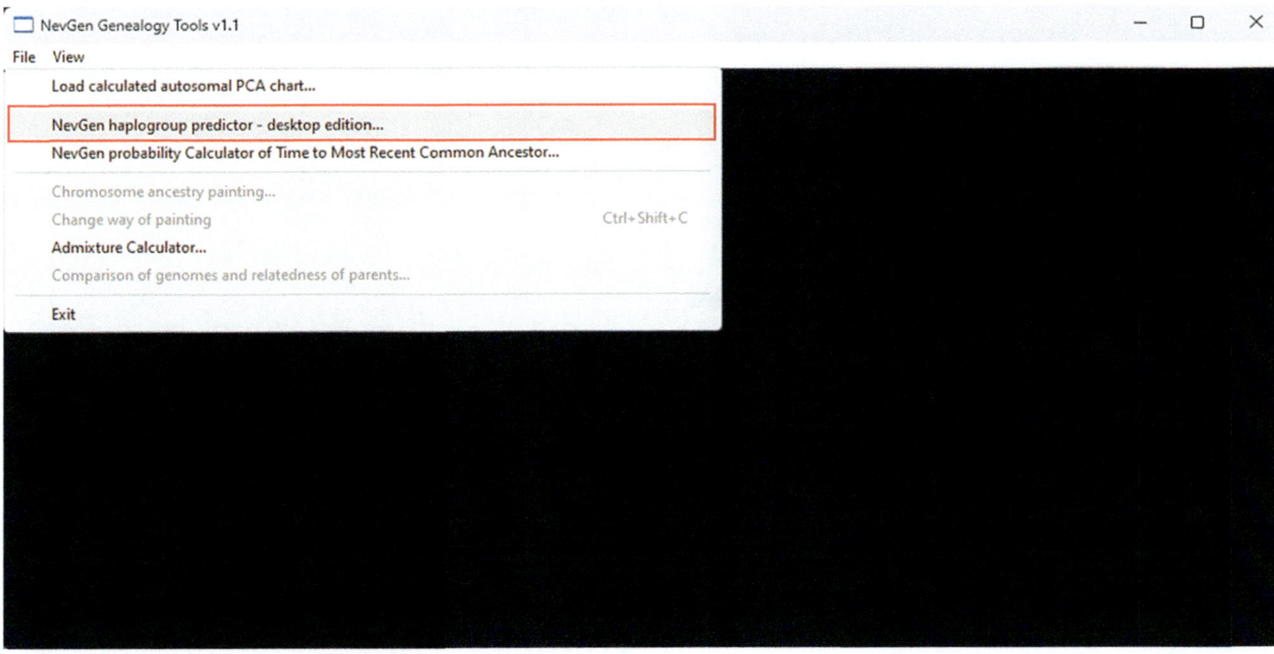

FIGURE 22.6 Panel to select the predictor function in NevGen Genealogy Tools. The desired function is indicated inside the red rectangle. Screenshot showing the graphical interface of the NevGen Haplogroup Predictor tool.
Screenshot taken from NevGen Genealogy Tools v1.1 (desktop edition).

haplotype followed by optional annotations submitted to the program (in this case, the individual ID). Then, further down on the third line, there is an estimate of the probability that this query haplotype is derived from an unsampled haplogroup. The remaining lines are the most probable haplogroups inferred from the Y-STR haplotype, arranged in decreasing order of probability. The output is presented without warnings when both fitness score and relative fitness, which is the value inside brackets, are high enough (Fitness > 30; Relative Fitness > 0.80): 30. ———————— 16 13 20 31 15 10 28 $ $ 11 14 20 22 21 12 11 13 13 17 $ $ 15 12; HGDP937 E Probability=100.00% Fitness=51.35 [1.18] E1b1a V38> >M4231.

Some comments are likely to follow the results. For example, when the fitness score values are low, there is a possibility of false positives for haplotypes that have little or no support or are indeed very rare. This message can accompany even results with 100% probability.

22.5.2.2 Whit Athey's Haplogroup Predictor (HAPEST5)

Available since October 2004, this widely used program allows users to input any number of markers from the Family Tree DNA 111-marker set and calculates a "goodness of fit" score for 21 haplogroups. The software is available at https://www.hprg.com/hapest5/.

22.5.2.2.1 Individual samples

To use the Y Haplogroup Prediction from Y-STR Values, also known as HAPEST5, you must visit their website at http://www.hprg.com/hapest5/ and select the desired method. There are models based on the prediction of 21 or 27 haplogroups, where you can enter the order of up to 86 Y-STR markers for each individual, either numerically or by following the FTDNA arrangement. There is also a third model that deals with 111 different Y-STR markers in the Y chromosome, although it returns a result with only 21 tested haplogroups. Nevertheless, you do not have to fill in the table to obtain a correct result with a good fitness score.

This is how to run the analysis by using HAPEST:

Step 1: You should define the model to conduct the analysis. The webpage http://www.hprg.com/hapest5/ will be used, and for this example, the model based on 27 haplogroups following the FTDNA order will be selected (Fig. 22.10):

A new tab will open, displaying the table with all the markers available for use. A table on the right side will show all the haplogroups supported by the program, and their probability values and fitness scores will be updated as the markers are entered.

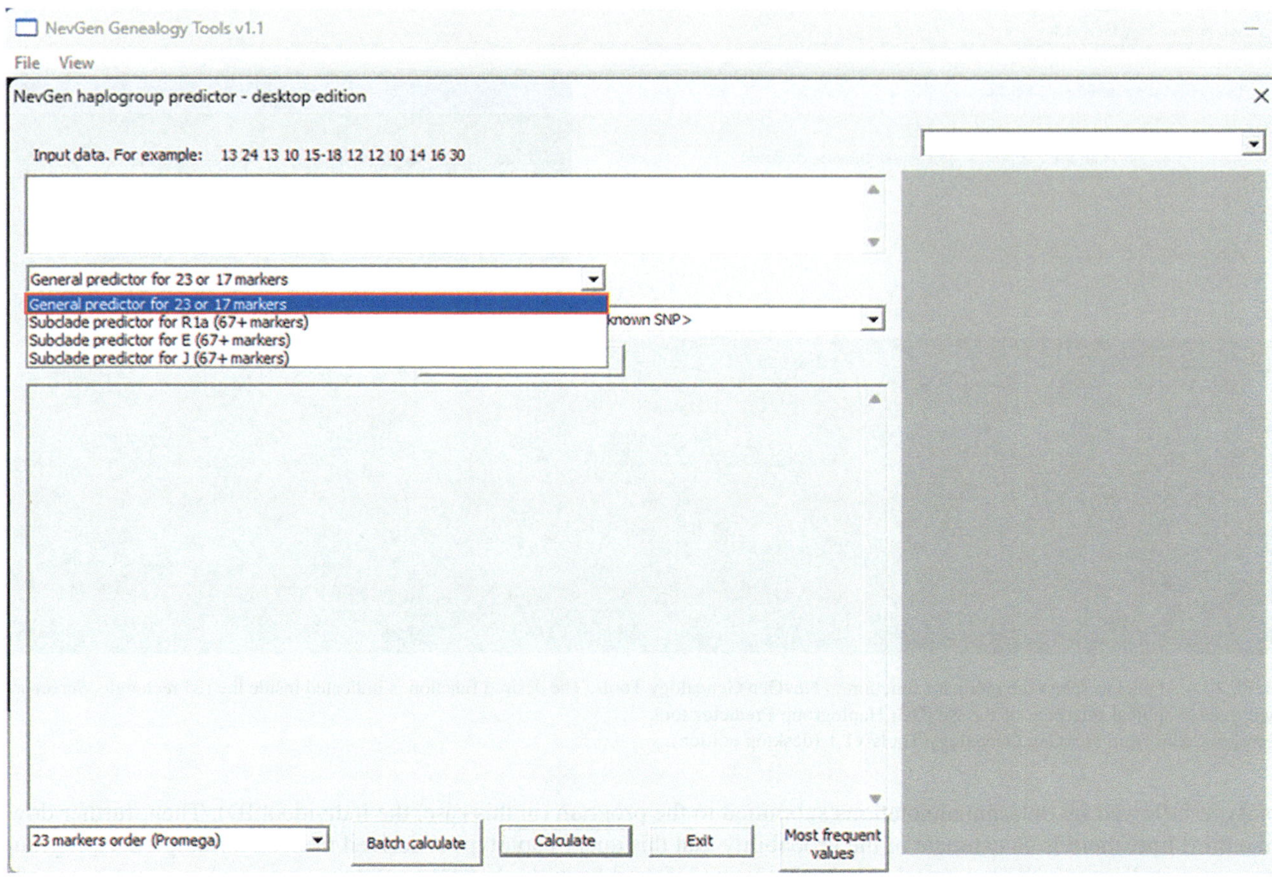

FIGURE 22.7 Panel to select the prediction method of the NevGen Haplogroup Predictor function. The desired method is indicated inside the red rectangle. Screenshot showing the graphical interface of the NevGen Haplogroup Predictor tool.
Screenshot taken from NevGen Genealogy Tools v1.1 (desktop edition).

Step 2: You can enter an individual allele for each Y-STR by using a drop-down menu, or all at once as a text line pasted in the lower portion of the page (Fig. 22.11).

For this example, we will use the Japanese haplotype (Table 22.1); we will use "0" as a character for undetected or unavailable Y-STR markers in the analysis. The input text line should be inserted as follows: 13,25,15,10,0,0,0,0,13,13,11,30,15,0,0,0,0,0,14,17,0,0,0,0,0,12,0,0,16,0,19,18,0,0,0,10,0,0,0,0-
,0,0,0,0,0,0,0,0,0,0,0,0,0,0,0,0,25,0,0,0,0,0,0,0,0,0,0,0,21,0,0,0,0,0,0,0,0,0,0,0,0,0,0,14.

The output should appear on the right side of the introduced Y-STR data (Fig. 22.12). The first column will display the possible haplogroups, the second column will show the fitness scores, and the last column will present the probability of the inference.

Ideally, haplogroup inference software based on Y-STRs should rely on a large Y-chromosome data repository, to grant high representativeness of the Y-STR profiles associated with each haplogroup, thereby resulting in highly reliable inferences. This means their databases should include the largest possible number of haplogroups, including less common haplogroups or those restricted to specific regions (Jannuzzi et al., 2020), as well as information regarding the corresponding mutation rates for some subclades (Petrejčíková et al., 2014). Despite their popularity, HAPEST5's and NevGen's databases differ, which culminates in potential variations in results and jeopardizes the accuracy of haplogroup assignments. Jannuzzi et al. (2020) mentioned that both tools incurred large error rates when it came to determining haplogroups for admixed African American ancestries. Núñez et al. (2012) observed that HAPEST5 required more reference data for haplogroups C, H, L, N, and Q (Núñez et al., 2012), while Muzzio et al. (2011) reported high prediction errors for HAPEST5 toward clade R (Muzzio et al., 2011). This bias appears to be subhaplogroup specific, given that Petrejčíková et al. (2014) controversially obtained adequate inferences for haplogroup R in their samples.

For instance, here we observed that the table of results of HAPEST5 (Figs. 22.11 and 22.12) does not include representative samples from Haplogroup D, mainly present in Central Asia, Southeast Asia, and Japan. This does not

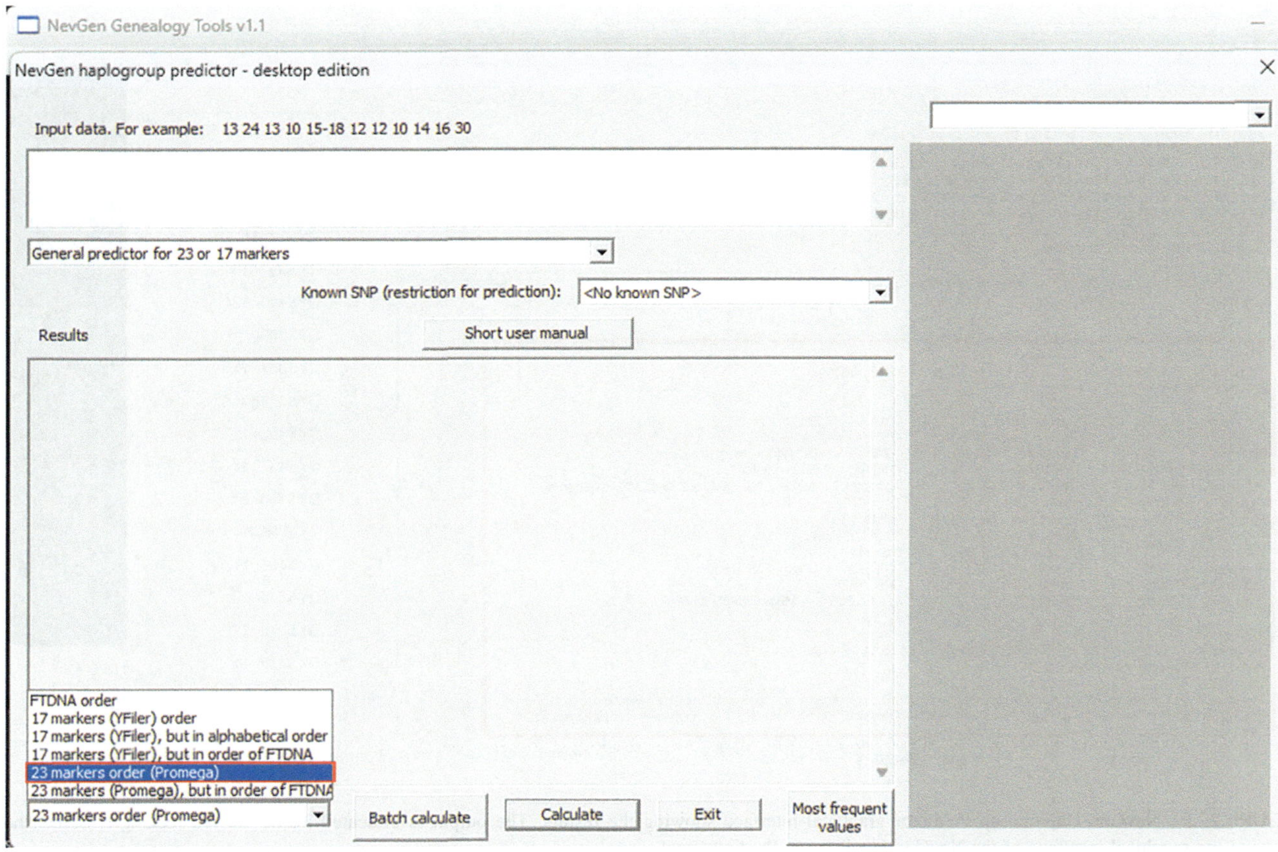

FIGURE 22.8 Panel to select the markers of the NevGen Haplogroup Predictor function. The desired markers are indicated inside the red rectangle. Screenshot showing the graphical interface of the NevGen Haplogroup Predictor tool.
Screenshot taken from NevGen Genealogy Tools v1.1 (desktop edition).

prevent accurate predictions for other well-represented haplogroups, but it compromises the analysis of individuals whose haplogroups are not sampled. For example, in the case of the Japanese individual analyzed in this tutorial, because HAPEST5 does not have a high representativeness of haplogroup D, the program will offer a different haplogroup prediction. As mentioned before, this may be due to a lack of samples from a specific region or to no record of a certain haplogroup in that region, as can be easily seen in Fig. 22.1. However, the low fitness score indicates the inference is uncertain in such situations.

Therefore, the fitness score value should always be considered in such analyses, and a flag suggesting potential variation in the results should be raised for low values. Thus, each inference must be carefully analyzed, especially in situations where the inference is provided without any prior knowledge of the individual's actual biogeographical origin. This also underscores the need for databases to be effectively representative of all regions or as many regions as possible.

To compare the accuracy of the two programs, we used the two tools described above based on haplotypes of 23 Y-STRs to reassign 31 admixed individuals from Latin America whose haplogroups had been determined by Y-SNPs. For the 10 African-American individuals, both programs achieved 90% accuracy. HAPEST5 had 10% errors, while NevGen did not achieve the minimum fitness score in 10% of the cases. Among 10 individuals with Amerindian haplogroups, HAPEST5 demonstrated 100% accuracy, whereas NevGen performed less efficiently—accuracy was 60%, and the minimum fitness score was not met in 40% of the cases. For 10 Brazilian individuals with European haplogroups, both tools correctly classified 80% of them, but HAPEST5 had a higher error rate (20%). In contrast, NevGen had a combination of errors (10%) and cases where the minimum fitness score was not achieved (10%). For the unique Asian haplogroup, both tools achieved 100% accuracy. Overall, HAPEST5 had a total accuracy of 92.5%, while NevGen achieved a lower accuracy of 82.5%. This comparison highlights that the tools have different performances, with HAPEST5 generally making fewer errors, while NevGen struggles more with cases that do not meet the minimum fitness score, particularly in the Amerindian and European groups. Nevertheless, we have been able to demonstrate the accuracy of both programs in predicting the haplogroups from admixed populations.

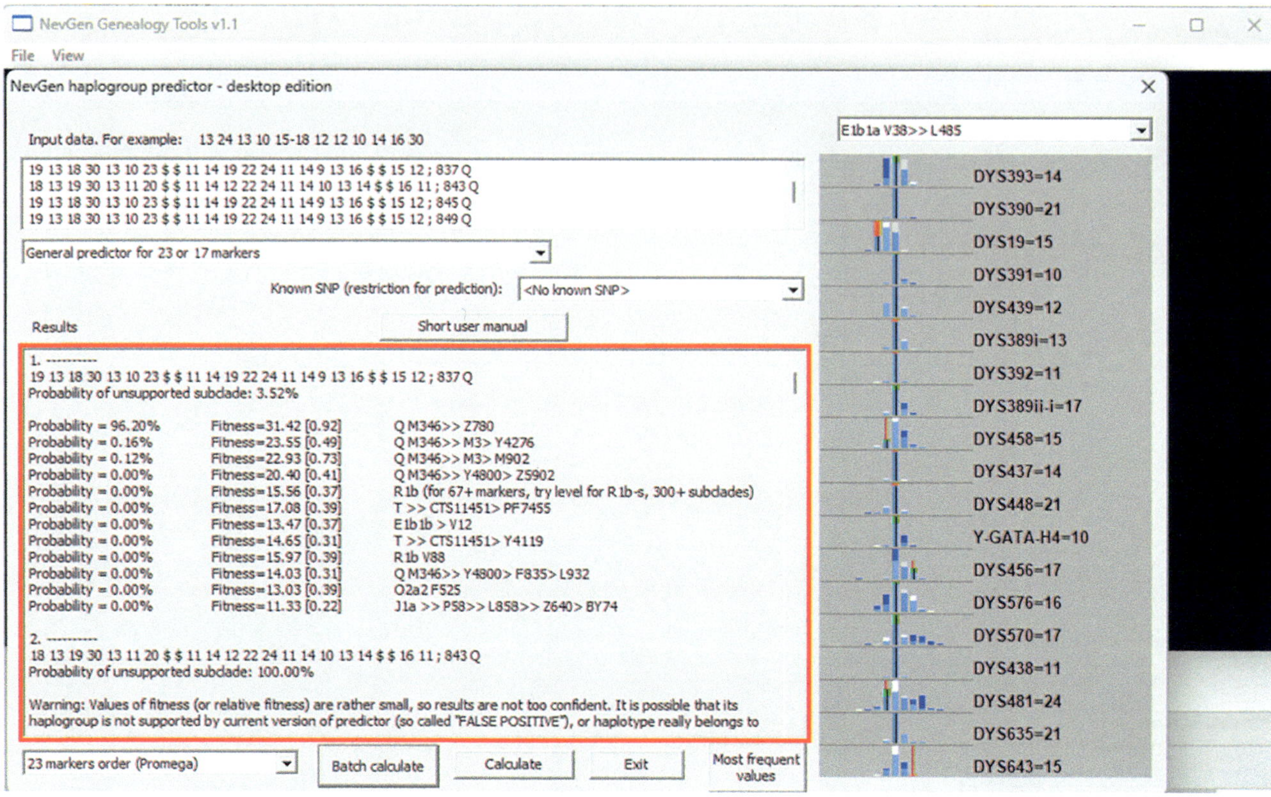

FIGURE 22.9 NevGen Haplogroup Predictor graphical interface showing the results. The output is indicated inside the red rectangle. Screenshot showing the graphical interface of the NevGen Haplogroup Predictor tool.
Screenshot taken from NevGen Genealogy Tools v1.1 (desktop edition).

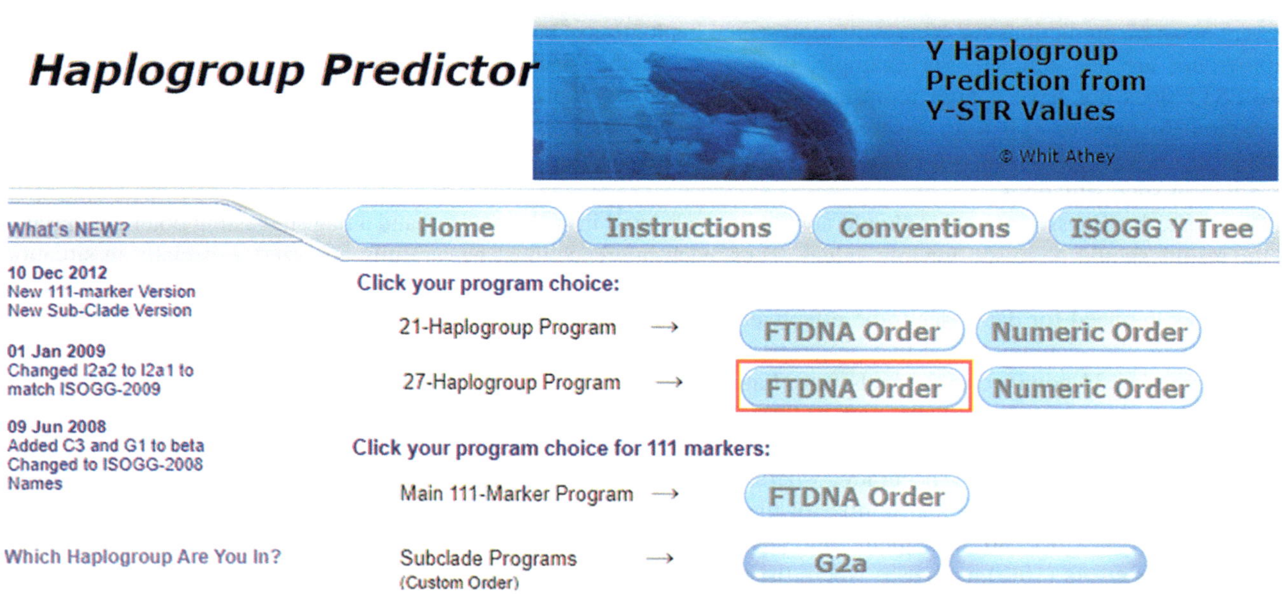

FIGURE 22.10 Haplogroup Predictor (HAPEST5) homepage. The red box indicates which option is being selected. Screenshot showing the graphical interface of the Haplogroup Predictor (HAPEST5) tool.
Screenshot taken from https://www.hprg.com/hapest5/index.html.

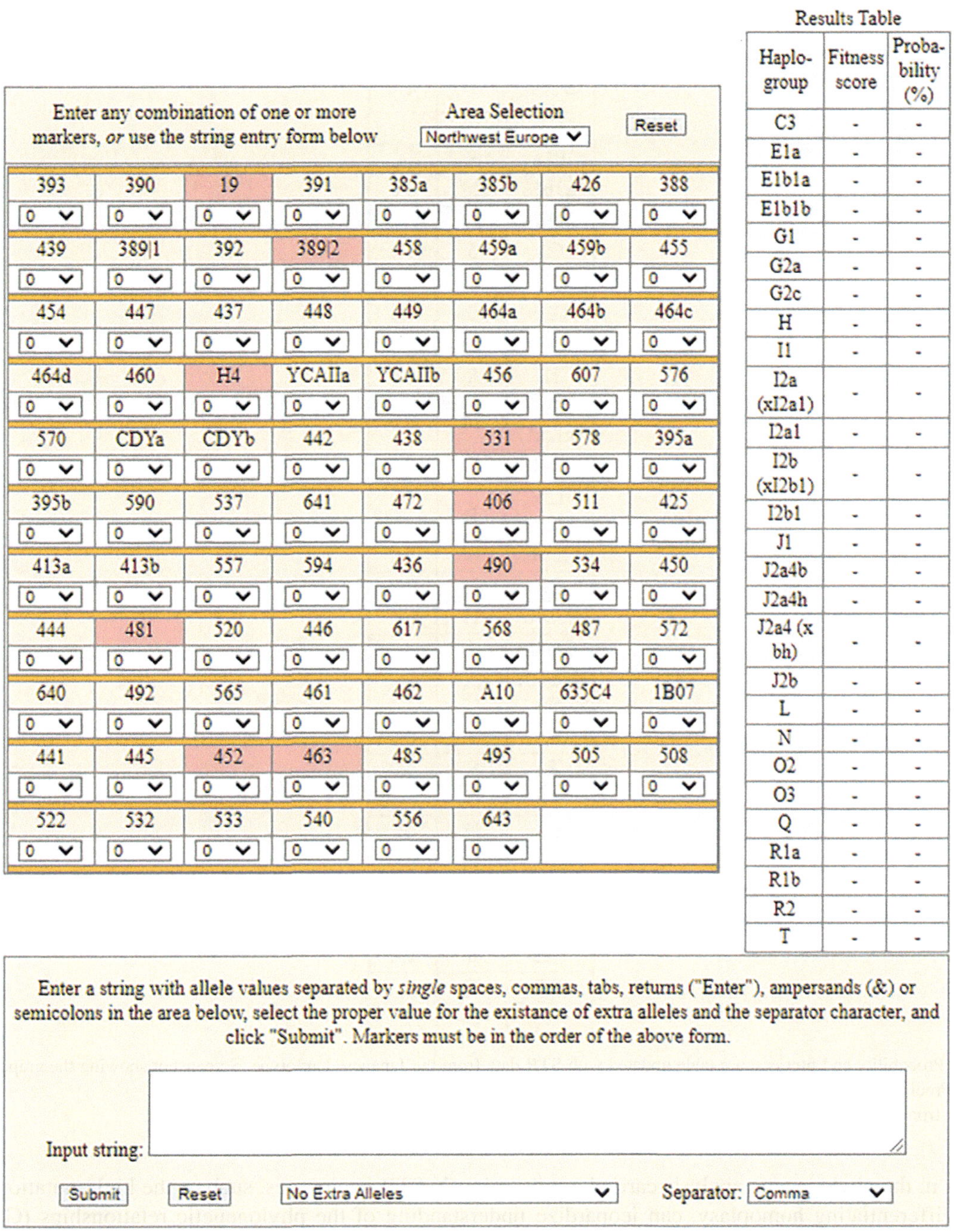

FIGURE 22.11 Allele input panel of the Haplogroup Predictor (HAPEST5) tool. Screenshot showing the graphical interface of the Haplogroup Predictor (HAPEST5) tool.
Screenshot taken from https://www.hprg.com/hapest5/index.html.

22.6 Final considerations

The tools we used to infer haplogroups through haplotypes proved easy and intuitive, but each algorithm for haplogroup classification employs distinct methods and has the potential for error. Combining their outputs has been suggested for making more robust decisions, which in turn should lead to highly accurate haplogroup assignments (Schlecht et al., 2008). Additionally, developments in the ML area are expected to be highly promising and could significantly advance haplogroup prediction methodologies (Bouakaze et al., 2020).

Results Table

Haplo-group	Fitness score	Proba-bility (%)
C3	18	-
E1a	8	-
E1b1a	20	0.0
E1b1b	29	51.8
G1	5	-
G2a	8	0.0
G2c	0	0.0
H	10	0.0
I1	8	0.0
I2a (xI2a1)	24	0.6
I2a1	7	0.0
I2b (xI2b1)	6	-
I2b1	12	0.0
J1	11	0.0
J2a4b	15	-
J2a4h	7	-
J2a4 (x bh)	10	-
J2b	3	0.0
L	9	0.0
N	5	-
O2	18	0.0
O3	14	0.0
Q	12	0.0
R1a	26	47.6
R1b	4	0.0
R2	7	-
T	19	-

FIGURE 22.12 Probability and fitness score table updated with STR data from the Japanese haplotype. Screenshot showing the graphical interface of the Haplogroup Predictor (HAPEST5) tool.
Screenshot taken from https://www.hprg.com/hapest5/index.html.

Limitations in the phylogenetic analysis carried out by using Y-STR haplotypes, such as the high mutation rate and the difficulty in differentiating homoplasy, can jeopardize understanding of the phylogenetic relationships (Gusmão et al., 2003) and reinforce that accurately defining phylogenetic branches of parental lineage still requires validation through Y-SNPs analysis (Petrejčíková et al., 2014). However, inferring haplogroups through Y-SNP markers is a more costly, time-consuming, and laborious process that requires multiple markers to define the haplogroup to which an individual belongs. In addition, the forensic routine does not include Y-SNP genotyping due to its limited informativeness for human identification purposes, low quantities of DNA obtained from traces, high costs, and restricted personal resources, among other factors (Babić Jordamović et al., 2021; Kayser, 2017).

Given that the relationship between Y-STRs and Y-chromosome haplogroups has been described to be strong enough to determine the population's geographical origin of a sampled chromosome based on haplotypes (Petrejčíková et al., 2014), Y-chromosome markers have proven indispensable in forensic genetics and offer invaluable insights in cases where autosomal markers may fall short. Looking ahead, integrating ML algorithms with Y-STR and Y-SNP data promises to enhance the precision of haplogroup assignments and forensic predictions.

Acknowledgments

We thank Cynthia Maria de Campos Prado Manso for her assistance with the English language. This study was financed in part by the Coordenação de Aperfeiçoamento de Pessoal de Nível Superior—Brasil (CAPES)—Finance Code 001 and Programa de Cooperação Acadêmica em Segurança Pública e Ciências Forenses (PROCAD—Segurança Pública e Ciências Forenses) Grants #88887.516236/2020-00 and #88881.516238/2020-01. C.T.M.J. (#310016/2022-3) is supported by a Research Fellowship from CNPq/Brazil.

References

Ali, S., & Hasnain, S. E. (2002). Molecular dissection of the human Y-chromosome. *Gene, 283*(1-2), 1–10. https://doi.org/10.1016/S0378-1119(01)00860-5.

Athey, T. W. (2005). Haplogroup prediction using an allele-frequency approach. *Journal of Genetic Genealogy, 1*(1), 1–7.

Babić Jordamović, N., Kojović, T., Dogan, S., Bešić, L., Salihefendić, L., Konjhodžić, R., Škaro, V., Projić, P., Hadžiavdić, V., Ašić, A., & Marjanović, D. (2021). Haplogroup prediction using Y-chromosomal short tandem repeats in the general population of Bosnia and Herzegovina. *Frontiers Media S.A., Bosnia and Herzegovina Frontiers in Genetics, 12*. https://doi.org/10.3389/fgene.2021.671467.

Bouakaze, C., Delehelle, F., Saenz-Oyhéréguy, N., Moreira, A., Schiavinato, S., Croze, M., Delon, S., Fortes-Lima, C., Gibert, M., Bujan, L., Huyghe, E., Bellis, G., Calderon, R., Hernández, C. L., Avendaño-Tamayo, E., Bedoya, G., Salas, A., Mazières, S., Charioni, J., ... Balaresque, P. (2020). Predicting haplogroups using a versatile machine learning program (PredYMaLe) on a new mutationally balanced 32 Y-STR multiplex (CombYplex): Unlocking the full potential of the human STR mutation rate spectrum to estimate forensic parameters. *Forensic Science International: Genetics, 48*, 102342. https://doi.org/10.1016/j.fsigen.2020.102342.

Bradbury, C., Köttgen, A., & Staubach, F. (2019). Off-target phenotypes in forensic DNA phenotyping and biogeographic ancestry inference: A resource. *Forensic Science International: Genetics, 38*, 93–104. https://doi.org/10.1016/j.fsigen.2018.10.010.

Butler, J. M. (2023). Recent advances in forensic biology and forensic DNA typing: INTERPOL review 2019–2022. *Forensic Science International: Synergy, 6*, 100311. https://doi.org/10.1016/j.fsisyn.2022.100311.

Canales Serrano, A. (2020). Forensic DNA phenotyping: a promising tool to aid forensic investigation. *Journal of Legal Medicine. 46*(4), 183–190. https://doi.org/10.1016/j.reml.2020.01.003, https://doi.org/10.1016/j.remle.2020.01.002.

Cetkovic Gentula, A., & Nevski, M. (2015). NEVGEN Y-DNA Haplogroup Predictor [Software]. http://www.nevgen.org/.

Chen, H., Lu, Y., Lu, D., & Xu, S. (2021). Y-LineageTracker: a high-throughput analysis framework for Y-chromosomal next-generation sequencing data. *BMC Bioinformatics, 22*(1). https://doi.org/10.1186/s12859-021-04057-z.

da Cruz, P. R. S., Ananina, G., Secolin, R., Gil-Da-Silva-Lopes, V. L., Lima, C. S. P., de França, P. H. C., Donatti, A., Lourenço, G. J., de Araujo, T. K., Simioni, M., Lopes-Cendes, I., Costa, F. F., & de Melo, M. B. (2022). Demographic history differences between Hispanics and Brazilians imprint haplotype features. *G3: Genes|Genomes|Genetics, 12*(7). https://doi.org/10.1093/g3journal/jkac111.

Doo, M., & Kim, Y. (2015). Obesity: Interactions of genome and nutrients intake. *Preventive Nutrition and Food Science, 20*(1), 1–7. https://doi.org/10.3746/pnf.2015.20.1.1.

Elgart, M., Goodman, M. O., Isasi, C., Chen, H., Morrison, A. C., de Vries, P. S., Xu, H., Manichaikul, A. W., Guo, X., Franceschini, N., Psaty, B. M., Rich, S. S., Rotter, J. I., Lloyd-Jones, D. M., Fornage, M., Correa, A., Heard-Costa, N. L., Vasan, R. S., Hernandez, R., ... Sofer, T. (2022). Correlations between complex human phenotypes vary by genetic background, gender, and environment. *Cell Reports Medicine, 3*(12), 100844. https://doi.org/10.1016/j.xcrm.2022.100844.

Emmerova, B., Ehler, E., Comas, D., Votrubova, J., & Vanek, D. (2017). Comparison of Y-chromosomal haplogroup predictors. *Forensic Science International: Genetics Supplement Series, 6*, e145. https://doi.org/10.1016/j.fsigss.2017.09.025.

Fan, H., & Chu, J.-Y. (2007). A brief review of short tandem repeat mutation. *Genomics, Proteomics and Bioinformatics, 5*(1), 7–14. https://doi.org/10.1016/S1672-0229(07)60009-6.

Fitzpatrick, C. (2022). Forensic genetic approaches for identification of human skeletal remains: Challenges, best practices, and emerging technologies. In: Ambers, A. (Ed.), *The emerging discipline of forensic genetic genealogy*. . Elsevier, United States. https://doi.org/10.1016/B978-0-12-815766-4.00022-4.

Fleckhaus, J., & Schneider, P. M. (2020). Novel multiplex strategy for DNA methylation-based age prediction from small amounts of DNA via pyrosequencing. *Forensic Science International: Genetics, 44*, 102189. https://doi.org/10.1016/j.fsigen.2019.102189.

Garcia, F. M., Bessa, B. G. O., dos Santos, E. V. W., Pereira, J. D. P., Alves, L. N. R., Vianna, L. A., Casotti, M. C., Trabach, R. S. R., Stange, V. S., Meira, D. D., & Louro, I. D. (2022). Forensic applications of markers present on the X chromosome. *Genes, 13*(9). https://doi.org/10.3390/genes13091597.

Gentile, F., Castoldi, E., Serventi, P., Colloca, D., Ciccotelli, R., & Marino, A. (2019). Early evaluation of five age-correlated DNA methylation markers in an Italian population sample. *Forensic Science International: Genetics Supplement Series, 7*(1), 424–426. https://doi.org/10.1016/j.fsigss.2019.10.037.

Gusmão, L., Sánchez-Diz, P., Alves, C., Beleza, S., Lopes, A., Carracedo, A., & Amorim, A. (2003). Grouping of Y-STR haplotypes discloses European geographic clines. *Forensic Science International, 134*(2-3), 172–179. https://doi.org/10.1016/S0379-0738(03)00160-9.

Gusmão, L., Brión, M., & Gomes, I. (2008). Handbook of Analytical SeparationsThe human Y chromosome male-specific polymorphisms and forensic genetics. *Handbook of Analytical Separations, 6*, 969–1000. https://doi.org/10.1016/S1567-7192(06)06034-7.

Gusmão, L., Butler, J. M., Linacre, A., Parson, W., Roewer, L., Schneider, P. M., & Carracedo, A. (2017). Revised guidelines for the publication of genetic population data. *Forensic Science International: Genetics, 30*, 160–163. https://doi.org/10.1016/j.fsigen.2017.06.007.

Hallast, P., Agdzhoyan, A., Balanovsky, O., Xue, Y., & Tyler-Smith, C. (2021). A Southeast Asian origin for present-day non-African human Y chromosomes. *GmbH, Estonia Human Genetics, 140*(2), 299–307. https://doi.org/10.1007/s00439-020-02204-9.

Hamczyk, M. R., Nevado, R. M., Barettino, A., Fuster, V., & Andrés, V. (2020). Biological versus chronological aging: JACC Focus Seminar. *Journal of the American College of Cardiology, 75*(8), 919–930. https://doi.org/10.1016/j.jacc.2019.11.062.

Heidegger, A., Pisarek, A., de la Puente, M., Niederstätter, H., Pośpiech, E., Woźniak, A., Schury, N., Unterländer, M., Sidstedt, M., Junker, K., Ventayol Garcia, M., Laurent, F. X., Ulus, A., Vannier, J., Bastisch, I., Hedman, J., Sijen, T., Branicki, W., Xavier, C., & Parson, W. (2022). VISAGE consortium, development and inter-laboratory validation of the VISAGE enhanced tool for age estimation from semen using quantitative DNA methylation analysis. *Forensic Science International. Genetics. 56*, 102596. https://doi.org/10.1016/j.fsigen.2021.102596.

Henn, B. M., Cavalli-Sforza, L. L., & Feldman, M. W. (2012). The great human expansion. *Proceedings of the National Academy of Sciences of the United States of America, 109*(44), 17758–17764. https://doi.org/10.1073/pnas.1212380109.

International Society of Genetic Genealogy 2019 2019 https://isogg.org/ Y-DNA Haplogroup Tree 2019-2020, version 15.73 (2020 Jul 11).

Jannuzzi, J., Ribeiro, J., Alho, C., de Oliveira Lázaro e Arão, G., Cicarelli, R., Simões Dutra Corrêa, H., Ferreira, S., Fridman, C., Gomes, V., Loiola, S., da Mota, M. F., Ribeiro-dos-Santos, Â., de Souza, C. A., de Sousa Azulay, R. S., Carvalho, E. F., & Gusmão, L. (2020). Male lineages in Brazilian populations and performance of haplogroup prediction tools. *Forensic Science International: Genetics, 44*, 102163. https://doi.org/10.1016/j.fsigen.2019.102163.

Jobling, M. A. (2012). The impact of recent events on human genetic diversity. *Philosophical Transactions of the Royal Society B: Biological Sciences, 367*(1590), 793–799. https://doi.org/10.1098/rstb.2011.0297.

Jobling, M. A. (2022). Forensic genetics through the lens of Lewontin: Population structure, ancestry and race. *Philosophical Transactions of the Royal Society B: Biological Sciences, 377*(1852). https://doi.org/10.1098/rstb.2020.0422.

Karafet, T. M., Fernando, L. M., Meilerman, Monica B., Underhill, P. A., Zegura, S. L., & Michael, F. H. (2008). New binary polymorphisms reshape and increase resolution of the human Y chromosomal haplogroup tree. *Genome Research, 18*(5), 830–838. https://doi.org/10.1101/gr.7172008.

Kayser, M. (2015). Forensic DNA Phenotyping: Predicting human appearance from crime scene material for investigative purposes. *Forensic Science International. Genetics, 18*, 33–48. https://doi.org/10.1016/j.fsigen.2015.02.003.

Kayser, M. (2017). Forensic use of Y-chromosome DNA: a general overview. *Human Genetics, 136*(5), 621–635. https://doi.org/10.1007/s00439-017-1776-9.

Kayser, M., Branicki, W., Parson, W., & Phillips, C. (2023). Recent advances in Forensic DNA phenotyping of appearance, ancestry and age. *Forensic Science International: Genetics, 65*, 102870. https://doi.org/10.1016/j.fsigen.2023.102870.

Kimura, M., & Ohta, T. (1978). Stepwise mutation model and distribution of allelic frequencies in a finite population. *Proceedings of the National Academy of Sciences of the United States of America, 75*(6), 2868–2872. https://doi.org/10.1073/pnas.75.6.2868.

de Knijff, P. (2000). Messages through bottlenecks: on the combined use of slow and fast evolving polymorphic markers on the human Y chromosome. *American Journal of Human Genetics, 67*(5), 1055–1061. https://doi.org/10.1016/S0002-9297(07)62935-8.

de Knijff, P. (2022). On the forensic use of Y-chromosome polymorphisms. *Genes, 13*(5), 898. https://doi.org/10.3390/genes13050898.

Llamas, B., Rada, X. R., & Collen, E. (2020). Ancient DNA helps trace the peopling of the world. *Biochemist, 42*(1), 18–22. https://doi.org/10.1042/BIO04201018.

Medina-Muñoz, S. G., Ortega-Del Vecchyo, D., Cruz-Hervert, L. P., Ferreyra-Reyes, L., García-García, L., Moreno-Estrada, A., & Ragsdale, A. P. (2023). Demographic modeling of admixed Latin American populations from whole genomes. *Journal of Human Genetics, 110*(10), 1804–1816. https://doi.org/10.1016/j.ajhg.2023.08.015.

Montesanto, A., D'Aquila, P., Lagani, V., Paparazzo, E., Geracitano, S., Formentini, L., Giacconi, R., Cardelli, M., Provinciali, M., Bellizzi, D., & Passarino, G. (2020). A new robust epigenetic model for forensic age prediction. *Journal of Forensic Sciences, 65*(5), 1424–1431. https://doi.org/10.1111/1556-4029.14460.

Moran, C. N., Scott, R. A., Adams, S. M., Warrington, S. J., Jobling, M. A., Wilson, R. H., Goodwin, W. H., Georgiades, E., Wolde, B., & Pitsiladis, Y. P. (2004). Y chromosome haplogroups of elite Ethiopian endurance runners. *Human Genetics, 115*(6), 492–497. https://doi.org/10.1007/s00439-004-1202-y.

Muzzio, M., Ramallo, V., Motti, J. M. B., Santos, M. R., López Camelo, J. S., & Bailliet, G. (2011). Software for Y-haplogroup predictions: A word of caution. *International Journal of Legal Medicine, 125*(1), 143–147. https://doi.org/10.1007/s00414-009-0404-1.

Myers, P. Z. (2007). Tandem repeats and morphological variation. *Nature Education. 1*(1). http://scienceblogs.com/pharyngula/2007/10/tandem_repeats_and_morphologic.php.

Nebel, A., Filon, D., Hohoff, C., Faerman, M., Brinkmann, B., & Oppenheim, A. (2001). Haplogroup-specific deviation from the stepwise mutation model at the microsatellite loci DYS388 and DYS392. *European Journal of Human Genetics, 9*(1), 22–26. https://doi.org/10.1038/sj.ejhg.5200577.

Nielsen, R., Akey, J. M., Jakobsson, M., Pritchard, J. K., Tishkoff, S., & Willerslev, E. (2017). Tracing the peopling of the world through genomics. *Nature, 541*(7637), 302–310. https://doi.org/10.1038/nature21347.

Núñez, C., Geppert, M., Baeta, M., Roewer, L., & Martínez-Jarreta, B. (2012). Y chromosome haplogroup diversity in a Mestizo population of Nicaragua. *Forensic Science International: Genetics, 6*(6), e192. https://doi.org/10.1016/j.fsigen.2012.06.011.

Pena, S. D. J., & Bortolini, M. C. (2004). Pode a genética definir quem deve se beneficiar das cotas universitárias e demais ações afirmativas? *Estudos Avançados, 18*(50), 31–50. https://doi.org/10.1590/S0103-40142004000100004.

Petrejčíková, E., Čarnogurská, J., Hronská, D., Bernasovská, J., Boroňová, I., Gabriková, D., Bôžiková, A., & Mačeková, S. (2014). Y-SNP analysis versus Y-haplogroup predictor in the Slovak population. *Anthropologischer Anzeiger. 71*(3), 275–285. https://doi.org/10.1127/0003-5548/2014/0368.

Pena, S. D. J., Pietro, G.di, Fuchshuber-Moraes, M., Pasqualini Genro, J., Hutz, M. H., Gomes Kehdy, F.de S., Kohlrausch, F., Alexandre Viana Magno, L., Montenegro, R. C., Moraes, M. O., Amaral de Moraes, M. E., de Moraes, M. R., Ojopi, É. B., Perini, J. A., Racciopi, C., Kely Campos Ribeiro-dos-Santos, Â., Rios-Santos, F., Romano-Silva, M. A., Sortica, V. A., & Suarez-Kurtz, G. (2011). The genomic ancestry of individuals from different geographical regions of Brazil is more uniform than expected. *PLoS ONE, 6*(2). https://doi.org/10.1371/journal.pone.0017063.

Phillips, C., & de la Puente, M. (2021). The analysis of ancestry with small-scale forensic panels of genetic markers. *Emerging Topics in Life Sciences, 5*(5), 443–453. https://doi.org/10.1042/ETLS20200327.

Phillips, C. (2015). Forensic genetic analysis of bio-geographical ancestry. *Forensic Science International: Genetics, 18*, 49–65. https://doi.org/10.1016/j.fsigen.2015.05.012.

Porras-Hurtado, L., Ruiz, Y., Santos, C., Phillips, C., Carracedo, Á., & Lareu, M. V. (2013). An overview of Structure: Applications, parameter settings, and supporting software. *Frontiers in Genetics, 4*. https://doi.org/10.3389/fgene.2013.00098.

Poznik, G. D. (2016). Identifying Y-chromosome haplogroups in arbitrarily large samples of sequenced or genotyped men. *BioRxiv*. 1–17. https://doi.org/10.1101/088716.

Quintana-Murci, L., Krausz, C., & McElreavey, K. (2001). The human Y chromosome: Function, evolution and disease. *Forensic Science International, 118*(2-3), 169–181. https://doi.org/10.1016/S0379-0738(01)00387-5.

Ralf, A., Montiel González, D., Zhong, K., & Kayser, M. (2018). Yleaf: Software for human Y-chromosomal haplogroup inference from next-generation sequencing data. *Molecular Biology and Evolution, 35*(5), 1291–1294. https://doi.org/10.1093/molbev/msy032.

Sahakyan, H., Margaryan, A., Saag, L., Karmin, M., Flores, R., Haber, M., Kushniarevich, A., Khachatryan, Z., Bahmanimehr, A., Parik, J., Karafet, T., Yunusbayev, B., Reisberg, T., Solnik, A., Metspalu, E., Hovhannisyan, A., Khusnutdinova, E. K., Behar, D. M., Metspalu, M., ... Villems, R. (2021). Origin and diffusion of human Y chromosome haplogroup J1-M267. *Scientific Reports, 11*(1). https://doi.org/10.1038/s41598-021-85883-2.

Schlecht, J., Kaplan, M. E., Barnard, K., Karafet, T., Hammer, M. F., & Merchant, N. C. (2008). Machine-learning approaches for classifying haplogroup from Y chromosome STR data. *PLoS computational biology, 4*(6). https://doi.org/10.1371/journal.pcbi.1000093.

Schneider, P. M., Prainsack, B., & Kayser, M. (2019). The use of forensic DNA phenotyping in predicting appearance and biogeographic ancestry. *Deutsches Arzteblatt International. 116*(51-52), 873–880. https://doi.org/10.3238/arztebl.2019.0873.

Suarez-Pajes, E., Díaz-De Usera, A., Marcelino-rodríguez, I., Guillen-guio, B., & Flores, C. (2021). Genetic ancestry inference and its application for the genetic mapping of human diseases. *International Journal of Molecular Sciences, 22*(13), 6962. https://doi.org/10.3390/ijms22136962.

Syndercombe Court, D. (2021). The Y chromosome and its use in forensic DNA analysis. *Emerging topics in life sciences, 5*(3), 427–441. https://doi.org/10.1042/ETLS20200339.

Tseng, B., Hwa, H.-L., Lin, C.-Y., Yu, Y.-J., Linacre, A., & Lee, J. C.-I. (2022). Y-SNP Haplogroup hierarchy finder: A web tool for Y-SNP haplogroup assignment. *Journal of Human Genetics, 67*(8), 487–493. https://doi.org/10.1038/s10038-022-01033-0.

Underhill, P. A., & Kivisild, T. (2007). Use of Y chromosome and mitochondrial DNA population structure in tracing human migrations. *Annual Review of Genetics, 41*, 539–564. https://doi.org/10.1146/annurev.genet.41.110306.130407.

Underhill, P. A., Poznik, G. D., Rootsi, S., Järve, M., Lin, A. A., Wang, J., Passarelli, Ben, Kanbar, J., Myres, N. M., King, R. J., Di Cristofaro, J., Sahakyan, H., Behar, D. M., Kushniarevich, A., Šarac, J., Šaric, T., Rudan, P., Pathak, A. K., ... Villems, R. (2015). The phylogenetic and geographic structure of Y-chromosome haplogroup R1a. *European Journal of Human Genetics, 23*(1), 124–131. https://doi.org/10.1038/ejhg.2014.50.

Van Oven, M., Van Geystelen, A., Kayser, M., Decorte, R., & Larmuseau, M. H. (2014). Seeing the wood for the trees: A minimal reference phylogeny for the human Y chromosome. *Human Mutation, 35*(2), 187–191. https://doi.org/10.1002/humu.22468.

Verdu, P., Pemberton, T. J., Laurent, R., Kemp, B. M., Gonzalez-Oliver, A., Gorodezky, C., Hughes, C. E., Shattuck, M. R., Petzelt, B., Mitchell, J., Harry, H., William, T., Worl, R., Cybulski, J. S., Rosenberg, N. A., & Malhi, R. S. (2014). Patterns of admixture and population structure in native populations of Northwest North America. *PLoS genetics, 10*(8). https://doi.org/10.1371/journal.pgen.1004530.

Whit Athey, T. (2006). Haplogroup prediction from Y-STR values using a Bayesian-allele-frequency approach. *Journal of Genetic Genealogy, 2*(2), 34–39.

Woźniak, A., Heidegger, A., Piniewska-Róg, D., Pośpiech, E., Xavier, C., Pisarek, A., Kartasińska, E., Boroń, M., Freire-Aradas, A., Wojtas, M., de la Puente, M., Niederstätter, H., Płoski, R., Spólnicka, M., Kayser, M., Phillips, C., Parson, W., & Branicki, W. (2021). Development of the VISAGE enhanced tool and statistical models for epigenetic age estimation in blood, buccal cells and bones. *Aging, 13*(5), 6459–6484. https://doi.org/10.18632/aging.202783.

Zieger, M., & Utz, S. (2019). A "forensic biobank" to establish comprehensive genetic frequency data for Switzerland. *Forensic Science International: Genetics, 40*, 46–51. https://doi.org/10.1016/j.fsigen.2019.02.004.

Chapter 23

Prediction of externally visible characteristics and its role in forensic investigation

Rym Kefi[1,2,3], Dora Becher[1,4], Nadia Kheriji[1,2], Mariem Hechmi[1,2,3], and Haifa Jmel[1,2]

[1]*Laboratory of Biomedical Genomics and Oncogenetics, Institut Pasteur de Tunis, Tunis, Tunisia,* [2]*University of Tunis El Manar, Tunis, Tunisia,* [3]*Genetic Typing Service, Institut Pasteur de Tunis, Tunis, Tunisia,* [4]*Directorate of Technical and Scientific Police, Sub-directorate of Forensic and Scientific Laboratories, Tunis, Tunisia*

23.1 Introduction

Human identification from biological material in forensic science heavily relies on the ability to analyze genetic polymorphisms in DNA (Parker et al., 2016). This method is a core component of forensic DNA typing, providing critical insights for criminal investigations and mass disaster identifications (Seo et al., 2013). In traditional forensic analysis, DNA is handled similarly to fingerprints, with short tandem repeats (STRs) being used to match a DNA sample to a suspect or a database (Alonso et al., 2018; Butler, 2011, 2014; Butler, 2015; Greytak & Armentrout, 2015). However, when biological samples such as hair, semen, or saliva are degraded or present in limited quantities, STR analysis becomes challenging (Chakravarty & Pandya, 2020; Marano & Fridman, 2019; Marrone et al., 2022; Sánchez-Diz, 2005; Seo et al., 2013). Moreover, if no suspect is identified and the database yields no matches, STR markers offer little to no additional information about the DNA recovered from a crime scene.

Numerous studies have highlighted the benefits of using single-nucleotide polymorphisms (SNPs) over STRs, noting their lower mutation rates, faster genotyping, abundance across the genome, and compatibility with high-throughput detection techniques. These studies have identified a range of SNPs in diverse populations that exhibit high heterozygosity and low-frequency variations, making them valuable for individual identification (Amorim & Pereira, 2005; Kidd et al., 2006).

SNPs, therefore, present a promising alternative for forensic purposes (Nievergelt et al., 2013). They have applications in individual identification, lineage tracing, ancestry determination, and forensic DNA phenotyping (FDP) (Shrivastava et al., 2020). FDP focuses on predicting externally visible characteristics (EVCs) such as eye, hair, and skin color of unknown persons based on biological samples (Montesanto et al., 2020). Recently, FDP has emerged as a rapidly advancing area in forensic genetics, aiding law enforcement in locating suspects and missing persons who cannot be identified through traditional DNA profiling, as well as in resolving cold cases (Chaitanya et al., 2018). Over the past decade, several research groups have developed highly effective individual identification SNP panels for various populations worldwide (Hou et al., 2014; Pakstis et al., 2010; Wei et al., 2014).

In the following sections, we start with an overview of the genetic markers for the prediction of traits such as eye color, hair color, and skin pigmentation. Next, we examine the genetic landscape of human identification markers linked to EVC among Mediterranean populations compared to other worldwide groups. Then we explore the application of FDP in casework, its methodologies, legal frameworks, and ethical considerations.

23.2 Externally visible characteristics prediction markers

A wealth of genetic information on EVC and human phenotypes has been made available through large-scale genome-wide association studies (GWAS). Variations such as SNP, INDEL, and copy number variants underlie phenotypic

Investigative and Predictive DNA Testing. DOI: https://doi.org/10.1016/B978-0-443-30128-5.00003-5

diversity. Among these, SNPs are the most widely used markers for phenotypic inference. Some common phenotypic traits related to appearance include eye color, hair color, and skin pigmentation (Becher et al., 2024).

23.2.1 Eye color prediction markers

Eye color is a highly heritable and complex genetic trait, with variations ranging from light to intermediate and dark shades. Eye pigmentation is determined by the amount of melanin stored in the melanocytes of the iris, with color variations depending on the type and amount of melanin produced. Numerous genes involved in melanin synthesis have been identified through various scientific studies. Since the beginning of research into external phenotypic traits, the *OCA2* gene has been shown to be a key factor in eye color variation, with 13 associated variants identified, including rs7495174, rs6497268, rs11855019, rs17566952, rs11638265, AY392134, rs1800411, rs1900758, rs1800404, rs1800407, rs749846, rs74653330, and rs121918166 (Marano & Fridman, 2019; Sturm & Frudakis, 2004; Sturm & Larsson, 2009; Dorgaleleh et al., 2020).

In 2008, a study by Kayser et al. confirmed through GWAS and linkage analysis that the *HERC2* gene also plays a role in eye color variation. Eye color has since been linked primarily to two genes on chromosome 15: *HERC2* and *OCA2*. Liu et al. proposed six additional SNPs across six pigmentation-related genes, achieving an area under the curve (AUC) of 0.91 for blue eyes, 0.93 for brown eyes, and 0.72 for intermediate colors in a study of 2300 Dutch Europeans. The six SNPs identified were: *HERC2* rs12913832, *OCA2* rs1800407, *IRF4* rs12203592, *TYR* rs1393350, *SLC45A2* rs16891982, and *SLC24A4* rs12896399 (Liu et al., 2009). In 2012, Walsh et al. developed the first blue/brown eye color prediction system, named IrisPlex, based on the six SNPs identified by Liu et al., which has been validated in European populations (Walsh et al., 2011).

Further studies, such as those by Ruiz et al., have confirmed that adding SNPs from *HERC2* and *OCA2* increases prediction accuracy, particularly for intermediate eye colors (Ruiz et al., 2013). However, discrepancies have arisen in its effectiveness across different populations. Some studies have validated its use in human identification, while others, particularly those focusing on the Iraqi and certain Mediterranean populations such as Italy and Spain, have questioned its accuracy (Martinez-Cadenas et al., 2013; Purps et al., 2011; Ruiz et al., 2013; Salvoro et al., 2019).

23.2.2 Hair color prediction markers

The primary differences in hair color result from the presence of two types of melanin: brown/black eumelanin and red/yellow pheomelanin (Gerstenblith et al., 2007; Koch et al., 2020). Individuals with red hair have a higher amount of pheomelanin compared to eumelanin, while darker hair is characterized by a higher concentration of eumelanin. In blonde hair, both types of melanin are present in lower amounts (Koch et al., 2020; Rees, 2003). Numerous studies have been conducted to investigate the genetic variations associated with hair color prediction. Indeed, research has shown that the genes *ASIP*, *SLC45A2*, *SLC24A5*, and *MC1R* are strongly associated with hair color variation, as they are involved in melanin biosynthesis (Van Daal, 2008). The agouti signaling protein (ASIP) is the second ligand for MC1R, functioning as an inverse agonist (Pavan & Sturm, 2019; Suzuki et al., 1997). The interaction between *MC1R* and *HERC2* influences the red/nonred hair color (Pospiech et al. 2014; Walsh et al., 2014). Another study highlighted a strong association between *OCA2* and *HERC2* with hair color variation, similar to their role in eye color (Andrade et al., 2017). According to Branicki et al. (2011), SNPs in *EXOC2, SLC45A2*, and *IRF4* were most significantly associated with black hair color (Branicki et al., 2011). SNPs in the *ASIP* gene were linked to red, dark blonde, and strawberry blonde hair, while the *OCA2* gene was significantly associated with brown hair color. Additionally, the *TYR* gene was found to be significantly associated with brown hair, and the *SLC24A4* gene with blonde and dark blonde hair (Valenzuela et al., 2010).

In 2013, the HIrisPlex system was developed, incorporating 18 hair color markers alongside the six preexisting IrisPlex SNPs (*KITLG* rs12821256, *SLC45A2* rs28777, *Tyrp1* rs683, *Tyr* rs1042602, *SLC24A4* rs2402130, and *EXOC2* rs4959270) and *MC1R* (rs1805005, rs11547464, rs1110400, rs1805007, rs1805006, rs1805009, rs201326893, rs2228479, and rs1805008), along with N29insA (rs312262906) (Branicki et al., 2011; Walsh et al., 2013). This system has proven effective for both European and non-European populations, unlike IrisPlex. A few years later, in 2015, Söchtig et al. conducted a study on SNPs affecting hair color prediction in Europe (Söchtig et al., 2015). They analyzed 605 Europeans using 63 SNPs, identifying 12 SNPs as excellent hair color markers, seven of which were already included in the HIrisPlex system (*OCA2* rs7495174 and rs4778138; *HERC2* SNPs: rs12931267, rs28777, rs12913832 and rs1129038; *TPCN2* rs35264875 and 5*MC1R* SNPs: rs1805006, rs11547464, rs1805007, rs1805009 and rs1805008).

Today, predicting hair color faces challenges due to the changes in hair color throughout life. Most studies rely on adult self-reports of childhood phenotypes, and predictions are primarily based on phenotypic information observed in adults (Kukla-Bartoszek et al., 2018).

23.2.3 Skin color prediction markers

It has been demonstrated that skin color is one of the most complex phenotypic traits to study (Jablonski & Chaplin, 2017). It is considered an adaptive trait, and melanin synthesis is a complex process involving multiple genes and other contributing factors, such as age, diseases, medications, and environmental influences (Spichenok et al., 2011). Skin pigmentation is crucial for protection against ultraviolet (UV) radiation (Jablonski, 2004; Parra, 2007; Rocha, 2020). A complex regulatory system controls melanin production, which determines skin color (Rocha, 2020; Slominski et al., 2004). Melanocytes in the epidermis produce melanin, which is deposited in melanosomes and then transferred to adjacent keratinocytes (Plonka et al., 2009). Skin color is strongly correlated with latitude due to variations in UV radiation intensity and differs across populations (Relethford, 1997; Rocha, 2020). Similar to other pigmentation traits, several genes are associated with skin color, such as variants in the *MC1R* gene (rs1805006, rs1805007, rs1805008, rs1805009, rs11547464, rs201326893, N29insA (InDel), and rs3212345) (Andersen et al., 2011; Neitzke-Montinelli et al., 2012), *HERC2* gene variants (rs1636232, rs1133496, and rs2238289) (Andrade et al., 2017), and *OCA2* gene variants (rs1800416, rs1800407, and rs1800404) for dark skin prediction. Other *OCA2* gene variants have been identified as being associated with skin type: rs7170989, rs4778138, rs1375164, rs1805007, and rs1805008 (Andrade et al., 2017; Zaumsegel et al., 2008). In Caucasian populations, dark skin prediction has been linked to *SLC45A2* gene variants rs16891982 and rs26722 (Pośpiech et al., 2014).

Maronas et al. investigated 29 SNPs that, according to their findings, are most correlated with skin color variation (Maroñas et al., 2014). They suggested six SNPs—*SLC45A2* rs16891982, *SLC24A5* rs1426654, *KITLG* rs10777129, *ASIP* rs6058017, *TYRP1* rs1408799, and *OCA2* rs1448484—for which they reported a classification success rate of 92.7% for black, 98.3% for white, and 83.7% for intermediate skin color. An extended set of 10 SNPs, including *SLC45A2* rs1329, *SLC24A4* rs2402130, *TPCN2* rs3829241, and *ASIP* rs6119471, was also highlighted, yielding AUC values of 0.999 for white, 0.966 for black, and 0.803 for intermediate skin color. These results were based on validation using 118 unrelated individuals of European and non-European ancestry (Americans, Africans, Eurasians, and Asians) (Maroñas et al., 2014).

23.3 Genetic landscape of externally visible characteristics markers

While significant progress has been made in identifying genetic variants associated with EVC, research analyzing the variation of phenotypic markers in Mediterranean populations remains limited. In this context, we conducted a comprehensive literature review to gather genes and variants previously linked to EVC. Our aim is to map the genetic landscape of human identification markers responsible for eye, hair, and skin color among Mediterranean populations, comparing these findings with those from other global groups (Becher et al., 2024).

Thus, we performed a systematic search for accessible peer-reviewed and full articles published from December 2000 to 2021, using PRISMA guidelines (Page et al., 2021) and the following search terms combinations of "Human identity markers, forensic, visual phenotypic traits, SNP involved in eye color, SNP involved in skin pigmentation, SNP involved in hair color and morphology.

Then, variants located on these genes were extracted from public genotypic data of Mediterranean, American, African, and East Asiatic populations. Genotyping data of 910 individuals from 17 populations were downloaded from the International 1000 Genome Project phase III (www.internationalgenome.org) and published data (Henn et al., 2012). The studied populations specifically included: individuals of European ancestry: Northwestern and Western European ancestry populations of Utah from the CEPH collection, Toscani people from Italy, South Spain, North Spain, North West Spain and Spain Basic populations; individuals of African American ancestry in the South Western USA; people of Mexican ancestry living in Los Angeles, California, USA; individuals from East Asian ancestry: Han Chinese in Beijing, China, the Chinese population of metropolitan Denver, Colorado, USA and Japanese in Tokyo, Japan; Individuals from North Africa: Algeria, Egypt, Libya, Tunisia Douiret, South Morocco and North Morocco. We extracted the chromosomal region plus 50 Kb of the aforementioned selected genes from this data.

The genetic landscape of these Human identification markers, their allelic distribution, and admixture analyses were determined using PLINK, R, and ADMIXTURE software. Our results indicate that Mediterranean populations are genetically closer to Mexican populations but distinct from sub-Saharan African populations residing in the USA and from East Asiatic populations (Figs. 23.1 and 23.2), (Becher et al., 2024).

We highlighted a total of 103,454 common variants shared among the studied populations, including 25 variants associated with eye, hair, and skin color (EVC). Notably the genotype frequency analysis revealed significant differences between the Mediterranean and Asiatic populations for the variants rs17646946, rs13016869, rs977588, rs1805008, and rs2240751 located in the *TCHH*, *PRKCE*, *OCA2*, *MC1R*, and *MFSD12* genes, respectively. Additionally, significant

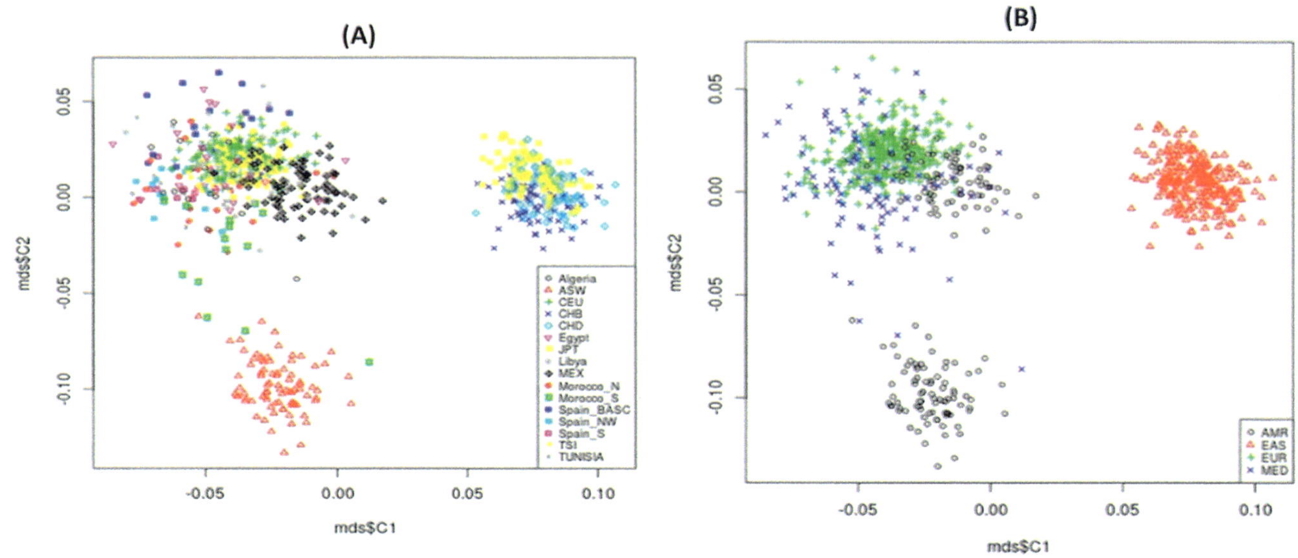

FIGURE 23.1 Multidimensional scaling (MDS) (A) of all populations analyzed in the present study. We used MDS to discriminate clusters of genetic variation within Mediterranean and other populations. MDS was applied using PLINK 2.0 software. The studied populations included; Algeria, Egypt, South Morocco (Morocco_S), North Morocco (Morocco_N), Tunisia, Libya, Toscani people of Italy (TSI), South Spain (Spain_S), North West of Spain (Spain_NW), Northwestern and Western European ancestry populations of Utah from the CEPH collection (CEU), Spain_BASC, Lebanon, USA (MEX), the sub-Saharan African population living in the USA (ASW), Individuals from East Asian ancestry; Han Chinese in Beijing, China (CHB), The Chinese population of metropolitan Denver, Colorado, USA (CHD) and Japanese in Tokyo, Japan; (JPT).(B) MDS of the three clusters: Mediterranean cluster, East Asiatic cluster, and American cluster. The composition of the clusters are the following Mediterranean (MED) group (Algeria, Egypt, South Morocco (Morocco_S), North Morocco (Morocco_N), Tunisia, Libya, Toscani people of Italy (TSI), South Spain (Spain_S), North West of Spain (Spain_NW), Lebanon, Northwestern and Western European ancestry populations of Utah from the CEPH collection (CEU)); East Asian (EAS) regroup (CHB, CHD, JPT) and American (AMR) group (ASW, MEX).

differences were found between the Mediterranean and American populations for the variants rs977589 and rs7179994 located in the *OCA2* gene, as well as rs12913832 and rs2240751 located in the *HERC2* and *MFSD12* genes, respectively. Our study provides a substantial set of EVC variants that may serve as a valuable resource for future forensic research field (Table 23.1) (Becher et al., 2024).

23.4 Role of externally visible characteristics markers in forensic investigation

FDP has revolutionized forensic science, becoming a vital tool in solving criminal and cold cases. Technological advancements now enable DNA evidence to play a pivotal role in narrowing down suspects and providing investigative leads. By predicting EVC such as skin, eye, and hair color, as well as facial morphology, FDP allows investigators to create physical profiles of unknown individuals based solely on their genetic material (Becher et al., 2024). This approach has proven particularly valuable in cases where no suspect or witness information is available (Elkins et al., 2023).

FDP has been instrumental in solving cold cases by offering physical profiles that help reduce suspect pools and generate new leads (Kaczmarek & Zysek, 2022). It complements traditional forensic techniques, such as fingerprinting and DNA matching, to build more comprehensive profiles in complex investigations (Canales, 2020). Unlike conventional DNA profiling, which requires a database match, FDP provides actionable insights even in the absence of a known suspect, making it an indispensable tool when other methods fall short (Pushpesh et al., 2023).

FDP has proven to be an invaluable asset in solving cold cases by providing physical profiles that help narrow down suspect pools or generate new leads (Kaczmarek & Zysek, 2022). This innovative approach complements traditional forensic techniques, such as fingerprinting and DNA matching, enabling investigators to create more comprehensive profiles in complex cases (Canales, 2020). Unlike conventional DNA profiling, which relies on a database match, FDP offers actionable insights even in the absence of a known suspect, making it particularly useful in situations where other methods may be inadequate (Pushpesh et al., 2023).

For instance, in 2017, DNA phenotyping played a crucial role in solving a criminal case involving the tragic murders of Candra Alston and her 3-year-old daughter, Malaysia Boykin, in Columbia, South Carolina, in 2011. Six years after the incident, investigators utilized DNA phenotyping to create a composite sketch of the suspect based solely on DNA

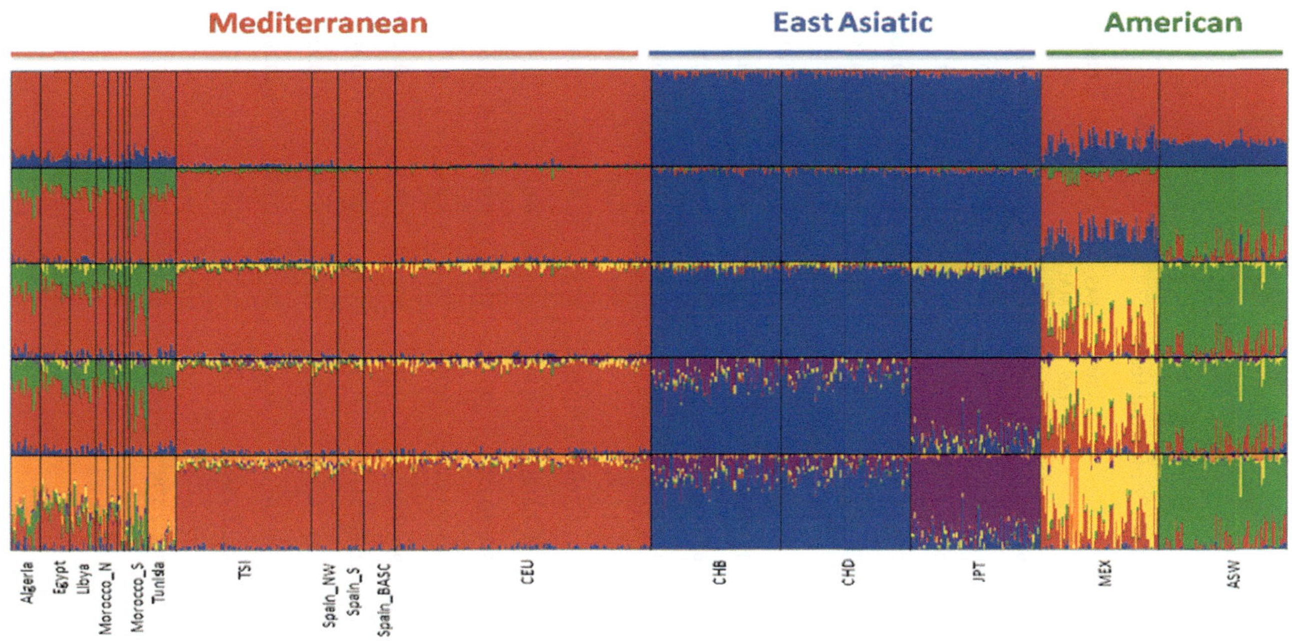

FIGURE 23.2 ADMIXTURE individual bar plots (K=2 to K=6). The plot shows the results of the ADMIXTURE analysis on the extent dataset. The ancestry proportion for each individual is estimated at k=6, displaying 17 populations and 30376 studied variants, resulting in cluster K2–10, with K referring to the hypothetical ancestry number. The K=6 is the number of groups that allows the model to have the best predictive accuracy. The ancestry proportions for each individual (columns) at K=6 are calculated by hypothesizing six hypothetical ancestral populations. Black vertical lines identify the population boundaries. The height extent of each color within an individual's color bar corresponds to the estimated membership of the individual in one of the clusters. For each cluster, we assigned a separate color. The bars with multiple colors can be interpreted as genetic admixture or as relative probabilities of belonging to the different clusters. The graphic of ADMIXTURE analysis demonstrates four components, with the predominance of African, European components in the Mediterranean population at the number of ancestors K=6. The studied population included; Algeria, Egypt, South Morocco (Morocco_S), North Morocco (Morocco_N), Tunisia, Libya, Lebanon, Toscani people of Italy (TSI), South Spain (Spain_S), Spain_BASC, North West of Spain (Spain_NW), Northwestern and Western European ancestry populations of Utah from the CEPH collection (CEU), USA (MEX), the sub Saharan African population living in the USA (ASW), Individuals from East Asian ancestry; Han Chinese in Beijing, China (CHB), the Chinese population of metropolitan Denver, Colorado, USA (CHD) and Japanese in Tokyo, Japan (JPT).

evidence collected at the scene. This sketch revealed essential physical characteristics, including skin tone, hair color, eye color, and facial features, which significantly narrowed the search and advanced the investigation toward resolving the case. This application of DNA phenotyping marked a pivotal milestone in forensic science, showcasing the technology's potential to aid in solving cases where traditional methods may have been insufficient (Hunter, 2021; Shih, 2020).

Another case occurred in 2012 in Reidsville, North Carolina, where a double murder took place involving a couple, Troy and LaDonna French. Their daughter, Whitley French, was the only survivor of the attack. The murderer left five drops of blood on the stairwell, but DNA testing could not link the samples to any suspects or individuals in open records. The crime remained unsolved for three years until early 2015, when FDP tools were employed. Using the DNA from the blood drops, the company predicted that the killer had fair skin, dark hair, and was of European and Latino descent. With these new leads, detectives focused on the family of Whitley's fiancé and discovered that the blood at the crime scene matched that of Whitley's future brother-in-law. He was arrested in June 2015, less than a month after serving as a groomsman at Whitley's wedding. After admitting his guilt, he is currently serving two life sentences (https://snapshot.parabon-nanolabs.com/snapshot-case-summary–rockingham-nc–french-homicides.html; Gannon, 2017).

In Glen Burnie, Maryland, in June 2017, a human skeleton was discovered and later identified through medical examination as that of a 20-year-old woman. DNA analysis did not yield conclusive results due to the lack of DNA samples from relatives for comparison. Fortunately, through FDP, investigators were able to determine the biogeographic origin, as well as the skin, eye, and hair color of the skeleton, revealing that the woman was of African American descent with brown skin, black hair, and brown eyes. These findings allowed for the establishment of a profile for the skeleton, and through parallel investigations and analyses, the body was identified as that of Shaquana Caldwell, a 26-year-old woman who had been missing since May 2017 (snapshot.parabon-nanolabs.https://snapshot.parabon-nanolabs.com/snapshot-case-summary–anne-arundel-county-md–shaquana-caldwell-murder.html).

TABLE 23.1 Allele frequency comparison of human identification traits among studied populations adjusted with Bonferroni test.

Gene	CHR	SNP	Associated Allele	Predicted phenotype	MED_ MAF	MED_ MAG	AMR_ MAF	AMR_ MAG	p.value	CI95 %	MAF_ EAS	EAS_ MAG	p.value	CI95 %
TCHH	1	rs17646946	G	Hair curl	0.209	79.04	0.162	83.73	0.466	[0.640–3.079]	0.005	99.40	3.759e-06	[4.009–1093.253]
LYST	1	rs3768056	A	Eye color	0.223	77.68	0.169	83.02	0.475	[0.642–2.982]	0.168	83.14	0.475	[0.642–2.982]
PRKCE	2	rs13016869	C	Eye color	0.257	74.27	0.325	67.47	0.352	[0.368–1.373]	0.035	96.45	1.503e-05	[2.730–34.371]
LPP	3	rs1464510	C	Vitiligo	0.432	56.71	0.443	55.66	1	[0.528–1.744]	0.445	55.49	0.886	[0.507–1.674]
DTNBP1	6	rs17470454	G	Light hair color	0.052	94.8	0.034	96.519	0.720	[0.320–11.235]	0.005	99.40	0.211	[0.564–248.582]
TYRP1	9	rs1325127	T	Eye color	0.391	60.82	0.487	51.25	0.199	[0.365–1.211]	0.269	73.03	0.097	[0.913–3.289]
BNC2	9	rs10756819	G	Skin	0.349	65.09	0.365	63.44	0.883	[0.493–1.701]	0.423	57.68	0.383	[0.403–1.369]
IL2RA	10	rs706779	C	Vitiligo	0.492	50.77	0.463	53.61	0.777	[0.623–2.040]	0.390	60.98	0.199	[0.825–2.739]
NTM	11	rs12421680	A	Skin	0.327	67.28	0.390	60.94	0.461	[0.414–1.430]	0.347	65.22	0.881	[0.488–1.712]
OCA2	15	rs1874835	G	Eye color	0.128	87.2	0.229	77.04	0.096	[0.217–1.115]	0.043	95.68	0.039	[1.049–15.571]
OCA2	15	rs977588	A	Eye color	0.464	53.56	0.471	52.81	1	[0.530–1.738]	0.181	81.87	3.497e-05	[1.954–7.854]
OCA2	15	rs977589	C	Eye color	0.374	62.54	0.140	85.94	0.0003045	[1.721–7.818]	0.196	80.31	0.011	[1.190–4.700]
OCA2	15	rs728405	C	Eye color	0.263	73.63	0.437	56.25	0.406	[0.539–7.395]	0.119	88.04	0.018	[1.301–432.516]
OCA2	15	rs1800407	C	Eye color	0.085	91.46	0.045	95.455	0.011	[0.235–0.845]	0.005	99.40	0.018	[1.154–5.989]
OCA2	15	rs2305252	C	Eye color	0.287	71.25	0.237	76.25	0.521	[0.656–2.557]	0.223	77.65	0.330	[0.727–2.902]
OCA2	15	rs1597196	T	Eye color	0.262	73.75	0.446	55.31	0.007602	[0.225–0.811]	0.221	77.84	0.619	[0.617–2.5237]
OCA2	15	rs7179994	A	Eye color	0.144	85.52	0.346	65.31	0.000892	[0.139–0.636]	0.244	75.59	0.103	[0.229–1.127]
OCA2	15	rs4778241	C	Eye color	0.288	71.19	0.434	56.56	0.055	[0.288–1.012]	0.207	79.22	0.252	[0.767–3.104]
HERC2	15	rs12913832	A	Eye, skin, hair	0.454	54.57	0.196	80.31	0.000259	[1.674–6.490]	-	-	-	-
HERC2	15	rs7183877	C	Eye color	0.106	89.33	0.125	87.5	1	[0.342–2.378]	0.302	69.76	0.001	[0.122–0.645]

MC1R	16	rs1805005	G	Red versus brown/ black hair color	0.103	89.64	0.026	97.368	0.081	[0.882–20.830]	0.005	99.40	0.009	[1.498–480.999]
MC1R	16	rs1805007	C	Red versus brown/ black hair color	0.079	92.03	0.012	98.734	0.034	[1.108–385.064]	0.005	99.40	0.034	[1.108–385.064]
MC1R	16	rs1805008	C	Hair, skin	0.09	91	0.022	97.799	0.058	[0.95946.940]	0.007	99.20	0.018	[1.301–432.516]
MFSD12	19	rs2240751	A	Skin	0.005	99.40	0.184	81.58	3.588e-05	[0.001–0.306]	0.348	65.16	3.457e-11	[0.0004–0.119]
SLC23A2	20	rs2748901	A	Eye color	0.490	50.91	0.346	65.31	0.062	[0.973–3.278]	0.405	59.45	0.319	[0.761–2.512]

Chr, chromosome; *MED_MAF*, derived allele frequency in the Mediterranean population; *AMR_MAF*, derived allele frequency in the American population; *EAS_MAF*, derived allele frequency in the East Asiatic population.
Note: The highlighted *p*-value indicates a statistically significant allelic difference between the MED and designated populations.
Source: Becher, D., Jmel, H., Kheriji, N., Sarno, S., Kefi, R. (2024). Genetic landscape of forensic DNA phenotyping markers among Mediterranean populations. *Forensic Science International*, 354111906. https://doi.org/10.1016/j.forsciint.2023.111906.

23.5 Steps to obtain external visible characteristic markers

Various steps must be followed to obtain the external visible characteristics of a suspect or victim from evidence collected at a crime scene using DNA phenotyping. The process begins with the collection of DNA evidence, such as blood, hair, or skin cells. Proper preservation techniques are essential, as samples can be easily contaminated or degraded, which would compromise the quality of the DNA.

Once collected, DNA is extracted from the biological material. It is crucial to handle the samples carefully during the extraction process to avoid contamination. The DNA extraction procedure consists of several key steps:

- Cell lysis.
- Lipid removal (achieved through detergents and centrifugation).
- Protein removal and denaturation using a protease such as proteinase K. Subsequently, the addition of RNase degrades RNA into ribonucleotides.
- Precipitation/aggregation and elution of the DNA.

Several typical methods of forensic DNA extraction may be employed:

Organic Extraction: This method uses organic solvents, such as phenol or a 1:1 mixture of phenol and chloroform, to remove proteins after denaturation. Proteins are precipitated by these solvents, while the polynucleotides (DNA and RNA) remain in solution. Although this technique is highly effective for obtaining high-quality DNA, it has disadvantages, including the use of hazardous organic solvents and the extended time required for extraction. Additionally, residues from phenol or chloroform can remain in the extracted DNA and inhibit downstream reactions such as PCR (Butler, 2010).
Silica-Based Extraction: This technique relies on the binding of DNA to silica in the presence of chaotropic salts, utilizing magnetic beads or centrifugation columns. After washing and elution, high-purity DNA is obtained. This method is favored in many forensic laboratories due to its effectiveness (Giacomazzi et al., 2005; Rudi et al., 1997).
Chelex Extraction: This method employs ion-exchange resin to bind metal ions that can interfere with DNA analysis. Human cells are lysed in a 5% suspension of Chelex and water at 100°C. This technique has been shown to protect DNA from degradation caused by high temperatures. Chelex extraction is often used for degraded samples but may yield lower-purity DNA compared to organic extraction (de Lamballerie et al., 1992).
FTA Paper Extraction: A Whatman FTA paper card is treated with chemicals that facilitate cell lysis and capture DNA, upon which a sample is placed. To recover the DNA, a specific area of the card is punched out and washed in a designated solution. The immobilized DNA is then ready for amplification. This method is practical for long-term storage and easy transport (Smith & Burgoyne, 2004).
Differential Extraction: This technique selectively isolates sperm cells from vaginal and epithelial cells, allowing for targeted DNA analysis in sexual assault cases. It involves enzymatic digestion and centrifugation to separate DNA from specific cell types (Elkins, 2013).

Each technique has its advantages, and the selection of a method depends on the nature of the sample, the integrity of the DNA, and the specific requirements of the forensic investigation. After extraction, the quantity and quality of the DNA are assessed to ensure that there is sufficient material for downstream analysis.

Once the DNA is prepared, sequencing is performed to analyze the markers associated with EVC. The first step involves defining all relevant genes and markers, followed by sequencing the various regions using either the traditional Sanger technique or advanced next-generation sequencing (NGS) methods. NGS allows the simultaneous sequencing of multiple SNPs and other genetic markers in parallel. Companies like Illumina, Thermo Fisher Scientific, Genomatix, Macrogen, and DNAVision are leaders in this field, providing the tools and expertise necessary for crime-solving.

Subsequently, bioinformatics software is essential for estimating the probable external traits of the individual based on the detected markers and generating a physical description of the person. To identify the individual, the created profile must be checked against available databases, including those containing information on victims and known suspects (Kayser & Schneider, 2009; Seo et al., 2013; Zidkova et al., 2013).

While this advancement holds significant potential for solving crimes, it also presents several ethical and legal challenges that must be carefully addressed.

23.6 Ethical aspects and legal considerations in forensic DNA investigation

The use of FDP raises ethical concerns related to privacy and consent, particularly when DNA is collected from crime scenes or individuals who are unaware of the analysis (Granja & Machado, 2023; Schneider et al., 2019). There is also the risk of exacerbating social pressures and discrimination against minority groups, although some argue that FDP may reduce such pressures compared to eyewitness testimonies (Granja & Machado, 2023; M'Charek, 2008). Therefore, legal safeguards must be implemented to prevent the misuse of FDP in ways that could reinforce racial profiling or discrimination. Laws should ensure that predictions of EVC are used responsibly, without perpetuating biases or targeting specific populations based solely on genetic traits (D'Amato et al., 2024).

However, legal frameworks regarding the collection and use of genetic material vary significantly across countries. In many jurisdictions, strict regulations govern how DNA samples can be obtained, stored, and utilized (Elhaik et al., 2021). The ethical challenge of consent arises when DNA is collected from individuals without their knowledge or when FDP is applied to large databases that may include innocent individuals (Pushpesh et al., 2023). Given that forensic investigations often involve cross-border collaboration, there is a critical need for international legal standards to regulate the use of FDP in investigations, ensuring that human rights are respected (Granja & Machado, 2023).

It is vital to consider the social implications of FDP applications, including the potential for misuse and the necessity for public discourse on their ethical consequences (Granja & Machado, 2023). Empirical research underscores the importance of protecting against discrimination and safeguarding privacy in the use of FDP (Schneider et al., 2019). The prediction of specific traits from DNA is not always accurate, particularly when environmental factors influence physical characteristics, which can lead to incomplete or misleading descriptions that may result in wrongful accusations or biased investigations (Canales, 2020; Kaczmarek & Zysek, 2022). Furthermore, methodological challenges arise from the polygenic nature of many traits, where multiple genetic variants contribute to a single characteristic, complicating the analysis significantly (Elkins et al., 2023).

In summary, while FDP has considerable potential to aid criminal investigations by predicting EVC from DNA samples, its implementation is fraught with ethical and legal challenges, including privacy concerns, the risk of discrimination, and the necessity for a robust legal framework. Addressing these issues through transparent and proportionate use, public discourse, and empirical research is essential for the responsible application of this promising forensic tool.

23.7 Future perspectives

As genetic research continues to evolve, future advancements in EVC are expected to enhance the accuracy of these predictions. This progress will not only encompass basic traits such as hair, eye, and skin color, but also more complex characteristics like facial structure, height, and even age-related changes. These improvements will render EVC prediction a more reliable tool for identifying unknown individuals.

In the future, the advancement of artificial intelligence (AI) and machine learning algorithms is likely to play a crucial role in refining the analysis of genetic data for EVC prediction. By leveraging large datasets, these technologies can uncover intricate patterns in DNA that correlate with specific traits, leading to faster and more precise predictions. AI could also facilitate the processing of vast amounts of forensic data, further supporting criminal investigations.

As technology progresses, the need for comprehensive ethical guidelines and international regulations will become increasingly urgent. Developing frameworks that ensure the responsible use of EVC prediction while safeguarding individual privacy and preventing misuse, such as racial or ethnic profiling, will be critical for the sustainable integration of this technology into forensic science. International collaboration will also be essential to standardize how EVC data is collected, utilized, and shared across borders.

As more cases are solved using EVC predictions, public trust in the technology will grow, provided it is applied ethically and transparently. Community engagement and education about the benefits and limitations of EVC prediction will be vital for its widespread acceptance within both the forensic and legal communities.

Beyond solving criminal cases, EVC prediction has the potential to extend to other fields, such as humanitarian efforts, disaster victim identification, and locating missing persons. In these contexts, predicting EVC from genetic data could assist in identifying individuals when no other information is available.

Currently, predicting complex traits from DNA, such as behavioral tendencies or disease predispositions, remains a significant challenge. Future research may address some of these limitations, but the potential ethical and privacy concerns related to predicting such traits must be carefully considered to prevent misuse in forensic contexts.

23.8 Conclusion

The prediction of EVC from DNA has emerged as a powerful and innovative tool in forensic investigations. By enabling law enforcement to create physical profiles of unknown individuals based solely on genetic information, EVC prediction has transformed crime-solving, particularly in cases where traditional DNA matching methods are insufficient. However, this advancing technology must be accompanied by robust ethical guidelines, transparent legal frameworks, and thoughtful public engagement to prevent misuse and ensure privacy protection.

Within the broader context of forensic science, EVC prediction serves as a complementary tool that must be employed responsibly and proportionately. It holds significant potential for solving complex cases, identifying unknown individuals, and supporting humanitarian efforts. Striking a balance between technological advancement and ethical responsibility will be essential for the sustainable and equitable use of this powerful technique in the future.

References

Alonso, A., Barrio, P. A., Müller, P., Köcher, S., Berger, B., Martin, P., Bodner, M., Willuweit, S., Parson, W., Roewer, L., & Budowle, B. (2018). Current state-of-art of STR sequencing in forensic genetics. *Electrophoresis, 39*(21), 2655–2668. https://doi.org/10.1002/elps.201800030, http://onlinelibrary.wiley.com/journal/10.1002/(ISSN)1522-2683.

Amorim, A., & Pereira, L. (2005). Pros and cons in the use of SNPs in forensic kinship investigation: A comparative analysis with STRs. *Forensic Science International, 150*(1), 17–21.

Andersen, J. D., Johansen, P., Mogensen, H. S., Børsting, C., & Morling, N. (2011). Eye colour and SNPs in Danes. *Forensic Science International: Genetics Supplement Series, 3*(1), e151. https://doi.org/10.1016/j.fsigss.2011.08.076.

Andrade, E. S., Fracasso, N. C. A., Strazza Júnior, P. S., Simões, A. L., & Mendes-Junior, C. T. (2017). Associations of OCA2-HERC2 SNPs and haplotypes with human pigmentation characteristics in the Brazilian population. *Legal Medicine, 24*, 78–83. https://doi.org/10.1016/j.legalmed.2016.12.003, www.elsevier.com/locate/legalmed.

Becher, D., Jmel, H., Kheriji, N., Sarno, S., & Kefi, R. (2024). Genetic landscape of forensic DNA phenotyping markers among Mediterranean populations. *Forensic Science International, 354*, 111906. https://doi.org/10.1016/j.forsciint.2023.111906.

Branicki, W., Liu, F., van Duijn, K., Draus-Barini, J., Pośpiech, E., Walsh, S., Kupiec, T., Wojas-Pelc, A., & Kayser, M. (2011). Model-based prediction of human hair color using DNA variants. *Human Genetics, 129*(4), 443–454. https://doi.org/10.1007/s00439-010-0939-8.

Butler, J. M. (2010). *Fundamentals of forensic DNA typing fundamentals of forensic DNA typing*. United States: Elsevier Inc. http://doi.org/10.1016/C2009-0-01945-X, http://www.sciencedirect.com/science/book/9780123749994.

Butler, J. M. (2011). *Advanced topics in forensic DNA typing: Methodology*. Academic press.

Butler, J. M. (2014). *Advanced topics in forensic DNA typing: Interpretation*. Academic Press.

Butler, J. M. (2015). The future of forensic DNA analysis. *Philosophical Transactions of the Royal Society B: Biological Sciences, 370*(1674), 20140252. https://doi.org/10.1098/rstb.2014.0252.

Canales, S. (2020). Aurora. Forensic DNA phenotyping: A promising tool to aid forensic investigation. Current situation. *Spanish Journal of Legal Medicine, 46*(4), 183–190. https://doi.org/10.1016/j.remle.2020.01.002.

Chaitanya, L., Breslin, K., Zuñiga, S., Wirken, L., Pośpiech, E., Kukla-Bartoszek, M., Sijen, T., Knijff, Pd, Liu, F., Branicki, W., Kayser, M., & Walsh, S. (2018). The HIrisPlex-S system for eye, hair and skin colour prediction from DNA: Introduction and forensic developmental validation. *Forensic Science International: Genetics, 35*, 123–135. https://doi.org/10.1016/j.fsigen.2018.04.004, http://www.elsevier.com.

Chakravarty, M., & Pandya, P. (2020). Introduction to forensic DNA typing and current trends. In *Handbook of DNA profiling*1–21.

D'Amato, M. E., Joly, Y., Lynch, V., Machado, H., Scudder, N., & Zieger, M. (2024). Ethical considerations for forensic genetic frequency databases: First report conception and development. *Forensic Science International: Genetics, 71*, 103053. https://doi.org/10.1016/j.fsigen.2024.103053.

de Lamballerie, X., Zandotti, C., Vignoli, C., Bollet, C., & de Micco, P. (1992). A one-step microbial DNA extraction method using "Chelex 100" suitable for gene amplification. *Research in Microbiology, 143*(8), 785–790. https://doi.org/10.1016/0923-2508(92)90107-y.

Dorgaleleh, S., Naghipoor, K., Barahouie, A., Dastaviz, F., & Oladnabi, M. (2020). Molecular and biochemical mechanisms of human iris color: a comprehensive review. *J. Cell. Physiol. 235*(12), 8972–8982. https://doi.org/10.1002/jcp.29824.

Elhaik, E., Ahsanuddin, S., Robinson, J. M., Foster, E. M., & Mason, C. E. (2021). The impact of cross-kingdom molecular forensics on genetic privacy. *Microbiome, 9*(1). https://doi.org/10.1186/s40168-021-01076-z.

Elkins, K. M. (2013). *DNA extraction*. Elsevier BV, 39–52. http://doi.org/10.1016/b978-0-12-394585-3.00004-3.

Elkins, K. M., Garloff, A. T., & Zeller, C. B. (2023). Additional predictions for forensic DNA phenotyping of externally visible characteristics using the ForenSeq and Imagen kits. *Journal of Forensic Sciences, 68*(2), 608–613. https://doi.org/10.1111/1556-4029.15215.

Gannon, M. (2017). Amazing DNA tool gives cops a new way to crack cold cases: DNA phenotyping can produce a sketch of the suspect. But is it ready for primetime? NBC News.

Gerstenblith, M. R., Goldstein, A. M., Fargnoli, M. C., Peris, K., & Landi, M. T. (2007). Comprehensive evaluation of allele frequency differences of MC1R variants across populations. *Human Mutation, 28*(5), 495–505. https://doi.org/10.1002/humu.20476.

Giacomazzi, S., Leroi, F., & Joffraud, J.-J. (2005). Comparison of three methods of DNA extraction from cold-smoked salmon and impact of physical treatments. *Journal of Applied Microbiology, 98*(5), 1230–1238. https://doi.org/10.1111/j.1365-2672.2005.02574.x.

Granja, R., & Machado, H. (2023). Forensic DNA phenotyping and its politics of legitimation and contestation: Views of forensic geneticists in Europe. *Social Studies of Science, 53*(6), 850–868. https://doi.org/10.1177/0306312720945033.

Greytak, E. M., & Armentrout, S. (2015). DNA phenotyping: Predicting ancestry and physical appearance from forensic DNA. *Proceedings of the 26th International Symposium on Human Identification.*

Henn, B. M., Botigué, L. R., Gravel, S., Wang, W., Brisbin, A., Byrnes, J. K., et al. (2012). Genomic ancestry of North Africans supports back-to-Africa migrations. *PLoS Genet, 8*(1), e1002397. https://doi.org/10.1371/journal.pgen.1002397.

Hou, G., Jiang, X., Yang, Y., Jia, F., Li, Q., Zhao, J., Guo, F., & Liu, L. (2014). A 21-locus autosomal SNP multiplex and its application in forensic science. *Journal of Forensic Sciences, 59*(1), 5–14. https://doi.org/10.1111/1556-4029.12259.

Hunter, P. (2021). Cold cases and ancient trade routes: DNA phenotyping and isotope analysis extend forensic science into new domains. *EMBO Reports, 22*(12), e54188.

Jablonski, N. G. (2004). The evolution of human skin and skin color. *Annual Review of Anthropology, 33*, 585–623. https://doi.org/10.1146/annurev.anthro.33.070203.143955.

Jablonski, N. G., & Chaplin, G. (2017). The colours of humanity: The evolution of pigmentation in the human lineage. *Philosophical Transactions of the Royal Society B: Biological Sciences, 372*(1724), 20160349. https://doi.org/10.1098/rstb.2016.0349.

Kaczmarek, M., & Zysek, W. (2022). DNA-based prediction of human phenotypic traits for the medico-legal and forensic purposes. *Postepy Biochemii, 68*(3), 264–287. https://doi.org/10.18388/pb.2021_449, https://postepybiochemii.ptbioch.edu.pl/index.php/PB/article/download/449/1061.

Kayser, M., & Schneider, P. M. (2009). DNA-based prediction of human externally visible characteristics in forensics: Motivations, scientific challenges, and ethical considerations. *Forensic Science International: Genetics, 3*(3), 154–161. https://doi.org/10.1016/j.fsigen.2009.01.012.

Kidd, K. K., Pakstis, A. J., Speed, W. C., Grigorenko, E. L., Kajuna, S. L. B., Karoma, N. J., Kungulilo, S., Kim, J. J., Lu, R. B., Odunsi, A., Okonofua, F., Parnas, J., Schulz, L. O., Zhukova, O. V., & Kidd, J. R. (2006). Developing a SNP panel for forensic identification of individuals. *Forensic Science International, 164*(1), 20–32. https://doi.org/10.1016/j.forsciint.2005.11.017, www.elsevier.com/locate/forsciint.

Koch, S. L., Tridico, S. R., Bernard, B. A., Shriver, M. D., & Jablonski, N. G. (2020). The biology of human hair: A multidisciplinary review. *American Journal of Human Biology, 32*(2). https://doi.org/10.1002/ajhb.23316, http://onlinelibrary.wiley.com/journal/10.1002/(ISSN)1520-6300.

Kukla-Bartoszek, M., Pośpiech, E., Spólnicka, M., Karłowska-Pik, J., Strapagiel, D., Żądzińska, E., Rosset, I., Sobalska-Kwapis, M., Słomka, M., Walsh, S., Kayser, M., Sitek, A., & Branicki, W. (2018). Investigating the impact of age-depended hair colour darkening during childhood on DNA-based hair colour prediction with the HIrisPlex system. *Forensic Science International: Genetics, 36*, 26–33. https://doi.org/10.1016/j.fsigen.2018.06.007.

Liu, F., van Duijn, K., Vingerling, J. R., Hofman, A., Uitterlinden, A. G., Janssens, A. C. J., et al. (2009). Eye color and the prediction of complex phenotypes from genotypes. *Curr. Biol. 19*(5), R192–R193. https://doi.org/10.1016/j.cub.2009.01.027.

Marano, L. A., & Fridman, C. (2019). DNA phenotyping: Current application in forensic science. *Research and Reports in Forensic Medical Science, 9*, 1–8.

Maroñas, O., Phillips, C., Söchtig, J., Gomez-Tato, A., Cruz, R., Alvarez-Dios, J., de Cal, M. C., Ruiz, Y., Fondevila, M., Carracedo, Á., & Lareu, M. V. (2014). Development of a forensic skin colour predictive test. *Forensic Science International: Genetics, 13*, 34–44. https://doi.org/10.1016/j.fsigen.2014.06.017.

Martinez-Cadenas, C., Pe˜na-Chilet, M., Ibarrola-Villava, M., & Ribas, G. (2013). Gender is a major factor explaining discrepancies in eye colour prediction based on HERC2/ OCA2 genotype and the IrisPlex model. *Forensic Sci. Int. Genet. 7*(4), 453–460. https://doi.org/10.1016/j.fsigen.2013.03.007.

Marrone, M., Tarantino, F., Stellacci, A., Baldassarra, S. L., Cazzato, G., Vinci, F., & Dell'Erba, A. (2022). Forensic analysis and identification processes in mass disasters: Explosion of gunpowder in the fireworks factory. *Molecules (Basel, Switzerland), 27*(1), 244. https://doi.org/10.3390/molecules27010244.

M'Charek, A. (2008). Silent witness, articulate collective: DNA evidence and the inference of visible traits. *Bioethics, 22*(9), 519–528. https://doi.org/10.1111/j.1467-8519.2008.00699.x.

Montesanto, A., D'Aquila, P., Lagani, V., Paparazzo, E., Geracitano, S., Formentini, L., Giacconi, R., Cardelli, M., Provinciali, M., & Bellizzi, D. (2020). A new robust epigenetic model for forensic age prediction. *Journal of Forensic Sciences.*

Neitzke-Montinelli, V., Urmenyi, T. P., Rondinelli, E., Cabello, P. H., Silva, R., & Moura-Neto, R. S. (2012). Polymorphisms upstream of the melanocortin-1 receptor coding region are associated with human pigmentation variation in a Brazilian population. *American Journal of Human Biology, 24*(6), 853–855. https://doi.org/10.1002/ajhb.22301.

Nievergelt, C. M., Maihofer, A. X., Shekhtman, T., Libiger, O., Wang, X., Kidd, K. K., & Kidd, J. R. (2013). Inference of human continental origin and admixture proportions using a highly discriminative ancestry informative 41-SNP panel. *Investigative Genetics, 4*(1), 13. https://doi.org/10.1186/2041-2223-4-13.

Page, M. J., McKenzie, J. E., Bossuyt, P. M., Boutron, I., Hoffmann, T. C., Mulrow, C. D., et al. (2021). The PRISMA 2020 statement: an updated guideline for reporting systematic reviews. *Int. J. Surg. 88*, 105906. https://doi.org/10.1136/bmj.n71.

Pakstis, A. J., Speed, W. C., Fang, R., Hyland, F. C., Furtado, M. R., Kidd, J. R., & Kidd, K. K. (2010). SNPs for a universal individual identification panel. *Human Genetics, 127*(3), 315–324. https://doi.org/10.1007/s00439-009-0771-1.

Parker, G. J., Leppert, T., Anex, D. S., Hilmer, J. K., Matsunami, N., Baird, L., Stevens, J., Parsawar, K., Durbin-Johnson, B. P., & Rocke, D. M. (2016). Demonstration of protein-based human identification using the hair shaft proteome. *PLoS One, 11*(9), e0160653. https://doi.org/10.1371/journal.pone.0160653.

Parra, E. J. (2007). Human pigmentation variation: Evolution, genetic basis, and implications for public health. *American Journal of Physical Anthropology: The Official Publication of the American Association of Physical Anthropologists, 134*(S45), 85–105.

Pavan, W. J., & Sturm, R. A. (2019). The genetics of human skin and hair pigmentation. *Annual Review of Genomics and Human Genetics, 20*, 41–72.

Plonka, P. M., Passeron, T., Brenner, M., Tobin, D. J., Shibahara, S., Thomas, A., et al. (2009). What are melanocytes really doing all day long? *Experimental Dermatology, 18*(9), 799–819.

Pośpiech, E., Wojas-Pelc, A., Walsh, S., Liu, F., Maeda, H., Ishikawa, T., et al. (2014). The common occurrence of epistasis in the determination of human pigmentation and its impact on DNA-based pigmentation phenotype prediction. *Forensic Science International: Genetics, 11*, 64–72. https://doi.org/10.1016/j.fsigen.2014.01.012.

Purps, J., Geppert, M., Nagy, M., & Roewer, L. (2011). Evaluation of the IrisPlex eye colour prediction tool in a German population sample. *Forensic Sci. Int. Genet. Suppl. Ser. 3*(1), e202–e203.

Pushpesh, K., Ahmed, U., & Rana, M. (2023). Forensic DNA phenotyping: A promising tool to predict human appearance for forensic purposes. *GSC Biological and Pharmaceutical Sciences, 23*(2), 034–041. https://doi.org/10.30574/gscbps.2023.23.2.0180.

Rees, J. L. (2003). Genetics of hair and skin color. *Annual Review of Genetics, 37*(1), 67–90. https://doi.org/10.1146/annurev.genet.37.110801.143233.

Relethford, J. H. (1997). Hemispheric difference in human skin color. *American Journal of Physical Anthropology: The Official Publication of the American Association of Physical Anthropologists, 104*(4), 449–457.

Rocha, J. (2020). The evolutionary history of human skin pigmentation. *Journal of Molecular Evolution, 88*(1), 77–87.

Rudi, K., Kroken, M., Dahlberg, O., Deggerdal, A., Jakobsen, K., & Larsen, F. (1997). Rapid, universal method to isolate PCR-ready DNA using magnetic beads. *Biotechniques, 22*(3), 506–511.

Ruiz, Y., Phillips, C., Gomez-Tato, A., Alvarez-Dios, J., De Cal, M. C., Cruz, R., et al. (2013). Further development of forensic eye color predictive tests. *Forensic Sci. Int.: Genet. 7*(1), 28–40. https://doi.org/10.1016/j.fsigen.2012.05.009.

Salvoro, C., Faccinetto, C., Zucchelli, L., Porto, M., Marino, A., Occhi, G., et al. (2019). Performance of four models for eye color prediction in an Italian population sample. *Forensic Sci. Int. Genet. 40*, 192–200. https://doi.org/10.1016/j.fsigen.2019.03.008.

Schneider, P. M., Prainsack, B., & Kayser, M. (2019). The use of forensic DNA phenotyping in predicting appearance and biogeographic ancestry. *Deutsches Arzteblatt International, 116*(51–52), 873–880. https://doi.org/10.3238/arztebl.2019.0873.

Seo, S. B., King, J. L., Warshauer, D. H., Davis, C. P., Ge, J., & Budowle, B. (2013). Single nucleotide polymorphism typing with massively parallel sequencing for human identification. *International Journal of Legal Medicine, 127*(6), 1079–1086.

Shih, Y. (2020). DNA phenotyping: from criminal investigations to surveillance. *Imperial Bioscience Review.*

Shrivastava, P., Dash, H. R., Lorente, J. A., & Imam, J. (2020). *Forensic DNA Typing: Principles, Applications and Advancements.* Springer.

Slominski, A., Tobin, D. J., Shibahara, S., & Wortsman, J. (2004). Melanin pigmentation in mammalian skin and its hormonal regulation. *Physiological Reviews, 84*(4), 1155–1228.

Smith, L., & Burgoyne, L. A. (2004). Collecting, archiving and processing DNA from wildlife samples using FTA® databasing paper. *BMC Ecology, 4*(1), 4.

Spichenok, O., Budimlija, Z. M., Mitchell, A. A., Jenny, A., Kovacevic, L., Marjanovic, D., et al. (2011). Prediction of eye and skin color in diverse populations using seven SNPs. *Forensic Science International: Genetics, 5*(5), 472–478. https://doi.org/10.1016/j.fsigen.2010.10.005.

Sturm, R. A., & Frudakis, T. N. (2004). Eye colour: portals into pigmentation genes and ancestry. *Trends Genet, 20*(8), 327–332. https://doi.org/10.1016/j.tig.2004.06.010.

Sturm, R. A., & Larsson, M. (2009). Genetics of human iris colour and patterns. *Pigment Cell Melanoma Res, 22*(5), 544–562. https://doi.org/10.1111/j.1755-148X.2009.00606.x.

Suzuki, I., Tada, A., Ollmann, M. M., Barsh, G. S., Im, S., Lamoreux, M. L., et al. (1997). Agouti signaling protein inhabits melanogenesis and the response of human melanocytes to α-melanotropin. *Journal of Investigative Dermatology, 108*(6).

Sánchez-Diz, A.C. (2005). Forensic DNA-Typing Technologies.

Söchtig, J., Phillips, C., Maroñas, O., Gómez-Tato, A., Cruz, R., Alvarez-Dios, J., et al. (2015). Exploration of SNP variants affecting hair colour prediction in Europeans. *International Journal of Legal Medicine, 129*(5), 963–975. https://doi.org/10.1007/s00414-015-1226-y.

Van Daal, A. (2008). The genetic basis of human pigmentation. *Forensic Science International: Genetics Supplement Series, 1*(1), 541–543.

Valenzuela, R. K., Henderson, M. S., Walsh, M. H., Garrison, N. A., Kelch, J. T., Cohen-Barak, O., et al. (2010). Predicting phenotype from genotype: Normal pigmentation. *Journal of Forensic Sciences, 55*(2), 315–322. https://doi.org/10.1111/j.1556-4029.2009.01317.x.

Walsh, S., Lindenbergh, A., Zuniga, S. B., Sijen, T., de Knijff, P., Kayser, M., et al. (2011). Developmental validation of the IrisPlex system: determination of blue and brown iris colour for forensic intelligence. *Forensic Sci. Int.: Genet. 5*(5), 464–471. https://doi.org/10.1016/j.fsigen.2010.09.008.

Walsh, S., Liu, F., Wollstein, A., Kovatsi, L., Ralf, A., Kosiniak-Kamysz, A., et al. (2013). The HIrisPlex system for simultaneous prediction of hair and eye colour from DNA. *Forensic Science International: Genetics, 7*(1), 98–115. https://doi.org/10.1016/j.fsigen.2018.04.004.

Walsh, S., Chaitanya, L., Clarisse, L., Wirken, L., Draus-Barini, J., Kovatsi, L., et al. (2014). Developmental validation of the HIrisPlex system: DNA-based eye and hair colour prediction for forensic and anthropological usage. *Forensic Science International: Genetics, 9*, 150–161. https://doi.org/10.1016/j.fsigen.2013.12.006.

Wei, Y. L., Qin, C. J., Liu, H. B., Jia, J., Hu, L., & Li, C. X. (2014). Validation of 58 autosomal individual identification SNPs in three Chinese populations. *Croatian Medical Journal, 55*(1), 10–13. https://doi.org/10.3325/cmj.2014.55.10.

Zaumsegel, D., Rothschild, M. A., & Schneider, P. M. (2008). SNPs for the analysis of human pigmentation genes—A comparative study. *Forensic Science International: Genetics Supplement Series, 1*(1), 544–546.

Zidkova, A., Horinek, A., Kebrdlova, V., & Korabecna, M. (2013). Application of the new insertion-deletion polymorphism kit for forensic identification and parentage testing on the Czech population. *International Journal of Legal Medicine, 127*(1) 7-1.

Chapter 24

The advent of investigative genetic genealogy

David Gurney, and Cairenn Binder

Ramapo College of New Jersey, Investigative Genetic Genealogy Center, Mahwah, NJ, United States

24.1 Introduction

Investigative genetic genealogy (IGG) uses genetic information and public records to provide leads for subjects in cases of violent crime, unidentified human remains (UHR), and wrongful convictions. The first attempts at using IGG were in 2011 and 2014, though neither resulted in a lead (Khatchadourian, 2021). The inflationary period of IGG began in April 2018, with the identification of the Golden State Killer as Joseph James DeAngelo and the Jane Doe, "Buckskin Girl," as Marcia King (Gurney, 2023).

Since 2018, IGG has helped resolve over 1000 cases (Dowdeswell, 2024). Most of the cases resolved with the use of IGG have been UHR, with violent crimes being a close second. Four individuals have been exonerated as a result of IGG (Dowdeswell, 2022; Taylor, 2024). That impressive record marks IGG as the most significant investigative tool since the advent of forensic DNA testing in the 1980s and 1990s.

This chapter provides an overview of IGG as an investigative method, briefly describes legal and ethical considerations for IGG, and demonstrates five specific IGG scenarios using case examples from the Ramapo College IGG Center's casework.

24.2 Investigative genetic genealogy—filling an investigative vacuum, and the forensic investigative genetic genealogy sandwich

IGG fills the investigative vacuum left by biological samples from which traditional short tandem repeat (STR) DNA profiles cannot be developed, and successful STR profiles produce no matches to known individuals in the Combined DNA Index System (CODIS), (Fig. 24.1). Whereas traditional DNA comparisons rely on STRs, IGG relies on genetic profiles made up of single-nucleotide polymorphisms (SNPs, pronounced "snips"). These SNP profiles can contain anywhere from several thousand to millions of markers along an individual's genome (Tillmar et al., 2020). This multiplicity of data points allows for statistical hypotheses about the degree of relatedness between individuals to be developed by comparing their SNP profiles (Bettinger, 2019).

In simple terms, the advent of IGG means that it is no longer necessary for a perpetrator of a violent crime to have been arrested for another crime and had his DNA uploaded to CODIS (or for a Jane Doe to have had her relatives submit reference samples to CODIS). With IGG, violent criminals and unidentified remains can be identified as long as a usable SNP profile can be developed from a biological sample and enough time and effort is expended by one or more IGG practitioners. Any killer or rapist who leaves behind even a trace amount of DNA at a crime scene can now, with enough effort, be identified—potentially before he is able to commit additional crimes. As CeCe Moore, one of the four pioneers of IGG put it, "I really believe we can stop serial killers from existing, stop serial rapists from existing" (Smith, 2023).

IGG is just one step in the process required to take a forensic sample and turn it into an investigative lead. IGG begins after an SNP profile has been uploaded to one of the two primary genetic genealogy databases that are accessible for IGG use: GEDmatch Pro and FamilyTreeDNA (Guerrini et al., 2021). Before an uploadable profile is generated, DNA must be extracted from a biological sample, and a laboratory must develop a SNP profile, either with whole genome sequencing technology or array genotyping. After genotyping, a bioinformatics practitioner must process the genomic data to generate

Investigative and Predictive DNA Testing. DOI: https://doi.org/10.1016/B978-0-443-30128-5.00028-X

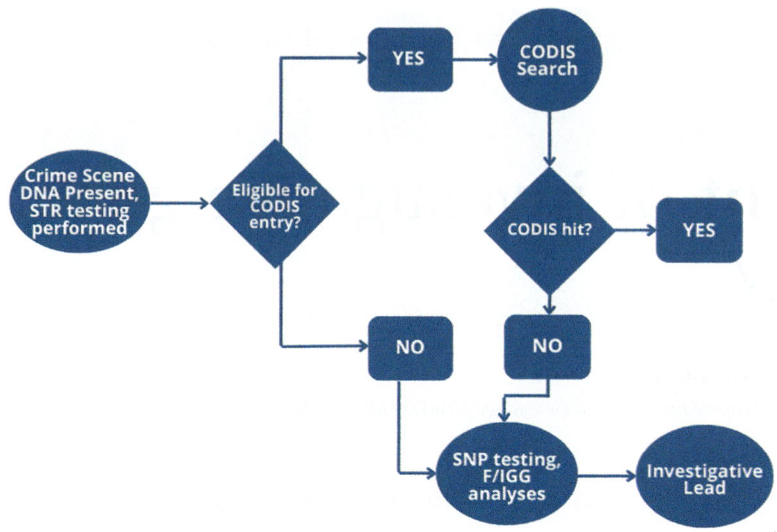

FIGURE 24.1 Investigative genetic genealogy (IGG) fills an investigative vacuum. IGG can provide leads in cases where a short tandem repeat (STR) profile cannot be generated from evidence or an STR profile has no match in the Combined DNA Index System.

a usable SNP profile that is compatible with the matching algorithms in GEDmatch Pro and FamilyTreeDNA (Greytak et al., 2024). These pre-IGG steps are performed by experts who are typically not IGG practitioners themselves.[1]

Once an IGG practitioner has turned over a lead to an investigating agency, it is up to that agency to confirm the lead through either a traditional one-to-one DNA comparison between the forensic sample and the candidate for the lead or other traditional investigative means. It is this confirmation that serves as evidence in a criminal case or proof of identification in the case of a John or Jane Doe (Gurney, 2023). These post-IGG steps are performed by law enforcement investigators, medical examiners, or attorneys working with one of the former.

This entire process of turning a forensic sample into an investigative lead is sometimes referred to as forensic investigative genetic genealogy (FIGG)—with IGG forming the middle of the process after a useable SNP profile has been uploaded to the genetic genealogy databases, and before the lead produced by IGG has been confirmed through traditional investigative means (Wickenheiser et al., 2023) (Fig. 24.2).

24.3 Legal and ethical considerations

When the Golden State Killer and "Buckskin Girl" cases were resolved with the help of IGG in 2018, there were no legal standards and few ethical standards to guide the field. While a small number of individuals had contemplated—and a few had even attempted—the technique before 2018, the sudden advent of IGG came largely as a surprise. The genetic genealogy databases that allowed uploads of third-party genetic profiles had not addressed law enforcement access to their databases in their terms of service at that time; no state had passed legislation regulating IGG; and there had been no public discussion about the ethical boundaries of IGG.

Much has changed since those early days. As of the writing of this chapter:

- Every genetic genealogy database addresses the use of IGG in their terms of service.
- Four states have passed laws that directly regulate at least some aspect of IGG work.
- The Department of Justice has provided guidelines for the practice of IGG.
- Two nonprofit organizations—the Investigative Genetic Genealogy Accreditation Board and the National Technology Validation and Implementation Collaborative—comprising IGG practitioners and IGG stakeholders, have organized to help guide the professionalization of IGG.
- Vigorous discussions of ethical and legal issues around IGG have taken place in scholarly works, popular media, and at academic and industry conferences.

1. There are some IGG practitioners who also work as forensic scientists and bioinformaticists

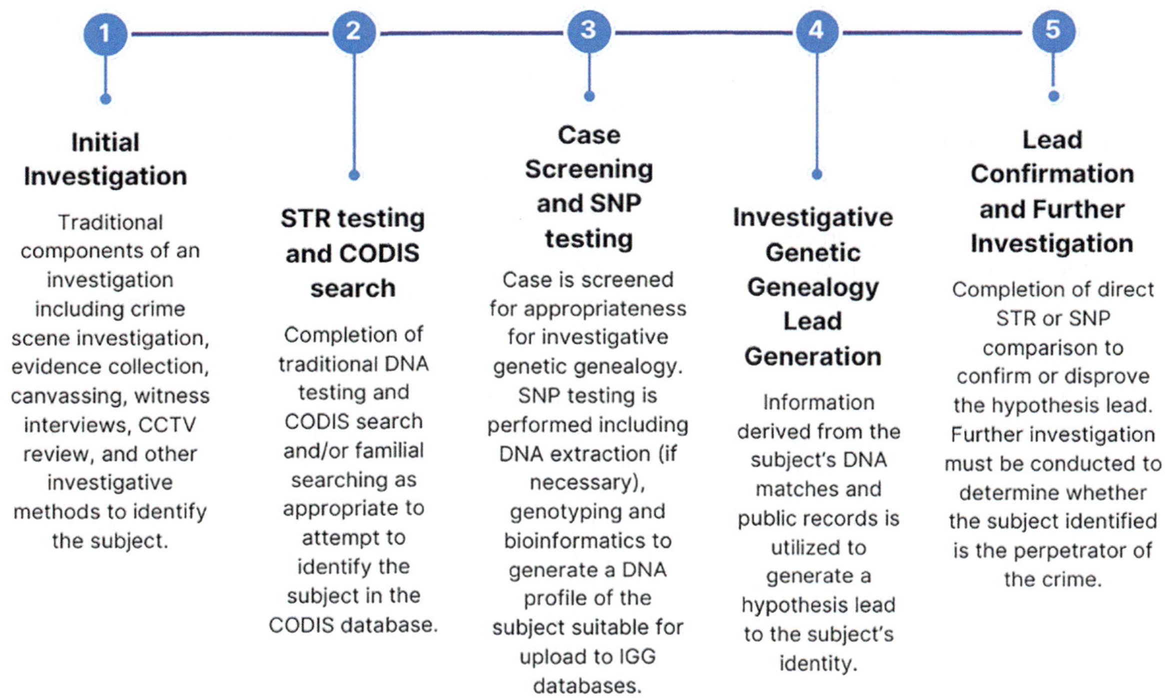

FIGURE 24.2 The forensic investigative genetic genealogy sandwich. Investigative genetic genealogy occurs after a genetic profile has been generated and before a lead has been confirmed.

This section provides an overview of the legal and ethical considerations that IGG practitioners and stakeholders must take into account as IGG continues its expansion from an upstart investigative method to the go-to tool for difficult DNA cases, whether cold or active.

24.3.1 Legal considerations

Four states have passed laws that directly address IGG in at least some manner. They are described here in descending order of their impact on the actual practice of IGG. Maryland has adopted comprehensive legislation that addresses nearly all aspects of IGG practice in that state (though a recent amendment exempts IGG performed on UHR from the provisions of the law) and includes criminal and civil penalties for failure to adhere to certain terms of the law (H.B. 240, 2021 Reg. Sess. Md., 2021). Utah's law is less detailed but still provides significant guidance to IGG practitioners, though it lacks meaningful penalties for violations (S.B. 156, 2023. Reg. Sess. Utah., 2023). Montana's two laws create a warrant requirement before IGG can begin in at least some circumstances (H.B. 602, 2021 Reg. Sess. Mt., 2021; S.B. 351, 2023 Reg. Sess. Mt., 2023). Finally, Florida's law does not impact the practice of IGG but limits what information from an IGG case can be disclosed through public records requests (H.B. 1327, 2023 Reg. Sess. Fla., 2023). IGG practitioners operating in these states should carefully review the laws and ensure that their practice accords.

Other legal considerations exist for IGG, but they apply less to IGG practitioners themselves and more to the agencies that have the authority to instigate IGG and use it to generate leads for criminal cases. Many courts have addressed constitutional challenges and requests for discovery related to IGG from the defense. The primary attack on IGG has been by way of the 4th Amendment or related state constitutional provisions. Thus far, all of those attacks have failed.

A recent example comes from the Minnesota Supreme Court's decision in *Minnesota v. Carbo* (Minnesota v. Carbo, 2024). In 2022, a Minnesota jury found Michael Allan Carbo guilty of the first-degree murder of Nancy Daugherty in 1986. Carbo was identified after IGG practitioners from Parabon Nanolabs produced a lead tying Carbo to DNA found at the crime scene, including semen, scrapings from under the victim's fingernails, and vomit found in the victim's yard. Direct comparisons between Carbo's DNA—taken by law enforcement from trash left outside his apartment—to the crime scene evidence confirmed the lead produced by IGG. At trial, Carbo argued that performing IGG on the evidence he left behind at the crime scene violated his privacy interest in the genetic information contained within. As such, he argued that the performance of IGG violated his Fourth Amendment rights, and any evidence derived from the IGG—including the direct confirmatory DNA match—should be excluded from trial. The trial court rejected Carbo's arguments (Minnesota v. Carbo, 2021). After his conviction, Carbo appealed to the Minnesota Supreme Court.

The Minnesota Supreme Court likewise rejected Carbo's Fourth Amendment challenge (Minnesota v. Carbo, 2024). Referencing the third-party doctrine of Fourth Amendment jurisprudence, the Court found that Carbo abandoned "*any* privacy interest in his genetic information" (emphasis in original) when he voluntarily—and admittedly—had unprotected sexual intercourse with the victim, leaving semen in the victim's body and skin under her fingernails. This deliberate abandonment of any privacy interest meant that Parabon's analysis of Carbo's genetic information did not even constitute a "search" for Fourth Amendment purposes.[2]

With respect to discovery, courts have been reluctant to grant sweeping defense requests to access detailed information about the IGG process, finding it mostly irrelevant to the defense case since it is the direct DNA comparison between the defendant and the crime-scene evidence—*not* the lead generated by IGG—that is introduced as evidence of guilt at trial (Minnesota v. Carbo, 2021). All the same, agencies should be aware of these defense tactics and be prepared to counter them. In the case of *Idaho v. Kohberger*, the district court ruled that at least some of the underlying IGG information beyond what had already been made available by the prosecution must be turned over to the defense (Idaho v. Kohberger, 2024). Much of that material was eventually made public when the court ruled on the defense's motion to suppress all genetic information in the case (Idaho v. Kohberger, 2025). The court's ruling, rejecting the defense's arguments on the same grounds as the court in *Carbo*, underscores the weakness of Fourth Amendment claims related to IGG.

As a final note, IGG practitioners who work on criminal cases should be prepared for the possibility of testifying, either in pretrial proceedings or, more rarely, in front of a jury. To date, a lead produced by IGG alone has never been used as evidence against a defendant. Instead, it is the direct STR DNA comparison between the defendant's DNA and crime-scene DNA that is presented to the jury. As a result of a recent ruling in California, this is likely to remain the case even in cases where a direct STR DNA comparison cannot be performed. In that ruling, the Kern County district court validated the use of SNP-to-SNP comparisons between a defendant and crime-scene evidence, providing an alternative avenue when direct STR DNA comparisons are impossible (Taylor, 2024). Several similar cases are pending in other state courts. Given these developments, it is unlikely that IGG practitioners will routinely have to testify in front of juries since the results of the IGG analysis will remain a lead, while the actual evidence of guilt presented to the jury will remain a direct DNA comparison between the defendant's DNA and the DNA taken from crime-scene evidence.

24.3.2 Ethical considerations

IGG involves probing into the lives and family connections of many individuals. While an individual who leaves his DNA at a crime scene forfeits any privacy interest in that material, the genetic matches that IGG practitioners rely on to begin a case, and the myriad of individuals connected to the family tree through genealogical work, are essentially innocent bystanders. IGG practitioners should take steps to ensure that any information gathered on these bystanders is kept as private as possible throughout the IGG process.

There are several steps IGG practitioners can take to protect the information gathered during their work on a case, which can be summarized as follows:

- Build any family trees on platforms that restrict who can view the information.
- Select the most stringent privacy settings possible for any databases accessed.
- Redact the names of DNA matches and genealogical relatives whenever possible in reports and other communications.
- Only contact reference testers when all other genealogical avenues have failed.

Concerning tree building, many IGG practitioners rely on Ancestry.com, while some use stand-alone platforms. Whatever the case, IGG practitioners can restrict who has access to the information contained in the tree. With Ancestry Trees, that requires making trees private and unsearchable and only sharing IGG trees with those who are working on the case or need to know the information. The latter applies to standalone platforms as well.

Most database searches leave breadcrumbs that can be discovered by other users if privacy settings are not adjusted. On Ancestry, the default setting alert a user when another user has viewed their tree. On Newspapers.com, the default setting attaches a username to information that is clipped. On LinkedIn, users are notified when another user views their profiles. On Facebook, repeatedly viewing another user's profile may create a "friend" suggestion based on the connection. IGG practitioners should be aware of these—and other—breadcrumbs and take steps to remove them whenever possible.

2. Although the Court rejected Carbo's Fourth Amendment arguments, it remanded the case back to the district court on other grounds

Finally, IGG reports for agencies point the agency to one or more leads. In those reports, IGG practitioners often include information about DNA matches and at least some genealogical relatives to help the agency see how the IGG practitioner arrived at their conclusion and present avenues for the agency to verify any leads. IGG practitioners should carefully consider which individuals' information it is necessary for the agency to know and redact the rest.

As one of the case studies below demonstrates, IGG practitioners sometimes rely on reference testers to narrow in on a lead. Contacting reference testers, whether it is done by the IGG practitioner or the agency, is inherently disruptive. It alerts the reference tester of their connection to a case and potentially introduces stress to their lives. Contacting reference testers should thus never be a first resort when the branches thin out. IGG practitioners should make every attempt to work the family tree before deciding to contact reference testers. When it does become necessary, care should be taken in deciding whom to contact. A scattershot approach should be avoided.

Along with protecting the privacy of DNA matches and genealogical relatives, IGG practitioners should only make use of direct DNA matches who have opted into law enforcement searching. The exception is when an IGG practitioner is acting under a warrant that provides access to additional databases and DNA matches.

Detailed ethical guidance for IGG practitioners is now available from the Investigative Genetic Genealogy Accreditation Board (Investigative Genetic Genealogy Accreditation Board, 2025).

24.4 A general approach for investigative genetic genealogy

There are many ways to conduct IGG, and each case has unique features that must be accommodated with appropriate methods. However, it is useful to describe a general approach to a typical IGG case by a typical IGG practitioner. One of the first clues available in an IGG case is information about a subject's bioancestral background, that is, which regions of the world the subject's ancestors came from. While this information is not definitive, it provides a starting point and can signal, upfront, whether a case will likely present difficulties. Cases involving individuals with Western European backgrounds are, on average, easier to work since the genetic genealogy databases are largely populated with individuals with that ancestral heritage. A case involving an African American or an individual of Asian descent, on the other hand, may pose challenges.

The genetic "match" lists in GEDmatch Pro and FamilyTreeDNA are the fulcrums of IGG work. The matches available to IGG practitioners have all agreed, per each genetic genealogy database's terms of service, to participate in IGG searching. IGG practitioners view these lists and begin by searching for matches that share the largest amount of DNA with the subject. In an average IGG case, the top matches may share between 60 and 100 centimorgans (cMs, a measure of shared genetic material) of DNA with the subject, though those numbers widely depend on the case. From the match lists, IGG practitioners can use a variety of tools to identify genetic clusters that represent individuals who share common ancestors with one another and the subject (Gurney, 2023).

Building out the family trees of these individuals to identify common ancestors forms the bulk of most IGG casework (Greytak et al., 2024). This step involves combing through public records—including old records such as censuses, and new records such as public social media posts—to build backward in time until common ancestors between matches are identified. It is a laborious process and requires careful attention to best practices in genealogical research. It is easy for amateurs to be misled by records and end up building erroneous branches in a family tree.

Once a critical mass of common ancestors has been identified, the genealogical work continues but in the opposite direction—moving forward in time from the common ancestors. The goal of this "downline" research is to identify connections between the genetic clusters, where separate family lines intermarried or produced children (Gurney, 2023). As the downlines approach the present day, the primary difficulty is in finding public records to prove relationships. Obituaries and social media are fertile ground for contemporary research.

Ultimately, the goal of a typical IGG case is to find a union couple—an intersection between genetic clusters—that represents the subject's maternal and paternal line coming together (Gurney et al., 2022). From the union couple will descend the subject. More than one candidate may arise if the union couple has several children. Context clues, such as where the children were living at the time of the crime in question, or if any of them have been reported missing in the case of a Doe, can help to narrow the possibilities.

24.5 Methods with case studies

The general approach described above works in many IGG cases. The first case study, described below, is illustrative. In other cases, additional methods must be brought to bear to produce a lead. Three of those additional methods, with accompanying case studies, are described in this section. Each of these case studies is based on a true UHR or criminal case, but names and key details have been modified for case privacy.

24.5.1 Average investigative genetic genealogy—finding the union couple

A typical IGG case may be described using the following research steps:

1. **Match selection**: The IGG practitioner selects a number of matches they believe will be relevant to the subject's family tree. These matches may be selected based on proximity to the subject, shared DNA with one another in addition to the subject ("genetic networks"), or other factors such as location or ethnicity.
2. **Ascendancy research:** The IGG practitioner performs genealogic research to build the family tree for the selected matches. This may lead to the identification of common ancestors between two or more genetic matches to the subject.
3. **Descendancy research:** The IGG practitioner researches the lines of descendancy or downlines from common ancestors. This may lead to the identification of one or more union couples who unite matches from multiple genetic networks.
4. Identification of the subject, based on estimated relationships to genetic matches as well as relevant context clues.

In the case of "Sally," this process occurred in precisely the order described. Sally, a Jane Doe, had thousands of genetic matches in the GEDmatch and FamilyTreeDNA databases, but a few key matches were critical to her identification using IGG (Fig. 24.3).

Match 1 was selected by the team working on her case because of their close proximity (800 cM) to Sally. Knowing that this match would likely share grandparents with Sally, the team quickly built Match 1's tree to the level of grandparents. Performing this step alone would have made it challenging to identify Sally, as the grandparents had many children and grandchildren.

Thankfully, other team members selected genetic Matches 2 and 3, who shared DNA with one another as well as sharing DNA with Sally. Upon building their family trees, the team quickly identified a common ancestor at the level of great-grandparent. They performed descendancy research and discovered that a grandchild of the common ancestor couple had married the uncle of Match 1, resulting in the identification of a union couple.

The IGG team then researched each of the union couple's children, discovering that they had one child who was unaccounted for and who lived in the region where Sally was found. This investigative lead was turned over to the investigating agency and, months later, the identity of Sally was officially confirmed.

24.5.2 Use of DNA inheritance patterns

In the aforementioned case, it was additionally helpful that Match 3 shared not only autosomal DNA with Sally, but also X-DNA. X-DNA follows a unique inheritance pattern: genotypic females (those with two X chromosomes) inherit DNA from both their mother and their father (Bettinger, 2019). Genotypic males, however, inherit X-DNA from their mother.

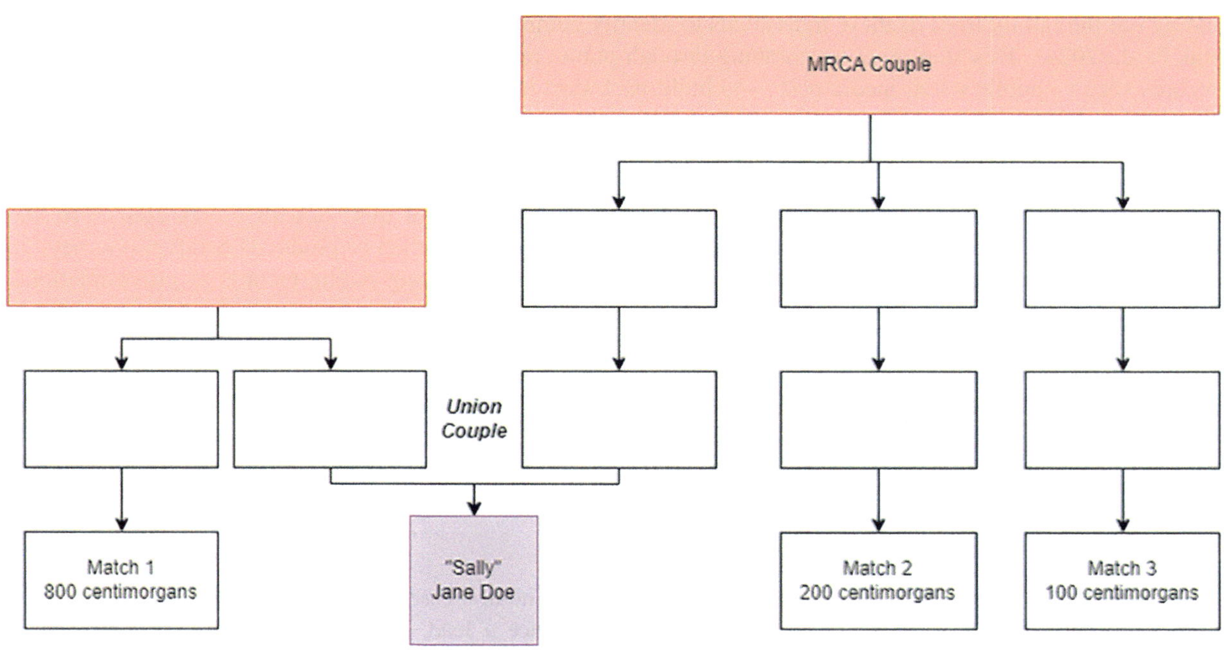

FIGURE 24.3 Family tree for "Sally." Sally was identified using three key matches.

This inheritance pattern is helpful for genetic genealogists, as X-DNA cannot be passed through two successive males in a familial line (Legacy Tree Genealogists, 2020). Sally and Match 3's relationship was one where it was reasonable for them to share X-DNA. This increased the team's confidence that they had correctly identified Sally.

Other DNA inheritance patterns can be helpful as well. For example, mitochondrial DNA is inherited via the direct maternal line (Bettinger, 2019). It can thus be assumed that any two individuals related to one another via the direct maternal line will share a common mitochondrial DNA haplotype.

Y-DNA is similar, but it is possessed only by genotypic males (those carrying a Y chromosome) and is inherited via the direct paternal line (Bettinger, 2019). Any two males directly related through the paternal line should thus share a common Y-DNA haplotype.

In addition to the case of Sally above, these two cases illustrate the power of DNA inheritance patterns and their utility in IGG cases.

First, the case of "Craig," a John Doe. Craig was believed to be Caucasian when his remains were discovered, but advanced DNA testing for IGG revealed that he was 50% Asian. Furthermore, his mitochondrial DNA haplogroup was associated with Asia, while his Y-DNA haplogroup was associated with Europe. This led the IGG team working on his case to determine it was likely that Craig's father would be of European descent, while his mother would be of Asian descent. After working on the available genetic matches, the team discovered a common ancestor couple of Irish ancestry who had many descendants in the area where Craig's remains were found. Further investigation determined that one of those descendants had married a Japanese woman, and together they had one son who had been missing for twenty years. Their son, Craig, was later determined to be the correct identity of John Doe.

In the case of "Tiffany," a Jane Doe, mitochondrial DNA haplogroups were critical to identifying which people in the family tree could *not* be Tiffany's mother. In this case, two genetic networks were united by a union couple, believed by the IGG team to be Tiffany's maternal great-grandparents. This couple had several children from whom Tiffany could be descended, and it was a challenge for the team to guess which child could be Tiffany's maternal grandparent.

The team eventually came to realize that a 2nd cousin once removed to Tiffany, Match 2, had listed her mitochondrial DNA haplotype, which differed from Tiffany's. This demonstrated that they could not be related matrilineally. Following the direct maternal ancestry, it was determined that Tiffany's mother could not be a daughter of any of the union couple's daughters because they would share Match 2's mitochondrial haplotype, H. Thus, a reference tester was contacted—a descendant of the couple's *sons*—and this tester was determined to be a half-aunt to Tiffany. Tiffany was found to be a grandchild of the same son (Fig. 24.4).

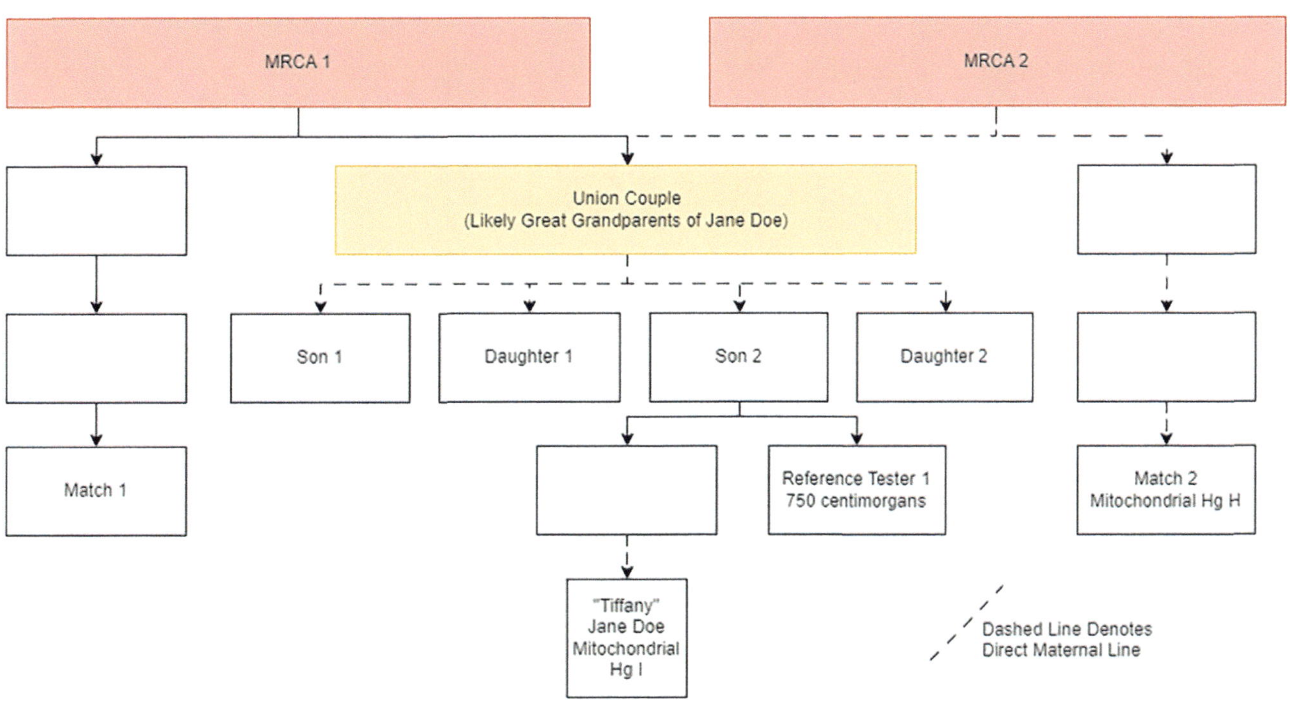

FIGURE 24.4 Family tree for "Tiffany." Tiffany was identified with the help of her mtDNA line.

X-DNA inheritance patterns were used in the case of "Belinda," a Jane Doe with British ancestry. Descending from a prolific family in England, the common ancestors identified among her genetic matches had thousands of descendants. From couples with 10–12 children each, it was difficult for the team to determine where Belinda might be descended from. The distance between her genetic matches presented an additional challenge. With a top match of 60 cM, it was anyone's guess how far back the common ancestors would be related to Jane Doe.

X-DNA inheritance patterns helped narrow the search. Match 2, who shared 30 cM of autosomal DNA with Jane Doe, also shared 20 cM of X-DNA. This match had a common ancestor with Match 1, the top match. It was determined that Belinda must be descended along an X-inheritance line from this common ancestor couple, who had 12 children. Sons of the couple's sons were eliminated from research, which reduced the number of lines of descendancy by one-third. Belinda was discovered, indeed, in an X-descendancy line at the level of third cousin once removed to Match 1 and fourth cousin once removed to Match 2 (Fig. 24.5).

24.5.3 Utilizing reference testers

In some cases, there is an absence of close genetic matches and genetic networks that could lead the team quickly to the dentification of the subject. In such cases, it may behoove the investigating team to utilize consenting reference testers to get closer to the subject's identity. An example of this can be found in the case of Jane Doe, "Layla."

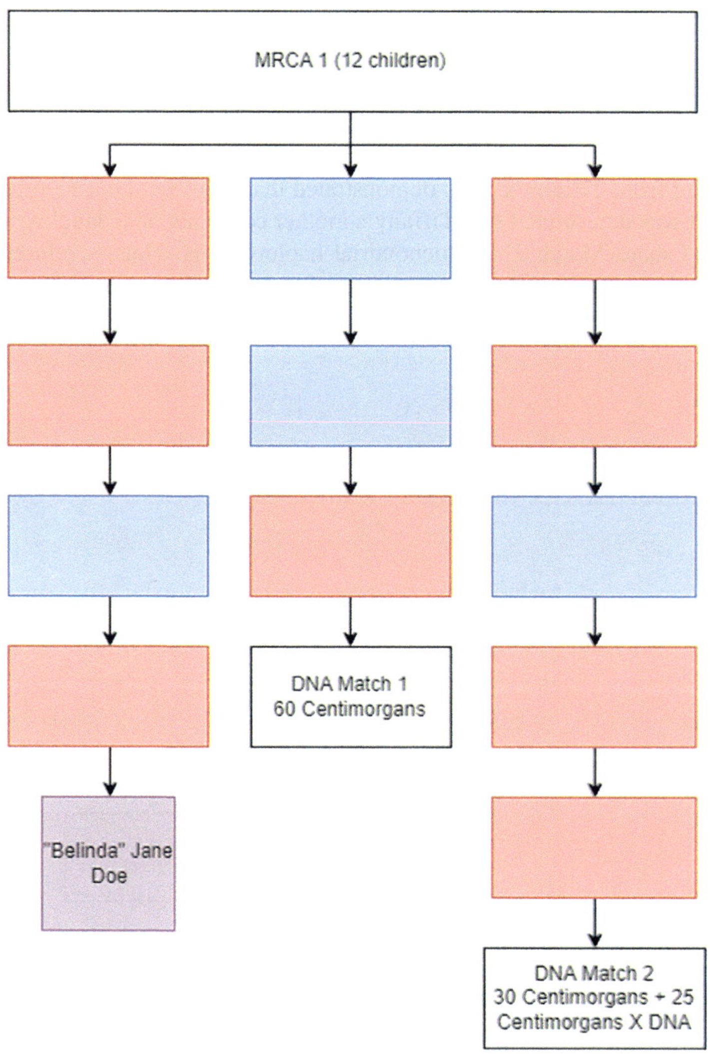

FIGURE 24.5 Family tree for "Belinda." Belinda was identified with the help of her X-DNA line.

Jane Doe, being of Mexican descent, had few genetic matches in the accessible DNA databases. Her closest match, Match 1, was estimated to be a 3rd cousin. However, there were no other genetic matches that shared a significant amount of DNA with either Layla or Match 1. The team proceeded by building the family tree for Match 1 to the level of 2nd-great-grandparents. It was discovered that Match 1 was half-American (of European descent) and, thus, his 2nd great-grandparent sets 1–4 were tentatively eliminated as ancestors of Layla (Fig. 24.6). It was believed that one of the remaining sets of 2nd great-grandparents was ancestors of Layla. All were of Mexican ancestry.

The team then identified 2nd and 3rd cousins of Match 1 who were related to Match 1 via 2nd great-grandparents 5, 6, 7, and 8 (Fig. 24.7). The law enforcement agency contacted these potential reference testers, and two consented to either take a DNA test or upload their existing direct-to-consumer DNA test to GEDmatch to be compared to Layla. The results made the research process clearer for the team working on the case.

Reference Tester 2, who shared 250 cM of DNA with Jane Doe, made it clear not only that the 2nd great-grandparent set 8 were ancestors to Layla, but also that Jane Doe was more closely related to Reference Tester 2 than to Match 1.

With Great Grandparent set 8 identified as common ancestors, the team performed downline research and provided a reference test to a second cousin of reference tester two. This individual, Reference Tester 3, a first cousin, became the critical match needed to identify Layla's family and place her in the tree (Fig. 24.8).

24.5.4 Application of context clues

On occasion, the uniqueness of context clues surrounding a case can be critical to the IGG research process. The cases of John Does "Clarence" and "Peter" were both dependent on data from a single genetic match, combined with circumstantial evidence that led to the identification of the decedents.

The remains of Clarence were found in a midwestern metropolitan area, and a crystal-clear image of his face was available. His case was predicted to be a challenge, as he was from a minority population (African American) that is underrepresented in genetic genealogy databases. The team was glad to find a genetic match at the level of first cousin upon viewing the match list.

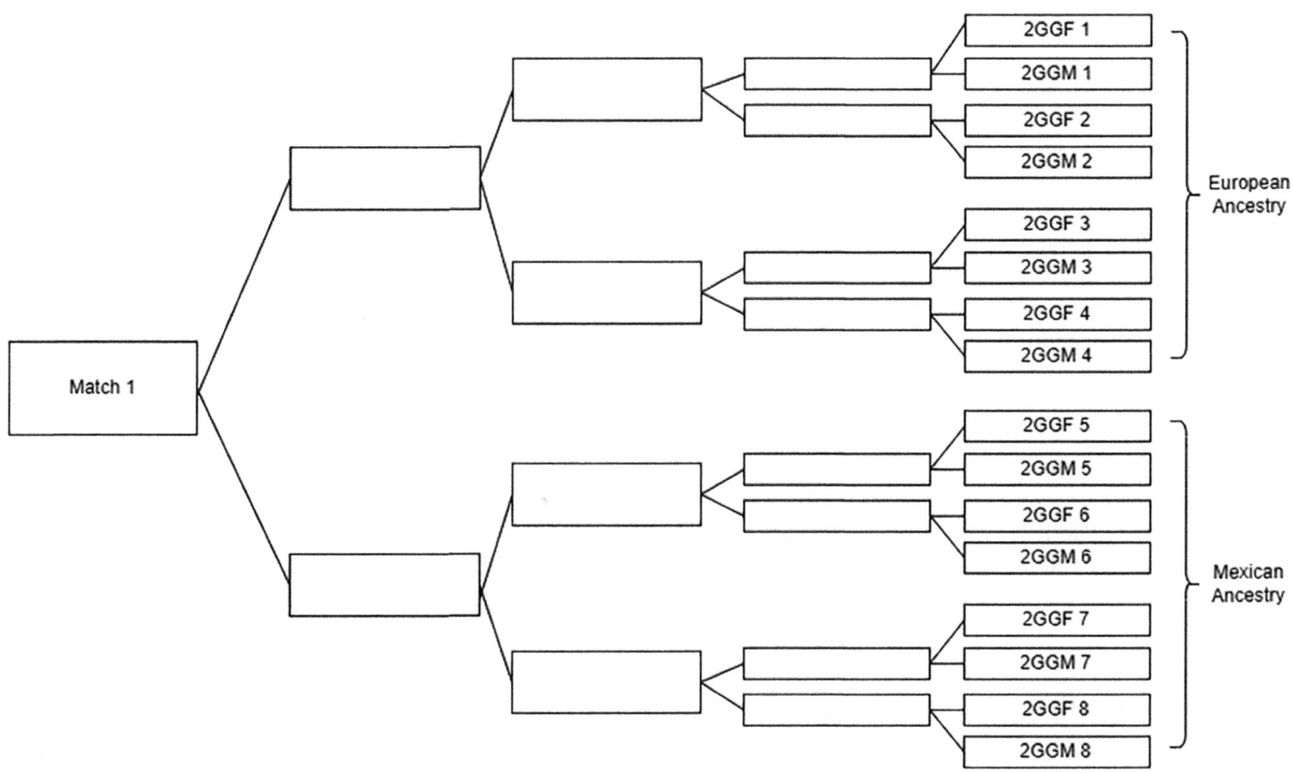

FIGURE 24.6 Match 1's family tree leading to European and Mexican ancestry. Match 1's paternal ancestry was European, while their maternal ancestry was Mexican.

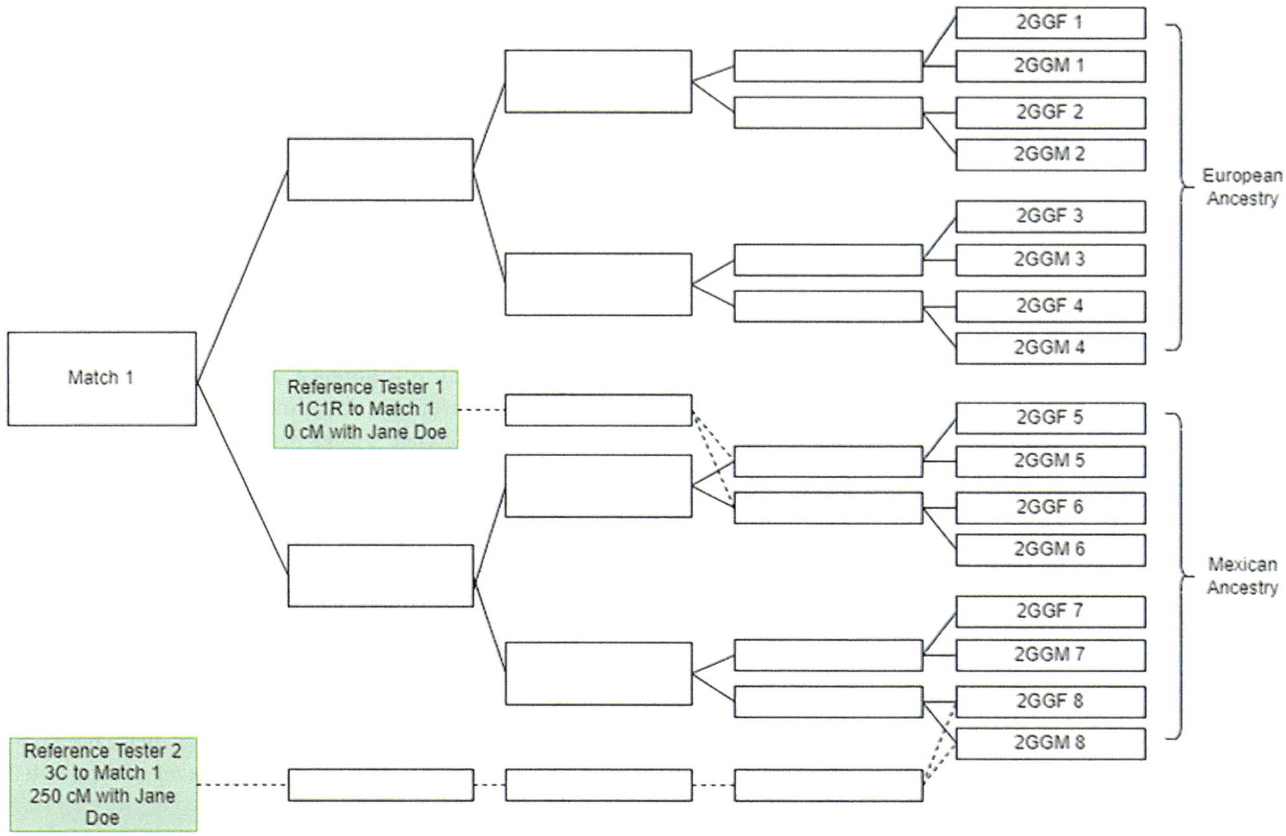

FIGURE 24.7 Match 1's family tree with the location of reference testers. Reference testers were identified from Match 1 maternal downlines.

It was quickly discovered that the top match was half African American and half Caucasian. Her mother was Caucasian, so the team could focus on the match's paternal line. Her paternal grandparents, likely Clarence's grandparents, had ten children who could potentially be one of his parents. At this point, a social media search was utilized to determine where the family members lived and if any of them were in close proximity to the location where Clarence's remains were found. Quickly, a social media post was discovered. One family member mentioned their missing nephew, Clarence, accompanied by a photo, an identical match to the face of John Doe. Weeks later, fingerprints confirmed Clarence's identity.

In the case of Peter, recent immigration on both sides of his family tree presented an enormous challenge to the IGG team. The closest genetic match was at the level of third to fifth cousin (50 cM), and the remaining genetic matches all shared less than 40 cM with Peter. The team dove deep into researching the various items carried by Peter at the time of his death: a t-shirt from a specific location, a pocket watch with initials, and a publication from a union.

The team made a guess, based on John Doe's bioancestral makeup, which of the top match's ancestors might be shared with John Doe. From there, they researched the third cousins of Peter and incredibly, found him: a man from the city the t-shirt displayed, who was a member of the union the publication was from, and whose father had the initials engraved on the pocket watch. The lead was bolstered when the team uncovered an unpaid tax lien and unclaimed funds associated with Peter from the year John Doe's remains were found.

The team's tenacity in researching these unusual clues paid off. Peter's identity was confirmed shortly after the lead was provided.

24.5.5 Investigative genetic genealogy for postconviction and defense work

Special considerations must be made for IGG cases worked on behalf of postconviction or defense attorneys. While these types of cases have made up only a fraction of the total cases resolved with the help of IGG, there is likely to be more postconviction and defense casework available as the field of IGG continues to expand its reach. The IGG Center cases of "Walter" and "Matthew" are illustrative of the unique issues present in these postconviction and defense work. Note that

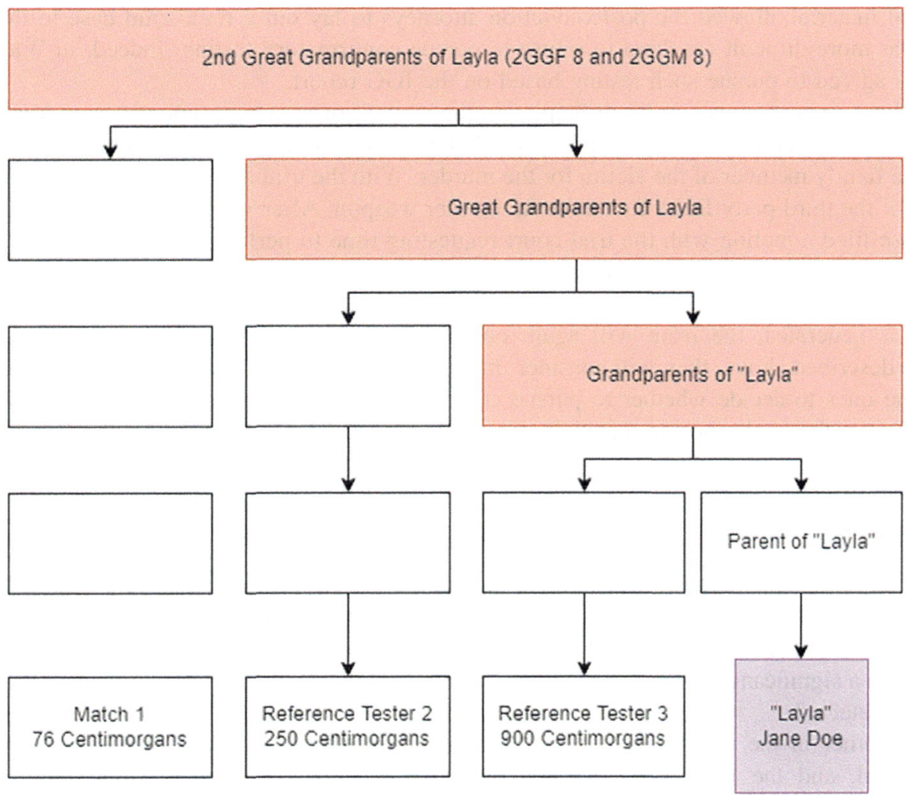

FIGURE 24.8 Family Tree for "Layla." Two reference testers were essential in identifying Layla.

the genealogical work does not differ in these case types. Rather, the difference is how cases get started and what IGG practitioners should consider including in final reports.

In these case examples, significant details have been changed to protect the cases.

Walter was an individual who left probative DNA on the victim of a homicide. Despite the presence of this probative, third-party DNA, another individual had been convicted of the crime in the 1970s. Postconviction attorneys wished to have IGG performed on the unknown DNA, with the hope that identifying the contributor would lead to the exoneration of their client. Yet, unlike in typical IGG cases worked with law enforcement or medical examiners, postconviction attorneys do not have direct access to the DNA in question, and they do not have the authority to order IGG to be performed. Postconviction attorneys must receive either an agreement from the law enforcement agency that controls the DNA or a court order before IGG can begin. In Walter's case, the postconviction attorneys reached out directly to the law enforcement agency in charge of the DNA and requested that the DNA be subjected to IGG. Here, the law enforcement agency agreed to IGG without the need for a court order.

Once the IGG team had developed a lead for the identity of Walter, they considered the context of postconviction when generating the final report. In a typical UHR or criminal case, IGG practitioners rarely have to convince the law enforcement agency or medical examiner to pursue confirmatory testing. This is because those agencies have the power to order additional testing and seek out IGG specifically to identify a criminal suspect or unidentified individual. In other words, nothing is holding them back from wanting an identification. Indeed, some agencies would prefer no final report from IGG practitioners—just a name, location, and date of birth.

Such is not the case for postconviction cases. When writing the final report for Walter, the team was acutely aware that the postconviction attorneys who would receive the report were not the ones who would be deciding whether to pursue confirmatory testing. Rather, the postconviction attorneys would have to use our report to convince either the law enforcement agency in charge of the DNA or a judge to move forward. With this in mind, the team included more detail in our final report than they would in a typical report to a law enforcement agency or medical examiner. To begin, the team provided an extended description of the IGG process for the postconviction attorneys to share with the law enforcement agency. Also, throughout the report, the team went into greater detail on the process that got us to the lead. Finally, the team included easy-to-follow hyperlinks to the most relevant genealogical and public records documents relied on in the case. This additional material allowed the postconviction attorneys to lay out a rock-solid case to the law enforcement agency, making it far more difficult for them to refuse to pursue confirmatory testing. Indeed, in Walter's case, the law enforcement agency agreed to pursue such testing based on the IGG report.

The case of Matthew demonstrates similar principles in the context of an active defense case. Matthew left probative DNA on the weapon used in a homicide. Without an identification for Matthew, law enforcement pursued other leads and eventually indicted a family member of the victim for the murder. With the trial approaching, the defense was desperate to identify the source of the third-party DNA found on the murder weapon. After vetting the IGG practitioners, the defense attorneys for the case filed a motion with the trial court requesting time to perform IGG and an order requiring the law enforcement agency to turn over the DNA evidence for testing. To convince the court, IGG team members were asked to provide resumes and a description of the their policies. With those in hand, the court ordered IGG and delayed the trial date. Once a lead is generated, the team will again consider the context when writing the final report. As in the postconviction case described above, they will consider that the recipient of our final report—the defense attorneys—will not ultimately be the ones to decide whether to pursue confirmatory testing. Because that decision will fall to the trial court, the team will tailor the final report to that audience, being sure to lay out the process of IGG and the steps taken to arrive at the lead in the clearest, simplest terms possible, with ample visual aids and links to relevant documents.

24.6 Conclusion

The advent of IGG represents the greatest investigative breakthrough since the introduction of forensic DNA forty years ago. With sufficient resources and personnel, the backlogs of cold cases—so long as they include biological material—can be cleared, and violent criminals can be stopped in their tracks after committing a single crime. Indeed, the publicity of IGG may, itself, create a significant deterrent for violent criminals, as they can now know, to a near certainty, that there is no way to remain undetected.

While still a newcomer in the investigative arena, IGG is in an active period of professionalization. Methods are becoming standardized, and the field has coalesced on a set of ethical principles while remaining open to new developments that will require additional careful consideration.

The methods described in this Chapter are a subset of those available to skilled IGG practitioners. Similarly, the legal and ethical considerations we have described here are not exhaustive, and developments in those areas occur regularly in this still-developing field. As such, investigating agencies who seek to hire IGG practitioners as contractors or in-house should carefully screen candidates to ensure that they are skilled in genealogy in general and IGG methods specifically, and that they adhere to the legal and ethical standards of the field.

References

Bettinger, B. T. (2019). *The family tree guide to DNA and genetic testing.* (2nd edition). Penguin Random House.

Dowdeswell, T. (2024). Forensic genetic genealogy project version December 2023. https://data.mendeley.com/datasets/cc5rh42mf9/1, https://doi.org/10.17632/cc5rh42mf9.1.

Dowdeswell, T. L. (2022). Forensic genetic genealogy: A profile of cases solved. *Forensic Science International: Genetics, 58*, 102679. https://doi.org/10.1016/j.fsigen.2022.102679, https://www.sciencedirect.com/science/article/pii/S1872497322000205.

Greytak, E., Wyatt, S., Cady, J., Moore, C. C., & Armentrout, S. (2024). Investigative genetic genealogy for human remains identification. *Journal of Forensic Sciences*. https://doi.org/10.1111/1556-4029.15469, https://onlinelibrary.wiley.com/doi/abs/10.1111/1556-4029.15469.

Guerrini, C. J., Wickenheiser, R. A., Bettinger, B., McGuire, A. L., & Fullerton, S. M. (2021). Four misconceptions about investigative genetic genealogy. *Journal of Law and the Biosciences, 8*(1). https://doi.org/10.1093/jlb/lsab001/37352563/lsab001.pdf, https://academic.oup.com/jlb/article-pdf/doi/.

Gurney, D. (2023). Investigative genetic genealogy and the First Amendment right to noninterference with receipt. *First Amendment Law Review, 21*, 1. https://heinonline.org/hol-cgi-bin/get_pdf.cgi?handle=hein.journals/falr21&section=5.

Gurney, D., Press, M., Moore, C. C., Rolnick, C., Hochreiter, A., & Bossert, B. (2022). The need for standards and certification for investigative genetic genealogy, and a notice of action. *Forensic Science International, 341*. https://doi.org/10.1016/j.forsciint.2022.111495, https://www.sciencedirect.com/science/article/abs/pii/S0379073822003255.

H.B. 1327, 2023 Reg. Sess. Fla. (2023).

H.B. 240, 2021 Reg. Sess. Md. (2021).

H.B. 602, 2021 Reg. Sess. Mt. (2021).

Idaho v. Kohberger. (2024). Public order for disclosure of IGG information, Idado, Second Judicial District, Latah County. No. CR29-22-2805 (Idaho Dist. Ct. January 11, 2024).

Idaho v. Kohberger. (2025). Order on defendant's motion to suppress re: Genetic information, Idaho, Fourth Judicial District, Ada County. No. CR01-24-31665 (Idaho Dist. Ct. February 19, 2025).

Investigative Genetic Genealogy Accreditation Board. (2025). IGG code of professional ethics.

Khatchadourian, R. (2021). How your family tree could catch a killer. *The New Yorker*. https://www.newyorker.com/magazine/2021/11/22/how-your-family-tree-could-catch-a-killer.

Legacy Tree Genealogists. (2020). DNA inheritance patterns: X-DNA and autosomal DNA. Legacy tree. Paul. https://www.legacytree.com/blog/x-dna-autosomal-dna-inheritance-paths.

Minnesota v. Carbo. (2021). Findings of fact, conclusions of law, and order. No. 69HI-CR-20-549 (Minn. Dist. Ct., November 10, 2021).

Minnesota v. Carbo. (2024). No.A22-1823 (Minn. May 8, 2024). https://cases.justia.com/minnesota/supreme-court/2024-a22-1823.pdf?ts=1715373350.

S.B. 156, 2023. Reg. Sess. Utah (2023).

S.B. 351, 2023 Reg. Sess. Mt. (2023).

Smith, J. (2023). Police are getting DNA data from people who think they opted out. The Intercept. Intercept: IGG DNA data. https://theintercept.com/2023/08/18/gedmatch-dna-police-forensic-genetic-genealogy/.

Taylor, M. (2024). Next generation sequencing accepted in court for first time first time. Forensicmag.com. https://www.forensicmag.com/610513-Next-Generation-Sequencing-Deemed-Admissible-in-Court-for-First-Time/.

Taylor, M. (2024). IGG helps exonerate brothers of 1987 murder. Forensicmag.com. https://www.forensicmag.com/3594-All-News/615248-IGG-Helps-Exonerate-Brothers-of-1987-Murder/

Tillmar, A., Sjölund, P., Lundqvist, B., Klippmark, T., Älgenäs, C., & Green, H. (2020). Whole-genome sequencing of human remains to enable genealogy DNA database searches – A case report. *Forensic Science International: Genetics, 46*, 102233. https://doi.org/10.1016/j.fsigen.2020.102233.

Wickenheiser, R.A., Naugle, J., Hoey, B., Nowlin, R., Kuman, S.A., Minton, A., Allen, L., & Glynn, C. (2023). National technology validation and implementation collaborative (NTVIC) policies and procedures for forensic investigative genetic genealogy (FIGG). https://www.sciencedirect.com/science/article/pii/S2589871X23000037.

Chapter 25

Application of next-generation sequencing techniques in generating investigation leads

Mahdi Haidar[1], Fatema Mousawi[1], and Mohammad H. Al-Hasan[2]

[1]*Kuwait Identification DNA Laboratory (KIDL), General Department of Criminal Evidence, Ministry of Interior, Kuwait City, Kuwait,* [2]*Environmental and Life Science Research Centre (ELSRC), Kuwait Institute for Scientific Research (KISR), Kuwait City, Kuwait*

25.1 Introduction

Providing intelligence on the origin or ancestry of unknown or unidentified samples is an invaluable tool in forensic cases, especially when there are no reliable eyewitness descriptions or if the DNA profile does not match any entries in a database. The primary goal of DNA analysis is to use DNA as an additional, or sometimes the only form of evidence, serving as a reliable "eyewitness" that offers data comparable to information gathered during police investigations. This type of biological intelligence can provide crucial leads, helping to narrow down suspect pools and guide investigative efforts when other evidence is lacking or inconclusive.

DNA profiling of biological evidence from crime scenes, mass disasters, or missing persons represents a significant challenge in forensic science. When suspects are identified, their DNA profiles are directly compared with the DNA profile of the evidence. However, if no suspects are identified, the profile is compared to a database to find a match. Both methods aim to identify the person who left the evidence.

However, there are instances where available DNA evidence does not provide sufficient information to identify or match a suspect; it happens when the DNA profile obtained does not match any existing profiles in the database, leaving investigators without a lead, which results in an information gap. Additionally, it can occur when there is insufficient or poor-quality DNA due to fragmentation, contamination, or minimal amounts of biological material, making the obtained profile unreliable. Incomplete genetic markers can also contribute to the gap; the genetic markers obtained might be incomplete or not sufficiently discriminatory to provide a definitive match, which could result from using less informative markers or obtaining partial profiles. Complex mixtures, where the sample contains DNA from multiple individuals, can complicate the analysis and make it difficult to separate and identify individual profiles accurately.

Over the years, DNA typing has increasingly used large sets of genetic markers. These markers, from various classes, can be analyzed either independently, where each marker is examined separately, or simultaneously, where multiple markers are analyzed at the same time from a single biological sample, even with a small amount of DNA, to maximize the information inferred from the genetic profile. The landscape of forensic DNA analysis has advanced significantly, with the development of additional tools and techniques to complement established methods. The progress has evolved from early methods like variable number of tandem repeats to the analysis of amplified short tandem repeat (STR) markers by capillary electrophoresis (CE). The analysis of STR markers using CE remains the foundation of forensic DNA analysis. Recent advancements in sequencing technologies, such as next-generation sequencing (NGS), also known as massively parallel sequencing (MPS), have been integrated into forensic science, offering extraordinary capabilities for comprehensive genetic analysis and significantly improving forensic investigations.

NGS facilitates the use of alternative markers, including autosomal and Y-chromosome, single-nucleotide polymorphisms (SNPs) and mitochondrial DNA (mtDNA), allowing for the genotyping of numerous STR loci and an expanding array of markers. These markers can provide additional information such as putative age, physical appearance, for

Investigative and Predictive DNA Testing. DOI: https://doi.org/10.1016/B978-0-443-30128-5.00031-X

example, hair, eye, and skin color, as well as biogeographical ancestry inference. By closing information gaps and providing detailed genetic insights, NGS technology improves the ability to generate investigative leads, thus improving the effectiveness of forensic investigations.

25.1.1 Next-generation sequencing platforms

NGS technologies, commercially introduced in 2005 (Margulies et al., 2005), have transformed DNA sequencing. Unlike traditional Sanger methods that sequence one DNA fragment at a time, NGS enables the simultaneous sequencing of numerous DNA fragments. NGS can conduct millions to billions of sequencing reactions simultaneously across multiple samples, allowing for the analysis of various genomic traits in a single run. While Sanger sequencing took up to 13 years and cost $2.7 billion to accomplish the first human genome sequence, NGS can now sequence a whole genome within 48 hours for less than $600. The high-throughput capacity, reduced costs, and high speed of NGS make it suitable for various applications, such as ancient DNA analysis (Poinar et al., 2006), disease diagnosis (Krier et al., 2016; McCarthy et al., 2013), human history (Marciniak & Perry, 2017), and forensics (Alvarez-Cubero et al., 2017; Ballard et al., 2020).

NGS platforms are categorized into two major types: second-generation and third-generation sequencing according to the read length, which refers to the number of base pairs (bp) sequenced from a DNA fragment. Second-generation sequencing includes Roche 454 pyrosequencing, Ion Torrent, and Illumina. Roche 454 pyrosequencing uses real-time sequencing based on detecting pyrophosphate release and light generation after nucleotide incorporation, making it faster and cheaper than Sanger sequencing. Ion Torrent, a nonoptical sequencing platform, uses a semiconductor ion to detect nucleotide incorporation. Protons released during DNA synthesis cause a small pH change detected by a sensor, with different chips enabling complete sequencing in less than two hours. Illumina, the most widely used sequencing technology, employs terminator molecules similar to those in Sanger sequencing, with each base labeled with a base-specific, cleavable fluorophore. The color emitted by total internal reflection fluorescence allows nucleotide identification. Illumina offers a range of platforms for whole-genome sequencing, from small benchtop instruments to large ultra-high-throughput instruments.

Third-generation sequencing, including PacBio sequencing, Oxford Nanopore, and VisiGen Biotechnology, aims to sequence longer DNA sequences. PacBio sequencing, also known as single-molecule real-time (SMRT) sequencing, allows the analysis of very long DNA fragments. DNA is sequenced in real-time as it progresses through a well in an SMRT flow cell, with imaging occurring when fluorescently labeled nucleotides are incorporated. The Oxford nanopore transports DNA through a nanopore membrane and detects changes in ion current as DNA passes through the pore. Each nucleotide produces distinct current levels, allowing the determination of the nucleotide sequence. VisiGen Biotechnology uses fluorescence resonance energy transfer (FRET) and a DNA polymerase to determine base sequences. FRET occurs when fluorescently labeled nucleotides are incorporated into a DNA strand, producing variable frequencies of FRET signal depending on the label incorporated.

Understanding the differences in principles, technologies, and applications of these sequencing methods is essential to select the appropriate method based on specific analytical needs.

25.1.2 Next-generation sequencing platforms for forensic use

Thermo Fisher Scientific has developed two primary NGS platforms customized for forensic applications: the Ion S5 System and the Ion GeneStudio S5 Series. These platforms were specifically designed to meet the requirements of forensic DNA analysis, offering high accuracy, flexibility, and rapid turnaround times.

The Ion S5 system is a versatile and cost-effective NGS platform and is well-suited for various forensic applications. It uses semiconductor sequencing technology to provide rapid sequencing capabilities with relatively low initial setup costs. The Ion S5 system is particularly advantageous for targeted sequencing applications, such as analyzing specific regions of interest, such as STRs, SNPs, and mtDNA. This system is known for its flexibility, speed, and scalability, making it an excellent choice for forensic laboratories that require reliable and efficient DNA sequencing solutions.

Building upon the capabilities of the Ion S5 System, the Ion GeneStudio S5 Series offers enhanced performance and higher throughput. This series includes multiple models, S5, S5 Plus, and S5 Prime, each designed to address different sequencing needs, from small-scale targeted sequencing to extensive genomic analyses. The Ion GeneStudio S5 Series maintains the core advantages of the Ion S5 System, such as rapid sequencing and cost-effectiveness, while providing additional features for higher throughput and more complex forensic applications. These systems can handle a wider range of samples, including degraded and low-quality samples commonly encountered in forensic cases.

The Ion S5 system and the Ion GeneStudio S5 Series share several key features that make them ideal for forensic DNA analysis. These platforms are adaptable to a wide range of sequencing applications, from targeted panels to comprehensive genomic analyses. They offer faster run times, enabling faster turnaround of sequencing results, which is critical for time-sensitive forensic investigations. Additionally, the different models and configurations allow forensic laboratories to scale their operations according to their throughput needs, whether for small or large projects.

Verogen is a company that specializes in advancing forensic science by developing and providing NGS kits and solutions specifically for forensic applications. Their partnership with Illumina, a global leader in sequencing and array-based technologies, has significantly impacted the field of forensic genomics. The partnership combines Verogen's forensic expertise with Illumina's advanced sequencing technology. This collaboration has led to innovative solutions that address the unique challenges of forensic DNA analysis. The critical outcome is the launch of the MiSeq FGx forensic genomics system, a groundbreaking NGS platform for forensic genomic applications. The system uses well-established Illumina sequencing technology, which uses a synthesis approach for labeling DNA. The MiSeq FGx system offers several advantages, such as higher throughput, improved sensitivity, and the ability to simultaneously analyze multiple markers, including STRs, SNPs, and other genomic regions of interest. These features make the platform particularly suitable for forensic genomics, where obtaining comprehensive and accurate data from challenging samples is often crucial.

25.2 Enhancing investigative leads with externally visible characteristics

Forensic DNA phenotyping (FDP) uses SNPs to predict externally visible characteristics (EVCs) such as eye color, hair color, skin color, and facial features (Haidar et al., 2024). These predictions are invaluable for identifying suspects and victims in criminal investigations, especially when conventional methods are inconclusive. Phenotypic-informative SNPs, which are genetic markers associated with specific physical traits, are highly reliable for forensic applications due to their stable inheritance and low mutation rates. FDP serves as a "biological eyewitness," providing critical information to support human identification when other methods fall short (Kayser, 2015; Valentine & Fitzgerald, 2016). Additionally, SNPs are particularly beneficial in cases involving degraded samples or low amounts of DNA, where STR genotyping may fail due to large allele fragmentation. Moreover, the simplicity of SNP alleles, without the repetitive structure found in STRs, reduces the likelihood of artefacts such as stutter peaks, making SNP analysis more straightforward and reliable. The 1000 Genomes Project has identified more than 80 million SNPs with minor allele frequencies greater than 1%, demonstrating their extensive variability and utility in forensic applications (Auton et al., 2015). It is estimated that 50 SNPs may be as informative as only 13 STRs, meaning that a single STR can match the identification power of 4–5 SNPs (Gill, 2001; Pakstis et al., 2010).

25.2.1 Assays for human pigmentation prediction

IrisPlex, the first pioneering sensitive genotyping assay released and validated by Walsh et al., has significantly advanced eye color prediction (Walsh et al., 2011). Eye color is a genetically complex trait influenced by the production and distribution of pigments, regulated by several genes and factors. The study of eye color genetics dates back to the early 20th century, when researchers first examined hereditary patterns (Davenport & Davenport, 1907). Understanding the genetics underlying eye color has notable implications for forensic investigations. The IrisPlex is an eye color prediction assay that includes six SNPs to predict blue, intermediate, and brown eye color with an accuracy rate of approximately 94% in correctly predicting an individual's eye color as blue or brown (Walsh et al., 2012).

A new forensic phenotyping assay, HIrisPlex, has been developed to enhance eye color prediction (Walsh et al., 2013). HIrisPlex comprises 24 forensically validated phenotypic markers in a single multiplex, including 18 new hair-color markers and the 6 IrisPlex SNPs. Hair pigmentation is one of the most visible phenotypic traits in human appearance, exhibiting significant variation between individuals, which is helpful in forensic contexts for identification purposes. The use of the HIrisPlex system is especially valuable in forensic cases involving skeletal remains and missing individuals, such as disaster victims (Chaitanya et al., 2017; Draus-Barini et al., 2013; Haeusler et al., 2016; King et al., 2014). Where traditional identification methods, such as dental records or fingerprints, may not be available, the HIrisPlex system can provide a valuable tool. Predicting physical hair and eye color can generate a more accurate biological profile in challenging forensic scenarios, aiding identification.

Validation studies have shown that the prediction precision of HIrisPlex is close to 70% for individuals with blond hair, 80% for brown hair, 80% for red hair, and 88% for black hair across various European regions (Walsh et al., 2014). In a broader range evaluation through the human genome diversity cell line panel (HGDP-CEPH), which includes 51 distinct ethnic groups, HIrisPlex can achieve an inference accuracy surpassing 85% in determining whether an individual with brown eyes and black hair has origins from non-European or European regions (Walsh et al., 2014). These results demonstrate the effectiveness of the HIrisPlex system in providing reliable phenotypic predictions, thus enhancing the capabilities of forensic investigations to identify unknown individuals based on genetic evidence. The HIrisPlex genotyping assay is highly sensitive and capable of producing complete profiles from as little as 63 picograms (pg) of DNA. It has proven precise and reliable when tested on various biological materials, including blood, semen, saliva, hair, and trace DNA (Walsh et al., 2014). This makes HIrisPlex particularly useful for identifying missing persons or disaster victims when traditional identification methods are unavailable. By predicting physical traits, the HIrisPlex can facilitate identification, making individuals available for further STR profiling and accurate identification.

In 2018, Chaitanya and his colleague developed the HIrisPlex-S as an advanced tool to its predecessor, the HIrisPlex system (Chaitanya et al., 2018). The system includes a novel 17 SNPs for skin color prediction and the previously developed assays for eye and hair color prediction. The system showed a strong correlation between the DNA-predicted skin color categories and the actual skin color phenotypes of the individuals. This demonstrates the effectiveness of the HIrisPlex-S system in accurately predicting skin color. The predictive model in the HIrisPlex-S system is based on the analysis of 41 HIrisPlex-S SNPs, which can be entered on the website (https://hirisplex.erasmusmc.nl/), to generate individual prediction probabilities for three eye colors, four hair colors, and five skin colors. By analyzing these genotypes, the model generates probabilities for different pigmentation categories (Chaitanya et al., 2018).

While the HIrisPlex-S assays have demonstrated robustness, especially with low quantity and quality DNA, they also have significant limitations. Despite their effectiveness, the SNaPshot single base primer extension and CE are limited in multiplex capacity, typically accommodating around 25 SNPs per assay (Breslin et al., 2019), leading to time, effort, sample, and cost consumption. For example, the HIrisPlex-S system requires two separate multiplex assays to analyze its 41 SNPs. Additionally, the SNaPshot-CE approach has difficulty separating DNA mixtures from multiple contributors because it cannot clearly distinguish between the different DNA fragments from various sources. The limited multiplex capacity also restricts the incorporation of other SNPs, such as ancestry-informative SNPs. Therefore, transitioning to NGS is necessary to take advantage of the growing knowledge of appearance and ancestry genetics.

NGS technologies dramatically increase multiplexing capacity compared to traditional DNA technologies, including the SNaPshot-CE system. The forensic field has begun to transition from CE to NGS, primarily for STRs and SNP sequencing, using technologies such as ThermoFisher and the Verogen-Illumina platforms. NGS allows the analysis of multiple sets of forensic markers, such as STRs and SNPs, in a single assay. In contrast, the CE system deals with each class of markers separately. This capability is invaluable in cases involving degraded, scarce, or mixed DNA samples, where traditional methods may not provide sufficient information. By simultaneously detecting a wide range of autosomal and sex STR loci as well as other classes of markers, NGS improves analysis efficiency and reduces costs.

25.2.2 Facial morphology prediction in forensic science

Facial morphology prediction has emerged as a significant tool in various scientific fields, including genetics, anthropology, and forensic science. This technology holds considerable promise for law enforcement agencies in identifying suspects, reconstructing victims' appearances, and aiding in missing person investigations. However, predicting facial traits using genetic markers is significantly more complex than predicting eye, hair, and skin pigmentation due to multiple genes that influence facial feature diversity. Consequently, achieving accurate and reliable facial morphology prediction remains a challenging task.

Recent advances in genomics have enabled the identification of genes and genetic variants associated with the development of facial characteristics. Specific SNPs within these genes have been analyzed for their impact on facial measurements, such as the distance between the eyes, the morphology of the nose, and the structure of the mandible (Boehringer et al., 2011). These advances stem from the integration of sequencing technologies and genome-wide association studies (GWAS). NGS has facilitated rapid and cost-effective sequencing of large populations, aiding in the discovery of genetic variants linked to facial features. Coupled with GWAS data, scientists have developed computational algorithms, statistical models, and high-resolution three-dimensional imaging, improving the ability to predict facial landmarks with approximate accuracy (Shaffer et al., 2016).

Early breakthroughs in understanding facial morphology came from two GWAS studies in Europeans, which identified significant associations between specific SNPs and nasal bridge morphology. These studies also discovered genetic variants close to genes that influence facial structure development (Adhikari et al., 2016). The same study also reported five genetic variants near genes involved in various developmental processes that affect nose shape. Another study has identified two novel genetic associations in individuals of European ancestry linked to central upper lip height (Lee et al., 2017).

These data indicate that facial traits undergo genetic regulation, creating an inherited, unique, and recognizable face crucial for physical identification. For instance, studies on different populations, such as African children and adolescents, and mouse models have highlighted the role of specific genes and genetic variants in facial morphogenesis (Cole et al., 2016). These findings suggest a link between facial traits and population genetic structure, providing strong evidence that ancestry can affect facial morphology. The importance of ancestry markers in determining facial features has been demonstrated by constructing facial feature models using these markers and additional SNPs, successfully predicting specific facial features such as eyebrow width and distance between the eyes (Claes et al., 2014, 2018). Additionally, a study in the Korean population identified significant genetic variants associated with facial morphology, showing that the nose shape is genetically inherited (Cha et al., 2018).

The Snapshot FDP system, developed by Parabon Nanolabs, is a DNA analysis tool that predicts facial morphology, sex, eye, hair, skin color, freckling, and biogeographic ancestry from DNA samples. Some law enforcement agencies have applied this system to solve cases. However, further research is needed to identify additional SNPs associated with facial features to improve the system's accuracy, which may vary depending on the DNA sample's quality and quantity.

25.2.3 Enhancing investigative leads with biogeographic ancestry

Biogeographic ancestry (BGA) differs from concepts like ethnic origin or race, which are influenced by many nongenetic factors such as culture, language, and religion. Biogeographic ancestry specifically relates to the geographical regions from which an individual's biological ancestors originated (Schneider et al., 2019). Since 2015, significant improvements have been made in the methods used to analyze BGA from forensic DNA, highlighting the ongoing evolution of forensic genomics (Phillips, 2015). Developing forensic BGA tools involves critical steps such as marker selection, genotyping multiplex design, and statistical analysis (Kayser et al., 2023). The selection of appropriate ancestry-informative markers (AIMs) is foundational to these tools, as they must accurately differentiate between populations and detect co-ancestry in admixed individuals (Kayser et al., 2023).

AIMs are transmitted through three principal mechanisms of inheritance. For example, autosomal markers are passed down from both parents, indicating the geographical origins of both parent lineages. Y-chromosomal markers, located on the nonrecombining portion of the Y-chromosome, are exclusively inherited from fathers to sons, mirroring the geographic ancestry of the paternal lineage. mtDNA markers, present in both males and females, are inherited solely from the mother, representing the geographic origins of the maternal lineage. While AIMs are diploid and account for maternal and paternal contributions, lineage markers are haploid and provide insights specific to maternal or paternal lines (McNevin, 2020). Autosomal SNPs that exhibit significant allele frequency differences between various populations are considered the best candidates for investigating ancestry (Phillips et al., 2007).

A well-established SNPforID 34-plex SNaPshot assay was developed by Phillips and his colleagues (Fondevila et al., 2013; Phillips et al., 2007). This assay can simultaneously genotype 34 ancestry SNPs, differentiating between Asian, African, and European populations. The SNPforID 34-plex was utilized to provide ancestry information for a crime scene sample that led to identifying the perpetrator in the Madrid train bombing (Phillips et al., 2009). Additionally, the SNPforID 34-plex assay has played a role in narrowing down the pool of suspects and aiding investigations, such as in Operation Minstead, described as Britain's largest investigation, where it found that the suspect was most likely of admixed African origins (Mehta et al., 2017). SNPforID 34-plex assay was also used to confirm the origin of the suspect in a murder in Madrid (Mehta et al., 2017). Moreover, Kidd's laboratory introduced a panel of 55 AIMS that are highly differentiated among 73 populations (Kidd et al., 2014). Additionally, Seldin's group identified a set of 128 SNPs that are effective in determining the continental origin of individuals and estimating their admixture proportions (Nassir et al., 2009). These two sets of ancestry SNPs are often combined and utilized in many commercial kits, as will be discussed in later sections.

25.3 Commercial externally visible characteristics and biogeographic ancestry predictions assays

Thermo Fisher Scientific has two main panels designed for forensic DNA analysis: the Ion AmpliSeq™ panels and Precision ID panels. Each of these panels offers different features. The Ion AmpliSeq™ panels are known for their high multiplexing capabilities, enabling the simultaneous amplification of numerous genetic targets in a single reaction. This makes them suitable for comprehensive genomic studies beyond forensic applications, including oncology and genetic disease research. However, specific Ion AmpliSeq™ panels are designed to address forensic needs.

25.3.1 Ion AmpliSeq™ DNA phenotyping panel

This panel targets 24 SNPs specifically chosen to determine hair and eye color (Walsh et al., 2013). The panel can predict eye color (blue, intermediate, or brown) and hair color (black, brown, red, or blond). This is particularly useful for creating descriptive profiles of unidentified individuals from DNA samples.

25.3.2 Ion AmpliSeq™ HID Y-SNP research panel v1

The Ion AmpliSeq™ HID Y-SNP research panel v1 focuses on 859 Y-SNPs, allowing the inference of 640 Y haplogroups. This panel is instrumental in tracing paternal lineage and inferring paternal biogeographical ancestry, offering high accuracy and sensitivity, particularly useful in cases involving low-template DNA samples. The validation study was conducted on various tests using the Ion Torrent S5 sequencing platform for Y-SNP genotyping (Ralf et al., 2019). Sensitivity tests revealed that performance declines with DNA inputs below 100–250 pg, but high-resolution haplogroups can still be determined with 100 pg despite a 44% marker dropout. DNA analysis of degraded samples showed reliable genotyping even after UV exposure and from long-stored samples. Direct amplification from biological samples produced good results, especially with diluted fresh blood and aged samples. Overall, using the Ion AmpliSeq™ HID Y-SNP research panel v1 with the Ion Torrent S5 platform demonstrated high concordance, precision, and accuracy in Y-SNP genotyping, effectively handling degraded DNA and enabling direct amplification from crude samples.

25.3.3 Ion AmpliSeq™ PhenoTrivium panel

Combining markers for ancestry, DNA phenotyping, and male lineage, the Ion AmpliSeq™ PhenoTrivium panel includes 41 phenotypic SNPs that correspond to the HIrisPlex-S panel and 163 AIM SNPs that overlap with the Precision ID ancestry panel, together with 120 lineage-specific Y-SNPs. The validation study showed that accurate genotyping was achieved down to 125 pg of DNA, with phenotype predictions remaining accurate up to 62 pg (Diepenbroek et al., 2020). The validation study included 17 casework samples and reliably predicted eye, hair, skin color, and biogeographical ancestry. Ancestry predictions using Converge Software provided precise ethnic background estimates (Diepenbroek et al., 2020). Y-haplogrouping analysis showed concordance with the Y-STR methods. Overall, the study demonstrated the robustness and accuracy of these tools in forensic investigations, aiding in identifying unknown individuals and assisting law enforcement.

25.3.4 Ion AmpliSeq™ VISAGE-basic tool research panel

The Visible Attributes through Genomics (VISAGE) project (https://www.visage-h2020.eu/) aims to enhance forensic DNA analysis by developing tools to predict physical appearance, age, and biogeographical ancestry from DNA traces found at crime scenes. The consortium consists of 13 partners from academic, police, and justice institutions in eight European countries. The project focuses on identifying new DNA predictors, developing prototype tools based on NGS technology, and ensuring responsible use within legal frameworks. The VISAGE AmpliSeq basic tool allows forensic investigators to predict appearance and ancestry from DNA samples. This initiative aims to generate investigative leads by providing composite sketches of unknown individuals, enhancing forensic investigations, and supporting law enforcement.

The first VISAGE appearance and ancestry tool development comprises 153 SNPs incorporated into a single assay, the Ion AmpliSeq™ VISAGE-basic tool research panel, which includes 153 SNPs aimed at FDP for eye, hair, and skin color prediction and markers for biogeographical ancestry inference. The assay demonstrated its ability to process low-level or degraded DNA and showed high genotyping concordance for casework-type samples (Xavier et al., 2020).

25.3.5 Precision ID ancestry panel

The Precision ID ancestry panel is designed to provide comprehensive ancestry inference by analyzing SNP markers, which is particularly valuable when determining the biogeographical origin of a DNA sample can offer valuable investigative leads. This panel genotypes 165 informative SNPs to provide reliable insights into an individual's ancestral background (Pereira et al., 2017). It is designed to work effectively with low quantities of DNA, making it suitable for challenging forensic samples that may be degraded or limited in quantity. Combining 55 Kidd lab AIMs with 128 SNPs from Seldin's panel (Kidd et al., 2014; Nassir et al., 2009), the Precision ID ancestry panel exemplifies the potential of integrating AIM markers to create a comprehensive ancestry prediction tool (Pakstis et al., 2019). This panel's inclusion of

165 ancestry-informative SNPs genotyped in a single MPS assay highlights its ability to provide detailed ancestry information crucial for distinguishing between population groups and detecting co-ancestry in admixed individuals.

Additionally, Verogen offers several kits designed to facilitate forensic DNA analysis using NGS. These kits further enhance the capabilities of forensic laboratories by enabling comprehensive and accurate analysis of DNA samples.

25.3.6 ForenSeq DNA signature prep kit

The ForenSeq DNA signature prep kit is an advanced solution for forensic DNA analysis, offering comprehensive capabilities for identity verification, phenotypic prediction, and ancestry inference. This advanced kit targets 231 genetic markers, ensuring high-resolution sequencing and detailed genetic insights. The ForenSeq DNA signature prep kit can analyze 153 identity markers simultaneously. These include amelogenin, a sex chromosome marker used to determine the sample's gender. Additionally, the kit targets 27 autosomal STRs, highly polymorphic markers widely used in forensic identification due to their variability between individuals. Including 24 Y-STRs allows paternal lineage tracing, while 7 X-STRs provide valuable information for complex kinship analysis. Moreover, the kit consists of 94 identity SNPs, which vary between individuals, enhancing the precision of genetic identification.

Beyond identity, the ForenSeq DNA signature prep kit targets 22 phenotypic SNPs identified by Walsh et al. (2013, 2014). These markers are associated with physical traits such as hair color, eye color, and skin pigmentation, providing crucial information on the phenotypic characteristics of individuals. This capability is particularly advantageous in forensic investigations, where physical descriptions of unknown individuals can significantly aid their identification.

In addition to identity and phenotypic markers, the kit includes the 56 ancestry Kidd panel SNPs (Kidd et al., 2014), which are used to infer individuals' biogeographical ancestry. These markers offer valuable information on genetic population background, assisting in determining an individual's ethnic and geographical origins. This comprehensive approach significantly enhances the resolution and precision of forensic investigations.

A considerable amount of literature has been published on the ForenSeq kit, which assesses the forensic performance of both the kit and the sequencing system. These studies include tests on sensitivity, reproducibility, concordance, and mixture analysis (Almalki et al., 2017; Caratti et al., 2015; Churchill et al., 2016; Iozzi et al., 2015; Just et al., 2017; Kocher et al., 2018; Moreno et al., 2018; Silvia et al., 2017; Xavier & Parson, 2017). Other research has focused on the applications of the ForenSeq kit in forensic DNA analysis, including its performance on degraded DNA (Calafell et al., 2016; Zhang et al., 2018), kinship testing (Kostrzewa et al., 2017; Ma et al., 2016), and ancestry prediction (King et al., 2018; Ramani et al., 2017). Furthermore, several studies have tested the performance of the ForenSeq markers on different populations (Casals et al., 2017; Churchill et al., 2017; Devesse et al., 2018; Kim et al., 2018; Novroski et al., 2016; Phillips et al., 2018; Salvador et al., 2018; Wendt et al., 2016, 2017).

25.3.7 ForenSeq Kintelligence kit

The ForenSeq Kintelligence kit enhances forensic DNA investigation by providing a multiplex of 10,230 SNPs (Antunes et al., 2024). It is designed to enable simultaneous sequencing of a wide range of SNPs for various applications, such as phenotypic prediction, ancestry determination, human identification, evaluation of kinship relationships, and analysis of X and Y-chromosome markers. The kit is intended to provide valuable investigation information, especially when dealing with highly degraded and challenging samples such as bones, teeth, saliva, and other casework samples. It supports long-range kinship associations, direct comparisons, and extended familial searching. The ForenSeq Kintelligence kit has demonstrated high performance on different types of samples, including highly degraded ones. It shows high allele and locus detection rates, indicating its ability to accurately identify genetic variations. This comprehensive approach makes it a reliable and effective tool for various forensic applications, significantly improving the efficiency and accuracy of forensic DNA analysis.

25.4 Epigenetics in forensic DNA analysis

Epigenetics is the inheritable modification of the expression of the gene and associated phenotype, activated by molecular mechanisms other than changes in the DNA sequence. These mechanisms include DNA base modifications such as cytosine/5'-CpG-3' methylation and post-translational histone modifications such as H3 histone methylation or acetylation (Bird, 2007). Epigenetics significantly affects gene expression in response to different environmental influences (Alegría-Torres et al., 2011; Mazzio & Soliman, 2012). Epigenetic regulation of gene expression mainly follows the assumption that methylation of a gene at its promoter region affects the binding affinity of transcription factors to that region, thereby inactivating gene expression. Epigenetic analysis offers valuable input for DNA forensic investigation. For example, it can be used to (1) determine the cellular origin of the DNA sample, (2) estimate the individual's age, and (3) discriminate monozygotic (MZ) twins (Vidaki & Kayser, 2018).

25.4.1 Determination of DNA origin

Identifying a DNA sample's cell or tissue type origin can be useful information for crime investigations, not only for reconstructing the events at the crime scene, but also for associating tissues with donors in mixed samples. For example, 150 tissue-specific markers were recognized to identify the origin of DNA from different biological samples such as blood, saliva, semen, vaginal fluid, and menstrual blood using Illumina Human Methylation 450 Beadchip microarray data (Forat et al., 2016). Another study identified a set of eight tissue-specific markers (CpG site) using Illumina 450 K data, which together with two control markers formed a 10-plex assay based on the methylation-specific restriction enzyme (MSRE)—PCR system (Lin et al., 2016). A new screening of semen-specific CpG markers (cg04382920 and cg11768416) screen was developed using highly sensitive singleplex pyrosequencing-based assays, which successfully analyzed 16-year-old semen stains (Vidaki et al., 2016). A different study used bisulfite pyrosequencing to develop a method for the identification of tissue-specific markers. The sensitivity and stability of the method were validated and confirmed using a low DNA input of 100 pg to identify epigenetic markers from different biological samples, including semen, saliva, and blood. Although the results showed that the identification of the markers tested was robust and reliable at low DNA concentration (100 pg), a minimum DNA input of 10 ng was recommended to avoid PCR bias (Silva et al., 2016).

25.4.2 Estimation of an individual's age

Estimating the age of the donor from an unknown DNA sample is forensically significant, as age can help identify unknown victims or perpetrators. Moreover, age estimation is important when combined with predicting EVCs, as certain traits, such as hair loss and hair graying, are highly age-dependent (Schneider et al., 2019). The most growing application of forensic epigenetics is the use of differential DNA methylation profiling to estimate the age of unknown stain donors (Vidaki & Kayser, 2018). The interest in DNA methylation profiling is based on two factors: (1) the well-established role of DNA methylation in the ageing process and (2) the reliability and robustness of DNA methylation profiling compared to other molecular approaches (Zubakov et al., 2016).

Hong et al. identified seven age-associated CpG markers in 54 saliva samples. They constructed a new 7-plex methylation SNaPshot system by combining saliva-identified markers with cell-type-specific markers for both blood and buccal cells. The system was tested with an independent set of 226 saliva samples and provided age prediction with a mean absolute deviation (MAD) of 3.2 years. Another study proposed different age prediction models using a bisulfite pyrosequencing approach to examine three previously reported genetic loci (*SCGN, DLX5*, and *KLF14*) in saliva. The MAD of the single- and dual-locus age-developed models was 8 and 7.1 years, respectively (Hong et al., 2017).

Hamano et al.used the methylation-sensitive high-resolution melting method to test gene promoters of two age-related genes (*ELOVL2* and *EDARADD*), in 250 saliva and cigarette butt samples. The MAD of age prediction in saliva samples was 6.25 and 7.65 years in cigarette butt samples (Hamano et al., 2017). The same research group in a previous study tested the promoter regions of *ELOVL2* and *FHL2* in postmortem blood samples, with similar results obtained from the blood of living individuals (Hamano et al., 2016).

Another study developed an age prediction algorithm for blood samples using a dual approach, NGS and random forest machine learning. Although MPS identified 13 CpG marker sites, the machine learning algorithm achieved an MAD of 3.16 years (Alghanim et al., 2017). Shi et al. investigated age-associated DNA methylation markers in Chinese children. They used a dual approach using the Illumina Human Methylation450 BeadChip system in 48 randomly selected children combined with skeletal and dental X-ray records of 48 randomly selected children aged 6–15 (Shi et al., 2018). Five age-related CpG markers were identified in four known genes (*DDO, PRPH2, DHX8*, and *ITGA2B*) and one unknown gene. Using such a combination, the study reported a high-accuracy mean absolute error (MAE) for boys and girls, 0.47 years and 0.33 years, respectively. The study emphasized the potential use of the developed models in forensic science and competitive sports in immigration cases where confirming the age of a child is necessary (Shi et al., 2018).

25.4.3 Discriminate monozygotic twins

Differentiating between MZ twins is one of the main challenges in forensic DNA identification, as they cannot be distinguished from each other using conventional DNA analysis techniques (Liu et al., 2023). MZ twins share roughly identical DNA sequences, so they have the same DNA profiles in general (Liu et al., 2023). In very isolated cases, variations are identified in forensic STR analysis due to individual-specific somatic mutations (Wang et al., 2015).

Similarly, these individual-specific somatic mutations in SNPs can be identified and used to discriminate MZ using ultra-deep whole genome sequencing (Weber-Lehmann et al., 2014). Nevertheless, these identification approaches are expensive. In addition, the validation process of identified SNPs is complicated and requires extra caution for interpretation due to tissue differences between the MZ twins (Krawczak et al., 2012). Validation also depends on the accessibility of reference and trace samples, which are not always available (Weber-Lehmann et al., 2014).

Vidaki et al. conducted two genome-wide methylation studies in whole blood and saliva samples from MZ twin pairs to identify markers that could be used to differentiate between them. They further validated these markers using qPCR analysis on trace-type samples such as bloodstains, saliva stains, and cigarettes Butts. These studies highlighted several critical technical and biological factors that affect the differentiation of MZ twins through DNA methylation profiling. These factors include method-to-method variations and differences in DNA methylation detection due to cell type composition. Additionally, the time difference between the collection of reference and trace samples can lead to changes in DNA methylation levels over time, an important biological factor that these studies could not investigate due to their design (Vidaki et al., 2017, 2018).

25.5 Conclusions

NGS has been developed for forensic genetics, offering several benefits over the conventional PCR-CE typing system for forensic STR and SNP markers. The main advantage of NGS is its ability to define the actual nucleotide sequence of the targeted region, rather than merely measuring the length of the STR fragment. The sequencing technology overcomes the limited multiplexing capacity of CE and increases the statistical power of forensic testing by combining hundreds to thousands of SNPs and STRs into a large panel. Sequencing a wide range of markers in one run reduces the need for multiple tests, saves time, reduces costs, increases throughput, and enhances the overall efficiency of forensic analysis.

Moreover, NGS has significantly enhanced DNA profiling, with the capability of generating detailed genetic information from a single biological sample, transforming it into a biologically reliable eyewitness. Therefore, it has become an invaluable investigative tool by predicting physical traits such as eye, hair, and skin color, as well as inferring the geographical background and predicting the age and biological source of the sample. These predictive capabilities offer leads that narrow investigations to specific groups of individuals, thereby aiding in individual identification.

The advantages of NGS have encouraged commercial companies to develop several sequencing platforms suitable for forensic applications, such as the Ion S5 System, the Ion GeneStudio S5 Series, and the MiSeq FGx. As a result, several NGS commercial kits were designed and validated for forensic use. These kits include hundreds of markers and can be tailored to users' needs. Recently, there has been significant growth in the forensic research field, particularly in the study of phenotypic and ancestry markers, which has attracted substantial attention. In response, several commercial companies have released kits accommodating phenotypic prediction SNPs and ancestry informative SNPs. Incorporating these SNPs into a single NGS kit can generate more detailed genetic information, providing more investigative leads. In order to achieve this, further exploration of the underlying genetic basis of complex physical traits, along with the development and validation of current markers, across a wide range of global populations, may enhance prediction accuracy.

Recent studies have shown that the addition of epigenetic analysis in forensic science can answer significant questions during investigations, such as determining the tissue source in biological samples, which is especially beneficial in cases of mixed samples, estimating the age of biological traces deposited at crime scenes, and differentiating between MZ twins. Models have been developed to analyze tissue-specific epigenetic markers in various biological samples such as blood, saliva, and semen. These models have proven efficient in identifying the source of the sample. Age estimation models based on DNA methylation profiling provide accurate age predictions, enhancing victim and suspect identification. Overall, studies have demonstrated that epigenetic markers have promising applications in addressing many investigative questions underlying biological samples.

The advancement of NGS in forensics has significantly improved the resolution and accuracy of DNA analysis from biological samples, particularly those often found in harsh conditions at crime scenes. This progress has introduced new tools and methodologies to generate investigative leads, making it a major focus of research in forensic science. As these technologies evolve, their integration into forensic practice will undoubtedly generate accurate and reliable predictions, ensuring that the most complex cases can be solved and the truth uncovered.

References

Adhikari, K., Fontanil, T., Cal, S., Mendoza-Revilla, J., Fuentes-Guajardo, M., Chacón-Duque, J.-C., Al-Saadi, F., Johansson, J. A., Quinto-Sanchez, M., Acuña-Alonzo, V., Jaramillo, C., Arias, W., Barquera Lozano, R., Macín Pérez, G., Gómez-Valdés, J., Villamil-Ramírez, H., Hunemeier, T.,

Ramallo, V., Silva de Cerqueira, C. C., ... Ruiz-Linares, A. (2016). A genome-wide association scan in admixed Latin Americans identifies loci influencing facial and scalp hair features. *Nature Communications, 7*(1). https://doi.org/10.1038/ncomms10815.

Alegría-Torres, J. A., Baccarelli, A., & Bollati, V. (2011). Epigenetics and lifestyle. *Epigenomics, 3*(3), 267–277. https://doi.org/10.2217/epi.11.22.

Alghanim, H., Antunes, J., Silva, D. S. B. S., Alho, C. S., Balamurugan, K., & McCord, B. (2017). Detection and evaluation of DNA methylation markers found at SCGN and KLF14 loci to estimate human age. *Forensic Science International: Genetics, 31*, 81–88. https://doi.org/10.1016/j.fsigen.2017.07.011, http://www.elsevier.com.

Almalki, N., Chow, H. Y., Sharma, V., Hart, K., Siegel, D., & Wurmbach, E. (2017). Systematic assessment of the performance of illumina's MiSeq FGx™ forensic genomics system. *Electrophoresis, 38*(6), 846–854. https://doi.org/10.1002/elps.201600511, http://onlinelibrary.wiley.com/journal/10.1002/(ISSN)1522-2683.

Alvarez-Cubero, M. J., Saiz, M., Martínez-García, B., Sayalero, S. M., Entrala, C., Lorente, J. A., & Martinez-Gonzalez, L. J. (2017). Next generation sequencing: An application in forensic sciences? *Annals of Human Biology, 44*(7), 581–592. https://doi.org/10.1080/03014460.2017.1375155.

Antunes, J., Walichiewicz, P., Forouzmand, E., Barta, R., Didier, M., Han, Y., Perez, J. C., Snedecor, J., Zlatkov, C., Padmabandu, G., Devesse, L., Radecke, S., Holt, C. L., Kumar, S. A., Budowle, B., & Stephens, K. M. (2024). Developmental validation of the ForenSeq® Kintelligence kit, MiSeq FGx® sequencing system and ForenSeq universal analysis software. *Forensic Science International: Genetics, 71*. https://doi.org/10.1016/j.fsigen.2024.103055, https://www.sciencedirect.com/science/journal/18724973.

Auton, A., Abecasis, G. R., Altshuler, D. M., Durbin, R. M., Bentley, D. R., Chakravarti, A., Clark, A. G., Donnelly, P., Eichler, E. E., Flicek, P., Gabriel, S. B., Gibbs, R. A., Green, E. D., Hurles, M. E., Knoppers, B. M., Korbel, J. O., Lander, E. S., Lee, C., Lehrach, H., ... Schloss, J. A. (2015). A global reference for human genetic variation. *Nature, 526*(7571), 68–74. https://doi.org/10.1038/nature15393, http://www.nature.com/nature/index.html.

Ballard, D., Winkler-Galicki, J., & Wesoły, J. (2020). Massive parallel sequencing in forensics: Advantages, issues, technicalities, and prospects. *International Journal of Legal Medicine, 134*(4), 1291–1303. https://doi.org/10.1007/s00414-020-02294-0, https://mdpi-res.com/d_attachment/genes/genes-11-01398/article_deploy/genes-11-01398-v2.pdf?version=1606373473.

Bird, A. (2007). Perceptions of epigenetics. *Nature, 447*(7143), 396–398. https://doi.org/10.1038/nature05913.

Boehringer, S., Van Der Lijn, F., Liu, F., Günther, M., Sinigerova, S., Nowak, S., Ludwig, K. U., Herberz, R., Klein, S., Hofman, A., Uitterlinden, A. G., Niessen, W. J., Breteler, M. M. B., Van Der Lugt, A., Würtz, R. P., Nöthen, M. M., Horsthemke, B., Wieczorek, D., Mangold, E., & Kayser, M. (2011). Genetic determination of human facial morphology: Links between cleft-lips and normal variation. *European Journal of Human Genetics, 19*(11), 1192–1197. https://doi.org/10.1038/ejhg.2011.110.

Breslin, K., Wills, B., Ralf, A., Ventayol Garcia, M., Kukla-Bartoszek, M., Pospiech, E., Freire-Aradas, A., Xavier, C., Ingold, S., de La Puente, M., van der Gaag, K. J., Herrick, N., Haas, C., Parson, W., Phillips, C., Sijen, T., Branicki, W., Walsh, S., & Kayser, M. (2019). HIrisPlex-S system for eye, hair, and skin color prediction from DNA: Massively parallel sequencing solutions for two common forensically used platforms. *Forensic Science International: Genetics, 43*. https://doi.org/10.1016/j.fsigen.2019.102152, http://www.elsevier.com.

Calafell, F., Anglada, R., Bonet, N., González-Ruiz, M., Prats-Muñoz, G., Rasal, R., Lalueza-Fox, C., Bertranpetit, J., Malgosa, A., & Casals, F. (2016). An assessment of a massively parallel sequencing approach for the identification of individuals from mass graves of the Spanish Civil War (1936–1939). *Electrophoresis, 37*(21), 2841–2847. https://doi.org/10.1002/elps.201600180, http://onlinelibrary.wiley.com/journal/10.1002/(ISSN)1522-2683.

Caratti, S., Turrina, S., Ferrian, M., Cosentino, E., & De Leo, D. (2015). MiSeq FGx sequencing system: A new platform for forensic genetics. *Forensic Science International: Genetics Supplement Series, 5*, e98. https://doi.org/10.1016/j.fsigss.2015.09.040.

Casals, F., Anglada, R., Bonet, N., Rasal, R., van der Gaag, K. J., Hoogenboom, J., Solé-Morata, N., Comas, D., & Calafell, F. (2017). Length and repeat-sequence variation in 58 STRs and 94 SNPs in two Spanish populations. *Forensic Science International: Genetics, 30*, 66–70. https://doi.org/10.1016/j.fsigen.2017.06.006, http://www.elsevier.com.

Cha, S., Lim, J. E., Park, A. Y., Do, J. H., Lee, S. W., Shin, C., Cho, N. H., Kang, J. O., Nam, J. M., Kim, J. S., Woo, K. M., Lee, S. H., Kim, J. Y., & Oh, B. (2018). Identification of five novel genetic loci related to facial morphology by genome-wide association studies. *BMC Genomics, 19*(1). https://doi.org/10.1186/s12864-018-4865-9, http://www.biomedcentral.com/bmcgenomics.

Chaitanya, L., Breslin, K., Zuñiga, S., Wirken, L., Pośpiech, E., Kukla-Bartoszek, M., Sijen, T., Knijff, Pd, Liu, F., Branicki, W., Kayser, M., & Walsh, S. (2018). The HIrisPlex-S system for eye, hair and skin colour prediction from DNA: Introduction and forensic developmental validation. *Forensic Science International: Genetics, 35*, 123–135. https://doi.org/10.1016/j.fsigen.2018.04.004, http://www.elsevier.com.

Chaitanya, L., Pajnič, I. Z., Walsh, S., Balažic, J., Zupanc, T., & Kayser, M. (2017). Bringing colour back after 70 years: Predicting eye and hair colour from skeletal remains of World War II victims using the HIrisPlex system. *Forensic Science International: Genetics, 26*, 48–57. https://doi.org/10.1016/j.fsigen.2016.10.004, http://www.elsevier.com.

Churchill, J. D., Novroski, N. M. M., King, J. L., Seah, L. H., & Budowle, B. (2017). Population and performance analyses of four major populations with Illumina's FGx forensic genomics system. *Forensic Science International: Genetics, 30*, 81–92. https://doi.org/10.1016/j.fsigen.2017.06.004, http://www.elsevier.com.

Churchill, J. D., Schmedes, S. E., King, J. L., & Budowle, B. (2016). Evaluation of the Illumina® Beta Version ForenSeq™ DNA signature prep kit for use in genetic profiling. *Forensic Science International: Genetics, 20*, 20–29. https://doi.org/10.1016/j.fsigen.2015.09.009, http://www.elsevier.com.

Claes, P., Liberton, D. K., Daniels, K., Rosana, K. M., Quillen, E. E., Pearson, L. N., McEvoy, B., Bauchet, M., Zaidi, A. A., Yao, W., Tang, H., Barsh, G. S., Absher, D. M., Puts, D. A., Rocha, J., Beleza, S., Pereira, R. W., Baynam, G., Suetens, P., ... Shriver, M. D. (2014). Modeling 3D facial shape from DNA. *PLoS Genetics, 10*(3). https://doi.org/10.1371/journal.pgen.1004224, http://www.plosgenetics.org/article/fetchObject.action?uri=info%3Adoi%2F10.1371%2Fjournal.pgen.1004224&representation=PDF.

Claes, P., Roosenboom, J., White, J. D., Swigut, T., Sero, D., Li, J., Lee, M. K., Zaidi, A., Mattern, B. C., Liebowitz, C., Pearson, L., González, T., Leslie, E. J., Carlson, J. C., Orlova, E., Suetens, P., Vandermeulen, D., Feingold, E., Marazita, M. L., ... Weinberg, S. M. (2018). Genome-wide mapping of global-to-local genetic effects on human facial shape. *Nature Genetics, 50*(3), 414–423. https://doi.org/10.1038/s41588-018-0057-4, http://www.nature.com/ng/index.html.

Cole, J. B., Manyama, M., Kimwaga, E., Mathayo, J., Larson, J. R., Liberton, D. K., Lukowiak, K., Ferrara, T. M., Riccardi, S. L., Li, M., Mio, W., Prochazkova, M., Williams, T., Li, H., Jones, K. L., Klein, O. D., Santorico, S. A., Hallgrimsson, B., & Spritz, R. A. (2016). Genomewide association study of African children identifies association of SCHIP1 and PDE8A with facial size and shape. *PLoS Genetics, 12*(8). https://doi.org/10.1371/journal.pgen.1006174, http://genetics.plosjournals.org/perlserv/?request=get-archive&issn=1553-7404.

Davenport, G. C., & Davenport, C. B. (1907). Heredity of eye-color in man. *Science, 26*(670), 589–592. https://doi.org/10.1126/science.26.670.589-b.

Devesse, L., Ballard, D., Davenport, L., Riethorst, I., Mason-Buck, G., & Syndercombe Court, D. (2018). Concordance of the ForenSeq™ system and characterisation of sequence-specific autosomal STR alleles across two major population groups. *Forensic Science International: Genetics, 34*, 57–61. https://doi.org/10.1016/j.fsigen.2017.10.012.

Diepenbroek, M., Bayer, B., Schwender, K., Schiller, R., Lim, J., Lagacé, R., & Anslinger, K. (2020). Evaluation of the ion Ampliseq™ phenotrivium panel: MPS-based assay for ancestry and phenotype predictions challenged by casework samples. *Genes, 11*(12), 1–24. https://doi.org/10.3390/genes11121398, https://www.mdpi.com/2073-4425/11/12/1398/pdf.

Draus-Barini, J., Walsh, S., Pośpiech, E., Kupiec, T., Głąb, H., Branicki, W., & Kayser, M. (2013). Bona fide colour: DNA prediction of human eye and hair colour from ancient and contemporary skeletal remains. *Investigative Genetics, 4*(1). https://doi.org/10.1186/2041-2223-4-3.

Fondevila, M., Phillips, C., Santos, C., Freire Aradas, A., Vallone, P. M., Butler, J. M., Lareu, M. V., & Carracedo, Á. (2013). Revision of the SNPforID 34-plex forensic ancestry test: Assay enhancements, standard reference sample genotypes and extended population studies. *Forensic Science International: Genetics, 7*(1), 63–74. https://doi.org/10.1016/j.fsigen.2012.06.007.

Forat, S., Huettel, B., Reinhardt, R., Fimmers, R., Haidl, G., Denschlag, D., Olek, K., & Kim, J.-W. (2016). Methylation markers for the identification of body fluids and tissues from forensic trace evidence. *PLoS One, 11*(2), e0147973. https://doi.org/10.1371/journal.pone.0147973.

Gill, P. (2001). An assessment of the utility of single nucleotide polymorphisms (SNPs) for forensic purposes. *International Journal of Legal Medicine, 114*(4-5), 204–210. https://doi.org/10.1007/s004149900117.

Haeusler, M., Haas, C., Lösch, S., Moghaddam, N., Villa, I. M., Walsh, S., Kayser, M., Seiler, R., Ruehli, F., Janosa, M., Papageorgopoulou, C., & Caramelli, D. (2016). Multidisciplinary identification of the controversial freedom fighter Jörg Jenatsch, assassinated 1639 in Chur, Switzerland. *PLoS One, 11*(12), e0168014. https://doi.org/10.1371/journal.pone.0168014.

Haidar, M., Mousawi, F., & Al-Matrouk, A. K. (2024). Chapter 15 - Forensic DNA phenotyping using next-generation sequencing. In H. R. Dash, K. M. Elkins, & N. R. Al-Snan (Eds.). *Next Generation Sequencing (NGS) Technology in DNA Analysis* (pp. 289–310). Academic Press. https://doi.org/10.1016/B978-0-323-99144-5.00014-7.

Hamano, Y., Manabe, S., Morimoto, C., Fujimoto, S., Ozeki, M., & Tamaki, K. (2016). Forensic age prediction for dead or living samples by use of methylation-sensitive high resolution melting. *Legal Medicine, 21*, 5–10. https://doi.org/10.1016/j.legalmed.2016.05.001.

Hamano, Y., Manabe, S., Morimoto, C., Fujimoto, S., & Tamaki, K. (2017). Forensic age prediction for saliva samples using methylation-sensitive high resolution melting: Exploratory application for cigarette butts. *Scientific Reports, 7*(1). https://doi.org/10.1038/s41598-017-10752-w.

Hong, S. R., Jung, S. E., Lee, E. H., Shin, K. J., Yang, W. I., & Lee, H. Y. (2017). DNA methylation-based age prediction from saliva: High age predictability by combination of 7 CpG markers. *Forensic Science International: Genetics, 29*, 118–125. https://doi.org/10.1016/j.fsigen.2017.04.006, http://www.elsevier.com.

Iozzi, S., Carboni, I., Contini, E., Pescucci, C., Frusconi, S., Nutini, A. L., Torricelli, F., & Ricci, U. (2015). Forensic genetics in NGS era: New frontiers for massively parallel typing. *Forensic Science International: Genetics Supplement Series, 5*, e418. https://doi.org/10.1016/j.fsigss.2015.09.166.

Just, R. S., Moreno, L. I., Smerick, J. B., & Irwin, J. A. (2017). Performance and concordance of the ForenSeq™ system for autosomal and Y chromosome short tandem repeat sequencing of reference-type specimens. *Forensic Science International: Genetics, 28*, 1–9. https://doi.org/10.1016/j.fsigen.2017.01.001, http://www.elsevier.com.

Kayser, M. (2015). Forensic DNA phenotyping: Predicting human appearance from crime scene material for investigative purposes. *Forensic Science International: Genetics, 18*, 33–48. https://doi.org/10.1016/j.fsigen.2015.02.003.

Kayser, M., Branicki, W., Parson, W., & Phillips, C. (2023). Recent advances in forensic DNA phenotyping of appearance, ancestry and age. *Forensic Science International: Genetics, 65*, 102870. https://doi.org/10.1016/j.fsigen.2023.102870.

Kidd, K. K., Speed, W. C., Pakstis, A. J., Furtado, M. R., Fang, R., Madbouly, A., Maiers, M., Middha, M., Friedlaender, F. R., & Kidd, J. R. (2014). Progress toward an efficient panel of SNPs for ancestry inference. *Forensic Science International: Genetics, 10*, 23–32. https://doi.org/10.1016/j.fsigen.2014.01.002.

Kim, S. Y., Lee, H. C., Chung, U., Ham, S. K., Lee, H. Y., Park, S. J., Roh, Y. J., & Lee, S. H. (2018). Massive parallel sequencing of short tandem repeats in the Korean population. *Electrophoresis, 39*(21), 2702–2707. https://doi.org/10.1002/elps.201800090, http://onlinelibrary.wiley.com/journal/10.1002/(ISSN)1522-2683.

King, J. L., Churchill, J. D., Novroski, N. M. M., Zeng, X., Warshauer, D. H., Seah, L. H., & Budowle, B. (2018). Increasing the discrimination power of ancestry- and identity-informative SNP loci within the ForenSeq™ DNA signature prep kit. *Forensic Science International: Genetics, 36*, 60–76. https://doi.org/10.1016/j.fsigen.2018.06.005, http://www.elsevier.com.

King, T. E., Fortes, G. G., Balaresque, P., Thomas, M. G., Balding, D., Delser, P. M., Neumann, R., Parson, W., Knapp, M., Walsh, S., Tonasso, L., Holt, J., Kayser, M., Appleby, J., Forster, P., Ekserdjian, D., Hofreiter, M., & Schürer, K. (2014). Identification of the remains of King Richard III. *Nature Communications, 5*. https://doi.org/10.1038/ncomms6631, http://www.nature.com/ncomms/index.html.

Kocher, S., Muller, P., Berger, B., Bodner, M., Parson, W., Roewer, L., Willuweit, S., & Consortium, D. N. (2018). Inter-laboratory validation study of the ForenSeq DNA signature prep kit. *Forensic Science International: Genetics, 36*, 77–85. https://doi.org/10.1016/j.fsigen.2018.05.007.

Kostrzewa, G., Konarzewska, M., & Pepiński, W. (2017). Application of massively parallel sequencing (MPS) in paternity testing – Case report. *Archives of Forensic Medicine and Criminology, 67*(1), 61–67. https://doi.org/10.5114/amsik.2017.70338.

Krawczak, M., Cooper, D. N., Fändrich, F., Engel, W., & Schmidtke, J. (2012). How to distinguish genetically between an alleged father and his monozygotic twin: A thought experiment. *Forensic Science International: Genetics, 6*(5), e129. https://doi.org/10.1016/j.fsigen.2011.11.003, http://www.elsevier.com.

Krier, J. B., Kalia, S. S., & Green, R. C. (2016). Genomic sequencing in clinical practice: Applications, challenges, and opportunities. *Dialogues in Clinical Neuroscience, 18*(3), 299–312. https://www.ncbi.nlm.nih.gov/pubmed/27757064, https://www.ncbi.nlm.nih.gov/pmc/PMC5067147/.

Lee, M. K., Shaffer, J. R., Leslie, E. J., Orlova, E., Carlson, J. C., Feingold, E., Marazita, M. L., & Weinberg, S. M. (2017). Genome-wide association study of facial morphology reveals novel associations with FREM1 and PARK2. *PLoS One, 12*(4). https://doi.org/10.1371/journal.pone.0176566, http://journals.plos.org/plosone/article/file?id=10.1371/journal.pone.0176566&type=printable.

Lin, Y. C., Tsai, L. C., Lee, J. C. I., Su, C. W., Tzen, J. T. C., Linacre, A., & Hsieh, H. M. (2016). Novel identification of biofluids using a multiplex methylation sensitive restriction enzyme-PCR system. *Forensic Science International: Genetics, 25*, 157–165. https://doi.org/10.1016/j.fsigen.2016.08.011, http://www.elsevier.com.

Liu, J., Zhang, S., Wen, Y., Su, Y., Jiang, L., Li, S., Shen, J., Zheng, X., Li, X., Chen, X., & Wang, Z. (2023). Exploring rare differences in mitochondrial genome between MZ twins using Ion Torrent semiconductor sequencing. *Forensic Science International, 348*, 111708. https://doi.org/10.1016/j.forsciint.2023.111708.

Ma, Y., Kuang, J. Z., Nie, T. G., Zhu, W., & Yang, Z. (2016). Next generation sequencing: Improved resolution for paternal/maternal duos analysis. *Forensic Science International: Genetics, 24*, 83–85. https://doi.org/10.1016/j.fsigen.2016.05.015, http://www.elsevier.com.

Marciniak, S., & Perry, G. H. (2017). Harnessing ancient genomes to study the history of human adaptation. *Nature Reviews Genetics, 18*(11), 659–674. https://doi.org/10.1038/nrg.2017.65, http://www.nature.com/reviews/genetics.

Margulies, M., Egholm, M., Altman, W. E., Attiya, S., Bader, J. S., Bemben, L. A., Berka, J., Braverman, M. S., Chen, Y. J., Chen, Z., Dewell, S. B., Du, L., Fierro, J. M., Gomes, X. V., Godwin, B. C., He, W., Helgesen, S., Ho, C. H., Irzyk, G. P., ... Rothberg, J. M. (2005). Genome sequencing in microfabricated high-density picolitre reactors. *Nature, 437*(7057), 376–380. https://doi.org/10.1038/nature03959.

Mazzio, E. A., & Soliman, K. F. A. (2012). Basic concepts of epigenetics impact of environmental signals on gene expression. *Epigenetics: Official Journal of the DNA Methylation Society, 7*(2), 119–130. https://doi.org/10.4161/epi.7.2.18764, http://www.landesbioscience.com/journals/epigenetics/MazzioEPI7-2-X.pdf.

McCarthy, J. J., McLeod, H. L., & Ginsburg, G. S. (2013). Genomic medicine: A decade of successes, challenges, and opportunities. *Science Translational Medicine, 5*(189). https://doi.org/10.1126/scitranslmed.3005785 189sr184-189sr184.

McNevin, D. (2020). Forensic inference of biogeographical ancestry from genotype: The genetic ancestry lab. *WIREs Forensic Science, 2*(2). https://doi.org/10.1002/wfs2.1356.

Mehta, B., Daniel, R., Phillips, C., & McNevin, D. (2017). Forensically relevant SNaPshot® assays for human DNA SNP analysis: A review. *International Journal of Legal Medicine, 131*(1), 21–37. https://doi.org/10.1007/s00414-016-1490-5.

Moreno, L. I., Galusha, M. B., & Just, R. (2018). A closer look at Verogen's Forenseq DNA signature prep kit autosomal and Y-STR data for streamlined analysis of routine reference samples. *Electrophoresis, 39*(21), 2685–2693. https://doi.org/10.1002/elps.201800087.

Nassir, R., Kosoy, R., Tian, C., White, P. A., Butler, L. M., Silva, G., Kittles, R., Alarcon-Riquelme, M. E., Gregersen, P. K., Belmont, J. W., De La Vega, F. M., & Seldin, M. F. (2009). An ancestry informative marker set for determining continental origin: Validation and extension using human genome diversity panels. *BMC Genetics, 10*. https://doi.org/10.1186/1471-2156-10-39.

Novroski, N. M. M., King, J. L., Churchill, J. D., Seah, L. H., & Budowle, B. (2016). Characterization of genetic sequence variation of 58 STR loci in four major population groups. *Forensic Science International: Genetics, 25*, 214–226. https://doi.org/10.1016/j.fsigen.2016.09.007, http://www.elsevier.com.

Pakstis, A. J., Speed, W. C., Fang, R., Hyland, F. C. L., Furtado, M. R., Kidd, J. R., & Kidd, K. K. (2010). SNPs for a universal individual identification panel. *Human Genetics, 127*(3), 315–324. https://doi.org/10.1007/s00439-009-0771-1.

Pakstis, A. J., Speed, W. C., Soundararajan, U., Rajeevan, H., Kidd, J. R., Li, H., & Kidd, K. K. (2019). Population relationships based on 170 ancestry SNPs from the combined Kidd and Seldin panels. *Scientific Reports, 9*(1). https://doi.org/10.1038/s41598-019-55175-x, www.nature.com/srep/index.html.

Pereira, V., Mogensen, H. S., Børsting, C., & Morling, N. (2017). Evaluation of the Precision ID ancestry panel for crime case work: A SNP typing assay developed for typing of 165 ancestral informative markers. *Forensic Science International: Genetics, 28*, 138–145. https://doi.org/10.1016/j.fsigen.2017.02.013, http://www.elsevier.com.

Phillips, C. (2015). Forensic genetic analysis of bio-geographical ancestry. *Forensic Science International: Genetics, 18*, 49–65. https://doi.org/10.1016/j.fsigen.2015.05.012.

Phillips, C., Devesse, L., Ballard, D., van Weert, L., de la Puente, M., Melis, S., Álvarez Iglesias, V., Freire-Aradas, A., Oldroyd, N., Holt, C., Syndercombe Court, D., Carracedo, Á., & Lareu, M. V. (2018). Global patterns of STR sequence variation: Sequencing the CEPH human genome diversity panel for 58 forensic STRs using the Illumina ForenSeq DNA signature prep kit. *Electrophoresis, 39*(21), 2708–2724. https://doi.org/10.1002/elps.201800117, http://onlinelibrary.wiley.com/journal/10.1002/(ISSN)1522-2683.

Phillips, C., Prieto, L., Fondevila, M., Salas, A., Gómez-Tato, A., Álvarez-Dios, J., Alonso, A., Blanco-Verea, A., Brión, M., Montesino, M., Carracedo, Á., Lareu, M. V., & Quintana-Murci, L. (2009). Ancestry analysis in the 11-M Madrid bomb attack investigation. *PLoS One, 4*(8), e6583. https://doi.org/10.1371/journal.pone.0006583.

Phillips, C., Salas, A., Sánchez, J. J., Fondevila, M., Gómez-Tato, A., Álvarez-Dios, J., Calaza, M., M.Casares de Cal, D., Ballard, M. V., Lareu, Á., & Carracedo (2007). Inferring ancestral origin using a single multiplex assay of ancestry-informative marker SNPs. *Forensic Science International: Genetics, 1*(3-4), 273–280. https://doi.org/10.1016/j.fsigen.2007.06.008.

Poinar, H. N., Schwarz, C., Qi, J., Shapiro, B., MacPhee, R. D. E., Buigues, B., Tikhonov, A., Huson, D. M., Tomsho, L. P., Auch, A., Rampp, M., Miller, W., & Schuster, S. C. (2006). Metagenomics to paleogenomics: Large-scale sequencing of mammoth DNA. *Science (New York, N.Y.), 311*(5759), 392–394. https://doi.org/10.1126/science.1123360.

Ralf, A., van Oven, M., Montiel González, D., de Knijff, P., van der Beek, K., Wootton, S., Lagacé, R., & Kayser, M. (2019). Forensic Y-SNP analysis beyond SNaPshot: High-resolution Y-chromosomal haplogrouping from low quality and quantity DNA using Ion AmpliSeq and targeted massively parallel sequencing. *Forensic Science International: Genetics, 41*, 93–106. https://doi.org/10.1016/j.fsigen.2019.04.001.

Ramani, Wong, Y., Tan, S. Z., Shue, B. H., & Syn, C. (2017). Ancestry prediction in Singapore population samples using the Illumina ForenSeq kit. *Forensic Science International: Genetics, 31*, 171–179. https://doi.org/10.1016/j.fsigen.2017.08.013, http://www.elsevier.com.

Salvador, J. M., Apaga, D. L. T., Delfin, F. C., Calacal, G. C., Dennis, S. E., & De Ungria, M. C. A. (2018). Filipino DNA variation at 12 X-chromosome short tandem repeat markers. *Forensic Science International: Genetics, 36*, e8. https://doi.org/10.1016/j.fsigen.2018.06.008, https://www.sciencedirect.com/science/journal/18724973.

Schneider, P. M., Prainsack, B., & Kayser, M. (2019). The use of forensic DNA phenotyping in predicting appearance and biogeographic ancestry. *Deutsches Arzteblatt International, 51-52*(51-52), 873–880. https://doi.org/10.3238/arztebl.2019.0873.

Shaffer, J. R., Orlova, E., Lee, M. K., Leslie, E. J., Raffensperger, Z. D., Heike, C. L., Cunningham, M. L., Hecht, J. T., Kau, C. H., Nidey, N. L., Moreno, L. M., Wehby, G. L., Murray, J. C., Laurie, C. A., Laurie, C. C., Cole, J., Ferrara, T., Santorico, S., Klein, O., ... Weinberg, S. M. (2016). Genome-wide association study reveals multiple loci influencing normal human facial morphology. *PLoS Genetics, 12*(8). https://doi.org/10.1371/journal.pgen.1006149, http://genetics.plosjournals.org/perlserv/?request=get-archive&issn=1553-7404.

Shi, L., Jiang, F., Ouyang, F., Zhang, J., Wang, Z., & Shen, X. (2018). DNA methylation markers in combination with skeletal and dental ages to improve age estimation in children. *Forensic Science International: Genetics, 33*, 1–9. https://doi.org/10.1016/j.fsigen.2017.11.005.

Silva, D. S. B. S., Antunes, J., Balamurugan, K., Duncan, G., Alho, C. S., & McCord, B. (2016). Developmental validation studies of epigenetic DNA methylation markers for the detection of blood, semen and saliva samples. *Forensic Science International: Genetics, 23*, 55–63. https://doi.org/10.1016/j.fsigen.2016.01.017, http://www.elsevier.com.

Silvia, A. L., Shugarts, N., & Smith, J. (2017). A preliminary assessment of the ForenSeq FGx system: next generation sequencing of an STR and SNP multiplex. *International Journal of Legal Medicine, 131*(1), 73–86. https://doi.org/10.1007/s00414-016-1457-6.

Valentine, T., & Fitzgerald, R. J. (2016). Identifying the culprit: An international perspective on the national academy of sciences report on eyewitness identification evidence. *Applied Cognitive Psychology, 30*(1), 135–138. https://doi.org/10.1002/acp.3164, http://onlinelibrary.wiley.com/journal/10.1002/(ISSN)1099-0720.

Vidaki, Díez López, C., Carnero-Montoro, E., Ralf, A., Ward, K., Spector, T., Bell, J. T., & Kayser, M. (2017). Epigenetic discrimination of identical twins from blood under the forensic scenario. *Forensic Science International: Genetics, 31*, 67–80. https://doi.org/10.1016/j.fsigen.2017.07.014, http://www.elsevier.com.

Vidaki, A., Giangasparo, F., & Syndercombe Court, D. (2016). Discovery of potential DNA methylation markers for forensic tissue identification using bisulphite pyrosequencing. *Electrophoresis, 37*(21), 2767–2779. https://doi.org/10.1002/elps.201600261.

Vidaki, Kalamara, V., Carnero-Montoro, E., Spector, T. D., Bell, J. T., & Kayser, M. (2018). Investigating the epigenetic discrimination of identical twins using buccal swabs, saliva, and cigarette butts in the forensic setting. *Genes, 9*(5). https://doi.org/10.3390/genes9050252, http://www.mdpi.com/2073-4425/9/5/252/pdf.

Vidaki, A., & Kayser, M. (2018). Recent progress, methods and perspectives in forensic epigenetics. *Forensic Science International: Genetics, 37*, 180–195. https://doi.org/10.1016/j.fsigen.2018.08.008.

Walsh, S., Chaitanya, L., Clarisse, L., Wirken, L., Draus-Barini, J., Kovatsi, L., Maeda, H., Ishikawa, T., Sijen, T., de Knijff, P., Branicki, W., Liu, F., & Kayser, M. (2014). Developmental validation of the HIrisPlex system: DNA-based eye and hair colour prediction for forensic and anthropological usage. *Forensic Science International: Genetics, 9*(1), 150–161. https://doi.org/10.1016/j.fsigen.2013.12.006.

Walsh, S., Liu, F., Ballantyne, K. N., Van Oven, M., Lao, O., & Kayser, M. (2011). IrisPlex: A sensitive DNA tool for accurate prediction of blue and brown eye colour in the absence of ancestry information. *Forensic Science International: Genetics, 5*(3), 170–180. https://doi.org/10.1016/j.fsigen.2010.02.004.

Walsh, S., Liu, F., Wollstein, A., Kovatsi, L., Ralf, A., Kosiniak-Kamysz, A., Branicki, W., & Kayser, M. (2013). The HIrisPlex system for simultaneous prediction of hair and eye colour from DNA. *Forensic Science International: Genetics, 7*(1), 98–115. https://doi.org/10.1016/j.fsigen.2012.07.005.

Walsh, S., Wollstein, A., Liu, F., Chakravarthy, U., Rahu, M., Seland, J. H., Soubrane, G., Tomazzoli, L., Topouzis, F., Vingerling, J. R., Vioque, J., Fletcher, A. E., Ballantyne, K. N., & Kayser, M. (2012). DNA-based eye colour prediction across Europe with the IrisPlex system. *Forensic Science International: Genetics, 6*(3), 330–340. https://doi.org/10.1016/j.fsigen.2011.07.009.

Wang, L. F., Yang, Y., Zhang, X. N., Quan, X. L., & Wu, Y. M. (2015). Tri-allelic pattern of short tandem repeats identifies the murderer among identical twins and suggests an embryonic mutational origin. *Forensic Science International: Genetics, 16*, 239–245. https://doi.org/10.1016/j.fsigen.2015.01.010, http://www.elsevier.com.

Weber-Lehmann, J., Schilling, E., Gradl, G., Richter, D. C., Wiehler, J., & Rolf, B. (2014). Finding the needle in the haystack: Differentiating "identical" twins in paternity testing and forensics by ultra-deep next generation sequencing. *Forensic Science International: Genetics, 9*(1), 42–46. https://doi.org/10.1016/j.fsigen.2013.10.015.

Wendt, F. R., Churchill, J. D., Novroski, N. M. M., King, J. L., Ng, J., Oldt, R. F., McCulloh, K. L., Weise, J. A., Smith, D. G., Kanthaswamy, S., & Budowle, B. (2016). Genetic analysis of the Yavapai Native Americans from West-Central Arizona using the Illumina MiSeq FGx™ forensic genomics system. *Forensic Science International: Genetics, 24*, 18–23. https://doi.org/10.1016/j.fsigen.2016.05.008, http://www.elsevier.com.

Wendt, F. R., King, J. L., Novroski, N. M. M., Churchill, J. D., Ng, J., Oldt, R. F., McCulloh, K. L., Weise, J. A., Smith, D. G., Kanthaswamy, S., & Budowle, B. (2017). Flanking region variation of ForenSeq™ DNA signature prep kit STR and SNP loci in Yavapai Native Americans. *Forensic Science International: Genetics, 28*, 146–154. https://doi.org/10.1016/j.fsigen.2017.02.014, http://www.elsevier.com.

Xavier, de la Puente, M., Mosquera-Miguel, A., Freire-Aradas, A., Kalamara, V., Vidaki, A., E. Gross, T., Revoir, A., Pośpiech, E., Kartasińska, E., Spólnicka, M., Branicki, W., E. Ames, C., M. Schneider, P., Hohoff, C., Kayser, M., Phillips, C., & Parson, W. (2020). Development and validation of the VISAGE AmpliSeq basic tool to predict appearance and ancestry from DNA. *Forensic Science International: Genetics, 48*. https://doi.org/10.1016/j.fsigen.2020.102336, http://www.elsevier.com.

Xavier, C., & Parson, W. (2017). Evaluation of the Illumina ForenSeq™ DNA signature prep kit – MPS forensic application for the MiSeq FGx™ benchtop sequencer. *Forensic Science International: Genetics, 28*, 188–194. https://doi.org/10.1016/j.fsigen.2017.02.018.

Zhang, Q., Zhou, Z., Liu, Q., Liu, L., Shao, L., Zhang, M., Ding, X., Gao, Y., & Wang, S. (2018). Evaluation of the performance of Illumina's ForenSeq™ system on serially degraded samples. *Electrophoresis, 39*(21), 2674–2684. https://doi.org/10.1002/elps.201800101.

Zubakov, Liu, F., Kokmeijer, I., Choi, Y., van Meurs, J. B. J., van IJcken, W. F. J., Uitterlinden, A. G., Hofman, A., Broer, L., van Duijn, C. M., Lewin, J., & Kayser, M. (2016). Human age estimation from blood using mRNA, DNA methylation, DNA rearrangement, and telomere length. *Forensic Science International: Genetics, 24*, 33–43. https://doi.org/10.1016/j.fsigen.2016.05.014, http://www.elsevier.com.

Part IV

Module IV: challenges and concerns in forensic DNA analysis

Chapter 26

Quality assurance and quality control in forensic science: from evidence collection to laboratory analysis

Abdullah Al-Matrouk[1], Abdullah Aldasem[2], Mahdi Haidar[3], and Fatema Mousawi[3]

[1]*General Department of Criminal Evidence, Quality Control Department, Ministry of Interior, Kuwait City, Kuwait,* [2]*University of Strathclyde, Glasgow, United Kingdom,* [3]*Kuwait Identification DNA Laboratory (KIDL), General Department of Criminal Evidence, Ministry of Interior, Kuwait City, Kuwait*

26.1 Introduction

Forensic science is a fundamental pillar of the criminal justice system, playing a critical role in solving crimes and aiding court proceedings by providing scientific evidence that can corroborate or challenge witness testimonies and case theories. The reliability and accuracy of forensic analyses are pivotal in determining the outcome of legal cases, making it essential that forensic laboratories operate under stringent quality management systems (Brown et al., 2020; Koehler et al., 2023). Quality assurance (QA) and quality control (QC) are key components of these systems, designed to safeguard the integrity of forensic evidence, from collection at the crime scene to its analysis and eventual presentation in court. These two processes, although often used interchangeably, serve distinct but complementary functions in forensic practice (Levy et al., 1999). QC is primarily concerned with the operational aspects of forensic work, ensuring that individual tasks such as DNA analysis, toxicology screening, or fingerprint comparison are conducted according to pre-defined standards. QC measures are embedded within daily forensic operations and are used to verify that each procedure produces accurate and consistent results (Page et al., 2019). This is achieved through the implementation of control samples, documentation of processes, and routine checks on the equipment used. For instance, running control samples alongside evidence samples in DNA profiling helps confirm that the analytical methods are functioning correctly, and any inconsistencies can be immediately identified and rectified before the results are finalized (Ansell, 2013). QC is a real-time measure, actively monitoring the quality of the forensic work as it is being conducted. On the other hand, QA encompasses a broader approach to ensuring the reliability of forensic findings. QA refers to the systematic actions taken by a forensic laboratory to monitor, evaluate, and document its overall performance. These actions include ensuring that the laboratory environment is conducive to producing reliable results, that the personnel involved in forensic analyses are adequately trained, and that the methods and equipment in use are validated and calibrated regularly (Tilstone, 2010). QA also involves periodic internal and external audits to verify that the laboratory complies with standard operating procedures (SOPs) and other regulatory requirements (Jul, n.d.). Unlike QC, which operates on a more granular level during specific tasks, QA focuses on the broader framework, ensuring that the forensic laboratory consistently produces high-quality results over time. The importance of QA and QC in forensic science cannot be overstated. Forensic evidence is often central to criminal investigations and court proceedings, and its accuracy can be the difference between a conviction and an acquittal. Errors in forensic analysis, whether due to faulty equipment, poorly trained personnel, or mishandled evidence, can lead to serious miscarriages of justice, such as wrongful convictions or the acquittal of guilty parties (Fan & Wang, 2021). Furthermore, the public's trust in the criminal justice system is heavily dependent on the reliability of forensic science. Therefore it is essential that forensic laboratories operate under stringent QA/QC protocols to prevent errors, uphold the integrity of the forensic process, and maintain public confidence in the justice system. One of the key challenges in forensic science is ensuring that evidence collected from crime scenes is handled properly throughout the entire forensic process. This is where QA becomes critically important. The chain of custody, a principle of evidence

Investigative and Predictive DNA Testing. DOI: https://doi.org/10.1016/B978-0-443-30128-5.00026-6

management that tracks the handling of evidence from the crime scene to the courtroom, is a crucial part of QA in forensic laboratories (Klesta & Bartz, 2018). Maintaining a well-documented chain of custody ensures that the evidence has not been tampered with, contaminated, or lost during transit. A breach in the chain of custody can result in evidence being deemed inadmissible in court, which can have serious consequences for the outcome of a case. QA measures ensure that every step of the evidence collection, transportation, analysis, and storage processes is carefully documented and followed, protecting the integrity of the forensic evidence. Another significant aspect of QA is the proficiency testing (PT) and competency assessment of forensic personnel. Forensic analysts must undergo regular PT to ensure they are capable of performing analyses accurately and reliably. This is especially important as forensic science is constantly evolving with new technologies and methodologies. Regular training and assessment ensure that forensic personnel stay updated with the latest developments in the field and are capable of applying new techniques effectively. Blind PT, in particular, offers an authentic evaluation of an analyst's day-to-day performance, as they are unaware that they are being tested. This prevents analysts from taking extra care or shortcuts during analysis, providing a more realistic assessment of their competencies (Neuteboom et al., 2024). Moreover, QA and QC also involve the regular calibration and maintenance of forensic laboratory equipment. Forensic analyses often rely on highly sophisticated instruments, such as mass spectrometers, DNA sequencers, and chromatography systems. The accuracy of these instruments directly affects the reliability of the forensic results, making regular calibration and maintenance essential. QA protocols require that these instruments be calibrated against known standards to ensure they are operating correctly. If any discrepancies are identified, corrective actions can be taken to recalibrate or repair the equipment before it is used for further analysis. Additionally, maintaining detailed logs of equipment performance and errors helps forensic laboratories identify trends and potential issues, allowing for proactive measures to ensure the reliability of forensic analyses (Tilstone, 2010). Forensic laboratories also seek to maintain accreditation from recognized international bodies, such as ISO/IEC 17025, which sets the standards for testing and calibration laboratories. Achieving and maintaining such accreditation demonstrates a laboratory's commitment to operating under stringent QA/QC protocols and adhering to international standards of scientific reliability. This is particularly important when forensic evidence is presented in court, as the accreditation status of the laboratory can be used to support the credibility of the forensic results (Malkoc & Neuteboom, 2007). In conclusion, QC and QA are indispensable elements of forensic science. They provide the necessary framework to ensure that forensic evidence is collected, handled, and analyzed in a manner that produces reliable, accurate, and legally defensible results. By implementing robust QA/QC systems, forensic laboratories can minimize errors, safeguard the chain of custody, and produce forensic findings that withstand scrutiny in legal settings. As forensic science continues to advance with new technologies and methodologies, maintaining rigorous QA/QC standards will remain essential to upholding the integrity of forensic evidence and ensuring justice is served.

26.2 Quality control in forensic laboratories

QC in forensic laboratories is a critical aspect of ensuring that the forensic analyses conducted are both accurate and reliable. QC encompasses the operational techniques and activities used to meet quality requirements throughout the laboratory processes. The importance of QC in forensic science cannot be overstated, as it directly affects the integrity of the results produced and, consequently, the outcomes of legal proceedings (Ansell, 2013).

26.2.1 Operational techniques for ensuring quality

To maintain high-quality forensic analyses, various operational techniques are implemented in laboratories.

26.2.1.1 Control samples

Control samples are a key element in QC protocols. These are known standards that are analyzed alongside actual evidence samples. The results obtained from control samples serve as a benchmark to ensure that the analytical methods and instruments are functioning correctly. For example, in DNA profiling, a control sample with a known genetic profile is run to validate that the test results for the evidence sample fall within expected ranges. If the control results do not match the expected outcomes, it indicates a potential issue with the analytical process, prompting immediate investigation and correction (Broadhurst et al., 2018).

26.2.1.2 Standard operating procedures

SOPs are essential for maintaining consistency and quality in laboratory operations. They outline the necessary steps to carry out specific forensic analyses and provide detailed instructions on how to handle samples, conduct tests, and document findings. By adhering to SOPs, forensic analysts can minimize variability in results and ensure that each procedure is performed systematically. Regular reviews and updates of SOPs are also necessary to incorporate advancements in techniques and technology (Ezzelle et al., 2008).

26.2.1.3 Documentation

Comprehensive documentation is a fundamental component of QC in forensic laboratories. Every step of the forensic process, from evidence collection to analysis and reporting, must be meticulously recorded. This includes maintaining logs of equipment performance, recording the results of control samples, and documenting any deviations from established procedures. Proper documentation not only provides a verifiable record of the processes but also aids in audits and reviews, ensuring transparency and accountability in forensic analyses (Olivares & Lopes, 2012; Tilstone, 2010).

26.3 Quality assurance programs

QA encompasses a wide range of systematic actions designed to monitor, evaluate, and enhance the performance of forensic laboratories. A robust QA program ensures that all aspects of the laboratory, from the handling of evidence to the qualifications of personnel, meet established quality standards (Levy et al., 1999).

26.3.1 Components of quality assurance systems

Several key components are integral to an effective QA system in forensic science:

26.3.1.1 Personnel training and competency

Competency of personnel is a critical factor in ensuring the quality of forensic services. Quality can only be achieved by competent forensic practitioners who work under the guidance of a QS and adhere to an appropriate philosophy and approach. Forensic experts are the key factor in a laboratory's accreditation, and as such, they should possess the appropriate scientific qualifications, typically a university degree, for the job they are performing. Given the rapidly evolving nature of forensic science, it is essential for employee working in the field to continue their education. This can be achieved by obtaining a higher degree in the field, taking courses, attending special training and seminars, and learning new techniques. Forensic personnel should also stay up to date with the professional scientific publications to remain aware of any developments in their work field. Each employee should have a written job description outlining their required responsibilities, duties, and skills. Managers are required to have several years of experience to effectively supervise and monitor working technicians. All staff should be audited both internally and externally by individuals inside and outside the laboratory. Good QA programs incorporate regular internal and external assessments. It is also important to train and prepare staff working in the forensic field to confidently defend their work during cross-examination while testifying in court before a judge and jury. This training should provide them with the main principles and details about forensic evidence processing (Grisso, 2006; Mahmood et al., 2005).

26.3.1.2 Regular audits

Audits are essential for evaluating the compliance of laboratory operations with established standards and procedures. Internal audits are conducted by laboratory personnel to assess adherence to SOPs and identify areas for improvement. External audits, performed by independent organizations or regulatory bodies, provide an unbiased evaluation of the laboratory's performance. Both types of audits are vital for maintaining high-quality standards and ensuring continuous improvement in laboratory practices (Ezzelle et al., 2008).

26.3.1.3 Corrective actions

When discrepancies are identified through audits or routine QC measures, corrective actions must be initiated to address the issues. This may involve revising SOPs, retraining personnel, or recalibrating equipment. Documenting these corrective actions is crucial for tracking improvements and ensuring that similar issues do not recur in the future (Westgard & Westgard, 2016).

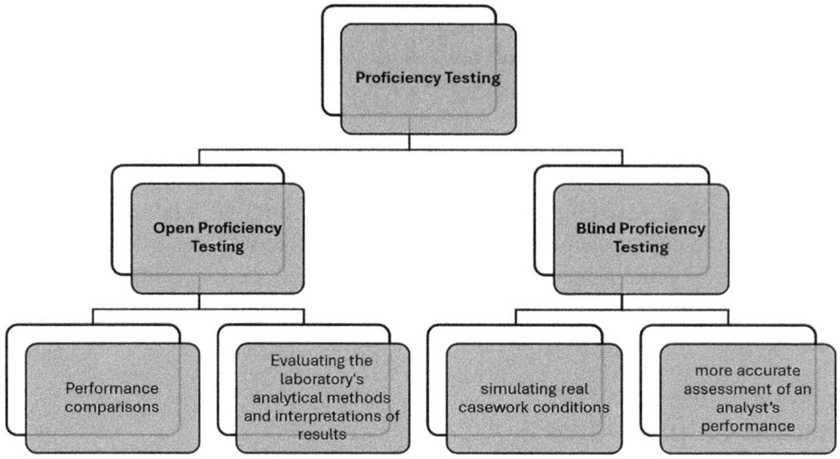

FIGURE 26.1 Proficiency testing (PT) and its essential components.

26.4 Proficiency testing and auditing

PT and auditing are essential components of QA in forensic laboratories, ensuring that analytical procedures are performed with a high degree of accuracy and reliability. These processes are designed to evaluate laboratory performance, identify areas for improvement, and foster a culture of continuous enhancement in forensic practices (Koehler, 2013; Fig. 26.1).

26.4.1 Proficiency testing

PT is an external assessment of a laboratory's ability to perform analyses at the required level of competence and quality. It evaluates not only the technical aspects of the analyses but also the laboratory's response to errors and areas for improvement. PT can be classified into two main types: open PT and blind PT (Peat et al., 1983).

26.4.1.1 Open proficiency testing

In open PT, analysts are aware that they are being tested. An external entity provides a set of samples in a mock case scenario, which the analysts are tasked with analyzing. This approach is beneficial for evaluating the laboratory's analytical methods and interpretations of results. It helps identify systematic errors due to equipment malfunctions, material contamination, or analyst misjudgment. Additionally, open PT allows for performance comparisons among different laboratories, facilitating inter-laboratory comparisons. This method of testing fosters a competitive environment that encourages laboratories to enhance their performance continuously (Koehler, 2013).

26.4.1.2 Blind proficiency testing

In contrast, blind PT involves analysts who are unaware that they are being evaluated. Blind PT can provide a more accurate assessment of an analyst's performance because they are likely to approach the analysis with the same level of care as they would for any regular case. This type of testing is often used to evaluate broader aspects of laboratory performance, from sample receipt through analysis to the final report generation. By simulating real casework conditions without the analysts' knowledge, blind PT provides insights into how routine laboratory operations perform under typical circumstances (Cowan & Koppl, 2011). PT serves multiple functions, including identifying training needs, verifying the effectiveness of analytical methods, and enhancing the credibility of the laboratory's results. Regular participation in PT is a hallmark of a quality laboratory, demonstrating its commitment to maintaining high standards of practice.

26.5 Auditing

Auditing is an independent review of laboratory operations aimed at determining whether the laboratory is performing according to defined standards. Audits are crucial for evaluating compliance with established protocols and identifying areas for improvement (Plebani, 2022). They can be categorized into two types: internal audits and external audits.

26.5.1 Internal audits

Internal audits are conducted by laboratory personnel or a designated internal team. These audits assess adherence to SOPs and other operational guidelines. Internal audits typically involve reviewing documentation, observing processes, and conducting interviews with personnel. The purpose of internal audits is to ensure that the laboratory is operating efficiently and in compliance with its own standards. It is essential that these audits are performed regularly, at least once a year to promote a culture of continuous improvement (Pramukti, 2024).

26.5.2 External audits

External audits are performed by independent organizations or regulatory bodies. These audits provide an objective evaluation of the laboratory's operations and ensure compliance with external standards, such as ISO/IEC 17025. External auditors review laboratory practices, assess the effectiveness of QA/QC measures, and verify the accuracy of results. The findings from external audits can lead to recommendations for improvement and may influence accreditation status (Plebani, 2022; Pramukti, 2024).

Auditing activities include reviewing SOPs, worksheets, laboratory notebooks, equipment calibration records, monitoring logs, and any other documents related to laboratory analysis processes. During audits, if errors or problems are identified, auditors will initiate corrective or preventive actions to ensure the continuous improvement of the quality system (QS). For example, if an audit reveals that certain SOPs are not being followed, the auditors might recommend retraining personnel or revising the procedures (Jul, n.d.).

The results of audits are documented and used to inform future auditing plans, defining the best approaches to address any issues identified. Regular audits not only help maintain high-quality standards but also foster transparency and accountability within the laboratory.

26.6 The forensic chain of custody

The forensic chain of custody is a fundamental principle in forensic science that ensures the integrity and reliability of evidence from the moment it is collected at the crime scene until it is presented in court. Properly managing the chain of custody is crucial for maintaining the admissibility of evidence in legal proceedings and for upholding public confidence in the forensic process (D'Anna et al., 2023).

26.6.1 Importance of chain of custody

The chain of custody refers to the chronological documentation of the handling of evidence. It provides a detailed record of every person who has handled the evidence, the time and date of each transfer, and the conditions under which the evidence was stored and transported (Prayudi & SN, 2015). The importance of maintaining a clear and accurate chain of custody includes (Fig. 26.2) the following:

26.6.1.1 Preventing tampering or contamination

A well-documented chain of custody helps ensure that evidence remains untampered and uncontaminated throughout the forensic process. Each transfer of evidence is recorded, providing a traceable path that can be audited at any point. This minimizes the risk of mishandling or accidental contamination that could compromise the integrity of the evidence (Ahmad et al., 2020).

26.6.1.2 Establishing authenticity and reliability

The chain of custody establishes the authenticity of the evidence, demonstrating that it has not been altered or tampered with since it was collected. This is critical for validating the evidence in court, as it shows that the evidence presented is a true representation of what was found at the crime scene (PILPG, n.d.).

26.6.1.3 Legal admissibility of evidence

Courts require that evidence be handled according to strict protocols to be admissible. A break in the chain of custody can lead to challenges regarding the reliability of the evidence, potentially resulting in its exclusion from court proceedings. Properly documenting the chain of custody ensures that evidence remains credible and that the laboratory can confidently assert its findings in court (Harmon, 2005).

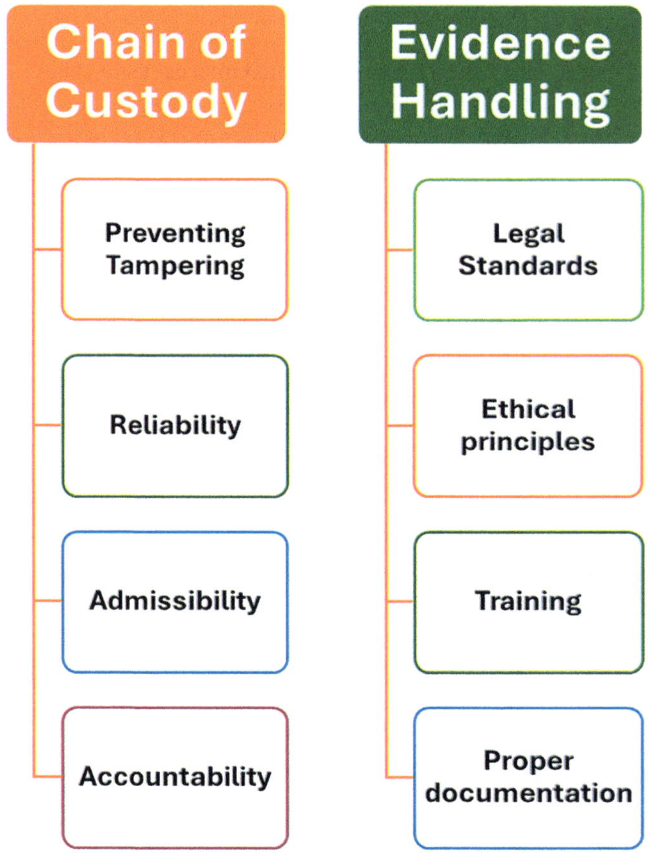

FIGURE 26.2 Overview of the forensic chain of custody process.

26.6.1.4 Accountability

A documented chain of custody holds individuals accountable for their handling of evidence. By clearly indicating who had access to the evidence and under what circumstances, it creates a transparent process that can be scrutinized during legal proceedings. This accountability is vital for maintaining ethical standards in forensic practice (D'Anna et al., 2023; Li et al., 2021).

26.6.2 Legal and ethical considerations in evidence handling

The legal and ethical considerations associated with the chain of custody are paramount in ensuring that forensic evidence is handled properly and in compliance with established laws and regulations (Maratsi et al., 2022).

26.6.2.1 Legal standards

Forensic laboratories must adhere to various legal standards when handling evidence. These include statutory requirements, professional guidelines, and specific legal precedents. For example, the Federal Rules of Evidence in the United States outline the standards for admitting forensic evidence, emphasizing the need for a reliable chain of custody. Laboratories must ensure compliance with these standards to avoid legal repercussions (Jonakait, 1991).

26.6.2.2 Ethical considerations

Ethical principles in forensic science dictate that evidence must be handled with the utmost care and integrity. Forensic professionals have a responsibility to uphold the principles of honesty, transparency, and objectivity in their work. Any breach of ethical standards, such as falsifying evidence or misrepresenting findings, undermines public trust in the forensic process and can have devastating consequences for individuals involved in legal proceedings (Hossain et al., 2024).

26.6.2.3 *Training and awareness*

Forensic personnel must be adequately trained in the legal and ethical implications of evidence handling. This includes understanding the importance of maintaining the chain of custody, as well as being aware of the legal consequences of mishandling evidence. Continuous education on legal standards and ethical practices should be integral to personnel training programs (Hossain et al., 2024).

26.6.2.4 *Documentation practices*

Proper documentation is a key element in ensuring compliance with legal and ethical standards. All actions related to evidence handling should be thoroughly recorded, including the time, date, individuals involved, and any observations made during the process. This documentation serves as a record that can be referenced in court and helps maintain accountability among forensic personnel (Almarzooqi, 2016).

26.7 Laboratory facilities, equipment, and supplies

The effectiveness of forensic analyses heavily relies on the quality of laboratory facilities, equipment, and supplies. Properly equipped laboratories are crucial for conducting accurate and reliable forensic investigations. This section outlines the essential equipment used in forensic analysis, the importance of calibration and maintenance, and the role of environmental controls and documentation in maintaining the integrity of forensic work (Pyrek, 2010).

26.7.1 Essential equipment for forensic analysis

Forensic laboratories utilize a wide array of sophisticated equipment to analyze various types of evidence. The choice of equipment depends on the specific needs of the laboratory and the types of analyses being conducted. Some of the essential equipment includes the following.

26.7.1.1 *Mass spectrometers*

Used for identifying and quantifying chemical substances, mass spectrometers analyze samples based on their mass-to-charge ratio. They are particularly important in toxicology and drug analysis, allowing forensic scientists to detect and measure drugs, poisons, and other chemical substances in biological samples (Brown et al., 2020).

26.7.1.2 *Gas chromatography and liquid chromatography*

These separation techniques are vital for analyzing complex mixtures. Gas chromatography is commonly used for volatile compounds, while liquid chromatography is more suited for nonvolatile substances. Both methods are instrumental in toxicology, environmental analysis, and trace evidence examination (Rawtani et al., 2020).

26.7.1.3 *DNA analyzers*

Essential for genetic profiling, DNA analyzers utilize various techniques, such as polymerase chain reaction and capillary electrophoresis, to analyze DNA samples. This equipment is crucial for identifying individuals in criminal cases and paternity testing (McKenna et al., 1994).

26.7.1.4 *Microscopes*

Various types of microscopes, including light microscopes, comparison microscopes, and scanning electron microscopes, are used in forensic laboratories for examining trace evidence, such as hair, fibers, and gunshot residue. Microscopy allows forensic scientists to observe details that are not visible to the naked eye, enhancing the accuracy of evidence analysis (Vaishnavi et al., 2024).

26.7.1.5 *Spectrophotometers*

Used for measuring the absorbance or transmittance of light by a sample, spectrophotometers are essential for analyzing substances based on their chemical properties. They are frequently used in toxicology and chemical analysis to identify and quantify compounds (Causin et al., 2005).

26.7.1.6 *Automated evidence processing systems*

These systems streamline the handling and analysis of evidence, reducing the potential for human error. Automated systems may include robotic sample handling, automated liquid handling, and software for data analysis and reporting (Jetten, 2021).

26.7.2 Calibration and maintenance of instruments

The accuracy and reliability of forensic analyses depend on the proper calibration and maintenance of laboratory instruments. Regular calibration ensures that equipment operates within specified parameters, providing accurate results (Tilstone, 2010; Kościelniak, 2023). Key aspects of calibration and maintenance include the following.

26.7.2.1 *Calibration procedures*

Calibration should be performed regularly according to established protocols. This involves comparing the instrument's output to known reference standards and making necessary adjustments. For example, mass spectrometers may be calibrated using calibration standards with known concentrations to ensure accurate measurements (Tilstone, 2010).

26.7.2.2 *Scheduled maintenance*

Routine maintenance is essential for preventing equipment malfunctions and ensuring optimal performance. Maintenance tasks may include cleaning, replacing worn parts, and conducting performance checks. Laboratories should establish a maintenance schedule for each piece of equipment and adhere to it strictly (Tilstone, 2010).

26.7.2.3 *Documentation of calibration and maintenance activities*

Keeping detailed records of calibration and maintenance activities is vital for maintaining QA. Documentation should include the date of calibration, the individual who performed the calibration, the results, and any corrective actions taken. This information is essential for audits and provides a verifiable history of equipment performance (Tilstone, 2010).

26.7.3 Environmental controls and documentation logs

Maintaining appropriate environmental conditions in forensic laboratories is critical for ensuring the reliability of test results. Several factors must be controlled, including temperature, humidity, and contamination risks (Morrison, 1999).

26.7.3.1 *Temperature and humidity control*

Forensic analyses can be sensitive to environmental conditions. Laboratories should monitor and control temperature and humidity levels to prevent degradation of samples and ensure accurate analyses. For example, DNA samples should be stored at low temperatures to preserve their integrity, while certain chemicals may require specific humidity levels to remain stable (Bonsu, 2022).

26.7.3.2 *Contamination prevention*

Implementing measures to prevent contamination is essential in forensic laboratories. This includes maintaining clean workspaces, using appropriate personal protective equipment, and employing sterile techniques when handling biological samples. Regular cleaning and decontamination protocols should be established to minimize the risk of cross-contamination (Nešić et al., 2018).

26.7.3.3 *Documentation logs*

Keeping comprehensive documentation logs is crucial for maintaining QA in forensic laboratories. These logs should include records of environmental monitoring (e.g., temperature and humidity readings), maintenance activities, calibration records, and cleaning schedules. Documentation logs serve as a reference for audits and provide a historical account of laboratory conditions and practices (Alshebel, 2020).

26.8 Competency for forensic personnel

The competency of personnel working in forensic laboratories is crucial for ensuring the quality and reliability of forensic analyses. Forensic analysts must possess the appropriate qualifications, skills, and training to conduct analyses accurately and effectively. This section highlights the importance of qualifications and continuous education for forensic scientists, along with their competency in providing courtroom testimony (Niskala, 1986).

26.8.1 Qualifications and continuous education for forensic scientists

Forensic science is a specialized field that requires a combination of formal education, practical skills, and ongoing professional development. Key aspects of qualifications and continuous education include the following.

26.8.1.1 Educational background

Forensic analysts typically hold advanced degrees in fields such as forensic science, chemistry, biology, or criminalistics. A strong educational foundation in the sciences is essential for understanding the principles underlying forensic techniques and methodologies. Many forensic scientists also possess specialized training in areas relevant to their specific roles, such as toxicology, DNA analysis, or trace evidence examination (Gaensslen, 2003).

26.8.1.2 Certification

Professional certification from recognized organizations enhances the credibility and competence of forensic analysts. Certification programs often require candidates to meet specific educational and experience criteria, as well as pass examinations. For example, the American Board of Criminalistics and the American Society of Crime Laboratory Directors offer certification programs for forensic professionals. Certification not only demonstrates expertize but also emphasizes a commitment to maintaining high standards of practice (Melbourn et al., 2019).

26.8.1.3 Continuing education

Given the rapidly evolving nature of forensic science, ongoing education is vital for forensic personnel. Continuous education ensures that analysts stay current with advancements in technology, techniques, and best practices. This may involve attending workshops, seminars, conferences, and online courses focused on new developments in forensic science. Organizations and laboratories should promote access to continuing education opportunities to encourage lifelong learning among their personnel (Gaensslen, 2003).

26.8.1.4 Competency assessments

Regular competency assessments are crucial for evaluating the skills of forensic analysts. This may include PT, where analysts are required to analyze blind samples without prior knowledge that they are being evaluated. These assessments help identify areas for improvement and reinforce the need for ongoing training (Grisso, 2006).

26.8.2 Courtroom testimony and legal competency

Forensic analysts often serve as expert witnesses in court, providing critical testimony regarding their analyses and findings. Their ability to communicate effectively in a legal setting is paramount to the credibility of forensic evidence (Bowers, 2013). Key aspects of courtroom testimony and legal competency include:

26.8.2.1 Understanding legal standards

Forensic analysts must be familiar with the legal standards governing the admissibility of forensic evidence. This includes knowledge of the Daubert standard in the United States, which requires that scientific evidence be based on reliable principles and methods. Understanding these legal requirements is essential for ensuring that forensic analyses are conducted in a manner that supports their admissibility in court (Kiely, 2006).

26.8.2.2 Effective communication

The ability to convey complex scientific concepts in a clear and understandable manner is vital for forensic analysts when testifying in court. Analysts must be able to explain their methodologies, findings, and the significance of their results to judges, juries, and attorneys who may not have a scientific background. Training in effective communication strategies, including public speaking and presentation skills, can enhance an analyst's ability to articulate their findings confidently (Hackman, 2021).

26.8.2.3 Preparation for cross-examination

Forensic analysts must be prepared to defend their findings during cross-examination by opposing counsel. This requires a thorough understanding of the analytical processes and methodologies used, as well as the ability to justify their conclusions. Preparation for courtroom appearances should include reviewing case materials, practicing responses to potential questions, and understanding the implications of their work in the context of the case (Thompson & Scurich, 2019).

26.8.2.4 Ethical considerations

Forensic analysts must adhere to ethical standards when providing testimony. This includes presenting findings objectively and without bias, maintaining transparency regarding any limitations or uncertainties in the analysis, and avoiding overstating the significance of their results. Ethical conduct in the courtroom is essential for preserving the integrity of the forensic process and maintaining public trust in the justice system (Downs & Swienton, 2012).

26.9 Managing nonconforming work

Despite the best QA/QC practices, errors can still occur. Nonconforming work must be addressed through corrective and preventive actions (CAPA). Therefore it is necessary to take prompt action upon the discovery of errors to correct them and ensure high-quality results, with high standards. One way to achieve this is by having well-defined SOPs and MSs. SOPs and MSs are written instructions that provide guidance on how to carry out a specific task using validated methods. These documents are essential for maintaining consistency and ensuring that unexpected results are addressed appropriately. SOPs and MSs should be regularly updated as new methods, equipment, and evidence emerge. Demonstrating to the judiciary system that the methods used to obtain results are validated and that the equipment is up to date and advanced can increase their confidence when making their decision. Moreover, nonconformities should be rectified and managed by applying appropriate preventive and corrective actions. Preventive actions are planned with the goal of preventing nonconformity from happening, while corrective actions are carried out after nonconformity has already occurred. Preventive actions aim to prevent occurrence, while corrective actions aim to prevent recurrence. Quality audits, whether by internal or external auditors, reviews, and commitment to a continuous improvement culture should also be applied and ensured by the forensic team (Doyle, 2024; Heavey et al., 2023).

To achieve this level of quality, forensic experts should take a scientific and objective approach in their work. This included being unbiased when analyzing results, repeating experiments when needed to confirm results, using control and reference samples for all experiments when required, and keeping accurate records of procedures and obtained results.

26.10 ISO/IEC 17025:2017 accreditation

ISO/IEC 17025:2017 is the international standard that specifies the requirements for the competence of testing and calibration laboratories. It provides a framework for laboratories to ensure they deliver accurate and reliable results. Accreditation to this standard demonstrates a commitment to quality and is essential for forensic laboratories aiming to uphold high standards of scientific practice (Miguel et al., 2021).

The key requirements of ISO/IEC 17025 include a management system, document control, internal audits, (CAPA), and technical requirements like method validation, measurement uncertainty, and traceability(International Organization for Standardization ISO, 2017). Accreditation to this standard can enhance the credibility and reliability of forensic results, improve the quality of forensic services, and increase confidence in the criminal justice system(no date a). Major accreditation bodies that offer ISO/IEC 17025 accreditation for forensic laboratories include the American National Standards Institute (ANSI) (National Accreditation Board) (https://anab.ansi.org/accreds/accreditation-for-forensic-service-providers/) and the United Kingdom Accreditation Service (International Laboratory Accreditation Cooperation ILAC, 2017). Maintaining accreditation requires ongoing efforts, such as regular internal audits, management reviews, PT, and continuous improvement initiatives. Crime scene investigation units, on the other hand, are accredited to ISO/IEC 17020:2012 "Conformity assessment- Requirements for the operation of various types of bodies performing inspection." This standard was determined to be the most appropriate for crime scene examination by the European Network of Forensic Science Institutes (ENFSI) in 2006, and later, ENFSI developed guidance for that standard in 2008, referred to as EA-5/03 [European Accreditation, 2008]. The International Standards Organization (ISO) is an independent, nongovernment organization with members from 162 standardization bodies (ISO, 2017). These standards provide specifications for products, services, and systems to ensure quality, safety, and efficiency. ISO/IES 17025, 2005 can be applied to all laboratories regardless of the number of personnel or the extent of the scope of testing and/or calibration activities.

The general requirements within ISO/IES 17025 are customized to forensic analysis by ILAC-G19:2002 guidance document "*Guidelines for Forensic Science Laboratories*" and other forensic criteria such as *ASCDL/LAB Supplemental Requirements*. In addition to this accreditation standard, forensic laboratories also use forensic-specific supplemental standards, scientific working group guidelines, as well as published materials on best laboratory practices and standard test methods, as guidance for their quality management systems.

ISO/IES 17025 standard has 15 management clauses and 10 technical clauses. The management clauses are associated with the performance and efficiency of the Quality Management System in the laboratory, and each clause requires clear procedures, policies, programs, and instructions to maintain the management system. The technical clauses determine the truth and reliability of testing and/or calibrations performed by the laboratory and are concerned with the competency of the employees, testing methodology, equipment, and the test and calibration results. However, some pitfalls and setbacks have been highlighted previously that are related to implementing and maintaining such QA systems. These include the time consumed to prepare for the requirements of the implementation of ISO/IEC 17025, especially when there is no previous experience. Moreover, applying such a system tends to expand bureaucracy as it requires management and regular updates to the records, which can be complicated, expensive, and demanding in terms of human resources. Furthermore, there are many cases and events that support the fact that being accredited with ISO/IEC 17025 and ISO/IEC 17020 is insufficient in itself to provide assurance of individual competence, and therefore, additional measures are required. Laboratory accreditation could solve many of the problems that would result in laboratory failures due to incompetence, sloppy procedures, fraud that could be presented by forged fingerprints, fake autopsies, and false laboratory reports. However, laboratory failure does not only occur in nonaccredited laboratories; it has also been shown in accredited laboratories due to incompetence of the forensic experts or inaccurate application of certain forensic techniques. Some of these laboratory techniques include hair comparisons, serological examination, drug analysis, fingerprint identification, and DNA profiling. In addition, the lack of QA protocols could result in difficulty distinguishing between malfeasance and misfeasance, which can go undetected for a long period of time (Miguel et al., 2021; no date a; no date b).

26.11 Case studies: laboratory failures

Several notable cases, such as those involving Fred Zain and Joyce Gilchrist, have demonstrated the devastating consequences of forensic misconduct. These examples underline the importance of stringent QA/QC measures in forensic laboratories.

26.11.1 Frederick Salem Zain and Joyce Gilchrist cases

The Frederick Salem Zain and Joyce Gilchrist cases are two examples of crime laboratory failures that led to wrongful convictions and highlighted the importance of strict QC procedures in forensic laboratories. Fred Salem Zain, the Chief Serologist in West Virginia State Police Crime Laboratory, was found guilty of falsifying test results in several cases (134 cases) from 1979 to 1989. The Supreme Court of Appeals of West Virginia reported that Zain's misconduct included overstating the strength of results, overstating the frequency of genetic matches, misreporting the number of tested evidence, and failing to report conflicting results. These actions led to the imprisonment of many defendants. Upon reviewing the cases that Zain processed, numerous charges were dismissed, and convictions were reversed for multiple cases. As a consequence, wrongfully convicted individuals due to Zain's misconduct raised lawsuits, causing West Virginia to pay a total of $6.5 million in settlements. Zain was charged with fraud and obtaining money by false pretenses, and the crime lab was revamped, establishing strict QC procedures. Due to such QC procedures, other inconsistencies were later discovered, committed by other lab members who were found to skip required tests and violate lab policies (Giannelli, 2010).

Similarly, Joyce Gilchrist, a forensic chemist working in the Oklahoma City Police Department crime laboratory in 1980, was accused of falsifying evidence in many cases to help prosecutors in Oklahoma. Concerns regarding her misconduct were first raised in the case of Jeffrey Todd Pierce, a landscaper who has been convicted of raping Sandra Burton in her apartment in 1985, despite a clean criminal record and good alibi. After Pierce's arrest, he voluntarily agreed to provide head and pubic hair exemplars, which Gilchrist examined. Gilchrist claimed that Pierce's hair samples were microscopically consistent with the hairs found at the crime scene, and as a result, Pierce was sentenced to 65e years in prison. However, later FBI investigations showed that Gilchrist had misidentified fiber in one case and hairs in six cases, including Pierce's case. The investigation results indicated that none of Pierce's hairs had similar microscopic characteristics to those taken from the crime scene. Consequently, the evidence was retested by another private DNA lab, and reexamination of the results cleared Pierce of the crime after he had served 15 years in prison (Giannelli, 2006).

These cases demonstrate the severe consequences of forensic misconduct and the importance of implementing and maintaining strict QC measures in crime laboratories. The actions of individuals like Zain and Gilchrist not only lead to wrongful convictions but also undermine the credibility of forensic science and the criminal justice system as a whole.

26.11.2 The Damilola Taylor case

The Damilola Taylor case is a tragic example of how errors in forensic examination can lead to significant delays in the administration of justice and cause immense pain for the victim's family. Damilola Taylor, a 10-year-old Nigerian schoolboy, was killed on November 27, 2000, in Peckham, South London. In 2004 four teenagers were tried for his murder, but all were found not guilty or had charges against them dropped due to some errors committed by the Forensic Science Service (FSS) team, an ISO/IEC 17025 accredited laboratory. The evidence was then reinvestigated by a private company called Forensic Alliance, which identified Damilola's blood spots on the trainer and jumper of the suspects, the Preddie brothers. As a result, new evidence was introduced in 2006, and the suspects, Hassan Jihad and the Preddie brothers, went on trial for the murder. The Preddie brothers were convicted of manslaughter (Innes, 2012).

In 2007 QC and Professor Brian Caddy conducted an independent review of the case and the mistakes made by the FSS team. Their report indicated that there were serious quality failures in the examination of items. The review concluded that there were no systematic failures in the FSS's operations but rather human failures in implementing the operating procedures, which consequently led to the failure to recover crucial evidence. The errors made by the forensic experts in this case resulted in years of delay and pain for Damilola's family. This case highlights the significant impact that the quality of the evidence generated and presented by forensic experts can have on court decisions. It also demonstrates that laboratory accreditation alone does not guarantee the highest quality of results.

The Damilola Taylor case underscores the importance of rigorous QC measures, ongoing training, and competency assessments for forensic practitioners. It is crucial for forensic laboratories to not only have well-defined SOPs but also to ensure that these procedures are followed consistently by all staff members. Regular audits, PT, and a culture of continuous improvement can help minimize the risk of human errors and ensure the accuracy and reliability of forensic evidence.

26.11.3 The R v T [2010] case

The R v T [2010] case is another example of how the quality of data generated by forensic experts can significantly affect court decisions. In this case, the English Court of Appeal overturned a murder conviction due to severe flaws in the prosecution's forensic evidence, including a lack of transparency in the expert's reasoning methodology and the absence of a sound scientific basis. In R v T [2010], T was convicted partly based on shoeprint comparison evidence provided by the FSS. The expert's opinion was based on the rarity of the sole pattern, shoe size, and indications of wear and damage. The expert concluded that there was a "moderate degree of scientific support" for the proposition that the shoeprints at the crime scene were made by T's Nike trainers. The expert used likelihood ratios to measure the strength of the evidence, in accordance with the FSS's SOP. The court found that the formula employed by the expert resulted in increasing the odds that T's trainers had made the footprint at the crime scene by 100-fold. Regarding transparency, the expert did not show any of the calculations used in his written report, nor did he present them during oral testimony at trial. The court held that it is simply wrong in principle for an expert to fail to set out the way in which he has reached his conclusion in the report.

The R v T [2010] case serves as a reminder of the importance of rigorous QC and transparency in forensic science. Experts must clearly communicate their reasoning and methodology, and their conclusions must be based on reliable scientific data and principles. Failure to adhere to these standards can have severe consequences, as demonstrated by the overturned conviction in this case.

26.12 Assessing forensic experts' competencies

Assessing Forensic Experts' Competencies is a crucial step in ensuring the reliability and credibility of forensic evidence. To address the issue of forensic experts' competencies, many countries have implemented different approaches and standards.

In the United States, the Forensic Specialties Accreditation Board (FSAB) requires accreditation to ISO/IES 17025:2012 "Conformity assessment- General requirements for bodies operating certification of persons." This standard is specifically designed for bodies assessing and certifying the competence of individual practitioners. By obtaining certification under ISO/IES 17025:2012, expert witnesses who are not employed by an organization accredited to ISO/IES 17025 or 17020 can demonstrate their competence as forensic science practitioners. Implementing this standard helps reduce the risk of releasing scientific information of insufficient quality, which could potentially lead to miscarriages of justice. Other countries have also developed their own approaches to assess the competencies of forensic experts. In Australia, the Australian Forensic Science Assessment Body (AFSAB) assesses and certifies crime scene, fingerprint, and toolmark/firearms examiners. In the United Kingdom, the Chartered Society of Forensic Sciences primarily awards certification to high-performing crime scene examiners.

However, it is important to note that adopting and applying different standards to the same activity across jurisdictions may not necessarily reflect competence. The use of different standards for the same activity can lead to risks of conflicting, duplicate, or overlapping standards, which goes against the principles of standardization as defined by ISO and the goal of harmonization. To address this issue, various bodies, including the Forensic Scientific Regulator of Forensic Science Institute (ENFSI), the ISO Technical Committee 272, and the National Institute of Forensic Science, should collaborate to achieve international harmonization. The aim should be to develop and implement a common standard that delivers forensic science optimally, regardless of jurisdiction.

In the United States, the National Commission on Forensic Science (NCFS) was created to improve forensic science practices and enhance their reliability. One of its objectives is enhancing the QA and QC in forensic laboratories. The NCFS provides recommendation and guidance on many aspects of forensic science, such as accreditation, PT, forensic investigations, scientific research, and training. These efforts are aimed at achieving harmonized practices and delivering high-quality forensic services (National Commission on Forensic Science, 2017).

26.13 Quality paradigm in forensic science

The quality paradigm consists of the implementation of standardization, certification, accreditation, and evaluation processes. This allows for the exchange of data between laboratories and the creation of databases due to standardized methods, which can ensure and maintain the integrity, accuracy, and reliability of the results throughout the forensic chain. Applying QA and QC systems in a forensic laboratory implies qualifying the lab, commonality of working procedures, defining used instruments along with their calibration, verification, and qualified working profile, and defining SOPs and standard methods (SMs), etc. (Milosevic et al., 2008). In addition, it defines the minimum skill requirements for those involved in a specific forensic science activity. For forensic findings to withstand court scrutiny, the QS must be assured at every level of the forensic analysis process and must be understood and implemented by every forensic employee from the beginning of the analysis until the end. Top management should also be committed to the highest level of testing quality. Lacking commitment from top management can result in flaws and systematic failures that will consequently affect the quality of the results, which can negatively impact the justice (Crispino et al., 2022).

In addition, since every human activity is associated with some risk of errors, some of which could have deleterious consequences, it is important for forensic analysts to perform their tasks with care, give attention to details, and use independent quality checks at all stages of the analytical process. Some potential sources of error include sample mishandling, data-recording errors, faulty reagents, equipment, controls, or technique, evidence contamination, and analyst bias(National Research Council US Committee on DNA Forensic Science: An Update, 1997).

In forensic laboratories, four broad categories can impact the quality of the forensic system. These include the forensic chain, competency of personnel, laboratory facilities, equipment, and supplies and managing nonconforming work. These categories will be discussed in detail subsequently.

26.14 Conclusion

QA and control are indispensable to forensic science. By implementing robust QA and QC systems, forensic laboratories can produce reliable and defensible results that support the legal process. Through adherence to ISO/IEC 17025:2017 accreditation standards, forensic laboratories establish a framework of excellence, maintaining high-quality outcomes through systematic PT, auditing, and continuous monitoring of environmental and technical conditions. The competency of forensic personnel is also pivotal, as well-trained and certified analysts contribute significantly to accurate analyses and credible courtroom testimony. Ultimately, QA/QC systems uphold the integrity of forensic investigations, ensuring that scientific evidence can withstand legal scrutiny and contribute to fair justice.

26.15 Future perspective

As forensic science continues to evolve with advancements in technology and analytical methods, the importance of quality management systems will grow. Future forensic laboratories will likely integrate more automation, artificial intelligence, and machine learning tools to enhance accuracy, minimize human error, and increase throughput. These advancements will necessitate updates to QA/QC protocols to ensure that emerging technologies comply with existing standards and maintain the reliability of forensic results. The development of new accreditation standards that address technological innovations, such as digital forensics and bioinformatics, will be crucial in shaping the future of forensic science. Furthermore, global collaboration on harmonizing accreditation and certification requirements across jurisdictions could help establish uniform standards of forensic practice worldwide.

26.16 Recommendations

To ensure the continued growth and improvement of forensic science, forensic laboratories should prioritize several key areas. First, continuous education and training programs for forensic analysts must be strengthened. As forensic techniques and technologies evolve, it is crucial that personnel stay up to date with the latest advancements. Offering specialized certifications and ongoing professional development will enhance the overall competency of forensic scientists. Additionally, forensic laboratories should embrace automation in their operations, such as robotic sample handling and AI-driven data analysis, which can significantly reduce human errors and improve efficiency. However, the implementation of these technologies must be accompanied by updates to QA/QC protocols to ensure they integrate seamlessly with current standards.

Acknowledgments

I extend my heartfelt gratitude to everyone who supported and contributed to this work.

Funding

Not applicable

Conflicts of Interest

The author declared no conflict of interest.

Data availability statement

Not applicable.

Author contributions

Abdullah Al-Matrouk, Abdullah Al-Dasem, Mahdi Haidar, Fatema Mousawi: Conceptualization and supervision, Methodology, interpretation, and writing original draft preparation. The author has read and agreed to the submitted version of the chapter.

References

Ahmad, L., Khanji, S., Iqbal, F., & Kamoun, F. (2020).Blockchain-based chain of custody: Towards real-time tamper-proof evidence management. In *ACM international conference proceeding series*, Association for Computing Machinery, United Arab Emirates. ISBN: 9781450388337. https://doi.org/10.1145/3407023.3409199, http://portal.acm.org/.

Almarzooqi, A. (2016). *Digital forensics practices: A road map for building digital forensics capability*.

Alshebel, A. K. S. (2020). *Standardization requirements for digital forensic laboratories: A document analysis and guideline*.

Ansell, R. (2013). Internal quality control in forensic DNA analysis. *Accreditation and Quality Assurance, 18*(4), 279–289. https://doi.org/10.1007/s00769-013-0968-9.

Bonsu, D. N. O. A. M. (2022). *Forensic trace DNA recovery and amplification from metal and metal-coated surfaces*.

Bowers, C. M. (2013). *Forensic testimony: Science, law and expert evidence forensic testimony: Science, law and expert evidence*. United States: Elsevier Inc, 1–263. http://www.sciencedirect.com/science/book/9780123970053, 10.1016/C2011-0-06823-0.

Broadhurst, D., Goodacre, R., Reinke, S. N., Kuligowski, J., Wilson, I. D., Lewis, M. R., & Dunn, W. B. (2018). Guidelines and considerations for the use of system suitability and quality control samples in mass spectrometry assays applied in untargeted clinical metabolomic studies. *Australia Metabolomics, 14*(6). https://doi.org/10.1007/s11306-018-1367-3, http://www.kluweronline.com/issn/1573-3890/.

Brown, H. M., McDaniel, T. J., Fedick, P. W., & Mulligan, C. C. (2020). The current role of mass spectrometry in forensics and future prospects. *Analytical Methods, 12*(32), 3974–3997. https://doi.org/10.1039/d0ay01113d, http://pubs.rsc.org/en/journals/journal/ay.

Causin, V., Marega, C., Guzzini, G., & Marigo, A. (2005). The effect of exposure to the elements on the forensic characterization by infrared spectroscopy of poly(ethylene terephthalate) fibers. *Journal of Forensic Sciences, 50*(4), 1–7. https://doi.org/10.1520/JFS2004519.

Cowan, E. J., & Koppl, R. (2011). An experimental study of blind proficiency tests in forensic science. *Review of Austrian Economics, 24*(3), 251–271. https://doi.org/10.1007/s11138-010-0130-4.

Crispino, F., Weyermann, C., Delémont, O., Roux, C., & Ribaux, O. (2022). Towards another paradigm for forensic science? *WIREs Forensic Science, 4*(3). https://doi.org/10.1002/wfs2.1441.

Downs, J. C. U., & Swienton, A. R. (2012). *Ethics in forensic science*. United States: Elsevier Inc. http://www.sciencedirect.com/science/book/9780123850195, 10.1016/C2010-0-64819-X.

Doyle, S. (2024). *QHFSS DNA laboratory–ISO/IEC 17025 conformance and accreditation*. Forensic Science International: Synergy.

D'Anna, T., Puntarello, M., Cannella, G., Scalzo, G., Buscemi, R., Zerbo, S., & Argo, A. (2023). The chain of custody in the era of modern forensics: From the classic procedures for gathering evidence to the new challenges related to digital data. *Healthcare, 11*(5), 634. https://doi.org/10.3390/healthcare11050634.

Ezzelle, J., Rodriguez-Chavez, I. R., Darden, J. M., Stirewalt, M., Kunwar, N., Hitchcock, R., Walter, T., & D'Souza, M. P. (2008). Guidelines on good clinical laboratory practice: Bridging operations between research and clinical research laboratories. *Journal of Pharmaceutical and Biomedical Analysis, 46*(1), 18–29. https://doi.org/10.1016/j.jpba.2007.10.010.

Fan, G., & Wang, Q. (2021). *Quality control and quality assurance clinical molecular diagnostics*. China: Springer Nature, 97–113. https://link.springer.com/book/9789811610370, 10.1007/978-981-16-1037-0_8.

Gaensslen, R. E. (2003). How do I become a forensic scientist? Educational pathways to forensic science careers. *Analytical and Bioanalytical Chemistry, 376*(8), 1151–1155. https://doi.org/10.1007/s00216-003-1834-0.

Giannelli, P. C. (2006). Alchemy, magic, and forensic science. *Criminal Justice, 21*.

Giannelli, P. C. (2010). Scientific fraud. *Criminal Law Bulletin*.

Grisso, T. (2006). *Evaluating competencies: Forensic assessments and instruments*. Springer Science & Business Media.

Hackman, L. (2021). Communication, forensic science, and the law. *WIREs Forensic Science, 3*(2). https://doi.org/10.1002/wfs2.1396.

Harmon, R. (2005). *Admissibility standards for scientific evidence microbial forensics*. United States: Elsevier, 381–392. https://www.sciencedirect.com/book/9780120884834, 10.1016/B978-012088483-4/50021-4.

Heavey, A. L., Turbett, G. R., Houck, M. M., & Lewis, S. W. (2023). Management and disclosure of quality issues in forensic science: A survey of current practice in Australia and New Zealand. *Forensic Science International: Synergy, 7*. https://doi.org/10.1016/j.fsisyn.2023.100339, https://www.journals.elsevier.com/forensic-science-international-synergy.

Hossain, M. Z., Kibria, H., & Johora, F. T. (2024). Ethical challenges in forensic accounting: Balancing professional responsibility and legal obligations. *Open Journal of Accounting, 13*(03), 57–73. https://doi.org/10.4236/ojacct.2024.133005.

Innes, M. (2012). Signal crimes': Detective work, mass media and constructing collective memory. In *Criminal visions*. Willan, 51–69.

International Laboratory Accreditation Cooperation (ILAC). (2017). Benefits of accreditation – ISO/IEC 17025:2017. https://ilac.org/latest_ilac_news/isoiec-170252017-published/

International Organization for Standardization (ISO)ISO/IEC 17025:2017 – General requirements for the competence of testing and calibration laboratories. ISO. https://www.iso.org/standard/66912.html.

JettenK.2021A contextual inquiry approach to improving DNA service delivery in sexual assault investigations.

Jonakait, R. N. (1991). Forensic science: The need for regulation. *Harvared Journal of Law & Technology. 4*.

Jul, T.S. (n.d.). Auditing and assessing the quality control laboratory.

Kiely, T. F. (2006). *Forensic evidence: Science and the criminal law*. CRC Press.

Klesta, E. J., & Bartz, J. K. (2018). Quality assurance and quality control methods of soil analysis. In *Part 3: Chemical methods*. United States: Wiley, 19–48. 10.2136/sssabookser5.3.c2, http://onlinelibrary.wiley.com/book/10.1002/9780891188667.

Koehler, J. J. (2013). Proficiency tests to estimate error rates in the forensic sciences. *Law, Probability and Risk, 12*(1), 89–98. https://doi.org/10.1093/lpr/mgs013.

Koehler, J. J., Mnookin, J. L., & Saks, M. J. (2023). The scientific reinvention of forensic science. *Proceedings of the National Academy of Sciences, 120*(41). https://doi.org/10.1073/pnas.2301840120.

Kościelniak, P. (2023). *Calibration in analytical science: Methods and procedures calibration in analytical science: Methods and procedures*. Poland: Wiley, 1–372. https://onlinelibrary.wiley.com/doi/book/10.1002/9783527831111, 10.1002/9783527831111.

Levy, S., Bergman, P., & Frank, A. (1999). Quality assurance in forensic science. *Accreditation and Quality Assurance, 4*(6), 253–255. https://doi.org/10.1007/s007690050361.

Li, M., Lal, C., Conti, M., & Hu, D. (2021). LEChain: A blockchain-based lawful evidence management scheme for digital forensics. *Future Generation Computer Systems, 115*, 406–420. https://doi.org/10.1016/j.future.2020.09.038, https://www.journals.elsevier.com/future-generation-computer-systems.

Mahmood, S., Lai, R., Kim, Y. S., Kim, J. H., Park, S. C., & Oh, H. S. (2005). A survey of component based system quality assurance and assessment. *Information and Software Technology, 47*(10), 693–707. https://doi.org/10.1016/j.infsof.2005.03.007, http://www.elsevier.com/wps/find/journaldescription.cws_home/525444/description#description.

Malkoc, E., & Neuteboom, W. (2007). The current status of forensic science laboratory accreditation in Europe. *Forensic Science International, 167*(2-3), 121–126. https://doi.org/10.1016/j.forsciint.2006.06.064, www.elsevier.com/locate/forsciint.

Maratsi, M.I., Popov, O., Alexopoulos, C., & Charalabidis, Y. (2022).Ethical and legal aspects of digital forensics algorithms: The Case of digital evidence acquisition. In *ACM international conference proceeding series,* Association for Computing Machinery, Sweden, pp. 32–40. 10.1145/3560107.3560114 9781450396356, http://portal.acm.org/.

McKenna, J. A., Cecil, J. S., & Coukos, P. (1994). Reference guide on forensic DNA evidence. Reference manual on scientific evidence. *Federal Judicial Center,* 273–330.

Melbourn, H., Smith, G., McFarland, J., Rogers, M., Wieland, K., DeWilde, D., Lighthart, S., Quinn, M., Baxter, A., & Quarino, L. (2019). Mandatory certification of forensic science practitioners in the United States: A supportive perspective. *Forensic Science International: Synergy, 1,* 161–169. https://doi.org/10.1016/j.fsisyn.2019.08.001, https://www.journals.elsevier.com/forensic-science-international-synergy.

Miguel, A. L. R., Moreiraa, R. P. L., & de Oliveira, A. F. (2021). ISO/IEC 17025: History and introduction of concepts. *Quimica Nova, 44*(6), 792–796. https://doi.org/10.21577/0100-4042.20170726, https://www.scielo.br/j/qn/a/3gwbfvVs8qydVd5ZstwXCph/?lang=en&format=pdf.

Morrison, R. D. (1999). Environmental forensics. *Environmental Claims Journal, 12*(2), 79–88. https://doi.org/10.1080/10406020009355116.

National Commission on Forensic Science (. (2017). .

National Research Council (US) Committee on DNA Forensic Science: An Update (1997).

Neuteboom, W., Ross, A., Bugeja, L., Willis, S., Roux, C., & Lothridge, K. (2024). Quality management in forensic science: A closer inspection. *Forensic Science International, 358,* 111779. https://doi.org/10.1016/j.forsciint.2023.111779.

Nešić, L., Lončar, M., & Inđić, J. (2018). Implementation of the new standard ISO/IEC 17025: 2017 and its impact on the quality of work in forensic laboratories. *Archibald Reiss Days.*

Niskala, H. (1986). Competencies and skills required by nurses working in forensic areas. *Western Journal of Nursing Research, 8*(4), 400–413. https://doi.org/10.1177/019394598600800402.

Olivares, I. R. B., & Lopes, F. A. (2012). Essential steps to providing reliable results using the analytical quality assurance cycle. *TrAC - Trends in Analytical Chemistry, 35,* 109–121. https://doi.org/10.1016/j.trac.2012.01.004, www.elsevier.com/locate/trac.

PILPG (n.d.). The engine room. Chain of custody. Documentation tools for human rights.

Page, H., Horsman, G., Sarna, A., & Foster, J. (2019). A review of quality procedures in the UK forensic sciences: What can the field of digital forensics learn? *Science and Justice, 59*(1), 83–92. https://doi.org/10.1016/j.scijus.2018.09.006, http://www.scienceandjusticejournal.com/contact.

Peat, M. A., Finnigan, J. S., & Finkle, B. S. (1983). Proficiency testing in forensic toxicology: A feasibility study. *Journal of Forensic Sciences, 28*(1), 139–158. https://doi.org/10.1520/jfs12247j.

Plebani, M. (2022). Internal quality control and external quality assurance: A great past opens the way to a bright future. *Advances in Laboratory Medicine, 3*(3), 215–217. https://doi.org/10.1515/almed-2022-0075, www.degruyter.com/journal/key/almed/html.

Pramukti, A. (2024). Internal audit versus external audit: A qualitative perspective. *Golden Ratio of Auditing Research, 4*(2), 78–88. https://doi.org/10.52970/grar.v4i2.392.

Prayudi, Y., & SN, A. (2015). Digital chain of custody: State of the art. *International Journal of Computer Applications, 114*(5), 1–9. https://doi.org/10.5120/19971-1856.

Pyrek, K. (2010). *Forensic science under siege: The challenges of forensic laboratories and the medico-legal investigation system.* Elsevier.

Rawtani, D., Mustansar Hussain, C., Pandey, G., Rawtani, D., & Mustansar Hussain, C. (2020). Chromatographic techniques for forensic investigations. In *Technology in forensic science: Sampling, analysis, data and regulations.* Wiley, 129–149. 10.1002/9783527827688.ch7.

Thompson, W. C., & Scurich, N. (2019). How cross-examination on subjectivity and bias affects jurors' evaluations of forensic science evidence. *Journal of Forensic Sciences, 64*(5), 1379–1388. https://doi.org/10.1111/1556-4029.14031, http://onlinelibrary.wiley.com/journal/10.1111/(ISSN)1556-4029.

Tilstone, W. J. (2010). *Quality in the forensic science laboratory.* Springer Science and Business Media LLC, 335–367. 10.1007/978-1-60761-872-0_12.

Vaishnavi, M. A., Venu, M., Kohli, R. K., & Puri, P. (2024). Microscopic techniques and their application in forensic science. In *Advances in analytical techniques for forensic investigation.* India: Wiley, 307–338. 10.1002/9781394167340.ch11, https://www.wiley.com/en-ie/Advances+in+Analytical+Techniques+for+Forensic+Investigation-p-9781394166312.

Westgard, J. O., & Westgard, S. A. (2016). Quality control review: Implementing a scientifically based quality control system. *Annals of Clinical Biochemistry, 53*(1), 32–50. https://doi.org/10.1177/0004563215597248, http://www.sagepub.com/journalsProdDesc.nav?prodId=Journal202177.

Chapter 27

Ethical issues in forensic DNA analysis

Prachilekha Sahoo
School of Law, Ramaiah University of Applied Sciences, Bengaluru, Karnataka, India

27.1 Introduction

Advances in scientific technology have transformed daily life in remarkable ways. While these developments have improved lifestyles and enhanced access to various services, they also present challenges for law enforcement agencies working to prevent technology misuse by criminals. Forensic science-the application of scientific methods to criminal investigations offers substantial scientific evidence through the analysis of evidence and exhibits. However, these technological advancements can also profoundly impact the administration of justice, sometimes posing complex challenges.

The judicious use of technology holds the potential to greatly enhance the criminal justice system (CJS). Research and case studies have shown that forensic methods have protected many individuals from wrongful convictions by providing objective evidence (Jayewardene, 1988; Sedley, 2005). Still, incorporating advanced technology into the justice system often requires adjustments to conventional legal principles. It remains a point of debate how best to utilize these advancements without compromising fundamental legal norms.

A careful balance must be struck to integrate scientific advancements while respecting traditional legal principles and constitutional rights. In particular, human rights such as the Right to Privacy and the Right Against Self-Incrimination must be safeguarded (Walsh, 2005). In criminal investigations, or when applying forensic evidence in court, these rights can sometimes conflict with technological practices. For example, using certain investigative technologies may create tensions between an individual's right to privacy and the need to gather evidence.

The discovery of deoxyribonucleic acid (DNA) has unlocked numerous possibilities in forensic genetics. DNA technology has proven to be a highly effective investigative tool, capable of confirming guilt or establishing innocence in legal proceedings (Dhikary, 2007). However, the legal framework for DNA evidence is an emerging issue in India, where specific legislation for its use in routine judicial processes remains absent. Presently, some existing laws provide guidance, though India has been working towards enacting comprehensive DNA legislation since 2003.

Another critical concern for policymakers is preventing the misuse of technology against public interests. Efforts should focus on implementing safeguards to ensure that technological advances are used responsibly, upholding justice without infringing on citizens' rights.

27.1.1 DNA analysis and its emergence in criminology

For nearly a century, the law enforcement community has sought a forensic "magic bullet"—a foolproof system to identify suspects with absolute certainty based on physical evidence left at crime scenes. Numerous methods have been explored over the years, each intended to track and identify past and future offenders. These methods have included fingerprinting, anthropometrics (systematic measurement of body parts), and voiceprints. However, over time, each of these techniques faced technical and practical limitations.

In 1984, forensic DNA analysis was developed in Britain by geneticist Alec Jeffreys. Judges and investigators hailed it as "the greatest advance in crime-fighting technology since fingerprints," with the media predicting it would "revolutionize law enforcement."[1] In many ways, it has. Over the past few decades, DNA evidence has been instrumental

1. New York and Wesley, 140 Misc. 2d 306 (1988).; Moss, D. C. (1988). DNA—The new fingerprints. ABA Journal, 74, 66.

Investigative and Predictive DNA Testing. DOI: https://doi.org/10.1016/B978-0-443-30128-5.00027-8

in solving countless violent crimes, leading to the incarceration of many rapists and murderers (Federal Bureau of Investigation, n.d.). The Federal Bureau Investigation's (FBI) national database, containing over four million criminal DNA profiles, has led to more than 45,400 matches since its inception in 1994 (Federal Bureau of Investigation, n.d.). Today, DNA analysis is widely considered the most effective forensic identification method ever developed. Even defense attorneys like Barry Scheck and Peter Neufeld, known for challenging the reliability of forensic DNA evidence in prominent cases (such as *People vs. Castro*, *United States vs. Yee*, and *People vs. Orenthal James Simpson*), now regard DNA testing as "a gold standard for truth-telling" (Scheck et al., 2000).

DNA analysis has extended far beyond the CJS, serving a wide range of identification needs. For instance, the New York City Medical Examiner's Office has used DNA analysis to identify victims of the September 11, 2001, terrorist attacks by matching small, charred remains to personal items like toothbrushes. DNA analysis also played a critical role in identifying the severely damaged remains of victims from the July 1996 TWA Flight 800 disaster (Aronson, 2007). Furthermore, DNA technology has been used to identify victims of human rights abuses and genocides in regions emerging from periods of violence and civil war.

Thanks to DNA analysis, cases previously shrouded in mystery are now being solved. For instance, it was DNA evidence that confirmed the identity of Michael J. Blassie, a Vietnam War veteran who had been interred in the Tomb of the Unknown Soldier (National Library of Medicine, 2014). With the establishment of DNA registries, the Department of Defense anticipates that unidentified soldiers may soon be a thing of the past, enriching history with newly uncovered identities and insights.

27.1.2 Foundation and breakthrough of DNA analysis

In 1984, Alec Jeffreys, a geneticist at the University of Leicester in England, developed the first practical form of DNA analysis. Following its introduction to the English legal system in 1985, this innovative method rapidly gained traction in the United States. Early on, the American DNA analysis market saw fierce competition between two biotechnology giants—Lifecodes Corporation and Cellmark Diagnostics—each launching variations of the technology within mere months of each other. Both companies were publicly traded multinational chemical corporations with annual revenues exceeding $20 billion, eager to capitalize on the biotech boom of the 1980s.

Both firms quickly filed patents on their techniques and tightly controlled access to their genetic markers—a decision that frustrated many academic scientists, who viewed such secrecy as antithetical to the transparency considered fundamental to scientific practice. The provision of forensic evidence by private companies was a novel development in law enforcement, as most scientific methods had previously been handled by public crime labs, especially the FBI. Although both companies initially focused on paternity testing, law enforcement soon sought their expertise to help solve complex criminal cases. Despite being a new and largely untested technology, DNA analysis was readily accepted in court, with the companies encountering little difficulty in convincing judges of its reliability as evidence.

Defense attorneys occasionally mounted weak attempts to challenge the accuracy and reliability of DNA analysis, but their limited understanding of genetics and molecular biology was no match for the robust teams of experts assembled by these private corporations. By the end of 1988, DNA analysis had been admitted into evidence without contest in nearly 200 cases in the United States.

27.1.3 The landmark DNA test: *Sarbah versus Home Office*

In a notable case, thirteen-year-old Andrew Sarbah expected a smooth re-entry through immigration when he arrived at London's Heathrow Airport. Though he spoke limited English, he was, after all, returning to the city where he was born. Born in London to Ghanaian parents, Andrew moved to Ghana with his father after his parents separated, spending seven years there. Wishing to return to the UK to live with his mother and siblings, he brought both a recently issued Ghanaian passport and his British passport from childhood, complete with a young photo of himself. However, his journey took a challenging turn at the immigration desk, where officials questioned his identity and citizenship, suspecting that the British passport was falsified and that he might actually be a cousin posing as Andrew. Pending investigation, he was allowed "temporary admission" while his case unfolded.

Seeking a resolution, the Sarbah family hired Sheona York, a London-based immigration lawyer. York promptly gathered photographic and testimonial evidence to support Andrew's case and also arranged for serological testing, which indicated a 98% likelihood that Andrew was indeed the biological son of his mother, Christina Sarbah. Despite this evidence, the Home Office remained skeptical, withholding Andrew's visa over concerns about possible tampering.

With limited options and mounting frustration, York filed a final appeal on Andrew's behalf. Around this time, a colleague presented her with a newspaper article from *The Guardian* that detailed a groundbreaking genetic technique developed by British geneticist Alec Jeffreys. This method, coined "genetic fingerprinting," was reported to be so precise that it could verify familial relationships, even within families with close genetic ties.

York reached out to Jeffreys to see if he would test his innovative DNA technique on Andrew's case. Despite his initial apprehension, he was worried that the test might fail and will yield no useful evidence. Jeffreys agreed to examine DNA samples from the Sarbah family. His findings provided clear proof of Andrew's relationship to his mother. The Home Office accepted the DNA evidence, allowing Andrew to reunite with his family in England after a prolonged two-year ordeal. This case marked a milestone, as the Home Office announced it would no longer contest immigration cases involving similar DNA evidence, underscoring the transformative power of DNA fingerprinting in legal and immigration matters.

27.1.4 Overview of DNA analysis methods

To understand Alec Jeffreys' groundbreaking identification tool, it is essential to explore some fundamentals of genetics and molecular biology. Jeffreys' method hinges on a core principle: except for identical twins, each individual possesses a unique genetic composition. Although humans share > 99.9% of their genome, there exist several differences, mostly in their non-coding region of the DNA, which creates a distinct DNA pattern. In fact, any two people chosen at random will differ in about one of every thousand DNA sites.

Some of these genetic variations can lead to serious health conditions (like sickle-cell anemia, Huntington's disease, Tay-Sachs disease, and cystic fibrosis), while others may have no noticeable impact or may even offer small advantages. Human bodies, made up of roughly a trillion cells, originate from a single fertilized egg. The genetic material, DNA, resides in the chromosomes within the cell nucleus. When the fertilized egg divides and develops, each resulting cell inherits the same DNA sequence, no matter whether it becomes part of the skin, blood, hair, or muscle tissue.

Chemically, DNA is a long chain composed of four types of nucleotides: adenine (A), cytosine (C), guanine (G), and thymine (T). Typically, DNA consists of two complementary strands held together by bonds between paired nucleotides: cytosine pairs with guanine, and adenine pairs with thymine.

Each bonded pair (either A-T or C-G) is called a base pair (bp), the basic unit of DNA length. While some DNA sequences, known as "genes," hold the instructions for producing proteins, organizing them into cells, and forming organs and body structures, other sequences, often called "junk DNA," play no known role but provide structural integrity to the chromosome.

Interestingly, it is in this "junk DNA" that the unique sequences used for DNA analysis are found. Within these non-coding areas of the genome, short DNA sequences—sometimes called "minisatellites" or "core sequences"—are repeated. Although scientists do not fully understand the reason behind these repetitions, they have observed that the number of repeats varies widely between individuals. Consequently, these segments are known as "variable number tandem repeats" (VNTRs). Each version of a repeated segment is called an "allele," a term in genetics that describes the variation at a specific physical location, or "locus," in the genome. At each locus where VNTRs occur, multiple alleles exist, each with a different number of repeated units, resulting in a distinct length. The specific pattern of alleles—variations of these repeat sequences—forms an individual's unique DNA profile.

As news of DNA fingerprinting spread, Jeffreys and his team at Leicester, home to the world's first lab using this method, were inundated with requests to assist in resolving immigration cases. His lab also began conducting paternity tests, and by mid-1986, courts had recognized DNA evidence as credible and trustworthy. In response to the growing demand for DNA testing, especially in immigration matters, Jeffreys and the Lister Foundation (which funded his research) partnered with Imperial Chemical Industries (ICI), the UK's largest chemical company, to commercialize this innovative technology. In 1987, ICI established a DNA testing center offering analyses for a fee of £105 per sample.

Official acknowledgment of DNA analysis as a legitimate method for determining paternity in immigration cases came with the Blood Test Regulations Amendment of 1989. Although formal recognition came later, the Home Office had already begun using DNA analysis in forensic casework, and it had been accepted by numerous courts.

Currently, Forensic DNA analysis relies on genetic markers such as short tandem repeats (STR) and single-nucleotide polymorphisms (SNPs) for applications including human identification, paternity testing, and criminal investigations. The process starts with the collection and proper preservation of biological samples, such as blood, saliva, hair, bone, and semen. To prevent DNA degradation and preserve sample quality, techniques like refrigeration or drying are employed, ensuring the integrity of the genetic material for subsequent analysis.

DNA is then extracted using chemical or enzymatic methods like organic extraction, Chelex extraction, or silica-based techniques to isolate and purify the DNA. Quantitative polymerase chain reaction (qPCR) measures the DNA concentration to ensure sufficient input for analysis.

In STR analysis, specific regions containing short repeating sequences of 4–5 base pairs are amplified using PCR. Multiple STR loci, typically 13–20, are amplified simultaneously in a multiplex reaction. Capillary electrophoresis (CE) separates DNA fragments based on size, producing distinct peaks for each STR allele. The resulting DNA profile is compared with suspects or database entries, and statistical validation, such as random match probability, determines match likelihood.

SNP analysis begins with genotyping to examine single-nucleotide variations. After DNA fragmentation, target regions are amplified using PCR. Detection methods include microarray chips, where DNA hybridizes with complementary probes, or next-generation sequencing (NGS), which identifies base sequences at SNP positions. SNPs are particularly useful for degraded samples due to their shorter DNA fragment requirements (Margoliash et al., 2023).

Statistical evaluation calculates match probabilities to verify results. STR profiles are stored in systems like the Combined DNA Index System (CODIS), while SNP profiles are often matched against ancestry databases. STR markers consist of repeating sequences with higher mutation rates, while SNP markers involve single-base changes with lower mutation rates. STRs analyze longer fragments (100–400 base pairs), whereas SNPs focus on shorter fragments (50–100 base pairs), making them ideal for degraded samples (Payseur & Cutter, 2006).

STR markers are primarily used for criminal cases, paternity testing, and disaster victim identification, while SNP markers assist in ancestry determination, DNA phenotyping for physical traits, and analyzing ancient or highly degraded samples. Both techniques provide accurate and reliable results, supporting forensic investigations and ensuring effective DNA profiling.

27.1.5 Validity and reliability of DNA technology: challenges of standardization and certification

Given the variations in DNA sequences among individuals, there is no scientific ambiguity that current DNA technologies accurately detect genetic differences. When conducted and interpreted correctly, a thorough examination of two DNA samples can determine whether the patterns match, and if they do, establish the likelihood that both samples originate from the same individual (except in cases of identical twins). In 1990, the US Congress Office of Technology Assessment reviewed DNA typing methods and reported that these techniques are valid and reliable when performed accurately. However, using DNA tests for forensic purposes, particularly in criminal investigations presents unique challenges that do not arise in basic research or clinical diagnostics.

Crime scene samples are often minute, potentially insufficient for testing, and may be degraded due to exposure to environmental factors. The primary concern in using DNA evidence is whether the detection methods are scientifically valid. Validity refers to a test's ability to correctly identify true matches and true non-matches. Initial concerns about the validity of DNA typing in forensic contexts focused on sample quality. Forensic samples are collected from a range of nonsterile materials (such as glass, wood, fabric, and soil) and are frequently exposed to sunlight, moisture, or dryness. These samples can also be contaminated by unknown biological materials, such as microbes or animal and plant residues. Therefore, extreme care must be taken when collecting forensic evidence to prevent contamination.

The second critical factor in forensic DNA testing is reliability, which involves several elements: the methods used, laboratory performance, record-keeping, quality control, and quality assurance. While DNA tests are indeed valid and reliable when conducted properly, there is a common misconception that DNA typing always produces a simple "yes" or "no" answer. A test that does not yield a definitive response is not necessarily inaccurate or unreliable. After the biochemical analysis, the key question becomes whether there is a "match" between the DNA profile of the forensic sample and that of a known sample. This issue is complex and requires expertise in molecular biology, population genetics, and statistics to interpret results accurately.

Errors in DNA testing can arise from various sources, such as sample mixing, contamination at the collection site or in the lab, and contamination from bacterial, viral, or other human or non-human DNA at the crime scene. A small or degraded sample can complicate the analysis and make further verification impossible, and the testing process itself can sometimes be prone to error (Singh, 2011).

In an April 1992 report,[2] the following recommendations were made:

- Current DNA typing procedures are scientifically sound.
- Each laboratory should implement a comprehensive quality assurance program.
- Laboratories should obtain proper accreditation.
- A national DNA profile database should be created, especially for convicted sex offenders and unidentified crime scene samples.

2. The National Academy of Sciences of the USA initiated a study by the National Research Council (NRC) on DNA typing methods.

The report also noted that interpreting DNA data often requires specialized expertise. For instance, samples from the same community or region may exhibit genetic similarities due to interbreeding, which could complicate the analysis. In such cases, a forensic genetics expert can provide critical guidance in interpreting the data accurately.

27.1.6 Ethics in forensic DNA are critical

Forensic DNA analysis plays a critical role in criminal investigations, but also raises complex legal and ethical challenges. It is essential to strike a balance between the advantages of DNA technology and the protection of individual rights, particularly privacy, consent, and ethical considerations. Rigorous regulation of DNA data management is necessary to prevent misuse and safeguard privacy. Technological advancements, such as probabilistic genotyping software, have improved the accuracy of DNA analysis, particularly in the interpretation of complex mixture samples. However, responsible use of DNA technology also requires careful management of emerging tools like NGS, while adhering to ethical standards.

To foster public trust, it is vital that transparency, accuracy, and ethical conduct are maintained throughout the analysis process. Regulatory oversight and accountability measures are essential for addressing errors and ensuring fairness in DNA data management. Ongoing dialogue among scientists, legal professionals, ethicists, and the public is crucial to maintaining scientific integrity and preserving confidence in forensic DNA analysis.

The responsible use of forensic DNA technology also involves careful consideration of emerging techniques, such as NGS and genetic phenotyping, which can provide detailed information about an individual's physical traits. While these technologies offer valuable insights, they must be used with a focus on ethical concerns to protect human privacy and uphold consent. The use of direct-to-consumer genetic testing for criminal investigations, particularly when identifying relatives, must be handled with caution, as it can have significant repercussions for both the individual and their extended family. Law enforcement's use of genealogy databases further underscores the potential implications of individuals' decisions regarding their genetic data.

DNA phenotyping, which reveals sensitive personal information, highlights the need to respect privacy while using genetic data. Personal identification is a fundamental human right, and each year, millions of people remain unidentified, preventing families from finding closure. Therefore, DNA technology is essential for identifying human remains, but it must be used with great care to balance investigative needs with the protection of individual rights and family privacy.

Establishing clear regulations and oversight mechanisms is critical to ensure that new DNA technologies are used ethically. These systems must emphasize transparency, accountability, and compliance with legal norms. Ongoing communication among scientists, policymakers, ethicists, and the public is vital to developing ethical practices that prioritize scientific integrity and human rights in forensic DNA analysis.

27.2 Privacy concerns in forensic DNA databases

27.2.1 Types of DNA databases

With the rise in cross-border crime, the international exchange of intelligence for policing has become increasingly essential. Forensic DNA data play a critical role in police investigations by aiding in the identification of suspects and victims, excluding individuals from criminal inquiries, and linking individuals to specific crimes. The effectiveness of forensic DNA data has been significantly enhanced through the establishment of DNA databases, which facilitate the identification of unknown and serial offenders by connecting different crimes.

Current developments in DNA databases include the expansion of national databases, innovative uses such as familial searching, and the analysis of offender mobility and crime patterns. Efforts are also being made to harmonize domestic and international DNA laws and policies while promoting the transnational exchange of DNA records.

Although these databases serve important public security interests, they raise substantial concerns regarding individual civil liberties. Key issues highlighted in existing literature include privacy threats, such as "function creep" and potential misuse of data, insufficient evidence on the overall effectiveness of DNA databases, and a significant public education gap regarding their purposes and implications.

27.2.2 Approaches to transnational DNA data exchange

The exchange of forensic DNA data across borders is facilitated through four primary approaches: (1) international DNA databases, (2) linked national DNA databases, (3) request-based data exchange, and (4) a hybrid combination of these methods. As detailed in subsequent sections, most nations engaged in forensic data sharing adopt a hybrid system (Amankwaa, 2019).

27.2.2.1 International DNA databases

International DNA databases are either global or regional in scope. A notable global initiative is the INTERPOL DNA Gateway, launched in 2002. By December 2017, this platform included 173,000 DNA profiles submitted by over 84 participating nations (INTERPOL, 2018a). The database contains profiles from convicted individuals, suspects, missing persons, unidentified remains, and crime scenes. However, it excludes personal data and is regulated by the domestic laws of the submitting country. For example, UK-submitted profiles adhere to the *Protection of Freedoms Act 2012* (Protection of Freedoms Act, 2012b). The INTERPOL database has proven effective in transnational crime investigations, such as Project Pink Panther, which led to the arrest of a network involved in jewelry theft between 2007 and 2016 (INTERPOL, 2018b).

At the regional level, the Europol Information System (EIS) serves as a significant repository for DNA and other biometric data within the European Union (EU). As of 2017, the EIS contained over 147,000 DNA profiles. Plans to expand its capabilities through partnerships, such as the EU Prüm framework, aim to bolster its transnational data exchange potential (Europol, 2013).

27.2.2.2 Linked national DNA databases

The EU Prüm framework represents a collaborative model that interlinks national DNA databases to facilitate seamless cross-border data sharing. Established in 2005 and incorporated into EU legislation by 2008, Prüm enables member states to exchange information on DNA, fingerprints, and vehicle registrations. Its streamlined two-stage process starts with a "hit/no-hit" query, which is followed by detailed information sharing if a match is identified. Unlike the INTERPOL system, Prüm achieves remarkable efficiency, enabling data exchanges in as little as 15 minutes (Home Office, 2015a).

A cornerstone of the Prüm regime is the requirement for each member state to establish a National Contact Point to coordinate and manage the exchange of intelligence. By 2018, 24 EU nations had actively engaged in DNA data exchanges under this framework, though some countries, including Croatia, Ireland, Italy, and the UK, were yet to fully implement their operational systems (Council of the European Union, 2017).

In the European context, the UK's National DNA Database (NDNAD) is particularly notable for its scale, containing over 6 million subject profiles and 590,000 crime scene profiles. Its vast size is largely attributed to earlier practices that permitted indefinite retention of profiles for minor offenses, a policy revised following the landmark *S. and Marper v. United Kingdom* case in 2008 (S et al., 2008). The subsequent legal reforms introduced more stringent retention guidelines, yet the UK remains a pivotal contributor to transnational DNA data sharing. In 2015, it formally joined the Prüm system, further solidifying its role in collaborative forensic intelligence.

27.2.2.3 Request-based DNA data exchange

Request-based DNA exchanges operate through bilateral agreements, granting conditional access to national DNA databases for law enforcement purposes. These exchanges emphasize the principles of necessity, relevance, and proportionality, with a primary focus on addressing serious crimes (Home Office, 2015b). While the approach mirrors the two-stage process of the Prüm framework, it is generally slower and less efficient (Wiles, 2017).

In the UK, the National Crime Agency oversees request-based DNA exchanges, often utilizing INTERPOL's I-24/7 network as a communication channel. Despite its utility, the system faces challenges, including delays in data sharing and a limited number of profile exchanges. For instance, in 2017, the UK shared only 23 subject profiles and 166 crime scene profiles internationally, resulting in a small number of positive matches (Wiles, 2018). Additionally, procedural lapses, such as unauthorized database searches and the improper sharing of personal data, have been identified but are reportedly being addressed through corrective measures.

The United States also employs the request-based model under bilateral agreements with countries such as Argentina, Australia, and several EU member states (Agreement Between the United States of America and Argentina, 2016). These agreements, inspired by the Prüm framework, prioritize serious criminal investigations while upholding the principle of reciprocity. However, the extent to which these agreements have been fully implemented by the US remains uncertain, leaving room for further evaluation and optimization.

27.2.2.4 Genetic genealogy databases

Genetic genealogy is the practice of combining DNA testing with conventional genealogical techniques to trace ancestry, determine biological relationships, and identify family connections. This process examines genetic markers to construct family trees, explore lineages, and uncover details about heritage. It serves as a valuable tool in forensic science, historical research, and personal ancestry investigations (Sharman, n.d.).

Investigative genetic genealogy (IGG) is an advanced forensic method that combines DNA profiling with genealogical research to identify individuals or resolve criminal cases by tracing familial connections. This approach involves analyzing DNA profiles obtained from crime scenes and comparing them against genetic data stored in publicly accessible databases, including those managed by direct-to-consumer testing services. By identifying genetic relatives and constructing family trees through genealogical investigation, authorities can narrow their focus to potential suspects. Data may be accessed from commercial companies with appropriate legal authorization, and databases such as GEDmatch facilitate cross-platform comparisons of genetic information. By 2019, over 26 million individuals worldwide had genetic profiles stored in one or more such databases (Fletcher, 2024).

In several countries, genetic data is classified as sensitive personal information and is protected under privacy and data protection regulations. When law enforcement agencies access such data from genealogy databases—where users initially uploaded it for a specific purpose—they may be stepping into a legal gray area. This issue raises concerns not only about the privacy of suspects but also about the relatives whose genetic data is accessed without their consent.

In the Swedish pilot study, authorities concluded that the benefits of removing a violent criminal from society outweighed the privacy concerns. A report following the study stated that the method proved effective enough to be used in certain circumstances, provided it is applied with "extreme care." However, the dilemma remains: where should the line be drawn between ensuring public safety and safeguarding personal privacy when it comes to genetic data?

In the UK and most European nations, there is still no clear regulation governing this practice, while in Germany, the use of IGG is explicitly prohibited (Samuel & Kennett, 2020). Furthermore, most genetic databases require law enforcement to obtain a warrant to access this sensitive information. In cases where a warrant is not necessary, individuals typically have the option to opt out of sharing their data with authorities. Under the GDPR, all individuals in the EU are automatically opted out. Consequently, IGG has seen far more widespread use in the United States, where regulations are more permissive.[3]

A major ethical concern with the use of genetic genealogy for criminal investigations is the potential invasion of privacy. Individuals typically upload their genetic data to genealogy databases for personal purposes, such as researching family history, but when law enforcement accesses this data, it raises important questions about whether users have given informed consent and whether they fully understood the potential future use of their genetic information. This issue extends to consent and autonomy, particularly when individuals or their family members may not have explicitly agreed to share their data with law enforcement (Takashima et al., 2018). Furthermore, genetic genealogy not only affects the person who uploads their data; relatives' genetic information can also be accessed, potentially violating their privacy, especially if they are not involved in any criminal activities. The security of genetic data is another pressing concern, as misuse or theft of this sensitive information could lead to identity theft, discrimination, or other unauthorized uses (Berkman et al., 2018). Additionally, the practice may unintentionally amplify biases or result in racial profiling, as certain genetic traits are more prevalent in specific ethnic groups, leading to unfair targeting during investigations. The lack of clear regulations in many regions only adds to the ethical concerns, as it allows for inconsistent and potentially unlawful use of genetic data by law enforcement (Ahmed, 2019). This absence of regulation may erode public trust in genetic databases, as individuals could be hesitant to share their information if they fear it may be used for criminal investigations without their consent. Finally, the practice may lead to "fishing expeditions," where law enforcement searches genetic data without a specific suspect, potentially expanding the scope of investigations unfairly and collecting data from innocent individuals.

27.2.2.5 Challenges and recommendations

The transnational exchange of DNA data is vital for tackling cross-border crimes but faces several challenges, including disparities in legislation, variations in database capabilities, and procedural inefficiencies. Studies recommend enhanced transparency, accountability, and follow-ups on confirmed DNA matches to improve the effectiveness and sustainability of these systems. For instance, analyzing confirmed DNA matches in the criminal justice process and improving stage-two operations in the Prüm system can foster greater trust and operational efficiency. Addressing these challenges will be crucial to leveraging the full potential of DNA data in transnational crime prevention.

27.2.3 Risks of privacy invasion

Unauthorized access, genetic privacy, and concerns about data misuse.

The battle for privacy in the 21st century extends far beyond the realm of government regulation and enforcement. It now encompasses private entities and commercial enterprises that collect, trade, and manipulate personal data—often

3. Id.

without the informed consent of individuals—for third-party use. These massive data repositories are frequently amassed under the guise of providing welfare services. Genetic privacy has emerged as a critical policy concern as personal genetic information becomes a valuable resource in developing goods and services. Policymakers must carefully evaluate restrictions on the future use of genetic data, especially as the risks of cybersecurity breaches, manipulation through targeted content delivery, and unforeseen consequences of machine learning applications continue to grow in our data-driven economy. The rising popularity of at-home genetic testing, coupled with the potential misuse of genetic information, heightens privacy concerns and raises pressing questions about the exploitation of personal genetic data.

27.2.3.1 Genetic discrimination, privacy, and its ramifications

The public availability of genetic data has profound implications, both positive and troubling. In May 2018, California law enforcement announced a breakthrough in the case of the "Golden State Killer," a notorious serial killer linked to 13 homicides (Chavez, 2018). Investigators identified the suspect, James DeAngelo, by leveraging a publicly accessible genetic database. DNA evidence from the crime scene was matched to a relative who had used GEDmatch, a genealogy website (Lussenhop, 2018). This case highlights a critical issue: when individuals make their genetic information public, they inadvertently expose their relatives' genetic data as well (Ferguson, 2017). This phenomenon creates a network effect, akin to how logging into Facebook can reveal information about other users and even non-users to third parties.

While the benefits of publicly available genetic databases are evident, such as aiding in criminal investigations, there are significant drawbacks. One alarming example is the potential misuse of genetic data in predictive policing. Predictive policing relies on algorithms that analyze crime data and police interactions to forecast future crimes and their perpetrators. However, the effectiveness of these models is only as reliable as the data they process, which often reflects biases and assumptions in its collection and programming.

For instance, over-policing in minority communities results in biased crime data, falsely identifying these groups as more likely to commit crimes (Gazzar, 2018). This creates a harmful feedback loop where minorities are disproportionately targeted based on attributes like their neighborhood or educational background, even if they have no criminal history. Reports from ProPublica reveal that CJSs increasingly use "risk-based assessments," where software predicts an individual's likelihood of reoffending (Angwin et al., 2016a). These scores influence sentencing and parole decisions, yet minorities consistently receive higher risk scores than white individuals, even when comparing white offenders to minority individuals without criminal records (Angwin et al., 2016b).

The inclusion of genetic data in such algorithms would exacerbate these disparities. Racial minorities, who are overrepresented in genetic databases due to frequent interactions with the CJS, are more vulnerable to false positives. This could lead to increased instances of wrongful accusations when genetic mismatches occur, deepening the systemic bias against minority communities.

Cathy O'Neill, in her book *Weapons of Math Destruction* (O'Neil, 2017), illustrates how unaccountable algorithms now determine critical life outcomes, such as health scores, credit ratings, and loan approvals, often relying on flawed metrics and biased assumptions. This uncritical adoption of algorithmic "science" lacks proper validation and oversight. Similarly, journalist Charles Seife, writing for *Scientific American*, highlights how companies like 23andMe (Mukherjee & Fry, 2018) analyze their genetic databases combined with user-submitted information to uncover genetic links to human traits (Seife, 2013).

This raises serious concerns about the misuse of genetic data in algorithms. For instance, if a genetic testing company identifies a supposed link between a genetic variant and violent behavior, this information could be presented as factual, similar to how ancestry companies often reduce identity to arbitrary percentages. Such "findings" might then be integrated into predictive policing algorithms, leading to a new form of genetic stereotyping. Individuals with particular genetic markers could face harsher treatment in the CJS, irrespective of their actual behavior.

This reductionist approach shifts the focus from broader societal, legal, and structural factors that influence violence to an individual's genetic makeup. Such genetic stereotyping could extend beyond criminal justice, influencing decisions on credit scores, mortgage approvals, and other life-altering outcomes, perpetuating discrimination under the guise of science. Addressing these issues requires a critical examination of the ethical, legal, and societal implications of genetic data usage, ensuring that technology serves justice rather than reinforcing prejudice.

27.2.3.2 Anonymity and genetic data

Anonymity is far from guaranteed when submitting your genetic data; it can still be used to identify you without your consent or knowledge. If people were fully aware of this possibility, many might reconsider using genetic testing services. The potential for breaches in genetic privacy is not hypothetical. Cyberattacks have already compromised customer data from major companies like Adidas, Sears, and Delta (Hanbury & Green, 2018). This raises serious concerns about genetic information stored in commercial databases being hacked. Stolen genetic data could enable third parties to identify individuals, posing far-reaching privacy and security risks.

The Cambridge Analytica scandal serves as a stark reminder of how companies, whether through negligence or deliberate actions, can violate their own privacy policies. Similarly, genetic testing companies are not immune to such breaches. The Federal Trade Commission (FTC) has cautioned consumers to carefully evaluate the privacy implications of home genetic tests, highlighting that even a small sample, such as a cheek swab, can reveal the intricate biological details that define a person (Maryland & King, n.d.).

Unlike financial breaches—where compromised credit card details can be replaced, and losses insured—genetic privacy breaches are irreversible. Once your genetic information becomes public, it remains exposed, potentially forever. This permanence underscores the urgent need for stringent safeguards and informed decision-making when dealing with genetic data.

27.2.3.3 Balancing privacy, policy, and the constitution

In a significant ruling, the US Supreme Court addressed the question of whether the Fourth Amendment permits DNA collection upon arrest (Maryland & King, n.d.). In a closely divided 5-4 decision, with Justice Breyer siding with the conservative bloc and Justice Scalia aligning with the liberals, the Court upheld the practice. While the decision aligns with certain public policy and scientific considerations, its legal reasoning has drawn criticism.

The case itself is straightforward. In 2009, Alonzo King was arrested for second-degree assault, a felony in Maryland. During booking, law enforcement collected a buccal swab of his DNA without a warrant, as permitted by Maryland's DNA Collection Act. King's genetic profile later matched evidence from a 2003 rape case, leading to his conviction. The key issue before the Court was whether the warrantless collection of King's DNA violated his Fourth Amendment rights.

The majority opinion centered on the reasonableness of the search, given that King was arrested for a serious crime. The Court found the swab minimally intrusive and justified by the circumstances, emphasizing its value as a tool for identification. However, while the policy outcomes of this decision may seem compelling, solving cold cases and potentially exonerating wrongfully convicted individuals—the legal analysis raised concerns.

From a public policy perspective, the benefits of DNA collection are evident. Genetic profiles, which consist of numerical data rather than full sequences, are less invasive than commonly feared. Buccal swabs are simple and nonintrusive, and laws generally prohibit misuses like familial searches. Moreover, the majority opinion demonstrated a reasonable understanding of the science, distinguishing between coding and non-coding regions and correctly explaining how CODIS profiles are constructed.

However, the legal foundation of the ruling is less solid. The Court's determination that buccal swabs are not unduly intrusive, and therefore reasonable, is based on their being less invasive than drawing blood, as noted in Missouri v. McNeely. While this distinction may seem logical at first glance, it creates inconsistencies within Fourth Amendment jurisprudence. The amendment's focus on bodily integrity suggests that the degree of intrusion is subjective and difficult to quantify. Lower courts now face the unenviable task of reconciling seemingly arbitrary distinctions between acceptable and unacceptable intrusions into personal privacy.

Additionally, the Court's reliance on probable cause to justify the swab complicates the broader legal landscape. While probable cause may validate the search of a person arrested for a serious crime, it contrasts sharply with the standard for Terry stop-and-frisks, which require less than probable cause and permit searches only for weapons. This inconsistency raises the question of why an individual's genetic material should be less protected than the contents of their pockets.

Most troubling is the Court's reasoning that the DNA collection was primarily for identification purposes. This argument rings hollow, as King's identity was never in question. The DNA sample was clearly intended to link him to unsolved crimes. The majority's historical appeal to outdated practices like Bertillonage—a discredited 19th-century method of criminal identification rooted in proto-eugenics further undermines its argument, casting an unfortunate shadow over its rationale.

Justice Scalia, in a sharp dissent, highlighted these flaws. He dismantled the majority's arguments, pointing out that the DNA samples were neither processed immediately nor used to confirm identity in the traditional sense. Instead, they were matched against unsolved crime databases, contrary to the stated purpose of identification. Scalia's critique underscores the contradictions in the majority's reasoning, questioning whether the Court's theory of identification holds any factual basis (Scalia, n.d.).

This case exposes a tension between public policy and constitutional rights, scientific precision and legal interpretation, and ideological divides on the Court. While the decision may align with the goals of law enforcement and justice, it leaves unresolved questions about the balance between privacy and state power in the age of genetic technology (Sherkow, 2013).

27.2.4 Balancing public safety with individual privacy rights

Ethics and morality are integral aspects of human existence, shaping not only personal behavior but also societal norms and laws. Ethics can be viewed as one side of a coin, with life on the other, forming a complex relationship. While ethics is often analyzed from a legal perspective, its relevance extends to technoscience, serving as a logical framework for examining actions and their alignment with standards of humanity. This framework incorporates mental elements and universal principles that should guide behavior.

The concept of ethics is intricate, reflecting a long-standing debate rooted in societal perspectives. Morality, in essence, stems from ethics, manifesting as external expressions of logical good or illogical bad behavior. Ethics, as described by Gowdy (2006), is a set of values that guides choices and actions, determining life's purpose and direction. Technological and legal issues, which are moral at their core, arise in diverse situations, often reflecting a conflict between technological advancements and respect for human dignity, as guaranteed under Article 21 of the Indian Constitution.

In many cases, technology prioritizes efficiency and utility over individual quality of life, even as it aims to serve the greater good. The central ethical challenge lies in striking a balance between an individual's right to privacy and societal interests. This requires carefully navigating the line between public and private interests, particularly as the issue of causation becomes relevant. Causation refers to the proximity and connection between an act and its consequences, helping to ascertain whether specific actions led to particular outcomes. Protecting the privacy of human life, including personal data and information, is fundamental to upholding moral values.

The ethical obligation to safeguard individuals is not new. It finds roots in ancient principles such as the Hippocratic Oath, which emphasized the physician's duty to avoid harm to patients. However, this oath has often been disregarded in practice, with human subjects treated as mere experiments. Advances in DNA analysis and collection, for instance, have the potential to revolutionize societal structures, particularly the judicial system, by improving evidence gathering, aiding research, and enhancing justice delivery.

Yet, these advancements are not without risk. The misuse of DNA analysis can lead to severe harm, particularly when it is weaponized against individuals or their families. Practices such as familial searches or framing vulnerable individuals to achieve specific outcomes highlight the dangers of unchecked technology. While DNA itself is impartial, human actions and intentions are not.

Ultimately, the ethical challenge is to balance the promise of DNA analysis as a tool for societal advancement with the moral imperative to protect individual rights and dignity. By addressing these concerns, society can ensure that technological progress does not come at the expense of humanity.

27.3 Consent and autonomy in DNA collection

The foundation of modern regulations governing informed consent in the United States can be traced back to the 1930s. During this period, Nazi physicians carried out heinous experiments on human subjects without their consent, particularly before and during World War II. The atrocities uncovered during the Nuremberg Trials led to the development of the Nuremberg Code—a landmark set of 10 ethical principles designed to guide physicians and researchers in conducting human experimentation. Central to these principles was the concept of informed consent.

However, informed consent in the US gained significant attention in the wake of the Tuskegee Syphilis Study, which was conducted between 1932 and 1972. In this government-led study, African American men were subjected to unethical experimentation, being deliberately denied treatment for syphilis without their informed consent. The public outcry following the exposure of these egregious actions spurred the establishment of the National Commission for the Protection of Human Subjects of Biomedical and Behavioral Research in 1974. Four years later, this commission released the Belmont Report, which underscored the critical importance of informed consent in research involving human subjects (National Human Genome Research Institute, n.d.).

Building on the Belmont Report's principles, the US Department of Health and Human Services formalized these ethical standards in the 1980s. They developed a comprehensive framework, now known as the "Common Rule," which outlines detailed guidelines for research ethics and mandates the requirements for informed consent.

Informed consent reflects the principle of Respect for Persons, a cornerstone of the Belmont Report. It recognizes individual autonomy and ensures that research participants are adequately informed about the study they are partaking in, including its purpose, risks, and potential benefits. This process is especially vital in research, as it distinguishes research objectives which often aim to generate broader societal benefits from clinical care, which focuses on individual patient welfare.

In DNA analysis, informed consent holds particular importance due to the inherently sensitive nature of genetic information. Even in situations where formal consent may not be legally mandated, researchers bear an ethical responsibility to ensure participants are fully informed, respecting their autonomy. It is crucial that participants comprehend the intended use of their DNA, whether it will be preserved for future studies, and the potential ramifications for privacy and familial relationships. This commitment to transparency is not merely a legal or ethical obligation but also a vital mechanism for building trust between researchers and participants. By prioritizing clear communication, researchers honor the fundamental principle of respecting individual rights in scientific exploration.

27.3.1 Legal implications

The evolution of DNA analysis has brought critical ethical concerns to the forefront, including autonomy, confidentiality, privacy, and equity. These concerns are heightened by the diverse contexts in which tests are conducted, the extensive range of potential genetic analyses, and the various institutions managing genetic data. Traditionally linked to reproductive health or newborn screening, genetic testing now extends to identifying predispositions, non-disease traits, and late-onset conditions like Huntington's disease.

Challenges arise as DNA analysis moves beyond the physician-patient relationship, involving research, public health, law enforcement, military, and even employers and insurers. Existing legal protections often fail to address the nuances of these contexts, leading to inconsistencies in respecting patient rights. Some healthcare providers conduct unauthorized tests, while others disclose genetic information without consent, highlighting disparities in upholding confidentiality and autonomy.

Further, institutions storing DNA samples and test results vary significantly in their practices, with some conducting unauthorized analyses or sharing data. Emerging technologies, such as medical information cards with potential genome storage, exacerbate privacy concerns, underscoring the urgent need for robust consent standards, safeguards, and ethical practices to manage future uses of genetic material responsibly.

27.4 Racial and socioeconomic bias in DNA databases

The debate over forensic DNA databases raises important ethical questions about genetic privacy and racial equity. While the government's power to compel DNA samples may be seen as an intrusion, the availability of genetic data from commercial services like 23andMe and Ancestry.com (Regalado, 2019) has shifted the conversation (Roberts, 2011). Familial DNA searches turn national databases into powerful tools for identifying potential relatives of individuals whose DNA is already in the system. However, these searches can disproportionately affect certain racial or ethnic communities if their representation in the database is skewed. This can lead to increased suspicion of specific groups, particularly if DNA profiles are retained indefinitely, compounding privacy and equity concerns over time. As future generations' profiles are added, the impact of such disparities grows, especially when familial searches remain a tool for law enforcement. Some argue that universal DNA databases could reduce racial bias in law enforcement, but critics contend that such databases won't address the deeper issue of racial disparities in policing. Even if the data exists elsewhere, concerns about racial targeting and the use of criminal law as a tool of oppression persist (Murphy & Tong, 2019).

27.5 Accuracy and reliability in DNA analysis

DNA evidence is often regarded as one of the most reliable tools in criminal investigations due to its high accuracy in identifying perpetrators. However, there are significant limitations that must be considered, particularly related to contamination, partial DNA profiles, and statistical interpretation. Contamination occurs when foreign DNA is introduced into a sample during any stage of handling, storage, or analysis. This introduces the possibility of false conclusions. Partial DNA profiles arise when only fragmented pieces of DNA are recovered, complicating their interpretation and potentially leading to ambiguous results that may sway the justice process.

In the 2003 People v. Puckett case[4] in California, partial DNA profiles were crucial to the defense's argument. The evidence from the crime scene did not fully match the defendant's DNA, but there were partial markers similar to Puckett's. While the prosecution suggested these markers indicated his involvement, the defense warned against drawing

4. People and Puckett, C040743 (Cal. Ct. App. November 5, 2003).

conclusions from incomplete data. This case highlights the risks of over-relying on partial DNA profiles, as their incomplete nature can easily lead to inaccurate conclusions about a suspect's involvement.

Another concern is the statistical interpretation of DNA evidence. Forensic scientists calculate the probability of a match using population databases, but these numbers can be influenced by the quality of the sample and the methods employed in analysis. When the DNA is partial or contaminated, the resulting probability may be misleading. A 1 in a million match might seem conclusive, but if the sample is compromised, the match may not be as reliable as it appears. Misleading statistical claims can create a false sense of certainty in court.

Ethically, these limitations raise concerns about wrongful convictions. DNA evidence is often regarded as definitive proof, which may lead courts to place undue reliance on it. When incomplete or tainted DNA evidence is presented as solid proof of guilt, there is a risk of wrongful conviction, undermining justice. This underscores the importance of courts carefully examining all available evidence and not relying solely on DNA analysis to make crucial decisions.

In conclusion, while DNA evidence is a powerful tool, its limitations—such as contamination, partial profiles, and statistical misinterpretation—must be considered to avoid miscarriages of justice. The People v. Puckett case serves as a reminder of the potential for misapplication and the ethical challenges in relying too heavily on DNA evidence in criminal cases. It is essential for the legal system to balance DNA evidence with other corroborating evidence to ensure fairness and accuracy in criminal justice proceedings.

27.6 Ethical issues in genetic privacy and future implications

The use of genetic data in forensic investigations presents growing concerns about health privacy. While designed to assist in solving crimes, forensic databases may also be repurposed for health-related purposes, raising privacy risks. Law enforcement's access to private genetic databases, such as those from ancestry services, exacerbates these concerns.

Genetic discrimination is a significant worry. Insurers or employers might misuse genetic data to deny coverage, increase premiums, or make biased decisions about employment, based on an individual's genetic predispositions. This potential for misuse underscores the need for stronger privacy protections.

The 2018 case involving the FBI's use of GEDmatch, an ancestry database, to solve cold cases exemplifies these issues. Individuals who voluntarily submitted their DNA for genealogical purposes were unaware that their data could be accessed for law enforcement. While this led to breakthroughs in solving crimes, it raised ethical concerns regarding consent and the repurposing of personal genetic information.

Advancements in DNA analysis heighten these concerns. Modern forensic techniques not only identify individuals but also uncover health-related genetic information, such as predispositions to diseases like cancer and Alzheimer's. This makes sensitive health data vulnerable to exposure without consent, presenting significant privacy risks.

The current legal landscape includes protections like the Genetic Information Non-discrimination Act (GINA), which prevents genetic discrimination in health insurance and employment. However, these laws do not fully address the broader risks, especially when genetic data is used outside of health contexts, such as in criminal investigations or by private companies. As genetic technology continues to evolve, it is essential to strengthen legal frameworks to ensure privacy, regulate usage, and protect individuals from genetic discrimination.

27.7 International legal and ethical standards in DNA forensics

Since the introduction of forensic DNA profiling and the creation of DNA databases, efforts have been made to enhance the effectiveness and utility of this technology. These efforts are driven by the belief that DNA analysis should be cost-effective while providing clear, tangible benefits. The practice of storing DNA profiles—whether from individuals involved in criminal investigations or recovered from crime scenes—has become widespread across the globe.

27.7.1 Comparison of laws: overview of DNA forensic laws across countries

27.7.1.1 United Kingdom

The UK's NDNAD, established in 1995, is one of the largest and oldest legal DNA databases globally. While it has led to higher conviction rates, its broader impact on public safety remains uncertain. Efforts to assess the effectiveness of the database suggest that while it aids in crime detection, its influence on overall public security goals is still unclear. Despite its extensive reach, the full benefits and long-term effectiveness of the NDNAD are still under scrutiny.

The UK was the first to pioneer the use of DNA profiling in 1988, initially for immigration disputes, later realizing its broader potential within the CJS. In 1995, the UK Home Ministry tasked the Forensic Science Service with establishing a national DNA criminal intelligence database, which became the largest in the world.[5] Initially, DNA databasing was limited to convicted individuals, with records of the unconvicted destroyed. However, after legal challenges arose due to unlawful database entries, the law was amended to include DNA profiles from arrested individuals, regardless of conviction. By 2001, the database was expanded to cover all arrestees, leading to concerns about privacy and ethics, particularly regarding the inclusion of innocent people. Despite the database's growth, its effectiveness was questioned, as only a small percentage of crime scenes yielded useful DNA evidence. In 2008, the European Court of Human Rights ruled the database's broad retention system was disproportionate. As a result, the Protection of Freedoms Act 2012 was enacted, mandating the destruction of DNA samples within six months, with retention allowed only during investigations or proceedings (Protection of Freedoms Act, 2012a).

Privacy concerns have been central in discussions about the effectiveness of DNA databases, especially regarding the retention of data from innocent individuals (McCartney, 2006). A successful DNA database should balance public security goals with the protection of human rights, particularly the right to privacy. Evaluating the effectiveness of such databases involves considering their ability to safeguard genetic privacy while serving legitimate public safety interests. Additionally, the cost and effort required to maintain DNA databases should be justified, with attention given to the broader resources needed for crime control, and the ethical and public trust aspects of their operation.

27.7.1.2 United States

Forensic DNA data banking began as a pilot project in the USA in 1990, serving fourteen states and local communities after its success in the UK. The FBI's National CODIS was created to link state DNA databanks and collect the DNA of convicted offenders and repeat sex offenders. Over time, however, the scope expanded to include individuals such as undocumented immigrants and misdemeanants who had neither been charged nor convicted of crimes, further broadening the range of individuals included in the system.

27.7.1.2.1 Fourth amendment and DNA databasing

Courts have consistently ruled that taking blood or DNA samples from a suspect without a warrant, based solely on the possibility of finding incriminating evidence, violates the Fourth Amendment. This amendment protects against unreasonable searches and seizures, requiring probable cause for any warrant. DNA collection, especially when suspicion is minimal and the intrusion significant, typically invokes Fourth Amendment protections. However, as suspicion rises and the method of collection becomes less intrusive, such as using a cheek swab instead of blood, the level of protection diminishes. Despite this, many jurisdictions allow the collection of DNA without a warrant, especially when law enforcement suspects a person may be involved in a crime. DNA evidence can exonerate or convict, with its power to exonerate often proving more impactful than its use for conviction.

27.7.1.2.2 Familial searching and Genetic Information Non-Discrimination Act

While genetic privacy has been increasingly protected in medical and employment contexts through the GINA of 2008, the CJS follows a different trend. Courts treat forensic DNA profiles similarly to fingerprints, viewing them primarily as a form of identification. DNA samples can now be collected with minimal intrusion, such as viacheek swabs, allowing police to gather DNA without a warrant in many cases.

GINA prevents insurance companies and employers from using genetic information for decisions about enrollment or hiring. However, DNA can reveal sensitive details, such as genetic disorders, family connections, and disease predispositions. Critics argue that if biological samples were destroyed after the DNA profile is obtained, many privacy concerns could be alleviated, as practiced in countries like Germany and Japan (Andrews et al., 1994).

Familial searching, a technique used to find close matches in DNA databases, raises significant privacy issues. It can lead to the identification of relatives of a person in the database, without reasonable suspicion. This practice has been criticized for disproportionately affecting certain communities, such as Black and Hispanic populations, which are overrepresented in criminal DNA databases. While some U.S. states allow familial searching, others, like Maryland, have banned it. This technique should ideally be restricted by a court order to safeguard privacy (Krimsky, 2014).

DNA dragnets, where DNA samples are voluntarily collected from people within a certain demographic, have been used in unsolved cases, though they rarely lead to the identification of suspects. A key concern is that individuals who are

5. Criminal Justice and Police Act 2001, Sect. 82. (2001). UK Legislation. Retrieved May 14, 2020, from http://www.legislation.gov.uk/ukpga/2001/16/contents.

excluded as suspects often remain in the database, with their DNA continuously monitored. Many states do not guarantee the removal of DNA profiles or biological samples from volunteers, raising further privacy issues. Ultimately, courts will likely continue to address unresolved questions regarding forensic DNA databases.

27.7.1.3 Australia

Australia has established three main DNA databases for law enforcement. The National Criminal Investigation DNA Database (NCIDD), launched in June 2001, supports both intra- and inter-jurisdictional DNA matching (Australian Law Reform Commission, 2010). The Disaster Victim Identification (DVI) Database, created in October 2002, aids in identifying victims of events like the Bali bombings. Additionally, the Australian Federal Police maintains its own DNA database (Australian Criminal Intelligence Commission, 2000). The CrimTrac Agency operates both the NCIDD and DVI databases under the Crimes Act 1914, with an inter-governmental agreement between all Australian police ministers.[6]

The NCIDD allows law enforcement to access DNA profiles from convicted individuals, facilitating the matching of crime scene samples. It connects all Australian states and territory' databases, enabling inter-jurisdictional sharing. However, there are concerns regarding data security during transfer and a lack of provisions for removing police officers' DNA profiles from the database. The database has grown substantially, holding 830,000 profiles as of 2014, with capabilities like familial searching (DNA databases in Australia and New Zealand, 2016). Despite its advantages, the system faces challenges with accuracy, contamination, and the de-identification of profiles. Each jurisdiction has specific legislation governing the handling of DNA samples.

27.7.2 Challenges in harmonization

Harmonization challenges in forensic DNA analysis stem from regional variations in ethical standards, resulting in inconsistencies in justice. These differences can undermine the reliability of evidence, procedures, and outcomes, demonstrating the need for standardized ethical guidelines and practices. Such uniformity would help ensure fairness and accuracy in criminal investigations, fostering greater trust in forensic DNA analysis as a tool for justice. Some of the key areas to focus are as discussed (Smith et al., 2024):

1. **Infrastructure**: The importance of establishing foundational infrastructure for DNA analysis laboratories, engaging key stakeholders such as law enforcement agencies, forensic science institutes, the justice system, and accreditation bodies. Key infrastructure needs include improving systematic investigations into unidentified human remains and missing persons, enhancing crime scene support, ensuring advanced laboratory analysis capabilities, developing databases for criminal intelligence and allele frequencies, refining the interpretation and communication of scientific evidence, and supporting ongoing research initiatives. These efforts aim to standardize and strengthen forensic DNA analysis across the region.
2. **Expertise**: Establishing a baseline level of expertise for practitioners to ensure universal recognition of qualifications across countries. Given the multidisciplinary nature of DNA analysis, this expertise should encompass formal education in relevant scientific fields, along with both general and specialized training in DNA analysis. It should also include competency and proficiency testing, as well as ongoing professional development to ensure practitioners maintain current knowledge and skills in the evolving field of forensic science.
3. **Quality Standards**: Few forensic DNA laboratories have obtained formal accreditation in the region. Delegates emphasized the need for established minimum quality standards and accreditation criteria, referencing ILAC G19 and ISO standards (such as ISO17020, ISO17025, ISO18385, and ISO21043). National and regional accreditation bodies should define these standards. While formal accreditation may not be widespread, delegates underscored that all forensic DNA activities must adhere to international standards. Additionally, a robust quality management system is essential to ensure the reliability of scientific evidence and prevent the misuse of misleading results.
4. **Legal Policy and Framework:** The DNA database policy panel and delegates observed limited transnational DNA data exchange in the region, mainly due to the lack of dedicated legislative frameworks and inconsistent policies. They recommended harmonizing laws and policies to foster regional cooperation by: (a) adopting research-based core STR/SNP markers in laboratories, (b) establishing a legislative framework for transnational forensic data exchange within the nations, and (c) developing future-proof national laws on forensic evidence that address data collection, retention, sharing, protection, and emerging technologies like forensic IGG.

---

6. Standing Committee of Attorneys-General, Canberra. (June 30, 2010). DNA forensic procedures: Further independent review of Part 1D of the Crimes Act 1914. Model Criminal Code Officers Committee, Final Draft: Model Forensic Procedures Bill and the Proposed National DNA Database.

5. **Operational Framework**: Researchers emphasized the importance of an independent oversight framework for forensic practices, particularly DNA analysis, to ensure transparency, fairness, and accountability. Since forensic services in Africa and a few other countries are predominantly managed or funded by the police, an independent framework can reduce the risk of miscarriages of justice and uphold the integrity of forensic science. It was underscored that regardless of who commissions the forensic work, scientists must maintain impartiality and prioritize their ultimate duty to justice, ensuring that their findings are free from bias.

27.8 Policy recommendations and future directions

27.8.1 Strengthening informed consent processes

Informed consent should be a dynamic and participant-centered practice, fostering empowerment and understanding. Researchers must provide participants with comprehensive yet clear explanations of the intended purpose, scope, and potential implications of DNA usage. Consent forms should allow individuals to specify how their genetic data can be used, including conditions for future research. Regular re-consent mechanisms are essential to keep participants informed about new developments in the handling or usage of their DNA. Integrating user-friendly digital tools can further enhance accessibility and engagement, making the consent process transparent and inclusive.

27.8.2 Improving oversight and regulation

Protecting the integrity of DNA databases necessitates a robust, multi-tiered regulatory structure. Policies should mandate the anonymization and encryption of genetic data to prevent misuse and breaches. Independent oversight bodies must conduct routine audits and enforce compliance through penalties. Detailed protocols should clearly define permissible sharing of genetic data, especially across sectors such as healthcare, research, and law enforcement. Given the global nature of genetic research, international treaties and cooperative frameworks are vital for ensuring consistency, accountability, and ethical standards in the management of DNA data.

27.8.3 Balancing innovation and ethics

Harmonizing scientific progress with ethical imperatives requires collaboration among stakeholders. Governments and institutions should incentivize ethical practices by linking research funding to adherence to privacy and rights-based standards. Independent ethical review boards must meticulously assess research proposals to ensure risks are minimized and benefits equitably distributed. Open and ongoing communication between researchers, policymakers, and the public is crucial for fostering trust and understanding. Transparent practices in data governance, such as publishing research findings and ethical implications, can help advance innovation responsibly while respecting individual dignity and privacy.

27.9 Conclusion

In the age of technoscience, human life has been reframed as a collection of data, driven by reductionist life sciences. DNA technology has unlocked incredible potential—curing diseases, exonerating the innocent, and solving crimes—while promising human welfare and progress. However, this immense growth challenges concepts like bodily integrity, as genetic material has become a lucrative "corporate heritage."

Ethical concerns in forensic DNA analysis include informed consent, privacy, data ownership, and misuse of genetic information. Innovations in profiling and predictive testing heighten dilemmas, necessitating safeguards against discrimination and surveillance. Justice in DNA forensics must balance innovation with ethical integrity through transparent governance, robust regulations, and public engagement, ensuring trust and fairness in an evolving technoscientific landscape.

References

Agreement between the United States of America and Argentina. (2016, March 23). Retrieved from https://www.state.gov/documents/organization/279149.pdf.

Ahmed, A. (2019). Ethical concerns of DNA databases used for crime control. The Petrie-Flom Center Staff, 7th Annual Health Law Year in Review, Genetics, Health Law Policy. https://petrieflom.law.harvard.edu (accessed 04.01.25).

Amankwaa, A. O. (2019). Trends in forensic DNA database: Transnational exchange of DNA data. *Forensic Science Research, 5*(1), 8–14. https://doi.org/10.1080/20961790.2019.1565651.

Andrews, L. B.,, Fullarton, J. E.,, & Holtzman, N. A., (Eds.). (1994). *Assessing genetic risks: Implications for health and social policy.* National Academies Press.

Angwin, J., Larson, J., Mattu, S., & Kirchner, L. (2016a, May 23). Machine bias. ProPublica. https://www.propublica.org/article/machine-bias-risk-assessments-in-criminal-sentencing (accessed 14.11.18)

Angwin, J., Larson, J., Mattu, S., & Kirchner, L. (2016b). *Machine bias.* ProPublica.

Aronson, J. D. (2007). *Genetic witness: Science, law, and controversy in the making of DNA profiling.* Rutgers University Press.

Australian Criminal Intelligence Commission. (2000, July 13). Agreement for the establishment and operation of 'CrimTrac', a national law enforcement information system for Australia's police services. www.crimtrac.gov.au/misc/ct_agreement.PDF (accessed 23.05.24).

Australian Law Reform Commission. (2010). DNA database systems (ALRC Report 96/43). https://www.alrc.gov.au/publication/essentially-yours-the-protection-of-human-genetic-information-in-australia-alrc-report-96/43-dna-database-systems/dna-database-systems/. (accessed 23.05.24).

Berkman, B. E., Miller, W. K., & Grady, C. (2018). Is it ethical to use genealogy data to solve crimes? *Annals of Internal Medicine,* 333. https://doi.org/10.7326/M18-1348 accessed 4 January 2025.

Chavez, N. (2018, June 2). DNA that led to Golden State Killer suspect's arrest was collected from his car while he shopped. CNN. Retrieved July 31, 2024, from https://edition.cnn.com/2018/06/02/us/golden-state-killer-unsealed-warrants (accessed 31.07.24).

Council of the European Union. (2017). Implementation of the provisions on information exchange of the "Prüm Decisions". Brussels, Belgium: Council of the European Union.

Dhikary, J. (2007). DNAo technology in administration of justice. Lexis Nexis Butterworths (A Division of Reed Elsevier India Pvt Ltd).

DNA databases in Australia and New Zealand. (2016, May). file:///C:/Users/HP/Downloads/Australia_and_New_Zealand.pdf.

Europol. (2013). Europol Information System (EIS) leaflet. https://www.europol.europa.eu/publications-documents/europol-information-system-eis-leaflet (accessed 06.06.24.

Federal Bureau of Investigation. (n.d.a). Measuring success. Combined DNA Index System (CODIS). http://www.fbi.gov/hq/lab/codis/success.htm. (accessed 02.03.24).

Federal Bureau of Investigation. (n.d.b). Measuring success. Combined DNA Index System (CODIS). http://www.fbi.gov/hq/lab/codis/success.htm (accessed 02.03.20).

Ferguson, A. G. (2017). Policing predictive policing. *Washington University Law Review, 95,* 1109–1189.

Fletcher, L. (2024, January 30). Genetic mysteries: Should your data be used to solve crimes? https://frontlinegenomics.com/genetic-mysteries-should-your-data-be-used-to-solve-crimes/#:~:text=This%20method%20involves%20comparing%20crime,can%20narrow%20down%20potential%20suspects (accessed 04.01.25).

Gazzar, B. (2018, February 4). Activists file lawsuit over LAPD's predictive policing program. Government Technology. http://www.govtech.com/public-safety/Activists-File-Lawsuit-Over-LAPDs-Predictive-Policing-Program.html (accessed 31.07.18).

Gowdy, L. N. (2006). *Definition of ethics, morals, virtue and quality* (1st ed.). Woven Strings Publishing.

Hanbury, D., & Green, M. (2018, July 20). If you shopped at these 15 stores in the last year, your data might have been stolen. Business Insider. http://uk.businessinsider.com/data-breaches-2018-4/#adidas-1 (accessed 31.07.18).

Home Office. (2015a). Government sets out case for joining Prüm. https://www.gov.uk/government/news/government-sets-out-case-for-joining-prum (accessed 05.07.24).

Home Office. (2015b). International DNA exchange policy for the United Kingdom. London, UK: Home Office.

INTERPOL. (2018a). Forensics: DNA. Retrieved March 23, 2024, from https://www.interpol.int/INTERPOL-expertise/Forensics/DNA.

INTERPOL. (2018b). Project Pink Panthers. https://www.interpol.int/Crime-areas/Organized-crime/Project-Pink-Panthers# (accessed 28.03.24).

Jayewardene, C. H. S. (1988). Forensic science and the law. *Forensic Science International, 36,* 41–45. https://doi.org/10.1016/0379-0738(88)90213-7.

Krimsky, S. (2014). Twenty years of DNA databanks in the U.S. Forensic Genetics Policy Initiative. http://dnapolicyinitiative.org/genewatch-forensic-dna/twenty-years-of-dna-databanks-in-the-u-s/ (accessed 20.05.24).

Lussenhop, J. (2018). Golden State Killer: The end of a 40-year hunt? *BBC News.* https://www.bbc.co.uk/news/world-us-canada-43915187.

Margoliash, J., Fuchs, S., Li, Y., Zhang, X., Massarat, A., Goren, A., & Gymrek, M. (2023). Polymorphic short tandem repeats make widespread contributions to blood and serum traits. *Cell Genomics,* 100458. https://doi.org/10.1016/j.xgen.2023.100458 accessed 4 January 2025.

Maryland and King. (n.d.). Oyez. https://www.oyez.org/cases/2012/12-207 (accessed 16.11.24).

McCartney, C. (2006). The DNA expansion programme and criminal investigation. British Journal of Criminology. https://www.researchgate.net/publication/31126414_The_DNA_Expansion_Programme_and_Criminal_Investigation (accessed 17.09.24).

Mukherjee, S., & Fry, E. (2018, March 19). Do DNA testing companies like 23andMe own your biological data? Fortune. http://fortune.com/2018/03/19/dna-testing-23andme-owns-biological-data (accessed 31.07.24).

Murphy, E.E., & Tong, J. (2019). The racial composition of forensic DNA databases. California Law Review, 108(Forthcoming). NYU School of Law, Public Law Research Paper No. 19-54. Available at SSRN: https://ssrn.com/abstract=3477974 or https://doi.org/10.2139/ssrn.3477974.

National Human Genome Research Institute. (n.d.). Why is informed consent required? National Institutes of Health. https://www.genome.gov/about-genomics/educational-resources/fact-sheets/why-is-informed-consent-required (accessed 16.11.24).

National Library of Medicine. (2014, December 3). Forensic views of the body. Visible Proofs. https://www.nlm.nih.gov/exhibition/visibleproofs/galleries/cases/blassie.html (accessed 03.05.24).

O'Neil, C. (2017). *Weapons of math destruction: How big data increases inequality and threatens democracy.* Harlow, England: Penguin Books.

Payseur, B. A., & Cutter, A. D. (2006). Integrating patterns of polymorphism at SNPs and STRs. *Trends in Genetics,* 424. https://doi.org/10.1016/j.tig.2006.06.009 accessed 4 January 2025.

Protection of Freedoms Act. (2012a). UK Legislation. http://www.legislation.gov.uk/ukpga/2012/9/contents/enacted (accessed 17.05.20).

Protection of Freedoms Act. (2012b). http://www.legislation.gov.uk/ukpga/2012/9/contents/enacted.

Regalado, A. (2019, February 11). More than 26 million people have taken an at-home ancestry test. *MIT Technology Review.* https://www.technologyreview.com/s/612880/more-than-26-million-people-have-taken-an-at-home-ancestry-test/.

Roberts, D. (2011). Collateral consequences, genetic surveillance, and the new biopolitics of race. *Howard Law Journal, 54*, 567–586.

S. and Marper v. The United Kingdom. (2008). http://www.bailii.org/eu/cases/ECHR/2008/1581.html. (accessed 11.03.24).

Samuel, G., & Kennett, D. (2020). The impact of investigative genetic genealogy: Perceptions of UK professional and public stakeholders. *Forensic Science International: Genetics,* 102366. https://doi.org/10.1016/j.fsigen.2020.102366 accessed 4 January 2025.

Scalia, J. (n.d.). Dissenting. Maryland v. King, slip op. at 12.

Scheck, B., Neufeld, P. J., & Dwyer, J. (2000). *Actual innocence: Five days to execution and other dispatches from the wrongly convicted.* Doubleday.

Sedley, S. (2005). Science & justice: DNA and the courts. *Science & Justice, 45*(2), 59–60. https://doi.org/10.1016/S1355-0306(05)71627-0.

Seife, C. (2013, November 27). 23andMe is terrifying, but not for the reasons the FDA thinks. Scientific American. https://www.scientificamerican.com/article/23andme-is-terrifying-but-not-for-the-reasons-the-fda-thinks (accessed 20.06.18).

Sharman, S. (n.d.). Genetic genealogy: DNA sleuthing to uncover familial relationships. (Everyday DNA Human Genomics, 30 March 2023) https://www.hudsonalpha.org/genetic-genealogy-dna-sleuthing-to-uncover-familial-relationships/ (accessed 04.01.25).

Sherkow, J. (2013, June 3). Maryland v. King: A mouthful of contradictions. Stanford Law School. https://law.stanford.edu/2013/06/03/lawandbiosciences-2013-06-03-maryland-v-king-a-mouthful-of-contradictions/.

Singh, S. C. (2011). DNA profiling and the forensic use of DNA evidence in criminal proceedings. *Journal of the Indian Law Institute, 53*(2). http://www.jstor.org/stable/43953503.

Smith, J. H., Lynch, V., Amankwaa, A. O., Budowle, B., Fonseca, S. A., Shako, S. K., & Molefe, I. (2024). 4th DNA Forensic Symposium: Challenges and future directions in Africa. *Forensic Science International: Synergy, 9*, 100555. https://doi.org/10.1016/j.fsisyn.2024.100555 Retrieved from https://www.sciencedirect.com/science/article/pii/S2589871X24001025.

Takashima, K., Maru, Y., Mori, S., et al. (2018). Ethical concerns on sharing genomic data including patients' family members. *BMC medical ethics,* 61. https://doi.org/10.1186/s12910-018-0310-5 accessed 4 January 2025.

Walsh, S. J. (2005). Legal perceptions of forensic DNA profiling: Part I: A review of the legal literature. *Forensic Science International, 155*, 51–60. https://doi.org/10.1016/j.forsciint.2004.11.001.

Wiles, P. (2017). *Annual report 2016: Commissioner for the retention and use of biometric material.* London, UK: Office of the Biometrics Commissioner.

Wiles, P. (2018). *Annual report 2017: Commissioner for the retention and use of biometric material.* London, UK: Office of the Biometrics Commissioner.

Chapter 28

DNA profiling and databases in criminal investigations: privacy-related issues and challenges

Oliver M. Tuazon[1], and Ray A. Wickenheiser[2]

[1]*Center for Law and Digital Technologies (eLaw), Institute for the Interdisciplinary Study of the Law, Leiden Law School, Leiden University, Leiden, The Netherlands,* [2]*New York State Police Crime Laboratory System, Forensic Investigation Center, (Retired during the course of writing this chapter; Ray Wickenheiser Forensic Consulting, Youngsville, LA), Albany, NY, United States*

28.1 Introduction

The first application of what was then called "DNA fingerprinting" in forensics was reported in 1985 (Gill et al., 1985). Forty years later, the use of DNA evidence in criminal investigations has become commonplace in many jurisdictions. It has displaced traditional fingerprinting as the "gold standard" in criminal identification techniques (Shelton, 2011).

Forensic DNA techniques have developed over time following the development and breakthroughs in DNA science and technology in general. Various DNA-based methods that are employed by law enforcement depend on the legal restrictions of their country, and at times, individual states of operation: short tandem repeat (STR) profile matching, familial DNA searching (FDS), forensic DNA phenotyping (FDP), mass DNA screening, Y-STR, mtDNA, and X chromosome searching (Wickenheiser, 2022). We have surveyed these techniques in relation to the relatively new technique, investigative forensic genetic genealogy (iFGG) (Tuazon et al., 2024). These methods are mainly based on two types of DNA profile data—STRs and single-nucleotide polymorphisms (SNPs)—and the corresponding DNA databases they utilize to help identify the source of the forensic sample: law enforcement databases and genetic genealogy databases.

This book chapter focuses on the issues and challenges that law enforcement agencies face in the use of STRs, SNPs, and their corresponding DNA databases in criminal investigations. More specifically, we identified technical, ethical and social, and legal and regulatory issues and challenges that revolve around data privacy issues. Given that an individual's DNA data is regarded to be "most intimate" (DNA Confidential, 2009) and that it is "uniquely identifiable" (McGuire, 2008), genetic data privacy has become a cause of great concern to many, including their access by law enforcement (Clayton et al., 2019). We present these privacy-related issues and challenges, and whenever applicable and relevant, we provide possible solutions.

We clarify from the outset the distinction among "forensic DNA samples" (those gathered at the crime scene from unknown donors), "abandoned DNA samples" (from known individuals who are unknowingly being investigated by the police as suspects) and "known DNA samples" (from known individuals whose DNA are knowingly collected by the police, say from arrestees and convicts). Forensic DNA samples have been "discarded" or left behind during the perpetration of a crime, and law enforcement has the full right—and in actuality, a duty—to interrogate these samples to determine who committed the crime, given that the suspect is at large and may commit new crimes on the same or new victims. We opine that the unknown donors of these forensic DNA samples have abandoned their right to privacy regarding their DNA samples by committing a crime (Wickenheiser, 2019, 2022). It is however possible that the DNA of an innocent individual be discovered in a forensic sample due to secondary DNA transfer, as in the case of Amanda Knox (Starr, 2016). It is for this reason that law enforcement should offer additional evidence outside of DNA profile matching data to secure the conviction of a suspect. As to abandoned DNA samples, we discussed the difference in approach as far as privacy is concerned between Europe and the United States in Section 28.4. Needless to say, we opine that the interrogation of known DNA samples outside criminal investigations invokes valid privacy concerns.

Investigative and Predictive DNA Testing. DOI: https://doi.org/10.1016/B978-0-443-30128-5.00033-3

The chapter consists of three major sections covering the three key issues and challenges surrounding the use of DNA profiling and databases in criminal investigations. Section 28.2 covers the technical issues and challenges, specifically the latest developments on STR research showing that they may also reveal individual characteristics, the current trend to remove nonmedical SNPs for identification purposes, and the development of professional organizations that set technical standards for these tests. Section 28.3 covers ethical and social issues and challenges, specifically those that concern genetic relatives who could be identified through the genetic data of their kin. Section 28.4 covers legal and regulatory issues, mainly comparing key laws and jurisprudence of major players in the field of forensic DNA technology, the United States and Europe.

28.2 Technical issues and challenges

There are two main technical issues and challenges that we discuss in this section. First, there is growing evidence showing that STR profile data may reveal individual characteristics previously unknown, thereby contradicting the initial claim that they are mainly "non-coding" (Section 28.2.1). Second, on the other hand, SNP profile data, which are known to reveal extensive personal information, may be reduced to a minimum set—sans medical data—and can still permit identification of an individual (Section 28.2.2). For a quick reference on the difference between STRs and SNPs and their corresponding DNA databases, please refer to Tables 1 and 2, which we updated from our previous publication (Tuazon et al., 2024). From the same publication, we also provide Fig. 28.1, which presents the interplay between the use of STRs and SNPs in criminal investigations. We close the section with a discussion on a possible solution to these technical issues: the creation of technical bodies to address these issues and ensure the protection of genetic data privacy, with a specific mention to the US-based National Technology Validation and Implementation Collaboration (NTVIC) (Section 28.3.3).

28.2.1 STR profile data and its failing "non-coding" promise

Law enforcement mainly uses STR profile direct matching as its first-level DNA-based method in criminal investigations. STRs are largely known to occur in the noncoding region of the genome. However, recent findings show that they may be associated with the expression of nearby genes (Fotsing et al., 2019) and may reveal some "trait-associated variants" (Link et al., 2023), which may compromise "medical privacy" (Bañuelos et al., 2022).

The debate as to whether STR profile data may reveal medical information has been ongoing for almost two decades now. In 2007, Professors Joh and Kaye took two different sides on the relevance of this type of DNA data, which was then known as "junk DNA," given the prevailing notion that they contain "useless" information, privacy-wise. The main contention was aptly summarized by Cole in this way: "that the information held in law enforcement genetic databases is

TABLE 1 Short tandem repeat (STR) versus single nucleotide polymorphism (SNP) profile data.

	STR	SNP
No. of markers	13-23	More than 600,000[a]
DNA detected	2-6 base pair repeats	Single nucleotide changes
Coverage	Mainly noncoding regions	Entire human genome[a]
Relatives detected	Mainly siblings, parents, aunts/ uncles	Up to 5th cousin and beyond
Data derived	Mainly identification markers	Identification markers, plus health, genetic predisposition and other trait markers[a]
Law enforcement activity	Directly performed	Usually outsourced
Database where stored	Law enforcement DNA databases	Third-party DNA databases

[a] *QIAGEN recently released* ForenSeq Kintelligence *(targeted amplicon sequencing) using only 10,230 SNP markers, none of which are known health-related SNPs (QIAGEN, 2024).*
Source: Originally published by Tuazon et al. (2024).

TABLE 2 Size of law enforcement (LE) and third-party (3P) DNA databases (arranged in descending order, as of September 15, 2023.

Database	Country of origin	Year launched	DNA markers	Size (min.)
AncestryDNA (3P)	United States	2012	SNP	25 M
National DNA Index (LE)	United States	1998	STR	22.4 M
23andMe (3P)	United States	2006	SNP	14.1 M
MyHeritage (3P)	Israel	2003	SNP	8.3 M
National DNA Database (LE)	United Kingdom	1995	STR	7.3 M
FamilyTreeDNA (3P)	United States	1999	SNP	2 M
GEDMatch (3P)	United States	2010	SNP	2 M

Sources for the US and UK forensic DNA databases: NDIS [https://le.fbi.gov/science-and-lab/biometrics-and-fingerprints/codis/codis-ndis-statistics]; NDNAD [www.gov.uk/government/statistics/national-dna-database-statistics]; Sources for the third-party DNA databases: the respective company websites. [last accessed 16 September 2024].
Source:

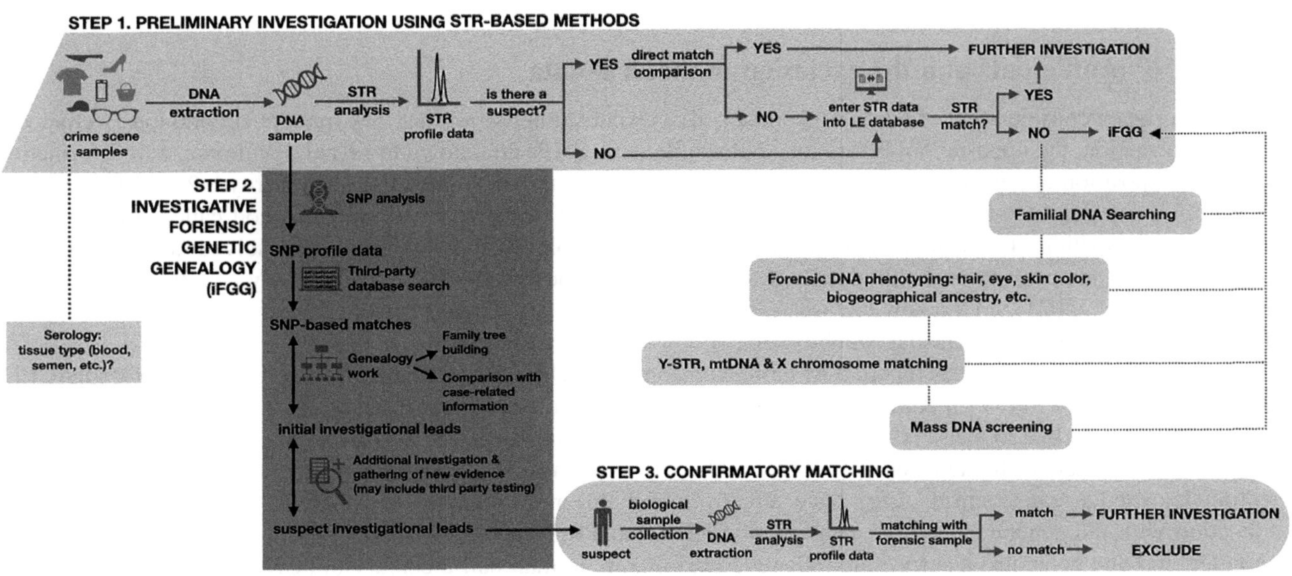

FIGURE 28.1 Flowchart detailing the forensic investigative process using STR-based methods and genetic geneal

innocuous from a privacy standpoint" (Cole, 2007a). According to Professor Joh, one cannot make use of the "junk DNA" defense given that "some markers now thought to be meaningless may be (and have been) found to contain predictive medical information as the science progresses" (Joh, 2007). On the other hand, Professor Kaye argued that "any claim that the DNA profiles currently used for identification constitute 'predictive medical information' is false" (Kaye, 2006). Professor Kaye himself took Cole to task on his statement that "[t]he privacy threat posed by forensic STRs may not be great," claiming that it had no basis (Cole, 2007b; Kaye, 2007). Later on Professor Kaye appeared to agree that it is possible that an STR may have "a dramatic effect on a trait," although it "may not be medically or socially significant" (Kaye, 2014).

More recently, a study has shown that "a large number of phenotypic traits" are associated with STR data although none of them were found to be "independently causative or predictive of disease" (Wyner et al., 2020). Another study reported that "information on gene expression levels may be revealed by Combined DNA Identification System (CODIS) profiles" and some of these are "connected to medical phenotypes" (Bañuelos et al., 2022). In yet another study, the authors found that "the regions around the CODIS loci are enriched for both known pathogenic variants" as well as for "trait-associated"

SNPs, although they did not claim to "establish or quantify associations between CODIS genotypes and phenotypes" (Link et al., 2023).

Be that as it may, the main reason for an increased interest in whether STR profile data reveal medical or other information aside from predicting the identity of the source of the forensic sample is its privacy implications. As some authors noted, "is there a duty to inform a DNA donor if a medical condition is discovered during routine analysis of their sample?" (Wyner et al., 2020). What is clear is that STR profile data can no longer be designated as "junk DNA," although the privacy implications of the other information—outside identity prediction—still need to be further elaborated (Scientific Working Group on DNA Analysis Methods SWGDAM, 2024).

We suggest that these developments should not hinder law enforcement in using STR profile data for criminal identification purposes for several reasons. First, these findings are still at their exploratory stage. Second, even if they have been validated convincingly and generally accepted by the scientific community, law enforcement does not have any business using them to elaborate on the medically related traits they purport to reveal; law enforcement is only concerned in using them for criminal identification purposes (cf. the concept of genometric data in Section 28.3.1 (Tuazon et al., 2025a)). Third, what law enforcement stores is DNA profile data expressed in numerical figures, not the DNA itself; corollary to the previous point, what can be done is to restrict the use of these DNA profile data for criminal identification purposes in the enabling law. Fourth, law enforcement has already built its own STR profile databases; hence, it is more economical and practical to continue building on them; the more profiles there are in these DNA databases, the more chances there are to make direct STR profile matches; they also facilitate the building up of universal forensic DNA databases (Tuazon, 2021). Fifth, direct STR profile matching continues to be the confirmatory step in probability-based DNA-based methods such as iFGG (Step 3, Fig. 28.1), hence, it is here to stay.

28.2.2 SNP profile data and the exclusion of medical data

If the main privacy consideration of STR profile data is that recent studies show that they may reveal previously unknown medical information, the opposite holds true for SNP profile data. SNPs are known to reveal a wide range of information such as predisposition to genetic diseases and external visible characteristics (Nolan & Ormondroyd, 2023; Tozzo et al., 2021). This is one reason why the use of SNPs by law enforcement in the relatively new method called iFGG has met considerable opposition. For example, although a 15-year-old cold case in Sweden was solved using iFGG, the government put it temporarily on hold, given unresolved doubts concerning genetic data privacy (Aili Fagerholm et al., 2021; Tillmar et al., 2021), until a new law was passed directly allowing iFGG use following specific safeguards (Ansell & Fagerholm, 2025; Sveriges Riksdag webb-tv, 2025).

An interesting development was reported by a life tech company, Qiagen, in late 2023 (Antunes et al., 2024; QIAGEN, 2024). They developed *ForenSeq Kintelligence System* to generate a more limited number of SNPs—from the usual over half a million to only 10,230 markers. What makes these markers promising from a data privacy perspective—aside from the fact that they include limited personal data—is that they claim to exclude medical information as far as current scientific knowledge can confirm. They have already tested this new method using iFGG with positive results. In conjunction with the genetic genealogy database of GEDmatch PRO, they have solved an unidentified human remains case in Oregon and generated a suspect investigative lead in a more than half a century old case in Georgia (*see* Step 2, Fig. 28.1 to see the process of generating a suspect investigative lead using iFGG).

This new development in SNP research, which focuses on using less privacy-sensitive markers, holds a great promise in law enforcement use of genetic genealogy databases in their criminal investigations. They avoid the unwarranted exposure of medically-related data while at the same time achieving the main purpose of a criminal investigation, that is, the identification of the source of the forensic sample or samples discovered at a crime scene. As this technology develops and fewer SNP markers will be required for identification purposes, the resistance to its use by police investigators may lessen. One day, it may no longer be considered law enforcement's method of last resort in identifying criminals.

We suggest the promulgation of specific iFGG legislation that allows the use of SNP indirect profile matching for criminal identification, replete with legal safeguards that protect genetic data privacy (more on this in Section 28.4) (Tuazon et al., 2025b). It is important to emphasize in this section that more research be conducted to continue reducing the amount of SNP data for criminal identification purposes, that is, only a small fraction is curated that do not reveal sensitive medical and other information. From a laboratory processing standpoint, there is no need to run two separate laboratory experiments with two different profiles, one with the entire SNP and another with SNP data sans the medically sensitive ones. A more practical and less costly solution would be to simply block the accessibility of medically sensitive information through the means of information technology in criminal investigations, and this must be supported by corresponding rules, regulations, and oversight as we discussed in Section 28.4.

28.2.3 A prospect in solving technical issues and challenges: the NTVIC model

Like any other science-based method, DNA profiling techniques develop over time, and their application to criminal investigations usually follow these breakthroughs. At the same time, law enforcement personnel are not always up to date with these scientific and technical developments, and there are frequently unique challenges to implementing new technology and applying it to forensic samples. There is then a need to set up professional organizations that evaluate and validate these developments and make them accessible to law enforcement. One such body that was recently founded in the United States is the National Technology Validation Implementation Collaborative (NTVIC).

The NTVIC was established in 2022 with a vision to collaborate nationally in the United States on validation, method development, and implementation (Gamette & Wickenheiser, 2023). Forensic science service providers (FSSPs) frequently duplicate validation studies, method development and performance verification as they seek to implement new technology. Working collaboratively, FSSPs can share their resources and at the same time strengthen policies, procedures, and scientific rigor by increasing the number of participants and skill sets, thereby further demonstrating that the technology at hand is fit for purpose. Once new technology is implemented—as multiple FSSPs are using the same set of methods and parameters—the learning of one FSSP becomes the learning of all. Outliers are also better identified and corrected earlier, benefiting and supporting the quality control system of participating FSSPs (Wickenheiser & Farrell, 2020).

FSSPs have learned that conducting these validation studies and technology implementation collaboratively has several benefits: (1) The most informed and brightest minds in the country work together in generating plans and procedures, and also in troubleshooting issues; (2) The existing resources at laboratories can be pooled to implement the solution more quickly and provide more robust data sets; (3) The validation work can be published as a collaborative effort for purposes of public scrutiny; and (4) The developmental validation work product can culminate in a technology transfer and implementation plan for instrument purchase, performance verification, and in a quick implementation in other laboratories interested in the technology (Gamette & Wickenheiser, 2023).

The first initiative undertaken by the NTVIC was iFGG, noting that the NTVIC had previously coined the term Forensic Investigative Genetic Genealogy (FIGG). Committees were established such as Policies and Procedures, Laboratory Technical Validation, and Training and Terminology. Following the NTVIC's goal of publication to provide open access, transparency and public scrutiny, the FIGG Policy and Procedures Group subsequently developed and published its first version of policies and procedures in February of 2023 in the open access journal Forensic Science International: Synergy (Wickenheiser et al., 2023a). Numerous suggestions and feedback were provided by forensic scientists, genealogists, attorneys, and bioethicists, which were incorporated into the next updated version published in November of 2023 in the same journal (Wickenheiser et al., 2023b). A third version of the NTVIC FIGG (iFGG) Guidelines was recently published in June of 2025, adding new recommendations on data retention, privacy, and discovery (Wickenheiser et al., 2025). With that updated version, no additional feedback was received, thereby achieving a balance of viewpoints to determine the use of best practices for the implementation and application of FIGG at its current state. This model of collaborative technology evaluation and implementation is now being applied to 3D Firearms Technology and Rapid DNA as similar committees have now been established in the NTVIC, which can be seen on the NTVIC website (National Technology Validation & Implementation Collaborative NTVIC, 2024a).

28.3 Ethical and social issues and challenges

A distinct feature of DNA profile data, whether in the form of STRs or SNPs, is that these data are shared by individuals with their genetic relatives. In criminal investigations, genetic similarity translates to the ability of law enforcement to help identify criminal suspects, whose DNA profiles are not stored in law enforcement DNA databases, through their next of kin whose STRs are enrolled in those databases via indirect matching. In the case of SNPs, the power to identify a genetic relative is even higher as was exhibited in the Golden State Killer who was identified through a second, third or a probable fourth cousin—as reports vary (Erlich et al., 2018; Wickenheiser, 2019; St. John, & 2020)—who was enrolled in a public genetic genealogy database. In this section, we will look into the ethical and social issues and challenges using DNA profiles in criminal investigations, specifically those that concern genetic relatives who could be identified through the genetic data of their kin. We will look into various concepts that have been developed to exemplify genetic relatedness: the dividuality of the human genome (Section 28.3.1), its relational aspect (Section 28.3.2), and social versus individual privacy concerns (Section 28.3.3). We close the section with our take on the future prospects of these ethical and social issues and challenges (Section 28.3.4).

28.3.1 Dividual nature of genetic data

The anthropological concept of "dividuality" was applied to genetic data and genomic research for the first time less than a decade ago (McGonigle, 2019; McGonigle & Shomron, 2016). In anthropology, as aptly explained by McGonigle, the "dividual self is a distributed entity, relationally constructed, partible, composite and essentially divisible" as opposed to the "individual self" (McGonigle, 2019). In ethnography, "the self is understood to exist only within the social networks and situated contexts that lend meaning and identity to the person" (McGonigle, 2019). Hence, he posited that the concept of the "dividual self" is a "good metaphor" to describe the "shared nature of genomic data" (McGonigle, 2019).

The concept of dividuality has repercussions on the way genetic data is perceived. It creates a certain awareness on the part of an individual that "when they disclose their personal genetic data publicly they also inadvertently share data about their biological kin" (McGonigle & Shomron, 2016). It also has an effect on the growing field of personalized medicine, which depends on "collective participation" (McGonigle, 2016). In the field of law enforcement, the dividual nature of genetic data concerns genetic relatives whose genetic data are enrolled in DNA databases and who may be considered by some people as "genetic informants" and "genetic witnesses" (Guerrini et al., 2021). However, genetic relatedness—following this dividual concept—is not in itself evidence of criminal activity. In the case of iFGG, for example, it may only be likened to a tip—well, a "scientific tip"—that helps narrow down the possible suspects for further investigation (Guerrini et al., 2021).

A more recent concept, *genometric data*, was proposed as a more privacy-sensitive solution for law enforcement use of DNA data in criminal investigations (Tuazon et al., 2025a). Differentiating it from the current concepts of genetic and biometric data as defined under the General Data Protection Regulation (GDPR) (European Union, 2016a), the Law Enforcement Directive (LED) (European Union, 2016b), and Convention 108+ (Modernized Data Protection Convention) (Council of Europe, 2018a, 2018b), genometric data is defined as "personal data resulting from specific technical processing relating to identification markers in the human genome, which allow or confirm the unique identification of an individual." In relation to that is the concept of *genometric data privacy* which is defined as "an interest that individuals and their genetic relatives have over their genetic data that can be used to identify them individually, whether directly or indirectly, within the bounds of data protection and privacy laws" (Tuazon et al., 2025a). It was argued that the "use of genometric data within the boundaries of genometric data privacy allows law enforcement to focus on their criminal investigative work without getting entangled with excessive data privacy considerations" (Tuazon et al., 2025a).

In any case, it is important not to exaggerate the concept of the dividual or shared nature of genetic data as far as its use by law enforcement in criminal identification is concerned. It is only a means for them to reach a criminal perpetrator who left biological samples at the crime scene. Law enforcement will have to gather additional evidence to link the identified suspect to the crime committed—such as their physical possibility of being present at the crime scene, among others—and conduct direct STR profile data matching between the DNA of the suspect and the one extracted from the forensic sample. Without that additional evidence, we suggest that the judge dismiss the case. We strongly opine that suspects should not be convicted with DNA matching data as sole evidence against them (Tuazon, 2021; Tuazon et al., 2025b). On the other hand, law enforcement must exclude a suspect in case of a non-match of probative evidence. The Innocence Project has helped reverse the convictions of more than 200 prisoners on death row (Innocence Project, 2024), thanks in large part to DNA non-match data, as well as a corresponding direct match to a new suspect.

28.3.2 Relational model of genetic privacy

Another concept related to dividuality is the "relational model of privacy" as applied to genetic data. Costello argued that the shared nature of genetic data—without using the world *dividual*—"challenges modern, individualistic conceptions of privacy" (Costello, 2022). Using the jurisprudence of the two European supranational courts—the European Court of Human Rights (ECtHR) and the Court of Justice of the European Union (CJEU)—she argued that the "collective impact of genetic data processing" can be evaluated using a relational model, given that the "revelatory capacity of genetic data" goes "beyond the immediately consenting party" (Costello, 2022).

The relational model of privacy—genetic data privacy, in this particular case—promises, among others, to provide protection to the biological group that shares the data being processed. This is in line with the developing concept of group privacy (Puri, 2021; Taylor et al., 2017). In this concept, individuals who share certain genetic data form a "genetic group." In a study of its application to EU's GDPR, the protection of these genetic groups using its current provisions was seen to be "problematic" and creates some "awkward problems," such as the application of the principle of data accuracy to all members of the genetic group (Hallinan & de Hert, 2017). In a later study, also on the application of genetic group protection within the GDPR regime, potential conflicts were identified—on the application of the right to information, right to access, right to erasure, and right to restriction of processing—and remain without clear solutions unless the legislation is reviewed and possibly updated (Kuru & Beriain, 2022).

The relational model of genetic privacy breaks into the current individualistic conception of genetic privacy laws. Although it has been indirectly acknowledged by the ECtHR in *S. and Marper v. the United Kingdom*—"that DNA profiles could be, and indeed had in some cases been, used for familial searching with a view to identifying a possible genetic relationship between individuals" (S & Marper v. the UK, 2008)—there has been no clear indication as to the protection of these genetic relatives. As Costello noted, this relational model is "not without complications" but at the same time it is "true to its origins" and it is "responsive to modern challenges" (Costello, 2022). We expect a heightened debate in this area in the coming decades as the general populace improves on their scientific culture and become more aware that sending their saliva to a direct-to-consumer genetic testing company may mean exposing not only their own genetic data but also that of their relatives.

28.3.3 Social value of genetic privacy

From the level of the biological or genetic group, there is yet another level that is being considered in terms of the shared nature of genetic data: the social level. De Groot argues that applying the concept of "privacy as a social value" to the law enforcement method making use of genetic genealogy databases to identify criminals—which we call iFGG (Tuazon et al., 2024)—"allows consideration of broader social issues, including informational and decisional interconnectedness, the risk of a 'tyranny of the minority', the involvement of multiple contexts, and the relationship between citizens and state" (de Groot, 2023).

Tackling it from a bioethical perspective, the same author earlier argued that in the case of shared genetic data such as that used in iFGG, the "decision of only a fraction of the population could have far-reaching consequences for the population as a whole" (de Groot et al., 2021). Hence, approaching it from a social value perspective can expose "more fundamental ethical, political, and social concerns" that go "beyond individual data control or consent" (de Groot, 2023).

Along the same line, a more explicit application of the bioethical principle of "solidarity" on the issue of processing shared genetic data in iFGG was proposed (Samuel, 2021). The author argued, among others, that assuming everyone's desire for an effective criminal investigation, "solidarity requires us all to accept this risk of harm because of the benefits that may come from it" (Samuel, 2021). Instead of creating a tension between the individual and the members of the genetic group, Samuel proposed that it be tackled as "both individualistic, through notions of individual privacy rights and data governance and collective" (Samuel, 2021).

This collective concept of genetic privacy is in line with the 1997 Universal Declaration on the Human Genome and Human Rights. It provides in Article 1 that "[t]he human genome underlies the fundamental unity of all members of the human family, as well as the recognition of their inherent dignity and diversity. In a symbolic sense, it is the heritage of humanity" (UNESCO, 2024). This claim is supported by a scientific study, following the Human Genome Project, claiming that human beings are 99.9% alike at the DNA level (Collins & Mansoura, 2001). We also foresee more discussion on this area in the coming decades: from the consciousness brought about by our genetic relatedness (as a biological family or an ethnic group) to its collective effect on society as a whole—that in the end, we all form one big human family.

28.3.4 Future prospects of these ethical and social issues and challenges

All the concepts presented in this section—the dividual nature of genetic data, the relational model of genetic privacy and its social value—all point to the scientific fact that individuals who share their genetic data may be potentially identified through the genetic data of one member of that genetic group. In law enforcement investigations, it would lead to the criminal identification of the forensic sample found at the crime scene. More practically, an individual does not have to be enrolled in a law enforcement DNA database—and to a more extensive degree, in a genetic genealogy database—to be capable of being identified in a criminal investigation, as they may be captured in family trees being built in the effort to develop an investigative lead through genealogical research and compiled publicly available data (Erlich et al., 2018; Greytak et al., 2018).

These concepts point to the need to protect not only individuals who are willing to share their genetic data but also their genetic relatives who can also be potentially identified through these data. Hence, these concepts point the direction from the current individualistic concept of genetic data protection to a more collective, relational, or dividual one. While it is true that these concepts provide some level of standing to a group of individuals connected in some way by genetics, they run counter to the bioethical principle of autonomy. Autonomy is one of the bedrock principles of medical bioethics, indicating that the individual has sole authority in the use of their body (Wickenheiser, 2019). This would include the informed consent provided by an individual to allow law enforcement access to their DNA profile in genealogy databases to help solve violent crimes, despite the obvious ability of DNA indirect comparison to not only implicate genetic relatives, but in theory even directly match to their identical twin.

We acknowledge the complexity of enforcing the right of genetic relatives—their right to genetic data privacy, which, in the regime of the European Convention on Human Rights (ECHR), falls under the right to respect for private life under Article 8 (Council of Europe, 1953). While it is true that its corresponding Court, the ECtHR, has affirmed that cellular samples "contain a unique genetic code of great relevance to both the individual and his relatives" (S & Marper v. the UK, 2008), it has not developed a legal procedure to protect these genetic relatives. For example, can genetic relatives file a case before a court because they felt that their privacy was unnecessarily exposed through the genetic data of one of their next of kin? Will they be given legal standing by a court, and to what extent can they claim damage over this alleged privacy intrusion? The current data protection and privacy laws do not provide such remedy at the moment because they are mainly based on the individual, who is the direct source of such data developed from biological evidence discarded at a crime scene. These are among the legal challenges that are a consequence of understanding the shared nature of genetic data that we expect to face in the coming decades. We tackle these legal and regulatory concerns further in Section 28.4.

28.4 Legal and regulatory issues and challenges

The legal and regulatory landscape governing DNA profiling and databases varies from place to place, or more technically, according to the legal traditions and dispositions of a particular state or country. At the same time, we argue that for law enforcement procedures involving scientific matters, such as the use of STRs and SNPs in identifying criminal perpetrators, there should be a more objective and unified set of laws and regulations given that they are not based on subjective cultural norms but rather on objective scientific principles, after due validation as mentioned in Section 28.2.3. This section provides the current legal landscape related to DNA profiling from major players in the field of forensic DNA technology, the United States, and Europe. We focus our discussion on two key areas where they differ in their approach: on DNA collection (Section 28.4.1) and DNA profiling for identification purposes (Section 28.4.2). We also present the current debate on the acceptability of universal forensic DNA databases vis-à-vis the use of SNPs and genetic genealogy databases in the technique, iFGG (Section 28.4.3). We end the section with a discussion on current legislation, specifically on iFGG, and the development of model legislation to provide a more coherent legal framework across the globe (Section 28.4.4).

28.4.1 On DNA collection in criminal investigations

Although the first-ever application of DNA evidence in a criminal investigation happened in Europe (Gill et al., 1985), recent breakthroughs come from the United States, such as the application of iFGG in identifying the Golden State Killer (Tuazon et al., 2024; Wickenheiser, 2019). This may be due to a more flexible approach in the United States in the handling of genetic data and the general attitude on genetic data privacy compared to Europe.

In a landmark case, the Supreme Court of the United States (SCOTUS) considers DNA collection and subsequent identification of arrestees as a "reasonable search that can be considered part of a routine booking procedure" (Maryland v. King, 2013). The Court clarified that in the case of a valid arrest, "expectations of privacy" are not offended by what it considers a "minor intrusion of a brief swab" on the cheeks of the arrestee. The disposition of the ECtHR on collection of DNA samples is rather different. In *Van der Velden v. the Netherlands*, the ECtHR considered the collection of cellular material through a mouth swab as an "intrusion on the applicant's privacy" and the systematic retention of the samples and DNA profiles were regarded as an interference with the right to respect for private life under Article 8 § 1 of the ECHR (Van der Velden v. the Netherlands, 2006). A couple of years later, in its Grand Chamber decision, *S and Marper v. the United Kingdom*, the ECtHR affirmed its ruling in *Van der Velden* and took precaution of future developments in DNA technology which may affect our privacy "in a manner which cannot be anticipated with precision today" (S & Marper v. the UK, 2008). Hence, the predictions on the acceptability of universal forensic DNA databases (Tuazon, 2021), FDP (Zieger, 2022), and iFGG (Tuazon et al., 2025b), within the ECtHR regime is not straightforward. In the United States, there appears to be no difficulty, for example, in using FDP by law enforcement in solving crimes as shown by Parabon Nanolab's law enforcement service (Arnold, 2020).

American legal scholars also do not find surreptitious DNA collection and testing by the police in their effort to solve crimes problematic, arguing that the biological material has been discarded or abandoned, and an innocent person should not feel stressed with that practice (Berkman et al., 2018; Guerrini et al., 2021; Wickenheiser, 2022). At the European level, however, the ECtHR in *S and Marper v. the United Kingdom* calls for the exercise of "careful scrutiny" when law enforcement uses personal data without consent in their effort to solve crimes (S & Marper v, 2008). For the ECtHR, personal data includes not only DNA profile data but also the cellular samples from which DNA profile data is extracted. This is extremely relevant because the mere fact of their collection triggers the application of the law on the right to respect for private life under Article 8 of the ECHR. However, the finding of an interference with the right to privacy does not necessarily mean a violation of Article 8 ECHR. The ECtHR still has to evaluate whether the interference is lawful, whether it serves a legitimate purpose, and whether it is necessary in a democratic society (*See* a summary of the four-fold privacy test employed by the ECtHR (Tuazon & Zwenne, 2021)).

28.4.2 On DNA profiling for criminal identification

The use of STR profiling for identification of forensic samples left at the crime scene is now a standard operating procedure in both American and European law enforcement regimes. The database system developed by the US Federal Bureau of Investigation, the CODIS, is widely used in Europe. The SCOTUS case law mentioned above, *Maryland v. King*, involves STR profile data (Maryland v. King, 2013). All the ECtHR case law involving law enforcement use of DNA data *to date*—from Van der Velden v. the Netherlands (2006) (Van der Velden v. the Netherlands, 2006) up to *Petrović v. Serbia* (2020) (Dragan Petrović v. Serbia, 2020)—presumably involved STR profiles, given that they involved law enforcement DNA databases which currently retain only STR and not SNP profile data. For the same reason, even the CJEU case law that mentions genetic data—*Ministerstvo na vatreshnite raboti* (2023) (Ministerstvo va vatreshnite raboti, 2023) and *NG v. Direktor na Glavna direktsia* (2024) (NG v. Direktor na Glavna direktsia, 2024)—also presumably involve STR profile data. In that regard, it is clear that the use of DNA profiling through STRs for criminal investigation by law enforcement is acceptable following the safeguards identified in the case law.

As to the use of the more "privacy-revealing" SNP profile data vis-à-vis genetic genealogy databases in iFGG, American law enforcement appears not to have much difficulty in using it for criminal investigations. It has been used even before the Golden State Killer case in 2018, although it was the latter that caught global attention (Wickenheiser, 2019). In Europe, iFGG was first used in 2019 to solve a double murder case in Linköping, Sweden (Tillmar et al., 2021). However, the Swedish data protection authority prohibited its use until its biometrics law was updated to include iFGG use following strict safeguards (Aili Fagerholm et al., 2021; Ansell & Fagerholm, 2025). This is one reason why we recommend that further use of iFGG in Europe be first supported by a solid legal foundation (Tuazon et al., 2025b; Tuazon, 2024a, 2024b). On the other hand, iFGG continues to be used in the United States, with some states passing specific legislations on iFGG use, such as Maryland, Montana, Utah and Florida (Dowdeswell, 2024; State of Maryland, 2025; State of Montana, 2025; State of Florida, 2025; State of Utah, 2025; Taylor, 2021).

The US Golden State Killer case actually came to light 10 years after ECtHR's decision in *S and Marper v. the United Kingdom* (S & Marper v. the UK, 2008). In *Marper*, the ECtHR acknowledged the sensitivity of genetic data and that genetic relatives could be identified through it. At that time, Familail DNA Searching (FDS) was used since only STR profile data were available for use by law enforcement. In the Golden State Killer case, it is SNP profile data that was involved, using iFGG. It is still unclear whether the ECtHR will apply the same legal principles it did in *Marper* and more recent cases, such as *Gaughran v. the United Kingdom*, in its assessment of iFGG. Given such uncertainty, the UK Biometrics and Forensics Ethics Group (BFEG) conducted a study where they recommended that the use of iFGG be "clearly establish[ed] whether it is in keeping with Article 8 of the ECHR or the right to respect for private life" (Biometrics & Forensics Ethics Group BFEG, 2024). A recent article claims that iFGG can be used within the Article 8 ECHR regime if law enforcement follows a set of safeguards that respect genetic data privacy (Tuazon et al., 2025b).

28.4.3 Universal forensic DNA databases or public genetic genealogy databases?

Hazel et al. (2018) published an article in Science with the title, *Is it time for a universal genetic forensic database?*, with an accompanying photo of the Golden State Killer during his arraignment in California. The authors argued that a universal database consisting of CODIS-STR profile data would be more advantageous and less privacy invasive compared to the use by law enforcement of genetic genealogy databases, which potentially reveal a lot of medical information. They acknowledged the ECtHR's decision in *Marper*, saying that it "spells doom for universal genetic databases" (Hazel et al., 2018).

In his analysis of ECtHR jurisprudence, Tuazon does not completely agree with Hazel et al.'s interpretation of *Marper* and the projected disposition of the said European court (Tuazon, 2021). He argued that it would be "extremely difficult, if not impossible" that universal forensic DNA databases be accepted under the ECHR regime. However, he also argued that there is a "glimmer of hope" should all the legal safeguards he indicated—following the fourfold privacy test (Tuazon & Zwenne, 2021)—be followed through an enabling universal DNA database law.

The debate on the acceptability of universal forensic DNA databases over iFGG will continue in the coming years (Hazel et al., 2018; Tuazon, 2021; Wickenheiser, 2022). It will be influenced by Court decisions on the current technology being employed—as mentioned in Section 28.2, non-medical information can now be used for iFGG, for example—and prevailing jurisprudence of various Courts around the world, not only in the United States and Europe. An important aspect of the ensuing debate will be the content of enabling laws, both for universal databases and iFGG, as they will be evaluated according to the sufficiency of legal safeguards they contain that protect genetic data privacy.

28.4.4 Prospects of legal and regulatory issues: developing model legislation

The difference in approach between the American and European—specifically, the European Union (EU)—legal systems on law enforcement access to data has been aptly expounded using both law and jurisprudence (Christakis & Terpan, 2021). What is relevant to our discussion is that the recent judgments of the CJEU, particularly in *Schrems* I and II (Data Protection Commissioner v. Schrems, 2020; Schrems v. Data Protection Commissioner, 2015), appear to show that current US law does not provide adequate protection on personal data compared to the European Union, as reflected for example in the GDPR (European Union, 2016a) and in its law enforcement application, the LED (European Union, 2016b). Added to this are the case law of the other European court, the ECtHR, as expounded *supra* which requires strict legal safeguards before DNA profile data can be used and retained by law enforcement. This difference will be a point of discussion in the coming years, and this chapter cannot possibly predict its outcome. What we aim to point out in this last subsection are possible solutions from the point of view of legislation. DNA profiling and databases will stand on more solid ground if they are backed up by clearly promulgated enabling laws.

As indicated *supra*, there appears to be no significant problem with the use of STR profile data and their corresponding law enforcement databases in criminal investigations. They have become customary in the field of law enforcement. What is more unclear is the use of iFGG given that it makes use of SNP profile data that, if utilized, could reveal more sensitive information such as medical data and predisposition to certain genetic diseases. In this regard, the US Department of Justice led the way by formulating an interim policy on what it called "forensic genetic genealogical DNA analysis and searching" (US Department of Justice, 2019). As explained in our paper on law enforcement use of genetic databases, the term iFGG is preferable (Tuazon et al., 2024). The said policy aimed to promote "reasoned exercise of investigative, scientific, and prosecutorial discretion" when iFGG is used by law enforcement (US Department of Justice, 2019). With respect to privacy, the policy also aims to develop practices that "protect reasonable interests in privacy" in the use of iFGG for specific purposes, like the identification of violent criminals (US Department of Justice, 2019).

The interim policy of the US DOJ may serve as a guide to other jurisdictions in formulating their own policies on iFGG (Tuazon et al., 2025b). Since the said policy provides that iFGG be consistent to the US Constitution, the policies that will be formulated in other jurisdictions should make sure that they are also consistent with the laws of their land. It shall indicate the type of crimes iFGG will be made applicable to, the prior techniques that should have been used prior to resorting to iFGG, legal safeguards on collection, processing and retention of samples, who has access to the data, possibility for review and oversight, among others.

In this regard, the NTVIC mentioned in Section 28.2.3 developed a model legislation for iFGG which can be adapted according to the legal requirements of a given jurisdiction (NTVIC, 2024b). These model legislations or "law templates" are specifically significant for science-based methods such as iFGG, as they provide a more coherent legal framework across the globe. In this way, data sharing across jurisdictions—following commonly acceptable legal safeguards—would be facilitated and the aim of law enforcement in protecting its people and reducing crimes would become more real than ideal.

In Europe, it would be advantageous to develop a common European-wide legislation on iFGG with minimum legal safeguards that will be applicable to all member states of both the European Union (EU) and the Council of Europe (COE), following the jurisprudence of their corresponding supranational courts, the CJEU and the ECtHR, respectively. Given the peculiarities of these supranational bodies and courts, each of them would obviously promulgate their own laws but with a certain harmony similar to what was done when introducing genetic data, among others, in the EU's GDPR (European Union, 2016a) and LED (European Union, 2016b) and COE's Convention 108+ (Council of Europe, 2018a).

28.5 Conclusions

DNA profile matching will continue to be the gold standard in criminal identification in the coming years, whether it be under its current form (STR profile matching in combination with law enforcement DNA databases) or under any of the new techniques that have been developed or will be developed following breakthroughs in DNA technology, such as FDP, mass DNA screening, Y-STR, mtDNA and X chromosome searching, and more recently, SNP profile matching in combination with public genetic genealogy databases through a technique called iFGG. The DNA's inherently personal and sensitive nature as well as personal data has brought with it some technical, ethical and social, and legal and regulatory issues and challenges.

We identified two main technical issues and challenges: (1) New studies show that STR profile data may reveal individual characteristics previously unknown, such as medical data, thereby making the initial claim that they are mainly "non-coding" doubtful. We opine that these developments should not hinder law enforcement use of STR profile matching given that they mainly use their numerical equivalents in the criminal identification process and are never interested in investigating personal information such as propensity to disease, for example; (2) A new technique using SNP profile data shows that it can permit identification of an individual without having to use medical data. This is a welcome development as it may help put iFGG at the forefront of criminal investigations and will no longer be called the method of last resort. To help address these technical issues and challenges, we discussed the need to set up professional organizations that evaluate and validate developments in DNA profiling technology and make them accessible to law enforcement. We provided as an example the recently founded NTVIC in the United States.

Our discussion of ethical and social issues and challenges revolved around the scientific fact that DNA data is shared among biological or genetic relatives, which is the scientific basis for law enforcement's use of familial DNA matching and iFGG. We looked into three concepts that have been developed to exemplify genetic relatedness: (1) The dividual nature of genetic data, that is, genetic data is not a merely personal (individual) concern but rather a legitimate concern of people who share such data and could be possibly identified through those data. We do not agree, however, that genetic relatives can be considered as "genetic informants" in a criminal investigation, but rather as leads that may help point to the criminal perpetrator, given that results of indirect matching cannot solely be used as evidence to convict an individual of a crime. The new concepts of genometric data and genometric data privacy are worth considering given their promise of being more privacy-sensitive for law enforcement use of DNA data in criminal investigations; (2) The relational model of genetic privacy, that is, the shared nature of genetic data implies with it a certain relationship to those individuals with whom it is shared. Hence, the concept of group privacy develops, where those who relate to each other genetically must be provided with similar protection. This concept is, however, difficult to implement in practice given that the current data protection laws are only geared towards the protection of individuals. We expect a heightened debate on this area in the coming years; (3) The social value of genetic privacy, that is, raising up further the genetic interconnectedness of individuals to the societal plane. This collective concept of genetic privacy is in line with Article 1 of the 1997 Universal Declaration on the Human Genome and Human Rights, which enshrined that the human genome "underlies the fundamental unity of all members of the human family" and that it is "the heritage of humanity." We also foresee more discussion on this area in the coming years as people's scientific culture improves.

Lastly, in our discussion on legal and regulatory issues and challenges, we acknowledged the current legal disposition that the regulatory landscape governing DNA profiling and databases vary from place to place, while at the same time arguing that they should be governed by a more objective and unified set of laws and regulations given that they are not based on subjective cultural norms but rather on objective scientific principles. We focus our discussion on two key areas where the two major players on the use of DNA profiling and databases, Europe and the United States, differ in their approach: (1) On DNA collection, the SCOTUS provides a more flexible approach, as it considers it as part of a "routine booking procedure," whereas the ECtHR considers it outrightly as an intrusion to privacy; (2) On the use of DNA profiling for criminal identification, both jurisdictions approve of STR profile data matching and currently have their respective law enforcement DNA databases. For SNP profile matching through iFGG, however, the technique is more commonly used in the United States. In Europe, its use was reported for the first time in Sweden but its further use was prohibited until an enabling-iFGG law was passed. A recent study shows that iFGG may be acceptable within the ECHR regime in the presence of safeguards that protect genetic data privacy. We also discussed the acceptability of universal forensic DNA databases compared to the use of public genealogy databases in criminal investigations, which we opine will be a subject of public debate in the coming years. Finally, we discussed the importance of developing model legislations or "law templates," which are useful for science-based methods such as iFGG, as they help provide a more coherent legal framework across the globe. It also facilitates data sharing across jurisdictions, with more robust legal safeguards, thereby contributing greatly to the main reason why law enforcement use DNA profiling and their corresponding DNA databases:

to protect the people under their care, reduce and prevent crimes, avoid wrongful accusations, and thereby create a better, safer and a more just, orderly, and peaceful society to live in.

References

Ansell, R., & Fagerholm, S. A. (2025). On the use of forensic investigative genetic genealogy in criminal cases in Sweden. *30th Congress of the International Society for Forensic Genetics Universidade de Santiago de Compostela, Spain,* 755–761. https://doi.org/10.15304/cc.2025.1869.

Antunes, J., et al. (2024). Developmental validation of the ForenSeq(R); Kintelligence kit, MiSeq FGx(R) sequencing system and ForenSeq Universal Analysis Software. *Forensic Science International: Genetics, 71,* 1–19. https://doi.org/10.1016/j.fsigen.2024.103055.

Arnold, C. (2020). The controversial company using DNA to sketch the faces of criminals. *Nature News Feature, 585*, 178–181. https://doi.org/10.1038/d41586-020-02545-5.

Bañuelos, M. M., et al. (2022). Associations between forensic loci and expression levels of neighboring genes may compromise medical privacy. *PNAS, 119*, 1–10. https://doi.org/10.1073/pnas.2121024119.

Berkman, B. E., Miller, W. K., & Grady, C. (2018). Is it ethical to use genealogy data to solve crimes? *Annals of Internal Medicine, 169*, 333–334. https://doi.org/10.7326/M18-1348.

Biometrics and Forensics Ethics Group (BFEG), (n.d.). *Should we be making use of genetic genealogy to assist in solving crime? A report on the feasibility of such methods in the UK.* https://www.gov.uk/government/publications/use-of-genetic-genealogy-techniques-to-assist-with-solving-crimes/should-we-be-making-use-of-genetic-genealogy-to-assist-in-solving-crime-a-report-on-the-feasibility-of-such-methods-in-the-uk-accessible-version.

Christakis, T., & Terpan, F. (2021). EU–US negotiations on law enforcement access to data: Divergences, challenges and EU law procedures and options. *International Data Privacy Law, 11*, 81–106. https://doi.org/10.1093/idpl/ipaa022.

Clayton, E. W., et al. (2019). The law of genetic privacy: Applications, implications, and limitations. *Journal of Law and the Biosciences, 6*, 1–36. https://doi.org/10.1093/jlb/lsz007.

Cole, S. A. (2007a). Coming clean about "junk DNA". *Northwestern University Law Review, 102*, 107–109.

Cole, S. A. (2007b). Is the 'junk' DNA designation bunk? *Northwestern University Law Review, 102*, 54–63.

Collins, F. S., & Mansoura, M. K. (2001). The Human Genome Project. Revealing the shared inheritance of all humankind. *Cancer, 91*, 221–225. https://doi.org/10.1002/1097-0142(20010101)91:1+<221::AID-CNCR8>3.0.CO;2-9.

Costello, R. A. (2022). Genetic data and the right to privacy: Towards a relational theory of privacy? *Human Rights Law Review, 22*, 1–23. https://doi.org/10.1093/hrlr/ngab031.

Council of Europe. (2018, May 18). *Convention 108 + Explanatory Report, 128th session of the Committee of Ministers.* http://rm.coe.int/convention-108-convention-for-the-protection-of-individuals-with-regar/16808b36f1.

Council of Europe. (2018a, May 18). Convention 108 +: *Convention for the protection of individuals with regard to the processing of personal data* (128th session of the Committee of Ministers). http://rm.coe.int/convention-108-convention-for-the-protection-of-individuals-with-regar/16808b36f1.

Council of Europe. (1953, September 3). *European Convention for the Protection of Human Rights and Fundamental Freedoms (European Convention on Human Rights)* (ETS 5, 213 UNTS 221). https://www.echr.coe.int/Documents/Convention_Eng.pdf.

Data Protection Commissioner v. Facebook Ireland Ltd, Maximillian Schrems (2020) C-311/18 [GC] (ECLI:EU:C:2020:559) CJEU paras. 1-203.

de Groot, N. F., van Beers, B. C., & Meynen, G. (2021). Commercial DNA tests and police investigations: A broad bioethical perspective. *Journal of Medical Ethics, 47*, 788–795. https://doi.org/10.1136/medethics-2021-107568.

de Groot, N. F. (2023). Commercial genetic information and criminal investigations: The case for social privacy. *Big Data & Society, 10*, 1–12. https://doi.org/10.1177/20539517231216957.

DNA confidential. (2009). *Nature Biotechnology*, *27*(9), 777. https://doi.org/10.1038/nbt0909-777.

Dowdeswell, T. L., (2023, June). *Forensic Genetic Genealogy Project v.JUN2023.* https://data.mendeley.com/datasets/82969bsmw4/3.

Dragan Petrović v. Serbia (2020) no. 75229/10 ECtHR paras. 1-98.

Erlich, Y., et al. (2018). Identity inference of genomic data using long-range familial searches. *Science (New York, N.Y.), 362*, 690–694. https://www.science.org/doi/10.1126/science.aau4832.

European Union. (2016a, April 27). *General Data Protection Regulation (GDPR): EU Regulation 2016a/679 on the protection of natural persons with regard to the processing of personal data and on the free movement of such data, and repealing Directive 95/46/EC* (OJ 2016 L 119/1). https://eur-lex.europa.eu/legal-content/EN/TXT/PDF/?uri=CELEX:32016R0679&from=EN.

European Union. (2016b, April 27). *Law Enforment Directive (LED): EU Directive 2016/680 of the European Parliament and of the Council of 27 April 2016 on the protection of natural persons with regard to the processing of personal data by competent authorities for the purposes of the prevention, investigation, detection or prosecution of criminal offences or the execution of criminal penalties, and on the free movement of such data, and repealing Council Framework Decision 2008/977/JHA* (OJ L 119/89). https://eur-lex.europa.eu/legal-content/EN/TXT/?uri=celex%3A32016L0680.

Fagerholm, S. A., et al. (2021, November), F*orensic DNA traces and genealogy: Use of investigative genetic genealogy in criminal investigations.* https://polisen.se/siteassets/dokument/forensik/forensic-dna-traces-and-genealogy.pdf.

Fotsing, S. F., et al. (2019). The impact of short tandem repeat variation on gene expression. *Nature Genetics, 51*(11), 1652–1659. https://doi.org/10.1038/s41588-019-0521-9.

Gamette, M. J., & Wickenheiser, R. A. (2023). Establishment of the National Technology Validation and Implementation Collaborative (NTVIC) and Forensic Investigative Genetic Genealogy Technology Validation Working Group (FIGG-TVWG). *Forensic Science International: Synergy, 6*, 1–3. https://doi.org/10.1016/j.fsisyn.2023.100317.

Gill, P., Jeffreys, A. J., & Werrett, D. J. (1985). Forensic application of DNA 'fingerprints. *Nature, 318*, 577–579. https://doi.org/10.1038/318577a0.

Greytak, E. M., Moore, C., & Armentrout, S. L. (2018, October 29). RE: Identity inference of genomic data using long-range familial searches, Erlich et al. *Science, eLetters.* https://www.science.org/doi/10.1126/science.aau4832.

Guerrini, C. J., et al. (2021). Four misconceptions about investigative genetic genealogy. *Journal of Law and the Biosciences, 8*, 1–18. https://doi.org/10.1093/jlb/lsab001.

Hallinan, D., & de Hert, P. (2017). Genetic classes and genetic categories: Protecting genetic groups through data protection law. In L. Taylor, L. Floridi, and, & B. van der Sloot (Eds.). *Genetic classes and genetic categories: Protecting genetic groups through data protection law, group privacy: New challenges of data technologies* (pp. 175–196). Springer International Publishing.

Hazel, J. W., et al. (2018). Is it time for a universal genetic forensic database? *Science (New York, N.Y.), 362*, 898–900. https://www.science.org/doi/10.1126/science.aav5475.

Innocence Project (n.d.). *Exonerate the Innocent.* https://www.innocenceproject.org/exonerate/.

Joh, E. A. (2007). Reclaiming "abandoned" DNA: The fourth amendment and genetic privacy. *Northwestern University Law Review, 102*, 857–884.

Kaye, D. H. (2014, May 28)., *"Open to dispute": CODIS STR loci as private medical information.* https://ssrn.com/abstract=2456300.

Kaye, D. H. (2007). Please, let's bury the junk: The CODIS loci and the revelation of private information. *Northwestern University Law Review, 102*, 70–81.

Kaye, D. H. (2006). Science fiction and shed DNA. *Northwestern University Law Review, 101*, 62–67.

Kuru, T., & Beriain, I. (2022). Your genetic data is my genetic data: Unveiling another enforcement issue of the GDPR. *Computer Law & Security Review, 47*, 1–15. https://doi.org/10.1016/j.clsr.2022.105752.

Link, V., et al. (2023). Microsatellites used in forensics are in regions enriched for trait-associated variants. *iScience, 26*, 1–12. https://doi.org/10.1016/j.isci.2023.107992.

Maryland V. K. 569 U.S. 435 (2013).

Maximillian Schrems v. Data Protection Commissioner. (2015). C-362/14 [GC] (ECLI:EU:C:2015:650) CJEU paras. 1-107.

McGonigle, I., & Shomron, N. (2016). Privacy, anonymity and subjectivity in genomic research. *Genetics Research (Camb.), 98*, e2. https://doi.org/10.1017/S0016672315000221.

McGonigle, I. (2019). Genomic data and the dividual self. *Genetics Research (Camb.), 101*, 1–4. https://doi.org/10.1017/S0016672319000107.

McGonigle, I. (2016). The collective nature of personalized medicine. *Genetics Research (Camb.), 98*, e3. https://doi.org/10.1017/S0016672315000270.

McGuire, A. L. (2008). Identifiability of DNA data: The need for consistent federal policy. *The American Journal of Bioethics AJOB, 8*, 75–76. https://doi.org/10.1080/15265160802478511.

Ministerstvo va vatreshnite raboti (2023) Case C-205/21 (ECLI:EU:C:2023:49) CJEU paras. 1-136.

National Technology Validation and Implementation Collaborative (NTVIC). https://sites.google.com/view/ntvic/home.

National Technology Validation and Implementation Collaborative (NTVIC). (n.d.). *Forensic Investigative Genetic Genealogy (FIGG) Model Legislation.* https://drive.google.com/file/d/1T8uUSaMse1G6EKoCHmN4dVm91ISZ62p6/view.

NG v. (2024). Direktor na Glavna direktsia 'Natsionalna politsia' pri Ministerstvo na vatreshnite raboti – Sofia. Case C 118/22 [GC] (ECLI:EU:C:2024:97) CJEU paras. 1–73.

Nolan, J. J., & Ormondroyd, E. (2023). Direct-to-consumer genetic tests providing health risk information: A systematic review of consequences for consumers and health services. *Clinical Genetics, 104*, 3–21. https://doi.org/10.1111/cge.14332.

Puri, A. (2021). A theory of group privacy. Cornell Journal of Law and Public Policy, *30*, 477–538.

QIAGEN. (2024). *DNA Labs International solves significant cold cases with QIAGEN workflow for forensic genetic genealogy.* https://corporate.qiagen.com/English/newsroom/press-releases/press-release-details/2023/DNA-Labs-International-solves-significant-cold-cases-with-QIAGEN-workflow-for-forensic-genetic-genealogy/default.aspx.

S and Marper v. (2008). The United Kingdom. nos. 30562/04 and 30566/04 [GC] ECtHR paras. 1–141.

Samuel, G. (2021). Investigative genetic genealogy: Can collective privacy and solidarity help? *Journal of Medical Ethics, 47*, 796–797. https://doi.org/10.1136/medethics-2021-107960.

Scientific Working Group on DNA Analysis Methods (SWGDAM). (n.d.). *Considerations for claims that the CODIS core loci are 'associated' with medical conditions/diseases.* http://media.wix.com/ugd/4344b0_5db9ecd570454b21a3dcbd6d165163bd.pdf.

Shelton, D. (2011). *Forensic science in court: Challenges in the twenty first century.* Rowman & Littlefield Publishers.

St. John, P. (2020, December 8). *The untold story of how the Golden State Killer was found: A covert operation and private DNA.* https://www.latimes.com/california/story/2020-12-08/man-in-the-window.

Starr, D. (2016). *Forensics gone wrong: When DNA snares the innocent.* https://www.science.org/content/article/forensics-gone-wrong-when-dna-snares-innocent.

State of Florida. (2023). *Committee Substitute for House Bill No. 1327, An act relating to public records; amending s. 119.071, F.S.; providing definitions; providing an exemption from public records requirements for investigative genetic genealogy information and materials.* http://laws.flrules.org/2023/235.

State of Maryland. (2021). *House Bill 240, Chapter 681, An act concerning criminal procedure – Forensic genetic genealogical DNA analysis, searching, regulation, and oversight.* https://mgaleg.maryland.gov/2021RS/chapters_noln/Ch_681_hb0240E.pdf.

State of Montana. (2021). *House Bill No. 602, An Act generally revising warrant requirements for DNA search results; Requiring a warrant for a search from a consumer DNA database; Requiring a warrant for familial DNA search or search results from partial matching from the state DNA identification index or a consumer DNA database; and providing definitions.* https://legiscan.com/MT/text/HB602/id/2384004/Montana-2021-HB602-Enrolled.pdf.

State of Utah. (2023). *Senate Bill 156: Investigative Genetic Genealogy Modifications.* https://le.utah.gov/~2023/bills/static/SB0156.html.

Sveriges riksdag (2025, 26 februari). *Beslut: Biometri i brottsbekämpningen.* https://www.riksdagen.se/sv/webb-tv/video/beslut/beslut-biometri-i-brottsbekampningen_hcc320250226juu18/.

Taylor, L., Floridi, L., & van der Sloot, B. (2017). Introduction: A new perspective on privacy. In L. Taylor, L. Floridi, & B. van der Sloot (Eds.). *Introduction: A new perspective on privacy, group privacy: New challenges of data technologies* (pp. 1–12). Springer International Publishing.

Taylor, M. (2021, June 30). *Maryland, Montana become first states to restrict genetic genealogy searches.* https://www.forensicmag.com/577298-Maryland-Montana-Become-First-States-to-Restrict-Genetic-Genealogy-Searches/.

Tillmar, A., et al. (2021). Getting the conclusive lead with investigative genetic genealogy – A successful case study of a 16 year old double murder in Sweden. *Forensic Science International: Genetics, 53*, 1–7. https://doi.org/10.1016/j.fsigen.2021.102525.

Tozzo, P., et al. (2021). External visible characteristics prediction through SNPs analysis in the forensic setting: A review. *FBL, 26*, 828–850. https://doi.org/10.52586/4991.

Tuazon, O. M., & Zwenne, G. J. (2021). DNA profile data and right to privacy. In C. Willemsen (Ed.). *DNA profile data and right to privacy, four decades of information technology and innovation: Experts and academics share their experience and knowledge* (pp. 19–27). Kindle Direct Publishing.

Tuazon, O. M., Custers, B., & Zwenne, G. J. (2025a). Genometric data privacy within the ECHR regime. *International Data Privacy Law,* 1–17. https://doi.org/10.1093/idpl/ipaf004.

Tuazon, O. M., Custers, B., & Zwenne, G. J. (2025b). Legitimacy of investigative forensic genetic genealogy under Art. 8 ECHR. *Forensic Science International: Synergy, 11*, 1–11. https://doi.org/10.1016/j.fsisyn.2025.100636.

Tuazon, O. M., et al. (2024). Law enforcement use of genetic genealogy databases in criminal investigations: Nomenclature, definition and scope. *Forensic Science International: Synergy, 8*, 1–12. https://doi.org/10.1016/j.fsisyn.2024.100460.

Tuazon, O. M. (2023, March 10). *Investigative forensic genetic genealogy (iFGG).* https://www.leidenlawblog.nl/articles/investigative-forensic-genetic-genealogy.

Tuazon, O. M. (2024, September 12). *Solving cold cases: Should the Netherlands use American genetic genealogy databases*? https://www.leidenlawblog.nl/articles/solving-cold-cases-should-the-netherlands-use-american-genetic-genealogy-databases.

Tuazon, O. M. (2021). Universal forensic DNA databases: Acceptable or illegal under the European Court of Human Rights regime? *Journal of Law and the Biosciences, 8*, 1–24. https://doi.org/10.1093/jlb/lsab022.

UNESCO. (1997, November 11). *Universal Declaration on the Human Genome and Human Rights.* https://www.unesco.org/en/legal-affairs/universal-declaration-human-genome-and-human-rights.

US Department of Justice, Office of Legal Policy (2019). *Interim policy: Forensic genetic genealogical DNA analysis and searching.* https://www.justice.gov/olp/page/file/1204386/download.

Van der Velden v. the Netherlands. (2006). no. 29514/05 [dec] EctHR.

Wickenheiser, R. A., & Farrell, L. (2020). Collaborative versus traditional method validation approach: Discussion and business case. *Forensic Science International: Synergy, 2*, 230–237. https://doi.org/10.1016/j.fsisyn.2020.08.003.

Wickenheiser, R. A., et al. (2023a). National Technology Validation and Implementation Collaborative (NTVIC) policies and procedures for Forensic Investigative Genetic Genealogy (FIGG). *Forensic Science International: Synergy, 6*, 1–5. https://doi.org/10.1016/j.fsisyn.2023.100316.

Wickenheiser, R. A., et al. (2023b). National Technology Validation and Implementation Collaborative (NTVIC): Guidelines for establishing Forensic Investigative Genetic Genealogy (FIGG) programs. *Forensic Science International: Synergy, 7*, 1–7. https://doi.org/10.1016/j.fsisyn.2023.100446.

Wickenheiser, R. A., et al. (2025). National Technology Validation and Implementation Collaborative (NTVIC): Updated guidelines for establishing Forensic Investigative Genetic Genealogy (FIGG) programs. *Forensic Science International: Synergy, 11*, 1–12. https://doi.org/10.1016/j.fsisyn.2025.100593.

Wickenheiser, R. A. (2022). Expanding DNA database effectiveness. *Forensic Science International, 4*, 1–17. https://doi.org/10.1016/j.fsisyn.2022.100226.

Wickenheiser, R. A. (2019). Forensic genealogy, bioethics and the Golden State Killer case. *Forensic Science International: Synergy, 1*, 114–125. https://doi.org/10.1016/j.fsisyn.2019.07.003.

Wyner, N., Barash, M., & McNevin, D. (2020). Forensic autosomal short tandem repeats and their potential association with phenotype. *Frontiers in Genetics, 11*, 1–7. https://doi.org/10.3389/fgene.2020.00884.

Zieger, M. (2022). Forensic DNA phenotyping in Europe: How far may it go? Journal of Law and the Biosciences, *9*, 1–22. https://doi.org/10.1093/jlb/lsac024.

Chapter 29

Challenges in interpretation of compromised cases by DNA profiling

Noora R. Al-Snan
Forensic Science Laboratory, Directorate of Forensic Science, General Directorate of Criminal Investigation and Forensic Science, Ministry of Interior, Manama, Bahrain

29.1 Introduction

DNA profiling, also known as DNA fingerprinting, has revolutionized forensic science and the criminal justice system over the past few decades. This powerful tool allows for the identification of individuals based on their unique genetic makeup. The process involves analyzing specific regions of DNA that vary greatly among individuals, known as short tandem repeats (STRs), which are highly polymorphic and provide a distinctive genetic profile for each person.

The development and refinement of DNA profiling techniques have led to significant advancements in solving crimes, identifying victims, and exonerating the wrongfully convicted. High-profile cases such as the O.J. Simpson trial and the exoneration of numerous individuals through organizations like the Innocence Project have brought DNA evidence into the public spotlight and highlighted its potential to transform forensic investigations (Dwyer et al., 2000).

However, the interpretation of DNA evidence is not without its challenges, particularly in cases where the DNA samples are compromised. Compromised DNA samples can result from various factors, including environmental exposure, contamination, degradation, inhibition, complex mixtures of DNA from multiple individuals, and low quantities of DNA. These factors can complicate the analysis and interpretation of DNA evidence, potentially leading to erroneous conclusions and miscarriages of justice. Table 29.1 illustrates the key aspects of DNA profiling.

One of the primary challenges in DNA profiling is contamination. Contamination can occur at any stage of the DNA analysis process, from sample collection to amplification and analysis. Contaminants can originate from the environment, forensic personnel, or other samples handled concurrently. The inclusion of extraneous DNA can lead to misleading or erroneous interpretations, as seen in the case of Amanda Knox, where questions about contamination played a significant role in the trial and subsequent appeals (Gill, 2014).

Degradation of DNA is another significant challenge, especially in cases involving old or improperly stored samples. DNA molecules can break down into smaller fragments due to exposure to environmental conditions such as heat, moisture, and UV radiation. This degradation can result in partial profiles that are difficult to interpret accurately, as demonstrated in the re-examination of evidence in the Boston Strangler case decades after the crimes were committed (Murphy, 2013).

Interpreting DNA mixtures, which contain genetic material from multiple individuals, adds another layer of complexity. Such mixtures are common in violent crimes involving multiple perpetrators or victims. The overlapping peaks in electropherograms make it challenging to distinguish between the contributors, necessitating sophisticated statistical methods and software to deconvolute the profiles. The case of R v Doheny and Adams (1997) highlights the difficulties and potential for error in analyzing complex DNA mixtures.

Low-template DNA, or low-quantity DNA, presents further challenges. When only minute amounts of DNA are available, the risk of stochastic effects, such as allelic drop-out and drop-in, increases. These effects can lead to incomplete or misleading profiles, complicating the interpretation process. The reliance on low-template DNA in the case of R v Broughton (2010) raised significant questions about the reliability of the evidence (Gill, 2014).

Moreover, interpretation biases pose a substantial risk in forensic DNA analysis. Cognitive factors, such as confirmation bias, can influence analysts' interpretation of DNA evidence, leading them to favor evidence that supports their

Investigative and Predictive DNA Testing. DOI: https://doi.org/10.1016/B978-0-443-30128-5.00025-4

TABLE 29.1 Key aspects of DNA profiling.

Aspect	Description
Definition	DNA profiling (or DNA fingerprinting) identifies individuals based on unique genetic markers.
Primary technique	Analysis of short tandem repeats (STRs), which are highly polymorphic regions of DNA.
Applications	Solving crimes, identifying victims, exonerating the wrongfully convicted.
High-profile cases	O.J. Simpson trial, Innocence Project exonerations.
Challenges	Contamination, degradation, inhibition, complex mixtures, low-template DNA, interpretation biases.

preconceived notions or the prosecution's theory of the case. Measures such as blind testing and independent review are essential to mitigate these biases and ensure objective analysis (Dror & Hampikian, 2011).

The legal and ethical implications of these challenges are profound. Misinterpretation of DNA evidence can lead to wrongful convictions, undermining the integrity of the judicial system and causing irreparable harm to innocent individuals. The ethical responsibilities of forensic scientists include ensuring the accuracy and reliability of their analyses, disclosing any limitations or uncertainties in the DNA evidence, and respecting individuals' privacy rights (Kaye et al., 2010).

Despite these challenges, ongoing advancements in forensic DNA analysis continue to improve the accuracy and reliability of DNA profiling. Technological innovations such as next-generation sequencing (NGS) and improved statistical models offer promising solutions to many of the current limitations. By embracing these innovations and adhering to high standards of practice, the forensic community can ensure that DNA evidence remains a powerful and reliable tool in the pursuit of justice.

This chapter delves into these challenges in detail, exploring the sources and impact of each issue on forensic investigations. By examining real-world cases and advancements in forensic DNA analysis, we aim to provide a comprehensive understanding of the complexities involved in interpreting compromised DNA evidence and highlight the importance of ongoing research and development in this critical field.

29.2 Contamination

Contamination of DNA samples is one of the most significant challenges in forensic DNA profiling. It can occur at various stages of the DNA analysis process, from the initial collection of samples, extraction, amplification, to their analysis process in the laboratory. Contaminants can originate from the environment, forensic personnel, other samples handled concurrently, or even from the reagents and equipment used in the analysis. Contamination can lead to the inclusion of extraneous DNA, which may result in misleading or erroneous interpretations.

29.2.1 Sources of contamination

29.2.1.1 Environmental contamination

One of the primary sources of contamination is the environment. DNA is ubiquitous, and microscopic amounts can easily transfer between surfaces and individuals. For example, secondary transfer of DNA can occur when an individual touches an object and then another person touches the same object, inadvertently transferring the first person's DNA. This issue was highlighted in the case of R v Reed & Reed; R v Garmson (2009), where the possibility of secondary transfer was a pivotal point in the defense argument.

Environmental factors such as wind, water, and other elements can also contribute to contamination. For instance, DNA from unrelated sources can be deposited on surfaces through airborne particles. This was demonstrated in studies showing how DNA can travel through the air and settle on objects, complicating forensic investigations (Cale et al., 2015).

A study by Goray, E. and van Oorschot, R. A. H. (2015) demonstrated that DNA can be transferred through secondary contact, where an intermediary surface or object acts as a vector. This secondary transfer of DNA can complicate the interpretation of forensic evidence, as it may not be immediately apparent that the DNA detected on an item of evidence originated from someone who never directly touched it (Goray & van Oorschot, 2015).

29.2.1.2 Personnel contamination

Forensic personnel can introduce contamination if proper protocols are not followed. This includes the use of contaminated gloves, tools, or work surfaces. The case of Amanda Knox is a notorious example, where DNA evidence was questioned due to potential contamination issues during evidence collection and processing (Gill, 2014).

Proper training and stringent adherence to contamination prevention protocols are essential. The use of personal protective equipment (PPE), regular decontamination of workspaces, and the implementation of strict chain-of-custody procedures can help minimize the risk of contamination. However, even with rigorous protocols, the risk of contamination cannot be entirely eliminated, underscoring the need for careful interpretation of DNA evidence.

29.3 Degradation

DNA degradation poses another formidable challenge in forensic analysis. Degradation occurs when the DNA molecules break down into smaller fragments, often due to exposure to environmental conditions such as heat, moisture, and UV radiation. Degraded DNA can result in partial profiles, which are more difficult to interpret accurately (Hall et al., 2016).

29.3.1 Factors contributing to DNA degradation

- **Heat**: High temperatures accelerate the breakdown of DNA molecules, making it challenging to obtain complete genetic profiles from samples exposed to heat. Studies have shown that prolonged exposure to heat can cause significant degradation, resulting in shorter and fragmented DNA strands.
- **Moisture**: Humid environments can promote the hydrolytic cleavage of DNA, leading to the formation of nicks and breaks in the DNA strands. This degradation is often exacerbated in forensic samples that have been exposed to moisture for extended periods.
- **UV Radiation**: Ultraviolet (UV) radiation from sunlight can cause thymine dimers and other photoproducts to form in the DNA, leading to strand breaks and cross-linking. UV exposure is particularly detrimental to DNA preserved in outdoor crime scenes, where it can lead to significant degradation over time.

29.3.2 Environmental impact on degradation

Degraded DNA often results in partial profiles, which are incomplete and can be challenging to interpret accurately. Partial profiles may lack some genetic markers, making it difficult to match the DNA to a specific individual conclusively. This can lead to challenges in forensic investigations, as incomplete profiles can either exclude potential suspects or fail to provide sufficient evidence for inclusion. For example, a study by Butler (2005) highlighted the difficulties in interpreting degraded DNA profiles, noting that the loss of alleles and the presence of artifacts can complicate the analysis (Butler, 2005).

Degradation is particularly problematic in cases involving old or improperly stored samples. The case of the Boston Strangler, where DNA evidence was revisited decades after the crimes, illustrates the difficulties associated with degraded DNA. In this case, the forensic team had to work with highly degraded samples to extract viable DNA profiles (Murphy, 2013).

Studies have shown that environmental factors play a crucial role in DNA degradation. For instance, exposure to high temperatures accelerates the breakdown of DNA molecules. Similarly, moisture and microbial activity can cause significant degradation, leading to partial or unusable DNA profiles (Daly et al., 2012).

29.3.3 Strategies to mitigate degradation

To mitigate the impact of degradation, forensic scientists use various strategies. One approach is the use of shorter DNA fragments, known as miniSTRs, which are more likely to be preserved in degraded samples. MiniSTRs target smaller regions of DNA, increasing the chances of obtaining a usable profile from degraded samples.

Another strategy involves the use of more robust preservation techniques. For example, samples can be stored at lower temperatures or in a desiccated environment to reduce the rate of degradation. Additionally, the use of specialized buffers and preservatives during the collection and storage of samples can help maintain DNA integrity over time.

29.4 Inhibition

DNA inhibition is a significant challenge in forensic DNA analysis, particularly when dealing with compromised samples. This issue arises when substances co-extracted with DNA interfere with the polymerase chain reaction (PCR) process, preventing the amplification of DNA and leading to false-negative results (Vajpayee et al., 2023).

29.4.1 Factors contributing to DNA inhibition

- **Environmental Contaminants**: Forensic samples often come from environments rich in contaminants, such as soil, water, or decaying organic matter. These contaminants can include humic acid, hematin, and other substances that inhibit PCR by binding to DNA or the polymerase enzyme (Hosbrough, 2020).
- **Sample Collection and Handling**: Improper collection, storage, or handling of forensic samples can introduce inhibitors. For instance, blood, tissue, and other biological samples might contain substances that impede PCR if not adequately purified (Hosbrough, 2020).
- **Forensic Processing**: During forensic analysis, materials and chemicals used in processing can also act as inhibitors if they are not fully removed before PCR amplification (Hosbrough, 2020).

29.4.2 Impact of DNA inhibition

- **False Negatives**: The primary impact of DNA inhibition is the potential for false-negative results, where the DNA present in a sample fails to amplify, leading to an erroneous conclusion that no DNA is present (Hsiao, 2019).
- **Partial Profiles**: Inhibited PCR reactions can produce partial DNA profiles that are difficult to interpret accurately. This is particularly problematic in forensic cases where complete profiles are essential for identification and comparison (Hsiao, 2019).
- **Increased Complexity in Interpretation**: DNA inhibition adds another layer of complexity to the interpretation of forensic DNA evidence. Analysts must distinguish between true allelic drop-out caused by inhibition and other forms of DNA degradation or low-template DNA effects (Hsiao, 2019).

29.5 Complex mixtures

DNA mixtures, which contain DNA from multiple individuals, are common in forensic cases, especially in violent crimes involving multiple perpetrators or victims. Interpreting these mixtures is inherently complex due to the overlapping peaks in electropherograms, which represent the combined genetic material of all contributors.

In electropherograms, the graphical representation of DNA fragments, peaks corresponding to alleles from different individuals can overlap. This overlap complicates the determination of which alleles belong to which contributor, making it difficult to interpret the DNA profile accurately. The challenge lies in distinguishing between true alleles and artifacts, such as stutter peaks, which are common in DNA mixture profiles. (Bieber et al., 2016).

This figure (Fig. 29.1) illustrates an example of electropherogram obtained from a bag containing RDX-C4 substance. The DNA mixture obtained for more than 3–4 contributors (Al-Snan, 2021).

29.5.1 Challenges in mixture interpretation

The interpretation of DNA mixtures requires sophisticated statistical methods and software to deconvolute the profiles. However, these methods are not foolproof and can lead to differing interpretations. In the case of People v. Puckett (2010), the presence of a complex DNA mixture led to significant debate over the inclusion of the defendant as a contributor, demonstrating the challenges and potential for error in mixture analysis.

Mixture interpretation is further complicated by the presence of low-level contributors. When DNA from a minor contributor is present in small quantities, it can be challenging to distinguish their profile from the background noise. This issue was evident in the investigation of the 2001 anthrax attacks, where the forensic team faced difficulties in interpreting mixtures containing trace amounts of DNA from multiple sources (Budowle et al., 2009).

Determining the number of contributors to a DNA mixture is another critical challenge. Variability in peak heights and areas in the electropherogram can make it difficult to accurately estimate the number of individuals contributing to the mixture (Butler & Kline, 2005).

FIGURE 29.1 Example of complex DNA mixture. An example of electropherogram for more than 3–4 contributors.

29.5.2 Statistical approaches to mixture analysis

To address the challenges of mixture interpretation, forensic scientists employ various statistical approaches. One widely used method is the combined probability of inclusion (CPI), which calculates the probability that a randomly selected individual would be included in the DNA mixture. However, CPI has limitations, particularly when dealing with complex mixtures containing multiple contributors (Gill et al., 2006).

More advanced statistical models, such as likelihood ratios (LRs) and probabilistic genotyping, offer improved accuracy in mixture interpretation. These models consider the probability of observing the DNA evidence under different hypotheses, providing a more robust framework for interpreting complex mixtures. Software tools like STRmix and TrueAllele have been developed to facilitate probabilistic genotyping, allowing forensic analysts to better handle mixtures (Perlin, 2015).

29.6 Low-quantity DNA

Low-quantity DNA, often referred to as low-template DNA (LT-DNA), is another challenge in forensic analysis. When only minute amounts of DNA are available, the risk of stochastic effects increases. These effects include allelic drop-out, where some alleles fail to amplify, and allelic drop-in, where extraneous alleles appear in the profile. Both phenomena can complicate the interpretation and reliability of the DNA profile.

29.6.1 Stochastic effects and their impact

The interpretation of low-template DNA (LT-DNA) profiles requires meticulous consideration due to the inherent challenges posed by such samples. Low-template DNA analysis often necessitates additional amplification cycles to generate sufficient genetic material for profiling. However, this increased amplification can exacerbate issues such as contamination and stochastic effects, which complicate the interpretation process. In the case of R v Broughton (2010), the reliance on low-template DNA raised questions about the reliability of the evidence, ultimately influencing the outcome of the case (Gill, 2014). The variability and potential for contamination associated with LT-DNA significantly influenced the outcome of the case, underscoring the need for rigorous interpretation protocols in forensic analysis.

29.6.1.1 *Stochastic effects in low-template DNA analysis*

Stochastic effects refer to the random variation in DNA profile components, which is particularly problematic in LT-DNA analysis. These effects can lead to significant variability in the profiles generated from such samples, making it difficult to obtain reproducible results. One major stochastic effect is allelic drop-out, where one allele of a heterozygous pair fails to amplify, resulting in incomplete profiles. This can potentially exclude true suspects or, conversely, falsely implicate innocent individuals. Stochastic effects can lead to significant variability in DNA profiles generated from low-template samples. This variability makes it challenging to obtain reproducible results, complicating the interpretation process. For instance, allelic drop-out can result in incomplete profiles that may wrongly exclude potential suspects or falsely implicate individuals.

Allelic Drop-Out: This occurs when some alleles fail to amplify during the PCR process, leading to incomplete DNA profiles. For example, if a heterozygous individual (having two different alleles at a locus) only shows one allele after amplification, this can result in incorrect exclusion of a potential suspect or failure to match a profile accurately. Allelic drop-out is a critical concern in LT-DNA analysis because it can lead to significant misinterpretations of genetic evidence (Taupin & Moira, 2013).

Allelic Drop-In: This phenomenon involves the appearance of extraneous alleles in the DNA profile, which were not present in the original sample. These additional alleles can arise from contamination or from the amplification of minute quantities of DNA from sources not related to the case. Butler (2015) highlights that allelic drop-in can falsely implicate individuals or create misleading links between samples (Butler, 2015).

29.6.1.2 *Quality control and validation in LT-DNA analysis*

To mitigate these challenges, implementing robust quality control measures and stringent validation protocols is crucial. These practices ensure the consistency and validity of the results, helping to account for the variability introduced by stochastic effects. Moreover, advancements in DNA profiling technologies, such as capillary electrophoresis and massively parallel sequencing (MPS), offer improved sensitivity and accuracy, further enhancing the reliability of LT-DNA analysis.

29.6.2 Techniques to enhance low-template DNA analysis

To improve the reliability of low-template DNA analysis, forensic scientists use several techniques. One approach is to increase the number of amplification cycles during the PCR process. While this can enhance the signal from low-template DNA, it also increases the risk of amplifying contaminants and introducing stochastic effects. Therefore, careful optimization of PCR conditions is essential (Gill, 2014).

Another technique involves the use of specialized DNA extraction methods that maximize the recovery of DNA from low-template samples. Methods such as micro-concentration and enhanced magnetic bead-based extraction can improve the yield of DNA from trace samples, increasing the chances of obtaining a usable profile (Budowle et al., 2009).

Utilizing advanced DNA profiling technologies such as capillary electrophoresis and MPS can further enhance the sensitivity and accuracy of low-template DNA analysis. These methods allow for the detection of alleles present in minute quantities, enabling forensic scientists to generate reliable genetic profiles from challenging samples. Additionally, implementing rigorous quality control measures and validation protocols is crucial to ensuring the consistency and validity of results in forensic DNA analysis (Young et al., 2017).

29.7 Interpretation bias

Interpretation bias is an inherent risk in forensic analysis, where subjective judgment can influence the interpretation of DNA evidence. Bias can arise from cognitive factors, such as confirmation bias, where analysts may subconsciously favor evidence that supports their preconceived notions or the prosecution's theory of the case. It is crucial for forensic analysts to be aware of their own biases and to follow rigorous protocols to minimize the impact of subjective judgment on their interpretations. By ensuring objectivity and maintaining a neutral stance, analysts can help ensure that the results of forensic analysis are based solely on the scientific evidence presented.

29.7.1 Sources of interpretation bias

Interpretation bias can stem from various sources, including the analyst's expectations, case context, and pressure to achieve a particular outcome. Studies have shown that analysts can be influenced by irrelevant contextual information, leading to biased interpretations. For example, knowing that a suspect has a prior criminal record can affect an analyst's judgment, increasing the likelihood of a biased conclusion (Dror & Charlton, 2006).

The risk of bias is particularly high in cases involving complex mixtures or low-template DNA, where the interpretation is inherently more challenging. In such cases, analysts may be more susceptible to confirmation bias, selectively interpreting ambiguous data in a manner that supports their expectations (Dror et al., 2011). This can potentially lead to inaccurate conclusions and wrongful convictions. Therefore, it is crucial for forensic analysts to be aware of their biases and take steps to mitigate them in order to ensure the integrity of the analysis and the justice system as a whole.

29.7.2 Mitigating interpretation bias

To mitigate interpretation bias, forensic laboratories implement various measures. One effective approach is blind testing, where analysts are unaware of the case context and the identities of the individuals involved. This helps reduce the influence of cognitive biases and promotes objective interpretation of the evidence (Koehler, 2008).

Independent review and verification of DNA profiles by multiple analysts can also help mitigate bias. By having different analysts independently interpret the same data, the likelihood of bias affecting the final conclusion is reduced. Additionally, the use of automated software for DNA profile interpretation can minimize human subjectivity and enhance the consistency of results. Furthermore, ongoing training and education for analysts are crucial in staying up to date with the latest forensic techniques and research findings. By continuously enhancing their skills and knowledge, analysts can better recognize and mitigate potential biases in their analyses. Regular proficiency testing and quality assurance programs can also help ensure the reliability and accuracy of forensic analyses, ultimately increasing confidence in the results produced by forensic laboratories. Table 29.2 summarizes the challenges in DNA profiling.

29.8 Legal and ethical considerations

The challenges in interpreting compromised DNA cases have significant legal and ethical implications. Misinterpretation of DNA evidence can lead to wrongful convictions, undermining the integrity of the judicial system and causing irreparable harm to innocent individuals. In addition, mishandling of compromised DNA cases can also prevent the true perpetrator from being held accountable, allowing them to remain free to commit further crimes. This highlights the crucial need for thorough investigation, proper training of forensic analysts, and strict adherence to protocols in handling and interpreting DNA evidence. Failure to do so can result in grave consequences for both the wrongly accused and society as a whole.

29.8.1 Wrongful convictions

Wrongful convictions based on faulty DNA evidence highlight the critical need for accurate and reliable forensic analysis. The case of Cameron Todd Willingham, who was executed based on forensic evidence later deemed unreliable, underscores the devastating impact of errors in forensic interpretation (Dror & Hampikian, 2011).

To prevent wrongful convictions, forensic laboratories must adhere to stringent standards and protocols. Continuous training and certification of forensic analysts, along with regular audits and proficiency testing, are essential to maintaining high standards of practice. Additionally, forensic evidence should be considered in conjunction with other forms of evidence, ensuring a holistic approach to criminal investigations. By ensuring thorough oversight and rigorous quality control measures within forensic laboratories, we can better safeguard against the risk of wrongful convictions. Collaboration across disciplines and transparency in the forensic process are also crucial in upholding the integrity of forensic evidence. Ultimately, a commitment to accuracy and reliability in forensic analysis is paramount in delivering justice and upholding the rights of the accused.

29.8.2 Ethical responsibilities

Forensic scientists have an ethical responsibility to ensure the accuracy and reliability of their analyses. This includes disclosing any limitations or uncertainties in the DNA evidence and avoiding overstatement of the evidence's significance. Transparency and honesty in reporting findings are crucial to upholding the integrity of forensic science (Kaye et al., 2010).

Furthermore, the use of DNA evidence must respect individuals' privacy rights. The collection, storage, and analysis of DNA samples should be conducted in accordance with legal and ethical guidelines to protect individuals' genetic information from misuse or unauthorized access. Forensic scientists must also consider the potential biases that may exist

TABLE 29.2 Challenges in DNA profiling.

Challenge	Description	Example case
Contamination	Extraneous DNA from the environment or personnel can lead to misleading interpretations.	Amanda Knox trial (Gill, 2014)
Degradation	Environmental factors like heat and UV radiation cause DNA to break into smaller fragments.	Boston Strangler case (Murphy, 2013)
Inhibition	Substances co-extracted with DNA interfere with the polymerase chain reaction (PCR) process	Amanda Knox trial (Gill, 2014)
Complex mixtures	DNA from multiple individuals complicates interpretation due to overlapping peaks in data.	People v. Puckett (2010)
Low-template DNA	Small quantities of DNA increase the risk of stochastic effects, leading to incomplete profiles.	R v Broughton (2010)
Interpretation biases	Cognitive biases can influence analysts, leading to skewed interpretations.	General issue in forensic analysis

in the collection, analysis, and interpretation of DNA evidence. It is essential to be aware of factors that could influence results, such as sample contamination, human error, or technological limitations. By remaining vigilant and thorough in their work, forensic scientists can minimize the risk of errors and ensure that justice is served accurately and fairly.

Table 29.3 explains the ethical and legal implications when handling DNA cases and the mitigation strategies and solutions to overcome the issues.

29.9 Advances in forensic DNA analysis

Despite the challenges, advances in forensic DNA analysis continue to improve the accuracy and reliability of DNA profiling. Technological innovations and methodological improvements offer new opportunities to address the limitations of traditional techniques. The integration of NGS technologies, the development of miniSTRs, and the utilization of probabilistic genotyping software have all played a significant role in enhancing the sensitivity and robustness of forensic DNA analysis. These advancements increase the likelihood of generating usable profiles from challenging samples, such as degraded or mixture samples, ultimately strengthening the criminal justice system and promoting justice for all. Fig. 29.2 illustrates the advancements in forensic DNA analysis, which can aid in solving DNA cases and minimize the challenges.

29.9.1 Next-generation sequencing

NGS is a cutting-edge technology that allows for the simultaneous analysis of multiple genetic markers. NGS can provide more detailed and comprehensive DNA profiles, improving the resolution of complex mixtures and degraded samples. By analyzing a broader range of genetic markers, NGS can enhance the discrimination power of DNA profiling and reduce the likelihood of ambiguous results (King et al., 2014).

NGS also enables the analysis of mitochondrial DNA (mtDNA) and Y-chromosome DNA, which can be particularly useful in cases where nuclear DNA is highly degraded or present in low quantities. These additional genetic markers provide valuable information that can complement traditional STR analysis, improving the overall reliability of forensic DNA evidence (Parson et al., 2018).

NGS technology has revolutionized the field of forensic genetics by offering unparalleled insights into DNA analysis. The ability to simultaneously examine multiple genetic markers with high sensitivity and accuracy has significantly advanced the capabilities of forensic investigations. This enhanced level of detail and precision not only aids in solving complex cases but also contributes to the overall reliability and credibility of DNA evidence presented in courts of law. With NGS paving the way for more in-depth genetic analysis, forensic scientists are better equipped than ever to unravel the mysteries hidden within DNA samples, ultimately serving the cause of justice with greater efficacy and confidence.

TABLE 29.3 Ethical and legal implications.

Implication	Description	Mitigation strategies
Wrongful convictions	Misinterpretation can lead to convicting innocent individuals.	Blind testing, independent review
Integrity of judicial system	Errors in DNA evidence can undermine public trust in the legal system.	Adherence to high standards of practice
Privacy rights	Unauthorized use or disclosure of genetic information can violate privacy rights.	Strict privacy protocols, informed consent
Ethical responsibilities	Forensic scientists must ensure accuracy, disclose limitations, and respect individuals' rights.	Ongoing training, ethical guidelines

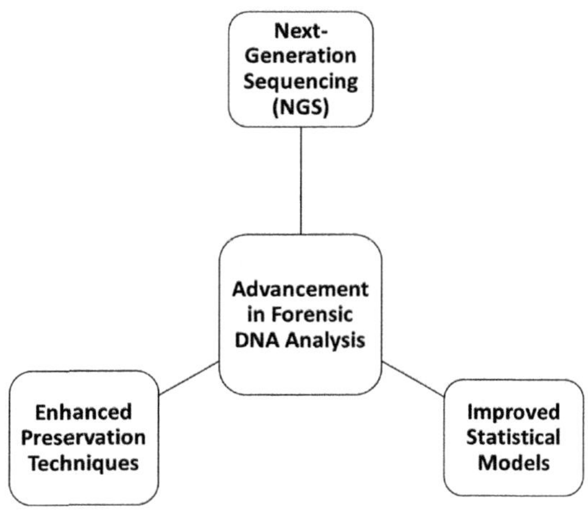

FIGURE 29.2 Advancement in forensic DNA analysis. Advancement in forensic DNA analysis, which can assist in lowering the challenges.

29.9.2 Improved statistical models

Advancements in statistical modeling and computational tools have significantly enhanced the interpretation of DNA evidence. Probabilistic genotyping software, such as STRmix and TrueAllele, utilizes sophisticated algorithms to analyze complex DNA mixtures and low-template samples. These tools provide more accurate and reproducible results, reducing the potential for subjective bias and interpretation errors (Perlin, 2015).

Probabilistic genotyping models incorporate various sources of uncertainty, such as stochastic effects and contamination, into their calculations. By accounting for these factors, the models can generate likelihood ratios that provide a quantitative measure of the strength of the DNA evidence. This approach offers a more rigorous and objective framework for interpreting forensic DNA profiles (Taylor et al., 2017).

The integration of statistical modeling and computational tools has revolutionized the field of forensic DNA analysis. These advancements have not only improved the accuracy and reliability of DNA evidence interpretation but have also allowed for a more transparent and robust analysis process. By leveraging sophisticated algorithms and probabilistic genotyping software, forensic scientists can now extract more meaningful information from complex DNA samples, ultimately enhancing the criminal justice system's ability to identify perpetrators and exonerate the innocent.

29.9.3 Enhanced preservation techniques

The development of enhanced preservation techniques has improved the stability and integrity of DNA samples. New preservation methods, such as the use of specialized buffers and desiccants, can significantly reduce the rate of DNA degradation, ensuring that samples remain viable for longer periods. Additionally, advancements in sample collection and storage protocols have minimized the risk of contamination, enhancing the overall quality of forensic DNA evidence. This has led to more accurate and reliable DNA analysis in forensic investigations, ultimately aiding in the resolution of criminal cases and providing closure for victims and their families. The continual improvement of preservation techniques will undoubtedly further strengthen the field of forensic DNA analysis, allowing for even more precise and impactful investigative work in the future.

29.10 Technological challenges and solutions

While technological advancements have significantly improved forensic DNA analysis, they also present new challenges and require ongoing research and development. One of the main challenges is keeping up with the evolving nature of DNA technology, as new techniques and tools are constantly being developed. Additionally, the increased sensitivity of these methods means that small traces of DNA can now be analyzed, opening up new possibilities for forensic investigations. Researchers must work diligently to ensure the accuracy and reliability of these cutting-edge technologies in order to uphold the integrity of DNA evidence in criminal cases.

29.10.1 Quality control and standardization

Ensuring the quality and consistency of forensic DNA analysis is paramount. Variability in laboratory practices and analytical methods can lead to discrepancies in DNA profiling results. Standardization of protocols and implementation of rigorous quality control measures are essential to maintain the reliability of forensic DNA evidence (Budowle et al., 2009). It is crucial for forensic laboratories to adhere to internationally recognized guidelines and accreditation standards to ensure the accuracy and validity of DNA analysis. Regular proficiency testing and continued education of forensic analysts are also vital components in promoting quality assurance and safeguarding the integrity of the criminal justice system. By upholding these rigorous standards, forensic DNA analysis can continue to serve as a powerful tool in solving crimes and ensuring justice is served.

29.10.2 Software and computational tools

The use of advanced software and computational tools, such as probabilistic genotyping, requires specialized training and expertise. The complexity of these tools can introduce new sources of error if not used correctly. Continuous training and certification of forensic analysts in the use of these tools are crucial to ensure accurate and reliable results (Perlin, 2015).

In addition to specialized training, it is important for forensic analysts to stay current with advancements in software and technology. Regular proficiency testing and proficiency demonstrations can help identify any areas that need improvement and ensure that analysts are using these tools effectively. Continuous learning and adaptation are essential in the field of forensic analysis to maintain the highest standards of accuracy and reliability.

29.10.3 Integration with traditional methods

Integrating new technologies, such as NGS, with traditional DNA profiling methods presents challenges in terms of data interpretation and compatibility. Ensuring that new technologies complement rather than replace existing methods is important for maintaining the robustness of forensic DNA analysis (King et al., 2014). As these advances continue to reshape the landscape of forensic science, it is crucial for the field to adapt and evolve in order to harness the full potential of these technologies. Embracing a harmonious integration of novel techniques with established practices will be essential in bolstering the reliability and accuracy of DNA analysis for investigative and legal purposes. By embracing the synergies between new and traditional methods, forensic scientists can enhance their ability to extract valuable insights from complex biological evidence, ultimately advancing the pursuit of justice.

29.11 Case studies

Examining real-world cases highlights the practical challenges and implications of interpreting compromised DNA evidence. For example, in the case of the Golden State Killer investigation, DNA evidence was crucial in identifying and capturing the perpetrator. However, due to the age of the evidence and the conditions in which it was stored, there were concerns about its integrity and reliability. This raised questions about the accuracy of the forensic analysis and the potential for contamination or degradation of the DNA samples. As a result, there were challenges in interpreting the compromised DNA evidence and ensuring that it was admissible in court. These challenges underscore the importance of proper handling and preservation of DNA evidence to maintain its integrity and validity for forensic analysis.

29.11.1 The Amanda Knox case

The Amanda Knox case is a prime example of the complexities involved in DNA evidence interpretation. Knox and her then-boyfriend Raffaele Sollecito were initially convicted of the murder of Meredith Kercher, largely based on DNA evidence. However, questions about the contamination of DNA samples and the handling of evidence led to their eventual acquittal. The case underscores the importance of proper evidence handling and the potential for contamination to compromise forensic analysis (Gill, 2014).

The Amanda Knox case serves as a poignant reminder of the intricate nature of DNA evidence evaluation. The initial convictions of Knox and Sollecito exemplify the significant impact DNA evidence can have on legal proceedings. However, uncertainties surrounding the integrity of DNA samples and the protocols for handling evidence cast doubt on the reliability of the conclusions drawn. It illuminates the critical significance of meticulous evidence management and the susceptibility of forensic analysis to potential contamination, highlighting the need for stringent procedures to safeguard the integrity of justice.

29.11.2 The Boston Strangler case

Revisiting the case of the Boston Strangler decades after the crimes demonstrated the challenges of working with degraded DNA samples. Advances in DNA analysis allowed forensic scientists to obtain a viable profile from highly degraded samples, ultimately confirming the identity of the perpetrator. This case highlights the importance of preserving evidence and the potential for technological advancements to solve cold cases (Murphy, 2013). In recent years, the field of forensic DNA analysis has seen remarkable progress, with improved techniques and tools enabling investigators to extract valuable information from even the most challenging samples. The ability to work with degraded DNA has opened new possibilities in solving previously unsolvable cases, shedding light on long-standing mysteries, and delivering justice to victims and their families. The relentless pursuit of truth through scientific advancements continues to drive the pursuit of closure and accountability in cases that once seemed destined to remain unsolved.

29.11.3 The 2001 anthrax attacks

The investigation of the 2001 anthrax attacks involved the analysis of complex DNA mixtures. The presence of trace amounts of DNA from multiple sources complicated the interpretation of the evidence. Probabilistic genotyping and advanced statistical methods were crucial in deconvoluting the mixtures and identifying the source of the anthrax spores. This case illustrates the importance of sophisticated analytical tools in interpreting complex DNA evidence (Budowle et al., 2005).

Dealing with mixed DNA samples requires precision and advanced techniques to extract meaningful information. It is evident from the anthrax investigation that probabilistic genotyping plays a key role in untangling complex genetic mixtures. As technology continues to evolve, forensic scientists are better equipped to handle challenging cases and deliver justice through DNA analysis.

29.12 Future directions

The field of forensic DNA analysis is continually evolving, with ongoing research aimed at addressing current challenges and improving the reliability of DNA evidence. One area of particular focus is the development of advanced techniques for analyzing degraded or mixed DNA samples, which can often present obstacles in criminal investigations. Researchers are also exploring ways to enhance the sensitivity and accuracy of DNA testing methods, ensuring that the results are as robust and dependable as possible. By staying at the forefront of scientific advancements, forensic DNA analysis continues to play a crucial role in modern law enforcement and justice systems.

29.12.1 Research and development

Continued research and development are essential to enhance the sensitivity and specificity of DNA profiling methods. Innovations in DNA extraction, amplification, and analysis techniques hold promise for overcoming the limitations of current methods and improving the accuracy of forensic DNA evidence (King et al., 2014). Efforts to explore novel technologies and methodologies in the field of DNA profiling are pivotal in the pursuit of heightened precision and reliability. Ongoing advancements in DNA extraction, amplification, and analytical strategies offer a pathway towards surmounting the existing boundaries and elevating the veracity of forensic DNA data.

29.12.2 Training and education

Investing in the training and education of forensic scientists is critical to ensure that they are equipped with the knowledge and skills required to use advanced technologies and interpret DNA evidence accurately. Ongoing professional development and certification programs can help maintain high standards of practice in forensic laboratories. Continuous investment in research and innovation in the field of forensic science is also essential to keep up with evolving technologies and criminal methods. By staying at the forefront of scientific advancements, forensic scientists can enhance their capabilities and stay ahead in solving complex cases and bringing criminals to justice.

29.12.3 Collaboration and standardization

Collaboration between forensic laboratories, academic institutions, and regulatory bodies is essential to develop standardized protocols and best practices. Standardization can help reduce variability in forensic DNA analysis and ensure the reliability of DNA evidence across different jurisdictions (Budowle et al., 2009).

Establishing consistent procedures and guidelines can also enhance the overall quality and accuracy of DNA testing results. By working together, these entities can share knowledge, resources, and expertise to continually improve the field of forensic DNA analysis. This collaboration ultimately serves to uphold the integrity of the criminal justice system and strengthen public trust in the use of DNA evidence.

29.13 Conclusion

The interpretation of DNA evidence in compromised cases presents numerous challenges that can significantly impact the outcome of forensic investigations. Contamination, degradation, complex mixtures, low quantity of DNA, and interpretation biases all contribute to the potential for errors and misinterpretations. Addressing these challenges requires stringent protocols, advanced analytical techniques, and a commitment to minimizing bias through blind testing and independent review.

As forensic science continues to evolve, ongoing research and development are essential to enhance the reliability and accuracy of DNA profiling in forensic contexts. Technological advancements, such as NGS and improved statistical models, offer promising solutions to many of the current challenges. By embracing these innovations and adhering to high standards of practice, the forensic community can ensure that DNA evidence remains a powerful and reliable tool in the pursuit of justice.

In addition to technological advancements, interdisciplinary collaboration between forensic scientists, geneticists, statisticians, and legal experts is crucial for advancing the field of forensic DNA analysis. By fostering a multidisciplinary approach, researchers can address complex issues and develop comprehensive solutions that meet the rigorous demands of the criminal justice system. Furthermore, continuous training and education are essential for forensic professionals to stay abreast of the latest developments and best practices in DNA analysis, ensuring the highest standards of quality and reliability in forensic investigations.

AI disclosure

During the preparation of this work, the author(s) used ScholarGPT in order to verify references, Language editing, and paraphrasing. After using this tool/service, the author(s) reviewed and edited the content as needed and take(s) full responsibility for the content of the publication.

References

Al-Snan, N. R. (2021). The recovery of touch DNA from RDX-C4 evidences. *International Journal of Legal Medicine, 135*(2), 393–397. https://doi.org/10.1007/s00414-020-02407-9, http://link.springer.de/link/service/journals/00414/index.htm.

Bieber, F. R., Buckleton, J. S., Budowle, B., Butler, J. M., & Coble, M. D. (2016). Evaluation of forensic DNA mixture evidence: Protocol for evaluation, interpretation, and statistical calculations using the combined probability of inclusion. *BMC Genetics, 17*(1). https://doi.org/10.1186/s12863-016-0429-7, http://www.biomedcentral.com/bmcgenet/.

Budowle, B., Eisenberg, A. J., & Van Daal, A. (2009). Validity of low copy number typing and applications to forensic science. *Croatian Medical Journal, 50*(3), 207–217. https://doi.org/10.3325/cmj.2009.50.207.

Budowle, B., Schutzer, S. E., Einseln, A., Kelley, L. C., Walsh, A. C., Smith, J. A. L., Marrone, B. L., Robertson, J., & Campos, J. (2005). Building microbial forensics as a response to bioterrorism. *Science, 301*(5641), 1852–1853. https://doi.org/10.1126/science.1090073.

Butler, J. (2015). *Advanced topics in forensic DNA typing: Interpretation*. Academic Press. https://doi.org/10.1016/B978-0-12-405213-0.00001-1.

Butler, J. M. (2005). *Forensic DNA typing: Biology, technology, and genetics of STR markers*. Elsevier.

Butler, J. M., & Kline, M. C. (2005). *NIST mixture interpretation interlaboratory study 2005 (MIX05*.

Cale, C. M., Earll, M. E., Latham, K. E., & Bush, G. L. (2015). The generation and interpretation of low quantity DNA. *Journal of Forensic Sciences, 60*(2), 396–401 2015.

Daly, D. J., Murphy, C., & McDermott, S. D. (2012). The transfer of touch DNA from hands to glass, fabric and wood. *Forensic Science International: Genetics, 6*(1), 41–46. https://doi.org/10.1016/j.fsigen.2010.12.016.

Dror, I. E., Thompson, W. C., Meissner, C. A., Kornfield, I., Krane, D., Saks, M., & Risinger, D. M. (2011). Context management toolbox: A linear sequential unmasking (LSU) approach for minimizing cognitive bias in forensic decision making. *Journal of Forensic Sciences, 60*(4), 1111–1112 2011.

Dror, I. E., & Charlton, D. (2006). Why experts make errors. *Journal of Forensic Identification, 56*(4), 600–616.

Dror, I. E., & Hampikian, G. (2011). Subjectivity and bias in forensic DNA mixture interpretation. *Science and Justice, 51*(4), 204–208. https://doi.org/10.1016/j.scijus.2011.08.004.

Dwyer, J., Neufeld, P. J., & Scheck, B. (2000). *Actual innocence: Five days to execution and other dispatches from the wrongly convicted*. Doubleday Books.

Gill, P., Brenner, C. H., Buckleton, J. S., Carracedo, A., Krawczak, M., Mayr, W. R., Morling, N., Prinz, M., Schneider, P. M., & Weir, B. S. (2006). DNA commission of the International Society of Forensic Genetics: Recommendations on the interpretation of mixtures. *Forensic Science International, 160*(2-3), 90–101. https://doi.org/10.1016/j.forsciint.2006.04.009.

Gill, P. (2014). *Misleading DNA evidence: Reasons for miscarriages of justice misleading DNA evidence: Reasons for miscarriages of justice*. Norway: Elsevier Inc, 1–174. http://www.sciencedirect.com/science/book/9780124172142, https://doi.org/10.1016/C2013-0-01382-5.

Goray, M., & van Oorschot, R. A. H. (2015). Secondary DNA transfer of biological substances under varying test conditions. *Forensic Science International: Genetics, 15*, 138–145. https://doi.org/10.1016/j.fsigen.2014.10.013.

Hall, A., Sims, L., Foster, A., & Ballantyne, J. (2016). DNA damage and repair in forensic science. In *Forensic science: A multidisciplinary approach*. United States: Wiley-VCH Verlag, 193–214. http://onlinelibrary.wiley.com/book/10.1002/9783527693535, https://doi.org/10.1002/9783527693535.ch10.

Hosbrough, M. K. (2020). A review of PCR inhibition and its mitigation in forensic DNA analysis (Master's thesis).

Hsiao, S. J. (2019). Sources of error in molecular diagnostic analysesIn *Accurate results in the clinical laboratory: A guide to error detection and correction* (Second Edition). United States: Elsevier, 337–347. https://www.sciencedirect.com/book/9780128137765, https://doi.org/10.1016/B978-0-12-813776-5.00021-2.

Kaye, D. H., Bieber, F. R., & Schwab, D. (2010). The new forensic science. *Michigan Law Review, 107*(4), 924–949 2010.

King, J. L., Churchill, J. D., Novroski, N. M. M., Zeng, X., Warshauer, D. H., & Budowle, B. (2014). Increasing the discrimination power of forensic DNA typing: Using next generation sequencing technologies. *Forensic Science International: Genetics, 18*, 20–29 2014.

Koehler, J. J. (2008). Fingerprint error rates and proficiency tests: What they are and why they matter. 5 2008/05 United States *Hastings Law Journal, 59*(5), 1077–1100 5 2008/05 United States.

Murphy, E. (2013). The forensic identification of the Boston Strangler: Rethinking the law's reliance on DNA. *Boston University Law Review, 93*, 130–159 2013.

Parson, W., Eduardoff, M., Ballard, D., & Budowle, B. (2018). Future potential of forensic applications of genomics: The role of mitochondrial DNA and single nucleotide polymorphisms. *Forensic Science International: Genetics, 34*, 153–160 2018.

Perlin, M. W. (2015). Forensic DNA mixture interpretation: Stratospheric science. *Philosophical Transactions of the Royal Society B: Biological Sciences, 370*(1674). https://doi.org/10.1098/rstb.2014.0256.

R v Broughton. (2010). EWCA Crim 549 (UK Court of Appeal).https://www.bailii.org/ew/cases/EWCA/Crim/2010/549.html

R v Doheny and Adams. (1997). 1 Cr App R 369 (UK Court of Appeal). https://www.bailii.org/ew/cases/EWCA/Crim/1996/728.html.

R v Reed & Reed; R v Garmson. (2009). EWCA Crim 2698 (UK Court of Appeal). https://www.bailii.org/ew/cases/EWCA/Crim/2009/2698.html.

Taupin, J., & Moira (2013). Introduction to forensic DNA evidence for criminal justice professionals. CRC Press.

Taylor, D., Bright, J. A., & Buckleton, J. (2017). The interpretation of single source and mixed DNA profiles. *Forensic Science International: Genetics, 24*, 21–25 2017.

Vajpayee, K., Dash, H. R., Parekh, P. B., & Shukla, R. K. (2023). PCR inhibitors and facilitators-Their role in forensic DNA analysis. *Forensic Science International.*

Young, B., King, J. L., Budowle, B., & Armogida, L. (2017). A technique for setting analytical thresholds in massively parallel sequencing-based forensic DNA analysis. *Plos One, 12*(5). https://doi.org/10.1371/journal.pone.0178005.

Chapter 30

Challenges of DNA fingerprinting in mass disaster victim investigations

Venusia Cortellini, Noemi Argirò, Antonio De Luca, and Andrea Verzeletti
Department of Medical and Surgical Specialities, Radiological Sciences and Public Health, Institute of Legal Medicine, University of Brescia, Brescia, Italy

30.1 Introduction

According to the World Health Organization (WHO), mass disasters refer to sudden catastrophic events of sufficient magnitude that require external assistance. These events can be caused by natural factors such as earthquakes, hurricanes, and tsunamis or anthropogenic like terrorist attacks, transportation accidents,[1] etc. In both instances, mass disasters result in personal injuries, mass fatalities, ecological disruption, and extensive damage to property and infrastructure (Alonso et al., 2005; Zięţkiewicz et al., 2012). Understanding the diverse nature of such catastrophes is crucial for successful disaster victim identification (DVI) operations. Natural disasters pose challenges due to their unpredictable nature and the widespread devastation they cause. On the other hand, the anthropogenic ones involve deliberate harm to human life and infrastructure, further complicating the identification process (Interpol, 2023c).

Another classification includes open and closed disasters. Open disasters are characterized by the absence of prior records or defining data available for victims. As a result, they require comprehensive search and recovery efforts to identify and name the deceased. On the other hand, closed disasters feature known, identifiable groups of victims, facilitating more targeted identification efforts such as passenger matching in airline accidents. This further distinction significantly influences DVI strategies since each incident has its characteristics and, in practice, requires different approaches (Caenazzo et al., 2013; Interpol, 2023c).

One of the most pressing challenges in the aftermath of such events is the identification of victims, which requires efficient investigation procedures (Alonso et al., 2005; Watherston et al., 2018). In the past, procedures mainly included the analysis of personal belongings, fingerprints, and dental data. However, with the advancements in molecular biology, DNA fingerprinting, a technique used to analyze specific regions of an individual's DNA, has emerged as the gold standard approach. Such an approach is valuable in cases where remains are degraded beyond recognition, as in mass casualty incidents (Alonso et al., 2005; De Boer et al., 2018; Watherston et al., 2018). In addition, to facilitate individual identification of disaster victims, it can also allow for the reassembly of heavily fragmented remains that would not be possible by any other means (Graham, 2006). Regardless, the amount and degradation of genetic material, combined with the nonhomogeneous nature of remains, often pose challenges to obtaining accurate results (De Boer et al., 2018; Mundorff et al., 2018; Zięţkiewicz et al., 2012). Although guidelines for post-mortem DNA sampling have been developed to overcome such limitations, challenges still need to be addressed (Mundorff et al., 2018; Watherston et al., 2018).

The following section will explore deeper into the challenges and obstacles associated with DNA fingerprinting when confronted with mass disasters. In addition, the guidelines and protocols that govern the process will also be explored.

1. Transportation accidents include different kinds of transport, such as road vehicles, planes, ships, and trains.

Investigative and Predictive DNA Testing. DOI: https://doi.org/10.1016/B978-0-443-30128-5.00029-1

30.2 Guidelines for disaster victim identification

Mass disasters have significantly affected society throughout history and have influenced the evolution of disaster response protocols. From natural calamities like the eruption of Vesuvius in 79 BC to the terrorist attack on September 11, 2001, these disasters have highlighted the importance of effective response planning and rapid victim identification protocols. This necessity has become even more crucial in recent years, primarily due to the escalating frequency of anthropogenic disasters, such as terrorist attacks resulting in a higher number of casualties (Fig. 30.1) and natural disasters (Fig. 30.2).

Since the beginning of 2010, several catastrophic events have occurred worldwide: the earthquake that severely affected Haiti in 2010, killing over 160,000 people. In the past decade, aeroplane crashes caused over 2700 fatalities, while terrorist attacks claimed more than 20,000 victims (Corach, 2017), as illustrated in Fig. 30.1. Over the last decade, the upward trend in both types of disasters underscores the pressing need for comprehensive guidelines and protocols to mitigate their impact and ensure rapid victim identification and response.

Interpol's (International Criminal Police Organization) resolution on DVI is currently the only internationally recognized legislation, and, therefore, the Guide to DVI is the globally accepted standard for DVI protocols. Initially produced in 1984, this guide has changed every five years to incorporate the experience gained by the international community of DVI disciplines. It was most recently published in 2023. It highlights the importance of applying international standards to promote a consistent and widely understood approach, especially in multinational DVI operations. The guide recommends that all 190 member countries adopt a common procedure for identifying victims in any disaster scenario, regardless of its cause or scale (Interpol, 2023c; See et al., 2016). The guide delineates and divides the victim identification process into four essential phases:

Phase 1: Scene Examination—During this phase, the scene is treated as a crime scene, adhering to protocols outlined by Crime Scene Examiners and DVI Specialist Teams. The scene examination and body recovery are conducted systematically, using temporary equipment and body removal methodologies based on the types of threats present. The remains are tagged, photographed, and transported, ensuring coordination with other scene-processing activities. This approach promotes efficiency and maintains the integrity of evidentiary material (Interpol, 2023c; See et al., 2016).

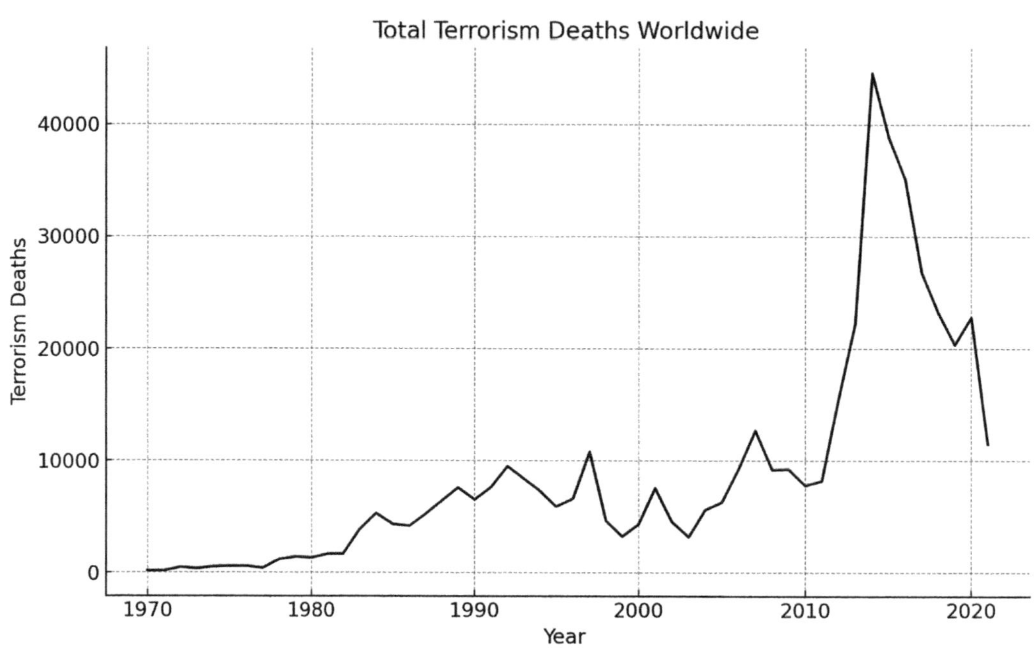

FIGURE 30.1 Total terrorism deaths worldwide. The line chart provides a graphical illustration of the number of fatalities caused by terrorism worldwide between 1970 and 2021. Terrorism Database defines a terrorist attack as an act of violence that seeks to achieve a political, economic, religious, and social goal through fear, intimidation, or coercion. Raw data derived from the Global Terrorism Database and processed using computational tools.
From START (National Consortium for the Study of Terrorism and Responses to Terrorism). (2022). Global Terrorism Database 1970–2020. Retrieved May 13, 2024. https://www.start.umd.edu/gtd.

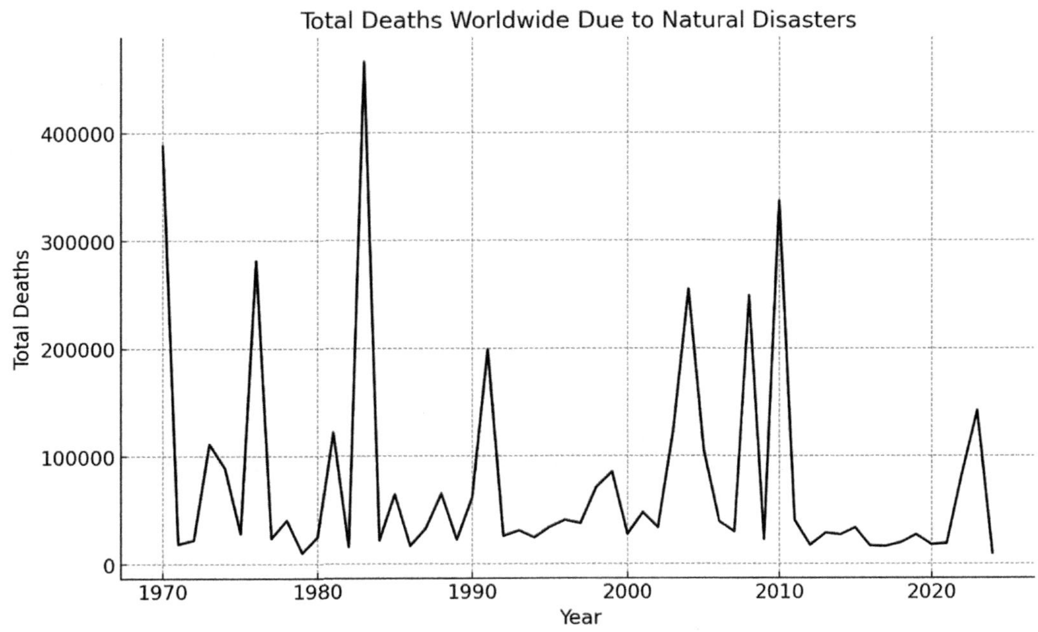

FIGURE 30.2 Total deaths worldwide due to natural disasters. The chart displays the total number of deaths due to natural calamities worldwide from 1970 to 2023, comprising various disasters. Raw data sourced from EM-DAT and processed using computational tools.
From EM-DAT, CRED / UCLouvain, Brussels, Belgium. The international disaster database. (2024). Retrieved August 25, 2024. https://www.emdat.be.

Phase 2: **Postmortem**—Human remains undergo processing, examination, and storage. The examination involves photography, ridge pattern analysis, medical imaging, odontology, DNA sampling, and post-mortem procedures (Interpol, 2023c). In this phase, the DNA fingerprinting takes place with the analysis of the different DNA sources collected.
Phase 3: Antemortem—An antemortem team, primarily composed of investigators, plays a key role in collecting identification data to facilitate the matching process. This phase requires an antemortem collection process to gather missing person information, which will be compared with the victim's data to identify them (Interpol, 2023c; See et al., 2016).
Phase 4: Reconciliation and Identification—At the end of the analysis, postmortem data is matched with antemortem data to name the deceased. This process involves comparing the information provided by families during the antemortem phase with the findings from the examination of bodies conducted during the postmortem phase. If sufficient reliable evidence, such as dental records, fingerprints, or DNA, is available, the individual can be identified (Interpol, 2023c; See et al., 2016).

These guidelines are in continuous improvement considering the experience gained from previous mass fatality incidents. They reinforce the need to take all necessary steps to guarantee DNA sample preservation and to use proper protocols for documenting the chain of custody[2] of DNA sampling and body tracking (Alonso et al., 2005; National Institute of Justice, 2023).

30.3 DNA fingerprinting

DNA fingerprinting, also known as DNA profiling, is a technique employed by forensic scientists to identify individuals or samples through their respective DNA profiles. Although more than 99.1% of the genome is the same throughout the

2. When gathering evidence at a crime scene, investigators establish a tracking record called "chain of custody". This record begins with detailed scene notes that describe where the evidence was received or collected. The process includes collection techniques, preservation, packaging, transportation, storage, and creation of an inventory list. By establishing the chain of custody, investigators ensure the evidence is handled correctly and can be relied upon (National Institute of Justice, 2023).

human population, the remaining 0.9% shows individual variations. These variable DNA sequences, known as polymorphic markers, can be used either to differentiate or correlate individuals. The first practical use of DNA fingerprinting was discovered in 1984 by British scientist Alec Jeffreys of the University of Leicester (Giardina, 2013).

This discovery changed the field of forensic investigations by enabling the possibility of precise human identification from biological samples and introducing new factual evidence to criminal investigations and court proceedings. The first application of DNA fingerprinting was to resolve an immigration dispute in 1985. The DNA technology developed by Jeffrey is now out of date for forensic use and has been overtaken by present-day techniques after continuous development (Shrivastava et al., 2016). The first technique used to type DNA polymorphisms was restriction fragment length polymorphism (RFLP), which relied on restriction enzyme digestion, followed by Southern blot analysis. However, the Southern blot technique was laborious and demanded substantial quantities of high-quality DNA to produce a profile. Currently, the most widely used method to type genetic markers relies on polymerase chain reaction (PCR), a technique capable of amplifying specific DNA fragments in a sample. The introduction of PCR led to the replacement of RFLP markers with short tandem repeats (STRs), which are smaller and more readily amplified. Additionally, the use of small-sized loci in PCR amplification has proven to be effective in cases involving degraded DNA samples, such as in case of mass disaster situations (Butler, 2012; Schneider et al., 2004; Whitaker et al., 1995). STRs are highly polymorphic DNA sequences with repeat units ranging from 2 to 6 base pairs in length, repeated in tandem along the DNA strand. Most of the loci -regions of the genome where the STR markers are located- currently used in forensic DNA analysis are tetranucleotide repeats composed of 4 bp repeat motifs. The number of repeats at each STR locus is highly variable among individuals, which makes them useful for human identification purposes, linkage analysis, and kinship relationships (Giardina, 2013). These STR loci are targeted with specific primers, amplified using PCR, and labeled with fluorescent dyes. Capillary electrophoresis is subsequently employed to separate, identify, and precisely measure the resulting DNA fragments. Advances in genotyping platforms and the study of the human genome sequence have resulted in the identification of several markers for forensic investigation.

Since its introduction, DNA typing has emerged as a primary method for DVI. It has been mainly prominent in cases such as the Spitsbergen aircraft disaster in 1996, where all 141 passengers and crew on board died. The extensive fragmentation of bodies required the application of DNA typing for identification and re-association of the remains (Graham, 2006). Similarly, in 1992, after the terrorist attack at the Israeli Embassy in Argentina, DNA markers were used to identify the victims despite the challenging conditions of the remains (Corach, 2017).

DNA technology offers at least two strategies for human identification:

- Direct comparison of DNA profiles from mass disaster remains with those obtained from reference samples, hair, or samples from personal effects of the victims.
- Samples provided by family members can be utilized for an indirect identification process through kinship analysis in situations where direct comparisons are not feasible.

One of the reasons why DNA typing is a primary methodology for disaster revolves around the possibility of not only identifying individual victims but also facilitating the reassembly of heavily fragmented remains. As such, forensic DNA profiling is increasingly recognized as a necessary tool in identifying individuals following mass disasters, underscoring the need to incorporate DNA sample collection and identification strategies into preparedness plans. However, the identification process presents specific challenges, requiring meticulous technical approaches and intensive labor (Caenazzo et al., 2013).

Efficient DNA extraction is a crucial step in DNA fingerprinting procedures. Over time, various protocols have been developed to ensure effectiveness and remove PCR inhibitors. Additionally, the use of STRs enhances the sensitivity and precision in forensic DNA analysis, in particular, the use of Multiplex STR analysis, capable of examining multiple markers simultaneously, has become a cornerstone in forensic DNA analysis. Standardised protocols for STR analysis have been established globally. Furthermore, advancements in automated detection have simplified DNA testing processes, meeting the growing demand for forensic casework (Shrivastava et al., 2016). In mass disaster cases, large numbers of biological samples and DNA profiles need to be managed, analyzed, and compared. These include mainly autosomal short tandem repeat (STR) profiles, and occasionally mitochondrial DNA (mtDNA) sequences and Y-Chromosome STR (Y-STR) haplotype data. These analyses require data visualization software and bioinformatic tools for DNA database searching according to different matching algorithms. Statistical programs are also needed to evaluate the significance of each match by likelihood ratio (LR) calculation (Alonso et al., 2005), such as LRmix and EuroForMix.

30.3.1 Preanalytical phase (source of DNA)

Any biological remains can be employed as a source of genetic material for DNA extraction or, more broadly, to obtain a genetic profile applicable to victim identification purposes. This method proves valuable in large-scale disasters, but it is crucial to abide by the Interpol Guide and The International Society for Forensic Genetics (ISFG) (Interpol, 2023a; Prinz et al., 2007). These institutions developed recommendations for gathering and preserving antemortem and post-mortem samples suitable for the DNA-based identification of victims based on experiences from recent large-scale disasters. According to the recommendation during sample collection at the disaster site, a forensic biologist or pathologist with basic knowledge of forensic DNA methods should be present to oversee the process. Attention should be paid to proper labeling and preservation of samples along with thorough documentation of the entire process, which are crucial. Factors including the time elapsed since death, the type of disaster (e.g., explosion, fire, shooting attack), and the environmental conditions that may accelerate the deterioration of biological remains can all affect the success and quality of DNA extraction (Interpol, 2023a). Different samples are collected depending on the body's condition and the mass disaster type (See Table 30.1).

Each scenario involving mass disaster also determines the most suitable sample for DNA analysis. Collecting samples from the least affected material is commonly recommended for minimizing exogenous contamination and body cross-contamination (Alonso et al., 2005; Interpol, 2023c).

The following are the most common sources that could be of assistance in mass disasters:

- Blood is the preferred sample for DNA analysis when the body is intact or mutilated but nondecomposed (Cortellini et al., 2022; Watherston et al., 2018). According to current Interpol Guidelines, blood should be collected on Flinders Technology Australia (FTA) paper (Whatman, GE Healthcare, Buckinghamshire, United Kingdom) or a swab. Usually, blood vessels yield better STR typing results than muscle tissue and are often less degraded than DNA from other sources, such as saliva (Scientific Working Group on DNA Analysis Methods SWGDAM, 2014; Watherston et al., 2018).
- Saliva is a suitable DNA source (Cortellini et al., 2022). Its collection is problematic once the production ceases. Buccal smears on FTA cards are recommended if the body is complete and nondecomposed (Watherston et al., 2018). Despite

TABLE 30.1 Illustration of DNA sources and applicability in different scenarios.

Body condition	Recommanded samples
Complete, nondecomposed	• Blood • Saliva
Mutilated, nondecomposed	• Blood • Deep-seated muscle tissue
Complete, decomposed	• Bones (long, compact bone) • Connective tissue (Achilles tendon)
Mutilated, decomposed	• Bones • Teeth
Fragmented body parts	• Bones • Teeth
Severely burnt	• Bladder • Deep-seated tissue
Reference	• Recommanded samples
From the victim	• Saliva (toothbrush) • Hair (brush)
From relatives	• Saliva

this limitation, saliva is primarily used as a reference sample, often obtained from the relatives of a missing person for kinship analysis (Hudson & Connon, 2023).

- Deep-seated red muscle tissue sampling is preferred if the remains are mutilated and incomplete. During decomposition, the body's soft tissues decompose much earlier than the hard tissues, along with DNA. When sampling soft tissue, skeletal muscle is recommended (Mundorff et al., 2018; Watherston et al., 2018). The muscle is recovered following an incision through the skin and then transferred to an FTA card. The FTA card is left to dry, before DNA profiling (Mundorff et al., 2018). In severely burnt corpses, samples from the bladder are a particularly effective alternate source of DNA (Brito et al., 2019).
- Organs and tissues offer an alternative source of DNA for decomposed remains (Watherston et al., 2018). Studies show that hyaline cartilage is superior to skeletal muscle in obtaining STR profiles due to its slower decomposition rate (Trindade-Filho et al., 2008). For instance, the Achilles tendon is a stable tissue and can also be used as a source of DNA in highly decomposed bodies due to the high quantity of DNA it yields (Watherston et al., 2018). A study examining different organs and skin samples indicated the brain, followed by the heart, as the most promising for DNA profiling of long-buried human remains, exhibiting the least degraded profiles (Van den Berge et al., 2016).
- Hairs are primarily used as an antemortem or reference sample; in fact, they are not a reliable source of DNA for identification after death, especially when separated from the body (Cortellini et al., 2022; Watherston et al., 2018). The structural composition of hair, predominantly keratinized proteins, poses challenges in obtaining sufficient DNA for usable profiles. The successful DNA profiling from hair samples depends on the presence of the root and adhering epithelial cells (Watherston et al., 2018).
- Nails are a valuable source of DNA due to their resistance to decay and preservation of DNA content. DNA, found both under and within the keratin structure of nails, remains well-preserved even under various environmental conditions (Allouche et al., 2008). Nail collection is simple, making it an option for urgent DNA-based identification. Even for exhumed and skeletonized bodies, as well as clippings stored for years, profiles have been successfully typed using STRs, providing results when other tissue failed (Watherston et al., 2018).
- Bone cells are reliable DNA sources, especially in adverse environmental conditions and for long-term sampling. Long, compact bones such as femurs are preferred for higher DNA recovery rates due to their unique composition, which is mainly made up of type I collagen and extracellular matrix proteins. Together with femurs, also tibias, pelvis, and metatarsals are identified as some of the most suitable skeletal elements for sample collection (International Commission on Missing Persons ICMP, 2015; Watherston et al., 2018). However, DNA analysis in bones may fail due to poor quality or quantity. In contrast, mtDNA is a viable alternative due to its higher resistance to degradation and abundance per cell (Watherston et al., 2018).
- Teeth, due to their composition and location in the jawbone, have additional protection against environmental and physical conditions that accelerate body decomposition and DNA deterioration. The total DNA content of teeth varies considerably between individuals and within the same individual. Teeth pulp and cementum are valuable sources of both nuclear DNA (nDNA) and mtDNA. Factors such as tooth type, health condition, and age influence the relative proportions of DNA present in teeth. Teeth with larger pulp volumes, such as molars, are preferred for DNA collection (Watherston et al., 2018).

30.3.2 Analytical phase

30.3.2.1 Extraction phase

After collecting biological material from victims of mass disasters, the first step is to extract DNA for subsequent analysis. The analytical approach varies based on the condition and type of the sample. Well-preserved samples will require different methods compared to those that are burned, decomposed, or otherwise compromised, as well as those already processed, such as saliva, bones, or hair. One of the critical factors in successful DNA extraction is addressing potential PCR inhibition, which is the most common cause of PCR failure, even when sufficient DNA is obtained. DNA extraction from postmortem samples often requires optimized conditions to enhance cell lysis and purification due to the progressive degradation of tissues after death (Butler, 2012; Cortellini et al., 2022). For instance, each biological substrate presents unique challenges depending on the expected number of nucleated cells, the tissue's biochemical composition, and the presence of inhibitory substances. For example, in bone DNA extraction, a demineralization process (EDTA solution) is often needed to access DNA, while in blood samples, hemoglobin can act as a potent inhibitor of PCR (Butler, 2005; Butler, 2012; Cortellini et al., 2022). To address these challenges, various DNA extraction methods and commercial kits are available. Today's forensic laboratories commonly rely on techniques such as solid-phase extraction, which include both silica-based (e.g., QIAGEN Investigator DNA Extraction kit [QIAGEN, Hilden, Germany], DNA IQ System

[Promega, Madison, WI, United States], PrepFiler [Applied Biosystem, Foster City, CA, United States]) and FTA paper approaches, as well as Chelex extraction (Butler, 2012; Walsh et al., 2013). Each of these methods offers advantages and are suited to different types of samples and conditions.

Solid-based extraction is a method in which DNA binds to a solid support, enabling the isolation of nucleic acids. This method can be employed with various materials:

Silica-based solid-phase extraction involves the use of silica as the solid support. DNA binds to the silica surface in the presence of chaotropic salts (e.g., guanidine hydrochloride), which disrupt protein structures and enable DNA to adhere to the silica (Connon, 2023). Kits like QIAGEN's QIAamp columns or DNA IQ use this method to efficiently extract DNA from a wide range of forensic samples, including blood, saliva, and bones (Connon, 2023; Cortellini et al., 2022). This technique is particularly effective for recovering high-quality DNA, even from degraded samples, as it removes inhibitors that can interfere with PCR amplification (Butler, 2012; Cortellini et al., 2022).

FTA paper, another form of solid-phase extraction, is a cellulose-based absorbent substrate that contains additives which lyse cells, protect DNA molecules from degradation by nucleases, as well as prevent the growth of microorganisms (Cortellini et al., 2022). DNA applied to FTA paper remains stable at room temperature for years. After a sample (e.g., blood or saliva) is dried, a small punch of the paper is washed to remove inhibitors. Then, the DNA is used directly in PCR. This method is useful for long-term storage and allows for automation but may not be as efficient for very degraded samples as silica-based methods (Connon, 2023; Cortellini et al., 2022).

Chelex is a reagent composed of styrene-divinylbenzene copolymers, and it contains paired iminodiacetate ions that act as chelators for polyvalent metal ions (Barbaro, 2022). The advantages of using Chelex-based protocols are that it is fast, simple, and inexpensive. The disadvantage is that it is not as effective in samples containing inhibitory substances, which are common in forensic cases (Cortellini et al., 2022).

Once extracted, DNA is typically stored at -20°C for short-term use, or at -80°C for long-term preservation to prevent degradation by nucleases (Butler, 2005).

30.3.2.2 DNA quantification

Following DNA extraction, Real-time PCR, or quantitative PCR (qPCR) is performed to assess both the quality and quantity of DNA in the sample. This technique uses fluorescent probes that bind to DNA during the analysis, allowing DNA amplification to be monitored in real-time. qPCR is particularly sensitive and allows the quantification of even very small quantities of DNA, ensuring that the DNA is of sufficient quality to proceed with subsequent analyses, such as genotyping. This provides also qualitative data, as unusual amplification profiles may indicate problems such as the presence of PCR inhibitors or DNA damage. This step is necessary to guide the decision on the appropriate amount of DNA template to include in the PCR reaction based on the kit being employed (Butler, 2005; Butler, 2012). In the forensic context, the kit usually adopted is the Quantifiler Trio DNA Quantification Kit (Applied Biosystems, Waltham, MA, United States), which allows to determine the quantity of human DNA present, the presence of a mixture of male and female DNA, as well as indicate the quality of the sample (Knight et al., 2023).

30.3.2.3 DNA amplification

After the quantification process, the PCR step takes place. The PCR is an enzymatic process in which a specific region of DNA is replicated to yield millions of copies of a particular sequence of interest and thus allows a minute amount of DNA to be examined. In the forensic context, Multiplex PCR is employed which allows to amplify simultaneously more than one region by simply adding more primer sets to the reaction mixture. This approach allows the analysis of different STRs regions, increasing the quality and reliability of the fingerprinting (Butler, 2005; Butler, 2012).

To facilitate the process and ensure global standardization, pre-mixed commercial kits, which contain all the reagents necessary for amplification, are used. These kits are designed to analyze a specific number of STR loci offering greater information power and optimal compatibility with international databases for profile exchange databases. The use of the same kits is useful for standardization procedures for global use, thus simplifying the process of victim identification and other forensic applications (Butler, 2005; Butler, 2012).

Forensic genetics employs various markers for identification, including autosomal STRs, Y-chromosome STRs (Y-STRs), X chromosome STRs (X-STRs), and mtDNA analysis. Depending on the response needs of the mass disaster, its complexity, and the workload, it may be useful to adopt alternative methods instead of the current gold standard, such as SNPs, InDel, and rapid DNA testing, which can complement traditional approaches.

30.3.2.3.1 Autosomal short tandem repeats

Autosomal STRs are highly polymorphic regions of the genome and offer a high rate of discrimination, making them extremely effective for distinguishing between individuals in forensic cases. Their high discriminative power is particularly valuable in mass fatality incidents. Modern multiplex systems, like PowerPlex Fusion 6 C (Promega Corporation, Madison, WI, United States) and AmpFISTR GlobalFiler (Life Technologies, Thermo Fisher Scientific, Waltham, WI, United States), allow for the simultaneous analysis of autosomal loci up to 27(Applied Biosystem, 2019; Promega Corporation, 2021), significantly enhancing the power of identification through the examination of additional genetic markers (Montelius & Lindblom, 2012; Watherston et al., 2018).

Therefore, to make sure data from different laboratories and jurisdictions can be compared reliably, a common set of standardized markers was needed. In 1997, 13 STR markers were selected to generate DNA profiles for inclusion in the US National DNA database, known as CODIS. The 13 CODIS markers are CSF1PO, FGA, TH01, TPOX, vWA, D3S1358, D5S818, D7S820, D8S1179, D13S317, D16S539, D18S51, and D21S11. European forensic DNA laboratories, however, do not share the same loci established for CODIS. In 1999, the European Standard Set (ESS) of STR loci was established, including FGA, TH01, vWA, D3S1358, D8S1179, D18S51, and D21S11. Later, 12 additional loci have been integrated into the ESS (Giardina, 2013). Today, commercially available kits use up to 27 For instance, both kits mentioned earlier include the suggested loci (Applied Biosystem, 2019; Promega Corporation, 2021).

30.3.2.3.2 Single nucleotide polymorphisms

Single-nucleotide polymorphisms (SNPs) are variations in the DNA sequence of a single base between individuals at a particular point in the genome. As the DNA fragments are very short, they are useful for degraded samples (Montelius & Lindblom, 2012; Watherston et al., 2018). They are classified based on their forensic application, such as identity-informative SNPs (IISNPs) for human identification, lineage-informative SNPs (LISNPs) for genealogy studies, phenotype-informative SNPs (PISNPs), and ancestry-informative SNPs (AISNPs). Phenotype-informative SNPs can reveal physical traits such as eye, hair, and skin color. These DNA tools, called forensic DNA phenotyping (FDP), have a high potential for missing persons and DVI cases. It can provide leads to potential victims or their relatives, offering new leads for investigation (Watherston et al., 2018). For these reasons, SNPs were successfully used to identify victims of the 9/11 World Trade Center attack (Montelius & Lindblom, 2012).

30.3.2.3.3 Y Chromosome short tandem repeats and X chromosome short tandem repeats

Y-STRs are useful in forensic investigations to establish paternal lineages as Y-STR profiles remain the same along the male line across generations. This makes them particularly useful for identifying male relatives. While Y-STRs alone offer lower discriminatory power compared to autosomal STRs, because the Y chromosome is passed from father to son with minimal variation, combining Y-STR analysis with autosomal STR markers can increase the overall accuracy of identification (Montelius & Lindblom, 2012; Watherston et al., 2018). In the context of mass DVI, where autosomal STRs may fail due to complex profile interpretation, Y-STRs can confirm familial relationships between male victims. Moreover, in mass disaster cases, Y-STRs can be combined with Y-SNPs to infer the biogeographical ancestry of unknown male victims. This combined approach can provide important clues regarding the ethnic or geographical origins of victims, thanks to a well-established Y-chromosome phylogenetic tree and extensive reference databases (Roewer, 2019).

X chromosome STRs represent another tool for genetic identity testing. The inheritance pattern of the X chromosome is used in certain forensic scenarios, such as complex kinship, missing persons, and DVI cases involving at least one female, especially in situations where other evidence is limited. It is important to note that these markers are primarily applied for population genetics analysis (Butler, 2012; Butler, 2014).

30.3.2.3.4 Mitochondrial DNA

Lineage markers include mtDNA haplotypes that are transferred directly from generation to generation from mother to child, and it is valuable in situations where victims of a disaster are suspected to be related (Butler, 2014). mtDNA is of great advantage in forensics due to its high copy number per cell making it useful when nuclear DNA is not available (compromised and degraded remains). Consequently, this analysis can either confirm or refute maternal relatedness between individuals, providing a tool for determining familial relationships (Watherston et al., 2018).

30.3.2.3.5 Insertion and deletion marker

InDels are small genetic variations where short sequences of DNA are inserted or deleted from the genome. Differences in the genome are used to discriminate between individuals. InDels, which account for 16%–25% of all genetic variations in the genome, are the most diffuse DNA polymorphisms after SNPs. Even though they are so common, studies with InDels are limited since commercial kit production for InDel loci is highly restricted, and forensic scientists often need to create their panels. They can potentially be used to identify disaster victims in mass fatality cases, such as aircraft accidents, terrorist attacks, or natural disasters, ancestry determination, evolutionary research, and molecular anthropology. More successful results can be achieved in identification by using InDel polymorphism with STR and SNP loci (Unsal Sapan, 2021).

30.3.2.3.6 Rapid DNA testing

Rapid DNA refers to the analysis that is performed with the use of a single-instrument profiling platform. This analysis, commonly called the "swab in– profile out" system combined the entire workflow into a single instrument without human intervention. The advantages are groundbreaking in the field application as it delivers results within 90 minutes and automates all the analytical steps without requiring an expert operator. Currently, the two main Rapid DNA instruments in forensic use are the ANDE 6 C (ANDE Corporation, Waltham, MA, United States) instrument and the RapidHIT200 and RapidHITID systems (Thermo Fisher Scientific, Waltham, MA, United States). All three have a similar principle. The user places the sample into a chamber then, within a closed system, the sample is subjected to a DNA extraction process, STR amplification, and separation/detection of fluorescently labeled STR products (Bowman et al., 2022).

30.3.2.4 DNA sequencing

DNA sequencing is a process that allows for the identification of individuals by analyzing specific regions of the genome. All the aforementioned markers, both STRs and sequence polymorphism undergo a sequencing step. Based on the different polymorphisms to investigate the technique employs a size separation technique (electrophoresis) of a sequence reading [Sanger sequencing, next-generation sequencing (NGS)] (Butler, 2012; Shrivastava et al., 2016).

30.3.2.4.1 Short tandem repeat sequencing

The gold standard for DNA fingerprinting is the STRs separation by capillary electrophoresis. To distinguish the various STR regions from one another, a separation step is required to pull the different-sized fragments apart. During electrophoresis, the amplified DNA fragments emit fluorescence when passing through a laser, creating an electropherogram where the peaks represent the specific STR alleles present in the sample. The size of each fragment, determined by comparing it to an allelic ladder which runs on the same plate, allows for the creation of a unique DNA profile for each individual (Butler, 2012; Shrivastava et al., 2016).

30.3.2.4.2 Single-nucleotide polymorphisms and mitochondrial DNA sequencing

In addition to STRs, SNPs,and mtDNA are sequenced using traditional methods such as Sanger sequencing. For both, the process involves PCR amplification of the region containing the SNP or the hypervariable regions, followed by sequencing using fluorescently labeled nucleotides to determine the base at the SNP location. The resulting data is analyzed through electrophoresis, where the sequence is read as fluorescent peaks (Watherston et al., 2018).

30.3.2.4.3 Next-generation sequencing

With the advent of NGS, called in the forensic field Massively Parallel Sequencing, DNA analysis has become more robust and efficient. NGS allows for the simultaneous sequencing of STRs, SNPs, and mtDNA in a single run, offering a view of an individual's genetic information. NGS is particularly beneficial in complex cases, such as mass disasters, where DNA mixtures or degraded samples are common. By providing greater depth and sensitivity, NGS enables the detection of even low-level contributors in mixed DNA samples, improving identification accuracy. Moreover, the integration of multiple marker types enhances the overall resolution of the genetic profile, making it a valuable tool for identifying closely related individuals, determining biogeographical ancestry, and even predicting phenotypic traits. In the context of DNA fingerprinting, DNA allows for STR sequencing, providing not only information about the number of repeats but also about potential internal mutations within the STR regions. These mutations can offer a more unique identification of an individual (Al-Snan, 2023).

30.3.3 Postanalytical phase (interpretation phase)

Once the laboratory workflow is completed, the post-analytical phase begins, focusing on the interpretation, comparison, and reporting of the genetic data. This phase transforms raw sequencing data into DNA profiles that can be used for identification. While various types of genetic information can be extracted from a sample, including SNPs and mtDNA, in this section, we will focus specifically on the comparison of STR profiles. As mentioned throughout the chapter, STRs are considered the gold standard in forensic DNA fingerprinting, and, for instance, STR analysis is the most developed and widely used method in forensic science, supported by comprehensive and well-established databases such as CODIS and ESS (Butler, 2014; Giardina, 2013). As mentioned previously, these sets of markers allow for the identification of a subject in a unique way.

In contrast, while SNPs and mtDNA provide important identification information, their use is currently limited by the lack of extensive databases and standardised comparison methods in forensic contexts. Thus, in this section, we will focus on the comparison of STR profiles and their challenges.

The DNA fragments are analyzed using specialized software, which produces an electrophrogram. An electropherogram is a graphical representation that displays peaks, each corresponding to specific STRs. Each peak on the graph corresponds to a specific DNA fragment, the height of the peak (expressed in RFU) indicates the amount of DNA present, while the position of the peak shows the number of STR repeats at that locus. These peaks are arranged according to the different and together they form a distinct pattern that is unique to each individual. This pattern is what we refer to as the person's DNA profile. The profile is essentially a "genetic fingerprint," which can then be used for further comparison and identification (Butler, 2014; Ziętkiewicz et al., 2012). Once we have the DNA profile, we can interpret it in three main ways:

1. **Direct comparison**: In this approach, the DNA profile is compared to a reference sample taken from the presumed victim (e.g., hair, piece of clothing, or a personal item that contains its DNA). The comparison between the two profiles is done using software like LRmix Studio. The software matches express a likelihood ratio (Caenazzo et al., 2013; Interpol, 2023c).
2. **Indirect comparison**: When a direct reference sample is not available, we compare the DNA profile to that of living relatives to determine kinship relationships. In this case, we use software like Familias, which helps trace family trees and assess whether the DNA from the victim matches that of their relatives, providing evidence of a familial connection (Caenazzo et al., 2013; Interpol, 2022; Interpol, 2023c).
3. **Database search**: If there is no reference sample or sufficient investigative information about the victim the DNA profile is compared against national or international DNA databases to look for potential matches. The profile is checked across thousands of profiles in the database until a match is found or further investigation is required (Giardina, 2013; Interpol, 2023c).

30.4 Mass disaster challenges

When confronted with the aftermath of a mass disaster, the process of DNA fingerprinting for victim identification encounters a multitude of challenges, each posing significant limitations in the investigation for accurate victim identification. These challenges range from DNA degradation due to environmental factors, large volume of samples, handling mixed profiles, and obtaining reliable reference samples for comparison. Furthermore, organizational challenges, such as the large volume of samples and lack of standardised protocols, add complexity to the forensic identification process (Caenazzo et al., 2013; Ziętkiewicz et al., 2012). Addressing these issues requires approaches and often calls for adjustments to traditional forensic protocols. Below, the challenges encountered in mass DVI are divided into three primary categories: intrinsic factors related to the samples, extrinsic factors affecting the samples, and organizational challenges.

30.4.1 Intrinsic challenges of the samples

30.4.1.1 DNA degradation

One of the primary challenges is the degradation of DNA extracted from human remains. DNA Degradation can occur due to various natural and environmental processes, such as extreme temperature (heat, fire), chemical interactions, and natural decomposition processes (Caenazzo et al., 2013). Upon death, DNA is attacked by cellular nucleases and is further degraded by bacterial, fungal, and insect activity, depending on the environmental conditions. All these factors accelerate the breakdown of biological tissues and, consequently, the degradation of DNA molecules themselves (Brenner & Weir,

2003; Bukyya et al., 2021). Degraded DNA often yields incomplete or unreliable profiles, making traditional DNA analysis techniques less effective. This not only slows down the identification process but also complicates the interpretation of genetic data, making it difficult to establish a positive match between victim samples and the reference samples provided (Brenner & Weir, 2003).

30.4.1.2 Mixed profiles interpretation

Another intrinsic challenge is the presence of mixed DNA profiles in samples collected from mass disaster scenes. These mixtures often arise from mixed remains belonging to different individuals or contaminated environments, making it difficult to separate and attribute specific DNA profiles to individual victims (Brenner & Weir, 2003). The presence of multiple individuals' DNA in a single sample complicates the identification process, necessitating advanced laboratory techniques and specialized software, such as STRmix or other statistical programs, to deconvolute mixed profiles. This challenge is further complicated by contamination, which can occur at various stages, including sample collection, handling, and processing (Bukyya et al., 2021).

30.4.2 Extrinsic factors affecting the samples

30.4.2.1 Environmental factors and inhibitors

Mass disaster scenarios often expose biological samples to harsh environmental conditions that degrade DNA and introduce inhibitors that interfere with DNA amplification. These PCR inhibitors may come from the surrounding environment, including substances like soil, chemicals, or materials from the disaster site (Interpol, 2023b). They bind to Taq DNA polymerase during PCR, preventing efficient DNA amplification. Additionally, samples exposed to extreme conditions like flooding, fire, or toxic chemicals may require specific extraction and amplification protocols to yield usable DNA profiles (Bukyya et al., 2021; Butler, 2012).

30.4.2.2 DNA contamination

Contamination poses a significant challenge in the context of mass DVI, especially because a disaster site is not a controlled environment. Contaminants from the environment, from handling by rescue workers, or during sample processing can introduce foreign DNA, complicating the accurate identification of victims. To mitigate this, strict contamination control protocols must be adhered to, including the use of protective equipment, sterile techniques, and clean laboratory spaces (Butler, 2005).

30.4.3 Organizational challenges

30.4.3.1 High volume of samples

The number of victims in mass disaster scenarios can overwhelm forensic laboratories and personnel, leading to delays in the processing and analysis of DNA samples (Saks & Koehler, 2005). For example, large-scale disasters such as plane crashes, earthquakes, or terrorist attacks generate high volumes of remains that need to be identified quickly. This logistical challenge often requires the coordination of multiple laboratories across different regions, each with varying levels of resources and expertise. Rapid DNA technology has been proposed as a potential solution, allowing for on-site analysis and eliminating the need to transport samples to centralized laboratories. However, as discussed later, Rapid DNA has limitations, particularly with highly degraded or mixed samples (Bowman et al., 2022).

30.4.3.2 Standardization of guidelines and protocols

Another critical organizational challenge is the lack of standardised guidelines for victim identification across different regions and forensic laboratories. While organizations such as Interpol have established general protocols, including DVI guidelines, variability in national-level protocols can lead to inconsistencies in identification results (Scientific Working Group on DNA Analysis Methods SWGDAM, 2014). Additionally, the lack of uniform training and the use of old/standard protocols by forensic personnel further complicates the identification process. There is a need for updated, universally accepted guidelines to ensure consistency and reliability in mass DVI efforts (Interpol, 2023c).

30.4.3.3 Collection of reference samples

A significant problem in victim identification is the collection of reliable reference samples. In mass disaster cases, it is often difficult to obtain appropriate reference material, especially if family members are not readily accessible or if victims are unidentified (Interpol, 2022; Interpol, 2023c). When possible, familial DNA matching is employed, using relatives' DNA to establish kinship relationships. However, in cases where no family members are available, forensic scientists must rely on national or international databases such as Interpol's I-Familia, a global platform designed for international DNA kinship matching. I-Familia has been recognized as a crucial tool in addressing the challenges of missing persons and unidentified victims in mass disaster cases (Interpol, 2022).

30.5 Challenges resolution

Successful resolution of challenges in mass disaster DNA identification requires a multi-faceted approach, involving advancements in technology, adaptation of forensic protocols, and improvements in organizational coordination. Mass disasters, by their nature, present a unique set of obstacles, and while technologies such as rapid DNA have the potential to accelerate the identification process, their limitations, particularly with degraded or complex samples, underscore the need for more comprehensive solutions. This section explores the strategies and methodologies for overcoming the intrinsic, extrinsic, and organizational challenges involved in DNA identification during mass disaster scenarios.

However, even after the highlight of the possible resolution, the technologies must be supported by more studies and research to become the new gold standard for mass DVI.

30.5.1 Intrinsic and extrinsic challenges

The challenges affecting the sample posed different solutions and precautions: degraded DNA samples can be addressed through several innovative approaches. Whole genome amplification and Massively Parallel Sequencing allow for the amplification and detection retrieval of low-quantity DNA, making analysis possible even from heavily degraded samples (Al-Snan, 2023). Methods combining silica-based extraction have proven particularly effective in isolating DNA from challenging materials (Interpol, 2023b). In cases where nuclear DNA is too degraded, mtDNA and SNPs offer alternatives for identification (Bhoyar et al., 2024). Since mtDNA is more resistant to degradation and present in higher copy numbers per cell, it is frequently used to trace maternal lineage, even in long-degraded remains. Moreover, NGS technologies now allow for the simultaneous analysis of multiple genetic markers, including STRs and SNPs, significantly improving identification success in disaster scenarios (Al-Snan, 2023; Butler, 2014; Montelius & Lindblom, 2012). More problematic is the resolution of mixed profile case, because while STRmix represents a valuable tool in profile deconvolution, the presence of mixed profile derived from contamination caused by mixed human remains cannot be eliminated or improved by laboratory approaches. Possible DNA contamination solutions could include strict adherence to contamination control protocols, such as using sterile equipment, protective clothing, and clean workspaces. Furthermore, internal positive and negative controls in PCR allow for the detection of contamination during the amplification process, enabling early intervention to mitigate its effects (Butler, 2012).

30.5.2 Organizational challenges

For organizational challenges, such as the lack of standardised guidelines, organizations like Interpol have spearheaded efforts to harmonize forensic protocols through initiatives like the DVI Guidelines, ensuring that best practices are followed consistently across regions. The I-Familia database is another important development, offering a global platform for kinship matching in missing person cases, which is essential when coordinating efforts across borders (Interpol, 2022; Interpol, 2023c). More countries must adopt the same protocol and establish a common protocol for each case scenario in order to help the forensic personnel working in a mass disaster scene, particularly for those who aren't experienced with the handling of a high sheer of samples. Regarding the handling of the high volume of samples in mass disasters can be addressed through automation and high-throughput processing technologies. Additionally, the use of Rapid DNA technology, such as the ANDE and RapidHIT systems, allows for on-site DNA profile results in 90 minutes, helping to expedite the identification process when time is critical, and prevent the workload of the laboratories (Bowman et al., 2022).

30.5.3 Rapid DNA as a potential solution?

Further research and development efforts are needed to enhance the robustness and reliability of DNA fingerprinting in mass disaster scenarios. This includes improving methods for DNA preservation, processing, and analysis, as well as integrating technologies like artificial intelligence and machine learning to assist in the identification process. One promising advancement is Rapid DNA testing, which allows DNA samples to be analyzed on-site, eliminating the need to transport them to centralized laboratories. This significantly reduces the time required for identification, improving the efficiency of forensic investigations (Saks & Koehler, 2005).

For example, a study using the ANDE 6 C instrument demonstrated its capability to produce DNA profiles from different tissues (buccal swabs, muscle, liver, and brain), within an exposure time of 4–6 days. The instrument was proven effective in producing full profiles from bone and tooth samples after 12 months of exposure, indicating the potential for its use with more degraded remains. However, challenges persist, particularly with samples damaged by water or collected from hurricane-related disasters, where Rapid DNA showed poor profiling success compared to traditional laboratory methodology (Bowman et al., 2022). Although Rapid DNA technology represents a significant advancement, it is not currently suitable for all types of samples, especially those from crime scenes, where samples are often irreplaceable and may contain low amounts of DNA or be degraded. This is, for example, the cases of terrorist attacks or mass disasters of human origin that could be classified as crime scenes and so persecuted. These samples require evaluation by trained forensic analysts to ensure the quality and accuracy of the DNA profiles entered into databases like CODIS, which must maintain public trust in forensic analysis (Scientific Working Group on DNA Analysis Methods SWGDAM, 2017).

References

Al-Snan, N. R. (2023). Transition of capillary electrophoresis to next generation sequencing for forensic DNA analysis: Need of the hour. In *Next Generation Sequencing (NGS) Technology in DNA Analysis*. Bahrain: Elsevier, 3–20. http://doi.org/10.1016/B978-0-323-99144-5.00023-8, https://www.sciencedirect.com/book/9780323991445.

Allouche, M., Hamdoum, M., Mangin, P., & Castella, V. (2008). Genetic identification of decomposed cadavers using nails as DNA source. *Forensic Science International: Genetics, 3*(1), 46–49. https://doi.org/10.1016/j.fsigen.2008.07.008.

Alonso, A., Martín, P., Albarrán, C., García, P., Fernández De Simón, L., Iturralde, M. J., Fernández-Rodríguez, A., Atienza, I., Capilla, J., García-Hirschfeld, J., Martínez, P., Vallejo, G., García, O., García, E., Real, P., Álvarez, D., León, A., & Sancho, M. (2005). Challenges of DNA profiling in mass disaster investigations. *Croatian Medical Journal, 46*(4), 540–548.

Applied Biosystem. (2019). GlobalFiler® and GlobalFiler® IQC PCR Amplification Kits - PCR Amplification and CE Manual. Life Technologies. Thermo Fisher Scientific, Thermo Fisher Scientific.

Barbaro, A. (2022). Challenges in DNA extraction from forensic samples. In *Handbook of DNA profiling*. Italy: Springer, 711–730. http://doi.org/10.1007/978-981-16-4318-7_36, https://link.springer.com/referencework/10.1007/978-981-15-9364-2.

Bhoyar, L., Mehar, P., & Chavali, K. (2024). An overview of DNA degradation and its implications in forensic caseworks. *Egyptian Journal of Forensic Sciences, 14*(1). https://doi.org/10.1186/s41935-024-00389-y.

Bowman, Z., Daniel, R., Gerostamoulos, D., Woodford, N., & Hartman, D. (2022). Rapid DNA from a disaster victim identification perspective: Is it a game changer? *Forensic Science International: Genetics, 58*, 102684. https://doi.org/10.1016/j.fsigen.2022.102684.

Brenner, C. H., & Weir, B. S. (2003). Issues and strategies in the DNA identification of World Trade Center victims. *Theoretical Population Biology, 63*(3), 173–178. https://doi.org/10.1016/s0040-5809(03)00008-x.

Brito, F. C. A., Nunes, M. R., Prata, D. R. B. M., Martha, S. F. P., Bottino, C., & Garrido, R. G. (2019). DNA extraction of urinary bladder swabs collected from carbonized and decomposing corpses: Possible application in disaster victim identification. *Legal Medicine, 37*, 15–17. https://doi.org/10.1016/j.legalmed.2018.12.002, https://www.elsevier.com/locate/legalmed.

Bukyya, J. L., Avinash Tejasvi, M. L., Avinash, A., P, C. H., Talwade, P., Afroz, M. M., Pokala, A., Neela, P. K., Shyamilee, T. K., & Srisha, V. (2021). DNA profiling in forensic science: A review. *Global Medical Genetics, 08*(04), 135–143. https://doi.org/10.1055/s-0041-1728689.

Butler, J. M. (2005). *Forensic DNA typing: Biology technology and genetics of STR markers*. Elsevier Academic Press.

Butler, J. M. (2012). *Advanced topics in forensic DNA typing: Methodology*. United States: Elsevier Inc. http://doi.org/10.1016/C2011-0-04189-3, http://www.sciencedirect.com/science/book/9780123745132.

Butler, J. M. (2014). *Advanced topics in forensic DNA typing: Interpretation*. United States: Elsevier Inc, 1–585. http://doi.org/10.1016/C2011-0-07649-4, http://www.sciencedirect.com/science/book/9780124052130.

Caenazzo, L., Tozzo, P., & Rodriguez, D. (2013). Ethical issues in DNA identification of human biological material from mass disasters. *Prehospital and Disaster Medicine, 28*(4), 393–396. https://doi.org/10.1017/s1049023x1300040x.

Connon, C. C. (2023). Manual silica-based DNA extractions. *Methods in Molecular Biology, 2685*, 35–51. https://doi.org/10.1007/978-1-0716-3295-6_3, https://www.springer.com/series/7651.

Corach, D. (2017). *New Perspectives in mass disaster victim identification assisted by DNA typing and forensic genomics*. Elsevier BV, 395–405. http://doi.org/10.1016/b978-0-12-802971-8.00022-5.

Cortellini, V., Franceschetti, L., Correa, H. S. D., & Verzeletti, A. (2022). DNA extraction in human bodies: From fresh to advanced stages of decomposition. In *Handbook of DNA profiling*. Italy: Springer, 731–753. http://doi.org/10.1007/978-981-16-4318-7_37, https://link.springer.com/referencework/10.1007/978-981-15-9364-2.

De Boer, H. H., Maat, G. J. R., Kadarmo, D. A., Widodo, P. T., Kloosterman, A. D., & Kal, A. J. (2018). DNA identification of human remains in disaster victim identification (DVI): An efficient sampling method for muscle, bone, bone marrow and teeth. *Forensic Science International, 289*, 253–259. https://doi.org/10.1016/j.forsciint.2018.05.044, http://www.elsevier.com/locate/forsciint.

Giardina, E. (2013). DNA Fingerprinting. In *Brenner's encyclopedia of genetics: Second edition*. Italy: Elsevier Inc, 356–359. http://doi.org/10.1016/B978-0-12-374984-0.00419-8, http://www.sciencedirect.com/science/book/9780080961569.

Graham, E. A. M. (2006). Disaster victim identification. *Forensic Science, Medicine, and Pathology, 2*(3), 203–207. https://doi.org/10.1007/s12024-006-0011-0, http://www.springerlink.com/content/120561/.

Hudson, B. C., & Connon, C. C. (2023). DNA purification from bloodstains and buccal cells/saliva on FTA® cards. *Methods in Molecular Biology, 2685*, 119–126. https://doi.org/10.1007/978-1-0716-3295-6_8, https://www.springer.com/series/7651.

International Commission on Missing Persons (ICMP). (2015). Standard operating procedure for sampling bone and tooth specimens from human remains for DNA testing at the ICMP.

Interpol. (2022). I-Familia. Identifying missing persons globally through family DNA matching.

Interpol. (2023a). Annexure 4 - Phase 2 > PM. Disaster victim identification guide.

Interpol. (2023b). Annexure 8 - Methods of identification. Disaster victim identification guide.

Interpol (2023c). Disaster victim identification guide.

Knight, K. L., Mauriello, A., & Williams, G. (2023). Quantitation of DNA Using the applied biosystems Quantifiler® Trio DNA quantification kit. *Methods in Molecular Biology, 2685*, 175–188. https://doi.org/10.1007/978-1-0716-3295-6_11, https://www.springer.com/series/7651.

Montelius, K., & Lindblom, B. (2012). DNA analysis in disaster victim identification. *Forensic Science, Medicine, and Pathology, 8*(2), 140–147. https://doi.org/10.1007/s12024-011-9276-z.

Mundorff, A. Z., Amory, S., Huel, R., Bilić, A., Scott, A. L., & Parsons, T. J. (2018). An economical and efficient method for postmortem DNA sampling in mass fatalities. *Forensic Science International: Genetics, 36*, 167–175. https://doi.org/10.1016/j.fsigen.2018.07.009, http://www.elsevier.com.

National Institute of Justice. (2023). Collecting DNA evidence at property crime scenes. Chain of custody.

Prinz, M., Carracedo, A., Mayr, W. R., Morling, N., Parsons, T. J., Sajantila, A., Scheithauer, R., Schmitter, H., & Schneider, P. M. (2007). DNA commission of the International Society for Forensic Genetics (ISFG): Recommendations regarding the role of forensic genetics for disaster victim identification (DVI). *Forensic Science International: Genetics, 1*(1), 3–12. https://doi.org/10.1016/j.fsigen.2006.10.003.

Promega Corporation. (2021). PowerPlex® 21 system for use on the Applied Biosystems® genetic analyzers, Technical manual.

Roewer, L. (2019). Y-chromosome short tandem repeats in forensics—Sexing, profiling, and matching male DNA. *WIREs Forensic Science, 1*(4). https://doi.org/10.1002/wfs2.1336.

Saks, M. J., & Koehler, J. J. (2005). The coming paradigm shift in forensic identification science. *Science (New York, N.Y.), 309*(5736), 892–895. https://doi.org/10.1126/science.1111565.

Schneider, P. M., Bender, K., Mayr, W. R., Parson, W., Hoste, B., Decorte, R., Cordonnier, J., Vanek, D., Morling, N., Karjalainen, M., Carlotti, C. M. P., Sabatier, M., Hohoff, C., Schmitter, H., Pflug, W., Wenzel, R., Patzelt, D., Lessig, R., Dobrowolski, P., ... Greenhalgh, M. (2004). STR analysis of artificially degraded DNA - Results of a collaborative European exercise. *Forensic Science International, 139*(2-3), 123–134. https://doi.org/10.1016/j.forsciint.2003.10.002, https://www.elsevier.com/locate/forsciint.

Scientific Working Group on DNA Analysis Methods (SWGDAM). (2014). Guidelines for Missing Persons Casework.

Scientific Working Group on DNA Analysis Methods (SWGDAM). (2017). Position statement on rapid DNA analysis.

See, K. L., Aziz, S., & Mahmood, M. S. (2016). Beyond DVI: Future identification, research and archiving. *Journal of Forensic Research, 07*(06). https://doi.org/10.4172/2157-7145.1000359.

Shrivastava, P., Jain, T., & Trivedi, V. (2016). DNA fingerprinting: A substantial and imperative aid to forensic investigation. *European Journal of Forensic Sciences, 3*(3), 23. https://doi.org/10.5455/ejfs.204929.

Trindade-Filho, A., Mendes, C., Ferreira, S., Oliveira, S., Vasconcelos, A., Maia, F., Pak, H., & Paula, K. (2008). DNA obtained from decomposed corpses cartilage: A comparison with skeleton muscle source. *Forensic Science International: Genetics Supplement Series, 1*(1), 459–461. https://doi.org/10.1016/j.fsigss.2007.10.210.

Unsal Sapan, T. (2021). *InDel loci in forensic DNA analysis*. Springer Science and Business Media LLC, 1–11. http://doi.org/10.1007/978-981-15-9364-2_49-1.

Van den Berge, M., Wiskerke, D., Gerretsen, R. R. R., Tabak, J., & Sijen, T. (2016). DNA and RNA profiling of excavated human remains with varying postmortem intervals. *International Journal of Legal Medicine, 130*(6), 1471–1480. https://doi.org/10.1007/s00414-016-1438-9.

Walsh, P. S., Metzger, D. A., & Higuchi, R. (2013). Chelex 100 as a medium for simple extraction of DNA for PCR-based typing from forensic material. *Biotechniques, 54*(3), 134–139. https://doi.org/10.2144/000114018.

Watherston, J., McNevin, D., Gahan, M. E., Bruce, D., & Ward, J. (2018). Current and emerging tools for the recovery of genetic information from post mortem samples: New directions for disaster victim identification. *Forensic Science International: Genetics, 37*, 270–282. https://doi.org/10.1016/j.fsigen.2018.08.016.

Whitaker, J. P., Clayton, T. M., Urquhart, A. J., Millican, E. S., Downes, T. J., Kimpton, C. P., & Gill, P. (1995). Short tandem repeat typing of bodies from a mass disaster: High success rate and characteristic amplification patterns in highly degraded samples. *Biotechniques, 18*(4), 670–676.

Ziętkiewicz, E., Witt, M., Daca, P., Żebracka-Gala, J., Goniewicz, M., Jarząb, B., & Witt, M. (2012). Current genetic methodologies in the identification of disaster victims and in forensic analysis. *Journal of Applied Genetics, 53*(1), 41–60. https://doi.org/10.1007/s13353-011-0068-7, http://www.springerlink.com/content/1234-1983/.

Chapter 31

Forensic implications of polymerase chain reaction inhibitors

Gagan Kumar Panigrahi[1], Sanjoy Majumder[2], and Annapurna Sahoo[3]
[1]*Department of Zoology, School of Applied Sciences, Centurion University of Technology and Management, Bhubaneswar, Odisha, India,* [2]*School of Biotechnology, Centurion University of Technology and Management, Bhubaneswar, Odisha, India,* [3]*Department of Zoology, Jatni College, Jatni, Odisha, India*

31.1 Introduction

Forensic deoxyribonucleic acid (DNA) analysis serves as a crucial instrument in criminal investigations and judicial processes, allowing for the accurate identification of individuals through unique genetic markers. By focusing on specific areas of DNA, commonly short tandem repeats (STRs) located in noncoding regions of the genome, scientists can create a genetic profile that is unique, with a level of accuracy that remains statistically significant even within large populations (Thanakiatkrai & Rerkamnuaychoke, 2017). Forensic DNA samples can be gathered from blood, saliva, hair, skin cells, or other bodily materials discovered at a crime scene. After collection, DNA is extracted, then amplified using polymerase chain reaction (PCR) and analyzed through methods such as electrophoresis or next-generation sequencing. PCR is an essential tool in forensic DNA analysis, facilitating the amplification of tiny amounts of DNA for examination (McDonald et al., 2024). The introduction of PCR has significantly improved the capacity to amplify and evaluate DNA from even compromised or minimal samples. However, one of the primary obstacles faced in forensic DNA testing is the presence of PCR inhibitors. These substances may arise from the sample surroundings or be inherent to the sample itself, and their interference with the PCR process can greatly affect the dependability of forensic evidence. PCR inhibitors found in forensic samples present a notable challenge. Such inhibitors can disrupt DNA amplification, resulting in false negatives or distorted results. This chapter discusses the characteristics of PCR inhibitors, their origins in forensic analysis of genetic materials, the mechanisms through which they inhibit the DNA amplification, and techniques for their identification and reduction, alongside the broader forensic consequences of their existence. Forensic DNA analysis has transformed criminal investigation practices, providing unmatched accuracy in identifying individuals associated with crimes (Oosthuizen & Howes, 2022).

31.2 Overview of forensic DNA analysis

Since its emergence in the mid-1980s, forensic DNA testing has significantly contributed to the criminal justice system by assisting in the conviction of the guilty and the clearing of the innocent (Alketbi, 2023; Butler, 2015; Chong et al., 2021). Remains of missing individuals and victims of mass disasters have been matched and identified by connecting reference samples to the recovered remains. Laboratories frequently adopt and validate new technologies to enhance their ability to obtain DNA results with greater sensitivity and detail. DNA profiling, often referred to as DNA fingerprinting, has emerged as one of the most groundbreaking advancements in forensic science (Butler, 2019; McCord et al., 2018; Odah, 2024). It enables the identification of individuals based on their distinct genetic characteristics. The capability to analyze genetic material has transformed the methods by which forensic evidence is gathered, examined, and interpreted, resulting in greater accuracy in solving criminal cases, paternity determinations, identifying disaster victims, and identifying missing individuals (Carey & Mitnik, 2002; McDonald & Lehman, 2012). Establishing an individual's identity is crucial in the criminal justice system, especially in cases of mass disasters and sexual assault. Traditional identification methods like fingerprint analysis, blood typing, and anthropometric assessments can be insufficient. Forensic DNA typing,

Investigative and Predictive DNA Testing. DOI: https://doi.org/10.1016/B978-0-443-30128-5.00030-8

however, offers a reliable and precise method for identification, making it the gold standard in forensic science (Behera et al., 2024; Butler, 2011; Steele & Balding, 2014).

The essential value of DNA profiling stems from its capacity to offer strong, objective evidence that connects people to crime scenes, other individuals, or particular occurrences (Das & Panigrahi, 2024; Finaughty et al., 2023; Haddrill, 2021; Jordan & Mills, 2021). The accuracy and precision of DNA profiling have notably strengthened the criminal justice system by boosting conviction rates and reducing instances of wrongful convictions (Smith & Singh, 2024). This method is so dependable that it can not only implicate suspects but also exonerate those who have been unjustly convicted of crimes, making it a vital instrument in maintaining justice (Majumder & Panigrahi, 2024a, 2024b; Wickenheiser, 2021).

DNA profiling has become a crucial component of forensic investigations mainly because every person (except identical twins) has a unique DNA sequence (Romeika & Yan, 2014). This distinctive genetic "fingerprint" can be extracted from various biological materials, such as blood, semen, saliva, skin cells, hair follicles, and other tissues found at crime scenes (Adamowicz et al., 2014; Barbour et al., 2023; Gray et al., 2023; Singh et al., 2024). By examining specific areas of DNA known to differ among individuals, forensic scientists can create a profile that can then be matched with DNA from known individuals or databases to determine identity or relationships.

The advent of PCR technology transformed forensic DNA testing by enabling forensic scientists to amplify minute amounts of DNA, even from severely degraded samples, into sufficient quantities for analysis (McDonald et al., 2024a). Two primary types of DNA markers are frequently utilized: STRs, which are highly variable regions of DNA that differ among individuals and mitochondrial DNA (mtDNA), inherited maternally and useful for identifying human remains when nuclear DNA is scarce (Balding & Buckleton, 2009; Connell et al., 2024; Panigrahi & Satapathy, 2020a, 2020b, 2020c). STR analysis, in particular, serves as the cornerstone of modern forensic DNA profiling, facilitating the creation of unique genetic fingerprints that can be correlated to individuals or stored in DNA databases such as Combined DNA Index System (CODIS) database, created and maintained by Federal Bureau of Investigation, United States (Panneerchelvam & Norazmi, 2003). Progress in forensic DNA testing has also led to innovations in predictive DNA testing, which seeks to anticipate traits like eye or hair color from DNA samples, providing valuable insights in criminal investigations where direct identification is not feasible. Although the precision and effectiveness of forensic DNA testing are unmatched, obstacles persist, especially in the presence of PCR inhibitors or contaminated samples, which can disrupt the process and result in inconclusive or incorrect outcomes. Nevertheless, forensic DNA testing has significantly influenced the justice system, providing a powerful resource for both criminal investigations and the resolution of civil matters, such as paternity issues (Meintjes-Van der Walt & Dhliwayo, 2021).

31.3 Basics of polymerase chain reaction and DNA amplification

The PCR is a technique used to amplify a specific segment of DNA, allowing researchers to produce millions of copies from a small initial sample (Gallup, 2008). Established by Kary Mullis in the 1980s, PCR has revolutionized genetic research and diagnostics due to its simplicity and effectiveness. PCR exponentially increases the quantity of DNA, making it essential for applications in genetic testing, medical diagnostics, forensic science, and research, where it assists in identifying genetic mutations, pathogens, or specific genes in a sample. The fundamental principle of PCR is to amplify specific DNA sequences, generating millions of copies of a target region from a small initial sample (Panigrahi & Satapathy, 2021; Panigrahi et al., 2024; Wang et al., 2024). This allows forensic scientists to utilize very small amounts of DNA, often obtained from trace materials like hair, blood, saliva, skin cells, or semen. The PCR process involves three main steps: denaturation, annealing, and extension, which are performed in a series of cycles (Ong et al., 2006; Shukurullaevna et al., 2024).

The first step involves heating the double-stranded DNA to about 94°C–98°C to disrupt the hydrogen bonds between the complementary strands, resulting in two single-stranded DNA molecules. This step ensures that the template DNA is available for subsequent binding with primers and polymerase (Jäger & Weiher, 2020; Lorenz, 2012). Following denaturation, the temperature of the reaction is decreased (generally to 50°C–65°C), allowing short DNA primers to bind or anneal to specific complementary sequences on the single-stranded DNA templates. These primers are designed to target the specific region within the DNA for amplification. After the primers have attached to the template DNA, the temperature is slightly increased (usually to 72°C), permitting DNA polymerase (most commonly Taq polymerase) to extend the primers by adding nucleotides to the developing DNA strand. The polymerase operates in a 5' to 3' direction, forming a new DNA strand complementary to the template strand. Each of these steps is repeated for 25–40 cycles, leading to exponential amplification of the target DNA segment. By the end of the process, millions or even billions of copies of the specific DNA region have been produced from the initial sample (Fig. 31.1).

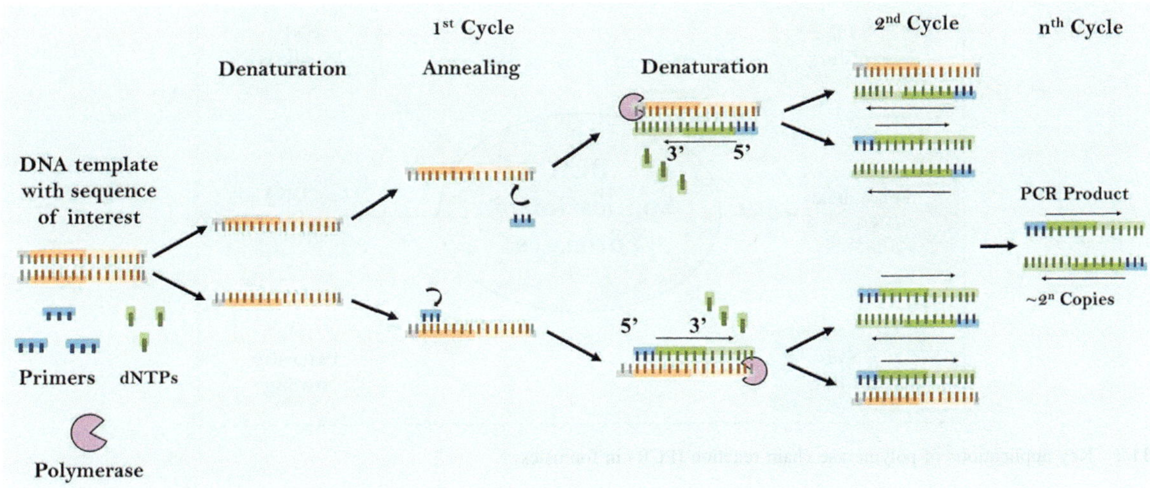

FIGURE 31.1 Schematic representation of the polymerase chain reaction process for the DNA sample.

The PCR has become a cornerstone technique across various scientific disciplines due to its remarkable ability to rapidly amplify specific DNA sequences (Primorac & Schanfield, 2023). In the field of medical diagnostics, PCR plays a crucial role by enabling the accurate identification of infectious agents, including a wide range of bacteria and viruses (Kojabad et al., 2021). This swift amplification process allows for timely and precise disease detection, which is essential in managing outbreaks and providing appropriate treatment. In addition to its applications in diagnostics, PCR is invaluable in genetic engineering and forensic science (Panigrahi et al., 2021a; 2021b; 2021c; Shewale & Liu, 2013). PCR has transformed forensic science by allowing for the amplification of tiny amounts of DNA from biological materials, making it possible to create strong DNA profiles even from severely degraded or contaminated specimens. By amplifying DNA quantity in these low DNA-containing samples, PCR enables investigators to create detailed genetic profiles. This process typically involves specific amplification of marker DNA segments, which can assist in solving crimes and identifying individuals with remarkable precision (Wang et al., 2024). It is a vital method in forensic investigations, facilitating the identification of individuals, their relationships, and linking evidence to crime scenes.

31.4 Applications of polymerase chain reaction in forensic science

PCR is utilized in various applications of forensics such as STR analysis, Y-STR profiling, mtDNA analysis, and single-nucleotide polymorphism (SNP) typing (Fig. 31.2). It plays a critical role in forensic DNA analysis, facilitating the examination of small or degraded samples like those often encountered at crime scenes. PCR in forensic science is revolutionizing the field by enabling the analysis of minute or degraded DNA samples that would otherwise be unusable for forensic analysis (Butler, 2012; Lynch & Fleming, 2019; Patro et al., 2023). This process facilitates the genetic analysis by increasing the quantity of DNA available for examination, even from a minimal sample. PCR allows forensic scientists to amplify specific regions of DNA exponentially, producing millions of copies from even the smallest biological trace left at a crime scene (Arsenault et al., 2024; Patro et al., 2024; Van Oorschot et al., 2010). This is crucial in forensic investigations where DNA may be present in very low quantities, such as in hair follicles, a few skin cells, or small amounts of blood, semen, or saliva. Without PCR, these samples might not yield enough DNA for conventional testing.

PCR is also essential in overcoming challenges posed by degraded DNA, which is common in forensic cases involving old, burned, or environmentally exposed samples (Alaeddini et al., 2010). By amplifying short regions of DNA, PCR can still generate reliable profiles even from fragmented genetic material. Furthermore, PCR is fast, sensitive, and adaptable, enabling real-time analysis through techniques like quantitative PCR (qPCR), which is useful for detecting the presence and quality of DNA before full profiling (Adams, 2020; Demeke & Jenkins, 2010; Ricci et al., 2024). The ability to amplify small and degraded samples has made PCR invaluable in cold case investigations, disaster victim identification, and even in noncriminal contexts like paternity testing. Additionally, advancements in PCR have led to specialized forms

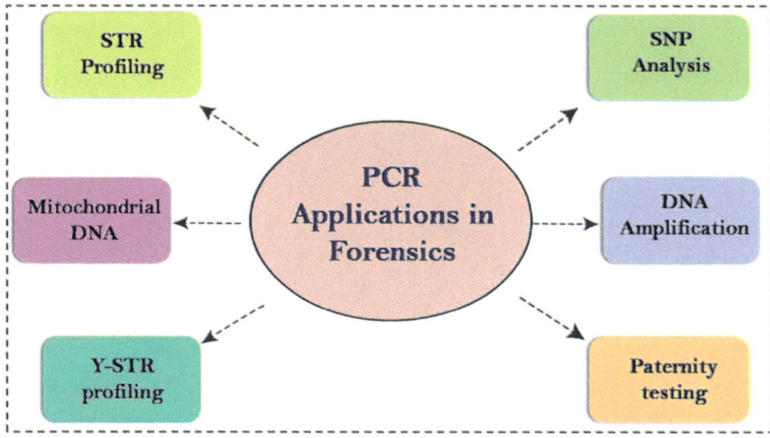

FIGURE 31.2 Key applications of polymerase chain reaction (PCR) in forensics.

such as multiplex PCR, where multiple DNA markers can be amplified simultaneously, increasing efficiency and throughput in forensic labs (Elnifro et al., 2000; Huang et al., 2024; Zhang et al., 2020). In essence, PCR has drastically expanded the scope and reliability of forensic DNA analysis, helping to ensure justice by providing accurate and robust genetic evidence. Forensic DNA profiling using STR typing has the capability to distinguish genetically between individuals. By facilitating STR analysis, mtDNA profiling, Y-STR, SNP analysis, and others, PCR forms the foundation for identifying individuals, determining familial connections, and resolving cold cases. PCR is frequently utilized to amplify DNA segments referred to as STRs, which show high variability among individuals (Sahoo & Satapathy, 2021; Tanudisastro et al., 2024). This makes PCR-based DNA profiling a critical tool in identifying suspects, victims, or missing persons (Chantzi & Georgakopoulos-Soares, 2024).

31.4.1 Short tandem repeat region profiling

STRs are specific segments of DNA characterized by the repetition of particular sequences, typically ranging from 2 to 6 base pairs in length, that occur in succession. This repetition creates a unique pattern that varies significantly among individuals, making STRs ideal markers for identity verification in various applications, especially in forensic science (Esslinger et al., 2004; Thanakiatkrai & Rerkamnuaychoke, 2017). In forensic analysis, PCR techniques are employed to amplify and target specific STR loci, distinct regions within the genome where these repetitive sequences are located. By examining multiple STR loci, scientists can generate a comprehensive DNA profile, ensuring a high degree of accuracy. The likelihood of two unrelated individuals sharing identical STR patterns across several loci is exceedingly low, which enhances the reliability of such analyses (Barbaro, 2020; Gill et al., 2015). One prominent application of STR profiling is the CODIS, which operates in the United States. This system utilizes a standardized set of 20 STR loci, allowing law enforcement agencies to create DNA profiles that can be compared against a vast database, aiding in criminal investigations and enhancing public safety (Frontanilla et al., 2022). PCR has transformed forensic DNA profiling by offering a strong, sensitive, and precise technique for amplifying STRs from forensic samples, which allows for dependable identification (Bornman et al., 2012). DNA profiles based on STR analysis generated through PCR have established themselves as fundamental in forensic science, aiding in connecting suspects to crime scenes, resolving cold cases, and supplying essential evidence in the legal system.

31.4.2 Mitochondrial DNA profiling

PCR enables the amplification of mtDNA, a process that is particularly advantageous when nuclear DNA is either inaccessible or damaged. mtDNA, which is uniquely inherited from the mother, exists in significant quantities within cells. This abundance makes mtDNA especially useful for forensic investigations involving challenging biological samples, such as strands of hair, fragments of bones, or remnants of teeth (Hofreiter et al., 2021; Xavier et al., 2021). The analysis of mtDNA serves a critical role in examining ancient or degraded specimens, including the remains of mass casualties, historical artifacts, or isolated hair strands (Orlando et al., 2021). This type of genetic examination is crucial for unraveling the mysteries of lineage and ancestry. Since mtDNA is transmitted maternally, it provides a reliable means to trace connections between maternal relatives. This unique characteristic aids in the identification of individuals, offering insights into familial relationships that might otherwise remain obscured in difficult cases.

31.4.3 Y-short tandem repeat profiling

The Y-chromosome is transmitted through the paternal lineage, and its STR regions can be amplified via PCR for forensic investigation. Y-STR profiling is especially beneficial in situations with multiple male suspects or when distinguishing male DNA in a mixed sample is necessary. Y-STR analysis assists in identifying male DNA in circumstances where female DNA is also present, like in sexual assault cases (Knijff, 2022). Ancestry and paternal lineage testing using Y-STR profiling can verify father-child relationships and trace male ancestry, aiding in the identification of remains (Roewer, 2019). Y-STR markers can offer insights into the history and structure of populations, supporting forensic anthropologists in cases involving unidentified remains.

31.4.4 Single-nucleotide polymorphism analysis

PCR is utilized to amplify areas that contain SNPs, which are variations involving a single base pair in the genome. While SNPs are less variable compared to STRs, they are particularly useful in degraded samples where STRs may not be identifiable. SNPs assist in deducing an individual's ancestral origins, offering investigative insights when a match is not available in DNA databases. In cases of highly degraded samples, SNP analysis proves effective, as traditional STR analysis might face limitations. Additionally, SNPs can provide predictions about certain phenotypic characteristics, like eye or hair color, which could be beneficial for suspect identification (Zupanič Pajnič et al., 2023).

31.4.5 Kinship and paternity testing

PCR-based DNA testing serves as a powerful tool for kinship analysis and paternity verification (Li et al., 2024). This method can play a crucial role in various contexts, including criminal investigations, where it helps in identifying victims or establishing familial relationships with suspects. For example, if a victim is found without identification, DNA samples can be compared to those of family members to confirm identity. In addition to its applications in criminal cases, PCR testing is also pertinent in civil matters, such as resolving disputes regarding parentage or inheritance. Overall, the versatility and accuracy of PCR-based DNA testing make it an essential resource in both criminal and civil contexts.

31.4.6 Trace DNA profiling

The remarkable sensitivity of PCR technology enables the analysis of minute traces of DNA, even from small or significantly degraded samples. This includes sensitive "touch DNA" that might be left behind on various surfaces. In situations involving extremely small amounts of DNA, the sensitivity of PCR allows for the extraction of DNA profiles from these minimal sources (Haarkötter et al., 2023; Xin & Chen, 2012; Xin et al., 2003). Techniques like low-copy-number PCR utilize a higher number of PCR cycles to enhance amplification of trace amounts, which allows forensic scientists to generate either partial or complete profiles from weak DNA evidence. This advancement opens new avenues for solving cases that would have otherwise remained unsolved.

The utilization of PCR in forensic DNA analysis has revolutionized criminal investigations, allowing forensic experts to examine DNA even from the most difficult samples. Whether for DNA profiling, sexual assault cases, disaster victim identification, or cold case resolutions, PCR serves as a flexible and highly sensitive technique for amplifying and analyzing genetic material. Its capability to handle degraded, mixed, or trace DNA samples renders it an essential asset in solving crimes, identifying victims, and achieving justice (Haarkötter Cardoso et al., 2024). As PCR technology continues to evolve, its significance in forensic science will only grow, providing new opportunities for DNA analysis and further improving the accuracy and dependability of forensic investigations (Rački et al., 2014).

31.5 Polymerase chain reaction inhibitors and their forensic relevance

Any substance that disrupts one or more of the processes involved in PCR, thus affecting the kinetics of the reaction and the detection threshold, is classified as a PCR inhibitor. Certain substances have been recognized as inhibitors of PCR and have been fully or partially characterized in terms of their molecular mechanisms of inhibiting PCR (Rossen et al., 1992; Sidstedt et al., 2019). As a result, categorizing PCR inhibitors is a challenging task. The precise mechanisms for most inhibitors remain unknown, which means that the typical classification is based on the specific targets they affect. They can be classified as (1) inhibitors of DNA polymerase, (2) inhibitors of nucleotides/nucleic acids, and (3) inhibitors of fluorescence (specific to qPCR) (Lance & Guan, 2020; Schrader et al., 2012; Thompson et al., 2014).

These inhibitors may naturally occur in biological samples gathered from crime scenes or could be introduced during the handling and processing of samples (Sessa et al., 2023; Vajpayee et al., 2023). Their presence can significantly impact DNA testing outcomes by decreasing the efficiency or completely halting DNA amplification, potentially leading to inaccurate results such as false negatives or incomplete DNA profiles. In forensic investigations, where evidence is frequently limited and compromised, PCR inhibitors pose a considerable challenge, requiring effective detection and mitigation strategies to ensure dependable DNA profiling.

PCR inhibitors can come from different sources, both intrinsic and extrinsic to the sample. Intrinsic inhibitors are mainly endogenous substances found within biological materials, including hemoglobin in the blood, bile salts in body fluids, or melanin in skin and hair. These substances may attach to either the DNA or the DNA polymerase enzyme, obstructing the PCR procedure. Conversely, extrinsic inhibitors are environmental contaminants that might mix with the sample at crime scenes, such as soil, fabric dyes, chemicals, or humic acids from soil and water (Alrabie et al., 2018; Smalla et al., 1993). Given that forensic samples are often collected from difficult environments like outdoor crime scenes, buried remains, or surfaces subjected to extreme conditions, the likelihood of encountering PCR inhibitors is considerably high.

The presence of PCR inhibitors presents multiple forensic challenges, particularly when the DNA evidence is essential for linking a suspect to a crime(Schrader et al., 2012). A false negative result, where no DNA profile is produced due to inhibition, may lead to the wrongful exclusion of a suspect or the inability to identify a victim. In cases where the DNA profile is partially inhibited, the incomplete genetic information might lead to ambiguous results or misinterpretations. For instance, in cold case investigations or mass disaster identifications, where samples are typically highly degraded or contaminated, the effects of PCR inhibitors can be particularly complicated in the generation of reliable DNA profiles suitable for comparison with known individuals or DNA databases. Moreover, the relevance of PCR inhibitors extends beyond the practical aspects of DNA profiling. Their presence can offer investigative leads. For example, identifying specific environmental contaminants like distinct chemicals, oils, or plant materials in a forensic sample might provide clues about the sample's origin or the circumstances surrounding the crime. In some instances, finding certain inhibitors could assist in establishing timelines or environmental conditions that a body or evidence has encountered, aiding in the reconstruction of the crime scene. Nonetheless, the presence of inhibitors can greatly impact PCR effectiveness, resulting in unsuccessful amplifications or diminished sensitivity (Sahoo et al., 2023a; 2023b; 2023c; Swari et al., 2023). Addressing these inhibitory effects is essential for achieving consistent DNA analysis.

The forensic implications of PCR inhibitors are crucial due to the characteristics of forensic evidence (Alaeddini, 2012). Unlike samples from clinical or research settings, forensic samples are frequently degraded, fragmented, or mixed with various substances that complicate the DNA extraction and amplification processes. These samples could be obtained from decayed human remains, surfaces tainted with dirt or chemicals, or materials exposed to harsh weather. Such conditions not only yield low-quality DNA but also heighten the risk of inhibitors being present, which can result in incomplete or undetectable DNA profiles. In criminal investigations, where every piece of evidence holds significant weight, the failure to obtain a trustworthy DNA profile due to PCR inhibition can hinder the identification of a suspect or victim, potentially obstructing justice.

31.6 Sources of polymerase chain reaction inhibitors in forensic DNA analysis

Contaminants represent a significant challenge in forensic DNA analysis, especially when DNA samples are obtained from crime scenes or outdoor settings (Fig. 31.3). The PCR is particularly sensitive to contaminants that can disrupt the amplification process, and these inhibitors may come from a variety of natural and artificial sources found in the environment (Cowell et al., 2015; Jäger, 2024; Sahoo et al., 2024). Understanding the characteristics and effects of these inhibitors is essential in the field of forensic science. This knowledge is vital for generating reliable DNA profiles, particularly when dealing with degraded samples. Effective detection and analysis of these inhibitors can significantly impact the accuracy and integrity of forensic investigations (Wilson, 1997). Environmental inhibitors can be categorized into several groups, such as organic compounds, inorganic materials, and biological substances, which interfere with the PCR process by either binding to DNA or directly impacting the enzymes (like Taq polymerase) that promote DNA amplification. Below is a comprehensive overview of the primary environmental sources of PCR inhibitors and their influence on forensic DNA testing. Soil samples can have a variety of substances, such as humic acids, metal ions, and organic compounds, that prevent PCR by attaching to DNA or disrupting the function of DNA polymerase(Bonsu et al., 2023; Seo et al., 2013). Contaminated water, particularly from rivers or lakes, might include heavy metals (like iron and

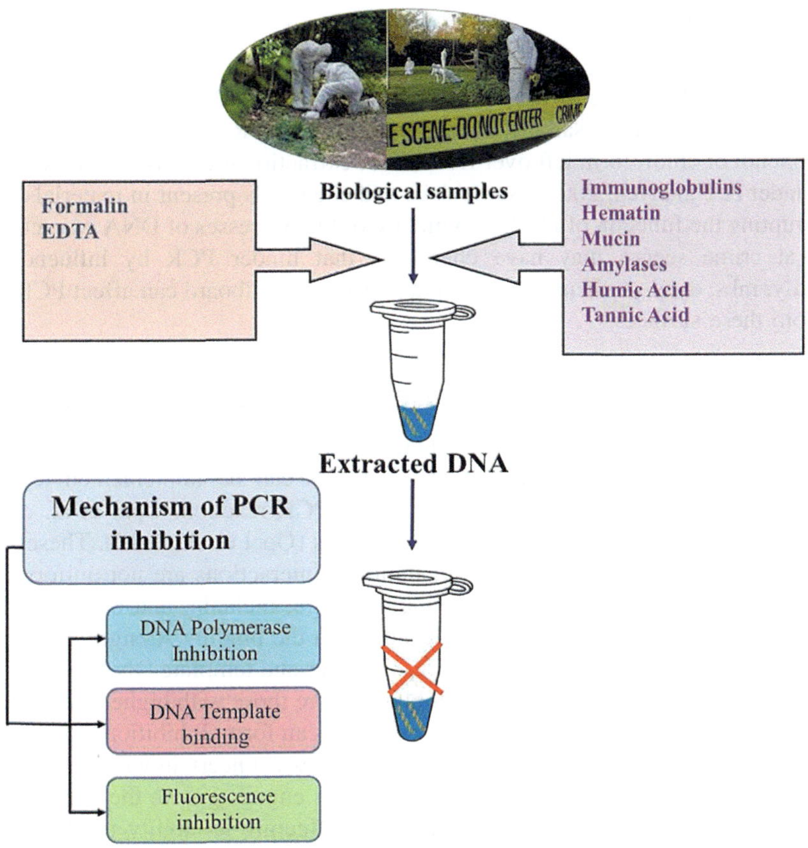

FIGURE 31.3 Diagrammatic representation of source of polymerase chain reaction (PCR) inhibitors and their mechanism of PCR process inhibition.

calcium) or organic compounds that hinder PCR reactions. Fine particulate matter found in dust and soil can introduce inhibitors that disrupt the enzymatic processes in PCR. Here are some common sources of PCR inhibitors in forensic DNA analysis.

31.6.1 Biological substances

Hemoglobin present in blood can significantly inhibit PCR. The iron in hemoglobin attaches to DNA, which obstructs the activity of polymerase. Mucins and various proteins found in saliva can hinder PCR by binding to or breaking down DNA or by interfering with the enzyme's function. Urea and other waste products in urine can disrupt enzyme activity, inhibiting PCR. DNA extracted from bones or teeth may become contaminated with calcium ions or collagen, both of which can impede PCR (D. B. & Joji, 2022).

31.6.2 Humic substances

Humic substances are organic materials that are frequently encountered in soil and plant debris, making them one of the most common PCR inhibitors faced in forensic DNA analysis. These compounds consist of complex mixtures of partially decomposed organic material, including humic acids and fulvic acids, which arise from the natural decay of plant and animal matter.

Humic substances hinder PCR by binding to DNA and obstructing the polymerase from accessing the template or by directly inhibiting the function of Taq polymerase (Sidstedt et al., 2015). Additionally, they can chelate magnesium ions (Mg^{2+}), which are crucial cofactors for the PCR reaction. The presence of humic substances can lead to diminished or nonexistent amplification, complicating the acquisition of DNA profiles from samples gathered in areas with high organic content, like soil or leaf litter.

31.6.3 Chemical substances

Detergents utilized during sample collection or DNA extraction (such as from swabs or cleaning tools) may obstruct PCR by denaturing the polymerase or disrupting DNA binding (Vitošević et al., 2018). Certain chemical preservatives, like formalin (formaldehyde), that are used for sample storage can inhibit PCR by cross-linking DNA, which complicates amplification. Residual phenol or chloroform left over from DNA extraction methods can carry into the PCR reaction and serve as inhibitors (Schander & Halanych, 2003). Some dyes and chemicals present in materials like clothing and leather can obstruct PCR by disrupting the function of DNA polymerase or the processes of DNA extraction. Adhesive items such as tape and bandages at crime scenes may have chemicals that hinder PCR by influencing DNA extraction or amplification. Additionally, inks, dyes, or chemicals found in paper and cardboard can affect PCR reactions when forensic samples are gathered from these surfaces.

31.7 Mechanisms of action by polymerase chain reaction inhibitors

One of the major challenges in tackling the issue of PCR inhibitors is understanding how each inhibitor interacts with the PCR process. Once the mechanism is identified, various strategies can be implemented to mitigate or remove the inhibitors, thereby enhancing the efficiency and sensitivity of the PCR reaction. Opel et al. determined that there are several key ways in which inhibitors can adversely affect the process (Opel et al., 2010). These ways include interfering with the polymerase or the template DNA. The effects of inhibitor interactions are not uniform and may vary between different loci. Chung observed that inhibitors such as hematin, indigo, melanin, and humic acids in blood, saliva, and tissue samples impacted different loci in various ways, depending on the binding strength of the primers to the template DNA or the interaction strength between the DNA polymerase, primer and template DNA (Eckhart et al., 2000; Hu et al., 2015). Loci with lower melting temperatures tend to be inhibited before those with higher melting temperatures due to the speed at which the DNA polymerase-primer-template DNA complex can form. Inhibition caused by collagen and calcium resulted in varying patterns of loci inhibition, although the researcher was uncertain why this phenomenon occurred. The presence of Collagen in the Sample results in a notable inhibition by engaging with the template DNA. This interaction disrupts the normal processivity of DNA polymerase, ultimately affecting its ability to effectively replicate the DNA (Sidstedt et al., 2020).

31.7.1 Binding of inhibitors to DNA polymerase enzyme

PCR inhibitors interfere with polymerase function via various mechanisms, such as directly hindering the enzyme, affecting cofactors, disrupting the annealing of primers, and obstructing the DNA template. These interferences diminish the effectiveness of DNA amplification and present notable difficulties in forensic DNA analysis, especially when dealing with tainted or degraded samples. Grasping these mechanisms is crucial for formulating approaches to counteract inhibition, including the use of inhibitor-resistant enzymes, improving extraction methods, or incorporating PCR enhancers that alleviate the impact of inhibitors (Dash et al., 2023). The PCR is an extremely sensitive method utilized to amplify DNA, which relies significantly on Taq polymerase, an enzyme responsible for synthesizing new DNA strands by adding nucleotides to a primer that has bonded to the DNA template. However, the effectiveness of PCR can be affected by various inhibitory substances that may directly or indirectly hinder the function of Taq polymerase. When these inhibitors are present, they can greatly decrease the DNA yield or potentially cause the amplification process to completely fail.

PCR inhibitors can disrupt various stages of a PCR analysis. Typically, different components of PCR, especially DNA, may adhere to polymer surfaces, such as the walls of tubes and vessels, during sample processing, extraction, or the PCR process itself (Rådström et al., 2003). Nucleases have the potential to break down template RNA or DNA. The DNA used as the template in PCR can undergo modification or degradation from nucleases and other agents. The annealing of primers to the DNA template may be affected by particular PCR inhibitors. This occurs because the inhibitor competes for binding with the template, but designing primers with higher melting points can mitigate this issue. Numerous PCR inhibitors directly or indirectly target DNA polymerase. Proteases or detergents present in the reaction can harm this enzyme. Urea and phenol, for instance, are known to damage DNA polymerases. Calcium, collagen, hematin, and tannic acid can all inhibit the activity of the polymerase. Melanin can form a reversible bond with the DNA polymerase, while polysaccharides may disrupt the enzymatic activity by mimicking the structure of nucleic acids. Humic acids can interact with both template DNA and the polymerase, thereby hindering the enzymatic process even at low concentrations. Other compounds may engage with the polymerase cofactors. High levels of calcium can lead to competitive binding by the DNA polymerase, displacing magnesium, while complexing agents like tannic acid can reduce magnesium availability.

In both scenarios, magnesium becomes unavailable as a cofactor for the polymerase, leading to decreased activity. In real-time PCR assays, the disruption of fluorescent probes or the increase in background fluorescence presents additional mechanisms through which PCR inhibitors lower sensitivity (Hedman & Rådström, 2013).

One of the most straightforward ways of inhibition involves the attachment of inhibitory substances to the active site or essential functional regions of the DNA polymerase enzyme. This attachment can hinder the enzyme from operating properly. Certain inhibitors, such as heme (found in blood), humic acids (from soil), and polysaccharides (from plants or tissues), exhibit strong affinities for the polymerase enzyme. When these substances attach to Taq polymerase, they can obstruct or modify the active site, which is vital for facilitating the addition of nucleotides to the expanding DNA strand. The direct engagement between the inhibitor and Taq polymerase obstructs the enzyme's ability to bind to the DNA template, diminishing its capacity to perform the elongation process (Sutlovic et al., 2008). Consequently, DNA amplification may be incomplete or completely halted, resulting in either a reduced yield or no amplification.

Heme, a component found in blood, can bind directly to Taq polymerase, which disrupts its crucial enzymatic functions and ultimately renders the enzyme inactive. Another example is humic acids, which are commonly found in environmental samples such as soil, which interact with Taq polymerase, leading to a reduction in its activity and resulting in inefficient amplification during processes like PCR (Sutlovic et al., 2005).

31.7.2 Chelation of essential cofactors

Taq polymerase necessitates magnesium ions (Mg^{2+}) as crucial cofactors for its catalytic function during PCR. These ions play a key role in stabilizing the interaction between the polymerase and the DNA template, as well as ensuring the proper alignment of nucleotides for inclusion into the growing DNA strand. PCR inhibitors can obstruct the availability of these cofactors, thereby indirectly hindering the activity of the polymerase. Agents like ethylenediaminetetraacetic acid (EDTA), which is a common chelating agent, can attach to and sequester Mg^{2+} ions within the PCR reaction mixture (Huggett et al., 2008). If there is an inadequate amount of magnesium ions, the polymerase cannot operate effectively. A deficiency of Mg^{2+} ions compromises the enzyme-template interaction and diminishes the polymerase's capacity to catalyze nucleotide incorporation. This leads to suboptimal amplification efficiency and may result in incomplete or failed PCR reactions. Even minimal concentrations of EDTA can significantly affect Taq polymerase activity because of its strong affinity for Mg^{2+}.

31.8 Impact of polymerase chain reaction inhibitors on forensic DNA profiling

Incomplete amplification of all loci due to PCR inhibition can produce partial DNA profiles. These profiles may miss some of the genetic markers commonly utilized for identification, which can complicate or render it impossible to link the sample to an individual in a DNA database.

In instances of significant inhibition, PCR may completely fail, resulting in no amplification products and thus no applicable DNA profile. PCR inhibition may lead to false negatives, where DNA exists in the sample, but cannot be amplified because of the inhibitors. In forensic situations, this could mean critical evidence is ignored or discarded, as it seems there is no DNA in the sample.

PCR inhibition can lead to allelic dropout, where specific alleles at a locus do not amplify, resulting in incomplete or incorrect profiles (Foley et al., 2024; Tvedebrink et al., 2012). This can cause misinterpretation in the context of mixed DNA samples, where identifying individual contributors is crucial. PCR inhibition can generate stochastic (random) effects during amplification, resulting in uneven amplification of alleles. This leads to peaks of different intensities in electropherograms, complicating the understanding of DNA profiles. Mixed DNA samples, often encountered in forensic cases like sexual assault, are especially vulnerable to PCR inhibitors. The presence of these inhibitors can hinder the ability to distinguish and identify individual contributors, particularly if they affect one contributor's DNA more than another's, resulting in imbalanced profiles.

The complete inhibition of PCR can give false negative results, where the DNA sample exists but fails to amplify due to the presence of inhibitors. In forensic situations, this could result in the incorrect assumption that no DNA is available at the crime scene, possibly missing crucial evidence. This issue is especially concerning in cold case investigations, where DNA may be degraded or tainted with environmental inhibitors, complicating the recovery of valuable genetic information (Bélec et al., 1998). False negatives can also influence the analysis of DNA evidence in cases where identifying a suspect's DNA is essential for proving their involvement in a crime.

31.9 Detection of polymerase chain reaction inhibitors

PCR inhibitors present in DNA samples can lead to diminished amplification efficiency or even complete reaction failure, particularly in cases involving degraded or very low quantities of templates such as ancient or forensic DNA (Al-Soud & Rådström, 2001; King et al., 2009; Straube et al., 2021). In these situations, it is essential to identify and accurately assess the extent of PCR inhibition. For qPCR, this can be achieved through two approaches: (1) measuring the variation in the quantification cycle (Cq) of an externally added internal positive control (IPC) DNA in the sample compared to control reactions, or (2) evaluating amplification efficiency via the slope of the amplification plots.

Identifying PCR inhibitors in forensic samples is vital for maintaining the trustworthiness and precision of DNA amplification methods. Considering the important function of PCR in forensic applications, from investigating crime scenes to conducting paternity tests, understanding and recognizing inhibitors is crucial. A variety of techniques have been created to identify PCR inhibitors, as their presence can greatly affect the success of amplification.

The impact of PCR inhibitory substances can be assessed indirectly by evaluating the quantity and quality of PCR products using various methods such as gel electrophoresis, dot blots, high-pressure liquid chromatography, and calorimetric assays (Al-Soud & Rådström, 2001). PCR inhibition can be detected through the amplification failures of internal standards and can be monitored semi-quantitatively by observing the PCR failures with different known quantities of the internal standard.

31.10 Strategies for overcoming polymerase chain reaction inhibitors

Given the significant effects of PCR inhibitors present in forensic samples, various strategies have been developed to counteract their effects (Cavanaugh & Bathrick, 2018; Martin & Linacre, 2020). A common method involves using DNA extraction techniques specifically designed to eliminate or neutralize inhibitors (Fig. 31.4). These methods often incorporate chemical purification strategies, such as utilizing silica-based columns or magnetic beads that selectively bind to DNA while discarding other contaminants (Liu et al., 2021; Makowski et al., 1995). Additionally, there are PCR additives, such as bovine serum albumin (BSA), which can help enhance the efficiency of the PCR process in the presence of inhibitors (Maher et al., 2001; Sur et al., 2010). This section explores various approaches to counteract PCR inhibition, emphasizing sample preparation, reagent optimization, and alternative techniques.

The process of eliminating PCR inhibitors can often be quite challenging and time-consuming. Several protocol modifications can be implemented to help mitigate the effects of these inhibitors (Baar et al., 2011; Rådström et al., 2004). One common approach is to dilute the sample, which can reduce the concentration of inhibitors, thereby enhancing the efficiency of the PCR amplification process.

In addition to dilution, the incorporation of amplification enhancers may also prove beneficial. These enhancers include substances such as BSA, sodium sulfate, dimethyl sulfoxide (DMSO), glycerol, and polyvinylpolypyrrolidone. Each of these compounds has unique properties that can improve the performance of PCR by stabilizing the DNA polymerase or by binding to inhibitors to decrease their detrimental effects on the reaction.

Another effective method for reducing the presence of inhibitors, particularly in water samples, is membrane filtration. This technique has been shown to effectively remove both organic and inorganic compounds that can interfere with PCR

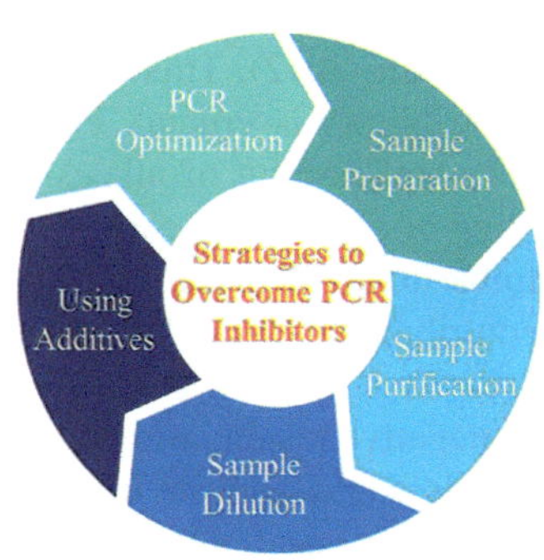

FIGURE 31.4 Schematic representation of strategies to overcome polymerase chain reaction (PCR) inhibitors.

amplification. By utilizing membrane filters, researchers can physically separate these contaminants from the samples before proceeding with the amplification steps, thus enhancing the overall yield and reliability of the PCR results.

31.10.1 Sample preparation technique

Effective sample preparation is crucial in reducing the effects of PCR inhibitors. Various methods can be utilized to improve DNA purity and lower concentration levels. Silica column extraction techniques are efficient for isolating DNA from samples that contain inhibitors. These approaches utilize the binding characteristics of silica in conjunction with chaotropic salts, aiding in the separation of DNA from proteins, cellular debris, and other impurities (Davalieva & Efremov, 2010; Liu et al., 2021). Phenol-chloroform extraction method can be used to isolate DNA from proteins and various organic substances. Although it is effective, it demands careful handling because of the hazardous risks associated with phenol and chloroform. Magnetic Bead-Based Methods employ magnetic beads that specifically bind to DNA, facilitating effective separation from inhibitors. This approach is gaining popularity due to its ease of use and potential for automation. In case of samples collected from soil, one of the most successful methods identified involved a pre-treatment with polyvinylpyrrolidone using a proteinase K extraction buffer, followed by purification using magnetic silica beads, which could include or exclude centrifugation (Fredricks & Relman, 1998; Howarth et al., 2022).

31.10.2 Sample dilution

Diluting DNA samples is a simple method to address inhibition before the PCR process. This can lower the concentration of inhibitors, enhancing the likelihood of successful amplification. However, it is crucial to find a balance in dilution to ensure that enough target DNA is present for effective amplification. Conducting serial dilutions can assist in determining the ideal dilution factor for amplification, providing a clearer insight into the inhibitory effects within the sample. When working with DNA extraction, the presence of detergents such as sodium dodecyl sulfate or Triton X-100 can interfere with subsequent processes like PCR. To mitigate this issue, incorporating an additional washing step during the DNA extraction process is helpful. This step effectively removes residual detergent, which may otherwise inhibit PCR efficiency (Bourke et al., 1999). Alternatively, there are specialized agents available that are specifically formulated to eliminate these detergents, or you might consider using extraction kits that are designed to target and neutralize such substances. These measures can help ensure the purity of the DNA sample and enhance the reliability of your PCR results. Centrifugation helps separate cellular debris from DNA by spinning the mixture at high speeds. This process allows the heavier debris to settle at the bottom, leaving the purified DNA on top. Filtration further improves the sample by removing larger particles and potential inhibitors that could affect results. Together, these methods ensure a cleaner and more effective DNA sample.

31.10.3 Using additives

BSA can bind to various PCR inhibitors, thereby stopping them from disrupting the function of DNA polymerase. It is especially proficient at countering inhibitors found in blood and environmental samples. Incorporating PCR enhancers and additives, like BSA or DMSO, can mitigate the effects of inhibitors and enhance the efficiency of the PCR reaction. These compounds can either bind to inhibitors or modify the reaction conditions to create a more conducive environment for Taq polymerase. Incorporating substances such as trehalose or betaine can enhance the stability of DNA polymerase and facilitate the amplification process, even when inhibitors are present (Frackman et al., 1998). Trehalose aids in safeguarding the enzyme from denaturation during the high-temperature cycles of PCR (Spiess et al., 2004). This stabilization enhances the polymerase's efficiency, even when inhibitors are present. By shielding the enzyme, trehalose contributes to more effective DNA amplification, resulting in clearer and more complete profiles in forensic or complex environmental samples. Additionally, trehalose can protect the DNA template from degradation, which is particularly advantageous for forensic samples that may be compromised or have low DNA concentrations. Betaine reduces the formation of secondary structures, such as hairpins, in GC-rich regions of the DNA template. This is particularly crucial for amplifying intricate sections of the genome that may be challenging to access under standard PCR conditions. By lowering the energy needed to unwind DNA, betaine improves primer-template hybridization, which increases amplification success even in samples containing PCR inhibitors. Betaine can alleviate the effects of inhibitors like salts, urea, and certain organic compounds by stabilizing the DNA polymerase and enhancing its performance in less-than-ideal conditions.

31.10.4 Optimization of polymerase chain reaction

Increasing the concentration of Mg^{2+} in the PCR mix can help mitigate the inhibitory impacts of elements such as humic acids or chelating agents. Employing inhibitor-resistant DNA polymerases, like modified versions of Taq polymerase that are more resilient to inhibition, can improve PCR outcomes when inhibitors are present (Abu Al-Soud & Rådström, 1998; Hedman et al., 2009). Mutational alteration of the polymerase can overcome the inhibition to the extent that no DNA purification is now required. An N-terminal deletion (Klentaq1) results in some 10–100-fold inhibition resistant to whole blood compared to full-length, wild-type (w.t.) Taq (Kermekchiev et al., 2009; Sidstedt et al., 2017). Hot-start PCR involves altering the reaction configuration so that the DNA polymerase only becomes active after the initial denaturation phase (Iyavoo, 2014; Karunanathie et al., 2022). This approach reduces the influence of inhibitors that might hinder the enzyme at lower temperatures. In certain instances, touchdown PCR can yield more effective amplification compared to conventional PCR, particularly in the presence of inhibitors (Baltrušis & Höglund, 2023; Hou et al., 2023; Salipante & Jerome, 2020; Sidstedt et al., 2018).

Hot-start PCR enhances both the specificity and yield of PCR when inhibitors are present. In hot-start PCR, the DNA polymerase remains inactive until the reaction attains a high temperature, minimizing nonspecific amplification and the chances of enzyme inhibition at lower temperatures (Kermekchiev et al., 2009). This activation method is crucial because it minimizes the enzyme's exposure to potential inhibitors that could interfere with its performance at lower temperatures. By delaying the activation of Taq polymerase, hot-start PCR enhances both the specificity and efficiency of the amplification process (Baltrušis & Höglund, 2023; Salipante & Jerome, 2020). This technique is particularly beneficial when working with samples that may contain background contaminants, ensuring that the desired DNA targets are amplified more accurately and effectively. DNA polymerases available commercially have been specially modified to stay inactive until they reach a designated temperature. Some utilize antibodies or chemical alterations to prevent premature activation of the enzyme. This method significantly boosts the efficiency and specificity of PCR, especially in samples with low DNA amounts or inhibitors that can hinder DNA polymerase function.

The enhancement of PCR to counteract inhibitors involves a variety of strategies, integrating improved DNA extraction techniques, unique PCR additives, specific enzyme choices, and modifications in cycling parameters (Chakrabarti & Schutt, 2001). By implementing the above-mentioned methods, forensic analysts can reduce the impact of inhibitors and boost the effectiveness of DNA profiling, even in difficult forensic specimens.

31.11 Conclusion

PCR inhibitors pose significant challenges in the realm of DNA profiling, undermining the accuracy, completeness, and clarity of DNA profiles crucial for forensic investigations. These inhibitors can lead to a range of unwanted outcomes, such as allelic dropout, where some genetic markers fail to amplify, resulting in incomplete profiles. Additionally, peak height discrepancies can arise, skewing the interpretation of results, while false negatives may prevent the identification of a suspect altogether. All of these factors can severely compromise the ability to pinpoint individuals or establish connections between suspects and crime scenes. To combat these issues, forensic scientists are continuously refining their methodologies. Improvements in sample purification techniques, for instance, help remove contaminants that contribute to PCR inhibition. The use of PCR enhancers can boost the efficiency of DNA amplification, while the introduction of inhibitor-resistant polymerases allows for more reliable reactions even in the presence of these problematic substances. However, the challenge remains particularly pronounced when dealing with environmental samples, which are often laden with inhibitors. To effectively harness PCR for these types of forensic analyses, innovative methods that bypass the effects of inhibition need to be developed. Once inhibition is detected in a sample, laboratories often implement a variety of troubleshooting strategies. Common approaches include significant dilution of samples, the addition of BSA to stabilize the reaction, and techniques like heat-soaked PCR and hot-start PCR to enhance performance. The incorporation of additional polymerase enzymes can also help in overcoming hurdles related to inhibition (Karunanathie et al., 2022). As forensic science continues to evolve, it becomes increasingly important to proactively address these obstacles. By doing so, the integrity and reliability of DNA evidence can be maintained, ultimately supporting the pursuit of justice.

References

Abu Al-Soud, W., & Rådström, P. (1998). Capacity of nine thermostable DNA polymerases to mediate DNA amplification in the presence of PCR-inhibiting samples. *Applied and Environmental Microbiology, 64*(10), 3748–3753. https://doi.org/10.1128/aem.64.10.3748-3753.1998, http://aem.asm.org/.

Adamowicz, M. S., Stasulli, D. M., Sobestanovich, E. M., & Bille, T. W. (2014). Evaluation of methods to improve the extraction and recovery of DNA from cotton swabs for forensic analysis. *PLoS One, 9*(12). https://doi.org/10.1371/journal.pone.0116351, http://www.plosone.org/article/fetchObject.action?uri=info%3Adoi%2F10.1371%2Fjournal.pone.0116351&representation=PDF.

Adams, G. (2020). A beginner's guide to RT-PCR, qPCR and RT-qPCR. *The Biochemist, 42*(3), 48–53. https://doi.org/10.1042/bio20200034.

Alaeddini, R. (2012). Forensic implications of PCR inhibition—A review. *Forensic Science International: Genetics, 6*(3), 297–305. https://doi.org/10.1016/j.fsigen.2011.08.006.

Alaeddini, R., Walsh, S. J., & Abbas, A. (2010). Forensic implications of genetic analyses from degraded DNA—A review. *Forensic Science International: Genetics, 4*(3), 148–157.

Alketbi, S. K. (2023). The role of DNA in forensic science: A comprehensive review. *SSRN*. https://doi.org/10.2139/ssrn.4550489, https://www.ssrn.com/index.cfm/en/.

Alrabie, S., Yaqoob, M., Turkiah, H., Salehi, J., Sanqoor, S., Lakshmi, V., & Nazir, M. (2018). Effect of fabric dyes and metal surfaces on DNA amplification. *International Journal of Sciences: Basic and Applied Research (IJSBAR, 38*(2), 152–159.

Al-Soud, W. A., & Rådström, P. (2001). Purification and characterization of PCR-inhibitory components in blood cells. *Journal of Clinical Microbiology, 39*(2), 485–493. https://doi.org/10.1128/JCM.39.2.485-493.2001.

Arsenault, H., Kuffel, A., Daeid, N. N., & Gray, A. (2024). Trace DNA and its persistence on various surfaces: A long term study investigating the influence of surface type and environmental conditions – Part one, metals. *Forensic Science International: Genetics, 70*. https://doi.org/10.1016/j.fsigen.2024.103011, https://www.sciencedirect.com/science/journal/18724973.

Baar, C., D'Abbadie, M., Vaisman, A., Arana, M. E., Hofreiter, M., Woodgate, R., Kunkel, T. A., & Holliger, P. (2011). Molecular breeding of polymerases for resistance to environmental inhibitors. *Nucleic Acids Research, 39*(8), e51. https://doi.org/10.1093/nar/gkq1360.

Balding, D. J., & Buckleton, J. (2009). Interpreting low template DNA profiles. *Forensic Science International: Genetics, 4*(1), 1–10. https://doi.org/10.1016/j.fsigen.2009.03.003.

Baltrušis, P., & Höglund, J. (2023). Digital PCR: Modern solution to parasite diagnostics and population trait genetics. *Parasites & Vectors, 16*(1). https://doi.org/10.1186/s13071-023-05756-7.

Barbaro (2020). STR typing and available multiplex kits including validation methods. In *Forensic DNA typing: Principles, applications and advancements*. Italy: Springer Singapore, 27–43. http://doi.org/10.1007/978-981-15-6655-4_2, https://link.springer.com/book/10.1007/978-981-15-6655-4.

Barbour, L.C., Tobe, S., Maker, G., & Baxter,F. (2023). Early-stage investigation into potential degradative effects of varied soils on human semen for the purpose of forensic investigation (Doctoral dissertation.

Behera, A., Panigrahi, G. K., & Sahoo, A. (2024). Nonsense-mediated mRNA decay in human health and diseases: Current understanding, regulatory mechanisms and future perspectives. *Molecular Biotechnology*, 1–15.

Bélec, L., Authier, J., Eliezer-Vanerot, M. C., Piedouillet, C., Mohamed, A. S., & Gherardi, R. K. (1998). Myoglobin as a polymerase chain reaction (PCR) inhibitor: A limitation for PCR from skeletal muscle tissue avoided by the use of Thermus thermophilus polymerase. *Muscle & Nerve, 21*, 1064–1067.

Bonsu, D. N. O., Higgins, D., Simon, C., Goodwin, C. S., Henry, J. M., & Austin, J. J. (2023). Quantitative PCR overestimation of DNA in samples contaminated with tin. *Journal of Forensic Sciences, 68*(4), 1302–1309. https://doi.org/10.1111/1556-4029.15312, http://onlinelibrary.wiley.com/journal/10.1111/(ISSN)1556-4029.

Bornman, D., Hester, M., Schuetter, J., Kasoji, M., Minard-Smith, A., Barden, C., Nelson, S., Godbold, G., Baker, C., Yang, B., Walther, J., Tornes, I., Yan, P., Rodriguez, B., Bundschuh, R., Dickens, M., Young, B., & Faith, S. (2012). Short-read, high-throughput sequencing technology for STR genotyping. *Biotechniques, 2012*. https://doi.org/10.2144/000113857.

Bourke, M. T., Scherczinger, C. A., Ladd, C., & Lee, H. C. (1999). NaOH treatment to neutralize inhibitors of Taq polymerase. *Journal of Forensic Sciences, 44*(5), 1046–1050. https://doi.org/10.1520/jfs12039j.

Butler, J. M. (2011). Forensic DNA testing. *Cold Spring Harbor Protocols, 2011*(12), pdb.top066928. https://doi.org/10.1101/pdb.top066928.

Butler, J. M. (2012). *PCR amplification*. Elsevier BV, 69–97. http://doi.org/10.1016/b978-0-12-374513-2.00004-x.

Butler, J. M. (2015). The future of forensic DNA analysis. *Philosophical Transactions of the Royal Society B: Biological Sciences, 370*(1674), 20140252. https://doi.org/10.1098/rstb.2014.0252.

Butler, J. M. (2019). Recent advances in forensic biology and forensic DNA typing: INTERPOL review. *Forensic Science International: Synergy, 6*.

Carey, L., & Mitnik, L. (2002). Trends in DNA forensic analysis. *Electrophoresis, 23*(10), 1386–1397. https://doi.org/10.1002/1522-2683(200205)23:10<1386::AID-ELPS1386>3.0.CO;2-M, http://onlinelibrary.wiley.com/journal/10.1002/(ISSN)1522-2683.

Cavanaugh, S. E., & Bathrick, A. S. (2018). Direct PCR amplification of forensic touch and other challenging DNA samples: A review. *Forensic Science International: Genetics, 32*, 40–49. https://doi.org/10.1016/j.fsigen.2017.10.005, http://www.elsevier.com.

Chakrabarti, R., & Schutt, C. E. (2001). The enhancement of PCR amplification by low molecular weight amides. *Nucleic Acids Research, 29*(11), 2377–2381. https://doi.org/10.1093/nar/29.11.2377, https://academic.oup.com/nar/issue.

Chantzi, N., & Georgakopoulos-Soares, I. (2024). The repertoire of short tandem repeats across the tree of life. *bioRxiv*. https://doi.org/10.1101/2024.08.08.607201, https://www.biorxiv.org.

Chong, K. W. Y., Thong, Z., & Syn, C. K. C. (2021). Recent trends and developments in forensic DNA extraction. *Wiley Interdisciplinary Reviews: Forensic Science, 3*(2), e1395.

Connell, J. R., Lea, R. A., Haupt, L. M., & Griffiths, L. R. (2024). Mitochondrial DNA analysis in population isolates: Challenges and implications for human identification. *Current Molecular Biology Reports, 10*(1), 1–8.

Cowell, R. G., Graversen, T., Lauritzen, S. L., & Mortera, J. (2015). Analysis of forensic DNA mixtures with artefacts. *Journal of the Royal Statistical Society Series C: Applied Statistics, 64*(1), 1–48. https://doi.org/10.1111/rssc.12071.

D. B., N., & Joji, J. G. (2022). Tooth as a source of DNA in forensic or investigative genetics: An overview. *Journal of Forensic Dental Sciences,* 186–196. https://doi.org/10.18311/jfds/12/3/2020.617.

Das, R., & Panigrahi, G. K. (2024). Messenger RNA surveillance: Current understanding, regulatory mechanisms, and future implications. *Molecular Biotechnology*. https://doi.org/10.1007/s12033-024-01062-4, https://www.springer.com/journal/12033.

Dash, H. R., Elkins, K. M., & Al-Snan, N. R. (2023). Technological advancements in DNA extraction and quantification of forensic samples. In *Advancements in forensic DNA analysis*. Singapore: Springer. https://doi.org/10.1007/978-981-99-6195-5_3.

Davalieva, K., & Efremov, G. D. (2010). Influence of salts and PCR inhibitors on the amplification capacity of three thermostable DNA polymerases. *Macedonian Journal of Chemistry and Chemical Engineering, 29*(1), 57–62. https://doi.org/10.20450/mjcce.2010.173, http://www.mjcce.org.mk/PDF/29_1_436.pdf.

Demeke, T., & Jenkins, G. R. (2010). Influence of DNA extraction methods, PCR inhibitors and quantification methods on real-time PCR assay of biotechnology-derived traits. *Analytical and Bioanalytical Chemistry, 396*(6), 1977–1990. https://doi.org/10.1007/s00216-009-3150-9.

Eckhart, L., Bach, J., Ban, J., & Tschachler, E. (2000). Melanin binds reversibly to thermostable DNA polymerase and inhibits its activity. *Biochemical and Biophysical Research Communications, 271*(3), 726–730. https://doi.org/10.1006/bbrc.2000.2716, http://www.sciencedirect.com/science/journal/0006291X.

Elnifro, E. M., Ashshi, A. M., Cooper, R. J., & Klapper, P. E. (2000). Multiplex PCR: Optimization and application in diagnostic virology. *Clinical Microbiology Reviews, 13*(4), 559–570. https://doi.org/10.1128/CMR.13.4.559-570.2000.

Esslinger, K. J., Siegel, J. A., Spillane, H., & Stallworth, S. (2004). Using STR analysis to detect human DNA from exploded pipe bomb devices. *Journal of Forensic Sciences, 49*(3), 481–484. https://doi.org/10.1520/jfs2003127, http://onlinelibrary.wiley.com/journal/10.1111/(ISSN)1556-4029.

Finaughty, C., Heathfield, L. J., Kemp, V., & Márquez-Grant, N. (2023). Forensic DNA extraction methods for human hard tissue: A systematic literature review and meta-analysis of technologies and sample type. *Forensic Science International: Genetics, 63*, 102818. https://doi.org/10.1016/j.fsigen.2022.102818.

Foley, M. M., Koehler, G., Fu, J., Allen, R., & Wagner, J. R. (2024). An exploratory view into allelic drop-out of sequenced autosomal STRs. *Journal of Forensic Sciences, 69*(3), 825–835. https://doi.org/10.1111/1556-4029.15504, http://onlinelibrary.wiley.com/journal/10.1111/(ISSN)1556-4029.

Frackman, S., Kobs, G., Simpson, D., & Storts, D. (1998). Beatine and DMSO: Enhancing agents for PCR. *Promega Notes, 65*, 27–30.

Fredricks, D. N., & Relman, D. A. (1998). Improved amplification of microbial DNA from blood cultures by removal of the PCR inhibitor sodium polyanetholesulfonate. *Journal of Clinical Microbiology, 36*(10), 2810–2816. https://doi.org/10.1128/jcm.36.10.2810-2816.1998, jcm.asm.org.

Frontanilla, T. S., Valle-Silva, G., Ayala, J., & Mendes-Junior, C. T. (2022). Open-access worldwide population STR database constructed using high-coverage massively parallel sequencing data obtained from the 1000 Genomes Project. *Genes, 13*(12). https://doi.org/10.3390/genes13122205, http://www.mdpi.com/journal/genes/.

Gallup, J.M. (2008). cPCR Inhibition and amplification of difficult templates. Reason.

Gill, P., Haned, H., Bleka, O., Hansson, O., Dørum, G., & Egeland, T. (2015). Genotyping and interpretation of STR-DNA: Low-template, mixtures and database matches—Twenty years of research and development. *Forensic Science International: Genetics, 18*, 100–117. https://doi.org/10.1016/j.fsigen.2015.03.014.

Gray, A., Kuffel, A., & Daeid, N. N. (2023). An improved rapid method for DNA recovery from cotton swabs. *Forensi c Science International: Genetics, 64*, 102848.

Haarkötter Cardoso, C., Sáiz Guinaldo, M., Gálvez Escolano, X., Vinueza Espinosa, D.C., Medina Lozano, M.I., Álvarez Merino, J.C., & Lorente Acosta, J.A. (2024). Evaluation of the usefulness of insertion-null markers in critical skeletal remains.

Haarkötter, C., Gálvez, X., Vinueza-Espinosa, D. C., Medina-Lozano, M. I., Saiz, M., Lorente, J. A., & Álvarez, J. C. (2023). A comparison of five DNA extraction methods from degraded human skeletal remains. *Forensic Science International, 348*. https://doi.org/10.1016/j.forsciint.2023.111730, https://www.elsevier.com/locate/forsciint.

Haddrill, P. R. (2021). Developments in forensic DNA analysis. *Emerging Topics in Life Sciences, 5*(3), 381–393. https://doi.org/10.1042/ETLS20200304, https://portlandpress.com/emergtoplifesci/article/5/3/381/228192/Developments-in-forensic-DNA-analysis.

Hedman, J., Nordgaard, A., Rasmusson, B., Ansell, R., & Rådström, P. (2009). Improved forensic DNA Analysis through the use of alternative DNA polymerases and statistical modeling of DNA profiles. *Biotechniques, 47*(5), 951–958. https://doi.org/10.2144/000113246.

Hedman, J., & Rådström, P. (2013). Overcoming inhibition in real-time diagnostic PCR. *Methods in Molecular Biology, 943*, 17–48. https://doi.org/10.1007/978-1-60327-353-4_2.

Hofreiter, M., Sneberger, J., Pospisek, M., & Vanek, D. (2021). Progress in forensic bone DNA analysis: Lessons learned from ancient DNA. *Forensic Science International: Genetics, 54*, 102538. https://doi.org/10.1016/j.fsigen.2021.102538.

Hou, Y., Chen, S., Zheng, Y., Zheng, X., & Lin, J.-M. (2023). Droplet-based digital PCR (ddPCR) and its applications. *TrAC Trends in Analytical Chemistry, 158*, 116897. https://doi.org/10.1016/j.trac.2022.116897.

Howarth, A., Drummond, B., Wasef, S., & Matheson, C. D. (2022). An assessment of DNA extraction methods from blood-stained soil in forensic science. *Forensic Science International, 341*, 111502. https://doi.org/10.1016/j.forsciint.2022.111502.

Hu, Q., Liu, Y., Yi, S., & Huang, D. (2015). A comparison of four methods for PCR inhibitor removal. *Forensic Science International: Genetics, 16*(1), 94–97. https://doi.org/10.1016/j.fsigen.2014.12.001.

Huang, Y., Wang, M., Liu, C., & He, G. (2024). Comprehensive landscape of non-CODIS STRs in global populations provides new insights into challenging DNA profiles. *Forensic Science International: Genetics, 70*, 103010. https://doi.org/10.1016/j.fsigen.2024.103010.

Huggett, J. F., Novak, T., Garson, J. A., Green, C., Morris-Jones, S. D., Miller, R. F., & Zumla, A. (2008). Differential susceptibility of PCR reactions to inhibitors: An important and unrecognised phenomenon. *BMC Research Notes, 1*. https://doi.org/10.1186/1756-0500-1-70.

Iyavoo, S. (2014). Development and application of a PCR multiplex to assess the quality and quantity of forensic DNA extracts (Doctoral dissertation.

Jäger, R. (2024). Genomic multicopy loci targeted by current forensic quantitative PCR assays. *Genes, 15*, 1299. https://doi.org/10.3390/genes15101299.

Jäger, R., & Weiher, H. (2020). Polymerase chain reaction. An introduction to molecular biotechnology: Fundamentals, methods and applications.

Jordan, D., & Mills, D. E. (2021). Past, present, and future of DNA typing for analyzing human and non-human forensic samples. *Frontiers in Ecology and Evolution, 9*. https://doi.org/10.3389/fevo.2021.646130.

Karunanathie, H., Kee, P. S., Ng, S. F., Kennedy, M. A., & Chua, E. W. (2022). PCR enhancers: Types, mechanisms, and applications in long-range PCR. *Biochimie, 197*, 130–143. https://doi.org/10.1016/j.biochi.2022.02.009, https://www.elsevier.com/locate/biochimie.

Kermekchiev, M. B., Kirilova, L. I., Vail, E. E., & Barnes, W. M. (2009). Mutants of Taq DNA polymerase resistant to PCR inhibitors allow DNA amplification from whole blood and crude soil samples. *Nucleic Acids Research, 37*(5), e40. https://doi.org/10.1093/nar/gkn1055.

King, C. E., Debruyne, R., Kuch, M., Schwarz, C., & Poinar, H. N. (2009). A quantitative approach to detect and overcome PCR inhibition in ancient DNA extracts. *Biotechniques, 47*(5), 941–949. https://doi.org/10.2144/000113244.

Knijff (2022). On the forensic use of Y-chromosome polymorphisms. *Genes, 13*(5).

Kojabad, A. A., Farzanehpour, M., Galeh, H. E. G., Dorostkar, R., Jafarpour, A., Bolandian, M., & Nodooshan, M. M. (2021). Droplet digital PCR of viral DNA/RNA, current progress, challenges, and future perspectives. *Journal of Medical Virology, 93*(7), 4182–4197. https://doi.org/10.1002/jmv.26846, http://onlinelibrary.wiley.com/journal/10.1002/(ISSN)1096-9071.

Lance, R. F., & Guan, X. (2020). Variation in inhibitor effects on qPCR assays and implications for eDNA surveys. *Canadian Journal of Fisheries and Aquatic Sciences, 77*(1), 23–33. https://doi.org/10.1139/cjfas-2018-0263, http://www.nrcresearchpress.com/journal/cjfas.

Li, H., Ren, H., Yang, F., Chen, M., Sun, W., Jiang, L., Gao, Z., Liu, Y., & Liu, X. (2024). Developmental validation of the microreader 23HS Plex ID system: A novel supplementary non-CODIS STR multiplex assay for forensic application. *Electrophoresis, 45*(21-22), 1976–1987. https://doi.org/10.1002/elps.202400106.

Liu, C., Huang, D., Yang, L., Wu, S., Shen, X., Pedersen-Bjergaard, S., & Huang, C. (2021). Removal of polymerase chain reaction inhibitors by electromembrane extraction. *Analytical Chemistry, 93*(33), 11488–11496. https://doi.org/10.1021/acs.analchem.1c01689.

Lorenz, T. C. (2012). Polymerase chain reaction: Basic protocol plus troubleshooting and optimization strategies. *Journal of Visualized Experiments*(63). https://doi.org/10.3791/3998, http://www.jove.com/pdf/3998/jove-protocol-3998-polymerase-chain-reaction-basic-protocol-plus-troubleshooting-and-optimization-strategies.

Lynch, C., & Fleming, R. (2019). A review of direct polymerase chain reaction of DNA and RNA for forensic purposes. *WIREs Forensic Science, 1*(4). https://doi.org/10.1002/wfs2.1335.

Maher, N., Dillon, H. K., Vermund, S. H., & Unnasch, T. R. (2001). Magnetic bead capture eliminates PCR inhibitors in samples collected from the airborne environment, permitting detection of Pneumocystis carinii DNA. *Applied and Environmental Microbiology, 67*(1), 449–452. https://doi.org/10.1128/AEM.67.1.449-452.2001.

Majumder, S., & Panigrahi, G. K. (2024a). Molecular docking and ADME evaluation of plant-based bioactive molecules targeting nonsense-mediated mRNA decay pathway factors to modulate tumorigenesis. *Plant Science Today, 11*(4), 1148–1155. https://doi.org/10.14719/pst.3737, https://horizonepublishing.com/journals/index.php/PST/article/view/3737.

Majumder, S., & Panigrahi, G. K. (2024b). Advancements in contemporary pharmacological innovation: Mechanistic insights and emerging trends in drug discovery and development. *Intelligent Pharmacy*. https://doi.org/10.1016/j.ipha.2024.10.001.

Makowski, G. S., Davis, E. L., Aslanzadeh, J., & Hopfer, S. M. (1995). Enhanced direct amplification of Guthrie card DNA following selective elution of PCR inhibitors. *Nucleic Acids Research, 23*(18), 3788–3789. https://doi.org/10.1093/nar/23.18.3788.

Martin, B., & Linacre, A. (2020). Direct PCR: A review of use and limitations. *Science & Justice, 60*(4), 303–310. https://doi.org/10.1016/j.scijus.2020.04.003.

McCord, B. R., Gauthier, Q., Cho, S., Roig, M. N., Gibson-Daw, G. C., Young, B., & Duncan, G. (2018). Forensic DNA analysis. *Analytical Chemistry, 91*(1), 673–688.

McDonald, J., & Lehman, D. C. (2012). Forensic DNA analysis. *American Society for Clinical Laboratory Science, 25*(2), 109–113. https://doi.org/10.29074/ascls.25.2.109.

McDonald, C., Taylor, D., & Linacre, A. (2024). PCR in forensic science: A critical review. *Genes, 15*(4), 438. https://doi.org/10.3390/genes15040438.

McDonald, C., Taylor, D., & Linacre, A. (2024a). Smart PCR leading to improved DNA profiles. *Australian Journal of Forensic Sciences, 56*(sup1), 49–51.

Meintjes-Van der Walt, L., & Dhliwayo, P. (2021). DNA evidence as the basis for conviction. *Potchefstroom Electronic Law Journal/Potchefstroomse Elektroniese Regsblad, 24*(1).

Odah, N. (2024). The double helix of justice: The crucial role of DNA in advancing criminal investigations.

Ong, Y. H. M., Cook, R., Silenieks, T., Henry, J., & Hefford, C. (2006). Evaluation of sampling techniques for trace DNA with robotic considerations. In *Biology report: R75. Evidence recovery and biology analytical groups*. Government of South Australia.

Oosthuizen, T., & Howes, L. M. (2022). The development of forensic DNA analysis: New debates on the issue of fundamental human rights. *Forensic Science International: Genetics, 56*, 102606. https://doi.org/10.1016/j.fsigen.2021.102606.

Opel, K. L., Chung, D., & McCord, B. R. (2010). A study of PCR inhibition mechanisms using real time PCR. *Journal of Forensic Sciences, 55*(1), 25–33.

Orlando, L., Allaby, R., Skoglund, P., Der Sarkissian, C., Stockhammer, P. W., Ávila-Arcos, M. C., Fu, Q., Krause, J., Willerslev, E., Stone, A. C., & Warinner, C. (2021). Ancient DNA analysis. *Nature Reviews Methods Primers, 1*(1). https://doi.org/10.1038/s43586-020-00011-0, https://www.nature.com/nrmp/journal-information.

Panigrahi, G. K., & Satapathy, K. B. (2020a). Arabidopsis DCP5, a decapping complex protein interacts with Ubiquitin-5 in the processing bodies. *Plant Archives, 20*(1), 2243–2247.

Panigrahi, G. K., & Satapathy, K. B. (2020b). Formation of Arabidopsis poly(A)-specific ribonuclease associated processing bodies in response to pathogenic infection. *Plant Archives, 20*(2), 4907–4912.

Panigrahi, G. K., & Satapathy, K. B. (2020c). Sacrificed surveillance process favours plant defense: A review. *Plant Archives, 20*(1), 2551–2559.

Panigrahi, G. K., Sahoo, S. K., Sahoo, A., Behera, S., Sahu, S. R., Dash, A., & Satapathy, K. B. (2021c). Bioactive molecules from plants: A prospective approach to combat SARS-Cov-2. *Advances in Traditional Medicine,* 1–14.

Panigrahi, G. K., & Satapathy, K. B. (2021). Pseudomonas syringae pv. syringae infection orchestrates the fate of the Arabidopsis J domain containing cochaperone and decapping protein factor 5. *Physiological and Molecular Plant Pathology, 113*. https://doi.org/10.1016/j.pmpp.2020.101598, http://www.elsevier.com/inca/publications/store/6/2/2/9/3/2/index.htt.

Panigrahi, G. K., Sahoo, A., & Satapathy, K. B. (2021a). Differential expression of selected Arabidopsis resistant genes under abiotic stress conditions. *Plant Science Today, 8*(4), 859–864.

Panigrahi, G. K., Sahoo, A., & Satapathy, K. B. (2021b). Insights to plant immunity: Defense signaling to epigenetics. *Physiological and Molecular Plant Pathology, 113*, 1–7.

Panigrahi, G. K., Sahoo, A., & Satapathy, K. B. (2024). The processing body component varicose plays a multiplayer role towards stress management in Arabidopsis. *Plant Physiology Reports,* 1–10.

Patro, I., Sahoo, A., Nayak, B. R., Das, R., Majumder, S., & Panigrahi, G. K. (2023). Nonsense-mediated mRNA decay: Mechanistic insights and physiological significance. *Molecular Biotechnology,* 1–15.

Patro, A. K., Sahoo, A., Satapathy, K. B., & Panigrahi, G. K. (2024). Nonsense-mediated mRNA decay: Physiological significance, mechanistic insights and future implications. *Pathology, Research and Practice, 264*, 155677.

Primorac, D., & Schanfield, M. (Eds.). (2023). *Forensic DNA applications: An interdisciplinary perspective*. CRC press.

Panneerchelvam, S., & Norazmi, M. N. (2003). Forensic DNA profiling and database. *Malaysian Journal of Medical Sciences, 10*(2), 20–26.

Rački, N., Dreo, T., Gutierrez-Aguirre, I., Blejec, A., & Ravnikar, M. (2014). Reverse transcriptase droplet digital PCR shows high resilience to PCR inhibitors from plant, soil and water samples. *Plant Methods, 10*(1). https://doi.org/10.1186/s13007-014-0042-6.

Rådström, P., Lövenklev, M., Wolffs, P., Löfström, C., & Knutsson, R. (2003). Pre-PCR processing strategies. In *PCR technology: Current innovations, second edition*. Sweden: CRC Press, 37–47. https://www.taylorfrancis.com/books/e/9781420040654.

Rådström, P., Knutsson, R., Wolffs, P., Lövenklev, M., & Löfström, C. (2004). Pre-PCR processing: Strategies to generate PCR-compatible samples. *Molecular Biotechnology, 26*(2), 133–146. https://doi.org/10.1385/MB:26:2:133.

Ricci, U., Ciappi, D., Carboni, I., Centrone, C., Giotti, I., Petti, M., Alice, B., & Pelo, E. (2024). Looking into the quantification of forensic samples with real-time PCR. *Genes, 15*(6), 759. https://doi.org/10.3390/genes15060759.

Roewer, L. (2019). Y-chromosome short tandem repeats in forensics—Sexing, profiling, and matching male DNA. *Wiley Interdisciplinary Reviews: Forensic Science, 1*(4).

Romeika, J. M., & Yan, F. (2014). Recent advances in forensic DNA analysis. *J. Forensic Res. s, 12*, 1–13. https://doi.org/10.4172/2157-7145.S12-001.

Rossen, L., Nørskov, P., Holmstrøm, K., & Rasmussen, O. F. (1992). Inhibition of PCR by components of food samples, microbial diagnostic assays and DNA-extraction solutions. *International Journal of Food Microbiology, 17*(1), 37–45. https://doi.org/10.1016/0168-1605(92)90017-W.

Sahoo, K. B., & Satapathy (2021). Differential expression of Arabidopsis EJC core proteins under short-day and long-day growth conditions. *Plant Science Today, 8*(4), 815–819. https://doi.org/10.14719/PST.2021.8.4.1214, https://horizonepublishing.com/journals/index.php/PST/article/view/1214.

Sahoo, A., Satapathy, K. B., & Panigrahi, G. K. (2023a). Ectopic expression of disease resistance protein promotes resistance against pathogen infection and drought stress in Arabidopsis. *Physiological and Molecular Plant Pathology, 124*(101949), 1–7.

Sahoo, A., Satapathy, K. B., Sahoo, S. K., & Panigrahi, G. K. (2023b). Microbased biorefinery for gold nanoparticle production: Recent advancements, applications and future aspects. *Preparative Biochemistry & Biotechnology, 53*(6), 579–590.

Sahoo, A., Sethi, J., Satapathy, K. B., Sahoo, S. K., & Panigrahi, G. K. (2023c). Nanotechnology for precision and sustainable agriculture: Recent advances, challenges and future implications. *Nanotechnology for Environmental Engineering, 8*(3), 775–787.

Sahoo, A., Satapathy, K. B., & Panigrahi, G. K. (2024). Security check: Plant immunity under temperature surveillance. *Journal of Plant Biochemistry and Biotechnology,* 1–4.

Salipante, S. J., & Jerome, K. R. (2020). Digital PCR—An emerging technology with broad applications in microbiology. *Clinical Chemistry, 66*(1), 117–123. https://doi.org/10.1373/clinchem.2019.304048, https://academic.oup.com/clinchem/article/66/1/117/5688828.

Singh, S. P., Panigrahi, G. K., & Sahoo, A. (2024). Optimizing nanoparticles use for growth promotion in zebrafish: Insights on concentration-dependent effects. *Nanotechnology for Environmental Engineering, 9*(4), 757–767.

Schander, C., & Halanych, K. M. (2003). DNA, PCR and formalinized animal tissue - A short review and protocols. *Organisms Diversity and Evolution, 3*(3), 195–205. https://doi.org/10.1078/1439-6092-00071, http://www.springer.com/life+sciences/ecology/journal/13127.

Schrader, C., Schielke, A., Ellerbroek, L., & Johne, R. (2012). PCR inhibitors - Occurrence, properties and removal. *Journal of Applied Microbiology, 113*(5), 1014–1026. https://doi.org/10.1111/j.1365-2672.2012.05384.x.

Seo, S. B., Jin, H. X., Lee, H. Y., Ge, J., King, J. L., Lyoo, S. H., Shin, D. H., & Lee, S. D. (2013). Improvement of short tandem repeat analysis of samples highly contaminated by humic acid. *Journal of Forensic and Legal Medicine, 20*(7), 922–928. https://doi.org/10.1016/j.jflm.2013.08.001.

Sessa, F., Pomara, C., Esposito, M., Grassi, P., Cocimano, G., & Salerno, M. (2023). Indirect DNA transfer and forensic implications: A literature review. *Genes, 14*(12), 2153. https://doi.org/10.3390/genes14122153.

Shewale, J. G., & Liu, R. H. (2013). *Forensic DNA analysis: Current practices and emerging technologies*. United States: Taylor and Francis, 1–414. http://doi.org/10.1201/b15361, https://www.routledge.com/products/9781466571266.

Shukurullaevna, B. S., Kamoliddinovna, N. N., & Khamidullayevna, T. M. (2024). The importance of PCR in the laboratory. *Tadqiqotlar. UZ, 48*(1), 54–60.

Sidstedt, M., Rådström, P., & Hedman, J. (2020). PCR inhibition in qPCR, dPCR and MPS-mechanisms and solutions. *Analytical and Bioanalytical Chemistry, 412*(9), 2009–2023. https://doi.org/10.1007/s00216-020-02490-2.

Smalla, K., Cresswell, N., Mendonca-Hagler, L. C., Wolters, A., & Elsas, J. V. (1993). Rapid DNA extraction protocol from soil for polymerase chain reaction-mediated amplification. *Journal of Applied Microbiology, 74*(1), 78–85.

Sutlovic, D., Definis, G. M., Andelinovic, S., Gugic, D., & Primorac, D. (2005). Taq polymerase reverses inhibition of quantitative real time polymerase chain reaction by humic acid. *Croatian Medical Journal, 46*, 556–562.

Sidstedt, M., Hedman, J., Romsos, E. L., Waitara, L., Wadsö, L., Steffen, C. R., Vallone, P. M., & Rådström, P. (2018). Inhibition mechanisms of hemoglobin, immunoglobulin G, and whole blood in digital and real-time PCR. *Analytical and Bioanalytical Chemistry, 410*(10), 2569–2583. https://doi.org/10.1007/s00216-018-0931-z, link.springer.de/link/service/journals/00216/index.htm.

Sidstedt, M., Jansson, L., Nilsson, E., Noppa, L., Forsman, M., Rådström, P., & Hedman, J. (2015). Humic substances cause fluorescence inhibition in real-time polymerase chain reaction. *Analytical Biochemistry, 487*, 30–37. https://doi.org/10.1016/j.ab.2015.07.002.

Sidstedt, M., Romsos, E. L., Hedell, R., Ansell, R., Steffen, C. R., Vallone, P. M., Rådström, P., & Hedman, J. (2017). Accurate digital polymerase chain reaction quantification of challenging samples applying inhibitor-tolerant DNA polymerases. *Analytical Chemistry, 89*(3), 1642–1649. https://doi.org/10.1021/acs.analchem.6b03746, http://pubs.acs.org/journal/ancham.

Sidstedt, M., Steffen, C. R., Kiesler, K. M., Vallone, P. M., Rådström, P., & Hedman, J. (2019). The impact of common PCR inhibitors on forensic MPS analysis. *Forensic Science International: Genetics, 40*, 182–191. https://doi.org/10.1016/j.fsigen.2019.03.001, http://www.elsevier.com.

Smith, J. H., & Singh, M. (2024). DNA forensic and forensic investigative leads. *Journal of Forensic Medicine, 9*.

Spiess, A.-N., Mueller, N., & Ivell, R. (2004). Trehalose Is a potent PCR Enhancer: Lowering of DNA melting temperature and thermal stabilization of Taq polymerase by the disaccharide trehalose. *Clinical Chemistry, 50*(7), 1256–1259. https://doi.org/10.1373/clinchem.2004.031336.

Steele, C. D., & Balding, D. J. (2014). Statistical evaluation of forensic DNA profile evidence. *Annual Review of Statistics and Its Application, 1*, 361–384. https://doi.org/10.1146/annurev-statistics-022513-115602, http://www.annualreviews.org/journal/statistics.

Straube, N., Lyra, M. L., Paijmans, J. L. A., Preick, M., Basler, N., Penner, J., Rödel, M. O., Westbury, M. V., Haddad, C. F. B., Barlow, A., & Hofreiter, M. (2021). Successful application of ancient DNA extraction and library construction protocols to museum wet collection specimens. *Molecular Ecology Resources, 21*(7), 2299–2315. https://doi.org/10.1111/1755-0998.13433, http://onlinelibrary.wiley.com/journal/10.1111/(ISSN)1755-0998.

Sur, K., McFall, S. M., Yeh, E. T., Jangam, S. R., Hayden, M. A., Stroupe, S. D., & Kelso, D. M. (2010). Immiscible phase nucleic acid purification eliminates PCR inhibitors with a single pass of paramagnetic particles through a hydrophobic liquid. *Journal of Molecular Diagnostics, 12*(5), 620–628. https://doi.org/10.2353/jmoldx.2010.090190, http://jmd.amjpathol.org/cgi/reprint/12/5/620.

Sutlovic, D., Gamulin, S., Definis-Gojanovic, M., Gugic, D., & Andjelinovic, S. (2008). Interaction of humic acids with human DNA: Proposed mechanisms and kinetics. *Electrophoresis, 29*(7), 1467–1472. https://doi.org/10.1002/elps.200700699, http://onlinelibrary.wiley.com/journal/10.1002/(ISSN)1522-2683.

Swari, F. A., Suharti, S., Susanti, E., Sanjaya, E. H., & Sumari, S. (2023). Study of inhibition DNA polymerase by phenolic compounds in traditional food spices on the polymerase chain reaction process. *Frontier Advances in Applied Science and Engineering, 1*(1), 1–13. https://doi.org/10.59535/faase.v1i1.108.

Tanudisastro, H. A., Deveson, I. W., Dashnow, H., & MacArthur, D. G. (2024). Sequencing and characterizing short tandem repeats in the human genome. *Nature Reviews. Genetics, 25*(7), 460–475. https://doi.org/10.1038/s41576-024-00692-3, https://www.nature.com/nrg/.

Thanakiatkrai, P., & Rerkamnuaychoke, B. (2017). Direct STR typing from bullet casings. *Forensic Science International: Genetics Supplement Series, 6*, e164–e166.

Thompson, R. E., Duncan, G., & McCord, B. R. (2014). An investigation of PCR inhibition using Plexor®-based quantitative PCR and short tandem repeat amplification. *Journal of Forensic Sciences, 59*(6), 1517–1529.

Tvedebrink, T., Eriksen, P. S., Mogensen, H. S., & Morling, N. (2012). Statistical model for degraded DNA samples and adjusted probabilities for allelic drop-out. *Forensic Science International: Genetics, 6*(1), 97–101. https://doi.org/10.1016/j.fsigen.2011.03.001.

Vajpayee, K., Dash, H. R., Parekh, P. B., & Shukla, R. K. (2023). PCR inhibitors and facilitators-Their role in forensic DNA analysis. *Forensic Science International*.

Van Oorschot, R. A., Ballantyne, K. N., & Mitchell, R. J. (2010). Forensic trace DNA: A review. *Investig Genet, 1*, 14. https://doi.org/10.1186/2041-2223-1-14.

Vitošević, K., Todorović, M., Varljen, T., Slović, Ž., Matić, S., & Todorović, D. (2018). Effect of formalin fixation on PCR amplification of DNA isolated from healthy autopsy tissues. *Acta Histochemica, 120*(8), 780–788. https://doi.org/10.1016/j.acthis.2018.09.005.

Wang, M., Cai, J., Chen, J., Liu, J., Geng, X., Yu, X., & Yang, J. (2024). *PCR techniques and their clinical applications*. IntechOpen. http://doi.org/10.5772/intechopen.110220.

Wickenheiser, R. A. (2021). The value of forensic DNA leads in preventing crime and eliminating the innocent. *Forensic Science International: Synergy, 3*, 100201. https://doi.org/10.1016/j.fsisyn.2021.100201.

Wilson, I. G. (1997). Inhibition and facilitation of nucleic acid amplification. *Applied and Environmental Microbiology, 63*(10), 3741–3751. https://doi.org/10.1128/aem.63.10.3741-3751.1997.

Xavier, C., Eduardoff, M., Bertoglio, B., Amory, C., Berger, C., Casas-Vargas, A., Pallua, J., & Parson, W. (2021). Evaluation of DNA extraction methods developed for forensic and ancient DNA applications using bone samples of different age. *Genes, 12*(2), 146. https://doi.org/10.3390/genes12020146.

Xin, Z., & Chen, J. (2012). A high throughput DNA extraction method with high yield and quality. *Plant Methods, 8*(1), 26. https://doi.org/10.1186/1746-4811-8-26.

Xin, Z., Velten, J. P., Oliver, M. J., & Burke, J. J. (2003). High-throughput DNA extraction method suitable for PCR. *Biotechniques, 34*(4), 820–826. https://doi.org/10.2144/03344rr04, https://www.future-science.com/journal/btn.

Zhang, J., Zhang, J., Tao, R., Jiang, L., Chen, L., Li, X., Li, C., & Zhang, S. (2020). A newly devised multiplex assay of novel polymorphic non-CODIS STRs as a valuable tool for forensic application. *Forensic Science International: Genetics, 48*, 102341. https://doi.org/10.1016/j.fsigen.2020.102341.

Zupanič Pajnič, I., Leskovar, T., & Črešnar, M. (2023). Improving kinship probability in analysis of ancient skeletons using identity SNPs and MPS technology. *International Journal of Legal Medicine, 137*(4), 1007–1015.

Chapter 32

Assessment of the reliability and reproducibility of forensic DNA profiling results

Heather Miller Coyle
Forensic Science Department, Henry C. Lee College of Criminal Justice & Forensic Sciences, University of New Haven, West Haven, CT, United States

32.1 Introduction

When one considers deoxyribonucleic acid (DNA), the molecule that holds the information for your unique blueprint for life, most consider it to be unique, testable, and infallible as a method for human identification. In general, this perception is very true. However, with all scientific methods, there exists the possibility for variation and error, and both are measurable. This chapter focuses on the reliability of the DNA test method called short tandem repeat (STR) analysis, where reliability refers to the reproducibility of a test method and the ability to achieve the same result using replicate testing strategies. Besides reliability, one must also prove that the results are accurate or conform to the true value. The difference between the test value and the true value would be defined as the error. For each step of the STR testing process for generating a unique DNA profile using DNA technology, the potential for error can be evaluated. These steps include DNA extraction, quantitation, polymerase chain reaction (PCR) amplification, capillary electrophoresis (CE) detection, software analysis, reporting, and testimony (Table 32.1). These are the standard steps for creating the DNA profile commonly used for human identification purposes in criminal and civil casework and immigration offices. This DNA profile or barcode is unique to everyone, to the best of our knowledge, except for identical twins who share the same DNA profile based on common genetics from birth.

32.2 Policy concerns

The National Academy of Sciences (NAS) in 2009 issued some recommendations regarding forensic method reliability (Giannelli, 2012; National Research Council, 2009). Recommendation 3 of the NAS report stated that there was a need for more forensic research "to address issues of accuracy, reliability, and validity in the forensic science disciplines." The report called for "quantifiable measures" of reliability and for research to be performed on relevant forensic casework situations. The casework analyst should know where the variance in the method should be, and research to quantify the variance should be published under the critical process of peer review. The report also called for a performance check (Recommendation 5) on forensic analysts to further define the variance in specific individuals as compared to the general practice in the forensic community. As part of the performance check, the NAS report suggests that human bias could be measured and that performance errors could be identified and further characterized to establish standardization in training and in laboratory procedures. Human bias in forensic evidence testing can often be characterized as confirmation bias and contextual bias (Starr, 2022). Confirmation bias is defined as the desire to identify facts or data that confirm prior beliefs, such as I believe Joe Smith committed the crime; therefore, I will seek information to fit that theory rather than trying to see all the options. This is also called by many the "tunnel vision" phenomenon. Contextual bias is defined as the ability to shift an individuals' perception of the data based on external nonrelevant information. An example of this phenomenon would be that the victim is the causative agent in the homicide due to dislike of the person and the prior behaviors rather than the true perpetrator of the crime. In both examples, bias can cloud the representation of the science, potentially, and forensic science training strives to reduce bias whenever possible in DNA testing, report writing, and testimony in court.

Investigative and Predictive DNA Testing. DOI: https://doi.org/10.1016/B978-0-443-30128-5.00024-2

TABLE 32.1 Steps in DNA profiling using short tandem repeat analysis in human identification.

DNA extraction	This step is for the lysis of the cell membrane and release of the DNA molecule into solution. Most extraction methods purify DNA away from other cellular components such as lipids, proteins, and carbohydrates. The extraction product consists of the nuclear DNA from the 23 pairs of chromosomes and the mitochondrial DNA combined in solution.
DNA quantitation	This step is designed to amplify specific regions of the nuclear DNA to estimate the quantity recovered from the extraction step. This step will also determine the ability of the DNA to be PCR amplified for the DNA profiling step based on purity and the level of degradation of the starting template.
Polymerase chain reaction (PCR) amplification	This step is to copy or amplify specific targeted regions of the human genome that are known to be highly variable between individuals. By generating test results from different chromosomal loci, a string of numbers representing varying alleles becomes the DNA profile.
Capillary electrophoresis detection	DNA is invisible and microscopic, so fluorescent dyes added in the PCR amplification step are detected by electrokinetic separation in microfluidic channels called capillaries. The size of the DNA fragment and the time for separation are plotted for different dye sets to generate an electropherogram.
Software analysis	The electropherogram or DNA peak profile is edited for PCR artifacts such as stutter, allele drop-in, baseline pull-up, peak saturation, etc., to assess the data and verify the allele calls made by the semi-expert allele calling software. All edits are traceable and reviewable by analysts, supervisors, and experts.
Reporting/testimony	Translating the data to verbal statements for reports and testimony is specified by individual forensic science laboratory protocols and guidelines. For this reason, verbiage of the reports or testimony may vary slightly in different jurisdictions, but the overall conclusions of exclusion, inclusion, or inconclusive should remain the same.

TABLE 32.2 President's Council of Advisors on Science and Technology report concerns regarding forensic DNA analysis.

Foundational validity	Used to show by empirical methods that the test procedure is repeatable, reproducible, and accurate. This term is used to show that the test method in principle is reliable and has been tested for the intended application.
Validity as applied	Used to show that the test method has been shown to be reliable in practice. Rule 702 in the Federal Rules of Evidence informs that the expert has reliably applied the principles and methods to the facts of the case.

The President's Council of Advisors on Science and Technology (PCAST) issued in 2016 some recommendations for strengthening the foundational validity of forensic science methods used in criminal courts (Table 32.2) (PCAST, 2016). Foundational validity is defined as the scientific standards that correspond to the legal standards that show reliability in methodology (Geddes, 2010). Another form of validity is called "validity as applied" meaning the scientific standard conforms to the legal standard for the reliable application of the science to the evidence method. It is possible to have a scientifically valid method but if used or applied incorrectly, it will not meet the standard for reliable acceptance in court (Catena and Judge, 2016; Ruling – The People of the State of New York versus Oral Nicholas Hillary (NY), 2017). Scientific validity means the method has been empirically determined to be accurate under specific sets of conditions and, therefore, approved for use in a court of law. To achieve scientific validity, one must test numerous samples, examine many conditions, determine error rates, and show experimental proof as validation studies for the courts in admissibility hearings. Empirical studies are used to support scientific validity by testing many forms of mock evidence under mock case scenarios. Blind studies are used to test analyst and examiner skills in an empirical study to assess variance and error in subjective analyses. The PCAST report specifically addresses concerns over statistical analyses of complex DNA mixtures and source attribution error (PCAST, 2016). The more individuals that you add to a DNA mixture, the more challenging it is to establish which DNA allele or peak goes to each contributor by visual examination of DNA electropherograms. One solution was to implement probabilistic genotyping (PG) software analysis of the complex DNA mixtures to generate a likelihood ratio (LR) value to give a probability estimate for inclusion, exclusion, or inconclusive conclusions as an aid to the forensic scientist. The software analysis must also be validated to establish proper use and standards for reporting as in the court decision for the *People of the State of New York v. Oral Nicholas Hillary*. As reliable and accurate as the computer analysis could be, and even with the improvement in detangling complex DNA mixture information, there were still circumstances where the software did not make adequate distinctions between DNA samples exhibiting a high degree of allele sharing or for minor contributor profiles with a high degree of allele drop-out.

32.3 Short tandem repeat validation

STR validation guidelines are included in the Scientific Working Group for DNA Analysis Methods (SWGDAM) Validation Guidelines for Forensic DNA Analysis Methods (https://www.swgdam.org/publications) (Coyle, 2022). Two forms of validation are generally recognized: (1) development validation by the manufacturer or developer of the equipment and/or methodology and (2) internal validation performed by the forensic laboratory and personnel to establish the benefits and limitations for use under their own work conditions. Both forms of validation require scientific and legal scrutiny and it is preferable to have the results of these studies publicly available and published in peer-reviewed scientific journals to establish forensic and legal general acceptance. The types of experimental data used to establish validity include the characterization of the genetic locations being tested; tests for the specificity of species detection; sensitivity studies to establish the appropriate quantities to use for the test method; stability studies for DNA that has been exposed to a variety of surfaces, chemistries and environmental conditions; population surveys for genetic variation; mock case sample testing; controlled DNA mixture studies, and reproducibility and accuracy studies.

32.4 Short tandem repeat method

32.4.1 Quantity effects

The quantity of DNA used in each test method is a variable that, when adjusted can yield variance in the test result. The optimal target quantity for generating a full DNA profile is approximately 500 - 1000 picograms (pg) for most STR profiling kits, and this is a good example of a manufacturer's recommendation based on developmental validation. This means the kit manufacturer recommends that to guarantee a full DNA profile, 500 to 1000 picograms of DNA template molecules are necessary in the PCR amplification step. However, with empirical validation studies, many times suboptimal DNA template can still provide a partial or full DNA profile result useful to the case. As an example, one forensic laboratory has empirically determined a STR profiling kit to be effective in their own hands for DNA template quantities of 525 to 37.5 pg based on their internal validation study that shows reliable data at suboptimal quantities (Office of the Chief Medical Examiner in New York City, https://www.nyc.gov/site/ocme/services/technical-manuals.page). Although different than the manufacturer's recommendation, this means that there is a wide range of input values for DNA quantity that could potentially yield valuable forensic information based on empirical testing. If the working parameters for the method are defined and provable using good scientific methods and principles, and there is substantial documentation to prove the claim, the technique may be allowed in court as good scientific practice based on the laboratory's own validation studies.

32.4.2 Quality effects

Besides quantity, the quality of the DNA profile can affect the interpretation of the data. A full DNA profile can yield sufficient information for a full identification. However, if a partial DNA profile is present, due to allele drop-out and degradation, an identification may not be possible. A term called stochastic variation is used to describe variance in PCR amplification efficiency of different allele peaks. If there is degradation of the DNA template, then it may not amplify very well, yielding peak imbalance. The PCR is an enzymatic replication process that occurs outside the cell that allows a scientist to study the molecular copying process of specific target regions of DNA from different chromosomes. However, DNA degradation can influence the final PCR result. The degradation pattern makes the data unreliable as far as human identification practice goes, if the degradation is sufficient to reduce the number of loci or points of comparison to below the minimum number allowed to make a comparison, and then that data is lost. Laboratory policy is designed to limit circumstances where too much coincidental matching may occur with other unrelated individuals in a population. The minimum number of required loci where test results may be presented in reports or testimony varies per forensic laboratory policy.

32.5 Equipment

32.5.1 Instrumentation differences

The PCR instrumentation contains a heat block that holds individual sample tubes, and this block should be temperature checked to be sure each sample well is reaching the expected temperature for the reaction. This quality control measure proves that the instrument is functioning properly and producing the best enzymatic replication result with optimal efficiency. The detection instrumentation for CE, the method used to detect fluorescence in DNA profiling, is a camera collection device associated with gel electrophoresis and runs by a charge-coupled device (CCD) camera chip. The manufacturing of the CCD camera chip introduces slight variability and hence, each chip is different in its sensitivity of detection. Therefore, calibration is required to equalize the detection limit for every CE instrument that it purchased. Most forensic science laboratories purchase multiple instruments and test run the instrumentation prior to casework to indicate comparability in detection limits for the same test sample. Instrumentation differences for the PCR amplification step may result in differences in PCR amplification efficiency and can be considered a source of variance. Likewise, the differences in the manufacturing of the CCD chip may be a source of variance in the CE detection step.

32.6 Short tandem repeat replicate testing

Most forensic science laboratories do not perform replicate testing on evidentiary samples to conserve the sample for later possible re-testing. The sample is tested once, and the DNA result is then reported. However, as a quality control measure, some forensic science laboratories will perform duplicate testing on DNA evidence. This is an expensive, time-consuming process to show that the data, when tested twice, yields approximately the same result. With high quantity samples (> 100 pg), the result should be consistent in replicate testing. However, when lower quantity samples are PCR amplified in duplicate (< 100 pg), low quantity and low-quality effects become magnified, and the results may be quite variable or stochastic. This results in some confusion in the courts as the replicates do not provide exact duplicate results but do corroborate similar amplification from the same DNA extract. In other words, a true sample switch could be detected using PCR duplicate reactions, but the stochastic nature of some samples may not give exact replicated data.

32.7 Quantifying variation

What is the best practice for measuring variation in DNA profiling tests? The simple measure is testing duplication and checking for concordance or similarity between results (Coyle, 2015). Duplicate or replicate testing is designed to test for a reproducible answer. The difference between each test result can be measured as testing variance (Table 32.3). Variance can be broken down into component parts to assess the testing process and determine if improvements can be made to reduce variance. For example, during CE, a DNA sample could be tested for the sizing of a band or peak, and the true value should be 150 base pairs in length for the DNA fragment. Every time the DNA fragment is processed through a Genetic Analyzer, the sizing can be measured, and it may vary by a fraction such as 150.25, 150.21 , and 150.18 bp. The difference between the expected value of 150 bp and the actual measured value, for example, 150.21 bp would be the size variance for the peak or 0.21 bp.

32.7.1 Polymerase chain reaction efficiency

PCR refers to the enzymatic replication of DNA outside of the cell in a sterile tube. All the components for PCR are also sterile and include water, buffer, free nucleotides, PCR primers (short sequences of DNA complementary to the target sequence), magnesium chloride, and Taq polymerase enzyme. The efficiency of the copying process can be measured and the factors that affect the efficiency include: (1) the G:C content of the target sequence, (2) the PCR primer complementation to the target sequence, (3) the exact concentrations of reagents (influenced by pipetting error, a major source of human error), (4) the exact temperatures for optimal replication, and (5) the presence of inhibitory molecules that were co-extracted during the DNA extraction process. As determined in biological modeling programs such as the probabilistic genotyping software STRmix, PCR amplification efficiency can vary but is measurable to establish if locus-by-locus the PCR conditions conform with the scientific expectation for the kit performance.

32.7.2 Contamination/allele drop-in and allele drop-out

The biological modeling of data is dependent on the number of contributors (NOCs) that can be established and is most often determined in forensic science by using the maximum allele count (MAC) method (Buckleton et al., 2007).

TABLE 32.3 Sources of variation in the DNA methodology.

DNA extraction	The sources of variation for DNA extraction method yields are derived from the differing methods, such as silica column and bead capture systems, Chelex extraction, phenol-chloroform extraction, etc. Some of these methods are stringent, and although the DNA yield is very pure, there may also be a reduction in the DNA quantity recovered.
DNA quantitation	The quantitation system yields generally reproducible DNA quantity estimates unless there is pipetting error or the presence of PCR inhibitors. There is variance in replicate testing due to these two factors, as well as sampling variation from the DNA extraction tube. Locus-to-locus variation is also evident due to differential PCR amplification efficiency between total and male DNA quantity estimate loci.
Polymerase chain reaction (PCR) amplification	The PCR amplification step does show variation in the PCR amplification efficiency per each test locus. In general, the kit manufacturer has tried to balance the PCR product from each locus to a similar scale but the product accumulation per locus is not exactly equal. PCR inhibitors can also reduce amplification efficiency. The optimal quantity template for the kit will yield the best DNA results. Suboptimal template quantity will introduce variance in the repeatability of the DNA profiles from test to test. Highly degraded or poor-quality DNA samples will yield less reliable results.
Capillary electrophoresis (CE) detection	The electrokinetic injection process can produce variable results due to sampling differences and varying injection times. The benefit of being able to alter the injection time is that low-level profiles can be enhanced; however, replicate testing may yield nonreproducible results for low quantity samples or minor contributors. The manufacturing of the CCD chip results in varying sensitivities of the CE equipment, and the chips must be calibrated to close equivalency but may not reach the point of exactness. The sizing of the peaks by the software shows slight variance due to differences in electrophoretic mobility from run to run.
Software analysis	Analytical and stochastic thresholds may vary per laboratory. Analysts are required to edit PCR artifacts, and may differ in their opinion on the editing. Laboratories set their own policies for editing and reporting, and there may be differences due to reporting strategies for peaks below the thresholds. The number of contributors can be challenging to assess consistently, depending on whether the peaks below the analytical threshold are considered.
Reporting/testimony	There are no national standards that forensic laboratories must follow in DNA reporting or testimony. There are, however, recommendations and guidelines to aid in providing some consistency from laboratory to laboratory.

The MAC method uses the maximum number of peaks detected at two or more loci in the DNA profile to establish the likely NOCs to a sample. For example, if six alleles are detected at a locus, then the user would divide by two and that would yield the answer of three contributors as the likely source of the DNA mixture. This is a reliable estimate for establishing the minimum number of individuals detected in the DNA mixture, but controlled laboratory studies have shown that with allele drop-out it should be considered a minimum estimate and that the true estimate could be more than what was detected in the data by using the MAC method. With evidentiary samples, there is no scientific way one can determine the true NOCs to a DNA mixture, only the detected NOCs. With allele drop-in, or the addition of a single stray allele, one may consider the possibility of an additional true contributor or the presence of environmental trace DNA contamination in a DNA sample. Once observed, the additional DNA peak must be accounted for as a possible contributor or sample "noise" and probabilistic genotyping systems now model this observation into the statistical assessments for LRs.

32.7.3 Injection effects/sampling variance

The CE detection instrument has an adjustable setting for the length of time that the sample can be injected into the capillary for further separation and detection of the allele peaks in the DNA profile. This is because the DNA sample can be adjusted by electrokinetic injection time to be more dilute or more concentrated for introduction into the sample injection port. This feature conserves the DNA sample that was prepared, as it allows for injection up to five times before a new sample preparation is required. Injection time can measure the variance in the low-level template by increasing the detection limit as more and more of the low-level molecules become "stacked" near the injection port to better visualize the minor components in a DNA mixture. The lower the DNA quantity of the starting template, the more likely stochastic or nonreproducible DNA test results may occur. Although the larger peaks in the dataset may be concordant with replicate test data, the minor components of a DNA mixture may be highly variable. This stochastic effect can explain the minimum estimate for the MAC methodology and the difficulty in establishing firmly the true NOCs to a DNA mixture (Buckleton et al., 2007; Krane, 2012).

32.8 Statistical methods and application

Statistics in forensic DNA analysis are designed to give an estimate of weight or certainty to match the evidentiary sample to a known reference sample (Weir, 2007). First and foremost, the numeric value is a probability estimate that is an approximation based on the uncertainties in the factors used to make the calculation. The forensic statistical calculations are based on the random sampling of human populations of different ancestry and, therefore, can only provide an estimate since not all individuals in that population would have been tested. Once sampling has been performed, the observed allele frequencies (estimates of how common or rare an allele may be present in the given population) are tabulated, and these observed frequencies are used to extrapolate out to the rest of that population for the estimated probability of a random chance match. If the allele is rare in a population and has never been observed, an estimated allele frequency can be calculated using the formula 5/2 N, where N is the total number of individuals in the database (Weir, 2007). For the estimation of rarity in a DNA match of the evidentiary DNA profile to the known reference sample of an individual, one must realize that the statistical calculation is an estimation only. In general, variance in statistical estimates between populations of differing ancestries is due to (1) the original population sampling, (2) inherent allele frequency differences in the founding population, and (3) mathematical calculations. NRC-II states the following: "It is probably safe to assume that within a race, the uncertainty of a value calculated from adequate databases (at least several hundred persons) by the product rule is within a factor of about 10 above and below the true value" (National Research Council (US), 1996; National Research Council (NRC-II), 1996; Weir et al., 1997).

32.8.1 Random match probability

The random match probability (RMP) is the estimated frequency for which one might expect a random chance match between the evidence and another person other than the matching defendant (Koehler et al., 1995; Thompson & Newman, 2015). This calculation can be used to show that the RMP in different populations would be anticipated to be rare in unrelated individuals. There is always the chance of a random match other than the matching defendant in partial DNA profiles, for genetic relatives, and for the identical twin scenario. This calculation could also be interpreted as the chance of a false positive match. This calculation is used most frequently for single-source DNA profiles or for an obvious major donor that can be interpreted as a single profile from a mixture. To calculate the RMP statistic, the allele frequencies at each locus must be acquired from an allele frequency table for each population of interest. Then, using the Hardy-Weinberg Equilibrium formula, the genotype per locus calculation is made. A heterozygous allele condition with two different forms of an allele would use $2pq$ as the formula, where p is the allele frequency for allele 1 and q is the allele frequency for allele 2 multiplied by 2 to yield the genotype frequency. A homozygous allele condition would use the formula p^2, where the allele is the same inherited from each parent. The allele frequency values per locus would be multiplied together for the genotype frequency. Once the genotype frequency has been calculated for all the tested loci, those values are multiplied together to yield the RMP value. The value of one divided by the RMP yields the LR value. The RMP value is a frequentist approach, which as the name implies, assesses the probability of a coincidental match within a population by extrapolating from a subset of the population. Sources of variance in these calculations include

TABLE 32.4 Variables in software statistical calculations.

Human error	Number of contributors, bias, analytical thresholds, stochastic thresholds, editing of polymerase chain reaction artifacts, report writing, and testimony
Computational error	Allele sharing, rounding decimals, coding issues, and choice of allele frequency database
Interpretational error	Choice of statistical method, choice of hypothesis

admixture, the original sampling to generate the allele frequency population reference database itself, and the number of decimal places used in the calculations (Table 32.4).

32.8.2 Likelihood ratio

The LR is a comparative hypothesis strategy that is designed to give a numeric value to indicate the strength of the hypothesis (Coyle, 2019). The numerator is designated Prosecutor's Hypothesis (Hp), the person of interest is more probable to be included, and the denominator is designated Defense Hypothesis (Hd), the person of interest is more probable not to be included. The numerator is 1 and is divided by all possible pair-wise comparisons of alleles per locus to yield the overall LR value. LR values can be small or large, as they are on an infinite scale in both directions from a zero point. The positive LR values are for support of Hp; the negative LR values are for support of Hd. Values that hover close to zero are uninformative due to the random chance coincidence of a low-level (a few alleles) match from unrelated individuals to the case. This range can be measured in validation studies by analyzing how often a noncontributor coincidentally matches with a positive LR value to a DNA mixture where the true contributors are known. Once defined, this false positive or false negative range can be designated as uninformative or inconclusive. The reporting for this form of statistic is, for example, "it is 1000 times more probable that Joe Smith is included if the hypothesis is true than if another unrelated random individual is the source" for positive LR values. The same type of verbiage is used for exclusions, but the statistic will favor the hypothesis that the data supports, that it is more probable that an unknown individual contributed to the DNA sample. This is the calculation that is performed by probabilistic genotyping software such as STRmix and TrueAllele for interpreting DNA mixtures. Some sources of error in this calculation could arise from the original hypothesis and who is being added to the numerator or denominator of the equation, the degree of allele sharing between the contributors, and whether conditioning is used on the victim or an elimination known reference sample for the calculation.

32.8.3 Observed frequency (counting method)

Some forms of DNA testing are lineage-based tests that show an association of the male Y-chromosome or the maternal mitochondrial genome with a family group. These types of tests are less able to individualize a person to a sample as the family group has shared genetics and the same haplotypes, a condition referred to as identity by descent or genetic relatedness. In addition, some unrelated family groups have by coincidence the same haplotypes, a condition known as identity by state. The way these data are presented is by an observed frequency from a database. The estimated probability of an individual with a male or female haplotype in any given population is based on the number of prior observations in the reference population database. These population databases are not searchable but are used for the observed frequency calculation, which can be expressed as X/N where X is the number of times the haplotype has been observed and N is the total number of observations in the database. These databases are ever-increasing in the number of observations, such that statistics can change or need to be updated over time. In the case of the paternal Y-chromosome haplotypes, many have prior observations, and so the discrimination power of this technique is very low; however, an exclusion is still informative. In the case of the maternal mitochondrial haplotype, many are still considered unique or have not ever been observed in the mitochondrial reference databases. These statistics are not as compelling or individualizing as the RMP or LR statistical calculations, but are still quite useful for establishing kinship.

32.8.4 Other statistics

There are other forensic genetic statistics that have been utilized and accepted in the courts. These statistics are less commonly used today, with a brief explanation for each form below.

32.8.4.1 2p rule

When variable number of tandem repeat technology was first identified, the 2p rule was recommended by the National Research Council (Buckleton & Triggs, 2006). This council also recommended that at least 200 alleles should be observed for the reference database for the allele frequency to be representative, and this data should be collected for each racial group (Black, Asian, Caucasian, Hispanic, Native American). The larger the population dataset, the more accurate the estimate would be for the allele's true representation in the dataset. The origin of the 2p rule is based on the migration of a band off the visual gel during electrophoresis and thus would make a heterozygote appear to be a false homozygote. The calculation was $2p_a$, where p_a is the probability of any given allele. It was designed to be an early statistical calculation to account for allele drop-out or lack of detection of a true heterozygote due to the technical issues with gel electrophoresis or highly degraded DNA samples. With STR technology, the precision is accurate to the point that this calculation is no longer required. Also, probabilistic genotyping software systems have developed to the point where allele drop-out can be modeled in complex DNA mixture data, and for these two reasons, the 2p rule has become defunct.

32.8.4.2 Random man not excluded/combined probability of inclusion/combined probability of exclusion

The random man not excluded (RMNE) calculation addresses the concept of the number of random matches to the evidence when allele drop-out has occurred. Related names for this calculation include the probability of inclusion (PI), the combined probability of inclusion (CPI), and the combined probability of exclusion (CPE). The CPI calculates the proportion of the population that may be included in a DNA mixture. No assumption is made about the NOCs to the DNA mixture. Only the loci where the peaks are clearly above the analytical and stochastic thresholds are used (where no allele drop-out is evident). The CPI is calculated by adding the frequencies of all the observed alleles and squaring the value to obtain the PI value per locus. All the PI values are multiplied together to obtain the CPI value. The CPE (or 1-CPI) calculates the proportion of the population that would have at least one allele not in common with the DNA sample. It is a conservative statistic, makes no assumption about the number of donors to the sample, and draws no conclusions about the mixture itself. The criticism of this general RMNE approach is that not all the data are used in the statistical calculations. It also assumes that the contributors are all from the same racial group and are genetically unrelated. These statistical calculations are becoming less commonly used in the forensic science community.

32.9 Summary

The purpose of the measurement of variance is to quantify the potential source of differences between replicate test measurements. Variance is the measurement of the data points that differ from the mean expected value. The cause of variance can be identified in this chapter and assessed for the final effect on the DNA data validity. Validity is defined as the quality of being logically or factually sound, and this is the condition that must be met before DNA data is presented in a court of law. Validation studies help define the steps within a scientific method that have identifiable variance and quantify the acceptable levels that do not significantly interfere with the data interpretation. The common areas of STR method variance are also the common points argued in a court of law, and this chapter has defined and discussed those main areas. The last half of this chapter spends some time discussing statistical methods as they have been applied to DNA test results, as they are still controversial and disputed, and have been so since originally discussed in NRC I and II. As the human identification kits and scientific methodology have evolved, admissibility hearings for the validity of the software used in DNA statistical calculations have also been held to assess software coding, the effect of rounding decimals, and the validity as applied on a case-by-case basis. In conclusion, variance can affect the reliability and reproducibility of DNA test results, but with proper developmental and internal validation studies to quantify and assess variance factors, DNA science can be reliable and trustworthy.

References

Buckleton, J., & Triggs, C. (2006). Is the 2p rule always conservative? *Forensic Science International, 159*(2-3), 206–209. https://doi.org/10.1016/j.forsciint.2005.08.004, https://www.elsevier.com/locate/forsciint.

Buckleton, J. S., Curran, J. M., & Gill, P. (2007). Towards understanding the effect of uncertainty in the number of contributors to DNA stains. *Forensic Science International: Genetics, 1*(1), 20–28. https://doi.org/10.1016/j.fsigen.2006.09.002.

Catena, F. (2016). Judge, decision and order -DNA analysis admissibility. The People of the State of New York v Oral Nicholas Hillary. County of St. Lawrence. https://johnbuckleton.wordpress.com/wp-content/uploads/2016/08/hillary-08-26-16-decision-and-order-dna-analysis-admissibility.pdf.

Coyle. (2015). Quality control and duplication for concordance in forensic DNA samples: implications for interpretation of mixtures. *International Research Journal of Computer Science, 2*(6), 16–18 2015.

Coyle. (2019). DNA mixture interpretation: effect of the hypothesis on the likelihood ratio. *International Research Journal of Computer Science, 6*(9), 672–675.

Coyle, H. M. (2022). Quality control measures in short tandem repeat (STR) analysis. In *Handbook of DNA profiling*. United States: Springer, 1107–1124. https://doi.org/10.1007/978-981-16-4318-7_53, https://link.springer.com/referencework/10.1007/978-981-15-9364-2.

Geddes, L. (2010). How DNA evidence creates victims of chance. New Scientist. https://www.newscientist.com/article/mg20727743-300-how-dna-evidence-creates-victims-of-chance/.

Giannelli, P. (2012). The 2009 NAS forensic science report: A literature review. *Criminal Law Bulletin, 48*, 2012.

Koehler, J. J., Chia, A., & Lindsey, S. (1995). The random match probability in DNA evidence: Irrelevant and prejudicial? *Jurimetrics, 35*(2), 201–219 1995.

Krane, D.E. (2012). Statistical weights of single source DNA profiles. https://corescholar.libraries.wright.edu/biology/246.

National Research Council. (2009). Strengthening forensic science in the United States: A path forward. https://www.ojp.gov/ncjrs/virtual-library/abstracts/strengthening-forensic-science-united-states-path-forward.

National Research Council (NRC-II). (1996). *The evaluation of forensic DNA evidence*. National Academy Press, Washington DC6.

National Research Council (US). (1996). Committee on DNA Forensic Science: Understanding of forensic statistics: An update. National Academies Press.

PCAST. (2016). The President's Council of Advisors on Science and Technology (PCAST) report on forensic science in criminal courts: Ensuring scientific validity of feature-comparison methods. https://obamawhitehouse.archives.gov/sites/default/files/microsites/ostp/PCAST/pcast_forensic_science_report_final.pdf.

Ruling – The People of the State of New York versus Oral Nicholas Hillary (NY). (2017). DNA evidence admissibility. https://www.strmix.com/news/ruling-the-people-of-the-state-of-new-york-versus-oral-nicholas-hillary-ny-dna-evidence-admissibility/.

Starr, D. (2022). The bias hunter. *Science, 376*(6594), 686–690. https://doi.org/10.1126/science.adc8720.

Thompson, W. C., & Newman, E. J. (2015). Lay understanding of forensic statistics: Evaluation of random match probabilities, likelihood ratios, and verbal equivalents. *Law and Human Behavior, 39*(4), 332–349. https://doi.org/10.1037/lhb0000134.

Weir, B. S. (2007). The rarity of DNA profiles. *The Annals of Applied Statistics, 1*(2). https://doi.org/10.1214/07-aoas128.

Weir, B. S., Triggs, C. M., Starling, L., Stowell, L. I., Walsh, K. A. J., & Buckleton, J. (1997). Interpreting DNA mixtures. *Journal of Forensic Sciences, 42*(2), 213–222. https://doi.org/10.1520/jfs14100j, http://onlinelibrary.wiley.com/journal/10.1111/(ISSN)1556-4029.

Chapter 33

Applications of digital polymerase chain reaction in environment and forensic chemistry

Sylvia Etim[1,*], Dipanwita Raha[1,*], George Duncan[2], and Bruce McCord[1]

[1]*Department of Chemistry and Biochemistry, Florida International University, Miami, FL, United States,* [2]*Halmos College of Natural Sciences and Oceanography, Nova Southeastern University, Dania Beach, FL, United States*

33.1 Introduction

This chapter reviews the application of digital polymerase chain reaction (dPCR) in the detection and analysis of environmental DNA (eDNA). The scope of the review covers the past two decades with a focus on the use of dPCR in the determination of different species and the evaluation of biodiversity in an ecological system (Handelsman, 2005). A second focus of the chapter involves applications within the realm of forensic science, notably wildlife trafficking and other legal issues.

The term "eDNA"was first coined in 1987 to describe a process for isolating the DNA of microorganism precipitates (Ogram et al., 1987). Yet, this phrase only became popular in the early 2000s, mostly among microbiologists (Handelsman, 2004). eDNA is defined as the genetic material isolated from substrates in the biosphere, including terrestrial, aquatic, and atmospheric environments, without first isolating the target organism. Animals expel DNA from the mitochondria or nucleus of their cells into the surrounding biosphere through shedding, death, and decay. The genomic material is extracted from aquatic substrates and typically exists in fragmented units of five hundred base pairs or less; however, a range of fragment sizes up to whole chromosomes may be present (Evans, 2019). The DNA can then be amplified by PCR for enhanced sensitivity and further analysis (Turner et al., 2015). eDNA is isolated from a blend of genetic material and consists of both intracellular DNA (from viable cells) and extracellular/cell-free DNA (Pietramellara et al., 2009) (from cellular decomposition or death).

The determination of eDNA involves a noninvasive sampling process that is not harmful to the organism or environment being studied. The results can be utilized to detect the presence and potential absence of fauna (micro-, meio-, and macrofauna) and flora. Initially, eDNA was used to examine and detect a variety of vertebrates and plants present in ancient sediment (Ariza et al., 2023; Ragot & Villemur, 2022). More recent work has demonstrated that eDNA can indicate the presence of a multiplicity of vertebrates, as shown in Fig. 33.1 below. eDNA is also useful as a biomonitoring tool due to its sensitivity and efficiency. It is time-saving and cost-effective when compared to traditional biological surveys and can alleviate the need for special permitting or invasive collection of samples. Simultaneously, it enables extensive comparison of biodiversity over extensive variations in time and place.

The application of eDNA technologies to environmental impact assessments, bio-surveillance, monitoring, and species inventories has enormous potential and interest since eDNA offers relatively inexpensive and credible biodiversity information. Effective land, water, mineral wealth, and natural resource management, as well as biodiversity and ecological productivity protection in an uncertain climate, necessitate an entire comprehension of the distribution and relative abundance of at-risk, sentinel, invasive, and pathogenic species. Advances in molecular technologies have greatly

* Etim and Raha share 1st author credit.

Investigative and Predictive DNA Testing. DOI: https://doi.org/10.1016/B978-0-443-30128-5.00032-1

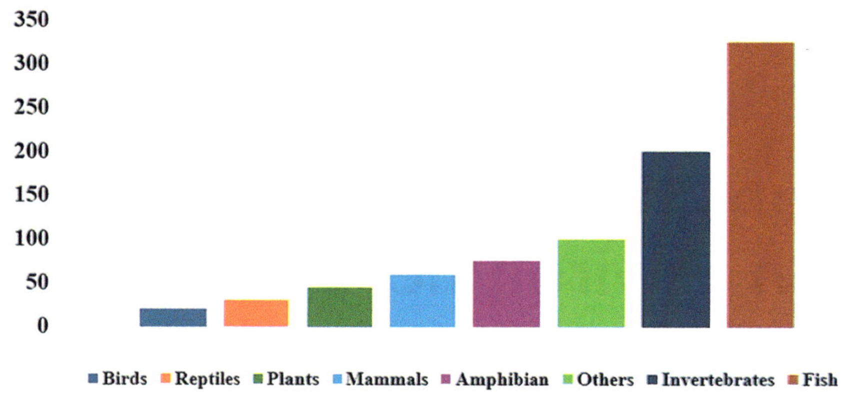

FIGURE 33.1 Investigations of environmental DNA in different groups of organisms (Sahu et al., 2023).

expanded the capacity to study biodiversity on Earth in natural ecosystems and anthropogenically affected areas (Barnes & Turner, 2016; Deiner et al., 2017; Goldberg et al., 2016; Jerde et al., 2011; Thomsen & Willerslev, 2015).

Furthermore, eDNA approaches are widely employed across the globe, including Africa (Elberri et al., 2020), Asia (Lee et al., 2020), and Europe (Brys et al., 2021). However, major variations in data quality, rigor, and reliability have negatively impacted confidence in eDNA applications, restricting regulatory and policy uptake (Nicholson et al., 2020). Specifically, false positives and negatives from field and analytical sources must be effectively mitigated (Nicholson et al., 2020; Thalinger et al., 2021). The design and performance components of targeted eDNA detection assays, which have been by far the most used in scientific literature, have received surprisingly little attention. A lot of work has gone into developing standardized frameworks and methods for eDNA field workflow (Baillie et al., 2019; Goldberg & Strickler, 2017; Hobbs et al., 2017; Laramie et al., 2015; The eDNA Society, 2019; US Fish and Wildlife Service, 2018).

The field of eDNA has the potential to transform archaeology, ecology, and wildlife forensics/conservation (Kelly et al., 2014). Body wastes, fluids, epidermal cells, and corpses are all sources of eDNA in the environment (Bohmann et al., 2014). There are a variety of methods that can be used to assess eDNA, including both targeted and nontargeted assays. Nontargeted methods include massively parallel sequencing and chip arrays. These procedures can be used as a survey tool for the detection of all fauna and flora in a particular sample or ecosystem, including soil, water, or sediments. Alternatively, targeted methods are utilized to detect and target specific species within an ecosystem (Rees et al., 2014). Here, methods such as quantitative and dPCR are used (Kudoh et al., 2020; Valentin et al., 2016; Wang et al., 2019). Sanger sequencing and electrophoresis/Amplified Fragment Length Polymorphism (AFLP) may also be applied (Jordan & Mills, 2021). Because of their lower cost and higher sensitivity, these methods can be particularly useful in the determination of rare or endangered species in the environment (Duarte et al., 2023) and in forensic investigations.

Targeted quantitative real-time polymerase chain reaction analysis, or quantitative PCR (qPCR), is the most widely available standards-development-friendly method for eDNA detection among the several techniques employed (Tsuji et al., 2019). Researchers embrace qPCR for its speed, sensitivity, specificity, and capability to detect single-nucleotide polymorphisms. Furthermore, dPCR can detect minor fold-change differences with 10% precision and mutation rates $<1\%$, in addition to providing absolute copy number quantification (Qiagen, 2023). It also has robust quantification, tolerance to PCR inhibitors, and is not impacted by fluctuations in PCR efficiency, unlike qPCR. dPCR uses endpoint detection. This review will primarily focus on applications of dPCR involving eDNA. This procedure has important advantages in this area, particularly in its resistance to PCR inhibition and its capability to perform quantitative analysis, often without the need for external standardization.

33.2 Digital polymerase chain reaction method and instruments

dPCR, a third-generation PCR diagnostic method, has recently been adopted by environmental microbiologists for the quantification of microorganisms in aquatic and terrestrial ecosystems (Borchardt et al., 2021). Its use has increased rapidly, due to its sensitivity, relevance in environmental monitoring, and unique advantage of partitioning individual DNA fragments into separate PCR reactions (Lei et al., 2021). The partitioning process, as shown in Fig. 33.2, generates hundreds of mini-reactions within the sample. Each partition undergoes PCR with fluorescent probes to confirm the presence or absence of a target gene in each partition. Finally, the target gene's absolute copy number can be estimated via the evaluation of positive and negative partitions based on the Poisson distribution (Basu, 2017; Quan et al., 2018; Shi et al., 2021).

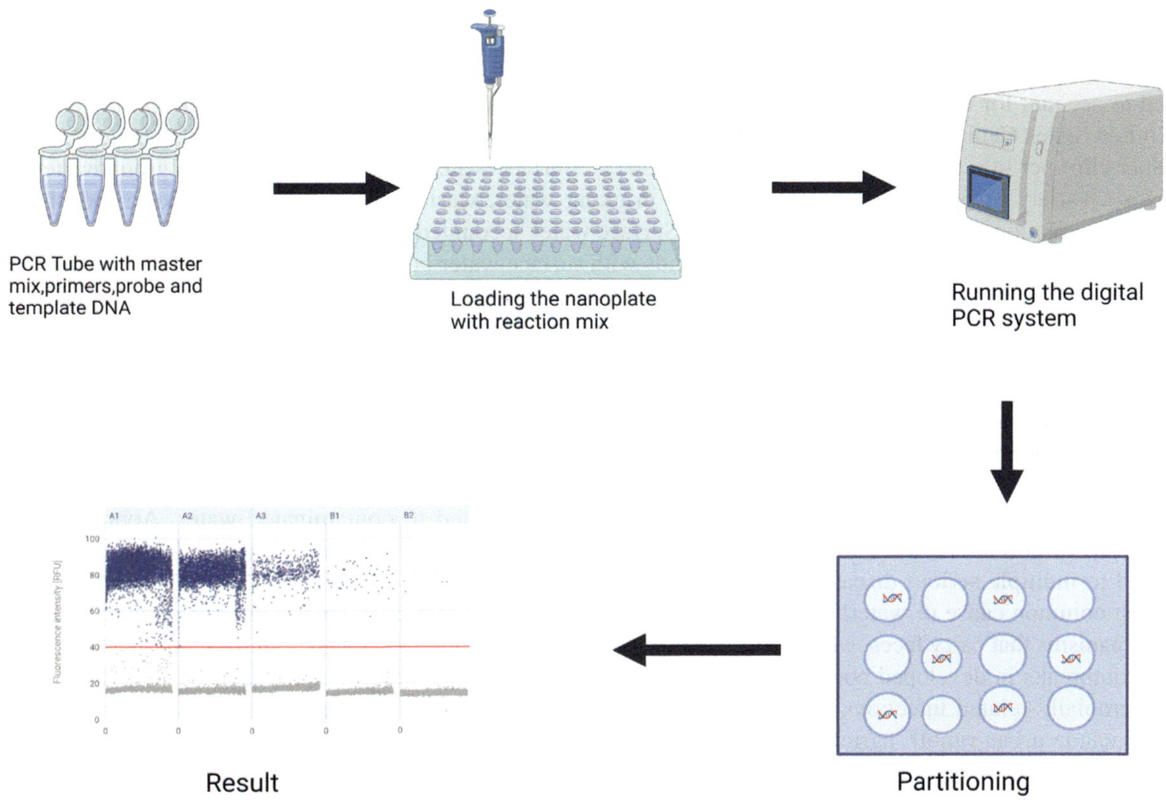

FIGURE 33.2 Digital polymerase chain reaction (PCR) workflow. The process involves preparing the reaction mix, loading the mix into nanoplate, and the instrument partitioning it into individual PCR reactions resulting in endpoint PCR amplification with precise quantification.
Created using BioRender.

33.3 Environmental DNA

eDNA research has gained popularity over the last decade, due to an increase in the number of studies that use DNA extracted from the environment. eDNA is a complex genomic DNA mixture from various kinds of organisms found in environmental samples (Taberlet et al., 2018). DNA extracted can be directly collected from samples such as water, air, soil, and sediments in the environment. It occurs in two forms as whole small organisms (e.g., microbes, single-cell algae, meiofauna) and zooplankton, or sloughed-off cellular material released into the environment by larger organisms through feces, urine, mucous, gametes, skin cells, and other particles (Pawlowski et al., 2020). Upon eDNA release into the environment, its survival can range from hours to weeks in temperate waters to several months or years in soil, caverns, permafrost, or sediments (Baillie et al., 2019). However, cellular material shed recently is still assumed to be the primary source of eDNA, and detection of the target sequence indicates the presence of the organism in the sampled system in most cases (Thomsen & Willerslev, 2015). eDNA is beneficial for aquatic and semiaquatic animals since it transports shed DNA away from the source, boosting detection ability.

To fully utilize eDNA-based biomonitoring technologies, it is important to understand the mechanisms by which DNA is released into the environment and its stability. Variables affecting stability include time, environment (e.g., soil, gut contents, water), UV levels, temperature, and microbial presence (Dejean et al., 2011). Rapid DNA degradation has the potential to generate false negative results (i.e., a visible absence of taxa that are present), which can confound biodiversity assessments and result in inaccurate characterization of community groups (Thomsen & Sigsgaard, 2019). Based on the number of taxa of interest, environmental samples can be targeted and amplified using either barcoding or metabarcoding.

DNA barcoding, also referred to as "targeted PCR," utilizes a single-species probe to detect a set of taxons within a sample set (Valentin et al., 2016). This technique can be used to evaluate the presence or absence of a small set of related species using qPCR and is often combined with Sanger sequencing. In contrast, metabarcoding amplifies multiple DNA fragments from mixed biological samples with "universal" primer sets (assays) to bind to a conserved homologous region of a gene shared by several species or groups of taxa via PCR and sequences them on a high-throughput screening platform (Sacco et al., 2022).

33.4 Surface water/disease detection

One of nature's essential assets that supports the existence of living organisms on our planet is water. Fresh water is estimated to account for roughly 2.5% of the Earth's water (Shiklomanov, 1993). Surface water is considered freshwater by the United States Geological Survey if the dissolved solids content is less than 1,000 milligrams (about the weight of a small paper clip) per liter (mg/L) (Surface-Water Use, 2019). Surface water is easier to access than groundwater; hence it is used for a variety of purposes such as sources of drinking water, wastewater treatment, livestock, industrial usage, hydropower, recreation, and irrigation on farmlands. Surface water accounted for about 80% of total water consumption in the United States in 2015 (Dieter et al., 2018). Wetlands with surface water provide essential habitat for aquatic plants and fauna.

Microbial pathogen contamination of water sources is a critical concern, with water quality. Pathogens such as *Escherichia coli* (*E. coli*), *Salmonella*, *Giardia, Norovirus, Rotavirus, Cryptosporidium*, and *Campylobacter* can infect water sources when sewage is left untreated, improperly treated, or accidentally released. Drinking water that has been contaminated by pathogenic microbes such as bacteria, fungi, viruses, and protozoa can have a serious negative impact on health and increase the risk of developing conditions like dermatitis, cholera, ulcers, hepatitis, typhoid fever, and nausea (Hanif et al., 2023). According to estimates from the World Health Organization, gastrointestinal diseases account for 485,000 deaths worldwide each year due to *E. coli* infections connected to contaminated water. Approximately 1400 different types of pollutants, including bacteria, viruses, parasitic protozoa, and a few fungal or helminth species, have been linked to multiple serious diseases (Hanif et al., 2023).

The most common cause of waterborne illnesses in people is either direct or indirect contact with water contaminated by microorganisms that carry feces, which is frequently the situation in underdeveloped countries. Additionally, connected to disease outbreaks in developed countries, especially in rural areas, are waterborne pathogens (Connelly & Baeumner, 2012). Microbially-related infections are the main cause of most waterborne disease outbreaks. Waterborne pathogenic pathways include urban runoff, agricultural wastewater, effluent discharges, and leaky sewage pipes. These sources can introduce enteric viruses, and poisonous water or foodborne pathogens can spread lethal aquatic diseases that are linked to both isolated cases and epidemics. Ensuring public health worldwide depends on the detection and identification of waterborne diseases. Consequently, it is crucial to have ongoing surveillance and efficient detection techniques.

There are several approaches for eDNA surface water detection. Major techniques rely on qPCR to detect a target species (BVNA, 2019). qPCR is a highly precise and sensitive DNA method that detects subnanogram amounts of DNA via repeated (up to 50) cycles of heat and cold that exponentially generate multiple copies of the target DNA; usually, a short sequence of DNA unique to the organism studied. The amplified target sequence is next visualized by a laboratory device that detects a fluorescent reporter dye that binds to the targeted DNA sequence (BVNA, 2019). It is important to note that the results of an assay of DNA are subjective. For example, a high level of eDNA may be due to many target species present, the rate at which species shed eDNA, or the proximity of sample collection to an individual species; hence, DNA levels do not necessarily provide a direct correlation with the number of DNA targets present in a sample (BVNA, 2019). Depending on environmental factors, such as temperature, microbial activity, and exposure to UV light, eDNA will remain in the environment for some time before degradation to the point of nondetection by qPCR assay (BVNA, 2019). When released from the organism into an aquatic environment, eDNA degrades within hours and may only be effectively detected by an eDNA assay 7 to 21 days later (Dejean et al., 2011). Therefore, while eDNA offers a current snapshot of a species' presence in an environment, the results cannot determine absolute quantity, nor can they determine if the source organism is alive or dead, or provide information about its age, size, gender, or reproductive state.

33.5 Bacterial contamination of foods

Bacterial contamination of food occurs when bacteria multiply in it and cause spoilage, resulting in foodborne diseases, commonly known as food poisoning. Some bacteria found in food products are harmless to consumers and could even provide some benefits, such as improved digestion and immune function, reduced inflammation, and improved nutrition. As a result, unless excessive growth or evident food spoilage is present, no attempt is made to eliminate such microbes from the processing environment (Liu et al., 2023). Microorganisms, however, are also subjected to a range of challenges in food processing environments, such as variations in moisture content, temperature, and cleaning methods, as well as effects due to periods of high nutrient levels and scarcity. Food-borne illness can be caused by chemicals, heavy metals, parasites, fungi, viruses, or bacteria (Wagner, 2008). Bacteria-related food poisoning is the most common, but only about 20 of several thousand different bacteria are responsible (Wagner, 2008). These include *Staphylococcus aureus*, *Salmonella*, *Clostridium perfringens*, *Campylobacter*, *Listeria monocytogenes* (*L. monocytogenes*), *Vibrio parahaemolyticus*, *Bacillus cereus*, and Enteropathogenic

E. coli, which can be found on many raw foods and cause more than 90% of food poisoning cases annually (Wagner, 2008). To produce disease, a significant number of pathogenic microorganisms are required.

Microorganisms produce biofilms, which attach to the surfaces of living or non-living things (Liu et al., 2023). Biofilm results when bacteria secrete extracellular polymers to form an extracellular matrix. These biological membranes are attached to the surfaces of complex microbial groups to produce a polymer matrix composed primarily of proteins, lipids, polysaccharides, and nucleic acids (Powell et al., 2018). Biofilms are composed of individual microbial species and may also include different types of protozoans, bacterial, archaeal, filamentous fungal, yeast, and algal species to form a complex three-dimensional microbiome or extracellular polymer structures such as flocs or granules (Raghupathi et al., 2017). Extracellular polymers protect biofilm residents by enhancing nutrient concentration, retaining moisture, and isolating biocides, heavy metals, and toxins.

The tendency of bacteria to form biofilms has been demonstrated to be an adaptive trait (Koczan et al., 2011). Currently, researchers agree that biofilms are the major means by which bacteria thrive and survive in natural environments. Scientists at the National Institutes of Health have discovered that biofilms are responsible for almost 80% of persistent bacterial infections (Worthington et al., 2012). Biofilm development and maturation is a complex, constant, and dynamic process influenced by the matrix, culture media, intrinsic features of cells, signaling molecules, cell metabolism, and gene regulation (Renner & Weibel, 2011). Adhesion is the initial reversible phase in biofilm formation; thereafter irreversible adhesion, also known as attachment (Vert et al., 2012). Proliferation of cells occurs after attachment to form unique extracellular polymeric substances (EPS) for the biofilm matrix. Major EPS constituents include carbohydrates, proteins, and extracellular DNA (Flemming & Wingender, 2010), which results in a highly diverse biofilm based on the microorganisms involved (Flemming et al., 2016). As the biofilm matures through the production of matrix and cell division, certain niches evolve within the biofilm (e.g., via nutrient supply and oxygen), and heterogeneous biofilms develop that can support a variety of specialized organisms. The final step in biofilm formation is dispersal and detachment, where single or clustered cells detach actively or passively and may adhere to a different surface to form a new biofilm.

Biofilms are a major concern in many food processing industries, like dairy, seafood, meat, beer, and fresh vegetables (Srey et al., 2013). *Lactobacillus*, *Listeria*, *Enterobacter*, *Micrococcus*, *Bacillus*, *Staphylococcus*, *Streptococcus*, and *Pseudomonas* are the most prevalent bacteria detected on surfaces in dairy facilities (Gutierrez et al., 2012), while *Staphylococcus aureus* (*S. aureus*), *Campylobacter*, *E. coli* O157:H7, *Salmonella*, and *L. monocytogenes* (Gutierrez et al., 2012) are common in the meat processing industry. The principal pathogenic bacteria to control in the fish industry are *E. coli*, *Vibrio*, *L. monocytogenes*, *Clostridium*, *S. enteritidis*, and *S. aureus* (Herrera et al., 2006). During infections, these foodborne pathogens frequently develop biofilms on both biotic and abiotic surfaces, causing cross-contamination, reduced product quality and shelf life, the possibility of disease transmission, and food safety concerns.

The issues of antimicrobial resistance, mechanical persistence, production of virulence factors, and other chemicals (Bridier et al., 2015) by bacteria in biofilms have received global attention, in the medicine and food industries. Overall, the techniques for the reduction of hazardous biofilms fit into three major categories, (Yin et al., 2021); (1) treatments such as regular disinfection, aseptic processing, and postuse equipment sterilization on abiotic surfaces; (2) regulation of signaling pathways to prevent biofilm development and promote diffusion; and (3) use of external pressures to eliminate biofilms.

A pioneering study by Klancnik et al. (2015) employed qPCR and dPCR to evaluate biofilm adherence or formation by *L. monocytogenes* in a 96-well polystyrene microtitre plate. A 200 μL of diluted bacterial culture in sterile TSB with a final concentration of 106 CFU/mL was used for inoculation and incubated at 25°C without shaking for 24 hours or 5 days. Template DNA was isolated and later quantified using QuantStudio™ 3D dPCR equipment. The results showed that qPCR had a detection range between 900,000 and 9 DNA copies, whereas dPCR had a 1000-fold dilution range of from 3.82 × 106 to 4.04 × 106 copies. These findings suggest that quantification of bacterial cells in a biofilm with qPCR and dPCR is a potential application that offers accurate results with a novel direction in adhesion and biofilm research.

33.6 Endangered species

eDNA metabarcoding is a noninvasive method for the detection of rare and endangered species in a variety of environments that relies on the retrieval of genetic traces released by animals. Biodiversity contributes significantly to ecosystem stability through critical functions such as nutrient recycling, climate regulation, erosion management, and waste processing (Dudgeon et al., 2006). Monitoring biodiversity is thus vital to evaluate ecosystem health, particularly to indicate a decline in endemic or endangered species population, early detection of invasive species, or changes in the range of native species in response to climate change. In terms of detecting (Navarro et al., 2017) biodiversity trends,

DNA obtained from the environment is revolutionary. Although the term "rare species" is commonly associated with indigenous endangered species, nonindigenous species can also be considered rare when population sizes are small early in the invasion process (Jerde et al., 2011).

An endangered species is a type of organism that is threatened by extinction. Species become endangered for two main reasons: loss of habitat and loss of genetic variation. The major emphasis on endangered species (also described as Endangered, critically endangered, vulnerable, threatened, imperiled, or of special concern) can be attributed to the widespread interest in mapping and understanding their distribution, which is critical for conservation management. Furthermore, most endangered species are rare, with relatively small populations and uneven distribution patterns, making them ideal targets for precise detection with eDNA-based assays. Most techniques based on DNA serve as an accurate means to determine diversity thresholds and monitor changes in the community due to actions taken to manage and characterize biodiversity (Langlois et al., 2021). The widespread adoption of molecular-based approaches that target eDNA has rapidly increased and changed how the Earth's biosphere is defined (Deiner et al., 2017). Molecular methods to detect invasive species are classified into two: (1) targeted methods, which require species-specific primers such as qPCR (Pochon et al., 2017; Sigsgaard et al., 2015); and (2) community-wide characterization methods, which employ universal primers to describe entire communities or groups of organisms, for example, metabarcoding (Bista et al., 2017; Hänfling et al., 2016).

qPCR assays are relatively sensitive, specific, and rapid (< 12 hours from DNA extraction to data), but detection of very low DNA concentrations can be subject to stochastic effects and inhibition (Ellison et al., 2006; Opel et al., 2010). Droplet ddPCR has the potential to significantly expedite the qPCR process since no inhibition assays, standard curves, or triplicate samples are required, and each result is an average of about 20,000 individual PCR reactions (Doi et al., 2015). Most studies use metabarcoding to analyze species diversity and detect rare or invasive species in environmental samples (Aylagas et al., 2016; Pochon et al., 2017; Zaiko et al., 2016). Unlike targeted techniques, metabarcoding allows for the simultaneous determination of multiple species, and based on the approaches utilized, the taxa may represent multiple trophic levels (Keeley et al., 2018). Although both strategies have been widely used in a variety of ecosystems (Evans et al., 2017), few researchers have directly compared the sensitivity (Hunter et al., 2017; Pochon et al., 2013).

In a study by Wood et al., (2019) to detect Mediterranean fanworm *Sabella spallanzanii* eDNA from 90 filtered seawater and 120 biofouling samples, qPCR, ddPCR, and metabarcoding were used for analysis. *S. spallanzanii* was found in 53% of water and 85% of biofouling samples via qPCR analysis; however, with ddPCR, 61% of water and 95% of biofouling samples were detected. Metabarcoding revealed an elevated detection rate in water samples when targeting the cytochrome c oxidase I (COI) (40%) compared to 18S rRNA (5.4%). Targeted techniques should be considered for studies that aim to detect invasive or rare species in environmental samples until an in-depth knowledge of how community, matrix complexity, as well as primer biases, affect metabarcoding data. Environmental changes like climate, pollution, biotic invasions, habitat destruction, and fragmentation, rapidly driven by humans, are major threats to global biodiversity and can have adverse effects on biodiversity and ecosystem function (Cardinale et al., 2012). Monitoring sensitive habitats and understanding the patterns and processes that regulate species distributions are critical to minimizing the effects of future changes in the environment.

33.7 Forensic investigations

dPCR has the potential to be extensively used in the fields of forensics by offering accurate and sensitive tools for DNA quantification, detection of microbes, and distinguishing between different body fluids, Fig. 33.3.

Conventional real-time qPCR requires a standard calibration curve that makes the quantification relative (Burns et al., 2010), whereas dPCR provides absolute quantification based on a method in which samples are distributed into multiple partitions, permitting a more direct detection technique. This procedure can improve the detection of low-level samples with less than 200 copies of DNA (Sanders et al., 2011; Tak et al., 2023). Because the sample is divided into partitions where each partition has its own PCR reaction (Gobert et al., 2018), susceptibility to PCR inhibitors is reduced when compared to real-time PCR (Dingle et al., 2013). Better precision and reproducibility can result in a lower false negative rate (Rene et al., 2019).

The sensitivity of dPCR may provide important information in situations where minute traces of biological material are the only evidence present and in case of mixed samples (Fu et al., 2024). Moreover, the robustness and precision of dPCR can offer advantages in situations in which samples may be degraded or at lower concentrations. The design of dPCR also reduces the impact of inhibitors (Dingle et al., 2013) commonly found in forensic samples. This is because dPCR utilizes endpoint detection. As a result, the potential for delayed amplification of inhibited samples has less impact on overall quantification, reducing the chance of negative or inaccurate results. However, it is important to note that similar to qPCR,

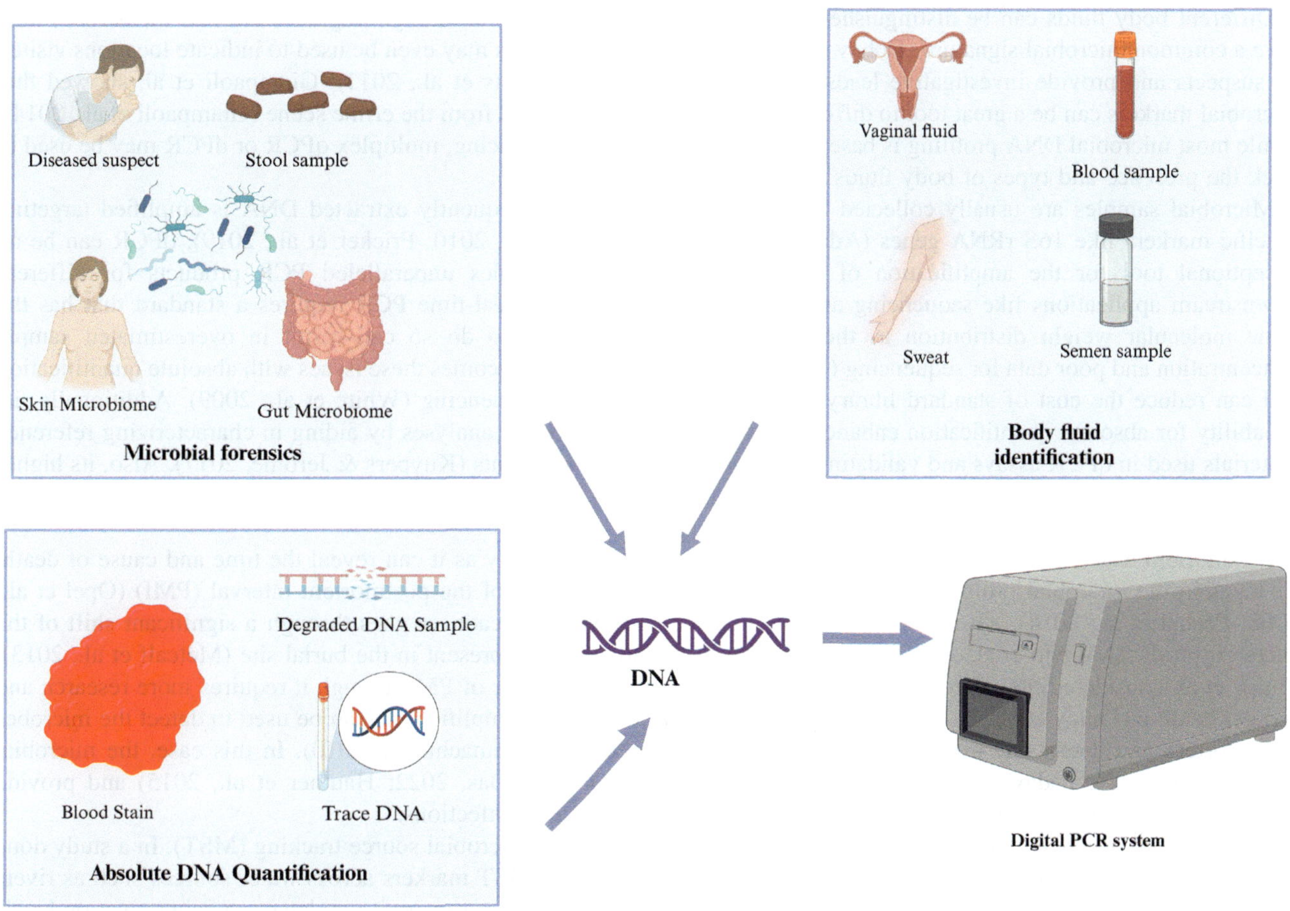

FIGURE 33.3 Applications of digital polymerase chain reaction in forensic chemistry. Created with BioRender.com.

dPCR can only detect and quantify accessible targets. For example, in a study evaluating the capability of dPCR for human quantification, Kline et al. explained that dPCR doesn't quantify the total number of PCR targets but instead can only amplify sequences that enzymes and reagents can access (Kline et al., 2016). For this reason, regardless of the procedure used, it is important to properly optimize and validate a procedure before use.

33.8 Microbial forensics

Microbial forensics is the cross-disciplinary link between microbiology and law enforcement. It aids in detecting and tracking microbial agents involved in bioterrorism, environmental crimes, and infectious disease outbreaks. Its significance lies in its ability to offer precise supporting data that can pinpoint the origin of a biological threat, aiding community surveillance and ultimately guarding public safety. Using different molecular methods, like qPCR, dPCR, and next-generation sequencing, forensic researchers can detect and quantify microorganisms, track their development, and find connections between evidence and potential suspects. Techniques like whole-genome sequencing and metagenomics can add to the reliability of investigations making microbial forensics essential in addressing and probing biosecurity breaches and bioweapons (Castro & De Ungria, 2022; Tripathi et al., 2024).

Previous research has shown that different factors like lifestyle, diet, health issues, personal hygiene, and environmental conditions directly impact the human gut microbial population (Clarke et al., 2017). These data can be used to detect differences in suspects and potentially detect the presence of certain individuals (Franceschetti et al., 2024). Microbial signature profiles can also differ between various parts of the body due to bodily function and differences in the number and types of sweat glands. (Schmedes et al., 2018). Such information can be useful in determining the types of body fluids (skin, semen, saliva, and vaginal secretions) detected at a crime scene (Hanssen et al., 2018; Wohlfahrt et al., 2023).

Different body fluids can be distinguished based on their bacterial profile, though vaginal fluid and menstrual blood have a common microbial signature (Dobay et al., 2019). Microbial profiles may even be used to indicate locations visited by suspects and provide investigative leads toward their identity (Knights et al., 2011). Giampaoli et al. showed that microbial markers can be a great tool to differentiate soil samples collected from the crime scene (Giampaoli et al., 2014). While most microbial DNA profiling is based on massively parallel sequencing, multiplex qPCR or dPCR may be used to track the presence and types of body fluids detected (Dass & Ghai, 2024).

Microbial samples are usually collected from crime scenes, and subsequently extracted DNA is amplified targeting specific markers like 16S rRNA genes (Adams et al., 2015; Fierer et al., 2010; Fricker et al., 2019). dPCR can be an exceptional tool for the amplification of extracted DNA, as it provides unparalleled PCR products for different downstream applications like sequencing and bioinformatics analysis. Real-time PCR requires a standard that has the same molecular weight distribution as the unknown sample. Failure to do so can result in overestimated sample concentration and poor data for sequencing (White et al., 2009). dPCR overcomes these issues with absolute quantification and can reduce the cost of standard library preparation required for sequencing (White et al., 2009). Additionally, its capability for absolute quantification enhances the credibility of microbial analyses by aiding in characterizing reference materials used in qPCR assays and validating forensic microbial assessments (Kuypers & Jerome, 2017). Also, its higher resistance against sample-associated inhibitors makes it useful in forensic settings in case of compromised or degraded samples (Strain et al., 2013)

Postmortem analysis is a significant area in microbial forensic chemistry as it can reveal the time and cause of death. Changes in the microbial profile after death can be used as an indicator of the postmortem interval (PMI) (Opel et al., 2008; Pechal et al., 2014). Also, it has been seen that the soil under the cadaver goes through a significant shift of the bacterial profile (Moreno et al., 2011), and Metcalf et al. found six phyla present in the burial site (Metcalf et al., 2013). Finley et al. considered the microbial communities as a biometric indicator of PMI, though it requires more research and standardization (Finley et al., 2015). It has been shown that nucleic acid amplification can be used to detect the microbes post-mortem and detect any associated disease (Wacharapluesadee & Hemachudha, 2010). In this case, the microbial communities in the cadavers can indicate the time of death (Dash & Das, 2022; Hauther et al., 2015) and provide information to assess whether the deceased had any previous disease or infections.

The potential of dPCR to reshape microbial forensics also extends to microbial source tracking (MST). In a study done by Staley et al. (2018), dPCR, along with qPCR, was used to evaluate MST markers across water sources such as rivers and beaches. While both methods were effective, dPCR demonstrated higher sensitivity in detecting contamination levels when compared to qPCR. In another study, results obtained by qPCR resulted in an underestimation of the target concentration when compared with dPCR. Resistance to interferences was also superior with dPCR (Cao et al., 2015), and the lack of a need for standards makes it particularly valuable for analyzing limited biomass samples commonly present in forensic investigations (Salipante & Jerome, 2020).

While dPCR presents advantages like enhanced accuracy due to direct calibration and better resistance to inhibition, it also comes with challenges that needs to be addressed. One drawback is its possible bias and need for normalization compared to qPCR systems, which could impact its use in clinical scenarios (Huggett et al., 2013; Sanders et al., 2013). Also, ensuring the compatibility of assays with the chosen dPCR platform through a selection of master mixes and primer/probe sets is crucial. It is essential to validate these assays according to established standards like the Minimum Information for Publication of Quantitative Digital PCR Experiments (MIQE) before applying them in forensic applications (Huggett et al., 2013).

Despite these obstacles, dPCR stands out for its precision in detecting low microbial loads (below 300 copies/106 cells), detecting rare sequences, and characterizing reference materials, where it outperforms qPCR in terms of accuracy and sensitivity (Kuypers & Jerome, 2017; Strain et al., 2013). For example, dPCR has been used to detect pathogens such as Mycobacterium tuberculosis from cell DNA in the bloodstream, presenting a fresh, noninvasive method for detecting bacterial presence in forensic samples (Devonshire et al., 2015; Ushio et al., 2016). Its exceptional sensitivity and ability to withstand inhibitors make dPCR ideal for analyzing microbial samples exposed to different environmental conditions. Standard qPCR may not work in some cases due to sample deterioration or higher susceptibility to PCR inhibitors (Salipante & Jerome, 2020). In such cases dPCR can be a better solution. The advantages of dPCR permit a more dependable and detailed microbial analysis of samples with highly diverse sequences and reduce the impact of sequence discrepancies and amplification inefficiencies. This quality makes the system well-suited for comparing microbial communities from crime scenes or deceased bodies, determining the cause of death, or estimating postmortem intervals (Salipante & Jerome, 2020).

Although dPCR presents some technical obstacles, like longer turnaround time, its capability to detect and quantify trace amounts of microbial DNA or RNA can assist in the determination of potential bioterrorism agents or pathogens during forensic investigations. The continuing development of dPCR technology may broaden its use in analyzing microbial communities, offering information on the microbial changes linked to human cadaver decomposition and estimating the postmortem interval (Liu et al., 2017). In short, dPCR is establishing its potential for use in microbial forensics, primarily based on its capability for absolute quantification and precision.

33.9 Quantification of DNA

Ensuring accurate DNA quantification is crucial in crime scene investigations, as it plays a major role in determining the relative amount of input DNA for downstream processes. Choosing amplification techniques is essential for accurate DNA quantification that guarantees that the obtained DNA profiles are both informative and precise, providing additional information like sex and degradation status (Fregel et al., 2011; Gray et al., 2014). DNA quantification in forensics is important as it affects the efficiency of subsequent downstream analysis procedures, such as short tandem repeat (STR) genotyping. dPCR assays have great potential in this field, as they can provide reliable DNA quantification.

dPCR is suitable for DNA quantification in mixed samples (Pacocha et al., 2019), which is often found in forensic scenarios. Precise quantification is imperative for analyzing low copy number and ancient forensic DNA samples (Fregel et al., 2011), which dPCR can provide. To detect copy number variations, duplex or multiplex dPCR assays can be designed for the analysis of valuable and limited samples (Whale et al., 2013). This approach boosts sensitivity and accuracy for quantification, allowing forensic analysts to manage degraded or mixed DNA samples efficiently.

Moreover, dPCR's ability to perform effectively when faced with PCR inhibitors in forensic samples ensures a more reliable quantification. Previous studies have shown that qPCR kits used in human DNA quantification are susceptible to higher levels of PCR inhibitors as well as calibration issues when used with standards prepared from cell lines or with badly inhibited DNA (Nielsen et al., 2008; Ricci et al., 2006). Furthermore, while real-time PCR has been successfully utilized in the quantitative analysis of forensic samples like hair samples, it can be challenged by highly degraded samples or inhibitors (Demeke & Jenkins, 2010). Such problems may be mitigated by using digital PCR (Geng & Mathies, 2015). dPCR has also been shown to be a useful tool for the analysis of paraffin-embedded and formalin-fixed tissue samples, and it showed efficacy in determining the DNA quality with enhanced sensitivity, enabling analysts to process better PCR products for targeted sequencing (Didelot et al., 2013).

Nevertheless, the integration of dPCR in DNA quantification comes with its set of challenges. One major hurdle is the method's intricacy and the necessity for standardization to guarantee reproducible outcomes. The variability stemming from different dPCR platforms and the unique characteristics of samples call for meticulous assay optimization and validation (Simoes Dutra Correa et al., 2020; Whale et al., 2013). Challenges like the "rain" effect during fluorescence detection and potential biases arising from sample preparation processes such as DNA fragmentation following bisulfite conversion need to be addressed (Kresse et al., 2023; Whale et al., 2013). Regardless, its ability to handle higher throughput using microarray chips and its resilience to variations in PCR efficiency make it a valuable choice (Shin et al., 2024). As forensic techniques progress, dPCR can emerge as a promising tool to enhance the accuracy and effectiveness of DNA analysis (Fregel et al., 2011; Shin et al., 2024).

33.10 Body fluids

In the field of forensic science, accurately detecting body fluids found at crime scenes is vital for connecting individuals to the actions, an important aspect of criminal investigations. dPCR has become an unprecedented technology in this area due to its sensitivity and precision, which significantly improves the ability to detect and measure DNA in forensic fluid samples. As mentioned above, the determination of specific microbial DNA species can be especially effective in distinguishing between types of body fluids.

Previous studies have shown the importance of differentiating body fluids. For example, the detection of saliva in bite marks can validate the origin of the DNA profiles detected at such locations (Chavez-Briones et al., 2015). The capability of dPCR was assessed in a research study where a multiplex PCR assay designed for dPCR targeted 16S microbial markers to detect and quantify bacterial species present in vaginal fluid and saliva. Important bacterial markers like *Lactobacillus crispatus* and *Streptococcus salivarius* were detected in both pure and mixed body fluid samples, demonstrating the reliability of dPCR in various scenarios (Dass & Ghai, 2024).

An additional advancement in utilizing dPCR within forensic science involves integrating it with methylation PCR (MSP-PCR) to create MSP-dPCR. This method emphasizes exploring DNA methylation patterns to distinguish different body fluids such as blood, semen, and saliva. MSP-dPCR has effectively been used to detect changes in methylation patterns of genes from different body fluids, connecting these alterations to health conditions or illnesses in samples of saliva and lung tissue (Yu et al., 2018). This demonstrates the versatility of dPCR not only in investigations but also in broader clinical diagnostics (Shahraki et al., 2023). Also, dPCR can be a powerful tool to assess the stability of different methylation markers in various age groups. Care must be taken to avoid complications with fluorescence detection due to issues with DNA fragmentation following bisulfite conversion (Kresse et al., 2023; Yu et al., 2018).

Additionally, applications have been developed utilizing methylation-sensitive restriction enzyme PCR techniques for the detection of microbial DNA in body fluids. Such methods have allowed for the detection of multiple bodily fluids in a single PCR test, significantly enhancing applications involving forensic body fluid analyses (Hanssen et al., 2017).

The significance of body fluid analysis was further highlighted in different research involving the sequencing of transcriptomes from various body fluids like semen, saliva, peripheral blood, and menstrual blood (Liu et al., 2024). This study disclosed a wide range of novel RNA molecules, and their genetic variations were characterized. dPCR assays can be designed to validate these novel markers and provide additional loci for targeted sequencing. Digital PCR has the potential to extend such assays to other biological fluids such as urine accurately providing microbial biomass and offering detailed microbial profiles which can be useful in forensic investigation (Moustafa et al., 2018).

33.11 Conclusions

Though dPCR comes with its own set of challenges, it shows great potential in the fields of environmental and forensic chemistry. Targeted dPCR assays can be designed with universal primers to detect a wide range of eDNA in a single reaction. The sensitivity of dPCR can be crucial in detecting and quantifying trace levels of microbial DNA coming from various sources, like foodborne and waterborne pathogens, as well as in the detection of potential bioterrorism agents. The robustness of the instrument plays a vital role in precise DNA quantification in a mixed sample to trace an extinct species, and gives enhanced efficacy in case of biological evidence like different body fluids. Future research should concentrate on the development of robust, portable, cost-effective, and user-friendly techniques for routine eDNA detection, to ultimately protect endangered species and public health, with artificial intelligence and data analysis serving essential roles in advancing these techniques for a safer environment.

Acknowledgments

The authors gratefully acknowledge follow-on funding under the Directorate Office of University Programs Summer Research Team Program for Minority Serving Institutions, administered by the Oak Ridge Institute for Science and Education (ORISE) through an interagency agreement between the US Department of Energy (DOE) and DHS. ORISE is managed by ORAU under DOE contract number DE-SC0014664. All opinions expressed in this paper are the author's and do not necessarily reflect the policies and views of DHS, DOE, or ORAU/ORISE. We also acknowledge technical support from QIAGEN.

References

Adams, R. I., Bateman, A. C., Bik, H. M., & Meadow, J. F. (2015). Microbiota of the indoor environment: A meta-analysis. *Microbiome, 3*, 49. https://doi.org/10.1186/s40168-015-0108-3.

Ariza, M., Fouks, B., Mauvisseau, Q., Halvorsen, R., Alsos, I. G., & de Boer, H. J. (2023). Plant biodiversity assessment through soil eDNA reflects temporal and local diversity. *Methods in Ecology and Evolution, 14*(2), 415–430. https://doi.org/10.1111/2041-210X.13865, http://onlinelibrary.wiley.com/journal/10.1111/(ISSN)2041-210X.

Aylagas, E., Borja, Á., Irigoien, X., & Rodríguez-Ezpeleta, N. (2016). Benchmarking DNA metabarcoding for biodiversity-based monitoring and assessment. *Frontiers in Marine Science, 3*. https://doi.org/10.3389/fmars.2016.00096.

Baillie, S.M., McGowan, C., May-McNally, S., Leggatt,R., Sutherland, B.J.G., & Robinson, S. (2019). Environmental DNA and its applications to Fisheries and Oceans Canada: National needs and priorities.

Barnes, M. A., & Turner, C. R. (2016). The ecology of environmental DNA and implications for conservation genetics. *Conservation Genetics, 17*(1), 1–17. https://doi.org/10.1007/s10592-015-0775-4, http://www.springerlink.com/content/1566-0621.

Basu, A. S. (2017). Digital assays part I: Partitioning statistics and digital PCR. *SLAS Technology, 22*(4), 369–386. https://doi.org/10.1177/2472630317705680, http://journals.sagepub.com/toc/jlad/current.

Bista, I., Carvalho, G. R., Walsh, K., Seymour, M., Hajibabaei, M., Lallias, D., Christmas, M., & Creer, S. (2017). Annual time-series analysis of aqueous eDNA reveals ecologically relevant dynamics of lake ecosystem biodiversity. *Nature Communications, 8*. https://doi.org/10.1038/ncomms14087, http://www.nature.com/ncomms/index.html.

Bohmann, K., Evans, A., Gilbert, M. T. P., Carvalho, G. R., Creer, S., Knapp, M., Yu, D. W., & de Bruyn, M. (2014). Environmental DNA for wildlife biology and biodiversity monitoring. *Trends in Ecology and Evolution, 29*(6), 358–367. https://doi.org/10.1016/j.tree.2014.04.003, https://www.elsevier.com/inca/publications/store/3/0/3/3/9.

Borchardt, M. A., Boehm, A. B., Salit, M., Spencer, S. K., Wigginton, K. R., & Noble, R. T. (2021). The environmental microbiology minimum information (EMMI) guidelines: QPCR and dPCR quality and reporting for environmental microbiology. *Environmental Science and Technology, 55*(15), 10210–10223. https://doi.org/10.1021/acs.est.1c01767, http://pubs.acs.org/journal/esthag.

Bridier, A., Sanchez-Vizuete, P., Guilbaud, M., Piard, J.-C., Naïtali, M., & Briandet, R. (2015). Biofilm-associated persistence of food-borne pathogens. *Food Microbiology, 45*(Pt B), 167–178. https://doi.org/10.1016/j.fm.2014.04.015.

Brys, R., Halfmaerten, D., Neyrinck, S., Mauvisseau, Q., Auwerx, J., Sweet, M., & Mergeay, J. (2021). Reliable eDNA detection and quantification of the European weather loach (*Misgurnus fossilis*). *Journal of Fish Biology, 98*(2), 399–414. https://doi.org/10.1111/jfb.14315.

Burns, M. J., Burrell, A. M., & Foy, C. A. (2010). The applicability of digital PCR for the assessment of detection limits in GMO analysis. *European Food Research and Technology, 231*(3), 353–362. https://doi.org/10.1007/s00217-010-1277-8.

BVNA. (2019). Understanding environmental DNA testing. Bureau Veritas North America. https://www.bvna.com/insight/understanding-environmental-dna-testing.

Cao, Y., Raith, M. R., & Griffith, J. F. (2015). Droplet digital PCR for simultaneous quantification of general and human-associated fecal indicators for water quality assessment. *Water Research, 70*, 337–349. https://doi.org/10.1016/j.watres.2014.12.008, https://www.elsevier.com/locate/watres.

Cardinale, B. J., Duffy, J. E., Gonzalez, A., Hooper, D. U., Perrings, C., Venail, P., Narwani, A., MacE, G. M., Tilman, D., Wardle, D. A., Kinzig, A. P., Daily, G. C., Loreau, M., Grace, J. B., Larigauderie, A., Srivastava, D. S., & Naeem, S. (2012). Biodiversity loss and its impact on humanity. *Nature, 486*(7401), 59–67. https://doi.org/10.1038/nature11148, http://www.nature.com/nature/index.html.

Castro, A. E., & De Ungria, M. C. A. (2022). Methods used in microbial forensics and epidemiological investigations for stronger health systems. *Forensic Sciences Research, 7*(4), 650–661. https://doi.org/10.1080/20961790.2021.2023272, https://www.tandfonline.com/action/journalInformation?journalCode=tfsr20.

Chavez-Briones, M. L., Hernandez-Cortes, R., Jaramillo-Rangel, G., & Ortega-Martinez, M. (2015). Relevance of sampling and DNA extraction techniques for the analysis of salivary evidence from bite marks: A case report. *Genetics and Molecular Research: GMR, 14*(3), 10165–10171. https://doi.org/10.4238/2015.August.21.23.

Clarke, T. H., Gomez, A., Singh, H., Nelson, K. E., & Brinkac, L. M. (2017). Integrating the microbiome as a resource in the forensics toolkit. *Forensic Science International: Genetics, 30*, 141–147. https://doi.org/10.1016/j.fsigen.2017.06.008, http://www.elsevier.com.

Connelly, J. T., & Baeumner, A. J. (2012). Biosensors for the detection of waterborne pathogens. *Analytical and Bioanalytical Chemistry, 402*(1), 117–127. https://doi.org/10.1007/s00216-011-5407-3.

Dash, H. R., & Das, S. (2022). Microbial community signatures for estimation of postmortem time intervals. *Advances in Applied Microbiology, 118*, 91–113. https://doi.org/10.1016/bs.aambs.2022.02.002, http://www.sciencedirect.com/science/journal/00652164.

Dass, M., & Ghai, M. (2024). Development of a multiplex PCR assay and quantification of microbial markers by ddPCR for identification of saliva and vaginal fluid. *Forensic Science International, 362*, 112147. https://doi.org/10.1016/j.forsciint.2024.112147.

Deiner, K., Bik, H. M., Mächler, E., Seymour, M., Lacoursière-Roussel, A., Altermatt, F., Creer, S., Bista, I., Lodge, D. M., de Vere, N., Pfrender, M. E., & Bernatchez, L. (2017). Environmental DNA metabarcoding: Transforming how we survey animal and plant communities. *Molecular Ecology, 26*(21), 5872–5895. https://doi.org/10.1111/mec.14350, http://www.blackwellpublishing.com/journals/MEC.

Dejean, T., Valentini, A., Duparc, A., Pellier-Cuit, S., Pompanon, F., Taberlet, P., Miaud, C., & Gilbert, J. A. (2011). Persistence of environmental DNA in freshwater ecosystems. *PLoS One, 6*(8), e23398. https://doi.org/10.1371/journal.pone.0023398.

Demeke, T., & Jenkins, G. R. (2010). Influence of DNA extraction methods, PCR inhibitors and quantification methods on real-time PCR assay of biotechnology-derived traits. *Analytical and Bioanalytical Chemistry, 396*(6), 1977–1990. https://doi.org/10.1007/s00216-009-3150-9.

Devonshire, A. S., Honeyborne, I., Gutteridge, A., Whale, A. S., Nixon, G., Wilson, P., Jones, G., McHugh, T. D., Foy, C. A., & Huggett, J. F. (2015). Highly reproducible absolute quantification of *Mycobacterium tuberculosis* complex by digital PCR. *Analytical Chemistry, 87*(7), 3706–3713. https://doi.org/10.1021/ac5041617, http://pubs.acs.org/journal/ancham.

Didelot, A., Kotsopoulos, S. K., Lupo, A., Pekin, D., Li, X., Atochin, I., Srinivasan, P., Zhong, Q., Olson, J., Link, D. R., Laurent-Puig, P., Blons, H., Hutchison, J. B., & Taly, V. (2013). Multiplex picoliter-droplet digital PCR for quantitative assessment of DNA integrity in clinical samples. *Clinical Chemistry, 59*(5), 815–823. https://doi.org/10.1373/clinchem.2012.193409France, http://www.clinchem.org/content/59/5/815.full.pdf+html.

Dieter, C.A., Maupin, M.A., Caldwell, R.R., Harris, M.A., Ivahnenko, T.I., Lovelace, J.K., Barber, N.L., & Linsey, K. (2018). Estimated use of water in the United States in 2015: U.S. geological survey circular 1441. doi: 10.3133/cir1441.

Dingle, T. C., Sedlak, R. H., Cook, L., & Jerome, K. R. (2013). Tolerance of droplet-digital PCR vs real-time quantitative PCR to inhibitory substances. *Clinical Chemistry, 59*(11), 1670–1672. https://doi.org/10.1373/clinchem.2013.211045, http://www.clinchem.org/content/59/11/1670.full.pdf+html.

Dobay, A., Haas, C., Fucile, G., Downey, N., Morrison, H. G., Kratzer, A., & Arora, N. (2019). Microbiome-based body fluid identification of samples exposed to indoor conditions. *Forensic Science International: Genetics, 40*, 105–113. https://doi.org/10.1016/j.fsigen.2019.02.010, http://www.elsevier.com.

Doi, H., Takahara, T., Minamoto, T., Matsuhashi, S., Uchii, K., & Yamanaka, H. (2015). Droplet digital polymerase chain reaction (PCR) outperforms real-time PCR in the detection of environmental DNA from an invasive fish species. *Environmental Science and Technology, 49*(9), 5601–5608. https://doi.org/10.1021/acs.est.5b00253, http://pubs.acs.org/journal/esthag.

Duarte, S., Simões, L., & Costa, F. O. (2023). Current status and topical issues on the use of eDNA-based targeted detection of rare animal species. *Science of the Total Environment, 904*. https://doi.org/10.1016/j.scitotenv.2023.166675, https://www.elsevier.com/locate/scitotenv.

Dudgeon, D., Arthington, A. H., Gessner, M. O., Kawabata, Z. I., Knowler, D. J., Lévêque, C., Naiman, R. J., Prieur-Richard, A. H., Soto, D., Stiassny, M. L. J., & Sullivan, C. A. (2006). Freshwater biodiversity: Importance, threats, status and conservation challenges. *Biological Reviews of the Cambridge Philosophical Society, 81*(2), 163–182. https://doi.org/10.1017/S1464793105006950.

Elberri, A. I., Galal-Khallaf, A., Gibreel, S. E., El-Sakhawy, S. F., El-Garawani, I., El-Sayed Hassab ElNabi, S., & Mohammed-Geba, K. (2020). DNA and eDNA-based tracking of the North African sharptooth catfish Clarias gariepinus. *Molecular and Cellular Probes, 51*. https://doi.org/10.1016/j.mcp.2020.101535, http://www.elsevier.com/inca/publications/store/6/2/2/9/1/8/index.htt.

Ellison, S. L. R., English, C. A., Burns, M. J., & Keer, J. T. (2006). Routes to improving the reliability of low level DNA analysis using real-time PCR. *BMC Biotechnology, 6*. https://doi.org/10.1186/1472-6750-6-33, http://www.biomedcentral.com/1472-6750/6/33.

Evans, A. R. (2019). *Exploring the biodiversity of the lakes of the Malay Archipelago using environmental DNA metabarcoding, Bangor University*. ProQuest Dissertations Publishing.

Evans, N. T., Li, Y., Renshaw, M. A., Olds, B. P., Deiner, K., Turner, C. R., Jerde, C. L., Lodge, D. M., Lamberti, G. A., & Pfrender, M. E. (2017). Fish community assessment with eDNA metabarcoding: Effects of sampling design and bioinformatic filtering. *Canadian Journal of Fisheries and Aquatic Sciences, 74*(9), 1362–1374. https://doi.org/10.1139/cjfas-2016-0306, http://pubs.nrc-cnrc.gc.ca/.

Fierer, N., Lauber, C. L., Zhou, N., McDonald, D., Costello, E. K., & Knight, R. (2010). Forensic identification using skin bacterial communities. *Proceedings of the National Academy of Sciences of the United States of America, 107*(14), 6477–6481. https://doi.org/10.1073/pnas.1000162107, http://www.pnas.org/content/107/14/6477.full.pdf.

Finley, S. J., Benbow, M. E., & Javan, G. T. (2015). Microbial communities associated with human decomposition and their potential use as postmortem clocks. *International Journal of Legal Medicine, 129*(3), 623–632. https://doi.org/10.1007/s00414-014-1059-0, link.springer.de/link/service/journals/00414/index.htm.

Flemming, H.-C., & Wingender, J. (2010). The biofilm matrix. *Nature Reviews. Microbiology, 8*(9), 623–633. https://doi.org/10.1038/nrmicro2415.

Flemming, H. C., Wingender, J., Szewzyk, U., Steinberg, P., Rice, S. A., & Kjelleberg, S. (2016). Biofilms: An emergent form of bacterial life. *Nature Reviews. Microbiology, 14*(9), 563–575. https://doi.org/10.1038/nrmicro.2016.94, http://www.nature.com/nrmicro/index.html.

Franceschetti, L., Lodetti, G., Blandino, A., Amadasi, A., & Bugelli, V. (2024). Exploring the role of the human microbiome in forensic identification: Opportunities and challenges. *International Journal of Legal Medicine, 138*(5), 1891–1905. https://doi.org/10.1007/s00414-024-03217-z.

Fregel, R., Almeida, M., Betancor, E., Suárez, N. M., & Pestano, J. (2011). Reliable nuclear and mitochondrial DNA quantification for low copy number and degraded forensic samples. *Forensic Science International: Genetics Supplement Series, 3*(1), e303. https://doi.org/10.1016/j.fsigss.2011.09.014.

Fricker, A. M., Podlesny, D., & Fricke, W. F. (2019). What is new and relevant for sequencing-based microbiome research? A mini-review. *Journal of Advanced Research, 19*, 105–112. https://doi.org/10.1016/j.jare.2019.03.006, http://www.elsevier.com/wps/find/journaldescription.cws_home/722881/description#description.

Fu, X., Li, X., Wang, Y., Xie, M., Wen, D., Tang, X., Wang, C., Jia, H., Liu, Y., Chen, S., Wang, Y., Zha, L., & Li, J. (2024). Discovery unbalanced DNA mixtures and evaluation mixing ratio via a droplet digital PCR method. *International Journal of Legal Medicine*. https://doi.org/10.1007/s00414-024-03306-z.

Geng, T., & Mathies, R. A. (2015). Minimizing inhibition of PCR-STR typing using digital agarose droplet microfluidics. *Forensic Science International: Genetics, 14*, 203–209. https://doi.org/10.1016/j.fsigen.2014.10.007, http://www.elsevier.com.

Giampaoli, S., Berti, A., Di Maggio, R. M., Pilli, E., Valentini, A., Valeriani, F., Gianfranceschi, G., Barni, F., Ripani, L., & Romano Spica, V. (2014). The environmental biological signature: NGS profiling for forensic comparison of soils. *Forensic Science International, 240*, 41–47. https://doi.org/10.1016/j.forsciint.2014.02.028.

Gobert, G., Cotillard, A., Fourmestraux, C., Pruvost, L., Miguet, J., & Boyer, M. (2018). Droplet digital PCR improves absolute quantification of viable lactic acid bacteria in faecal samples. *Journal of Microbiological Methods, 148*, 64–73. https://doi.org/10.1016/j.mimet.2018.03.004.

Goldberg, C., & Strickler, K. (2017). eDNA protocol sample collection.

Goldberg, C. S., Turner, C. R., Deiner, K., Klymus, K. E., Thomsen, P. F., Murphy, M. A., Spear, S. F., McKee, A., Oyler-McCance, S. J., Cornman, R. S., Laramie, M. B., Mahon, A. R., Lance, R. F., Pilliod, D. S., Strickler, K. M., Waits, L. P., Fremier, A. K., Takahara, T., Herder, J. E., & Taberlet, P. (2016). Critical considerations for the application of environmental DNA methods to detect aquatic species. *Methods in Ecology and Evolution, 7*(11), 1299–1307. https://doi.org/10.1111/2041-210X.12595, http://onlinelibrary.wiley.com/journal/10.1111/(ISSN)2041-210X.

Gray, K., Crowle, D., & Scott, P. (2014). Direct amplification of casework bloodstains using the Promega PowerPlex® 21 PCR Amplification System. *Forensic Science International: Genetics, 12*, 86–92. https://doi.org/10.1016/j.fsigen.2014.05.003.

Gutierrez, D., Delgado, S., Vazquez-Sanchez, D., Martinez, B., Cabo, M. L., Rodriguez, A., Herrera, J. J., & Garcia, P. (2012). Incidence of *Staphylococcus aureus* and analysis of associated bacterial communities on food industry surfaces. *Applied and Environmental Microbiology, 78*(24), 8547–8554. https://doi.org/10.1128/AEM.02045-12.

Handelsman, J. (2004). Metagenomics: Application of genomics to uncultured microorganisms. *Microbiology and Molecular Biology Reviews, 68*(4), 669–685. https://doi.org/10.1128/mmbr.68.4.669-685.2004.

Handelsman, J. (2005). Metagenomics or megagenomics? *Nature Reviews. Microbiology, 3*(6), 457–458. https://doi.org/10.1038/nrmicro1185.

Hanif, K., Muhammad, R., & Iqra, Z. (2023). Pathogens detection and identification in drinking water. *International Journal of Chemical and Biochemical Sciences*.

Hanssen, E. N., Avershina, E., Rudi, K., Gill, P., & Snipen, L. (2017). Body fluid prediction from microbial patterns for forensic application. *Forensic Science International: Genetics, 30*, 10–17. https://doi.org/10.1016/j.fsigen.2017.05.009, http://www.elsevier.com.

Hanssen, E. N., Liland, K. H., Gill, P., & Snipen, L. (2018). Optimizing body fluid recognition from microbial taxonomic profiles. *Forensic Science International: Genetics, 37*, 13–20. https://doi.org/10.1016/j.fsigen.2018.07.012, http://www.elsevier.com.

Hauther, K. A., Cobaugh, K. L., Jantz, L. M., Sparer, T. E., & Debruyn, J. M. (2015). Estimating time since death from postmortem human gut microbial communities. *Journal of Forensic Sciences, 60*(5), 1234–1240. https://doi.org/10.1111/1556-4029.12828, http://onlinelibrary.wiley.com/journal/10.1111/(ISSN)1556-4029.

Herrera, F. C., Santos, J. A., Otero, A., & García-López, M. L. (2006). Occurrence of foodborne pathogenic bacteria in retail prepackaged portions of marine fish in Spain. *Journal of Applied Microbiology, 100*(3), 527–536. https://doi.org/10.1111/j.1365-2672.2005.02848.x, https://academic.oup.com/jambio.

Hobbs, J., Goldberg, C.S., Helbing, C.C., & Veldhoen, N. (2017). Environmental DNA protocol for freshwater aquatic ecosystems Version 2.

Huggett, J. F., Foy, C. A., Benes, V., Emslie, K., Garson, J. A., Haynes, R., Hellemans, J., Kubista, M., Mueller, R. D., Nolan, T., Pfaffl, M. W., Shipley, G. L., Vandesompele, J., Wittwer, C. T., & Bustin, S. A. (2013). The digital MIQE guidelines: Minimum information for publication of quantitative digital PCR experiments. *Clinical Chemistry, 59*(6), 892–902. https://doi.org/10.1373/clinchem.2013.206375, http://www.clinchem.org/content/59/6/892.full.pdf+html.

Hunter, M. E., Dorazio, R. M., Butterfield, J. S. S., Meigs-Friend, G., Nico, L. G., & Ferrante, J. A. (2017). Detection limits of quantitative and digital PCR assays and their influence in presence–absence surveys of environmental DNA. *Molecular Ecology Resources, 17*(2), 221–229. https://doi.org/10.1111/1755-0998.12619, http://onlinelibrary.wiley.com/journal/10.1111/(ISSN)1755-0998.

Hänfling, B., Handley, L. L., Read, D. S., Hahn, C., Li, J., Nichols, P., Blackman, R. C., Oliver, A., & Winfield, I. J. (2016). Environmental DNA metabarcoding of lake fish communities reflects long-term data from established survey methods. *Molecular Ecology, 25*(13), 3101–3119. https://doi.org/10.1111/mec.13660, http://www.blackwellpublishing.com/journals/MEC.

Jerde, C. L., Mahon, A. R., Chadderton, W. L., & Lodge, D. M. (2011). "Sight-unseen" detection of rare aquatic species using environmental DNA. *Conservation Letters, 4*(2), 150–157. https://doi.org/10.1111/j.1755-263X.2010.00158.x, http://onlinelibrary.wiley.com/journal/10.1111/(ISSN)1755-263X.

Jordan, D., & Mills, D. E. (2021). Past, present, and future of DNA typing for analyzing human and non-human forensic samples. *Frontiers in Ecology and Evolution, 9*. https://doi.org/10.3389/fevo.2021.646130.

Keeley, N., Wood, S. A., & Pochon, X. (2018). Development and preliminary validation of a multi-trophic metabarcoding biotic index for monitoring benthic organic enrichment. *Ecological Indicators, 85*, 1044–1057. https://doi.org/10.1016/j.ecolind.2017.11.014, http://www.elsevier.com/locate/ecolind.

Kelly, R. P., Port, J. A., Yamahara, K. M., Martone, R. G., Lowell, N., Thomsen, P. F., Mach, M. E., Bennett, M., Prahler, E., Caldwell, M. R., & Crowder, L. B. (2014). Harnessing DNA to improve environmental management. *Science, 344*(6191), 1455–1456. https://doi.org/10.1126/science.1251156, http://www.sciencemag.org/content/344/6191/1455.full.pdf.

Klancnik, A., Toplak, N., Kovac, M., Marquis, H., & Jersek, B. (2015). Quantification of *Listeria monocytogenes* cells with digital PCR and their biofilm cells with real-time PCR. *Journal of Microbiological Methods, 118*, 37–41. https://doi.org/10.1016/j.mimet.2015.08.012.

Kline, M. C., Romsos, E. L., & Duewer, D. L. (2016). Evaluating digital PCR for the quantification of human genomic DNA: Accessible amplifiable targets. *Analytical Chemistry, 88*(4), 2132–2139. https://doi.org/10.1021/acs.analchem.5b03692, http://pubs.acs.org/journal/ancham.

Knights, D., Parfrey, L. W., Zaneveld, J., Lozupone, C., & Knight, R. (2011). Human-associated microbial signatures: Examining their predictive value. *Cell Host and Microbe, 10*(4), 292–296. https://doi.org/10.1016/j.chom.2011.09.003, http://www.elsevier.com.

Koczan, J. M., Lenneman, B. R., McGrath, M. J., & Sundin, G. W. (2011). Cell surface attachment structures contribute to biofilm formation and xylem colonization by *Erwinia amylovora*. *Applied and Environmental Microbiology, 77*(19), 7031–7039. https://doi.org/10.1128/AEM.05138-11, http://aem.asm.org/content/77/19/7031.full.pdf.

Kresse, S. H., Brandt-Winge, S., Pharo, H., Flatin, B. T. B., Jeanmougin, M., Vedeld, H. M., & Lind, G. E. (2023). Evaluation of commercial kits for isolation and bisulfite conversion of circulating cell-free tumor DNA from blood. *Clinical Epigenetics, 15*(1). https://doi.org/10.1186/s13148-023-01563-0, https://clinicalepigeneticsjournal.biomedcentral.com/.

Kudoh, A., Minamoto, T., & Yamamoto, S. (2020). Detection of herbivory: eDNA detection from feeding marks on leaves. *Environmental DNA, 2*(4), 627–634. https://doi.org/10.1002/edn3.113.

Kuypers, J., & Jerome, K. R. (2017). Applications of digital PCR for clinical microbiology. *Journal of Clinical Microbiology, 55*(6), 1621–1628. https://doi.org/10.1128/JCM.00211-17, http://jcm.asm.org/content/55/6/1621.full.pdf.

Langlois, V. S., Allison, M. J., Bergman, L. C., To, T. A., & Helbing, C. C. (2021). The need for robust qPCR-based eDNA detection assays in environmental monitoring and species inventories. *Environmental DNA, 3*(3), 519–527. https://doi.org/10.1002/edn3.164, onlinelibrary.wiley.com/journal/26374943.

Laramie, M. B., Pilliod, D. S., Goldberg, C. S., & Strickler, K. M. (2015). Environmental DNA sampling protocol— Filtering water to capture DNA from aquatic organisms. *Geological Survey Techniques and Methods*.

Lee, A. H., Lee, J., Hong, S., Kwon, B. O., Xie, Y., Giesy, J. P., Zhang, X., & Khim, J. S. (2020). Integrated assessment of west coast of South Korea by use of benthic bacterial community structure as determined by eDNA, concentrations of contaminants, and in vitro bioassays. *Environment International, 137*. https://doi.org/10.1016/j.envint.2020.105569, https://www.elsevier.com/locate/envint.

Lei, S., Chen, S., & Zhong, Q. (2021). Digital PCR for accurate quantification of pathogens: Principles, applications, challenges and future prospects. *International Journal of Biological Macromolecules, 184*, 750–759. https://doi.org/10.1016/j.ijbiomac.2021.06.132.

Liu, X., Yao, H., Zhao, X., & Ge, C. (2023). Biofilm formation and control of foodborne pathogenic bacteria. *Molecules, 28*(6), 2432. https://doi.org/10.3390/molecules28062432.

Liu, Y., Cathcart, A. L., Delaney, W. E., & Kitrinos, K. M. (2017). Development of a digital droplet PCR assay to measure HBV DNA in patients receiving long-term TDF treatment. *Journal of Virological Methods, 249*, 189–193. https://doi.org/10.1016/j.jviromet.2017.09.015, https://www.elsevier.com/locate/jviromet.

Liu, Z., Yang, J., Wang, N., Liu, J., Geng, J., Zhu, J., Cong, B., Sun, H., & Wu, R. (2024). Integrative lncRNA, circRNA, and mRNA analysis reveals expression profiles of six forensic body fluids/tissue. *International Journal of Legal Medicine, 138*(3), 731–742. https://doi.org/10.1007/s00414-023-03131-w.

Metcalf, J. L., Parfrey, L. W., Gonzalez, A., Lauber, C. L., Knights, D., Ackermann, G., Humphrey, G. C., Gebert, M. J., Van Treuren, W., Berg-Lyons, D., Keepers, K., Guo, Y., Bullard, J., Fierer, N., Carter, D. O., & Knight, R. (2013). A microbial clock provides an accurate estimate of the postmortem interval in a mouse model system. *eLife, 2013*(2). https://doi.org/10.7554/eLife.01104.001, http://elife.elifesciences.org/content/elife/2/e01104.full.pdf.

Moreno, L. I., Mills, D. E., Fetscher, J., John-Williams, K., Meadows-Jantz, L., & McCord, B. (2011). The application of amplicon length heterogeneity PCR (LH-PCR) for monitoring the dynamics of soil microbial communities associated with cadaver decomposition. *Journal of Microbiological Methods, 84*(3), 388–393. https://doi.org/10.1016/j.mimet.2010.11.023.

Moustafa, A., Li, W., Singh, H., Moncera, K. J., Torralba, M. G., Yu, Y., Manuel, O., Biggs, W., Venter, J. C., Nelson, K. E., Pieper, R., & Telenti, A. (2018). Microbial metagenome of urinary tract infection. *Scientific Reports, 8*(1), 4333. https://doi.org/10.1038/s41598-018-22660-8.

Navarro, L. M., Fernández, N., Guerra, C., Guralnick, R., Kissling, W. D., Londoño, M. C., Muller-Karger, F., Turak, E., Balvanera, P., Costello, M. J., Delavaud, A., El Serafy, G. Y., Ferrier, S., Geijzendorffer, I., Geller, G. N., Jetz, W., Kim, E. S., Kim, H. J., Martin, C. S., ... Pereira, H. M. (2017). Monitoring biodiversity change through effective global coordination. *Current Opinion in Environmental Sustainability, 29*, 158–169. https://doi.org/10.1016/j.cosust.2018.02.005, http://www.elsevier.com/wps/find/journaldescription.cws_home/718675/description#description.

Nicholson, A., McIsaac, D., MacDonald, C., Gec, P., Mason, B. E., Rein, W., Wrobel, J., de Boer, M., Milián-García, Y., & Hanner, R. H. (2020). An analysis of metadata reporting in freshwater environmental DNA research calls for the development of best practice guidelines. *Environmental DNA, 2*(3), 343–349. https://doi.org/10.1002/edn3.81, onlinelibrary.wiley.com/journal/26374943.

Nielsen, K., Mogensen, H. S., Hedman, J., Niederstätter, H., Parson, W., & Morling, N. (2008). Comparison of five DNA quantification methods. *Forensic Science International: Genetics, 2*(3), 226–230. https://doi.org/10.1016/j.fsigen.2008.02.008, http://www.elsevier.com.

Ogram, A., Sayler, G. S., & Barkay, T. (1987). The extraction and purification of microbial DNA from sediments. *Journal of Microbiological Methods, 7*(2-3), 57–66. https://doi.org/10.1016/0167-7012(87)90025-X.

Opel, K. L., Chung, D., & McCord, B. R. (2010). A study of PCR inhibition mechanisms using real time PCR. *Journal of Forensic Sciences, 55*(1), 25–33. https://doi.org/10.1111/j.1556-4029.2009.01245.x.

Opel, K. L., Fleishaker, E. L., Nicklas, J. A., Buel, E., & McCord, B. R. (2008). Evaluation and quantification of nuclear DNA from human telogen hairs. *Journal of Forensic Sciences, 53*(4), 853–857. https://doi.org/10.1111/j.1556-4029.2008.00777.x.

Pacocha, N., Scheler, O., Nowak, M. M., Derzsi, L., Cichy, J., & Garstecki, P. (2019). Direct droplet digital PCR (dddPCR) for species specific, accurate and precise quantification of bacteria in mixed samples. *Analytical Methods, 11*(44), 5730–5735. https://doi.org/10.1039/c9ay01874c.

Pawlowski, J., Apothéloz-Perret-Gentil, L., & Altermatt, F. (2020). Environmental DNA: What's behind the term? Clarifying the terminology and recommendations for its future use in biomonitoring. *Molecular Ecology, 29*(22), 4258–4264. https://doi.org/10.1111/mec.15643.

Pechal, J. L., Crippen, T. L., Benbow, M. E., Tarone, A. M., Dowd, S., & Tomberlin, J. K. (2014). The potential use of bacterial community succession in forensics as described by high throughput metagenomic sequencing. *International Journal of Legal Medicine, 128*(1), 193–205. https://doi.org/10.1007/s00414-013-0872-1.

Pietramellara, G., Ascher, J., Borgogni, F., Ceccherini, M. T., Guerri, G., & Nannipieri, P. (2009). Extracellular DNA in soil and sediment: Fate and ecological relevance. *Biology and Fertility of Soils, 45*(3), 219–235. https://doi.org/10.1007/s00374-008-0345-8.

Pochon, X., Bott, N. J., Smith, K. F., & Wood, S. A. (2013). Evaluating detection limits of next-generation sequencing for the surveillance and monitoring of international marine pests. *PLoS One, 8*(9). https://doi.org/10.1371/journal.pone.0073935, http://www.plosone.org/article/fetchObjectAttachment.action;jsessionid=0F589D2B027F92343C11F367F8560B06?uri=info%3Adoi%2F10.1371%2Fjournal.pone.0073935&representation=PDF.

Pochon, X., Zaiko, A., Fletcher, L. M., Laroche, O., & Wood, S. A. (2017). Wanted dead or alive? Using metabarcoding of environmental DNA and RNA to distinguish living assemblages for biosecurity applications. *PLoS One, 12*(11). https://doi.org/10.1371/journal.pone.0187636, http://journals.plos.org/plosone/article/file?id=10.1371/journal.pone.0187636&type=printable.

Powell, L. C., Pritchard, M. F., Ferguson, E. L., Powell, K. A., Patel, S. U., Rye, P. D., Sakellakou, S. M., Buurma, N. J., Brilliant, C. D., Copping, J. M., Menzies, G. E., Lewis, P. D., Hill, K. E., & Thomas, D. W. (2018). Targeted disruption of the extracellular polymeric network of *Pseudomonas aeruginosa* biofilms by alginate oligosaccharides. *npj Biofilms and Microbiomes, 4*(1). https://doi.org/10.1038/s41522-018-0056-3, https://www.nature.com/npjbiofilms/.

Qiagen (2023). Assessment of the precision and sensitivity of QIAcuity nanoplate digital PCR vs qPCR.

Quan, P.-L., Sauzade, M., & Brouzes, E. (2018). dPCR: A technology review. *Sensors, 18*(4), 1271. https://doi.org/10.3390/s18041271.

Raghupathi, P. K., Liu, W., Sabbe, K., Houf, K., Burmolle, M., & Sorensen, S. J. (2017). Synergistic interactions within a multispecies biofilm enhance individual species protection against grazing by a pelagic protozoan. *Front Microbiol, 8*, 2649. https://doi.org/10.3389/fmicb.2017.02649, https://doi.org/10.1007/s10661-021-09668-w.

Ragot, R., & Villemur, R. (2022). eDNA profiling of mammals, birds, and fish of surface waters by mitochondrial metagenomics: Application for source tracking of fecal contamination in surface waters. *Environmental Monitoring and Assessment, 194*(2). https://doi.org/10.1007/s10661-021-09668-w.

Rees, H. C., Maddison, B. C., Middleditch, D. J., Patmore, J. R. M., & Gough, K. C. (2014). The detection of aquatic animal species using environmental DNA – A review of eDNA as a survey tool in ecology. *Journal of Applied Ecology, 51*(5), 1450–1459. https://doi.org/10.1111/1365-2664.12306, http://onlinelibrary.wiley.com/journal/10.1111/(ISSN)1365-2664.

Rene, K. A., Ganeshan, F., van Velsen, S., Weber, J., Schmid, C., Graf, & Hochegger, R. (2019). Digital duplex versus real-time PCR for the determination of meat proportions from sausages containing pork and beef. *European Food Research Technology,* 151–157. https://doi.org/10.1007/s00217-018-3147-8.

Renner, L. D., & Weibel, D. B. (2011). Physicochemical regulation of biofilm formation. *MRS Bulletin, 36*(5), 347–355. https://doi.org/10.1557/mrs.2011.65.

Ricci, U., Marchi, C., Previderè, C., & Fattorini, P. (2006). Quantification of human DNA by real-time PCR in forensic casework. *International Congress Series, 1288*, 750–752. https://doi.org/10.1016/j.ics.2005.11.135.

Sacco, M., Guzik, M. T., van der Heyde, M., Nevill, P., Cooper, S. J. B., Austin, A. D., Coates, P. J., Allentoft, M. E., & White, N. E. (2022). eDNA in subterranean ecosystems: Applications, technical aspects, and future prospects. *The Science of the Total Environment, 820*, 153223. https://doi.org/10.1016/j.scitotenv.2022.153223.

Sahu, A., Kumar, N., Pal Singh, C., & Singh, M. (2023). Environmental DNA (eDNA): Powerful technique for biodiversity conservation. *Journal for Nature Conservation, 71*, 126325. https://doi.org/10.1016/j.jnc.2022.126325.

Salipante, S. J., & Jerome, K. R. (2020). Digital PCR—An emerging technology with broad applications in microbiology. *Clinical Chemistry, 66*(1), 117–123. https://doi.org/10.1373/clinchem.2019.304048, https://academic.oup.com/clinchem/article/66/1/117/5688828.

Sanders, R., Huggett, J. F., Bushell, C. A., Cowen, S., Scott, D. J., & Foy, C. A. (2011). Evaluation of digital PCR for absolute DNA quantification. *Analytical Chemistry, 83*(17), 6474–6484. https://doi.org/10.1021/ac103230c.

Sanders, R., Mason, D. J., Foy, C. A., & Huggett, J. F. (2013). Evaluation of digital PCR for absolute RNA quantification. *PLoS One, 8*(9). https://doi.org/10.1371/journal.pone.0075296, http://www.plosone.org/article/fetchObject.action;jsessionid=3BA3A00AFDD3FAA0EE6684E95CF22E1D?uri=info%3Adoi%2F10.1371%2Fjournal.pone.0075296&representation=PDF.

Schmedes, S. E., Woerner, A. E., Novroski, N. M. M., Wendt, F. R., King, J. L., Stephens, K. M., & Budowle, B. (2018). Targeted sequencing of clade-specific markers from skin microbiomes for forensic human identification. *Forensic Science International: Genetics, 32*, 50–61. https://doi.org/10.1016/j.fsigen.2017.10.004.

Shahraki, K., Shahraki, K., Ghasemi Boroumand, P., & Sheervalilou, R. (2023). Promotor methylation in ocular surface squamous neoplasia development: Epigenetics implications in molecular diagnosis. *Expert Review of Molecular Diagnostics, 23*(9), 753–769. https://doi.org/10.1080/14737159.2023.2240238.

Shi, X., Liu, G., Shi, L., Chen, M., Wu, X., Lu, Y., & Zhao, J. (2021). The detection efficiency of digital PCR for the virulence genes of waterborne pathogenic bacteria. *Water Supply, 21*(5), 2285–2297. https://doi.org/10.2166/ws.2021.056, https://watermark.silverchair.com/ws021052285.pdf?.

Shiklomanov, L. A. (1993). *Water in crisis: A guide to world's freshwater resources*. Oxford University Press.

Shin, J. C., Jeong, J. Y., Son, S. G., Choi, S. H., Nam, H. C., Yoon, T. H., Kim, H. J., Choi, D. G., Lee, H., Lee, U., Yang, S. M., Kang, I., Jung, D. Y., Lee, H. W., Lee, M. K., Lee, T. J., Kim, G., Park, H. O., & Lee, S. W. (2024). Developing centrifugal force real-time digital PCR for detecting extremely low DNA concentration. *Scientific Reports, 14*(1). https://doi.org/10.1038/s41598-024-62199-5, https://www.nature.com/srep/.

Sigsgaard, E. E., Carl, H., Møller, P. R., & Thomsen, P. F. (2015). Monitoring the near-extinct European weather loach in Denmark based on environmental DNA from water samples. *Biological Conservation, 183*, 46–52. https://doi.org/10.1016/j.biocon.2014.11.023, https://www.elsevier.com/inca/publications/store/4/0/5/8/5/3.

Simoes Dutra Correa, H., Brescia, G., Cortellini, V., Cerri, N., & Verzeletti, A. (2020). DNA quantitation and degradation assessment: A quantitative PCR protocol designed for small forensic genetics laboratories. *Electrophoresis, 41*(9), 714–719. https://doi.org/10.1002/elps.201900360.

Srey, S., Jahid, I. K., & Ha, S. D. (2013). Biofilm formation in food industries: A food safety concern. *Food Control, 31*(2), 572–585. https://doi.org/10.1016/j.foodcont.2012.12.001.

Staley, Z. R., Boyd, R. J., Shum, P., & Edge, T. A. (2018). Microbial source tracking using quantitative and digital PCR to identify sources of fecal contamination in stormwater, river water, and beach water in a Great Lakes area of concern. *Applied and Environmental Microbiology, 84*(20). https://doi.org/10.1128/AEM.01634-18, https://aem.asm.org/content/aem/84/20/e01634-18.full.pdf.

Strain, M. C., Lada, S. M., Luong, T., Rought, S. E., Gianella, S., Terry, V. H., Spina, C. A., Woelk, C. H., & Richman, D. D. (2013). Highly precise measurement of HIV DNA by droplet digital PCR. *PLoS One, 8*(4). https://doi.org/10.1371/journal.pone.0055943, http://www.plosone.org/article/fetchObjectAttachment.action?uri=info%3Adoi%2F10.1371%2Fjournal.pone.0055943&representation=PDF.

Surface-Water Use. (2019).

Taberlet, P., Bonin, A., Zinger, L., & Coissac, E. (2018). *Environmental DNA: For biodiversity research and monitoring*. France: Oxford University Press, 1–253. http://www.oxfordscholarship.com/view/10.1093/oso/9780198767220.001.0001/oso-9780198767220 10.1093/oso/9780198767220.001.0001.

Tak, L. J., Shin, M. K., Yoo, J. I., Cho, M. C., & Kim, W. (2023). Development of droplet digital PCR-based detection of bacterial pathogens in prosthetic joint infection: A preliminary study using a synthesized model plasmid. *Frontiers in Cellular and Infection Microbiology, 13*. https://doi.org/10.3389/fcimb.2023.1301446, http://www.frontiersin.org/Cellular_and_Infection_Microbiology/archive.

Thalinger, B., Deiner, K., Harper, L. R., Rees, H. C., Blackman, R. C., Sint, D., Traugott, M., Goldberg, C. S., & Bruce, K. (2021). A validation scale to determine the readiness of environmental DNA assays for routine species monitoring. *Environmental DNA, 3*(4), 823–836. https://doi.org/10.1002/edn3.189, onlinelibrary.wiley.com/journal/26374943.

The eDNA Society (2019). In environmental DNA sampling and experiment manual. Version 2.1.

Thomsen, P. F., & Sigsgaard, E. E. (2019). Environmental DNA metabarcoding of wild flowers reveals diverse communities of terrestrial arthropods. *Ecology and Evolution, 9*(4), 1665–1679. https://doi.org/10.1002/ece3.4809, http://onlinelibrary.wiley.com/journal/10.1002/(ISSN)2045-7758.

Thomsen, P. F., & Willerslev, E. (2015). Environmental DNA – An emerging tool in conservation for monitoring past and present biodiversity. *Biological Conservation, 183*, 4–18. https://doi.org/10.1016/j.biocon.2014.11.019, https://www.elsevier.com/inca/publications/store/4/0/5/8/5/3.

Tripathi, P., Render, R., Nidhi, S., & Tripathi, V. (2024). Microbial genomics: A potential toolkit for forensic investigations. *Forensic Science, Medicine and Pathology*. https://doi.org/10.1007/s12024-024-00830-7.

Tsuji, S., Takahara, T., Doi, H., Shibata, N., & Yamanaka, H. (2019). The detection of aquatic macroorganisms using environmental DNA analysis—A review of methods for collection, extraction, and detection. *Environmental DNA, 1*(2), 99–108. https://doi.org/10.1002/edn3.21, onlinelibrary.wiley.com/journal/26374943.

Turner, C. R., Uy, K. L., & Everhart, R. C. (2015). Fish environmental DNA is more concentrated in aquatic sediments than surface water. *Biological Conservation, 183*, 93–102. https://doi.org/10.1016/j.biocon.2014.11.017, https://www.elsevier.com/inca/publications/store/4/0/5/8/5/3.

US Fish and Wildlife Service (2018). Quality assurance project plan eDNA monitoring of bighead and silver carps (Bloomington, MN: US Fish and Wildlife Service, USFWS Midwest Region). 168 pp. https://www.fws.gov/midwest/fisheries/eDNA/documents/QAPP.pdf.

Ushio, R., Yamamoto, M., Nakashima, K., Watanabe, H., Nagai, K., Shibata, Y., Tashiro, K., Tsukahara, T., Nagakura, H., Horita, N., Sato, T., Shinkai, M., Kudo, M., Ueda, A., & Kaneko, T. (2016). Digital PCR assay detection of circulating *Mycobacterium tuberculosis* DNA in pulmonary tuberculosis patient plasma. *Tuberculosis, 99*, 47–53. https://doi.org/10.1016/j.tube.2016.04.004.

Valentin, R. E., Maslo, B., Lockwood, J. L., Pote, J., & Fonseca, D. M. (2016). Real-time PCR assay to detect brown marmorated stink bug, *Halyomorpha halys* (Stål), in environmental DNA. *Pest Management Science, 72*(10), 1854–1861. https://doi.org/10.1002/ps.4217, http://onlinelibrary.wiley.com/journal/10.1002/(ISSN)1526-4998.

Vert, M., Doi, Y., Hellwich, K. H., Hess, M., Hodge, P., Kubisa, P., Rinaudo, M., & Schué, F. (2012). Terminology for biorelated polymers and applications (IUPAC recommendations 2012). *Pure and Applied Chemistry, 84*(2), 377–410. https://doi.org/10.1351/PAC-REC-10-12-04, http://www.iupac.org/publications/pac/84/2/0335/pdf/.

Wacharapluesadee, S., & Hemachudha, T. (2010). Ante- and post-mortem diagnosis of rabies using nucleic acid-amplification tests. *Expert Review of Molecular Diagnostics, 10*(2), 207–218. https://doi.org/10.1586/erm.09.85.

Wagner, A.B. Jr. (2008). Bacterial food poisoning. Texas A&M AgriLife Extension, Food Technology & Processing. https://aggie-horticulture.tamu.edu/food-technology/bacterial-food-poisoning/.

Wang, P., Yan, Z., Yang, S., Wang, S., Zheng, X., Fan, J., & Zhang, T. (2019). Environmental DNA: An emerging tool in ecological assessment. *Bulletin of Environmental Contamination and Toxicology, 103*(5), 651–656. https://doi.org/10.1007/s00128-019-02720-z.

Whale, A. S., Cowen, S., Foy, C. A., & Huggett, J. F. (2013). Methods for applying accurate digital PCR analysis on low copy DNA samples. *PLoS One, 8*(3). https://doi.org/10.1371/journal.pone.0058177, http://www.plosone.org/article/fetchObjectAttachment.action?uri=info%3Adoi%2F10.1371%2Fjournal.pone.0058177&representation=PDF.

White, R. A., Blainey, P. C., Fan, H. C., & Quake, S. R. (2009). Digital PCR provides sensitive and absolute calibration for high throughput sequencing. *BMC Genomics, 10*. https://doi.org/10.1186/1471-2164-10-116.

Wohlfahrt, D., Tan-Torres, A. L., Green, R., Brim, K., Bradley, N., Brand, A., Abshier, E., Nogales, F., Babcock, K., Brooks, J. P., Seashols-Williams, S., & Singh, B. (2023). A bacterial signature-based method for the identification of seven forensically relevant human body fluids. *Forensic Science International: Genetics, 65*. https://doi.org/10.1016/j.fsigen.2023.102865, http://www.elsevier.com.

Wood, S. A., Pochon, X., Laroche, O., von Ammon, U., Adamson, J., & Zaiko, A. (2019). A comparison of droplet digital polymerase chain reaction (PCR), quantitative PCR and metabarcoding for species-specific detection in environmental DNA. *Molecular Ecology Resources, 19*(6), 1407–1419.

Worthington, R. J., Richards, J. J., & Melander, C. (2012). Small molecule control of bacterial biofilms. *Organic and Biomolecular Chemistry, 10*(37), 7457–7474. https://doi.org/10.1039/c2ob25835h.

Yin, W., Xu, S., Wang, Y., Zhang, Y., Chou, S. H., Galperin, M. Y., & He, J. (2021). Ways to control harmful biofilms: Prevention, inhibition, and eradication. *Critical Reviews in Microbiology, 47*(1), 57–78. https://doi.org/10.1080/1040841X.2020.1842325, https://www.tandfonline.com/loi/imby20.

Yu, M., Heinzerling, T. J., & Grady, W. M. (2018). *DNA methylation analysis using droplet digital PCR, 1768*. Springer Nature, 363–383 10.1007/978-1-4939-7778-9_21.

Zaiko, A., Schimanski, K., Pochon, X., Hopkins, G. A., Goldstien, S., Floerl, O., & Wood, S. A. (2016). Metabarcoding improves detection of eukaryotes from early biofouling communities: Implications for pest monitoring and pathway management. *Biofouling, 32*(6), 671–684. https://doi.org/10.1080/08927014.2016.1186165, https://www.tandf.co.uk/journals/titles/08927014.asp.

Index

Note: Page numbers followed by "*f*" and "*t*" refer to figures and tables, respectively.

G

O

P

Q

R

S

T

U

V

W

X

Y

Z

9780443301285